Modeling Groundwater Flow and Contaminant Transport

Theory and Applications of Transport in Porous Media

Series Editor:
Jacob Bear: *Department of of Civil and Environmental Engineering, Technion-Israel Institute of Technology, Haifa, and School of Engineering, Kinneret College on the Sea of Galilee, Israel*

Volume 23

For further volumes: http://www.springer.com/series/6612

Modeling Groundwater Flow and Contaminant Transport

by

Jacob Bear
Department of Civil and Environmental Engineering, Technion-Israel Institute of Technology, Haifa, and School of Engineering, Kinneret College on the Sea of Galilee, Israel

and

Alexander H.-D. Cheng
Department of Civil Engineering, University of Mississippi, Oxford, MS, USA

Springer

Prof. Dr. Jacob Bear
Technion-Israel Institute of Technology
Dept. Civil & Environmental Engineering
32000 Haifa
Israel
cvrbear@tx.technion.ac.il

Dr. Alexander H.-D. Cheng
University of Mississippi
Dept. Civil Engineering
P.O.Box 1848
University MS 38677-1848
USA
acheng@olemiss.edu

ISBN 978-94-024-0477-7 ISBN 978-1-4020-6682-5 (eBook)
DOI 10.1007/978-1-4020-6682-5
Springer Dordrecht Heidelberg London New York

Printed on acid-free paper

Springer is part of Springer Science+Business Media (www.springer.com)

Contents

Preface

Groundwater, extracted from deep geological formations (called aquifers) through pumping wells, constitutes an important component of many water resource systems. A spring constitutes an outlet for groundwater from an underlying aquifer to ground surface; its discharge rate may be strongly affected by pumping from the same aquifer in the vicinity of the spring. Water can be injected through specially designed wells into an aquifer, say, for storage purposes. A water table aquifer can also be artificially recharged through infiltration ponds. These are just a few examples of factor that may affect the management of a groundwater system. Decisions associated with such management include, for example,

- The volume that can be safely withdrawn annually from the aquifer.
- The location of pumping and artificial recharge wells, and their rates.
- The quality of the water to be maintained in the aquifer, and/or to be pumped from it.

In fact, in the management of water resources, the quantity and quality problems cannot be separated from each other. In many parts of the world, as a result of increased withdrawal of groundwater, often beyond permissible limits, the quality of groundwater has been continuously deteriorating, causing much concern to both suppliers and users. The quality deterioration may manifest itself in the form of an increase in the total salinity, or as increased concentrations of nitrates and other undesirable chemical species, or as increased concentrations of harmful pathogens.

Traditionally, hydrogeologist dealt with flow in aquifers, and with certain water quality aspects, e.g., salinization. Soil physicist and agronomists, in connection with agricultural activities, have modeled the movement of water and chemicals (e.g., fertilizers) in the unsaturated zone. The hydrogeologist, whose primary interest has been water in aquifers, regarded the unsaturated zone only as the domain through which water from precipitation passes on its way to replenish an underlying water table aquifer. The details of the actual movement of water through the unsaturated zone have been of little or no interest. The situation has completely changed with the rising interest in subsurface contamination. Clearly, interdisciplinary efforts, straddling numerous areas of expertise, are now required to model and solve contamination problems of practical importance.

In recent years, in addition to the general groundwater quality aspects mentioned above, public attention has been focused on groundwater contamination by hazardous industrial wastes, on leachate from landfills and spills of oil and other toxic liquids, on agricultural activities, such as the use of fertilizers, pesticides, and herbicides, and on radioactive waste in repositories located in deep geological formations, to mention some of the more acute contamination sources. Although originating at ground surface, these contaminants (e.g., spilled oil, pesticides applied to an orchard, or leachate from a landfill), soon infiltrate through ground surface, percolate through the unsaturated zone, and find their way to groundwater in an underlying aquifer. Once reaching the aquifer, contaminants are transported with the moving groundwater, eventually reaching pumping wells, streams and lakes.

Sometimes, a toxic chemical may constitute a separate, nonaqueous, liquid phase, e.g., oil, that occupies part of the void space. Components of such toxic liquids may dissolve in percolating water, thus constituting a source for groundwater contamination. The volatile contaminant components may evaporate to become components of the gaseous phase (air) present in the void space. In the subsurface, while being transported with the water, the various contaminants undergo complex physical, chemical, and biological transformations. Chemical species carried by the water often interact with each other and with the soil, especially with the clay and organic fractions of the latter. Phenomena such as adsorption, ion exchange, chemical reactions, dissolution, volatilization, and biological decay, continuously affect the concentration of the chemical constituents present in the percolating water.

Data obtained by monitoring concentrations of hazardous contaminants in the subsurface, often call for remedial action. Regulations on quality standards may require cleanup of the contaminated aquifer and the unsaturated zone. The latter may be visualized as a huge physical-chemical-biological reactor in which many processes occur simultaneously among species present in the water, in nonaqueous fluids, and on the solid matrix. The biota present on the soil and in the fluids may also play an important role. Because of the way fluids behave in the unsaturated zone, cleaning that zone is often very complicated and costly, as it often requires sophisticated in situ chemical and biological methods.

Any plan of mitigation, cleanup operations, or control measures, once contamination has been detected in the subsurface, requires the prediction of pathways and fate of the contaminants, in response to certain planned remediation activities. Similarly, any monitoring or observation network must be based on the anticipated behavior of the system.

Management means making decisions to achieve goals, without violating specified constraints. Therefore, good management requires information on the response of the managed system to proposed activities. This information enables the planner, or the decision-maker, to compare alternative actions, to select the best one, and to ensure that constraints are not violated. All such predictions can be obtained, within the framework of a considered management problem, by constructing and solving mathematical models of the investigated domain, and of the flow and solute transport phenomena that take place in it.

Using mathematical models for making prediction requires input data. The major role of the data is to enable model validation and calibration. The more accurate and complete are the collected field data, the more reliable are the values of model parameters obtained by model calibration, and, hence, also of the model's predictions. However, data gathering activities face the reality of uncertainty: subsurface geological formations are far too heterogenous to provide the accurate detailed information required for their complete description. Nevertheless, since there is no other way but to use models in order to predict the future behavior of an investigated system, using whatever data that are available for model calibration cannot be avoided, in spite of the associated uncertainty. Various tools are available for coping with this uncertainty.

For most practical problems, because of the heterogeneity of the considered domain and the irregular shape of its boundaries, it is not possible to solve the mathematical models analytically. Instead, the mathematical model is transformed into a numerical one that can be solved by means of computer programs. Indeed, excellent computer programs are available for this purpose. Unfortunately, too often, practitioners use such available computer programs without really grasping the theory and assumptions underlying the models that they are solving. Our purpose in this book is to present not only the models that describe phenomena of flow and solute transport in aquifers, but also to emphasize the theoretical foundation and the various assumptions that simplify the complex reality to the extent that it can be described by rather simple and solvable models.

With this background, the major objectives of this book are:

To construct conceptual and mathematical models that can provide the information required for making decisions associated with the management of groundwater resources, and the remediation of contaminated aquifers.

More specifically,

- To describe the mechanisms that govern the movement of fluids and contaminants in aquifers and in the unsaturated zone.
- To construct well-posed mathematical models of saturated flow in three-dimensional porous medium domains and in aquifers, and of single and multiphase flow in the unsaturated zone.
- To construct well-posed mathematical models of transport of single and multiple chemical species in the unsaturated zone and in aquifers.

Three additional topics, strongly related to use of models for predicting flow and transport regimes in aquifers, within the framework of management, are discussed:

- The use of numerical models and computer codes as practical tools for solving the mathematical models.
- The issues of uncertainty associated with modeling.
- Certain mathematical tools for groundwater management.

With these objectives in mind, the book is aimed at practitioners, modelers, water resources managers, scientists, and researchers, who face the need to build and solve models of flow and contaminant transport in the subsurface. It is also suitable for graduate and upper level undergraduate students who are interested in such topics as groundwater, water resources, and environmental engineering. The basic scientific background needed is the concepts and terminologies of hydrology and hydraulics.

Finally, we wish to acknowledge the following colleagues who have provided useful advice for various parts of the book: Yunwei Sun and Walt McNab of Lawrence Livermore National Laboratory; Quanlin Zhou of Lawrence Berkeley National Laboratory; Peter Lichtner of Los Alamos National Laboratory; Vicky Freeman of Pacific Northwest National Laboratory; Ajit Sadana of the University of Mississippi; Randy Gentry of the University of Tennessee; Shlomo Neuman of the University of Arizona; Prabhakar Clement of Auburn University; T.N. Narasimhan of the University of California, Berkeley; Shlomo Orr of MRDS, Inc.; Shaul Sorek of Ben-Gurion University, Israel; Uri Shavit and Leonid Fel of Technion, Israel; Jacob Bensabat of Environmental and Water Resources Engineering Inc., Israel; Dalila Loudyi of Hassan II University, Morocco; and Don Nield of University of Auckland, New Zealand.

Jacob Bear Haifa, Israel
Alexander H.-D. Cheng Oxford, Mississippi, USA

2009

List of Main Symbols

a Dispersivity of a porous medium (fourth rank tensor).
a_{ijkl} Component of a.
a_L, a_T Longitudinal and transversal dispersivities (isotropic medium), respectively.
B Thickness of a confined aquifer.
B' Thickness of a semipervious layer (aquitard).
$\mathcal{B}$ Mass balance operator.
c_{ab}, c_{aa} Covariance between a and b, and autocovariance of a.
c_r Hydraulic resistance of a semipervious layer.
c_α^γ Mass concentration of a γ-species in an α-phase.
$[c_\alpha^\gamma]$ Molar concentration of a γ-species in an α-phase.
C Normalized salt mass fraction.
C_v Consolidation coefficient.
C_w Moisture (water) capacity.
d Effective grain diameter.
Microscopic length characterizing the void space.
d^* Length characterizing macroscopic heterogeneity.
Da Darcy number.
Dm Damköhler number.
$\mathbf{D}_\alpha^\gamma$ Coefficient of dispersion of the mass of a γ-species in an α-phase.
$\mathbf{D}_{\alpha h}^\gamma$ Coefficient of hydrodynamic dispersion of a γ-species in an α-phase ($= \boldsymbol{\mathcal{D}}_\alpha^{*\gamma} + \mathbf{D}_\alpha^\gamma$).
D_L, D_T Longitudinal and transversal dispersion coefficients.
$\mathcal{D}_\alpha^\gamma$ Coefficient of molecular diffusion of a γ-species in an α-phase.
$\boldsymbol{\mathcal{D}}_\alpha^{*\gamma}$ Coefficient of molecular diffusion of a γ-species in an α-phase in a porous medium ($= \mathcal{D}_\alpha^\gamma \mathbf{T}^*$).
$\mathbf{D}_w$ Moisture diffusivity.
e An intensive quantity, density of an extensive quantity, E.
Void ratio ($= \mathcal{U}_{ov}/\mathcal{U}_{os}$).
E An extensive quantity.
e_α^γ Density of E_α^γ ($= E$ of γ in α per unit volume of α-phase).
E_α^γ An extensive quantity, E, of a γ-species in an α-phase.
$f_{\alpha\rightarrow\beta}^E$ Rate of transfer of E from an α-phase to a β-phase, across their

	common microscopic interface, per unit volume of porous medium.
$\mathbf{f}_\alpha$	Fractional flow rate of an α-phase in two-phase flow.
F	Concentration of a species adsorbed on a solid ($= m_s^\gamma/m_s$).
	Function representing a surface.
$\mathbf{F}$	Body force.
$\mathcal{F}$	Faraday's constant.
Fo	Fourier number.
Fr	Froude number.
g	Gravity acceleration.
G	Gibbs free energy.
	Green's function.
h	Piezometric head.
h^*	Hubbert's potential.
h_c	Height of capillary fringe.
h_r	Relative humidity.
H	Height of water table in phreatic aquifer.
	Heaviside unit step function.
$\mathcal{H}$	Henry's coefficient.
H	Hessian matrix.
I	Rate of infiltration.
	Ionic strength of solution.
	Integral scale.
I	Unit tensor.
$\mathbf{j}^\gamma$	Microscopic diffusive mass flux of a γ-species, relative to the mass weighted velocity ($= \rho^\gamma(\mathbf{V}^\gamma - \mathbf{V})$).
$\mathbf{j}^{tE_\alpha^\gamma}$	Total microscopic flux of E_α^γ ($= e_\alpha^\gamma \mathbf{V}^{E_\alpha^\gamma}$).
$\mathbf{J}_\alpha^\gamma$	Macroscopic diffusive flux of a γ-species in an α-phase ($\equiv \overline{\mathbf{j}^\gamma}^\alpha$).
$\mathbf{J}^{\star E}$	Dispersive flux of E.
$\mathbf{J}^{tE_\alpha^\gamma}$	Total macroscopic flux of E_α^γ.
$\mathbf{J}_{\alpha h}^E$	Sum of diffusive and dispersive fluxes of E in an α-phase.
$\mathcal{J}$	Hydraulic gradient.
k	Permeability (scalar).
$\mathbf{k}$	Permeability (second rank tensor).
$\mathbf{k}_\alpha$	Effective permeability of an α-phase.
$\mathbf{k}_{r\alpha}$	Relative permeability of an α-phase, in an isotropic porous medium.
k_α^γ	Degradation rate constant of a γ-species in an α-phase.
K	Hydraulic conductivity (scalar).
$\mathbf{K}$	Hydraulic conductivity (second rank tensor).
K'	Hydraulic conductivity of semi-pervious layer.
K_d	Partitioning coefficient.
K_{eq}	Equilibrium coefficient.
K_{sp}	Solubility product.
ℓ	Characteristic size (diameter) of REV.
L	Length of a column.

L^*	Characteristic size of domain.
$L_c^{(v)}$	Length over which a significant change takes place in the value v.
$\mathcal{L}$	Partial differential operator.
m	Mass.
m_α^γ	Mass of a γ-species of an α-phase.
$\hat{m}^{\mathrm{A}}$	Molality (moles of species A per kilogram of solvent).
M	Molecular mass.
M^γ	Molecular mass of a γ-species.
$\mathbf{M}_\alpha$	Momentum of an α-phase.
$\mathbf{n}_\alpha$	Outward unit vector to a surface (e.g., to $\mathcal{S}_{\alpha\beta}$).
$\mathbf{N}$	Rate of accretion on a phreatic surface.
N_i	The ith basis (shape) function.
N_α^γ	Number of moles of a γ-component in an α-phase.
$\mathcal{N}$	Partial differential operator for natural boundary condition.
p	Pressure.
p_c	Capillary pressure, macroscopic.
p'_c	Capillary pressure, microscopic.
p_α	Pressure in an α-phase.
p_α^γ	Partial pressure of a γ-species in an α-phase.
p^v	Vapor pressure.
P	Precipitation recharge rate.
	Pumping rate in an aquifer.
P	Primary species.
Pe	Peclet number.
$\mathbf{q}_\alpha$	Specific discharge of an α-phase $(= \phi\mathbf{V}_\alpha)$.
$\mathbf{q}_{\alpha r}$	Specific discharge of an α-phase, relative to the solid matrix.
q_v	Vertical leakage into an aquifer.
Q	Discharge (dims. L^3/T).
Q_w	Pumping well discharge (dims. L^3/T).
Q'	Discharge per unit width of aquifer (dims. L^2/T).
Q	Secondary species.
r	Radial distance.
r_w	Well radius.
r_{ab}	Correlation of a and b.
r_{aa}	Autocorrelation of a.
R_{ab}	Correlation coefficient of a and b.
R_{aa}	Autocorrelation coefficient of a.
R	Universal gas constant.
	Rate of recharge of an aquifer (volume per unit area per unit time).
	Residual error residual in numerical method.
$\mathcal{R}$	Reliability (= probability that a system does not fail).
R_α^γ	Solubility of a γ-species in an α-phase.
R_d, R_v	Partitioning coefficient.
R_r	Reaction rate.
Re	Reynolds number.

s	Drawdown in aquifer.
	Length along streamline.
S	Aquifer storativity.
S_o	Specific storativity of a porous medium.
S_y	Specific yield.
S_r	Specific retention.
S_α	Saturation of an α-phase.
$S_{\alpha r}$	Irreducible, or residual saturation of an α-phase.
$\tilde{S}_w$, $\tilde{S}_{at}$	effective water saturation and effective trapped air saturation.
$\widehat{S}_w$	Apparent water saturation.
S_a	Power spectral density of quantity a.
St	Strouhal number.
S_i	Slack variable in linear programming.
$\mathcal{S}$	Partial differential operator for essential boundary condition.
$\mathcal{S}_o$, $\mathcal{S}_{o\alpha}$	Area of surfaces surrounding $\mathcal{U}_o$, and $\mathcal{U}_{o\alpha}$, respectively.
$\mathcal{S}_{\alpha\beta}$	Area of surface of contact of α-phase with all other (β) phases within $\mathcal{U}_o$.
t	Time.
T	Temperature.
T	Aquifer transmissivity (scalar).
$\mathbf{T}$	Aquifer transmissivity (second rank tensor).
$\mathbf{T}^*$	Tortuosity of a porous medium (second rank tensor).
$\mathbf{u}_\alpha$	Velocity of a surface (e.g., of $\mathcal{S}_{\alpha\beta}$).
$\mathcal{U}_o$	Volume of REV.
$\mathcal{U}_{o,s}$, $\mathcal{U}_{o,v}$	Volume of solid and void, respectively.
$\mathcal{U}_\alpha$	Volume of α-phase.
$\mathcal{U}^E$	Volume occupied by a quantity E.
v	Specific volume of mass ($= 1/\rho$).
v^E	Specific volume of E ($= 1/e$).
$\mathbf{V}$	Velocity of a fluid phase.
$\mathbf{V}^m$	Mass weighted velocity of a fluid phase.
$\mathbf{V}_s$	Velocity of a solid phase.
$\mathbf{V}^{E_\alpha^\gamma}$	Velocity of an E_α^γ-continuum.
$\mathbf{V}_\alpha^\gamma$	Velocity of a γ-species in an α-phase.
w	Weighting function.
$\mathbf{w}$	Displacement.
W	Well function.
$\mathbf{x}$	Position vector.
$\mathring{\mathbf{x}}$	$= \mathbf{x} - \mathbf{x}_o$.
$\mathbf{x}'$	Position vector of point at the microscopic level.
$\mathbf{x}_o$	Position vector of the centroid of an REV.
X_α^γ	Mole fraction of a γ-species in an α-phase.
Y_s, Y_v	Volume of solid and void, respectively, in a Y-periodic cell in a homogenization process.
$\mathbf{y}$	Coordinate system used in the period cell (Y-cell) in a

	homogenization process.
z	Vertical coordinate (positive upward).
z^γ	Electrical charge of ion of γ species.
Z	Objective function for optimization.

Greek letters

α	Symbol for an α-phase.
	Coefficient of soil compressibility.
α'	Coefficient of rock compressibility.
$\alpha^{\star E}$	Transfer coefficient of E.
β_p, β_T	Fluid compressibilities at constant p and constant T, respectively.
γ	Symbol denoting a γ-species.
	Euler constant ($= 0.5772157\ldots$).
	Specific weight.
	Semivariogram.
$\gamma(\mathbf{x})$	Characteristic function of the void space.
γ_α	Unit weight of an α-phase ($= \rho_\alpha g$).
$\gamma_{\alpha\beta}$	Surface tension between α- and β-phases.
$\Gamma^{E_\alpha^\gamma}$	Rate of production of E_α^γ, per unit mass of an α-phase.
δ	Dirac delta function.
δ_{ij}	Components of Kronecker delta.
Δ	Characteristic distance from solid surface to the fluid in an REV.
Δ_α	Hydraulic radius of an α-phase.
ε	Volumetric strain.
	Small parameter in perturbation and homogenization.
$\boldsymbol{\varepsilon}$	Strain tensor (Components ε_{ij}).
$\dot{\boldsymbol{\varepsilon}}$	Rate of strain.
ε_{sk}	Dilatations of the solid matrix (skeleton).
ζ	Saltwater-freshwater interface location.
θ	Weighting factor for Crank-Nicolson scheme.
θ_{LG}	Contact angle (between liquid and gas).
θ_α	Volumetric fraction of an α-phase ($\equiv \mathcal{U}_\alpha/\mathcal{U}_o = \phi S_\alpha$).
$\theta_{\alpha r}$	Irreducible, or residual volumetric fraction of an α-phase.
λ	Coefficient of radioactive decay ($= 1/t_{1/2}$, where $t_{1/2}$ = mean half life).
	Viscosity associated with fluid compressibility.
	Leakage factor.
λ_s''	Lamé constant of an elastic solid matrix.
μ_s'	Lamé constant of an elastic solid matrix.
μ_α	Dynamic viscosity of an α-phase.
μ_α^γ	Chemical potential of a γ-species of an α-phase.
ν_α	Kinematic viscosity of an α-phase ($\equiv \mu_\alpha/\rho_\alpha$).
ν^γ	Stoichiometric coefficients of γ species.
ξ	Location of particle in a Lagrangian system.
π_α	Equivalent pressure head in an α-phase.

ρ_b	Bulk mass density of soil.
ρ_α	Mass density of an α-phase.
ρ_α^γ	Mass density of a γ-species in an α-phase.
σ	Standard deviation.
σ^2	Variance.
$\boldsymbol{\sigma}$	Stress tensor.
$\boldsymbol{\sigma}'_s$	Effective stress ($\equiv \overline{\boldsymbol{\sigma}'_s}$).
$\Sigma_{\alpha\beta}$	Specific area of $\mathcal{S}_{\alpha\beta}$.
$\boldsymbol{\tau}$	Shear stress.
ϕ, ϕ_{eff}	Porosity and effective porosity, respectively.
φ_α	Piezometric head in an α-phase, in terms of a reference α-fluid.
φ	Surface potential.
	Strack's potential.
$\chi(\theta)$	Function of moisture content.
χ^γ	Stoichiometric coefficients of γ species.
ψ	Suction, or matric suction.
ψ_{ab}, ψ_{aa}	Spatial (temporal) covariance of a and b, and autocovariance of a.
Ψ_{ab}, Ψ_{aa}	Spatial (temporal) correlation of a and b, and autocorrelation of a.
Ψ	Stream function.
Ψ_g, Ψ_m	Gravity and matric potentials, respectively.
Ψ_p, Ψ_{sol}	Pressure and solute (osmotic) potentials, respectively.
Ψ_{sw}, Ψ_T	Soil-water and thermal potentials, respectively.
Ψ_{total}	Total potential.
ω_α^γ	Mass fraction of a γ-species in an α-phase.
Ω	Porous medium domain.
$\partial\Omega$	Boundary of porous medium domain Ω.

Subscripts

a	Air.
c	Characteristic value.
g	Gas.
f	Fluid.
i	Intermediate wetting phase.
im	Immobile phase.
ℓ	Liquid.
m	Mobile phase.
n	Nonwetting phase.
N	Nonaqueous liquid phase (NAPL).
o	Organic liquid. Oil.
pm	Porous medium.
s	Solid.
v	Void space.
w	Water.
	Wetting phase.

α	α-phase.
β	β-phase. Also, a symbol for all other phases, except α.

Superscripts

a	Air, as a species.
H	Heat.
T	Transpose of a matrix.
v	Vapor.
w	Water, as a component.
γ	γ-component.
ε	Quantities dependent on the small parameter ε in homogenization.

Special symbols

$\overline{(..)}$	Average, volume average, or phase average of (..) ($=\frac{1}{\mathcal{U}_o}\int_{\mathcal{U}_o}(..)\,d\mathcal{U}$). Ensemble average.
$\overline{(..)}^{\alpha}$	Intrinsic phase average of (..) ($=\frac{1}{\mathcal{U}_{o\alpha}}\int_{\mathcal{U}_o}(..)\,d\mathcal{U}$).
$\mathring{G}$	Deviation of G from its intrinsic phase average, $\overline{G}^{\alpha}$, over an REV.
$\widetilde{(..)}^{\alpha\beta}$	Average of (..) over the $\mathcal{S}_{\alpha\beta}$-surface.
$\overline{\overline{(..)}}$	Average of (..) over the thickness of thin domain.
$\nabla\cdot\mathbf{A}$	Divergence of a vector $\mathbf{A}$ ($\equiv$ div $\mathbf{A}$).
∇A	Gradient of a scalar A ($\equiv$ grad A).
$\frac{\mathrm{D}_E(..)}{\mathrm{D}t}$	Material derivative of (..), as observed by the E-continuum.
$\langle..\rangle$	Temporal or spatial average.
$[\![A]\!]_{\alpha,\beta}$	Jump in A across an α-β-interface ($\equiv A\vert_\alpha - A\vert_\beta$).
$\{A\}$	Activity of a species A.

Chapter 1

INTRODUCTION

The objective of this book is to present the methodology and procedure for constructing complete *conceptual and mathematical models* of two types of problems: (1) groundwater flow, and (2) groundwater contaminant transport, both in the saturated and unsaturated zones. The construction of such models, however, is not the ultimate goal. In fact, these models are used as essential tools for the planning and management for sustainable use of groundwater resources. This use of models is based on our belief that the physical reality can be represented by mathematical models, albeit with acceptable approximations. In this chapter, we examine the concept, process, and limitations of modeling.

Accordingly, this chapter discusses the role of groundwater in the hydrological cycle, and presents aquifers as a part of water resources systems. The objectives of water resources management are then presented. In view of the complex hydrological, environmental, and economic constraints, these objectives may be in conflict with each other. Alternative management schemes often have to be evaluated and compared by solving models that can simulate the various scenarios. The evaluation of water resources systems, particularly groundwater systems with complex objectives, is possible today, primarily due to the advancement of computer systems since the 1970s. The advent of new computer technologies and capabilities has allowed for the development of more complex mathematical models.

An aquifer is a *porous medium domain*. Hence, flow through the void space of a porous medium is involved in all (physical and, hence, mathematical) models of groundwater systems. Since it is neither feasible nor required to model the detailed flow inside the pore space, we shall show and discuss how this flow and other phenomena of transport (e.g., solute transport) are modeled without information on the details of the pore space geometry. This is accomplished by introducing the concept of a *continuum*.

In this chapter, we may use terms and concepts that will be defined and explained later in the book in more details. This is based on the assumption that the reader is somewhat familiar with these concepts and terminology from earlier studies.

J. Bear, A.H.-D. Cheng, *Modeling Groundwater Flow and Contaminant Transport*, Theory and Applications of Transport in Porous Media 23,
DOI 10.1007/978-1-4020-6682-5_1, © Springer Science+Business Media B.V. 2010

1.1 Groundwater in Water Resources Systems

1.1.1 Hydrological cycle

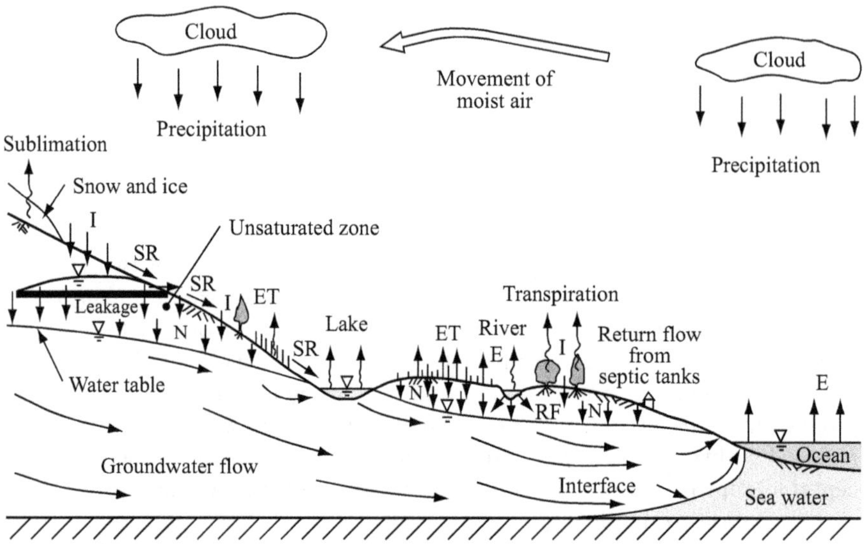

Figure 1.1.1: The hydrological cycle (SR = surface runoff, E = Evaporation, ET = evapotranspiration, I = infiltration, RF = return flow, N = natural replenishment).

The *hydrological cycle* is the term used to describe the cyclic movement of water (in all its states) in nature. As shown by the schematic description in Fig. 1.1.1, the hydrological cycle includes the following components:

- Evaporation from the world's oceans and lakes.
- Transport of water vapor by atmospheric circulation.
- Condensation and precipitation over land and oceans.
- Impoundment in lakes.
- Movement of water back to the oceans as groundwater and river flow.

Groundwater is a term used to denote all the waters found beneath ground surface, as a part of the hydrological cycle. However, often this definition is reserved for the term *subsurface water*, while the term *groundwater* is associated primarily with that part of the water in the hydrological cycle that occurs only in the zone of saturation. Most groundwater hydrologists employ the latter definition, although water in the unsaturated zone is an important part of the hydrological cycle. In addition, understanding what happens in the unsaturated zone is essential, especially in the connection with subsurface contamination.

The role of the unsaturated zone is clearly depicted in Fig. 1.1.1. A portion of the precipitation infiltrates through land surface, and then percolates through the unsaturated zone to an underlying groundwater reservoir. On its way, part of it is taken up by plants and transpires to the atmosphere. Another part may evaporate back to the atmosphere through the ground surface. Water in the unsaturated zone, often referred to as *soil moisture*, amounts to only about 1.6% of the generally accessible subsurface water (at less than half a mile deep), on a global basis. This amount is of little significance as a source of water. However, the moisture in the unsaturated zone plays a very important role in providing nutrients for plants, as well as transporting, transforming, and *temporarily* storing dissolved substances originating at ground surface and carried downward by the water.

1.1.2 Surface water versus groundwater

In order to discuss the role that groundwater may play in the management of regional water resources, let us assume that both surface water and groundwater are present in relatively significant quantities in a region.

Actually, surface water (in lakes and streams) and groundwater (in aquifers) are not necessarily independent water resources. Consider, for example, the interrelations between a river (or a lake) and an adjacent aquifer, or a river passing through a region in which a phreatic aquifer exists. If the river (or lake) bed is not completely clogged, water will flow through it from the river into the aquifer when water levels in the former are higher than in the latter, and vice versa. Base flow in streams is nothing more than groundwater emerging at ground surface. In this way, rivers and lakes in direct, continuous hydraulic contact with adjacent, or underlying aquifers serve as boundaries to the flow domain in the latter. By controlling water levels on such boundaries, we can control the flow of water through them into or out of an aquifer.

Discharge from a spring is another example of groundwater emerging, under certain conditions, at ground surface and becoming surface runoff. By controlling groundwater levels in the vicinity of a spring, its discharge is controlled, or even stopped completely.

The above considerations apply not only to water quantity, but also to water quality, defined, for example, by some chemical species or microorganisms carried with the water. Polluted surface water may easily reach and pollute groundwater and vice versa.

Thus, it is obvious that the management of regional water resources should always include both resources, incorporating each of them in the overall system according to its individual features. In one way or another, any control of one resource will eventually, if not immediately, affect the other. The possible time lag may be due to storage and/or the relatively slow movement of groundwater and of pollutants carried by it. One should note, however, that the *water divide* delineating a groundwater basin and that delineating a surface one are not necessarily geographically identical; in fact, generally, they are not. Depending on the geographical boundaries of any such region, management may include transfer of water from one basin to another within the framework of regional conjunctive use.

Although it seems obvious that groundwater, when present in a region, should be used conjunctively with surface water within the framework of any development and management scheme, in many parts of the world, one finds a certain degree of reluctance to include groundwater in such schemes. Perhaps, in part, this attitude stems from the fact that unlike water in streams and lakes, one cannot actually see groundwater in aquifers. In trying to rationalize this attitude, the following reasons were given (Wiener, 1972):

- Exploitation of groundwater is energy consuming and expensive, especially when the water table, or the piezometric surface, is deep.
- Planning the development of groundwater resources requires long-term data, which, usually, are not available.
- Evaluation and planning groundwater resources requires highly trained personnel that are not available.
- It is difficult to predict the response of an aquifer (in terms of both water quantity and quality) to proposed management schemes.
- Groundwater projects are usually single purpose ones, namely, to supply pumped water, whereas most surface water projects have multiple purposes, e.g., to supply water, produce energy, and provide recreation.

Obviously, in order to examine these arguments and to compare surface water with groundwater, one must know the local conditions. In general, however, it seems that at least in part, these arguments are based on lack of knowledge. For example:

- It is true that when pumping heads are large, energy costs may be significant (whereas energy may be produced from surface water). However, if one includes in the annual expenditures also the relatively high initial investments required for hydraulic structures, such as storage dams, diversion structures, canals, and pipes, the overall economic picture may sway in the direction favoring groundwater.
- Because of the large storage and slow motion involved, groundwater levels at any instant reflect the accumulated effect of a rather long period of time; changes are relatively small and slow, in comparison with those of surface water. Hence, in general, shorter groundwater records give sufficient information for planning purposes, whereas much longer records are required in order to obtain a complete picture of the more frequent and rapid fluctuating behavior of surface water.
- It is true that a certain amount of knowledge is required, but this knowledge has been developed to the extent that nowadays this knowledge is included in most training programs of hydrologists and civil and environmental engineers, or in special courses of continuing education. Most of

the necessary theory is also included in the present text, as a contribution to the dissemination of knowledge on groundwater. Consequently, the lack of skilled personnel can easily be overcome, even in regions where this subject has been neglected in the past.

- With modern hydrological tools, there is no difficulty in modeling the behavior of a groundwater system and forecasting its response (both quantity- and quality-wise). In general, the forecasts are reliable. In most cases, because of the system's complexity, digital computers and computer programs have to be used.
- Certainly, unlike surface water, groundwater cannot be used for recreation. Nevertheless, and perhaps to a more limited extent, groundwater projects may also serve multiple purposes. For example, in addition to water supply, drainage and reclamation of land may be achieved. We have already mentioned above that by controlling groundwater levels in adjacent aquifers, we can control base flow in streams, and this, in turn, has water quality as well as ecological consequences. Using the purifying and mixing properties of an aquifer, artificial recharge can be used for the disposal of reclaimed sewage water, thus augmenting groundwater quantity.

1.1.3 Characteristics of groundwater

Our main purpose in bringing these arguments is not to show that groundwater utilization is always superior and more advantageous, but to emphasize, again, that whenever the two resources are present, they should be used conjunctively, according to their individual features. Following are some of the main characteristics of groundwater (Wiener, 1972).

Location. Surface water flows along fixed curved paths. Their utilization usually requires the construction of regulative facilities that will make the water available only along certain portions of their path. On the other hand, groundwater, when present, underlies extended areas. If these coincide also with demand areas, there is almost no need for a surface distribution system, as the aquifer acts also as a conduit, and consumers can pump their share directly from the aquifer. This feature is of special interest in regions where development is gradual. More wells are sunk whenever an increase in pumping is required. Often control structures for surface water (e.g., dams, or diversions) cannot be built in stages, and the one-time investment is large.

Flow and availability. Fluctuations in surface flow may be significant. Minimum flows, including zero flow, occur often during the season of highest demand for water. On the other hand, climatic fluctuations in groundwater levels are usually small relative to the aquifer's thickness, so that the large volume of water stored in the aquifer may serve as a buffer that can supply water in periods of drought. Whereas the regulation of surface flow requires hydraulic structures, which are often rather costly, the regulation of groundwater flow is incorporated in the implementation of management schemes, namely through an appropriate areal and temporal distribution of pumping and artificial recharge.

The regulation of groundwater flow is, therefore, in general, much less expensive. Base flow in streams and spring flow (including the drying up of streams, which means transforming surface flow into groundwater flow) can be regulated by controlling groundwater levels in their vicinity.

Annual and seasonal variability. Annual and seasonal fluctuations are much more pronounced in surface than in groundwater flow. In surface flow, this means large losses of water by spillage in periods of excess water, or the need for expensive regulatory structures (e.g., dams). In groundwater flow, storage is provided by the aquifer itself; spillage due to very high water levels near an outlet is relatively small and can easily be avoided by manipulating water levels through pumping.

Energy. Energy must be expended in order to lift groundwater to the ground surface. In general, capital investment in wells are low, but operating costs (i.e., cost of electricity or fuel) are relatively high. On the other hand, surface water, stored in reservoirs, often serve also as sources of energy.

Quality of water. Both surface water, e.g., in rivers, and groundwater are susceptible to man-made pollution, which usually requires costly remediation and treatment operations for its removal. In certain formations, pollutants may travel large distances in an aquifer without being attenuated. As for mineral contents, although the range of concentrations encountered is very large, in general one may observe that groundwater is more liable to picking up minerals in solution. The removal of such minerals is usually very expensive. When groundwater does get polluted (e.g., by polluting solutes, e.g., leachate from landfills carried down with the water from the ground surface, or by intrusion of groundwater of inferior quality into an aquifer), the restoration of quality and the removal of pollutants, by mixing with and leaching by clean water, is a very slow, hence, lengthy, and, sometimes, practically impossible, process; the process is often practically irreversible. This is due to the very slow movement of groundwater, especially in layers of very fine material, imbedded in formations of higher permeability, and to adsorption and ion-exchange phenomena on the surface of the solid matrix. These phenomena are especially significant when fine grained materials, e.g., clay, are present in an aquifer. Adsorbed species continue to be fed into the groundwater flow for prolonged periods.

On the other hand, for certain polluting elements carried with the water, such as treated waste water, the above processes of adsorption and ion exchange are an advantage as they remove them from the water. The aquifer then plays the role of a filter and a purifier, taking advantage of the adsorptive capacity of the solid matrix.

In general, there is always the trend of salinization of groundwater by solutes brought down from the surface. Under natural conditions, an equilibrium is reached by the fact that water leaving the formation carries solutes with it. However, when a management program calls for a reduction of outflow (i.e., increased pumping) and/or the introduction of more solutes (e.g., when the aquifer is artificially recharged with water of inferior quality, or when more soluble polluting sources are introduced on the ground surface), this equilibrium is destroyed and we observe an inevitable rise in groundwater solute concentration, sometimes beyond permissible limits.

Impact on drainage problems. The lowering of the phreatic surface by pumping may solve drainage problems in areas where the problems are produced by a high water table. The application of surface water in such cases may require a drainage system to maintain the water table at the desired depth. In the case of marshes or of a water table which is close to the ground surface, the lowering of the water table will also reduce evapotranspiration, thus making more water available for beneficial use. When artificial recharge (Sec. 3.4) is implemented, one should make sure that the rising water table of a phreatic aquifer will not create drainage problems.

Land subsidence. When water is pumped out of a confined aquifer, the intergranular stress in the solid matrix is increased even without changing the load at ground surface. When relatively soft layers (e.g., clay, or silt) are present within the aquifer, they are compressed and we observe land subsidence. In certain areas this subsidence is very significant and can limit, or even force the stoppage, of pumping.

Data. The main source for information about the movement and accumulation in an aquifer of water and solutes carried with the water, are measurements of water levels and solute concentrations in observation wells. Spring discharge and base flow are also sources of data. Even with a relatively small amount of information, preliminary conclusions as to the feasibility of groundwater development can be drawn; they can be modified when more data becomes available as development proceeds. The construction of more refined models of aquifer behavior requires more data, well distributed over space and time.

Staged and gradual development. The fact that groundwater is withdrawn through wells, often located at the actual area of consumption, with each well adding an increment to annual withdrawal, makes it easy to develop groundwater stage by stage as needs arise, or according to a development plan. Only a relatively small investment is required at each incremental development. Surface water projects must, usually, be large in order to be economical. The economy of scale, important in surface flow regulative structures, is not a constraint in groundwater projects. In addition, a large storage dam may not be fully utilized for a long period after its construction, depending on the rate of consumption growth.

Legal and institutional aspects. Because large scale groundwater developments are relatively recent, legal and institutional framework regulating these developments is generally inadequate or nonexistent in many regions. This is so even in regions where this framework is well established for the management of surface water. Because of the interrelations between surface water and groundwater, and the recommended conjunctive use wherever and whenever called for, these two kinds of water should be incorporated within a single, unified legal and administrative framework.

In establishing the legal and administrative framework for the exploitation of groundwater, it is important to consider some of its basic features. The entire aquifer may be regarded as a single basin from the point of view of its water balance. In the long run, all consumers together cannot withdraw more than what is made available by the water balance, which takes into account all inputs (natural and artificial) and all outputs. Temporarily, excess of outflow over inflow can be allowed by reducing storage, i.e., decline of water levels. By pumping, each well produces a drawdown also in its vicinity and may affect the pumping of neighboring wells. The aquifer is also a basin with a certain internal flow pattern established by the pattern of pumping and recharge, both natural and artificial. Pollutants reaching the aquifer will be transported according to this flow pattern, and may reach areas at large distances from where they were originally introduced. In this way, many wells downstream of a source of pollution may be affected. All these factors call for basin-wide regulation of groundwater withdrawal, if aquifer sustainability is set as a goal.

Finally, the problems of groundwater quantity and quality cannot be handled separately, as they are interrelated. For example, seawater intrusion depends on the rate of freshwater flow from the aquifer to the sea, and the movement of contaminants depends on the flow pattern.

All these considerations lead to the conclusion that the management of an aquifer should be centralized and that it requires an appropriate legal and institutional framework. One cannot leave individual land owners to pump according to their needs, or to allow them to dump pollutants on their land.

1.1.4 Functions of aquifers

From the discussion presented above, it becomes obvious that beyond serving as just a source for water, an aquifer is a *system* that should be managed and operated as a unit to achieve various objectives. For example:

Source for water. This is the more obvious function. When an aquifer contains only water stored in it from the far past, usually under different climatic conditions, this water should be considered a *nonrenewable resource*. However, in general, an aquifer is replenished annually from precipitation over the region overlying it, or over its intake region (if it is confined). Thus, in general, groundwater is a *renewable resource*. Obviously only a certain part of the precipitation, depending on the distribution of storms, land topography and cover, permeability of soil, etc., infiltrates through the ground surface and replenishes the underlying phreatic aquifer. Aquifers can also be replenished from streams (with permeable beds) and floods. In many arid regions, aquifers in the low lands are replenished during a very short period once in several years, from flash floods originating in the mountains. Under natural conditions, a quasi-equilibrium situation is maintained with inflow equal to outflow. Part of the replenishment can be intercepted by pumping, thus reducing the outflow and establishing a new quasi-equilibrium. In this way the aquifer serves as a source for water.

Storage reservoir. Every water supply system requires storage, especially when replenishment is intermittent and is subject to random fluctuations. Demand may also fluctuate, e.g., seasonally. A large volume of storage is available in the void space of a phreatic (water table) aquifer. Just to give a rough idea, we can store 15×10^6 m^3 of water in a portion of an aquifer of 10 km $\times$ 10 km, with a storativity of 15%, raising the water table by 1 m. Using the technique of artificial recharge, large quantities of water can, thus, be stored in a phreatic aquifer. By doing so, water levels rise and outward (from the recharge area) gradients are established. These cause the stored water to spread over ever increasing areas and/or to leave the aquifer through its boundaries. In this way, if not used, the stored water is gradually lost. Nevertheless, due to the slow movement of groundwater, and with appropriate management, these losses can be minimized. A possible management procedure is first to lower the water table by pumping in excess of natural replenishment, withdrawing the volume of water as a *one time reserve* stored in the aquifer between the initial and planned water levels, and then use the dewatered void space for storage. In this way, storage is provided without raising the water table to excessive elevations. Sometimes, we even start by producing a crater in the groundwater table and then filling it up for storage. Again, losses from storage are minimized (recalling always that losses exist also for surface storage by evaporation and seepage).

An aquifer can be used for long term storage, e.g., from a wet sequence of years to a dry one, for seasonal storage, from the rainy season to the dry one, or even for shorter periods. The selection of the type of storage, the right combination of surface and underground storage, etc., depends on local conditions, economics, etc.

Conduit. Using the techniques of artificial recharge, water can be introduced into an aquifer at one point and be withdrawn by pumping at another (or at several other points). The injected water will flow through the aquifer from regions of high water levels, produced by the recharge, to region of low water levels, produced by the pumping. The rate of flow will depend also on the aquifer's transmissivity. In this way, large distribution systems, say to individual consumers spread out over large areas, may be avoided. Obviously, there is a limit to the permissible rise in water levels, as well as to permissible drawdown; these impose a limit on the use of an aquifer as a conduit.

Filter plant. Using the techniques of artificial recharge, an aquifer may serve as a filter and purifier for water of inferior quality injected into it. This may take several forms:

- By recharging an aquifer (through infiltration ponds) with surface water containing fine suspended load, we remove the fine suspension by the time the water reaches the water table. The bottom of the pond and the soil column act as a filter.
- The subsurface in the unsaturated zone and the aquifer material act as huge *chemical reactors*. Various chemical species may be removed from the flowing water by chemical reactions, by adsorption and by ion exchange phenomena on the solid surface of the porous matrix, especially when clay colloids are present. Obviously, by these reactions, the chemical species are removed from the water, but they remain on the solid. If these chemical species are considered as contaminants, the fact that they remain in the aquifer means that clean-up (remediation) activities will still be required in order to remove them from the aquifer. Certain chemical species will undergo natural attenuation and decay as they move through the aquifer. Often, such attenuation is associated with and enhanced by biological activities. Of special interest is the reduction in organic matter content as well as the removal of taste, or bacteria and viruses, especially if the flow is of sufficient length and duration. Sometimes, minerals are added to the water by dissolution.
- Mixing of injected water with the indigenous water of an aquifer is achieved by their simultaneous movement in the aquifer due to both the mechanism of hydrodynamic dispersion and the geometry of the flow pattern.
- Pumping near a river induces recharge from the latter into the aquifer. Filtering of the river water and purification are achieved by the flow through the aquifer material from the river to the wells.

In each case, the ability of an aquifer to improve the quality of water depends on the chemical and physical properties of the aquifer material and on the type of mineral and organic impurities contained in the water, taking into account that a significant part of the removal of impurities takes place at the phase of entry into the soil material (that is, bottom of an infiltration pond, or vicinity of a recharge well).

Control of base flow. This can be achieved in springs and streams by controlling water levels in the aquifer supplying water to them.

Water mine. We have already mentioned above the possibility of mining a *one time reserve* stored in an aquifer between some initial phreatic surface and a planned ultimate one. The same is true (using some coefficient of efficiency of mining, due to hydrodynamic dispersion) in the case of the advancement of saltwater in a coastal aquifer toward its planned position.

In general, the yield of an aquifer is a long-term average of part of its replenishment (renewable resource). However, under certain conditions, albeit very rarely, we may plan to completely mine an aquifer (like any other nonrenewable resource), not worrying about what will happen once this source has been depleted. In this case mining is, usually, based only on economic considerations.

We have, thus, summarized the roles that groundwater can play in the management of regional water resources. We have also suggested that the aquifer be considered as a system which can perform different functions to achieve desired goals, and we have analyzed these functions. More about aquifer management, will be presented in Chap. 11.

1.1.5 Subsurface contamination

Liquid contaminants, whether as a toxic liquid or as a solution of a toxic chemical species dissolved in water, are sometimes spilled at ground surface, intentionally or by accident. Once released, a liquid contaminant will percolate downward through the unsaturated zone, eventually reaching an underlying water table. Various biological and chemical processes may take place along its pathway through the unsaturated zone, prior to reaching the underlying aquifer. Once reaching the aquifer, the liquid contaminant will be transported through the aquifer to the latter's outlets. Along its way towards the aquifer's outlets (rivers, lakes, springs, or pumping wells), the contaminant's concentration will be gradually reduced by various processes, biological, chemical and physical (see Chap. 7). In many cases such contaminants may render groundwater useless for most purposes. Hence, management of aquifers must include both management of water quantity as well as of water quality. Obviously, the best strategy is to make every possible effort, e.g., via appropriate technological means, legislation and education, to prevent contamination. However, when pollution does occur, unfortunately too often, remediation of the polluted subsurface is called for. Some major remediation technologies are presented in Sec. 7.10. In what follows, we shall review various sources of subsurface pollution.

Anthropogenic materials are ones that are introduced into the environment primarily, or exclusively, by human activities. They include inorganic and organic chemicals used for agricultural, industrial and domestic purposes. Such materials usually dissolve in the aqueous phase—the water. They may also enter the soil from leaky storage tanks, pipes, and sewers, from landfills, and from evaporation ponds. Some materials, such as fluoride and arsenic, occur naturally in the aqueous phase in certain regions. Heavy metals may originate in wastes from industrial processes, such as metal plating or from 'spent' organic liquids used for cleaning metallic products. Altogether, infiltrating water from precipitation is seldom 'pure'; it contains various chemicals that are present in the atmospheric air. Chemicals present above ground surface, e.g., in an open air storage facility, may also serve as sources of dissolved matter in the infiltrating water. Surface runoff comes in contact, dissolves and carries various chemicals present on ground surface. Part of this runoff infiltrates through ground surface, thus contaminating water in underlying aquifers.

In addition to water (= aqueous liquid phase), certain nonaqueous phase liquids (NAPLs), e.g., hydrocarbons and organic solvents, may, sometimes, be present in the unsaturated zone. Most often, the source is a spill, leakage from pipes and storage reservoirs at ground surface, or from improper waste disposal facilities. Although small quantities of such materials do dissolve in the aqueous phase, these two phases are usually assumed to be *immiscible.* However, since the acceptable concentrations of these chemicals in drinking water is often less than 5 parts per billion (ppb), or even less than 10 parts per trillion (ppt), even very low concentration may be significant as far as water contamination is concerned.

Chemical substances present in the water may react with the soil solids, or with other chemicals present in the indigenous water in the pores. This may result in the precipitation of various chemical substances, with the possibility of changing the physical nature of the soil.

A. Sources of groundwater contamination

Contaminants in groundwater may take different forms. For example:

- **Pathogenic organisms**, originating from poorly constructed septic tanks, or from improper disposal of waste from hospitals.
- **Inorganic contaminants**, e.g., as increased levels of chloride, sulphate, nitrate, and sodium ions, originating from landfills of domestic waste. Cyanide, arsenic and heavy metals may also be present in the leachate from landfills, when containing waste from sources other than domestic. We may include here also radionuclides in uranium tailings, and nuclear waste in storage sites and in processing plants.
- **Organic contaminants**, which originate mainly from industrial wastes and spills. Examples are chlorinated hydrocarbons (e.g., chloroform, carbon tetrachloride, trichloroethylene), and aromatic hydrocarbons (e.g., benzene, naphthalene).

Our objective in this section is to describe and discuss major sources for subsurface contamination. Both hazardous and non-hazardous wastes, as well as non-waste substances released to the subsurface, constitute actual or potential sources of subsurface contamination. In some cases, sources are the consequence of installations that are designed specifically for the purpose of discharging substances to the subsurface. Unfortunately, too often, waste and non-waste materials are released to the subsurface unintentionally.

B. Classification of contamination sources

The Office of Technology Assessment (OTA) of the US Congress, has categorized (OTA, 1984) sources known to have contaminated the subsurface, according to the nature of their release characteristics:

Category I: Sources designed to release substances. These include engineered structures for subsurface disposal of waste by percolation (e.g., septic tanks, and cesspools), injection wells (for disposal of hazardous and non-hazardous materials, e.g., brine), and surface application (e.g., disposal of wastewater by surface irrigation, or dumping of wastewater sludge). Solution mining and in-situ mining activities are also included in this category.

Category II: Sources designed to store, treat, and/or dispose of substances, as well as discharge resulting from unplanned release. Here the OTA report lists, primarily, *landfills* for municipal, and industrial (hazardous and non-hazardous) waste. However, included in this category are also open (legal and illegal) waste dumps, surface impoundments of hazardous and non-hazardous liquid wastes, waste tailings, (non-waste) material stockpiles, graveyards and animal burial sites, above-ground storage tanks for hazardous and non-hazardous waste materials, containers for all kinds of non-waste materials, and disposal sites for radioactive materials.

Category III: Sources that retain substances during transportation or transmission. Pipelines and material transportation and transfer operations of hazardous and non-hazardous materials, as well as non-waste materials are included in this category.

Category IV: Sources that discharge substances to the environment as part of various planned activities. Examples are: irrigation practices, pesticide and fertilizer applications, animal feeding operations, application of de-icing salts, urban runoff, structures for infiltrating storm water precipitation (carrying atmospheric pollutants), surface and subsurface mining, and mine-drainage operations.

Category V: Wells and construction excavation. Included here are (oil and gas) production wells, geothermal and heat recovery wells, water supply wells, monitoring wells, and exploration wells.

Category VI: Naturally occurring sources whose discharge is created or exacerbated by human activities. This category includes water infiltrating from precipitation and carrying atmospheric pollutants, natural leaching, saltwater intrusion, and encroachment of poor quality water as a result of man-made changes in the flow regime in an aquifer.

Another way of classifying sources is according to their geometry:

Point sources. Here, 'point' has to be interpreted as 'of small areal extent', relative to the subsurface domain under consideration. For example, a large landfill may be considered as a 'point source' relative to an underlying aquifer contaminated by the leachate from the landfill, once the plume of contaminants has reached a distance which is much larger than the dimensions of the landfill itself. An oil spill from a ruptured pipeline, leaks from an above-ground, or buried tank, a radioactive waste repository, and a septic tank, may serve as additional examples of point sources. However, if we have a large number of septic tanks within a certain area, we may average (or integrate) their effect, and regard the source created by them as a distributed, rather than point, source.

Distributed sources, also called **non-point** or **diffuse sources**. Here a source extendsover a large horizontal area relative to that of the contaminated subsurface. The application of pesticides, herbicides, and fertilizers in agriculture are examples. Another example is the infiltration of rainwater carrying atmospheric pollutants (e.g., acid rain).

As already emphasized in Subs. 1.1.1, liquid-borne contaminants originating from sources at or near ground surface travel first through the *unsaturated zone* ($\equiv$ vadose zone), (primarily) downward to an underlying aquifer. The rate at which contaminants are transported to the aquifer depends on the quantity of the contaminant, and on the quantity and flux of the percolating water passing through the contaminated soil volume, dissolving and carrying the contaminants.

Of special interest is the case of a contaminant that constitutes a *non-aqueous liquid phase*, i.e., a liquid phase different from water. This case is discussed in detail in Subs. 1.1.5E.

In some cases, the moving contaminated liquid, once in the subsurface, may spread out horizontally, beyond the relatively small spreading produced by capillarity, because of *anisotropy* in soil permeability and/or because of the presence of horizontal lenses and strata of relatively low permeability.

Altogether, the issue of whether to regard a source as a 'point' source or a 'non-point' one depends on the scale of the problem. This scale may evolve in time, such that after a certain period, a plume originating from a diffuse source may be approximated as originating from a point one.

Sources may also be classified according to the temporal variation in their rate of release, e.g., a one-time, short duration spill, a slug over some time period, occurring once or repetitively, or a continuous release over an extended period of time. Some specific sources are described below.

C. Point sources

Following are examples of some of the more commonly encountered pollution point sources.

Septic tanks. These are used as a means of disposal of domestic sewage in many (especially rural) areas. Waste water enters first the septic tank, where scum, grease and heavier than water solids are removed by gravity segregation. The clarified liquid proceeds to the subsurface infiltration system, where it is discharged to the soil. In general, a properly designed and maintained septic tank should be regarded as an efficient and economical means of domestic sewage disposal. However, even when each unit in itself is well designed and maintained, a high density of septic tanks may exceed the natural ability of the subsurface environment to absorb and purify effluents, thus causing a degradation of groundwater quality due to release of bacteria, organic contaminants and nitrates. Also, because non-domestic sewage is diverted to them in many cases, and because of unfavorable soil and climatic conditions, septic tanks are considered a potential source for groundwater contamination.

Raw sewage contains biological contaminants (bacteria and viruses), inorganic contaminants (phosphorus, nitrogen, and metals), and organic ones (synthetic organics, pesticides, and hydrocarbons, which are compounds of hydrogen and carbon). Bacteria tend to be removed from percolating water through the physical process of straining and/or the chemical process of adsorption onto soil particles. Phosphorus movement is retarded by chemical transformations and adsorption. Ammoniacal-nitrogen is retarded primarily by adsorption, but can also be subject to cation exchange or incorporation into microbial biomass. Nitrate-nitrogen tends to be highly mobile, moving essentially with the water. Nitrates may also be removed through plant uptake and microbiological denitrification. Metals in soils can be rendered immobile, primarily by adsorption, but also through ion exchange, chemical precipitation, or complexation with organic substances. The transport and fate of organic contaminants is affected by volatilization, adsorption, incorporation into plants and microbial biomass, and bacterial degradation.

Storage tanks. In terms of the number of incidents (e.g., gasoline tanks and service stations), this is probably the major source of subsurface contamination in the USA and other industrialized countries. The tanks may contain hydrocarbons, organic compounds, and inorganic liquid chemicals. The main cause of leakage from steel tanks is corrosion.

Gasoline, an example of a hydrocarbon, actually contains dozens of different hydrocarbon compounds. It contains *aliphatic* compounds, such as *pentane* and *butane*, which are characterized by carbon atoms linked to an open chain, and *aromatic* compounds such as *benzene* and *toluene*, which are characterized by a ring structure of carbon atoms. Being lighter than water, gasoline, spilled in a sufficiently large quantity, will tend to create a lens which will float and spread out on the water table. Depending on the pressure and temperature prevailing in the subsurface, certain chemical species in gasoline may volatilize, and then diffuse within the gaseous phase. Because diffusion in a gaseous phase is much faster than in a liquid, the vapor of a volatile species may spread extensively. As such vapor spreads in the gaseous phase, it may reenter the water phase, in an attempt to achieve equilibrium between the two phases. Chemical species, such as *benzene*, *toluene*, *xylene* (or BTX), dissolve in water. These species may also be adsorb onto soil particles. Although the solubility level may be low, concentrations above permissible levels are often reached.

In many cases, the leakage is not from the tank itself, but from the inlet and outlet pipes and valves at loading and unloading facilities, and from trucks or tank cars while they are being cleaned.

Landfills (or **sanitary landfills**). If properly designed, maintained, and managed (including enforcement of their designation as solely for domestic refuse), landfills should not pose any threat to the environment. In practice, however, they do pose a serious threat to groundwater resources, due to the actual way in which they are operated and the sheer number of such facilities.

Leachate is generated by the liquids usually contained in the refuse, as well as by infiltrating rainwater, if the landfill is not properly capped. As infiltrating water percolates through a landfill, it removes and carries dissolved waste species. The chemical composition of the leachate depends on the nature of the (domestic and industrial) waste, on the rate of leaching, and on the age of the landfill. Pressure, temperature, and moisture conditions within the landfill may be favorable, or unfavorable to certain chemical reactions and biological activities, thus affecting the nature of the resulting leachate. The actual amount of leachate leaving the landfill depends primarily on the rate of infiltration from precipitation; it is, thus, higher in humid areas. Well designed landfills are constructed with an impervious base, say, a clay layer, and/or liners, which are designed to prevent infiltration of leachate from the landfill into the subsurface. Above this base, a specially designed drainage system collects the leachate, and conveys it to where it is properly treated and disposed. Capping a landfill with impervious material reduces infiltration. Unfortunately, even some modern landfills are not designed and operated according to these criteria. Over time, caps crack and allow infiltration. Liners, weathered by the presence of unanticipated chemicals in the water, become permeable. Synthetic liners that were stressed during construction develop leaks, and leachate collection systems can become clogged.

Uncontrolled hazardous waste dumps. Numerous such sites, many of them long abandoned, are scattered all over the USA and, indeed, the world. They contain a mixture of many kinds of refuse, as well as solid and liquid toxic waste materials. Many sites are operating illegally. In most cases, the nature and the quantity of the disposed materials are unknown. Sometimes, sites are abandoned, making the determination of ownership and responsibility difficult and, often, impossible. Extensive exploration programs are usually required in order to identify such contamination sources and to plan cleanup operations.

In recent years, developing nations have legally accepted hazardous waste materials from industrialized nations, or from large corporations, to spur their economies. Sooner or later, they will constitute severe hazard to the environment and to human health.

Surface impoundments of liquids. Referred to as pits, ponds, lagoons, and basins, these are natural or man-made depressions, lined or unlined, from a few feet in diameter to hundreds of acres in size, that are used for impounding various kinds of liquids. One example is the impoundment of municipal wastewater, including *waste stabilization ponds*, and *aerobic, anaerobic*, and *facultative* lagoons.

Surface impoundment, often called 'evaporation pits', are encountered as a part of the production process in many industries, including food and poultry processing, refining of hydrocarbons, petrochemical and chemical production, mining, and oil and gas production. They become sources of contaminants when the impounded liquids leak into the subsurface. In the case of sewage, the contaminants may be ammonia, nitrates, nitrites, phosphates, and biological components, such as bacteria and viruses.

Abandoned wells. In many areas, oil and gas wells are abandoned when production becomes uneconomical. Improperly sealed abandoned wells may serve as avenues for contaminants. In addition, over the years, many water supply, waste injection, and mineral resource extraction wells have been abandoned without proper sealing. With time, the casing of such a well will corrode, thus providing connections between formations with clean water and ones with low quality or polluted water. Leakage of contaminated or highly mineralized water through abandoned wells and unplugged exploration boreholes has also led to groundwater pollution.

Disposal of drilling mud. Wells are often installed in holes created by drilling. A slurry (or mud) composed of water, clay, and certain chemical additives is used in rotary and some other drilling technologies. Various chemical additives in the slurry may be pollutants. If drilling is taking place through such a polluted formation, the mud may also pick up contaminants and may endanger groundwater resources.

Additives of potential concern from the pollution point of view are, usually, *ferochrome, lingosulfonate*, and *lead compounds*. In addition, the sediments in the mud consist of fine solid particles and interstitial water of very high ionic strength. The water may contain such constituents as chromium, barium, lead and arsenic.

Transportation accidents. Spills may take place as a result of accidents that occur in the process of transporting toxic liquids by trucks or trains. Usually, the quantity associated with a single vehicle is relatively small, although, in train accidents, the spilled quantity may be much larger. The infiltrates and migrates downward (and laterally by capillarity). Even if the quantity is not large, and spilled liquid migration stops after a certain volume of soil has been contaminated, subsequent percolating water will dissolve and carry the contaminants to underlying groundwater, unless the contaminated soil has been removed. If the chemical is volatile, its vapor will diffuse and may contaminate an increasingly larger soil volume.

Storage of hazardous solid chemicals. If not properly stored and handled (including loading and unloading), and especially if exposed to precipitation, such facilities may constitute potential sources of pollution.

Tailings of coal, copper, uranium and other mines. These are waste mounds of excavated material from mining operations, with no economic value. When infiltrated by water from precipitation, and exposed to atmospheric conditions, this material produces leachate that contains toxic and hazardous compounds that may contaminate surface water and groundwater.

D. Distributed sources

Agriculture. Many agricultural activities produce potential sources of groundwater contamination. Among such sources, we may mention pesticides and herbicides, fertilizers, animal feed and waste, irrigation, and plant residues. The first three are associated with nitrate (NO_3^-), which is the most common environmental form of nitrogen, because it is the end product of the aerobic biological process called *nitrification.*

Fertilizers (chemical and manure) constitute a serious danger to the subsurface, both at the handling stage (transportation and storage) and when applied in the field. Irrigation, especially excessive irrigation, may leach and transport significant quantities of nitrogen fertilizers, in the form of nitrate, to underlying groundwater.

Animal raising activities produce contaminants that include nitrogen compounds, phosphates, chlorides, bacteria, and sometimes heavy metals (Ba, Cd, Cr, Cu, Pb, Ag, and Zn).

Irrigation water often contains high levels of total dissolved solids (TDS), especially salts. By taking up water, plants raise the concentration of TDS in the water remaining in the soil within the root zone. As this concentration rises, soil productivity is reduced. When the TDS concentration becomes too high to be flushed by infiltration from precipitation, the salinity can be flushed by excess irrigation. Obviously, the flushed salts end up in the underlying groundwater. Also, as part of the usual irrigation practice, the amount of water applied is often in excess of *field capacity* (Subs. 6.1.9), so that part of it (called 'return flow from irrigation') continues to migrate downward, carrying whatever salinity is contained in the water.

Acid precipitation. *Acid rain* and other airborne contaminants, constitute a source for groundwater contamination. The danger is due not only to the mere fact that certain polluting constituents, e.g., sulfates contained in acid rain, are introduced into groundwater; as the acid water travels through the vadose zone, it will increase leaching and mobilization of other chemical species. It is often found that sulfate and aluminum content increases in acidic groundwater. The mobility of heavy metals may also be increased.

E. Non-Aqueous Liquid Phase (NAPL)

Too often, a contaminant in the form of a third fluid phase—a liquid that is *practically* immiscible in water—is introduced into the air-water system of the unsaturated zone through ground surface. Organic liquids, e.g., hydrocarbons and various organic solvents, such as TCE (*Trichloroethylene*), used in industry, may serve as an examples. Another example is pesticides, such as DBCP, which are highly toxic. They are introduced into the unsaturated zone as a result of spills, leaks from faulty storage tanks or pipes, leakage from corroded drums, and burst of pipelines.

Referring to water as the *aqueous phase*, this third fluid is often referred to as a *nonaqueous phase liquid*, abbreviated, NAPL. It is sometimes referred to as *free product.* In this book, we shall sometimes use 'oil' as a typical example of NAPL. The term 'oil' will then be used generically, without any implied meaning regarding its actual composition. Thus, when a NAPL is introduced into the unsaturated zone, the latter becomes a three-phase flow domain. Some brief comments on three fluid phases are presented in Sec. 6.5.

When its quantity is small, the NAPL will move and spread out in the unsaturated zone, occupying part of the void space within a certain domain, jointly with water and air. If the quantity of NAPL is sufficiently large, this soil domain will continue to expand, primarily downward, until reaching an underlying water table.

In the subsurface, certain species of an organic liquid may dissolve in the water moving in the void space (e.g., from infiltration). In this way, these species will reach and contaminate an underlying aquifer. Volatile species may evaporate and spread out by diffusion in the gaseous phase occupying part of the void space.

Sharp (microscopic) interfaces are maintained between a NAPL and both the gaseous and the aqueous phases within each pore. Some NAPLs, like gasoline, are less dense than water, while others, such as *chlorinated solvents*, are denser than water. When the NAPL is less dense than water, it is called 'light NAPL', or LNAPL, and, conversely, when it is denser than water, it is called 'dense NAPL', or DNAPL. NAPLs may be pure organic compounds, or, like gasoline, complex mixtures of a large number of compounds. The various NAPL components may dissolve in the aqueous phase in small quantities, each according to its own water solubility. Thus, the term 'immiscible' is used here in the sense that the water and the NAPL are separated (inside the void space) by a sharp physical interface, despite the small amount of transfer of components between the phases. Unfortunately, some chemical species in drinking water may have deleterious effects on human beings, even at concentrations as low as a few parts per billion (ppb). Because the water solubilities of commonly encountered NAPL components are high, relative to those concentrations (e.g., TCE is soluble to about 1.1×10^6 ppb at 20°C), the solubility of NAPLs in water is an important factor in such subsurface contamination problems.

If a small quantity of NAPL is spilled on ground surface, it will percolate through the vadose zone, primarily downward. As the NAPL travels downward, it leaves behind isolated blobs and *ganglia* of NAPL. If the volume of the spill is not too large, and the water table is at a large depth, after some distance the movement of all the spilled NAPL will be arrested, creating a zone containing immobile NAPL at residual NAPL saturation. When percolating water (from precipitation) passes through this zone, a certain quantity of the NAPL dissolves in the water, and is carried away. In this way, a secondary source of NAPL pollution is created.

When the quantity of spilled LNAPL is larger, the downward moving LNAPL may reach the capillary fringe and form a lens floating and moving according to the prevailing hydraulic gradient. The moving lens leaves

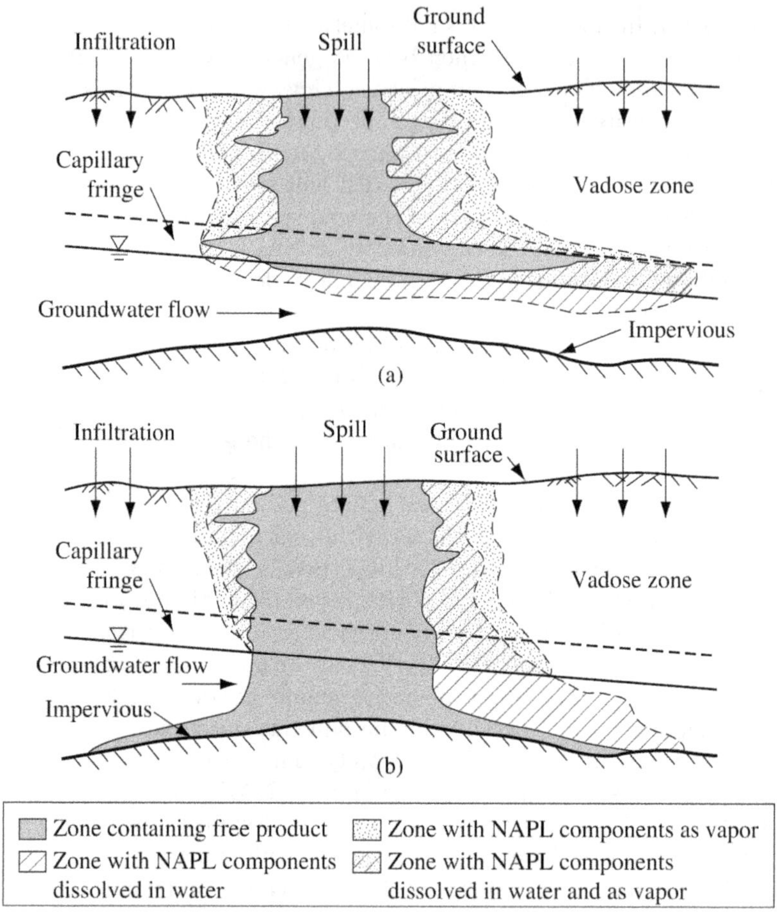

Figure 1.1.2: NAPL spills in large volume: (a) LNAPL, (b) DNAPL.

behind a trail of ganglia at residual LNAPL saturation. Eventually the floating LNAPL lens may become immobile at residual LNAPL saturation. Water passing through this domain of immobile water will be contaminated. Again, the remaining LNAPL acts as an immobile source for a long time, until all LNAPL has been dissolved.

Figure 1.1.2a illustrates an example of a contamination event by LNAPL, e.g., gasoline. From its point of infiltration into the subsurface, the LNAPL flows essentially downward through the unsaturated zone simultaneously with water. Due to heterogeneities, such as layering and the occurrence of low permeability lenses, lateral movement of the NAPL often occurs. Some of the LNAPL will be retained along its flow path by adsorptive and capillary forces at the *residual saturation* (Subs. 6.2.3). As long as part of the LNAPL is at a saturation that exceeds this critical value, it will continue to move until it reaches a zone of complete water saturation, leaving behind porous medium subdomains at the residual LNAPL saturation. For a deep water table, and a relatively small LNAPL quantity, the entire volume of LNAPL may, at some stage, become immobile at the residual saturation, occupying a certain domain within the vadose zone.

If the LNAPL's quantity is sufficiently large, it will continue to move downward until a water table is reached. Being lighter than water, the LNAPL will form a 'pool' or 'lens' that floats on top of the zone of full water saturation. Actually, LNAPL and water can coexist within the lens. In this zone, both fluids are moving in response to pressure gradients and to gravity.

Some components of the LNAPL are volatile, while others may be soluble in water to varying degrees. Hence, within the pore space, LNAPL components will evaporate into the gaseous phase (air) and dissolve into any adjacent water phase. Of major interest is the dissolution of NAPL components in water percolating, essentially downward, from precipitation. LNAPL components will also dissolve in the water present within the lens. Chemical and biological transformations of the LNAPL may also occur.

When a LNAPL lens floats on the water-saturated zone, it flows predominantly by gravity along the (sloping) water table. Remedial schemes take advantage of this fact. By pumping water from the water-saturated zone, a drawdown cone is formed, into which the LNAPL also flows.

The case of DNAPL is shown in Fig. 1.1.2b. The transport of the DNAPL through the unsaturated zone, and the dissolution and volatilization of components of it along its path, are similar to those of the LNAPL as described above. Residual DNAPL will be retained in the vadose zone following a small quantity spill. However, for a large DNAPL spill, because its density exceeds that of the water, as the DNAPL reaches the water table, it will continue to percolate downward through the water saturated zone, under two-phase (water-DNAPL) flow conditions, leaving behind residual DNAPL saturation in the form of isolated ganglia. As in the case of LNAPL, these ganglia, due the dissolution of DNAPL components in water, constitute a source of pollution both for the water percolating through the vadose zone, and for the essentially horizontally moving groundwater in the saturated zone.

If there exists a sufficient amount of DNAPL, it will continue to percolate downward until it encounters an impervious formation, e.g., the impervious bottom of the aquifer. Upon reaching such surface, it will accumulate above it in the form of a lens, or pool. The lens will move along the impervious surface, primarily in the direction of the downward slope of the surface, due to gravity. A pool of DNAPL may also be formed at the interface between two soils having different *entry pressure values* to the DNAPL (Subs. 6.2.3). If the entry pressure value of a non-invaded zone (where no DNAPL has yet entered) is zero, then pooling will occur until a pressure sufficient to cause entry is developed, or until the miscible components of the DNAPL dissolve at the interface, leading to reduced interfacial tension and, hence, a reduction in this pressure barrier to invasion by the DNAPL.

1.1.6 Sustainable yield

In what follows, we shall discuss the concepts of *safe yield*, and *sustainable yield*, and try to highlight issues and difficulties associated with their determination. The presented discussion should be sufficient for their determination.

Two preliminary questions may be asked:

(a) Is the considered aquifer an isolated source of water for a well-defined population, such that this is their sole source of water? Or, do other sources of water exist within the region, or within a reasonable distance, e.g., surface water, that can be managed conjunctively with the aquifer with the objective of supplying the population's demand? In addition to water for domestic use, demand should also take into account water required for the conservation of nature and landscape, as well as water for economic activities in industry, agriculture, commerce, tourism and recreation.

(b) Does the management of this aquifer include the option of artificial recharge (Sec. 3.4), i.e., the option of importing water from sources external to the region, or even surface water within the region (in fact, also desalinized or treated water), and using it to recharge the aquifer?

For the sake of simplicity, let us continue under the assumption that (1) the considered aquifer is an isolated source of water and (2) the extracted water is supplied only to the local population (i.e., not for export to other regions, as an economic commodity). An additional assumption is that the aquifer is managed centrally. This means, for example, that the volume of water extracted by individual well owners is controlled according to some agreed (by the population) aquifer management scheme.

Given such a scenario, the question 'what is the yield of the aquifer?', or 'what is the maximum quantity of water that can be extracted, say, annually, from that aquifer?' is always raised, as the answer is essential for the aquifer's management. Obviously, it is impossible to manage an aquifer, as a source of water, without knowing its yield. We should add that the yield concerns not only water quantity, but also water quality and reliability of supplying the promised quantity and quality, with the latter complying with prevailing standards.

Next we ask: 'do we make the fundamental assumption that the aquifer should be maintained as a viable, or *sustainable* source of water *forever*'. The opposite option is to *mine* water from the aquifer for a number of years, like an oil reservoir, until the source is fully depleted (either quantity- or quality-wise). In what follows, we shall not consider this option (but partial/temporary mining, while passing from one steady state regime to another, without destroying the aquifer as a sustainable source, is considered acceptable).

Another preliminary question is: 'should the yield be a constant value, independent of the fluctuating nature of the aquifer's natural replenishment,

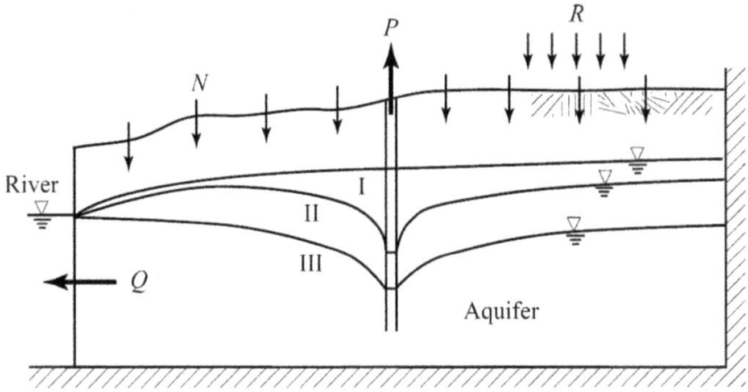

Figure 1.1.3: Yield of an aquifer.

or should we adjust it annually, say in response to changes in water table elevations?'. Obviously, to answer this question, we have also to consider (1) the nature of the consumers (e.g., can they tolerate fluctuations in the supply, or is their demand rigid), and (2) is the aquifer (and the consumers) part of a larger system, such that reduction of supply from one source, say the aquifer, can be compensated by increased supply from another source.

A starting point for this discussion is the understanding that the *yield of an aquifer* is, actually, a *decision variable*, i.e., its value may vary (at least, up to a certain upper limit) according to some management scheme that involves various objectives and constraints. As a simple example, consider steady state flow in an aquifer, isolated on the right side and connected to a river on its left side (Fig. 1.1.3). The aquifer is naturally replenished at an annual rate N, and artificially at a rate R. With no pumping and no artificial recharge of the aquifer, i.e., $P = 0$ and $R = 0$, so called 'virgin conditions', the discharge, Q, leaving the aquifer to an outlet, say a river, is $Q = N$, and the water table is in position I (Fig. 1.1.3). As pumping is increased, say to P_1, Q decreases to $Q = N - P_1$ and the water table drops to position II. Recall that *a water balance must always be satisfied*; in steady state, there is no change in storage in the aquifer. If we further increase P, the water table continues to drop, until, at some value of $P = P_2 = N$, we will have $Q = 0$. Beyond that P-value, say, $P = P_3 > N$, the water table will drop to below river level, as in position III, and Q will reverse direction. This shows that the extraction rate can exceed natural replenishment, still with a steady state maintained. However, if the water in the river is polluted, or there is a requirement to maintain a certain base flow to the river, these will serve as constraints on the permissible P. Pumping yield can also be constrained by restriction on water table elevations, e.g., due to legal considerations. So, what is the permissible yield? Altogether the allowed maximum abstraction will depend on the constraints imposed on the system, based on various hydrological and socio-economic considerations. Another example is the case of the coastal aquifer discussed in Chap. 9, and especially in Subs. 9.2.5A, where the rate of pumping is shown to affect the extent of seawater intrusion, thus making the latter a decision variable.

However, constraints need not be related only to water levels. For example, all aquifers, especially water table aquifers, are continuously threatened by pollution originating from human activities at ground surface, reaching the aquifer after passing through the unsaturated zone. The passage through the unsaturated zone may take some time, often a long time, depending on the depth of the water table and on soil properties. To mitigate groundwater contamination from such sources, we should allow a certain volume of water, $Q_{\min}$, to continuously leave the aquifer, carrying with it (= flushing) pollutants (assuming that there is enough mixing in the aquifer). This means that the yield is also constrained by the need to maintain a certain minimum value of outflow, $Q_{\min}$, to ensure water quality. Obviously, another option that will achieve the same goal is to undertake a comprehensive approach that protects the land above the aquifer against pollution, as a part of a comprehensive approach to regional management, incorporating the management of land use with that of the underlying water resources.

One should note that if we aim at a constant annual yield, the latter is not simply equal to the long term average of the natural replenishment minus the long term average drainage required for flushing purposes. The reason is that it is not *a priori* obvious that the aquifer can store all excess water of rainy years for use in dry years; this has to be tested and proven, say, by appropriate modeling.

It may be interesting to ask: 'if a decision to increase annual pumping will (eventually) lead to a new lower steady state water table, where did the water that occupied the pore space between the initial water table and the new lower water table go?' Obviously, it was drained to the outlet (e.g., river, or sea) during the transition from the initial water table to the new one. If there exists no prior obligation/constraint to discharge this water volume to the river, why not pump it (in excess of a yield that is based on natural replenishment alone)? We refer to this scheme as 'pumping a one-time reserve'.

Let us now allow the import of water from some external source (and/or surface water in the same area) and use it for artificially recharging the aquifer at an annual rate R (obviously, if a management decision is made to recharge and then pump, rather than supplying the imported water directly to consumers). To maintain a water balance, we can, in principle, increase pumping (above the yield) by the amount of recharge, without lowering water table elevations. On the other hand, if we recharge the aquifer, but we do not increase the pumping rate by the same amount, water levels will rise and so will Q. Thus, artificial recharge can be a tool for increasing the yield of an aquifer, only if it is properly incorporated in management considerations. Obviously, the pumping and the recharge (e.g., timing, spatial distribution, and water quality) must be coordinated. Just as an example, if we recharge an aquifer, but do not pump in excess of the yield that corresponds to no recharge, then the recharged water will gradually be drained out of the aquifer, and we shall gain nothing by the fact that we have artificially recharged the aquifer. Similarly, water quality of pumped water is generally some average of the quality of the recharged water and that of the indigenous water in the aquifer. Obviously, in this preliminary discussion, no attention has been given to economic considerations and constraints, although these may play a significant role.

In addition to induced recharge (Sec. 3.5) of polluted water from a river, there may be other sources of inferior water quality (e.g., high salinity) within an aquifer, or along its boundaries with adjacent aquifers, such that this inferior quality water may be mobilized by lowering water levels, and eventually pollute the aquifer and the pumped water. Obviously, this may affect the aquifer's safe yield.

So far, we have discussed the relatively simple case of a steady state flow. However, how can we determine the yield in view of the fact that natural replenishment varies from year to year? One way is to take some long term average replenishment as a base for estimating aquifer yield; this is allowed, provided we are sure that the aquifer itself will store the water from rainy years for use in drier years. To examine this point, as part of a management procedure, we run models (Sec. 1.2) in order to discover the highest value possible for a constant aquifer yield, say, over a design period of 30 years. Or, we can consider a yield that varies, say, from year to year, as a function of the volume of water stored in the aquifer (as manifested by water table elevations), and run the model for the same period.

Up to this point, the discussion has been based on the assumption that the aquifer should be preserved as a source of water for any foreseeable future. This underlying guideline should be compared, *using socio-economic criteria*, with (the opposite) approach that regards the aquifer as a *mine of water*, such that water, similar to oil, can be extracted until the aquifer is 'empty', or destroyed, say, by the invasion by water of inferior quality, e.g., seawater in a coastal aquifer.

At this stage, let us define 'sustainable yield' of an aquifer (again, in fact, of any water resources system) as

> *the maximum volume of water that can be annually withdrawn from the aquifer, for an indefinite period, without causing any undesired effects.*

In the past, hydrologists used the term 'safe yield' to indicate that the yield should be determined such that the aquifer as a source of water will be sustained forever. Its determination is based on regional water balances for the considered aquifer, using as input from precipitation a synthetic sequence (or synthetic sequences of various probability of occurrence) based on the statistics of precipitation. A design period of 30 years is usually used.

To emphasize the underlying (and indisputable) requirement that an aquifer should serve as a source of water that must be sustained for an indefinite period of time, we shall, henceforth, use the term 'sustainable yield'. We can include the 'impairment of sustainability' as one of the undesired effects.

This brings us to the issue of 'undesired effects'. Obviously, as stated above, a primary undesired effect is that the aquifer will not continue to serve as a source of water for the needs of the population (and nature). The requirement of satisfactory water quality is always included. Among 'undesired effects' ($\equiv$ constraints), we may include hydrological, social, economic, legal, and political criteria. Following are a few examples:

- Water levels everywhere, or at specific locations, should not rise above specified maximum elevations.
- Water levels everywhere, or at specific locations, should not drop below specified minimum elevations.
- The discharge of a spring, or base flow in a stream fed by groundwater emerging from the aquifer, should not drop below a specified minimum.
- The concentrations of certain chemical or biological species in the water, pumped at specific locations, should not exceed specified threshold values.
- Land subsidence (caused by lowering water levels) should not exceed specified values.
- The length of the seawater wedge intruding into the (coastal) aquifer should not exceed a specified value.
- If artificial recharge is implemented, the residence time for recharge water in the aquifer, before being pumped, should exceed a specified minimum period.
- If parts of the same aquifer belong to different political entities, there may be a constraint on the volume of water pumped by each of them, or on the drawdown produced in one region by pumping in the other.
- The cost of pumping (or supplying) water should not exceed a specified value.
- The total volume of water pumped from the aquifer should be such that certain portions of it will be pumped from certain portions of the aquifer.

Terms like 'optimal yield' or 'operational yield' are also often used. The term 'optimal yield' is used for that value of the annual yield, whether constant or variable, which, when pumped, achieves a certain objective. For example, the optimal yield could be such that it maximizes the total net benefits from operating the system (or present worth of the costs and benefits, if timing is taken into account). Other examples are the minimization of the cost of a supplied unit volume of water, or the minimization of the amount of energy consumed by pumping. Operational yield refers to the annual volume to be pumped in a specific year/season, taking into account the current water levels and pollution concentration in the aquifer, the possibility of importing water for artificial recharge, population (possibly growing) demand, costs of various operations, and, perhaps, other economic factors. It is an approach which, perhaps, is more suitable for a large system, with many (both surface and groundwater) sources, with different kinds of consumers, etc.

Altogether, we accept the concept of sustainable (annual or seasonal) yield, whether in the form of a constant value or as an algorithm that will provide the yield as it varies, say annually, in conjunction with various parameters, and constraints.

Recall that up to this point, we have been discussing the yield of an aquifer as a whole, overlooking the fact that eventually, the implementation must be through spatially distributed wells, although, to some extent, the spatial distribution of wells may also affect the resulting yield.

It is often emphasized that socio-economic considerations must be taken into account when determining the yield of an aquifer. In fact, the sustainability of the aquifer may be regarded as the major socio-economic constraint. However, many other socio-economic considerations can also be expressed in the form of constraints. If we accept the notion of seeking an optimal value for the sustainable yield, then such considerations can also be expressed within the framework of management objective functions (Sec. 11.1).

With the above consideration, how do we determine the sustainable yield of a considered aquifer? We continue to focus attention on a single aquifer or part of an aquifer. As discussed above, the primary tool (one may call this a constraint) is that no matter what we decide as the yield, the water balance of the considered aquifer must always be satisfied. Using hydrological information and data about the considered aquifer, we can express the water balance in the simple form (3.10.1). This balance equation is written for the entire aquifer. Note that all terms are in volume per year. As presented throughout this book, the balance equation can also be written in the form of a partial differential equation for a point in the aquifer.

The various terms appearing in this equation are discussed in detail in Chap. 3. Let us refer to them briefly:

- *Groundwater inflow and groundwater outflow* are the rates of inflow and outflow through aquifer boundaries. This information can be obtained from contour maps and information on aquifer permeability or transmissivity. Estimates of local values of these parameter can be obtained from pumping tests. However, since the aquifer, and the subsurface in general, are usually very heterogeneous, values of transmissivity and their spatial variations are obtained through the process of model calibration (Step 7 in Subs. 1.2.2 and Sec. 11.3).
- *Natural replenishment* (N) is that part of the precipitation that infiltrates through ground surface, percolates through the unsaturated zone and reaches an underlying water table. There is no way to actually measure this value over large areas. Instead, there exist various kinds of models that estimate this value from information on precipitation. Because of the uncertainty involved in these estimates, the value of N, or the coefficients that relate precipitation to N, are obtained as calibration parameters.

- *Return flow* is a term used for that part of the water used for irrigation that is not consumed by the vegetation, but infiltrates and reaches the water table. Again, we can relate return flow to the (measurable) volume used for irrigation through certain parameters and obtain estimates of these parameters in the process of model calibration.
- *Artificial recharge*, *inflow from lakes and streams* and *spring discharge* require no further explanation. Artificial recharge is a measured quantity. Inflow from lakes and streams has to be estimated as function of the difference in water levels, and coefficients (= calibration parameters) that express soil properties and geometry.
- *Evapotranspiration* is the term used for water lost to the atmosphere by evaporation from the water table and transpiration through the vegetation. This loss, which is of significance only in the case of a shallow water table, has to be estimated.
- As a consequence of excess of all inflows over outflows, the water table (or the piezometric head) will rise, thus storing this excess. Here, the relevant coefficient is aquifer storativity (S or S_y), again an unknown coefficient that has to be estimated in the process of model calibration.

Altogether, on the basis of water level observations during a sufficiently large number of years, we obtain a set of annual balance equations and use them to derive the unknown parameters and coefficients for the considered aquifer. Although there exist a number of software packages that help in performing this process of parameter estimation, often, in practice, calibration is performed by a process of *trial and error*, in which, basically, we assume a set of coefficients and examine whether, by using them in a model, we can reconstruct ('predict' the observed past) the measured water levels.

The main outcome of such process is information on the various parameters that enable us to estimate the replenishment and its relationship to precipitation (on which much more data are available), and aquifer coefficients that are required for modeling the flow regime in the aquifer.

Once we have a calibrated model of the considered aquifer, we can run models for various values of aquifer yield, predict water levels (which enable us to estimate groundwater flow and storage changes) and examine whether or not that yield causes the violation of imposed constraints. In this way, we can derive (or estimate) the sustainable yield of the aquifer. If quality of water is a constraint, we also have to run calibrated solute transport models. Altogether, by using appropriate flow and solute transport models, we can also derive the sustainable yield as a variable that depends on the varying elevations of the water table in the aquifer. We can also investigate artificial recharge (using imported water) as a tool for increasing the sustainable yield.

1.2 Modeling

The management of a groundwater resources system, which can be an aquifer, or a system of aquifers, alone or conjunctively with surface water sources, aims at achieving certain goals through a set of decisions concerning the development and/or operation of the system. A more detailed discussion on the management of groundwater resources, particularly the quantitative analysis involved, is presented in Chap. 11. The remarks in the present section are aimed at providing the background needed for the discussion on modeling of the physical and chemical behavior of water in aquifers, presented throughout this book. Particularly, we wish to emphasize the need for employing the various models introduced in this book as essential management tools.

The modeling concept and process are elaborated in the following sections.

1.2.1 Modeling concepts

A model may be defined as a *selected simplified version of a real system and phenomena that take place within it, which approximately simulates the system's excitation-response relationships that are of interest.* For example, a groundwater system may be 'excited' by pumping, by the introduction of a contaminant, by artificial recharge, or by changing a boundary condition. Its 'response' takes the form of spatial and temporal changes in water levels and in contaminant concentrations.

With the above definition, we emphasize that all that a model can do is to predict the future behavior of an investigated, say aquifer, system. However, this information may be used in different contexts and for different purposes. Modeling activities may be conducted to achieve any of the following objectives:

(a) To predict the behavior of a system, say, an aquifer, in response to excitations that stem from the implementation of management decisions.
(b) To obtain a better understanding of a system from the geological, hydrological, and chemical points of view.
(c) To provide information required in order to comply with regulations.
(d) To provide information for the design of a monitoring network, by predicting the system's future behavior.
(e) To provide information for the design of field experiments.

In principle, for both (a) and (b), we need to write a flow and, sometimes, also a solute transport model, but the requirements for details and accuracy may be significantly different. This difference may strongly affect the amount of (geological, hydrological and geochemical) data needed and its accuracy requirements.

Figure 1.2.1 shows how modeling activities are incorporated in the management process of a given management problem. We always start from the given management problem, identifying what we wish to achieve in order to

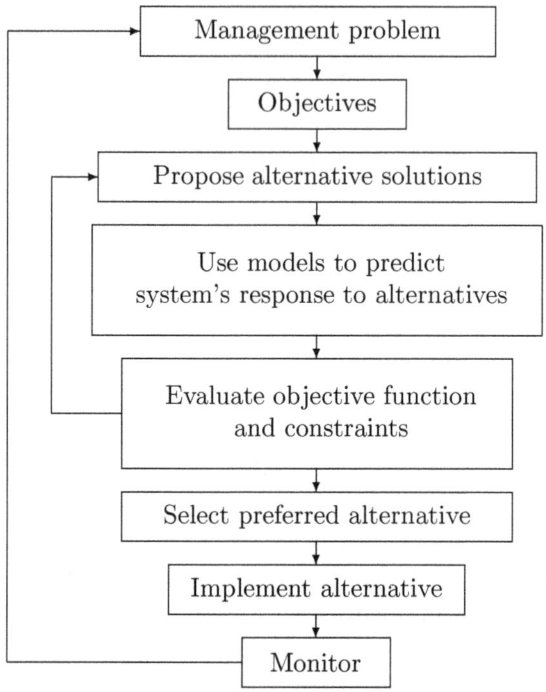

Figure 1.2.1: The role of modeling in the decision-making process.

solve it. We express what we wish to achieve in terms of an *objective function* (Sec. 11.1). For example, suppose we wish to cleanup a contaminated domain within a certain time interval at minimum costs. The objective function expresses the cost of each alternative solution, and we wish to minimize this cost. In this example, the prescribed time interval is a constraint that the solution must satisfy. Obviously, the cost is associated with the behavior of the system. We then identify a number of alternative solutions, say, various well locations and pumping rates, with the intention of selecting the one that will be the 'best' (or the 'optimal') from the point of view of the selected objective function. We determine the information that is required in order to enable this selection, e.g., information on how will the system respond to the implementation of the various management alternatives. The 'response' is then expressed in terms of 'cost'. The model is the tool that will provide the information on the system's response. We then construct the model (and that includes model validation, calibration, etc., to be discussed below), and use it to obtain the required information. The latter is then used to select the most desired alternative. The subject of aquifer management is discussed in detail in Chap. 11. As a solution is being implemented, additional data is continuously being assembled and used to reevaluate the problem, to update

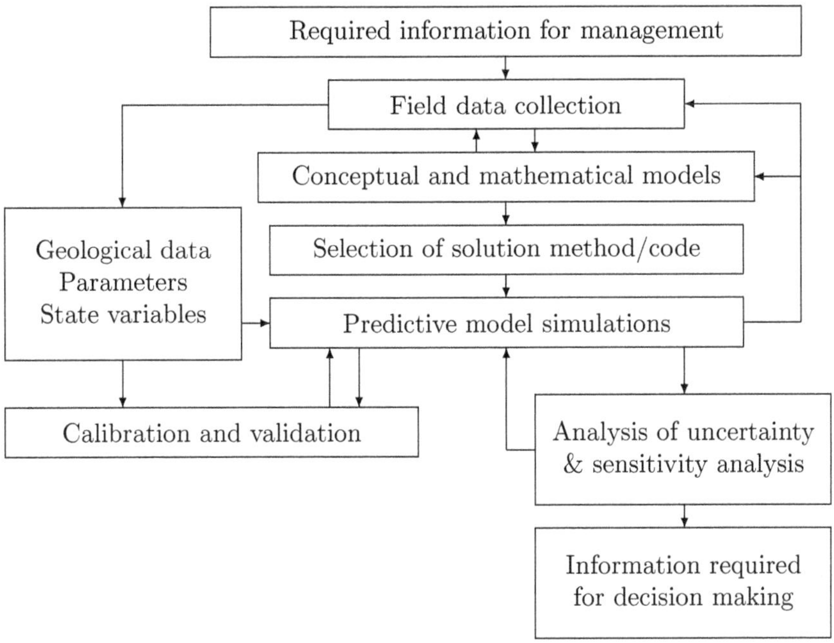

Figure 1.2.2: The modeling process.

the model and its predictions, and, perhaps, to modify the solution. This additional data may be used to continuously improve the model.

In the following subsection, we present the methodology of constructing models that are required for solving problems of flow and contaminant transport in the subsurface. We start by introducing some fundamental modeling concepts, then discuss the model as a tool for forecasting the future behavior of an investigated physical domain, suggest a procedure for constructing models, outline methods of solution, and discuss how models are used.

We wish to emphasize from the outset that although a model is eventually expressed in mathematical terms, the mathematical notation is used merely as a compact way of describing the physical and chemical phenomena that are relevant to the problem. We shall make an effort to present the physical interpretation of each term that appears in the model's equations. Obviously, a mathematical model is needed in order to obtain a prediction of the system's behavior.

1.2.2 Modeling process

The modeling process reviewed in this subsection is summarized in Fig. 1.2.2.

Step 1: Identification of the information required for making management decisions. As emphasized above, models are needed in order to predict the outcome of implementing management decisions. In practice, the same project objectives may be achieved by more than one plan of operation. In order to select the most desirable plan, the decision maker needs to assign values to a set of *measures*, that are used to evaluate the success of a selected plan in achieving the desired objectives. The cost of a project, the period required for remediation, the quality of water reaching an underlying aquifer, and the quantity and quality of water that can be pumped from a considered aquifer, may serve as examples of such measures. The actual values of the various measures depend on the managed system and on its response to the implementation of a proposed plan. Accordingly, we need information on such parameters as future water levels and concentrations of relevant chemical species, spring discharge, quality of pumped water, etc. These, in turn, are used to evaluate the criteria, or objective function, employed for selecting the preferred management alternative. Information of this kind may also be required in order to ensure that a proposed alternative does not violate constraints imposed on the managed system. Examples of constraints are regulatory limits on contaminant concentrations, and on the leakage to an underlying aquifer.

Step 2: Development of a conceptual model. The real system and its behavior may be very complicated, depending on the amount of details we wish, or need to include in describing them. For example, is the configuration of each individual soil grain needed in the description of flow in a porous medium domain? Should we include every clay lens in a sandy formation? Should we include every detail of the domain's stratigraphy? It is intuitively obvious that no practical management decision can depend on an excessively detailed description of a system and its behavior. On the other hand, certain features may be significant in governing those aspects that are important from the management point of view, say the advance of a contaminant plume. We should also bear in mind that gathering information is always costly, so that a balance should be sought between additional information and the benefits to be gained from it. *The art of modeling is to simplify the description of the system and its behavior to a degree that will be useful for the purpose of planning and making management decisions in specific cases.*

The simplifications are introduced in the form of a *set of assumptions* that expresses our understanding of the system and its behavior. Because the model is a *simplified* version of the real system, no unique model exists to describe it. Different sets of simplifying assumptions will result in different models, each approximating the domain and its behavior in a different way.

Thus, the second step in the modeling process is the construction of a *conceptual model of the problem and of the relevant domain.* The conceptual model consists of a set of assumptions that reduce the real problem and the real domain to simplified versions that are satisfactory in view of the modeling objectives, the associated management problem, and the available data.

Assumptions should be related to such characteristics as:

- The domain's hydrogeology, stratigraphy, etc.
- The dimensionality of the model (one, two, or three dimensions), and the geometry of the boundary of the domain of interest.
- The behavior of the system: steady state or time-dependent.
- The kind of soil and rock materials comprising the domain, as well as inhomogeneity, anisotropy, and deformability of these materials.
- The number and kinds of fluid phases (e.g., water, air, NAPL), and the relevant chemical species.
- The extensive quantities transported within the domain.
- The relevant material properties of the fluid phases (density, viscosity, compressibility, presence of solutes).
- The relevant transport mechanisms within the domain.
- The possibility of phase change and exchange of chemical species between adjacent phases.
- The relevant chemical, physical, and biological processes that take place in the domain.
- The flow regimes of the involved fluids (e.g., laminar or non-laminar).
- The existence of nonisothermal conditions and their influence on fluid and solid properties and on chemical–biological processes.
- The presence of assumed sharp macroscopic fluid-fluid boundaries, such as a phreatic surface.
- The relevant state variables, and the areas or volumes over which averages of such variables should be taken.
- The presence of sources and sinks of fluids and contaminants within the domain, and their nature (spatial distribution and temporal variation).
- The initial conditions within the domain, and conditions on its boundaries.

Obviously, more items may be included in the conceptual model of any specific case.

The set of assumptions comprising the conceptual model is expressed *in words*. It is recommended that these assumptions be numbered, say [A1], [A2], etc., as is done for equations, so that they can be referred to.

The selection of the appropriate model for a particular case depends on three main factors:

- The objective(s) of the investigations, i.e., what kind of information is the model required to provide for the purpose of making management decisions. Here we may include rough preliminary estimates vs. more detailed predictions.
- The available resources required to construct and solve the model. Included here are the availability of expertise, skilled personnel, and computers. Also included is the ability to describe processes that take place, and the availability of field data required to validate the model and determine the numerical values of its coefficients.
- The legal and regulatory framework which pertains to the considered case.

The objectives of an investigation dictate which features of the domain and its behavior should be represented in the model, and to what degree of accuracy and detail.

The constraints imposed by limited resources are very real and, although we shall not dwell on them in this book, cannot be overlooked. For example, a more detailed model is, obviously, more costly and requires more skilled modelers, more sophisticated computer codes, and larger computers. As we shall emphasize below, it is important to understand that a more detailed model requires more field measurements in order to calibrate it and to determine its coefficients. Data acquisition is usually more expensive than code and computer costs.

Selecting the appropriate conceptual model for a given problem is the most important step in the modeling process. If we oversimplify, we may not produce the required information. If we undersimplify, we may have neither the information required for model calibration (see below) and coefficient determination, nor the resources to solve it. If we select inappropriate or wrong assumptions, our model may not represent those features of the system behavior that are relevant to the management problem on hand. The set of assumptions serves as a 'prescription label' of the model. One should not use a model developed for a different problem, unless one examines its 'label' to ensure that it fits one's problem.

Step 3: Development of a mathematical model. In this step, the conceptual model is expressed in the form of a *mathematical model*. The *continuum type of model*, discussed in Sec. 1.3, is usually (but not always) employed. The mathematical model consists of:

- A definition of the *geometry* of the surfaces that bound the domain.
- *Equations* that express the *balances* of the relevant extensive quantities (e.g., mass of fluids, mass of chemical species, energy).
- *Flux equations* that relate the fluxes of the extensive quantities to the relevant state variables of the problem (e.g., Fick's law for the diffusive mass flux of a chemical species in a fluid phase).
- *Constitutive equations* that define the behavior of the particular phases and chemical species involved (e.g., dependence of density and viscosity on pressure, temperature, and solute concentration).
- *Sources and sinks*, often referred to as *forcing functions*, of the relevant extensive quantities.
- *Initial conditions* that describe the *known* state of the system at some initial time.
- *Boundary conditions* that describe the interaction of the domain with its environment (i.e., outside the delineated domain) across their common boundaries.

Mathematical models of flow nd solute transport are discussed in Chaps. 5 and 7, respectively.

In a continuum model (Sec. 1.3), a partial differential balance equation describes the behavior *at every point within the domain.* However, sometimes we are not interested in what happens *at every point.* Instead, we need information on the lumped, or averaged, behavior of an entire domain, or of parts of it. Another kind of model is called for in such cases, referred to as a *lumped parameter model*, a *compartmental model*, a *multi-cell model*, or an *input-output* one (Subs. 7.2.1). In such a model, balances of the relevant extensive quantities are stated for 'cells' of different shapes and sizes of the domain; state variables are averages over these cells. Sometimes, the heterogeneity of a domain is such that an appropriate representative elementary volume (Sec. 1.3) cannot be found for it and the continuum approach cannot be applied. A lumped parameter model may be required. In this book we focus our attention mainly on continuum models.

The core of any model of transport on an extensive quantity is the balance equation of that quantity. In this book, we deal with models of transport of two extensive quantities: mass of a fluid phase, and mass of a component of a fluid phase. The extensive quantity of momentum of a phase is not referred to directly, as we replace it by an approximation that takes the form of Darcy's law. We do, however, refer to the momentum of a solid phase when we deal with porous medium deformation.

In passing from a model at the *microscopic scale* to the *macroscopic scale* model (these terms are defined in Subs. 1.3.1), by some process of averaging, various *coefficients of transport, transformation*, and *storage* of the extensive quantities are introduced. The permeability of a porous medium (Subs. 4.1.3), moisture diffusivity (Subs. 6.2.2), and dispersivity (Subs. 7.1.6) are examples of such coefficients. Permeability and dispersivity of a phase are examples of coefficients that express the macroscopic effects of the microscopic configuration of the boundaries between a considered phase and all other phases present in a representative elementary volume (REV) of the medium.

It is important to realize that the coefficients derived by field experiments, in which (in principle) we compare measurements of certain state variables with the corresponding values predicted by a model, actually correspond to *that particular model.* If possible, one should not employ coefficients derived for one model in another one. When the coefficients developed for one model are used in another model, errors may result. The magnitude of the error will depend on the differences between the two models.

In principle, to employ a particular model for a particular domain, the values of the coefficients that appear in the model should be determined *from field experiments* conducted in the domain. Typically, such an experiment involves a comparison between actual field observations and predictions made by the model, employing some parameter identification technique (Sec. 11.3).

Step 4: Development of a numerical model and code. Having constructed a mathematical model, in terms of relevant state variables, it has to be solved for cases of interest. The preferred method of solution is the *analytical* one, as it provides a general solution (for a given domain geometry) that can be applicable to various sets of domain and fluid parameters. However, because of the complexity of most problems of practical interest (shape of domain's boundaries, heterogeneity, nonlinearity, irregular source functions, etc.), it is impossible, in general, to derive analytical solutions for them. Numerical methods are usually employed for solving the mathematical model. This means that various methods are used in order to transform the mathematical model into a numerical one, in which the partial differential equations are represented by their numerical counterparts. A computer program, or a *code* is required in order to solve the numerical model. Numerical methods are discussed in Chap. 8.

Very often, there is no need to develop a code that will solve a given problem, as such code is readily available, either as a public domain code, or a commercial one that has to be purchased. In most cases, it is much cheaper to purchase a code than to develop one. Codes are available not only for flow problems, but also for most problems of contaminant transport in single phase flow, some with chemical and biological reactions and transformations.

Step 5: Code verification. When a new numerical model and a code are developed for solving a mathematical model, the code is not considered ready for use unless it undergoes a proper verification procedure. Here, *verification* means checking that the code does what it proclaims to do, namely, to solve the mathematical model. Verification involves comparing solutions obtained by using the code with those obtained by analytical methods, whenever such solutions are possible. This is usually done for some simplified domain geometry, homogeneous materials, etc. In many cases, analytical solutions cannot be derived. The only procedure, then, is to compare code solutions with solutions obtained by other codes.

Good codes, especially commercial ones, should have documented code verification.

Step 6: Model validation. Once a model has been selected for a *particular problem* at a *particular site*, the model must be *validated.* Model validation is the process of making sure that the model correctly describes all the relevant processes that affect the excitation-response relations of interest to an acceptable degree of accuracy. The only way to validate a model is an *experiment.* Although it is desirable to perform the model validation for the actual site of interest, we often validate the model *in principle*, i.e., ensuring that it represents the phenomena, by conducting controlled field or laboratory experiments. Unlike laboratory experiments, many features encountered in field experiments, such as heterogeneity and anisotropy, cannot be controlled or identified. Unfortunately, in many cases they dominate the system's behavior.

If model validation cannot be implemented, it is sometimes combined with *model calibration* (see below).

Step 7: Model calibration and parameter estimation. Obviously, no model can be employed in any particular case of interest, unless *numerical* *values* are assigned to all the coefficients (e.g., permeability, transmissivity, storativity) that appear in it. Natural replenishment, and the location and type of boundaries should be included in the list of model coefficients and parameters that have to be identified. We refer to the activity of identifying the values of these model coefficients as the *identification problem*, the *inverse problem*, or the *parameter estimation problem.*

We use the term *model calibration* for the activity that combines *model validation* and *parameter estimation* at a specific site of interest. These activities are actually executed simultaneously. Thus, in the procedure of calibration, the values of model coefficients for a site are determined by solving an inverse problem, using field data from that site.

In principle, the only way to obtain the values of these coefficients for a model of a considered domain, e.g., an aquifer, is to investigate the domain itself. Historical data are reviewed to find a period in the past for which information is available on (a) initial conditions of the system, (b) excitations of the system, say, in the form of pumping, natural replenishment (infiltration from precipitation), introduction of contaminants, or changes in boundary conditions, and (c) observations of the response of the system, say, in the form of temporal and spatial distributions of state variables, e.g., moisture content and solute concentrations. If such a period (or periods) are found, we impose the known initial conditions and the known excitations of the real system on the model, and derive the response of the latter to these excitations. Obviously, in order to derive the model's response, we have to assume some trial values of the sought coefficients. We then compare the observed and modeled responses. The sought values of the coefficients are those that would make the two sets of values of state variables identical. The measured response is compared with model predictions. However, because the model is only an approximation of the real system, we should never expect these two sets of values to be really identical. Instead, we search for a 'best fit' between them, according to some criteria. It is also possible to conduct *field tests*, or *pumping tests* (e.g., Bear, 1979), in which the system is excited artificially (see discussion on pumping tests in Subs. 11.3.1).

Various techniques exist for determining the 'best', or 'optimal', values of these coefficients, i.e., values that will make the predicted values and the measured ones sufficiently (or acceptably) close to each other. Obviously, the values of the coefficients eventually accepted as 'best' for a considered model depend on the criteria selected for 'goodness of fit' between the observed and predicted values of the state variables. These, in turn, depend on the objective of the modeling. Some techniques use the basic trial-and-error approach described above, while others employ more sophisticated optimization methods. In some methods, *a priori* estimates of values to be expected for the coefficients, as well as information about lower and upper bounds, are introduced. In addition to the question of selecting the appropriate criteria, the question of the conditions for which this inverse problem will result in a *unique* solution still remains. The inverse problem, as described above is *ill-posed*, and there is no reason to expect a unique solution. The inverse problem is discussed in Sec. 11.3.

When conditions for determining coefficients, as described above, do not exist, they can be created as a *laboratory experiment*, usually on a *soil core* taken from the considered formation, or as a *field test*. In such experiments, we create a situation (in the laboratory, or in the field), in which state variables can be observed and for which a solution (preferably, analytical) of the model can be determined. It is important to emphasize again that coefficients for any particular site must be determined by making use of data assembled from that particular location.

In addition to model calibration, or solving the inverse problem for the considered domain, model coefficients and parameters can also be obtained (better: estimated) from:

- **Literature survey.** In many cases, we can find in the literature values of coefficients and parameters that have been derived and used in modeling similar situations, similar soils, fluid phases and chemical species, etc. One should be careful in using this information, at least as far as soils are concerned, as soils at different sites, even when belonging to the same class in some standard classification, seldom behave identically. Nevertheless, the values found in the literature may be employed as *first estimates* in a model calibration process, or in sensitivity analysis runs. Such runs are conducted in the process of planning field experiments aimed at obtaining the site specific values for such coefficients.
- **Laboratory experiments.** Laboratory experiments provide good information on parameters that are independent of soil characteristics. As for soil dependent coefficients, we should recall that our modeling is at the field scale, which always involves heterogeneities, etc. Laboratory experiments are usually carried out on relatively small samples ('cores'), and not always under undisturbed conditions. Hence, at best, the values obtained in such experiments should be considered as *first estimates* to be used in model calibration runs.
- **Small scale field experiments.** These include both standard tests, like pumping tests, slug tests, etc., as well as specially designed ones. If properly designed, they could serve for model validation as well as provide site-specific information on values of coefficients.

Step 8: Model application. Once a model has been calibrated for a considered problem (and this includes all the required site-specific coefficients and parameters), the model is ready for use. Computer runs are then conducted to provide the required forecasts.

Step 9: Sensitivity analysis. The term *sensitivity analysis* is used here to describe tools that help the modeler evaluate the impact of uncertainty, say, in the values of model coefficients, on the results predicted by the model. Briefly, we want to know how sensitive are the predicted values to changes in the values of model coefficients. If these effects are not significant (from the point of view of the decision maker, who makes use of these predictions in the management process), we can accept the predicted values and make decisions. If, however, the predicted values are sensitive to changes in parameter values, we must reduce the range of uncertainty in the values of these parameters. In most cases, this means that we must invest more resources in order to acquire more and more accurate data.

Sensitivity analysis can also be used to assess the reliability of parameters determined in the calibration, or parameter estimation procedure described in Step 7 above. Typically, a *residual error* is usually expressed in the calibration process, e.g., as the sum of the squared differences between the measured and the predicted water levels. The optimal set of model parameters is the one that minimizes this error. This subject is discussed in Chap. 10. There, we show also how to enable a more accurate determination of model parameters and how to asses the reliability of the obtained parameter values. In Subs. 11.3.1, we also present the sensitivity analysis in connection with the analysis of pumping tests used for determining aquifer parameters. Altogether, if a small change in a parameter causes a large change in the residual error, we may say that the residual error is *sensitive* to that parameter. Hence, our information on that parameter is of high quality, or reliable. Conversely, if a large change in the parameter causes only a small change in the residual error, we say that the calibration process is *insensitive* to that parameter; hence, our information on that parameter is of low quality, or less reliable. Poor quality information means that more data is required in order to improve calibration results, and vice versa.

When the quantity of available data (e.g, water levels at observation wells) is insufficient for performing regional model calibration, we use information on the domains stratigraphy, which should be available, to *estimate* values of model coefficients. Then, it is certainly of interest to study the sensitivity of the model's predictions to the uncertainty in these input parameters. More information on how such analysis can be performed is presented in Chap. 10. Again, a high sensitivity indicates that it may be worthwhile spending extra efforts and resources, such as commissioning a geological or geophysical survey, in order to improve our information concerning that feature. Sensitivity analysis is discussed in more details in Subs. 10.2.2.

Altogether, the result of a sensitivity analysis can either increase the modeler's confidence in the model, or, conversely, reveal the deficiency of the model. In the latter case, the sensitivity analysis can be used to help the manager to efficiently allocate resources to expand data acquisition, thus improving model prediction.

Step 10: Stochastic analysis. The sensitivity analysis discussed in Step 9 provides a *qualitative* description of uncertainty. It addresses questions like 'what if there exists 20% uncertainty in this parameter or that condition'; it does not express the range of uncertainty in either the input or the output in terms of statistical measures, such as mean and standard deviation.

A *stochastic analysis* not only takes into account the simple statistical measures of mean and standard deviation of the input data, such as the hydraulic conductivity and the natural replenishment, but it also examines the temporal and spatial correlations of these data. Following are some typical questions that can be asked :

- What is the predicted mean piezometric head at a given time and location?
- What is the standard deviation of that prediction?
- If a 90% reliability is needed in the manager's recommendation, what is the allowable pumping rate?
- If a head is observed at a certain location at a certain time, what is the probability that the value will stay correlated a certain distance away or after a certain time period?

In Chap. 10 we shall present and discuss the stochastic analysis technique that can provide answers to the above typical questions.

Step 11: Summary, conclusions, and reporting. The summary and conclusions should include the information that the model was expected to provide, including additional information concerning the accuracy of the information, the uncertainty involved, and suggested follow-up work. The report on the modeling activities may be part of the report on solving the management problem, say as an appendix, or as a separate report.

1.2.3 Model use

Admittedly, the subsurface is highly heterogeneous, and we seldom have enough data to calibrate the model of an aquifer domain in a way that will accurately describe its heterogeneity. We are also uncertain about model boundaries and conditions occurring on them, and even less certain about what will occur in the future. In contamination problems, we *never* have sufficient knowledge and data concerning *all* the chemicals and biological transformations that may take place. Usually, we have insufficient knowledge of the thermodynamic information required for modeling the complex case of multiple multicomponent phases, which may be under nonisothermal conditions.

What, then, is the use of a model? Can the model be expected to predict future states to any desired, or acceptable, degree of accuracy? In some cases, such as in the case of a repository for radioactive waste located at depth in the unsaturated zone of a highly heterogeneous fractured rock, the required extrapolation in time is for hundreds and thousands of years, and the domain of interest extends for many kilometers. Can we validate a model under laboratory, or small scale field conditions and then apply it to such a large scale and long term problem? The answer is, generally, not.

An argument for the use of models is that, in most cases, *decisions will be made anyway*, with or without sufficient information. The use of a model can, at least, provide some information, even if it lacks the desirable accuracy.

Hence, modeling still plays an essential role in decision making. However, in view of all the uncertainties, we should not claim that the model's usefulness rests only in its ability to *accurately* predict the system's response. The model should also be used for enhancing and organizing our understanding of the phenomena, say, of flow and contaminant transport that take place in it as a result of applied excitations.

By running the model of a given problem in a given domain, under various assumed conditions, we gain insight into the roles that various processes play in producing the system's response. A sensitivity analysis (Step 9 above and Subs. 10.2.2), in which various parameters and coefficient values are systematically changed, within specified ranges, will indicate the more significant ones. A sensitivity analysis may also indicate ranges of outcomes that may be expected. When combined with methods of stochastic analysis (Step 10 above and Sec. 10.3), probability distributions of outcomes may be derived. Such analysis will guide the acquisition of additional field data. A model may, thus, be used for two additional objectives:

- design of observation or monitoring networks; and
- design of field experiments.

By running the model, and learning what to expect in the future, observation and early-warning networks can be designed. Altogether, the use of models will aid in making informed decisions, even in the absence of model validation and accurate parameter identification in the strict sense of these terms.

Finally, a few warnings concerning the use of models, from the establishment of the conceptual model through the use of a computer code to derive a solution:.

- Before running a computer program in order to obtain a solution (= prediction), make sure that one knows what to expect as the answer (obviously, approximately, or at least the trend of it). A small error somewhere in the construction of the model, in the computer program, and in data input, may lead to erroneous results. If one does not know *a priori* what to expect, one may make decisions based on erroneous results.
- Make sure that the computer program has been properly verified.
- Make sure that the assumptions that constitute the conceptual model are justified. A wrong assumption may lead to a model that does not represent the physical reality.
- Finish the modeling efforts by an appropriate sensitivity analysis, so that one knows where more data is needed to reduce uncertainties. If possible, use stochastic approaches to overcome uncertainties. Such model will take uncertainties in the input, say the proper values of model coefficients, into account, and provide a prediction with stated reliability level to assist the manager in decision making.

1.3 Continuum Approach to Transport in Porous Media

As will be shown throughout this book, the core of any flow and solute transport model consists of a partial differential equation (abbreviated PDE) that represents the physical reality of the subsurface porous medium and the phenomena of flow and solute transport that occur there. The construction of such equation is based on the assumption that the porous medium domains of interest may be described as *continua*; the models so constructed are referred to as 'continuum models'. Accordingly, in this section, we shall start by examining the fundamental question of 'what is a continuum model?'

The subsurface—both the saturated and unsaturated (or vadose) zones—is a domain occupied by a *porous medium*. Soil, sand, fissured rocks, sandstone and karstic limestone, are examples of porous media. Common to all these examples is the presence of both a *solid matrix* and a *void space* within such domains. The void space is occupied by one or more fluid phases (e.g., water and air). Our interest is in the mathematical representation, or model, that describes the movement of water and contaminants through the void space. Once such a mathematical model has been obtained for a given case, its solution, analytically or numerically, can provide information on the future behavior of the fluids and contaminants within the considered domains.

Because the flow and contaminant transport in the subsurface involves multiple phases and often multiple multicomponent fluid phases, we shall start the discussion by defining phases, chemical species, and chemical components.

1.3.1 Phases, chemical species and components

A *phase* is a portion of space occupied by a material such that a *single* set of *constitutive relations*, e.g., the relationship between density, pressure, and temperature, describes the behavior everywhere within that material. Another definition of a phase is a portion of space that is separated from other such portions by a definite physical boundary (*interface*, or *interphase* boundary). According to the second definition, globules of NAPL surrounded by water are considered separate phases (unless we regard them as a single multiply connected phase). There can be only one gaseous phase in a system, as all gaseous phases are completely *miscible* and do not maintain a distinct boundary between them. We may, however, have more than one liquid phase in a system. Such liquid phases are referred to as *immiscible fluids*.

In these definitions, a domain occupied by a phase is regarded as a *continuum* at the microscopic level (or scale of description), if values of state variables, and of material parameters or coefficients of the phase, can be assigned to *every* point within the domain. Examples of state variables are pressure and solute concentration, and of material coefficients are fluid viscosity and coefficient of compressibility. These coefficients are obtained by averaging the behavior that takes place at the *molecular level* over a microscopic representative elementary volume (microREV), denoted as μREV, around every point within the domain occupied by the phase. Thus, a 'fluid phase' and a 'solid phase' are *continuum concepts*, obtained from the molecular level by volume averaging over appropriate representative elementary volumes. This idea will be further explained below.

A phase may be composed of a large number of different chemical species. A *chemical species* is an atom, a molecule or an ion distinguishable from the rest of the phase due to its chemical composition. Often, the number of species in a phase is very large and in a liquid they may also interact chemically. However, under conditions of *chemical equilibrium*, the *minimum* number of independent chemical species necessary to completely describe the composition of a given phase may be much smaller. We use the term *component* to denote a chemical species that belongs to the smallest set of such species that is required in order to completely define the chemical composition of a phase under equilibrium conditions. As a simple example, consider liquid water in contact with its vapor. The water is composed of oxygen and hydrogen, with these two elements being *always* present in fixed and definite proportions. Therefore, the system is composed of two chemical species, but only a single component. When chemical equilibrium **is not** assumed, all species are defined to be components. In this book, we use the terms chemical species and component interchangeably, unless we wish to emphasize that we are referring specifically only to one of them.

Often, only a subset of the species within a phase is in equilibrium. Then, the set of components consists of the components of the system that are in equilibrium, together with the species that are not in equilibrium.

For the sake of simplicity, we shall often use the term 'component' also to denote the mixture of a number of independent chemical species in a liquid or a gas. The selection of components is not unique, in the sense that different chemical species, may be selected as components of a given phase.

1.3.2 Need for continuum approach

A spatial domain is said to behave as a *continuum* if state variables and properties that describe the behavior of the material occupying it can be assigned to *every* point within it. With this definition in mind, if we consider a *porous medium domain*, containing a solid matrix and at least one fluid phase, we can identify within this domain the subdomains occupied, respectively, by the solid and by each fluid phase. The domain occupied by water behaves as a continuum with respect to water, and so is the domain occupied by solid. As defined earlier, we refer to the level of description of phenomena *within* each phase as a description at the *microscopic level*.

As an example, let us consider a porous medium domain whose void space is entirely occupied by water. The domain occupied by solid behaves as a continuum with respect to the solid. For example, we can define solid density at *every* point within the solid subdomain. Similarly, we can describe certain properties or behavior variables at every point in the domain occupied by water, e.g., water density, pressure, and velocity. As described in the preceding section, at this microscopic level of description, the water phase is regarded as a continuum within the domain it occupies, and so is the solid. For example, to obtain the velocity distribution within the water, regarded as a Newtonian fluid, we could make use of the *Navier-Stokes equation* (which is a PDE that describes the linear momentum balance of a Newtonian fluid), and solve it within the water occupied domain, subject to boundary conditions on the solid-liquid interface that bounds this domain. Obviously, for flow through porous medium problems, this approach is, usually, impractical, due to our inability to describe the complex configuration of the solid-liquid boundary over large porous medium domains. Moreover, even if we could solve for values of state variables, e.g., pressure, it would be impractical to verify the solution by measurements at this level.

To circumvent these difficulties, associated with trying to solve problems at the microscopic level, another level of description is introduced, referred to as the *macroscopic level.* At this level, properties are defined at *every point in the porous medium domain*, such that the knowledge of the complex interphase geometry is not needed.

The continuum approach, described in the preceding subsection for the transport of a species *within* a phase, is obtained by averaging the behavior of the phase at the molecular level (Bear, 1972, p. 17). We referred to this description as one at the *microscopic level.* This approach of averaging can be extended to a multiphase system such as a porous medium domain, where the various phases are separated from each other by abrupt interfaces. To achieve this goal, the real porous medium domain, containing two, or more phases (each of which is already regarded as a continuum at the microscopic level that occupies a certain portion of space that together completely occupy the porous medium domain), is replaced by a *model*, in which each of the phases is assumed to behave as a continuum that fills up the *entire* porous medium domain. We then speak of '*overlapping continua*', each continuum corresponding to one of the phases present in the domain. If the individual phases interact with each other, so will these continua. The space occupied by these overlapping continua will be referred to as the *macroscopic space.* For each point within this macroscopic space, average values of the variables that describe the behavior of a phase are taken over *elementary volumes*, centered at the point, regardless of whether, in the real domain, this point falls within that phase, or outside it. The averaged values are referred to as *macroscopic values* of the considered variables. By traversing the entire porous medium domain with a moving *elementary volumes*, thus assigning averaged values to *every* point, we obtain *fields* of macroscopic variables, which are differentiable functions of the spatial coordinates. In this way, we have turned the porous medium domain into a continuum; the behavior of each of the phases within this continuum is described at every point by the averaged values of variables and material properties.

The advantages of the continuum model of a porous medium are:

- It circumvents the need to specify the exact configuration of the interphase boundaries, the knowledge of which is, anyway, not available.
- It describes processes occurring in porous media in terms of *differentiable quantities*, thus enabling the solution of problems by employing methods of mathematical analysis.
- The (macroscopic) quantities mentioned above are *measurable*, and can, therefore, be useful in solving field problems of practical interest.

These advantages are at the expense of the loss of information concerning (1) the microscopic distributions of variables *within* each phase, and (2) the microscopic configuration of the interphase boundaries. However, the macroscopic effects of this configuration are retained in the form of *coefficients* that are created in the process of averaging. The structure and relationship of these coefficients to the statistical properties of the void space (or phase) configuration within the *elementary volumes* can be determined. Examples of resulting coefficients are the porosity, permeability, and dispersivity. For a specific porous medium, *the numerical values of these coefficients must be determined experimentally, in the laboratory, or in the field.*

In the following subsection, we shall discuss the procedure for passing from the microscopic level to the macroscopic, or continuum, one, by averaging over an REV. In Subs. 1.3.5, we shall introduce another approach for smoothing out heterogeneities.

1.3.3 Representative elementary volume and averages

A basic feature, which is common to the porous materials that occupy the subsurface, in fact, common to *all* porous media, is that both the solid matrix and the void space are distributed *throughout* the porous medium domain. This implies that samples of a *sufficiently large* volume, taken at different locations within the domain, will *always* contain both a solid phase and a void space. How large should such volume be? A porosity (= ratio between the volume of void space to the sample's volume) can be defined for the sample. Obviously, there is no meaning to porosity at the microscopic level of description. For the time being, let us refer to the volume of such a sample as a *representative elementary volume* (abbreviated REV). We shall discuss the size of an REV in Subs. 1.3.4A.

A. Definition of representative elementary volume

Based on this brief discussion, we may now define a *porous medium* as a portion of space (1) that is occupied by a number of phases, *at least one of which is a solid*, and (2) for which an REV can be found. This means that a porous medium is not merely 'a domain containing a void space and a solid phase'. Our definition implies that if a *common* REV cannot be found for *all* points of a domain, that domain *cannot* qualify as a porous medium.

Specifically, suppose we consider the density of the solid phase, ρ_s, comprising the solid matrix within a porous medium domain. At *every* point, $\mathbf{x}$, within the domain, we place an REV of volume $\mathcal{U}_o$, with the point $\mathbf{x}$ as its centroid, and determine the volume average of the solid density, $\overline{\rho_s}$, over the REV. We then assign the calculated average density to the centroid, $\mathbf{x}$, of the REV. We repeat this procedure for *all* points within the domain, whether the centroid falls within the solid matrix or in the void space. We can do so because the definition of the REV states that the size of the REV must be such that it will *always* contain *both* a solid matrix and a void space. By following this (conceptual) procedure for all points in the domain, we obtain a continuous function $\overline{\rho_s} = \overline{\rho_s}(\mathbf{x})$ within the porous medium domain. We say that the (entire) porous medium domain behaves as a continuum with respect to the average density, $\overline{\rho_s}$. In this way, the porous medium domain, comprised of a solid phase and a void space containing one or more fluid phases, each occupying only a portion of the spatial domain, is replaced by a *model* (Sec. 1.2) that visualizes the domain as a continuum. In this continuum, the value of $\overline{\rho_s}(\mathbf{x})$ is assigned to *every* point (not only to points inside the solid matrix). We can repeat this procedure for any material property and state variable of interest of the solid and of the fluid(s) occupying the void space. We refer to the averaged values as *macroscopic values.*

In fact, what we have obtained is a description of the porous medium domain as *multiple overlapping continua*, one for every variable of interest, e.g., $\overline{\rho_s} = \overline{\rho_s}(\mathbf{x})$, for ρ_s and $\overline{\rho_f} = \overline{\rho_f}(\mathbf{x})$. The macroscopic (continuum) description of flow, or of any other phenomenon of transport, is, thus, derived by averaging the microscopic description over an REV. The macroscopic model (of the 'real' porous medium) obtained in this way describes the flow in terms of *macroscopic*, or averaged quantities. **In this book, our objective is to learn how to write models of flow and contaminant transport at the macroscopic level.** Following are the mathematical definitions of averaging and for REV size.

B. Representative elementary volume averaging

Let the symbol E_α denote the amount of an *extensive quantity*, E, in a phase denoted by the subscript α. An extensive quantity is an additive quantity, like mass, momentum and energy. The corresponding intensive quantity, or density of E_α, is denoted by e_α,

$$e_\alpha = \frac{E_\alpha}{\mathcal{U}_\alpha}, \tag{1.3.1}$$

where $\mathcal{U}_\alpha$ indicates the volume of the α-phase.

Three kinds of average values of e_α can be defined (Fig. 1.3.1):

Intrinsic phase average. The intrinsic phase average of e_α, taken over the domain of the REV of volume $\mathcal{U}_o$, centered at the point $\mathbf{x}$, is defined as

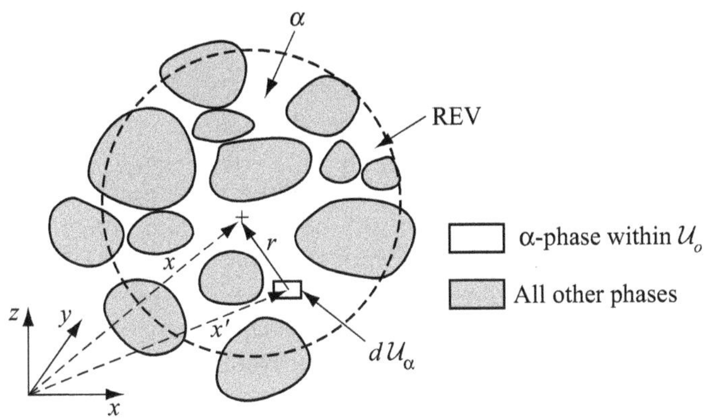

Figure 1.3.1: Nomenclature for averaging over an REV.

$$\overline{e_\alpha}^\alpha(\mathbf{x},t) = \frac{1}{\mathcal{U}_{o\alpha}(\mathbf{x},t)} \int_{\mathcal{U}_{o\alpha}(\mathbf{x},t)} e_\alpha(\mathbf{x}',t;\mathbf{x})\, d\mathcal{U}_\alpha(\mathbf{x}'), \tag{1.3.2}$$

where $\mathcal{U}_{o\alpha}$ is the volume of the α-phase within $\mathcal{U}_o$, $\mathbf{x}'$ is a point in α-phase within the REV centered at $\mathbf{x}$, and t denotes time. The *intrinsic phase average* is, thus, an average of E_α per unit volume of the considered phase.

Phase average. The phase average of e_α is defined as

$$\overline{e_\alpha}(\mathbf{x},t) = \frac{1}{\mathcal{U}_o} \int_{\mathcal{U}_{o\alpha}(\mathbf{x},t)} e_\alpha(\mathbf{x}',t;\mathbf{x})\, d\mathcal{U}_\alpha(\mathbf{x}'). \tag{1.3.3}$$

This is an average of E_α per unit volume of porous medium.

Mass average. Let $e'_\alpha(\mathbf{x}',t) = dE_\alpha/dm_\alpha \equiv e_\alpha/\rho_\alpha$ denote the quantity of E of an α-phase per unit mass of that phase, with $\rho_\alpha = dm_\alpha/d\mathcal{U}_\alpha$. The mass average of e'_α is defined as

$$\begin{aligned} < e'_\alpha >^\alpha (\mathbf{x},t) &= \frac{1}{m_{o\alpha}(\mathbf{x},t)} \int_{m_{o\alpha}} e'_\alpha\, dm_\alpha \\ &= \frac{1}{\overline{\rho_\alpha}^\alpha(\mathbf{x},t)\mathcal{U}_{o\alpha}(\mathbf{x},t)} \int_{\mathcal{U}_{o\alpha}} \rho_\alpha(\mathbf{x}',t;\mathbf{x}) e'_\alpha(\mathbf{x}',t;\mathbf{x})\, d\mathcal{U}_\alpha(\mathbf{x}') \\ &= \frac{1}{\overline{\rho_\alpha}^\alpha} \overline{\rho_\alpha e'_\alpha}^\alpha = \frac{\overline{e_\alpha}^\alpha(\mathbf{x},t)}{\overline{\rho_\alpha}^\alpha(\mathbf{x},t)}, \end{aligned} \tag{1.3.4}$$

where $\overline{\rho_\alpha}^\alpha(\mathbf{x},t)$ is the *intrinsic phase average mass density* of the α-phase.

The first two kinds of averages are related to each other by

$$\overline{e_\alpha} = \theta_\alpha \overline{e_\alpha}^\alpha, \tag{1.3.5}$$

where

$$\theta_\alpha = \frac{\mathcal{U}_{o\alpha}}{\mathcal{U}_o} \tag{1.3.6}$$

is the *volumetric fraction* of the α-phase within $\mathcal{U}_o$.

The kind of average to be used in each case depends on the way the averaged quantity is actually measured in the field. For example, if, at a point, we take a liquid sample out of a porous medium domain, in order to determine the concentration of a solute in it, the latter is an intrinsic phase average, as it is taken only over the liquid phase. The measuring device (e.g., screened portion of a piezometer, or porous cup of a tensiometer (Fig. 6.1.9)) of an instrument designed to measure an averaged, macroscopic quantity, must also be of the size of an REV, in order to yield observations compatible with the averaged values calculated by a macroscopic model for that point.

Throughout this book, it is assumed that the porous medium comprising the subsurface may be considered as a continuum in the sense explained above. Accordingly, the phenomena of fluid mass transport, solute transport, and heat transport (not considered in his book) are described (modeled) at the macroscopic level. This is the level at which engineers and hydrologists make predictions and measure state variables in the field. However, certain phenomena (e.g., capillary pressure and dispersion of a solute) cannot be understood unless we consider and understand them first at the *microscopic level*. Once these phenomena are understood and described at the microscopic level, they are averaged to yield their macroscopic description.

1.3.4 Scale of heterogeneity in continuum models

A. Size of representative elementary volume

We must still select the appropriate size and shape of the averaging volume, earlier referred to as REV. Usually, this size is selected such that (1) the average value of any geometrical characteristic of the microstructure of the void space, at any point in a porous medium, will be *a unique function (or almost so, within an acceptable error) of the location of that point only*, and (2) the measured averaged value should be *independent of small perturbations in the size of the* REV. This means that the average value should remain more or less constant over a range of REV volumes that correspond to the range of variation in the size of the sample or instrument that measures that average.

Denoting the characteristic dimension of an REV by ℓ, e.g., diameter of a spherical REV, and the length characterizing the microscopic structure (or, heterogeneity) of the void space by d (say, the typical size of grain or pore, or the *hydraulic radius*, which is proportional to the reciprocal of the specific surface area of the solids within an REV), a necessary condition for obtaining non-random estimates of the geometrical characteristics of the void space at any point within a porous medium domain is

$$\ell \gg d. \tag{1.3.7}$$

Another condition that sets an upper limit to the size of the REV is

$$\ell \ll \ell_{\max}, \tag{1.3.8}$$

where $\ell_{\max}$ is the distance beyond which the spatial distribution of the relevant macroscopic coefficients that characterize the configuration of the void space (e.g., porosity, permeability) deviates from the linear one by not more than some acceptable value (Bear and Bachmat, 1990, p. 22). The selection of the size of the REV is also constrained by the requirement that

$$\ell \ll L, \tag{1.3.9}$$

where L is a characteristic length of the porous medium domain. We shall discuss these limits in greater details in Subs. 1.3.4B, together with the general question of *scales* and their corresponding elementary volumes.

Although we have shown here how the macroscopic level of description is obtained from the microscopic one by *volume averaging*, other techniques that lead to macroscopic flow and solute transport models are also presented in the literature. Among them, we may mention another volume averaging approach proposed by Whitaker and coworkers (e.g., Hassanizadeh, 1986; Whitaker, 1967, 1986a, 1986b; Plumb and Whitaker, 1990) and the *homogenization* technique (e.g., Sanchez-Palencia, 1980; Hornung, 1997), which aims at smoothing out the heterogeneity at the pore scale, as well as at other scales. The homogenization technique is discussed in Subs. 1.3.5.

In the previous section, we have discussed two levels of description of phenomena: the microscopic level, obtained by averaging over a μREV (read: microREV), to smooth out spatial variations at the molecular level, and the macroscopic level, obtained by averaging over an REV, to smooth out variations at the microscopic level, resulting from the presence of a solid matrix and a void space within a porous medium domain. With the above in mind, we can extend this smoothing out approach to other scales of heterogeneity. Altogether, we shall have:

- **molecular scale**, when we consider the behavior of individual molecules (obviously, impractical).
- **microscopic scale**, at which we describe what happens at points within every phase present in the void space.
- **macroscopic scale**, which is the usual scale of describing phenomena in porous media.
- **megascopic scale**, which is the scale at which we describe phenomena of transport in the field. The latter is, usually, very large and very heterogeneous in a random fashion.

By smoothing out heterogeneity at one level we obtain a model at the higher level. The smoothing operation introduces 'coefficients' that represent the effect of the smoothed out properties. Note that the spatial distribution of the coefficients at the higher level may still be heterogeneous. Heterogeneity of a continuum domain with respect to a given property at the macroscopic level will be defined in Sec. 2.5.

B. Averaging over microscopic heterogeneity

For the purpose of the discussion here, let us refer to a given domain (regarded as a continuum) as *homogeneous* (= opposite of *heterogeneous*, or *inhomogeneous*) with respect to a given property, if that property has the *same* value at all points of the domain. For example, if the property of interest is 'the presence of solid' at a point, then, by the very definition of a porous medium (Subs. 1.3.3A), this domain is always heterogeneous with respect to that property.

As an example, let us compare two porous medium domains, one filled with large gravel, say, 1 cm in mean radius, and the other with sand, say, 1 mm in radius. Both contain a solid matrix and void space, and both are heterogeneous at the microscopic level. If the granules are spherical and in cubical arrangement, both will have the same porosity. What is the difference in heterogeneity between these two porous media? If we pick a point inside a solid grain, we have a good chance of finding solid at points within a distance of 1 cm in the first case, and within a distance of 1 mm in the second. We can say that their 'scales of heterogeneity' are 1 cm and 1 mm, respectively. Extending the above example to a porous medium with random grain arrangement, we can define a distance along which a selected property (here the 'presence of solid') at a point is *strongly correlated* to that at another point.

We continue to discuss what happens when we consider flow and transport in large field domains, say, in an aquifer or in the unsaturated zone, which are always *highly heterogeneous* with respect to (macroscopic) flow and solute transport coefficients. For example, we consider the permeability (in a macroscopic level continuum), and ask: 'given the permeability at a point A, how far from A is the permeability still correlated to that at A?' This question makes sense, because the heterogeneity in permeability in the field is a consequence of geological processes that produced it in the first place. Repeating this question for a large number of points at various distances around A, and for various points A within the considered domain, we shall find a certain distance that we refer to as the *scale of heterogeneity* of the given domain, with respect to the considered property, here, permeability; within that distance the permeability is still strongly (or sufficiently) correlated to that at A.

It is natural to assume that the *scale of heterogeneity* introduced above is related to the size of the considered domain, with larger domains exhibiting higher heterogeneity; so when we consider domains with characteristic lengths of tens, hundreds and thousands of meters, the scale of heterogeneity will also gradually grow. However, we usually assume that at some field size, this scale will level off at some value.

Returning to the microscopic level, let us use porosity as an example to illustrate the concept of scale of heterogeneity introduced above.

Porosity, ϕ, which, obviously, is a macroscopic property, is defined as *the ratio between the volume of void space and the sample's volume.* We can also define porosity as the volume average of a *characteristic function of the void space*, γ, which is defined in the following way. Let $\mathbf{x}'$ denote the position vector of a point within a small porous medium element of volume $\mathcal{U}$ centered at $\mathbf{x}$. The volume $\mathcal{U}$ is composed of two portions: $\mathcal{U}_v$, denoting the volume of void space, and $\mathcal{U}_s$ ($\equiv \mathcal{U} - \mathcal{U}_v$), denoting that of the solid matrix. The function $\gamma(\mathbf{x}')$ is defined by

$$\gamma(\mathbf{x}') = \begin{cases} 1, \text{ for } \mathbf{x}' \text{ within } \mathcal{U}_v, \\ 0, \text{ for } \mathbf{x}' \text{ within } \mathcal{U}_s. \end{cases} \tag{1.3.10}$$

For this function, an average, $\overline{\gamma}(\mathbf{x})$, may be defined by (see Fig. 1.3.1)

$$\overline{\gamma}(\mathbf{x}) = \frac{1}{\mathcal{U}} \int_{\mathcal{U}(\mathbf{x})} \gamma(\mathbf{x}';\mathbf{x})\, d\mathcal{U}(\mathbf{x}') = \frac{1}{\mathcal{U}} \int_{\mathcal{U}_v(\mathbf{x})} d\mathcal{U}_v(\mathbf{x}') = \left.\frac{\mathcal{U}_v}{\mathcal{U}}\right|_{\mathbf{x}}, \tag{1.3.11}$$

and a deviation from the average, $\mathring{\gamma}(\mathbf{x}';\mathbf{x})$, defined as

$$\mathring{\gamma}(\mathbf{x}';\mathbf{x}) = \gamma(\mathbf{x}';\mathbf{x}) - \overline{\gamma}(\mathbf{x}), \tag{1.3.12}$$

where the $\mathbf{x}'$ in $\mathring{\gamma}(\mathbf{x}';\mathbf{x})$ and in $\gamma(\mathbf{x}';\mathbf{x})$ indicates that these values correspond to a point $\mathbf{x}'$ that is located within a domain $\mathcal{U}$ centered at $\mathbf{x}$. From (1.3.11) and (1.3.12), it follows that

$$\overline{\gamma}(\mathbf{x}) = \left.\frac{\mathcal{U}_v}{\mathcal{U}}\right|_{\mathbf{x}} = \phi(\mathbf{x}) \quad \text{and} \quad \overline{\mathring{\gamma}}(\mathbf{x}) = 0. \tag{1.3.13}$$

In this way, we have defined the porosity, ϕ, as the *volume average of the characteristic function* $\gamma(\mathbf{x}')$. However, this immediately raises the question 'what size of sample should be selected in order to represent the porosity at a point in a porous medium domain?'

How small or large should $\mathcal{U}$ be in order to qualify as a *representative elementary volume*? In other words, 'how large should $\mathcal{U}$ be so that ϕ will represent the porosity at a point that serves as the centroid of an REV?'

Let us consider a small porous medium element of volume $\mathcal{U}_1$ centered at a point $\mathbf{x}$ within a given porous medium domain. For this volume, we determine the ratio $\mathcal{U}_v/\mathcal{U}$. We then consider a sequence of volumes of increasing sizes $\mathcal{U}_1 < \mathcal{U}_2 < \mathcal{U}_3 < \ldots$, up to that of many times the (typical) pore sizes. For each such volume, we calculate the ratio $\mathcal{U}_v/\mathcal{U}$ ($\equiv \overline{\gamma}$, where the average is taken over $\mathcal{U}$). Drawing this ratio (Fig. 1.3.2) as a function of the size of the volume of the element, $\mathcal{U}$, we observe that, the ratio $\mathcal{U}_v/\mathcal{U}$ starts from 0 or 1, depending on whether the point $\mathbf{x}$ falls in the void space or in the solid matrix. As the volume increases, we note oscillations with a decreasing

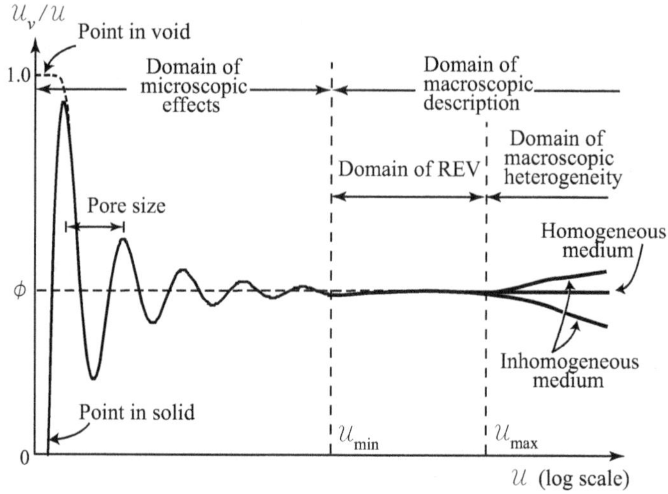

Figure 1.3.2: Variation of $\mathcal{U}_v/\mathcal{U}$ in the neighborhood of a point as a function of the averaging volume.

amplitude. At some volume, $\mathcal{U}_{\min}$, we observe approximately a *plateau* in the ratio of $\mathcal{U}_v/\mathcal{U}$. For a homogeneous porous medium, the plateau in the ratio $\mathcal{U}_v/\mathcal{U}$ stays at the same level. This means that we have achieved two goals: we found the value of $\mathcal{U}_v/\mathcal{U}$ which is independent of the volume of averaging, as long as $\mathcal{U} > \mathcal{U}_{\min}$, and we have determined the porosity at that point, ϕ $(= \mathcal{U}_v/\mathcal{U})$. Obviously, in order for this average to represent the porosity at the point, we should take the smallest possible value, i.e., $\mathcal{U} = \mathcal{U}_{\min}$. This volume, $\mathcal{U}_{\min}$, at which the porosity value stabilizes, is estimated to be a radius of about 50 times the pore radius (Dagan, 1989). This is the lower bound for the size of an REV.

If, however, the porous medium domain is inhomogeneous with respect to porosity, beyond a $\mathcal{U}_{\max}$, the value of the ratio $\mathcal{U}_v/\mathcal{U}$ will rise or drop, depending on the kind of heterogeneity that exists in the vicinity of $\mathbf{x}$. The volume of the REV, $\mathcal{U}_o$, should be selected in the range, $\mathcal{U}_{\min} \leq \mathcal{U}_o \leq \mathcal{U}_{\max}$.

When the size of $\mathcal{U}$ is further increased, the value of the ratio $\mathcal{U}_v/\mathcal{U}$ may stabilize, or may again rise and drop, similar to what is observed at the microscopic level, but at a larger scale, until, at some value $\mathcal{U}_{\text{meg}}$, a new *plateau* is reached. The corresponding porosity at the megascopic scale will then be ϕ_{meg}.

In principle, the averaging process (see Subs. 1.3.3B) should be repeated for any density ($\equiv$ extensive quantity per unit volume, also referred to as *intensive quantity*) relevant to a considered problem, hoping that the same REV can be determined for all of them. If, indeed, such a common REV can be found, then the considered domain is regarded as a porous medium, in which the behavior of each phase and component can be described as a continuum at the macroscopic level.

As we shall see throughout this book, the core of any model that describes the transport of an extensive property (e.g., mass, mass of a component, energy and momentum) in a considered domain, is the balance of that quantity *at a point*, meaning within a small volume centered at the point. When written per unit volume of the domain, this balance takes the form of a partial differential equation (PDE). At the microscopic level, i.e., within a phase, this volume is the μREV, and the PDE describes the (instantaneous) balance of that quantity as it accumulates within the μREV. This accumulation is a consequence of the net inflow of the considered quantity (by advection and diffusion) and the rate of its production, per unit volume of the μREV centered at the point. Since averaging is actually an integration operation (dividing the result by the volume over which the integration has been performed), and averaging a sum is equal to the sum of averages, the averaged equation is obtained by averaging each term in the equation over the μREV centered at the point.

Once we have the balance equation at the microscopic level, we can average it to obtain its macroscopic counterpart. Bear and Bachmat (1990) describe the methodology of averaging a partial differential equation.

An important example of deriving an averaged equation is the derivation of the averaged momentum balance equation for a fluid that moves in the void space of a porous medium. The movement of a Newtonian fluid is described by the Navier-Stokes equation (see most textbook on Fluid Mechanics). This description of motion is at the microscopic level. As stated earlier, solving problems of flow in porous medium domains at this level is impractical, as we cannot describe the fluid-solid interface that bounds this flow domain. The construction of a continuum model at the macroscopic level is called for. Accordingly, the averaging procedure is applied to the Navier-Stokes equation (e.g., Bear and Bachmat, 1990), in order to find its macroscopic scale counterpart. Often, the averaged expressions become rather complicated, and various simplifying assumptions have to be introduced in order to eventually reach a relatively simple form of the averaged equation. For example, *subject to certain simplifying assumptions, the averaged Navier-Stokes equation is reduced to Darcy's law.* The averaged equations always contain coefficients that incorporate the information on the detailed configuration of the solid-fluid microscopic boundaries that we have been trying to avoid. For example, in Darcy's law the permeability is such a coefficient. Although these coefficients have rigorous definitions and structure, their numerical values for particular porous media can be obtained only experimentally. In Subs. 1.3.5, the homogenization technique is utilized to derive some of the above-mentioned results.

C. Macroscopic heterogeneity

We have already introduced the option of smoothing out heterogeneity at the macroscopic level, in order to obtain the description of phenomena at the megascopic level (which may still be heterogeneous). In this case, the averaging is performed over a representative macroscopic volume (abbreviated RMV). The characteristic size, ℓ^*, of this volume, is constrained by

$$d^* \ll \ell^* \ll L, \tag{1.3.14}$$

where d^* is a length characterizing the macroscopic heterogeneity that we wish to smooth out, and L is a length characterizing the porous medium domain. In fact, the features of the REV listed in Subs. 1.3.4B, as well as the constraints imposed on its size, may, at least in principle, be repeated also here, replacing the terms 'microscopic' and 'macroscopic' by the terms 'macroscopic' and 'megascopic', respectively. Obviously, the length scale of heterogeneity at the megascopic level will be much larger than that corresponding to the macroscopic one.

Similar to what happens at the microscopic-to-macroscopic smoothing, here also, the information about the heterogeneity at the macroscopic level appears at the megascopic one in the form of various coefficients that reflect the effect of the actual spatial distribution of the (geometrical) parameters at the macroscopic level on various phenomena of transport. In practice, when we consider field-scale problems, we use the same models as we use for the description of phenomena of transport (e.g., flow of water) at the macroscopic level, but we *interpret* the coefficients and the variables as average values over an RMV. The only measurable quantities is piezometric head and concentrations. Eventually, we use this data with an inverse approach to determine the values of field coefficients and their spatial variability.

Consider the case of lenses of one material embedded in a domain composed of another material (Fig. 1.3.3); silt lenses in a sandy domain may serve as an example. Let the lenses be of dimensions that are much smaller than the characteristic length of the domain, L, i.e., $L_1, L_2 \ll L$, and be randomly distributed in space. Since we do not know the detailed spatial distribution of the lenses, we wish to replace the real heterogeneous domain by a homogeneous one without the lenses. We achieve this goal by defining a *representative macroscopic volume* (abbreviated as RMV), which is much larger than the scale of heterogeneity at the macroscopic level (i.e., that of variations in permeability, say, spacing between lenses) and averaging over it. By doing so, the permeability of the (macroscopic level) lenticular structure of heterogeneity is replaced by that of an equivalent fictitious homogeneous material having some average, or effective, permeability that is anisotropic at the megascopic scale, with higher permeability in the direction parallel to the lenses. The effective permeability has to be determined by an appropriate field experiment. In Sec. 7.5, in connection with contaminant transport, we shall discuss how such a domain can be modeled as a 'dual continuum'

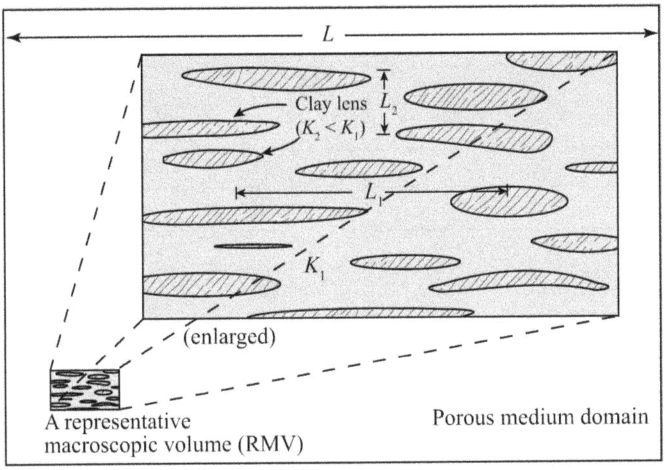

Figure 1.3.3: An inhomogeneous aquifer treated as an equivalent homogeneous one using Representative Macroscopic Volume averaging.

(or 'double porosity') model. If the lenses are non-uniformly distributed, averaging will lead to an equivalent heterogeneous domain at the megascopic scale.

1.3.5 Homogenization

Although the volume averaging technique discussed above has been widely used (mostly conceptually) for the passage from the microscopic level to the macroscopic one, and from the macroscopic to the megascopic one, let us introduce another technique that is generally acknowledged to be more appropriate for handling multiple-scale heterogeneity. This technique is known as the *mathematical theory of homogenization*, a term coined by Babuška (1975, 1976-1977). Since the 1970s, this technique has been applied to a range of physical problems that involve composite materials, heterogeneous geological media, and porous media (Bensoussan *et al.*, 1978; Sanchez-Palencia, 1974, 1980; Lions, 1981; Bakhvalov and Panasenko, 1989; Jikov *et al.*, 1994; Mikelic, 2000).

Briefly, homogenization is a mathematical technique applied to differential equations that describe physical phenomena associated with a domain exhibiting heterogeneities and/or geometrical features at two scales or more. Figure 1.3.4 shows an illustration of such case. The microscopic (pore) scale, and the macroscopic scale may serve as an example. By homogenization, we obtain a domain which is more homogeneous, at least locally. The coefficients, which characterize this 'homogenized' medium, are referred to as 'homogenized', 'equivalent', or 'effective'. In the process of homogenization,

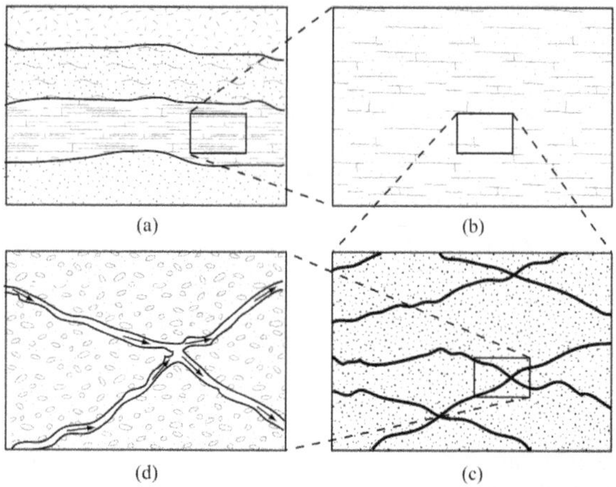

Figure 1.3.4: A porous medium with multiple scales. (a) Scale 1: effective porous medium of a layered formation; (b) Scale 2: a homogeneous porous medium; (c) Scale 3: a dual porosity medium; (d) Scale 4: viscous flow in fracture.

each equation is separated into a number of equations, each addressing the dominant physical phenomena at one of the scales.

Suppose we wish to investigate (1) the microscopic flow of a viscous fluid in a given pore geometry, described by Navier-Stokes equations, (2) the flow at the *macroscopic* scale, where Darcy's law is valid, without being interested in the detailed flow within the pores, or, (3) a layered porous medium, and our interest lies in finding an 'equivalent hydraulic conductivity' at the megascopic scale. These are three problems at three different scales. In principle, this process of *upscaling* may continue indefinitely for whatever scale-dependent physical features exist in the porous medium domain. Different differential equation systems are set up to model the dominant flow at each scale. These constructed physical/mathematical models are correlated—whatever happens at the smaller scale is carried over to the larger scale in the form of lumped coefficients that appear in the differential equations corresponding to that scale. In this way, the *homogenization technique permits the simultaneous examination of processes taking place at multiple scales.* This is the strength of the homogenization method. On the other hand, in most cases, we are not interested in what is happening at the smaller scale. In such case, similar to the volume averaging approach, homogenization is applied to smooth out the smaller scale features that, anyway, are of no practical interest. Once the macroscopic model is constructed, the microscopic model is discarded; only the (single scale) macroscopic (or megascopic) model is used for applications.

A key to the application of the homogenization theory is the existence of a *periodic* structure. This condition is needed for a rigorous mathematical proof of the existence and uniqueness of the solution. Physically, this means a struc-

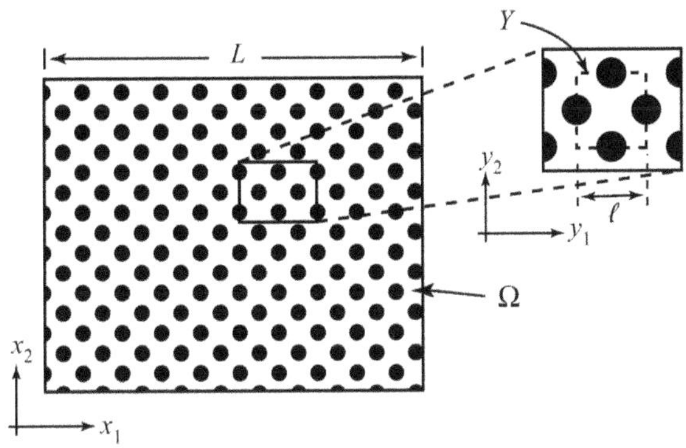

Figure 1.3.5: A domain Ω containing periodic cells Y with period ℓ; $\ell/L \ll 1$.

ture based on a repeated pattern, e.g., a certain form of stacking of grains, or of alternating layers of different materials. Figure 1.3.5 illustrates a domain of size L containing a periodic pattern, with ℓ denoting the period. This pattern could be the packing of spheres in a porous media, or the embedding of particles or fibers in a composite material. For the homogenization theory to be applicable, it is necessary that the ratio $\varepsilon(=\ell/L)$ be a small parameter. In fact, the homogenization seeks the *asymptotic solution* in the limit $\varepsilon \to 0$. For multiple scale domains, such as the one shown in Fig. 1.3.4, we can define several scales $L \gg \ell_1 \gg \ell_2 \gg \dots$, with $\varepsilon_1(=\ell_1/L)$, $\varepsilon_2(=\ell_2/\ell_1)$, $\dots \ll 1$.

In what follows, we shall present an example of homogenization in one spatial dimension, involving an ordinary differential equation exhibiting a two-scale characteristic. The considered (elliptic-type) differential equation, with a variable coefficient, is

$$\frac{d}{dx}\left[a^\varepsilon(x)\frac{du^\varepsilon(x)}{dx}\right] = 0, \quad 0 \leq x \leq L. \tag{1.3.15}$$

It is subject to the boundary conditions

$$u^\varepsilon(0) = 0, \quad \text{and} \quad u^\varepsilon(L) = 1. \tag{1.3.16}$$

Here, L is the length of the domain of interest, considered as the large scale. Within the considered domain, there exist repeated small scale features of size ℓ, introduced through the functional relation of the coefficient a^ε, and $\varepsilon - \frac{\ell}{L} \ll 1$.

As a consequence, the solution u^ε depends also on ε; the superscript is used to emphasize this fact. We shall demonstrate below that even if the variations in the coefficient a^ε are large, expressed in terms of large amplitude fluctuations, their effect on the solution of the differential equation is small; it is of the order $\mathcal{O}(\varepsilon)$ only! In fact, the purpose of homogenization is to seek

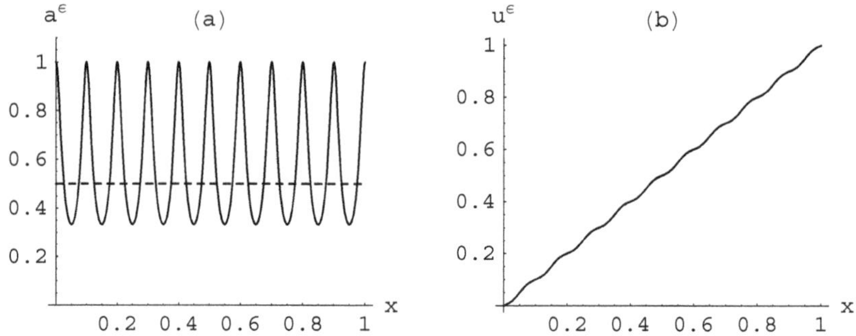

Figure 1.3.6: Solution of (1.3.15) with a rapidly fluctuating coefficient: (a) a plot of a^ε as (1.3.18), with $\varepsilon = 0.1$, in which the dashed line indicates the 'effective coefficient' a_o in (1.3.39); and (b) the solution (1.3.19) for u^ε.

the asymptotic solution as $\varepsilon \to 0$,

$$u(x) = \lim_{\varepsilon \to 0} u^\varepsilon(x). \tag{1.3.17}$$

To demonstrate this feature, and to validate the result of homogenization, we shall use an example for which we can obtain an exact solution.

In the present example, without loss of generality, we shall assume $L = 1$, and hence $\ell = \varepsilon$. Let us assume that a^ϵ is a periodic function, with period ε,

$$a^\varepsilon(x) = \frac{1}{1 + 2\sin^2 \frac{\pi x}{\varepsilon}}. \tag{1.3.18}$$

For a small ε, say, $\varepsilon = 0.1$, this coefficient is plotted as Fig. 1.3.6a, where we observe 'rapid' fluctuations (small period) with a large amplitude.

Before applying the homogenization technique, it would be instructional to examine the exact solution of the problem represented by (1.3.15) and (1.3.16). For $1/\varepsilon$ an integer, this solution is

$$u^\varepsilon(x) = \frac{\int_0^x \frac{dx}{a^\varepsilon(x)}}{\int_0^1 \frac{dx}{a^\varepsilon(x)}} = \frac{4\pi x - \varepsilon \sin \frac{2\pi x}{\varepsilon}}{4\pi - \varepsilon \sin \frac{2\pi}{\varepsilon}} = x - \frac{\varepsilon}{4\pi} \sin \frac{2\pi x}{\varepsilon}, \tag{1.3.19}$$

plotted as Fig. 1.3.6b for $\varepsilon = 0.1$. As observed in the figure, and also in (1.3.19), the solution consists of two parts: a slowly varying part (linear), with rapidly fluctuating, small amplitude 'ripples' superposed on top. Indeed, the magnitude of the 'disturbance' in the solution, caused by the fluctuating coefficient a^ε, is controlled not by the coefficient's amplitude, but by its period, which is small.

As emphasized earlier, homogenization requires the existence of *two scales.* At the larger scale, we denote the domain of size L ($\equiv 1$ in the current case) as

Ω, and use the coordinate system $0 \le x \le 1$. At the small scale, characterized by the periodic cells of size ℓ ($\equiv \varepsilon$) (see Fig. 1.3.5 for a two-dimensional conceptualization), we denote the repeated domain as Y, and use the scaled coordinate

$$y = \frac{x}{\varepsilon}, \tag{1.3.20}$$

such that $0 \le y \le 1$ in a Y-cell.

With the above definition, we now express the coefficient a^ε, defined in (1.3.18), as

$$a^\varepsilon(x) = a(y) = \frac{1}{1 + 2\sin^2 \pi y}. \tag{1.3.21}$$

For $u^\varepsilon(x)$, we can express it as a two-scale function, $u(x,y)$, and expand it into a power series in terms of the small parameter, ε,

$$u^\varepsilon(x) = u(x,y) = u^{(o)}(x,y) + \varepsilon u^{(1)}(x,y) + \varepsilon^2 u^{(2)}(x,y) + \dots \tag{1.3.22}$$

This is known as the *perturbation technique* (Nayfeh, 2000). Substituting the above expression into (1.3.15), and applying the chain rule

$$\frac{d}{dx} = \frac{\partial}{\partial x} + \frac{1}{\varepsilon}\frac{\partial}{\partial y}, \tag{1.3.23}$$

to the two-scale functions, we can expand and separate the resulting equation into several equations, each corresponding to the same power of ε:

$$\mathcal{O}(\varepsilon^{-2}): \quad \frac{\partial}{\partial y}\left[a(y)\frac{\partial u^{(o)}(x,y)}{\partial y}\right] = 0, \tag{1.3.24}$$

$$\begin{aligned}\mathcal{O}(\varepsilon^{-1}): \quad & \frac{\partial}{\partial x}\left[a(y)\frac{\partial u^{(o)}(x,y)}{\partial y}\right] + \frac{\partial}{\partial y}\left[a(y)\frac{\partial u^{(o)}(x,y)}{\partial x}\right] \\ & + \frac{\partial}{\partial y}\left[a(y)\frac{\partial u^{(1)}(x,y)}{\partial y}\right] = 0,\end{aligned} \tag{1.3.25}$$

$$\begin{aligned}\mathcal{O}(\varepsilon^{0}): \quad & \frac{\partial}{\partial x}\left[a(y)\frac{\partial u^{(o)}(x,y)}{\partial x}\right] + \frac{\partial}{\partial x}\left[a(y)\frac{\partial u^{(1)}(x,y)}{\partial y}\right] \\ & + \frac{\partial}{\partial y}\left[a(y)\frac{\partial u^{(1)}(x,y)}{\partial x}\right] + \frac{\partial}{\partial y}\left[a(y)\frac{\partial u^{(2)}(x,y)}{\partial y}\right] = 0,\end{aligned} \tag{1.3.26}$$

and higher order equations. The boundary conditions (1.3.16), are assigned to the leading terms, such that the higher order terms take the null boundary conditions:

$$\begin{aligned} u^{(o)}(0,y) = 0, \quad u^{(o)}(1,y) = 1; \quad u^{(1)}(0,y) = u^{(1)}(1,y) = 0; \\ u^{(2)}(0,y) = u^{(2)}(1,y) = 0; \quad \dots \end{aligned} \tag{1.3.27}$$

Also, the periodicity of the Y-cells requires that

$$
\begin{aligned}
& u^{(1)}(x,0) = u^{(1)}(x,1); \quad u^{(2)}(x,0) = u^{(2)}(x,1); \quad \dots \\
& \left.\frac{\partial u^{(1)}(x,y)}{\partial y}\right|_{y=0} = \left.\frac{\partial u^{(1)}(x,y)}{\partial y}\right|_{y=1}; \\
& \left.\frac{\partial u^{(2)}(x,y)}{\partial y}\right|_{y=0} = \left.\frac{\partial u^{(2)}(x,y)}{\partial y}\right|_{y=1}; \quad \dots
\end{aligned}
\tag{1.3.28}
$$

A quick inspection of the $\mathcal{O}(\varepsilon^{-2})$-equation, (1.3.24), shows that

$$
u^{(o)}(x,y) = u^{(o)}(x) \tag{1.3.29}
$$

is an admissible solution. Based on a theorem of existence and uniqueness presented in Subs. 4.2.3, as shown in (4.2.67) and the bounding condition below it, (1.3.29) is, in fact, the *only* admissible (unique) solution of (1.3.24).

With the condition expressed by (1.3.29), the $\mathcal{O}(\varepsilon^{-1})$-equation, (1.3.25), can be simplified to

$$
\frac{\partial a(y)}{\partial y}\frac{\partial u^{(o)}(x)}{\partial x} + \frac{\partial}{\partial y}\left[a(y)\frac{\partial u^{(1)}(x,y)}{\partial y}\right] = 0. \tag{1.3.30}
$$

To solve this equation, we assume the existence of a Y-periodic function $w(y)$, satisfying

$$
\frac{\partial a(y)}{\partial y} + \frac{\partial}{\partial y}\left[a(y)\frac{\partial w(y)}{\partial y}\right] = 0. \tag{1.3.31}
$$

Comparing (1.3.31) with (1.3.30), and realizing that $\partial u^{(o)}/\partial x$ is not a function of y, it can easily be shown that

$$
u^{(1)}(x,y) = w(y)\frac{\partial u^{(o)}(x)}{\partial x} + f(x), \tag{1.3.32}
$$

where $f(x)$ is an arbitrary function of x. Equation (1.3.32) can be differentiated with respect to y, yielding

$$
\frac{\partial u^{(1)}(x,y)}{\partial y} = \frac{\partial w(y)}{\partial y}\frac{\partial u^{(o)}(x)}{\partial x}. \tag{1.3.33}
$$

We can solve (1.3.31) with the periodicity condition $w(0) = w(1)$, to obtain

$$
w(y) = \frac{\int_0^y \frac{dy}{a(y)}}{\int_0^1 \frac{dy}{a(y)}} - y + c, \tag{1.3.34}
$$

where c is an arbitrary additive constant.

We now turn to the $\mathcal{O}(\varepsilon^0)$-equation, (1.3.26). Integrating it with respect to y over a Y-cell, i.e., from $y = 0$ to 1, we observe that due to the periodicity property, (1.3.28), the integration of the last two terms in (1.3.26) vanishes:

$$\int_0^1 \frac{\partial}{\partial y}\left[a(y)\frac{\partial u^{(1)}(x,y)}{\partial x}\right] + \frac{\partial}{\partial y}\left[a(y)\frac{\partial u^{(2)}(x,y)}{\partial y}\right] dy$$
$$= a(y)\frac{\partial u^{(1)}(x,y)}{\partial x}\bigg|_{y=0}^{1} + a(y)\frac{\partial u^{(2)}(x,y)}{\partial y}\bigg|_{y=0}^{1} = 0. \qquad (1.3.35)$$

We substitute (1.3.33) for the first two terms in (1.3.26), to obtain

$$\frac{\partial}{\partial x}\left[a(y)\frac{\partial u^{(o)}(x)}{\partial x}\right] + \frac{\partial}{\partial x}\left[a(y)\frac{\partial u^{(1)}(x,y)}{\partial y}\right]$$
$$= \frac{\partial}{\partial x}\left[a(y)\left(1+\frac{\partial w(y)}{\partial y}\right)\frac{\partial u^{(o)}(x)}{\partial x}\right]. \qquad (1.3.36)$$

After integrating the right hand side of the above expression over the Y-cell, (1.3.26) becomes

$$\frac{d}{dx}\left[a_o\frac{du^{(o)}(x)}{dx}\right] = 0, \qquad a_o = \int_0^1 a(y)\left(1+\frac{\partial w(y)}{\partial y}\right) dy. \qquad (1.3.37)$$

With the substitution of (1.3.34), we obtain

$$a_o = \frac{1}{\int_0^1 \frac{dy}{a(y)}}. \qquad (1.3.38)$$

Hence a_o lumps the details of the periodic cell; it is a new coefficient for the 'homogenized' differential equation (1.3.37). By inserting the actual expression for $a(y)$ appearing in (1.3.21) into the above equation, we obtain

$$a_o = \frac{1}{2}. \qquad (1.3.39)$$

This value is plotted as the dashed line in Fig. 1.3.6a.

We are now ready to solve the homogenized equation, (1.3.37). With the boundary condition $u^{(o)}(0) = 0$ and $u^{(o)}(1) = 1$, as given in (1.3.27), equation (1.3.37) can be solved to yield

$$u^{(o)}(x) = x. \qquad (1.3.40)$$

As observed in Fig. 1.3.6b, the above linear term is exactly the anticipated large scale behavior.

Due to the simplicity of the one-dimensional problem, this result may look trivial. However, we can examine the next quantity of interest, the 'flux',

$$q^\varepsilon = -a^\varepsilon\frac{du^\varepsilon}{dx}. \qquad (1.3.41)$$

This term often has a physical meaning, e.g., the heat flux in the case of thermal conduction, or the specific discharge in porous medium flow. Using (1.3.18) and (1.3.19) in (1.3.41), we find that

$$q^\varepsilon = -\frac{1}{2}. \tag{1.3.42}$$

This behavior is not obvious from the coefficient a^ε (see also Fig. 1.3.6a). The homogenization process, however, correctly captures this behavior by providing the effective coefficient a_o as in (1.3.38), and in (1.3.39), such that

$$q_o = -a_o \frac{du^{(o)}}{dx} = -\frac{1}{2}. \tag{1.3.43}$$

Just as expected, the solutions (1.3.40) and (1.3.43) are independent of ε, because these are the asymptotic solutions of u^ε and q^ε as $\varepsilon \to 0$.

Following the same procedure, we can find the solution of higher order terms. The first order asymptotic expansion term, $u^{(1)}$, is obtained from (1.3.32) and (1.3.34), together with the boundary conditions (1.3.27) and (1.3.28), in the form

$$u^{(1)}(y) = -\frac{1}{4\pi} \sin(2\pi y). \tag{1.3.44}$$

This term can be compared with the second term of the exact solution (1.3.19). Hence the term $\varepsilon u^{(1)}$ in (1.3.22) is a rapidly fluctuating term, with period $\varepsilon = \ell/L$, and amplitude of order $\mathcal{O}(\varepsilon)$. This term is needed only if ε is of intermediate scale. Its effect vanished as $\varepsilon \to 0$.

The above analysis, carried out for a one-dimensional problem, presents the essentials of the *homogenization* procedure. Let us reiterate a few important points and conclusions.

- Homogenization models require the assumption of periodicity at the smaller scale (Fig. 1.3.6a). Although this requirement may be viewed as a restriction, as natural materials are not periodic, This assumption provides the necessary boundary condition for a rigorous mathematical analysis, ensuring the existence and uniqueness of the solution. This theoretical *closure* (≡ existence and uniqueness theorem) of the mathematical problem is the strength of the homogenization theory.
- We have demonstrated that in a problem that involves two (or multiple) scales, the asymptotic expansion procedure, which is the basic tool of the homogenization technique, allows us to separate the governing equation into a number of scales, each governing the process at a specified scale. The periodic assumption allows the smaller scale effects to be averaged (= integrated) to produce lumped (or effective) coefficients.
- The small scale feature is unlikely to be truly periodic; the periodicity, introduced in order to facilitate homogenization, should be regarded as a *conceptual model.*

In Chapter 4, the homogenization technique will be used to derive the (macroscopic) Darcy's law from the (microscopic) law governing viscous flow in the void space (Subs. 4.2.2), and the equivalent anisotropic hydraulic conductivity at the megascopic scale for a layered formation (Subs. 4.2.3).

1.4 Scope and Organization

At the beginning of this chapter, we introduced the 'problems' that are of interest to groundwater hydrologists and managers: water flow, and contaminant transport in both the saturated and the unsaturated zones. In both cases, models provide management with information that is essential to decision making, e.g., water levels and solute concentrations, to be expected if certain management decisions will be implemented. We have limited the presentation to isothermal conditions, although, under certain circumstances, (man-made, or naturally occurring) temperature variations may have a significant influence on the flow and solute transport regimes in the subsurface.

With the above in mind, Chapter 1 introduces groundwater within the hydrological cycle, and outlines the role of aquifers within the framework of a water resources system. Since the models discussed in this book visualize the porous medium as a continuum, we also explain how the complex solid-fluid(s) domain, called 'a porous medium', is transformed into a continuum.

Chapter 2 Presents the definition and classification of aquifers and introduces the hydraulic approach, based on the assumption of 'essentially horizontal flow', as an important and practical mode of modeling employed in many cases in practice.

Chapter 3 reviews the regional groundwater balance and its components. Both natural and man-introduced components are considered. In each case, both the quantity and the quality of the water are discussed.

Chapter 4 is devoted to the basic equation of groundwater motion—Darcy's law. First, the equation for three-dimensional flow is introduced. Then, the integrated equations for flow in confined and phreatic aquifers are developed and discussed. In doing so, the transmissivity, as an aquifer parameter, and the Dupuit assumption, as a good approximation, are introduced in connection with the hydraulic approach to flow in aquifers. The basic motion (or Darcy's) equation is extended to inhomogeneous fluids and to inhomogeneous and anisotropic aquifers.

Chapter 5 leads to complete saturated flow models. It introduces the definitions of specific storativity and aquifer storativity, and employs these coefficients to construct mass balance equations for flow in three-dimensional domains. Then, models based on the concept of 'essentially horizontal flow' are developed for confined, phreatic, and leaky aquifers by integrating the three-dimensional flow model over the vertical thickness of the aquifer. It is shown how, by doing so, the phreatic and other upper and lower boundary conditions are incorporated as source terms in the equations. Following a discussion of boundary and initial conditions, the structure of the complete mathematical statement of any groundwater flow problem is presented.

Upon reaching this point, the reader should be able to state, correctly and completely, any problem of groundwater flow in an aquifer, in terms of a partial differential equation and appropriate initial and boundary conditions.

Chapter 6 presents flow in the unsaturated zone. This is an important subject as aquifer replenishment takes the form of unsaturated downward flow, and because of the transport of pollutants with this downward flow. The discussion leads to complete well-posed flow models.

Chapter 7 presents a comprehensive discussion on the transport and accumulation of chemicals in the subsurface. The main new feature here is the phenomenon of hydrodynamic dispersion. The general mass balance equation for a chemical species, often a contaminant, is developed for both saturated and unsaturated flow. Following a discussion on boundary and initial conditions, the complete, mathematical statement of the problem of movement and accumulation of a chemical species is presented. The discussion includes the effects of sources and sinks of chemical species, due to such phenomena as adsorption, dissolution, volatilization, and chemical reactions, under both chemical equilibrium and kinetic conditions.

Chapter 8 reviews numerical solution techniques for solving the flow and transport models presented in Chaps. 5, 6 and 7. The objective is to present an introduction to the fundamentals and methodologies of constructing numerical flow and transport models and solving them by using computer programs ('codes'). A brief introduction is presented to some of the more commonly used codes.

Chapter 9 deals with the important problem of seawater intrusion into coastal aquifers, as a consequence of over-exploitation. The discussion leads to complete models that describe this phenomenon. Two approaches are discussed: one that visualizes seawater and freshwater as two immiscible fluids separated by a sharp interface, and another one that regards seawater and freshwater as a single liquid with variable concentrations of dissolved salts and, thus, having a variable density. Some comments are made on the issue of management of a coastal aquifer.

Chapter 10 reviews the issues of uncertainty associated with modeling. Sources of uncertainty are examined. The basic definitions of a stochastic process and tools for its analysis, such as sensitivity analysis, kriging, Monte Carlo and perturbation methods, are introduced.

Chapter 11 serves as an introduction to the management of groundwater. It discusses the management issues involved, and then introduces tools that can be used by managers for quantitative predictions. Several optimization methodologies, including the constrained linear programming, the unconstrained gradient search, the genetic algorithm, optimization under uncertainty, and multiobjective optimization, are introduced. Inverse problems, particularly problems of parameter estimation, are also presented.

Chapter 2

GROUNDWATER AND AQUIFERS

2.1 Definitions of Aquifers

Subsurface water, or **groundwater**, is a term used to denote all the waters found beneath ground surface. However, groundwater hydrologists, who are primarily (but not exclusively) concerned with the water contained in the zone of saturation (Subs. 1.1.1), often use the term 'groundwater' to denote water in only this zone. In this book, we adhere to this definition, using the term *subsurface water* to denote *all* the water below ground surface. Practically, all groundwater can be regarded as part of the hydrological cycle (Fig. 1.1.1; or see any textbook on Hydrology). Very small amounts, however, may enter the cycle from other sources (e.g., magmatic water).

Aquifer. Todd (1959) traces the term 'aquifer' to its Latin origin: *aqui* comes from *aqua*, meaning "water", and *fer*, from *ferre*, meaning "to bear". Thus, an aquifer is a term used for a porous geological formation that (1) contains water at full saturation (i.e., the entire interconnected void space is filled with water), and (2) permits water to move through it under ordinary field conditions. This means that whether a geological formation can be designated as an aquifer, or not, depends on its ability to store and transmit water *relative* to other formations in the vicinity. Other terms often used are: groundwater reservoir (or basin) and water bearing zone (or formation).

Aquitard. This is a semipervious geologic formation that transmits water at a very low rate compared to an aquifer. However, the term should not be used just for any low permeability formation. Instead, the term is restricted to describe a semipervious layer which (1) is thin relative to the thickness of the aquifer underlying or overlying it, (2) has a permeability that is much smaller than that of such an aquifer, and (3) extends over relatively large horizontal areas. In spite of its relatively low permeability, over such large (horizontal) areas, it may permit the passage of large quantities of water between the aquifers that are separated from each other by it. It is often referred to as a *semipervious formation* or a *leaky formation*.

Aquifuge. This is a rarely used term that describes an impervious formation, which neither contains nor transmits water.

J. Bear, A.H.-D. Cheng, *Modeling Groundwater Flow and Contaminant Transport*,
Theory and Applications of Transport in Porous Media 23,
DOI 10.1007/978-1-4020-6682-5_2,

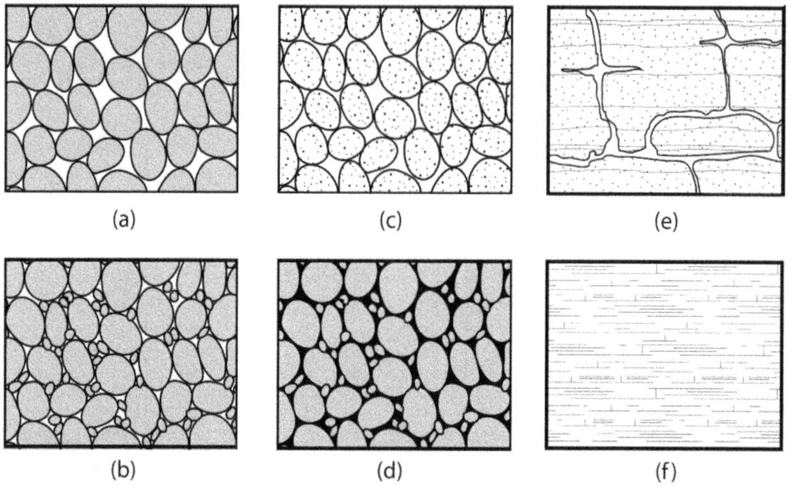

Figure 2.1.1: Diagram showing types of rock interstices: (a) Well-sorted sedimentary deposit having high porosity; (b) Poorly sorted sedimentary deposit having low porosity; (c) Well-sorted sedimentary deposit consisting of porous pebbles, so that the deposit as a whole has a very high porosity; (d) Well-sorted sedimentary deposit whose porosity has been diminished by the deposition of mineral matter in the interstices; (e) Rock rendered porous by solution; and (f) Rock rendered porous by fracturing (after Meaner, 1942).

The solid portion of a rock formation is called the *solid matrix*. The portion of the rock formation which is not occupied by solid matter is the *void space* (or *pore space*) (Sec. 1.3). In general, the void space may contain in part a liquid phase (water), and in part a gaseous phase (air). Only connected interstices can act as elementary conduits within the formation. Figure 2.1.1 shows several types of rock interstices. These may range in size from huge limestone caverns to minute subcapillary openings in which water is held primarily by adhesive forces. The interstices of a rock formation can be grouped in two classes: original interstices (mainly in sedimentary and igneous rocks) created by geological processes at the time the rock was formed, and secondary interstices, mainly in the form of fissures, joints, and solution passages developed after the rock was formed.

2.2 Moisture Distribution in Vertical Soil Profile

The subsurface is a *porous medium* (see definition in Sec. 1.3) comprised of a *solid matrix* and a *void space*. The latter is filled with one or more fluids. The term *subsurface water* has been used to designate all water present in the soil below ground surface. Usually, this term refers only to liquid water.

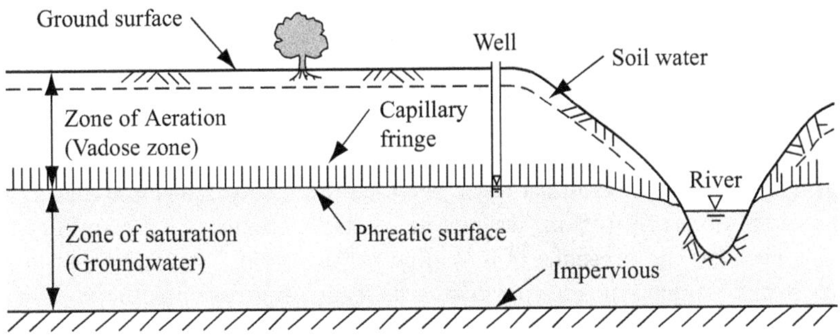

Figure 2.2.1: Subsurface moisture zones (after Bear, 1979).

When we are interested in the movement of fluids and contaminants, the term *subsurface fluids* will, usually, be used to refer to fluids (liquids and gases) in the interconnected part of the void space, through which fluid motion can take place.

The soil beneath ground surface may be divided vertically into a number of essentially horizontal zones, according to the proportion of the void space that is filled with water. Figure 2.2.1 provides a convenient schematic description for a homogeneous soil. We note two major zones:

- The *saturated zone*, in which the entire void space is occupied by water.
- The *unsaturated zone* (also referred to as the *zone of aeration*, or the *vadose zone*), where only part of the void space is occupied by water, the remainder being occupied by a gaseous phase, air. The behavior of water in this zone is discussed in Chap. 6.

A nonaqueous liquid phase may occupy part of the void space in both zones (see Subs. 1.1.5E).

Water (e.g., from precipitation and/or irrigation) infiltrates through ground surface, moves downwards, primarily under the influence of gravity, and accumulates above some impervious bedrock, completely saturating all the interstices of the rock formation. This is the saturated zone. Wells, springs, and effluent streams act as outlets of water from this zone. The saturated zone in Fig. 2.2.1 is bounded from above by a *water table*, or *phreatic surface*. We shall see below that under different circumstances, the upper boundary of a saturated zone can be an impervious one. The phreatic surface will be discussed in detail in Sec. 4.5. At this time, let us define it as the *imaginary surface at all points of which the pressure is atmospheric*. In Fig. 2.2.1, this surface is revealed by the level at which water stands in a well just penetrating the aquifer. When the flow in an aquifer is essentially horizontal, the depth of the observation well's screen below the water table is immaterial. Actually, as will be shown in Subs. 6.1.8, full saturation (or almost so) extends a certain distance above the water table, depending on the type of soil. The unsaturated zone can further be divided into three subzones: the soil water zone, or the *root zone*, close to ground surface, the intermediate zone, and the capillary zone (or capillary fringe), immediately above the water table.

The upper part of the unsaturated zone is called *soil water zone*. It extends downward through the root zone. Vegetation depends on water in this zone, as plants require both air and water. The moisture distribution in this zone is strongly affected by conditions at ground surface, i.e., seasonal and diurnal fluctuations of precipitation, irrigation, air temperature, and humidity. It is also affected by the presence of a shallow water table. When the water table of the aquifer is deep, it does not affect the moisture distribution in this zone. Water in this zone moves downward during infiltration (e.g., from precipitation, flooding of ground surface or irrigation), and upward by evaporation and plant transpiration. Temporarily, during a short period of excessive infiltration, the soil in this zone may be almost completely saturated.

After an extended period of gravity drainage, without additional supply of water at ground surface, the amount of moisture remaining in the soil is called *field capacity* (see Subs. 6.1.9). Below field capacity, the soil contains water in the form of continuous thin films on the soil particles and menisci between them, held by surface tension. Water in these films is moved by capillary action and is available to plants. Below some moisture content, called the *hygroscopic coefficient* (= maximum moisture which an initially dry soil will adsorb when brought in contact with an atmosphere of 50% *relative humidity* at 20°C), the water in the soil is called *hygroscopic water*. It also forms very thin films of moisture on the surface of soil particles, but the adhesive forces are very strong, so that this water is unavailable to plants.

The *intermediate zone* extends from the lower edge of the soil water zone to the upper limit of the capillary fringe (see below). Its thickness depends on the depth of the water table below ground surface; it does not exist when the water table is high, in which case the capillary fringe may extend into the soil water zone, or even to ground surface. Temporarily, during replenishment periods, water moves downward through this zone as gravitational water.

The *capillary fringe* (Subs. 6.1.8) extends from the water table up to the limit of the capillary rise of water. Its thickness depends on the soil properties and on the homogeneity of the soil, mainly on the pore size distribution. The capillary rise ranges from practically nothing in coarse material, to as much as 2 m to 3 m and more in fine materials (e.g., clay). Within the capillary fringe there is a gradual decrease in moisture content with height above the water table. Just above the water table, the pores are practically saturated. Moving higher, only the smaller connected pores contain water. Still higher, only the smallest connected pores are filled with water. Hence, the upper limit of the capillary fringe has an irregular shape. For practical purposes, some average smooth surface is taken as the upper limit of the capillary fringe, such that below it the soil is *assumed* practically saturated (say > 75%).

In most regional groundwater studies, the capillary fringe is much thinner than the thickness of the saturated zone and is, therefore, disregarded. The phreatic surface then serves as the upper bound of the saturated zone.

Within a homogeneous unsaturated zone, in the absence of infiltration or evaporation, the moisture content generally decreases gradually with height above the phreatic surface. Infiltration will cause saturation to rise temporarily as ground surface is approached. In many cases, the spatial variability in soil properties is the dominant factor in determining the distribution of moisture content, with regions of fine soils having a higher moisture content, while those of coarse materials having a low moisture content.

Numerous complications are introduced into the schematic moisture distribution described here by the large variability in pore sizes, by the presence of layers of different permeability, and by the temporary movement of infiltrating water. All these subjects will be discussed in detail in Chap. 6

Note that although, traditionally, water in the intermediate zone was referred to as 'vadose water', in recent years, the term 'vadose zone' is often used to denote the 'unsaturated' zone, i.e., the zone between ground surface and the underlying phreatic surface.

2.3 Classification of Aquifers

The term *aquifer* was introduced in Sec. 2.1. Let us now introduce the definitions of specific aquifer types.

The *piezometric head* and the *piezometric surface* will be defined in Subs. 4.1.1. At this point, the former will be defined as the water elevation at a well that has a screened portion at a point within an aquifer. By measuring the piezometric heads at a number of spatially distributed observation wells, tapping the same aquifer, especially with essentially horizontal flow, we obtain a contour map that defines a surface called the *piezometric surface*. The elevation of this surface at a point in the horizontal plane gives the piezometric head in the aquifer at that point. The phreatic surface is also a piezometric surface (just think of observation wells with screens just below the water table).

In what follows, the location of the piezometric surface will be used in the classification of aquifers.

A **confined aquifer** (Fig. 2.3.1a) is one that (1) is bounded from above and from below by impervious formations, and (2) the water pressure in it is such that the level of water in a well that is open in it will be at, or will rise above the upper impervious bounding surface. In other words, the piezometric surface of a confined aquifer is above the latter's impervious ceiling.

An aquifer that is bounded from above by a phreatic surface (as in Fig. 2.3.1b) is called a **phreatic aquifer**, or an **unconfined aquifer**. For an aquifer with essentially horizontal flow, the water table is also the piezometric surface of the aquifer. A special case of a phreatic aquifer is the **perched aquifer** (Fig. 2.3.1c). This is a phreatic aquifer of limited areal extent, formed on a semipervious, or impervious, layer that is present between the persistent water table of a phreatic aquifer and ground surface. A perched aquifer may

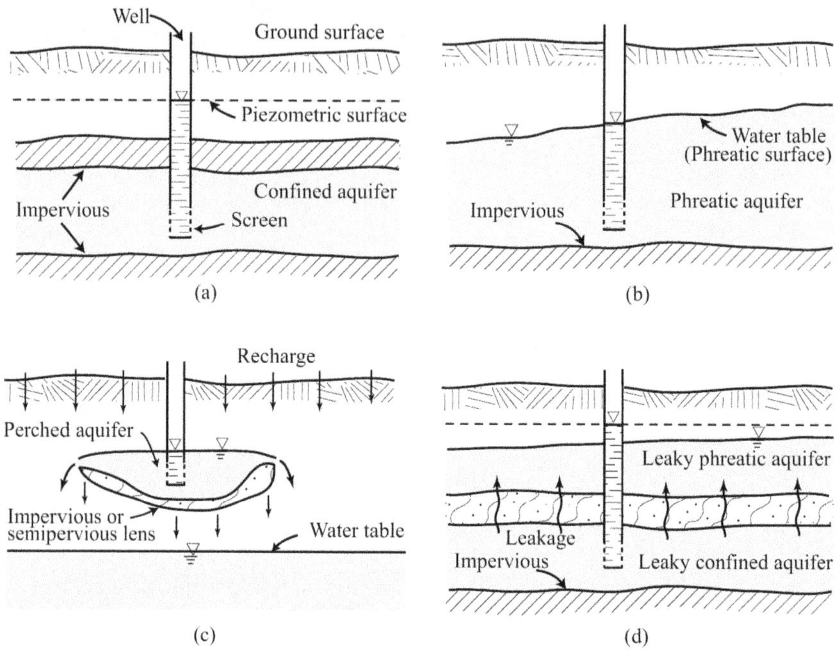

Figure 2.3.1: (a) Confined aquifer. (b) Phreatic aquifer. (c) Perched aquifer. (d) Leaky aquifer.

exist only for a limited period of time, seasonal or ephemeral, as its water drains into the underlying phreatic aquifer.

A **leaky phreatic aquifer**, shown in Fig. 2.3.1d, is a phreatic aquifer that is bounded from below by a semipervious layer, usually referred to as an *aquitard.* The latter was defined in Sec. 2.1.

A **leaky confined aquifer** is a confined aquifer, except that one or both confining layers are aquitards, through which leakage may take place. Figure 2.3.1d shows the two kinds of a leaky aquifer.

Figure 2.3.2 shows several aquifers and observation wells. Also indicated are the phreatic surface of aquifer A and the piezometric surfaces of aquifers B and C. The upper phreatic aquifer, A, is underlain by two confined aquifers, (B and C). In the recharge area, aquifer B becomes phreatic. Portions of aquifers A, B, and C are leaky, with directions and rates of leakage determined by the elevations of the piezometric surface in each of these aquifers. The boundaries between the various confined and unconfined portions may vary with time, as a result of fluctuations in the piezometric surfaces.

Because, in zones with non-horizontal flow, the piezometric head may vary along the vertical, it is important that the screened portion of a piezometer be made relatively short. A well with an elongated screen, will provide a piezometric head that cannot be assigned to a specific point. In fact, when

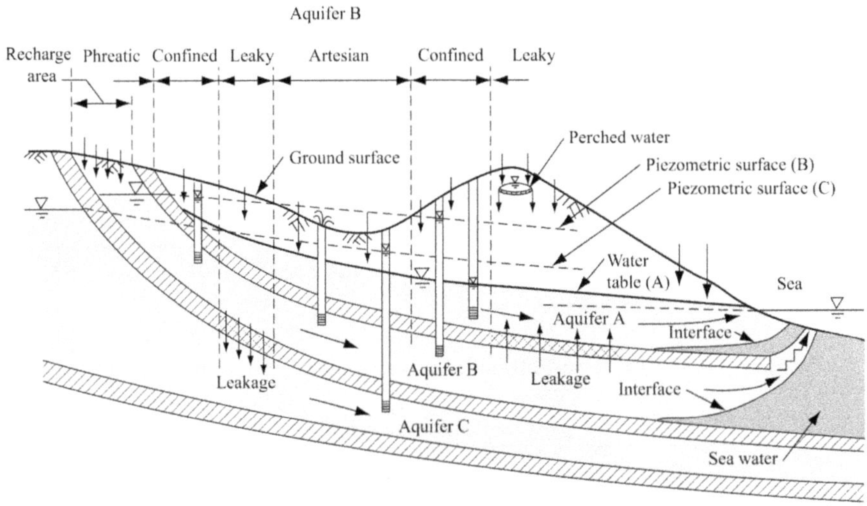

Figure 2.3.2: Types of aquifers.

different screens on the same casing are located at points that have different piezometric heads, flow will take place *through the casing* from high to low piezometric heads, thus creating a new flow regime. When such screens are located within different aquifers, they constitute conduits that enable flow between these aquifers.

2.4 Solid Matrix Properties

2.4.1 Soil classification based on grain size distribution

Soils are composed of mineral particles, or grains, that may have a broad range of sizes. The *grain size distribution* is an important characteristic of a soil, as it affects soil properties such as permeability to fluids, and retention of fluids in the case of multiple phases. It also affects various chemical properties that depend on the surface area of the mineral particles. Soil particles are divided into classes referred to as *size separates.* A number of organizations have defined various standards for size-separate classes. Figure 2.4.1 shows the classification used by the American Society of Testing Materials (ASTM), the US Department of Agriculture (USDA), and the International Soil Science Society (ISSS). In all systems, the primary classes are: cobbles, gravel, sand, silt, and clay. The upper limit of the clay-size separate is 0.002 mm in both the USDA and the ISSS systems, while the range of sand size is between 0.05 and 2.0 mm.

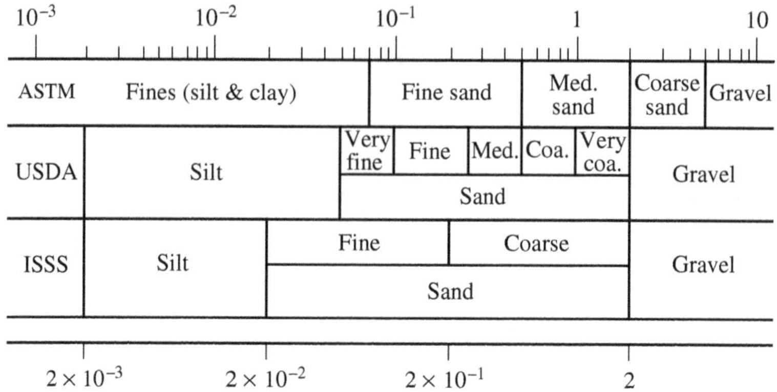

Figure 2.4.1: Major soil classifications according to particle size (mm).

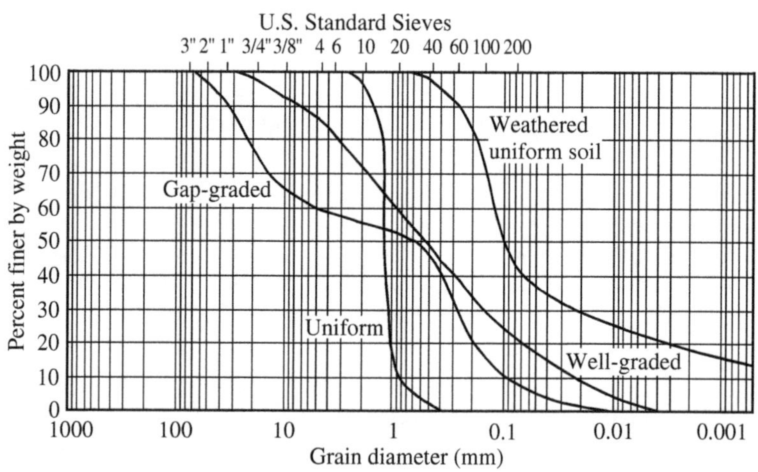

Figure 2.4.2: Typical cumulative grain size distribution curves.

Laboratory measurements, involving sieve fractionation techniques for coarse particles (sand and gravel) and sedimentation methods for fine particles (silt and clay), may be performed to determine the mass fractions in the various size separates. The particle size distribution is often represented as a plot of the cumulative mass fraction of the soil finer than a given size (Fig. 2.4.2). From such a plot, it is possible to determine the median grain size (denoted as d_{50}), or the grain size for which a certain fraction of the soil mass is finer than (e.g., 10% of the soil mass is finer than the diameter d_{10}). Such values are often useful for characterizing the soil. In many cases,

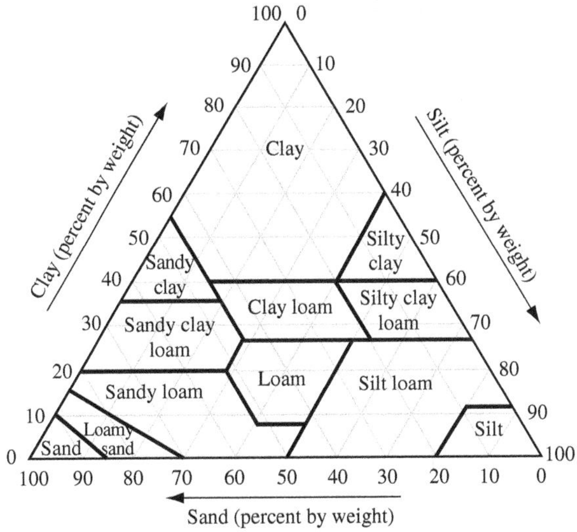

Figure 2.4.3: USDA soil textural triangle.

they have been found to correlate well with various important physical soil properties, such as permeability.

Various soil classification systems, based solely or partly on soil particle size distribution, have been proposed and used by various organizations. The USDA classification system is based only on the size distribution of particles finer than 2 mm (i.e., sand, silt, and clay). Depending on the fraction of these three size separates, the soil textural class is assigned as shown in Fig. 2.4.3. Another system, commonly used by geotechnical and geological engineers, is the *unified* soil classification (USC) method originally proposed by A. Casagrande in 1942, and revised in 1952 by the Corps of Engineers and the Bureau of Reclamation (Das, 1983, p. 35).

2.4.2 Porosity and void ratio

Porosity, ϕ, at a point in a porous medium domain, is defined as the volume of void space per unit volume of porous medium at that point

$$\phi(\mathbf{x}, t) = \frac{\mathcal{U}_{ov}(\mathbf{x}, t)}{\mathcal{U}_o(\mathbf{x})}, \tag{2.4.1}$$

where $\mathcal{U}_o$ and $\mathcal{U}_{ov}$ are the volume of the REV centered at point $\mathbf{x}$, and the void space within the REV, respectively. The porosity depends on the texture and structure of the soil. Soil porosity varies over a wide range of values, from less than 30% to over 90%. Table 2.4.1 gives typical porosity values for a range of natural materials.

Material	Porosity
Peat soil	0.6–0.8
Soils	0.5–0.6
Clay	0.45–0.55
Silt	0.4–0.5
Medium to coarse mixed sand	0.35–0.4
Uniform sand	0.3–0.4
Fine to medium mixed sand	0.3–0.35
Gravel	0.3–0.4
Gravel and sand	0.3–0.35
Glacial till	0.1–0.2
Sandstone	0.1–0.2
Shale	0.01–0.1
Limestone	0.01–0.1
Carbonate mud	0.4-0.7
Dolomite	0.001–0.15
Chalk	0.15–0.45
Fractured igneous rock	0.01–0.1
Karst limestone	0.05–0.5
Basalt	0.01-0.25

Table 2.4.1: Typical porosity of natural materials.

Sometimes, the void space is made up of two parts: an *interconnected* portion, through which a fluid can move from any point to any other point within this portion, and a *non-interconnected* portion. A fluid present in the latter portion of the void space cannot leave it, except by crossing solid phase boundaries. Because we are interested primarily in the transport of mass of fluid phases and chemical species within the void space, unless otherwise specified, we shall use the term porosity to indicate only the interconnected portion of the void space. The terms *interconnected porosity* and *effective porosity* are sometimes used for this purpose.

One should be particularly careful in using the term 'effective porosity', because this term has a number of additional interpretations. For example, in some porous media, the configuration of the (interconnected) void space is such that most of the flow of a fluid phase takes place through only part of the interconnected void space, with only a small fraction of the total flux taking place through the remaining portion of the void space. We often approximate the situation by assuming that in the latter portion of the void space, the fluid is (practically) stationary, or *immobile*. This may happen, for example, when pores have the shape of a *dead-ends*, or when very small pores, say between very small grains, are mixed with very large ones. We then assume that the fluid is practically immobile within the dead-end pores and in the very small ones. The term 'effective porosity' is used to describe that part of the total void space through which (most of the) flow takes place.

The term *void ratio*, e, defined as

$$e(\mathbf{x},t) = \frac{\mathcal{U}_{ov}(\mathbf{x},t)}{\mathcal{U}_{os}(\mathbf{x},t)} = \frac{\phi}{1-\phi}, \tag{2.4.2}$$

is used mainly in soil mechanics, with $\mathcal{U}_{os}$ denoting the volume of the solid matrix within an REV.

The *bulk density* of the soil, ρ_b (= mass of the solid per unit volume of porous medium), is defined as

$$\rho_b = \frac{m_s}{\mathcal{U}_o} = \rho_s \frac{\mathcal{U}_s}{\mathcal{U}_o} = \rho_s(1-\phi). \tag{2.4.3}$$

2.4.3 Specific surface

The specific surface area, $\Sigma_{sv} = S_{sv}/m_s$, where S_{sv} is the total surface area of the solid matrix (= solid-void interface) and m_s is the mass of the solid matrix, is defined as the surface area of the solid matrix per unit mass of soil. A typical unit is m^2/g. It is a very important soil characteristic, especially in connection with surface phenomena such as adsorption and ion-exchange (Sec. 7.3). Fine soils, e.g., clay, are characterized by a huge value of Σ_{sv}.

To estimate Σ_{sv}, consider a soil made up of spherical particles of diameter d. For such spheres, the area per unit mass is given by $6/\rho_s d$, where ρ_s is the mass density of the solid. For a soil composed of a number of fractions of particle sizes, with m_i denoting the mass of solid in the ith fraction, we have

$$\Sigma_{sv} = \frac{6}{\rho_s} \sum_{(i)} \frac{m_i}{m_s} \frac{1}{d_i}. \tag{2.4.4}$$

For soil particles in the form of platelets $\ell \times \ell \times b$, the specific area is

$$\Sigma_{sv} = \frac{2(\ell + 2b)}{\rho_s \, \ell \, b}, \tag{2.4.5}$$

where b is the platelet thickness. For very thin platelets, $\Sigma_{sv} \approx 2/\rho_s b$.

For soils made of spherical or cubical particles, with a distribution of sizes, we can make use of the relationship

$$\Sigma_{sv} = \frac{3(C_u + 7)}{4\rho_s \, d_{50}}, \tag{2.4.6}$$

where d_{50} is soil particle size of 50% passing (50% of the cumulative distribution curve), and $C_u = d_{60}/d_{10}$ is the soil uniformity coefficient. The above relation assumes that the soil cumulative grain size distribution can be approximated by a straight line on a standard semi-logarithmic plot. For clayey, plate-like particles, the above equation can be modified to

$$\Sigma_{sv} = \frac{(2+\beta)(C_u + 7)}{4\rho_s \, d_{50}}, \tag{2.4.7}$$

Soil	Specific Surface (m^2/g)
Montmorillonite	100–1,000
Attapulgite	140–170
Zeolite	40–140
Illite	70–100
Kaolinite	6–100
Halloysite	60–80
Chlorite	10–30
Bituminous shale	16
Coal	8
Sandstone	1–4
Limestone	1–2
Feldspar	0.1

Table 2.4.2: Specific surfaces for some soils and rocks (compiled from Kleineidam *et al.*, 1999; Santamarina *et al.*, 2002; Yukselen and Kaya, 2006).

where $\beta = \ell/b$ is the slenderness of particle (Santamarina *et al.*, 2002).

Specific surfaces are normally measured by utilizing the particle's ability to adsorb molecules onto its surface in mono- or multiple-layers. The amount of adsorbed substance gives an indication of the amount of surface area. The technique can involve gas adsorption, such as nitrogen, or the adsorption of polar liquid, such as ethylene glycol, methylene blue, etc. The specific surface areas measured by using different techniques may not be consistent with each other; a certain conversion may be needed (Yukselen and Kaya, 2006). Table 2.4.2 shows some specific surfaces of different soils and rocks.

2.5 Inhomogeneity and Anisotropy

In this book, unless otherwise specified, homogeneity and isotropy of a porous medium refer to its permeability, k. In general, these terms may be applied to other porous medium properties.

A porous medium domain is said to be *homogeneous* if its permeability is the same *at all its points*. Otherwise, the domain is said to be *heterogeneous* (or *inhomogeneous*). **Most subsurface domains are highly heterogeneous.** The heterogeneity of the subsurface is a consequence of the way the latter has been shaped and reshaped over millions of years by geological processes.

When the permeability at a considered point is independent of direction, the medium is said to be *isotropic at that point.* Similar considerations apply to the hydraulic conductivity, K, and to the transmissivity, T, of an aquifer; in the latter case, the considered directions are only in the xy plane.

In many cases, aquifers are anisotropic, with the vertical permeability being higher than the horizontal one. This may happen, for example, when the sediments comprising the aquifer are such (e.g., flat shaped mica particles) that when deposited, the resulting porous medium has a higher permeability in one direction (usually the horizontal one, unless later tilting of the formation occurs) than in other directions. Both sedimentation and pressure of overlying material cause flat particles to be oriented with their longest dimension parallel to the plane on which they settle. Later this produces flow channels parallel to the bedding plane, which are different from those oriented normal to this plane, thus rendering the medium anisotropic. If the flow through a formation takes place in some predominant direction over prolonged periods of time, it may produce a more developed network of channels parallel to that direction by removing fine material. In carbonate rocks, flowing water may dissolve the rock, producing solution channels beginning as very thin fissures in the direction of the predominant flow. In some soils, structural fissures develop more readily in one direction than in others, and the soil will exhibit anisotropy. In certain rocks, fractures produce a very high permeability in the direction of the fractures. All these circumstances anisotropy in permeability.

In general, we may distinguish two types of inhomogeneous aquifers:

(a) **Type 1.** A gradual change in transmissivity; the variable transmissivity may be expressed as a function of the space coordinates, $\mathbf{T} = \mathbf{T}(x, y)$.
(b) **Type 2.** Abrupt changes across well-defined surfaces of discontinuity. Each subdomain, enclosed by boundaries of discontinuity, is homogeneous in itself, or is heterogeneous of Type 1, discussed above, and should be treated as such. Across the boundary of discontinuity, there is a jump either in $\mathbf{T}$, or in its derivative normal to the boundary.

Under certain conditions, an inhomogeneous domain of Type 2 may be regarded as equivalent in its overall behavior to a (fictitious) homogeneous one. We may treat the aquifer domain shown in Fig. 1.3.3 as homogeneous, as long as we are interested in phenomena, say head drop between two points, the length scale of which is much larger than lengths (L_1, and L_2) that characterize the heterogeneity of the aquifer (Fig. 1.3.3). Homogeneity of a porous medium is, thus, judged by comparing the length scale of the phenomenon of interest with that of inhomogeneity, say in permeability, of the porous medium.

An inhomogeneous material composed of alternating layers of different textures, say low permeability silt and high permeability sand, is equivalent in its overall behavior to a homogeneous anisotropic porous medium in which the permeability parallel to the layers is larger than that perpendicular to them (Bear, 1972, p. 155).

In a similar way, if an inhomogeneous medium of Type 2 is non-repetitive, but the characteristic length scale of its inhomogeneity is much smaller than that of some phenomenon of interest, the medium may be considered as being equivalent in its behavior to a domain of Type 1 inhomogeneity. Figure 2.5.1 shows how a layered aquifer (Type 1 inhomogeneity) may be considered as an inhomogeneous aquifer with a gradually varying hydraulic conductivity.

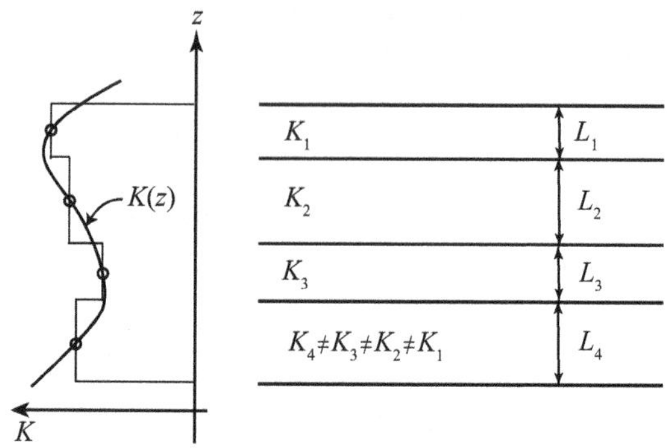

Figure 2.5.1: A layered aquifer.

An inhomogeneous material of Type 2, composed of alternating layers of different textures, is equivalent in its behavior to an homogeneous anisotropic medium (Bear, 1972, p. 155). However, in order for a stratified formation of this kind to be considered as an equivalent homogeneous anisotropic aquifer, the thickness of the individual layers must be much smaller than lengths of interest. For example, it is meaningless to determine the equivalent permeability of such a formation from a core whose size is smaller than the thickness of a single layer.

2.6 Hydraulic Approach to Flow in Aquifers

In general, flow through the subsurface, e.g., an aquifer, is three-dimensional. Also, the piezometric head, h, defined in Subs. 4.1.1, usually varies in space, i.e., $h = h(x, y, z, t)$. However, since the geometry of most aquifers, is such that they are much thinner relative to their horizontal dimensions (e.g., tens or hundreds of meters as compared to thousands of meters), a simpler approach is often employed. According to this approach, we *assume* that the flow in the aquifer is everywhere *essentially horizontal* (often referred to as *aquifer-type flow*), or that it may be approximated as such, neglecting vertical flow components. This is strictly true (not just an assumption) for flow in a horizontal, homogeneous, isotropic, confined aquifer of constant thickness and with fully penetrating wells. Nevertheless, the approximation is still a good one when the thickness of the aquifer varies, but in such a way that the variations are much smaller than the average thickness (Fig. 2.5.2).

Whenever justified on the basis of the geometry, i.e., thickness versus horizontal length, and the flow pattern, the assumption of horizontal flow, which is equivalent to assuming vertical equipotentials (= surfaces of constant value of piezometric head), $h = h(x, y, t)$, greatly simplifies the mathematical analysis of the flow in the aquifer. The error introduced by this assumption is

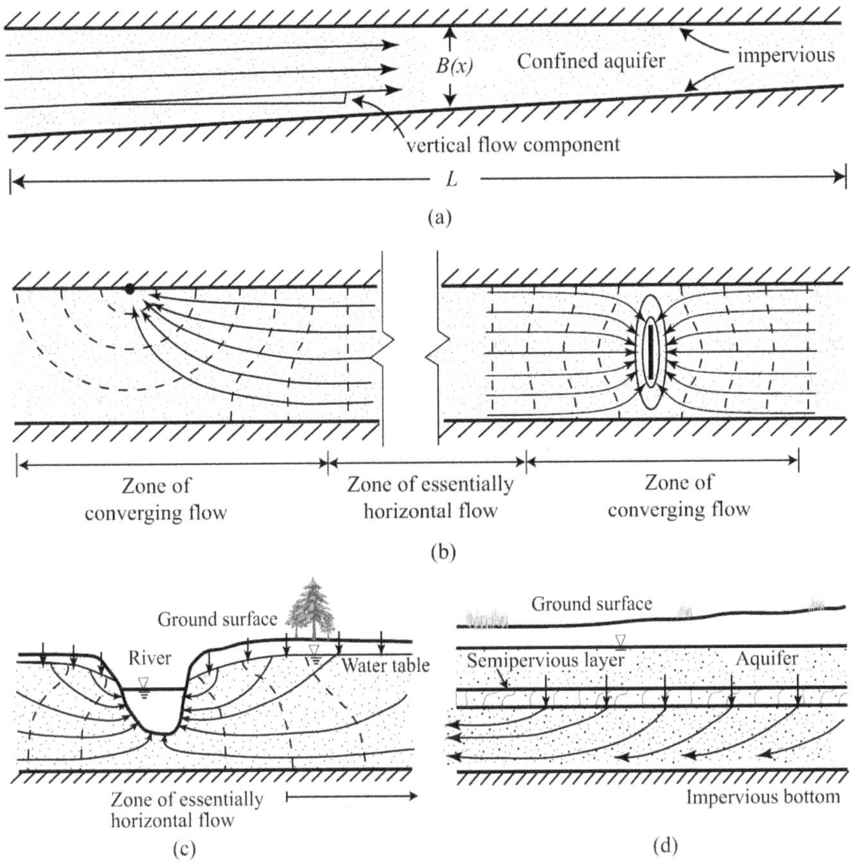

Figure 2.5.2: The hydraulic approach to flow in aquifers: (a) Flow in a confined aquifer with variable thickness, $B(x) \ll L$, (b) Flow in a confined aquifer with partially penetrating wells, (c) Flow in a phreatic aquifer with accretion, (d) Flow in a leaky-confined aquifer.

small in most cases of practical interest. We shall return to this subject in Sec. 4.4.

The assumption of essentially horizontal flow is applicable also to leaky aquifers (Fig. 2.5.2d). When the hydraulic conductivity of the aquifer is much larger than that of the semipermeable layer, and the thickness of the first is much larger than that of the latter, it follows from the law of refraction of streamlines (e.g., Bear, 1972, p. 26) that the flow in the aquifer is essentially horizontal, while it is essentially vertical in the semipermeable layer. These assumptions greatly simplify the analysis of flow in leaky aquifers and, in cases of practical interest, introduce very small errors.

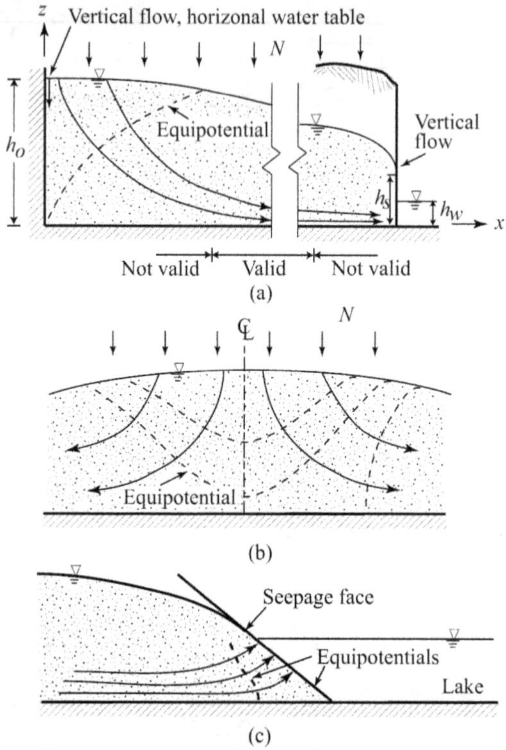

Figure 2.6.1: Regions where the Dupuit assumption fail.

In phreatic aquifers the *essentially horizontal flow* approximation is the basis for the *Dupuit assumption* presented in Sec. 4.5 (Fig. 2.5.2c). In Sec. 5.4, we shall present aquifer flow models, for both confined, leaky and phreatic aquifers, based on the concept of essentially horizontal flow. In most cases, it will be shown how the aquifer flow equations are derived by averaging the basic, three-dimensional flow equations along the thickness of the aquifer, using the assumption of vertical equipotential surfaces. This procedure is called *the hydraulic approach.*

The assumption of essentially horizontal flow fails in regions where the flow has a large vertical component as, for example, in the vicinity of partially penetrating wells, or outlets in the form of springs or rivers (Figs. 2.5.2 and 2.6.1). However even in these cases, at some distance from the source, sink, or the special feature, the assumption of essentially horizontal flow is, again, valid. As a simple rule, one may assume that at distances larger than 1.5 to 2 times the thickness of the aquifer away from these special features, aquifer-type flow occurs. At smaller distances, equipotentials are no more vertical, the flow is three-dimensional and should be treated as such.

Chapter 3

REGIONAL GROUNDWATER BALANCE

The groundwater part of the hydrological cycle is presented in Fig. 1.1.1. In the management of groundwater resources, man intervenes in the this cycle in order to achieve beneficial goals. This intervention takes the form of modifications imposed on the various components of the water balance, for example by pumping, by artificial recharge or by affecting natural replenishment. Another, unfortunately detrimental effect is the contamination of groundwater by human activities at ground surface.

Water and contaminants carried with it may enter an aquifer, or a portion of it, in the following ways:

- Groundwater inflow through aquifer boundaries and leakage from overlying or underlying aquifers.
- Natural replenishment (infiltration) from precipitation over the area.
- Return flow from irrigation and septic tanks (or similar structures, including faulty water supply and sewage networks).
- Artificial recharge.
- Seepage from influent streams and lakes.

Water and pollutants carried with the water may leave an aquifer in the following ways:

- Groundwater outflow through boundaries and leakage out of the considered aquifer into underlying or overlying strata.
- Pumping and drainage.
- Seepage into effluent streams and lakes.
- Spring discharge.
- Evapotranspiration from a shallow water table.

The difference between total inflow and total outflow of water volume and of total mass of contaminants during any period of time is stored in the aquifer, causing a rise in water levels and in the concentration of contaminants, respectively.

In this chapter, we shall review these components, which comprise the groundwater balance of a region, in order to facilitate the more detailed discussion in the following chapters. The objective is to construct models

J. Bear, A.H.-D. Cheng, *Modeling Groundwater Flow and Contaminant Transport*, Theory and Applications of Transport in Porous Media 23,
DOI 10.1007/978-1-4020-6682-5_3,

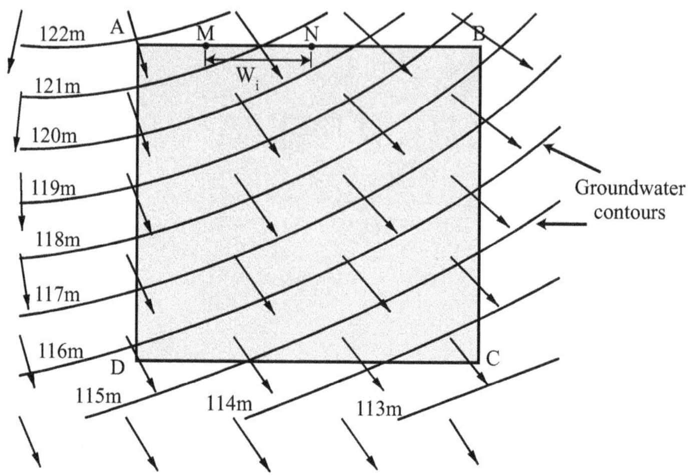

Figure 3.1.1: Inflow and outflow through aquifer boundaries.

that will enable the forecast of the aquifer's response to all these man-made (in fact, also natural) changes in components of the hydrological cycle. The region's boundaries may be the natural boundaries of a groundwater basin (e.g., an impervious boundary, a water divide, or a river fully penetrating an aquifer), or any closed boundary drawn on a map (say, for administrative reasons).

In the present chapter, we shall use some of the aquifer concepts and definitions introduced in later chapters (especially in Chaps. 4 and 5), assuming that the reader is familiar with them, at least in a general way, from previous studies and reading.

The 'balance', or 'budget', discussed here is only for groundwater in the saturated zone. It is obviously possible to discuss a water balance for the entire subsurface, including water in the unsaturated zone (Chap. 6), or even a regional water balance which will also include surface water.

3.1 Groundwater Flow and Leakage

3.1.1 Inflow and outflow through aquifer boundaries

When a boundary of a considered aquifer domain—an entre aquifer or a portion of it—is pervious, groundwater may enter or leave the aquifer through it. The flow across such boundaries is governed by the gradient of the water table, or of the piezometric head (Subs. 4.1.1), across the boundary.

Figure 3.1.1 shows a contour map (Subs. 5.4.6) and a portion of an aquifer, ABCD, for which a water balance is being established. Groundwater enters the aquifer through the boundary DAB and leaves the aquifer through BCD.

Because the hydraulic gradient (Subs. 4.1.1) varies in magnitude and direction along the boundary, and streamlines are, in general, not perpendicular to it, we divide the boundary into segments (like MN in Fig. 3.1.1) of length W_i, $i = 1, 2, \ldots, N$, each. Segments are chosen such that along each of them, the normal component of the hydraulic gradient can be satisfactorily represented by an average value $\overline{\mathcal{J}_{ni}}$, similarly, the transmissivity (Subs. 4.4) along the segment is represented by an average value $\overline{\mathsf{T}}_i$. The inflow through W_i is then given by the expression $W_i\overline{\mathsf{T}}_i\overline{\mathcal{J}_{ni}}$. We may then sum such expressions for the entire boundary with a positive value of $\overline{\mathcal{J}_{ni}}$ for inflow and a negative one for outflow, and obtain $\sum_i W_i\overline{\mathsf{T}}_i\overline{\mathcal{J}_{ni}}$ for the instantaneous net inflow through aquifer boundaries. Explanations of the above calculations are presented in Sec. 4.4. In general, the instantaneous net inflow may vary with time. With Q denoting the average net inflow during a water balance period, Δt, obtained, for example, by taking half the sum of net inflows computed at the beginning and at the end of Δt, the volume added to the aquifer by net inflow of groundwater is given by $Q\Delta t$.

Groundwater entering a balance area carries dissolved solutes (e.g., pollutants) present in the formation on the external side of the boundary. Groundwater leaving the balance area carries with it solutes present in the formation on the inner side of the boundary. The difference between solute inflow and outflow is stored in the considered balance domain. Often, especially if the balance area is not too large, it is assumed that complete 'mixing' of groundwater takes place in the balance area so that the water leaving it carries the average concentration in the balance area (Subs. 7.2.1).

3.1.2 Leakage

This subject is discussed in Sec. 5.4. The leakage, q_v, (volume of water per unit area and per unit time) into an aquifer, in which the piezometric head is h (Subs. 4.1.1), from an overlying (or underlying) aquifer, in which the piezometric head is h_{ext}, through a relatively thin semipermeable layer (= aquitard) (Sec. 2.1), is given by

$$q_v = \mathsf{K}' \frac{h_{\text{ext}} - h}{B'}, \tag{3.1.1}$$

where K' and B' are the hydraulic conductivity and the thickness, respectively, of the semipervious layer. If q_v, as calculated by (3.1.1), is negative, we have leakage out of the aquifer. As leakage may vary from point to point, we may divide the relevant area of the semipervious layer into elementary areas, A_j, through each we calculate an average leakage, q_j, during Δt. Then, $\Delta t \sum_j q_j$ gives the total net inflow into the aquifer by leakage during Δt. As we lower the piezometric head in a pumped aquifer, the leakage may reverse its direction, e.g., from outflow to inflow.

The remarks given above with respect to inflow and outflow of solutes and polluting elements carried with the water, are valid also with respect to the leakage, which may carry contaminants across aquitards.

3.2 Natural Replenishment from Precipitation

Phreatic aquifers are replenished from above by precipitation that falls directly over the ground surface overlying the aquifer, provided the ground surface is sufficiently pervious. The fraction of precipitation that infiltrates depends, among other factors, on the type of precipitation, the climatic conditions, especially evapotranspiration, the moisture of the soil prior to storm, storm characteristics (duration, intensity, peak intensity), topography of ground surface, slope and perviousness of ground surface and vegetation cover. Part of the area may be completely impervious (houses, streets, parking areas, and highways, or an impervious rock or top soil, which is practically impervious) and does not contribute to the natural replenishment of the aquifer beneath it. In fact, in certain countries, the reduction in natural replenishment, caused by impervious areas at ground surface, as a result of urbanization, has reduced natural replenishment to the extent that special laws are introduced in order to encourage (or enforce) the practice of 'water conserving construction', or 'Rainwater harvesting techniques'. These techniques are aimed at collecting surface runoff over urban impervious areas, directing it to artificial recharge facilities, thus recharging underlying aquifers.

A confined aquifers is replenished by groundwater inflow from an adjacent phreatic aquifer, which, in turn, is replenished from precipitation (e.g., aquifers B and C in Fig. 2.3.2).

In principle, infiltration is downward unsaturated flow from the ground surface to the water table, and, hence, the theory presented in Chap. 6 is applicable. From this theory, it follows, for example, that the instantaneous rate of infiltration from precipitation (e.g., hourly or daily rate) is strongly affected by the soil moisture conditions just below ground surface (as produced by antecedent rains). However, in most groundwater investigations, the use of this theory is not regarded as a practical way to determine the natural replenishment of an aquifer, as it requires detailed information on soil characteristics along the vertical column, on storm details, etc. Moreover, in general, for the purpose of management of a groundwater system, and in view of the buffer effect of the large volume of water in storage in the aquifer at any time, the groundwater hydrologist is not really interested in the variability in infiltration during any individual storm and not even that resulting from storms during the year, taking each storm as an instantaneous pulse. For most regional management purposes. we are interested in annual or seasonal replenishment. Within the framework of management models, we often assume that the natural replenishment is uniformly distributed throughout the year, or throughout the rainy season. In certain cases, where more details are required, monthly averages are used.

A number of methods are available for estimating natural replenishment from annual or seasonal precipitation (Walton, 1970; Simmers, 1988; Sophocleous, 1991; Healy and Cook, 2002). For example, we may regard replenishment as an aquifer parameter, rather than relate it to precipitation. We can then estimate this parameter by employing any of the parameter identification techniques described in Sec. 11.3.

Except for precipitation, which varies from one year to the next, all other factors affecting replenishment are constant in time, or vary only gradually (e.g., due to changes in land use). Hence, rather than refer to annual replenishment as an unknown variable, the natural replenishment is often related to precipitation, for which a much larger amount of data is usually available. One possible such relationship is

$$\begin{aligned} N &= \alpha(P - P_o), \qquad & P > P_o, \\ N &= 0, & P \le P_o, \end{aligned} \tag{3.2.1}$$

where N is annual natural replenishment, α is a coefficient, P is the annual precipitation, and P_o denotes threshold precipitation. For example, for $\alpha = 0.9$ and $P_o = 200$ mm/year, we obtain $N = 405$ mm/year for $P = 650$ mm/year. In this way, the number of unknown variables defining natural replenishment is reduced to only two: α and P_o. They may vary from one part of a considered aquifer to the next. These are then regarded as parameters of the aquifer model; they have to be determined as part of the parameter estimation process.

Another method often used for estimating natural replenishment, when detailed data on precipitation are available, is the use of any of the so called Watershed Models, e.g., the Stanford Watershed Model (SWM), developed by Crawford and Linsley (1966), and the Hydrocomp Simulation Program (Hydrocomp, 1968). Many such models (e.g., the Hydrological Simulation Program (HSPF) (USEPA, 1997), the Storm Water Management Model (SWMM) (Metcalf and Eddy, *et al.*, 1971; USEPA, 1994), and the Système Hydrologique Européen (SHE) hydrological modeling system (Abbott *et al.*, 1986a, 1986b) have been developed and published since the pioneering work of Crawford and Linsley (1966). A survey of such models is given by Fleming (1975, p. 190). A comprehensive review of computer models is given by Singh (1995). Like the pioneering Stanford model, most models of this kind simulate the hydrologic cycle, using a moisture accounting procedure of one form or another. A system of equations describes the interrelationships among the various elements of the model. During the simulation, a running record is maintained of all moisture entering the basin (or the considered part of it), stored in it, and leaving it as evapotranspiration, surface runoff and groundwater. The latter is the natural replenishment considered in the present section. Some of the models, e.g., SHE, the Institute of Hydrology Distributed

Model (IHDM) (Beven *et al.*, 1987), and Simulator for Water Resources in Rural Basins (SWRRB) (Williams *et al.*, 1985), consider surface runoff and groundwater flow, simultaneously.

Many parameters are included in the more sophisticated models of this kind. Obviously, calibration (i.e., identification of all model parameters) is always required before such a model can be used.

When we wish to predict future water levels for a relatively short period of time, say two to three years, we introduce, as input, (assumed) values of future natural replenishment, say in the form of monthly averaged values. However, in general, when a forecasting problem is solved as part of a management one, we are interested in a much longer period, say a planning horizon of 15 or 20 years. Then, we usually use annual or seasonal (averaged) values. However, we have to take into account the fact that, like precipitation, annual natural replenishment is a *random phenomenon.* In order to take care of this feature, we use one of the models, which uses available past data to generate *synthetic sequences* of values of that data, in order to generate a number of *possible sequences* of values of annual natural replenishment, each with some probability of occurrence. These sequences, in turn, are introduced as time-dependent (deterministic) input in the forecasting and management models.

A commonly used model for the generation of synthetic sequence is the *autoregressive* (AR) method. For hydrological data with a yearly cycle due to seasonal variations, e.g., precipitation and streamflow records, we use the *periodic autoregressive* (PAR) method. Another example is the monthly averaged streamflow discharge of a stream. We are given a historical discharge record that ends at year i and month $j-1$, and we would like to successively generate the next flow rate in order to create a time series. In order to mimic nature's process, these newly generated data should be random, but not arbitrary, as they need to be generated based on the statistics of the past record. The PAR formula takes the form (Salas, 1993):

$$Q_{i,j} = \overline{Q}_j + \sum_{k=1}^{p} b_{k,j} \left(Q_{i,j-k} - \overline{Q}_{j-k}\right) + \varepsilon_{i,j}, \tag{3.2.2}$$

where $Q_{i,j}$ denotes the monthly averaged streamflow at year i and j, $\overline{Q}_j$ is the mean value of the jth month flow, obtained as the ensemble average of all jth month flow of the previous years, $b_{k,j}$ are the autoregressive parameters, $\varepsilon_{i,j}$ are normally distributed, uncorrelated random variables with zero mean, and p is an integer denoting the order of the method. In the above formula, since the month index can only take a value between 1 and 12, when the subscript $j-k \leq 0$, it is necessary to replace $\overline{Q}_{j-k}$ by $\overline{Q}_{12+j-k}$, and $Q_{i,j-k}$ by $Q_{i-1,12+j-k}$.

The simplest PAR model is the first order one that corresponds to $p=1$; it leads to the well-known Thomas-Fiering model (Thomas and Fiering, 1962):

$$Q_{i,j} = \overline{Q}_j + b_j \left(Q_{i,j-1} - \overline{Q}_{j-1}\right) + s_j \left(1 - r_j^2\right)^{1/2} t, \tag{3.2.3}$$

where b_j is a regression coefficient for estimating the flow rate in month j from information on month $j-1$, s_j is the standard deviation of the jth monthly flow, r_j is the correlation coefficient between the previous $(j-1)$ and the current (j) month, (see Sec. 10.1.3 for basic definitions of statistical quantities), and t is a random number generator with Gaussian distribution of zero mean and unit variance. For other models, see Matalas and Wallis (1976), and Salas (1993).

The main advantages of the procedure for generating synthetic sequences of events (natural replenishment in the case considered here) are: (1) the possibility of creating records which are longer than the historical ones, and (2) that different sequences with different probabilities of occurrence can be generated and used as inputs to models. This is especially important for management purposes (Chap. 11). It is also possible to generate synthetic sequences that take into account future climatic changes. Such sequences will enable the investigation of the effects of climate changes on various hydrological components of the water cycle.

The portion of precipitation that becomes natural replenishment, does not reach the aquifer immediately. The time lag depends on the depth of the water table and the hydraulic properties of the soil. However, the mechanism can be visualized approximately as one of displacement (at some more or less constant degree of saturation) whereby water is continuously added at the top of a soil column (i.e., at ground surface) and removed from the column at the bottom (i.e., at the water table). In this way, if we think of water as labeled, say by some contaminant, we may have a considerable time lag (even tens of years) before the contaminant reaches the water table. However, from the point of view of water quantity, the actual replenishment practically does not lag behind the precipitation producing it.

This phenomenon is of special importance when water quality is being considered. In addition, as water passes through the soil column, from ground surface to the water table, changes may take place in the quality of the water. Precipitation water is not distilled water. Depending on the location, the air pollution conditions, and the distance from the sea, precipitation water usually contains dissolved matter. For example, observations close to the coast of Israel show up to 25 ppm Cl^- in rainwater (with a fast reduction farther inland). Rainwater will further pick up salinity, and, in general, contaminants, at ground surface and upon passing through the upper soil layer. Some of the dissolved species (e.g. Cl^-) undergo no changes as the water percolates downward. Others may undergo changes as a result of interaction with the solid matrix (e.g., adsorption, ion exchange). For example, due to adsorption, the downward movement of some heavy metals will be slowed down many times with respect to the movement of the Cl^-. Their arrival at the water table may be delayed for many years. The occurrence of clay layers may appreciably affect adsorption and ion exchange phenomena along the column. This subject is of special importance in the case of artificial recharge of reclaimed sewage through infiltration ponds.

Infiltration may take place also from rivers and perennial and intermittent streams that serve as the drainage system of the watershed that overlies the aquifer. In some areas, this may be the only source for groundwater replenishment, with the rain occurring in mountain areas that are far away, in the upstream portions of the watershed. In semi-arid areas, we may encounter short duration flash floods over large areas, with infiltration from the flooded area as the only source for aquifer replenishment.

Briefly, groundwater recharge can be estimated by one of the following methods:

- Hydrological water budget, based on mass balance. In such a balance, groundwater recharge is obtained as the difference between of precipitation and the sum of interception, evapotranspiration, runoff, and other components expressing loss.
- Infiltration rate based on precipitation and an infiltration coefficient that depends on soil type and on soil saturation history.
- Methods based on water table fluctuations—for short duration and for a shallow water table.
- Methods based on the use of tracers—dissolved, natural and artificial tracers (Allison *et al.*, 1994; Ekwurzel *et al.*, 1994), and on the use of heat as a tracer.
- Methods based on the calibration of groundwater computer models, using inverse technique.

The U.S. Geological Survey provides two public domain computer programs for estimating groundwater recharge/discharge interaction with stream systems. PULSE (Rutledge, 1997) gives model-estimated groundwater recharge and hydrographs of groundwater discharge to a stream. The model is applicable to a groundwater flow system that is driven by areally uniform recharge to the water table, and in which groundwater discharges to a gaining stream. RORA (Rutledge, 1998) uses the recession-curve-displacement method to estimate groundwater discharge to a stream.

3.3 Return Flow from Irrigation and Sewage

Even in efficient irrigation practices, a certain portion of the water applied to an area is not used up by the plants as consumptive use. Instead, it infiltrates, eventually reaching the water table. We refer to this contribution to the aquifer's replenishment as *return flow from irrigation*. It includes also seepage from open channels and leakage from faulty pipes. It may amount to as much as 20-40% of the volume of water used for irrigation, depending on the irrigation technique. As the irrigation becomes more efficient, this percentage is reduced. The water used for irrigation may be that pumped from an underlying aquifer (hence the term 'return flow'), surface water, or water imported from other regions. Obviously, return flow carries with it salts and other dissolved matter: those contained in the original irrigation water, augmented by evaporation, and those picked up upon passage through ground surface and through the root zone.

Sometimes, return flow is created on purpose, in order to leach salts from the root zone, often overlooking the fact that in the absence of adsorption and other attenuating phenomena, the leached salts, eventually, reach the underlying aquifer and pollute it. For the sake of simplicity, we use here the term 'salts', but this should be understood to include also dissolved fertilizers, pesticides, and other kinds of (potential) groundwater pollutants present in the root zone and on ground surface.

Denoting the concentration of salts in irrigation water by c_i and the maximum permissible salt concentration in the soil solution (without causing undesirable losses in production) by $c_{\max}(> c_i)$, then out of the volume of irrigation water, $\mathcal{U}_i$, the minimum volume of water, $\mathcal{U}_L$, required for leaching the soil, such that equilibrium will be maintained at $c_{\max}$, is

$$\mathcal{U}_L = \mathcal{U}_i c_i / c_{\max}. \tag{3.3.1}$$

Hence, the amount of salt continuously added to the aquifer is $\mathcal{U}_L c_{\max} (= \mathcal{U}_i c_i)$. Actually, a volume larger than $\mathcal{U}_L$, as defined by (3.3.1) is required, because of the inefficiency of the leaching process. If we increase $\mathcal{U}_i$, consumptive use will practically remain unchanged, $\mathcal{U}_L$ will increase, causing a reduction in $c_{\max}$. Obviously, we have to take into account that part of the leaching is also takes place during rainy seasons by natural replenishment from precipitation that, in general, has a relatively low salt content.

The quality problem associated with return flow and leaching should be more carefully studied when reclaimed sewage (or other kinds of contaminated water) is used for irrigation.

3.4 Artificial Recharge

Artificial recharge may be defined as man-made operations aimed at transferring water from ground surface into an underlying aquifer. This is in contradistinction to natural replenishment (or natural recharge), considered in Sec. 3.1 above, whereby water from precipitation and surface runoff reaches the aquifer without man's intervention. Whereas natural replenishment is an uncontrolled (by man) input to the a groundwater system, artificial recharge is a controlled input. The quantity, quality, location, and time of artificial recharge are decision variables, the values of which are determined as part of the management of a considered groundwater system.

3.4.1 Objectives

Artificial recharge may be practiced in order to achieve various objectives. Among them, we may list the following:

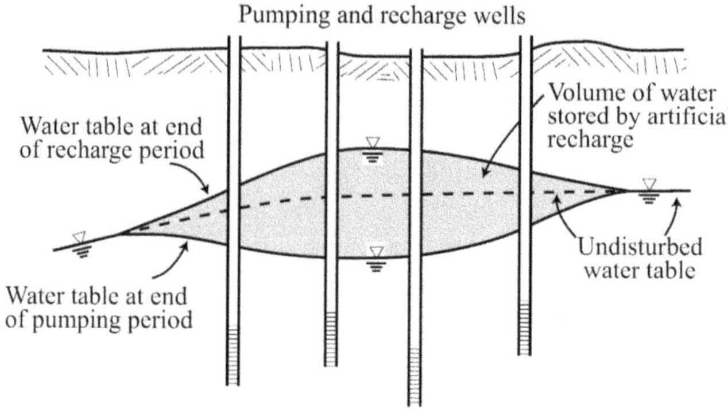

Figure 3.4.1: Storage of water in a phreatic aquifer.

Control of regional hydrological regime. By artificially recharging an aquifer, water levels, or piezometric heads, are raised. By manipulating these levels (obviously, taking into account also the effect of pumping), we can control the rate and direction of flow in an aquifer, control the movement of water bodies of inferior quality, e.g., resulting from seawater intrusion, control spring discharge, and control seepage to or out of adjacent water bodies (rivers and lakes). Because of the very low value of storativity of a confined aquifer, a relatively small volume of injected water is required in order to produce a large rise in piezometric head elevations in such an aquifer. On the other hand, a very large pressure or piezometric head is needed in order to inject this volume of water. This fact is used, for example, to control seawater intrusion in coastal confined aquifers discussed in Sec. 9.3.

Storage of water. Water can be stored in an aquifer, to be pumped at a later time. Phreatic aquifers, because of their large storativity (= specific yield), relative to that of confined aquifers, may serve as very large storage reservoirs. Water can be stored in the void space of such aquifers, to be pumped at a later time. For example, a portion of a phreatic aquifer of (horizontal) area of 100 km^2 and storativity (actually, specific yield, discussed in Subs. 6.1.9) of 15% can store as much as 150×10^6 m^3 of water, if water levels are raised by 10 m. Figure 3.4.1 shows the volume of water that can be stored in a phreatic aquifer by recharge and pumping. However, as groundwater is never at rest, if no use is made of the stored water, i.e., no portion of the stored water is pumped, it will gradually leave the area as groundwater outflow. Fortunately, due to the relatively slow movement of water, most of the stored water can be recovered by appropriate management.

Both long-term and short-term storage in aquifers may be practiced. In years with excess surface runoff, water may be diverted from streams and lakes to be stored in aquifers for use in dryer years. Short-term storage may be practiced in order to make a more efficient use of the water supply lines. Water may be delivered to a demand area at a constant rate throughout the year, to be stored in the aquifer when supply exceeds demand, and pumped by local wells to supplement demand in excess of direct supply.

Obviously, storage of water in any quantity of economical and hydrological significance is possible only in phreatic aquifers, where the storativity (actually, the specific yield) is related to the porosity (or at least to the effective one) and not, as in a confined aquifer, to the elastic properties of the water and the solid skeleton of the aquifer (see Subs. 5.1.3).

In each particular case, a comparison should be made between storage in aquifers and storage in surface reservoirs. Among the points to be considered in such comparison, we could mention:

Storage in aquifers:

- Cost of recharge wells, including land for wells;
- Cost for additional pumping wells, in excess of existing pumping;
- Cost associated with the loss of water by evapotranspiration (if water levels are close to ground surface);
- Cost of additional energy;
- Cost associated with water that cannot be recovered due to groundwater flow.

Storage in surface reservoirs:

- Cost and availability of (geologically) appropriate site;
- Cost of dams and other diversion, conveyance (by pipeline and open channel), and regulation structures;
- Benefit from the recovery of energy by the production of hydroelectric power;
- Benefit from recreation;
- Loss of water by evaporation and infiltration;
- Cost associated with damage resulting from the contamination of stored water.

In many instances, especially in arid and semiarid regions, storage in aquifers has been proven to be more economical than surface storage.

Control of water quality. As the water introduced into an aquifer and the indigenous water in the aquifer move, they 'mix' as a result of hydrodynamic dispersion (Subs. 7.1.4). Mixing is also achieved by wells that pump simultaneously from the two kinds of water, when the latter are of different qualities, in terms of concentration of dissolved matter. We can control the quality of the pumped water by manipulating pumping and artificial recharge, thus controlling the movement of the water bodies introduced into the aquifer by artificial recharge and the mixing that takes place in the aquifer and in the pumping wells (Bear, 1979, p. 282–292).

Water used for artificial recharge may be either water of a quality higher than that of the indigenous water of the aquifer, or of an inferior quality.

In the former case, we improve the quality of the pumped water. In the latter case, we lower the quality. Yet, if the resulting quality is still within the permissible range, we may upgrade the efficiency of the entire water resource system by increasing the total pumped quantity of water, and, perhaps, making use of water that may otherwise be unacceptable from the quality point of view.

Due to the very slow movement of water in an aquifer, a period of years, sometimes many years, may elapse between the time water is introduced into an aquifer and the time it is pumped. During that time, phenomena such as chemical reactions among constituents present in the water, interaction with the solid skeleton (adsorption and ion exchange), decay (e.g., radioactive), and filtering may take place. In this way, the aquifer acts to improve the quality of the injected water. It is for that last reason that, very often, reuse of reclaimed sewage water is implemented in conjunction with artificial recharge.

Suspended fine material in surface water used for artificial recharge can be removed by the filtering that takes place as the water percolates through the bottom of infiltration basins and the soil underlying them on its downward way to the aquifer. Of special interest is the improvement of water quality (e.g., removal and destruction of microorganisms) as the recharge water percolates through the unsaturated zone.

In addition to these major objectives, we may also mention:

- Supplementing the difference between the demand for groundwater and the natural replenishment of an aquifer.
- Disposal of liquid waste into deep formations (often, containing brine), where it will stay or move very slowly (sometimes for thousands of years) towards outlets. It is always important to verify by thorough hydrogeological investigations that, indeed, there exists no possibility of contact between the injected waste and groundwater that will, eventually, be used.
- It is possible to create a flow pattern within the aquifer from the area of artificial recharge to that of withdrawal by pumping, with the aquifer serving as a conduit. Wells distributed over an area may withdraw water for local use, thus avoiding the need for an above-surface distribution system.
- Maintenance of high water levels (or heads) to prevent land subsidence, or other undesirable phenomena, which result from lowered water levels (e.g., damage to foundations).
- Conservation of water. For example, water used only for cooling can be re-circulated by injecting the warm water back into the aquifer from which it is pumped.
- Energy storage. Warmer surface water (or hot water from the cooling systems of power plants) can be pumped into an aquifer during summer and recovered during winter at higher than normal temperature for certain water usage.

In most cases, artificial recharge is implemented to achieve a number of goals and in conjunction with the utilization of surface water. However, in spite of many advantages, in each case, one should carefully examine the danger of permanently (or at least for a long time) damaging an aquifer's water quality by recharging it with water containing toxic or non-degradable pollutants.

3.4.2 Methods

Artificial recharge can be implemented by several methods. In each particular case, the selected method depends on the source of water, the quality of the water, the type of aquifer, the availability of land, the topographical and geological conditions, the type of soil, and the economic conditions.

Methods for enhancing infiltration. In these methods, the objective is to increase infiltration by various agro-techniques, which affect ground surface roughness, slope, vegetation cover, etc. The purpose is to extend the time and area through which infiltration from precipitation and surface runoff takes place. Both the slopes of the watershed and the drainage channel network can be treated to achieve this goal. For example, small (rock and wire) check dams in the natural channels will cause water to spread over a larger area, or delay its movement.

Surface spreading methods. Here water is diverted to specially constructed ponds, basins, or trenches, dug along ground surface contours, and allowed to infiltrate through their pervious bottom.

Figure 3.4.2 shows a typical scheme of a project in which water is diverted from a perennial or an intermittent stream to a settling basin, where most of the fine material is removed, and then to infiltration basins. Usually the peak of large floods, carrying large quantities of silt and debris, are not diverted. Wells are located at some distance from the infiltration basins to allow for a certain minimum retention time, say, one year, before pumping.

Two objectives are achieved by the project shown in Fig. 3.4.2: (1) storage, say, if water in the river is available in winter and is needed for irrigation in summer, and (2) improvement of water quality, at least from the removal of fines in the settling basins, and also through the soil layer just beneath the infiltration basins.

Ditches and furrows are sometimes used instead of basins. Excess irrigation, especially during non-irrigation seasons, can also be used as a method for artificially recharging the underlying aquifer.

Obviously, surface spreading techniques should be implemented only when the recharged aquifer is a phreatic one, and when no impervious layer of significant areal extent is present between the bottom of the infiltration basins and the water table.

The economy of artificial recharge by surface spreading techniques depends to a large extent upon the availability of land and the maintenance of high infiltration rates. Depending on the type of soil, rates of 3–15 m/day (that is, 15 m^3/m^2/day) have been observed in gravel, up to 3 m/day in gravel and sand, up to 2 m/day in fine sand and sandstone, and up to 0.5 m/day

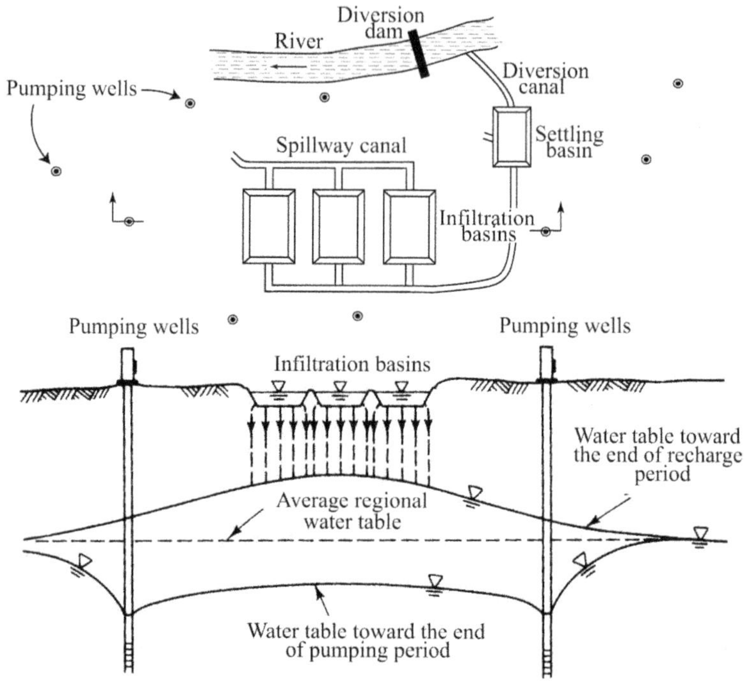

Figure 3.4.2: Schematic plan and cross-section of artificial recharge by means of infiltration basins.

in sand and silt. Values lower than 0.5 m/day have also been reported. All these rates represent initial values, because as infiltration continues through an infiltration basin, its bottom becomes gradually clogged. A typical curve showing the infiltration rate is presented in Fig. 3.4.3; the initial reduction in infiltration rate is caused by dispersion and swelling of soil particles after wetting. The subsequent increase results from elimination of entrapped air by dissolution in the water. The following reduction in infiltration rate, which has a more or less exponential form, is due to the clogging of the soil pores at the bottom of the basin and just beneath it. This clogging is due to the retention of suspended solids (when present in the recharge water, as, for example, when water is diverted from flash floods), growth of algae and bacteria (when nutrients are present in the water), entrained or dissolved gases released from the water, and precipitation of dissolved solids and chemical reactions between dissolved solids and the soil particles and/or the native water present in the void space.

When the infiltration rate of a basin drops below some design value, its use as a recharge basin is discontinued. By drying it, cleaning, and sometimes scraping the top 2–5 cm of bottom material, the infiltration rate is brought

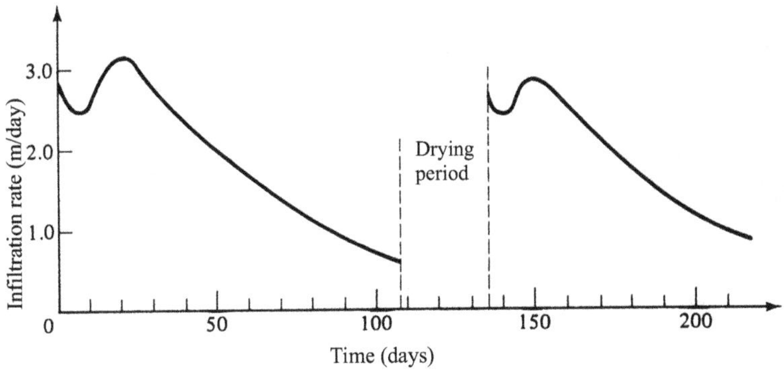

Figure 3.4.3: Typical reduction in infiltration rate.

back almost to its initial value, and the basin can be returned to use. The frequency of cleaning depends, of course, on the local conditions (type of water and soil) and may be as often as every three months or once in 2–3 years. Sometimes it is found economical to pre-treat the water (chlorination, adding polyelectrolytes, etc.) to increase settlement in the settling basin, thus prolonging the period of effective operation.

Artificial recharge through wells. Artificial recharge can be carried out through ordinary pumping wells, or through specially constructed recharging (= injection) wells. It is also possible to design a dual purpose well. Figure 3.4.4 shows somc typical recharging wells in use in Israel (Harpaz, 1971), where artificial recharge is an intrinsic part of the operation of the national water resources system.

The phenomenon of clogging, causing a reduction in injection rate, occurs also in wells. The reasons are similar to those causing clogging of infiltration basins. However, clogging in wells is usually under *anaerobic conditions*. Because the velocity of the injected water decreases as water travels away from a well, the deposition of fines will occur at some distance from a recharging well, making cleaning more difficult.

Artificial recharge through wells is practiced (a) for recharging confined aquifers, (b) when extended impervious layers are present between ground surface and an underlying phreatic aquifer, and (c) when land is expensive, or unavailable. Because clogging is more severe in wells, artificial recharge through wells is mostly implemented with high quality water, often drinking water quality. Pre-treatment (chlorination, sedimentation, filtration, etc.) is the rule rather than the exception. In spite of all these precautions, clogging does take place.

When the injection rate drops to below some design value, renovation by various chemical treatment techniques, e.g., acidation, oxygen supply, or enzymes, is required. One possible technique, called 'backwashing', is implemented by pumping at a high rate.

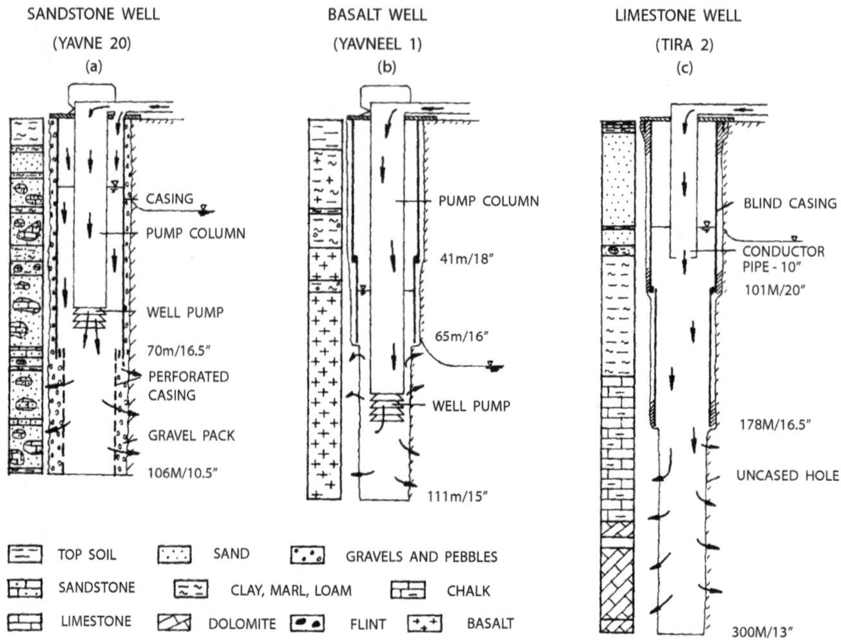

Figure 3.4.4: Typical recharging wells, (a) In sandstone, (b) In basalt, (c) In limestone (Harpaz, 1971).

Induced recharge. This term is used for cases in which withdrawal installations, in the form of a gallery or an array of shallow wells, are located at a relatively small distance from a river, or a lake, and parallel to it. By withdrawing water through these installations, the groundwater table is lowered in their vicinity, thus inducing the movement of water from the river, or the lake, into the aquifer, and to the wells, provided, of course, that the river bed is not completely clogged. Figure 3.4.5 shows a typical cross-section with induced recharge from a river. The gallery intercepts water originally drained to the river, but also water from the river.

Two goals can be achieved by induced recharge:

- The aquifer is recharged by river water, which, in turn, is pumped for beneficial use, without constructing any recharge installations (the aquifer itself is used as a conduit).
- The river water is filtered, and fines are removed, as it travels through the aquifer towards the abstraction installations. The aquifer acts as a large slow sand filter. Detention time of 2–3 months may serve as a typical example.

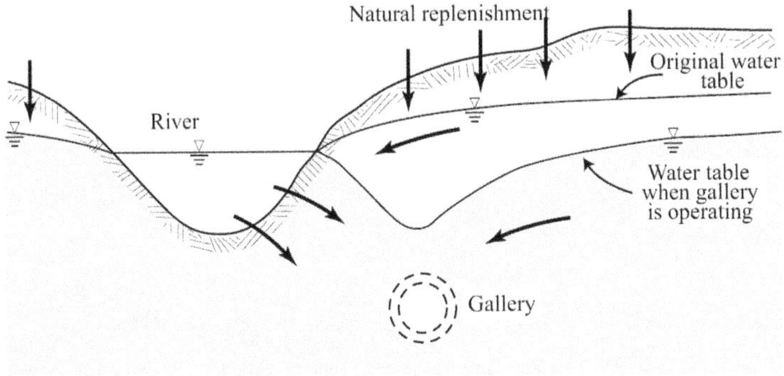

Figure 3.4.5: Induced recharge.

The brief review of artificial recharge presented above is by no means a full coverage of this important subject for those who manage groundwater resources. It is presented here both as an element in the water balance, and also as an example of a technique that requires model investigations, based on the knowledge of most of the material presented in this book for the design of their implementation. For example, investigations should be used for:

- Determining travel time of water injected into aquifers (Chap. 4).
- Forecasting water level changes caused by artificial recharge in different aquifers (Chap. 5).
- Studying the movement of the water infiltrating from spreading basins through the unsaturated zone (Chap. 6).
- Determining the changes of quality that take place both in the aquifer and in the unsaturated zone (Chap. 7).
- Determining the movement of an (assumed) abrupt front between injected and indigenous water.
- Controlling seawater intrusion by maintaining a piezometric head barrier (Chap. 3).

A vast amount of literature is available on artificial recharge. Among summaries available on this subject we may mention the proceedings of the International Symposium on Artificial Recharge of Groundwater (Johnson and Finlayson, 1988; Johnson and Pyne, 1994; Peters, 1998; Dillon, 2002), practicing guideline (Environmental and Water Resources Institute, ASCE, 2002), USGS reports (Todd, 1959; Aiken and Kuniansky, 2002) and other sources (Pyne, 1995).

3.5 River-Aquifer Interrelationships

Rivers passing through a region underlain by a phreatic aquifer may either contribute water to the aquifer, or serve as its drain. Much of the low water

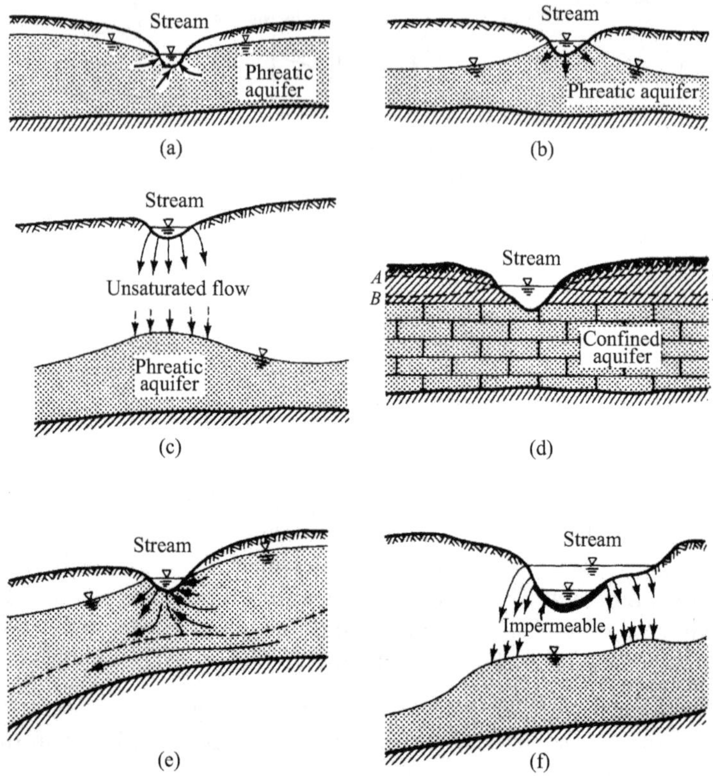

Figure 3.5.1: River-aquifer relationships. (a) Effluent stream. Groundwater drains into stream. (b) Influent stream. River contributes to groundwater flow. (c) Influent stream (deep water table). River contributes to groundwater. (d) Influent stream (piezometric surface B), or effluent one (piezometric surface A) intersecting a confined aquifer. (e) A stream which is both influent and effluent. (f) A partly clogged influent stream.

flow in streams (base flow) is derived from groundwater whose water table elevations in the vicinity of a stream are higher than in the stream. Such streams are called *effluent streams* (Fig. 3.5.1a). On the other hand, when the water level in a stream is higher than the water level in an adjacent (or underlying) aquifer, water will flow from the river to the aquifer. The river is then called an *influent river* (Figs. 3.5.1b and c). When a stream cuts through an impervious layer, establishing a direct contact with an underlying confined aquifer, the stream may be either an influent one or an effluent one, depending on whether the piezometric head in the aquifer is above or below the water level in the stream (Fig. 3.5.1d). The same stream can be an influent one along one river stretch and an effluent one along another. Or, it can be both influent and effluent at the same point, as shown in Fig. 3.5.1e. Obviously, the entire discussion presented here is based on the assumption that the river bed is not completely clogged and that water can flow freely through it. Otherwise, there is no hydraulic contact between the water in the river and in the aquifer and no relationship exists between the two. It is possible that the profile of a stream is such that its deeper part, accommodating for low flows, is completely clogged, while above a certain level, the river bed is pervious (Fig. 3.5.1f). When the water table under an influent stream is sufficiently deep, a mound is formed in the former by the percolating water (Figs. 3.5.1c and f).

The volume of water contributed to an aquifer by streamflow (or drained into a stream from an aquifer), is part of the regional water balance. The rate of flow in either direction, also when the stream bed is partly clogged, can be calculated.

In view of the different possible situations shown in Fig. 3.5.1, a river may play several roles when solving a groundwater problem:

- The river may act as a boundary of specified head to the flow domain in the adjacent aquifer (Figs. 3.5.1a, b, d, and e). We have in mind an *aquifer-type flow*, which is based on the assumption of essentially horizontal flow in the aquifer, so that the shape of this boundary is a curve in the xy-plane. This is a good approximation, overlooking the details of the flow-net under the stream. A somewhat different boundary condition should be employed when the river bed is semipervious (Sec. 5.2.3c).
- A river may serve as a source, contributing water to the aquifer. The rate of leakage depends on the depth of water in the river and on the permeability of the river-bed. This can be a line (actually curve) source, or a strip of some width, when the river is sufficiently wide. The main point is that the rate of seepage is independent of the water levels in the aquifer (Figs. 3.5.1c and f).

Thus, depending on the elevation of the water table, both situations (a) and (b) above are possible at different periods of time for the same stretch of river (and certainly for different stretches of the same river).

Finally, we note that when a river is sufficiently large, in terms of the rate of flow, the exchange of water between it and an aquifer practically does not affect the its flow, and, hence, the depth, of flow in it. However, in small streams, the leakage itself may lower the water level in the stream, and even completely dry it up. We may encounter a passage from condition (a) to (b) as defined above, at some *a priori* unknown time.

Often, when solving a forecasting problem, the type of situation to be realized is not known *a priori*, and some trial and error, or iteration technique is required.

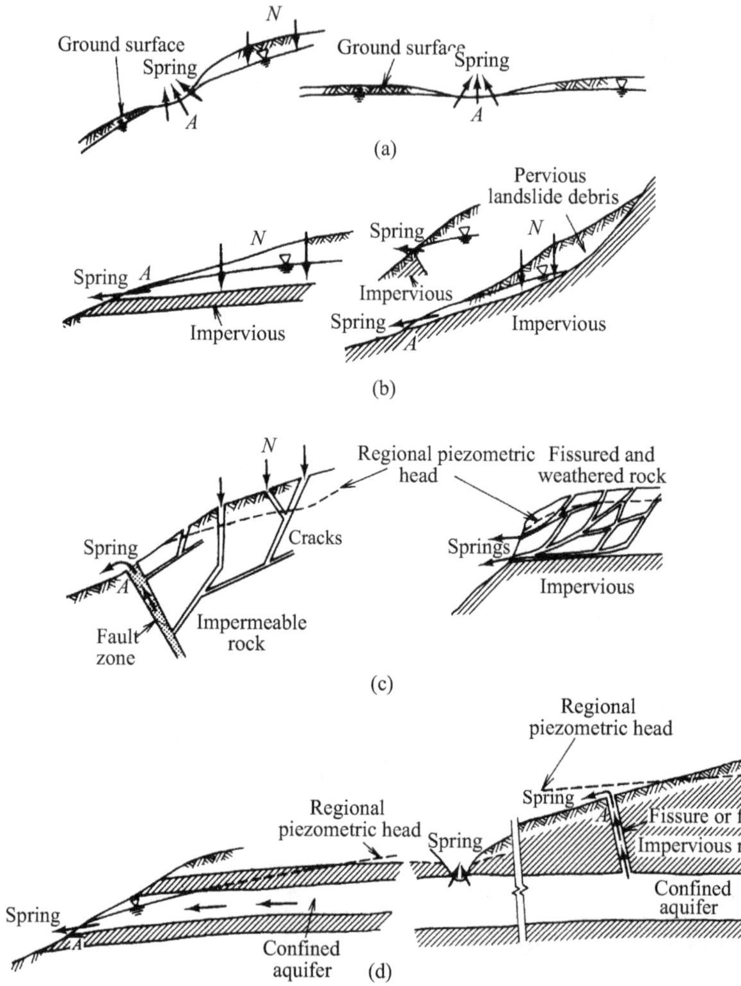

Figure 3.6.1: Types of springs: (a) Depression springs; (b) Perched springs; (c) Springs in cracked, impermeable rock; (d) Springs in a confined aquifer.

3.6 Springs

A spring is a point (or a small area) through which groundwater emerges from an underlying aquifer to ground surface. The discharge of some springs is small and of no significance in the groundwater balance; however, some are very large and dominate the groundwater flow pattern in their vicinity.

Figure 3.6.1 shows several types of springs. A *depression spring* (Fig. 3.6.1a) occurs when a water table intersects ground surface. A *perched spring* (Fig. 3.6.1b) occurs when an impervious layer, which serves as the bottom

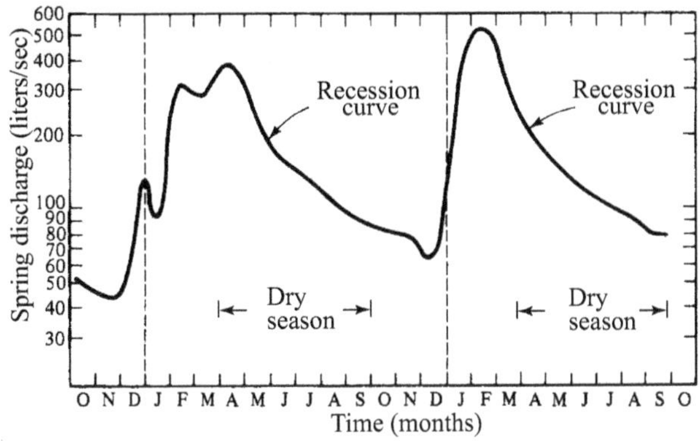

Figure 3.6.2: A typical spring hydrograph with significant seasonal fluctuations.

of a phreatic aquifer, intersects ground surface. A confined aquifer can be drained in the form of an *artesian* spring, either through a pervious fault, or fissure, reaching the ground surface, or where it becomes exposed at ground surface. The driving force for spring discharge is the piezometric head in the aquifer.

The instantaneous rate of discharge of a spring depends on the difference between the elevations of the water table (or piezometric head) in the aquifer in the vicinity of the spring, and the elevation of the spring's threshold (point A in Fig. 3.6.1a). As the spring's discharge is derived from water stored in the aquifer, during the dry season, water levels in the aquifer will gradually decline and with it the spring's discharge.

A spring may completely dry up when water levels fall below its threshold. Thus, the relationship between the rate of decline of a spring's discharge depends on the storage characteristics of the aquifer (storativity and geometry of aquifer, e.g., areal extent). Figure 3.6.2 shows a typical portion of a spring's hydrograph; the recession (or depletion) portions of this hydrograph corresponds to the dry seasons. On a semi-log paper (with time on the linear scale) the recession curve usually plots as a straight line.

Using the simple model of a spring draining an aquifer, envisioned as the reservoir shown in Fig. 3.6.3a, with $Q = \alpha_1 h$, $\alpha_1 =$ constant, we express the change in storage in the form

$$Q\,dt \equiv \alpha_1 h\,dt = -S_y\,A\,dh, \tag{3.6.1}$$

where S_y denotes the specific yield of the phreatic aquifer (Sec. 5.4.1), α_1 is a coefficient, and A denotes the (horizontal) area of the aquifer. By solving the above equation, subject to $h = h_o$ at $t = t_o$, and $Q = Q_o = \alpha_1 h_o$, we obtain

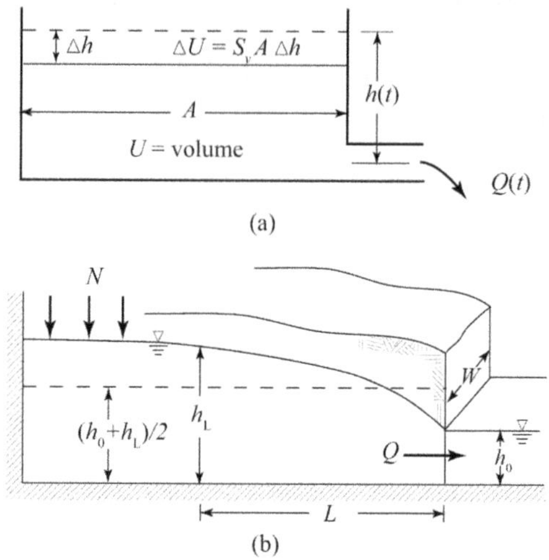

Figure 3.6.3: Simple models of a spring.

$$Q(t) = Q_o \exp\left[-\frac{\alpha_1}{S_y A}(t - t_o)\right], \tag{3.6.2}$$

which plots as a straight line on a semi-log paper (Q on the logarithmic scale).

To obtain an interpretation of α_1, consider another simple model of steady flow to a spring, shown in Fig. 3.6.3b. As we shall learn below (Sec. 4.5), the steady rate of flow, Q, in this model is given by

$$Q = WK\frac{h_L^2 - h_o^2}{2L} = WK\frac{h_L + h_o}{2}\frac{h_L - h_o}{L}, \tag{3.6.3}$$

where W is the width of the spring. The product $K(h_L+h_o)/2 \equiv \mathsf{T}$ represents an average *transmissivity* of the aquifer (Sec. 4.4), and $h_L - h_o$ represents the difference in head above the spring's threshold. Thus, with $Q = W\mathsf{T}(h_L - h_o)/L$, we may express α_1 by $\alpha_1 = (W/L)\mathsf{T}$, where L is a characteristic length of the aquifer. Or we may replace the expression α_1/S_yA by $\alpha_2\mathsf{T}/S_y$, where $\alpha_2(= W/LA)$ is a coefficient (dims. L^{-2}) that represents the geometry of the aquifer.

We may conclude this discussion by expressing the spring's *recession curve* by the formula

$$Q = Q_o \exp\left[-\beta(t - t_o)\right], \tag{3.6.4}$$

where we have some indication as to the nature of the coefficient β, since $\beta \equiv \alpha_2\mathsf{T}/S$. Often the aquifer contributing to the spring is made of several separate subregions, each with its own characteristic coefficient β.

Since in one form or another, the coefficient, or coefficients, appearing in the expression describing a spring's recession curve are related to the aquifer's geometry, transmissivity, and storativity, it is possible to investigate these properties of an aquifer by analyzing the hydrograph of the spring's discharge. We should note that in the above models, no pumping or recharge takes place during the analyzed aquifer depletion period. It is obvious that when wells operate in the region, they affect the water levels, which, in turn, affect the spring's discharge. In other words, spring discharge can be controlled (and that includes drying-up a spring if it is so indicated by the optimal management scheme) by controlling water levels in its vicinity.

In a regional aquifer model, a spring serves as a boundary condition. Usually it is considered a fixed head (= elevation of physical threshold) boundary condition. However, we have to watch for the possibility that at some point in time, which is *a priori* unknown, water levels in the vicinity of the spring may drop to below the threshold: the spring then dries up and ceases to act as a boundary of the flow domain. As water levels rise, it may return to its role as boundary of the flow domain. Another point to watch for is the possibility that as the rate of flow increases, a layer of water of a certain thickness covers the spring, thus making the water level at the spring a function of the discharge rate.

In a management problem a constraint of minimum spring discharge (say to supply downstream consumers or to maintain wildlife) is sometimes imposed.

3.7 Evapotranspiration

Evapotranspiration is another mechanism by which groundwater may leave an aquifer. Evaporation is the net transfer of water from the liquid phase to the vapor one. Transpiration is the process by which plants remove moisture from the soil and release it to the atmosphere as vapor. *Evapotranspiration*, a combination of the above two processes, is the term used to describe the total water removal from an area, partly covered by vegetation, by transpiration, evaporation from soil (actually from the water present in the void space of unsaturated soil), from snow, and from open water surfaces (lakes, streams, and reservoirs).

The amount of energy required to evaporate 1 cm^3 of water is 597 calories. The sun is the source of energy for the process of evapotranspiration in the hydrological cycle. However, the actual amount of energy available for evapotranspiration depends on the type of surface and the degree of cloudiness. Given this amount of energy, the actual evapotranspiration also depends on temperature, air pressure, wind, salinity of water, and the curvature of the air water interface through which evaporation takes place.

Obviously, evapotranspiration requires the availability of water. The term *potential evapotranspiration* is used to define the (say, annual) rate of evapotranspiration that would occur were there an adequate supply of soil moisture at all times. Actual evapotranspiration is less than, or at most equal to potential evapotranspiration. The latter is affected mainly by meteorological factors, whereas the former depends on plant and soil conditions.

Various methods are available for determining actual and potential evapotranspiration. Among the better known methods are the Thornthwaite formula (Thornthwaite, 1948; Thornthwaite and Hare, 1965), which estimates potential evaporation of open water, based on air temperature; the Hargreaves formula (Hargreaves, 1975), which considers the effect of solar radiation; the Blaney and Criddle formula (Blaney and Criddle, 1950), which is a temperature based method, but includes the effect of vegetation cover; the Priestley-Taylor formula (Priestley and Taylor, 1972), based on energy budget method; the Penman (Penman, 1948) and the Penman-Monteith (Monteith, 1965) model, which are methods that combine the energy budget that includes air and water temperature, and solar radiation effects, together with the aerodynamic effect of wind; the McNaughton-Black model (McNaughton and Black, 1973), which excludes the radiation budget; and the Shuttleworth-Wallace model (Shuttleworth and Wallace, 1985), which adds a soil layer to the Penman-Monteith model. For determining actual evapotranspiration we have the water balance method (based on lysimeter studies, water level fluctuations, or soil moisture balances), the use of evaporimeter (evaporation pan), and the moisture flux method (e.g., Rider, 1957). More details about evaporation and transpiration theory and measurement can be found in Brutsaert (1982), Hydrology Handbook (ASCE, 1996), Shuttleworth (1993), Verhoff and Campbell (2005), Allen (2005), and Roberts (2005).

Unless the water table is within 1–1.5 m from ground surface, evaporation from groundwater is negligible. When the water table is near ground surface, evapotranspiration may constitute a significant factor in the water balance. Certain plants have very deep roots and take up water by transpiration even from a rather deep water table (e.g., 15–20 m). White (1932) reports 10% of pan evaporation at a depth of 1 m. We do have some evaporation from water in the unsaturated zone, but this loss does not enter a water balance, when the latter is formulated only for the saturated zone.

In a large scale field experiment, Chen *et al.* (1996) compared four land surface evaporation models. In the *bucket evaporation model* (Manabe, 1969; Robock *et al.*, 1995), a simple mass balance approach is used to estimate the soil bucket water content (the available soil moisture for evaporation in the uppermost 1 m of soil). The *simple water balance (SWB) model* (Schaake *et al.*, 1996), is a two-layer (upper and lower 'buckets') water balance model that accounts for the spatial heterogeneity of rainfall, soil moisture, and runoff. In the *Oregon State University (OSU) model* (Pan and Mahrt, 1987; Ek and Mahrt, 1991), the soil hydrology (mass balance and unsaturated flow, see Sec. 6.3), and the soil thermodynamics (energy balance and heat flux), as well as the interaction with the vegetation, are modeled in two to three soil layer geometry. The fourth model, the *simplified simple biosphere (SSiB) model* (Xue *et al.*, 1991), is a biophysically based model of land-surface-atmosphere interaction, having three hydrological (unsaturated flow) soil layers, one thermodynamic soil slab, with a total of eight prognostic variables (soil wetness in three layers, temperature at the canopy, ground surface, and bottom boundary, water stored in the canopy, and snow stored on the ground). After applying these four models to the long term (5 months) and large study area (15 km × 15 km), it was concluded that the two simpler models, the bucket model and the SWB model, respectively, overestimated the soil evaporation during the wet periods and underestimated the evaporation during the dry periods. The two more complex models, the OSU model and the SSiB model, both were able to simulate the observed diurnal and seasonal variations in evaporation, soil moisture, sensible heat flux, and surface skin temperature.

3.8 Pumping and Drainage

Water can be withdrawn from an aquifer for beneficial usages by means of shallow dug wells, tubular deep wells, horizontal wells (also known as *radial collector wells*), kanats, and galleries. The reader is referred to the literature for details on well design and construction (e.g., American Water Work Assoc., 1967; Driscoll, 1986; Lehr *et al.*, 1988; Harlan *et al.*, 1989; Detay, 1997).

Often, wells penetrate an aquifer only partially. In three-dimensional flows, they are line sinks of finite length, producing in their vicinity a converging flow pattern toward them. Under the assumption of 'essentially (2-D) horizontal flow' in an aquifer, they are point sinks, producing a radially converging flow in their vicinity (to be superimposed on whatever other flow pattern exists in the aquifer). Very seldom should the fact that the well has actually a finite diameter be taken into account.

In consolidated materials, wells are often completed as uncased holes. In unconsolidated materials, a gravel pack filter is placed around the well screen to prevent sand from entering the well in order to ensure the production of cleaner water and to prevent damage to the formation by the removal of fines. The permeability of the gravel pack is higher than that of the formation. Figure 3.8.1 illustrates a typical construction of a deep well.

While a pumping well is a point sink in the flow domain in an aquifer, a gallery and a kanat (qanat) (Fig. 3.9.1) are line sinks of finite length. Actually, in the vicinity of a gallery (also surrounded by a gravel pack), the flow is two-dimensional in the vertical plane. However, in view of the assumption of essentially horizontal flow in an aquifer, in regional studies we regard it as a line sink in the xy-plane. A well can pump water as long as the water table at its location is higher than the elevation of the bottom of the suction pipe installed in it. For water to enter a gallery, the water table should be above its bottom.

In a regional water balance, we are often interested only in the total withdrawal by pumpage during the balance period. In a detailed forecasting problem, the areal distribution of pumpage is important.

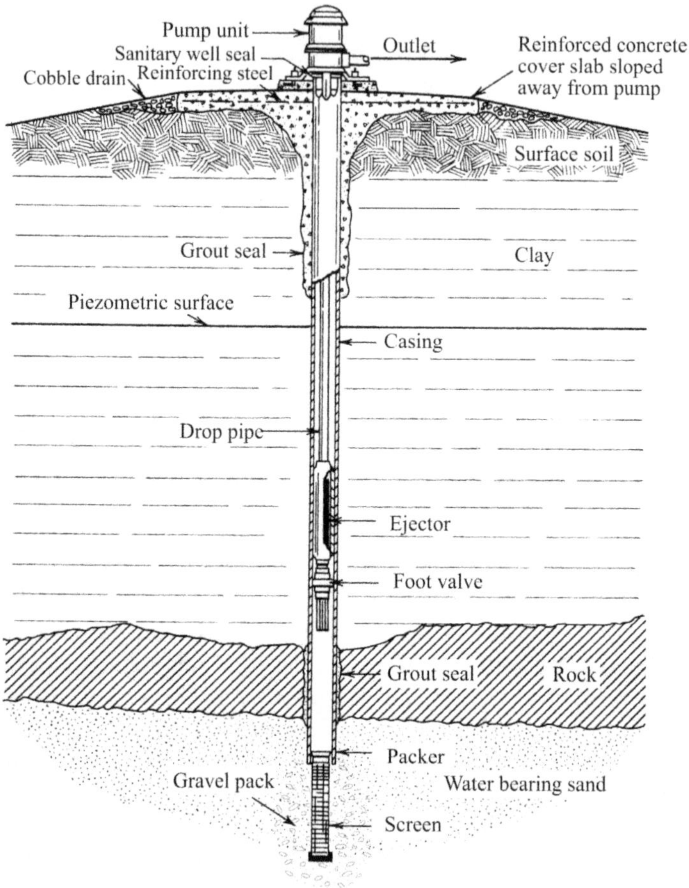

Figure 3.8.1: A drilled deep well (after US EPA, 2001).

A drainage system (open channels, or buried drains) is usually installed in order to control the elevation of the water table (say, to maintain water levels below the root zone). Groundwater will then leave the aquifer through this system (say, to a nearby stream) whenever the water table is higher than the drains. The overall behavior is similar to that of a depression spring (Sec. 3.6.1). The volume of water drained out of an aquifer in this way should not be left out of the water balance.

Usually, a second important objective of a drainage system is the removal (with the drainage water) of salts flushed down to the water table. In fact, every advanced irrigation system is always supplemented by a drainage system. Salts drained out of an aquifer in this way should be taken into account in a *regional salt balance.*

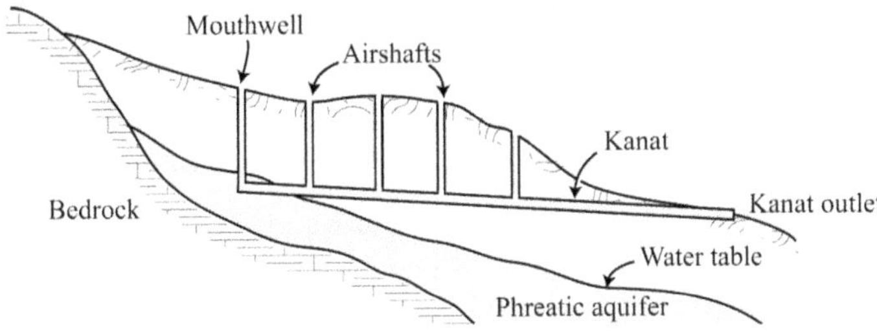

Figure 3.9.1: Schematic sketch of a kanat.

3.9 Change in Storage

The difference between all inflows and outflows during a balance period accumulates in the considered aquifer region. In a phreatic aquifer, water is stored in the void space (i.e., in that portion of the void space not occupied already by water or from which air can readily be displaced by water). In a confined aquifer, water is stored on account of water and solid matrix compressibility. In the first case, increased storage is followed by a rise of the water table. In the second case, by a rise in the piezometric head. These concepts are discussed in Sec. 5.4.1. At this point, we shall introduce (with no further explanations) the definition of aquifer storativity, S, associated with essentially horizontal flow in an aquifer, as the volume of water added to a unit horizontal area of aquifer, per unit rise in the water table elevation. Over an area A, a volume of water $\mathcal{U}_w$ stored in an aquifer causes the water table to rise by Δh,

$$\mathcal{U}_w = S \times A \times \Delta h. \tag{3.9.1}$$

If the rise is not uniform and the storativity varies from point to point, we can always divide the balance area into N sub-areas, such that

$$\mathcal{U}_w = \sum_{j=1}^{N} S_j \times A_j \times (\Delta h)_j. \tag{3.9.2}$$

Obviously, excess of outflow over inflow produces a drop in the water table, or in the piezometric surface.

3.10 Regional Groundwater Balance

We can now summarize the regional groundwater balance by the following equation

$$\left\{\begin{matrix} Groundwater \\ inflow \end{matrix}\right\} - \left\{\begin{matrix} Groundwater \\ outflow \end{matrix}\right\} + \left\{\begin{matrix} Natural \\ replenishment \end{matrix}\right\} + \left\{\begin{matrix} Return \\ flow \end{matrix}\right\}$$

$$+\left\{\begin{matrix} Artificial \\ recharge \end{matrix}\right\} + \left\{\begin{matrix} Inflow\ from \\ streams\ and\ lakes \end{matrix}\right\} - \left\{\begin{matrix} Spring \\ discharge \end{matrix}\right\}$$

$$-\left\{\begin{matrix} Evapo- \\ transpiration \end{matrix}\right\} - \left\{\begin{matrix} Pumpage\ and \\ drainage \end{matrix}\right\} = \left\{\begin{matrix} Increased\ volume \\ stored\ in\ aquifer \end{matrix}\right\}, \quad (3.10.1)$$

where all terms are expressed as volume of water during the balance period. Obviously, not all terms appear in all cases.

The various terms appearing in (3.10.1) can be calculated (better, estimated) for a specified aquifer domain as follows:

- *Groundwater inflow and outflow through the domains boundaries*: This subject is discussed in Subs. 3.1.1. We need values of aquifer transmissivity. Either this information is available (say from pumping tests), or the value of this coefficient is an unknown, the value of which should be determined as part of the calibration process.
- *Natural replenishment*: This term is discussed in Sec. 3.2. Often, (3.2.1) is employed, with P taken from precipitation records, and P_o and α regarded as unknown coefficients to be determined by the calibration process.
- *Return flow (from irrigation)*: This subject is discussed in Sec. 3.3. We need data on the annual rate of irrigation over the area, and an estimate on the fraction of it that is not consumed by the vegetation. Again, this fraction can be estimated, or considered a calibration factor.
- *Artificial recharge*: This subject is discussed in Sec. 3.4. Here we use actual data of the volume of water recharging the aquifer through wells and infiltration ponds.
- *Inflow from streams and lakes*: This subject is discussed in Sec. 3.5. In general, this value has to be estimated.
- *Spring discharge*: This subject is discussed in Sec. 3.6. Here, data is available from actual measurements.
- *Evapotranspiration*: This subject is discussed in Sec. 3.6. Estimated evapotranspiration values can be obtained from data on climatic conditions, vegetation, etc.
- *Pumping and drainage*: Data on pumping is usually available from direct measurements, or estimated from information on crops and irrigation practices. Information on drainage has to be estimated.
- *Increased volume stored in aquifer*: This subject is discussed in Sec. 3.9. The coefficient of aquifer storativity, is either known or has to be estimated by calibration.

Chapter 4

GROUNDWATER MOTION

As part of the hydrological cycle (Subs. 1.1.1), water from precipitation *infiltrates* through ground surface and *percolates*, primarily downward, through the unsaturated zone, or *vadose zone*, until it reaches a water table. The source of the infiltrating water may also be irrigation, or infiltration ponds for the purpose of artificially recharging an underlying aquifer.

The objective of this chapter is to present the basic laws that govern the flux of fluids in the saturated zone under isothermal conditions. The discussion on the movement of fluids in the unsaturated zone is presented in Chap. 6. As everywhere in this book, the entire discussion in this chapter, except in Subs. 4.2.2, is at the *macroscopic level* introduced in Sec. 1.3. Thus, all variables and parameters in this chapter are assumed to be macroscopic quantities defined everywhere in a porous medium domain regarded as a continuum. No special symbol is used to indicate this fact. We shall start from the empirical law suggested by *Henri Darcy* (1803-1858) in 1856, and then extend this law to three dimensions and to inhomogeneous and anisotropic porous media. The range of validity of Darcy's law is also discussed. Although Darcy's law was originally introduced by Darcy (1856) as an empirical (macroscopic) law, we shall also briefly mention its physical origin as a momentum equation at the microscopic level, and the derivation of its macroscopic counterpart by volume averaging. We shall then implement the mathematically more rigorous homogenization technique to develop the macroscopic flux law. Only the case of constant fluid density is considered in this chapter. The case of density-dependent fluid is discussed in Subs. 9.3.1.

4.1 Darcy's Law

4.1.1 Empirical law

Henri Darcy, the water engineer of Dijon, a city in the southern part of France, investigated the flow of water through vertical, saturated, homogeneous sand filters (columns) for the water for the city's fountains. Figure 4.1.1 shows the original experimental setup used by Darcy to reach his *empirical* law. From his experiments (Darcy, 1856), Darcy concluded that the rate of flow

J. Bear, A.H.-D. Cheng, *Modeling Groundwater Flow and Contaminant Transport*, Theory and Applications of Transport in Porous Media 23,
DOI 10.1007/978-1-4020-6682-5_4, © Springer Science+Business Media B.V. 2010

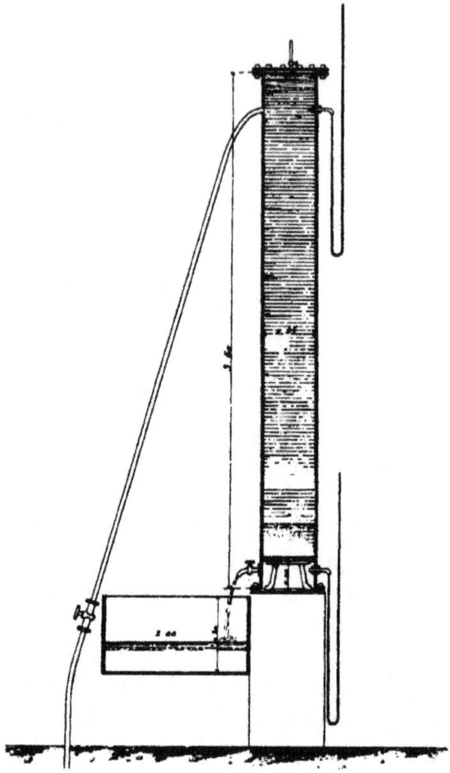

Figure 4.1.1: Darcy's column experiment. (After Darcy, 1856)

(= volume of water passing per unit time), Q, through a homogeneous sand column of constant cross-sectional area is:

- proportional to the cross-sectional area of the column, $\mathcal{A}$,
- proportional to the difference in water level elevations, $h^{(1)}$ and $h^{(2)}$, at the inflow and outflow reservoirs of the column, respectively, and
- inversely proportional to the column's length, L.

When combined, these conclusions give the famous *Darcy's formula* (or *law*)[1]

$$Q = \mathsf{K}\mathcal{A}\frac{h^{(1)} - h^{(2)}}{L}, \tag{4.1.1}$$

[1] Prof. T.N. Narasimhan of University of California, Berkeley, comments (Private communication) that Henri Darcy's fundamental contribution was to extend the one-dimensional linear flux law, already known for viscous flow in tubes (e.g., in the form of Poiseuille's (1799-1869) law) for viscous flow in capillary tubes, to natural earth materials, and to account for gravity in addition to pressure. He realized that the conductivity (or, resistance) for natural earth materials, must be obtained empirically. (See also Narasimhan (1998).)

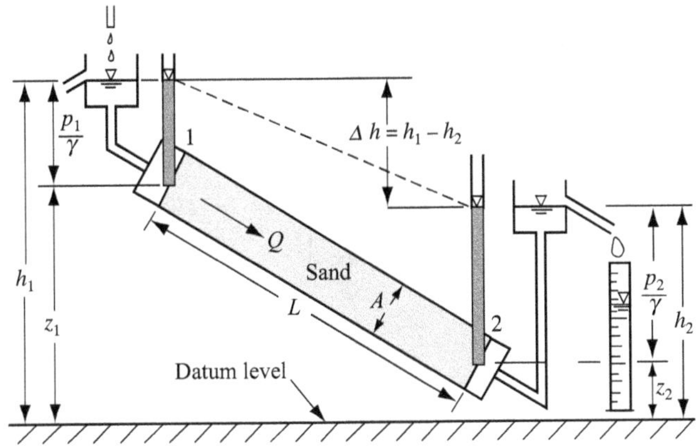

Figure 4.1.2: Flow in an inclined porous medium column.

where the coefficient of proportionality, K, called *hydraulic conductivity* (dims. L/T), is discussed in detail in Subs. 4.1.3. Introducing the definition of *specific discharge*, defined as the volume of water passing through a unit area of porous medium cross-section, in the direction normal to the latter, per unit time,

$$q = \frac{Q}{A}, \tag{4.1.2}$$

we can rewrite (4.1.1) in the form:

$$q = \mathsf{K}\frac{h_{\text{inflow}} - h_{\text{outflow}}}{L}. \tag{4.1.3}$$

Although Darcy's law was derived from experiments on a vertical column, we can easily extend the basic principles to the case of one-dimensional flow in the inclined column of saturated, homogeneous porous medium shown in Fig. 4.1.2. In this figure, the elevations $h^{(1)}$ and $h^{(2)}$ represent also the *piezometric heads* (dims. L) in the respective reservoirs, defined as

$$h = z + \frac{p}{\rho g}, \qquad \rho g = \gamma, \tag{4.1.4}$$

where z is the elevation of the point at which the piezometric head is being considered, above some datum level, p, ρ and γ are the fluid's pressure, mass density, and specific weight, respectively, and g is the gravity acceleration.

The *total mechanical energy per unit weight* of fluid at a point within a porous medium domain, h_{total}, often referred to as *total head*, is expressed by

$$h_{\text{total}} = \underset{\text{(a)}}{z} + \underset{\text{(b)}}{\frac{p}{\rho g}} + \underset{\text{(c)}}{\beta\frac{V^2}{2g}}, \tag{4.1.5}$$

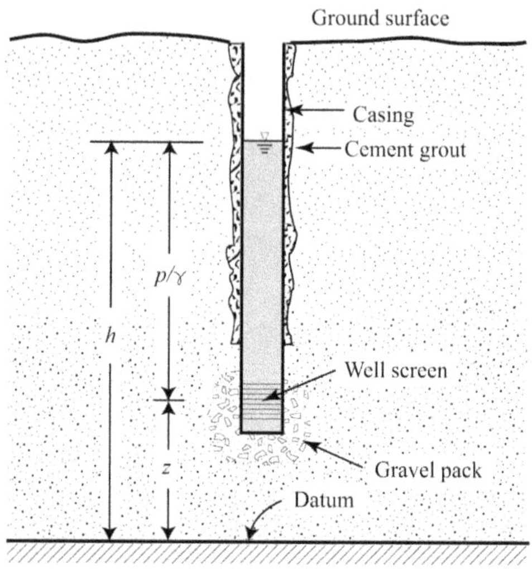

Figure 4.1.3: The piezometric head measured in an observation well.

where V is the magnitude of the fluid's (intrinsic phase average) velocity, and β is a coefficient introduced by the averaging process. Thus, the mechanical energy (per unit weight of fluid) is made up of three parts:

(a) **Potential energy** per unit weight (dims. L), also referred to as *elevation head*, z, resulting from the elevation of the point above some datum level.
(b) **Pressure energy** per unit weight, or **pressure head**, $p/\rho g$ (dims. L), resulting from the pressure in the fluid.
(c) **Kinetic energy** per unit weight, $\beta V^2/2g$ (dims. L), associated with the fluid's velocity. In most cases of flow through porous media, the kinetic energy head is much smaller than the pressure one, due to the very low velocity of the fluid.

Figure 4.1.3 shows the piezometric head, h, measured in an observation well, with respect to the same datum level as the elevation, z. It expresses the sum of the potential energy and the pressure energy per unit weight of fluid.

An *observation*, or *monitoring, well* is a device used for measuring the *piezometric head*, defined in (4.1.4), at a point within an aquifer. It is also called a *piezometer*. It can be briefly described as a vertical pipe (= *casing*) inserted down to the point where the piezometric head measurement is required. The portion of the pipe that is located at the elevation at which this measurement is required, is slotted, perforated, or screened, and is often surrounded by a *gravel pack* to prevent clogging. In this way, we enable a good hydraulic connection between water in the pipe (= well) and water in the formation. The elevation of the water surface inside the well gives the piezometric head at the location of the screen.

By measuring the piezometric heads at a number of spatially distributed observation wells tapping the same aquifer, a contour map can be drawn of a surface called the *piezometric surface*. The elevation of this surface at a point in the horizontal plane gives the piezometric head in the aquifer at that point. The phreatic surface is also a piezometric surface. In a given flow domain, the surface composed of all the points at which the piezometric head has the same value is called a *equipotential surface*.

To complete the picture, in a compressible fluid, $\rho = \rho(p)$, and the pressure head is expressed by

$$\int_{p_o}^{p} \frac{dp}{g\rho(p)},$$

where p_o is some reference pressure. This expression indicates that the pressure energy stored in the fluid per unit weight of fluid is obtained from the work done in compressing the fluid. For such a fluid, it is common to define a piezometric head, h^*, often called *Hubbert's potential* (Hubbert, 1940):

$$h^* = h^*(\mathbf{x}, t) = z + \int_{p_o}^{p} \frac{dp}{g\rho(p)}. \tag{4.1.6}$$

In Sec. 4.2, we shall make use of Hubbert's potential to introduce Darcy's law for a compressible fluid.

Note that although the piezometric head, h, is often referred to by groundwater hydrologists as a 'potential', or a 'groundwater potential', it is actually not a potential in the mathematical sense. A function Φ^E is said to be a 'flux potential of E' if the flux of E, $\mathbf{q}^E$, at every point within a considered (here, porous medium) domain, is given by $\mathbf{q}^E = -\nabla\Phi^E$ (Bear and Bachmat, 1990, p. 67). When this relationship is valid, the flow of E is *irrotational*, i.e., $\nabla \times \Phi^E = 0$, and Φ^E is a *harmonic* function, i.e., satisfying the Laplace equation, $\nabla^2\Phi^E = 0$. Also the integral $\oint \Phi^E\, ds = 0.$, for any closed curve within the domain.

In a homogeneous isotropic porous medium domain, with constant fluid density, where Darcy's law is $\mathbf{q} = -\mathsf{K}\nabla h \equiv -\nabla(\mathsf{K}h)$ (see Subs. 4.1.2), we may refer to $\mathsf{K}h$ as a potential (or to h as a *pseudo-potential*). Then Hubbert's potential, where $\rho = \rho(p)$, is also a (pseudo)potential. However, in the general case of $\rho = \rho(p, c, T)$, where c is solute concentration and T is temperature, $z + p/g\rho$ is *not a potential.*

With the definition of h as energy per unit weight of water, the *energy loss*, $h^{(1)} - h^{(2)}$, in (4.1.1), is due to friction in the flow through the narrow tortuous paths of the porous medium; changes in the kinetic energy have been neglected as being much smaller than those in piezometric head.

Although, originally, Darcy's law in the form of (4.1.1), or (4.1.3), was derived from experiments on a finite length column, we can extend Darcy's conclusion to what happens *at a point* along a column. To achieve this goal, consider flow in a segment of a stream-tube in three-dimensional space aligned in a direction indicated by the unit vector $\mathbf{1s}$ (Fig. 4.1.4). The piezometric

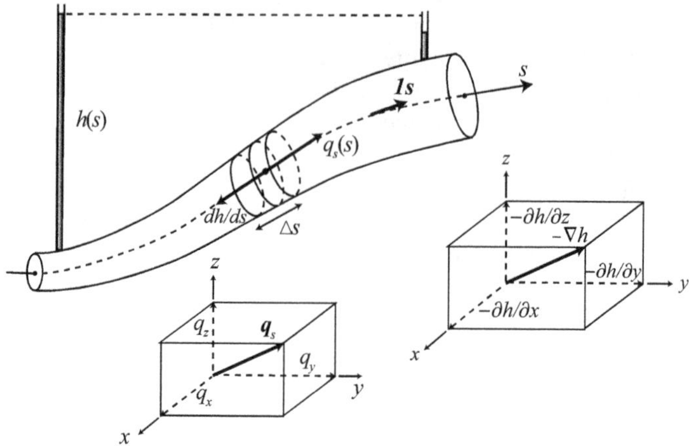

Figure 4.1.4: A stream-tube in three-dimensional space.

head varies along the stream-tube (≡ 'column'), i.e., $h = h(s)$. Let us consider a segment of this column of length Δs along the s-axis, between the coordinates $s - \frac{\Delta s}{2}$, and $s + \frac{\Delta s}{2}$. For this case, (4.1.3) takes the form:

$$q_s(s) = \mathsf{K}\frac{h\big|_{s-\frac{\Delta s}{2}} - h\big|_{s+\frac{\Delta s}{2}}}{\Delta s}, \tag{4.1.7}$$

where the subscript s in q_s indicates that the flow is in the s-direction. In the limit, as $\Delta s \to 0$, we obtain

$$\lim_{\Delta s \to 0} \frac{h\big|_{s-\frac{\Delta s}{2}} - h\big|_{s+\frac{\Delta s}{2}}}{\Delta s} = -\frac{dh}{ds}, \tag{4.1.8}$$

and (4.1.7) reduces to

$$q_s = -\mathsf{K}\frac{dh}{ds} \equiv \mathsf{K}\mathcal{J}_s, \tag{4.1.9}$$

where q_s is considered positive in the positive direction of the s-axis, and $\mathcal{J}_s$ is defined by

$$\mathcal{J}_s = -\frac{dh}{ds}. \tag{4.1.10}$$

In the one-dimensional flow considered so far, the derivative dh/ds expresses the slope of the *piezometric line*, $h = h(s)$, with a positive value indicating a rising function, $h = h(s)$, and a negative value indicating that h decreases with s. We refer to $\mathcal{J}$ as the *hydraulic gradient.*

Equation (4.1.9) states that the flow takes place from a higher piezometric head to a lower one, and *not* necessarily from a higher to a lower pressure. For example, in the case shown in Fig. 4.1.2, $p^{(1)} < p^{(2)}$, i.e., the flow is in

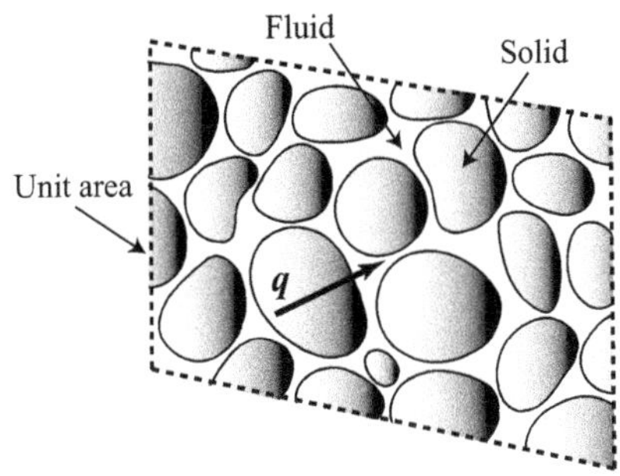

Figure 4.1.5: A porous medium cross-section.

the direction of *increasing* pressure; however, it is the direction of *decreasing* piezometric head. It is only in the special case of horizontal flow, where $z_1 = z_2$, that we may write

$$Q = \mathsf{K}\,\mathcal{A}\frac{p_1 - p_2}{\rho g L}. \qquad (4.1.11)$$

It should be emphasized that Darcy's law (4.1.9), expressed in terms of the piezometric head, h, is valid only for a fluid of constant density. When the fluid's density varies, because of variations in pressure, concentration of dissolved matter, or temperature, the hydraulic gradient, $\mathcal{J}$, should not be used as a driving force. Instead, as we shall suggest in Sub. 9.3.1, the state variable to be used in the motion equation is the pressure.

So far, we have been discussing the fluid's specific discharge. However, when considering the transport of a solute dissolved in the moving fluid (Chap. 7), we need to know the fluid's *velocity*. Actually, as shown in Fig. 4.1.5, at the microscopic level, flow takes place only through part of the cross-sectional area of the porous medium, the remaining part being occupied by the *solid matrix*, or *solid skeleton*, of the porous medium. Because it can be shown (Delesse, 1848; Bear and Bachmat, 1990 p. 37) in all but exceptional cases, that the *average areal porosity equals the volumetric one*, ϕ, the portion of the area $\mathcal{A}$ available to flow is $\phi\mathcal{A}$, where ϕ denotes the porosity (= volume of voids per unit volume of a porous medium sample). Accordingly, the *average velocity*, V, of a fluid flowing through a porous medium is given by

$$V = \frac{Q}{\phi\mathcal{A}} = \frac{q}{\phi}. \qquad (4.1.12)$$

This velocity is sometimes called the *seepage velocity*. As explained in Subs. 4.1.5, the velocity as defined above is the *mass-averaged velocity* of the fluid.

Sometime, part of the void space is unavailable to fluid flow, or almost so, due to dead-end pores in which the fluid is (practically) immobile. We then define an *effective porosity*, ϕ_{eff}, and use it to determine the velocity,

$$V = \frac{Q}{\phi_{\text{eff}}\mathcal{A}} = \frac{q}{\phi_{\text{eff}}}. \tag{4.1.13}$$

Note how we make a very clear distinction between **specific discharge** and **velocity**.

4.1.2 Extension to three-dimensional space

In the preceding section, we have extended Darcy's law from the original vertical column (Fig. 4.1.1) to an inclined straight column (Fig. 4.1.2), and then to a short segment of a curved stream-tube (Fig. 4.1.4) in a three-dimensional space. Now, we are ready to extend the law to the general three-dimensional case.

Consider flow in a three-dimensional porous medium domain. We focus our attention on a (possibly curved) streamline and identify a stream-tube of infinitesimal cross-section around it (Fig. 4.1.4). The stream-tube serves just like the 'column' in the figure. The piezometric head, h, varies with distance, s, along the stream-tube, i.e., $h = h(s)$, with a derivative dh/ds, in the s-direction. In this way we have established the relationship $h = h(s)$, and its derivative, dh/ds, in the s-direction in a three-dimensional context. The specific discharge in the s-direction, q_s, which is now marked as a vector $\mathbf{q}_s$, takes place in the direction of the hydraulic gradient, i.e., opposite to the direction of the gradient of h ($\equiv \text{grad}\, h$).

As demonstrated in Fig. 4.1.4, let $\mathbf{1s}$ denote a vector in the three-dimensional Cartesian space in x, y and z coordinates. We can replace the (scalar) gradient ($\mathcal{J}_s \equiv -dh(s)/ds$) by the *hydraulic gradient vector*, $\boldsymbol{\mathcal{J}}$, which can be decomposed into three components: $-\partial h/\partial x$, $-\partial h/\partial y$, and $-\partial h/\partial z$. In other words, $\boldsymbol{\mathcal{J}}$, is equal to the negative of the *gradient* vector, $\text{grad}\, h$ ($\equiv \nabla h$). Similarly, the specific discharge $\mathbf{q}_s$ in the $\mathbf{1s}$ direction can be decomposed into three components q_x, q_y and q_z.

Given a spatial distribution $h = h(x, y, z)$, the vector ∇h at a point (x, y, z), indicates, *in direction and magnitude*, the steepest ascent of the function $h = h(x, y, z)$ at that point. It is a vector that is everywhere normal to the surfaces $h =$ constant. Figure 4.1.6 gives an illustration of a two-dimensional field $h(x, y)$; its magnitude is plotted as the surface in Fig. 4.1.6a. The arrow indicates the direction of *steepest ascent*. The flow will take place in the opposite direction, i.e., 'downhill', in the direction of *steepest descent*. Figure 4.1.6b presents the same idea in a two dimensional contour plot of constant h values. The arrows indicate both the magnitude and the direction of the gradient, which must be perpendicular to the constant head line (or surface in three dimensions).

With the above illustration, we are now ready to write Darcy's law in three dimension. In fact, we can extend the above discussion to include also time, with $h = h(x, y, z, t)$. At any point in a three-dimensional porous medium

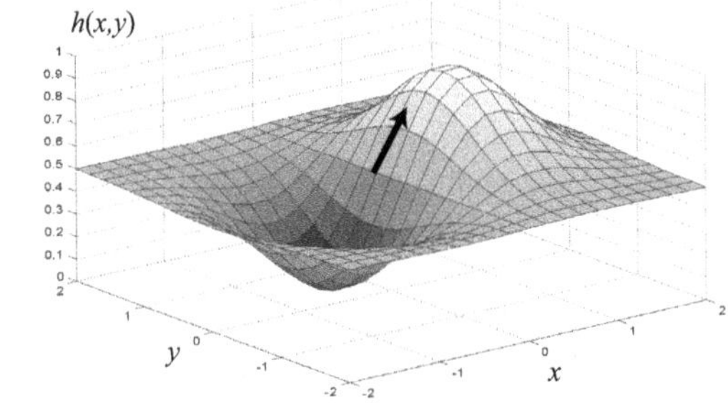

(a)

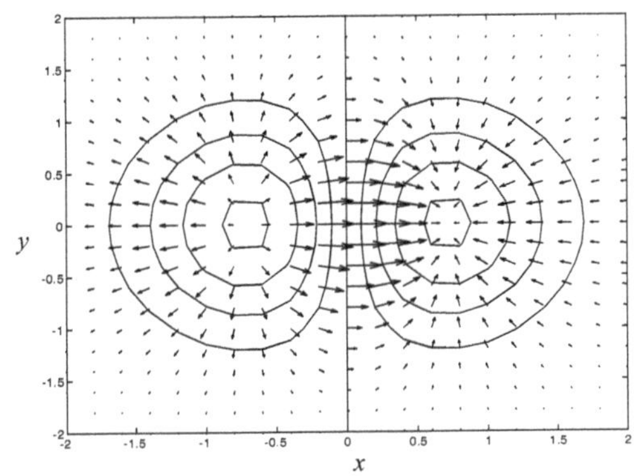

(b)

Figure 4.1.6: The concept of gradient: Plot of piezometric head $h(x, y)$: (a) As surface, (b) As contour lines; arrows indicate direction of steepest ascent.

domain, at any time t, given the value of $\mathsf{K}(x, y, z)$ *at that point* and the spatial distribution of the piezometric head, $h = h(x, y, z, t)$, the extended Darcy's law, can be written in the form:

$$q_x = -\mathsf{K}(x, y, z)\frac{\partial h}{\partial x}, \quad q_y = -\mathsf{K}(x, y, z)\frac{\partial h}{\partial y}, \quad q_z = -\mathsf{K}(x, y, z)\frac{\partial h}{\partial z}. \tag{4.1.14}$$

Note that we have emphasized that we take the value of K *at* the consider point, (x, y, z). In this way, we have, actually, extended Darcy's law also to the case of an inhomogeneous porous medium domain, where $\mathsf{K} = \mathsf{K}(x, y, z)$. In the compact *vcctor form*:

$$\mathbf{q} = -\mathsf{K}(x, y, z)\nabla h \equiv -\mathsf{K}(x, y, z)\boldsymbol{\mathcal{J}}. \tag{4.1.15}$$

Darcy's law (4.1.15), expressed in terms of the piezometric head, h, is valid only for a fluid of constant density. Variable density is discussed in Subs. 9.3.1). The mass flux is expressed by $\rho\mathbf{q}$.

4.1.3 Hydraulic conductivity

The coefficient of proportionality, K, appearing in Darcy's law (4.1.1) is called the *hydraulic conductivity* of the porous medium. In an *isotropic* porous medium (Subs. 4.1.4), this equation may be used to define it as the *specific discharge per unit hydraulic gradient.* It is a scalar that expresses the ease with which a fluid flows through the tortuous void space. It is, therefore, a coefficient that depends on both matrix and fluid properties. As demonstrated in Subs. 4.2.2, the relevant fluid properties are the density, ρ, and the dynamic viscosity, μ (or in the combined form of the *kinematic viscosity,* ν $(= \mu/\rho)$). For water at 15°C, $\mu = 1.139 \times 10^{-6}\mathrm{m}^2/s = 1.139$ centipoise. The relevant solid matrix properties are, for example, grain- or pore-size distribution, shape of grains or pores, tortuosity of passages, specific surface, and porosity. Bear and Bachmat (1990) show that in saturated flow, the *hydraulic radius* of the fluid filled void space is the characteristic length that determines the hydraulic conductivity. The effects of the various solid matrix (actually, void space) features are combined in the form of a coefficient called *permeability.*

The hydraulic conductivity, K, can then be expressed as

$$\mathsf{K} = \mathsf{k}\frac{\rho g}{\mu} = \frac{\mathsf{k}g}{\nu}, \tag{4.1.16}$$

where g is the gravity acceleration, μ and ν are the dynamic and kinematic viscosities of the fluid, respectively, and k (dims. L^2) is the *permeability*, or *intrinsic permeability,* of the porous medium. It is a coefficient that depends solely on the properties of the configuration of the void space and not on those of the fluid.

Various units are used in practice for the hydraulic conductivity, K. Some hydrologists prefer the unit m/d (meters per day). Soil scientists and geotechnical engineers often use cm/s (centimeters per second). In SI units, m/s (meters per second) is used. In the United States and other countries that use the English system of units, hydrologists use ft/d (feet per day) and gal/d-ft^2 (gallons per day per square foot). Representative values of hydraulic conductivity are given in Fig. 4.1.7.

The permeability, k, is measured in the metric system in cm^2, or in m^2. Petroleum engineers use the unit *darcy,* defined by

$$1 \text{ darcy} = \frac{1 \text{ cm}^3/\text{s}/\text{cm}^2 \ \times \ 1 \text{ centipoise}}{1 \text{ atmosphere/cm}}, \tag{4.1.17}$$

with 1 darcy $= 9.87 \times 10^{-9}\mathrm{cm}^2$. This is obtained from Darcy's law for horizontal flow: $q = -(k/\mu)[(\Delta p)/L]$. Figure 4.1.7 includes also permeability values.

Numerous formulæ that relate permeability to various geometric properties of the solid matrix are presented in the literature. Some are purely

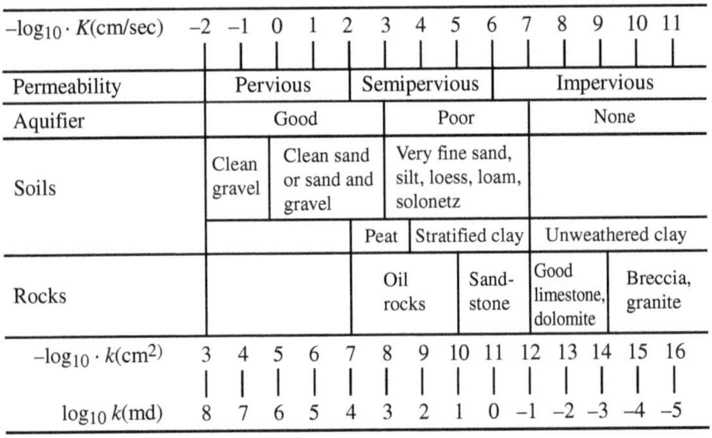

Figure 4.1.7: Representative values of hydraulic conductivity (for water at $20°C$) and permeability for selected soils (Bear *et al.*, 1968).

empirical, as, for example,

$$\mathsf{k} = Cd^2, \tag{4.1.18}$$

where C is a dimensionless coefficient and d is an effective grain diameter, say, d_{10} (i.e., 10% of the grains by weight are smaller than this diameter). Krumbein and Monk (1943) suggested $C = 6.17 \times 10^{-4}$. Although this is an empirical formula, the dependence on the square of a characteristic length of the pore space can be justified by a theoretical analysis (e.g., Bear and Bachmat, 1990).

Another example is the Fair and Hatch (1933) formula, developed from dimensional considerations, and verified experimentally,

$$\mathsf{k} = \frac{1}{\beta}\left[\frac{(1-\phi)^2}{\phi^3}\left(\frac{\alpha}{100}\sum_{(m)}\frac{P_m}{d_m}\right)^2\right]^{-1}, \tag{4.1.19}$$

where β is a *packing factor*, found experimentally to be 5, α is a sand *shape factor*, varying from 6 for spherical grains to 7.7 for angular ones, P_m is the weight percentage of sand held between adjacent sieves, and d_m is the geometric mean diameter of the adjacent sieves.

An often used formula for permeability is the Kozeny-Carman equation:

$$\mathsf{k} = C_o\frac{\phi^3}{(1-\phi)^2(\Sigma_{vs})^2}, \tag{4.1.20}$$

where Σ_{vs} is the specific surface area of the solid (defined per unit volume of solid matrix), and C_o is a coefficient for which Carman (1937) suggested the value 0.2. Often, $1/(\Sigma_{vs})^2$ is replaced by d^2, with d = mean grain size.

Under certain conditions, the permeability may vary with time. Such a change may be caused by compaction of a layer due to external loads. Clogging by precipitation or dissolution of minerals, filtration of fine-grained solids, or swelling of clay may also produce changes in the structure and texture of the solid matrix. When a soil contains argillaceous material, drying of the soil may shrink the clay, especially bentonite, causing the permeability to air of the dried soil to be higher than to water before drying. Freshwater replacing saltwater in a soil sample may cause the clay to swell, thereby reducing the permeability. Biological activity in the void space may produce biomass that tends to fill the void space, thus reducing the permeability with time.

4.1.4 Extension to anisotropic porous media

When, *at a given point*, the permeability is independent of direction, the porous medium at that point is said to be *isotropic*. Otherwise, the porous medium is referred to as *anisotropic* with respect to permeability.

For an anisotropic porous medium, Darcy's law takes the form:

$$\begin{aligned} q_x &= \mathsf{K}_{xx}\mathcal{J}_x + \mathsf{K}_{xy}\mathcal{J}_y + \mathsf{K}_{xz}\mathcal{J}_z, \\ q_y &= \mathsf{K}_{yx}\mathcal{J}_x + \mathsf{K}_{yy}\mathcal{J}_y + \mathsf{K}_{yz}\mathcal{J}_z, \\ q_z &= \mathsf{K}_{zx}\mathcal{J}_x + \mathsf{K}_{zy}\mathcal{J}_y + \mathsf{K}_{zz}\mathcal{J}_z. \end{aligned} \tag{4.1.21}$$

Equation (4.1.21) expresses a linear relationship between the $x-$, $y-$, and $z-$components, q_x, q_y, q_z, of the specific discharge vector, $\mathbf{q}$, and the components $\mathcal{J}_x$, $\mathcal{J}_y$, $\mathcal{J}_z$, of the hydraulic gradient, $\boldsymbol{\mathcal{J}}$. These two sets of components are related to each other by *nine* coefficients, K_{xx}, K_{xy}, etc. Together, these coefficients represent the hydraulic conductivity of the porous medium. In a heterogeneous medium, each of these coefficients may vary in space.

One may regard (4.1.21) just as the most general *linear* relationship between the 'driving force', which is the hydraulic gradient, $\boldsymbol{\mathcal{J}}$, and the resulting fluid flux, $\mathbf{q}$. However, it is also possible to obtain (4.1.21) from first principles, as discussed in Sec. 4.2

The component K_{ij} may be interpreted as the contribution to the specific discharge in the ith direction, q_i, produced by a unit component of the hydraulic gradient in the jth direction, $\mathcal{J}_j$. The total specific discharge in any direction is the sum of the partial specific discharges in that direction caused by $\mathcal{J}_x$, $\mathcal{J}_y$, and $\mathcal{J}_z$. This is a consequence of the microscopic structure of the tortuous pore space (Bear, 1972, p. 111).

The nine K-coefficients appearing in (4.1.21) are components of the *second rank tensor of hydraulic conductivity* of an anisotropic porous medium. A detailed discussion on the nature and use of second rank tensors is beyond the scope of this book. We'll try to minimize the use of the theory of tensor analysis. The reader is referred to texts on tensor analysis, such as Morse and Feshbach (1953) and Aris (1962).

Nevertheless, let us say a few words about tensors. In Civil and Mechanical Engineering, we encounter the concept of *stress*, defined as *force per unit area*. However, dividing the vector *force* by the vector *area*, which has a magnitude and a direction perpendicular to the tangent plane of the surface, is not permitted. Instead, we may say that a *force vector*, $\mathbf{F}$, is obtained by the product of *stress*, $\boldsymbol{\sigma}$, and the area vector $\boldsymbol{\mathcal{A}}$

$$\mathbf{F} = \boldsymbol{\sigma} \cdot \boldsymbol{\mathcal{A}}. \tag{4.1.22}$$

The *stress*, $\boldsymbol{\sigma}$, is then said to be a *second rank symmetric tensor.* In fact, whenever two vectors (e.g., $\mathbf{F}$ and $\boldsymbol{\mathcal{A}}$ above) are related to each other by a relationship like (4.1.22), in which the dot ($\cdot$) denotes a product (*contraction*), the term represented by $\boldsymbol{\sigma}$ is a second rank tensor; this includes a scalar regarded as a zeroth rank tensor.

Another (more mathematical) way to explain what is a tensor, is to refer to its behavior under rotation of the coordinate system. We recall that when we consider a force (vector) in a three-dimensional space, and rotate the coordinates, the force itself is not changed by the fact that we have rotated the coordinate system. Only the *components* of the vector vary as a consequence of this rotation. In mathematics, a certain rule expresses the relation between the magnitude of a vector component before and after rotation of the coordinate system. We may now turn this statement around and state that any three numbers that, under rotation of coordinates, behave according to this mathematical rule are components of a vector. Similarly, any 9 numbers that, under rotation of the coordinate system, behave according to the rule that defines the change in magnitude of the components of a second rank tensor, are components of such a tensor.

A vector (three components in a three-dimensional space) may be considered as a first rank tensor, while a scalar (one component) is a zeroth rank tensor. Then, in a three-dimensional space, a second rank tensor has 9 components.

In vector notation, we use the *sans serif* boldface symbol, here, $\mathsf{\mathbf{K}}$, to denote the second rank tensor of hydraulic conductivity. We use the regular *sans serif* symbol to denote the scalar components of a tensor and the scalar value of a (second rank) tensorial property of an isotropic porous medium. The components of $\mathsf{\mathbf{K}}$ in a three-dimensional space, can be written in the matrix form:

$$\mathsf{\mathbf{K}} = \begin{bmatrix} \mathsf{K}_{xx} & \mathsf{K}_{xy} & \mathsf{K}_{xz} \\ \mathsf{K}_{yx} & \mathsf{K}_{yy} & \mathsf{K}_{yz} \\ \mathsf{K}_{zx} & \mathsf{K}_{zy} & \mathsf{K}_{zz} \end{bmatrix}, \tag{4.1.23}$$

and in a two-dimensional space as:

$$\mathsf{\mathbf{K}} = \begin{bmatrix} \mathsf{K}_{xx} & \mathsf{K}_{xy} \\ \mathsf{K}_{yx} & \mathsf{K}_{yy} \end{bmatrix}. \tag{4.1.24}$$

The hydraulic conductivity tensor, K_{ij}, is *symmetric*, i.e., $\mathsf{K}_{ij} = \mathsf{K}_{ji}$, and *positive definite*. This means that actually only *six distinct components* are needed to fully define it in a three-dimensional domain. If the principal direction are known, then only three distinct coefficients are required to fully define the hydraulic conductivity tensor in a three-dimensional space, as three angles are required to define the principal directions (two angles to define one direction, one additional angle for the second direction, and the third direction is automatically determined). In two dimensions, three coefficients are needed, but only two if we know the directions of the principal axes (only one angle is needed to define the two orthogonal directions in two-dimensional space). The fact that the tensor K_{ij} that appears in (4.1.21) is symmetric is a consequence of the *Onsager reciprocal relationship* (e.g., de Groot and Mazur, 1962), which is applicable to cases in which a force that produces a flux of one extensive quantity produces also fluxes of other extensive quantities. Here, each of the 3 components of the hydraulic gradient, which is the driving force, contributes to each of the three flux components. It is interesting to note that we may have cases with $\mathsf{K}_{ij} \leq 0$ for $i \neq j$. This is a consequence of the positiveness of the principal minor: in 2-D we have the following relationships

$$\mathsf{K}_{11}\mathsf{K}_{22} - \mathsf{K}_{12}^2 \geq 0, \quad \Rightarrow \quad -\sqrt{\mathsf{K}_{11}\mathsf{K}_{22}} \leq \mathsf{K}_{12} \leq \sqrt{\mathsf{K}_{11}\mathsf{K}_{22}}. \tag{4.1.25}$$

Note that in an anisotropic porous medium, the information on the six independent coefficients that define the hydraulic conductivity tensor can be transformed into three principal directions and three K-values (= *principal values*) in these directions.

Equations (4.1.21) may be written in several compact forms. For example, they may be written in the compact vector form:

$$\mathbf{q} = \mathbf{K} \cdot \mathcal{J} \equiv -\mathbf{K} \cdot \nabla h, \tag{4.1.26}$$

i.e., expressing $\mathbf{q}$ as a *scalar* (or *dot*) *product* of the hydraulic conductivity tensor, $\mathbf{K}$, and the hydraulic gradient vector, $\mathcal{J}$ ($\equiv -\nabla h$), or, in *indicial notation* for an orthogonal coordinate system as:

$$q_i = \mathsf{K}_{ij}\mathcal{J}_j \equiv -\mathsf{K}_{ij}\frac{\partial h}{\partial x_j}, \tag{4.1.27}$$

where the subscripts i and j indicate x_i and x_j, respectively, with $x_1 \equiv x$, $x_2 \equiv y$, and $x_3 \equiv z$. In writing (4.1.27), we have made use of *Einstein's summation convention* (or *double index convention*) according to which a subscript (or superscript) repeated twice and only twice in any product or quotient of factors is summed over the entire range of values of that subscript (or superscript), i.e., $i, j = 1, 2, 3$ and $1, 2$ for three- and two-dimensional spaces, respectively. Thus, in three dimensions,

$$\mathsf{K}_{ij}\mathcal{J}_j \equiv \sum_{j=1}^{3} \mathsf{K}_{ij}\mathcal{J}_j = \mathsf{K}_{i1}\mathcal{J}_1 + \mathsf{K}_{i2}\mathcal{J}_2 + \mathsf{K}_{i3}\mathcal{J}_3.$$

In this book, we shall often write equations in their compact vector form. Although only scalars are used when actually making calculations, the advantage of writing the more compact vector equations (in addition to their conciseness) is that given a problem, they can easily be rewritten in the coordinate system, say cartesian, radial, cylindrical, etc., that is suitable for that problem. Note that although the physical nature of the hydraulic conductivity tensor, $\mathbf{K}$, at a point within an anisotropic porous medium is independent of the coordinate system used, its mathematical representation in terms of the value of tensor components, K_{ij}, *does* depend on the chosen coordinate system.

Textbooks on tensor analysis (Aris, 1962) give the rules for transforming these components from one coordinate system to another: the components K_{ij}, $i,j = 1,2,3$ of the second rank tensor, $\mathbf{K}$, in the coordinate system (x_1, x_2, x_3), are transformed into K'_{ij} in the coordinate system (x'_1, x'_2, x'_3), according to the transformation rule

$$\mathsf{K}'_{ij} = \sum_{(m)} \sum_{(n)} \mathsf{K}_{mn} \cos\alpha_{mi} \cos\alpha_{nj}, \tag{4.1.28}$$

in which α_{mi} denotes the angle between the axes $0x_m$ and $0x'_i$. They also show that it is always possible to find *three mutually orthogonal directions* in space such that when these directions are chosen as the coordinate system for expressing the components K_{ij}, we find that $\mathsf{K}_{ij} = 0$ for all $i \neq j$ and $\mathsf{K}_{ij} \geq 0$ for $i = j$. These directions in space are called the *principal directions of the hydraulic conductivity* of the anisotropic porous medium. In a heterogeneous porous medium domain, the principal directions may vary from point to point.

Thus, for example, in two dimensions, the transformation from any coordinate system, (x, y) to the principal coordinate system, (x', y'), is given by the relationship

$$\begin{matrix} \mathsf{K}'_{x'x'} \\ \mathsf{K}'_{y'y'} \end{matrix} = \frac{\mathsf{K}_{xx} + \mathsf{K}_{yy}}{2} \pm \left[\left(\frac{\mathsf{K}_{xx} - \mathsf{K}_{yy}}{2} \right)^2 + \mathsf{K}_{xy}^2 \right]^{1/2}. \tag{4.1.29}$$

The angle of rotation needed to reach the principle axes is given by

$$\alpha = \frac{1}{2} \tan^{-1} \frac{2\mathsf{K}_{xy}}{\mathsf{K}_{xx} - \mathsf{K}_{yy}}. \tag{4.1.30}$$

When $\mathsf{K}'_{x'x'}$ and $\mathsf{K}'_{y'y'}$, and xy are the cartesian coordinates rotated clockwise by an angle α, with respect to $x'y'$, we obtain:

$$\begin{aligned} \begin{matrix} \mathsf{K}_{xx} \\ \mathsf{K}_{yy} \end{matrix} &= \frac{\mathsf{K}'_{x'x'} + \mathsf{K}'_{y'y'}}{2} \pm \frac{\mathsf{K}'_{x'x'} - \mathsf{K}'_{y'y'}}{2} \cos 2\alpha, \\ \mathsf{K}_{xy} &= -\frac{\mathsf{K}'_{x'x'} - \mathsf{K}'_{y'y'}}{2} \sin 2\alpha. \end{aligned} \tag{4.1.31}$$

When the principal directions are aligned with a selected coordinate system, (4.1.23) and (4.1.24) may be represented in matrix form as:

$$[\mathsf{K}] = \begin{bmatrix} \mathsf{K}_{xx} & 0 & 0 \\ 0 & \mathsf{K}_{yy} & 0 \\ 0 & 0 & \mathsf{K}_{zz} \end{bmatrix}, \quad \text{and} \quad [\mathsf{K}] = \begin{bmatrix} \mathsf{K}_{xx} & 0 \\ 0 & \mathsf{K}_{yy} \end{bmatrix}, \tag{4.1.32}$$

respectively, so that (4.1.21) reduces to

$$q_x = \mathsf{K}_{xx}\mathcal{J}_x, \qquad q_y = \mathsf{K}_{yy}\mathcal{J}_y, \qquad q_z = \mathsf{K}_{zz}\mathcal{J}_z. \tag{4.1.33}$$

Because the hydraulic conductivity is related to the permeability by the scalar factor $\rho_w g/\mu_w$, the permeability of an anisotropic porous medium is also a second rank tensor, **k**.

4.2 Darcy's Law as Momentum Balance Equation

Darcy's law is a fundamental building block of modeling flow and solute transport through porous media. As presented in Sec. 4.1, originally, Henry Darcy (1856) suggested this law on the basis of sand column experiments aimed at determining the relationship between the flux through a sand column and the difference in head imposed across that column. However, over the years, and especially in the last few decades, a number of researchers have noted that Darcy's law is actually a *motion equation* for a fluid in the void space. Hence, following the practice in hydrodynamics, it should be expressed as a *momentum balance equation.* In hydrodynamics, the momentum balance equation is considered at the *microscopic* level (Sec. 1.3), i.e., *at points inside a fluid continuum.* In the case of a porous medium, the hydrodynamic approach means considering what happens *at points inside the fluid that occupies the void space, or part of it.* of course, this is not a practical approach for modeling flow through a porous medium domain, as we have no way of describing the geometry of the fluid occupied domain within the void space. To obtain a practical flux law for a porous medium, the microscopic motion equation needs to be transformed into a *macroscopic* one by some averaging technique.

As discussed in Chap. 1, there are three major approaches that can be used to transform the microscopic physical laws, e.g., balance equations of extensive quantities, such as mass, momentum and energy, into their macroscopic counterparts for practical application purposes: (1) the representative elementary volume (REV) averaging approach (Subs. 1.3.3), (2) the homogenization approach (Subs. 1.3.5), and (3) the mixture-theory approach (de Boer, 2000). A variety of simplifying assumptions and constraints are introduced in each of these three approaches; hence, the resulting macroscopic laws depend on the selected set of underlying assumptions. A complete review of these three approaches is beyond the scope of this book. See, for example, Bear (1972), Bear and Bachmat (1986, 1990), Bowen (1984), and Hassanizadeh (1986) for the REV averaging approach, Ene and Poliševski (1987), Mei and Auriault (1989), Hornung (1997), and Chen *et al.* (2001), for the homogenization approach, and Bowen (1980) and de Boer (2000) for the mixture-theory approach. In this section, we shall consider only the momentum balance equation, leading to Darcy's law. We shall start (next subsection) by introducing the momentum balance equation at the microscopic level and lead to Darcy's law by volume averaging. Then in Subs. 4.2.2, we shall derive Darcy's law by homogenization. Furthermore, in Subs. 4.2.3, we shall make use of the development of the macroscopic mass balance equation for the fluid in order to derive an expression for the equivalent, or effective hydraulic conductivity of a heterogeneous (with respect to hydraulic conductivity) porous medium domain.

All three techniques—volume averaging, homogenization, and mixture-theory—have a common objective, namely to smooth out the effects of the (geometrical) heterogeneity at the microscopic level (produced by the presence of solids and pores) on the description of flow and other phenomena of transport, by passing to a the macroscopic level of description. Let us start with a brief presentation of Darcy's law as derived by volume averaging. More about volume averaging can be found, for example, in Bear (1972), Bowen (1984), Hassanizadeh (1986), Bear and Bachmat (1986, 1990), Mei and Auriault (1989), and Hornung (1997).

4.2.1 Darcy's law by volume averaging

The starting point is always the microscopic *differential balance equation of linear momentum of a phase*, written in the (vector) form as:

$$\underset{\text{(a)}}{\frac{\partial}{\partial t}\rho\mathbf{V}} = \underset{\text{(b)}}{-\nabla\cdot\rho\mathbf{V}\mathbf{V}} + \underset{\text{(c)}}{\nabla\cdot\boldsymbol{\sigma}} + \underset{\text{(d)}}{\rho\mathbf{F}}, \tag{4.2.1}$$

where ρ is the fluid's density, $\mathbf{F}$ is the body force, usually, due to gravity, and $\boldsymbol{\sigma}$ denotes stress (= minus the diffusive flux of momentum). Thus, by comparison with the law of momentum conservation, we recognize that

(a) = rate of accumulation of momentum,
(b) = rate of momentum gained by advection,
(c) = rate of momentum gained by diffusive momentum transfer,
(d) = rate of supply of momentum by the body force,

and all terms are *per unit volume of the phase.*

By averaging this momentum balance equation over the fluid volume within an REV (e.g., Bear and Bachmat, 1990, p. 162), and introducing the assumptions:

- The fluid is Newtonian.
- Inertial forces are negligible relative to the viscous ones.
- The effect of momentum transfer within the fluid, as a result of (microscopic) velocity gradients, is negligible, in comparison with the drag produced at the fluid-solid interface,

a simplified form is obtained for the averaged momentum balance equation. When the fluid occupies the entire void space (i.e., saturated flow), this expression takes the form:

$$\mathbf{V} - \mathbf{V}_s = -\frac{\mathsf{k}(x, y, z)}{\phi\mu} \cdot (\nabla p + \rho g \nabla z), \tag{4.2.2}$$

where $\mathbf{V}$, p, ρ, and μ denote the (intrinsic phase) average velocity, pressure, density, and viscosity of the fluid, respectively, $\mathbf{V}_s$ denotes the (intrinsic phase) average velocity of the solid, z is the elevation, and k is the permeability tensor, which reduces to a scalar, k, in an isotropic porous medium. In indicial notation and Cartesian coordinates, (4.2.2) takes the form:

$$V_i - V_{si} = -\frac{\mathsf{k}_{ij}(x, y, z)}{\phi\mu}\left(\frac{\partial p}{\partial x_j} + \rho g \frac{\partial z}{\partial x_j}\right), \qquad i, j = 1, 2, 3. \tag{4.2.3}$$

Equation (4.2.2) is the general motion equation (we still refer to it as Darcy's law) for saturated flow of a single Newtonian fluid in an anisotropic, heterogeneous, deformable porous medium.

Note that the fluid's velocity, $\mathbf{V}$, in (4.2.2) is a *mass-weighted* one, $\rho\mathbf{V} = \sum_\gamma \rho^\gamma \mathbf{V}^\gamma$, where superscript γ denotes a component in the fluid. This fact is significant when we consider a multicomponent fluid with spatial variations in component concentrations that produce density variations and fluxes by molecular diffusion.

In (4.2.2), the term $-(\nabla p + \rho g \nabla z)$ represents the macroscopic driving force, per unit volume of fluid, which produces fluid motion: the term $-\nabla p$ represents a force due to a pressure gradient, while $-\rho g \nabla z$ is a force due to gravity, recalling that ∇z represents a unit vector directed vertically upward. The (vector) sum of these two forces (Bear, 1972, p. 160) is balanced by the drag, or resistance to flow at the solid-fluid interface, which is expressed by $-\phi\mu\mathsf{k}^{-1} \cdot (\mathbf{V} - \mathbf{V}_s)$. Written in this form, Darcy's law states that the so-called *Stokes drag* is proportional to the fluid's velocity relative to the solid skeleton, proportional to the fluid viscosity, and inversely proportional to the macroscopic coefficient k/ϕ, which represents (at the macroscopic level) the effect of the microscopic configuration of the void space of the porous medium on the saturated flow.

We wish to reiterate that the motion equation for a fluid phase, e.g., (4.2.2), is nothing but a simplified form of the macroscopic momentum balance equation for that phase. It expresses the exchange of momentum between the fluid and the solid phases. When the above simplifying assumptions that underlie Darcy's law are not applicable, other ('non-Darcy') forms of the averaged momentum balance equation are obtained; these are mentioned in Subs. 4.3.2.

Introducing the definition of *specific discharge relative to the solid*

$$\mathbf{q}_r \equiv \phi(\mathbf{V} - \mathbf{V}_s) \equiv \phi\mathbf{V}_r, \tag{4.2.4}$$

equation (4.2.2) becomes

$$\mathbf{q}_r = -\frac{\mathsf{k}}{\mu} \cdot (\nabla p + \rho g \nabla z). \tag{4.2.5}$$

In this equation, the fluid's density, ρ, may depend on pressure, concentration of components and temperature.

For a compressible fluid, $\rho = \rho(p)$, the right hand side of (4.2.5) reduces to

$$\mathbf{q}_r = -\mathsf{K} \cdot \nabla h^*, \tag{4.2.6}$$

where $\mathsf{K} = \mathsf{k}\rho g/\mu$, and h^* is Hubbert's potential defined by (4.1.6). For a constant ρ, the right-hand side reduces to $-\mathsf{K} \cdot \nabla h$.

When a gaseous phase flows at a low pressure through very small pores, the *mean free path* of the molecules is of the order of magnitude of the pore size, and the *no slip* condition (at the microscopic fluid-solid interface) is no longer valid (Knudsen, 1934). Klinkenberg (1941) suggested the following expression for gas permeability under such conditions:

$$\mathsf{k}_g = \mathsf{k}_\ell(1 + 4c\lambda/r) = \mathsf{k}_\ell(1 + b/p), \tag{4.2.7}$$

where k_g and k_ℓ ($< \mathsf{k}_g$) denote gas and liquid (or high density gas) permeabilities, respectively, λ denotes the mean free path of the gas molecules under the pressure p, $c\,(\approx 1)$ is a proportionality factor, r is mean pore size, and $b = b(\lambda, r)$ is a coefficient. At high p, $b/p \to 0$.

When $\mathbf{V}_s = 0$ (or approximately so), we denote $\phi\mathbf{V}$ by $\mathbf{q}$ to obtain

$$\mathbf{q} = -\frac{\mathsf{k}}{\mu} \cdot \Big(\nabla p + \rho g \nabla z\Big). \tag{4.2.8}$$

Thus, in principle, we should always use $\mathbf{q}_r$ in the flux equation (with $\mathbf{q}_r \equiv \phi(\mathbf{V} - \mathbf{V}_s)$, referred to as 'relative specific discharge'). However, as long as it does not lead to misunderstandings, we shall continue to use the symbol $\mathbf{q}$ for specific flux whenever $\mathbf{V}_s = 0$, or approximated as such; under such conditions, $\mathbf{q} \equiv \phi\mathbf{V}$.

4.2.2 Darcy's law by homogenization

The *differential balance equation of linear momentum of a phase*, e.g., in the (vector) form (4.2.1), is also the starting point for the derivation of Darcy's law by the method of homogenization.

We introduce the assumptions:

- *Locally*, i.e., at every point within the domain, the fluid motion is at steady state (i.e., the local acceleration term is negligible). We introduce this assumption because our limited objective here is to demonstrate the development of Darcy's original law from the fundamental momentum balance equation.
- The body force is due to gravity only.
- The solid phase is non-deformable and stationary.
- The fluid is a single-component, compressible Newtonian one, with the viscous stress tensor described by the constitutive relation

$$\boldsymbol{\tau} = \mu\left(\nabla\mathbf{V} + (\nabla\mathbf{V})^T\right) + \lambda(\nabla\cdot\mathbf{V})\,\mathbf{I}, \quad \boldsymbol{\tau} = \boldsymbol{\sigma} - p\mathbf{I}, \tag{4.2.9}$$

  in which μ is the fluid's dynamic viscosity, and λ is the viscosity associated with the fluid's compressibility.

With these assumptions, the simplified momentum balance equation is the steady-state Navier-Stokes equation, written in vector notation as

$$\begin{aligned}\rho(\mathbf{x})\left(\mathbf{V}(\mathbf{x})\cdot\nabla\right)\mathbf{V}(\mathbf{x}) &= -\nabla p(\mathbf{x}) + \mu\nabla^2\mathbf{V}(\mathbf{x}) \\ &\quad + (\lambda+\mu)\nabla\left(\nabla\cdot\mathbf{V}(\mathbf{x})\right) + \rho(\mathbf{x})\mathbf{g}, \quad \mathbf{x}\in\Omega_v, \end{aligned}\tag{4.2.10}$$

where p is the pressure, $\mathbf{g}$ is the gravity acceleration vector, and ∇, and ∇^2 are, respectively, the gradient, and the Laplacian operators.

Equation (4.2.10) is applicable only to the Ω_v-space occupied by the fluid phase; the fluid quantities $\mathbf{V}$, p, ρ, etc., are not defined in Ω_s. Because of the existence of the two subdomains, Ω_v and Ω_s, the values of these three fluid-variables fluctuate over the Ω-space, particularly $\mathbf{V}$. Some averaging, or 'smoothing', is needed to produce a continuous distribution of these variables within Ω. Such a distribution is required if we wish to apply (4.2.10) to Ω, within the framework of a *continuum* model, in which the averaged, macroscopic variables are used and can be measured and coefficients can be determined. In the subsection below, the homogenization approach (see Subs. 1.3.5) is used to average the fluctuating (microscopic) behavior of fluids within the porous medium domain as a whole. It will be shown that the resulting (smoother) macroscopic model is Darcy's law.

Consider the porous medium domain, Ω, characterized by the length scale L, shown in Fig. 4.2.1. A portion of the domain, Ω_s, is occupied by a solid, while the remaining portion, Ω_v, is occupied by a void space, with $\Omega = \Omega_s + \Omega_v$. The interface between Ω_s and Ω_v is denoted by Γ_i. The domain is visualized as a *periodic* structure composed of a large number of identical

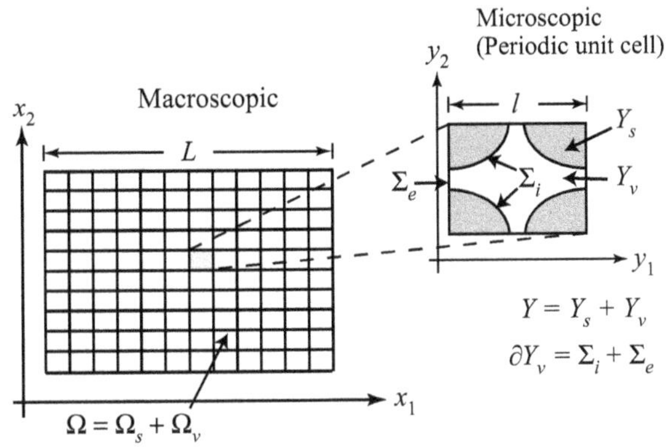

Figure 4.2.1: Periodic cells in the homogenization theory.

elementary *cells*, Y, each characterized by the length scale ℓ, with the ratio between the two scales, ε, being much smaller than 1:

$$\varepsilon \equiv \frac{\ell}{L} \ll 1. \tag{4.2.11}$$

Each Y-cell contains two parts: a solid portion, Y_s, and a void space, Y_v, such that $Y = Y_s + Y_v$. The boundary of the void space domain, denoted as ∂Y_v, consists of two parts: a solid-void space interface, Σ_i, and the exterior surface of Y, Σ_e, with $\partial Y_v = \Sigma_i + \Sigma_e$ (see Fig. 4.2.1).

As in Subs. 1.3.5, the small parameter, ε, is used as the *asymptotic expansion parameter* in the homogenization process. At the macroscopic scale, i.e., at the scale associated with the length L, we define a coordinate system $\mathbf{x} = (x_1, x_2, x_3)$. At the microscopic scale, i.e., the scale associated with the length ℓ, we introduce a *local system*, $\mathbf{y} = (y_1, y_2, y_3)$ (Fig. 4.2.1), with

$$\mathbf{y} = \frac{\mathbf{x}}{\varepsilon}, \tag{4.2.12}$$

such that the Y-cell is scaled to unity in the $\mathbf{y}$-coordinate system.

We shall examine the problem of a viscous fluid that flows through the void space, Ω_v (and Y_v). For a Newtonian fluid, this flow is governed by the Navier-Stokes equation (4.2.10), rewritten here with a superscript ε to emphasize the solution's dependence on the small parameter ε:

$$\begin{aligned}\rho^\varepsilon(\mathbf{x})\left[\mathbf{V}^\varepsilon(\mathbf{x})\cdot\nabla\right]\mathbf{V}^\varepsilon(\mathbf{x}) &= -\nabla p^\varepsilon(\mathbf{x}) + \mu\nabla^2\mathbf{V}^\varepsilon(\mathbf{x})\\ &+(\lambda+\mu)\nabla\left[\nabla\cdot\mathbf{V}^\varepsilon(\mathbf{x})\right] + \rho^\varepsilon(\mathbf{x})\mathbf{g}; \quad \mathbf{x}\in\Omega_v. \end{aligned} \tag{4.2.13}$$

Note that because our objective is derive Darcy's law, we are starting from the Navier-Stokes equation for steady flow. The above equation is subject to the no-slip boundary condition on Γ_i (i.e., the fluid sticks to the stationary solid surface),

$$\mathbf{V}^\varepsilon(\mathbf{x}) = 0, \quad \mathbf{x} \in \Gamma_i. \tag{4.2.14}$$

To obtain a solution for $\mathbf{V}$, p and ρ, we need one more equation and a constitutive relation for ρ. Accordingly, we introduce the fluid's mass balance equation (see any textbook on fluid mechanics, or Sec. 5.1, with reference to a porous medium):

$$\nabla \cdot [\rho^\varepsilon(\mathbf{x})\, \mathbf{V}^\varepsilon(\mathbf{x})] = 0, \quad \mathbf{x} \in \Omega_v. \tag{4.2.15}$$

To be consistent with (4.2.13), the above equation describes steady state.

Next, we introduce an equation of state for the fluid's density. We shall assume that the density is a function of pressure only, obeying the linear constitutive equation

$$\rho^\varepsilon = \rho^\varepsilon(p^\varepsilon) = \rho^\varepsilon(p_o)[1 + \beta(p^\varepsilon - p_o)], \tag{4.2.16}$$

where $\rho^\varepsilon(p_o)$ is the reference density corresponding to the reference pressure p_o, and β is the compressibility coefficient. Altogether, we have to solve (4.2.13), (4.2.15), and (4.2.16) for the three variables: $\mathbf{V}^\varepsilon$, p^ε, and ρ^ε.

With the above system in mind, our goal is to find the (differential) equation, with corresponding coefficients, that describes the relation between the *average fluid velocity* and *average pressure* in the porous medium domain, taking into account the dependence of average density on average pressure.

We assume that each dependent variable in the above flow model can be expanded into an asymptotic series in ε (see discussion in Subs. 1.3.5),

$$p^\varepsilon(\mathbf{x}) = p^{(o)}(\mathbf{x},\mathbf{y}) + \varepsilon p^{(1)}(\mathbf{x},\mathbf{y}) + \varepsilon^2 p^{(2)}(\mathbf{x},\mathbf{y}) + \dots, \tag{4.2.17}$$

$$\mathbf{V}^\varepsilon(\mathbf{x}) = \varepsilon^2\mathbf{V}^{(o)}(\mathbf{x},\mathbf{y}) + \varepsilon^3\mathbf{V}^{(1)}(\mathbf{x},\mathbf{y}) + \varepsilon^4\mathbf{V}^{(2)}(\mathbf{x},\mathbf{y}) + \dots, \tag{4.2.18}$$

$$\rho^\varepsilon(\mathbf{x}) = \rho^{(o)}(\mathbf{x},\mathbf{y}) + \varepsilon \rho^{(1)}(\mathbf{x},\mathbf{y}) + \varepsilon^2 \rho^{(2)}(\mathbf{x},\mathbf{y}) + \dots, \tag{4.2.19}$$

where $\mathbf{x} \in \Omega$ and $\mathbf{y} \in Y_v$. Here, in anticipation of a two-scale solution—at the microscopic level, and at the macroscopic one—the expansion terms have been expressed as functions of both $\mathbf{x}$ and $\mathbf{y}$. All quantities related to the $\mathbf{y}$-coordinates are assumed to be Y-periodic. We notice that we have formally extended the definition of the dependent variables, $\mathbf{V}^\varepsilon$, p^ε and ρ^ε, to the *entire domain*, $\mathbf{x} \in \Omega$, and not just to the void space Ω_v.

In the expansions (4.2.17)–(4.2.19), we notice that by starting from different orders of ε-terms, the physical quantities, particularly the velocity, are scaled differently. The selection of the scale is important as it will affect the outcome of the analysis—different scales in the asymptotic expansion lead to different laws (Ene and Poliševski, 1987). For example, the selection of the leading term of $\mathbf{V}^\varepsilon$ to be of the order ε^0 or ε^1 will cause $\mathbf{V}^{(o)}$ to appear in the highest order equation, giving the result that $\mathbf{V}^{(o)}$ is independent of the microscopic pore geometry, which is inconsistent with Darcy's law. On the other hand, different selections of the order of the leading term will lead to nonlinear laws, such as the *Forchheimer law* (Woodie and Levy, 1991; Chen *et al.*, 2001), and *Brinkman law* (Allaire, 1991, 1997a).

By inserting the perturbed expressions (4.2.17) and (4.2.19) into the equation of state, (4.2.16), and comparing terms of the same order in ε, we obtain

$$\rho^{(o)} = \rho(p_o)[1 + \beta(p^{(o)} - p_o)]; \quad \rho^{(1)} = \rho(p_o)\beta p^{(1)}; \quad \rho^{(2)} = \rho(p_o)\beta p^{(2)}; \quad \dots \tag{4.2.20}$$

Next, we substitute (4.2.17)–(4.2.19) into the Navier-Stokes and the continuity equations, (4.2.10) and (4.2.15). We then make use of the two-scale differentiation rule

$$\nabla = \nabla_{\mathbf{x}} + \frac{1}{\varepsilon}\nabla_{\mathbf{y}}, \tag{4.2.21}$$

in which the subscripts indicate that the differentiation is conducted with respect to the **x**- or to the **y**-coordinate. For example,

$$\nabla_{\mathbf{x}} = \frac{\partial}{\partial x_1}\mathbf{i} + \frac{\partial}{\partial x_2}\mathbf{j} + \frac{\partial}{\partial x_3}\mathbf{k}. \tag{4.2.22}$$

With $O(\varepsilon^i)$ denoting 'order of ε to the power i', we obtain from the Navier-Stokes equation (4.2.10):
$O(\varepsilon^{-1})$:

$$\nabla_{\mathbf{y}}\, p^{(o)}(\mathbf{x},\mathbf{y}) = 0, \tag{4.2.23}$$

$O(\varepsilon^0)$:

$$\begin{aligned}\mu\nabla^2_{\mathbf{y}}\,\mathbf{V}^{(o)}(\mathbf{x},\mathbf{y}) &+ (\lambda+\mu)\nabla_{\mathbf{y}}\left[\nabla_{\mathbf{y}}\cdot\mathbf{V}^{(o)}(\mathbf{x},\mathbf{y})\right]\\ &= \nabla_{\mathbf{y}}\, p^{(1)}(\mathbf{x},\mathbf{y}) + \nabla_{\mathbf{x}}\, p^{(o)}(\mathbf{x},\mathbf{y}) - \rho^{(o)}(\mathbf{x},\mathbf{y})\mathbf{g},\end{aligned} \tag{4.2.24}$$

where we recall that $\mathbf{x}\in\Omega$ and $\mathbf{y}\in Y_v$. From the continuity equation (4.2.15), we find
$O(\varepsilon^1)$:

$$\nabla_{\mathbf{y}}\cdot\left[\rho^{(o)}(\mathbf{x},\mathbf{y})\mathbf{V}^{(o)}(\mathbf{x},\mathbf{y})\right] = 0, \tag{4.2.25}$$

$O(\varepsilon^2)$:

$$\begin{aligned}\nabla_{\mathbf{x}}\cdot\left[\rho^{(o)}(\mathbf{x},\mathbf{y})\mathbf{V}^{(o)}(\mathbf{x},\mathbf{y})\right] &+ \nabla_{\mathbf{y}}\cdot\left[\rho^{(o)}(\mathbf{x},\mathbf{y})\mathbf{V}^{(1)}(\mathbf{x},\mathbf{y})\right]\\ &+\nabla_{\mathbf{y}}\cdot\left[\rho^{(1)}(\mathbf{x},\mathbf{y})\mathbf{V}^{(o)}(\mathbf{x},\mathbf{y})\right] = 0.\end{aligned} \tag{4.2.26}$$

The boundary condition (4.2.14) gives

$$\mathbf{V}^{(o)}(\mathbf{x},\mathbf{y}) = \mathbf{V}^{(1)}(\mathbf{x},\mathbf{y}) = \mathbf{V}^{(2)}(\mathbf{x},\mathbf{y}) = \ldots = 0, \quad \mathbf{x}\in\Gamma_i, \quad \mathbf{y}\in\Sigma_i. \tag{4.2.27}$$

We are now ready to solve the above asymptotic expansion equations. First, from (4.2.23), we conclude that

$$p^{(o)}(\mathbf{x},\mathbf{y}) = p^{(o)}(\mathbf{x}). \tag{4.2.28}$$

From (4.2.20), we conclude that

$$\rho^{(o)}(\mathbf{x},\mathbf{y}) = \rho^{(o)}(\mathbf{x}). \tag{4.2.29}$$

Equations (4.2.28) and (4.2.29) show that the *zeroth order* terms (meaning the most dominant terms in the asymptotic expansions of pressure and density, (4.2.17)–(4.2.19), are macroscopic variables. Their values change very little within the microscopic Y-cell, to the extent that they may be considered constant, i.e., independent of $\mathbf{y}$ within the Y-cell. This conclusion is not trivial, as we shall observe below that the zeroth order velocity term depends on *both* $\mathbf{x}$ and $\mathbf{y}$. The reason is that at the microscopic scale, the velocity changes from some finite value in the flow channels (voids) to zero at the solid wall, due to the no-slip condition mentioned earlier. In fact, these large changes in velocity at the small (microscopic) scale are the reason for performing homogenization, with the objective of producing a smoothed, macroscopic velocity field.

Introducing the above relationships into the $O(\varepsilon^1)$ continuity equation (4.2.25), we find

$$\nabla_{\mathbf{y}} \cdot \mathbf{V}^{(o)}(\mathbf{x},\mathbf{y}) = 0. \tag{4.2.30}$$

The $O(\varepsilon^2)$ continuity equation, (4.2.26), becomes

$$\nabla_{\mathbf{x}} \cdot \left[\rho^{(o)}(\mathbf{x})\mathbf{V}^{(o)}(\mathbf{x},\mathbf{y})\right] + \nabla_{\mathbf{y}} \cdot \left[\rho^{(o)}(\mathbf{x})\mathbf{V}^{(1)}(\mathbf{x},\mathbf{y})\right] + \nabla_{\mathbf{y}} \cdot \left[\rho^{(1)}(\mathbf{x},\mathbf{y})\mathbf{V}^{(o)}(\mathbf{x},\mathbf{y})\right] = 0. \tag{4.2.31}$$

With (4.2.28)–(4.2.30), the $O(\varepsilon^0)$ Navier-Stokes equation, (4.2.24), reduces to

$$\mu\nabla^2_{\mathbf{y}}\mathbf{V}^{(o)}(\mathbf{x},\mathbf{y}) = \nabla_{\mathbf{y}}\, p^{(1)}(\mathbf{x},\mathbf{y}) + \nabla_{\mathbf{x}}\, p^{(o)}(\mathbf{x}) - \rho^{(o)}(\mathbf{x})\mathbf{g}. \tag{4.2.32}$$

The last two equations need to be averaged over the Y-cell in order to smooth out the microscopic variations and to produce equations in terms of the *macroscopic* $\mathbf{x}$-coordinate only. This is accomplished by integration.

For example, when the integration is applied to the velocity term, $\mathbf{V}^{(o)}$, we obtain

$$\mathbf{q}_o(\mathbf{x}) = \frac{1}{|Y|}\int_{Y_v} \mathbf{V}^{(o)}(\mathbf{x},\mathbf{y})\, d\mathcal{U}_v(\mathbf{y}), \tag{4.2.33}$$

where $d\mathcal{U}_v(\mathbf{y})$ denotes an infinitesimal volume element of Y_v (the void space) with respect to the $\mathbf{y}$-coordinate, and the integration is performed only over the void domain, Y_v. The symbol $|Y|$ denotes the volume of Y, and $\mathbf{q}$ is the 'homogenized' velocity. In groundwater flow, as everywhere in this book, $\mathbf{q}$ is referred to as the 'specific discharge' (= fluid volume passing through a unit area of porous medium, per unit time).

Applying the same integration to the second term in (4.2.31), and utilizing the *Gauss (divergence) theorem* (see any textbook on vector analysis) to convert the volume integral to a surface one, we obtain:

$$
\begin{aligned}
\int_{Y_v} \nabla_{\mathbf{y}} \cdot \left[\rho^{(o)}(\mathbf{x})\mathbf{V}^{(1)}(\mathbf{x},\mathbf{y})\right] \, d\mathcal{U}(\mathbf{y}) &= \rho^{(o)}(\mathbf{x}) \int_{\partial Y_v} \mathbf{n} \cdot \mathbf{V}^{(1)}(\mathbf{x},\mathbf{y}) \, d\mathcal{S}(\mathbf{y}) \\
&= \rho^{(o)}(\mathbf{x}) \int_{\Sigma_i} \mathbf{n} \cdot \mathbf{V}^{(1)}(\mathbf{x},\mathbf{y}) \, d\mathcal{S}(\mathbf{y}) + \rho^{(o)}(\mathbf{x}) \int_{\Sigma_e} \mathbf{n} \cdot \mathbf{V}^{(1)}(\mathbf{x},\mathbf{y}) \, d\mathcal{S}(\mathbf{y}) \\
&= 0,
\end{aligned} \tag{4.2.34}
$$

where $d\mathcal{S}$ denotes a surface element, $\mathbf{n}$ is the outward unit vector normal to the surface, and $\partial Y_v = \Sigma_i + \Sigma_e$ is the boundary of Y_v, which consists of two parts: the solid-fluid interface, Σ_i, and a part that is exposed to the external surface of Y, Σ_e (see Fig. 4.2.1). The integration over Σ_i vanishes because of the no-slip condition (4.2.27); the integration over Σ_e also vanishes because of the Y-periodicity condition. Similarly, it can be shown that

$$
\int_{Y_v} \nabla_{\mathbf{y}} \cdot \left[\rho^{(1)}(\mathbf{x},\mathbf{y})\mathbf{V}^{(o)}(\mathbf{x},\mathbf{y})\right] \, d\mathcal{U}(\mathbf{y}) = 0. \tag{4.2.35}
$$

From (4.2.33)–(4.2.35), we obtain the average of (4.2.31) in the form

$$
\nabla_{\mathbf{x}} \cdot \left[\rho^{(o)}(\mathbf{x})\mathbf{q}_o(\mathbf{x})\right] = 0. \tag{4.2.36}
$$

This is the *macroscopic mass balance equation*, or *continuity equation* for steady flow in a porous medium domain. Although we have introduced here the microscopic fluid's mass balance equation, and obtained, as a by-product, its macroscopic counterpart, the mass balance equation will be discussed in detail in Chap. 5.

Next, we handle the $O(\varepsilon^0)$ Navier-Stokes equation (4.2.32). It is solved together with (4.2.30), subject to the boundary condition (4.2.27). To solve these equations, we first create an auxiliary problem defined for the same geometry in the Y-cell, and with the same boundary conditions:

$$
\nabla^2_{\mathbf{y}} \, \mathsf{w}(\mathbf{y}) = \nabla_{\mathbf{y}} \, \mathbf{s}(\mathbf{y}) - \mathsf{I}, \quad \mathbf{y} \in Y_v, \tag{4.2.37}
$$

$$
\nabla_{\mathbf{y}} \cdot \mathsf{w}(\mathbf{y}) = 0, \quad \mathbf{y} \in Y_v, \tag{4.2.38}
$$

$$
\mathsf{w}(\mathbf{y}) = 0, \quad \mathbf{y} \in \Sigma_i, \tag{4.2.39}
$$

where w is a second rank tensor, $\mathbf{s}$ is a vector, and I is the second rank unit, or identity tensor, used here as a unit *forcing function.* With the prescribed pore domain Y_v in the periodic cell, and the periodic boundary condition on Σ_e, this is a well-posed boundary value problem that can be solved for w and $\mathbf{s}$, using certain analytical or numerical solution techniques, depending on the geometrical complexity of the pore domain.

We can obtain the solution for $\mathbf{V}^{(o)}$, as appearing in (4.2.30) and (4.2.32), by comparing these equations, together with the boundary condition (4.2.27), with the system (4.2.37)–(4.2.39). We observe that the two systems are governed by the same differential operators, with the exception that the 'forcing term' in (4.2.32), $-(1/\mu)\left[\nabla_{\mathbf{x}} p^{(o)}(\mathbf{x}) - \rho^{(o)}(\mathbf{x})\mathbf{g}\right]$ (a vector), is replaced by the second rank unit tensor I in (4.2.37). The replacement of a vector by a tensor also expands the dependent variable $\mathbf{V}^{(o)}$ (a vector) in (4.2.32) to w (a second rank tensor) in (4.2.37), and $p^{(1)}$ (a scalar) to $\mathbf{s}$ (a vector). Since the partial differential equation system is linear, the solution is proportional to the imposed forcing function. Hence, the solution for $\mathbf{V}^{(o)}$ is obtained by the following dot (or scalar) product (which leads to the contraction of the tensor rank),

$$\mathbf{V}^{(o)}(\mathbf{x},\mathbf{y}) = -\frac{1}{\mu}\mathsf{w}(\mathbf{y}) \cdot \left[\nabla_{\mathbf{x}} p^{(o)}(\mathbf{x}) - \rho^{(o)}(\mathbf{x})\mathbf{g}\right], \quad \mathbf{g} \equiv -g\nabla_{\mathbf{x}}\, z, \qquad (4.2.40)$$

where g is the gravity acceleration, $z\ (= x_3)$ is the coordinate pointing in the opposite direction of gravity (i.e., $\mathbf{g} = 0\,\mathbf{e}_1 + 0\,\mathbf{e}_2 - g\,\mathbf{e}_3$).

Next,we integrate (4.2.40) with respect to the Y-cell, as in (4.2.33), to obtain the specific discharge

$$\mathbf{q}_o(\mathbf{x}) = -\frac{\mathsf{k}}{\mu} \cdot \left[\nabla_{\mathbf{x}} p^{(o)}(\mathbf{x}) - \rho^{(o)}(\mathbf{x})\mathbf{g}\right], \qquad (4.2.41)$$

in which

$$\mathsf{k} = \frac{1}{|Y|}\int_{Y_v} \mathsf{w}(\mathbf{y})\, d\mathcal{U}(\mathbf{y}), \qquad (4.2.42)$$

is the *(intrinsic) permeability tensor.* For the convenience of presentation, from this point on, we shall drop the subscripts representing the zeroth order term in (4.2.41), as well as the subscript $\mathbf{x}$ in the gradient operator, and express (4.2.41) in the form

$$\mathbf{q} = -\frac{\mathsf{k}}{\mu} \cdot (\nabla p + \rho g \nabla z). \qquad (4.2.43)$$

Equation (4.2.43) is *Darcy's Law* derived by averaging the microscopic momentum equation, using the homogenization theory.

We have, thus, shown that the motion equation for the fluid phase, (4.2.43), is nothing but a simplified form of the macroscopic momentum balance equation for that phase. It expresses the exchange of momentum between the fluid and the solid. When the simplifying assumptions that have been introduced during the above development of Darcy's law are not applicable, other forms of the averaged momentum balance equation are obtained; examples of such equations are presented in Subs. 4.3.2.

Let us discuss some implications of the fluid's macroscopic momentum equation, (4.2.43). So far, we have assumed that the fluid is *incompressible*;

this is a good assumption for most groundwater flow problems, although we do take into account the fluid's compressibility in determining the aquifer's storativity (Sec. 5.1.3). Thus, with $\rho = \text{const.}$, we rewrite (4.2.43) as:

$$\mathbf{q}(\mathbf{x}) = -\frac{\rho g \mathbf{k}}{\mu} \cdot \nabla \left(\frac{p}{\rho g} + z \right) \equiv -\mathbf{K} \cdot \nabla h. \tag{4.2.44}$$

This equation can be compared to the generalized Darcy's law, (4.1.26) for a three-dimensional anisotropic domain. In Darcy's law (4.2.44), we observe that the *piezometric head*, h, is related to the homogenized pressure in the form

$$h = \frac{p}{\rho g} + z, \tag{4.2.45}$$

which is identical to (4.1.4). We also note that the *hydraulic conductivity*, $\mathbf{K}$ (dims. L/T), appearing in Darcy's law is expressed in the form presented earlier as (4.1.16).

From the definition of $\mathbf{k}$ in (4.2.42), and, in turn, in the definition of $\mathbf{w}$ in the boundary value problem (4.2.37)–(4.2.39), it is clear that $\mathbf{k}$ is a property of the *pore geometry only*, as no physical fluid parameters, e.g., viscosity and density, appear in these equations.

Furthermore, $\mathbf{w}$, and hence $\mathbf{k}$, is a second rank tensor by virtue of the introduction of the second rank unit tensor, $\mathbf{I}$, as the forcing function in (4.2.37). Hence, $\mathbf{k}$ is subject to the coordinate transformation rules described in Subs. 4.1.4.

We also recognize that the differential operator acting on $\mathbf{V}^{(o)}$ (as in (4.2.32) and (4.2.30)) and $\mathbf{w}$ (as in (4.2.37) and (4.2.38)) are the same. This operator, denoted by $\mathcal{L}$, is *self-adjoint*, meaning that the *inner product* defined as

$$\langle \mathbf{w}, \mathbf{V}^{(o)} \rangle = \int_{Y_v} \mathbf{w} \cdot \mathcal{L} \left\{ \mathbf{V}^{(o)} \right\} \, d\mathcal{U}, \tag{4.2.46}$$

has the *reciprocity property* (see any advanced calculus book, e.g., Greenberg, 1998)

$$\langle \mathbf{w}, \mathbf{V}^{(o)} \rangle = \langle \mathbf{V}^{(o)}, \mathbf{w} \rangle. \tag{4.2.47}$$

An important consequence of the above statement is that it can be proven (Hornung, 1997) that the tensor $\mathbf{k}$ (based on $\mathbf{w}$) is *symmetric*, with its components $k_{ij} = k_{ji}$ (see also discussion in Subs. 4.1.4 pertaining to the symmetry of the permeability tensor $\mathbf{k}$).

Another important property of $\mathbf{k}$ is that it is a *positive definite matrix*, meaning that the principal permeabilities are all positive. This condition prevents the physically unacceptable result of fluid moving in the direction of increasing head gradient, violating the laws of thermodynamics.

Next, we examine the dependence of the permeability, $\mathbf{k}$, on the microscopic scale ℓ. As demonstrated in Subs. 1.3.5, in the example of homogenization of a differential equation with a periodic variable coefficient, the influence of ℓ on the effective coefficient in the homogenized, large scale model vanishes in the limit as $\varepsilon = \ell/L \to 0$. In other words, the effective coefficient a_o is independent of the size of periodic cell, ℓ. This however, is not the case in the derivation of Darcy's law. As defined by the boundary value problem, (4.2.37)–(4.2.39), and (4.2.42), we observe that $\mathbf{k} \to 0$ as $\ell \to 0$. Hence the homogenization solution is an *asymptotic* solution, and not a *limiting* solution. The macroscopic coefficient $\mathbf{k}$ does depend on the actual size of the microscopic cell.

We can explore the explicit dependence of $\mathbf{k}$ on ℓ by observing that the system (4.2.37)–(4.2.39) can be expressed in the non-dimensional form

$$\nabla^2_{\mathbf{y}^*} \mathbf{w}^*(\mathbf{y}^*) = \nabla_{\mathbf{y}^*} \mathbf{s}^*(\mathbf{y}^*) - \mathbf{I}, \quad \mathbf{y}^* \in Y_v^*, \tag{4.2.48}$$

$$\nabla_{\mathbf{y}^*} \cdot \mathbf{w}^*(\mathbf{y}^*) = 0, \quad \mathbf{y}^* \in Y_v^*, \tag{4.2.49}$$

$$\mathbf{w}^*(\mathbf{y}^*) = 0, \quad \mathbf{y}^* \in \Sigma_i^*, \tag{4.2.50}$$

where $\mathbf{y}^* = \mathbf{y}/\ell$, $\mathbf{w}^* = \mathbf{w}/\ell^2$, $\mathbf{s}^* = \mathbf{s}/\ell$, and the domain Y is mapped to Y^* of unit size (say, a cube of $1 \times 1 \times 1$), with corresponding partial volumes and interfaces, Y_v^* and Σ_i^*, etc. It is apparent that, based on the above system, $\mathbf{w}^*$ is scale *independent.* Its solution depends only on the geometric pattern of the pore space, Y_v^*, within the unit cell Y^*. We can define a *dimensionless intrinsic permeability* $\mathbf{k}^*$ similar to (4.2.42), in the form

$$\mathbf{k}^* = \frac{1}{|Y^*|} \int_{Y_v^*} \mathbf{w}^*(\mathbf{y}^*)\, d\mathcal{U}(\mathbf{y}^*). \tag{4.2.51}$$

Thus, it becomes clear that $\mathbf{k} = \mathbf{k}^* \ell^2$, and (4.1.16) can be expressed as

$$\mathbf{K} = \frac{\rho g \ell^2 \mathbf{k}^*}{\mu}. \tag{4.2.52}$$

This relation shows the dependence of the hydraulic conductivity on the square of a microscopic characteristic length, ℓ. At the same time, $\mathbf{k}^*$ is a constant that depends on the pore or grain shape and packing pattern, e.g., cubic or a rhombohedral, and not on size. In a random porous medium, $\mathbf{k}^*$ can be empirically related to some lumped geometrical parameters such as porosity, grain (or pore) shape factor, and tortuosity. On the other hand, the length scale ℓ can be associated with quantities such as mean grain size, hydraulic radius, or the inverse of the specific surface area. These ideas were discussed from the empirical point of view in Subs. 4.1.3.

As demonstrated above, the homogenization theory provides a rigorous basis for a number of empirical observations used in practice. In addition, it substantiates the generalization of the concept of permeability as a symmetric second rank tensor. The coordinate transformation rules, the existence of principal directions, and the assurance of physically meaningful positive permeability coefficients, follow as rigorous consequences from the mathematical analysis.

The above comparison with Darcy's law was based on the assumption that the fluid is incompressible. For a compressible fluid, we have to return to (4.2.43). The question is, whether there exists a 'potential', such that the flux $\mathbf{q}$ is proportional to its gradient, i.e., proportionate to the gradient of that potential, analogous to the piezometric head, h that appears in Darcy's law, (4.2.44), derived for $\rho =$ const. In general, it is impossible to express the sum $\nabla p + \rho g \nabla z$, appearing in (4.2.43), as a gradient of a single potential. However, for a single component compressible fluid, under isothermal conditions, i.e., for $\rho = \rho(p)$, it can be shown (Hubbert, 1940) that such a potential does exist. In this case we can write (4.2.43) as

$$\mathbf{q} = -\mathbf{K} \cdot \nabla h^*, \tag{4.2.53}$$

where

$$h^* = \int_{p_o}^{p} \frac{dp}{g\rho(p)} + z \tag{4.2.54}$$

is called *Hubbert's potential*, and p_o is a reference pressure (see Subs. 4.1.1 and equation (4.1.6)). By applying the chain rule to the first term on the right hand side, it is easy to prove that

$$\nabla h^* = \nabla \left(\int_{p_o}^{p} \frac{dp}{g\rho(p)} + z \right) = \frac{1}{\rho g} \nabla p + \nabla z. \tag{4.2.55}$$

Hence, the concept of Hubbert's potential, (4.2.54), and the extension of Darcy's law to a compressible fluid, (4.2.53), are firmly established on the foundation of the microscopic momentum balance.

We shall conclude this subsection with a discussion on whether the macroscopic (or average) velocity of the fluid, or the specific discharge, should be based on an area averaged velocity, or on a volume averaged one. As demonstrated in this section, the origin of Darcy's law is the momentum balance equation in which the (microscopic mass weighted) fluid velocity is used to express (the microscopic value of) the fluid's momentum per unit volume ($= \rho \mathbf{V}$). Let us focus on the derivation of Darcy's law by homogenization. We notice that according to (4.2.33), as well as in the continuity equation, (4.2.36), the specific discharge, $\mathbf{q}$, appearing in the derivation of Darcy's law, (4.2.43), is a *volume* average of the velocity. Based on this interpretation, the specific discharge should be obtained by volume averaging.

However, in actual measurement, for example, in a laboratory porous medium column, the specific discharge is measured as the volume of water passing through a *unit cross-sectional area* of the column, per unit time. The hydraulic conductivity is then interpreted on the basis of such area averaged measurement. Also, we notice that the piezometric head is often observed by a piezometer, or a tensiometer, in the case of partially saturated porous medium. The pressure measured in a piezometer is likely to be an average over the well's screen area, and that for a tensiometer, an average over the surface of the ceramic cup. So how can we reconcile the difference between the theoretically derived Darcy's law (volume average) and the practice (area average)? Bear and Bachmat (1990, p. 34) discussed this issue and showed that the average over an REV of the area average is equal to the volume average over the REV. Then, they justified the use of the volume averaged velocity in transport models by showing that the volumetric intrinsic phase average and the areal intrinsic phase average of the flux of an extensive quantity (i.e., $e_\alpha \mathbf{V}^{E_\alpha}$) of a phase at a point are identical.

In what follows, we shall demonstrate that the difference between the two kinds of averages, volume and area, can be reconciled under the periodicity assumption that underlies the homogenization theory, thus bridging the gap between the two definitions of macroscopic fluid velocity.

Assuming isochoric fluid flow at the microscopic level (see (4.2.30)), i.e., $\nabla_y \cdot \mathbf{V} = 0$ (or $\partial V_j/\partial y_j = 0$), it is easily shown that the integrand $\mathbf{V}^{(o)}(\mathbf{x}, \mathbf{y})$ in (4.2.33), can be obtained from the following relationship, written in indicial notation, making use of Einstein's summation convention (Subs. 4.1.4):

$$\frac{\partial(y_i V_j)}{\partial y_j} = \frac{\partial y_i}{\partial y_j} V_j + y_i \frac{\partial V_j}{\partial y_j} \equiv \delta_{ij} V_j = V_i, \tag{4.2.56}$$

in which δ_{ij} is the Kronecker delta ($\equiv \partial y_i/\partial y_j$). Hence, we can apply the *divergence theorem* to (4.2.33) in the form

$$\begin{aligned} q_i(\mathbf{x}) &= \frac{1}{|Y|} \int_{Y_v} V_i(\mathbf{x}, \mathbf{y})\, d\mathcal{U}_v(\mathbf{y}) = \frac{1}{|Y|} \int_{Y_v} \frac{\partial[y_i V_j(\mathbf{x}, \mathbf{y})]}{\partial y_j}\, d\mathcal{U}_v(\mathbf{y}) \\ &= \frac{1}{|Y|} \int_{\partial Y_v} y_i V_j(\mathbf{x}, \mathbf{y}) n_j(\mathbf{y})\, d\mathcal{S}_v(\mathbf{y}) \\ &= \frac{1}{|Y|} \int_{\Sigma_e} y_i V_j(\mathbf{x}, \mathbf{y}) n_j(\mathbf{y})\, d\mathcal{S}_v(\mathbf{y}), \end{aligned} \tag{4.2.57}$$

where n_i is the ith component of the outward normal unit vector $\mathbf{n}$. We recall that ∂Y_v ($\equiv \Sigma_i + \Sigma_e$) is the boundary of Y_v. Also, in the transition to the last line of (4.2.57), the integration over the internal surface, Σ_i, vanishes due to the vanishing velocity components (no-slip condition), such that the only integration needed is that over the external void surface, Σ_e.

For simplicity, yet without loss of generality, we shall present the following proof using only one component of the specific discharge, q_1. For this component, we can express (4.2.57) as

$$q_1(\mathbf{x}) = \frac{1}{\ell_1 \ell_2 \ell_3} \int_{\Sigma_1 + \ldots + \Sigma_6} y_1 \left(V_1 n_1 + V_2 n_2 + V_3 n_3\right) d\mathcal{S}_v(\mathbf{y}), \tag{4.2.58}$$

in which we have replaced $|Y|$ by the volume $\ell_1 \ell_2 \ell_3$, and the external void surface Σ_e is decomposed into six parts, $\Sigma_1, \ldots, \Sigma_6$, corresponding to the six sides of the cube (see Fig. 4.2.2). Utilizing the periodicity condition, it is easy

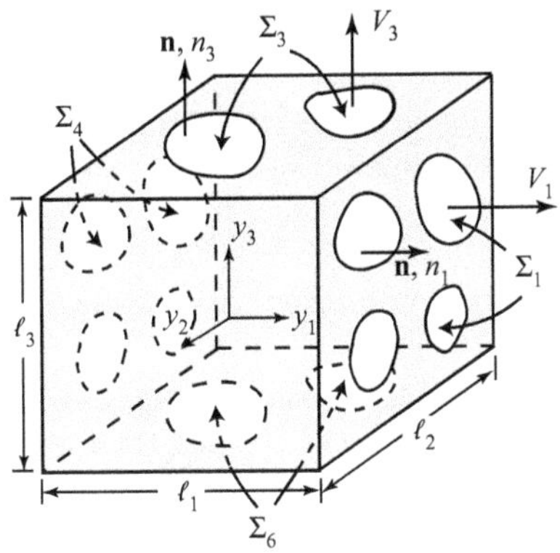

Figure 4.2.2: Illustration of the surface integral resulting from the divergence theorem.

to show that the integral over the two surfaces perpendicular to the y_3-axis, $\Sigma_3 + \Sigma_6$, vanishes, due to the following conditions: for $\mathbf{n}$ on these surfaces, only the n_3 component exists; n_3 takes opposite values ($+1$ and -1) on these surfaces; $y_1 V_3$ has the same value on the opposing surfaces; and Σ_3 and Σ_6 have identical shapes. The same is true for the integral over the two surfaces perpendicular to y_2-axis. Hence, we are left with

$$\begin{aligned} q_1(\mathbf{x}) &= \frac{1}{\ell_1\ell_2\ell_3}\int_{\Sigma_1+\Sigma_4} y_1 V_1 n_1 \, d\mathcal{S}_v(\mathbf{y}) \\ &= \frac{1}{\ell_1\ell_2\ell_3}\left[\int_{\Sigma_1}\frac{\ell_1}{2}V_1\cdot(+1)\,d\mathcal{S}_v(\mathbf{y}) + \int_{\Sigma_4} -\frac{\ell_1}{2}V_1\cdot(-1)\,d\mathcal{S}_v(\mathbf{y})\right] \\ &= \frac{1}{\ell_2\ell_3}\int_{\Sigma_1} V_1(\mathbf{x},\mathbf{y})\,d\mathcal{S}_v(\mathbf{y}). \end{aligned} \tag{4.2.59}$$

Again, the periodicity condition (V_1 on $\Sigma_1 = V_1$ on Σ_4, and $\Sigma_1 = \Sigma_4$) is used. Equation (4.2.59) is indeed a surface average. The same procedure can be applied to the other two specific discharge components. Thus, we have proven that

$$q_i(\mathbf{x}) = \frac{1}{|Y|}\int_{Y_v} V_i(\mathbf{x},\mathbf{y})\,d\mathcal{U}_v(\mathbf{y}) = \frac{1}{|\partial Y_i|}\int_{\Sigma_i} V_i(\mathbf{x},\mathbf{y})\,d\mathcal{S}_v(\mathbf{y}), \tag{4.2.60}$$

where ∂Y_i is the (solid plus void) surface area of the cube perpendicular to the y_i axis. The inconsistency between the volume and the area averaged specific discharge has thus been resolved.

4.2.3 *Effective hydraulic conductivity by homogenization*

Usually, as extensively discussed in Subs. 1.3.4, there exist many scales of heterogeneity in natural porous medium domains. In the preceding subsection, we have derived the flux laws at the macroscopic scale. A typical example of macroscopic scale is *laboratory scale*, say, 1 to 2 meter. However, groundwater domains of practical interest are, often, on a scale that is much larger than the macroscopic one.

At the *field scale*, or *megascopic scale*, porous media are highly heterogeneous. The permeability may vary significantly over relatively short distances, such as meters or tens of meters, in fields that extend over kilometers, tens of kilometers, or even more. In large scale cases, we are faced with a situation that is similar to the one we have faced at the microscopic scale, namely, that we are not interested in modeling the (relatively) small scale fluctuations, the solution of which requires a detailed knowledge that we are unable to obtain anyway. On the other hand, there may exist large scale variations that are of interest, and, hence, need to be modeled. The purpose of homogenization is to smooth out small scale heterogeneities, while preserving the larger scale trends. In what follows, we shall use the homogenization technique in order to transform the macroscopic scale motion equation and its coefficient of hydraulic conductivity into their counterparts at the *megascopic scale*.

For simplicity, we shall consider an incompressible fluid, $\rho =$ const. The macroscopic mass balance equation (4.2.36) then takes the form

$$\nabla \cdot \mathbf{q}(\mathbf{x}) = 0, \quad \mathbf{x} \in \Omega. \tag{4.2.61}$$

By combining the above equation with Darcy's law, (4.2.44), we obtain

$$\nabla \cdot [\mathsf{K}(\mathbf{x}) \cdot \nabla h(\mathbf{x})] = 0, \quad \mathbf{x} \in \Omega, \tag{4.2.62}$$

where Ω is the solution domain. Let (4.2.62) be subject to the boundary condition on the boundary, $\partial\Omega$, of Ω,

$$h(\mathbf{x}) = h_D(\mathbf{x}), \quad \mathbf{x} \in \partial\Omega. \tag{4.2.63}$$

Figure 4.2.3 illustrates the case of an aquifer containing interspersed clay lenses of low hydraulic conductivity $\mathsf{K}(\mathbf{x})$. The scale of heterogeneity of such domain is characterized by the length of a clay lens, or the distance between lens centers. The piezometric head distribution, $h(\mathbf{x})$, across the field, however, exhibits only small fluctuations; this is a lesson learned from the homogenization example in Subs. 1.3.5. Our goal is to remove these small amplitude head fluctuations at the macroscopic scale, and to derive a smooth, megascopic (field) scale head distribution.

For the purpose of homogenization, it is necessary to assume that the pattern of the inhomogeneity is *periodic* in the local scale ℓ, as illustrated in the inset of Fig. 4.2.3, with $\ell \ll L$, where L is the scale of the considered do-

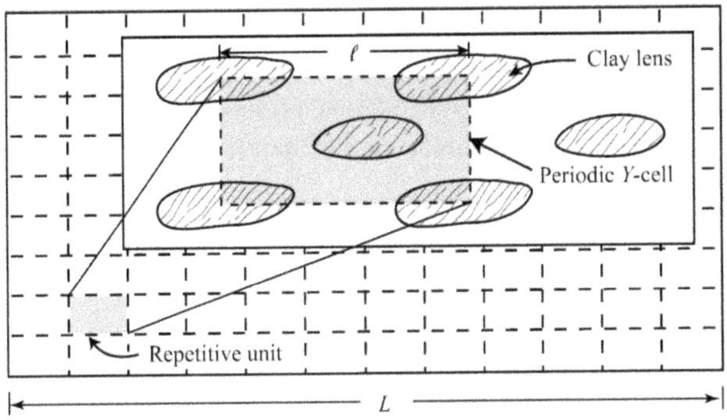

Figure 4.2.3: From macroscopic to megascopic scale.

main, or the megascopic scale. To emphasize the dependence of the hydraulic conductivity tensor on the small parameter $\varepsilon = \ell/L \ll 1$, we shall denote it as $\mathsf{K}^\varepsilon(\mathbf{x})$. As in (4.2.12), we shall introduce the local, small scale coordinate system $\mathbf{y} = \mathbf{x}/\varepsilon$. We then Next, we express this dependence, explicitly, at the two scales as

$$\mathsf{K}^\varepsilon(\mathbf{x}) = \mathsf{K}(\mathbf{x}, \mathbf{y}). \tag{4.2.64}$$

Here, the $\mathbf{y}$-dependence implies the periodic nature of the function at the local scale, while the $\mathbf{x}$ dependence indicates the large scale heterogeneity trend. For a field that is homogeneous (no trend) at a large scale, we can write $\mathsf{K}^\varepsilon(\mathbf{x}) = \mathsf{K}(\mathbf{y})$.

The piezometric head, h^ε, governed by (4.2.62), is also a function of both $\mathbf{x}$ and $\mathbf{y}$. The expansion of h^ε into an asymptotic series with respect to the small parameter ε, takes the form

$$h^\varepsilon(\mathbf{x}) = h^{(o)}(\mathbf{x}, \mathbf{y}) + \varepsilon h^{(1)}(\mathbf{x}, \mathbf{y}) + \varepsilon^2 h^{(2)}(\mathbf{x}, \mathbf{y}) + \dots \tag{4.2.65}$$

Using the above equation in (4.2.62), following the differentiation rule (4.2.21), and sorting out terms of the same order in ε, we obtain the equation of the lowest order in ε, $O(\varepsilon^{-2})$, in the form

$$\nabla_{\mathbf{y}} \cdot [\mathsf{K}(\mathbf{x}, \mathbf{y}) \cdot \nabla_{\mathbf{y}} h^{(o)}(\mathbf{x}, \mathbf{y})] = 0; \quad \text{for } \mathbf{x} \in \Omega, \quad \mathbf{y} \in Y. \tag{4.2.66}$$

Higher order equations have a similar form; they will be introduced later.

To solve equations of this type, we make use of an important theorem in homogenization theory that states that given the elliptic partial differential equation

$$\frac{\partial}{\partial y_i}\left(a_{ij}\frac{\partial \varphi}{\partial y_j}\right) = f \quad \text{in } Y, \tag{4.2.67}$$

where the coefficients a_{ij} are symmetric and positive definite, and a_{ij} and f are Y-periodic, the Y-periodic solution for φ exists and is unique up to an additive constant (see, for example, Cioranescu and Donato, 1999, Theorem 4.26, or Allaire, 1997b, p. 234). Furthermore, the norm of φ is bounded by the right hand side, f, as $\|\varphi\| \leq C\,\|f\|$. A consequence of the above theorem is that the only admissible solution of (4.2.66) is

$$h^{(o)}(\mathbf{x}, \mathbf{y}) = h^{(o)}(\mathbf{x}), \tag{4.2.68}$$

because the null right hand side of (4.2.66) bounds the solution to be a function of x only (with $h^{(o)}(\mathbf{x})$ as the additive constant).

Incorporating the functional relationship of (4.2.68), the $O(\varepsilon^{-1})$-equation is simplified to the following form

$$\nabla_{\mathbf{y}} \cdot [\mathsf{K}(\mathbf{x}, \mathbf{y}) \cdot \nabla_{\mathbf{y}} h^{(1)}(\mathbf{x}, \mathbf{y})] = -[\nabla_{\mathbf{y}} \cdot \mathsf{K}(\mathbf{x}, \mathbf{y})] \cdot [\nabla_{\mathbf{x}} h^{(o)}(\mathbf{x})], \tag{4.2.69}$$

while the $O(\varepsilon^{0})$-equation becomes

$$\begin{aligned} \nabla_{\mathbf{x}} \cdot [\mathsf{K}(\mathbf{x}, \mathbf{y}) \cdot \nabla_{\mathbf{x}} h^{(o)}(\mathbf{x})] &= -\nabla_{\mathbf{y}} \cdot [\mathsf{K}(\mathbf{x}, \mathbf{y}) \cdot \nabla_{\mathbf{x}} h^{(1)}(\mathbf{x}, \mathbf{y})] \\ -\nabla_{\mathbf{y}} \cdot [\mathsf{K}(\mathbf{x}, \mathbf{y}) \cdot \nabla_{\mathbf{y}} h^{(2)}(\mathbf{x}, \mathbf{y})] &- \nabla_{\mathbf{x}} \cdot [\mathsf{K}(\mathbf{x}, \mathbf{y}) \cdot \nabla_{\mathbf{y}} h^{(1)}(\mathbf{x}, \mathbf{y})]. \end{aligned} \tag{4.2.70}$$

To solve (4.2.69), we start with the following definitions. Let $\mathbf{u}$ (a vector) be the Y-periodic solution of the problem

$$\nabla_{\mathbf{y}} \cdot [\mathsf{K}(\mathbf{x}, \mathbf{y}) \cdot \nabla_{\mathbf{y}} \mathbf{u}(\mathbf{x}, \mathbf{y})] = -\nabla_{\mathbf{y}} \cdot \mathsf{K}(\mathbf{x}, \mathbf{y}); \quad \mathbf{y} \in Y. \tag{4.2.71}$$

By comparing (4.2.71) with (4.2.69), we conclude that the solution for $h^{(1)}$ is

$$h^{(1)}(\mathbf{x}, \mathbf{y}) = \mathbf{u}(\mathbf{x}, \mathbf{y}) \cdot \nabla_{\mathbf{x}} h^{(o)}(\mathbf{x}) + f(\mathbf{x}), \tag{4.2.72}$$

where $f(\mathbf{x})$ is an arbitrary function of $\mathbf{x}$. By applying the gradient operator, the above equation can be differentiated, to yield

$$\nabla_{\mathbf{y}} h^{(1)}(\mathbf{x}, \mathbf{y}) = \nabla_{\mathbf{y}} \mathbf{u}(\mathbf{x}, \mathbf{y}) \cdot \nabla_{\mathbf{x}} h^{(o)}(\mathbf{x}). \tag{4.2.73}$$

Next, we turn to the $O(\varepsilon^{0})$-equation. Equation (4.2.70) can be integrated over the Y-cell to yield a volume average. For the first two terms on the right hand side, we apply the *divergence theorem* (e.g., Greenberg, 1998), obtaining

$$\begin{aligned} &\int_{Y} \nabla_{\mathbf{y}} \cdot [\mathsf{K}(\mathbf{x}, \mathbf{y})\, \nabla_{\mathbf{x}} h^{(1)}(\mathbf{x}, \mathbf{y}) + \mathsf{K}(\mathbf{x}, \mathbf{y})\, \nabla_{\mathbf{y}} h^{(2)}(\mathbf{x}, \mathbf{y})]\, d\mathcal{U}(\mathbf{y}) \\ &= \int_{\partial Y} \mathbf{n}(\mathbf{y}) \cdot [\mathsf{K}(\mathbf{x}, \mathbf{y})\, \nabla_{\mathbf{x}} h^{(1)}(\mathbf{x}, \mathbf{y}) + \mathsf{K}(\mathbf{x}, \mathbf{y})\, \nabla_{\mathbf{y}} h^{(2)}(\mathbf{x}, \mathbf{y})]\, d\mathcal{S}(\mathbf{y}) \\ &= 0, \end{aligned} \tag{4.2.74}$$

where $d\mathcal{U}$ and $d\mathcal{S}$ denote a volume and a surface elements, respectively, ∂Y is the boundary of the Y-cell, and $\mathbf{n}$ is the outward unit vector normal to ∂Y. We notice that the above integral vanishes because of the periodicity condition: the values of K, $h^{(1)}$ and $h^{(2)}$ and their derivatives are the same on both the left and the right sides of the Y-cell, as well as on the top and bottom sides, thus canceling each other. The remainder of (4.2.70) gives

$$\int_Y \nabla_{\mathbf{x}} \cdot [\mathsf{K}(\mathbf{x},\mathbf{y}) \cdot \nabla_{\mathbf{x}} h^{(o)}(\mathbf{x})] \, d\mathcal{U}(\mathbf{y}) + \int_Y \nabla_{\mathbf{x}} \cdot [\mathsf{K}(\mathbf{x},\mathbf{y}) \cdot \nabla_{\mathbf{y}} h^{(1)}(\mathbf{x},\mathbf{y})] \, d\mathcal{U}(\mathbf{y})$$
$$= \nabla_{\mathbf{x}} \cdot \left[\int_Y \mathsf{K}(\mathbf{x},\mathbf{y}) \, d\mathcal{U}(\mathbf{y}) \cdot \nabla_{\mathbf{x}} h^{(o)}(\mathbf{x}) \right]$$
$$+ \nabla_{\mathbf{x}} \cdot \left[\int_Y \mathsf{K}(\mathbf{x},\mathbf{y}) \cdot \nabla_{\mathbf{y}} \mathbf{u}(\mathbf{x},\mathbf{y}) \, d\mathcal{U}(\mathbf{y}) \cdot \nabla_{\mathbf{x}} h^{(o)}(\mathbf{x}) \right] = 0, \quad (4.2.75)$$

where we have made a substitution, using (4.2.73). The above equation can be expressed as

$$\nabla_{\mathbf{x}} \cdot [\mathsf{K}^{\text{eq}}(\mathbf{x}) \cdot \nabla_{\mathbf{x}} h^{(o)}(\mathbf{x})] = 0, \quad (4.2.76)$$

where K^{eq} is the *effective hydraulic conductivity* given by

$$\mathsf{K}^{\text{eq}}(\mathbf{x}) = \frac{1}{|Y|} \int_Y \mathsf{K}(\mathbf{x},\mathbf{y}) \, d\mathcal{U}(\mathbf{y}) + \frac{1}{|Y|} \int_Y \mathsf{K}(\mathbf{x},\mathbf{y}) \cdot \nabla_{\mathbf{y}} \mathbf{u}(\mathbf{x},\mathbf{y}) \, d\mathcal{U}(\mathbf{y}). \quad (4.2.77)$$

Equation (4.2.77) can be rewritten in indicial notation, as

$$\mathsf{K}^{\text{eq}}_{ij}(\mathbf{x}) = \frac{1}{|Y|} \int_Y \mathsf{K}_{ij}(\mathbf{x},\mathbf{y}) \, d\mathcal{U}(\mathbf{y}) + \frac{1}{|Y|} \int_Y \mathsf{K}_{ik}(\mathbf{x},\mathbf{y}) \frac{\partial u_j(\mathbf{x},\mathbf{y})}{\partial y_k} \, d\mathcal{U}(\mathbf{y}), \quad (4.2.78)$$

in which *Einstein's summation convention*, introduced after (4.1.27), is applicable. We note that K^{eq} is a symmetric, positive definite, second rank tensor by virtue of the observation that the differential operators involved are self-adjoint, and that K is symmetric and positive definite, a fact proven earlier.

Equation (4.2.76) is, thus, the homogenized equation at the megascopic scale; it replaces (4.2.62), which is valid at the macroscopic scale. The coefficient K^{eq}, defined in (4.2.77), is a function of the large scale coordinates only; hence, it is free from small scale fluctuations. The asymptotic piezometric head at the megascopic scale, $h^{(o)}(\mathbf{x})$, is also smooth and without fluctuation.

It is of interest to examine the special case of a layered aquifer, which is often encountered in geological formations of sedimentary origin. If all layers together are considered as a single aquifer unit at the field (megascopic) scale, that aquifer as a whole has a higher ability to transmit water in the direction parallel to the layers than in the direction perpendicular to them. Our purpose is to find the equivalent homogeneous anisotropic conductivity of such an aquifer.

We shall assume that each layer consists of a homogeneous isotropic material, but the material varies from layer to layer (Fig. 4.2.4). With the isotropy assumption, the tensor $\mathsf{K}(\mathbf{y})$ reduces to a scalar, $\mathsf{K}_{ij}(\mathbf{y}) = \delta_{ij}\mathsf{K}(\mathbf{y})$, where δ_{ij} is the Kronecker delta. Here, we have ignored the large scale trend, that is, the

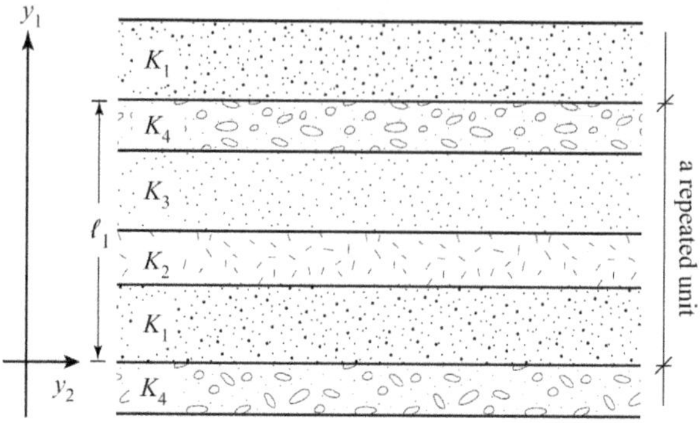

Figure 4.2.4: Equivalent hydraulic conductivity of a layered formation. Each layer is homogeneous and isotropic.

$\mathbf{x}$ dependence of $\mathbf{K}$. A layered formation means that the hydraulic conductivity varies only in one direction, say in the direction y_1; hence we may further conclude that $\mathsf{K}(\mathbf{y}) = \mathsf{K}(y_1)$. Recall that we always use the $\mathbf{y}$-coordinates for the smaller scale domain, and $\mathbf{x}$ for the larger, homogenized one. The homogenization theory and the use of the $\mathbf{y}$-coordinates also implies that the layers form repeated units of thickness ℓ_1, as illustrated in Fig. 4.2.4. In this special case, it is possible to solve $\mathbf{u}$ in (4.2.71), obtaining explicit result.

Equation (4.2.71) can be expressed in component form, incorporating the above assumptions, in the form

$$\frac{\partial}{\partial y_1}\left[\mathsf{K}(y_1)\frac{\partial u_1(\mathbf{y})}{\partial y_1}\right] + \frac{\partial}{\partial y_2}\left[\mathsf{K}(y_1)\frac{\partial u_1(\mathbf{y})}{\partial y_2}\right] + \frac{\partial}{\partial y_3}\left[\mathsf{K}(y_1)\frac{\partial u_1(\mathbf{y})}{\partial y_3}\right] = -\frac{\partial \mathsf{K}(y_1)}{\partial y_1}, \tag{4.2.79}$$

$$\frac{\partial}{\partial y_1}\left[\mathsf{K}(y_1)\frac{\partial u_2(\mathbf{y})}{\partial y_1}\right] + \frac{\partial}{\partial y_2}\left[\mathsf{K}(y_1)\frac{\partial u_2(\mathbf{y})}{\partial y_2}\right] + \frac{\partial}{\partial y_3}\left[\mathsf{K}(y_1)\frac{\partial u_2(\mathbf{y})}{\partial y_3}\right] = 0, \tag{4.2.80}$$

$$\frac{\partial}{\partial y_1}\left[\mathsf{K}(y_1)\frac{\partial u_3(\mathbf{y})}{\partial y_1}\right] + \frac{\partial}{\partial y_2}\left[\mathsf{K}(y_1)\frac{\partial u_3(\mathbf{y})}{\partial y_2}\right] + \frac{\partial}{\partial y_3}\left[\mathsf{K}(y_1)\frac{\partial u_3(\mathbf{y})}{\partial y_3}\right] = 0. \tag{4.2.81}$$

Studying (4.2.80) and (4.2.81), we conclude that the only admissible periodic solution of these equations is $u_2 = u_3 = 0$, ignoring an arbitrary additive constant. For (4.2.79), we observe that the right hand side is a function of y_1 only; hence, the periodic solution of this equation is restricted to the form of $u_1 = u_1(y_1)$. This simplifies (4.2.79) to

$$\frac{\partial}{\partial y_1}\left[\mathsf{K}(y_1)\frac{\partial u_1(y_1)}{\partial y_1}\right] = -\frac{\partial \mathsf{K}(y_1)}{\partial y_1}. \tag{4.2.82}$$

The solution of the above equation, with the periodic condition of $u_1(0) = u_1(\ell_1)$, is

$$u_1(y_1) = \ell_1 \frac{\int_0^{y_1} \frac{dy_1}{\mathsf{K}(y_1)}}{\int_0^{\ell_1} \frac{dy_1}{\mathsf{K}(y_1)}} - y_1, \tag{4.2.83}$$

in which, again, we ignored an arbitrary additive constant. The solution $\mathbf{u}$ is then used in (4.2.78) to find the equivalent hydraulic conductivity. For the tensor component $\mathsf{K}_{11}^{\text{eq}}$, we find

$$\mathsf{K}_{11}^{\text{eq}} = \frac{1}{\ell_1}\int_0^{\ell_1} \mathsf{K}(y_1)\, dy_1 + \frac{1}{\ell_1}\int_0^{\ell_1} \mathsf{K}(y_1)\frac{\partial u_1(y_1)}{\partial y_1}\, dy_1 = \frac{\ell_1}{\int_0^{\ell_1} \frac{dy_1}{\mathsf{K}(y_1)}}. \tag{4.2.84}$$

For the component $\mathsf{K}_{22}^{\text{eq}}$, we obtain

$$\mathsf{K}_{22}^{\text{eq}} = \frac{1}{\ell_1}\int_0^{\ell_1} \mathsf{K}(y_1)\, dy_1. \tag{4.2.85}$$

Compiling all components, the effective hydraulic conductivity tensor becomes

$$\mathbf{K}^{\text{eq}} = \begin{bmatrix} \ell_1 \Big/ \int_0^{\ell_1} \frac{dy_1}{\mathsf{K}(y_1)} & 0 & 0 \\ 0 & \int_0^{\ell_1} \mathsf{K}(y_1)\, dy_1 \Big/ \ell_1 & 0 \\ 0 & 0 & \int_0^{\ell_1} \mathsf{K}(y_1)\, dy_1 \Big/ \ell_1 \end{bmatrix}. \tag{4.2.86}$$

The above result shows that the hydraulic conductivity, K, is *anisotropic*, with its principal axes perpendicular and parallel to the layers. In the direction normal to the layers (x_1-direction), the effective hydraulic conductivity is the *harmonic mean* of the individual hydraulic conductivities, while in the directions parallel to the layers (x_2- and x_3-direction), it is the algebraic mean. Bear (1972, Subs. 5.8.1) shows the analogy to the equivalent resistance of resistances connected in series and in parallel.

4.3 Non-Darcy Laws

4.3.1 Range of validity of Darcy's law

Column experiments, similar to those conducted by Darcy, indicate that as the specific discharge increases, its relationship to the hydraulic gradient gradually deviates from the linear relationship expressed by Darcy's law (4.1.9). Figure 4.3.1a shows this deviation.

In fluid mechanics, when considering flow through pressurized conduits, the (dimensionless) *Reynolds number*, Re, is used as a criterion for distinguishing between *laminar flow* occurring at low velocities and turbulent flow

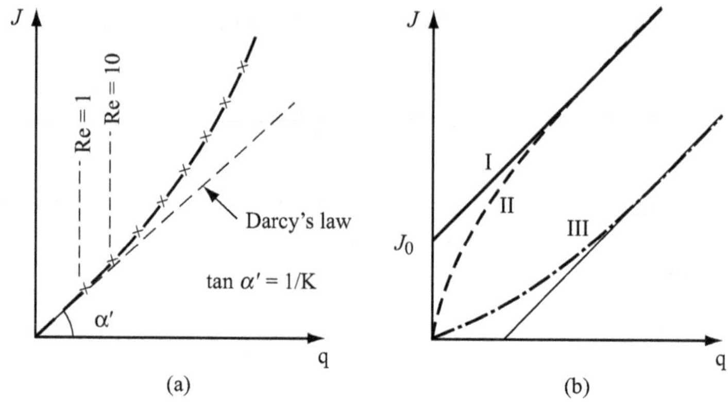

Figure 4.3.1: Experimental relationship between specific discharge, q, and hydraulic gradient, $\mathcal{J}$, in (a) sand, and (b) clay.

occurring at higher ones (Bear and Bachmat, 1990, p. 276, and any textbook on fluid mechanics, e.g., Munson *et al*, 2005). This number expresses the ratio between inertial and viscous forces acting on a moving fluid. In pipes, the critical value of Re, indicating the transition from laminar to turbulent flow, is usually about 2000, although widely varying values may apply under special conditions. By analogy, a Reynolds number is defined for flow through porous media, as

$$\mathrm{Re} = \frac{q\,d}{\nu}, \tag{4.3.1}$$

where d is some representative (microscopic) length characterizing the void space and ν is the fluid's kinematic viscosity. Although, by analogy with pipe flow, d should be a length characterizing the cross-section of an elementary channel of the porous medium, it is customary, for unconsolidated porous media, to employ for d some characteristic length of the grains, probably because it is more easily measured.

Often, the mean grain diameter is used for d in (4.3.1). Sometimes, d_{10} is mentioned in the literature as the representative grain diameter to be used for d. Collins (1961) suggested $d = (\mathsf{k}/\phi)^{\frac{1}{2}}$, where k is the permeability and ϕ is the porosity, as the representative length. Bear and Bachmat (1990), on the basis of theoretical analysis, suggested the *hydraulic radius* of the void space (defined as the ratio of the volume of the void space to the area of solid-fluid interface) as the characteristic length. In their analysis, they define a Reynolds number and a *Darcy number*

$$\mathrm{Re} = \frac{V\varDelta}{\nu}, \quad \mathrm{Da} = \left(\frac{\varDelta}{L^{(v)}}\right)^2, \quad \varDelta = \sqrt{\frac{\mathsf{k}}{\phi T^*}}, \tag{4.3.2}$$

where Δ denotes the characteristic hydraulic radius, T^* is the *tortuosity* (Subs. 4.1.3), and $L^{(V)}$ is the length characterizing spatial variations in the fluid's velocity, V. They suggest that Darcy's law be used for $\mathrm{ReDa}^{1/2} \ll 1$. Following their analysis, we suggest replacing (4.3.1) by

$$\mathrm{Re} = \frac{V\sqrt{\mathsf{k}/\phi T^*}}{\nu}. \tag{4.3.3}$$

In spite of the various definitions for the characteristic length used in (4.3.1), practically, all evidence indicates that *Darcy's law is valid as long as* Re *does not exceed a value of about* 1 (but sometimes as high as 10). Most saturated groundwater flows occur in this range, except in the very close vicinity of high-rate pumping or recharging wells, or large (point) springs. High Reynolds number flows may also be observed in very porous aquifers, such as cavernous limestone, where the hydraulic radius is large.

4.3.2 Non-Darcian motion equations

In the previous subsection, we showed how Darcy's law can be derived from first principles, as an average of the momentum balance equation, subject to certain simplifying assumptions. By relaxing some of these assumptions, primarily by leaving the inertial terms in the averaged equation, we obtain other forms of the motion equation, which are valid for larger values of the Reynolds number.

Bear and Bachmat (1990, p. 172) used volume averaging and a certain set of simplifying assumptions to derive the following general macroscopic momentum balance (≡ flow) equation:

$$\rho\left\{\frac{\partial q_i}{\partial t} + \frac{\partial}{\partial x_j}\left(\frac{q_i q_j}{\phi}\right)\right\} = -\phi\left(\frac{\partial p}{\partial x_j} + \rho g\frac{\partial z}{\partial x_j}\right)T^*_{ji} + \mu\frac{\partial^2 q_{ri}}{\partial x_j \partial x_j} - \mu\alpha_{ij}\frac{C_f}{\Delta^2}q_{rj}, \tag{4.3.4}$$

where T^*_{ij} and α_{ij} are two (2nd order) tensorial properties of the configuration of the $\mathcal{S}_{fs}$-surface in saturated, single phase flow, Δ was defined in (4.3.2), and all variables and coefficients are macroscopic. The two tensorial coefficients reflect at the macroscopic level the effect of the microscopic configuration of the fluid-solid interface within the REV. The first, T^*_{ji}, transforms the local body force into a macroscopic one. The second, α_{ij}, introduces the effect of the configuration of the solid-fluid surface in the term that transforms part of the force resisting the flow at a point to an averaged resistance force at the fluid-solid interface.

When, in addition, $\mathbf{V}_s \equiv 0$, and the fluid is incompressible, $\mathbf{q} \equiv \mathbf{q}_r$ and $\nabla \cdot \mathbf{q} = 0$. Then, (4.3.4) reduces to

$$\rho\left(\frac{\partial q_i}{\partial t} + q_j\frac{\partial (q_i/\phi)}{\partial x_j}\right) = -\phi\left(\frac{\partial p}{\partial x_j} + \rho g\frac{\partial z}{\partial x_j}\right)T^*_{ji} + \mu\frac{\partial^2 q_i}{\partial x_j \partial x_j} - \mu\alpha_{ij}\frac{C_f}{\Delta^2}q_j. \tag{4.3.5}$$

Equation (4.3.4) represents an approximation of the macroscopic momentum balance equation for a fluid phase that fully occupies the void space of a porous medium domain.

The first term on the r.h.s. of (4.3.4), represents the resultant force acting on the fluid, due to gravity and pressure gradient, per unit volume of porous medium.

The second term on the r.h.s. of (4.3.4) represents the force acting on the fluid, due to the viscous resistance to its flow inside the fluid, per unit volume of porous medium. Actually, this force is exerted on the fluid within the REV, across the fluid-fluid portion of the surface bounding the REV.

The last term on the r.h.s. of (4.3.4), expresses the viscous resistance, or viscous drag force exerted by the solid phase on the flowing fluid at their contact surfaces within the REV, per unit volume of porous medium.

Often, in a given problem, one of these forces is much smaller with respect to the remaining ones and, therefore, may be deleted from the momentum balance equation. Hence, following the discussion on the deletion of non-dominant effects presented in Sec. 7.7, let us consider some simplified cases of (4.3.4).

A. Darcy-Forchheimer equation

In the range of validity of Darcy's law, i.e., Re < 1–10, the viscous forces that resist flow are predominant. As the flow velocity increases, a gradual transition is observed (Fig. 4.3.1a) from (microscopically) laminar flow, where viscous forces are predominant, to, still, laminar flow, but with inertial forces gradually taking over. Often, the value of Re = 100 is mentioned as the upper limit of this transition region in which Darcy's linear law is no longer valid. The reason for this deviation from the linear law is that *at the microscopic level*, as velocities increase, local separations of flow from the solid surfaces of the solid matrix occur at an increasing number of places, where the flow curves or diverges. Local vortices and countercurrent flow regimes are caused by inertial and viscous forces along portions of the solid.

For a rigid porous medium, neglecting the effects of inertia *at the macroscopic level*, (4.3.4) reduces to

$$q_i = -\frac{\mathsf{k}_{ij}}{\mu}\left(\frac{\partial p}{\partial x_j} + \rho g\frac{\partial z}{\partial x_j}\right) - \frac{\rho\beta_{ij}}{\mu}qq_j, \tag{4.3.6}$$

where $q = |\mathbf{q}|$, and the β_{ij}'s are coefficients that are related to the configuration of the void space. This motion equation is known as the *Darcy-Forchheimer equation* (Forchheimer, 1901).

At low Re, the second term on the right-hand side, which expresses the average of the *microscopic inertial effects*, becomes negligible.

B. Brinkman equation

When all inertial effects are negligible, but we do not neglect the dissipation of energy *within* the fluid, the motion equation takes the form known as *Brinkman equation* (Brinkman, 1948; Gray and O'Neill, 1976). Bear and Bachmat (1990, p. 177) generalized this equation to the form:

$$q_i = -\frac{\mathsf{k}_{ij}}{\mu}\left(\frac{\partial p}{\partial x_j} + \rho g\frac{\partial z}{\partial x_j}\right) + \frac{\mathsf{k}_{mj}(\mathsf{T}^*_{im})^{-1}}{\phi}\frac{\partial^2 q_j}{\partial x_k \partial x_k}. \tag{4.3.7}$$

In deriving this equation, the viscosity causing energy dissipation at the solid-fluid microscopic interfaces is assumed to be the same as that *within* the fluid.

A detailed discussion on the Dupuit-Forchheimer and the Brinkman equations is presented in Nield and Bejan (1998). Many authors have suggested expressions for the motion equation, or fluid flux, when $\text{Re} \ll 1$ (e.g., Pedras and de Lemos (2001)).

C. Rapid velocity variations

When local acceleration may not be neglected, especially at the onset of flow and in oscillatory flows, but the advective acceleration and the internal friction in the fluid may be neglected, the motion equation takes the form:

$$q_i = -\frac{\mathsf{k}_{ij}}{\mu}\left(\frac{\partial p}{\partial x_j} + \rho g\frac{\partial z}{\partial x_j}\right) - \frac{\rho}{\mu}(\mathsf{T}^*_{im})^{-1}\mathsf{k}_{mj}\frac{\partial V_j}{\partial t}. \tag{4.3.8}$$

Although, for the sake of completeness, we have introduced the above equations, applicable to cases in which Darcy's law is not valid, conditions that justify their application are seldom encountered in problems of flow and contaminant transport in the subsurface.

Like all other macroscopic equations, the non-Darcy equations can also be obtained by homogenization techniques (e.g., Allaire, 1997a; Nield, 2000; Chen *et al.*, 2001; Auriault *et al.*, 2005).

4.4 Aquifer Transmissivity

The subject of 'essentially horizontal flow' in confined and phreatic aquifers has already been introduced in Sec. 2.6. Consider the flow through the confined aquifer of thickness B shown in Fig. 4.4.1. If the aquifer is homogeneous and isotropic, with hydraulic conductivity K, and we assume *essentially horizontal flow*, the total discharge in the $+x$ direction, Q_x, through the area $W \times B$, which is normal to the direction of flow, is given by Darcy's law, written, for example for the x-direction, in the form

$$Q_x = -\mathsf{K}BW\frac{\partial h}{\partial x} \equiv \mathsf{K}BW\mathcal{J}_x, \quad \mathcal{J}_x \equiv -\frac{\partial h}{\partial x}, \quad \boldsymbol{\mathcal{J}} = -\nabla h, \tag{4.4.1}$$

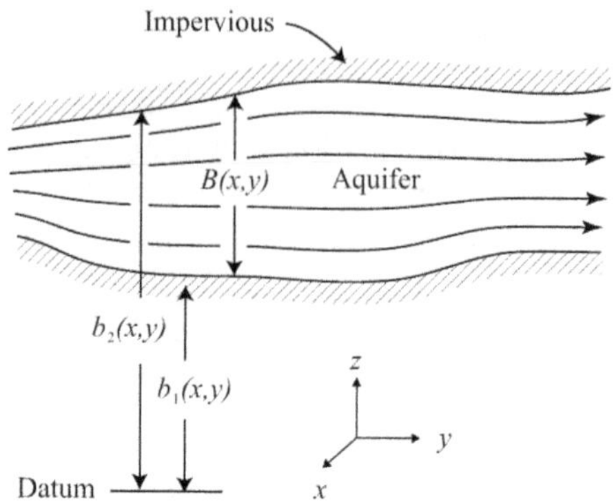

Figure 4.4.1: Flow through a confined aquifer.

in which the vector $\boldsymbol{\mathcal{J}} \equiv -\nabla h$ is the *hydraulic gradient.*

The discharge per unit width of aquifer, Q'_x, normal to the direction of the flow, $+x$, is

$$Q'_x = -\frac{Q_x}{W} = \mathsf{K}B\mathcal{J}_x, \equiv \mathsf{T}\mathcal{J}_x, \quad \mathsf{T} = \mathsf{K}B = \frac{Q_x}{\mathcal{J}_x}. \tag{4.4.2}$$

A similar expression can be written for flow in the y-direction. In vector form, we may write

$$\mathbf{Q}' = -\mathsf{T}\nabla' h = \mathsf{T}\boldsymbol{\mathcal{J}}', \quad \boldsymbol{\mathcal{J}}' = -\nabla' h, \quad \mathsf{T} = \mathsf{K}B = \frac{Q_x{}'}{\mathcal{J}_x}, \tag{4.4.3}$$

where the piezometric head, h, denotes the average value over the thickness of the aquifer, i.e., $h = h(x, y, t)$, and the prime symbol indicates that the operation is only in the xy-plane,

$$\nabla'(..) = \frac{\partial(..)}{\partial x}\mathbf{1x} + \frac{\partial(..)}{\partial y}\mathbf{1y},$$

with $\mathbf{1x}$ and $\mathbf{1y}$ denoting unit vectors in the respective directions. The product $\mathsf{K}B$, denoted by T, which appears whenever flow through the entire thickness of a confined aquifer is being considered, is called *transmissivity*. It is an aquifer property, which is defined by the rate of flow per unit width through the entire thickness of an aquifer, per unit hydraulic gradient. The concept is valid only in aquifer-type, two-dimensional flow. In three-dimensional flow through porous media, the concept of transmissivity is meaningless.

Let us present a more rigorous definition of the transmissivity concept. Consider a confined aquifer of variable thickness, $B = B(x, y) = b_2(x, y) - b_1(x, y)$, where $b_1(x, y)$ and $b_2(x, y)$ are the elevations of the fixed bottom and ceiling of the aquifer. Let $h = h(x, y, z, t)$ denote the piezometric head in this aquifer. The total discharge through the aquifer can then be expressed by

$$Q'_x = \int_{b_1(x,y)}^{b_2(x,y)} q_x\,dz = -\int_{b_1(x,y)}^{b_2(x,y)} \mathsf{K}(x,y,z)\frac{\partial h}{\partial x}\,dz,$$
$$Q'_y = \int_{b_1(x,y)}^{b_2(x,y)} q_y\,dz = -\int_{b_1(x,y)}^{b_2(x,y)} \mathsf{K}(x,y,z)\frac{\partial h}{\partial y}\,dz. \tag{4.4.4}$$

To integrate a derivative, or to differentiate an integral, we make use of *Leibnitz' rule* (see any textbook on advanced calculus, e.g., Greenberg, 1998)

$$\frac{\partial}{\partial t}\int_{b(x)}^{a(x)} f(x,t)\,dt = \int_{b(x)}^{a(x)} \frac{\partial f(x,t)}{\partial x}\,dt + f(x,a)\frac{\partial a}{\partial x} - f(x,b)\frac{\partial b}{\partial x}. \tag{4.4.5}$$

For the sake of simplicity, we shall consider a homogeneous isotropic confined aquifer. Applying this rule to (4.4.4), we obtain

$$Q'_x = -\mathsf{K}B\frac{\partial \tilde{h}}{\partial x} - \mathsf{K}\left[\tilde{h}\frac{\partial B}{\partial x} - h(x,y,b_2)\frac{\partial b_2}{\partial x} + h(x,y,b_1)\frac{\partial b_1}{\partial x}\right],$$
$$Q'_y = -\mathsf{K}B\frac{\partial \tilde{h}}{\partial y} - \mathsf{K}\left[\tilde{h}\frac{\partial B}{\partial y} - h(x,y,b_2)\frac{\partial b_2}{\partial y} + h(x,y,b_1)\frac{\partial b_1}{\partial y}\right], \tag{4.4.6}$$

or, in vector notation,

$$\mathbf{Q}' = -\mathsf{K}B\nabla\tilde{h} - \mathsf{K}\left[\tilde{h}\nabla' B - h(x,y,b_2)\nabla' b_2 + h(x,y,b_1)\nabla' b_1\right], \tag{4.4.7}$$

in which

$$\tilde{h}(x,y,t) = \frac{1}{B}\int_{b_1(x,y)}^{b_2(x,y)} h(x,y,z,t)\,dz \tag{4.4.8}$$

is the average piezometric head along the vertical at point (x,y) at time t.

If we now assume 'essentially horizontal flow', that is, vertical equipotentials, (4.4.8) may be approximated by introducing $\tilde{h} \cong h(x,y,b_2) \cong h(x,y,b_1)$. Then, (4.4.8) is approximated as

$$\mathbf{Q}' = -\mathsf{T}(x,y)\nabla'\tilde{h}, \qquad \mathsf{T}(x,y) = \mathsf{K}(x,y)B(x,y). \tag{4.4.9}$$

The error resulting from employing (4.4.9), based on the assumption of 'essentially horizontal flow', is given by the second term on the right-hand side of (4.4.9). We could reduce the error by expressing $\mathbf{Q}'$ as

$$\mathbf{Q}' = -\mathsf{K}\nabla'(B\tilde{h}). \tag{4.4.10}$$

In a heterogeneous aquifer, $\mathsf{K} = \mathsf{K}(x,y,z)$, we also start from the assumption of 'essentially horizontal flow', that is, $h = h(x,y,t)$ in (4.4.4). Then, the latter equation leads to

$$\mathbf{Q}' = -\mathsf{T}(x,y)\nabla' h, \quad \mathsf{T}(x,y) = \int_{b_1}^{b_2} \mathsf{K}(x,y,z)\,dz = \widetilde{\mathsf{K}}B, \tag{4.4.11}$$

where we used the tilde symbol, $\tilde{(.)}$, to denote the average over the thickness, just like 4.4.8. In general, we have

$$\nabla'\tilde{h} = \widetilde{\nabla' h} - \left(\tilde{h}/B\right)\nabla' B + \left[h(x,y,b_2)\nabla' b_2 - h(x,y,b_1)\nabla' b_1\right]/B. \tag{4.4.12}$$

When $h(x,y,b_1) \cong h(x,y,b_2) \cong \tilde{h}(x,y)$, then $\nabla'\tilde{h} = \widetilde{\nabla' h}$, i.e., the gradient of the average head is equal to the average of the head gradient. Equations (4.4.3), (4.4.9), and (4.4.11) are identical if h is understood as mean $\tilde{h}$.

The above discussion serves as a justification for employing the concepts of *aquifer flow* and *aquifer transmissivity,* also in cases of inhomogeneous hydraulic conductivity and variable thickness.

When a confined aquifer is composed of N distinct homogeneous layers, each of thickness B_i and hydraulic conductivity K_i, (4.4.12) reduces to

$$\mathbf{Q}' = -\mathsf{T}(x,y)\nabla' h, \quad \mathsf{T}(x,y) = \sum_{i=1}^{i=N} B_i \mathsf{K}_i. \tag{4.4.13}$$

Following the discussion above, (4.4.13) is valid as an approximation also when $B_i = B_i(x,y)$ and $\mathsf{K}_i = \mathsf{K}_i(x,y)$.

As indicated in Sec. 2.6, the assumption of 'essentially horizontal flow' can be extended also to leaky aquifers (Fig. 2.5.2), with the concept of transmissivity defined by (4.4.13). This point is further discussed in Sec. 5.4.

4.5 Dupuit Assumption for Phreatic Aquifer

A phreatic aquifer is defined in Sec. 2.3 as one in which a water table ($\equiv$ phreatic surface) serves as its upper boundary. In Sec. 2.2, we have introduced the fact that, actually, above the phreatic surface, which is defined as *an imaginary surface at all points of which the pressure is atmospheric,* moisture does occupy at least part of the void space (Fig. 2.2.1). The *capillary fringe* was introduced as an approximation of the actual distribution of moisture in the soil above the phreatic surface. The details of what happens in the unsaturated zone is discussed in Chap. 6. At this stage, we wish only to emphasize that moisture exists, whether stationary or mobile, also in the unsaturated zone.

Figure 4.5.1 shows a typical distribution of immobile water in the unsaturated zone. As an approximation, we often replace this distribution by assuming the existence of a fully saturated capillary fringe up to an elevation h_c immediately above the water table, with water at a constant saturation, equal to the *irreducible water saturation,* above h_c. Within the capillary fringe, water is at a pressure which is *less than atmospheric.* The capillary

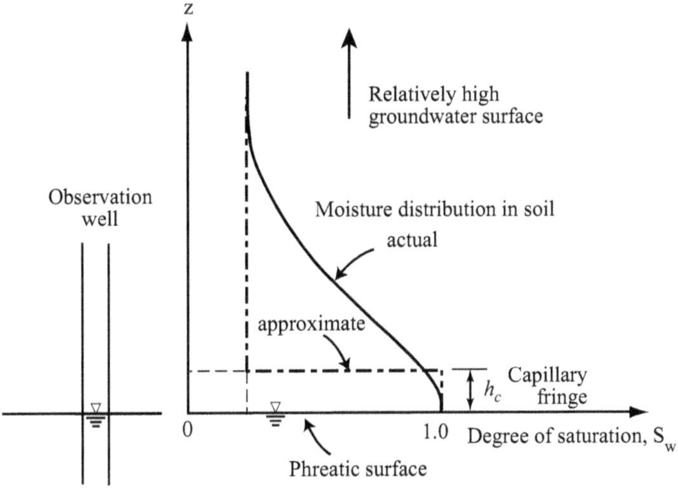

Figure 4.5.1: A phreatic surface and a capillary fringe.

fringe concept better approximates reality in the case of a poorly graded soil than in a well-graded soil (Sec. 4.5.1). Often it is assumed that the existence of a capillary fringe may be neglected i.e., that no moisture is present in the soil above the water table. Obviously, this approximation is justified only when the thickness of the capillary fringe thus defined is much smaller than the thickness of the (saturated) aquifer below the water table.

Sometimes, we assume that no moisture exists in the void space above h_c. With this approximation, the upper end of the capillary fringe serves as a boundary to the saturated zone. However, when h_c is much smaller than the thickness of the underlying phreatic aquifer, and this is, indeed, the situation encountered in most aquifers, hydrologists often neglect the existence of the capillary fringe, assuming full saturation up to the phreatic surface, and no moisture in the void space above it.

An estimate of the value of h_c can be obtained, for example, from (Mavis and Tsui, 1939)

$$h_c = \frac{2.2}{d_H}\left(\frac{1-\phi}{\phi}\right)^{2/3}, \tag{4.5.1}$$

where h_c is in mm, d_H is the harmonic mean grain diameter also in mm, and ϕ is the porosity. Another expression is (Polubarinova-Kochina, 1951, 1962)

$$h_c = \frac{0.45}{d_{10}}\left(\frac{1-\phi}{\phi}\right), \tag{4.5.2}$$

where both h_c and the effective particle diameter, d_{10}, are in cm. Silin-Bekchurin (1958) suggested a capillary rise of 2–5 cm in coarse sand, 12–35

Material	Grain size (mm)	Capillary rise (cm)
Fine gravel	2–5	2.5
Very coarse sand	1–2	6.5
Coarse sand	0.5–1	13.5
Medium sand	0.2–0.5	24.6
Fine sand	0.1–0.2	42.8
Silt	0.05–0.1	105.5
Silt	0.02–0.05	200

Table 4.5.1: Capillary rise in samples after 72 days. Samples have virtually the same porosity, 41% (From Lohman, 1972, cited in Todd, 1980).

cm in sand, 35–70 cm in fine sand, 70–150 cm in silt, and 2–4 m and more in clay. Equations (4.5.1) and (4.5.2) can be compared with the rise of water in a capillary tube of radius r: $h_c = 2\gamma_{aw}/(\rho g)r$, where γ_{aw} is the (air-water) surface tension. For water at $20°C$, $\gamma_{wa} = 0.073$ g/cm, $\rho g = 1$ g/cm^3, so that the estimated capillary rise becomes $h_c \simeq 0.15/r$ (r and h_c in cm). Table 4.5.1 gives the capillary rise in unconsolidated soil sample with the same porosity but different grain sizes The observed values are fairly consistent with the above formula.

We continue under the assumption that when considering flow in a phreatic aquifer, the moisture in the void space above the water table can be neglected. Below the water table, both the piezometric head, h, and the specific discharge, $\mathbf{q}$, vary from point to point within the aquifer. In order to obtain the specific discharge $\mathbf{q} = \mathbf{q}(x, y, z, t)$ at every point, we have to know $h = h(x, y, z, t)$, either by field measurements, of by solving an appropriate model. This topic is discussed in Chap. 5. However, as we shall see there, in the case of a three-dimensional flow domain (or a vertical two-dimensional one) for which a (possibly moving) phreatic surface is a boundary, the solution is possible only numerically, because the forecasting model becomes nonlinear.

Hence, let us examine the possibility of assuming that the flow in the aquifer is *essentially horizontal*, as we did in the case of a confined aquifer. Basically, the phreatic surface is *never horizontal*, except for special cases, like when the water is immobile. This observation is valid even in the case of an aquifer with a horizontal impervious bottom; equipotentials (surfaces of constant h) are never vertical (Fig. 4.5.2a).

In spite of this statement, in practice, except in very limited special domains, *the slope of the water table is very small.*, i.e., close to horizontal. With such observation, Dupuit (1863) suggested that for most cases of practical interest, we may assume that flow in a phreatic aquifer is *practically horizontal.* Equivalently, this assumption states that surfaces of equal piezometric heads ($\equiv$ equipotential surfaces) are vertical. This also means that *the pressure distribution in the aquifer is hydrostatic.* This assumption is known as the 'Dupuit assumption'. It is a most powerful tool for treating flow in

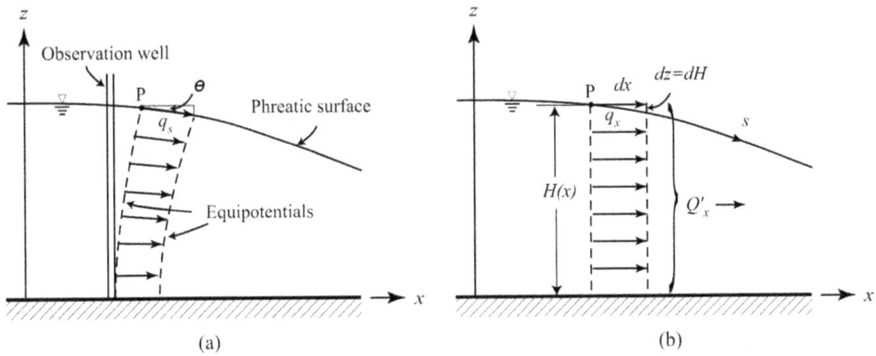

Figure 4.5.2: Dupuit assumption.

unconfined aquifers. In fact, it is the only simple tool available to most engineers and hydrologists for solving such problems. Dupuit (1863) based his assumption on the observation that in most groundwater flows, the slope of the phreatic surface is very small. Slopes of 1/1000 and 10/1000 are commonly encountered.

Consider steady flow without accretion in the vertical two-dimensional xz-plane shown in Fig. 4.5.2a. The assumptions of steady flow and a horizontal bottom are intended only to simplify the discussion. Under these conditions, the phreatic surface is a streamline. At every point P along this streamline, $p = 0$ and $h = H$, where h denotes the piezometric head. We have introduced the symbol H to indicate the elevation of P above the horizontal impervious bottom that serves as a datum level. We note that, in the case considered here, $h = h(x, z)$ and $H = H(x)$. The specific discharge at P, which is in the direction of the tangent to the streamline, is given by Darcy's law

$$q_s = -\mathsf{K}\frac{dh}{ds} = -\mathsf{K}\frac{dH}{ds} = -\mathsf{K}\sin\theta, \tag{4.5.3}$$

where θ is the angle between the tangent of the phreatic surface and the horizontal (Fig. 4.5.2). Dupuit suggested that for small θ, $\sin\theta$ in (4.5.3) can be replaced by the slope, $\tan\theta = dH/dx$, and $\cos\theta \approx 1$, such that

$$q_x = q_s \cos\theta \approx -\mathsf{K}\frac{dH}{dx}. \tag{4.5.4}$$

The assumption of small θ is equivalent to assuming that (1) equipotential surfaces are vertical, i.e., $h = h(x)(= H(x))$, rather than $h = h(x, z)$, and (2) the flow is essentially horizontal.

For unsteady flow in a three-dimensional domain, the Dupuit assumption presented above is extended to $h = h(x, y, z, t)$ and $H = H(x, y, t)$. Then, the Dupuit assumption leads to the specific discharge expressed by

$$q_x = -\mathsf{K}\frac{\partial H}{\partial x}, \quad q_y = -\mathsf{K}\frac{\partial H}{\partial y}. \tag{4.5.5}$$

Since, by making use of the Dupuit assumption, $\mathbf{q}$ is independent of elevation z, the corresponding total discharge through a vertical surface of unit width (normal to the direction of flow) can be expressed in the vector form:

$$\mathbf{Q}' = -\mathsf{K}H\nabla'H, \quad \text{or} \quad \mathbf{Q}' = -\mathsf{K}\nabla'(H^2/2). \tag{4.5.6}$$

Recall that the aquifer's bottom is horizontal and that it serves as a datum level for H.

The important advantage gained by employing the Dupuit assumption is that in a considered three-dimensional flow domain with a phreatic surface, the variable $h = h(x, y, z, t)$ has been replaced by $H = H(x, y, t)$, that is, z does not appear as an independent variable.

In order to obtain a better understanding of the approximation involved in the Dupuit assumption, let us determine the exact expression for flow in the horizontal direction in a phreatic aquifer, with $h = h(x, y, z, t)$. We obtain this expression by integrating along the vertical from the bottom of the aquifer, $\eta = \eta(x, y)$, which needs not be horizontal, to the phreatic surface at elevation $H = H(x, y, t)$. For flow in the $+x$ direction, assuming $\mathsf{K} = \text{const.}$, we obtain

$$\begin{aligned}
Q'_x &= \int_{\eta(x,y)}^{H(x,y,t)} q_x\,dz = -\mathsf{K}\int_{\eta(x,y)}^{H(x,y,t)} \frac{\partial h}{\partial x}dz \\
&= -\mathsf{K}\left\{\frac{\partial}{\partial x}\int_{\eta}^{H} h\,dz - h|_H\frac{\partial H}{\partial x} + h|_\eta\frac{\partial \eta}{\partial x}\right\} \\
&= -\mathsf{K}\left\{\frac{\partial}{\partial x}\left[(H-\eta)\tilde{h}\right] - h|_H\frac{\partial H}{\partial x} + h|_\eta\frac{\partial \eta}{\partial x}\right\} \\
&= -\mathsf{K}\left\{(H-\eta)\frac{\partial \tilde{h}}{\partial x} + \tilde{h}\frac{\partial H-\eta}{\partial x} - h|_H\frac{\partial H}{\partial x} + h|_\eta\frac{\partial \eta}{\partial x}\right\},
\end{aligned} \tag{4.5.7}$$

where $\tilde{h}$ denotes the average (over the vertical) head,

$$\tilde{h} = \frac{1}{H-\eta}\int_{\eta}^{H} h\,dz, \qquad h|_H = H. \tag{4.5.8}$$

Equation (4.5.7) involves no approximation. If we now assume vertical equipotentials, i.e., $\tilde{h} \simeq h|_\eta \simeq h|_H (= H)$, then (4.5.7) reduces to

$$\mathbf{Q}' = -\mathsf{K}(H-\eta)\nabla'H, \tag{4.5.9}$$

which is the same as (4.5.6), written for a non-horizontal bottom. Or, without the Dupuit approximation, for a horizontal bottom, $\eta = 0$, equation (4.5.7) can be written as

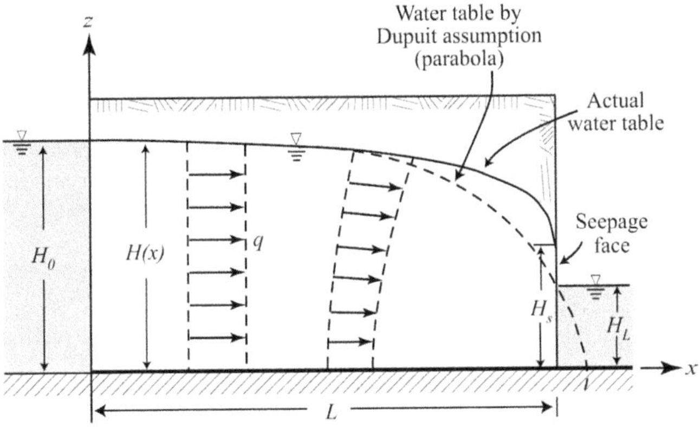

Figure 4.5.3: Dupuit-Forchheimer discharge formula.

$$Q'_x = -\mathsf{K}\frac{\partial}{\partial x}\left(H\tilde{h} - \frac{H^2}{2}\right), \quad \text{or} \quad \mathbf{Q}' = -\mathsf{K}\nabla'\left(H\tilde{h} - \frac{H^2}{2}\right). \tag{4.5.10}$$

By comparing (4.5.10) with (4.5.6), we observe that we have replaced $H\tilde{h} - H^2/2$ by $H^2/2$ in the approximate expression based on the Dupuit assumption. The error reduces to zero as $\tilde{h} \to H$. Bear (1972, p. 363) gives an estimate of the error involved in replacing $H\tilde{h} - H^2/2$ by $H^2/2$ in (4.5.6),

$$0 < \frac{H^2/2 - (H\tilde{h} - H^2/2)}{H^2/2} < \frac{i^2}{1+i^2}, \quad i = \frac{dH}{dx}, \tag{4.5.11}$$

so that the error is small as long as $i^2 \ll 1$, where i denotes the slope of the phreatic surface. When the medium is anisotropic, with $\mathsf{K}_x \neq \mathsf{K}_z$ (x, z principal directions), i^2 in (4.5.11) should be replaced by $(\mathsf{K}_x/\mathsf{K}_z)i^2$.

As a simple example of the application of (4.5.4), consider the case of steady unconfined flow through a homogeneous formation between two reservoirs with vertical faces (Fig. 4.5.3). Following the Dupuit assumption, the total discharge in the x direction, per unit width, through a vertical cross section of height $H(x)$ is given by (4.5.6)

$$Q'_x \equiv Q' = -\mathsf{K}H\frac{dH}{dx} = \text{const.} \quad \Rightarrow \quad Q'\,dx = -\mathsf{K}H\,dH. \tag{4.5.12}$$

By integrating this expression between the boundary at $x = 0$, where $H = H_o$, and any distance x, where $H = H(x)$, we obtain

$$Q'\int_{x=0}^{x} dx = -\mathsf{K}\int_{H=H_o}^{H(x)} H\,dH, \quad \Rightarrow \quad Q' = \mathsf{K}\frac{H_o^2 - H^2}{2x}. \tag{4.5.13}$$

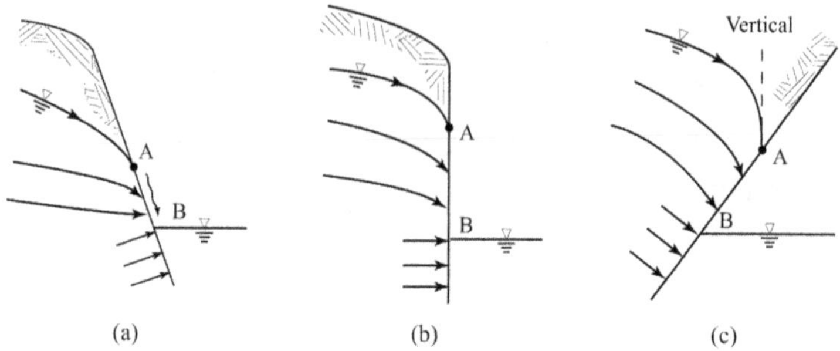

Figure 4.5.4: Seepage face.

Equation (4.5.13) describes a water table, $H = H(x)$, which has the shape of a parabola passing through $x = 0, H = H_o$. If we know H_o and $H(x)$ at some distance x, or H at any two x-values, we can use the above equation to derive Q' (obviously, if K is known). However, the boundary condition at the other end, $x = L$, requires special attention.

As a rule, whenever a phreatic surface approaches a downstream reservoir that serves as the external boundary of the flow domain, it always terminates on such boundary at a point that is some distance above the water table of the body of open water. Points A in Figs. 4.5.4a, b, and c, are such points.

The segment AB of the boundary above the water table and below the phreatic surface is called the *seepage face.* Along the seepage face, water emerges from the soil, trickling down along the seepage face. In Figs. 4.5.4a and b, the phreatic surface is tangent to the external boundary at A; in Fig. 4.5.4c, it is tangent to a vertical line at A (Bear, 1972, p. 260).

Because of the presence of a seepage face, which terminates at an unknown point on the phreatic surface, H_s in Fig. 4.5.3 is unknown. However, when the Dupuit assumption is employed, we approximate the situation by overlooking the presence of the seepage face and *assume* that the water table at $x = L$ passes through $H = H_L$. Using this as the downstream boundary condition, we obtain from (4.5.13)

$$Q' = \mathsf{K}\frac{H_o^2 - H_L^2}{2L}, \tag{4.5.14}$$

known as the *Dupuit-Forchheimer discharge formula.* The parabolic water table is shown as a dashed line in Fig. 4.5.3. The discrepancy exists mainly at the boundaries, i.e., at $x = 0$, where the water table should be tangent to the horizontal line, whereas the parabola has a slope of $dH/dx|_{x=0} = -Q'/\mathsf{K}H_o$, and at $x = L$, where the seepage face has been neglected. Otherwise, the discrepancy between the curves derived by the exact theory of the phreatic surface boundary, and by the Dupuit approximation is negligible. A simple rule is that a solution based on the Dupuit assumption is sufficiently accurate for practical purposes at a distance from the downstream end that is larger than 1.5–2 times the thickness of the flow domain. Moreover, it can be shown

(Bear, 1972; p. 367) that (4.5.14) is accurate as far as the rate of discharge is concerned, although (4.5.13) does not give the accurate water table elevations, $H = H(x)$.

The Dupuit assumption is not applicable in regions where the vertical flow component is not negligible (e.g., Fig. 2.6.1). Note that in cases with accretion, a horizontal water table (or almost so) is not sufficient to justify the application of the Dupuit assumption. One must verify that vertical flow components are indeed negligible, before applying the Dupuit assumption.

Another case to which the Dupuit assumption should be applied with care is that of unsteady flow in a decaying phreatic surface mound. Although no accretion takes place, yet, at the crest and in its vicinity, the flow is vertically downward. At a distance of say 1.5 to 2 times the thickness of the flow, the approximation of vertical equipotentials is again valid.

In spite of the cautionary comments above, in regional groundwater studies, the Dupuit assumption is usually applied everywhere, even to those (relatively small) parts of an investigated region where it is not strictly applicable. This is because of its simplicity and the relatively small error involved.

Chapter 5

WATER BALANCE AND COMPLETE FLOW MODEL

Each of the motion equations discussed in Chap. 4 involves two dependent variables: the flux and the pressure (or piezometric head, or suction). For example, (4.1.26) involves $\mathbf{q}$ and h, while (4.2.43) involves $\mathbf{q}$ and p. Therefore, to obtain a solution, we need one additional equation. This is the *mass balance equation* of the fluid phase. This equation is considered in this chapter.

The density is not included as an additional variable because we can always use an appropriate equation of state to express the fluid's density, ρ, in terms of the pressure, and, if relevant, the temperature and the fluid's composition. However, when temperature and fluid composition vary in space and time, they become additional state variables of the problem, and additional equations are required in order to obtain a solution. In this chapter, we assume that the spatial and temporal changes in the temperature and composition of each phase are sufficiently small so as not to affect the fluid's density. In Chap. 7 we shall consider changes in phase composition, including the case of composition-dependent density.

The justification presented above is a simplified one. More rigorously, the mathematical model that completely describes the movement of a fluid, at the continuum level, consists of balance equations for the mass, the momentum, and the energy of that fluid. By solving these three equations simultaneously, we obtain the pressure, velocity, and temperature of the fluid, as functions of space and time. Together, these three variables define the complete behavior of the fluid. This statement is valid for the description of fluid flow at the microscopic level, as well as at the macroscopic, or averaged one. Under isothermal conditions, the energy balance equation is not required.

We begin this chapter with the *mass balance equations* for saturated flow, i.e., where a single fluid phase occupies the entire void space. In Chap. 6, we shall extend the discussion to unsaturated flow domains. Deformation of the solid matrix will be taken into account. For a given fluid phase, by combining the mass balance equation with an appropriate motion equation, a *flow equation* is obtained. The complete flow model requires, in addition, specification of appropriate *initial* and *boundary conditions*. The various boundary conditions that may be encountered are discussed in detail. The numerical values of all the coefficients that appear in the model equations have also

J. Bear, A.H.-D. Cheng, *Modeling Groundwater Flow and Contaminant Transport*, Theory and Applications of Transport in Porous Media 23,
DOI 10.1007/978-1-4020-6682-5_5, © Springer Science+Business Media B.V. 2010

to be specified. All this supplemental information must be specified for the considered domain and fluids in order to obtain solutions that pertain to the particular flow problem under consideration.

Unless otherwise specified, the entire presentation will be at the macroscopic level. We shall omit the symbol that denotes an average value.

5.1 Mass Balance Equations

5.1.1 Fundamental mass balance equations

A. General considerations

We start by considering the mass balance equation at the microscopic level. The corresponding equation at the macroscopic level can be obtained, for example, by averaging the microscopic level equation over an REV.

The basic idea of a balance is straightforward. A balance can be written for any extensive quantity, E, within *a specified spatial domain*, and for a *specified period* of time. Let Δt denote the time interval, and $\mathcal{U}_o$, bounded by a surface $\mathcal{S}$, denote the volume of a phase domain (i.e., at the microscopic level) for which the balance is written. The balance is then stated as:

$$\left\{\begin{array}{c}\text{Quantity of } E\\ \text{accumulating}\\ \text{in } \mathcal{U}_o\\ \text{during } \Delta t\end{array}\right\} = \left\{\begin{array}{c}\text{Net quantity of}\\ E \text{ entering } \mathcal{U}_o\\ \text{through } \mathcal{S}\\ \text{during } \Delta t\end{array}\right\} + \left\{\begin{array}{c}\text{Net production}\\ \text{of } E\\ \text{in } \mathcal{U}_o\\ \text{during } \Delta t\end{array}\right\}. \tag{5.1.1}$$

When this balance is written for a small volume around a point, and we let both volume and time interval shrink to zero, the balance will be represented by a *partial differential equation* (abbreviated PDE). This PDE can, then, be interpreted as 'a statement of balance at a point', i.e., for a small volume about the point.

The total flux, $\mathbf{j}^{tE}$, of any extensive quantity, E, having a density e, is expressed by the product $e\mathbf{V}^E$, with $\mathbf{V}^E$ denoting the velocity of the E-continuum. At this time, let us introduce the mathematical term 'divergence of a flux', and its physical interpretation.

Consider the parallelepiped *control volume* with dimensions Δx, Δy, Δz, shown in Figure 5.1.1. During a period Δt, the excess of inflow over outflow of the extensive quantity through the surfaces that bound the control volume, is expressed by

$$\begin{aligned}\Delta t \Big[&\left(j_x^{tE}\big|_{x-\frac{1}{2}\Delta x,y,z} - j_x^{tE}\big|_{x+\frac{1}{2}\Delta x,y,z}\right) \Delta y\, \Delta z \\ &+ \left(j_y^{tE}\big|_{x,y-\frac{1}{2}\Delta y,z} - j_y^{tE}\big|_{x,y+\frac{1}{2}\Delta y,z}\right) \Delta x\, \Delta z \\ &+ \left(j_z^{tE}\big|_{x,y,z-\frac{1}{2}\Delta z} - j_z^{tE}\big|_{x,y,z+\frac{1}{2}\Delta z}\right) \Delta x\, \Delta y \Big].\end{aligned}$$

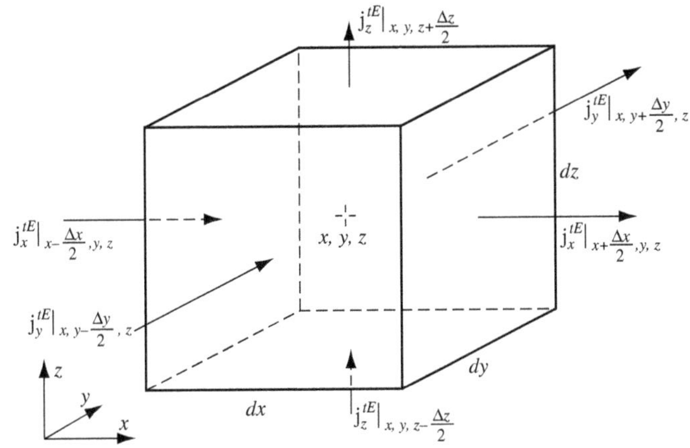

Figure 5.1.1: A control volume.

The excess of inflow over outflow, per unit volume and per unit time, is obtained by dividing the above expression by $\Delta x\,\Delta y\,\Delta z\,\Delta t$, yielding

$$\frac{j_x^{tE}\big|_{x-\frac{1}{2}\Delta x,y,z} - j_x^{tE}\big|_{x+\frac{1}{2}\Delta x,y,z}}{\Delta x} + \frac{j_y^{tE}\big|_{x,y-\frac{1}{2}\Delta y,z} - j_y^{tE}\big|_{x,y+\frac{1}{2}\Delta y,z}}{\Delta y}$$
$$+\frac{j_z^{tE}\big|_{x,y,z-\frac{1}{2}\Delta z} - j_z^{tE}\big|_{x,y,z+\frac{1}{2}\Delta z}}{\Delta z}.$$

In order to obtain the above excess of inflow over outflow of the extensive quantity *at a point* (that is, per unit volume of an REV centered at the point), we let $\Delta x,\ \Delta y,\ \Delta z \to 0$, leading to the expression

$$-\left(\frac{\partial j_x^{tE}}{\partial x} + \frac{\partial j_y^{tE}}{\partial y} + \frac{\partial j_z^{tE}}{\partial z}\right), \quad \text{or, equivalently} \quad -\nabla{\cdot}\mathbf{j}^{tE}.$$

The right-hand side of this expression (written, equivalently, in the form $-\text{div}\ \mathbf{j}^{tE} \equiv \nabla\cdot\mathbf{j}^{tE}$) is read as 'minus divergence of the vector $\mathbf{j}^{tE}$'.

Thus, the physical interpretation of *minus the divergence of the total flux of any extensive quantity* is *the excess of inflow over outflow of that quantity, per unit volume of fluid phase and per unit time* at a point. This excess flow is a consequence of the spatial variations of the flux.

With this interpretation of 'divergence of a flux', and noting that $\partial e/\partial t$ is interpreted as *the rate of increase in e* (*i.e., in the amount of E per unit volume of a phase*), the (microscopic) differential balance of E of a phase is given by

$$\frac{\partial e}{\partial t} = -\nabla{\cdot}e\mathbf{V}^E + \rho\Gamma^E, \tag{5.1.2}$$

where $\mathbf{j}^{tE} \equiv e\mathbf{V}^{E}$, and Γ^{E} is the rate of production of E per unit mass of the phase.

Another way to express the total flux of an extensive quantity, E, is by making use of the *diffusive flux*, $\mathbf{j}^{E}$ of E. To understand the definition of the diffusive flux of an extensive quantity E, we note that the total flux of E, $e\mathbf{V}^{E}$, may be expressed as a sum

$$e\mathbf{V}^{E} \equiv e\mathbf{V} + e(\mathbf{V}^{E} - \mathbf{V}) \equiv e\mathbf{V}^{E} + \mathbf{j}^{E}, \tag{5.1.3}$$

i.e., the sum of an advective flux, $e\mathbf{V}$, carried by the fluid as it moves at the mass-weighted velocity, $\mathbf{V}$, defined in (7.1.13), and a diffusive flux, $\mathbf{j}^{E}$, which expresses the flux of E relative to the fluid's motion. The latter flux is defined in (7.1.11) as the *diffusive flux* of E.

Thus, expressing the total E-flux as a sum of advective and diffusive fluxes, (5.1.2) may also be written in the form:

$$\frac{\partial e}{\partial t} = -\nabla\cdot(e\mathbf{V} + \mathbf{j}^{E}) + \rho\Gamma^{E}. \tag{5.1.4}$$

This is the most general balance equation for any extensive quantity, E, of a phase at the microscopic level.

Each of the terms in this balance equation represents an added quantity of E per unit time and per unit volume of the considered fluid phase. The net added quantity, represented by the term on the left-hand side, is due to the right-hand side terms:

- An excess of inflow over outflow of E, per unit time and per unit volume of the fluid phase, due to spatial variations in the total flux of E.
- A quantity of E produced *within* the phase, per unit time and per unit volume of the fluid phase.

Equation (5.1.4) describes the (microscopic) balance of E at any point within an α-phase. According to our methodology, to obtain a macroscopic balance equation, we average (5.1.4) over the domain $\mathcal{U}_{o\alpha}$ occupied by an α-phase within an REV, centered at any point within $\mathcal{U}_{o\alpha}$, using the intrinsic phase average defined in (1.3.2). We obtain (Bear and Bachmat, 1990)

$$\begin{aligned}\frac{\partial\theta\overline{e}^{\alpha}}{\partial t} = &- \nabla\cdot\theta(\overline{e}^{\alpha}\overline{\mathbf{V}}^{\alpha} + \overline{\mathring{e}\mathring{\mathbf{V}}}^{\alpha} + \overline{\mathbf{j}^{E}}^{\alpha}) \\ &- \frac{1}{\mathcal{U}_o}\int_{\mathcal{S}_{\alpha\beta}} [e(\mathbf{V}-\mathbf{u}) + \mathbf{j}^{E}]\cdot\mathbf{n}\,d\mathcal{S} + \theta\overline{\rho\Gamma^{E}}^{\alpha},\end{aligned} \tag{5.1.5}$$

where θ denotes the volumetric fraction of the considered phase, $\mathbf{n}$ denotes the unit vector normal to $\mathcal{S}_{\alpha\beta}$, and we have made use of the decomposition of the (intrinsic phase) averaged advective flux of E within the α-phase, $\overline{e\mathbf{V}}^{\alpha}$, into two fluxes: a macroscopic advective flux $\overline{e}^{\alpha}\overline{\mathbf{V}}^{\alpha}$, and another macroscopic flux, $\overline{\mathring{e}\mathring{\mathbf{V}}}^{\alpha}$. In the latter expression, $\mathring{e}$ and $\mathring{\mathbf{V}}$ are deviations of e and $\mathbf{V}$ from their averages over the α-phase within the REV, $\overline{e}^\alpha$ and $\overline{\mathbf{V}}^\alpha$, respectively. Note that (1) we have made use of the average symbol, as we wish to make a distinction between microscopic and macroscopic level variables, and (2) we have introduced here the volumetric fraction of a fluid, θ, defined as the volume of the fluid phase per unit volume of the REV; in saturated flow, it is equal to the porosity, ϕ.

Equation (5.1.5) is the *general (macroscopic) differential balance equation* of an extensive quantity, E, of a phase. By comparing this equation with the microscopic one, (5.1.4), we note that the former contains two additional terms, introduced as a result of the averaging process:

- A flux, $\overline{\mathring{e}\mathring{\mathbf{V}}}^\alpha$, which is the flux of E in excess of the average advection of E by the phase. This flux is discussed in detail in Subs. 7.1.4 and 7.1.5.
- A term
$$\frac{1}{\mathcal{U}_o}\int_{\mathcal{S}_{\alpha\beta}}[e(\mathbf{V}^E-\mathbf{u})+\mathbf{j}^E]\cdot\mathbf{n}\,dS,$$
which expresses the flux of E across the $\mathcal{S}_{\alpha\beta}$-surface that separates the considered phase from all other phases within $\mathcal{U}_o$.

Note that the second term is a sum over *all* other phases, β, when such phases are present in the system.

The (microscopic) flux normal to the $\mathcal{S}_{\alpha\beta}$-surface, appearing in the integrand, is made up of advection of e, relative to the surface (possibly moving at a velocity $\mathbf{u}$), and diffusion. When the $\mathcal{S}_{\alpha\beta}$-surface is a *material surface* with respect to the E-quantity, the term $e(\mathbf{V}^E-\mathbf{u})\cdot\mathbf{n}$ vanishes, and the flux through $\mathcal{S}_{\alpha\beta}$ is due to diffusion only. It is of interest to note that by the averaging process, the boundary conditions on the interphase boundaries, $\mathcal{S}_{\alpha\beta}$, became a source term in the macroscopic equation, (5.1.5).

Each of the four terms in the macroscopic balance equation represents a quantity of E added per unit time to a unit volume of porous medium. The net added quantity, represented by the term on the left hand side is due to the three right hand side terms:

- A net influx of E, per unit time and per unit volume of porous medium, due to spatial variations in the total E-flux.
- An outflow of E that leaves the phase through the $\mathcal{S}_{\alpha\beta}$-surface that bounds the phase within the REV (of volume $\mathcal{U}_o$), per unit time and per unit volume of porous medium, and
- A quantity of E produced *within* the considered phase, per unit time and per volume of porous medium.

We note that the total macroscopic flux of E, per unit area of porous medium, is made up of three parts:

- An advective flux, $\theta\overline{e}^\alpha\overline{\mathbf{V}}^\alpha$.
- A dispersive flux, $\theta\overline{\mathring{e}\mathring{\mathbf{V}}}^\alpha$.
- A diffusive flux, $\theta\overline{\mathbf{j}^E}^\alpha$.

B. Mass balance equation

Let us now return to the particular case considered in this section, viz., the extensive quantity under consideration is the *mass of a fluid phase.* For this case, we replace E by m, and e by ρ, representing the mass density of the fluid (= mass per unit volume of fluid phase), and $\mathbf{V}^E$ by $\mathbf{V}$. Equation (5.1.2) takes then the form

$$\frac{\partial \rho}{\partial t} = -\nabla \cdot \rho \mathbf{V} + \rho \Gamma'^m. \tag{5.1.6}$$

We note that by definition, *the diffusive flux of the mass is identically zero.* The term $\rho\Gamma'^m$ represents a source (= a negative sink) of phase mass. Actually, at the microscopic level in a three-dimensional domain, such a source does not exist, unless there are chemical, biological, or nuclear reactions. Nevertheless, we shall leave this term here, as often point sources or sinks of fluid mass are represented, symbolically, by a function $\Gamma'^m = \Gamma'^m(x, y, z, t)$. This topic is further discussed in Subs. 5.1.4.

The averaged, or macroscopic mass balance equation for a fluid phase that occupies the entire void space (porosity, ϕ), can be obtained by averaging the microscopic mass balance equation (5.1.6), or by applying (5.1.5) to the case $e = \rho$, with θ replaced by ϕ.

Assuming that the dispersive flux (discussed in Subs. 7.1.4 and 7.1.5) of the mass is much smaller than its advective one, i.e., $|\overline{\rho}^\alpha \overline{\mathbf{V}}^\alpha| \gg |\overline{\mathring{\rho}\mathring{\mathbf{V}}}^\alpha|$, we obtain the *macroscopic mass balance equation for a fluid phase,*

$$\frac{\partial}{\partial t}(\phi \rho) = -\nabla \cdot (\rho \mathbf{q}) + \rho \Gamma^m, \tag{5.1.7}$$

where $\mathbf{q}(= \phi \mathbf{V})$ denotes the *specific discharge* (= volume of fluid per unit area of porous medium per unit time) of the fluid-phase, and the symbol Γ^m denotes a source of fluid (= added volume of fluid phase per unit volume of porous medium, per unit time), other than through the (microscopic) solid-fluid interface. We regard a sink as a negative source. To obtain (5.1.7), we have assumed that the microscopic solid-fluid boundaries are *material interfaces* with respect to fluid mass, so that $(\mathbf{V} - \mathbf{u}) \cdot \mathbf{n} = 0$ in (5.1.5). Recall that when a velocity (here, $(\mathbf{V} - \mathbf{u})$) is multiplied ('scalar product') by a unit vector normal to an area (here, $\mathbf{n}$), we obtain the component of that velocity vector normal to the surface. The case in which we do not neglect the dispersive mass flux is presented in the discussion on variable density flow and transport in Subs. 9.3.1.

Note that we have omitted the average symbol over ρ in (5.1.7), because it is obvious that this is a macroscopic equation (as it involves ϕ and $\mathbf{q}$). In Subs. 4.2.2, we have derived this macroscopic mass balance equation also as (4.2.36) (for steady state), by using the *homogenization approach.*

Equation (5.1.7) is the second equation that we mentioned at the beginning of this chapter. We note that we also need information on ϕ and Γ^m. We have also to relate h to the pressure, p, and p to ρ (through the constitutive relationship $\rho = \rho(p)$).

For steady flow of a constant density fluid, in the absence of sources and sinks within the flow domain, the mass balance equation (5.1.7) reduces to

$$\nabla \cdot \mathbf{q} = 0. \tag{5.1.8}$$

When using (4.1.15) to express $\mathbf{q}$, and assuming K = const., we obtain the *Laplace equation*[1]

$$\nabla^2 h = 0, \quad \nabla^2 \equiv \frac{\partial^2}{\partial x^2} + \frac{\partial^2}{\partial y^2} + \frac{\partial^2}{\partial z^2}, \tag{5.1.9}$$

to be solved for $h = h(x, y, z)$.

5.1.2 Deformable porous medium

In general, the porous medium comprising an aquifer is *deformable*, i.e., it deforms under applied stresses. Obviously, this subject is of interest in soil mechanics and geotechnical engineering, especially in connection with consolidation produced by construction. As we shall show below, in groundwater hydrology, the subject is of interest because of two reasons: (1) the deformation of the solid matrix affects the storage of water in the void space, as water pressure changes, and (2) the deformation may reach such an extent that it manifests itself as *land subsidence*. We shall deal with both aspects in the current subsection, in Subs. 5.1.3 and in Sec. 5.5.

In a deformable porous medium, we have: (1) a time-dependent porosity, i.e., $\partial\phi/\partial t \neq 0$, and (2) a moving solid matrix, i.e., $\mathbf{V}_s \neq 0$, where $\mathbf{V}_s$ $(\equiv \overline{\mathbf{V}_s}^s)$ is the macroscopic velocity of the solid matrix. We recall (Sec. 4.2) that Darcy's law, e.g., (4.2.2), expresses the fluid flux relative to the (possibly moving) solid. Hence, when considering flow of water (w) through a deformable porous medium, we have to take into account also the solid's (s) velocity, i.e.,

$$\mathbf{q}_r = \phi(\mathbf{V}_w - \mathbf{V}_s). \tag{5.1.10}$$

Because the above expression involves the solid's velocity as an additional variable, we have to consider also the solid's mass balance equation. To do so, we make use of the general balance equation (5.1.5), with $e = \rho_s$, $\theta \Rightarrow (1 - \phi)$, $\mathbf{V} = \mathbf{V}_s$, $\mathbf{j}^E = 0$, $\Gamma^E = 0$, assuming that at the microscopic level, $(\mathbf{V}_s - \mathbf{u}) \cdot \mathbf{n} = 0$, i.e., the solid-fluid interface is a *material surface* with respect to the solid mass. Under these conditions, (5.1.5) reduces to

[1] Prof. T.N. Narasimhan of the University of California, Berkeley, comments (Private communication) that placing a linear flux law in the context of a balance equation that takes the form of a partial differential equation describing multidimensional diffusion, or diffusion-like phenomena, was attributed to Fourier (1822), Ohm (1827), and Fick (1855). In earth sciences, it was probably Dupuit (1863) who inserted Darcy's one-dimensional law into the mass balance equation to obtain a description of regional groundwater movement.

$$\frac{\partial}{\partial t}(1-\phi)\rho_s = -\nabla\cdot[(1-\phi)\rho_s\mathbf{V}_s]. \tag{5.1.11}$$

Recalling the physical meaning of the time derivative and of the 'divergence of a flux', this equation states that the rate of added solid mass, per unit volume of porous medium, is equal to the net mass of moving solid entering that volume, per unit volume and unit time, through the boundaries of that volume. Introducing the *material* (or *total*, or *substantial*) *time derivative* for the solid phase, defined by

$$\frac{\mathrm{D}_s(..)}{\mathrm{D}t} = \frac{\partial(..)}{\partial t} + \mathbf{V}_s\cdot\nabla(..). \tag{5.1.12}$$

Equation (5.1.11) can be rewritten in the form

$$\frac{1}{1-\phi}\frac{\mathrm{D}_s(1-\phi)}{\mathrm{D}t} + \frac{1}{\rho_s}\frac{\mathrm{D}_s\rho_s}{\mathrm{D}t} = -\nabla\cdot\mathbf{V}_s. \tag{5.1.13}$$

The deformation of the solid phase (not the solid matrix!) is usually assumed to be *volume preserving*. This means that at the *microscopic* level, $\nabla\cdot\mathbf{V}_s = 0$, and $\mathrm{D}_s\rho_s/\mathrm{D}t = 0$. Hence, at the macroscopic level, $\mathrm{D}_s\rho_s/\mathrm{D}t$ vanishes, and (5.1.13) reduces to

$$\frac{1}{1-\phi}\frac{\mathrm{D}_s}{\mathrm{D}t}(1-\phi) = -\nabla\cdot\mathbf{V}_s. \tag{5.1.14}$$

The left-hand side of (5.1.14) may be interpreted as 'the relative rate of expansion of the volume occupied by the solid phase'.

We now rewrite the (macroscopic) water (w) mass balance equation (5.1.7) in the form

$$\begin{aligned}\frac{\partial\phi\rho_w}{\partial t} &= -\nabla\cdot\phi\rho_w(\mathbf{V}_w-\mathbf{V}_s) - \nabla\cdot(\phi\rho_w\mathbf{V}_s) + \rho_w\Gamma^w \\ &= -\nabla\cdot(\rho_w\mathbf{q}_r) - \mathbf{V}_s\cdot\nabla(\phi\rho_w) - \phi\rho_w\nabla\cdot\mathbf{V}_s + \rho_w\Gamma^w \\ &= -\nabla\cdot(\rho_w\mathbf{q}_r) - \mathbf{V}_s\cdot\nabla(\phi\rho_w) + \phi\rho_w\frac{1}{1-\phi}\frac{\mathrm{D}_s(1-\phi)}{\mathrm{D}t} + \rho_w\Gamma^w,\end{aligned} \tag{5.1.15}$$

or,

$$\phi\frac{\mathrm{D}_s\rho_w}{\mathrm{D}t} + \rho_w\frac{1}{1-\phi}\frac{\mathrm{D}_s\phi}{\mathrm{D}t} = -\nabla\cdot(\rho_w\mathbf{q}_r) + \rho_w\Gamma^w, \tag{5.1.16}$$

or, in terms of the material derivative with respect to the water phase,

$$\phi\frac{\mathrm{D}_w\rho_w}{\mathrm{D}t} + \rho_w\frac{1}{1-\phi}\frac{\mathrm{D}_s\phi}{\mathrm{D}t} = -\rho_w\nabla\cdot\mathbf{q}_r + \rho_w\Gamma^w, \tag{5.1.17}$$

where

$$\frac{\mathrm{D}_w(..)}{\mathrm{D}t} = \frac{\partial(..)}{\partial t} + \mathbf{V}_w\cdot\nabla(..). \tag{5.1.18}$$

For a *stationary* ($\mathbf{V}_s = 0$) *nondeformable solid matrix*, $\mathrm{D}_s(1-\phi)/\mathrm{D}t = -\mathrm{D}_s\phi/\mathrm{D}t = 0$, equation (5.1.16) reduces to (5.1.7).

Assuming that in a deformable porous medium

$$\left|\frac{\partial \rho_w}{\partial t}\right| \gg \left|\mathbf{V}_s \cdot \nabla \rho_w\right|, \quad \left|\frac{\partial \phi}{\partial t}\right| \gg \left|\mathbf{V}_s \cdot \nabla \phi\right|, \tag{5.1.19}$$

i.e., assuming that the spatial variations are much smaller than the corresponding temporal ones, (5.1.16) reduces to mass balance equation for the water,

$$\phi\frac{\partial \rho_w}{\partial t} + \rho_w \frac{1}{1-\phi}\frac{\partial \phi}{\partial t} = -\nabla\cdot(\rho_w \mathbf{q}_r) + \rho_w \Gamma^w. \tag{5.1.20}$$

A detailed analysis of soil deformability requires the introduction of the soil (macroscopic) *volumetric strain*, or *dilatation*, ε_{sk}. Denoting the (macroscopic) displacement of the soil's solid skeleton by the vector $\mathbf{w}_s$, the soil volumetric strain, a second rank symmetric tensor, is expressed by

$$\varepsilon_{sk} = \nabla\cdot\mathbf{w}_s. \tag{5.1.21}$$

Then, with the assumption $|\partial \mathbf{w}_s/\partial t| \gg |\mathbf{V}_s \cdot \nabla \mathbf{w}_s|$, we obtain

$$\mathbf{V}_s \equiv \frac{\mathrm{D}_s \mathbf{w}_s}{\mathrm{D}t} \approx \frac{\partial \mathbf{w}_s}{\partial t}. \tag{5.1.22}$$

Equation (5.1.14) becomes

$$\frac{\partial \varepsilon_{sk}}{\partial t} = -\frac{1}{1-\phi}\frac{\mathrm{D}_s(1-\phi)}{\mathrm{D}t}, \tag{5.1.23}$$

and the mass balance equation (5.1.16) is replaced by

$$\phi\frac{\mathrm{D}_s \rho_w}{\mathrm{D}t} + \rho_w \frac{\partial \varepsilon_{sk}}{\partial t} = -\nabla\cdot(\rho_w \mathbf{q}_r) + \rho_w \Gamma^w. \tag{5.1.24}$$

The mass balance equation (5.1.20) can be rewritten as

$$\phi\frac{\partial \rho_w}{\partial t} + \rho_w \frac{\partial \varepsilon_{sk}}{\partial t} = -\nabla\cdot(\rho_w \mathbf{q}_r) + \rho_w \Gamma^w. \tag{5.1.25}$$

Finally, if we assume that

$$\left|\phi\frac{\partial \rho_w}{\partial t}\right| \gg \left|\mathbf{q}_r \cdot \nabla \rho_w\right|, \tag{5.1.26}$$

which may be interpreted as stating that the temporal rate of density change at a point is much larger than the spatial one, we may approximate $\nabla\cdot(\rho_w \mathbf{q}_r)$ in (5.1.20), (5.1.24), and (5.1.25) by $\rho_w \nabla\cdot\mathbf{q}_r$. For example, the mass balance equation for water, (5.1.20), then reduces to

$$\phi \frac{\partial \rho_w}{\partial t} + \rho_w \frac{1}{1-\phi} \frac{\partial \phi}{\partial t} = -\rho_w \nabla \cdot \mathbf{q}_r + \rho_w \Gamma^m. \tag{5.1.27}$$

For a compressible fluid, $\rho_w = \rho_w(p)$, the above equation takes the form

$$\phi \beta \frac{\partial p}{\partial t} + \frac{\partial \varepsilon_{sk}}{\partial t} = -\nabla \cdot \mathbf{q}_r + \Gamma^w, \tag{5.1.28}$$

in which β is the coefficient of fluid compressibility, defined by

$$\beta = \frac{1}{\rho_w} \frac{d\rho_w}{dp}. \tag{5.1.29}$$

Expressing $\mathbf{q}_r$ by (4.2.2), we obtain a single equation in the three variables p, ϕ and ε_{sk}. Actually, we have shown earlier that changes in ε_{sk} are associated with changes in ϕ. The second term on the left-hand side of the above equation expresses the temporal rate of change in the *volume strain* of the solid skeleton. It has to be expressed in terms of the variable(s) of the problem, e.g., in terms of the rate of change in water pressure. To achieve this goal, we shall introduce the concept of effective stress in the next subsection. Then we shall define the specific storativity of a deformable saturated porous medium, which leads to the derivation (end of Subs. 5.1.3) of a model that describes saturated flow and deformation in a three-dimensional porous medium domain.

Finally, in Subs. 5.1.4 we shall present a different form of (5.1.28), which describes the mass balance for a fluid in a deformable porous medium—an equation written in terms of a single variable, pressure or piezometric head.

5.1.3 Specific storativity

Geotechnical engineers regard the soil in the subsurface as a deformable material. They usually apply to such material the concepts and terminology of solid mechanics. For example, they use stress and strain in order to investigate deformation, e.g., as manifested by the phenomenon of consolidation. Hydrogeologists have introduced the concept of specific storativity to express the changes in the mass of water stored in a formation as a consequence of pressure changes, taking into account the compressibility of the water and the deformability of the soil.

The term on the left-hand side of (5.1.7) represents the *mass of water added to a unit volume of porous medium per unit time*. This term can also be written as

$$\frac{\partial}{\partial t}(\phi \rho_w) = \phi \frac{\partial \rho_w}{\partial t} + \rho_w \frac{\partial \phi}{\partial t}. \tag{5.1.30}$$

We note that two effects contribute to the added water mass: water compressibility and porous medium deformability.

In general, the density of water depends on pressure, p, solute concentration, c, and temperature, T, i.e., $\rho_w = \rho_w(p, c, T)$. Thus,

$$\frac{\partial \rho_w}{\partial t} = \frac{\partial \rho_w}{\partial p}\frac{\partial p}{\partial t} + \frac{\partial \rho_w}{\partial c}\frac{\partial c}{\partial t} + \frac{\partial \rho_w}{\partial T}\frac{\partial T}{\partial t}. \tag{5.1.31}$$

In this section, we shall restrict the discussion to the case in which the density of water depends on pressure only, i.e., $\rho_w = \rho_w(p)$. Then,

$$\frac{\partial \rho_w}{\partial t} = \frac{\partial \rho_w}{\partial p}\frac{\partial p}{\partial t} = \rho_w \beta \frac{\partial p}{\partial t}, \tag{5.1.32}$$

where β is *water compressibility*, defined in (5.1.29).

To develop the second term on the right-hand side of (5.1.30), we start from the assumption (already introduced earlier) that the density of the solid, ρ_s, *not of the solid matrix*, remains unchanged as the porosity undergoes changes. Given a fixed mass of solid matrix, m_s, this means that $\partial \mathcal{U}_s/\partial t = 0$, where $\mathcal{U}_s$ denotes the volume of solid. Recall that the entire discussion here is at the macroscopic level.

To facilitate the discussion, we have to introduce here the concept of *effective stress*, commonly used in geotechnical engineering.

By averaging the momentum balance equations for the solid and fluid phases within an REV, neglecting the inertial terms and terms that express friction *within* the fluids, and summing up the equations for the individual phases (in order to eliminate terms representing momentum exchange across fluid-solid interfaces), we obtain (Bear and Bachmat, 1990) the *equilibrium equation*

$$\nabla \cdot \overline{\boldsymbol{\sigma}} + \overline{\rho \mathbf{F}} = 0, \quad \overline{\rho \mathbf{F}} = \rho_w \phi + \rho_s (1-\phi), \tag{5.1.33}$$

in which, $\overline{\rho \mathbf{F}}$ $(= -\overline{\rho} g \nabla z)$ denotes the *phase average* (Sec. 1.3) gravity force acting on the total mass within the REV, and $\overline{\boldsymbol{\sigma}}$ is the phase averaged *total stress* at a point (in a three-dimensional domain), given by

$$\overline{\boldsymbol{\sigma}} = \frac{1}{\mathcal{U}_o}\int_{\mathcal{U}_o} \boldsymbol{\sigma}\, d\mathcal{U} = \frac{1}{\mathcal{U}_o}\sum_{\alpha=w,s}\int_{\mathcal{U}_{o\alpha}} \boldsymbol{\sigma}_\alpha\, d\mathcal{U}_\alpha = \overline{\boldsymbol{\sigma}_w} + \overline{\boldsymbol{\sigma}_s}. \tag{5.1.34}$$

All stresses are second rank symmetric tensors.

For a saturated porous medium, neglecting the shear stress in the fluid, the total stress in (5.1.33) takes the form

$$\overline{\boldsymbol{\sigma}} = (1-\phi)\overline{\boldsymbol{\sigma}_s}^s - \phi \overline{p_w}^w \mathbf{I}. \tag{5.1.35}$$

Our objective is to determine the stress that produces the strain of the solid matrix in a porous medium domain. The knowledge of this strain is required in order to deal with the deformation of a porous medium.

In order to deal with the deformation of a porous medium, we use the concept of *effective stress* introduced in soil mechanics by Terzaghi (1923). Essentially, this concept assumes that in a granular porous medium, the pressure in the water (or in the fluids, in a multiphase system) that *almost* completely surrounds each solid grain, produces in the latter a stress of equal

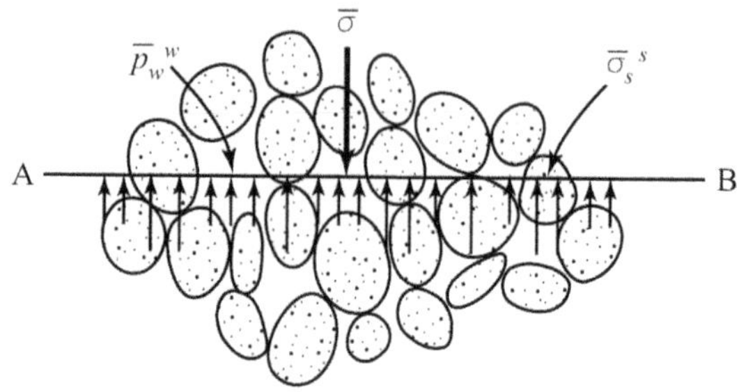

Figure 5.1.2: Nomenclature for the definition of Terzaghi's effective stress.

magnitude, without contributing to the deformation of each grain. Instead, deformation occurs mainly due to the forces at the points of contact between the grains. At these points, concentrated normal and shear forces are transmitted from grain to grain. Thus, the (macroscopic) *strain-producing stress*, or *intergranular stress*, or *effective stress*, is obtained by subtracting the water pressure (or, in a multiphase system, the average pressure of the fluids in the void space) from the stress in the solid material (all stresses and pressures being averaged ones).

The deformation of the solid matrix is produced mainly by the rearrangement of the grains, with localized slipping and rolling. This deformation is, largely, irreversible. A change in water pressure, with an equal change in total stress, produces no deformation and, hence, should produce no change in the effective stress. In considering porous medium deformation, the deformation of the solid itself is neglected

To illustrate the concept of effective stress in a simple way, let us limit the discussion for the moment to *vertical forces only*, and consider the vertical cross-section through a saturated porous medium domain and the horizontal unit area, AB, shown in Fig. 5.1.2.

In this section we shall make use of the symbols that indicate average, or macroscopic values, in order to emphasize the meaning of these averaged values.

At every instant, the load acting on the upper side of AB is due to the weight of soil, water, and whatever load exists at ground surface. This load, which produces a (macroscopic) stress, $\overline{\sigma}$ (= total force per unit area of porous medium), must be in equilibrium with two stresses acting on AB from below: a stress, $\phi\overline{p_w}^w$, resulting from the (average) pressure in the water, $\overline{p_w}^w$, acting on the water portion of AB, and a stress $(1-\phi)\overline{\sigma_s}^s$, resulting from the (average) stress, $\overline{\sigma_s}^s$, in the solid skeleton, acting on the solid portion of AB. Both $\overline{p_w}^w$ and $\overline{\sigma_s}^s$ are intrinsic phase averages, while $\overline{\sigma}$ is a volume average of σ. It can be shown (e.g., Bear and Bachmat, 1990) that these intrinsic phase averages are equal to intrinsic areal averages, that is, $\overline{p_w}^w$ and $\overline{\sigma_s}^s$ also express force per unit area of water and per unit area of solid, respectively. The shear stress in the water has been neglected.

With the stresses $\overline{\sigma}$ and $\overline{\sigma_s}^s$ taken as positive for tension, while pressure in the water taken as positive for compression (as is common in fluid mechanics), the above statement of force (actually, stress) equilibrium can be expressed in the form:

$$\overline{\sigma} = (1-\phi)\overline{\sigma_s}^s - \phi\overline{p_w}^w. \tag{5.1.36}$$

In order to express the above equation in terms of Terzaghi's effective stress, $\overline{\sigma'_s}$, we add $\overline{p_w}^w$ to both sides of (5.1.36), obtaining

$$\overline{\sigma'_s} = (1-\phi)(\overline{\sigma_s}^s + \overline{p_w}^w), \tag{5.1.37}$$

where

$$\overline{\sigma'_s} = \overline{\sigma} + \overline{p_w}^w. \tag{5.1.38}$$

Although (5.1.36) through (5.1.38) are based on the *simplification of vertical stress only*, they can easily be extended to three-dimensional domains in the following manner. The (phase average) stress in the water may be decomposed into two parts, using the definition

$$\overline{\boldsymbol{\sigma}_w}^w = \overline{p_w}^w\mathbf{I} + \overline{\boldsymbol{\tau}_w}^w, \tag{5.1.39}$$

where $\overline{\boldsymbol{\tau}_w}^w$, is the (intrinsic phase average) shear (or deviatoric) stress in the water. Here, $\overline{\boldsymbol{\sigma}_w}^w$ and $\overline{\boldsymbol{\tau}_w}^w$ are second rank symmetric tensors, and $\mathbf{I}$ is the unit tensor.

Neglecting $\overline{\boldsymbol{\tau}_w}^w$, and defining the effective stress as

$$\overline{\boldsymbol{\sigma}'_s} = (1-\phi)(\overline{\boldsymbol{\sigma}_s}^s + \overline{p_w}^w\mathbf{I}), \tag{5.1.40}$$

the total stress in a three-dimensional saturated medium is expressed by

$$\overline{\boldsymbol{\sigma}} = \overline{\boldsymbol{\sigma}'_s} - \overline{p_w}^w\mathbf{I}, \tag{5.1.41}$$

where $\overline{\boldsymbol{\sigma}}$ and $\overline{\boldsymbol{\sigma}'_s}$ are second rank symmetric tensors.

When changes take place, either in the external load (producing changes in the total stress distribution, $\overline{\boldsymbol{\sigma}}$), or in the water pressure, $\overline{p_w}^w$, e.g., by pumping, we have

$$d\overline{\boldsymbol{\sigma}} = d\overline{\boldsymbol{\sigma}'_s} - d\overline{p_w}^w\mathbf{I}. \tag{5.1.42}$$

We can now return to the discussion on specific storativity.

In the case of *vertical stresses only*, considered above, omitting the overbar symbol for an average, the statement that the solid is not deformable, is expressed by

$$\frac{\partial \mathcal{U}_s}{\partial \sigma'_s} = 0, \tag{5.1.43}$$

where σ'_s is the vertical effective stress. With $\mathcal{U}_{\rm pm}$ $(= \mathcal{U}_s/(1-\phi))$ denoting the porous medium volume containing $\mathcal{U}_s$, we can rewrite (5.1.43) as

$$\frac{\partial \mathcal{U}_s}{\partial \sigma'_s} \equiv (1-\phi)\frac{\partial \mathcal{U}_{\rm pm}}{\partial \sigma'_s} + \mathcal{U}_{\rm pm}\frac{\partial (1-\phi)}{\partial \sigma'_s} = 0. \tag{5.1.44}$$

Hence, in view of (5.1.38), written for the vertical direction only, and *assuming no change in the total stress*, i.e., $d\sigma = 0$, and $d\sigma'_s = dp$, we may write

$$\frac{1}{\mathcal{U}_{\rm pm}}\frac{\partial \mathcal{U}_{\rm pm}}{\partial \sigma'_s} = \frac{1}{1-\phi}\frac{\partial \phi}{\partial \sigma'_s} = \frac{1}{1-\phi}\frac{\partial \phi}{\partial p}. \tag{5.1.45}$$

At this point, we assume that we deal with relatively small volume changes, and that the soil may be assumed to behave as an *elastic material.* The *coefficient of soil compressibility*, α, is defined for this case of vertical stresses only, as

$$\alpha = \frac{1}{\mathcal{U}_{\rm pm}}\frac{\partial \mathcal{U}_{\rm pm}}{\partial \sigma'_s} = \frac{1}{1-\phi}\frac{\partial \phi}{\partial p}. \tag{5.1.46}$$

The coefficient α can be determined in a laboratory experiment with a fixed mass of soil, and a *representative volume* of soil. Sometimes, soil compressibility is expressed by a *coefficient of rock compressibility*, α', such that

$$\phi = \phi_o\left[1 + \alpha'(p - p_o)\right], \tag{5.1.47}$$

where p_o and ϕ_o are reference values of pressure and porosity, respectively.

We now return to the second term on the right-hand side of (5.1.30). Making use of (5.1.46), we obtain

$$\frac{\partial \phi}{\partial t} = (1-\phi)\alpha\frac{\partial p}{\partial t}. \tag{5.1.48}$$

With the above equation, we can now rewrite (5.1.30) in the form

$$\frac{\partial}{\partial t}(\phi\rho_w) = \rho_w\left[\phi\beta + (1-\phi)\alpha\right]\frac{\partial p}{\partial t} \equiv S^{m*}_{op}\frac{\partial p}{\partial t}. \tag{5.1.49}$$

Recalling the physical interpretation of the left-hand side of the above equation, S^{m*}_{op} can be interpreted as the *specific mass storativity*, here for a saturated porous medium. It is defined as the *mass of water released from (or added to) storage in a unit volume of a deformable porous medium per unit decline (or rise) in water pressure.* We have used the superscript m and subscript p to indicate that this is a specific *mass* storativity associated with *pressure* changes, as several other types of storativities will be defined below.

One may easily define a specific storativity with respect to changes in piezometric head (actually, in terms of Hubbert's potential, h^*, defined in (4.1.6)), by making use of the relationship (6.3.22)

$$\frac{\partial h^*}{\partial t} = \frac{1}{\rho_w(p)g}\frac{\partial p}{\partial t}.$$

Thus, groundwater hydrologists define a *specific storativity* (for saturated flow),

$$S_o^* \equiv gS_{op}^{m*} = \rho_w g[\phi\beta + (1-\phi)\alpha], \tag{5.1.50}$$

as the *volume of water released from (or added to) storage in a unit volume of porous medium, per unit decline (or rise) in the piezometric head* (e.g., Bear, 1972),

$$S_o^* = \frac{\Delta \mathcal{U}_w}{\mathcal{U}_{\text{pm}}\Delta h}. \tag{5.1.51}$$

Here,

$$\rho_w S_o^* \frac{\partial h^*}{\partial t} = S_{op}^{m*}\frac{\partial p}{\partial t}. \tag{5.1.52}$$

Again, the above discussion can be extended to a three-dimensional domain.

Following the above discussion, we may now rewrite the mass balance equation (5.1.7) in the form

$$S_{op}^m \frac{\partial p}{\partial t} = -\nabla\cdot(\rho_w \mathbf{q}_r) + \rho_w \Gamma^w, \quad S_{op}^m = \rho_w(\phi\beta + \alpha). \tag{5.1.53}$$

We can now write the following mass balance equations for water in saturated flow, in terms of the piezometric head, h, in the form:

$$\rho_w S_o \frac{\partial h}{\partial t} = -\nabla\cdot(\rho_w \mathbf{q}_r) + \rho_w \Gamma^w, \quad S_o = \rho_w g(\phi\beta + \alpha), \tag{5.1.54}$$

in which S_o is another form for the *specific (volume) storativity.* Note that by expressing $\mathbf{q}_r$ in terms of h, the above equation contains only a single variable, h, to be solved for.

We note the difference between the expressions for S_o and S_o^* defined in (5.1.50). This difference is explained by the difference between $\mathbf{q}_r$ and $\mathbf{q}_w$ appearing in the divergence term in the mass balance equation. In soil mechanics, an *undrained test* (on a saturated soil sample) is one in which water is (practically) stationary relative to the solids, i.e., $\mathbf{q}_r = 0$. Then, any added water goes only into storage in the sample, raising the pressure of the water in the sample. Accordingly, S_o $(= \rho_w(\alpha + \phi\beta))$ may be considered to be another definition for specific storativity, this time under conditions equivalent to those prevailing in an undrained test. Thus, S_o is applicable to a coordinate system that moves with the solid phase, while S_o^* is appropriate for a reference frame in which solid and fluids move about freely.

When we invoke assumption (5.1.26), i.e., $|\phi\partial\rho_w/\partial t| \gg |\mathbf{q}_r\cdot\nabla\rho_w|$, (5.1.54) is replaced by the simpler form

$$S_o \frac{\partial h}{\partial t} = -\nabla\cdot\mathbf{q}_r + \Gamma^w. \tag{5.1.55}$$

This equation, with specific storativity, S_o, defined by (5.1.51), i.e., a verbal definition identical to that of S_o^*, is the one commonly used to describe the movement of water under saturated flow conditions. Actually, hydrologists use (5.1.55) with S_o defined by the right-hand side of (5.1.50), and with $\mathbf{q}_r$ replaced by $\mathbf{q}_w$.

We wish to emphasize again that underlying (5.1.55) is the assumption that water density is assumed constant, *except* in the expression for the specific storativity, S_o, where we do take into account water compressibility.

Altogether, we have achieved our goal of expressing the mass balance equation for a compressible fluid in a deformable porous medium in terms of a single equation written in terms p and $\mathbf{q}_r$. In the next subsection we shall use Darcy's law, thus leading to a single equation in a single variable.

Before moving to the next subsection, let us present a three-dimensional model of saturated flow and deformation.

Following Verruijt (1969), we separate the total stress, $\boldsymbol{\sigma}$, the effective stress, $\boldsymbol{\sigma}'_s$, the pressure p, and the body force $\mathbf{f}(\equiv \overline{\rho \mathbf{F}})$ into initial steady-state values, $\boldsymbol{\sigma}^o$, ${\boldsymbol{\sigma}'_s}^o$, p^o and $\mathbf{f}^o$, and deformation-producing increments, $\boldsymbol{\sigma}^e$, ${\boldsymbol{\sigma}'_s}^e$, p^e and $\mathbf{f}^e$, with

$$\boldsymbol{\sigma}^o = {\boldsymbol{\sigma}'_s}^o - p^o\mathbf{I}, \qquad \text{and} \qquad \boldsymbol{\sigma}^e = {\boldsymbol{\sigma}'_s}^e - p^e\mathbf{I}. \tag{5.1.56}$$

We assume, as a good approximation, that the body force, $\mathbf{f}$, remains unchanged, although ϕ and ρ_w do vary, i.e., $\mathbf{f}^e = 0$. Then, the equilibrium equation for the initial steady state, is

$$\nabla \cdot {\boldsymbol{\sigma}'_s}^o + \mathbf{f}^o - \nabla p^o = 0. \tag{5.1.57}$$

For the incremental (deformation producing) effective stress and pressure, we have

$$\nabla \cdot \boldsymbol{\sigma}^e \equiv \nabla \cdot {\boldsymbol{\sigma}'_s}^e - \nabla p^e = 0. \tag{5.1.58}$$

We now make the assumptions that the solid matrix is isotropic and, for the relatively small excess effective stresses considered here, is made of a perfectly elastic material that obeys the macroscopic strain-stress relationship

$${\boldsymbol{\sigma}'_s}^e = \mu'_s\{\nabla \mathbf{w} + (\nabla \mathbf{w})^T\} + \lambda''_s(\nabla \cdot \mathbf{w})\mathbf{I}, \tag{5.1.59}$$

or, in indicial notation

$$({\sigma'_s}^e)_{ij} = \mu'_s\left(\frac{\partial w_i}{\partial x_j} + \frac{\partial w_j}{\partial x_i}\right) + \lambda''_s\left(\frac{\partial w_k}{\partial x_k}\right)\delta_{ij} = 2\mu'_s\varepsilon_{ij} + \lambda''_s\varepsilon\delta_{ij}, \tag{5.1.60}$$

where ε_{ij} denotes a component of the strain tensor, $\boldsymbol{\varepsilon}$, $\varepsilon \equiv \varepsilon_{sk}$, and μ'_s and λ''_s are macroscopic constant coefficients called the *Lamé's coefficients of the solid skeleton* (see any text on theory of elasticity). Their values must be obtained experimentally. Only the incremental effective stress causes displacement. Here, $\mathbf{w}$ denotes the macroscopic displacement vector $\overline{\mathbf{w}_s}^s$.

By inserting (5.1.60) into (5.1.58), we obtain

$$\frac{\partial}{\partial x_i}\left[\mu_s'\left(\frac{\partial w_i}{\partial x_j}+\frac{\partial w_j}{\partial x_i}\right)+\lambda_s''\frac{\partial w_k}{\partial x_k}\delta_{ij}\right]-\frac{\partial p^e}{\partial x_i}\delta_{ij}=0, \tag{5.1.61}$$

to be used for determining the *displacement vector*, $\mathbf{w}$.

The mass balance equation (5.1.25) may also be rewritten as two balance equations, one representing the initial steady state (with variables denoted by superscript o), and the other, involving increments of pressure (with variables denoted by superscript e), that produce displacements. Thus, the second equation may be written in the form

$$\nabla\cdot(\rho_w\mathbf{q}_r^e)+\phi\rho\beta\frac{\partial p^e}{\partial t}+\rho\frac{\partial\varepsilon_{sk}}{\partial t}=0, \tag{5.1.62}$$

where $\varepsilon_{sk}^e \equiv \varepsilon_{sk}$ since $\varepsilon_{sk}^o \equiv 0$, and, for the isotropic porous medium considered here,

$$\mathbf{q}_r^e=-\frac{\mathsf{k}}{\mu}(\nabla p^e+\rho_w g\nabla z). \tag{5.1.63}$$

In writing (5.1.62) and (5.1.63), we have introduced the approximations

$$\rho_w=\rho_w^o+\rho_w^e\simeq\rho_w^o,\quad \rho_w^e\ll\rho_w^o,\quad \mu^e\ll\mu^o,\quad \phi=\phi^o+\phi^e\simeq\phi^o,\quad \phi^e\ll\phi^o.$$

We also assume that the permeability k remains unchanged in spite of the deformation that takes place.

By inserting the expression for $\mathbf{q}_r^e$ into (5.1.62), we obtain the mass balance equation for a compressible fluid phase in a deformable, isotropic and linearly elastic porous medium, in the form

$$-\nabla\cdot\left[\rho_w\frac{\mathsf{k}}{\mu}(\nabla p^e+\rho_w g\nabla z)\right]+\phi\rho_w\beta\frac{\partial p^e}{\partial t}+\rho_w\frac{\partial\varepsilon_{sk}}{\partial t}=0. \tag{5.1.64}$$

This is a single equation in two variables p^e and ε_{sk}.

The complete set of equations describing the flow of a single compressible Newtonian fluid (ρ,μ) in a deformable isotropic porous medium consists now of the equations and relationships summarized in Table 5.1.1. From this table, it follows that we have a sufficient number of equations to solve for the various dependent variables involved. In principle, this is the model introduced by Biot (1941). We note that this model also yields the displacement vector, $\mathbf{w}$. It can be used, for example, for determining soil consolidation and land subsidence.

As an example, consider a homogeneous isotropic porous medium, with λ_s'', $\mu_s'=$ const. We rewrite (5.1.61) in the form of the three equations

$$\mu_s'\frac{\partial^2 w_i}{\partial x_j\partial x_j}+(\lambda_s''+\mu_s')\frac{\partial\varepsilon_{sk}}{\partial x_i}-\frac{\partial p^e}{\partial x_i}=0,\qquad i=1,2,3. \tag{5.1.65}$$

Equations		Additional Dependent Variables
Mass balance equation for the fluid, (5.1.25)–(1 eqn.)	$\nabla \cdot \rho_w \mathbf{q}_r^e + \phi \frac{\partial \rho_w}{\partial t} + \rho_w \frac{\partial \varepsilon_{sk}}{\partial t} = 0$	ρ_w, ϕ, $\mathbf{q}_r^e$, ε_{sk} (6 vars.)
Equation of motion for the fluid (5.1.63)–(3 eqns.)	$\mathbf{q}_r^e = -\frac{\mathsf{k}}{\mu}(\nabla p^e + \rho_w g \nabla z)$	p^e (1 var.)
Equilibrium relationships (5.1.58)–(3 eqns.)	$\nabla \cdot \boldsymbol{\sigma}_s'^{\,e} - \nabla p^e = 0$	$\boldsymbol{\sigma}_s'^{\,e}$ (6 vars.)
Stress-strain relationships for the solid matrix (5.1.59)–(6 eqns.)	$\boldsymbol{\sigma}_s'^{\,e} = \overline{\mu_s'}\left[\nabla \mathbf{w} + (\nabla \mathbf{w})^T\right] + \overline{\lambda_s''}(\nabla \cdot \mathbf{w})\mathbf{I}$	$\mathbf{w}$ (3 vars.)
Dilatation-displacement relations (5.1.21)–(1 eqn.)	$\varepsilon_{sk} = \nabla \cdot \mathbf{w}$	(none)
Equation of state for the fluid–(1 eqn.)	$\rho_w = \rho_w(p)$	(none)
Dilatation-porosity relation (5.1.23)–(1 eqn.)	$\dot{\varepsilon}_{sk} = \frac{1}{1-\phi}\dot{\phi}$	(none)
Total: 16 equations		16 (scalar) variables

Table 5.1.1: Summary of balance equations and constitutive relations for Darcian, saturated flow of a compressible Newtonian fluid in an isotropic linearly elastic porous medium (Bear and Bachmat, 1990).

By differentiating each of these equations with respect to the corresponding x_i, and adding the resulting three equations, we obtain the single equation (Verruijt, 1969)

$$(\lambda_s'' + 2\mu_s')\nabla^2 \varepsilon_{sk} - \nabla^2 p^e = 0, \tag{5.1.66}$$

which, together with (5.1.28) and (4.2.2), often simplified for a homogeneous isotropic porous medium to the form

$$-\frac{k}{\mu}\nabla^2 p^e + \phi\beta \frac{\partial p^e}{\partial t} + \frac{\partial \varepsilon_{sk}}{\partial t} = 0, \tag{5.1.67}$$

constitute two equations in the variables p^e and ε_{sk}.

Following Verruijt (1969), we integrate (5.1.66), and obtain

$$(\widetilde{\lambda_s''} + 2\widetilde{\mu_s'})\widetilde{\varepsilon_{sk}} = \widetilde{p^e} + \Pi(\mathbf{x}, t), \tag{5.1.68}$$

where Π is a function of position and time that for every value of time, t, satisfies

$$\nabla^2 \Pi = 0. \tag{5.1.69}$$

When $\Pi \equiv 0$ (see below), we may insert

$$\widetilde{\varepsilon_{sk}} = \frac{\widetilde{p^e}}{\widetilde{\lambda_s''} + 2\widetilde{\mu_s'}} \tag{5.1.70}$$

in (5.1.67), and obtain (omitting the averaging symbol)

$$\frac{k}{\mu}\nabla^2 p^e = \left(n\beta + \frac{1}{\lambda_s'' + 2\mu_s'}\right)\frac{\partial p^e}{\partial t} \equiv (n\beta + \alpha)\frac{\partial p^e}{\partial t}, \tag{5.1.71}$$

which is a simple (diffusion-type) mass balance equation commonly employed in hydraulics of groundwater for determining the pressure distribution. In the above, we have

$$\alpha = \frac{1}{\lambda_s'' + 2\mu_s'}, \tag{5.1.72}$$

which may be interpreted as a *coefficient of porous medium compressibility.* This is the same coefficient α defined in (5.1.46) that appears in the definition of specific storativity, (5.1.50).

As pointed out by Verruijt (1969, p. 348), the function Π 'describes the deviation of the simplified Terzaghi-Jacob theory from the Biot theory', where the former assumes vertical consolidation only, while the latter takes into account the three-dimensional nature of consolidation. Here, Π expresses the deviation in the integrated approach to aquifer consolidation. In principle, however, horizontal displacements do take place. Their effect in hydrology may be negligible, but as part of consolidation, their damage may be significant. A discussion on modeling land subsidence is presented in Sec. 5.5.

5.1.4 Flow equations

We use the term *flow equation* when referring to a mass balance equation for a fluid phase, combined with the appropriate form of Darcy's law (= motion equation). The objective is to obtain a single equation for that phase, written in terms of a single state variable, such as pressure or piezometric head. As examples, we shall use the mass balance equation for three-dimensional saturated flow of water in the forms of (5.1.53) or (5.1.54).

The equation of motion (Darcy's law) for saturated flow is presented in Chap. 4. In particular, we shall use two forms of Darcy's law: (4.2.2) in terms of pressure, and (4.2.53) in terms of Hubbert's potential. Usually, we assume a constant fluid density, and replace Hubbert's potential, h^*, by the piezometric head, h.

With the above in mind, we may start from the mass balance equation (5.1.53), recalling all the assumptions that underlie it, and rewrite it for the flow of a compressible fluid, $\rho_w = \rho_w(p)$, in a deformable porous medium, in the form:

$$S_{op}^m \frac{\partial p}{\partial t} = \nabla \cdot \left[\rho_w \frac{\mathbf{k}}{\mu} \cdot (\nabla p + \rho_w g \nabla z) \right] + \rho_w \Gamma^w, \tag{5.1.73}$$

where $\mathbf{k}$ is the permeability (tensor), Γ^w represents a possible water source, and the *specific mass storativity*, S_{op}^m, is defined in (5.1.53), recalling that superscript m and subscript p indicate that this is a specific *mass* storativity associated with *pressure* changes. The pressure, $p = p(\mathbf{x}, t)$, is the variable for which a solution is sought by solving this equation.

If we invoke assumption (5.1.26), which is usually justified in practice, (5.1.73) reduces to

$$S_{op} \frac{\partial p}{\partial t} = \nabla \cdot \left[\frac{\mathbf{k}}{\mu} \cdot (\nabla p + \rho_w g \nabla z) \right] + \Gamma^w, \quad S_{op} = (\phi \beta + \alpha). \tag{5.1.74}$$

This is a single equation in terms of pressure, p, as a single variable (although we have to make use also of the constitutive equation $\rho = \rho(p)$).

In modeling flow in aquifers, it is convenient to use the piezometric head, h, instead of pressure, as the state variable. Then, defining *specific (volume) storativity*

$$S_o \equiv g S_{op}^m = \rho_w g \, (\phi \beta + \alpha), \tag{5.1.75}$$

(dims. L^{-1}) as the *volume of water released from (or added to) storage per unit volume of porous medium per unit decline (or rise) in the piezometric head*, we obtain

$$S_o \frac{\partial h}{\partial t} = \nabla \cdot (\mathbf{K} \cdot \nabla h) + \Gamma^w. \tag{5.1.76}$$

This is the most commonly used flow equation for describing three-dimensional, saturated flow in terms of the single variable, piezometric head, $h = h(\mathbf{x}, t)$. In a homogeneous domain, the quotient $\mathbf{K}/S_o$ is sometimes referred to as the *(flow) diffusivity of the porous medium* (in analogy to the coefficient that appears in the balance equation for molecular diffusion).

Note that, actually, we do not really consider here the deformation of the porous medium. Instead, as is common in groundwater hydrology, we make the assumption that the fluid compressibility and the effect of solid deformation can be incorporated in the scalar coefficient of specific storativity.

In (5.1.76), Γ^w (dims. T^{-1}) represents the strength of fluid sources. When these sources take the form of point sources of strength $Q^i(\mathbf{x}, t)$ (dims. L^3/T) located at points $\mathbf{x}_i$, we use the *Dirac delta-function* (or the *Dirac distribution*, or the *unit impulse*), $\delta(\mathbf{x} - \mathbf{x}_i)$ (dims. L^{-3}), defined formally by

$$\delta(\mathbf{x}-\mathbf{x}_i) = \lim_{a\to 0}\begin{cases} 1/a^3 & |\mathbf{x}-\mathbf{x}_i| < a, \\ 0 & \text{elsewhere,} \end{cases} \tag{5.1.77}$$

to express the source term in the form:

$$\Gamma^w = \sum_{(i)} Q^i(\mathbf{x}_i, t)\delta(\mathbf{x}-\mathbf{x}_i). \tag{5.1.78}$$

Actually, Q^i may represent both pumping wells (with $Q_i < 0$) and recharge wells (with $Q^i > 0$).

For an artesian well, the rate of discharge is dictated by the drawdown. Specifically, the rate for a well at (x_m, y_m, z_m) can be expressed as $Q_m = \alpha_m[h(x_m, y_m, z_m, t) - \zeta^*_m]$ in which α_m is a coefficient that relates drawdown to discharge in the artesian well (e.g., Bear, 1979, p. 326), ζ^*_m denotes the elevation of ground surface, or of the well's outlet, and $Q_m = 0$ when $h(x_m, y_m, z_m, t) \leq \zeta^*_m$.

In Subs. 9.3.1C, we shall return to the discussion of point sources. There, in connection with variable density flow, we shall make a distinction between the density of pumped water and injected water.

Additional forms of the fundamental mass balance equation

In indicial notation, (5.1.76) takes the form

$$S_o\frac{\partial h}{\partial t} = \frac{\partial}{\partial x_i}\left(\mathsf{K}_{ij}\frac{\partial h}{\partial x_j}\right) + \Gamma^w. \tag{5.1.79}$$

For an homogeneous isotropic porous medium, this equation becomes

$$S_o\frac{\partial h}{\partial t} = \mathsf{K}\left(\frac{\partial^2 h}{\partial x_i \partial x_i}\right) + \Gamma^w. \tag{5.1.80}$$

Finally, for steady state, or when we may neglect the specific storativity, $S_o \approx 0$, and in the absence of sources, this equation reduces to the *Laplace equation* (5.1.9), i.e.,

$$\frac{\partial^2 h}{\partial x_i \partial x_i} = 0, \quad \text{or} \quad \frac{\partial^2 h}{\partial x^2} + \frac{\partial^2 h}{\partial y^2} + \frac{\partial^2 h}{\partial z^2} \equiv \nabla^2 h = 0. \tag{5.1.81}$$

5.2 Initial and Boundary Conditions

Each of the (macroscopic) partial differential equations presented in the previous subsection describes the mass balance of water, either in terms of p or in terms of h. In subsequent chapters, we shall see additional equations, describing balances of mass of water in the unsaturated zone and of chemical components in the water. Often, the balance equation contains more than a single variable. For example, as already mentioned earlier, the mass balance equation contains also the mass density, and we have to make use of the constitutive equation $\rho = \rho(p)$. The flow equation contains also the soil permeability, and the fluid's density and viscosity

All these balance equations contain no information related to any *particular* case of water flow or of contaminant transport, because a balance equation itself contains no information on the shape of the boundaries of the problem domain, nor on the behavior of the specific materials (solid matrix and fluids) involved.

Accordingly, for a balance equation to describe (and, thus, to be solved for) a *particular* case of interest, it has to be supplemented by the following information:

- *Constitutive equations* that provide information on the behavior of the solid and fluid phases involved in the considered case (here, $\rho = \rho(p)$, and in a deformable porous medium, a constitutive equation for solid matrix deformation). We have tried to avoid all these by introducing the specific storativity.
- Functions that represent the rate of (positive or negative) production (= pumping or injection) of water.
- Information on the numerical values of all the coefficients that appear in the constitutive and balance equations and in the source terms.

When all of this information is put together, we obtain a *closed set of equations*, i.e., a set in which the number of equations equals that of the variables to be solved for. In the case considered here, this set contains a single partial differential equation and a number of algebraic equations. The solution provides the future spatial distributions of the value(s) of the state variable(s) within the domain.

However, this closed set of equations generally has an infinite number of solutions. To obtain a unique one that corresponds to a particular case of interest, it is necessary to provide supplementary information:

(a) The configuration of the boundaries of the domain within which the flow takes place.
(b) For unsteady flow, a description of the initial state of the domain (= *initial conditions*) in terms of the state variables.
(c) A description of the interaction of the fluid within the domain with its environment, i.e., conditions on the boundaries specified in (a). These conditions are referred to as *boundary conditions.*

5.2.1 Boundary surface

Any closed surface may serve as a boundary of an investigated domain, provided we can state the conditions that prevail on it. It is, therefore, convenient, but not mandatory, to select natural boundaries for a problem domain, e.g., an impervious geological formation, a lake, or a river.

In the strict continuum sense, at the macroscopic level, *sharp boundaries* that separate a porous medium domain from its environment, or that delineate subdomains of different media, do not exist. By taking averages over REV's in the vicinity, and across any boundary between two different media, e.g., between two porous media (with $\phi_1 \neq \phi_2$), between a porous medium and an adjacent domain of solid without voids ($\phi = 0$), or between a porous medium and a body of fluid ($\phi = 1$), we obtain a gradual transition in the averaged solid matrix properties (see discussion in Bear and Bachmat, 1990, p. 232). Usually, no information is available on how the averaged values of a considered property vary within this transition zone. However, we recall that in defining a porous medium in Sec. 1.3, we required: (a) the existence of an REV the size of which is much smaller than the size of the domain, and (b) that the variation of any macroscopic quantity (e.g., porosity) over the REV be *linear*, or approximately so. If these conditions are satisfied, the actual variation in porosity across the transition zone may be approximated as an *idealized boundary in the form of a surface across which an abrupt change in porosity takes place.* The sharp boundary may be arbitrarily located at any point within the transition region. For convenience, however, we usually locate this surface at the point corresponding to the mean value of the property between the two adjacent regions. These considerations are applicable also to an impervious boundary, i.e., when the external domain is impervious ($\phi = 0$).

The sharp boundary surfaces introduced in this way, divide the entire space into subdomains; the continuum approach is applicable to each of them. Across the boundaries, we *assume* the existence of a jump in porosity and in other macroscopic solid matrix properties. On the two sides of each such a boundary, the values of these properties are obtained by extrapolating the spatial linear trend in the averaged property values as the boundary is approached from within each subdomain.

Rigorously, the behavior close to an impervious boundary, say, within a boundary layer of thickness equal to half the size of an REV, *cannot* be described by the continuum approach, as, *within* such layer, we do not have the REV required to obtain averaged values. For example, within the framework of the continuum approach as described here, a porosity, *phi* cannot be defined within this boundary layer. In most cases, we extrapolate the value of ϕ from the interior of the domain. According to the definition of an REV, the width of the boundary domain must be much smaller than the size of the domain itself, so that the effect (on the solution) of the error resulting from extrapolating the value of porosity from the interior to the boundary should be negligible. The only way to study what happens *within* the boundary layer is to do so at the *microscopic level.*

So far, we have considered boundaries which are either some arbitrary surfaces, or surfaces of (hypothetical) discontinuities in solid matrix properties. However, a sharp boundary may be introduced *as an approximation* in two additional cases.

A fictitious abrupt boundary between two miscible fluids. Here, the 'two miscible fluids' may be the same fluid, but with significantly different concentrations of certain components. The transport of these components is considered in detail in Chap. 7. At this point, it is sufficient to note that, usually, in reality, a transition zone is created between two adjacent domains with different solute concentration in the fluid that occupies the void space. The concentration varies gradually across this transition zone. When the latter is narrow, relative to the dimensions of the two domains of interest, we may *approximate* it as a sharp boundary between two fluids, across which the concentration changes abruptly from that of one fluid to that of the other.

A boundary between two immiscible fluids. Here, due to capillary effects (Subs. 6.1.3), the saturation of each fluid varies gradually across a transition zone. If this zone is narrow relative to the domains of interest on its two sides, it may be *approximated* as a sharp boundary across which, as an approximation, we *stipulate* a jump in the saturation of the fluids. The phreatic surface (Subs. 4.5) may serve as an example; the two fluids are air and water, and we assume that only water is present in the void space below this surface, while only air, plus water at the irreducible water saturation, is present in the void space above it.

Because of the approximation involved, whenever sharp boundaries are introduction to replace transition zones, measurements *within* the latter should not be expected to compare well with predictions obtained by solving the mathematical models that include such (hypothetical sharp) boundaries.

In general, a boundary surface may be stationary or moving. It may also be material or non-material with respect to any extensive quantity. In this book, the extensive quantities of interest are the mass of a fluid and the mass of a component of a fluid.

Let $F(\mathbf{x},t) \equiv F(x,y,z,t) = 0$ represent the equation of a possibly moving (macroscopic) boundary surface between two fluids. This means that fluid particles on that surface stay on it, but molecules of dissolved species in the fluids may (and, in fact, do) cross this interface. The speed of displacement of this boundary, $\mathbf{u}$, should not be mixed up with the velocities of the fluids present on both sides of the surface. As the surface moves, its shape may change, but its equation, $F(\mathbf{x},t) = 0$, remains unchanged. The quantity F is, thus, a *conservative property* of the points on the surface, for which the total derivative vanishes, i.e.,

$$\frac{\mathrm{D}F}{\mathrm{D}t} \equiv \frac{\partial F}{\partial t} + \mathbf{u} \cdot \nabla F = 0. \tag{5.2.1}$$

By definition,

$$\mathbf{n} = \frac{\nabla F}{|\nabla F|}, \tag{5.2.2}$$

where $\mathbf{n}$ denotes the unit vector normal to the surface $F = 0$.

From (5.2.1), we obtain

$$\mathbf{u} \cdot \nabla F = -\frac{\partial F}{\partial t}. \tag{5.2.3}$$

The component of $\mathbf{u}$ normal to the surface is then given by

$$u_n \equiv \mathbf{u} \cdot \mathbf{n} = -\frac{\partial F/\partial t}{|\nabla F|} = -\frac{\partial F/\partial t}{\partial F/\partial s_n}, \tag{5.2.4}$$

where s_n is a distance measured along $\mathbf{n}$.

In this section, we shall continue to consider flow in a three-dimensional domain. In Sec. 5.4 we shall focus on horizontal two-dimensional flow domains.

5.2.2 Initial and general boundary conditions

A. Initial conditions

Initial conditions specify the value of the (macroscopic) dependent variable, e.g., p, or h, at all points within the modeled domain at some initial time, usually taken as $t = 0$. For example, in terms of h, initial conditions may take the form

$$h(x, y, z, 0) = f(x, y, z), \tag{5.2.5}$$

where $f(x, y, z)$ is a known function.

B. General boundary condition

In general, there exist *two* kinds of boundary conditions:

- Continuity in the (macroscopic) value of intensive quantities, $\overline{e}^\alpha$, across the boundary,

  $$[\![\, \overline{e}^\alpha \,]\!]_{1,2} = 0. \tag{5.2.6}$$

  where the symbol $[\![\, (..) \,]\!]$ denotes the jump in (..), and the subscripts 1 and 2 mark the two adjacent domains. This is a consequence of the continuity in the microscopic value, e, as any microscopic boundary is crossed. A jump in $\overline{e}^\alpha$ would lead to an infinite gradient which, in turn, would create an infinite diffusive flux that will instantly eliminate the jump.
- In the absence of sources and sinks on a boundary, the total amount of any extensive quantity that is transferred by all phases present in the porous medium domain must be conserved as it is being transported across a boundary. This condition arises from the balance of that quantity as it is transported across a boundary.

Because, in the general case of a multi-fluid phase system, we, usually, need information regarding the transport of the various extensive quantities within each phase separately, we could assume that conservation of an extensive quantity (e.g., mass of a phase, mass of a component of a phase) is maintained separately for each phase. However, the quantity that is being transported across a boundary within any phase would have to enter *only that same phase* on the other side of the boundary. This assumption violates our conceptualization of the boundary as representing an abrupt change in porosity. This implies that, in addition to a portion of the boundary across which phases remain continuous, surfaces of contact must also exist on the boundary between any given phase and all other phases. Across such surfaces of contact, extensive quantities may be transported from one phase to another. For example, if porosity varies across such boundary, some solid surface portion of this boundary must face a liquid. In a heat transport problem, heat must be transferred from one phase to the other by conduction across such boundary.

With the above considerations in mind, for any extensive quantity, E, in the absence of sources and sinks of E on the boundary, the boundary condition may be stated mathematically in the form (Bear and Bachmat, 1990):

$$\sum_{(\alpha)} [\![\, \theta_\alpha \overline{e}^\alpha (\overline{\mathbf{V}_\alpha}^\alpha - \mathbf{u}) + \theta_\alpha \overline{\mathbf{j}^E}^\alpha \,]\!]_{1,2} \cdot \mathbf{n} = 0, \tag{5.2.7}$$

where α denotes a phase within an REV, with $\sum_{(\alpha)} \theta_\alpha = 1$, $\overline{\mathbf{j}^E}^\alpha$ denotes the sum of the dispersive and (macroscopic) diffusive fluxes of E, and $\mathbf{u}$ denotes the velocity of the boundary. The macroscopic dispersive and diffusive fluxes are discussed in Sec. 7.1. Equation (5.2.7) represents the *general macroscopic boundary condition for any extensive* quantity, E, *in a porous medium.* It is often referred to as the *no-jump* condition, meaning no-jump in the normal component of the total flux across the boundary. We note that it expresses the notion that E does not accumulate on the boundary. We also note that the total flux consists of advection relative to the boundary (possibly moving at velocity $\mathbf{u}$), dispersion, and diffusion in *all* phases. To be used as a boundary condition in a transport problem, *we must know the value of the flux on the external side of the boundary.* By using this information, the no-jump condition becomes a boundary condition to the transport problem.

Thermodynamic scalar variables, such as pressure, temperature, and concentration, must also be continuous across the boundary.

In practice, the boundary conditions used in models of flow and transport in porous medium domains may not necessarily take the exact form of the no-jump condition (5.2.7). Instead, they may be reduced to forms that specify values of variables, or of their derivatives, on the boundary. However, their origin is still the fundamental no-jump condition in fluid mass and in fluid components.

The kind of PDE describing the mass balance, e.g., (5.1.54), requires only *one* condition on each boundary segment, and we should prefer a condition based on flux continuity, if such information is available. If not, we shall base the condition on available values of the scalar variables. Sometimes, approximations concerning the continuity in fluxes produce a jump in the values of the variables; we have to accept this consequence.

In the next subsection, we present boundary conditions for saturated flow models, without presenting the details of developing them from the general boundary condition stated above. The type of boundary condition to be used in any particular case depends on the available data concerning the actual or anticipated behavior in the field.

The partial differential equation (PDE) that we wish to solve—any of the flow equations presented in the previous subsection—describes the mass balance equation for water, combined with Darcy's law. By inserting Darcy's law into the mass balance equation, we obtain a single linear second order PDE in terms of either pressure (p), or piezometric head (h). A well-posed problem involving this equation requires a *single* boundary condition on each segment of the boundary. However, in Subs. 5.2.3I, we shall introduce a case in which the motion equation takes the form of Brinkman's equation (Subs. 4.3.2B), rather than Darcy's law. The reason for the need to use Brinkman's equation is that the momentum transfer due to the fluid's velocity gradient cannot be neglected, at least in part of the considered domain. Although the use of Brinkman's equation is not common in groundwater flow, we believe that it is an interesting example of a case encountered under certain conditions when dealing with phenomena of transport in porous media.

5.2.3 Particular boundary conditions

Following are some of the more commonly encountered boundary conditions for saturated flow. In each case, the boundary condition should to be stated in terms of the relevant state variable of the problem. The boundary surface is described by $F(\mathbf{x}, t) = 0$ (Subs. 5.2.1). The discussion is at the macroscopic level, and we shall use the symbol $\mathcal{B}$ to denote a boundary segment.

A. Boundary of prescribed pressure, or piezometric head

In this case, the boundary condition takes the form

$$p = f_1(x, y, z, t), \quad \text{or} \quad h = f_2(x, y, z, t), \quad \text{on} \quad \mathcal{B}, \tag{5.2.8}$$

where f_1 and f_2 are known functions.

Actually, the value of p is seldom known on the boundary, except when a porous medium domain is bounded by a body of water (e.g., a pond). In such case, the pressure along the pond's bottom is dictated by the depth of water in the pond. Whenever the density, ρ_w, is constant, the piezometric head, h, may also be prescribed on such a boundary.

A boundary condition that specifies the value of a state variable (here, p, or h) along a boundary segment is called *boundary condition of the first type*, or *Dirichlet boundary condition*.

B. Boundary of prescribed water flux

This case occurs, when water, at a *known flux*, enters a domain through its boundary. This includes the case of no-flow through such a boundary. We shall assume that such boundary is a *material surface* with respect to the solid, i.e.,

$$(\mathbf{V}_s - \mathbf{u})\big|_{\text{side 1}} \cdot \mathbf{n} = (\mathbf{V}_s - \mathbf{u})\big|_{\text{side 2}} \cdot \mathbf{n} = 0. \tag{5.2.9}$$

Assuming $[\![\, \rho_w \,]\!]_{1,2} = 0$, the general boundary condition, (5.2.7), for such a surface takes the form:

$$[\![\, \phi(\mathbf{V} - \mathbf{u}) \,]\!]_{1,2} \cdot \mathbf{n} = 0. \tag{5.2.10}$$

With (5.2.9), for water, (w), equation (5.2.10) reduces to the form:

$$[\![\, \mathbf{q}_r \,]\!]_{1,2} \cdot \mathbf{n} = 0, \quad \text{or} \quad \mathbf{q}_r\big|_1 \cdot \mathbf{n} = \mathbf{q}_r\big|_2 \cdot \mathbf{n}. \tag{5.2.11}$$

To serve as a boundary condition, information must be available on what happens on the external side of the boundary, say, side 2. Obviously, the relative specific flux, $\mathbf{q}_r$, has to be expressed by an appropriate motion equation, written in terms of p, or h.

For an *impervious boundary*, say, a pervious side 1 and an impervious side 2, equation (5.2.11) reduces to

$$\mathbf{q}_r \cdot \mathbf{n} = 0. \tag{5.2.12}$$

Note that this equation constrains only the normal component of the flux. The tangential components may take on any value; we may have *slippage* along such a boundary.

Let $\mathbf{N}$ denote the prescribed flux on the external side of a stationary boundary ($\mathbf{u} = 0$) described by $F = F(\mathbf{x})$, with $\mathbf{n} \equiv \nabla F/|\nabla F|$ denoting the unit outward normal vector to it. We assume that the water density, $[\![\, \rho_w \,]\!]_{1,2} = 0$, and that ρ_w is a constant. Then, the prescribed flux boundary condition takes the form:

$$\mathbf{q}_r \cdot \nabla F = N|\nabla F|, \qquad N = \mathbf{N} \cdot \mathbf{n}. \tag{5.2.13}$$

In this equation, $\mathbf{q}_r$ can be expressed by any of the motion equations presented in Chap. 4. For example, we can rewrite (5.2.13) in the form:

$$-(\mathsf{K} \cdot \nabla h) \cdot \mathbf{n} = \mathbf{N} \cdot \mathbf{n}, \tag{5.2.14}$$

where K denotes the hydraulic conductivity.

A condition that specifies the gradient of a scalar variable on the boundary is called a *boundary condition of the second kind* or a *Neumann boundary condition*.

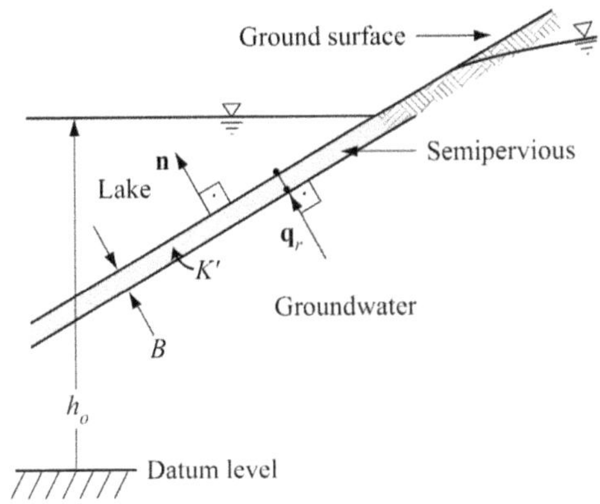

Figure 5.2.1: A semipervious boundary.

The condition of prescribed flux provides no explicit information on the values of the state variables, say, p or h, at (i.e., just inside) the boundary. These values will adjust themselves to accommodate the specified rate of flow through the boundary.

C. Semipervious boundary

Let us assume that just below ground surface we have a relatively thin soil layer which behaves as a semipervious membrane that resists the downward movement of water through it. A layer of fine sediments on the bottom of a pond is another example of such case (Fig. 5.2.1). We assume that this 'membrane' is saturated when present at the bottom of an active water pond, and whenever infiltration occurs through it.

Let us assume that water is ponded on the upper side of this 'membrane', possibly with a zero thickness of ponding, such that a piezometric head h_o is specified there. Let us denote the *resistance* of the semipervious membrane by c_r (= thickness of the membrane, B, divided by its hydraulic conductivity, K', i.e., the reciprocal of the leakance), and the piezometric head on the lower side of the membrane by h. Then, the flux through the membrane is expressed by

$$\mathbf{q}_r \cdot \mathbf{n} = \frac{h - h_o}{c_r}, \quad c_r = \frac{B}{\mathsf{K}'}, \tag{5.2.15}$$

where $\mathbf{q}_r$ may be expressed by any of the flux equations.

This is a *boundary condition of the third type*, or a *Robin boundary condition.*

D. Boundary between different porous media

Figure 5.2.2 shows a boundary between two regions of different hydraulic conductivities. In principle, we should avoid the modeling of discontinuities

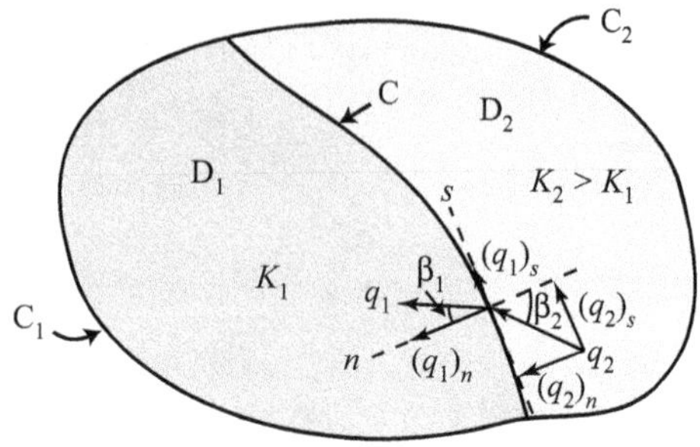

Figure 5.2.2: Boundary between regions of different hydraulic conductivities.

within a modeled domain, e.g., discontinuity in the values of coefficients. When an investigated domain does includes such discontinuities, it is useful to divide the domain into subdomains along the surfaces of discontinuity, in order to obtain subdomains without discontinuities. We then write a model for each subdomain. On each of the common boundary segments, we need *two* boundary conditions—one for each side: these are the continuity of flux and the continuity of pressure. Because both flux and pressure are unknown *a priori*, we have to write these conditions in terms of the state variables for both sides, and solve for all the subdomains simultaneously.

As the boundary is approached from within each side, the continuity of pressure (or piezometric head), is expressed as

$$p\Big|_{\text{side 1}} = p\Big|_{\text{side 2}}, \quad \text{or} \quad h\Big|_{\text{side 1}} = h\Big|_{\text{side 2}}, \tag{5.2.16}$$

and the continuity of flux, following the discussion leading to (5.2.11), takes the form:

$$\mathbf{q}_r\Big|_{\text{side 1}} \cdot \mathbf{n} = \mathbf{q}_r\Big|_{\text{side 2}} \cdot \mathbf{n}. \tag{5.2.17}$$

In the above equation, we have expressed the boundary condition in terms of the relative flux, $\mathbf{q}_r$, expressed by Darcy's law. However, in most cases of practical interest, as henceforth in this section, we assume $\mathbf{V}_s \equiv 0$, so that $\mathbf{q}_r = \mathbf{q}$.

Although Figure 5.2.2 is presented in two-dimension, (5.2.16) and (5.2.17) are valid also on a boundary in a three-dimensional domain. The explicit expression (5.2.17), in terms of h_1 in D_1, and h_2 in D_2, depends on the nature of the materials occupying the two subdomains (also, with respect to isotropy or anisotropy). For example, if K_1 and K_2 are both isotropic, (5.2.17) in two-dimensional flow reduces to the condition

$$\mathsf{K}_1 \frac{\partial h_1}{\partial s_n} = \mathsf{K}_2 \frac{\partial h_2}{\partial s_n}, \quad \text{on C}, \tag{5.2.18}$$

where s_n is distance measured along the normal.

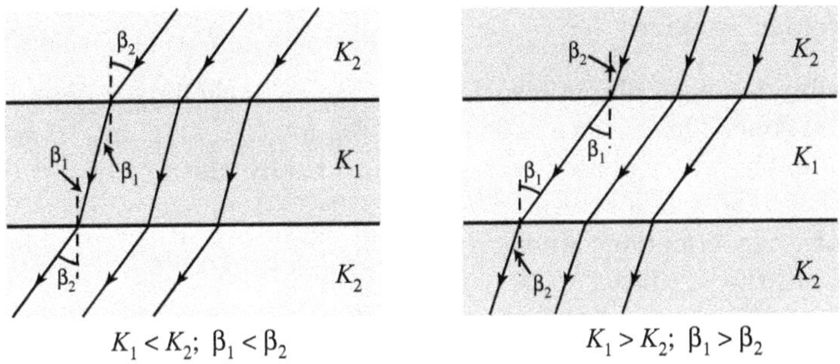

Figure 5.2.3: Refraction of streamlines at an interface between different hydraulic conductivities.

Thus, for the case shown in Fig. 5.2.2, the two boundary conditions to be satisfied on C are (5.2.16) and (5.2.17). Since each of these equations includes both h_1 and h_2, the two problems (for D_1 and D_2) must be solved simultaneously.

From (5.2.16), it follows that $\partial h_1/\partial s = \partial h_2/\partial s$, where s is a distance measured along the tangent to C (in Fig. 5.2.2). This can also be expressed as

$$\frac{(q_s)_1}{\mathsf{K}_1} = \frac{(q_s)_2}{\mathsf{K}_2}, \tag{5.2.19}$$

where both K_1, and K_2 are isotropic. By combining (5.2.17) with (5.2.19), we obtain

$$\frac{\mathsf{K}_1}{\tan\beta_1} = \frac{\mathsf{K}_2}{\tan\beta_2}, \quad \tan\beta_1 = \frac{(q_s)_1}{(q_n)_1}, \quad \tan\beta_2 = \frac{(q_s)_2}{(q_n)_2}, \tag{5.2.20}$$

where β_1 and β_2 are the angles which $\mathbf{q}_1$ and $\mathbf{q}_2$ make with the normal to the boundary C. This means that along such a boundary, the incident streamline is refracted. Equation (5.2.20) is the *law of refraction of streamlines for two-dimensional flow*, when both subdomains are occupied by isotropic media.

Bear (1972, p. 263) discusses the laws of refraction of streamlines and of equipotentials also for three-dimensional flows and for cases where the two subdomains are occupied by anisotropic media.

From (5.2.20), it follows that when $\mathsf{K}_1 \gg \mathsf{K}_2$, then $\beta_1 \gg \beta_2$, and the refracted streamline approaches the normal to the common boundary upon passing from a more pervious to a less pervious medium. When $\mathsf{K}_1 \ll \mathsf{K}_2$, then $\beta_1 \ll \beta_2$, and the refracted streamline tends to become almost parallel to the common boundary upon passing from a less pervious (e.g., semipervious) to a more pervious medium. This justifies the assumption of 'essentially horizontal flow' in a leaky aquifer. The two cases mentioned here are shown in Fig. 5.2.3.

E. Phreatic surface

A (possibly moving) phreatic surface may serve as the upper boundary of a saturated zone. This surface is defined as the locus of all points at which the pressure in the water phase is atmospheric, usually taken to be $p = 0$. Below this surface, the soil is saturated, with $p > 0$.

As for every boundary, we have to specify both the *shape* of the boundary surface and the *condition* to be satisfied on it.

Usually, the shape of the phreatic surface, $F(x, y, z, t) = 0$, is *a priori* unknown. In fact, as we have already emphasized earlier, in many flow problems, determining the shape and (possibly time-dependent) position of this surface is the very objective of model investigations. However, once we have a solution, say, in the form of $p = p(x, y, z, t)$, whether in the unsaturated flow domain, or in the saturated one underlying it, since on the phreatic surface

$$p\Big|_{\text{sat}} = p\Big|_{\text{unsat}} = 0, \tag{5.2.21}$$

the shape of the phreatic surface boundary is given by

$$F(x, y, z, t) \equiv p(x, y, z, t) = 0. \tag{5.2.22}$$

When the water density, ρ_w, remains unchanged, since on the phreatic surface,

$$h\Big|_{\text{sat}} = h\Big|_{\text{unsat}} = z, \tag{5.2.23}$$

we may also define the shape of the phreatic surface as

$$F(x, y, z, t) \equiv h(x, y, z, t) - z = 0. \tag{5.2.24}$$

The condition on the phreatic surface boundary is that of continuity of the normal water flux across it. Usually, when we consider a phreatic surface as a boundary, the underlying assumption is that the moisture content above this surface is at its irreducible level, θ_{rw}. The concept and definition of irreducible moisture content is discussed in Subs. 6.1.4. This assumption is valid as long as the thickness of the capillary fringe is small relative to either the thickness of the unsaturated zone, or of the saturated one. If this condition is not satisfied, then the use of the sharp interface approach is questionable

The condition to be satisfied on the phreatic surface, assuming no change in density as water crosses this surface, is expressed in the form

$$\phi(\mathbf{V}_w - \mathbf{u})\Big|_{\text{sat}} \cdot \mathbf{n} = \theta_{rw}(\mathbf{V}_w - \mathbf{u})\Big|_{\text{unsat}} \cdot \mathbf{n}, \tag{5.2.25}$$

where $\mathbf{u}$ is the speed of the moving phreatic surface, and $\mathbf{n}$ denotes the unit vector normal to that surface, pointing away from the saturated zone. They are related to the shape of the surface, $F(x, y, z, t) = 0$, by (5.2.2) and (5.2.3), respectively.

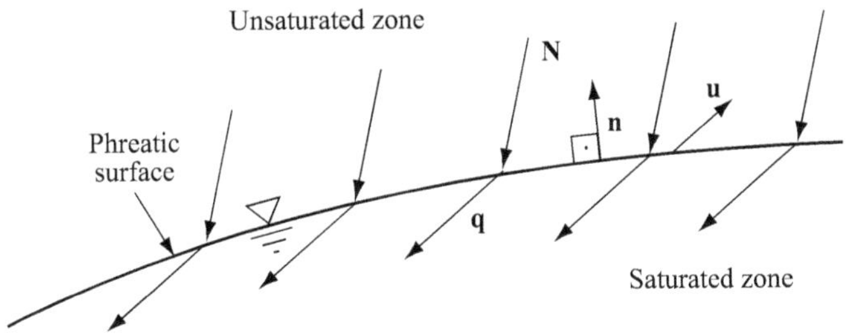

Figure 5.2.4: Phreatic surface with accretion.

Let us consider the case of flow in a phreatic aquifer, encountered in groundwater hydrology. The details of flow in the unsaturated zone are of no interest. Instead, we assume that accretion, $\mathbf{N}$ (e.g., from precipitation), takes place on the upper side *of the phreatic surface* (Fig. 5.2.4). The rate at which water moves from the unsaturated zone to the saturated one through the phreatic surface may be expressed in the form:

$$\theta_{rw}(\mathbf{V}_w - \mathbf{u})\big|_{\text{unsat}} \cdot \mathbf{n} \equiv (\mathbf{N} - \theta_{rw}\mathbf{u})\big|_{\text{unsat}} \cdot \mathbf{n}, \tag{5.2.26}$$

where $\mathbf{N} = \theta_{rw}\mathbf{V}_w\big|_{\text{unsat}}$. For a vertically downward accretion at a rate N, we use $\mathbf{N} = -N\nabla z$. Equation (5.2.26) is the sought boundary condition. Let us rewrite it in a number of equivalent forms.

We have used the term ‘accretion’ to denote the rate at which water is added to, or removed from the phreatic surface, independent of the movement of the latter and of any moisture (if present) in the void space above it. However, it should be emphasized that $\mathbf{N} \cdot \mathbf{n}$ is not the rate at which water actually crosses the phreatic surface and augments (or reduces) the quantity of water in the saturated zone. This net rate depends also on the movement of the phreatic surface.

We can rewrite (5.2.25), or (5.2.26), in the form:

$$(\mathbf{q}_w\big|_{\text{sat}} - \phi\mathbf{u}) \cdot \mathbf{n} = (\mathbf{N} - \theta_{rw}\mathbf{u}) \cdot \mathbf{n}, \tag{5.2.27}$$

or

$$(\mathbf{q}_w\big|_{\text{sat}} - \mathbf{N}) \cdot \mathbf{n} = (\phi - \theta_{rw})\mathbf{u} \cdot \mathbf{n}. \tag{5.2.28}$$

In view of (5.2.3) and (5.2.4), we may rewrite (5.2.28) as

$$(\mathbf{q}_w\big|_{\text{sat}} - \mathbf{N}) \cdot \nabla F = -(\phi - \theta_{rw})\frac{\partial F}{\partial t}. \tag{5.2.29}$$

Making use of (5.2.24), we may rewrite (5.2.29) in terms of the piezometric head, h, in the form:

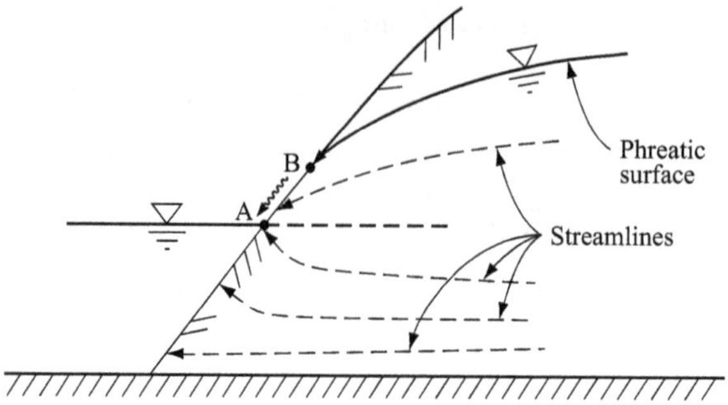

Figure 5.2.5: The seepage face, AB.

$$(\mathbf{q}_w\big|_{\text{sat}} - \mathbf{N}) \cdot \nabla(h - z) = -(\phi - \theta_{rw})\frac{\partial h}{\partial t}. \tag{5.2.30}$$

By inserting $\mathbf{q}_w\big|_{\text{sat}} \equiv \mathbf{q}_r = -\mathbf{K} \cdot \nabla h$ into this equation, we obtain

$$(\mathbf{K}_w \cdot \nabla h + \mathbf{N}) \cdot \nabla(h - z) = \phi_{\text{eff}}\frac{\partial h}{\partial t}, \tag{5.2.31}$$

where $\phi_{\text{eff}} \equiv \phi - \theta_{rw}$. We wish to reiterate that this equation expresses nothing but the continuity of fluid flux across the phreatic surface boundary.

Although we have presented here the boundary condition of a phreatic surface, assuming that the rate of accretion, $\mathbf{N}$ is known, in most cases of practical interest, this rate is actually unknown. It is certainly not the rate of rainfall; among other factors, its value depends both on the rainfall and on the moisture conditions of the soil. We shall discuss this issue in detail in Subs. 6.3.2.

F. Seepage face

This kind of boundary appears when a phreatic surface approaches a body of open water, a river or a lake, which serves as part of the boundary of a flow domain (Fig. 5.2.5). In such cases, the phreatic surface will *always* terminate on that (known) boundary at a point (Point B in Fig. 5.2.5) located at some elevation *above* the water surface of the body of open water (Point A). The segment AB is called the *seepage face.* Through it, water seeps out of the porous medium domain.

The reason for the existence of a seepage face is that otherwise (i.e., if point B would coincide with A), the velocity at that point would be infinite. This is an impossible situation (Muskat, 1937; Bear, 1972).

Since on a seepage face, which is exposed to the atmosphere, the pressure in the water is $p = 0$ (assuming atmospheric pressure is $p_a = 0$), the boundary condition is

$$p(\mathbf{x}, t) = 0, \quad \text{or} \quad h(\mathbf{x}, t) = z, \tag{5.2.32}$$

i.e., the head at every point of the seepage face is specified to be equal to its *known* elevation. The geometry of the seepage face is known, except for the location of its end point, B, which is also a point on the (*a priori* unknown) phreatic surface.

G. Well

In Sec. 5.4, we shall see how pumping and injection wells are represented as *point* sinks and sources in regional models that describe two-dimension flow in aquifers. In Fig. 3.8.1, we see a complete well equipped with an ejector and a discharge pipe. Water enters the casing (or leaves it in the case of an injection well) through its screened, or slotted portion. Without pumping, the well's casing acts a piezometer; water in the casing rises to a level that corresponds to the piezometric head in the aquifer outside the screen. During pumping, water is first removed from the casing (by a pump discharging water to ground surface through a pipe), causing a drop in the water level in the casing. Then, under the difference in head between the exterior and interior of the screen, water enters the casing. If the pumping rate is constant, eventually, the water level in the casing will reach some constant level lower than the initial one. During injection, the water level in the casing rises, and water leaves the casing. For a constant injection rate, eventually, the water level in the casing will reach a constant level that is higher than the initial one. Head ($\equiv$ energy) is lost as water moves through the screen.

In an *artesian well* (in an artesian aquifer), a pump and a discharge pipe are not required; this is a consequence of the fact that the piezometric head at the screen is higher than ground surface. This causes water in the casing to rise and discharge freely at ground surface. To account for head loss in the pipe, the piezometric head at the screen should be higher than ground surface by an amount equal to the head loss along the pipe. *Hydraulics of wells* (e.g., Bear, 1979; Boonstra, 1998) is a term used to describe the study of drawdown and build-up in pumping and injection wells, respectively, and in their vicinity in an aquifer.

In the present section, we consider three-dimensional flow models, and, therefore, let us focus on the conditions that prevail on the circumference of a well's screen in three-dimensional flow in an aquifer. For the purpose of this discussion, we shall assume that the screened (or slotted) portion of a well's casing has the shape of a vertical cylinder of finite length through which water can be injected into or extracted from the aquifer. We shall overlook the detailed structure of the well (casing, gravel pack, etc.; Fig. 3.8.1).

The specific discharge, or flux, through the screen is not uniformly distributed along the screen (e.g., Bear, 1979, p. 346). It is higher at the upper and lower portions of the screen, close to the edges, than in the central portion. In addition, depending on the conditions in the vicinity of the screen, at every elevation, the flux around the screen is also, usually, not uniform. Nevertheless, for the sake of simplicity, we often assume, as a good approximation, that the specific discharge is uniformly distributed over the entire screen area. Moreover, it is always taken to be in the radial direction. In a flow model, solved for $h = h(r,t)$, the screen serves as a boundary to the (porous medium) flow domain and a boundary condition has to be specified on it. We shall consider a few examples.

A pumping well with a known specific discharge through the screen, q_{well} (positive for pumping), neglecting head loss through the screen. The condition is

$$q_{well} = K\frac{\partial h(r,t)}{\partial r}. \tag{5.2.33}$$

A pumping well with specified head inside the well, h_{well}, and the head $h(r,t)|_{r=r_{well}}$ outside the well. The boundary condition is

$$h_{well} - h(r,t)|_{r=r_{well}} = -\alpha_1\frac{\partial h(r,t)}{\partial r}, \tag{5.2.34}$$

in which α_1 is a coefficient that depends on the resistance of the screen and the permeability of the aquifer. Another option is to specify the known piezometric head as a first kind boundary condition on the screen. It is also possible to take into account the head loss through the screen, which is proportional to the square of the specific discharge through the latter.

An artesian, or flowing well. In such a well, when active, the piezometric head within the well (i.e., the screened portion), is known; it is equal to the elevation of ground surface. This head is then assigned as specified head boundary condition. Actually, the head is known only *at ground surface*, or at the point of the well's outlet, while inside the well it is slightly higher; however, we usually neglect the head loss along the well's casing, from the screen to ground surface (or point of water outlet). It is, of course, possible to take into account this head loss, which is proportional to the square of the well's discharge.

H. Spring

In a phreatic aquifer, a spring occurs where groundwater emerges at ground surface. This occurs wherever the water table elevation is above ground surface over a small area (Sec. 3.6). Often, this happens where a local depression exists in ground surface elevations (Fig. 3.6.1a). A spring may also occur in the case of a confined aquifer, when a fissure connects the aquifer with ground surface through the upper impervious layer (Fig. 3.6.1d). This latter case is similar to the case of an *artesian*, or *flowing well.* We shall examine these two cases separately.

A spring in a phreatic aquifer

A spring in a phreatic aquifer (Fig. 5.2.6a) may be treated in one of the following two ways:

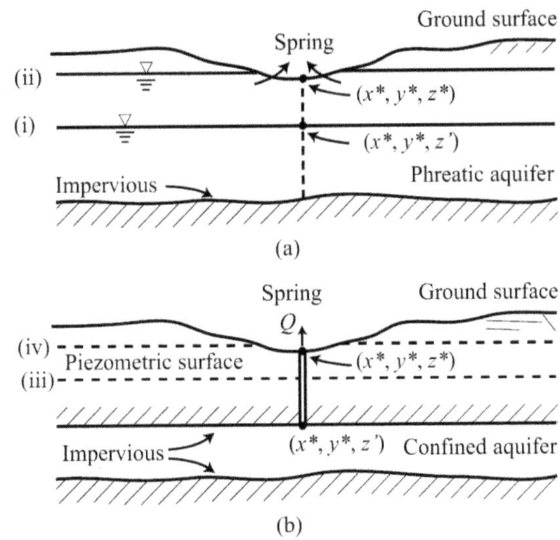

Figure 5.2.6: Springs in (a) a phreatic aquifer, and (b) a confined aquifer.

- We can start by applying the phreatic surface boundary condition (Subs. 5.2.3E) on the water table boundary, and solve the flow problem without a spring. If the calculated phreatic surface elevation, z', satisfies $z' \leq z^*$ (see surface (i) in Fig. 5.2.6a), i.e., it is below ground surface elevation, the spring does not exist, and no further action is needed (for that time step). If $z' > z^*$ (surface (ii) in Fig. 5.2.6a), the spring exists. In that case, the upper boundary at the spring's location (or an area in its vicinity, or at a node in a numerical solution) is adjusted to ground surface elevation, z^*, and the condition (Subs. 5.2.3F), $h(x^*, y^*, z') = z^*$, is applied. The solution for that time step is repeated.
- Or, we assume that the spring exists, and use ground surface as the upper boundary to apply the $h(x^*, y^*, z') = z^*$ condition. Once the solution is obtained, the flux to the spring area is determined. If the flux is towards the spring, the initial assumption was correct and we can continue to the next time step. If the flux is away from the spring, the latter does not exist, and we have to solve for that time step for a phreatic surface condition without the presence of a spring.

A spring in a confined aquifer

Figure 5.2.6b gives a schematic sketch of a spring in a confined aquifer. Two situations may occur:

- We can apply the impermeable condition on the upper boundary of the confined aquifer, without considering a spring, and solve the flow problem. If the resulting piezometric head at the spring's origin, (x^*, y^*, z'), is below ground surface elevation, i.e., $h(x^*, y^*, z', t) \leq z^*$ (see surface (iii) in

Fig. 5.2.6b), the spring is dry. The impermeable boundary condition was correctly applied, and no further action is required for that time step.

- Apply the same impermeable condition described above. If the piezometric head at the spring's origin is above ground surface, i.e., $h(x^*, y^*, z', t) > z^*$ (see surface (iv) in Fig. 5.2.6b), a spring will emerge. In such case, the impermeable condition at and around (x^*, y^*, z') needs to be replaced by a third type boundary condition described below, and the solution is repeated for that time step.

A few comments are appropriate here.

- Unlike the case of a phreatic aquifer, a spring originating in a confined aquifer will *not* occur anywhere at ground surface, even when the piezometric surface is above ground surface elevation at that point (see (iv) in Fig. 5.2.6b). Instead, the spring emerges *only* at locations where a fracture system exists to serve as a conduit. This is illustrated, schematically, in Fig. 5.2.6b as a pipe, which is analogous to the case of an artesian well. Furthermore, the piezometric head at (x^*, y^*, z') should not be made equal to ground surface elevation, i.e., $h(x^*, y^*, z') \neq z^*$. Instead, in order to overcome the resistance of the fracture system and maintain a spring discharge, the condition $h(x^*, y^*, z') > z^*$ should be satisfied. The actual head, $h(x^*, y^*, z')$, however, is not known *a priori*; hence, a Dirichlet boundary condition is not applicable.
- One way to model a spring resulting from a fracture system is to assume that the specific discharge in the fracture system is proportional to the difference in piezometric head between the top of the confined aquifer, z', and ground surface elevation, z^*, i.e., $h(x^*, y^*, z', t) - z^*$. It is also inversely proportional to the fracture system's resistance, c_{sp} (dims. T), which, in turn, depends on the length of the fracture zone, its cross-sectional area, and fracture surface roughness, among other factors. This discharge, equated to that at the aquifer top, leads to a third type boundary condition, similar to the semipervious layer boundary condition

$$-\mathsf{K}\left.\frac{\partial h}{\partial z}\right|_{x^*, y^*, z', t} = \frac{h(x^*, y^*, z', t) - z^*}{c_{sp}}. \tag{5.2.35}$$

The above boundary condition should be applied on the top aquifer boundary, over the area where the fracture system exists.

- A comment concerning the assignment of a first type boundary condition, $h = z^*$, or the third type boundary condition, (5.2.35), is needed. Unlike a pumping well, which could be modeled as a point sink on the boundary, the first and third type boundary conditions need to be applied to a *finite area*, in fact, to the actual area where the spring occurs. As the flux across the boundary is finite, the total spring discharge is proportional to the area where such a boundary condition is applied. Note that for the case of a point sink on the aquifer boundary, the area is zero, and the specific discharge is infinite, such that a finite discharge can result.

Assigning a first type or third type boundary condition to a point on the upper aquifer boundary (impermeable ceiling) will not only result in a zero total discharge, but also violate the principle of existence of a well-posed boundary value problem.

In a numerical implementation, using the finite difference or finite element methods (Chap. 8), assigning a first or third type boundary condition to a single node has the effect of spreading its interpolated values to adjacent nodes, or to the area represented by an element, so that a solution can be found. As this procedure is somewhat arbitrary, it is desirable to have a refined mesh near the spring, such that the actual spring area can be better represented. In reality, however, it is unlikely that the spring area and its resistance coefficient can be observed and known to any precision. In order to obtain reliable prediction of spring discharge, model calibration is needed. In a calibration procedure, the third type boundary condition can be applied to an area which may not be an accurate representation of the true spring area. The resistance coefficient, c_{sp}, is considered as a calibration coefficient; its value needs to be adjusted such that the numerically simulated total spring discharge matches field observed value.

I. Boundary with a body of flowing water

In Subs. 5.2.3A, we discussed the boundary condition of a porous medium in contact with a static body of water; in Subs. 5.2.3C, the presence of a semipervious layer separating open water and a porous medium domain was discussed. In this subsection, we shall examine the conditions on the boundary between a porous medium domain and a body of *flowing water*.

In a body of water, the flow is governed by the Navier-Stokes equation (= momentum balance equation for a Newtonian fluid), or by some simplified version of it, together with the mass balance equation. To have a well-posed boundary value problem, either the velocity components, or the stresses, but not both, have to be specified on all parts of the boundary (Ladyzhenskaya, 1963). For example, for the two-dimensional flow shown in Fig. 5.2.7a, the boundary condition on the solid surface is that the fluid must adhere there to the solid; this is referred to as the 'no-slip' condition. In the case of a stationary solid surface, the boundary conditions are $V_x = 0$ and $V_z = 0$. On the other hand, on a free surface, the shear stress and pressure are specified as $\tau_{zx} = 0$ and $p = 0$.

In principle, flows in a porous medium, and in a fluid continuum, are governed by the same physical laws—momentum balance and mass balance (and solute balance and heat balance, in the cases of solute and heat transport). However, for flow in a porous medium domain, as a consequence of the process of averaging or homogenization, these two fundamental equations are often expressed in different forms, containing *coefficients*, such as permeability; and the microscopic geometry of the void space no longer appears explicitly.

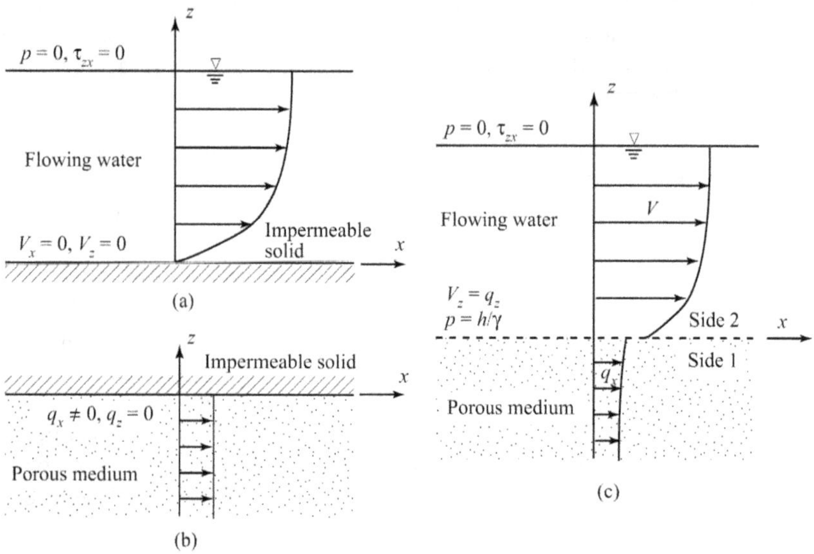

Figure 5.2.7: Boundary conditions for: (a) Flowing water with impermeable boundary, (b) Porous medium with impermeable boundary, (c) Common boundary between flowing water and porous medium.

Furthermore, the porous medium equations are often simplified, because the effects of certain terms appearing in these equations are negligible.

Let us first assume that the flow in the porous medium domain is governed by Darcy's law, as discussed in Subs. 4.2.2. This law is a simplified form of the homogenized Navier-Stokes equation. When Darcy's law is combined with the mass balance equation, we obtain a single governing, or flow, equation, expressed in terms of a single scalar variable—the piezometric head, h. As a consequence, one and only one boundary condition, either the normal flux or the head, needs to be specified on any part of the domain's boundary. As illustrated in Fig. 5.2.7b, $q_z = 0$ on a horizontal impervious boundary. Particularly, we notice that q_x cannot be specified and must be solved for; hence, $q_x \neq 0$ on the impervious boundary, i.e., we have a 'slip'. We recall that $\mathbf{q}$ is the macroscopic average of the microscopic velocity over the REV, and that the microscopic velocity *does not slip* on a solid surface.

The discussion above serves to illustrate that although the physical principles need to be obeyed, the averaging process makes the two sets of equations, one based on Navier-Stokes equation in the free-flowing water, and the other based on the homogenized Darcy's law in the porous medium, *incompatible* on a shared boundary. Hence, a coupled solution of the two domains is not possible.

This incompatibility between governing equations may be resolved if we assume that flow in the porous medium domain is governed by *Brinkman's equation* (4.3.7). When the two flow domains are in contact with each other, that is, the flowing (viscous) fluid is bounded from below by a porous medium saturated with the same fluid (Fig. 5.2.7c), we need to consider the conditions on the common boundary carefully, in order to determine the set of necessary and sufficient boundary conditions that will ensure the existence of a solution of the problem.

First, consider the interface between two immiscible flowing fluids with different densities and viscosities. A total of four conditions are needed on the interface (for two-dimensional flow): (1) velocity continuity, $V_x|_{z=0^+} = V_x|_{z=0^-}$; (2) $V_z|_{z=0^+} = V_z|_{z=0^-}$; (3) pressure continuity, $p|_{z=0^+} = p|_{z=0^-}$; and (4) shear stress continuity $\tau_{zx}|_{z=0^+} = \tau_{zx}|_{z=0^-}$. A similar situation exists at the interface between a free-flowing fluid and a porous medium (Kohr and Sekhar, 2007). Two of these interface conditions are obvious: (1) velocity continuity normal to the interface (mass conservation), $V_z|_{z=0^+} = q_z|_{z=0^-}$, and pressure continuity $p|_{z=0^+} = \gamma h|_{z=0^-}$. The other two conditions: the relation between the horizontal velocity of the flowing fluid, $V_x|_{z=0^+}$ and the flux $q_x|_{z=0^-}$, and between the two shear stresses, are not so obvious, as the quantities in the porous medium are homogenized ones.

Bear and Bachmat (1990, p. 245) considered the case of free flowing water overlaying a porous medium (Fig. 5.2.7c) containing an incompressible, Newtonian fluid. They showed that *when we assume no jump in pressure, and no jump in effective stress, there should not be a jump in the normal (to the common interface) component of the shear stress, $\boldsymbol{\tau}$, across the interface*, i.e.,

$$[\![\ \boldsymbol{\tau}_f\ ]\!]_{1,2} \cdot \boldsymbol{n} = 0, \tag{5.2.36}$$

where $\boldsymbol{n}$ is the unit normal vector. In order to express (5.2.36) in terms of fluid velocities, we need an appropriate constitutive relation. For $\boldsymbol{\tau}_f|_2$, i.e., in the free-flowing fluid, we use the constitutive relationship for an incompressible single component Newtonian fluid,

$$\tau_{ij} = \mu\left(\frac{\partial V_i}{\partial x_j} + \frac{\partial V_j}{\partial x_i}\right). \tag{5.2.37}$$

We assume that this relationship is valid also for the porous medium at the macroscopic level, but with an apparent viscosity μ^* that takes into consideration the added porous medium resistance (Shavit *et al.*, 2004). Different studies have derived and used different values of μ^*, $\mu^* >, =, < \mu$ (Nield and Bejan, 2006; Koplik *et al*, 1983; Kim and Russel, 1985). In the case under consideration here, we shall neglect the velocity gradient terms $\partial V_z/\partial x|_{z=0^+}$ and $\partial q_z/\partial x|_{z=0^-}$ in (5.2.37), to obtain the condition for shear stress continuity on the interface,

$$\left.\frac{\mu^*}{\phi}\frac{\partial q_x}{\partial z}\right|_{z=0-} = \left.\mu\frac{\partial V}{\partial z}\right|_{z=0+}. \tag{5.2.38}$$

This can be used as the condition on the interface between the two domains in the coupled boundary value problems. A more thorough examination of the condition of shear stress compatibility can be found in Kubik (2004), who suggested that the horizontal velocities tangential to the interface, $V|_{z=z^+}$ and $q_x|_{z=0^-}$ are not continuous. Their relation should be determined from the continuity of both momentum and energy near the interface. More discussion about the interface condition between free-flowing fluid and porous medium can be found in Rosenzweig and Shavit (2007).

Finally, let us examine a well-known condition at the interface between a free flowing water domain and a saturated porous medium domain, known as the *Beavers-Joseph condition* (Beavers and Joseph, 1967). This condition was motivated by the observation that in open channel flow, with a porous channel bottom, the discharge tends to be slightly greater than the one bounded by an impermeable bottom. The reason is that the porous interface condition allows the velocity at the channel bottom to *slip*, as shown in Fig. 5.2.7c. The Beavers-Joseph condition approximates the velocity gradient on the left side of (5.2.38) in the form

$$\frac{\alpha^M}{\phi}\left(V\big|_{z=0^+} - q_o\right) = \left.\frac{\partial V}{\partial z}\right|_{z=0^+}, \tag{5.2.39}$$

where α^M (dims. L^{-1}) is a momentum transfer coefficient that depends only on porous medium properties, such as permeability and porosity, and q_o is the uniform specific discharge in the porous medium starting from a certain distance away from the 'velocity boundary layer' in the vicinity of the interface. Beavers and Joseph (1967) proposed $\alpha^M = C^M\phi/\sqrt{\mathsf{k}}$, where the dimensionless coefficient, C^M, which depends only on ϕ and k, has to be determined experimentally. We note that (5.2.39) is a third type boundary condition for the free-flowing water, with empirical coefficients α^M and q_o. As the equation does not contain head or specific discharge information, it is not a boundary condition for the porous medium flow. More discussion on the Beavers-Joseph type boundary condition can be found in Ochoa-Tapia and Whitaker (1995), Nield and Bejan (2006), and Jager and Mikelic (2000).

J. Artificial boundary

The need to delineate boundaries for a modeled flow domain has already been presented in Subs. 5.2.1. The domain's boundaries have to be selected such that conditions with respect to the relevant state variables of the problem are known on them. However, for a numerical solution involving an *unbounded* domain, or practically so (e.g., a domain which is much larger than the domain of interest), it is often necessary to truncate the modeled domain to a finite size, as the computer cannot handle a discrete system with an infinite number of unknowns (or it is uneconomical to handle a very large number of nodes). The truncation of the domain requires the introduction of an *artificial boundary*, on which the boundary condition is unknown; a certain approximation of it is called for. Obviously, the most prudent thing to do is to make sure that the artificially introduced boundary is sufficiently far away from the region of interest, i.e., the region where significant piezometric head changes occur in a flow problem. An often used condition on such a boundary is to fix and maintain the piezometric head on it equal to the initial head there (in some computer codes this is referred to as to 'clamp' the condition).

As simulation time increases, the zone of major activities may expand and the selected location of the artificial boundary may no longer be appropriate; the boundary may have to be moved farther away. A trial-and-error approach may be required.

5.3 Complete 3-D Mathematical Flow Model

We now have all the elements required in order to formulate the complete mathematical model of a problem of forecasting the flow of a single fluid phase (saturated flow) in a porous medium domain. The objective of this section is to review the standard content of any such model.

5.3.1 Well-posed problem

The solution of a mathematical model of a problem takes the form of temporal and spatial distributions of the state variables of interest within the problem's prescribed time and space domains.

From the mathematical point of view, given a model composed of one or more partial differential equations, not every set of conditions imposed on the boundaries of the problem domain is satisfactory. This is even more so because, often, we have to resort to estimates of coefficients and simplifications of the mathematical models in specifying the boundary conditions for a given problem.

A mathematical model that represents a *physical reality* (and only such cases are considered in this book) is said to be *well-posed* if it satisfies the following requirements (e.g., Courant and Hilbert, 1962):

- A solution of the problem exists (*existence*).
- The solution is unique (*uniqueness*).
- The solution is stable (*stability*).

The first requirement simply states that at least one solution exists. The second one stipulates completeness of the problem statement, with no ambiguity. The third requirement means that small variations in data (e.g., initial and boundary conditions, and/or values of model coefficients) should lead to small changes in the resulting solution. If small errors in the data do not lead to correspondingly small errors in the solution, then the mathematical model is *ill-posed.* This last requirement is of particular interest, as all our observations have always some measurement errors. A model will be meaningless if these small errors will significantly affect the solution, i.e., the prediction obtained by the model.

Thus, once a complete mathematical model has been stated, the next step is to ensure that it is well-posed. Only then should a solution be sought.

The models developed and presented in this book, since they are based on a thorough analysis of a physical reality and on the description of this reality, albeit with certain simplifying assumptions, are implicitly *assumed* to be always well-posed. Therefore, they should provide unique, stable solutions. We shall not go into the mathematical analysis of whether a model developed here is well-posed, or not, although, as stated above, such an analysis is an essential step in the modeling process. The techniques used for such analysis can be found in appropriate mathematical texts on partial differential equations.

5.3.2 Conceptual model

As emphasized in the discussion on the modeling process (Subs. 1.2.2), after identifying the information required for making decisions, the first step in the modeling process is the establishment of a *conceptual model.* This model, in the form of a list of assumptions, indicates how the real, often complicated, problem is simplified for the purpose of providing the required specific information.

In order to establish the conceptual model, investigations are often undertaken, including the analysis of available data, field and laboratory work, etc. Such investigations should provide answers to questions related to the domain, the fluids involved, the modes of fluid transport, fluid sources, etc. Whenever definite information is not available, assumptions (or even educated guesses) are introduced, subject to *a-posteriori* validation. The investigations and simplifications are guided by the kind of information which the model is expected to provide for the purpose of making management decisions.

The conceptual model usually includes items such as:

- The domain's hydrogeology, stratigraphy, etc.
- The dimensionality of the model (one, two, or three dimensions), and the geometry of the boundary of the domain of interest.
- The behavior of the system: steady state or time-dependent.
- The kind of soil and rock materials comprising the domain, as well as their inhomogeneity, anisotropy, and deformability.
- The relevant properties of the fluid phase (density, viscosity, compressibility, presence of solutes).
- The flow regimes of the involved fluids (e.g., laminar or non-laminar).
- The presence of assumed sharp macroscopic fluid-fluid boundaries, such as a phreatic surface.
- The relevant state variables, and the areas or volumes over which averages of such variables should be taken.

- The presence of fluid sources and sinks, and their spatial distribution and temporal variation.
- The initial conditions within the domain, and conditions on its boundaries.

Once a conceptual model has been established, a well-posed mathematical model can be constructed. In the present chapter, we consider only isothermal saturated flow models. The effects of solutes are discussed in Chaps. 7.

5.3.3 Standard content of flow model

Based on the verbal statements and definitions included in the conceptual model, the complete mathematical model consists of the following items:

(a) Definition of the boundaries of the flow domain. The boundary surface must form a closed surface.
(b) A list of the variables that describe the state of the system, e.g., the piezometric head, or pressure.
(c) Partial differential flow equation for the water. We recall that the flow equation is obtained by inserting the appropriate motion equation (= Darcy's law) into the mass balance equation.
(d) Constitutive equations (and equations of state) for the phases involved, including, if necessary, the solid matrix. For the case of saturated flow, we may need the relationships between density and pressure, and between porosity and effective stress.
(e) Information on the various sources and sinks of water mass. Sometimes, these take the form of functions of the problem's state variables.
(f) Formulation of the conditions that prevail everywhere within the considered domain at some initial time, in terms of the problem's state variables.
(g) Formulation of the conditions that prevail on the domain boundaries, specified in item (a) above, during the period of interest. In many cases, the delineation of a boundary segment and the conditions on it have to be considered simultaneously, i.e., we select boundaries such that we can specify the conditions on them.
(h) Numerical values, or functional relations, of all the coefficients and parameters that appear in the equations mentioned in items (c) through (g) above.

The set of equations (mass balance equations, motion equations, and constitutive relations) must constitute a *closed* one, i.e., it should contain a sufficient number of equations to permit the simultaneous solution for all state variables of the problem.

After writing the closed set of equations, we use the methodology discussed in Subs. 7.9.4 to determine the number of *primary variables* of the problem and to select the most convenient ones. We then identify an equal number of (partial differential) balance equations that have to be solved in order to determine the values of these variables (whether they appear explicitly in the equations or not). All the remaining equations and relationships, including any partial differential equations used for defining variables, are then employed in order to determine the remaining variables. Initial and boundary conditions are specified *only* for the partial differential equations that have to be solved.

The situation described above becomes much simpler in the case of saturated flow considered in this chapter, for which we have to solve only a single partial differential equation—the flow equation—in the single variable pressure, or piezometric head. We have presented the discussion above in a somewhat more generalized form in order to facilitate the discussion in Chaps. 6 and 7, where we shall be discussing more than one fluid phase and more than one dissolved chemical species.

All the information included in the conceptual model must be available in order to construct and solve the model that will provide the required prediction. In practice, however, this is not always possible, primarily because of the lack of sufficient data, or the lack of resources to acquire and interpret the necessary data. Nevertheless, our objective should still be to provide management with the best prediction, approximate as it might be, under the prevailing circumstances. For example, the problem domain may not have natural boundaries (e.g., a lake, river, or an impervious geological formation) on which conditions are known, or such boundaries may be at large distances from the primary region of interest, or measurements may not be available. Since we still have to delineate some boundaries to the domain of interest, an estimate (even a guess based on our experience and prior knowledge) of what conditions prevail on them, or are anticipated to prevail on them in the future, may be used in the model.

Obviously, we have to make sure that errors in our estimates do not affect the results significantly. We do so by performing a *sensitivity analysis*, discussed in Sec. 10.2.2. The impacts of a number of possible alternative boundary locations and a range of possible conditions that may prevail on them, are evaluated to ensure that their effects are minimal. If, on the other hand, we find that the location of the boundary or conditions on it may significantly affect model predictions, we may conclude that more investment to acquire additional data are justified. Recall that data is required in order to solve an inverse (or parameter estimation) problem, aimed at estimating values of coefficients appearing in the model.

We take the same approach with respect to model parameters and source or sink functions. Measurements should be combined with our best judgement and available experience to produce estimated values. Then, a sensitivity analysis should be performed to determine to what extent the predicted values of the state variables are affected by these estimates. When the analysis indicates that a certain parameter does affect the predicted values significantly, it may mean that a greater investment in determining the value of that parameter is justified.

Although the preferred method of solution is the *analytical* one, it is, generally, impossible to derive analytical solutions for most cases of practical interest, and numerical methods have to be employed. Chapter 8 is devoted to numerical methods and computer codes.

5.4 Modeling 2-D Flow in Aquifers

In principle, flow always takes place in a three-dimensional domain. However, as discussed in Sec. 2.6, the main feature of an aquifer is that it is an essentially horizontal flow domain, characterized by a thickness that is much smaller than its horizontal extent of interest; hence, the vertical variations in piezometric head are, usually, much smaller than the horizontal ones. Under such circumstances, flow in an aquifer may be *conceptually* modeled (albeit *as an approximation*) as taking place in a horizontal two-dimensional domain. We referred to this approximation as 'the essentially horizontal flow' approximation, or 'the hydraulic approach'. The transformation of the three-dimensional mathematical model into a horizontal two-dimensional one is performed by integrating (or averaging) the former along the vertical coordinate axis. We have already demonstrated this approach in Sec. 4.4, where we have introduced the concept of 'aquifer transmissivity' for a confined aquifer. When we employ the hydraulic approach, the flow in the aquifer is described in terms of the average piezometric head, taken over its thickness.

In this section, we shall present the derivation of the mass balance equation for 2-D flow in aquifers in two ways. First, we shall develop the equation, rigorously, by vertical integration over the aquifer thickness. In a subsequent subsection, for the benefit of those who may have difficulties with the mathematics, we shall develop *the same equations* in a simpler way.

5.4.1 Deriving 2-D balance equations by integration

Consider, for example, the piezometric head, $h = h(x, y, z, t)$. Its average over the vertical thickness, $B(x, y, t)$, of an aquifer is defined by

$$\tilde{h}(x, y, t) = \frac{1}{B(x, y, t)} \int_{B(x,y,t)} h(x, y, z, t)\, dz. \tag{5.4.1}$$

In terms of this averaged variable, $\tilde{h}$, which is a function of x, y and t only, the flow equation is reduced to a two-dimensional one in the horizontal, xy-plane.

In addition to the mathematical advantage achieved by reducing the model from three to two dimensions, a two-dimensional flow model requires less data about the spatial distributions of the various model coefficients. This also means that fewer field observations may be needed in order to evaluate these coefficients. However, it should be emphasized again that the hydraulic approach may be employed only when the vertical variations of the relevant state variables, in comparison with their respective averages, are much smaller (or less important to the modeler) along the vertical axis than along the horizontal ones.

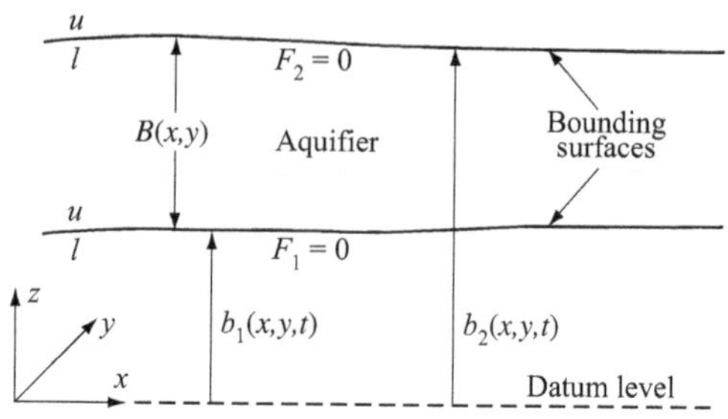

Figure 5.4.1: Nomenclature for integration over the thickness of an aquifer.

We recall that the assumption of 'essentially horizontal flow', usually referred to as the *Dupuit assumption* (Sec. 2.6), introduced by Dupuit (1863) in connection with phreatic aquifers, means that equipotentials are vertical and (equivalently) the vertical pressure distribution is hydrostatic. Although we shall apply here the Dupuit assumption to water flow in an aquifer, the presented material may be extended to the transport of any extensive quantity in any horizontal thin domain (Bear and Bachmat, 1990, p. 481).

We shall start by developing the *integrated balance equation* for *any* extensive quantity, E, having a (microscopic) density e (= amount of E per unit volume of the phase). When such a quantity is being transported by the fluid phase that occupies the entire void space in an aquifer, the general *macroscopic balance equation* is given by (5.1.5), rewritten here in the form:

$$\frac{\partial}{\partial t}(\phi e) + \nabla \cdot (e\mathbf{q} + \phi\,\mathbf{J}_h^E) - \Gamma'' = 0, \tag{5.4.2}$$

where ϕ denotes the porosity, $\mathbf{q}$ denotes the specific discharge of the phase, $\phi\mathbf{J}_h^E$ denotes the sum of dispersive and diffusive fluxes (Sec. 7.1) of the extensive quantity, per unit area of porous medium, and Γ'' denotes the total source (= rate of production) of E, due to both internal production and influx across the (microscopic) surface that bounds the phase, per unit volume of porous medium.

The methodology of the hydraulic approach calls for the integration of (5.4.2) along the vertical (possibly varying) thickness of the aquifer. Let the aquifer be bounded from above and below by (possibly moving) surfaces whose elevations are at $z = b_1(x,y,t)$ and $z = b_2(x,y,t)$, respectively, with $b_2 - b_1 = B$. Another way of expressing the geometry of these boundary surfaces is by (Fig. 5.4.1):

$$\begin{aligned} F_1 &\equiv F_1(x,y,z,t) = z - b_1(x,y,t) = 0, \\ F_2 &\equiv F_2(x,y,z,t) = z - b_2(x,y,t) = 0, \end{aligned} \tag{5.4.3}$$

where $F_i(\mathbf{x},t) = 0$ represents the equation of a boundary surface, or a segment of it. Time is introduced to allow for the possibility of a moving boundary, e.g., a phreatic surface, with $\mathbf{u}$ denoting the speed of displacement of such a boundary. This speed should not be confused with the velocity of the fluid present on both sides of the surface. As explained in Subs. 5.2.1, when a surface moves, its shape may change, but its equation, $F(\mathbf{x}, t) = 0$, remains unchanged. The quantity F is, thus, a conservative quantity of the points on the surface for which the total derivative vanishes. This leads to

$$\mathbf{n} = \frac{\nabla F}{|\nabla F|}, \qquad u_n = \mathbf{u} \cdot \mathbf{n} = -\frac{\partial F/\partial t}{|\nabla F|} = -\frac{\partial F/\partial t}{\partial F/\partial s_n}, \tag{5.4.4}$$

where $\mathbf{n}$ denotes the unit vector normal to the surface $F = 0$, and s_n is a distance measured along $\mathbf{n}$.

With the above definitions of the F-surfaces, we have from (5.4.3):

$$\nabla F_i = \nabla(z - b_i) \qquad \text{and} \qquad \frac{\partial F_i}{\partial t} = -\frac{\partial b_i}{\partial t}, \qquad i = 1, 2. \tag{5.4.5}$$

For example, for a horizontal F_1-surface, ∇F_1 is directed upward, normal to the surface. Also, for any surface $F_i = F_i(x, y, z, t)$, we have:

$$\frac{\partial F_i}{\partial t} + \mathbf{u} \cdot \nabla F_i = 0, \quad \text{or} \quad \frac{\partial b_i}{\partial t} - \mathbf{u} \cdot \nabla(z - b_i) = 0, \qquad i = 1, 2. \tag{5.4.6}$$

For a stationary boundary

$$b_i = b_i(x, y), \quad \frac{\partial F_i}{\partial t} = 0, \quad i = 1, 2. \tag{5.4.7}$$

By integrating (5.4.2) along the thickness, B, we obtain

$$\int_{b_1}^{b_2} \frac{\partial \phi e}{\partial t}\, dz + \int_{b_1}^{b_2} \nabla \cdot (e\mathbf{q} + \phi \mathbf{J}_h^E)\, dz - \int_{b_1}^{b_2} \Gamma''\, dz = 0. \tag{5.4.8}$$

Since we have here integrals of derivatives, with integration boundaries that are space- and possibly time-dependent, we have to introduce a certain rule for taking integrals of derivatives. This rule is based on *Leibnitz' rule* for a derivative of an integral with respect to a variable upon which the boundaries of the latter depend. This rule was introduced as (4.4.5) in Sec. 4.4. We rewrite this rule here in the form:

$$\frac{\partial}{\partial r} \int_{b_1}^{b_2} \mathbf{A}\, dz = \int_{b_1}^{b_2} \frac{\partial \mathbf{A}}{\partial r}\, dz + \mathbf{A}\Big|_{b_2} \frac{\partial b_2}{\partial r} - \mathbf{A}\Big|_{b_1} \frac{\partial b_1}{\partial r}, \tag{5.4.9}$$

where $\mathbf{A} = \mathbf{A}(x, y, z, t)$ is any tensor field, and r stands for x, y, z, or t.

Let us define the symbol $\widetilde{\mathbf{A}}$ as

$$\widetilde{\mathbf{A}}(x,y,t) = \frac{1}{B(x,y,t)} \int_{b_1(x,y,t)}^{b_2(x,y,t)} \mathbf{A}(x,y,z,t)\, dz, \tag{5.4.10}$$

in which the prime symbol denotes a vector (or vector operator) in the two-dimensional (xy)-plane only, viz.,

$$\mathbf{A}' = A_x \mathbf{1x} + A_y \mathbf{1y}, \qquad \nabla'(..) = \frac{\partial}{\partial x}(..)\, \mathbf{1x} + \frac{\partial}{\partial y}(..)\, \mathbf{1y},$$

with $\mathbf{1x}$ and $\mathbf{1y}$ denoting unit vectors in the x- and y-directions.

Making use of Leibnitz rule, we may write for any vector, $\mathbf{A}$:

$$\begin{aligned}
\int_{b_1(x,y,t)}^{b_2(x,y,t)} \nabla \cdot \mathbf{A}\, dz &= \int_{b_1}^{b_2} \left(\nabla' \cdot \mathbf{A}' + \frac{\partial A_z}{\partial z} \right) dz \\
&= \nabla' \cdot \int_{b_1}^{b_2} \mathbf{A}'\, dz - \mathbf{A}'\big|_{b_2} \cdot \nabla' b_2 + \mathbf{A}'\big|_{b_1} \cdot \nabla' b_1 + A_z\big|_{b_2} - A_z\big|_{b_1} \\
&= \nabla' \cdot B\widetilde{\mathbf{A}'} + \mathbf{A}\big|_{b_2} \cdot \nabla(z - b_2) - \mathbf{A}\big|_{b_1} \cdot \nabla(z - b_1) \\
&= \nabla' \cdot B\widetilde{\mathbf{A}'} + \mathbf{A}\big|_{F_2} \cdot \nabla F_2 - \mathbf{A}\big|_{F_1} \cdot \nabla F_1.
\end{aligned} \tag{5.4.11}$$

Here, and henceforth, $\big|_{F_i}$ stands for $\big|_{F_i=0}$.

For any scalar, $A(x,y,z,t)$, with $b_1 = b_1(x,y,t)$, $b_2 = b_2(x,y,t)$, we have

$$\begin{aligned}
\int_{b_1}^{b_2} \frac{\partial A}{\partial t}\, dz &= \frac{\partial}{\partial t} \int_{b_1}^{b_2} A\, dz - A\big|_{b_2} \frac{\partial b_2}{\partial t} + A\big|_{b_1} \frac{\partial b_1}{\partial t} \\
&= \frac{\partial}{\partial t} B\tilde{A} + A\big|_{F_2} \frac{\partial F_2}{\partial t} - A\big|_{F_1} \frac{\partial F_1}{\partial t}.
\end{aligned} \tag{5.4.12}$$

By applying (5.4.6), (5.4.12), (5.4.11), and (5.4.8), we obtain

$$\begin{aligned}
&\frac{\partial}{\partial t} B\widetilde{\phi e} + \nabla' \cdot B\left(\widetilde{e\,\mathbf{q}'} + \widetilde{\phi\, \mathbf{J}'^E_h} \right) + [\phi e(\mathbf{V} - \mathbf{u}) + \phi\, \mathbf{J}^E_h]\big|_{F_2} \cdot \nabla F_2 \\
&\qquad - [\phi e(\mathbf{V} - \mathbf{u}) + \phi\, \mathbf{J}^E_h]\big|_{F_1} \cdot \nabla F_1 - B\widetilde{\Gamma''} = 0,
\end{aligned} \tag{5.4.13}$$

in which $\mathbf{q}'$ denotes the vector of specific discharge in the horizontal xy-plane. This is the averaged, two-dimensional (in the horizontal plane) balance equation for any E in an aquifer. The dependent variables and fluxes, $\widetilde{\phi e}$, $\widetilde{e\mathbf{q}'}$, and $\widetilde{\phi \mathbf{J}'^E_h}$, are functions of x, y, and t only.

In (5.4.13), the terms

$$[\phi e(\mathbf{V} - \mathbf{u}) + \phi \mathbf{J}^E_h]\Big|_{F_2} \cdot \nabla F_2 \quad \text{and} \quad [\phi e(\mathbf{V} - \mathbf{u}) + \phi \mathbf{J}^E_h]\Big|_{F_1} \cdot \nabla F_1$$

represent the total flux of E through the (possibly moving) boundaries $F_2 = 0$ and $F_1 = 0$, which bound the aquifer from above and below, respectively. In other words, these terms represent *boundary conditions* on these surfaces. We note that while these terms are boundary conditions for the three-dimensional balance equation, (5.4.2), they appear as *source terms* in the averaged, two-dimensional equation (5.4.13). Our next task is to express these conditions.

The general condition that must be satisfied at any point on a boundary $F = 0$, for all α-phases present in a system, in the absence of sources or sinks of the quantity E on the latter, is the continuity of the normal component of the total flux of E in all phases. Using subscripts *ext* and *int* to denote the *external* and *internal* sides of the boundary $F = 0$, respectively, we may rewrite this condition in the form:

$$\sum_{(\alpha=s,f)} [\![\, \theta_\alpha \left\{ e_\alpha(\mathbf{V}_\alpha - \mathbf{u}) + \mathbf{j}^{E_\alpha}_{h\alpha} \right\}]\!]_{\text{ext,int}} \cdot \mathbf{n} = 0, \tag{5.4.14}$$

where the symbols f and s denote the fluid and solid phases, respectively, and $[\![\, (..) \,]\!]$ denotes the jump in (..) from one side (here, external) of the boundary to the other (here, internal). When the microscopic interphase boundary is a *material boundary* with respect to the considered quantity, i.e., there is no exchange of that quantity among the phases, the *no-jump* condition (5.4.14) may be written separately for each phase.

In what follows, we shall assume that the top and bottom surfaces that bound a confined or a leaky aquifer are material surfaces with respect to the solid mass. Hence, on these surfaces, $(\mathbf{V}_s - \mathbf{u}) \cdot \mathbf{n} = 0$, and, therefore,

$$\begin{aligned} \phi(\mathbf{V}_f - \mathbf{u}) \cdot \mathbf{n} &= \phi(\mathbf{V}_f - \mathbf{u}) \cdot \mathbf{n} - \phi(\mathbf{V}_s - \mathbf{u}) \cdot \mathbf{n} \\ &= \phi(\mathbf{V}_f - \mathbf{V}_s) \cdot \mathbf{n} \equiv \mathbf{q}_r \cdot \mathbf{n}. \end{aligned} \tag{5.4.15}$$

Recall that $\mathbf{q}_r$ denotes the specific discharge of the fluid relative to the solid; it is expressed by Darcy's law.

We can now use (5.4.14) to replace the terms in (5.4.13) that express the flux conditions on the 'internal sides' of the boundaries by terms that involve (*known*) information on the corresponding 'external sides.'

Let us develop the condition for an upper boundary, $F_2 = 0$, and for the specific case of the mass of water phase, $e = \rho$, in saturated flow. Since we have assumed that the boundary is a material surface with respect to the solid, we have on it

$$\rho_s(\mathbf{V}_s - \mathbf{u}) \cdot \nabla F_2 = 0. \tag{5.4.16}$$

In addition, for the F_2-surface, (5.4.14) reduces to

$$[\phi\rho(\mathbf{V}_f - \mathbf{u})]\big|_{\text{ext}} \cdot \nabla F_2 = [\phi\rho(\mathbf{V}_f - \mathbf{u})]\big|_{\text{int}} \cdot \nabla F_2, \tag{5.4.17}$$

where $\rho \equiv \rho_f$, or

$$(\rho\mathbf{q}_r)\big|_{\text{ext}} \cdot \nabla F_2 = (\rho\mathbf{q}_r)\big|_{\text{int}} \cdot \nabla F_2. \tag{5.4.18}$$

For an impervious boundary, $\mathbf{q}_r\big|_{\text{ext}} \cdot \nabla F_2 = \mathbf{q}_r\big|_{\text{ext}} \cdot \mathbf{n} = 0$, so that (5.4.18) reduces to

$$(\rho \mathbf{q}_r)\big|_{\text{int}} \cdot \nabla F_2 = 0. \tag{5.4.19}$$

At a leaky boundary, i.e., a boundary through which fluid mass can enter or leave the aquifer at a *known* rate, $\rho \mathbf{q}_{\text{leak}}$, the condition is

$$(\rho \mathbf{q}_r)\big|_{\text{int}} \cdot \nabla F_2 = (\rho \mathbf{q}_{\text{leak}})\big|_{\text{ext}} \cdot \nabla F_2. \tag{5.4.20}$$

The term $\mathbf{q}_{\text{leak}}\big|_{\text{ext}}$ represents the leakage into (or out of) the aquifer on the *external* side of the latter. It can now be expressed in terms of the state variables of a considered problem.

Let the surface F_2 serve as the upper boundary for the saturated domain of a phreatic aquifer with accretion. The condition on such a boundary takes the form

$$[\phi \rho (\mathbf{V}_f - \mathbf{u})]\big|_{\text{int}} \cdot \nabla F_2 = \rho_N (\mathbf{N} - \theta_{rf} \mathbf{u})\big|_{\text{ext}} \cdot \nabla F_2, \tag{5.4.21}$$

or, by rearranging terms, and using (5.2.3),

$$(\rho_N \mathbf{N} - \rho \mathbf{q}) \cdot \nabla F_2 = (\phi \rho - \theta_{rf} \rho_N) \frac{\partial F_2}{\partial t}. \tag{5.4.22}$$

Here, $\mathbf{N}$ denotes the rate of accretion of water of density ρ_N, and θ_{rf} denotes the irreducible moisture content that is assumed to prevail above the phreatic surface.

For downward accretion at a rate N, we introduce $\mathbf{N} = -N \nabla z$, and (5.4.22) becomes:

$$(\rho_N N \nabla z + \rho \mathbf{q}) \cdot \nabla F_2 = -(\phi \rho - \theta_{rf} \rho_N) \frac{\partial F_2}{\partial t}. \tag{5.4.23}$$

Let us now rewrite the mass balance equation (5.4.13), making use of the following approximations:

- The macrodispersive flux (Subs. 7.1.8) of the total mass, $\widetilde{\hat{\rho}\hat{\mathbf{q}}'}$, due to vertical variations in the horizontal flux, $\mathbf{q}'$, and in ρ, may be neglected, i.e.,

$$\widetilde{\rho \mathbf{q}'} = \widetilde{\rho}\, \widetilde{\mathbf{q}'} + \widetilde{\hat{\rho}\hat{\mathbf{q}}'} \approx \widetilde{\rho}\, \widetilde{\mathbf{q}'}, \tag{5.4.24}$$

  where the symbol $\hat{(..)}$ denotes deviation of (..) from its average, $\widetilde{(..)}$, over the vertical, B. Note that in (7.1.67), the macrodispersive flux, denoted by the double overbar, was obtained by a volume average (over an RMV), while here it is obtained by integration over the vertical, and denoted by the *tilde* symbol.
- The average of the sum of the components of the dispersive and diffusive fluxes of the total mass is much smaller than the advective mass flux at the averaged level, i.e.,

$$|\widetilde{\phi \mathbf{J}^{m\prime}_{fh}}| \ll |\widetilde{\rho}\,\widetilde{\mathbf{q}'}|.$$

With these approximations, (5.4.13) can be rewritten in the form:

$$\frac{\partial}{\partial t}(B\widetilde{\phi\rho}) + \nabla' \cdot (B\widetilde{\rho}\,\widetilde{\mathbf{q}'}) + \left[\rho(\mathbf{q} - \phi\mathbf{u})\right]\big|_{F_2} \cdot \nabla F_2$$
$$- \left[\rho(\mathbf{q} - \phi\mathbf{u})\right]\big|_{F_1} \cdot \nabla F_1 - B\widetilde{\Gamma''} = 0. \qquad (5.4.25)$$

Finally, let us rewrite the last balance equation for each type of aquifer, separately.

A. Confined aquifer

For this case, both the upper and lower bounding surfaces are impervious boundaries. On such boundaries, $\mathbf{u} \cdot \nabla F = \mathbf{V}_s \cdot \nabla F$; hence, two conditions prevail: (5.4.19) and a similar one for $F_1 = 0$. By inserting these conditions into (5.4.25), we obtain:

$$\frac{\partial(B\widetilde{\phi\rho})}{\partial t} + \nabla' \cdot (B\widetilde{\rho}\,\widetilde{\mathbf{q}'}) - B\widetilde{\Gamma''} = 0. \qquad (5.4.26)$$

This is the *(integrated) balance equation for flow in a confined aquifer.* We note in it the specific discharge, $\widetilde{\mathbf{q}'}$, rather than the specific discharge relative to the moving solids, $\widetilde{\mathbf{q}'_r}$. Pumping and artificial recharge may serve as examples of distributed sinks and sources expressed by $B\widetilde{\Gamma''}$.

In order to express the mass balance equation (5.4.26), which applies to a confined aquifer with a compressible fluid, in terms of a single state variable, $\widetilde{h}$, we make use of (5.4.12), and introduce the approximations

$$\frac{\partial(B\widetilde{\phi\rho})}{\partial t} \approx B\frac{\partial(\widetilde{\phi\rho})}{\partial t} \approx \widetilde{\rho}BS_o\frac{\partial\widetilde{h^*}}{\partial t}, \qquad (5.4.27)$$

$$(\phi\rho)\big|_{F_1} \approx (\phi\rho)\big|_{F_2} \approx \widetilde{\phi\rho}, \qquad (5.4.28)$$

where h^* is Hubbert potential defined by (4.1.6), and neglect averages of products of fluctuations over the thickness. We also assume that

$$h^*\big|_{F_1} \approx h^*\big|_{F_2} \approx \widetilde{h^*} \approx \widetilde{h} \equiv h, \qquad (5.4.29)$$

$$\nabla' \cdot (B\widetilde{\rho}\,\widetilde{\mathbf{q}'}) \approx -\nabla' \cdot (\widetilde{\rho}B\widetilde{\mathbf{K}'} \cdot \nabla'\widetilde{h^*}), \qquad (5.4.30)$$

$$\mathbf{q} \approx \mathbf{q}_r, \qquad (5.4.31)$$

and

$$\left|B\widetilde{\phi}\frac{\partial\widetilde{\rho}}{\partial t}\right| \gg \left|B\widetilde{\mathbf{q}'} \cdot \nabla'\widetilde{\rho}\right|. \qquad (5.4.32)$$

These approximations lead to the averaged mass balance equation

$$\widetilde{\rho} S \frac{\partial h}{\partial t} = \nabla' \cdot (\widetilde{\rho} \mathbf{T} \cdot \nabla' h) + B\widetilde{\Gamma''}, \tag{5.4.33}$$

in which the *aquifer storativity* or storage coefficient, S, is defined by

$$S = \int_{b_1}^{b_2} S_o \, dz, \qquad S = B\widetilde{S_o}, \tag{5.4.34}$$

where $\widetilde{S_o}$ represents the average value of S_o along B. In words, S is defined as the *volume of water release from (or added to) storage in a confined aquifer per unit area of aquifer per unit decline (or rise) the piezometric head.*

The term $-\mathbf{T} \cdot \nabla' \widetilde{h}$, which we denote as $\mathbf{Q}'$, expresses the total discharge through the entire thickness of the aquifer per unit width, with

$$\mathbf{T} = \int_{b_1}^{b_2} \mathbf{K} \, dz = B\widetilde{\mathbf{K}'} \tag{5.4.35}$$

denoting *aquifer transmissivity.* The same definition is presented in (4.4.12).

If, now, we invoke the Dupuit assumption that equipotentials can be approximated as vertical, i.e., $\widetilde{h} \approx h(x, y, b_1) \approx h(x, y, b_2)$, then (5.4.35) reduces to

$$\mathbf{Q}' = -\mathbf{T} \cdot \nabla' \widetilde{h}, \qquad \mathbf{T} = \mathbf{K}(x, y) B(x, y). \tag{5.4.36}$$

In (5.4.33), $B\widetilde{\Gamma''}/\widetilde{\rho}$ denotes a water source (dims. $L^3/T/L^2$). For wells, we express the sources by using the *Dirac delta function*, $\delta(x - x^i, y - y^i)$, modified to two dimensions from the three-dimensional definition (5.1.77).

In groundwater hydrology, it is often assumed that $\widetilde{\rho} =$ const. in (5.4.33). Thus, for water of constant density, with pumping wells at points (x^m, y^m), with pumping rates, P^m, we obtain the confined aquifer flow equation

$$S \frac{\partial h}{\partial t} = \nabla' \cdot (\mathbf{T} \cdot \nabla' h) - \sum_{(m)} P^m \delta(x - x^m, y - y^m). \tag{5.4.37}$$

The quotient $\mathbf{T}/S$ is often referred to as the *aquifer diffusivity.*

B. Leaky aquifer

In this case, (5.4.20) serves as a boundary condition on the upper bounding surface (a similar expression can be written for the lower surface). We assume that the upper semipervious boundary can be approximated as a thin membrane through which water leaks out of the aquifer into an overlying aquifer. Then, the rate of leakage through the upper surface of the aquifer can be expressed by

$$\mathbf{q}_{\text{leak}}\big|_{F_2} \cdot \nabla F_2 = \mathsf{K}_\ell \frac{\widetilde{h} - h\big|_{\text{ext}}}{B_\ell} |\nabla F_2|, \tag{5.4.38}$$

where $h|_{\text{ext}}$ denotes the piezometric head above the upper semipervious layer, and K_ℓ and B_ℓ denote the hydraulic conductivity and thickness of the semipervious layer. This term is sometimes called *leakance*. The term B_ℓ/K_ℓ is referred to as the *resistance* (dims. T) of the semipervious layer. In writing (5.4.38), we have also assumed that the water density is constant, and is, thus, the same on both sides of the semipervious layer.

We now use (5.4.13), making the same approximations used to derive (5.4.33) and (5.4.36) for the confined aquifer, defining $q_{\text{leak}}|_{F_i} \equiv \mathbf{q}_{\text{leak}}|_{F_i} \cdot \nabla F_i$ $i = 1, 2$, to obtain the following relationships from (5.4.20):

$$\mathbf{q}|_{F_1} \cdot \frac{\nabla F_1}{|\nabla F_1|} = q_{\text{leak}}|_{F_1} \qquad \text{and} \qquad \mathbf{q}|_{F_2} \cdot \frac{\nabla F_2}{|\nabla F_2|} = q_{\text{leak}}|_{F_2},$$

which represent the conditions on the top and the bottom boundaries. With $R(x, y, t) \equiv B\widetilde{\Gamma''}/\rho$, we then obtain

$$\nabla' \cdot (\mathsf{T} \cdot \nabla' \widetilde{h}) + R(x, y, t) - q_{\text{leak}}|_{F_2} |\nabla F_2| + q_{\text{leak}}|_{F_1} |\nabla F_1| = \widetilde{S} \frac{\partial \widetilde{h}}{\partial t}. \quad (5.4.39)$$

This is the *(integrated) mass balance equation for a leaky aquifer*. The leakage terms express (possible) loss of water to the overlying and underlying aquifers.

We now use (5.4.13), making the same approximations used to derive (5.4.33) and (5.4.36) for the confined aquifer, defining $q_{\text{leak}}|_{F_i} \equiv \mathbf{q}_{\text{leak}}|_{F_i} \cdot \nabla F_i$ $i = 1, 2$, to obtain the following relationships from (5.4.20):

$$\mathbf{q}|_{F_1} \cdot \frac{\nabla F_1}{|\nabla F_1|} = q_{\text{leak}}|_{F_1} \qquad \text{and} \qquad \mathbf{q}|_{F_2} \cdot \frac{\nabla F_2}{|\nabla F_2|} = q_{\text{leak}}|_{F_2},$$

which represent the conditions on the top and the bottom boundaries. With $R(x, y, t) \equiv B\widetilde{\Gamma''}/\rho$, we then obtain

$$\nabla' \cdot (\mathsf{T} \cdot \nabla' \widetilde{h}) + R(x, y, t) - q_{\text{leak}}|_{F_2} |\nabla F_2| + q_{\text{leak}}|_{F_1} |\nabla F_1| = \widetilde{S} \frac{\partial \widetilde{h}}{\partial t}. \quad (5.4.40)$$

This is the *(integrated) mass balance equation for a leaky aquifer*. The leakage terms express (possible) loss of water to the overlying and underlying aquifers. The quotient $\mathsf{T}/\widetilde{S}$ is often referred to as the *aquifer diffusivity*. We can also define a characteristic time for the aquifer as $t_c = S_c L_c^2/\mathsf{T}_c$, where subscript c denotes characteristic values, and a corresponding *Fourier number* for an aquifer, following the discussion on dominance of effects, in Sec. 7.7.

C. Phreatic aquifer

Let the lower boundary be impervious, so that (5.4.19) is valid there. Then, using (5.4.21) with (5.2.3), and employing the same assumptions as introduced above, including constant water density, (5.4.25) becomes

$$\frac{\partial(B\widetilde{\phi})}{\partial t} + \nabla' \cdot (B\widetilde{\mathbf{q}'}) + \left(\mathbf{N} \cdot \nabla F_2 + \theta_{rw}\frac{\partial F_2}{\partial t}\right) - B\frac{\widetilde{\Gamma''}}{\rho} = 0, \tag{5.4.41}$$

where we have assumed $b_2 \approx \widetilde{h}$ so that $B = \widetilde{h} - b_1$. This is the *(integrated) mass balance equation for a phreatic aquifer.*

Assuming that $|B(\partial\widetilde{\phi}/\partial t)| \ll |(\phi|_{\widetilde{h}}\, \partial\widetilde{h}/\partial t)|$, and since $F_1 = z - b_1(x,y)$, and $F_2 = z - \widetilde{h}(x,y,t)$, the first term of the left-hand side of (5.4.41) becomes:

$$\begin{aligned}\frac{\partial B\widetilde{\phi}}{\partial t} &\equiv \frac{\partial}{\partial t}\int_{b_1}^{\widetilde{h}} \phi\, dz = \int_{b_1}^{\widetilde{h}} \frac{\partial \phi}{\partial t}\, dz + \phi\bigg|_{\widetilde{h}} \frac{\partial \widetilde{h}}{\partial t} \\ &= B\frac{\partial\widetilde{\phi}}{\partial t} + \phi\bigg|_{\widetilde{h}}\frac{\partial\widetilde{h}}{\partial t} \approx \phi\bigg|_{\widetilde{h}}\frac{\partial\widetilde{h}}{\partial t}.\end{aligned} \tag{5.4.42}$$

With $\mathbf{N} = -N\nabla z$, by inserting (5.4.42) into (5.4.41), we now obtain

$$S_y\frac{\partial\widetilde{h}}{\partial t} + \nabla' \cdot [(\widetilde{h} - b_1)\widetilde{\mathbf{q}'}] - N - R + P = 0, \tag{5.4.43}$$

in which we have assumed that $\nabla F_2 \approx -\nabla z$, i.e., the water table is approximately horizontal with respect to $\mathbf{N}$, and the net withdrawal from the aquifer, i.e., pumping minus artificial recharge, is denoted by $P - R \equiv -(\widetilde{h} - b_1)\widetilde{\Gamma''}/\rho_w$.

The symbol $S_y \equiv \phi_{\text{eff}} = \phi - \theta_{rw}$ is the *specific yield,* which is equivalent to the storativity of a phreatic aquifer. It is defined as the *volume of water released from storage in a phreatic aquifer, per unit area and per unit decline in the water table elevation,*

$$S_y = \frac{\Delta\mathcal{U}_w}{\mathcal{A}\Delta h}. \tag{5.4.44}$$

In this definition of S_y, we have assumed that $(\widetilde{h} - b_1)S_o \ll \phi_{\text{eff}}$, i.e., the effect of elastic storativity is negligible. The specific yield is further discussed in Subs. 6.1.9. In that discussion we shall emphasize that during to the delay in drainage from storage, the specific yield is actually time dependent.

Figure 5.4.2 shows the dependence of the specific yield, S_y, on grain- (actually pore-) size. We note that for clay, S_y is very small, although the porosity is relatively large. This behavior stems from the fact that the size of the pores in clays is very small, so that capillary forces are very large, and so is the irreducible moisture content, θ_{wr}.

Equation (5.4.43) is the balance equation commonly employed for a phreatic aquifer. It is often referred to by groundwater hydrologists as the *Boussinesq equation.* In (5.4.43), $\mathbf{Q}'$ $(\equiv (\widetilde{h} - b_1)\widetilde{\mathbf{q}'})$ denotes the total discharge through the saturated thickness, $\widetilde{h} - b_1$, per unit width of aquifer.

Let us use vertical integration to derive an expression for $\mathbf{Q}'$ in terms of $\widetilde{h}$. Assuming that $\mathsf{K} = \mathsf{K}(x,y)$, we obtain

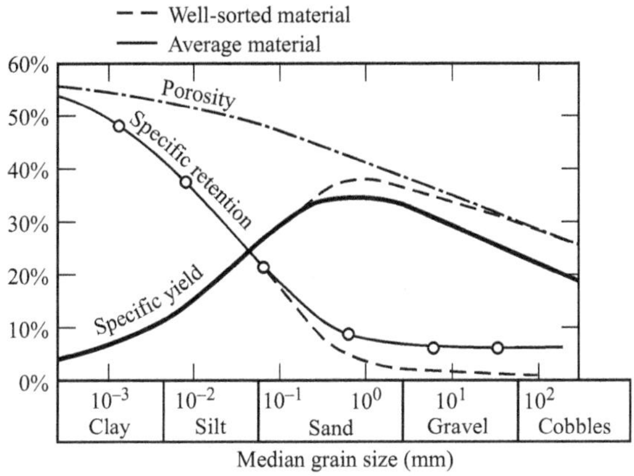

Figure 5.4.2: Relation between specific yield and grain size (from Conkling *et al.*, 1934, as modified by Davis and DeWiest, 1966).

$$\begin{aligned}\mathbf{Q}' &= \int_{b_1(x,y)}^{\widetilde{h}(x,y,t)} \mathbf{q}\, dz = -\mathbf{K} \cdot \int_{b_1}^{\widetilde{h}} \nabla h \, dz \\ &= -\mathbf{K} \cdot \left\{ \nabla' \left[(\widetilde{h} - b_1)\widetilde{h} \right] - h\Big|_{x,y,z=\widetilde{h}} \nabla' \widetilde{h} + \widetilde{h}\Big|_{x,y,z=b_1} \nabla' b_1 \right\}, \end{aligned} \tag{5.4.45}$$

where $h\big|_{(x,y,z=h)} = \widetilde{h}$.

By invoking now the Dupuit assumption, i.e., $\widetilde{h} \approx h\big|_{(x,y,z=h)} \approx h\big|_{(x,y,z=b_1)}$, we obtain

$$\mathbf{Q}' = -(\widetilde{h} - b_1)\mathbf{K} \cdot \nabla' \widetilde{h}. \tag{5.4.46}$$

Equation (5.4.43) can then be written as

$$S_y \frac{\partial \widetilde{h}}{\partial t} + \nabla \cdot [(\widetilde{h} - b_1)\mathbf{K} \cdot \nabla' \widetilde{h}] - N - R + P = 0. \tag{5.4.47}$$

When recharge and pumping is implemented through wells ($\equiv$ point sources and sinks), we may, symbolically, use the notation

$$\begin{aligned} &R(x, y, t) - P(x, y, t) \equiv \\ &\sum_{(i)} R^i(t)\delta(x - x^i, y - y^i) - \sum_{(j)} P^j(t)\delta(x - x^j, y - y^j), \end{aligned} \tag{5.4.48}$$

where $\delta(x - x^i, y - y^i)$ is the *Dirac delta function* at (x^i, y^i).

In (5.4.47), the product $(\widetilde{h} - b_1)\mathbf{K}$ plays the role of *transmissivity* of a phreatic aquifer. However, here the transmissivity may be time dependent, because $\widetilde{h} = \widetilde{h}(x, y, t)$. As a result, the equation for flow in a phreatic aquifer is *nonlinear*.

In principle, the non-linear balance equation (5.4.47) can be solved numerically. However, often this equation is approximated by linearization prior to being solved numerically. Commonly, linearization is achieved by replacing h in the product $(\widetilde{h} - b_1)\mathbf{K}$, which represents the aquifer transmissivity, by some mean (in time!) value $\widehat{\widetilde{h}}$, assuming $|\widetilde{h} - \widehat{\widetilde{h}}| \ll \widehat{\widetilde{h}}$. Equation (5.4.47) then becomes

$$S_y \frac{\partial \widetilde{h}}{\partial t} + \nabla \cdot [(\widehat{\widetilde{h}} - b_1)\mathbf{K} \cdot \nabla' \widetilde{h}] - N - R + P = 0, \tag{5.4.49}$$

which is now linear in $\widetilde{h} = \widetilde{h}(x, y, t)$. The introduction of an average thickness of the saturated zone is justified whenever fluctuations in the water table elevations are much smaller than the thickness itself.

5.4.2 Another derivation of 2-D balance equations

The macroscopic mass balance equations for essentially horizontal water flow in an aquifer can also be obtained directly from the discussion in Subs. 5.1.1. Again, the basic idea of a balance is simple and straight forward. We write a balance for the mass of water within *a specified spatial domain*, and for a *specified period* of time. Because we assume essentially horizontal flow, we consider a balance for a volume with a horizontal area A_o and a thickness equal to the aquifer thickness, B, of the aquifer. Let Δt denote the time interval, and $\mathcal{S}$ denote the area of the vertical surface that bounds the volume. The mass balance is then stated as

$$\left\{\begin{array}{c}\text{Mass of water}\\ \text{accumulating}\\ \text{in } A_o \times B\\ \text{during } \Delta t\end{array}\right\} = \left\{\begin{array}{c}\text{Net mass of w.}\\ \text{entering } A_o \times B\\ \text{through } \mathcal{S}\\ \text{during } \Delta t\end{array}\right\} + \left\{\begin{array}{c}\text{Net rate of}\\ \text{w. production}\\ \text{in } A_o \times B\\ \text{during } \Delta t\end{array}\right\}. \tag{5.4.50}$$

Dividing by Δt and by A_o, we obtain the mass balance in the form:

$$\left\{\begin{array}{c}\text{Rate of w. mass}\\ \text{accumulating in}\\ \text{a unit area}\end{array}\right\} = \left\{\begin{array}{c}\text{Net rate of w.}\\ \text{mass entering}\\ \text{a unit area}\end{array}\right\} + \left\{\begin{array}{c}\text{Net rate of w.}\\ \text{production in}\\ \text{a unit area}\end{array}\right\}. \tag{5.4.51}$$

From the discussion in Subs. 5.1.1, it follows that

- $\partial B\phi\rho/\partial t =$ The rate of accumulation of water mass in the specified volume, per unit area.
- $-\nabla\cdot\rho B\mathbf{q}' =$ The net rate of mass influx, per unit area, with $\mathbf{Q}' = B\mathbf{q}'$.

Using the symbol Γ'' to denote the net rate of water production per unit volume, we can replace (5.4.51) by

$$\frac{\partial B\phi\rho}{\partial t} = -\nabla\cdot\rho\mathbf{Q}' + B\Gamma'', \tag{5.4.52}$$

to be compared with (5.4.26).

5.4.3 Complete aquifer flow models

The discussion presented in the Sec. 5.3, on the content of a 3-D mathematical flow model and on the requirement that the mathematical model be *well-posed*, is valid also here, and, therefore, need not be repeated. We shall focus only on the mass balance equations and on the initial and boundary conditions.

A. Summary of mass balance equations

Let us summarize the flow equations for flow in an aquifer domain $\Omega(x,y)$, bounded by segments $\mathcal{B} = \mathcal{B}(x,y)$ that together surround the entire aquifer domain. The symbol h will denote the average piezometric head, $\tilde{h}$. The aquifer transmissivity is defied by (5.4.35); the aquifer storativity is defined by (5.4.34). We shall assume that the water density is constant, except that we do take into account the contribution of water compressibility to aquifer storativity.

A confined aquifer. The aquifer is inhomogeneous and anisotropic, with a transmissivity $\mathbf{T} = \mathbf{T}(x,y)$, and storativity $S = S(x,y)$. The flow equation is:

$$S\frac{\partial h}{\partial t} = \nabla' \cdot (\mathbf{T} \cdot \nabla' h) - P(x,y,t) + R(x,y,t), \qquad (5.4.53)$$

in which $\nabla'\cdot$ and ∇' denote the divergence and the gradient operators in the xy-plane, and $P = P(x,y,t)$, and $R = R(x,y,t)$ denote, symbolically, pumping and artificial recharge, respectively (dims. LT^{-1}). When such sinks and sources take the form of point wells, these symbols can be replaced by

$$R(x,y,t) - P(x,y,t) = \sum_{(i)} R^i(t)\delta(\mathbf{x}-\mathbf{x}^i) - \sum_{(j)} P^j(t)\delta(\mathbf{x}-\mathbf{x}^j), \qquad (5.4.54)$$

in which $\delta(\mathbf{x}-\mathbf{x}^i)$ denotes the Dirac function for two dimensions (dims. L^{-2}) defined by (5.1.77), but with a^3 replaced by a^2.

Note that, as almost everywhere in this book, the balance equations are written in *vector form*; they can easily be rewritten in *indicial notation*, in terms of the selected coordinate system.

For steady flow ($\partial h/\partial t = 0$) in a homogeneous aquifer ($\mathbf{T} =$ const.) and in the absence of sources and sinks ($R = P = 0$), the balance equation (5.4.53) reduces to the *Laplace equation*

$$\nabla' \cdot \nabla' h = 0, \quad \text{or} \quad \nabla'^2 h \left(\equiv \frac{\partial^2 h}{\partial x^2} + \frac{\partial^2 h}{\partial y^2} \right) = 0, \quad h = h(x,y). \qquad (5.4.55)$$

A phreatic aquifer. The aquifer is inhomogeneous and anisotropic, with average (over the vertical) hydraulic conductivity $\mathbf{K} = \mathbf{K}(x,y)$, and specific

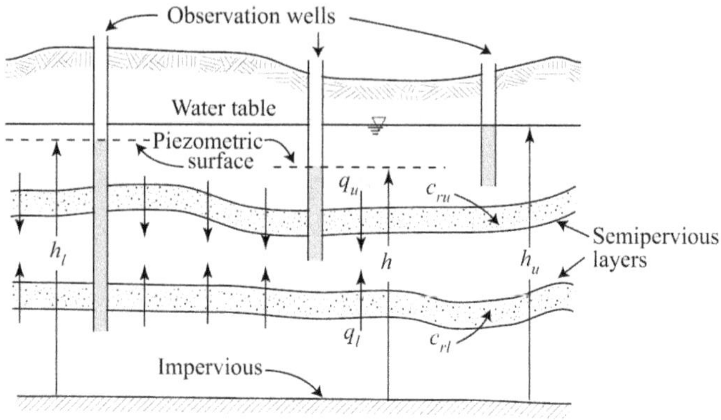

Figure 5.4.3: Multiple leaky aquifers.

yield, $S_y = S_y(x, y)$, defined in (5.4.44). Elastic storativity is neglected. The aquifer's bottom is at an elevation $\eta = \eta(x, y)$. The flow equation is

$$S_y \frac{\partial h}{\partial t} = \nabla' \cdot [(h - \eta)\mathbf{K} \cdot \nabla' h] + N(x, y, t) - P(x, y, t) + R(x, y, t), \quad (5.4.56)$$

in which $N = N(x, y, t)$ denotes the rate of natural replenishment of the aquifer, i.e., rate at which water reaches the water table from precipitation. One can include in this rate also water reaching the water table from excess irrigation.

A leaky confined aquifer. Consider the middle aquifer shown in Fig. 5.4.3. The piezometric head in this aquifer is denoted by $h_u = h_u(x, y, t)$. The aquifer is separated from an overlying (leaky phreatic, but could also be a leaky-confined) aquifer by a semipervious layer (= aquitard) having the resistance c_{ru}, and from an underlying (leaky confined) aquifer by a semipervious layer of resistance $c_{r\ell}$. These resistances are defined in (5.2.15). We denote the piezometric heads in the underlying leaky confined aquifer by $h_\ell = h_\ell(x, y, t)$, and that in the overlying phreatic leaky aquifer by $h_u = h_u(x, y, t)$. The flow equation in the considered leaky confined aquifer is

$$S \frac{\partial h}{\partial t} = \nabla' \cdot (\mathsf{T} \cdot \nabla' h) + \frac{h_u - h}{c_{ru}} + \frac{h_\ell - h}{c_{r\ell}} - P(x, y, t) + R(x, y, t). \quad (5.4.57)$$

For a homogeneous isotropic aquifer, the above equation reduces to

$$\frac{S}{\mathsf{T}} \frac{\partial h}{\partial t} = \frac{\partial^2 h}{\partial x^2} + \frac{\partial^2 h}{\partial y^2} + \frac{h_u - h}{\lambda_u^2} - \frac{h - h_\ell}{\lambda_\ell^2} + \frac{R - P}{\mathsf{T}}, \quad (5.4.58)$$

where $\lambda_i^2 = Tc_{ri}$, $i = \ell, u$, is another leaky aquifer coefficient called *leakage factor*, which determines the areal distribution of leakage.

To solve (5.4.57) for $h = h(x, y, t)$, we need, in addition to values of all relevant coefficients, also information, say from measurements, on $h_u = h_u(x, y, t)$ and $h_\ell = h_\ell(x, y, t)$. When this information is not available, we have three variables to solve for, and we need two additional equations:

For the lower leaky confined aquifer:

$$S_\ell \frac{\partial h_\ell}{\partial t} = \nabla' \cdot (\mathbf{T}_\ell \cdot \nabla' h_\ell) - \frac{h_\ell - h}{c_{r\ell}} - P_\ell(x, y, t) + R_\ell(x, y, t). \tag{5.4.59}$$

For the leaky phreatic aquifer:

$$S_y \frac{\partial h_u}{\partial t} = \nabla' \cdot (\mathbf{T}_u \cdot \nabla' h_u) - \frac{h_u - h}{c_{ru}} + N(x, y, t) - P_u(x, y, t) + R_u(x, y, t). \tag{5.4.60}$$

Obviously, the three PDEs have to be solved simultaneously, with appropriate initial and boundary conditions.

B. Initial and boundary conditions

In order to solve any of the aquifer flow equations presented above, or the set of equations describing the flow in the case of multiple leaky aquifers, we need to state appropriate initial and boundary conditions for every flow equation. We recall that we are considering here flow equations based on the assumption of essentially horizontal flow. Hence, the flow domain, $\Omega(x, y)$, which is in the horizontal plane, is bounded by a *closed boundary* $\mathcal{B}$, composed of straight line segments and curves, with $F_i = F_i(x, y, t) = 0$ representing the equation of the i-th segment. Unless otherwise stated, we assume that the boundaries are stationary; we shall regard h as the variable appearing in the flow equation.

Initial conditions Initial conditions take the form

$$h = h(x, y, t) = f(x, y), \quad \text{on} \quad \Omega, \tag{5.4.61}$$

where $f = f(x, y)$ is a known function.

Boundary conditions Several types of boundary conditions may be encountered.

(a) *Boundary of prescribed piezometric head.* This condition takes the form

$$h = f_1(x, y, t), \quad \text{on} \quad \mathcal{B}_1, \tag{5.4.62}$$

where $f_1 = f_1(x, y, t)$ is a known function.

A special case of this kind of boundary is the equipotential boundary, $f_1(x, y, t) = \text{constant}$, or $f_1(x, y, t) = f_1^*(t)$, where $f_1^*(t)$ is a known function. A specified piezometric head boundary is encountered whenever the aquifer is in direct contact with a lake or a river, in which the water level is known.

Another special case of this kind of boundary is a *spring* through which groundwater emerges to ground surface. The spring's outlet threshold is at a fixed elevation. Water emerges from the aquifer into the atmosphere (say, at p=0) at that fixed elevation, and, hence, this is a boundary of a specified piezometric head. Sometime, a layer of water exists above the threshold, which may vary with the rate of flow. However, when the piezometric head in the aquifer in the vicinity of the spring is lower than this threshold, the spring dries up and ceases to act as a boundary to the flow domain. It is thus a boundary of specified (known) head only as long as the water heads in the vicinity of the spring are higher than the spring's outlet. When a spring emerges at the bottom of a lake, the specified head at the spring is dictated by the water level in the lake. This kind of boundary conditions was discussed in detail in Subs. 5.2.3.

(b) *Boundary of prescribed flux.* Along such a boundary

$$Q'_n = \mathbf{Q}' \cdot \mathbf{n} = f_2(x, y, t), \quad \text{on} \quad \mathcal{B}_2, \tag{5.4.63}$$

where $f_2 = f_2(x, y, t)$ is a known function. For a confined aquifer, $\mathbf{Q}' = -\mathsf{T} \cdot \nabla' h$. For a phreatic aquifer, $\mathbf{Q}' = -(h - \eta)\mathsf{K} \cdot \nabla' h$.

For an isotropic porous medium, (5.4.63) can be reduced to the form

$$\partial h / \partial s_n = f'_2(x, y, t), \quad \text{on} \quad \mathcal{B}_2, \tag{5.4.64}$$

where s_n denotes the distance measured along the normal to the boundary, and $f'_2(x, y, t)$ is a known function.

For an impervious boundary, $f_2(x, y, t) = 0$ in (5.4.63) and $f'_2(x, y, t) = 0$ in (5.4.64). We recall that a *streamline* and a *water divide* behave as an impervious boundary.

(c) *Semipervious Boundary.* As in three-dimensional flow, this kind of boundary condition occurs when a partly clogged river-bed (e.g., by a thin layer of silt or clay) serves as a boundary of a flow domain. Because of the resistance to the flow offered by the semipervious layer, the piezometric head in the aquifer, next to this layer, is different from that on its external side (which is dictated by the water level in the river (Fig. 5.4.4).

For the phreatic aquifer shown (Fig. 5.4.4), since the flow is assumed horizontal, continuity of flux through the entire thickness of the aquifer requires that:

$$(Q'_n \equiv \mathbf{Q}' \cdot \mathbf{n} =) - (\mathsf{K} h \cdot \nabla' h) \cdot \mathbf{n} = h \frac{h_o - h(x, y, t)}{c_r}, \tag{5.4.65}$$

where $c_r = \mathsf{K}'/B'$, with K' and B' denoting the hydraulic conductivity and the thickness of the semipervious layer, respectively. For a confined aquifer, we replace $\mathsf{K}h$ by the aquifer transmissivity, T, and the thickness, h, on the r.h.s. by the aquifer thickness, B.

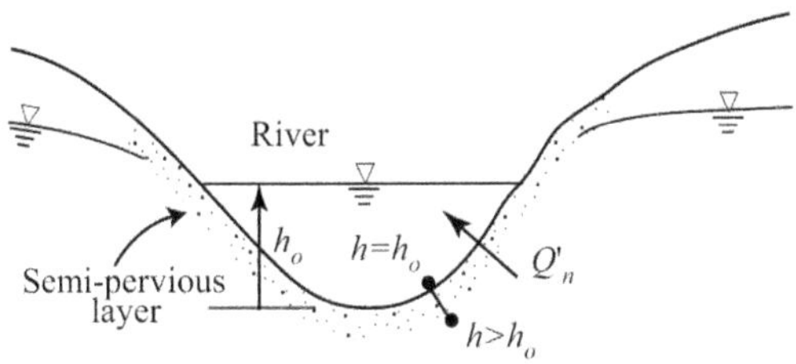

Figure 5.4.4: Semipervious boundary.

C. Complete model statement

Similar to the discussion on three-dimensional flow models, recalling that we are modeling the flow in an aquifer under the assumption of 'essentially horizontal flow', say, in the xy-plane, the standard content of a model should include the following items:

- Delineation of the closed curve that bounds the problem area. This means that all boundaries are really vertical surfaces extending through the entire thickness of the aquifer.
- Specification of the state variable, usually the average piezometric head h, for which a solution is sought.
- Statement of the partial differential equation that represents the mass balance of water in the aquifer. In the case of multiple leaky aquifers, we need a variable of state and a mass balance equation (PDE) for each aquifer.
- Specification of all the (storage and transport) coefficients that appear in the balance equation(s).
- Statement of initial conditions that the state variables have to satisfy.
- Statement of boundary conditions for each balance equation.

The investigated domain need not extend to natural boundaries, such as a river, or an impervious fault. Often, arbitrary boundaries are introduced. This is appropriate, as long as we know (or assume or even guess) the conditions on these boundaries.

The case of a boundary between two porous media domains was discussed in Sec. 5.2.3 and need not be repeated here, as the extension to two-dimensional models is obvious.

5.4.4 Effect of storage changes in aquitard

In the discussion on leaky confined aquifers (Subs. 5.4.3A), we assumed that the semipervious layers—aquitards—have zero storativity and, hence, any change in the piezometric head in the overlying and/or underlying aquifers propagates instantaneously within these aquitards. Therefore, we could as-

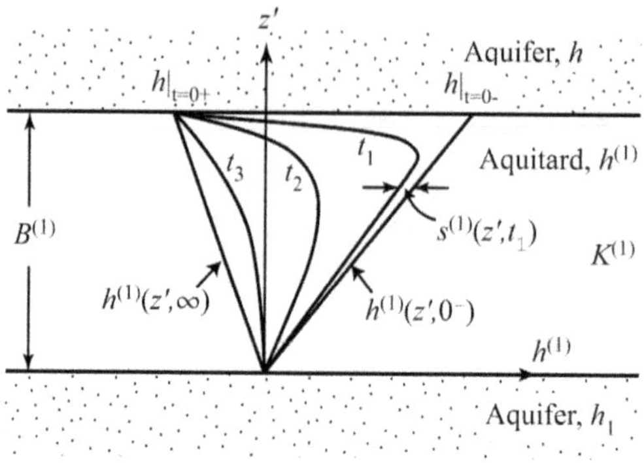

Figure 5.4.5: Changes in piezometric head within an aquitard.

sume that a linear distribution of head always exists in these layers. In reality, changes in water storage do take place in semipervious layers.

In unsteady flow, as the piezometric heads in the aquifers above and below the semipervious layer vary, continuous changes are produced in the piezometric head distribution in that layer. Figure 5.4.5 shows how an instantaneous step drop in the piezometric head in the upper aquifer, from $h|_{t=0^-}$ to $h|_{t=0^+}$, produces a gradual change in the piezometric head distribution, $h^{(1)}(z',t)$, in the aquitard. We note that the gradient $\partial h^{(1)}/\partial z'$ varies both with time and along the aquitard thickness, $B^{(1)}$. At the same time, the quantity of water stored (elastically) within each unit volume of porous medium along the thickness of the aquitard also varies as $h^{(1)}(z',t)$ varies.

In Fig. 5.4.5, we note that, whereas, initially, the flow throughout $B^{(1)}$ is everywhere downward, instantly, the flow moves in both directions, upward and downward. After some time, say t_3 (assuming that $h_{t=0^+}$ does not vary with time), the flow in the lower part of the aquitard also changes its direction. Eventually, the flow will be entirely upward throughout the aquitard. Employing the definition of specific storativity, S_o, the total volume of water, $\Delta\mathcal{U}_w$, released from storage in the aquitard (per unit horizontal area) during t is given by

$$\Delta\mathcal{U}_w = \int_0^{B^{(1)}} S_o^{(1)}[h^{(1)}(x,y,z',t) - h^{(1)}(x,y,z',0)]\, dz'. \qquad (5.4.66)$$

Depending on the aquitard permeability, $\mathsf{K}^{(1)}$ (actually on the ratio $S_o^{(1)}/\mathsf{K}^{(1)}$), it will take some time for this volume of water to be released from storage in the aquitard. Accordingly, the distribution of the piezometric head within the aquitard cannot respond instantaneously to head changes in the adjacent aquifers. We refer to this phenomenon as *delayed storage*. The entire picture is more complicated when the piezometric heads in the adjacent aquifers vary continuously.

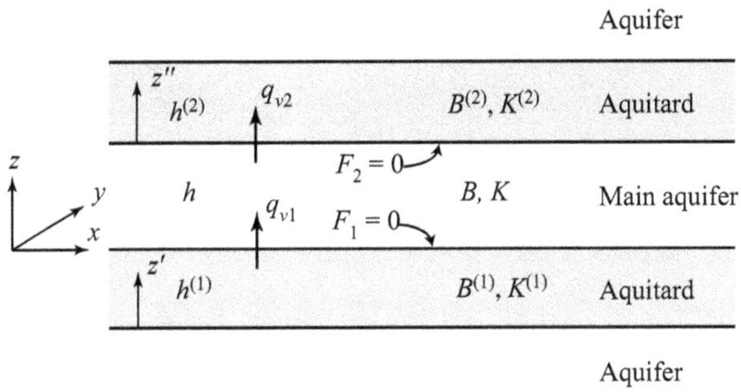

Figure 5.4.6: Nomenclature for an aquifer-aquitard system.

For the aquifer above the aquitard (see Figure 5.4.6), the rates of leakage entering and leaving the aquifer are expressed by

$$q_{v1}|_{F_1} = -\mathsf{K}^{(1)} \frac{\partial h^{(1)}}{\partial z'}\bigg|_{F_1}, \qquad q_{v2}|_{F_2} = -\mathsf{K}^{(2)} \frac{\partial h^{(2)}}{\partial z''}\bigg|_{F_2}, \tag{5.4.67}$$

where $h^{(1)}(z',t)$ and $h^{(2)}(z'',t)$, respectively, denote the piezometric heads in the upper and in the lower aquitards, and the gradients are taken at the interfaces between the main aquifer and the top and bottom aquitards. With (5.4.67), the flow equation for the leaky confined aquifer, (5.4.57), becomes

$$S\frac{\partial h}{\partial t} = \nabla' \cdot (\mathbf{T} \cdot \nabla' h) - \mathsf{K}^{(1)} \frac{\partial h^{(1)}}{\partial z'}\bigg|_{F_1} + \mathsf{K}^{(2)} \frac{\partial h^{(2)}}{\partial z''}\bigg|_{F_2} - P(x,y,t) + R(x,y,t). \tag{5.4.68}$$

We note that in the above equation, we now have two additional dependent variables $h^{(1)}(z',t)$ and $h^{(2)}(z'',t)$. To solve for these variables, we have to construct a model that describes the (vertical) flow within each aquitard.

Referring to Fig. 5.4.6, we assume that $\mathsf{K}^{(1)}$, $\mathsf{K}^{(2)} \ll \mathsf{K}$, so that the flow in the semipervious layer is *essentially vertical.* Let a steady flow be established through the layer and then assume that a stepwise reduction of head is produced in the main aquifer by pumping. After a sufficiently long time, a new steady state will be established, with a linear head distribution, as demonstrated in Fig. 5.4.5. During this period, the reduction of head in the semipervious layer will lag behind that corresponding to the new steady state. The problem of determining $h^{(1)}(x,y,z',t)$ for $\mathsf{K}^{(1)} =$ constant, is stated in the domain $0 \le z' \le B^{(1)}$, by the following partial differential equation and initial and boundary conditions, written in terms of the drawdown variable, $s^{(1)}(x,y,z',t) = h^{(1)}(x,y,z',0) - h^{(1)}(x,y,z',t)$,

$$S_o^{(1)} \frac{\partial s^{(1)}}{\partial t} = \mathsf{K}^{(1)} \frac{\partial^2 s^{(1)}}{\partial z'^2},$$

$$
\begin{aligned}
s^{(1)}(x,y,z',0) &= 0, \quad 0 \le z' \le B^{(1)}, \\
s^{(1)}(x,y,0,t) &= 0, \quad t > 0, \\
s^{(1)}(x,y,B^{(1)},t) &= \begin{cases} 0, & t \le 0, \\ H_o, & t > 0, \end{cases}
\end{aligned} \tag{5.4.69}
$$

Carslaw and Jaeger (1959, p. 310) provide a solution of this problem, in the form

$$
\begin{aligned}
&\frac{s^{(1)}(x,y,z',t)}{H_o} \\
&= \sum_{n=0}^{\infty} \left\{ \operatorname{erfc}\left[\frac{(2n+1)B^{(1)} - z'}{2\left(\mathsf{K}^{(1)}t/S_o^{(1)}\right)^{1/2}}\right] - \operatorname{erfc}\left[\frac{(2n+1)B^{(1)} + z'}{2\left(\mathsf{K}^{(1)}t/S_o^{(1)}\right)^{1/2}}\right]\right\}.
\end{aligned} \tag{5.4.70}
$$

Figure 5.4.7 shows this solution graphically. From $s^{(1)} = s^{(1)}(x,y,z',t)$, one can determine the increase in the rate of flow, q_{v1}, into the pumped aquifer, produced by the stepwise reduction in head,

$$
\begin{aligned}
\varDelta q_{v1} = \mathsf{K}^{(1)} \left.\frac{\partial s^{(1)}}{\partial z'}\right|_{z'=B^{(1)}} &= \frac{\mathsf{K}^{(1)} H_o}{B^{(1)}\left(\pi \mathsf{K}^{(1)} t / B^{(1)2} S_o^{(1)}\right)^{1/2}} \\
&\times \left[1 + 2\sum_{n=1}^{\infty} \exp\left(-\frac{n^2}{\mathsf{K}^{(1)} t / B^{(1)2} S_o^{(1)}}\right)\right].
\end{aligned} \tag{5.4.71}
$$

This flow is plotted in Fig. 5.4.7b. From this figure, it follows that a long time may elapse before steady flow is re-established in the semipervious layer.

The above solution for the aquitard assumes that we know the head change in the aquifer. In fact, as indicated in (5.4.68), the head in the aquifer, h, is coupled with that in the aquitards, $h^{(1)}$ and $h^{(2)}$, and hence, it cannot be solved alone. In the next subsection, we shall give a more complete treatment of a multilayered aquifer-aquitard system.

5.4.5 Multilayered aquifer-aquitard system

Figure 5.4.8 give a schematic illustration of a multilayered system with the more permeable layers (aquifers) separated by the less permeable ones (aquitards). As discussed in Subs. 5.4.4, the flow in the aquifers can be considered as essentially horizontal, while flow in the aquitards is assumed to be essentially vertical, due to the contrast in hydraulic conductivity. As suggested by Neuman (1968) (see also Cheng, 2000, p. 116) the hydraulic conductivity contrast should be at least 1:10 for these assumptions to be valid.

Following (5.4.66), for aquifer i, bounded from below and from above by aquitards $i-1$ and i (Fig. 5.4.8), respectively, the flow equation takes the form

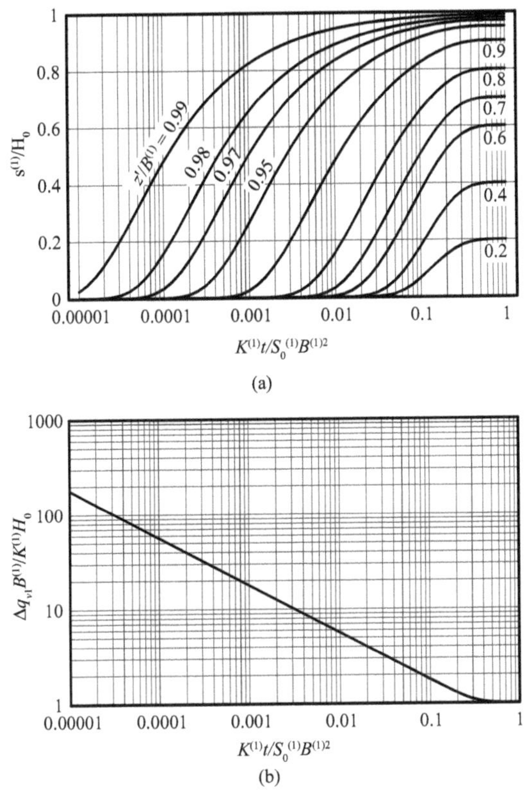

Figure 5.4.7: Graphical representation of equations (a) (5.4.70), and (b) (5.4.71).

$$S\frac{\partial s_i}{\partial t} = \nabla' \cdot (\mathbf{T} \cdot \nabla' s_i) - \mathsf{K}^{(i-1)} \left. \frac{\partial s^{(i-1)}}{\partial z} \right|_{z=z^t_{i-1}} + \mathsf{K}^{(i)} \left. \frac{\partial s^{(i)}}{\partial z} \right|_{z=z^b_i} + P; \quad i = 1, \ldots, n, \tag{5.4.72}$$

where n is the number of aquifer layers. In (5.4.72), we have used drawdown, s, instead of head, as the variable, and ignored the recharge term. The set of equations, (5.4.72), cannot be solved independently; they need to be coupled with the aquitard flow equations. Following (5.4.69), we can express the aquitards' flow equations,

$$\mathsf{K}^{(i)} \frac{\partial^2 s^{(i)}}{\partial z^2} = S_o^{(i)} \frac{\partial s^{(i)}}{\partial t}; \quad i = 0, \ldots, n. \tag{5.4.73}$$

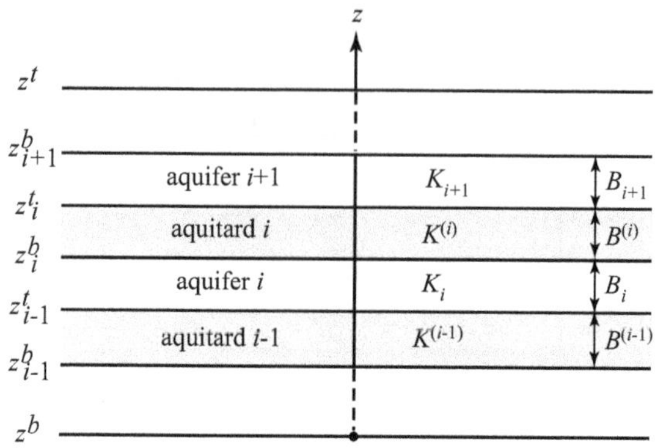

Figure 5.4.8: A multilayered aquifer-aquitard system.

Combining (5.4.72) and (5.4.73), together with the proper initial and boundary conditions, we have enough equations to solve for the aquifer and aquitard drawdowns, s_i and $s^{(i)}$, *simultaneously.* The above formulation for a two-aquifer system is attributed to Neuman and Witherspoon (1969a, 1969b).

For a numerical solution, (5.4.72) and (5.4.73) have to be discretized (Chap. 8). We observe that the aquifer equations, (5.4.72), are two-dimensional in the x-y plane; hence, a two-dimensional mesh is needed for each aquifer layer. On the other hand, the aquitard equations, (5.4.73), are one-dimensional, in the z-direction. However, we note that the aquitard drawdown, $s^{(i)}$, is in the form of $s^{(i)}(x, y, z, t)$, and that only one such equation is needed for each (x, y) location. The solution meshes for the aquitards are three-dimensional.

As demonstrated in Subs. 5.4.4, and particularly in (5.4.7), the aquitard drawdown, $s^{(i)}$, can be derived analytically from the one-dimensional equation (5.4.73), provided the drawdowns in the aquifer just above and just below, s^{i+1} and s^i (Figure 5.4.8), are known. Utilizing the analytical solutions for aquitards, Herrera and Figueroa (1969) and Herrera (1970) present an integrodifferential equation formulation that eliminates the aquitard drawdown from the governing equations. The resulting solution system can be discretized, using two-dimensional meshes representing only the aquifers. In this way, the size of the required mesh is significantly reduced. Cheng (2000) provides a detailed treatment of multilayered aquifer-aquitard systems with pumping wells.

5.4.6 Groundwater maps and streamlines

A. Hydrological maps

A groundwater map is a compact visual form of presenting the results of both field and model investigations of the *instantaneous* flow regime in an aquifer. It expresses the elevations $h = h(x, y, t_{\text{map}})$ of the water table in a phreatic

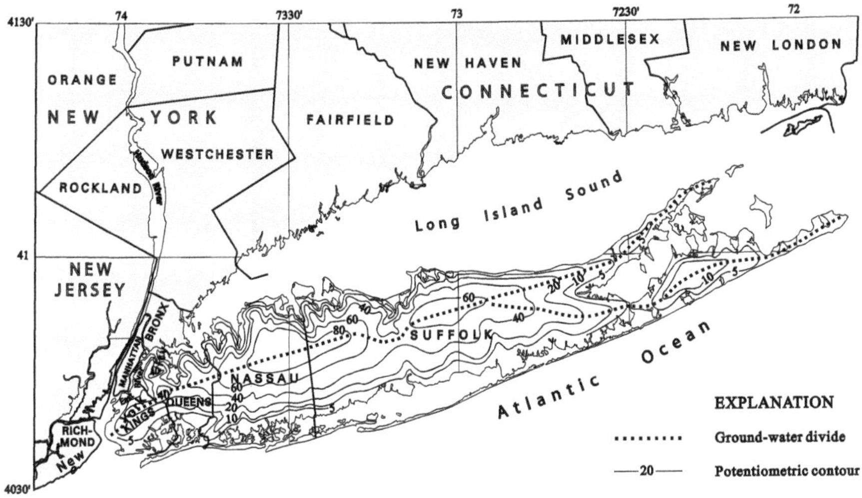

Figure 5.4.9: Groundwater contour map of Long Island, NY, prior to any large scale withdrawal. The potentiometric surface shows a smoothly sloping surface to the north and south in the western part of the island, indicating freshwater outflow to the sea. (After Franke and McClymonds, 1972)

aquifer, or of the piezometric surface in a confined, or a leaky one, at time $t = t_{\text{map}}$, as measured, or determined by a model. Underlying the drawing of a hydrologic map and its utilization in practice is the assumption that flow in the aquifer is essentially horizontal and, therefore, equipotential surfaces are vertical (Sec. 2.6). The hydrologic map is drawn for a specific aquifer and for a specified instant of time. UNESCO (1977), jointly with WMO, has published a summary on hydrologic maps. A detailed discussion of contour maps and their interpretation is also given by Dalton *et al.* (1991). Using water levels measured simultaneously at a certain time in all observation wells in a given aquifer (and one should be careful not to mix observations in wells that tap different aquifers), we employ an interpolation technique of one kind or another (e.g., linear) to draw contours of the water table, or of the piezometric surface at desired times. In the latter case, the term *equipotentials* is often used.

Figure 5.4.9 shows an example of a groundwater contour map (= hydrologic map, potentiometric map) obtained in this way. When the objective of a map is to provide a *regional* picture of the flow regime in an aquifer, it should exclude local cones of depression that occur around pumping wells operating in the area. To eliminate these local effects, pumping wells (and recharging ones) are shut off for a sufficient period of time (often 24–48 h) prior to measuring the water levels in them (if they are used as observation wells) and/or in observation wells in their vicinity. The period of time should be sufficient for complete recovery of water levels, so that the measured ones

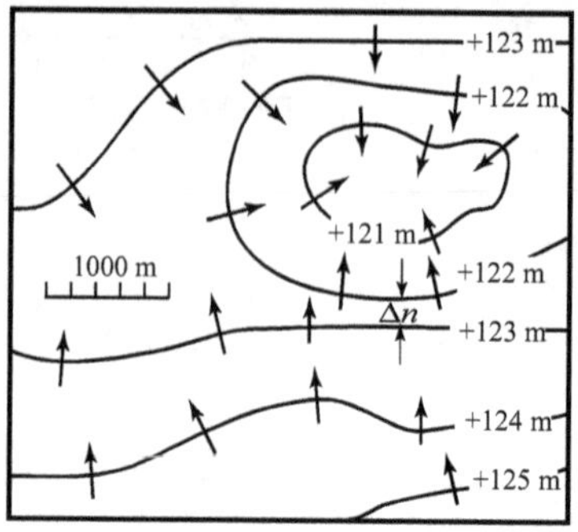

Figure 5.4.10: A groundwater contour map with arrows indicating flow directions.

correspond to the regional water table (or piezometric surface). Public domain and commercial computer programs are available for drawing contour maps by computers.

Forecasts of water levels by solving flow models (Sec. 5.4), analytically or numerically, may also be represented in the form of contour maps. The flow regime in an aquifer can be investigated by using contour maps in the following ways:

(a) Assuming that the aquifer is isotropic, the direction of flow (of water and pollutants carried with the water) at every point is *perpendicular to the equipotentials*. An example is shown in Fig. 5.4.10. We shall explain below why, in general, we draw only short arrows on a contour map to indicate flow direction and not complete streamlines.

(b) For an isotropic aquifer, $\mathsf{T}_x = \mathsf{T}_y = \mathsf{T}$, the rate of flow $\mathbf{Q}'$ is given by $\mathbf{Q}' = -\mathsf{T}\nabla h \equiv \mathsf{T}\boldsymbol{\mathcal{J}}$, $Q'_n = \mathsf{T}\mathcal{J}_n$. The gradient vector, $\boldsymbol{\mathcal{J}}$, and its direction (normal to the equipotential) can be determined from the map, as shown in Fig. 5.4.10, using $\mathcal{J}_n = \Delta h/\Delta n$, where n is measured normal to the contours. In a phreatic aquifer, we make use the Dupuit assumption and the *Dupuit-Forchheimer discharge formula* (4.5.14) for determining $\mathbf{Q}'$. In this case, information on the elevations of the impervious bottom of the aquifer is also needed.

(c) In an anisotropic aquifer, with principal directions x and y, i.e., $\mathsf{T}_x \neq \mathsf{T}_y$, the flow is no more perpendicular to the equipotentials. The discharge, $\mathbf{Q}'$ is calculated from $\mathbf{Q}'_i = \mathsf{T}_i\mathcal{J}_i$, $i = x, y$, with $\mathcal{J}_x = \Delta h/\Delta x$, $\mathcal{J}_y = \Delta h/\Delta y$ that can be read from the map.

(d) We can use two superimposed contour maps of an aquifer at two different times to draw a map of water table changes. Given aquifer storativity, this information is used to determine changes in the volume of water stored in the aquifer during the time interval between the dates for which the two maps were drawn.

In Sec. 3.9, we have shown how the above results are used for setting up a regional water balance. By analyzing the pattern of equipotentials, one learns about the general features of the flow pattern in a given aquifer, as shown by the following examples.

(a) In the absence of sources and sinks, and under steady state flow condition, a contour map cannot show a minimum and/or a maximum anywhere inside a flow domain. These may occur only along the boundaries of the domain. If a minimum in water levels, indicated by a closed contour curve, does occur in a flow domain, this means that a sink (e.g., due to intensive pumping, a spring, or marshland losing water by evapotranspiration) occurs in that area (Fig. 5.4.11a). Similarly a maximum (also indicated by a closed contour) indicates a zone of recharge (Fig. 5.4.11b), both natural and artificial. One should note that sources and sinks are not only those that are visible at ground surface (e.g., pumping, natural and artificial replenishment, springs), but may include hidden underground ones, such as a zone of leakage into or out of an aquifer from underlying or overlying aquifers through semipervious strata, fissures, or local direct contact with an adjacent aquifer. In unsteady flow, maxima and minima may (temporarily) exist inside a flow domain. In this case the interpretation is different, namely, that water is taken out of storage in the aquifer (say in the case of a decaying mound, following a period of artificial recharge), or added to it (as where an area of depression in the water table is being filled up once pumping has stopped).

(b) Contours can indicate river-aquifer relationships (Figs. 5.4.11c and d). Note that the streams in Figs. 5.4.11c and d are not equipotentials.

(c) Gradients may increase (Fig. 5.4.11e) or decrease in the direction of flow. In the absence of sources or sinks, this is caused by either a reduction in aquifer thickness and/or its hydraulic conductivity. (See discussion below on flow nets.)

(d) Impervious zones or lines (e.g. faults) force the flow to change direction. These obstacles may not be visible at ground surface, but they are indicated by studying the contour map (Fig. 5.4.11f).

(e) Zones of very high transmissivity may behave practically as equipotential (or nearly so) open water bodies (Fig. 5.4.11g). Old buried rivers may produce in their vicinity flow patterns similar to those shown in Fig. 5.4.11c and d, although no stream is visible at ground surface.

(f) Different water levels may occur across an impervious fault (Fig. 5.4.11h). One should remember that an impervious boundary is also a streamline. If a fault zone is semipervious, water may flow through it, indicating a local loss of head across the fault.

(g) A groundwater divide is a curve in the horizontal plane that separates the flow domain into subdomains, such that all groundwater from a subdomain will eventually drain out through a separate outlet (a lake, spring, river, etc.). Figure 5.4.11i shows a segment of a groundwater divide. This is an important feature in pollution studies, as a spring, a lake, or a river

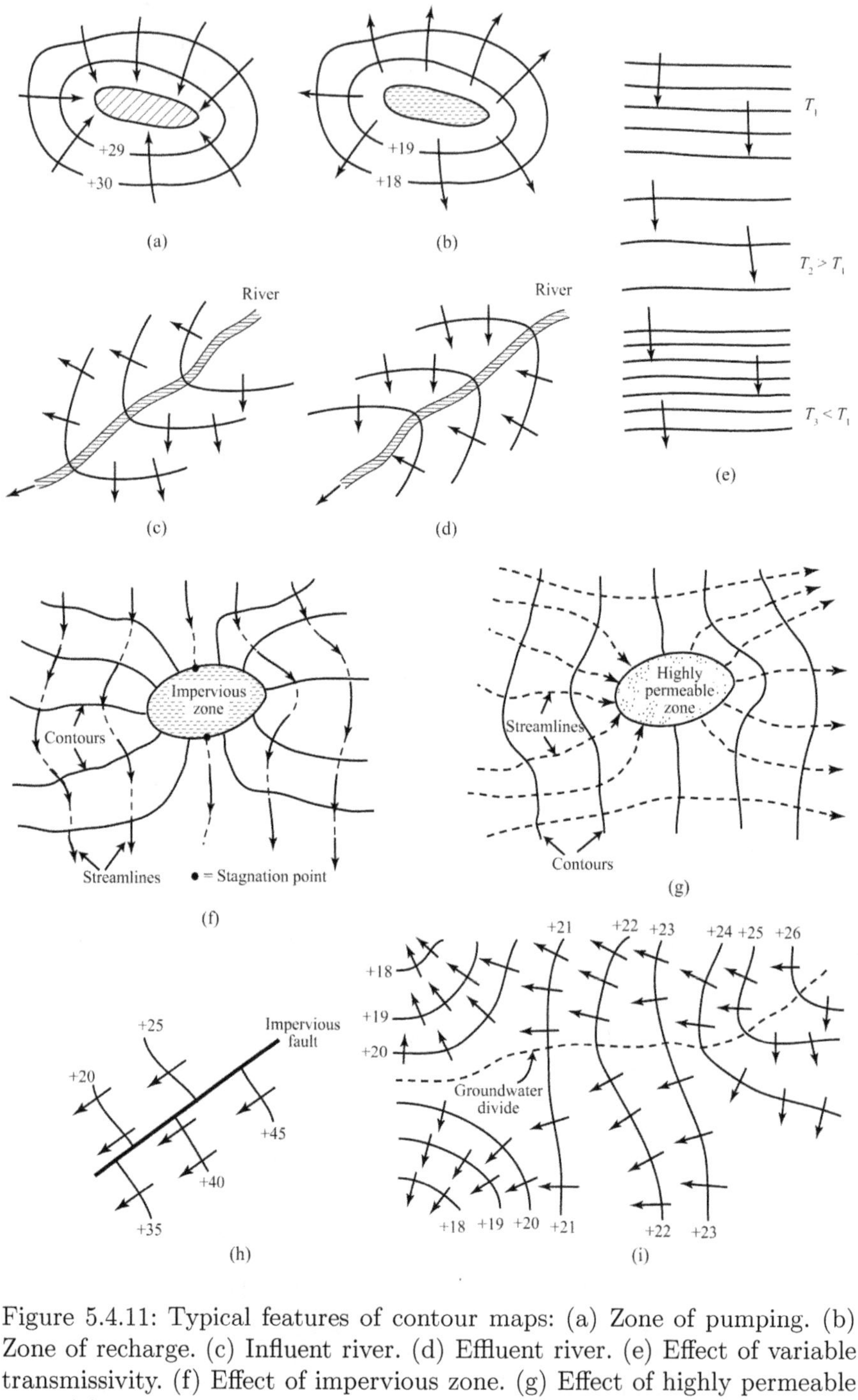

Figure 5.4.11: Typical features of contour maps: (a) Zone of pumping. (b) Zone of recharge. (c) Influent river. (d) Effluent river. (e) Effect of variable transmissivity. (f) Effect of impervious zone. (g) Effect of highly permeable zone. (h) Difference in water level across a semipervious fault. (i) Groundwater divide.

can only be polluted by polluted groundwater originating in the subdomain drained into it. It is of interest to note that unlike a topographic water divide, the groundwater one may shift with time. It can be made to shift by controlling groundwater levels.

An important conclusion from the above examples (and this is by no means a complete list of the possible special features observed on contour maps) is that serious mistakes in a hydrologic map may be made if contours are constructed by mechanical interpolation of observed water levels only, without taking into account all available geological information on faults, variations in aquifer thickness and hydraulic conductivity, depth of water table below ground surface (which may indicate area of groundwater loss in the form of a spring, or evapotranspiration from marshes, or from a water table which is very close to ground surface), etc. We have already mentioned above that another source of errors is the mixing of data on water levels observed at wells tapping different aquifers.

B. Flow nets

In two-dimensional flow, a *streamline* is a curve that is everywhere tangent to the specific discharge vector, **q**. Thus, a streamline indicates the direction of flow at every point along it. A stream-tube is formed by adjacent streamlines. Since, by definition, no flow can cross a streamline, the flow rate along a stream-tube is constant. This statement is true provided the flow is steady (i.e., no water is taken out of, or added to storage) and no distributed sources and sinks exist in the flow domain. Point sources and sinks (e.g., wells) may exist, but they can be excluded from the flow domain, leaving the latter with no sources and sinks. By the above definition, the equation of a streamline is

$$\mathbf{q} \times \mathbf{ds} = 0, \tag{5.4.74}$$

where $\times$ denotes a *cross product* and **ds** is an element of length along the streamline (Fig. 5.4.12a). In Cartesian coordinates, (5.4.74) is equivalent to

$$q_y dx - q_x dy = 0, \quad \frac{q_x}{dx} = \frac{q_y}{dy}. \tag{5.4.75}$$

We now define a function $\Psi = \Psi(x, y)$, which is a constant along a streamline (see below). Hence, along a streamline

$$d\Psi \equiv \frac{\partial \Psi}{\partial x} dx + \frac{\partial \Psi}{\partial y} dy = 0. \tag{5.4.76}$$

In fact, the function $\Psi = \Psi(x, y)$ is a solution of (5.4.75); it describes the geometry (in steady flow) of the streamlines. For the condition in (5.4.75) to be an exact differential of some function $\Psi(x, y)$, it is required that $\partial q_x/\partial x + \partial q_y/\partial y = 0$; this is nothing but the mass balance equation, (5.1.8), for a constant ρ.

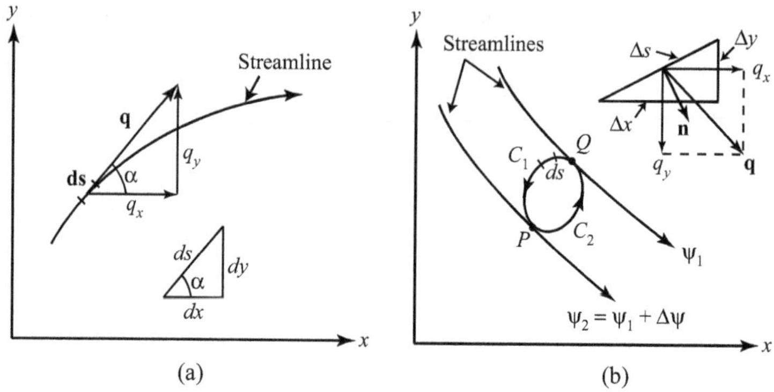

Figure 5.4.12: Streamlines and specific discharge in plane flow.

By comparing (5.4.75) and (5.4.76), we find that

$$q_x = -\frac{\partial \Psi}{\partial y}, \qquad q_y = -\frac{\partial \Psi}{\partial x}. \tag{5.4.77}$$

The function Ψ is called a *stream function* (dims. L^2/T), or *Lagrange stream function*. The physical interpretation of the generating function Ψ (as, according to (5.4.77) it can be used to generate the specific discharge), may be obtained by considering the integral of Ψ between two points, say P and Q in Fig. 5.4.12b.

Before determining the value of this integral, let us consider the two integrals

$$\begin{aligned} \int_P^Q \rho \mathbf{q} \cdot \mathbf{n}\, ds &= \text{mass flow rate across } C_1, \\ \int_Q^P \rho \mathbf{q} \cdot \mathbf{n}\, ds &= \text{mass flow rate across } C_2, \end{aligned} \tag{5.4.78}$$

where $\mathbf{n}$ is unit normal as shown in Fig. 5.4.12b. If the fluid and medium are incompressible (i.e., steady flow) and no sources and sinks are present within the area bounded by the curves C_1, and C_2, the mass of fluid in the area bounded by these curves remains constant and, therefore,

$$\int_P^Q \rho \mathbf{q} \cdot \mathbf{n}\, ds\Big|_{\text{along } C_1} + \int_Q^P \rho \mathbf{q} \cdot \mathbf{n}\, ds\Big|_{\text{along } C_2} = 0. \tag{5.4.79}$$

From the above, it follows that

$$\int_P^Q \rho \mathbf{q} \cdot \mathbf{n}\, ds\Big|_{\text{along } C_1} = \int_P^Q \rho \mathbf{q} \cdot \mathbf{n}\, ds\Big|_{\text{along } C_2}, \tag{5.4.80}$$

i.e. the result is independent of the curve between points P and Q on the two streamlines selected as the path of integration. Since the differential of $\Psi_P - \Psi_Q$ ($\equiv \Psi_2 - \Psi_1$) depends only on the endpoints of the integration, we obtain

$$\int_{\Psi_Q}^{\Psi_P} d\Psi = \Psi_P - \Psi_Q \, (\equiv \Psi_2 - \Psi_1). \tag{5.4.81}$$

On the other hand, the total flow through the stream-tube is given by

$$\begin{aligned} Q_{QP} &= \int_Q^P \mathbf{q} \cdot \mathbf{n} \, ds = \int_Q^P (q_x \, dy - q_y \, dx) = \int_Q^P \left(-\frac{\partial \Psi}{\partial y} \, dy - \frac{\partial \Psi}{\partial x} \, dx \right) \\ &= -\int_Q^P d\Psi = \Psi_Q - \Psi_P. \end{aligned} \tag{5.4.82}$$

Hence, the total discharge (in terms of volume per unit width normal to the xy-plane, per unit time) between two streamlines is equal to the difference in the values of the stream function corresponding to these lines. Note that according to our sign convention here (where $\nabla\Psi$ is obtained from ∇h by a counterclockwise rotation), $\Psi_P > \Psi_Q$. If points Q and P are on the same streamline, $Q_{QP} = 0$, $d\Psi = 0$, and $\Psi =$ constant. We must emphasize again that this discussion is valid for an incompressible fluid and in the absence of sources and sinks.

For an isotropic domain, we have the *Cauchy-Riemann conditions*

$$q_x = -\mathsf{K}\frac{\partial h}{\partial x} = -\frac{\partial \Psi}{\partial y}, \qquad q_y = -\mathsf{K}\frac{\partial h}{\partial y} = -\frac{\partial \Psi}{\partial x}. \tag{5.4.83}$$

Hence. by multiplying the two equations, we obtain

$$\frac{\partial h}{\partial x}\frac{\partial \Psi}{\partial x} + \frac{\partial h}{\partial y}\frac{\partial \Psi}{\partial y} \equiv \nabla' h \cdot \nabla' \Psi = 0, \tag{5.4.84}$$

that is, the family of equipotentials, $h =$ constant, is perpendicular everywhere to the family of streamlines, $\Psi =$ constant. The two families of curves are not orthogonal to each other when the medium is anisotropic.

Finally, it can be shown (Bear, 1972, p. 230), that Ψ satisfies the following partial differential equation:

- For a homogeneous isotropic medium

$$\nabla^2 \Psi(x, y) = 0, \quad \nabla^2 = \frac{\partial^2}{\partial x^2} + \frac{\partial^2}{\partial y^2}, \tag{5.4.85}$$

  which is the *Laplace equation* (in terms of Ψ). This equation, in terms of $h(x, y, z)$, has already been presented as (5.1.81).
- For a homogeneous anisotropic medium, where K is a tensor, with principal directions x and y,

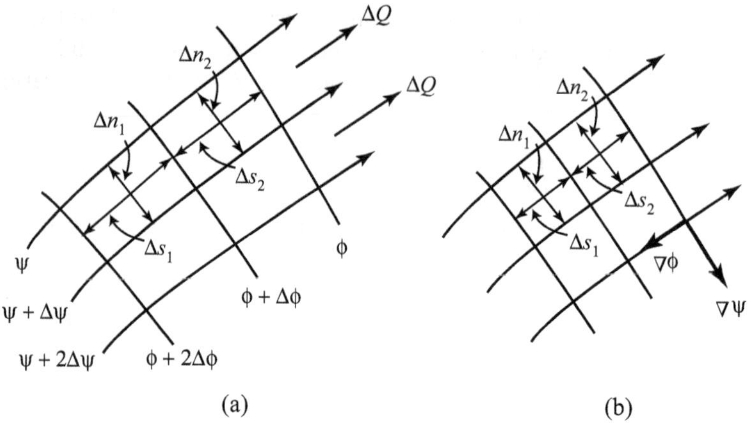

Figure 5.4.13: Flow net.

$$\mathsf{K}_x \frac{\partial^2 \Psi}{\partial x^2} + \mathsf{K}_y \frac{\partial^2 \Psi}{\partial y^2} = 0. \tag{5.4.86}$$

- For a nonhomogeneous, isotropic medium

$$\mathsf{K} \nabla^2 \Psi - \nabla \mathsf{K} \cdot \nabla \Psi = 0. \tag{5.4.87}$$

- For a nonhomogeneous anisotropic medium (x, y principal directions)

$$\mathsf{K}_x \frac{\partial^2 \Psi}{\partial x^2} + \mathsf{K}_y \frac{\partial^2 \Psi}{\partial y^2} - \frac{\mathsf{K}_x}{\mathsf{K}_y} \frac{\partial \mathsf{K}_y}{\partial x} \frac{\partial \Psi}{\partial x} - \frac{\mathsf{K}_y}{\mathsf{K}_x} \frac{\partial \mathsf{K}_x}{\partial y} \frac{\partial \Psi}{\partial y} = 0. \tag{5.4.88}$$

These equations, with appropriate boundary and initial conditions on Ψ, can be used to determine $\Psi = \Psi(x, y)$ in the given domain. Of special interest is (5.4.85), which is the same as the Laplace equation (5.1.9), satisfied by $h = h(x, y)$ in steady flow in an isotropic homogeneous domain, in the absence of distributed sources and sinks. In two-dimensional plane flow, a plot of the two families of curves: equipotentials $h = h(x, y) =$ const. and streamlines $\Psi = \Psi(x, y) =$ const., is called a *flow net.* In an isotropic domain, the two families are *orthogonal to each other.*

The use of Ψ is of practical value only in steady flow and in the absence of distributed sources and sinks within the flow domain. This, in fact, is the reason for not plotting streamlines on hydrological maps. Figure 5.4.13 shows a portion of a flow net in a homogeneous isotropic medium. It is customary to draw the flow net such that the difference Δh between any two adjacent equipotentials is constant. The same is true for $\Delta \Psi$ between adjacent streamlines. For each stream-tube, we have

$$\Delta Q' = \mathsf{K} \Delta n_1 \frac{\Delta h}{\Delta s_1} = \mathsf{K} \Delta n_2 \frac{\Delta h}{\Delta s_2}, \tag{5.4.89}$$

where Q' is the discharge per unit horizontal width, per unit thickness of aquifer. We may replace K by the transmissivity T and then Q' is the flow through the entire thickness of the aquifer. From (5.4.89) we obtain

$$\frac{\Delta n_1}{\Delta s_1} = \frac{\Delta n_2}{\Delta s_2}, \tag{5.4.90}$$

i.e., in a homogeneous medium, the ratio $\Delta n/\Delta s$ must remain constant throughout the flow net.

For an inhomogeneous medium

$$\mathsf{K}_1 \Delta n_1 \frac{\Delta h}{\Delta s_1} = \mathsf{K}_2 \Delta n_2 \frac{\Delta h}{\Delta s_2}, \qquad \mathsf{K}_1 \frac{\Delta n_1}{\Delta s_1} = \mathsf{K}_s \frac{\Delta n_2}{\Delta s_2}, \tag{5.4.91}$$

i.e., the ratio $\Delta n/\Delta s$ varies. When streamlines are approximately parallel ($\Delta n_1 \approx \Delta n_2$), we have $(\mathsf{K}_1/\mathsf{K}_2) \cong (\Delta h/\Delta s_2)/(\Delta h/\Delta s_1)$, that is, the hydraulic conductivity is inversely proportional to the hydraulic gradient; contours will be closely spaced in regions of low hydraulic conductivity.

It is convenient to draw the flow net for a homogeneous isotropic medium so that approximate curvilinear squares are formed (Fig. 5.4.13b). For such case $\Delta s = \Delta n$, and $Q' = \mathsf{K}\Delta h$. In certain cases, however, it is more convenient to draw the flow net such that we have m stream-tubes, each carrying the same discharge $\Delta Q' = Q'_{\text{total}}/m$, with n equal drops in piezometric head, $\Delta h = (h_{\max} - h_{\min})/n$. Then

$$Q'_{\text{total}} = m\Delta Q' = m\mathsf{K}\Delta n \frac{\Delta h}{\Delta s} = m\mathsf{K}\Delta n \frac{h_{\max} - h_{\min}}{n\Delta s} = \frac{m}{n}\frac{\Delta n}{\Delta s}\mathsf{K}(h_{\max} - h_{\min}). \tag{5.4.92}$$

5.5 Land Subsidence

Land subsidence is the settling of ground surface over large areas, due to the *compaction* of certain subsurface materials, primarily clay layers or lenses, usually as a consequence of pumping from underlying aquifers. The term *consolidation* is used for the same phenomenon, when it occurs over relatively small areas, especially as a result of loading ground surface, e.g., by new structures. In this section, we shall focus on land subsidence produced by pumping, as this is the subject that interests hydrogeologists. In addition to downward settling of ground surface, horizontal displacement may also occur.

It is important to emphasize that significant land subsidence does not always occur directly on top of overpumped aquifers. It occurs when a large amount of groundwater is squeezed out of fine-grained sediments, like clay; the pumping itself may take place some distance away.

The basic cause for subsidence has already been discussed in Subs. 5.1.3. We have shown there that pumping produces a reduction in water pressure and, therefore, simultaneously, an increase in the effective stress in the solid matrix. This increase produces compaction that manifests itself as consolidation or land subsidence. The degree and rate of (vertical) consolidation and horizontal displacement depend on the stress-strain relationship of the solid matrix comprising the aquifer.

There exist many examples of subsidence over large areas as a result of pumping groundwater (Poland, 1984). Perhaps, the most spectacular one in an urban area is in Mexico City, where almost the entire metropolitan area has subsided more than 3 m (with up to 8 m at some locations). As a result of excessive withdrawal of groundwater for irrigation, the San Joaquin Valley in California has experienced subsidence at a rapid rate of 30–40 cm per year, reaching a total subsidence of 9 m at some locations. Another famous place is Venice, Italy, where all pumping has recently been stopped in an effort to prevent further land subsidence. Significant subsidence has occurred also in the Taipei Basin, Tokyo, the Texas Gulf Coast, London, and Bangkok. In almost all these cases, the aquifers (or a sequence of confined aquifers separated from each other) contains clay strata.

In general, two main approaches exist for the analysis of both consolidation and land subsidence.

- The approach introduced by Biot (1941) (see also Detournay and Cheng, 1993) which regards consolidation as an elasticity problem with interacting pore fluid. Stress, strain, and fluid flow are considered in a three-dimensional and time-dependent setting. The considered physical phenomena include the pore pressure change induced by skeleton deformation, the (Darcy's) flow produced by pressure gradient, the effective stress on the skeleton produced by the modified pore pressure, and the three-dimensional strain compatibility of elastic materials. A simultaneous solution is sought for two sets of dependent variables: pressure in the water and displacements of the solid.
- A generalization of Terzaghi's (1923) theory (Subs. 5.1.3) to include three-dimensional dissipation of water pressure, without considering the state of strain compatibility in the solid skeleton at the same time. The redistribution of total stress and its effect on water pressure is ignored when the latter is solved from a single PDE in terms of pressure only. Once the pressure distribution is known, effective stress, accompanying strain and vertical settlement, can be determined.

In the practice of geotechnical engineering, the second approach is commonly employed for determining consolidation; it is considered a fair approximation of the corresponding Biot solution. Hydrogeologists use the same geotechnical engineering approach in developing the concept of specific storativity (Subs. 5.1.3) and for dealing with deformable porous media.

Basically, consolidation is a three-dimensional phenomenon. De Josselin de Jong (1963), Verruijt (1969, 1995), Gambolati *et al.* (1973, 1974), Bear (1972, p. 208), among others, presented theories and examples of three-dimensional consolidation. On the other hand, as will be shown below, land subsidence can be visualized and approximated as two-dimensional in the horizontal plane. Bear and Corapcioglu (1981a, b) presented a regional, two-dimensional model of land subsidence with horizontal displacement due only to pumping. Parts of their presentation will be followed below.

Actually, in most consolidation analyses carried out in soil mechanics, it is assumed that both flow and soil deformation occur mainly in the vertical direction, ignoring lateral deformation. The problem then reduces to one-dimensional consolidation in the vertical direction. In such case, the two approaches coincide. The subject of consolidation is discussed in standard texts on soil mechanics (e.g., Terzaghi, 1943; Scott, 1963; Harr, 1966). Of special interest should be the Proceedings of the UNESCO/IAHS Symposium on Land Subsidence (Tison, 1969; Rodda, 1976; Johnson *et al.*, 1984; Johnson, 1991; Barends *et al.*, 1995; and Poland, 1984).

Like water levels and solute concentrations, land subsidence is part of the aquifers response to pumping. Maximum permissible land subsidence may serve as a *constraint* in groundwater management. Therefore, whenever appreciable subsidence is anticipated, modeling of subsidence is called for. In the discussion here, we shall follow the second approach, which is, usually, sufficient for predicting land subsidence due to large scale pumping.

5.5.1 Integrated water mass balance equation

Our objective is to construct a model that describes land subsidence, say, in response to pumping from a confined aquifer of thickness $B = B(x, y, t)$. We start from (5.1.28) that expresses the mass balance for a compressible fluid in a deformable porous medium in three dimensions,

$$\nabla\cdot \mathbf{q}_r + \rho g\phi\beta\frac{\partial h^*}{\partial t} + \frac{\partial \varepsilon_{sk}}{\partial t} + P(x, y, z, t) = 0, \tag{5.5.1}$$

in which $\rho(\equiv \rho_f)$ denotes fluid's density, h^* is Hubbert's potential, defined by (4.1.6), $P(x, y, z, t)$ represents distributed pumping (as volume of water extracted per unit volume of soil, per unit time).

To obtain a two-dimensional model, following the methodology presented in Subs. 5.4.1, we integrate (5.5.1) over the thickness, $B(x, y, t)$, obtaining

$$\begin{aligned}
&\int_{b_1(x,y,t)}^{b_2(x,y,t)} \left(\nabla\cdot \mathbf{q}_r + \rho g\phi\beta\frac{\partial h^*}{\partial t} + \frac{\partial \varepsilon_{sk}}{\partial t} + P\right)\,dz \\
&= \nabla'\cdot B\widetilde{\mathbf{q}_r} + \mathbf{q}_r|_{F_2}\cdot\nabla F_2 - \mathbf{q}_r|_{F_1}\cdot\nabla F_1 + B\frac{\partial\widetilde{\varepsilon_{sk}}}{\partial t} + \tilde{\rho} g\tilde{\phi}\beta B\frac{\partial\widetilde{h^*}}{\partial t} \\
&\quad + \tilde{\rho} g\tilde{\phi}\beta\left(h^*\frac{\partial B}{\partial t} + h^*\big|_{F_2}\frac{\partial F_2}{\partial t} - h^*\big|_{F_1}\frac{\partial F_1}{\partial t}\right) + B\widetilde{P} = 0,
\end{aligned} \tag{5.5.2}$$

where $F_i = F_i(x, y, z, t) = z - b_i(x, y, t) = 0$, $i = 1, 2$, describes the bottom and top surfaces bounding the aquifer, the prime symbol over an operator indicates that the operator is in the xy-plane only, and we have made use of the approximation

$$\int_{b_1}^{b_2} \rho g \phi \beta \frac{\partial h^*}{\partial t}\, dz \simeq \tilde{\rho} g \tilde{\phi} \beta \int_{b_1}^{b_2} \frac{\partial h^*}{\partial t}\, dz, \qquad \widetilde{(\,)} = \frac{1}{B} \int_{b_1}^{b_2} (\,)\, dz. \tag{5.5.3}$$

Obviously, all averaged terms are functions of x, y and possibly t.

For the impervious top and bottom bounding surfaces considered here, we use the boundary condition (5.2.12), rewritten in the form

$$\mathbf{q}_r|_{F_1} \cdot \nabla F_1 = 0, \qquad \mathbf{q}_r|_{F_2} \cdot \nabla F_2 = 0. \tag{5.5.4}$$

Adding the assumption that equipotentials are essentially vertical, i.e., $h^*|_{F_1} \simeq h^*|_{F_2} = \widetilde{h^*}$, equation (5.5.2) reduces to

$$\nabla' \cdot B\widetilde{\mathbf{q}_r} + B\frac{\partial \widetilde{\varepsilon_{sk}}}{\partial t} + \tilde{\rho} g \tilde{\phi} \beta B \frac{\partial \widetilde{h^*}}{\partial t} + B(x,y,t)\widetilde{P}(x,y,t) = 0, \tag{5.5.5}$$

where $\widetilde{P}$ represents volume of water withdrawn from the aquifer per unit horizontal area, per unit time, and $B\widetilde{\mathbf{q}_r} = -B\widetilde{\mathbf{K}} \cdot \nabla' \widetilde{h^*}$ represents integrated horizontal flux.

We note that in (5.5.5)

$$\frac{\partial \widetilde{h^*}}{\partial t} \simeq \frac{1}{\tilde{\rho} g} \frac{\partial \tilde{p}}{\partial t} + \frac{\partial \tilde{z}}{\partial t}, \qquad \nabla' \widetilde{h^*} \simeq +\frac{1}{\tilde{\rho} g} \nabla' \tilde{p} + \nabla' \tilde{z}, \tag{5.5.6}$$

in which $\tilde{z} = (b_1 + b_2)/2$ is the elevation of the midpoint of the aquifer.

From (5.1.21) and (5.1.22), we obtain

$$B\frac{\partial \widetilde{\varepsilon_{sk}}}{\partial t} = B\widetilde{\nabla \cdot \mathbf{V}_s} = \nabla' \cdot B\widetilde{\mathbf{V}'_s} + \mathbf{V}_s\big|_{F_2} \cdot \nabla F_2 - \mathbf{V}_s\big|_{F_1} \cdot \nabla F_1. \tag{5.5.7}$$

Since the top and bottom surfaces of the aquifer are assumed to be material surfaces with respect to the solid, following (5.2.9), we have on them

$$(\mathbf{V}_s - \mathbf{u})\big|_{F_1} \cdot \nabla F_1 = 0, \qquad (\mathbf{V}_s - \mathbf{u})\big|_{F_2} \cdot \nabla F_2 = 0. \tag{5.5.8}$$

or, in view of (5.2.3)

$$\mathbf{V}_s\big|_{F_1} \cdot \nabla F_1 = -\frac{\partial F_1}{\partial t}, \qquad \mathbf{V}_s|_{F_2} \cdot \nabla F_2 = -\frac{\partial F_2}{\partial t}. \tag{5.5.9}$$

Hence, (5.5.7) becomes

$$B\frac{\partial \widetilde{\varepsilon_{sk}}}{\partial t} = \nabla' \cdot B\widetilde{\mathbf{V}'_s} - \frac{\partial (F_2 - F_1)}{\partial t} = \nabla' \cdot B\widetilde{\mathbf{V}'_s} + \frac{\partial B}{\partial t}. \tag{5.5.10}$$

With the solid velocity, $\mathbf{V}_s$, related to the displacement vector, $\mathbf{w}(\equiv \mathbf{w}_s)$ by (5.1.14), noting the approximation included in this equation, we obtain

$$
\begin{aligned}
B\widetilde{\mathbf{V}'_s} &= \int_{(B)} \mathbf{V}'_s\, dz - \int_{(B)} \frac{\partial \mathbf{w}}{\partial t}\, dz = \frac{\partial}{\partial t}(B\widetilde{\mathbf{w}'}) + \widetilde{\mathbf{w}'}|_{F_2}\frac{\partial F_2}{\partial t} - \widetilde{\mathbf{w}'}|_{F_1}\frac{\partial F_1}{\partial t} \\
&= B\frac{\partial \widetilde{\mathbf{w}'}}{\partial t} + \left(\widetilde{\mathbf{w}'}\frac{\partial B}{\partial t} + \mathbf{w}'|_{F_2}\frac{\partial F_2}{\partial t} - \mathbf{w}'|_{F_1}\frac{\partial F_1}{\partial t}\right).
\end{aligned}
\tag{5.5.11}
$$

At this point, we need information on $\mathbf{w}'|_{F_2}$ and $\mathbf{w}'|_{F_1}$, which are the displacement boundary conditions on F_1 and F_2, respectively. This information is not available. We circumvent this difficulty by introducing the simplifying assumption that *the horizontal displacement is constant along the vertical*, i.e.,

$$
\mathbf{w}'|_{F_2} = \mathbf{w}'|_{F_1} = \widetilde{\mathbf{w}'}. \tag{5.5.12}
$$

Then,

$$
B\frac{\partial \widetilde{\varepsilon_{sk}}}{\partial t} = \nabla' \cdot B\frac{\partial \widetilde{\mathbf{w}'}}{\partial t} + \frac{\partial B}{\partial t}. \tag{5.5.13}
$$

Following Verruijt (1969), we now express the specific discharge and the piezometric head as a sum of initial steady state values and deviations that express excess above the latter. By averaging these expressions, we obtain

$$
\begin{aligned}
\widetilde{h^*}(x,y,t) &= \widetilde{h^{*o}}(x,y) + \widetilde{h^{*e}}(x,y,t), \\
\widetilde{\mathbf{q}'_r}(x,y,t) &= \widetilde{\mathbf{q}'^{\,o}_r}(x,y) + \widetilde{\mathbf{q}'^{\,e}_r}(x,y,t), \\
\widetilde{P}(x,y,t) &= \widetilde{P^o}(x,y) + \widetilde{P^e}(x,y,t).
\end{aligned}
\tag{5.5.14}
$$

In terms of these variables, the averaged mass balance equation (5.5.5) is separated into two equations: a steady state mass balance equation

$$
\nabla' \cdot B\widetilde{\mathbf{q}'^{\,o}_r} + \widetilde{P^o} = 0, \tag{5.5.15}
$$

and an unsteady one

$$
\nabla' \cdot B\widetilde{\mathbf{q}'^{\,e}_r} + \nabla' \cdot \left(B\frac{\partial \widetilde{\mathbf{w}'}}{\partial t}\right) + \frac{\partial B}{\partial t} + \tilde{\rho} g \tilde{\phi} \beta B \frac{\partial \widetilde{h^{*e}}}{\partial t} + B\widetilde{P^e} = 0. \tag{5.5.16}
$$

The last equation can be linearized by introducing

$$
\begin{aligned}
B(x,y,t) &= b_2(x,y,t) - b_1(x,y,t) \\
&= (b_2^o(x,y) + w_z|_{F_2}) - (b_1^o(x,y) + w_z|_{F_1}) = B^o(x,y) + \Delta_z, \\
\Delta_z &= w_z|_{F_2} - w_z|_{F_1} \ll B^o.
\end{aligned}
\tag{5.5.17}
$$

We note that Δ_z, denoting compaction, is positive in the $+z$-direction. Neglecting the effect of compaction on permeability, this approximation also leads to the flux equation

$$B\widetilde{\mathbf{q}_r'^{\,e}} = -B^o\widetilde{\mathbf{K}'} \cdot \nabla'\widetilde{h^{*e}}. \tag{5.5.18}$$

Altogether, by substituting (5.5.18) in (5.5.16), we obtain a single equation in the variables $\widetilde{h^{*e}}$, B, and $\widetilde{\mathbf{w}'}$. Our next step is to make use of the equilibrium equation.

5.5.2 Integrated equilibrium equation

The total stress tensor, $\boldsymbol{\sigma}$, at a point within an aquifer, satisfies the *equilibrium equation* (5.1.33), rewritten here for convenience in the form

$$\nabla \cdot \boldsymbol{\sigma} + \rho\mathbf{F} = 0, \tag{5.5.19}$$

where we have omitted the average symbol; the body force acting on the porous medium, $\rho\mathbf{F}$, defined in (5.1.33), is assumed to remain unchanged by the compaction of the porous medium, i.e., $(\rho\mathbf{F})^e = 0$, where the superscript e denotes the increment. In (5.5.19), as in (5.1.33), $\rho = \phi\rho_f + (1-\phi)\rho_s$, represents the combined density of fluid and solid matrix.

We start from (5.1.58), which involves the incremental effective stress and effective pressure. We repeat it here for convenience as

$$\nabla \cdot \boldsymbol{\sigma}_s'^{\,e} - \nabla p^e = 0. \tag{5.5.20}$$

Note that in this equation, p is positive for compression.

We then assume that the solid matrix comprising the aquifer behaves like an isotropic and (for the relatively small displacements considered here) perfectly elastic body, for which the stress-strain relationship (5.1.60) is valid. By integrating (5.5.20) (see nomenclature in Fig. 5.1.2), we obtain

$$\int_{b_1}^{b_2} \left(\nabla \cdot \boldsymbol{\sigma}_s{}^e - \nabla p^e\right) dz = \nabla' \cdot B\widetilde{\boldsymbol{\sigma}_s'^{\,e}} + \left.(\boldsymbol{\sigma}_s'^{\,e} - p^e\mathbf{I})\right|_{F_2} \cdot \nabla F_2$$
$$-\left.(\boldsymbol{\sigma}_s'^{\,e} - p^e\mathbf{I})\right|_{F_1} \cdot \nabla F_1 - \nabla' B\widetilde{p^e} = 0. \tag{5.5.21}$$

We assume that the porous medium on both sides of a boundary surface, say the upper one, $F_2 = 0$, is deformable. On such boundary, we have to maintain the *condition of no-jump in the total stress*, i.e.,

$$[\![\, \boldsymbol{\sigma} \,]\!]_{u,\ell} \cdot \nabla F_2 = 0, \qquad [\![\, \boldsymbol{\sigma}_s' - p\mathbf{I} \,]\!]_{u,\ell} \cdot \nabla F_2 = 0, \tag{5.5.22}$$

where u and ℓ, respectively, denote the upper and lower sides. Following the methodology introduced earlier, this condition leads to an analogous condition related to the incremental effective stress and pressure,

$$[\![\, \boldsymbol{\sigma}_s'^{\,e} - p^e\mathbf{I} \,]\!]_{u,\ell} \cdot \nabla F_2 = 0. \tag{5.5.23}$$

When the excess stress and pressure in an aquifer are due only to pumping, and not to changes in the overburden load, say, by excavation, the total stress on the upper side of the boundary, $\boldsymbol{\sigma}\big|_u$, remains unchanged. Hence, from (5.5.22), we obtain

$$\boldsymbol{\sigma}\big|_u \cdot \nabla F_2 = \boldsymbol{\sigma}^o\big|_u \cdot \nabla F_2 = \left(\boldsymbol{\sigma}'^{\,o}_s\big|_\ell - p^o\big|_\ell \mathbf{I}\right) \cdot \nabla F_2, \tag{5.5.24}$$

and

$$\boldsymbol{\sigma}^e\big|_u \cdot \nabla F_2 = \left(\boldsymbol{\sigma}'^{\,e}_s\big|_\ell - p^e\big|_\ell \mathbf{I}\right) \cdot \nabla F_2 = 0. \tag{5.5.25}$$

In view of the boundary conditions (5.5.25), equation (5.5.21) reduces to

$$\nabla' \cdot B(\widetilde{\boldsymbol{\sigma}'^{\,e}_s}) - \nabla' B\widetilde{p^e} = 0, \tag{5.5.26}$$

in which averaged values are functions of x, y and t only.

Let us rewrite (5.5.26) in the form

$$\frac{\partial}{\partial x} B(\widetilde{\sigma'^{\,e}_s})_{xx} + \frac{\partial}{\partial y} B(\widetilde{\sigma'^{\,e}_s})_{xy} - \frac{\partial}{\partial x} B\widetilde{p^e} = 0, \tag{5.5.27}$$

and two analogous equations in the y and z directions (Bear and Bachmat, 1990, p. 509). We then express the averaged excess effective stress tensor in terms of averaged displacements, making use of (5.1.21) and (5.1.60), and the assumption expressed by (5.5.12). We obtain

$$\begin{aligned} \widetilde{\varepsilon_{sk}} &= \widetilde{\varepsilon}_{xx} + \widetilde{\varepsilon}_{yy} + \widetilde{\varepsilon}_{zz} \\ &= \frac{\partial \widetilde{w_x}}{\partial x} + \frac{\partial \widetilde{w_y}}{\partial y} + \frac{\partial \widetilde{w_z}}{\partial z} = \frac{\partial \widetilde{w_x}}{\partial x} + \frac{\partial \widetilde{w_y}}{\partial y} + \frac{\Delta_z}{B}, \end{aligned} \tag{5.5.28}$$

$$\begin{aligned} (\widetilde{\sigma'^{\,e}_s})_{xx} &= \lambda''_s\, \widetilde{\varepsilon_{sk}} + 2\mu'_s\, (\widetilde{\varepsilon_{sk}})_{xx} \\ &= (\lambda''_s + 2\mu'_s)\frac{\partial \widetilde{w_x}}{\partial x} + \lambda''_s\left(\frac{\partial \widetilde{w_y}}{\partial y} + \frac{\Delta_z}{B}\right), \end{aligned} \tag{5.5.29}$$

and additional analogous equations for $(\widetilde{\sigma'^{\,e}_s})_{yy}$, $(\widetilde{\sigma'^{\,e}_s})_{xy} = (\widetilde{\sigma'^{\,e}_s})_{yx}$, $(\widetilde{\sigma'^{\,e}_s})_{zx} = (\widetilde{\sigma'^{\,e}_s})_{xz}$, $(\widetilde{\sigma'^{\,e}_s})_{yz} = (\widetilde{\sigma'^{\,e}_s})_{zy}$, and $(\widetilde{\sigma'^{\,e}_s})_{zz}$. For example (Bear and Bachmat, 1990, p. 510), we obtain the linearized equation

$$(\widetilde{\boldsymbol{\sigma}'^{\,e}_s})_{zz} = \overline{\lambda''_s}\, \widetilde{\varepsilon_{sk}} + 2\overline{\mu'_s}\frac{\partial \widetilde{w_z}}{\partial z} = \overline{\lambda''_s}\left(\frac{\partial \widetilde{w_x}}{\partial x} + \frac{\partial \widetilde{w_y}}{\partial y}\right) + (\overline{\lambda''_s} + 2\overline{\mu'_s})\frac{\Delta_z}{B}. \tag{5.5.30}$$

By inserting these expressions into (5.5.27) and the additional, not shown, equations, and making use of (5.5.12), we obtain three equations in the four averaged variables $\widetilde{p^e}$, $\widetilde{w_x}$, $\widetilde{w_y}$, and $\widetilde{w_z}$, all functions of x, y and t only,

$$\frac{\partial}{\partial x}\left\{B\left[(\lambda''_s + 2\mu'_s)\frac{\partial \widetilde{w_x}}{\partial x} + \lambda''_s\left(\frac{\partial \widetilde{w_y}}{\partial y} + \frac{\Delta_z}{B}\right)\right]\right\}$$

$$+\frac{\partial}{\partial y}\left[B\mu'_s\left(\frac{\partial\widetilde{w_x}}{\partial y}+\frac{\partial\widetilde{w_y}}{\partial x}\right)\right]-\frac{\partial}{\partial x}B\widetilde{p^e}=0, \qquad (5.5.31)$$

$$\frac{\partial}{\partial x}\left[B\mu'_s\left(\frac{\partial\widetilde{w_y}}{\partial x}+\frac{\partial\widetilde{w_x}}{\partial y}\right)\right]+\frac{\partial}{\partial y}\left\{B\left[\lambda''_s\frac{\partial\widetilde{w_x}}{\partial x}\right.\right.$$
$$\left.\left.+(\lambda''_s+2\mu'_s)\frac{\partial\widetilde{w_y}}{\partial y}+\lambda''_s\frac{\Delta_z}{B}\right]\right\}-\frac{\partial}{\partial y}B\widetilde{p^e}=0, \qquad (5.5.32)$$

$$\frac{\partial}{\partial x}\left[B\mu'_s\frac{\partial\widetilde{w_z}}{\partial x}+\mu'_s\left(\widetilde{w_z}+w_z\big|_{F_2}\frac{\partial F_2}{\partial x}-w_z\big|_{F_1}\frac{\partial F_1}{\partial x}\right)\right]+\frac{\partial}{\partial y}\left[B\mu'_s\frac{\partial\widetilde{w_z}}{\partial y}\right.$$
$$\left.+\mu'_s\left(\widetilde{w_z}+w_z\Big|_{F_2}\frac{\partial F_2}{\partial y}-w_z\Big|_{F_1}\frac{\partial F_1}{\partial y}\right)\right]=0. \qquad (5.5.33)$$

For constant λ''_s and μ'_s, and with

$$B(x,y,t)=B^o(x,y)+\Delta_z(x,y,t), \qquad \Delta_z\ll B^o, \qquad (5.5.34)$$

we obtain the linearized forms of (5.5.31) and (5.5.32)

$$\mu'_s\nabla'^2\widetilde{w_x}+(\lambda''_s+\mu'_s)\left(\frac{\partial\widetilde{w_x}}{\partial x}+\frac{\partial\widetilde{w_y}}{\partial y}\right)+\lambda''_s\frac{\partial(\Delta_z/B^o)}{\partial x}-\frac{\partial\widetilde{p^e}}{\partial x}=0, \quad (5.5.35)$$

$$\mu'_s\nabla'^2\widetilde{w_y}+(\lambda''_s+\mu'_s)\left(\frac{\partial\widetilde{w_x}}{\partial x}+\frac{\partial\widetilde{w_y}}{\partial y}\right)+\lambda''_s\frac{\partial(\Delta_z/B^o)}{\partial y}-\frac{\partial\widetilde{p^e}}{\partial y}=0. \quad (5.5.36)$$

With the same linearization, and assuming $|\widetilde{\mathbf{V}'_s}\cdot\nabla'B|\ll|\partial B/\partial t|$, the second and third terms in (5.5.16) reduce to $B^o\partial\widetilde{\varepsilon_{sk}}/\partial t$. Thus, we may approximate the volume balance equation, (5.5.16), by (Bear and Bachmat, 1990, p. 511)

$$\nabla'\cdot B^o\widetilde{\mathsf{K}}\left(\frac{1}{\widetilde{\rho}g}\nabla'\widetilde{p^e}+\nabla'z\right)+B^o\frac{\partial\widetilde{\varepsilon_{sk}}}{\partial t}+\widetilde{\phi}\beta B^o\frac{\partial\widetilde{p^e}}{\partial t}+B^o\widetilde{P^e}=0. \quad (5.5.37)$$

In principle, (5.5.28), (5.5.33), (5.5.35), (5.5.36) and (5.5.37) are five equations in the variables $\widetilde{p^e}$, $\widetilde{w_x}$, $\widetilde{w_y}$, Δ_z and $\widetilde{\varepsilon_{sk}}$. However, in (5.5.33) we still have the terms

$$w_z\big|_{F_1}, \quad \text{and} \quad w_z\big|_{F_2}, \quad \text{with} \quad \Delta_z\equiv w_z\big|_{F_2}-w_z\big|_{F_1}, \quad \text{and} \quad B=B^o+\Delta_z,$$

which are actually conditions on the surfaces $F_1=0$ and $F_2=0$, for which we have no information. In fact, in most subsidence problems, the land subsidence, as expressed by $w_z\big|_{F_2}$ is the very state variable for which a solution is sought.

At this point we may continue by introducing certain simplifying assumptions, as a substitute for the missing information. For example, we may assume that the bottom of the aquifer is fixed, i.e., $w_z\big|_{F_1} = 0$ and that w_z varies linearly with elevation, i.e., $\widetilde{w_z} = \frac{1}{2}w_z\big|_{F_2} = \Delta_z/2$, where $-\Delta_z$ denotes land subsidence (positive downward). We then end up with equations for $\widetilde{w_x}, \widetilde{w_y}, \Delta z$ and $\widetilde{p^e}$. In this way, we have achieved our goal of determining the land subsidence $\Delta_z(x, y, t)$. In fact, we have solved, simultaneously, for the horizontal displacement, $\widetilde{w_x}$, as well as for the pressure in the aquifer, $\widetilde{p^e}$.

Verruijt (1969, p. 347) suggested an approach based on the assumption that consolidation occurs under *conditions of planar incremental total stress*,

$$\sigma^e_{zz} = 0, \qquad \sigma^e_{xz} = \sigma^e_{zx} = 0, \qquad \sigma^e_{yz} = \sigma^e_{zy} = 0. \tag{5.5.38}$$

This is a consequence of the assumption that displacements occur in the vertical direction only, i.e., $w_z \neq 0$, $w_x = w_y = 0$, while the total stress remains unchanged, i.e., $\boldsymbol{\sigma} \equiv \boldsymbol{\sigma}^o$ and $\boldsymbol{\sigma}^e = 0$. This assumption is justified when the aquifer is located between two soft confining layers (e.g., clay) which cannot resist shear stress. Furthermore, this assumption also justifies (5.5.12), since in a relatively thin aquifer, as implied by the planar stress assumption, lateral deformation is, more or less, uniform throughout the relatively small thickness of the layer.

From (5.5.38), it follows that the equilibrium equation (5.5.20) reduces to

$$\nabla' \cdot \boldsymbol{\sigma}'^{\,e}_s - \nabla' p^e = 0, \tag{5.5.39}$$

with the boundary condition (5.5.25), and a similar one for $F_1 = 0$, also written in the xy-coordinates only.

Following the integration procedure, which led above to (5.5.27) and to analogous equations in y and z, we now obtain only (5.5.27) and an analogous equations in y; the z-equation has been eliminated.

Accordingly, we now have to solve (5.5.16), or any equivalent form of it, (5.5.35) and (5.5.36), for $\widetilde{p^e}$, $\widetilde{w_x}$, $\widetilde{w_y}$ and Δ_z. The required fourth equation is now obtained from the first condition in (5.5.38), which leads to

$$(\boldsymbol{\sigma}'^{\,e}_s)_{zz} = p^e. \tag{5.5.40}$$

From (5.5.30) and (5.5.40), we now obtain

$$\begin{aligned}\widetilde{p^e} &= \overline{\lambda''_s}\left(\frac{\partial \widetilde{w_x}}{\partial x} + \frac{\partial \widetilde{w_y}}{\partial y}\right) + (\overline{\lambda''_s} + 2\overline{\mu'_s})\frac{\Delta_z}{B}\\ &= \overline{\lambda''_s}\,\widetilde{\varepsilon_{sk}} + 2\overline{\mu'_s}\,\frac{\Delta_z}{B}.\end{aligned} \tag{5.5.41}$$

This completes the formulation of the mathematical model for land subsidence. Usually we assume that $w_z\big|_{F_1} = 0$, and that $-\Delta_z = -w_z\big|_{F_2}$ expresses land subsidence.

5.5.3 *Terzaghi-Jacob vs. Biot approaches*

By differentiating (5.5.35) with respect to x, equation (5.5.36) with respect to y, linearizing both equations and then adding them, assuming constant $\overline{\lambda''_s}$, $\overline{\mu'_s}$, and B^o, we obtain

$$\nabla'^2\{(\overline{\lambda''_s} + 2\overline{\mu'_s})\nabla' \cdot \widetilde{\mathbf{w}'} + \overline{\lambda''_s}\,\frac{\Delta_z}{B^o} - \widetilde{p^e}\} = 0. \tag{5.5.42}$$

Following Verruijt (1969), we integrate (5.5.42), obtaining

$$\begin{aligned}(\overline{\lambda''_s} + 2\overline{\mu'_s})\nabla' \cdot \widetilde{\mathbf{w}'} + \overline{\lambda''_s}\frac{\Delta_z}{B^o} &= (\overline{\lambda''_s} + 2\overline{\mu'_s})\widetilde{\varepsilon_{sk}} - 2\overline{\mu'_s}\frac{\Delta_z}{B^o}\\ &= \widetilde{p^e} + \Pi'(x, y, t),\end{aligned} \tag{5.5.43}$$

where Π' satisfies

$$\nabla^2\Pi' = 0, \qquad \text{for every} \quad t.$$

The case $\Pi' = 0$, is presented after (5.1.69).

By comparing (5.5.43) with (5.5.41), obtained by introducing the *planar stress assumption*, we find that

$$\Pi' = 2\overline{\mu'_s}\left(\nabla' \cdot \widetilde{\mathbf{w}'} - \frac{\Delta_z}{B^o}\right) \simeq 2\overline{\mu'_s}\left(\widetilde{\varepsilon_{sk}} - 2\frac{\Delta_z}{B^o}\right), \tag{5.5.44}$$

where $\widetilde{\varepsilon_{sk}}$ is defined by (5.5.28).

If we assume $\widetilde{P^e} = 0$, and *no horizontal displacement*, i.e., $\widetilde{\mathbf{w}'} \equiv 0$, equation (5.5.16) reduces to

$$\nabla' \cdot B\widetilde{\mathbf{q}'_r{}^e} + \frac{\partial B}{\partial t} + \widetilde{\rho} g\widetilde{\phi}\beta B\frac{\partial \widetilde{h^*}}{\partial t} = 0, \tag{5.5.45}$$

where $B = B^o + \Delta_z$. Under the same conditions, (5.5.41) reduces to

$$\widetilde{p^e} = (\overline{\lambda''_s} + 2\overline{\mu'_s})\frac{\Delta_z}{B}. \tag{5.5.46}$$

Together, (5.5.45) and (5.5.46) can now be solved for $\widetilde{p^e}$ and Δ_z.

By combining the two equations, and assuming $\Delta_z \ll B$, we obtain

$$\nabla' \cdot B\widetilde{\mathbf{q}'_r{}^e} + B\left(\frac{1}{\overline{\lambda''_s} + 2\overline{\mu'_s}} + \widetilde{\phi}\beta\right)\frac{\partial \widetilde{p^e}}{\partial t} = 0. \tag{5.5.47}$$

By comparing (5.5.47) with (5.1.71), we may conclude that we could have obtained the last equation by *assuming, from the onset, that only vertical compressibility prevails*, with a coefficient of vertical compressibility

$$\alpha = \frac{1}{\overline{\lambda''_s} + 2\overline{\mu'_s}} \tag{5.5.48}$$

as in (5.1.71). Furthermore, by comparing (5.5.44), with $\widetilde{\mathbf{w}'} = 0$, with (5.5.41), obtained by assuming (i) planar stress, and (ii) no horizontal displacement, we obtain

$$\Pi' = -2\overline{\mu'_s}\frac{\Delta_z}{B}. \tag{5.5.49}$$

It is of interest to return at this point to the end of Subs. 5.1.3, where a comparison is made between the Terzaghi-Jacob and the Biot approaches.

5.5.4 Land subsidence produced by pumping

As an example for the use of the land subsidence model developed above, consider the case of land subsidence presented by Bear and Corapcioglu (1981b). In this example, both vertical and horizontal displacements, produced by pumping from a single well in a homogeneous confined or phreatic aquifer, are considered.

With $\widetilde{P^e}$ denoting the constant pumping rate from a well of radius r_w in a confined aquifer, a land subsidence model is constructed in terms of the four variables: $\delta(r,t)(\equiv \Delta_z)$, $\widetilde{w_r}(r,t)$, $-\widetilde{p^e}$, and $\widetilde{\varepsilon^e}$, denoting (vertical) subsidence, horizontal displacement, pressure drop, and strain, or *volume dilatation*, respectively.

Based on certain simplifying assumptions, e.g., $\partial B/\partial r \ll B/r$, and $(1/\Delta_z)\partial\Delta_z/\partial t \ll (1/B)\partial B/\partial t$, the Bear and Corapcioglu (1981b) land subsidence model is composed of the following four equations:

- **(Linearized) mass balance equation:**

$$-\frac{1}{r}\frac{\partial}{\partial r}\left(r\frac{\widetilde{\mathsf{k}^o}}{\widetilde{\mu^o}}\frac{\partial\widetilde{p^e}}{\partial r}\right) + \frac{\partial\widetilde{\varepsilon}}{\partial t} + \widetilde{\phi^o}\beta\frac{\partial\widetilde{p^e}}{\partial t} = 0. \tag{5.5.50}$$

- **Definition of dilatation**, averaged over the vertical:

$$\widetilde{\varepsilon^e} = \frac{\partial\widetilde{w_r}}{\partial t} + \frac{\widetilde{w_r}}{r} + \frac{\Delta z}{B^o}. \tag{5.5.51}$$

- **A combination of averaged (= integrated over the vertical) equilibrium equation, combined with the constitutive equations:**

$$\left(2\overline{\mu'_s} + \overline{\lambda''_s}\right)\widetilde{\varepsilon^e} - 2\overline{\mu'_s}\frac{\Delta_z}{B^o} = \widetilde{p^e} + 2g(t), \tag{5.5.52}$$

  where $g(t)$ is an arbitrary function of t.

- **Averaged constitutive relation, combined with the assumption of plane incremental total stress, suggested by Verruijt (1969):**

$$\widetilde{p^e} = 2\overline{\mu'_s}\frac{\Delta_z}{B^o} + \overline{\lambda''_s}\widetilde{\varepsilon^e}. \tag{5.5.53}$$

These four equations, in the variables: $\widetilde{p^e}$, $\widetilde{\varepsilon^e}$, $\widetilde{w_r}$, and Δ_z, are solved for the boundary and initial conditions,

$$\begin{aligned}
t \le 0,\ r \ge r_e, &\quad \widetilde{p^e}, \widetilde{\varepsilon^e}, \widetilde{w_r}, \Delta_z = 0\\
t > 0,\ r = r_w, &\quad \frac{\partial \widetilde{p^e}}{\partial r} = \frac{Q_w \widetilde{\mu^o}}{2\pi r_w B^o \widetilde{\mathsf{k}^o}}\\
r = r_w, &\quad \widetilde{w_r} = 0\\
r \to \infty, &\quad \widetilde{p^e},\ \widetilde{w_r},\ \widetilde{\varepsilon^e},\ \Delta_z = 0.
\end{aligned} \tag{5.5.54}$$

Subject to certain simplifying assumptions, the solutions derived by Bear and Corapcioglu (1981b) for the excess pressure, expressed in terms of change in piezometric head, is:

$$\frac{\widetilde{p^e}}{\widetilde{\rho^o} g} = \Delta h = -\frac{Q_w}{4\pi \mathsf{T}} W(u), \qquad u = \frac{r^2}{4C_v t} = \frac{Sr^2}{4\mathsf{T}t}, \tag{5.5.55}$$

which is the usual equation describing drawdown in a confined aquifer, as a result of pumping from a well (e.g., Bear, 1979, p. 321). In this equation, $C_v (= 1/(\overline{\mu'_s} + \overline{\lambda''_s}) \equiv \mathsf{T}/S)$ is a consolidation coefficient, T and S are the aquifer transmissivity and storativity, respectively, and $W(u)$ is the *well function of a confined aquifer* ($\equiv$ the *exponential integral*, see (11.3.8)). The vertical displacement is

$$\delta(\equiv -\Delta_z) = \frac{Q_w}{8\pi C_v} W(u) = -\frac{S}{2}\Delta h. \tag{5.5.56}$$

This is half the value obtained by assuming that only vertical consolidation takes place (Bear and Corapcioglu, 1981a). The horizontal displacement is

$$\widetilde{w_r} = -\frac{Q_w r}{16\pi C_v B^o}\left[W(u) + \frac{1 - e^{-u}}{u}\right], \tag{5.5.57}$$

with a maximum value at

$$r|_{w_r,\max} \approx 1.1367(C_v t)^{1/2} = 1.1367(\mathsf{T}t/S)^{1/2}. \tag{5.5.58}$$

Figure 5.5.1 shows these results in graphical form.

In may be of interest to note that in the four-variables model presented here, the integrated flow equation, equilibrium equation and constitutive relationship are *coupled*. A simpler approach would be to solve for the pressure drop, assuming no soil deformation, and then to estimate soil compaction, or/and subsidence from

$$\delta(x, y, t) = \int_B \frac{\partial w_z}{\partial z}\, dz = \int_B \varepsilon_{sk}(x, y, z, t)\, dz = \int_B \frac{\widetilde{p^e}}{\widetilde{\lambda''_s} + 2\widetilde{\mu'_s}}\, dz, \tag{5.5.59}$$

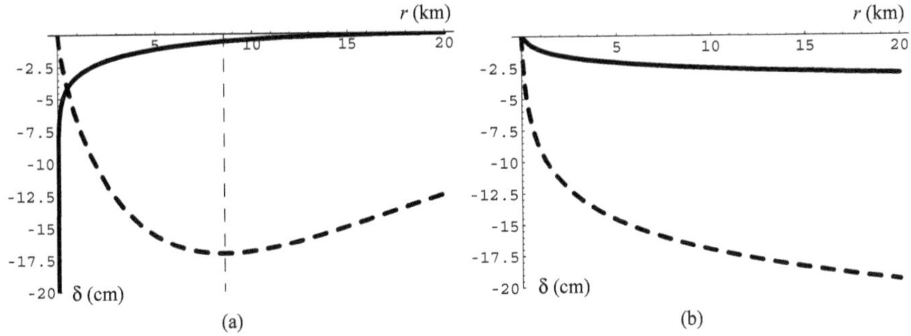

Figure 5.5.1: Land subsidence (δ) and horizontal displacement ($\widetilde{w_r}$) produced by a single pumping well in a confined aquifer ($Q_w = 50\,\ell/\mathrm{s}$, $C_v = 6 \times 10^3$ cm^2/s, $T = 95$ cm^2/sec, $B^o = 142$ m). (a) Subsidence after 3 years of pumping versus radial distance r; (b) Subsidence at 3 km from pumping well versus time. (Dashed line: $\widetilde{w_r}$; Solid line: $\varDelta_z (\equiv -\delta)$; Vertical dashed line marks $r|_{w_r,\max} = 8.54$ km.

where B denotes the thickness of the considered layer, and we have made use of (5.1.70) to express ε_{sk} in terms of the pressure $\widetilde{p^e}$. The solution for $\widetilde{p^e}$ can be obtained by solving (5.5.47), with $\phi \approx \tilde{\phi}$.

Some researchers (e.g., Gambolati *et al.*, 1973, 1974; Corapcioglu and Brutsaert, 1977) have presented subsidence models that take into account the time lag between measured changes in piezometric head and the observed resulting compaction. This time lag is an indication that the purely elastic constitutive relations is not appropriate for clay and silt lenses.

Chapter 6

UNSATURATED FLOW MODELS

The previous two chapters dealt with the modeling of groundwater flow in aquifers, i.e., in the saturated zone. However, as emphasized on several occasions already in Chaps. 1 and 2, certain flow processes, which take place in the unsaturated zone, or zone of aeration (Fig. 2.2.1), are highly important also from the regional point of view and should be incorporated in our modeling considerations. The first example is the infiltration process. A phreatic aquifer is replenished from above by water from various sources: precipitation, irrigation, artificial recharge by surface spreading techniques, etc. In all these cases, water moves downward, from ground surface to the water table, through the unsaturated zone. The understanding of, and, consequently, the ability to calculate and predict the movement of water in the unsaturated zone is, therefore, essential when we wish to determine the replenishment of a phreatic aquifer.

A second example is related to the contamination of groundwater from sources at ground surface. Contaminants from such sources dissolve in water applied to ground surface. The infiltrating water will then carry the dissolved contaminants as it moves downward towards the water table. As contaminants travel downward with the infiltrating water, various phenomena, e.g., dispersion and adsorption, take place. These affect the concentration of pollutants in the water, which, eventually, reaches the water table. The ability to forecast the movement and accumulation of contaminants in the unsaturated zone is required if we wish to clean the subsurface from these contaminants, or to determine the rate at which they will reach the water table. However, one cannot study the movement of contaminants carried by the water without information on the movement of the water itself. In this chapter we shall discuss the modeling of flow in unsaturated flow.

It is interesting to note (Bear *et al.*, 1968; Narasimhan, 2005) that Buckingham (1907) was probably the first to present a detailed analysis of unsaturated flow. He assumed that capillary attraction is a conservative force field, and related water mass flow to the gradient of the *capillary potential* which he implicitly assumed to depend on the underpressure of pure water. His definition of *pressure* is equivalent to the instrumental pressure determined by a

J. Bear, A.H.-D. Cheng, *Modeling Groundwater Flow and Contaminant Transport*, Theory and Applications of Transport in Porous Media 23,
DOI 10.1007/978-1-4020-6682-5_6, © Springer Science+Business Media B.V. 2010

porous plate or cup (Subs. 6.1.5). Later, Richards (1931) and Childs (1936) introduced the flux and the gradient of the capillary pressure.

With the background presented in Chaps. 4 and 5, we can, easily, present the modeling of flow in the unsaturated zone in this chapter. The modeling of contaminant transport in both the saturated and unsaturated zones will be presented in Chap. 7.

More information on flow in the unsaturated zone can be found in many reviews and books, e.g., Stallman (1967), Childs (1969), Swartzendruber (1969), Philip (1970), Bear (1972, Chap. 9), Kirkham and Powers (1972), Morel-Seytoux (1973), Kool *et al.* (1987), Wilson *et al.* (1994), Stephens (1996), Parlange and Hopmans (1999), Tindall and Kunkel (1999), and Feddes *et al.* (2007). Information on field and laboratory measurement techniques and instrumentation concerning the unsaturated zone, e.g., sampling techniques, properties of the solid matrix and the phase or phases occupying the void space, can be found in Dane and Topp (2002).

6.1 Statics of Fluids in Unsaturated Zone

6.1.1 Water content

In the unsaturated zone, the void space is partly filled by air (subscript a) and partly by water (subscript w). Two variables may be used to define the relative quantity of water at a certain time in the vicinity of a point in a porous medium domain (i.e., in an REV for which this point is a centroid):

A. Volumetric fraction of an α-phase:

$$\theta_\alpha = \frac{\text{Volume of } \alpha\text{-phase in REV}}{\text{Bulk volume of REV}}, \qquad \alpha = w, a, \qquad 0 < \theta_\alpha \leq \phi, \quad (6.1.1)$$

where ϕ denotes the porosity.

B. Saturation of an α-phase:

$$S_\alpha = \frac{\text{Volume of } \alpha\text{-phase in REV}}{\text{Volume of void space in REV}}, \qquad \alpha = w, a, \qquad 0 < S_\alpha \leq 1, \quad (6.1.2)$$

with the two definitions related to each other by

$$\theta_w + \theta_a = \phi, \qquad S_\alpha + S_a = 1. \quad (6.1.3)$$

6.1.2 Surface tension

A liquid that partially fills a container forms a surface with any other immiscible fluid—a liquid or a gas—present in the same container. This 'surface' is actually a very thin zone of transition between the two phases. Close to this surface, say, within a distance of a few molecules on either side of the transition zone, the fluid's properties differ significantly from those within the body of either fluid. To understand the reason for the difference in behavior, we have to refer to what happens in the fluids at the *molecular level.*

Molecules of a fluid are attracted to each other by an attractive force. As an example, let us focus on the interface between a liquid domain and another domain occupied by the liquid's vapor. Because a molecule in the interior of the liquid body is surrounded by liquid molecules of the same kind, having a similar mean spacing, it is attracted, on the average, equally in all directions, and the resultant attractive force acting on it vanishes. The same is true for a molecule in the interior of the vapor domain. However, the situation changes as we approach the surface from either side. A molecule belonging to the surface is subjected to a stronger resultant attractive force towards the interior of the liquid body. As a consequence of the pull towards the liquid's interior, *work* must be performed in order to increase the surface of the interface by bringing liquid molecules from the interior of the liquid body to the interface. Left alone, the surface will tend to assume, spontaneously, the shape that corresponds to a state of minimum energy under the prevailing conditions. Thus, the surface of the liquid always tends to contract to the smallest area possible under the prevailing circumstances. The same phenomenon takes place at the interface between a liquid and a gas, or between any two immiscible liquids.

Because the molecules in the transition zone behave differently from those in the interior of the respective fluids, we replace the transition zone, *conceptually*, or as an approximation, by a sharp *interface* that separates the two fluids (Gibbs, 1906). The exact position of this surface within the transition zone is arbitrary. Although molecules are continuously joining and leaving it, its identity is retained as a distinct surface separating the two fluid phases. Extensive quantities (e.g., heat or mass of a chemical species) may cross it.

In order to increase the area of the interface between two immiscible fluids, molecules from the interior of the two fluid bodies must be brought into the surface. This requires that work be done against the net *cohesive force* among the molecules in the two fluids. On the other hand, energy is gained when the area of an interface is reduced. The work required in order to increase the surface of an interface by one unit area is called *surface* (or *interfacial*) *free energy.*

The tendency of a surface to contract may be regarded as a manifestation of the surface free energy. The molecules at the surface behave *as if* they belong to a thin, skin-like elastic layer, or *membrane*, under tension, that adjusts its geometry to attain the smallest possible surface area under the prevailing conditions. Obviously, the 'membrane' is only a *model* of the behavior of the interfacial boundary surface, and no such membrane actually exists. This property of interfaces causes a liquid droplet to assume a spherical shape (which has the smallest surface area for a given volume), in the absence of any other forces.

We have to be careful with the analogy to a 'stretched membrane'. The tension in the latter, generally, increases with increased surface area, whereas

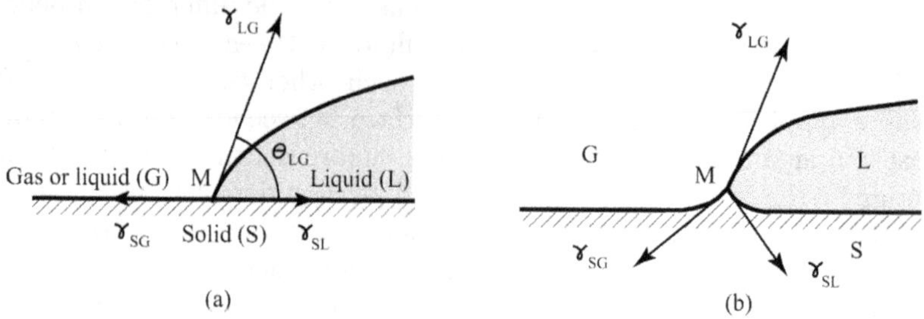

Figure 6.1.1: Interfacial tension.

the surface tension is independent of area. Furthermore, contrary to the case of an interface, molecules are not added to a true membrane as it is being stretched.

The interfacial free energy manifests itself as an *interfacial tension* (inside the apparent 'membrane'), measured as energy per unit area. For a pair of fluid phases, α and β, the interfacial tension, $\gamma_{\alpha\beta}$, is defined as *the amount of work that must be performed in order to separate a unit area of phase α from phase β, or, equivalently, to increase their interface by a unit area.* For air (a) and water (w) at $20°C$, $\gamma_{wa} = 72.88$ erg/cm^2 ($\equiv 72.88 \times 10^{-3}$ J/m^2). Equivalently, the interfacial tension can also be expressed as force per unit distance along the membrane's surface, i.e., $\gamma_{\alpha\beta} = 72.88$ dyne/cm ($\equiv 72.88 \times 10^{-3}$ N/m). For comparison, for water and benzaldehyde the interfacial tension is 15.5 dyne/cm. The interfacial tension between an α-phase and its own vapor is called *surface tension*, γ_α. Often, the term *surface tension* is also used to indicate the *interfacial tension* associated with the interface between two immiscible liquids, or between any liquid and a gas. Henceforth, we shall also use the term 'surface tension' to indicate 'interfacial tension'.

Surface tension depends on temperature; it decreases by approximately 5.5×10^{-2} dyne/cm/°C for a crude oil-water interface. It is strongly affected by surface active agents (called *surfactants*), by gas in solution, and by pH (Schowalter, 1979).

Figure 6.1.1a shows two immiscible fluids in contact with a plane solid surface (S). The point M in the figure is the trace of the line (perpendicular to the figure) along which the three phases are in contact with each other. Due to the surface tension, three forces act at this line, each being directed along the tangent to the interface between adjacent phases. The magnitude of each force, per unit length of the contact line, is equal to the corresponding interfacial tension: γ_{SG}, γ_{SL}, and γ_{LG}.

The angle θ_{LG}, called *contact angle*, or *wetting angle*, denotes the angle between the solid surface and the fluid-fluid interface, measured through the denser fluid. It depends on the properties of the two fluids, and expresses the affinity of the fluids for the solid. At equilibrium,

$$\gamma_{LG} \cos \theta_{LG} + \gamma_{SL} = \gamma_{SG}, \quad \text{or} \quad \cos \theta_{LG} = \frac{\gamma_{SG} - \gamma_{SL}}{\gamma_{LG}}. \tag{6.1.4}$$

Equation (6.1.4), called *Young's*, or *Dupré equation*, states that $\cos \theta_{LG}$ is the ratio of the work required to change a unit area of the S-G-interface into a unit area of S-L-interface to the work required to form a unit area between the L- and the G-phases.

Actually, Young's equation considers only equilibrium of the force components along the tangent to the solid surface. Requiring equilibrium of force components also along the normal to the surface would mean that, in principle, we must have the situation shown in Fig. 6.1.1b. We note that the solid surface is not planar, although the actual deviation from a plane may be very small.

The fluid for which $\theta_{LG} < 90°$ (e.g., L in Fig. 6.1.1a), is said to *wet* the solid and is called the *wetting fluid.* When $\theta_{LG} > 90°$, the fluid (G in Fig. 6.1.1a) is called a *nonwetting fluid.* In any system similar to that shown in Fig. 6.1.1a, it is possible to have either a L-fluid-wet, or a G-fluid-wet solid surface, depending on the chemical composition of the two fluids and of the solid. In the unsaturated (air-water) zone in the soil, water is, usually, the wetting phase, while air is the nonwetting one. Sometimes, due to the heterogeneity of the mineral composition of the solid matrix, we encounter *fractional wettability*, defined as the fraction of the total surface area that is preferentially wet by one of the phases (e.g., Anderson, 1987; Demond *et al.*, 1994). This phenomenon may strongly affect the transport of fluid phases, and of dissolved chemical species.

Additives, called *surfactants*, or *surface active agents*, tend to accumulate in the liquid close to and at the interface. We say that they 'adsorb' on the interface. They reduce the interfacial tension, sometimes significantly, and may alter the contact angle, mainly due to modifications of solid surface properties. The presence of surfactants, even in minute quantities, may significantly change the capillary behavior of water in a soil.

Let us now focus on the interaction at the solid-fluid interface, in particular between an aqueous solution and a non-reactive mineral solid.

Figure 6.1.2 shows, schematically, the magnitude of the forces of molecular attraction between a solid and an adjacent wetting fluid. These forces decrease rapidly with the distance from the solid surface. Various explanations have been given for the resulting 'adsorbed water'. One hypothesis is that positively charged ions, which are surrounded by water molecules, are attracted to the mineral surfaces, which, as is well known, are usually negatively charged. Other forces that attract water molecules to a solid surface are the *van der Waals force*, and interactions between the electric field produced by the solid and water dipoles. The result is a layer of water, perhaps only a few molecules thick, with thermodynamic properties, such as density and viscosity, which are different from those of the bulk water at the same pressure and temperature. Low (1976) observed higher viscosity and higher

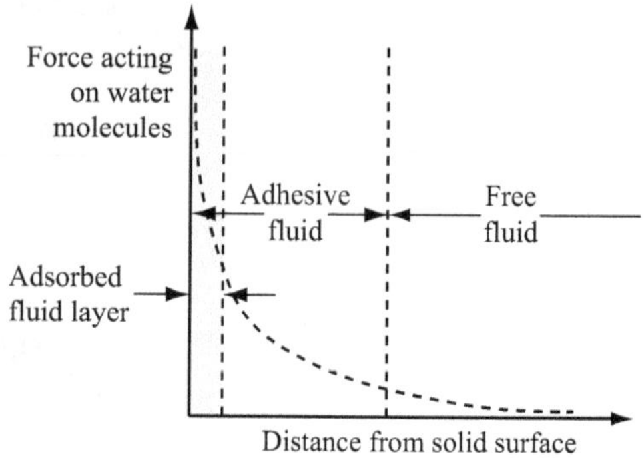

Figure 6.1.2: Schematic diagram of adhesive fluid near a solid surface.

density at distances of up to 16 molecular layers from clay surfaces. The term *adhesive fluid* is used to denote the fluid layer in which the above forces are significant.

In two-phase flow, e.g., water and air in the unsaturated zone, molecular forces prevent the complete drainage of the wetting fluid—water—from the void space. A thin *film* of adsorbed wetting fluid will always remain on the solid. As we shall see below, some wetting fluid may also remain in the void space for other reasons.

As an *illustration* of the effect of surface tension, consider a soap (subscript b) bubble of radius r, with gas (= air, subscript a) on both sides. Actually, in a soap bubble, we have two interfaces: a soap-(internal) air interface, and a soap-(external) air one. With γ ($= 2\gamma_{ab}$) denoting the surface tension in the film, measured as energy per unit area, the total energy in the film surface is $4\pi r^2 \gamma$. When the radius is increased by dr, the added energy will amount to $8\pi r\gamma dr$. This increase in film area is produced by increasing the pressure difference, Δp, say, by increasing the inner pressure, p_{in}, more than the outer one, p_{out}. The added energy is due to the work of Δp. Thus,

$$\Delta p \times (4\pi r^2) \times dr = 8\pi r\gamma dr, \quad \text{or} \quad \Delta p = \frac{2\gamma}{r}. \tag{6.1.5}$$

Since $\gamma > 0$, and $r > 0$, we must have $\Delta p > 0$, or $p_{\text{in}} > p_{\text{out}}$.

As in the soap bubble example presented above, a discontinuity in fluid stress exists across a *curved* interface that separates any two immiscible fluids (say, air and water). The jump in the normal stress, or pressure, is a consequence of the interfacial tension which exists at every point of such an interface. The difference between the pressure in the fluid that occupies the *concave* side of an interface, p_{concave}, and the pressure that occupies the fluid on the *convex* side of the interface, p_{convex}, is called *capillary pressure*, p'_c,

$$p'_c = p_{\text{concave}} - p_{\text{convex}}. \tag{6.1.6}$$

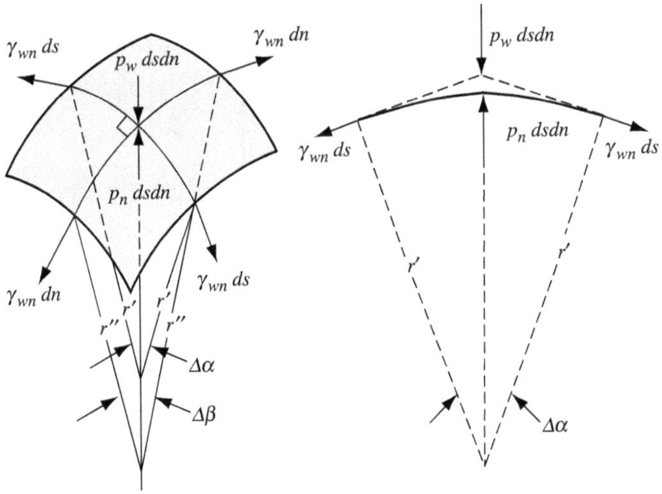

Figure 6.1.3: Force balance at a curved interface.

In this equation, the pressures are taken as the interface is approached within the appropriate phase. We have used the prime symbol to indicate that p'_c is the capillary pressure *at the microscopic level*, i.e., at a point on the interface.

The magnitude of the pressure difference at a point on an interface depends on the radius of curvature of the interface, as well as on the surface tension at that point. The soap-bubble discussed above has a spherical surface.

Let us now consider the relationship for a general curved surface. Figure 6.1.3 shows an infinitesimal element of a curved interface between a w-fluid, which occupies the convex side of the interface, and an n-fluid, which occupies the concave side. The reason for choosing these symbols will become clear below. The figure shows the various forces acting on this element. Assuming the interfacial tension between these two fluids, γ_{wn}, to be constant, a balance of force components normal to this element requires (Adamson, 1982) that at equilibrium

$$p'_c = p_n - p_w = \gamma_{wn}\left(\frac{1}{r'} + \frac{1}{r''}\right) = \frac{2}{r^*}\gamma_{wn}. \qquad (6.1.7)$$

In this expression, r' and r'' are the two *principal radii* of curvature of the surface, with a radius considered positive when it lies within the n-fluid. The radius r^* denotes the *mean radius of curvature*, defined by $2/r^* - 1/r' + 1/r''$. Equation (6.1.7) is known as the *Laplace*, or *Young-Laplace formula* for capillary pressure. Because γ_{nw} is positive when both radii are positive, the pressure is greater in the n-fluid, for which the surface is convex.

When any two fluids are in contact with a solid, one of them will behave like a *wetting* fluid with respect to the solid, while the other will behave as

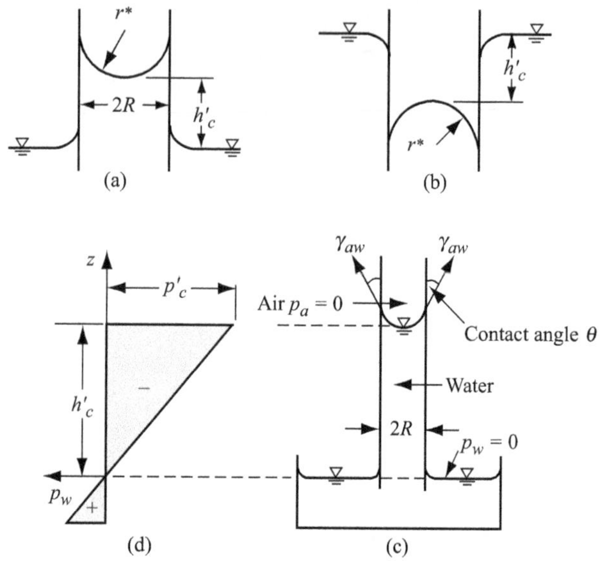

Figure 6.1.4: Two immiscible fluids in a capillary tube, with (a) capillary rise, $\theta = 0°$, (b) capillary depression, $\theta = 180°$, and (c) capillary rise, $\theta \neq 0°$.

a nonwetting one. The wetting fluid tends to be as close as possible to the solid, while the nonwetting fluid tends to shy away from the solid.

An example of two fluids in a capillary tube is shown in Fig. 6.1.4. Similarly, when two fluid phases occupy the void space, one will be identified as the wetting fluid and the other will be the nonwetting one. The definition of capillary pressure presented above still holds, but now we interpret the subscripts w and n as 'wetting' and 'nonwetting' fluids, respectively.

Bear and Bachmat (1990, p. 252) derive the Laplace formula by considering the boundary condition for momentum transfer across a microscopic interphase boundary. They show that this condition reduces to the Laplace formula when the two fluids are assumed stationary, or when the effects of shear and spatial variations in surface tension (e.g., due to temperature variations) are neglected. For moving fluids, the Laplace equation expresses the difference between the normal stresses on both sides of the interface.

In the general case of a wetting liquid, $\theta_{LG} \neq 0°$, we have the situation shown in Fig. 6.1.4c. If the capillary tube has a circular cross-section, with a radius R that is not too large, the curved interface (= meniscus) will be approximately in the shape of a hemisphere. In this case, $r' = r''$, and $r^* = R/\cos\theta_{LG}$. Then, in a small diameter circular tube, with an approximately spherical meniscus, the pressure difference given by (6.1.7) can be written in terms of the tube's radius, R, in the form

$$p'_c = \frac{2\gamma_{wn} \cos \theta_{LG}}{R}. \tag{6.1.8}$$

In a porous medium, one may visualize the narrow passage between grains (often referred to as 'pore throat') as a capillary tube, with R representing some equivalent radius of such passage (or a pore size).

6.1.3 Capillary pressure

In the previous section, we have discussed the behavior of an interface at the microscopic level, i.e., at a point on the curved interface (= meniscus) between two immiscible fluids within a pore. Equation (6.1.7) describes what happens at a point on such an interface *at equilibrium*. It expresses a microscopic relationship satisfied by the pressures within the two fluids as the interface is approached from both sides. In the interior of each fluid, the pressure varies from point to point according to whether the fluid is stationary or moving. For example, if the fluid is stationary, the pressure in it varies *hydrostatically*. At the microscopic level, the interface serves as a common boundary to the two fluid domains.

In the macroscopic approach, however, the microscopic pressure does not appear as a state variable. Instead, we need information only on the *average pressure* within each of the two fluid phases. For this purpose, we define a *macroscopic capillary pressure*, p_c, as the difference in the average pressures, viz.,

$$p_c = \overline{p_n}^n - \overline{p_w}^w, \tag{6.1.9}$$

in analogy to the definition of capillary pressure at the microscopic level. Actually, we should derive (6.1.9) by averaging (6.1.7) over the microscopic interface *surface area*, S_{wn}, within an REV. Note the use of the averaging symbol here. Later we shall drop its usage whenever it is obvious that we are referring to averaged values.

In the unsaturated zone, the wetting fluid is water and the nonwetting one is air. It is often *assumed* that the air is at a constant atmospheric pressure, taken as zero datum, i.e., $\overline{p_n}^n = \overline{p_a}^a = 0$. Then,

$$p_c = -\overline{p_w}^w. \tag{6.1.10}$$

Since $p_c > 0$, except on a planar surface, the (average) gauge pressure, $\overline{p_w}^w$, in the water present in the void space in the unsaturated soil, with gas at atmospheric pressure, is always negative. Under such conditions, we often introduce the definition of (macroscopic) *capillary pressure head*, ψ (dims. L), also called (macroscopic) *suction*, or *tension*, or *matric suction*,

$$\psi = -\frac{\overline{p_w}^w}{g\overline{\rho_w}^w}. \tag{6.1.11}$$

Thus, when the pressure in the water is negative, suction is positive. In fact, this is the reason for introducing suction as a variable, rather than working with water pressure, or pressure head. Note that ψ should be employed only when the (macroscopic) water density, $\overline{\rho_w}^w$, is constant.

In soil science, the unit pF is defined as the logarithm of the negative pressure head in the water, measured in cm. Thus, pF $= 1$ means a suction of 10 cm, while pF $= 4$ indicates a suction of 10^4 cm (of water).

Although introduced here for the case $\overline{p_a}^a = 0$, the same definition is sometimes extended to cases where $\overline{p_a}^a \neq 0$, viz.,

$$\psi = \frac{\overline{p_a}^a - \overline{p_w}^w}{g\overline{\rho_w}^w} = \frac{p_c}{g\overline{\rho_w}^w}. \tag{6.1.12}$$

The symbol h_c is often used interchangeably with ψ. There is no advantage in using the concept of suction (rather than capillary pressure head) when $\overline{p_a}^a \neq 0$. Note that the definition for ψ is not $p_a/\rho_a g - p_w/\rho_w g$, i.e., it is not defined as the difference between the pressure heads in air and in water. Instead, in (6.1.12), we have used the density of water as a *reference density* for defining *equivalent pressure heads.*

In the remaining part of this chapter, the averaging symbol, $\overline{(..)}^\alpha$, will be omitted when referring to the average of (..). Unless otherwise stated, the term capillary pressure, and the symbol p_c will be used for the difference between the macroscopic pressures defined by (6.1.9).

The pressure distribution within a stationary α-fluid that occupies part of the void space of a porous medium domain, is *hydrostatic.* This means that differences in potential (= gravitational) energy balance pressure differences within the fluid, i.e.,

$$\nabla p_\alpha + \rho_\alpha g \nabla z = 0, \qquad \nabla(..) \equiv (\partial(..)/\partial x_i)\mathbf{1x}_i, \tag{6.1.13}$$

where Einstein's summation convention is applicable, $\nabla(..)$ denotes the gradient of (..), p_α is the (average) pressure in the α-phase, $\mathbf{1x}_i$ denotes the component in the i direction of the unit vector $\mathbf{1x}$, g is gravity acceleration, ρ_α is the (average) density of that α-phase, and z is elevation relative to an arbitrary datum.

For a stationary homogeneous fluid ($\nabla\rho_\alpha = 0$), we can write

$$\nabla h_\alpha \equiv \nabla\left(\frac{p_\alpha}{\rho_\alpha g} + z\right) = 0, \qquad h_\alpha = z + \frac{p_\alpha}{\rho_\alpha g}, \tag{6.1.14}$$

where, as in single-phase flow, h_α denotes the (macroscopic) *piezometric head* (dims. L) in the α-phase. Equations (6.1.13) and (6.1.14) are equivalent.

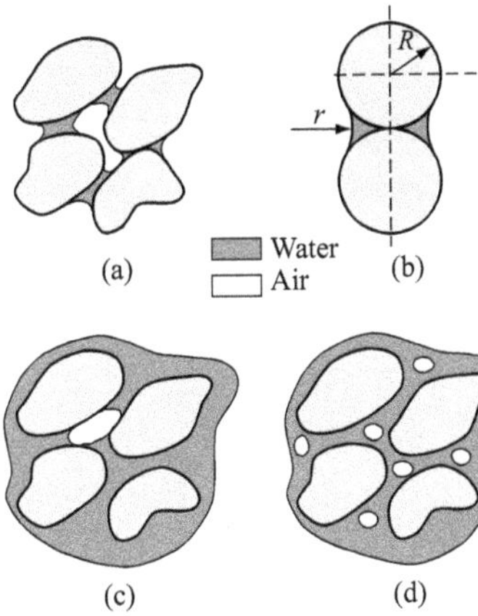

Figure 6.1.5: Air and water distributions at various saturations: (a) Pendular saturation; (b) Pendular ring between two spheres; (c) Funicular water saturation; (d) Insular air saturation.

6.1.4 Retention curve

In (6.1.7), we note the relationship between the capillary pressure and the radius of the meniscus. It seems reasonable to assume that for the many menisci within an REV, the average capillary pressure will depend on some average radius of the menisci, and, hence, on the saturation. Our objective in this subsection is to present the relationship between the (average) capillary pressure and the saturations of two fluid phases that occupy the void space in the unsaturated zone.

We start by considering the distributions of the two phases, air (= nonwetting phase) and water (= wetting phase), within the void space. At low saturation, as a result of the surface tension phenomenon discussed above, water occupies domains in the form of rings around the grain contact points (Fig. 6.1.5a). These are called *pendular rings*. The air-water interface has the shape of a saddle. A number of adjacent pendular rings may coalesce. We observe how water touches the solid at the *contact angle*. When the grains are close, but not touching each other, the water takes the form of a 'bridge', connecting the two grains.

At this low saturation, the pendular rings are isolated and do not form a continuous water phase, except for the very thin film of adsorbed water on the solid surfaces. Figure 6.1.5b shows a pendular ring between two spheres.

As water saturation increases, the pendular rings expand and coalesce, until a continuous water phase is formed. Above this critical saturation, the 'bulk' water forms a continuous phase, and its saturation is called *funicular*; flow of 'bulk' water is possible (Fig. 6.1.5c). Both the water and the air phases are continuous. As water saturation increases, a situation develops in which

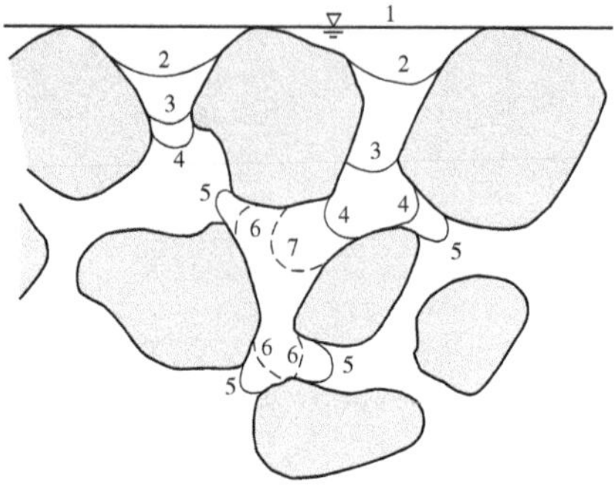

Figure 6.1.6: Gradual drainage and rewetting.

the air is no longer a continuous phase; it breaks into individual bubbles (globules, blobs, ganglia) lodged in the larger pores (Fig. 6.1.5d). The air is then said to be in a state of *insular air saturation*. An air globule can move only if a pressure difference is applied to it by the surrounding water that is sufficient to squeeze it through the constriction. In the absence of air in the void space, we have complete water saturation.

Depending on the pore size distribution, the above stages, say from pendular to funicular water, do not have to occur simultaneously across the entire unsaturated domain. It is possible to envision that at low saturations, part of the water forms a continuous phase, while the remaining part is still in a pendular state, with a gradual transition as saturations increase. A similar transition may occur as saturations are reduced. This may give rise to *models* in which part of the water in the porous medium is mobile, while the other part is *immobile*.

In the course of time, the volume of air at insular saturation may decrease due to air solubility in water. Similarly, the volume of water in pendular rings may decrease with time due to evaporation.

With these definitions, we may now follow the changes in water saturation as an initially saturated porous medium sample is gradually drained from its bottom, with air introduced at its top. Figure 6.1.6 shows several successive stages of drainage (stages 1 through 5) and rewetting (stages 6 and 7). Each state corresponds to a certain volume of air occupying a certain portion of the void space at a corresponding saturation. As water drains, water-air interfaces (menisci) are formed. The radius of curvature at every point on a meniscus depends on the (local, microscopic) capillary pressure, i.e., on the pressure jump across the meniscus, with the radius becoming smaller as the capillary pressure increases.

At every stage, the largest capillary pressure that can be maintained by a local interface corresponds to the smallest radius of curvature that can be accommodated in a pore, or channel, through which the interface is being withdrawn. Therefore, the smallest radius of curvature occurs in the narrowest pores that correspond to the prevailing air volume (e.g., interface 3 in Fig. 6.1.6).

In general, pores have different dimensions and shapes. Therefore, they will not all empty at the same capillary pressure. The large pores (or those with larger channels, or *throats* of entry) will empty at low capillary pressures, while those with narrow channels of entry, supporting interfaces of smaller radius of curvature, will empty at higher capillary pressures.

As water drainage progresses, the water-air interface retreats into channels which support a curvature of still smaller radius (e.g., interface 4). The wetting fluid will continue to retreat until the local interfaces have taken up positions of equilibrium in channels which are sufficiently narrow to support interfaces with smaller radius of curvature. Obviously, if all channels are equal and large, at a given capillary pressure, no equilibrium can be maintained any longer, and a sudden, almost complete, drainage of water from the entire porous medium sample will be observed. We say 'almost', because some water will always remain as isolated pendular rings and as a film adsorbed to the solid surface. Within the pendular rings, the pressure is independent of that in the remaining, continuous water phase in the void space. However, the pressure there is related to pressure in the gaseous phase (which is a mixture of air and water vapor) by the capillary pressure relationship. As water evaporates, the volume of water in a pendular ring decreases, the radius of curvature of the meniscus decreases, and the capillary pressure increases.

If at some point in time, the drainage of the wetting fluid at the bottom of a sample is stopped, overlooking evaporation and air solubility processes, an equilibrium will be established, with no further motion of either fluid. The pressure distribution within each fluid will be hydrostatic, while satisfying the pressure jump condition at every point of the (microscopic) interfaces. At every such point, the pressure jump will correspond to the radius of curvature of the interface at that point. In this way, equilibrium is established between surface tension and gravity.

From the above description, it is obvious that at each stage of a drainage process, the quantity of water remaining in the void space, say, within an REV centered at a point, takes on a certain (microscopic) configuration. The latter is related to the distribution of the (microscopic) interface geometry within that REV. As a consequence, the quantity of water remaining in the void space depends on the (macroscopic) capillary pressure defined by (6.1.9). The capillary pressure increases as the water saturation decreases, which, in turn, corresponds to a decrease in interfacial surface area.

Let us now reverse the process and begin to refill the pore space with water. In Fig. 6.1.6, this is pictured as transition from stage 5 to stages 6 and 7. The interfaces' radii of curvature become progressively larger.

Equation (6.1.7) relates the capillary pressure at a point on the water-air interface to the mean radius of curvature of the latter. In Fig. 6.1.6 we note how the capillary pressure in a tube is related to the radius of curvature of the meniscus, which, in turn, is related to the radius of the tube. Visualizing a porous medium domain as a random assembly of tubes of various radii, the above relations can be interpreted as stating which of these tubes will drain at any given capillary pressure (as long as there is a continuous passage for the drained water to the external boundary of the sample). We may conclude that *a strong relationship exists between the macroscopic capillary pressure and water saturation.*

The relationship between the quantity of water present in the void space (within an REV), in terms of its saturation, to the prevailing capillary pressure, is recorded as a *capillary pressure curve*, $p_c = p_c(S_w)$, or as a *suction curve*, $\psi = \psi(S_w)$. In soil science, these curves are called *retention curves*, as they show how much water is retained in the soil by the capillary pressure. Some authors refer to the drainage retention curve as a *desorption curve* and to the wetting, or *imbibition*, curve as a *sorption* curve.

We recall that at every microscopic point on a meniscus, the capillary pressure depends on the surface tension between the wetting and nonwetting fluids, γ_{wn}. As explained when the concept of surface tension was introduced in Subs. 6.1.1, γ_{wn} depends on the temperature, T_{wn}, and on the concentrations, c_{wn}^{γ}, of the γ-components at the interface. Under thermodynamic equilibrium, the temperature is the same as in the two adjacent phases, i.e., $T_w = T_n = T_{wn}$, and concentrations are a function of those on either one of the phases, i.e., $c_{wn} = f_w(c_w) = f_n(c_n)$.

Therefore, the capillary pressure will also depend on these factors, viz.,

$$p_c = p_c(S_w, \gamma_{wn}(T_{wn}, c_{wn})),$$

or, more concisely,

$$p_c = p_c(S_w, T_{wn}, c_{wn}).$$

For the sake of simplicity, in this chapter we shall continue to assume that for a given pair of fluids, $p_c = p_c(S_w)$ only.

Figure 6.1.7 shows typical examples of capillary pressure head curves, $h_c = h_c(\theta_w)$, during drainage. Point A in Fig. 6.1.7 indicates the *threshold capillary pressure head*, h_c^{cr}, corresponding to the largest pore size. If we start from a sample that is fully saturated by water, we can produce the capillary head, h_c^{cr}, by draining a very small quantity of water. Practically, no air will penetrate the sample, until the critical capillary head is reached. The corresponding pressure, p_c^{cr}, is called the *critical pressure*, or *threshold pressure*. When expressed in terms of pressure, the critical value is also called the *bubbling pressure*, or *air entry pressure*. As the magnitude of the capillary pressure head, h_c, is increased, an initial small reduction in θ_w, associated with the retreat of the air-water menisci into the pores at the external surface of the sample, is observed. Then, at the critical pressure head value h_c^{cr}, air enters the larger pores and they begin to drain.

The shape of the capillary pressure curve, and, hence, also the threshold pressure, depends on the distributions of pore sizes and shapes. Curves 1 and

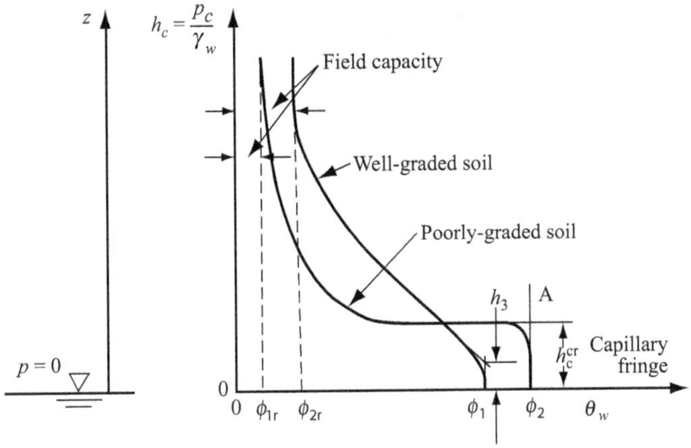

Figure 6.1.7: Typical capillary pressure curves during drainage.

2 in Fig. 6.1.7 correspond to *well graded* and *poorly graded* granular porous media, respectively.

As water drainage progresses, we observe that a certain quantity of water remains in the void space even at very high capillary pressures, in the form of isolated pendular rings and relatively immobile thin films. The corresponding value of water saturation, denoted by S_{wr}, is called the *irreducible water (or moisture) saturation.* In terms of the volumetric fraction, θ_w $(= \phi S_w)$, it is denoted by θ_{wr}, and called *irreducible water (or moisture) content.* When the air is present in the form of isolated bubbles only, the air saturation, S_a, is referred to as *residual air saturation*, S_{ar}. Note that, as is customary, we have used here the term 'irreducible' for the wetting phase, while for the nonwetting phase we have used the term 'residual'. Further reduction in these limiting values is possible by water evaporation and by film flow. At these low saturations, the concept of *matric potential* should be used, rather than capillary pressure (Subs. 6.1.6).

The capillary pressure curve is an expression of the pore-size distribution of the soil. The *Laplace formula* at the macroscopic level, as an analog to (6.1.7) and (6.1.8), is

$$p_c \equiv p_n - p_w = \frac{2}{r^*}\gamma_{wn}, \quad \text{or} \quad p_c \equiv p_n - p_w = \frac{2\gamma_{wn}\cos\theta_{LG}}{R}. \tag{6.1.15}$$

Here r^* is the mean radius of curvature of the microscopic interfaces between the wetting and nonwetting fluids (in our case, water and air) inside an REV, and θ_{LG} is the contact angle. Note that if we assume equilibrium and no gravity effects, all air-water menisci at a (macroscopic) point, i.e., within the REV centered at that point, must have exactly the same radius of curvature. Under such conditions, the microscopic value of capillary pressure and its macroscopic counterpart become identical. Recalling that in (6.1.8) we have replaced (the local) r^* by $R/\cos\theta_{LG}$, with R regarded as a measure of the size of the pores occupied by the wetting fluid. For water on mineral grain surfaces, it is known that $\theta_{LG} \approx 0$.

For a given rigid soil, neglecting any effect of fluid composition on the structure of the pores, the effect of the pore- or grain-size distribution on capillary pressure curves is the same, regardless of the nature of the two fluids. The effect of fluid properties may be stated by the general expression

$$S_w(p_c|_{\text{fluids } n_1,w_1}) = S_w(\beta_{12} p_c|_{\text{fluids } n_2,w_2}), \tag{6.1.16}$$

where

$$\beta_{12} = \frac{\gamma_{n_1w_1}\cos\theta_{n_1w_1}}{\gamma_{n_2w_2}\cos\theta_{n_2w_2}} \tag{6.1.17}$$

is a *scaling factor*, and $\{n_1, w_1\}$ and $\{n_2, w_2\}$ represent pairs of nonwetting and wetting fluids. When the contact angles remain unchanged, then

$$\beta_{12} = \frac{\gamma_{n_1w_1}}{\gamma_{n_2w_2}}. \tag{6.1.18}$$

The pairs of immiscible fluids may be different, or they may be the same pair of fluids under different temperature and slightly different chemical compositions. Therefore, using fluids n_1 and w_1 as reference fluids, we can use the above relationships to obtain the capillary pressure curve for any other pair of immiscible fluids.

6.1.5 *Experimental determination of retention curve*

A variety of methods are employed for determining the capillary pressure relationship. A direct approach is to obtain them from laboratory measured static saturation-capillary pressure data, obtained from soil cores, or from simultaneous field measurements of water content and matric potential.

A typical experimental apparatus for determining capillary pressure relationships of soil or rock cores is shown schematically in Fig. 6.1.8. A porous medium sample is placed in a cell in contact with a 'capillary barrier', or 'porous plate', which itself is a porous material, such as ceramic, sintered metal, or fritted glass. The grain (or pore) size of this porous material is selected such that it is sufficiently small so as to prevent air from invading the sample under the capillary pressures to be applied during the experiment. To achieve this goal, the *air entry pressure* of the porous plate should exceed the range of capillary pressures that are planned for the experiment. At the same time, the pores in the barrier should not be too small, as, otherwise, a long time will be required for equilibrium to be reached at every stage of the experiment. The sample (= core) is initially saturated with water, with a zero capillary head maintained by adjusting the water level in the tube connected to the bottom of the cell. The top of the sample is in contact

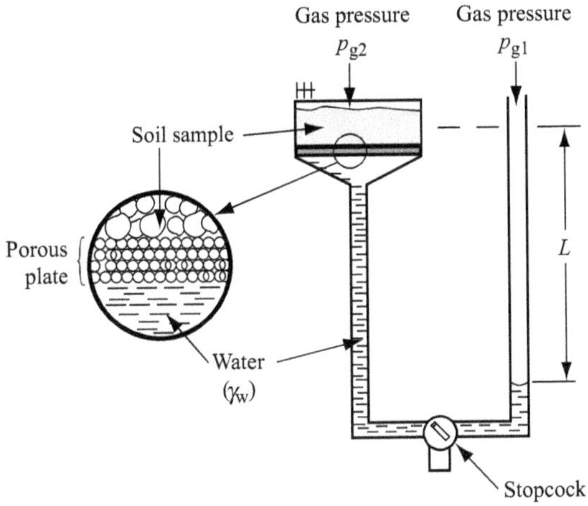

Figure 6.1.8: Laboratory determining the capillary pressure curve.

with gas at atmospheric pressure. The capillary pressure in the sample is increased incrementally, by producing a negative (gauge) water pressure at the bottom of the core (e.g., by lowering the tube), or by increasing the gas pressure in contact with the top of the core. Either way, the capillary pressure is increased and water will leave the core through the capillary barrier. After equilibrium has been attained, following each incremental change in the capillary pressure, the volume of water outflow is measured and the new water saturation computed. This procedure is repeated step-wise to generate points on the drainage capillary pressure curve.

In this experiment, water can be removed as long as the remaining water, at least in part, constitutes a continuous phase (i.e., above the 'irreducible liquid saturation', S_{wr}). When the liquid saturation level is dropped to S_{wr}, it becomes discontinuous everywhere, in the form of isolated pendular rings, ganglia, or very thin liquid films on the solid surface in pores, from which most of the liquid has been evacuated. Under such conditions, the liquid's effective permeability vanishes, and further drainage by *liquid flow* cannot be produced by a pressure gradient and gravity. The above statement is not completely accurate as some water flow may take place even under saturations below S_{wr} in the form of film flow (Dullien, 1992). The experiment is terminated when the air entry value of the porous plate is approached. An imbibition capillary pressure curve is determined by reversing the process.

Common techniques for measuring water content in the field include simple gravimetric determination from soil core samples, calculation from downhole neutron attenuation measurements, inference from calibrated electrical resistance (or dielectric) cells, and time-domain reflectometry (TDR). The most common method for measuring negative (gauge) water pressures in the field utilizes devices known as *tensiometers* (a name introduced by Richards

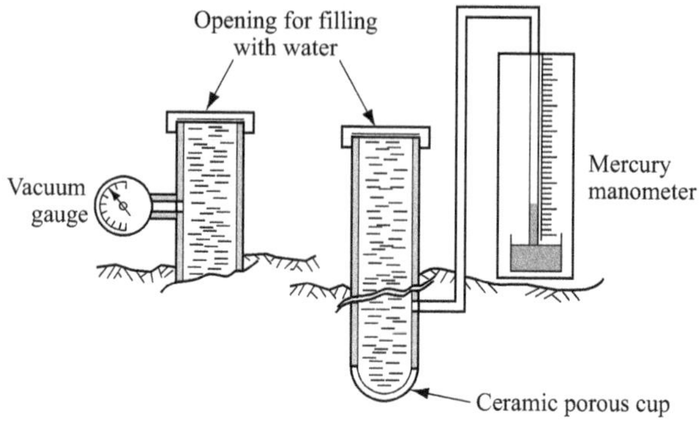

Figure 6.1.9: Tensiometer for determining water suction.

and Gardner (1936)), shown schematically in Fig. 6.1.9. The tensiometer is made up of a rigid tube filled with water and fitted with a manometer and a vacuum gauge, or a transducer, to measure the pressure in the tube. The contact between the water in the soil and in the tensiometer is established through a 'porous cup' made of ceramic (or other rigid and inert porous materials). When inserted into the soil, the water saturating the rigid and tightly sealed pipe is initially at atmospheric pressure (approximately). The water in the soil is at sub-atmospheric pressure. The suction applied through the wall of the porous cup draws out a small volume of water, producing the same sub-atmospheric pressure in the tube. This pressure is indicated by the measuring device. Before determining pressure and water content at every step, one has to make sure that equilibrium between the water pressure in the soil and in the tensiometer has indeed been reached. This may take a very long time, depending on the hydraulic resistances of the porous cup and the surrounding soil. The time to establish equilibrium can be reduced by selecting a cup with higher permeability, but then we encounter the possibility of air being sucked into the tensiometer. The range of such a device is, thus, limited by the *bubbling pressure* of the tensiometer, and, ultimately, by the cavitation pressure of water. This limits the use of tensiometers to capillary pressures of less than about 0.8 atm.

Common to the above laboratory and field methods is the fact that in all of them, measurements are taken once the system has reached (or has been assumed to have reached) equilibrium, or no-flow conditions. This statement questions the use of the resulting capillary pressure curves in *flow models*.

Another approach for estimating saturation-capillary pressure relations is to conduct a *dynamic* flow experiment in the laboratory (say, in a soil column), or in the field. A typical experiment involves some sequence of water infiltration, redistribution, drainage, and/or evaporation events. The capillary pressure curve is estimated by matching observed water contents, pressures, and/or fluxes to results of simulating the corresponding transient flow initial/boundary value problem. Since the solution of such model depends on the hydraulic properties of the soil (e.g., saturation-capillary pressure relations and permeability relations; see Subs. 6.1.4), it may be possible to infer these relationships by matching observed data with those predicted by the model (history matching). The process of determining soil (or aquifer) properties by comparing measured data with values predicted by solving a mathematical model for the corresponding boundary-value problem is referred to as the *inverse problem*, or the *parameter estimation problem*, or the *calibration problem*. This topic is discussed as Step 7 in Subs. 1.2.2 and in Sec. 11.3.

Inverse methods have the advantage that they yield 'effective parameters' that are directly relevant to dynamic experimental conditions. They may, therefore, require less experimental time or effort than conventional laboratory or field methods. Their primary disadvantage is that solutions to inverse problems are not always *unique*, i.e., different values of an hydraulic parameter may yield similar model predictions.

Over the years, methods have also been developed for estimating soil hydraulic properties from grain size distribution data. Such methods are appealing, since grain size distribution data can more easily be obtained than static or dynamic hydraulic data. Some methods for estimating parameters that appear in saturation-capillary pressure models are based on statistical analyses of tabulated data (e.g., Campbell, 1974; McCuen *et al.*, 1981; Rawls and Brakensiek, 1982; Carsel and Parrish, 1988; and Russo and Bouton, 1992). Other models employ quasi-physical models (Arya and Paris, 1981; Mishra *et al.*, 1989; Celia *et al.*, 1995; Schaap *et al.*, 1998; Arya *et al.*, 1999a,b; Wösten, 2001; and Pan *et al.*, 2001). The latter are based on the proposition that the pore size distribution may be inferred from the grain size distribution by assuming some packing geometry, and that the saturation-capillary pressure curve may be inferred from the pore size distribution function via the Laplace capillary formula, (6.1.7).

Various authors have proposed analytical expressions for the general shape of fluid retention curves. These expressions involve coefficients that must be determined by solving the inverse problem, i.e., by fitting the analytical expression to measured experimental data. Following are a number of such expressions for an air-water system. The symbol $\tilde{S}_w$, defined by

$$\tilde{S}_w = \frac{S_w - S_{wr}}{1 - S_{wr}}, \tag{6.1.19}$$

is called the *effective*, or *reduced water saturation*. In it, the *irreducible water saturation*, S_{wr}, is a fitting parameter whose definition may depend on the conditions under which the saturation-capillary pressure data are measured.

- Brooks and Corey (1966) proposed the relationship

$$\tilde{S}_w = \begin{cases} \left(\dfrac{p_b}{p_c}\right)^{\lambda} & \text{for} \quad p_c \geq p_b, \\ 1 & \text{for} \quad p_c < p_b, \end{cases} \tag{6.1.20}$$

where λ is called *pore size distribution index*, and p_b is the *bubbling pressure.* This is, approximately, the minimum value of p_c on a drainage capillary pressure curve at which a continuous air phase exists in the void space.

- Brutsaert (1966) proposed the relationship

$$\tilde{S}_w = \begin{cases} \dfrac{1}{1+(A\psi)^B} & \text{for} \quad \psi \geq 0, \\ 1 & \text{for} \quad \psi < 0, \end{cases} \tag{6.1.21}$$

where A and B are positive curve fitting coefficients.

- Vauclin *et al.* (1979) introduced the relationship

$$\tilde{S}_w = \begin{cases} \dfrac{1}{1+(A\ln\psi)^B} & \text{for} \quad \psi \geq 1\,\text{cm}, \\ 1 & \text{for} \quad \psi < 1\,\text{cm}, \end{cases} \tag{6.1.22}$$

where A and B are positive curve fitting coefficients.

- Van Genuchten (1980) proposed the relationship

$$\tilde{S}'_w \equiv \frac{S_w - S_{wr}}{1 - S_{ar} - S_{wr}} = \begin{cases} \left[\dfrac{1}{1+(A\psi)^B}\right]^C & \text{for} \quad \psi \geq 0, \\ 1 & \text{for} \quad \psi < 0, \end{cases} \tag{6.1.23}$$

where A, B and C $(= 1 - 1/B)$ are positive curve fitting coefficients.

6.1.6 Matric and other potentials

Thus far, the retention curve has been presented as a relationship between (macroscopic) capillary pressure and saturation. We have explained that the microscopic notion of capillary pressure is associated with the phenomenon of surface tension at the fluid-fluid interface. Although we have repeatedly mentioned the presence of films, and the interaction between the fluid and the solid at their common interface, we have not shown how these phenomena affect the capillary pressure curve. The effects of these and of other phenomena will be discussed in this subsection. We shall include a brief discussion on what is a 'potential' and define the matric and other kinds of potentials.

The *potential* is well established concept in thermodynamics, where, in our language, it is discussed 'at the microscopic level'. Nitao and Bear (1996) presented a rigorous discussion on potentials at the macroscopic level. As in the theory of thermodynamics, the potentials were defined at the *microscopic* level, and then their *macroscopic* counterparts were obtained by averaging.

Any chemical species in an aqueous phase within the void space, including water (H_2O) itself, as a species, is acted upon by various forces. Specifically, water molecules are acted upon by forces that arise from gravity, from the interaction with other water molecules, from the interaction with molecules of other chemical species present in the water, and by interaction with molecules of the solid matrix. Although this statement is valid for any fluid, we shall continue to use water as an example, because it is the main fluid of interest in the unsaturated zone. *Work* has to be expended in order to change the state of a considered system in a direction that is opposite to the resultant of the forces mentioned above (which in themselves may act in different directions). The *potential* is a concept that facilitates the discussion of such changes. It expresses *the reversible work that has to be expended in order to transform a given system from some specified reference state to its current one.* This amount of work is equivalent to the increase in the system's energy. Because of the different nature of the various forces, some being non-mechanical, it is more convenient to define a number of potentials, each corresponding to a specific force, or a combination of forces. The potential is expressed as an *intensive quantity*, either per unit volume, per unit mass, or per unit weight of the considered system—here the water—within the void space.

A. Total potential

The concept of *total potential* was discussed already by Buckingham (1907). A detailed description of Buckingham's work on flow in the unsaturated zone is presented by Narasimhan (2005).

According to the Terminology Committee of Commission *I* of the International Soil Science Society (1976),

> "...the *total potential*, Ψ^w_{total}, of water, H_2O, as a chemical species in the soil solution, at a point in the soil, is defined as the amount of work that must be done, per unit mass of pure water, in order to transfer reversibly and isothermally to the soil water at that point, an infinitesimal quantity of pure water from a pool that contains pure water. The pool is at a specified elevation, and with the same temperature and external gas pressure as at the considered point."

Here 'pure water' refers to the water (H_2O) as a *substance*, or a *species*, while the 'soil water' is the liquid (aqueous) phase (or soil solution) in the pore space, which contains both water and dissolved matter as components.

The above definition refers to 'soil water at a considered point', where the 'point' is within a phase, i.e., at the *microscopic level.* A macroscopic potential is defined as the average of the microscopic one over all (microscopic) points within a representative elementary volume centered at a (macroscopic) point (= centroid of an REV) within a porous medium domain. The same extension from microscopic to macroscopic level may be applied to other types of potentials.

The total potential of a fluid phase is composed of the following parts:

(a) A *matric potential*, associated with the adsorption of water to soil surfaces.
(b) A *solute potential* (or *osmotic potential*), associated with the concentration of species in solution.
(c) A *gravity potential*, which expresses the change in the potential energy associated with the elevation of the considered point above a reference reservoir.
(d) A *thermal potential*, which expresses the change in the free energy associated with the temperature of the considered point above that prevailing in the reference reservoir.

In Soil Physics, the sum of (a) and (b) is called *soil water potential.* Note that the kinetic energy, which is proportional to the water velocity squared, is not mentioned here, because, in soils, it is negligible due to the very small water velocity.

B. Matric potential

The *matric potential*, Ψ_m^w, at a point within a fluid phase (occupying part of the void space in the soil) is defined as (Commission *I*, ISSS, 1976)

> "...the amount of work that must be done, per unit mass of pure water (as a component, w) in order to transfer reversibly and isothermally to the soil water at a considered point, an infinitesimal quantity of pure water from a reference pool. The latter is at the same elevation, temperature, and external gas pressure as at the considered point, and contains water (= soil solution) identical in composition to that present in the soil at the considered point."

Again, the matric potential as defined above refers to a *component* (or chemical species) at a point, i.e., at the microscopic level, while our interest is really (a) in the soil solution, or fluid phase, and (b) *within* the void space, i.e., at the macroscopic level.

The matric potential can also be defined for a phase as a whole. At a point, it is defined as the sum of the (reversible) work that has to be expended in order to move each of the individual components comprising the phase, from a reservoir at the same elevation, temperature, external gas pressure, and composition, to the considered point.

The macroscopic matric potential of a phase is the average over the REV of the microscopic one defined above. This matric potential, often used in soil science when dealing with the unsaturated (air-water) zone of the soil, is a consequence of two phenomena:

- Unbalanced forces across water-air interfaces (menisci), manifested as *surface tension.*
- Attraction of molecules in the phase to the solid surface. This effect is significant only in the vicinity of the solid surface, and, hence, this effect

is manifested as a thin film that coats the solid (see discussion on surface potential below).

The presence of air-water interfaces gives rise to the phenomenon of capillary pressure, viz., the jump in pressure across the microscopic water-air interfaces inside the void space. The microscopic capillary pressure is then averaged to obtain its macroscopic counterpart. In the simplified model of a curved meniscus, with a sufficiently thick fluid layer on both sides, the concept of surface tension is valid, and so is the resulting *Laplace formula*, (6.1.7), that relates the radius of curvature of the meniscus to the (microscopic) capillary pressure. However, in developing an expression for the capillary pressure, say, the Laplace formula, the presence of a film of adsorbed water on the solid surface, and its effect on the relationship between water and gas pressures, was overlooked. This approach is not justified at low saturations, as the portion of the void space from which water has been drained always contains some water in *pendular rings* and as a *thin film* that coats the solid, and, therefore, the effect of the solid surface in the unsaturated zone cannot be ignored at low saturations.

The definition of matric potential incorporates also the effect of the attractive forces acting on these films (Nitao and Bear, 1996). The matric potential is, thus, not identical to the *capillary potential*, which is associated with capillary forces only. The surface effects at the liquid-gas interface becomes negligible at sufficiently high saturations, as the distance from the interface to the solid surface increases. Consequently, the effect of the adsorbed water films becomes negligible. The matric potential is, then, essentially equal to the capillary potential. If we define the matric potential *per unit volume of water*, then, in this range of saturations, it is identical to the capillary pressure.

The effect of the films becomes more significant as the soil is drained, and water saturation approaches the *irreducible* value. As water is further removed by evaporation to *below* the irreducible saturation, the effect of adsorbed water films, and its contribution to the matric potential, become even more significant, as the only water left occurs as films. At some low water saturation, water in the void space can be present only as films that coat the soil surface (and pendular rings at points of contact between grains). They become thinner as water is removed by evaporation. In this range, the behavior is dominated not by the air-water surface tension, but by the interaction between water and solid molecules.

When expressed per unit volume of water, the matric potential at sufficiently high saturations is equal to the pressure in the water (as a liquid phase), $p_w - p_o$, with respect to a datum pressure, p_o, in the reservoir. It is positive (i.e., above the reference atmospheric pressure) in the saturated zone, and negative (i.e., below the reference atmospheric pressure) in the unsaturated zone above it. When the gaseous phase in the void space (primarily air) is at the pressure $p_g = p_o$, the matric potential per unit volume of water

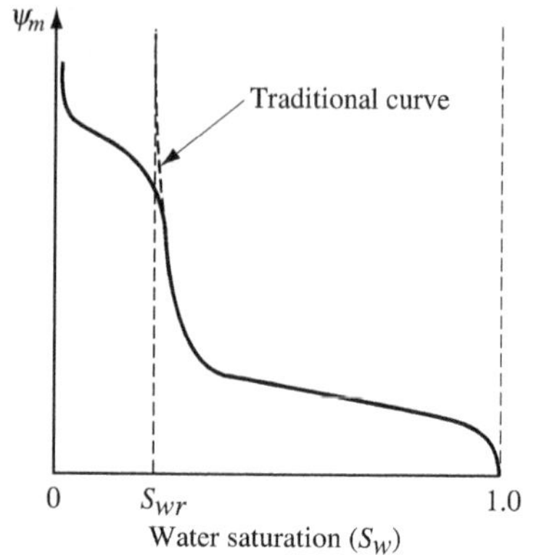

Figure 6.1.10: A typical matric potential curve, including the effect of drying.

is expressed by the difference $p_w - p_g$, or by $-p_c$. The matric potential per unit weight of water is, thus, expressed by $(p_w - p_g)/\rho_w g$.

For a given soil, the relationship between the matric potential and saturation has to be derived *experimentally*, because it depends on the pore size distribution and on the complex geometry of the pore space. For sufficiently coarse soils, at the irreducible water saturation, $S_w \gg S_{wr}$, the $\Psi_m^w(S_w)$-curve (with Ψ_m^w expressed per unit volume of water) is essentially that for the capillary pressure. At full saturation, the matric potential becomes zero.

The matric potential is the more appropriate thermodynamic quantity to be used for describing phenomena in the unsaturated zone than the capillary pressure, as it incorporates both the effect of capillarity, and that of fluid-solid interaction. The latter is significant mainly at very low saturations, where water is primarily in the form of films. To obtain the relationship between the matric potential and the saturation, let us conduct an experiment in which water is removed from a sample, not by gravity drainage, but by evaporation. In this experiment, conducted with pure water in the sample, the vapor content in the gaseous phase and liquid water in the pore space are brought to equilibrium at different levels of water saturation. Kelvin's law (6.1.24) is used to determine the matric potential at each stage. We obtain the typical matric potential curve shown in Fig. 6.1.10. This curve raises the question: "what is the meaning of the term 'irreducible water saturation?". This question is justified as water can be further removed from the sample by evaporation, and by flow along surface films (albeit, at a much lower rate than 'bulk' liquid flow). We have, therefore, to interpret this term as indicating the saturation at which the liquid water becomes discontinuous, such that pressure cannot be transmitted through it to produce motion (e.g., as described by Darcy's law).

Experiments show that the value of irreducible saturation is asymptotically approached as water is (gravitationally) drained from a sample. However, if evaporation and/or film flow are allowed to continue to remove water from the sample, saturation can be further reduced, until it is practically zero. Marshall and Holmes (1979) describe such a drying technique and refer to it as the *vapor sorption method.* Thus, by controlling the vapor content (or vapor pressure) in the gas, waiting at every step for equilibrium to be reached between the vapor pressure in the gas and the pressure in the liquid, the latter's saturation (determined, say, by weighing the sample, or by a mass balance) will gradually decrease, with liquid remaining in isolated smaller pores as pendular rings and as a thin film on the solid. At some saturation level, it will remain only in the latter form. The gradually increasing (volume averaged) matric potential in the water as the soil dries up can, then, be calculated by the relationship

$$p'_c = -\frac{\rho_w RT}{M^w} \ln \frac{p^v}{p^{v_o}}, \quad \text{or} \quad \psi = \frac{RT}{gM^w} \ln h_r, \tag{6.1.24}$$

known as *Kelvin's law,* and drawn as a function of the decreasing saturation. In this equation, R is the universal gas constant, T is the temperature, M_w is the molecular weight of water, $p^v/p^{v_o} \leq 1$ is the ratio between the partial pressure of water in the gas phase at a curved meniscus, to the partial pressure of water in the absence of interfacial effects, and h_r is the *relative vapor pressure,* or *relative humidity.* In this way, the retention curve can be extended to zero liquid saturation. The above process leads to the definition of *surface potential* discussed below.

C. Surface potential

We have seen above, (1) that the amount of energy required for removing water from the soil increases with decreasing saturation, and (2) that this is due primarily to the need to overcome capillary and surface forces. However, capillarity alone cannot account for the energy required to remove water from the soil *at very low saturations,* when water is present in the void space in the form of small pendular rings and thin water films in aerated pores. The water in such films is affected by Van der Waals and electrostatic forces acting on the water (and its dissolved components) by the solid surfaces within the porous medium.

At the microscopic level, the surface potential, φ^w, defines the contribution to the matric potential by forces other than capillary, i.e.,

$$\varphi^w = \Psi^w_m - \Psi^w_p, \tag{6.1.25}$$

where Ψ^w_m is the matric potential, and Ψ^w_p is the *pressure potential* at a point, defined as the work that is required in order to move a unit quantity of water (solution) from a reference reservoir at location (x_o, y_o, z_o) to a point in the porous medium at location (x', y', z_o), with the two points having the same elevation, temperature, and chemical composition.

The pressure potential in a saturated porous medium can be expressed in term of the *pressure energy* per unit weight, resulting from the pressure in the fluid. Furthermore, when the fluid is compressible, the pressure energy stored in the fluid, per unit fluid weight, is obtained from the work done by compressing the fluid. For a compressible fluid, the pressure energy is commonly expressed by *Hubbert's potential* (Hubbert, 1940) defined by (4.1.6).

Consider now forces acting at the microscopic level on a small water particle (infinitesimal quantity) at some point $\boldsymbol{\xi}$, which is close to a solid surface. We assume that the net force acting on this particle by the solid surfaces is *conservative*, i.e., we can define for it a potential energy, $\varphi^w(\boldsymbol{\xi})$, at $\boldsymbol{\xi}$, as the work needed to move the particle from a reference point sufficiently removed form the solid surface, so that there is no force acting on the water there, to the point $\boldsymbol{\xi}$. In a similar way, we may define a surface potential, $\varphi^\gamma(\boldsymbol{\xi})$, for any γ-component in the fluid. In general, these *surface potentials* will depend also on temperature, pressure, and the concentrations of the various components.

The concept of *surface potential*, as used here, is an idealization. Actually, the exact nature of a surface potential is quite complex, and not truly conservative, because the very introduction of the water particle will disturb the motion of molecules and ions, thus causing changes in the potential field. In addition, fluid motion may partially destroy the ordering of (the dipolar) water molecules near a surface, resulting in a change in potential energy.

To obtain a macroscopic surface potential, say of a wetting α-phase, occupying the volume $\mathcal{U}_{o\alpha}$ within an REV of volume $\mathcal{U}_o$, we have to average the microscopic potential, φ, over the REV. Because the surface potential depends on the distance from the solid surface, to facilitate the averaging, $\mathcal{U}_{o\alpha}$ may, in some cases, be divided into three 'apparent phases':

- An **immobile adsorbed layer**, which is next to the solid surface. In it, the fluid-solid forces are so strong such that no advective fluid movement can take place. Structural changes in the fluid may occur in this layer.
- A **mobile adsorbed layer**. The fluid in this layer is mobile, although it is still within the range in which surface forces may lead to attraction or repulsion of components in the fluid. Structural changes in the fluid may also occur in this layer.
- The remaining **bulk phase**. The fluid here is at a sufficiently large distance from the solid surface such that the effect of surface potential is insignificant and there are no structural changes.

Altogether, we may summarize that (1) at sufficiently high saturation, the surface potential, which is part of the matric potential, may be neglected, but that (2) at very low saturations, when moisture takes the form of very this films, the matric potential is due only to surface potential.

D. Solute potential

The solute potential, Ψ^w_{sol}, is also referred to as *osmotic potential.* The liquid in the soil, referred to by soil scientists as 'soil solution', usually contains dissolved matter. The concentration of the latter affects both the surface tension and the forces that attract water molecules to solid surfaces. It also affects the energy relationships that determine the equilibrium among phases and components. Thus, the solute potential at a point in the soil (i.e., microscopic level) obeys a definition similar to that of the matric potential, except that the reference pool contains pure water at the same pressure, temperature and elevation, as the considered point, while the void space at the point contains a soil solution. We wish to emphasize that this potential is defined for the water *as a component* and not as a *liquid phase.* For the latter, we have to sum over all components, and the pool has to contain a dilute solution.

E. Soil water potential

This potential Ψ_{sw}^w combines the work required to overcome the forces due to both surface effect, pressure, and concentration differences between the reference pool and the considered point. The soil-water potential is, thus, the sum of the matric and the solute potentials.

Taking into account the effect of dissolved matter, and following a derivation similar to that for Kelvin's equation, (6.1.24), we may write:

$$\Psi_{sw}^w = -\frac{\rho_w^w RT}{M^w} \ln \frac{p^v}{p^{v_o}}, \tag{6.1.26}$$

in which Ψ_{sw}^w is per unit volume of water in the void space. Note that superscript w refers to water as a component, while subscript w refers to water as a liquid phase, which consists primarily of water, but may contain other components. For the sake of clarity, we could have used subscript ℓ instead of w to denote the liquid phase. The above equation is valid at both the microscopic and the macroscopic levels.

F. Gravity potential

This potential Ψ_g^w expresses the change in the potential energy associated with the elevation of the considered point above the reference pool. Thus, we can use the same definition as that of the matric potential, except that the considered point and the reference pool are at different elevations, and both reservoirs have flat interfaces. When expressed per unit weight, the gravity potential is equal to the elevation of the point, $z - z_o$. When expressed per unit volume, the gravity potential for the w-component is given by $\rho_w^w g(z - z_o)$, assuming that ρ_w^w does not change appreciably from z_o to z.

G. Thermal potential

This potential Ψ_T^w expresses the change in the free energy associated with the temperature of the considered point above that prevailing in the reference pool. Thus, we can use the same definition as that of the matric potential,

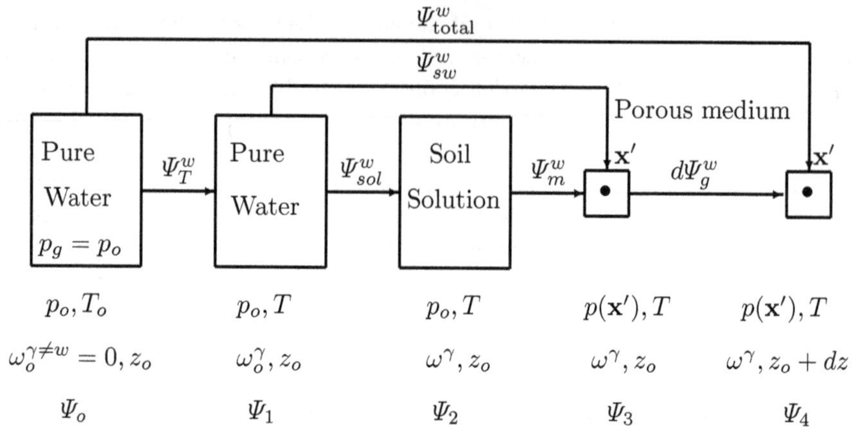

Figure 6.1.11: Definition of various soil potentials (Nitao and Bear, 1996).

except that the considered point and the reference pure water reservoir are at different temperatures.

H. Total potential

Altogether, the total potential, Ψ^w_{total}, is given by

$$\Psi^w_{\text{total}} = \Psi^w_m + \Psi^w_{sol} + \Psi^w_g + \Psi^w_T = \Psi^w_{sw} + \Psi^w_g + \Psi^w_T. \tag{6.1.27}$$

Many of these potentials depend on the liquid's saturation. The total potential is nothing but the *chemical potential* of water, as a component in the soil, per unit volume of the water phase. The relationship among the various potentials discussed above is shown in Fig. 6.1.11. In this figure,

$$\begin{aligned} \Psi^w_T &= \Psi_1 - \Psi_2, & \Psi^w_{sol} &= \Psi_2 - \Psi_1, \\ \Psi^w_m &= \Psi_3 - \Psi_2, & \Psi^w_g &= \Psi_4 - \Psi_3, \\ \Psi^w_{\text{total}} &= \Psi_4 - \Psi_0, & \Psi^w_{sw} &= \Psi_3 - \Psi_1, \end{aligned} \tag{6.1.28}$$

and $d\Psi^w_g$ is defined under the condition that dz is sufficiently small so that pressure and mass fraction of γ do not change appreciably.

6.1.7 Hysteresis

Upon rewetting (or *imbibition*), we observe (Fig. 6.1.12) that the capillary pressure curve, $p_c = p_c(S_w)$, for the drainage of a soil sample, differs from that obtained during wetting. The figure shows one drainage curve—main drainage curve—and three wetting curves.

This phenomenon, the dependence of the capillary pressure curve on the direction and history of drainage and wetting of a sample, is called *hysteresis*.

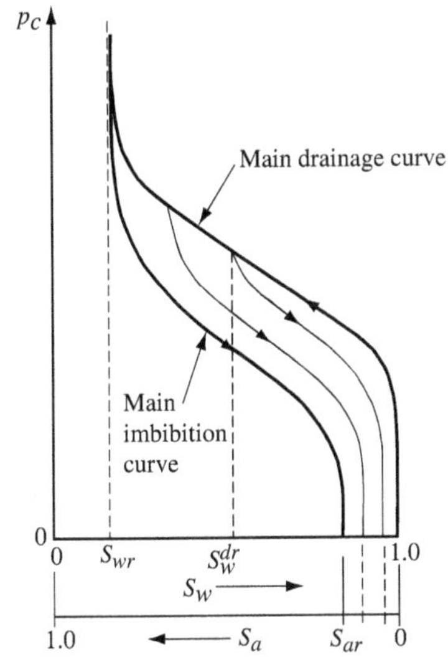

Figure 6.1.12: Hysteresis in capillary pressure curves.

It is attributed to a number of causes. The first, called the *ink-bottle effect*, results from the shape of the pore space, with interchanging narrow (throats) and wide passages (Fig. 6.1.13a). During drainage and rewetting, menisci having the same radius of curvature occur at different elevations, thus yielding different water saturations for the same capillary pressure, or suction, ψ $(= 2\gamma_{wa}/r)$. As water is drained, the radius of curvature of the air-water meniscus diminishes. At the narrowest part of the throat, the curvature of the meniscus cannot continue to increase gradually; instead, the meniscus abruptly retreats to a nearby throat. This sudden change is called *Haines jump* (Haines, 1930). A similar phenomenon occurs during wetting. Altogether, the drying curve depends on the narrow throats (small radii of meniscus curvature), while wetting depends on the maximum diameter of the large pores. In Fig. 6.1.13a, the pore is abruptly drained as the suction exceeds $\psi = 2\gamma_{wa}/r_1$. For the same pore to abruptly fill-up, suction must decrease to below $\psi = 2\gamma_{wa}/r_3, r_3 > r_1$. The hysteresis effect is more significant in coarse-textured soil, in the low-suction range, where pores may empty at an appreciable higher suction than that at which they fill-up (Hillel, 1980).

A second effect, called the *raindrop effect* (Fig. 6.1.13b), is due to the fact that the contact angle is larger at the advancing trace of a water-air interface on a solid than at the receding one, because of impurities and possible variability in the minerals that compose the surface, because of the roughness of the solid, and because of gravity. Also, when a fluid is *polar*, as is water, the contact angle depends on whether the solid surface has been previously wetted by the fluid, or not.

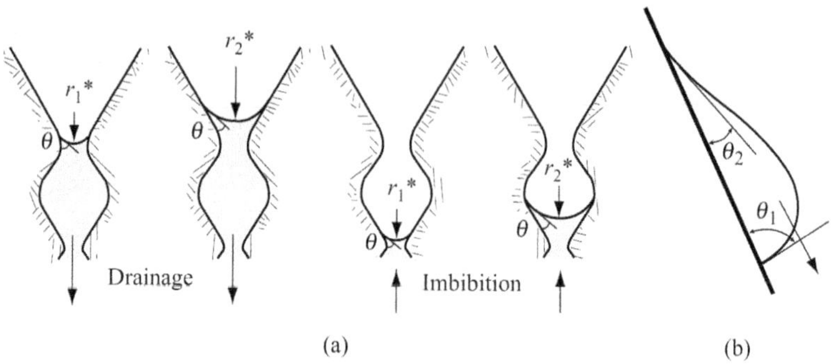

Figure 6.1.13: Hysteresis in capillary pressure curve due to (a) the ink-bottle effect, and (b) the raindrop effect.

A third cause for hysteresis is the entrapment of air, as an initially saturated sample is drained and then rewetted.

Finally, consolidation, swelling, and shrinkage of the solid matrix as it is dried and wetted may also contribute to hysteresis in the capillary pressure curve, especially in fine, unconsolidated porous media.

Figure 6.1.14 shows the effect of entrapped air on the capillary pressure relationship, $p_c = p_c(S_w)$. Starting from any point on the boundary drainage curve, it is possible to follow the wetting process and observe a *wetting scanning curve. Drying scanning curves* are obtained by draining the sample from an initial point on the boundary wetting curve. The scanning curves are shown as dashed lines on the figure. In this way, the macroscopic relationship between capillary pressure and saturation, expressed by the capillary pressure curve, depends also on the wetting-drying history of the particular sample under consideration. For a given capillary pressure, a higher wetting fluid saturation is obtained when a sample is being drained than during imbibition.

As a sample is being rewetted to zero capillary pressure, air at *residual air saturation*, S_{ar}, remains in the sample. We note that the value of S_{ar} depends also on the drainage-imbibition history. For example, when a saturated sample is drained and at some point imbibition begins, continuing up to zero capillary pressure, the amount of entrapped air is smaller than when the imbibition begins from a dryer state. The curve beginning at 100% water saturation and proceeding to the residual water saturation at a high capillary pressure is referred to as the *main drainage curve*. The curve beginning at residual water saturation and proceeding in the direction of higher saturations, ending at zero capillary pressure is called the *main imbibition curve*. The two curves define the *hysteresis envelope* which encompasses an infinite number of possible *scanning curves*. An imbibition curve that begins at a *reversal point*, where the saturation change reverses direction along the main drainage curve, is referred to as a *primary imbibition scanning curve*.

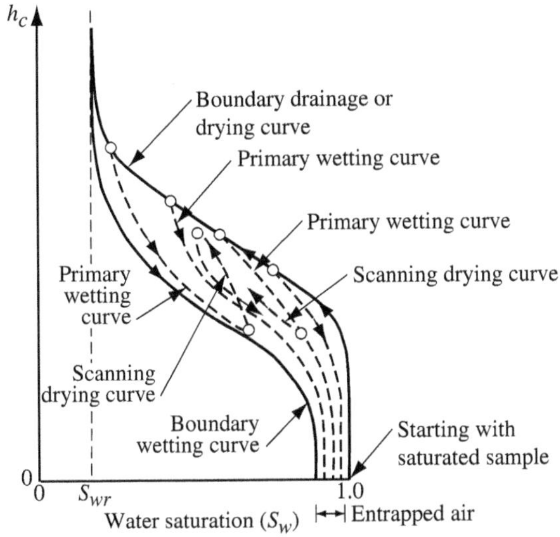

Figure 6.1.14: Hysteresis and entrapped air in a capillary pressure curve.

A drainage curve starting along the main imbibition branch is referred to as a *primary drainage scanning curve*. Secondary scanning curves may be initiated at reversal points on a primary scanning path. Similarly, higher order scanning curves may be defined.

To account for the effects of air entrapment during water imbibition, we may employ the concept of *apparent saturation* as a macroscopic surrogate for tracking the position of the interfaces between continuous portions of the air and water phases. We define the *apparent water saturation*, $\widehat{S}_w$, as

$$\widehat{S}_w = \tilde{S}_w + \tilde{S}_{at}, \tag{6.1.29}$$

where $\tilde{S}_w$ is the *effective water saturation* defined by

$$\tilde{S}_w = \frac{S_w - S_{wr} - S_{ta}}{1 - S_{wr}}, \tag{6.1.30}$$

and $\tilde{S}_{at}$ is the *effective trapped air saturation* defined as

$$\tilde{S}_{at} = \frac{S_{ta}}{1 - S_{wr}}, \tag{6.1.31}$$

in which S_{ta} is the actual trapped air saturation. Note that for the main drainage branch, the apparent and effective water saturations are equal because there is no trapped air along this path.

After water has imbibed, starting from some saturation, S_w^{dr}, on the main drainage path to an apparently saturated condition, i.e., to zero capillary pressure, the residual air saturation, S_{ar}, may be estimated by the empirical relation developed by Land (1968)

$$\tilde{S}_{ar} = \frac{1 - \tilde{S}_w^{dr}}{1 + R_{aw}\left(1 - \tilde{S}_w^{dr}\right)}, \tag{6.1.32}$$

$$R_{aw} = \left[(1 - S_{wr})/\tilde{S}_{ar}^{\text{im}} - 1\right], \tag{6.1.33}$$

where $\tilde{S}_{ar} = S_{ar}/(1 - S_{wr})$ is the *effective residual air saturation*, $\tilde{S}_w^{dr}$ is the effective water saturation at S_w^{dr}, and $\tilde{S}_{ar}^{\text{im}}$ is the residual air saturation for the main imbibition branch (i.e., the branch starting at $\tilde{S}_w^{dr} = 0$), as illustrated in Fig. 6.1.12. Since S_{ar} represents air which is occluded within the wetting phase, it is sometimes referred to as the *insular residual air saturation*.

The residual air saturation corresponds to the amount of trapped air at zero capillary pressure for a given imbibition saturation path. The effective trapped air saturation, $\tilde{S}_{at}$, at an arbitrary water saturation along an imbibition path, may be approximated by linearly interpolating between the reversal point on the main drainage curve and zero capillary pressure. A more accurate representation that follows from (6.1.32) is

$$\tilde{S}_{at} = \begin{cases} \tilde{S}_{ar}\left(\tilde{S}_w^{dr}\right) - \tilde{S}_{ar}\left(\widehat{S}_w\right) & \text{for} \quad \widehat{S}_w > \tilde{S}_w^{dr}, \\ 0 & \text{for} \quad \widehat{S}_w \leq \tilde{S}_w^{dr}, \end{cases} \tag{6.1.34}$$

where $\widehat{S}_w$ designates the current apparent water saturation, $\tilde{S}_w^{dr}$ is the effective saturation at the reversal point on the main drainage curve, and $\tilde{S}_{ar}$ is the function of $\tilde{S}_w^{dr}$ defined by (6.1.32).

Theoretical analyses of hysteresis in the air-water capillary pressure curve have been presented by Poulovassilis (1962), Topp (1969, 1971), Mualem (1974, 1976, 1984), Kool and Parker (1987), Luckner *et al.* (1989), and others. Hysteresis in oil-water and gas-oil systems have also been studied by numerous researchers (e.g., Naar and Henderson, 1961; Snell, 1962; Land, 1968; Schneider and Owens, 1970; Parker and Lenhard, 1987).

Nowadays, the most commonly used theory that explains and describes soil water hysteresis is the *independent domain theory* (Poulovassilis, 1962). Mualem (1973) suggested a similarity hypothesis, according to which the bivariate domain density distribution function is represented as a product of two univariate distribution functions. The resulting model significantly reduced the amount of data required for calibration. In subsequent years, Mualem (1974, 1977) introduced the *universal hysteresis model* based on a non-dimensional formulation (Mualem, 1979). Extension of the *domain theory* to the prediction of hysteresis in unsaturated hydraulic conductivity has been successfully initiated by Mualem (1976).

6.1.8 Saturation distribution along vertical

Let us consider a homogeneous soil occupied, simultaneously, by water and air, each at a constant density, and a water table that is very deep below ground surface. At equilibrium, with no flow occurring within the domain, the piezometric head within each of the fluid phases, h_α, $\alpha = w, a$, is constant. Consider two points, $m = 1$ and $m = 2$, at elevations z_1 and z_2, respectively. Using ρ_w and ρ_a as reference densities for the piezometric heads for water and for air, respectively, we can write (see Subs. 3.2.1) for these two points:

$$h_{wm} = z_m + \frac{p_{wm}}{g\rho_w}, \quad \text{and} \quad h_{am} = z_m + \frac{p_{am}}{g\rho_a}, \quad m = 1, 2. \tag{6.1.35}$$

Since no flow takes place, we write $h_{w1} = h_{w2}$, and $h_{a1} = h_{a2}$; hence,

$$z_2 - z_1 = \frac{p_{c2} - p_{c1}}{g(\rho_w - \rho_a)}. \tag{6.1.36}$$

Choosing the datum level such that $S_w = 1.0$ at $z_1 = 0$, we have $S_{w1} = 1$, and $p_{c1} = 0$; hence,

$$z_2 = \frac{p_{c2}}{g(\rho_w - \rho_a)} \approx \frac{p_{c2}}{g\rho_w} = h_{c2}, \tag{6.1.37}$$

where the approximation stems from $\rho_a \ll \rho_w$. Thus, the capillary pressure head curve, $h_c = h_c(S_w)$, defines the distribution of water saturation as a function of the distance above the surface at which $p_w = 0$, and $S_w = 1.0$. As the elevation, z, above this surface increases, so does h_c, while S_w decreases with elevation.

The surface at which $p_w = p_a \equiv p_{\text{atm}}$ (often taken as $p_a = 0$) is the *phreatic surface*, or *water table*, introduced in Subs. 5.2.1E. The zone immediately above the phreatic surface is referred to as the *capillary fringe*. It extends up to an elevation of about h_c^{cr} above the phreatic surface (Fig. 6.1.7). This zone is practically saturated, while the (gauge) pressure in the water within it is negative. From the above discussion on $S_w = S_w(z)$, it follows that the residual water saturation occurs only at a point that is located sufficiently high above the water table.

So far, we have been assuming a homogeneous soil profile. Let us now assume that the soil is layered, with an assumed sharp macroscopic boundary between two soil domains, *I* and *II* (say, a fine and a coarse sand). The assumption of local equilibrium prohibits any jump in pressure across such a boundary in both the water and the air phases. This means that

$$p_{\alpha\text{I}} = p_{\alpha\text{II}}, \quad \text{for} \quad \alpha = w, a, \quad \text{and hence} \quad p_{c\text{I}} = p_{c\text{II}}. \tag{6.1.38}$$

The condition of equality of pressures at the interface between different porous media remains valid also when the fluids are moving.

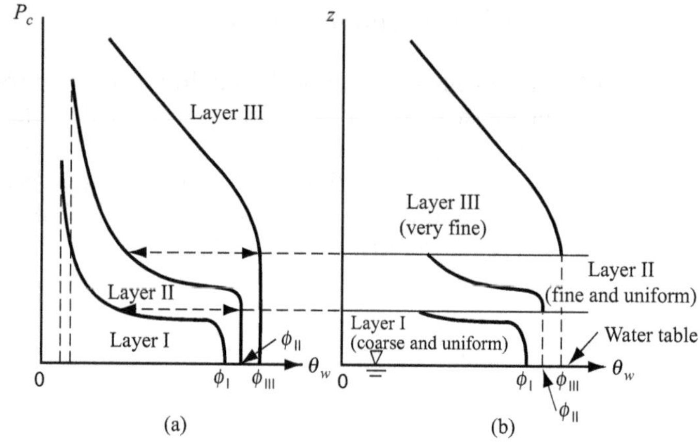

Figure 6.1.15: Saturation discontinuities at boundaries in a layered soil. (a) Capillary pressure curves for three soils. (b) Equilibrium moisture distribution in a three-layered system.

Figure 6.1.15a shows three typical capillary pressure curves for three soils in the unsaturated zone. Figure 6.1.15b shows the moisture distribution in a three-layered, unsaturated soil, with the same soils, under *equilibrium conditions* (i.e., no flow). In this figure we observe $\theta_{wII} > \theta_{wI}$ at the interface between layers *I* and *II*. Similarly, $\theta_{wIII} > \theta_{wII}$ at the boundary between layers *II* and *III*. Thus, as in (6.1.38), the conditions $p_{cI} = p_{cII}$ and $p_{cII} = p_{cIII}$ lead to *jumps in water saturation* between layers *I* and *II* and between layers *II* and *III* as the respective boundaries are crossed.

An interesting phenomenon occurs at the interface between two soils. Let us consider a horizontal layer of unsaturated fine-textured soil that overlies a coarse-textured one, which is practically dry. The capillary pressure is continuous at the interface between the two soils. The large capillary pressure of the fine soil, even at high saturations, is sufficient to maintain, at the interface, the moisture content of the coarse soil essentially at the residual value.

Another case of interest occurs when air tries to move from a coarse-textured soil into an overlying, practically saturated, fine-textured one. The air cannot penetrate the fine textured soil until its pressure is equal to the *threshold*, or *bubbling pressure* that corresponds to the fine-textured soil. Thus, the saturated fine-textured soil acts as a barrier to the upward movement of air.

The concept of *threshold pressure* introduced above for an air-water system is also valid for any two-phase (wetting-non-wetting) system. In the latter, the threshold pressure is the pressure in the nonwetting phase required in order to overcome the capillary pressure and penetrate into a wetting fluid (water) saturated domain. Thus, a saturated layer with small pore sizes serves as a *capillary barrier*. The density of the non-wetting fluid, ρ_n, has to accumulate

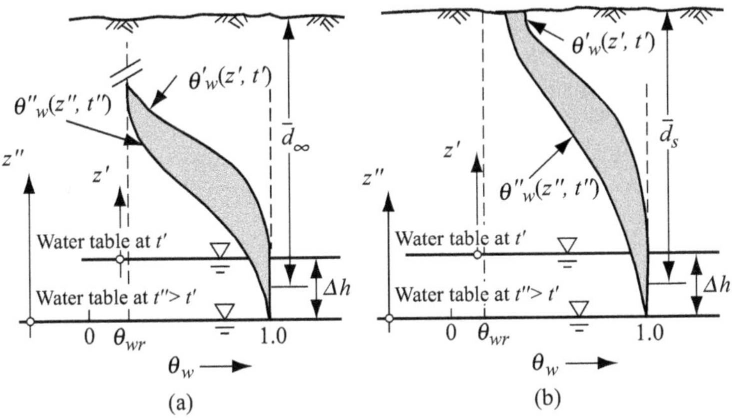

Figure 6.1.16: Specific yield for (a) A deep ($\overline{d}_\infty$); (b) A shallow ($\overline{d}_s$) phreatic surface.

to a thickness d_n, expressed by

$$d_n = \frac{2\gamma_{wn} \cos \theta_{LG}}{gR(\rho_n - \rho_w)}, \tag{6.1.39}$$

before it can penetrate into a barrier with pores of size R.

6.1.9 Specific yield and field capacity

The concept of *specific yield*, S_y, related to the unsaturated zone, was introduced in Subs. 5.4.1C. It is often used in modeling drainage of agricultural lands and drawdown of the water table in phreatic aquifers, assuming 'essentially horizontal flow'. It is defined as *the volume of water drained from a soil column of a unit horizontal cross-sectional area, extending from the soil surface down to the underlying phreatic surface, per unit lowering of the phreatic surface's elevation.* Because it takes time to complete the drainage process and to establish a new moisture distribution, following a rapid change in water table elevation, the specific yield is a *time-dependent quantity.* However, after a sufficiently long time (depending on the type of soil), a new equilibrium moisture distribution will be reached. It is common to assume that such conditions have been attained when employing the term 'specific yield' as defined above.

Following the discussion in Subs. 6.1.8, Fig. 6.1.16a shows the moisture content distribution in a homogeneous soil profile, corresponding to two water table depths, d' at time t', and d'' $(= d' + \Delta h)$ at time t'' $(= t' + \Delta t)$. The volume of water drained is indicated by the shaded area in the figure. When the initial and final water tables are sufficiently deep below ground surface, and a sufficient time has elapsed, so that a new equilibrium moisture distribution has been reached, the curves θ'_w and θ''_w are identical in shape, with one being shifted vertically with respect to the other.

The above definition of specific yield is also valid when the rate of water table drawdown is very small.

With the nomenclature of this figure, and the above definition of S_y, we have

$$S_y = \frac{1}{\Delta h}\left(\phi\Delta h + \int_0^{d'} \theta'_w(z',t')\,dz' - \int_0^{d''} \theta''_w(z'',t'')\,dz''\right). \qquad (6.1.40)$$

When the water table is at a shallow depth below ground surface (Fig. 6.1.16b), the specific yield, S_y, is a function of both water table depths, d' and d''. The same is true when the soil is inhomogeneous (e.g., layered).

When changes in the water table elevations are slow, the corresponding changes in moisture distribution have sufficient time to adjust continuously, and the lag between the lowering of the water table and the total volume of water drained practically vanishes.

Let us supplement the definition of specific yield presented above by requiring that the volume of water be drained during a sufficiently long time following the change in water table elevations, such that drainage may be assumed to be complete (or practically so). Then the two equilibrium moisture profiles have the same shape, i.e., $\theta''_w(z) = \theta'_w(z)$. Equation (6.1.40) can then be written as

$$S_y = \phi - \frac{1}{\Delta h}\int_{d'}^{d''} \theta'_w(z')\,dz'. \qquad (6.1.41)$$

In the limit, as $\Delta h \to 0$, we obtain

$$S_y \equiv \phi - \theta_{wr} \equiv \phi(1 - S_{wr}), \qquad (6.1.42)$$

where $\theta_{wr} = \phi S_{wr}$. The volume of water that will drain from a soil column of unit cross-sectional area is thus $\Delta h \times \phi(1 - S_{wr})$. Note that in view of the above discussion, the *specific yield* is a property of a horizontal two-dimensional model of a phreatic aquifer, or an integrated property of the vadose zone, and not one at a point in the vadose zone (Subs. 5.4.1C).

In reality, equilibrium conditions are rarely, if ever, reached, or approximated, even after prolonged periods, unless the soil is very coarse. The reason is that at low saturations, the low effective permeability to water (Subs. 6.2.1) will impede the approach to equilibrium.

Thus, in general, the value of θ_{wr} (or S_{wr}) relevant to (6.1.42) under 'quasi-static' conditions will exceed corresponding values obtained from laboratory measurements under conditions that are closer to equilibrium.

The minimum water content attained in practice during drainage under the influence of gravity is often referred to in soil science as *field capacity*; it is approximately equal to θ_{wr}. The water content may be reduced below this value by evaporation. However, this definition is valid only at points that are sufficiently high above the water table. Water contents below the field capacity may be produced by evaporation. Closer to the water table, the saturation retained against gravity depends on the elevation of the considered point above the water table. Note that the specific yield has been expressed above as a function of moisture content. Sometimes, it is expressed in terms of the corresponding water saturation.

Field capacity is usually defined as the water content remaining in a unit volume of soil after gravity drainage has ceased, or practically so, after a period of rain or irrigation. A difficulty inherent in this definition is that there is no quantitative specification given for what is meant by 'practically so' (some authors add 'after 2–3 days'). Although, according to this definition, field capacity is a property of the soil, depending on structure, grain-size distribution, etc., it is obvious from our previous discussion that the moisture distribution depends also on the elevation of the considered point above the water table. Also, close to ground surface, equilibrium conditions are seldom reached, because of infiltration or evaporation there. Usually, this definition is useful only if it relates to a unit volume of highly permeable soil, which is located sufficiently high above a water table, and at a sufficient depth below ground surface.

To gain some insight into the concepts of water retention and field capacity introduced above, consider the following case. Following a long period of no infiltration, water is applied at ground surface, say, by irrigation, such that infiltration takes place during a finite period. Prior to the application of irrigation, the moisture in the soil, say above an underlying water table, has reached equilibrium conditions. As explained above, under such conditions, the moisture is vertically distributed according to the retention curve.

At any depth below ground surface, water will move, primarily downward, as long as the water saturation at that point is (1) above the irreducible water saturation (below which point the permeability to water vanishes), and (2) above the value that corresponds to the elevation of the point on the retention curve. Let us focus our attention on a volume of coarse soil at some depth below ground surface that is sufficiently high above the underlying water table, such that the water there is at the irreducible moisture content. At this saturation, the immobile water remaining in the soil takes the form of pendular rings and isolated blobs, in addition to water in films. If the volume of water applied at ground surface is sufficiently large, some time after infiltration begins, we should observe a rise in saturation within the considered soil volume (and flow will take place). The gradual increase in saturation will reach some peak, and then saturations will decrease, until the irreducible moisture content is re-established.

Figure 6.1.17a shows an individual water blob within the void space. It is bounded by air-water menisci and by (water-wet) solid surfaces (overlooking the presence of water films). We focus our attention on the uppermost meniscus, and the lowest one. Within the stationary water blob, the pressure distribution is hydrostatic. Hence, with r_u and r_ℓ denoting the radii of up-

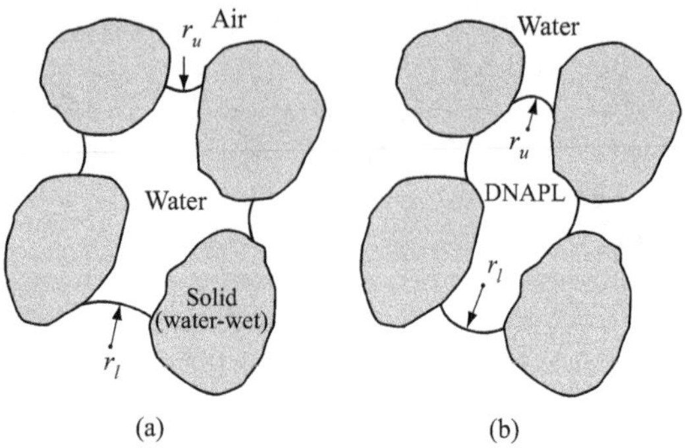

Figure 6.1.17: (a) A water blob, and (b) A DNAPL ganglion.

permost and lowest menisci, respectively, and assuming air to be at constant atmospheric pressure, we have

$$p_a - p_{w,u} \equiv p_{c,u} = \frac{2\gamma_{wa}}{r_u}, \qquad p_a - p_{w,\ell} \equiv p_{c,\ell} = \frac{2\gamma_{wa}}{r_\ell}. \tag{6.1.43}$$

Hence, for a blob of vertical length L_b, we have:

$$p_{w,\ell} - p_{w,u} = 2\gamma_{wa}\left(\frac{1}{r_u} - \frac{1}{r_\ell}\right) = \rho_w g L_b, \quad L_b = \frac{2\gamma_{wa}}{\rho_w g}\left(\frac{1}{r_u} - \frac{1}{r_\ell}\right), \tag{6.1.44}$$

where $r_u < r_\ell$. The radii of the intermediate menisci will correspond to the varying pressure along the vertical. As drainage takes place, these menisci will occur in pores that have the appropriate sizes. Obviously, for a blob to be formed, the presence of an appropriate pore size distribution is required. In the above expressions, we may replace the meniscus radius by an appropriate $R/\cos\theta_{LG}$, where R represents a throat or pore size.

Immobile DNAPL ganglia are formed when the quantity of a DNAPL spill at ground surface is sufficiently large, so that when, eventually, the DNAPL becomes immobile at residual DNAPL saturation, part of it may be located below the water table. Following the same considerations as presented above in the air-water system, but now in a water-DNAPL one, we have (Fig. 6.1.17b)

$$L_b = \frac{2\gamma_{Nw}}{(\rho_N - \rho_w)g}\left(\frac{1}{r_\ell} - \frac{1}{r_u}\right). \tag{6.1.45}$$

Since $\rho_N > \rho_w$, $r_\ell > r_u$, $L_{b,\max}$ occurs when $r_\ell \to r_{\min}$ and $r_u \gg r_{\ell,\min}$. Hunt *et al.* (1988) present some examples of the length of DNAPL ganglia.

6.2 Motion Equations

6.2.1 Coupling between the phases

The two fluid phases, water and air, that occupy the void space in the unsaturated zone, can be in motion, simultaneously. At every instant, each of these fluids occupies a well defined subdomain within the void space, with the two subdomains being separated from each other by microscopic air-water interfaces. A detailed discussion of the spatial distributions of air and water within the void space was presented in Sec. 6.1.

In this section, we shall limit the discussion to the movement of air and water phases of fixed composition, overlooking the possible transfer of chemical components between phases. We shall also limit the discussion to cases in which pressure gradients and gravity are the only driving forces. The effects of concentration of dissolved matter are introduced in Chap. 7.

As in the case of saturated flow, we are interested in *macroscopic* flux laws that will express the specific discharge of each of the fluid phases in terms of macroscopic quantities, such as saturation and pressure.

In Sec. 4.1, Darcy's law for a single fluid that occupies the entire void space was first presented as an *empirical* law. Buckingham (1907), Richards (1931), and Childs (1967, 1969) *assumed* that Darcy's empirical law is also applicable to the flow of a fluid phase in an isotropic unsaturated zone, with the hydraulic conductivity being a function of the saturation. Later, in Sec. 4.1, we mentioned that Darcy's law, say in the form of (4.2.5), can also be *derived* from fundamental principles, by taking an average (over an REV) of the microscopic linear momentum balance equation, written for a point within a fluid phase, and making certain simplifying assumptions. The same procedure of averaging can also be applied to a fluid phase that occupies only part of the void space, the remaining part being occupied by one or more additional fluid phases (e.g., Bear and Bachmat, 1990).

For the sake of simplicity, let us accept here the conceptual model in which each fluid (in our case, either air or water) occupies (instantaneously) a distinct portion of the void space. *Water, the wetting phase, occupies primarily the smaller pores, while air, the nonwetting phase, occupies primarily the larger ones.* Although we know that a thin film of water covers the solid face in the air-occupied pores, we assume that this film, due to its special structure (Subs. 6.1.2), behaves as if it were an extended part of the solid in transmitting momentum from the air to the solid. Thus, we assume that each fluid has a microscopic interface with the solid, in addition to an interface with the other fluid. Let $\mathcal{S}_{ws}$, $\mathcal{S}_{wa}$, and $\mathcal{S}_{as}$ represent the water-solid, water-air and air-solid interfaces within an REV. Momentum can then be transferred across each of these internal surfaces.

Consider the water phase. When this phase occupies the entire void space, the averaged momentum balance equation contains the term

$$-\frac{1}{\mathcal{U}_o}\int_{\mathcal{S}_{ws}} \boldsymbol{\tau}\cdot\mathbf{n}\,dS,$$

where $\boldsymbol{\tau}$ is the microscopic viscous shear in the fluid, $\mathcal{U}_o$ is the volume of the REV, $\mathcal{S}_{ws}$ is the water-solid surface within $\mathcal{U}_o$, and $\mathbf{n}$ is the outward unit vector (from the water occupied domain) on $\mathcal{S}_{ws}$. This term expresses the momentum transfer across the entire water-solid interface within the averaging volume (REV). In fact, it is this term that leads to Darcy's law for single-phase flow. When the same procedure is applied to air and water flow in the unsaturated zone, the total rate of transfer of momentum from the water to its surroundings is made up of two parts:

$$-\frac{1}{\mathcal{U}_o}\int_{\mathcal{S}_{ws}} \boldsymbol{\tau}\cdot\mathbf{n}\,dS, \quad \text{and} \quad -\frac{1}{\mathcal{U}_o}\int_{\mathcal{S}_{wa}} \boldsymbol{\tau}\cdot\mathbf{n}\,dS.$$

The first term expresses the resistance to water flow by momentum transfer across the water-solid interface, and the second one represents the drag at the water-air interface. In this way, at least in principle, the motion in each of the two phases is *coupled* to that of the other; a pressure gradient in one fluid produces *also* flow in the other fluid.

One may express this viscous coupling, say between the flux of a wetting phase (w) and of a nonwetting one (n), in the form:

$$\mathbf{q}_{rw} = -\frac{\mathsf{k}^w_w(S_w)}{\mu_w}\cdot(\nabla p_w + \rho_w g\nabla z) - \frac{\mathsf{k}^n_w(S_n)}{\mu_n}\cdot(\nabla p_n + \rho_n g\nabla z), \quad (6.2.1)$$

$$\mathbf{q}_{rn} = -\frac{\mathsf{k}^w_n(S_w)}{\mu_w}\cdot(\nabla p_w + \rho_w g\nabla z) - \frac{\mathsf{k}^n_n(S_n)}{\mu_n}\cdot(\nabla p_n + \rho_n g\nabla z), \quad (6.2.2)$$

in which we note the coupling coefficients. Coupled two-phase flow in homogeneous, isotropic porous media has been studied by many authors (e.g., Rose (1972, 1988), Sanchez-Palencia (1980), Whitaker (1986a), Kalaydjian (1987), Auriault *et al.* (1989)). The significance of this coupling has been also extensively debated in the literature, starting in the 1950's (e.g., Yuster,1951; Odeh, 1959; Bentsen and Manai, 1993; Goode and Ramakrishnan, 1993; and Lasseux *et al.*, 1996). Rose (1972, 1988, 1990, 1997) and Rose and Rose (2005) suggested a relationship between the two cross-permeability coefficients. Avroam and Payatakes (1995) discuss this topic and report on experimental investigations. Unfortunately, relatively few experiments have been conducted to determine the significance of coupling, due to momentum transfer across fluid-fluid interfaces, that takes place in multiphase flows (e.g., Liang and Lohrenz (1994), Dullien and Dong (1996)).

Bear and Bachmat (1990) also add another term to each of the above equations, due to gradients in (averaged) surface tension between the two fluid phases. Such gradients may be produced by temperature and concentration gradients.

By neglecting the momentum exchange across the (microscopic) interfaces, e.g., the air-water interface, the resulting averaged momentum balance equation, written separately for each fluid phase, is *identical in form* to that written for that phase when it occupies the entire void space. However, since the shape and size of the solid-fluid surfaces and of the volumes occupied by these phases within an REV vary with the saturation of the considered phase, the resistance to the flow of each fluid phase also depends on its saturation. Bear and Bachmat (1986, 1990) show that the permeability of a saturated porous medium depends on (1) certain geometric features of the fluid-solid interface, (2) a length (hydraulic radius) that characterizes the distance between the interior of the volume occupied by the considered phase (within the REV) and the fluid-solid interface, and (3) the porosity. When a fluid occupies only part of the void space, the geometrical features become functions of the fluid's saturation, while the porosity is replaced by the volumetric fraction of the void space occupied by the fluid. The conclusion is that *the permeability of a considered fluid phase is a function of the saturation of that fluid.*

Thus, when the momentum transfer across the water-air interface is much smaller than across the fluid-solid interface, the motion equation (4.2.5) can be used also as a good approximation for the flow of a fluid phase in a multiphase system, with the permeability being a function of the fluid's saturation.

By the above brief discussion, we have, actually, based the motion equation for a fluid in the unsaturated zone on first principles, as we did in Subs. 4.2.1 for a single fluid that occupies the entire void space.

6.2.2 Darcy's law for unsaturated flow

With the above considerations in mind, we can now write the macroscopic equations that describe the simultaneous motion of water and air, each occupying part of the void space, in the form

$$\mathbf{q}_{rw} = -\frac{\mathbf{k}_w(S_w)}{\mu_w}\cdot(\nabla p_w + \rho_w g \nabla z), \tag{6.2.3}$$

$$\mathbf{q}_{ra} = -\frac{\mathbf{k}_a(S_a)}{\mu_a}\cdot(\nabla p_a + \rho_a g \nabla z). \tag{6.2.4}$$

In these equations, ∇z denotes a unit vector directed upward, while $\mathbf{q}_{rw} \equiv \phi S_w(\mathbf{V}_w - \mathbf{V}_s)$ and $\mathbf{q}_{ra} \equiv \phi S_a(\mathbf{V}_a - \mathbf{V}_s)$ are the specific discharges of the water and the air, respectively, relative to the solid, $\mathbf{V}_s$ is the velocity of the (possibly moving) solid, p_w and p_a are the pressures in the water and in the air, respectively, ρ_w and ρ_a are the respective fluid densities, ϕ is porosity, and S_w and S_a are the respective fluid saturations. We recall that $\mathbf{V}_w$, $\mathbf{V}_a$, $\mathbf{V}_s$, p_w, p_a, ρ_w, and ρ_a denote intrinsic phase averaged quantities, as defined in Sec. 1.3. When the solid matrix is stationary and nondeformable, i.e., ϕ is a constant and $\mathbf{V}_s = 0$, we may replace $\mathbf{q}_{rw}$ by $\mathbf{q}_w$, and $\mathbf{q}_{ra}$ by $\mathbf{q}_a$. Obviously, (6.2.3) and (6.2.4) are also valid for any pair of wetting and nonwetting fluids.

The (tensor) coefficients k_w and k_a denote the *effective permeabilities* to the water and to the air phases, respectively, discussed in Subs. 6.2.3. Equations (6.2.3) and (6.2.4) are written for the general case of an *anisotropic* porous medium and for fluid densities that may depend on pressure, concentrations of components, and temperature. These two equations are not independent of each other. They are linked by the condition $S_a + S_w = 1$, and by the relationship between the saturation and the (macroscopic) capillary pressure, $p_c = p_c(S_w)$, discussed in Subs. 6.1.3.

When water and air densities remain unchanged (or may be assumed to be practically so), we may rewrite (6.2.3) and (6.2.4) in terms of the piezometric heads in the water, h_w, and in the air, h_a, in the form:

$$\mathbf{q}_{rw} = -\mathsf{K}_w(S_w) \cdot \nabla h_w, \tag{6.2.5}$$

$$\mathbf{q}_{ra} = -\mathsf{K}_a(S_a) \cdot \nabla h_a, \tag{6.2.6}$$

where, following (4.1.4),

$$h_w = z + \frac{p_w}{\rho_w g}, \qquad h_a = z + \frac{p_a}{\rho_a g}. \tag{6.2.7}$$

Here,

$$\mathsf{K}_w(S_w) = \frac{\mathsf{k}_w(S_w)\rho_w g}{\mu_w}, \qquad \text{and} \qquad \mathsf{K}_a(S_a) = \frac{\mathsf{k}_a(S_a)\rho_a g}{\mu_a}, \tag{6.2.8}$$

are the *effective hydraulic conductivities* to water and to air, respectively.

Because ρ_a is very small, the (interconnected) air-phase is assumed to be everywhere under (practically) atmospheric pressure, usually taken as $p_a = 0$. The equation of motion for the water phase may then take the form:

$$\mathbf{q}_{rw} = \mathsf{K}_w(\psi) \cdot \nabla(\psi - z), \tag{6.2.9}$$

where ψ is the *suction* (or *suction head* discussed in Subs. 6.1.3). However, sometimes, a significant resistance to air flow may exist, e.g. when a fine grained soil becomes water-logged. Then, the assumption of 'water flow only' is not justified, and a two-phase flow model must be used.

Another form of the motion equation is often used in soil science for an isotropic soil and constant ρ_w. It is based on the assumption that the relations $p_c = p_c(S_w)$ and $\psi = \psi(S_w)$ are unique, single-valued functions. Starting with (6.2.9), we write Darcy's law in terms of S_w, in the form:

$$\mathbf{q}_{rw} = \mathsf{K}_w(S_w) \cdot \left(\frac{d\psi}{dS_w} \nabla S_w - \nabla z \right), \tag{6.2.10}$$

or

$$\mathbf{q}_{rw} = -\mathbf{D}_w(S_w) \cdot \nabla S_w - \mathbf{K}_w(S_w) \cdot \nabla z, \tag{6.2.11}$$

where

$$\mathbf{D}_w(S_w) \equiv -\frac{\mathbf{K}_w(S_w)}{(dS_w/d\psi)} \tag{6.2.12}$$

is called the *moisture* (or *capillary*) *diffusivity* (dims. L^2T^{-1}).

For horizontal flow, (6.2.11) reduces to

$$\mathbf{q}_{rw} = -\mathbf{D}_w(S_w) \cdot \nabla S_w, \tag{6.2.13}$$

which, by comparison with Fick's law of molecular diffusion, explains the origin of the term 'diffusivity' assigned to $\mathbf{D}_w(S_w)$.

When the solid matrix is stationary and nondeformable, it is possible to define another moisture diffusivity

$$\mathbf{D}'_w(\theta_w) = -\frac{\mathbf{K}_w(\theta_w)}{(d\theta_w/d\psi)}, \qquad \theta \equiv S_w\phi, \tag{6.2.14}$$

and rewrite (6.2.11) in terms of the moisture content, θ_w, in the form:

$$\mathbf{q}_w = -\mathbf{D}'_w(\theta_w) \cdot \nabla \theta_w - \mathbf{K}_w(\theta_w) \cdot \nabla z. \tag{6.2.15}$$

By examining any of the motion equations presented above, we note that they are all *nonlinear*, due to the dependence of the effective permeability on saturation (and, hence, on fluid pressure, or on water suction). This is a fundamental feature of multiphase flow in porous media, and is one of the principal reasons that unsaturated flow problems are more difficult to solve than saturated ones. Another difficulty is that the two functional relationships, $\mathsf{K}_w = \mathsf{K}_w(S_w)$ and $p_c = p_c(S_w)$, which are needed in order to solve unsaturated flow problems, are really not unique, cannot be easily measured, and are subject to hysteresis (Subs. 6.1.7).

6.2.3 Effective permeability

Bear and Bachmat (1990) showed that the (macroscopic) coefficients $\mathbf{k}_w$ and $\mathbf{k}_a$ are related to (microscopic) properties of the geometrical configuration of the portion of void space occupied by each fluid phase. Since, for each phase, this configuration depends on the phase saturation, the effective permeabilities also depend on the phase saturations, i.e.,

$$\mathbf{k}_w = \mathbf{k}_w(S_w) \qquad \text{and} \qquad \mathbf{k}_a = \mathbf{k}_a(S_a).$$

Another important feature is that for an isotropic porous medium, each of these effective permeabilities is a second rank symmetric tensor. In component notation, we write $\mathsf{k}_{wij}(S_w)$ and $\mathsf{k}_{aij}(S_a)$ to emphasize that each of the ij-components, of either $\mathbf{k}_w$, or $\mathbf{k}_a$, may have a *different functional relation-*

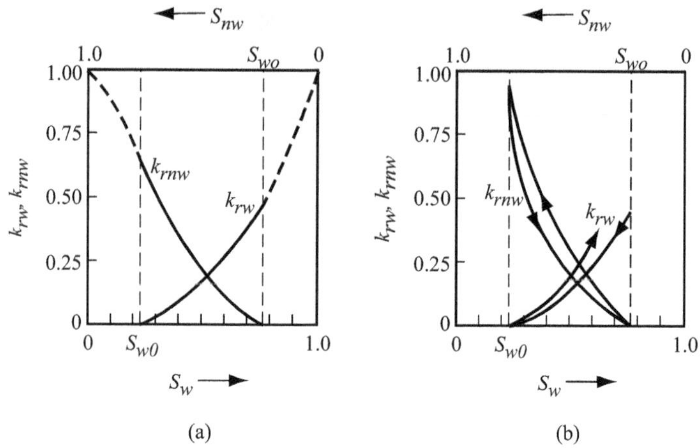

Figure 6.2.1: Typical relative permeability curves (a) Without hysteresis, (b) With hysteresis.

ship to saturation (Bear *et al.*, 1987; Stephens and Heermann, 1988; McCord *et al.*, 1991; and Friedman and Seaton, 1996).

For an isotropic porous medium, and only for such a medium, *relative permeabilities* to water and to air may be defined by (Bear *et al.*, 1987)

$$\mathsf{k}_{rw} = \frac{\mathsf{k}_w(S_w)}{\mathsf{k}_{\text{sat}}} \qquad \text{and} \qquad \mathsf{k}_{ra} = \frac{\mathsf{k}_a(S_a)}{\mathsf{k}_{\text{sat}}}, \tag{6.2.16}$$

where k_{sat} is the permeability at full saturation ($S_w = 1$). The relative permeability is a convenient and commonly used concept for describing permeability for multiphase flow in an isotropic porous medium. Note that $0 \leq \mathsf{k}_{rw} \leq 1$, and $0 \leq \mathsf{k}_{ra} \leq 1$.

Figure 6.2.1 shows typical relative permeability curves for a wetting phase and for a non-wetting one. Starting the drainage of a soil sample from full water saturation, we note a rapid decline in k_{rw} as the larger pores are drained first, and the flow of water takes place through the smaller pores. This means that a smaller cross-sectional area is available for flow and that the flow paths of the water become increasingly more and more tortuous. When the water saturation is below the *irreducible water saturation* S_{wr} (Subs. 6.1.4), the water remaining in the soil is in the form of isolated *pendular rings* (Subs. 6.1.4), and very thin films that cover the solid surface in the larger pores from which water has already been drained. In this form, the water constitutes a discontinuous, immobile phase that cannot transmit pressure. Thus, $\mathsf{k}_{rw} = 0$ for $S_w \leq S_{wr}$. In reality, given enough time, the wetting phase will continue to drain by gravity in the form of films, reducing the saturation to below the irreducible saturation (Dullien, 1992). For $S_w = 1$, temporarily overlooking the meaning of the dashed portion of the k_{rw}-curve, we have $k_{rw} = 1$, i.e., $k_{rw}(S_w)\big|_{S_w=1} \equiv k_{sat}$.

When a *nonwetting* fluid is drained as it is being displaced by a wetting fluid to below a critical saturation value, S_{nr}, referred to as the *residual nonwetting fluid saturation*, the latter breaks down into isolated blobs, or *globules*. Usually, these remain immobile under the pressure gradient that drives the wetting fluid. The value of S_{nr} is determined by properties of the nonwetting fluid and of the solid. This phenomenon can easily be explained by the concept of capillary pressure discussed in Subs. 6.1.3. The globules establish their shapes within the void space in response to capillary forces. These forces establish a pressure gradient within each globule that opposes that in the mobile fluid around it. Following the Laplace equation (6.1.7), the capillary pressure is of the order $2\gamma_{wn}/r$, where r denotes some characteristic radius of a pore. Thus, in an immobile globule, the menisci configurations adjust themselves to maintain a pressure equilibrium with the mobile wetting fluid, as long as the pressure gradient in that fluid is not too high. As this pressure gradient increases, at some point, equilibrium can no longer be maintained, and the globule will be displaced, until equilibrium is re-established.

This explains why the nonwetting fluid becomes immobile below the critical saturation S_{nr}. This phenomenon is of major significance in reservoir (petroleum) engineering and contaminant hydrology, as large quantities of oil remain immobile in the form of globules; they cannot be mobilized by the displacing water (which is the wetting fluid). The same phenomenon also occurs in an air-water system, as water displaces air.

If, at the some value of S_w, the drainage process is stopped and wetting of the sample begins, the latter cannot be brought back to full saturation. The air saturation of the sample cannot be lowered to below the *residual air saturation*, S_{ar}, because of *entrapped air* (Subs. 6.1.4). The amount of entrapped air, in the form of air bubbles and air-filled portion of the void space completely surrounded by water in the larger pores, grows gradually as the sample is rewetted. Sometimes, this means that effective permeability never rises back to more than 0.5 k_{sat}, especially in view of the fact that for many soils, the slope of the effective permeability-saturation-curve becomes steeper as full saturation is approached.

The relative permeability curve for air is also shown in Fig. 6.2.1. Again, we note that $k_{ra} = 0$ for $S_a < S_{ar}$, and that due to the irreducible water saturation, k_{ra} cannot rise above $k_{ra}\big|_{S_a=1-S_{wr}}$, unless we start by wetting an initially dry sample at $S_a = 1$.

We emphasize that relative permeability curves have to be determined experimentally for each particular soil. However, various investigators have suggested analytical expressions for the relationship between relative permeability (or relative hydraulic conductivity) and saturation. These expressions were usually obtained by analyzing simplified models of porous media, such as a bundle of parallel capillary tubes, or a network of such tubes. The results, while highlighting the main features of the sought relationship, always contain numerical coefficients that characterize the considered model. They cannot be used for soils that have a much more complicated irregular structure. For a particular soil, the numerical values of the coefficients have to be determined by fitting the analytical expression to experimental curves. Analytical expressions (as compared with tables of experimental results) have the advantages that they can be used in analytical or semi-analytical solutions, and can more easily be used as input to numerical models.

Following are a few examples, all for isotropic porous media.

- Gardner (1958) suggested the expression

$$\mathsf{K}_w = \frac{a}{b + |\psi|^m}, \tag{6.2.17}$$

where a, b and m are constants, with $m \approx 2$ for heavy clay soil, and $m \approx 4$ for sand. He also suggested the exponential model

$$\mathsf{K}_w = \mathsf{K}_{\text{sat}}\, e^{-\alpha\psi}, \tag{6.2.18}$$

where α is a soil index parameter, related to pore size distribution, or, according to Raats (1976), to the reciprocal of a macroscopic capillary length scale.

- Childs and Collis-George (1950), for the flow of water in the unsaturated zone, introduced the expression

$$\mathsf{K}_w = B\frac{\theta_w^3}{\Sigma_{vs}}, \tag{6.2.19}$$

where Σ_{vs} is the specific surface area of the soil, and B is a coefficient.

- Irmay (1954) suggested

$$\mathsf{K}_w(S_w) = \mathsf{K}_{\text{sat}}\left(\tilde{S}_w\right)^3 \equiv \mathsf{K}_{\text{sat}}\left(\frac{S_w - S_{wr}}{1 - S_{wr}}\right)^3, \tag{6.2.20}$$

where $\tilde{S}_w$ is the *effective water saturation* defined in Subs. 6.1.7.

- Corey (1957) suggested a relationship proportional to $\left(\tilde{S}_w\right)^4$.
- Brooks and Corey (1964, 1966) suggested

$$\mathsf{k}_w = \begin{cases} \mathsf{k}_{\text{sat}} & \text{for} \quad p_c < p_b, \\ \mathsf{k}_{\text{sat}}\left(\dfrac{p_b}{p_c}\right)^{\frac{2+\lambda}{\lambda}} & \text{for} \quad p_c \geq p_b, \end{cases} \tag{6.2.21}$$

where p_b is the *bubbling pressure*, or *air entry pressure*, related to the largest pore size forming a continuous network of water occupied channels within the porous medium, and λ is an index of the pore-size distribution of the porous medium. In this equation, $\mathsf{k}_{\rm sat}$ is the permeability at $\tilde{S}_w = 1$, and not at $S_w = 1$. When combined with (6.1.20), Brooks and Corey (1964) obtained

$$\mathsf{k}_w(\tilde{S}_w) = \mathsf{k}_{\rm sat}\left(\tilde{S}_w\right)^{\epsilon}, \qquad \epsilon = \frac{2+3\lambda}{\lambda}, \tag{6.2.22}$$

and

$$\mathsf{k}_a(\tilde{S}_w) = \mathsf{k}_{\rm sat}(1-\tilde{S}_w)^2\left[1-\left(\tilde{S}_w\right)^{\gamma}\right], \quad \gamma = \frac{2+\lambda}{\lambda}, \tag{6.2.23}$$

where $\tilde{S}_w$ is less than some maximum value (usually ≈ 0.85) at which k_a still exists.

- Van Genuchten (1980) suggested the relationship

$$\mathsf{K}_w = \mathsf{K}_{\rm sat}\,\tilde{S}_w^{\frac{1}{2}}\left[1-\left(1-\tilde{S}_w^{\frac{1}{m}}\right)^m\right]^2, \tag{6.2.24}$$

where m is the same as the coefficient C that appears in (6.1.23).

Effective permeability, or hydraulic conductivity, may also be presented as a function of the pressure head, ψ. However, the relationship $\mathbf{k}_w(\psi)$ shows much more hysteresis than $\mathbf{k}_w(S_w)$, probably due to the large hysteresis in the function $\psi(S_w)$. Hysteresis in $\mathbf{k}_w(S_w)$ is generally ignored because the function $p_c(S_w)$ usually exhibits far greater hysteretic effects and because the values of the parameters required to describe hysteresis in $\mathbf{k}_w(S_w)$ are highly uncertain.

6.3 Mass Balance Equation and Complete Model

In Sec. 5.1, we presented the fundamental concepts underlying the mass balance equation. We started from the mass balance equation for a fluid continuum, and developed the corresponding equation for saturated flow. Here, we wish to extend the same basic ideas to the unsaturated zone, where the void space is occupied by two fluid phases. The motion equations for these two phases are presented in the previous section. We shall limit the discussion to the case of a non-deformable porous medium.

6.3.1 Mass balance equations

A. Fundamental equations

We can start from the microscopic mass balance equation (5.1.6), rewritten here for convenience, as

$$\frac{\partial \rho}{\partial t} = -\nabla\cdot\rho\mathbf{V} + \rho\Gamma'^m, \tag{6.3.1}$$

noting the comment concerning the source term $\rho\Gamma'^m$ that follows (5.1.6). As suggested in the discussion in Sec. 5.1.1B, the averaged, or macroscopic mass balance equation for a fluid phase that occupies part of void space (volumetric fraction, θ_α), can be obtained by averaging the microscopic mass balance equation (5.1.6) over an REV.

As in the case of saturated flow, assuming that the dispersive flux of the fluid's mass is much smaller than its advective one (Subs. 7.1.5), i.e., $|\overline{\rho}^\alpha \overline{\mathbf{V}}^\alpha| \gg |\overline{\mathring{\rho}\mathring{\mathbf{V}}}^\alpha|$, we obtain the averaged mass balance equation for any α-phase, in a multiphase system, in the form

$$\frac{\partial}{\partial t}(\theta_\alpha \rho_\alpha) = -\nabla \cdot (\rho_\alpha \mathbf{q}_\alpha) + f_{\alpha \to \beta} + \rho_\alpha \Gamma'^\alpha, \tag{6.3.2}$$

where $\mathbf{q}_\alpha = \theta_\alpha \mathbf{V}_\alpha$ denotes the *specific discharge* (= flux) of the α-phase, $f_{\alpha\to\beta}$ denotes the transfer of α-phase mass into the β-phase, across their common (microscopic) $\mathcal{S}_{w\beta}$-interface, say by dissolution, and the symbol Γ'^α denotes a source of α-fluid (= added volume of α-phase per unit volume of porous medium, per unit time), other than through the (microscopic) $\mathcal{S}_{\alpha\beta}$-interface (e.g., evaporation, in the air-water case).

Let the two fluid phases that occupy the void space in the unsaturated zone be air and water. In the absence of evaporation or air dissolution, i.e., $f_{w\to a} = 0$, the macroscopic mass balance equation for the water can be obtained from (5.1.5) in the form

$$\frac{\partial}{\partial t}(\phi S_w \rho_w) = -\nabla \cdot (\rho_w \mathbf{q}_w) + \rho_w \Gamma'^w. \tag{6.3.3}$$

Although, as explained above for the microscopic equation, in the rigorous sense, there cannot be a distributed source of mass in a three-dimensional space, we have, symbolically, introduced such a source here to denote an *external* source of water mass that takes the form of individual points at which water is injected. Sometimes, such point sources are approximated as a distributed source. A negative value means withdrawal of water from the void space. For example, water imbibed by distributed vegetation roots may be sufficiently well approximated as a distributed water sink.

Under the same conditions, the equation of mass balance for the air is

$$\frac{\partial}{\partial t}(\phi S_a \rho_a) = -\nabla \cdot (\rho_a \mathbf{q}_a) + \rho_a \Gamma'^a, \tag{6.3.4}$$

where Γ'^a represents the volume of air added from external sources per unit volume of porous medium per unit time.

For constant ρ_w and ρ_a, the mass balance equations, (6.3.3) and (6.3.4), reduce to

$$\frac{\partial(\phi S_w)}{\partial t} = -\nabla \cdot \mathbf{q}_w + \Gamma'^w, \tag{6.3.5}$$

$$\frac{\partial(\phi S_a)}{\partial t} = -\nabla \cdot \mathbf{q}_a + \Gamma'^a. \tag{6.3.6}$$

The complete set of equations that describe the simultaneous flow of air and water, considered as compressible fluids, in a stationary rigid soil (i.e., $\mathbf{V}_s = 0$, $\partial\phi/\partial t = 0$, $\mathbf{q} \equiv \mathbf{q}_r$), neglecting interphase mass transfers and external sources, are summarized below:

Mass balances for water and air:

$$\phi\frac{\partial S_w \rho_w}{\partial t} = -\nabla \cdot (\rho_w \mathbf{q}_w), \qquad \phi\frac{\partial S_a \rho_a}{\partial t} = -\nabla \cdot (\rho_a \mathbf{q}_a). \tag{6.3.7}$$

Darcy's law for water and air:

$$\mathbf{q}_w = -\frac{\mathbf{k}_w}{\mu_w} \cdot (\nabla p_w + \rho_w g \nabla z), \qquad \mathbf{q}_a = -\frac{\mathbf{k}_a}{\mu_a} \cdot (\nabla p_a + \rho_a g \nabla z). \tag{6.3.8}$$

Capillary pressure and density relationships:

$$p_c = p_a - p_w = p_c(S_w), \qquad \rho_w = \rho_w(p_w), \qquad \rho_a = \rho_a(p_a). \tag{6.3.9}$$

Sum of saturations:

$$S_w + S_a = 1. \tag{6.3.10}$$

Altogether eight equations that must be solved simultaneously for the eight variables:

$$p_w, \; p_a, \; S_w, \; S_a, \; \rho_w, \; \rho_a, \; \mathbf{q}_w, \; \mathbf{q}_a.$$

In three-dimensional domains, these are twelve scalar variables and twelve scalar equations.

We wish to emphasize that we have here only two partial differential equations to be solved for two (properly selected) variables. The remaining variables can be determined from them. The concept of *primary variables* is introduced in Subs. 7.9.4, where we consider the general case of multiple multicomponent phases under nonisothermal conditions in a deformable porous medium.

Although not always justified, very often, in practice, only the flow of the water is considered, overlooking air flow. For example, it is certainly not justified when air flow is produced by air injection and/or extraction as part of contaminant cleanup operations. The assumption underlying the 'water flow only' model is that the resistance to flow in the air phase is everywhere negligible, so that we have hydrostatic pressure distribution in the air, i.e., $\nabla h_a = 0$. We certainly have stationary air in steady flow of water. Because air density is very small, the air-phase is assumed to be everywhere under (practically) atmospheric pressure, usually taken as zero pressure. Of course, if this assumption cannot be made, like when both fluids are moving, the 'water flow only' model cannot be used, and we have to add and solve a mass balance equation for the air.

B. Deformable porous medium

In a deformable porous medium, $\partial\phi/\partial t \neq 0$. The case of saturated flow in a deformable porous medium was discussed in Sec. 5.1.2. Let us extend this discussion to unsaturated flow, following the same approach. We shall focus on the mass balance equation for the water. As in the case of saturated flow, we start by analyzing the term on the l.h.s. of (6.3.3), which expresses the rate at which mass of water is accumulated in a unit volume of porous medium as a consequence of flow and of water sources. Following the presentation in Sec. 5.1.2, we can rewrite this term in the form

$$\frac{\partial}{\partial t}\phi S_w \rho_w = \phi S_w \frac{\partial \rho_w}{\partial t} + S_w \rho_w \frac{\partial \phi}{\partial t} + \phi \rho_w \frac{\partial S_w}{\partial t}. \qquad (6.3.11)$$

We note the three effects that contribute to the added water mass: fluid compressibility, solid matrix compressibility, and saturation change. In general, $\rho_w = \rho_w(p_w, c_w, T)$, but here we focus on $\rho_w = \rho_w(p_w)$. The case of variable density as a function of concentration is considered in Subs. 9.3.1.

The first term on the r.h.s. can be expressed, similar to the saturated case, by making use of the relationship (5.1.29). For the second term, we follow the discussion on *effective stress* that, for saturated flow, leads to the relationship (5.1.41). This relationship can be extended to unsaturated flow by replacing $\overline{p_w}^w$ by some average fluid pressure in the void-space, $\overline{p_{\text{void}}}^v$. For example, for the air-water system in the unsaturated zone, the average pressure in the water, $\overline{p_w}^w$, may be replaced by $\overline{p_{\text{void}}}^v$, defined by

$$\overline{p_{\text{void}}}^v = S_w \overline{p_w}^w + S_a \overline{p_a}^a, \qquad (6.3.12)$$

i.e., using fluid saturations as weights in determining the mean pressure. Aitchison and Donald (1956) suggested the relationship

$$\overline{p_{\text{void}}}^v = \chi(S_w)\overline{p_w}^w, \qquad (6.3.13)$$

where $\chi(S_w)$ is some function of the saturation. We recall that for $p_a = 0$, we have $S_w = S_w(\overline{p_w}^w)$. With (6.3.13), equation (5.1.41) becomes

$$\overline{\boldsymbol{\sigma}} = \overline{\boldsymbol{\sigma}'_s} - \chi(S_w)\overline{p_w}^w \mathbf{I}. \qquad (6.3.14)$$

Bear *et al.* (1984) used (6.3.14) with $\chi(S_w) = S_w$.

With the introduction of S_w and $\chi(S_w)$, we can extend (5.1.48) for two-phase flow to the form

$$\frac{\partial \phi}{\partial t} = (1-\phi)\alpha\left(\chi(S_w)\frac{\partial p_w}{\partial t} + p_w \frac{\partial \chi}{\partial S_w}\frac{\partial S_w}{\partial t}\right). \qquad (6.3.15)$$

To obtain an expression for the third term on the r.h.s. of (6.3.11), we apply the *chain rule of differentiation* to the term $\partial S_w/\partial t$, and noting that $S_w =$

$S_w(p_c)$, we obtain

$$\frac{\partial S_w}{\partial t} = \frac{\mathrm{d}S_w}{\mathrm{d}p_c}\frac{\partial p_c}{\partial t} = \frac{\mathrm{d}S_w}{\mathrm{d}p_c}\left(\frac{\partial p_a}{\partial t} - \frac{\partial p_w}{\partial t}\right). \tag{6.3.16}$$

We define the *water (moisture) capacity*, C_w, by

$$C_w = -\phi\frac{\mathrm{d}S_w}{\mathrm{d}p_c}. \tag{6.3.17}$$

Then,

$$\frac{\partial S_w}{\partial t} = \frac{C_w}{\phi}\left(\frac{\partial p_w}{\partial t} - \frac{\partial p_a}{\partial t}\right). \tag{6.3.18}$$

If p_a is constant, $\partial p_a/\partial t = 0$, and

$$\frac{\partial S_w}{\partial t} = \frac{C_w}{\phi}\frac{\partial p_w}{\partial t}. \tag{6.3.19}$$

In this case, we may also write $S_w = S_w(p_w)$; therefore,

$$\frac{\partial S_w}{\partial t} = \frac{dS_w}{dp_w}\frac{\partial p_w}{\partial t}, \quad \text{and} \quad C_w = \phi\frac{\mathrm{d}S_w}{\mathrm{d}p_w}. \tag{6.3.20}$$

Altogether, we obtain

$$\begin{aligned}\frac{\partial}{\partial t}(\phi S_w \rho_w) = \rho_w\Bigg\{&\phi S_w \beta_w + \phi\frac{\mathrm{d}S_w}{\mathrm{d}p_w} \\ &+ S_w(1-\phi)\alpha\left[\chi(S_w) + p_w\frac{\mathrm{d}\chi}{\mathrm{d}S_w}\frac{\mathrm{d}S_w}{\mathrm{d}p_w}\right]\Bigg\}\frac{\partial p_w}{\partial t},\end{aligned} \tag{6.3.21}$$

in which α is defined by (5.1.46), and $\chi(S_w)$ is defined in (6.3.13).

Equation (6.3.21) can be rewritten in a form analogous to (5.1.49), with a specific mass storativity that takes into account also the change in saturation. For the saturated zone, $S_w = 1$, $\chi(S_w) = 1$, and (6.3.21) reduces to (5.1.49).

One may easily define a specific storativity with respect to changes in piezometric head (actually, in terms of Hubbert's potential, h^*, defined in (4.1.6)), by making use of the relationship

$$\frac{\partial h^*}{\partial t} = \frac{1}{\rho_w(p)g}\frac{\partial p_w}{\partial t}. \tag{6.3.22}$$

Thus, groundwater hydrologists define a *specific storativity*, S_o^*, defined as the volume of water released from storage in a unit volume of porous medium, per unit decline in the piezometric head (e.g., Bear, 1972). Here, this definition is expressed by (5.1.52), repeated here for convenience

$$\rho_w S_o^* \frac{\partial h^*}{\partial t} = S_{op}^{m*} \frac{\partial p_w}{\partial t}. \tag{6.3.23}$$

By assuming $\chi(S_w) \equiv 1$ for saturated flow, we have overlooked the possibility of entrapped air that may fill up part of the void space. We also note that, while the specific storativity of a saturated zone depends on water pressure only, through $\phi = \phi(p_w)$, the specific mass storativity in the unsaturated zone varies also with p_a, through $S_w(p_c)$.

With the above developments, the mass balance equation for water in a deformable porous medium, (6.3.3), takes the form

$$\rho_w \left\{ \phi S_w \beta_w + \phi \frac{\mathrm{d}S_w}{\mathrm{d}p_w} + S_w \alpha \left[p_w \frac{\mathrm{d}\chi}{\mathrm{d}S_w} \frac{\mathrm{d}S_w}{\mathrm{d}p_w} + \chi(S_w) \right] \right\} \frac{\partial p_w}{\partial t}$$
$$= -\nabla \cdot (\rho_w \mathbf{q}_{rw}) + \rho_w \Gamma'^w. \tag{6.3.24}$$

We can write this equation also in the form (5.1.53), repeated here for convenience:

$$S_{op}^m \frac{\partial p_w}{\partial t} = -\nabla \cdot (\rho_w \mathbf{q}_{rw}) + \rho_w \Gamma'^w, \tag{6.3.25}$$

in which the coefficient is defined by (6.3.24). Note that S_{op}^m is different from the coefficient S_{op}^{m*} defined by (6.3.21), and that in the divergence term on the r.h.s. we have $\mathbf{q}_{rw}$ rather than $\mathbf{q}_w$. The explanation is given following (5.1.54). There should not be any difficulty in writing the above equation in terms of the piezometric head instead of pressure.

C. Flow equations

As in saturated flow, the term *flow equation* is used for the mass balance equation for a fluid phase, combined with the appropriate form of the motion equation, e.g., Darcy's law. In two-phase flow, the objective is to obtain a single equation for each phase, written in terms of a single state variable, e.g., pressure, piezometric head, or saturation. Actually, in most two- or three-phase cases, the phases are strongly coupled and this goal cannot be achieved because of the non-analytical form of the relevant constitutive equations. It is, therefore, often preferable to leave the mathematical model in its original form (of balance, motion, and constitutive equations), and handle the complexity through the algorithm of the numerical solution, rather than attempt to eliminate variables (and equations).

In the following, we shall assume that hysteresis in C_w may be neglected (Subs. 6.1.7).

In terms of pressure

First, we focus on the balance equation for water in the unsaturated zone. Any of the balance equations: (6.3.3), (6.3.5), and (6.3.25), may be used as a point of departure.

Let us start by assuming that water is a compressible fluid, with $\rho_w = \rho_w(p_w)$, so that the motion equation (= Darcy's law) takes the form of (6.2.3). Furthermore, we assume that changes in porosity, due to changes in fluid pressures, may be written in the form (6.3.15), i.e., as an explicit function of p_w and S_w, and of changes in these quantities only. This assumption implies (6.3.25), using (6.3.18). Then, with no interphase mass transfers, we obtain the following flow equation for water in a deformable porous medium:

$$S_{op}^{m*}\frac{\partial p_w}{\partial t} - \rho_e C_w \frac{\partial p_a}{\partial t} = \nabla\cdot\left[\rho_w \frac{\mathbf{k}_w(S_w)}{\mu_w}\cdot(\nabla p_w + \rho_w g \nabla z)\right] + \phi S_w \rho_w \Gamma'^w, \tag{6.3.26}$$

where the *specific mass storativity*, S_{op}^{m*}, is defined by

$$S_{op}^{m*} = \rho_w\left[\phi S_w \beta_w - \phi\frac{dS_w}{dp_c} + \alpha S_w\left(\chi - p_w \frac{d\chi}{dS_w}\frac{dS_w}{dp_c}\right)\right]. \tag{6.3.27}$$

Using the definition (6.3.17) for water capacity, C_w, we also have

$$S_{op}^{m*} = \rho_w\left\{\phi S_w \beta_w + C_w(S_w) + \alpha S_w\left[\frac{p_w}{\phi}\frac{\mathrm{d}\chi}{\mathrm{d}S_w}C_w(S_w) + \chi(S_w)\right]\right\}, \tag{6.3.28}$$

in which $\rho_w = \rho_w(p_w)$, and $p_c \equiv p_a - p_w = p_c(S_w)$. Although our objective was to write an equation in terms of only a single variable, p_w, p_a also appears, because of the relationship between S_w and the capillary pressure, $p_a - p_w$. If $p_a = 0$ (or a constant), then (6.3.26) is an equation for p_w only.

To solve (6.3.26), a companion equation for the air phase is required (e.g., $p_a = 0$). We also need information on Γ'^w, $\mathbf{k}_w = \mathbf{k}_w(S_w)$, $\rho_w = \rho_w(p_w)$, ϕ, β_w, $\chi = \chi(S_w)$, and $S_w(p_c)$. In a deformable porous medium, the porosity, ϕ, also varies with water (or, generally, the fluid) pressure, and we need information on the coefficient α.

Rather than continue with this quite complicated case of a deformable porous medium in the unsaturated zone, let us assume here that under the pressure changes that take place in the unsaturated zone, the solid matrix may be considered nondeformable. This is the more common case in practice.

Thus, for a nondeformable, stationary porous medium, i.e., $\partial\phi/\partial t = 0$, $\mathbf{V}_s = 0$, $\mathbf{q}_r \equiv \mathbf{q}$, the flow equation for water takes the form:

$$\phi\frac{\partial \rho_w S_w}{\partial t} = \nabla\cdot\left[\rho_w \frac{\mathbf{k}_w(S_w)}{\mu_w}\cdot(\nabla p_w + \rho_w g \nabla z)\right] + \phi S_w \rho_w \Gamma'^w. \tag{6.3.29}$$

Alternatively, with $C_w = C_w(S_w) = \phi\,\mathrm{d}S_w/\mathrm{d}p_c$, we have

$$\begin{aligned}\rho_w\left[\phi S_w \beta_w + C_w\right]\frac{\partial p_w}{\partial t} &- \rho_w C_w \frac{\partial p_a}{\partial t}\\ &= \nabla\cdot\left[\rho_w \frac{\mathbf{k}_w(S_w)}{\mu_w}\cdot(\nabla p_w + \rho_w g \nabla z)\right] + \phi S_w \rho_w \Gamma'^w.\end{aligned} \tag{6.3.30}$$

An analogous equation can be written for the gaseous phase.

If we invoke assumption (5.1.26), which is usually justified in practice also in the unsaturated zone, (6.3.30) reduces to

$$(\phi S_w \beta_w + C_w) \frac{\partial p_w}{\partial t} - C_w \frac{\partial p_a}{\partial t} = \nabla \cdot \left[\frac{\mathbf{k}_w(S_w)}{\mu_w} \cdot (\nabla p_w + \rho_w g \nabla z) \right] + \phi S_w \Gamma'^w. \tag{6.3.31}$$

Under the same conditions, although less justified, we obtain for air

$$(\phi S_a \beta_a + C_w) \frac{\partial p_a}{\partial t} - C_w \frac{\partial p_w}{\partial t} = \nabla \cdot \left[\frac{\mathbf{k}_a(S_a)}{\mu_a} \cdot (\nabla p_a + \rho_a g \nabla z) \right] + \phi S_a \Gamma'^a. \tag{6.3.32}$$

The partial differential (flow) equations, (6.3.31) and (6.3.32), the constitutive relations, $\rho_w = \rho_w(p_w)$ and $\rho_a = \rho_w(p_a)$, the capillary pressure relationship, $p_a - p_w = p_c(S_w)$, and the constraint, $S_w + S_a = 1$, constitute a set of six equations to be solved simultaneously for the six state variables: p_a, p_w, ρ_w, ρ_a, S_w, and S_a. We note that only two of the six equations are partial differential equations that have to be solved, say for p_w and S_a (or any other selected pair of *independent* variables). For example, we may select p_w and p_a, and solve (6.3.31) and (6.3.32), with appropriate initial and boundary conditions related to the flow of the respective fluid phases. The concept of *primary variables* is introduced in Subs. 7.9.4.

Once these variables have been determined, the other four can be derived from the remaining four equations. The two partial differential equations are coupled by the capillary pressure relationship, and, therefore, have to be solved simultaneously. A knowledge of the saturations will also provide information on $C_w(S_w)$, $\mathbf{k}_w(S_w)$ and $\mathbf{k}_a(S_a)$. The source terms, $\phi S_w \rho_w \Gamma'^w$, and Γ'^a, may also depend on these state variables.

In terms of piezometric head

When $\rho_w, \rho_a =$ const., we may write the flow equations for water and air in terms of the *piezometric heads*, $h_w (= z + p_w/\rho_w g)$ and $h_a (= z + p_a/\rho_a g)$. For example, for a nondeformable, stationary porous medium ($\partial \phi/\partial t = 0$, $\mathbf{q}_r \equiv \mathbf{q}$), we make use of the motion equations (6.2.5) and (6.2.6), to obtain for the water:

$$C'_w(S_w) \left(\frac{\partial h_w}{\partial t} - \frac{\rho_a}{\rho_w} \frac{\partial h_a}{\partial t} \right) = -\nabla \cdot [\mathbf{K}_w(S_w) \cdot \nabla h_w], \quad C'_w = \rho_w g C_w. \tag{6.3.33}$$

where, for the sake of simplicity, we have omitted the source terms, and $C'_w(S_w)$ expresses another *moisture capacity*. For air, the flow equation is

$$C'_w(S_w)\left(\frac{\partial h_a}{\partial t}-\frac{\rho_w}{\rho_a}\frac{\partial h_w}{\partial t}\right)=-\nabla\cdot[\mathbf{K}_a(S_a)\cdot\nabla h_a]. \tag{6.3.34}$$

In terms of suction

When only water flow (at ρ_w = const.) is being considered, assuming that pressure in the air is approximately atmospheric, viz., $p_a \approx 0$, we often express the water balance equation in terms of the *suction*, or *suction head*, ψ_w $(= -p_w/\rho g)$, introduced in Subs. 6.1.3. From (6.3.33), we then obtain

$$C'_w(\psi_w)\frac{\partial \psi_w}{\partial t}=\nabla\cdot[\mathbf{K}_w(\psi_w)\cdot\nabla(z-\psi_w)]. \tag{6.3.35}$$

In terms of moisture content—Richards' equation

We continue to limit the discussion to the case of a stationary, nondeformable porous medium, ρ_w = const., and constant air pressure. Making use of the motion equation (6.2.11), and omitting source terms, (5.1.20) becomes

$$\frac{\partial \theta_w}{\partial t}=\nabla\cdot[\mathbf{D}_w(\theta_w)\cdot\nabla\theta_w]+\nabla\cdot[\mathbf{K}_w(\theta_w)\cdot\nabla z], \tag{6.3.36}$$

known as *Richards' equation.*

If the effect of gravity, represented by the second term on the right-hand side of (6.3.36), is neglected, or the flow is horizontal, (6.3.36) reduces to

$$\frac{\partial \theta_w}{\partial t}=\nabla\cdot[\mathbf{D}_w(\theta_w)\cdot\nabla\theta_w], \tag{6.3.37}$$

known as *moisture diffusivity equation.*

When the porous medium is deformable, $\partial\phi/\partial t \neq 0$, and the flow equation should be written in terms of saturation as a state variable.

The flow equation in the form of (6.3.37) has been proven useful, especially in the development of analytical and quasi-analytical solutions to problems of unsaturated flow (Irmay, 1968; Brutsaert, 1968; Philip, 1969; Parlange, 1971, 1972; Braester, 1973; Broadbridge and White, 1988).

D. Flow with interphase mass transfer

In many cases of multiphase flow, phase components are exchanged among the fluid phases. Dissolution of air in water and evaporation of water in the unsaturated zone, may serve as examples. In fact, a term, $f_{\alpha\rightarrow\beta}$, that expresses the transfer of phase mass, from one phase (α) to the other (β), appears in the fundamental mass balance equation (6.3.2). However, immediately after introducing this equation, we have assumed that none of these exchange processes takes place, and developed mass balance equations without these source/sink terms. Here, we shall construct models that take into account processes of phase change under isothermal conditions.

Although we have referred above to 'mass of a phase' that crosses interphase boundaries, it is actually mass of certain phase components that do so. The transport of fluid phase components is discussed in details in Subs. 7.9.3. There, we consider three fluid phases, and the emphasis is on the various mechanisms of transport of individual components. Our objective here is to introduce an example that involves the phenomenon of interphase mass exchange in the case of two fluid phases, with its effect on fluid density.

We consider a case in which the wetting phase is an aqueous liquid denoted by subscript ℓ, and the nonwetting phase is a gas, denoted by subscript g. The liquid is assumed to be made up of two components: (primarily) pure water ($= H_2O$), denoted by superscript w, and dissolved air, denoted by superscript a. The gas is also made up of two components: (primarily) 'dry' air, superscript a, and water vapor, denoted by superscript w. Accordingly, c_ℓ^w and c_ℓ^a will denote the mass of pure water and of dissolved air, respectively, both per unit volume of the liquid phase, while c_g^a and c_g^w will denote the mass of (dry) air and of water vapor, respectively, per unit volume of the gas.

Because we are considering here components of phases, we have to take into account the flux of these components due to hydrodynamic dispersion (discussed in Subs. 7.1.4 and 7.1.5). Thus,

$$\mathbf{J}_{h\alpha}^{\gamma}(=\mathbf{J}_{\alpha}^{*\gamma}+\mathbf{J}_{\alpha}^{\gamma}),$$

denotes the sum of dispersive and diffusive fluxes of the γ-component in the α-phase, per unit area of the latter (Subs. 7.1.7).

Four phase change phenomena occur: evaporation of water, condensation of water vapor, dissolution of air in the liquid, and release of dissolved air into the gaseous phase. In the balance equation for the mass of a γ-component, the symbols $f_{\ell\to g}^{\gamma}$ and $f_{g\to\ell}^{\gamma}$ will indicate, the rate of transfer from liquid to gas and from gas to liquid, respectively, of a γ-component, both in terms of mass of component per unit volume of porous medium. For example, $f_{\ell\to g}^{a}$ is the rate at which mass of dry air is added to the gaseous phase from the liquid phase (i.e., evaporation). We note that $f_{\ell\to g}^{a}=-f_{g\to\ell}^{a}$ and $f_{\ell\to g}^{w}=-f_{g\to\ell}^{w}$.

With these symbols, and making use of the component mass balance equation (5.1.5), in which we replace $\overline{e}^{\alpha}$ by c_{α}^{γ}, and with the appropriate f-symbol replacing the surface integral, we can now write one mass balance equation for each of the four combinations of phases and components.

Mass balance for 'pure' water in the liquid phase:

$$\frac{\partial\theta_\ell c_\ell^w}{\partial t}=-\nabla\cdot\theta_\ell(c_\ell^w\mathbf{V}_\ell+\mathbf{J}_{h\ell}^w)+f_{g\to\ell}^w+\theta_\ell\rho_\ell\Gamma_\ell^w, \tag{6.3.38}$$

where $\theta_\ell c_\ell^w\mathbf{V}_\ell$ denotes the advective flux of the w-component in the ℓ-phase and $\theta_\ell\rho_\ell\Gamma_\ell^w$ denotes an external source of the w-component ($=$ mass of water per unit volume of porous medium).

For dissolved air in the liquid phase:

$$\frac{\partial \theta_\ell c_\ell^a}{\partial t} = -\nabla \cdot \theta_\ell (c_\ell^a \mathbf{V}_\ell + \mathbf{J}_{h\ell}^a) + f_{g\to\ell}^a + \theta_\ell \rho_\ell \Gamma_\ell^a. \tag{6.3.39}$$

For 'dry' air in the gaseous phase:

$$\frac{\partial \theta_g c_g^a}{\partial t} = -\nabla \cdot \theta_g (c_g^a \mathbf{V}_g + \mathbf{J}_{hg}^a) + f_{\ell\to g}^a + \theta_a \rho_a \Gamma_g^a. \tag{6.3.40}$$

For water vapor in the gaseous phase:

$$\frac{\partial \theta_g c_g^w}{\partial t} = -\nabla \cdot \theta_g (c_g^w \mathbf{V}_g + \mathbf{J}_{hg}^w) + f_{\ell\to g}^w + \theta_g \rho_g \Gamma_g^w. \tag{6.3.41}$$

The expressions for the advective liquid- and gas-phase fluxes, to be inserted in these component balance equations, are

$$\mathbf{q}_\ell \equiv \theta_\ell \mathbf{V}_\ell = -\frac{\mathsf{k}_\ell}{\mu_\ell} \cdot (\nabla p_\ell + \rho_\ell g \nabla z), \tag{6.3.42}$$

and

$$\mathbf{q}_g \equiv \theta_g \mathbf{V}_g = -\frac{\mathsf{k}_g}{\mu_g} \cdot (\nabla p_g + \rho_g g \nabla z), \tag{6.3.43}$$

in which we regard the effective permeabilities as known functions of the moisture content, θ_ℓ.

The capillary pressure,

$$p_g - p_\ell = p_c(\theta_\ell), \tag{6.3.44}$$

in which $p_c(\theta_\ell)$ is assumed to be a known function, relates the pressures in the two phases to the moisture content. In addition, we have

$$\theta_\ell + \theta_g = \phi. \tag{6.3.45}$$

Concentrations of components (= partial densities) are related to phase densities by

$$\rho_\ell = c_\ell^w + c_\ell^a, \tag{6.3.46}$$

$$\rho_g = c_g^a + c_g^w. \tag{6.3.47}$$

Underlying the above two equations is the assumption (reasonable in the case considered here) that the volume of water is not affected by the considered phase exchange phenomena (compare with Subs. 9.3.1).

At this point, we have 14 scalar equations in terms of 18 scalar variables: c_ℓ^w, c_ℓ^a, c_g^a, c_g^w, $\mathbf{V}_\ell$, $\mathbf{V}_g$, p_ℓ, p_g, θ_ℓ, θ_g, ρ_ℓ, ρ_g, and the two rates of exchange, $f_{\ell\to g}^w$ $(= -f_{g\to\ell}^w)$, and $f_{g\to\ell}^a$ $(= -f_{\ell\to g}^a)$. The $\mathbf{J}_h$'s have not been counted as variables, because they can be related to the advective phase fluxes and to concentration gradients (Subs. 7.1.7). The source functions, Γ_ℓ^a, Γ_g^a, etc., are also assumed known, as are the constitutive relations $\mu_\ell(p_\ell, c_\ell^w, \ldots)$, and $\mu_g(p_g, c_g^a, \ldots)$.

To eliminate the rates of phase change from the component balance equations, we sum up, for each component, the corresponding balance equations for the two phases. The result is a single balance equation for the considered component in the porous medium.

The two mass balance equations for the components for the porous medium as a whole, are:

Mass balance equation for pure water, obtained by summing (6.3.38) and (6.3.41), is

$$\frac{\partial}{\partial t}(\theta_\ell c_\ell^w + \theta_g c_g^w) = -\nabla\cdot(\theta_\ell c_\ell^w \mathbf{V}_\ell + \theta_g c_g^w \mathbf{V}_g) \\ -\nabla\cdot(\theta_\ell \mathbf{J}_{h\ell}^w + \theta_g \mathbf{J}_{hg}^w) + \theta_\ell \rho_\ell \Gamma_\ell^w + \theta_g \rho_g \Gamma_g^w. \tag{6.3.48}$$

Mass balance equation for dry air, obtained by summing (6.3.39) and (6.3.40), is

$$\frac{\partial}{\partial t}(\theta_\ell c_\ell^a + \theta_g c_g^a) = -\nabla\cdot(\theta_\ell c_\ell^a \mathbf{V}_\ell + \theta_g c_g^a \mathbf{V}_g) \\ -\nabla\cdot(\theta_\ell \mathbf{J}_{h\ell}^a + \theta_g \mathbf{J}_{hg}^a) + \theta_g \rho_g \Gamma_g^a + \theta_\ell \rho_\ell \Gamma_\ell^a. \tag{6.3.49}$$

In this way, we have eliminated the rates of phase change, but now each of the fluid phase balance equations involves component concentrations in both phases.

Altogether, we now have 12 scalar equations for the 16 scalar variables. The required additional equations must express thermodynamic relationships between components in the two phases. This calls for the introduction of two additional variables, the partial pressures for the gaseous phase, p_g^a and p_g^w, with the relationship

$$p_g^a + p_g^w = p_g. \tag{6.3.50}$$

Now we have 13 equations and 18 state variables to be solved for.

As an example of the relations between component concentrations, densities, and pressures, we may use the symbolic expression

$$c_\ell^a = c_\ell^a(p_g, p_\ell), \tag{6.3.51}$$

that relates air solubility in liquid water to air and water pressures. Alternatively, we may use *Henry's law*, which relates the mole fraction of a gas component (= the solute) in a dilute liquid solution (= the solvent) to its partial pressure in the solution. In addition, assuming that the water vapor and the dry air components in the gaseous phase behave as *ideal gases*, we can write for the gaseous (air) phase

$$p_g^w = \frac{RT}{M^w} c_g^w, \qquad p_g^a = \frac{RT}{M^a} c_g^a, \tag{6.3.52}$$

where R (= 8.1347 Joule/mole°K) is the *universal gas constant*, T is the absolute (°K) temperature, and M^w (= 18 gr/mole) and M^a (= 29 gr/mole) are the molar masses of 'pure' water and 'dry' air, respectively.

The *relative humidity* in the soil, h_r $(= c_g^w / c_g^w|_{\text{sat}})$, in which $c_g^w|_{\text{sat}}$ is the vapor's concentration (= density) at saturation, is given by (Edelfsen and Anderson, 1943)

$$\left.\frac{c_g^w}{(c_g^w)}\right|_{\text{sat}} = \exp\left\{-\frac{p_g - p_\ell}{\rho_\ell}\frac{M^w}{RT}\right\}. \tag{6.3.53}$$

Finally, the mass density of the liquid phase is related to its pressure and to the amount of air dissolved in it by

$$\rho_\ell = \rho_\ell(p_\ell, c_\ell^a). \tag{6.3.54}$$

Altogether we now have 18 scalar equations in 18 scalar variables. In principle, with appropriate initial and boundary conditions, a solution can be obtained.

E. Combined saturated-unsaturated flow model

A coupled saturated-unsaturated model, with a moving phreatic surface separating the two zones, sounds rather complicated. However, by comparing the flow equation for the unsaturated zone, (6.3.26), with no interphase mass transfer and with $p_a = 0$, with (5.1.73) for the saturated zone, we note that the two are identical in form. For the former, the specific storativity is expressed by (6.3.28), while for the latter, it can be expressed by the same equation, with $S_w = 1$ and $\chi = 1$. Hence, we may treat the combined saturated-unsaturated domain as a single one, in which the pressure in the water is the *only* state variable that satisfies (6.3.26), as long as the pressure in saturated regions of the domain stays above zero. In fact, this should not be surprising, as each of the two equations expresses nothing but the mass balance of the water in the porous medium domain. The advantage of this (combined) approach is that we have removed the phreatic surface as a boundary. The two domains constitute a single continuous domain with pressure as a variable that varies in a continuous way throughout the domain. In computer codes that are used for solving for the flow in such a domain, we insert the constraints:

Saturated zone

$$\begin{aligned} p_w > 0, \quad S_w = 1, \\ \mathbf{k}_w = \mathbf{k}_w(\mathbf{x}). \end{aligned} \tag{6.3.55}$$

Unsaturated zone

$$p_w < 0, \quad S_w = S_w(p_w), \quad S_{wr} \leq S_w \leq 1,$$

$$\mathbf{k}_w = \begin{cases} \mathbf{k}_w(S_w(\mathbf{x},t)), & S_w > S_{wr}, \\ \mathbf{k}_w = 0, & S_w \leq S_{wr}. \end{cases} \tag{6.3.56}$$

6.3.2 Initial and boundary conditions

The need for initial and boundary conditions and their role in models has already been presented and discussed in Sec. 5.2, in connection with saturated flow. There, we have also discussed the general concept and representation of boundary surfaces, and the principles that serve as the basis for determining boundary conditions. In connection with two phase flow in the unsaturated zone, it may be of interest to note the discussion in Subs. 5.2.1. Here, we are discussing the simultaneous flow of two (assumed immiscible) fluids, like water and air, with no sharp (macroscopic) interface between them. Instead, there always exists a zone across which the saturation varies in space and time. Finally, in Sec. 5.2, we have presented a number of boundary conditions that are commonly encountered in saturated flow. Here, we shall discuss conditions that are usually encountered when modeling flow in the unsaturated zone.

Needless to say that when we model the simultaneous flow of two (assumed) immiscible fluids, we have two degrees of freedom, i.e., two independent variables to solve for, and we have two partial differential balance equations to be solved simultaneously. The kind of PDE describing the mass balance requires only *one* condition on each boundary segment. We should prefer a condition based on flux continuity, if such information is available. If not, we shall base the condition on available information on values of a scalar variable, e.g., pressure or saturation. Sometimes, approximations concerning the continuity in fluxes produce a jump in the values of the variables; we have to accept this consequence.

A. Initial conditions

Initial conditions specify the values of the two selected (macroscopic) dependent variable, e.g., p_w, p_a, S_w, θ_w, or h_w, at all points within the modeled domain at some initial time, usually taken as $t = 0$. For example, in terms of p_w, initial conditions may take the form (5.2.5), in which we replace p by p_w.

B. General boundary condition

The discussion in Subs. 5.2.1B is valid also here. In fact, we note that (5.2.7) is already written for the general case of multiple fluid phases. However, we have to be careful in the case of a boundary of discontinuity between two different porous media. The pressure in each fluid must undergo no jump, but, following the discussion in Subs. 6.1.8, *there is always a jump in saturation*, because it is a non-thermodynamic variable.

Following are some of the more commonly encountered boundary conditions for unsaturated flow. In each case, the boundary condition has to be stated in terms of the relevant state variable of the problem. The boundary surface is defined by the equation $F(\mathbf{x}, t) = 0$ discussed in Subs. 5.2.1.

C. Boundary of prescribed saturation, or moisture content

In this case, the external domain imposes a certain saturation on the domain's boundary. In practice, this kind of boundary seldom occurs, except in the case of full saturation, $S_w = 1$, such as when the considered domain is in contact with a body of water (a lake, a river, or a pond). For example, $S_w = 1$ is prescribed on the bottom of a water pond, dictating there a surface at full saturation (even in the limiting case, when a very thin layer of water is present in the pond). Similarly, under the assumption of completely dry soil, the condition $S_a = 1$ is prescribed at ground surface that serves as a boundary to the unsaturated domain.

Another possibility is to prescribe the moisture content, θ_w ($\equiv \phi S_w$). For a deformable porous medium, $\partial\phi/\partial t \neq 0$, it is better to prescribe the boundary conditions in terms of S_w.

When ground surface without ponding serves as the upper boundary for the unsaturated zone, neither the water pressure on it, nor the water saturation are known. The only information that we have is the rate of water infiltration through such boundary (including the case of no-infiltration). This information is then used to specify the boundary condition (see below).

D. Boundary of prescribed pressure, head, or suction

As in the case of prescribed saturation, the value of p_w is seldom known on a boundary, except when a porous medium domain is bounded by a body of water (e.g., a pond). In the latter case, the pressure along the pond's bottom is dictated by the depth of water in the pond. Whenever the density, ρ_w, is constant, the piezometric head, h_w, or the suction, ψ ($\equiv -p_w/\rho_w g$), may also be prescribed on such a boundary.

In air-water flow, atmospheric conditions at ground surface are employed. The condition is specified air pressure, $p_a = p_{\text{atm}}$, or specified matric suction, $\psi = \psi^o$, where Kelvin's law,

$$\psi^o = \frac{RT\rho_w}{M^w} \ln h_r, \tag{6.3.57}$$

is used to relate suction to *relative humidity*, h_r, in the atmosphere. In this equation, R is the universal gas constant, T is the temperature, and M_w is the molecular weight of water.

Another type of boundary condition at ground surface, for air-water flow in the unsaturated zone, is $p_a = p_{atm}$ for the air phase, and a known flux for the water phase. Yet, another type of boundary condition, appropriate under a pond, is known water pressure, and no flux of air phase.

The boundary condition that specifies the value of S_w, p_w, ψ, or h_w along a boundary segment is a *boundary condition of the first type*, or *Dirichlet boundary condition.*

E. Boundary of prescribed water flux

This case occurs, for example, when water (e.g., from rainfall or irrigation by sprinklers) infiltrates *at a known rate* through ground surface, which serves as a boundary to the unsaturated zone. This includes the case of no-flow through such a boundary.

Ground surface is a *material surface* with respect to the solid, and hence, (5.2.9) is applicable, i.e.,

$$(\mathbf{V}_s - \mathbf{u})\big|_{\text{side 1}} \cdot \mathbf{n} = (\mathbf{V}_s - \mathbf{u})\big|_{\text{side 2}} \cdot \mathbf{n} = 0. \tag{6.3.58}$$

Since the microscopic water-solid and air-solid interfaces are material with respect to fluid mass, (5.2.7) can be written separately for every fluid phase.

With $[\![\, \rho_f \,]\!]_{1,2} = 0$, assuming no dispersive flux of the total mass of a phase, and replacing $(\mathbf{V} - \mathbf{u})$ by $(\mathbf{V} - \mathbf{V}_s) + (\mathbf{V}_s - \mathbf{u})$, the general boundary condition, (5.2.7), takes for such a surface the form:

$$[\![\, \theta_\alpha(\mathbf{V}_\alpha - \mathbf{u}) \,]\!]_{1,2} \cdot \mathbf{n} = 0. \tag{6.3.59}$$

Thus, with (5.2.9), for an α-fluid phase, $\alpha = w, a$, the last equation reduces to the form of (5.2.11), repeated here for convenience as

$$[\![\, \mathbf{q}_{r\alpha} \,]\!]_{1,2} \cdot \mathbf{n} = 0, \quad \text{or} \quad \mathbf{q}_{r\alpha}\big|_1 \cdot \mathbf{n} = \mathbf{q}_{r\alpha}\big|_2 \cdot \mathbf{n}, \tag{6.3.60}$$

in which one of the sides, say side 2, is the external (atmospheric) one. The relative specific flux, $\mathbf{q}_{r\alpha}$, is expressed by an appropriate motion equation.

For an *impervious boundary*, say, a pervious side 1 and an impervious side 2, equation (5.2.11) reduces to (5.2.12). Note that this equation constrains only the normal component of the flux. The tangential components may take on any value, meaning that we may have *slip* along the boundary.

Let $\mathbf{N}$ denote the prescribed flux (say, upward for evaporation and downward for infiltration) on the external side of a stationary ($\mathbf{u} = 0$) ground surface, described by $F = F(\mathbf{x})$, with $\mathbf{n} \equiv \nabla F / |\nabla F|$ denoting the unit outward normal vector to it. We assume that the water density, ρ_w, on the external side is the same as within the unsaturated zone, and that it is a constant. Then, the prescribed flux boundary condition takes the form of (5.2.13), repeated here for convenience as

$$\mathbf{q}_{r\alpha} \cdot \nabla F = N |\nabla F|, \qquad N = \mathbf{N} \cdot \mathbf{n}. \tag{6.3.61}$$

In this equation, $\mathbf{q}_{r\alpha}$ can be expressed by any of the motion equations presented in Chap. 4. For example, in terms of ψ, we use (6.2.9), and (5.2.13) becomes

$$[\mathsf{K}_w(\psi) \cdot \nabla(\psi - z)] \cdot \mathbf{n} = \mathbf{N} \cdot \mathbf{n}. \tag{6.3.62}$$

As this condition specifies the gradient of a scalar variable on the boundary, it is a *boundary condition of the second kind*, or a *Neumann boundary condition.*

For a horizontal ground surface that serves as the upper boundary of the unsaturated zone, $\mathbf{n} \equiv \nabla z$. If the soil is isotropic and the infiltration is vertically downward, i.e., $\mathbf{N} = -N\nabla z$, we obtain from (6.3.62):

$$N = -\mathsf{K}_w(\psi)\frac{\partial \psi}{\partial z} + \mathsf{K}_w(\psi), \tag{6.3.63}$$

or, in terms of θ_w and using (6.2.15), we obtain

$$N = D_w(\theta_w)\frac{\partial \theta_w}{\partial z} + \mathsf{K}_w(\theta_w). \tag{6.3.64}$$

The boundary condition (6.3.63) or (6.3.64) specifies a constraint that is a combination of ψ (or θ_w) and $\nabla\psi$ (or $\nabla\theta_w$). This is a *boundary condition of the third type*, or a *Robin boundary condition.*

The condition of prescribed flux provides no explicit information on the values of the state variables, say θ_w or ψ, at (i.e., just inside) the boundary. These values will adjust themselves (thus modifying also the values of the effective hydraulic conductivity, $\mathsf{K}_w(\psi)$) to accommodate the specified rate of flow through the boundary. The flux through a ground surface that serves as the upper boundary of the unsaturated zone, requires special attention.

F. Infiltration and evaporation at ground surface

This type of boundary always occurs when ground surface serves as the upper boundary of a modeled subsurface unsaturated domain (= vadose zone). For the air, we can specify atmospheric pressure as a known value. However, for the water, neither the pressure nor the saturation is known. The boundary condition to be used in such a case is that of specified flux (due to infiltration or evaporation), with the special case of zero flux when no infiltration from precipitation or from irrigation takes place. There are two approaches that one can take: (1) specify the net infiltration flux (= precipitation minus evaporation), if it is known, or (2) model the evaporation by specifying the relative humidity, h_r, in the air close to ground surface, and use Kelvin's law in (6.3.57) to specify the resulting suction in the water that occupies the void space close to ground surface. The specified suction is then used as a first type boundary condition. The second approach usually requires that thermal effects be also modeled.

To gain some understanding about what is involved in determining infiltration, let us start with a few remarks on infiltration. More information on this subject may be found in standard texts on hydrology (e.g., Chow, 1964; Maidment, 1993; Bras, 1990).

Depending on the local conditions during a storm, e.g., rate of precipitation, type of soil, vegetation cover, surface topography, climatic conditions, antecedent soil moisture, etc., part of the precipitation reaching ground surface *infiltrates* through the latter and continues to percolate downward towards an underlying water table. The remaining part will either pond above ground surface, or become *surface runoff*. Infiltration also takes place from water storage ponds and infiltration basins used for artificially recharging the underlying aquifer (Subs. 3.4). When the supply of water to ground surface exceeds the rate of infiltration, which is commensurate with the properties of the soil and its moisture content, the infiltration rate is the maximum possible one, under the prevailing conditions. This rate is referred to as *infiltration capacity*, I_c.

The processes of infiltration and percolation are accompanied by changes in saturation within the unsaturated zone. The redistribution of moisture in the subsurface will continue for some time after cessation of infiltration. This process, its relationship to the rate of infiltration, and to the rate at which water is applied at ground surface, are of major interest in the design of irrigation systems. Much attention has, therefore, been devoted to these topics by soil scientists. Traditionally, however, hydrologists, who only need an estimate of the *annual or seasonal natural replenishment* of an aquifer from precipitation (and, perhaps, from return flow of excess irrigation) in order to determine its *sustainable yield* (Subs. 1.1.6), have not been interested in infiltration from individual storm events. They obtained estimates of annual or seasonal aquifer *natural replenishment*, by considering the aquifer's water balance. To them, the time scale of interest is the season, or the year, and they are interested only in the average *net* rate at which water from precipitation reaches the water table and replenishes an aquifer. In irrigation, the time scale is much shorter; sometimes hours, days, or weeks during an irrigation event, or between successive irrigation events, as they are interested in variations in moisture content only in the root zone.

This situation has completely changed as attention has focused on contaminant transport, especially in the vadose zone. When dealing with the downward movement of contaminants through this zone, we need information on the rate at which water infiltrates, acting as the carrier of the contaminants. Usually, an estimate of this movement can be obtained by considering the annual (or seasonal) average rate of infiltration. However, in a semi-arid region, most of the infiltration that recharges an underlying aquifer may be produced by a small number of relatively large storms, with long dry periods between them. Sometimes, a large storm occurs once in a number of years. During a dry period, evaporation (and evapotranspiration) may remove all or a large portion of the moisture that has infiltrated during the antecedent storms. Therefore, in such a region, a significant difference may exist between the estimate of contaminant movement as obtained by using a long term average annual infiltration rate, as is usually done for estimating the aquifer's sustainable yield, and the one which takes into account the effect of individual short duration storms. Thus, we may have to consider the random occurrence of annual rains, with certain statistical features, and the statistics of individual storms within the year. The question we face is the effect of such rainfall events, followed by dry periods, on the estimation of infiltration (which, in turn, affects contaminant transport).

In principle, the rate of (actual) infiltration, $I = I(t)$, should be derived by solving a model of flow in the unsaturated zone, subject to appropriate boundary conditions, especially at ground surface. The properties of the soil and the precipitation characteristics will be represented in such a model. However, the rate of infiltration to be used as a boundary condition *is not known*, as it seldom equals the rate at which water is applied at ground surface. This fact, combined with the difficulties inherent in solving such a problem, has made this approach impractical, at least until recent years. Instead, in the practice of hydrology, various empirical formulas have been employed. Such formulas include coefficients that describe the type of soil and the moisture conditions that prevail at the beginning of a storm event. For such conditions, the empirical formulas suggest some initial infiltration capacity rate, I_o (that depends on the kind of soil and on the initial soil moisture) and dictate how this rate decreases as the storm continues and the soil's saturation increases. Obviously, such formulas are valid for a recharge rate that is higher than I_o. The formulas suggest a gradual reduction in the *infiltration capacity*, $I_c(t)$, as time progresses, until the latter levels off at a value, $I_{c\infty}$, that corresponds to a saturated soil, at least close to ground surface. Usually, the rate of attenuation in I_c, is related also to soil surface conditions, e.g., the vegetation, and to rainfall intensity.

An exact analytical solution of a model of flow in a vertical soil column, subject to recharge conditions at ground surface is not possible, due to its nonlinearity. In a series of papers, Philip (1957a–e, 1958a,b; 1969) proposed an approximate solution for the moisture distribution in the soil in the form of an infinite series (Subs. 6.4.1). Based on this solution, the infiltration capacity, $I_c(t)$, may be estimated from the formula

$$I_c = I_{c\infty} + \frac{s}{2\sqrt{t}}, \tag{6.3.65}$$

in which s is a coefficient that depends on the initial soil moisture. We note that initially, $I_c = \infty$. In the limit, as $t \to \infty$, $I_c(t) \to I_{c\infty}$. Figure 6.3.1 shows the effect of the initial soil moisture on the infiltration capacity. It can easily be shown that as infiltration proceeds, in the absence of ponding and air entrapment, the soil close to ground surface becomes saturated, and $I_c(t) \to \mathsf{K}|_{\text{sat}}$, i.e., approaches the hydraulic conductivity at saturation. At the same time, the saturation gradient at ground surface vanishes asymptotically. The combination of these two processes leaves gravity as the only driving force, with the result that the flux becomes equal to $\mathsf{K}|_{\text{sat}}$. At shorter times, $I_c(t) > \mathsf{K}|_{\text{sat}}$, because of an additional force due to capillarity.

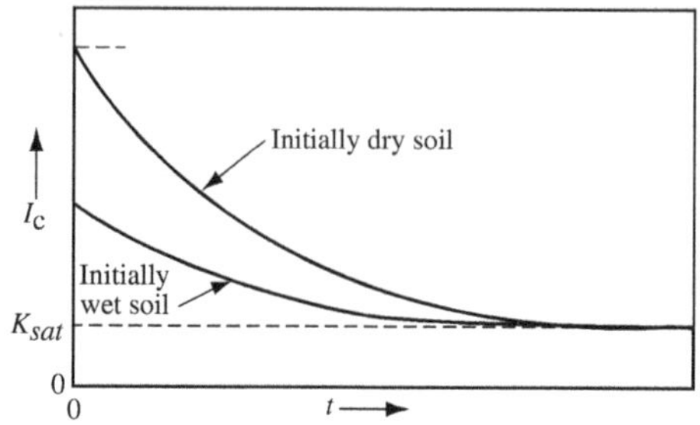

Figure 6.3.1: Effect of initial soil moisture on the rate of infiltration.

One of the better known formulas for estimating infiltration is the *Horton infiltration equation* (Horton, 1940)

$$I_c(t) = I_{c\infty} + (I_o - I_{c\infty})e^{-\alpha t}, \tag{6.3.66}$$

in which α is a parameter, which, together with I_o and $I_{c\infty}$, has to be estimated from observed data by some calibration procedure.

Figure 6.3.2 shows a typical infiltration capacity curve for a rainfall of variable intensity. From this figure, it follows that, at first, actual infiltration (shaded area) equals the rate of rainfall, $R(t)$. At some point, $t = t_i$, at which rainfall exceeds infiltration, only part of the rainfall infiltrates. Eventually, as rainfall continues, the rate of infiltration approaches the value of the infiltration capacity asymptotically. Obviously, if rainfall reduces to below the infiltration capacity, the actual infiltration must equal the rainfall.

The Green-Ampt model (Green and Ampt, 1911) assumes 'piston' flow, with a sharp wetting front between the infiltration zone and the zone of soil at the initial water content. The wetted zone increases in length as infiltration progresses. Additional infiltration models can be found in the literature, for example, Morel-Seytoux (1973) and Espinoza (2006).

Our intention here is not to discuss these models in detail, but to indicate that knowing the rate of rainfall, or irrigation, does not imply that the rate of infiltration is also known, and can, therefore, be used as a specified flux boundary condition at ground surface. As we shall see below, conditions at ground surface that control infiltration may vary to the extent that boundary conditions in a model may have to be switched from one type to another as infiltration progresses.

With this introduction in mind, we can now proceed to discuss the infiltration boundary condition.

Let $R(\mathbf{x}, t)$ (≥ 0) denote the rate at which water is applied in a downward direction to ground surface by precipitation, or irrigation, with $\mathbf{x}$ denoting points on ground surface. We start by assuming that this rate is also equal to the rate of (vertically downward) infiltration, i.e., $I(\mathbf{x}, t) = R(\mathbf{x}, t)$. We are faced with two important questions:

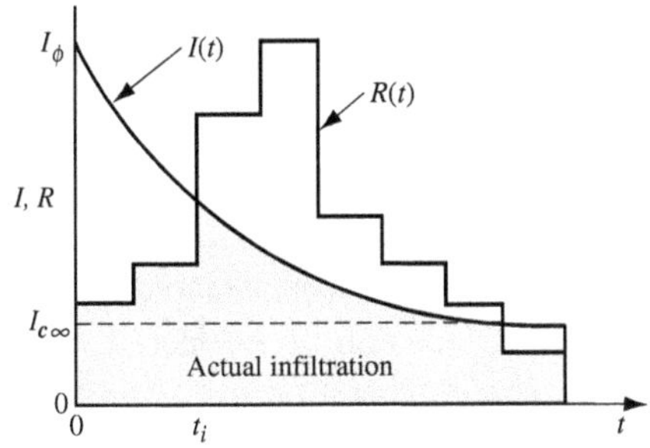

Figure 6.3.2: Infiltration capacity as rainfall exceeds infiltration capacity values.

- Under what conditions is this assumption valid?
- What do we do when it is not valid, viz., we know R, but not I?

Consider the case of rainfall at a rate R over a horizontal ground surface, resulting in infiltration at a rate I; the soil is assumed to be isotropic. The boundary condition, say (6.3.63), may then be rewritten as:

$$I = -\mathsf{K}_w(\psi)\frac{\partial \psi}{\partial z} + \mathsf{K}_w(\psi), \tag{6.3.67}$$

or, in terms of θ_w:

$$I = D_w(\theta_w)\frac{\partial \theta_w}{\partial z} + \mathsf{K}_w(\theta_w). \tag{6.3.68}$$

From our discussion so far, it follows that at any instant, depending on soil properties and on the prevailing moisture content and distribution, the soil just below ground surface can transmit only a certain flux of water, provided such water quantity is applied at ground surface. If the rate of application is higher, the difference will pond on ground surface, or produce surface runoff. The behavior above and below ground surface is, thus, coupled by the common condition at ground surface. To avoid modeling what happens above ground surface, the verbal constraint is often added that 'no ponding of water is allowed to take place above ground surface'. In reality, because of ground surface roughness, some ponding may take place before surface runoff actually begins. The no-ponding constraint limits the rate of infiltration at every instant to what the soil can transmit at the prevailing saturation and saturation gradient conditions. This constraint has to be incorporated in the statement of the boundary conditions at ground surface.

From (6.3.68), it follows that as the soil just below ground surface approaches full saturation, overlooking entrapped air, $\theta_w \to \phi$ (or $\theta_w \to \phi - \theta_{ao}$, if entrapped air is considered), and $\partial\theta_w/\partial z \to 0$, the first term on the right-hand side vanishes, while the second one approaches the value of the saturated hydraulic conductivity, $\mathsf{K}\big|_{\text{sat}}$ (or, with entrapped air, $\mathsf{K}\big|_{(\phi-\theta_{ao})}$). In an anisotropic porous medium, it will approach the value of the vertical component, $\mathsf{K}_z\big|_{\text{sat}}$. In the current discussion on infiltration, for brevity, we shall use $\mathsf{K}_z\big|_{\text{sat}}$ to mean $\mathsf{K}\big|_{(\phi-\theta_{ao})}$, when entrapped air may be present.

Let us consider two situations. In both cases, we shall assume that R is constant, and that initially the soil is relatively dry, say at *field capacity*. We shall assume that the water table is sufficiently deep so that its influence on soil moisture near ground surface can be neglected.

CASE A. Precipitation is applied at a constant rate $R > \mathsf{K}_z\big|_{\text{sat}}$. At first, even at low saturations, the soil can absorb the incoming water at a very high rate, as the gradient in moisture content that is produced at ground surface is very high. Theoretically, this rate is infinite at the initial time. Thus, for a certain period, we have $I = R$. We use this value in the boundary condition (6.3.68). During this initial period, as a wetting front advances downward, the infiltrating water produces two phenomena:

- At ground surface, and just below it, the water content in the soil gradually increases. In the limit, full saturation may be reached (i.e., water content equals porosity). The increase in water content is accompanied by an increase in effective hydraulic conductivity, up to the limiting value of $\mathsf{K}_z\big|_{\text{sat}}$, corresponding to a saturated soil.
- As water percolates downward, the gradient in the soil's water content close to ground surface decreases with time. In the limit, this gradient at ground surface approaches the value of zero, so that gravity remains the only driving force there.

The initial period continues until the soil at ground surface reaches a point at which the combination of saturation (dictating the moisture diffusivity and effective hydraulic conductivity) and saturation gradient are such that the soil can no longer transmit water at the rate applied at ground surface. This occurs when the pressure in the water occupying the pore space at ground surface approaches atmospheric pressure. With $p_w = p_a\big|_{\text{atm}}$, we also have full water saturation (or practically so) at ground surface. During this period, the infiltration rate remains constant, equal to the (assumed constant) rate of accretion.

Actually, because of the (discontinuous, i.e., composed of isolated air bubbles) *entrapped air* that remains in the void space, the soil will never reach full saturation. Although the volumetric fraction of entrapped air depends on the wetting-drainage history, we shall approximate it as a constant, θ_{ao}. Thus, upon rewetting, the moisture content of water cannot exceed $\phi - \theta_{ao}$.

In our model, the initial period continues as long as $\theta_w < \phi - \theta_{ao}$, or, equivalently, as long as $p_w < p_a\big|_{\text{atm}}$. Once full saturation is reached ($\theta_w = \phi - \theta_{ao}$), we have to replace (6.3.68) by the first type boundary condition

$$p_w = p_a\big|_{\text{atm}}, \tag{6.3.69}$$

(where we often assume $p_a\big|_{\rm atm} = 0$ as the datum) corresponding to a condition just below a state of zero ponding depth.

Next, we calculate the rate of infiltration, $I(t)(< R)$ by substituting the solutions for $\theta_w(t)$ in (6.3.68). We would then observe a gradual reduction in $I(t)$, approaching the limiting value of $\mathsf{K}\big|_{(\phi-\theta_{ao})}$.

Let us now consider the time-varying case $R = R(t)$. The discussion presented so far remains valid. With the resulting value of $I(t)$, we should keep track of whether the calculated rate of infiltration, I, is less than that of application, $R(t)$. As soon as we reach the situation of $I \geq R$, we should switch back to the condition (6.3.68), with $I = R$.

It is important to emphasize again that initially, and for some time (which may be significant when considering irrigation, or an individual storm event), the rate of infiltration will exceed the limiting value of $\mathsf{K}_z\big|_{\rm sat}$. The latter is approached *from above*, provided the rate of application of water to ground surface remains larger than the rate of infiltration.

CASE B. Precipitation is applied at a constant rate $R < \mathsf{K}_z\big|_{\rm sat}$. Initially, the saturation gradient at ground surface will be very high, and the soil will absorb all incoming water. However, rather rapidly, as infiltration continues, and saturation at ground surface increases, the saturation gradient there will decrease. Eventually, asymptotically, a saturation level is reached with a zero saturation gradient, so that the rate of infiltration (which equals to the rate of application) becomes equal to the effective hydraulic conductivity at the prevailing saturation. Under such conditions, the only force driving infiltration at ground surface is gravity. Capillarity still plays a role in the part of the wetting front that is ahead of the (practically) saturated zone.

For $R = R(t) < \mathsf{K}_z\big|_{\rm sat}$, the same phenomena, as described above, will occur, viz., the rate of infiltration will equal that of accretion, but saturation at ground surface will vary, without leveling off at any asymptotic value.

These phenomena have been known to hydrologists for many years. Here, we have expressed them as constraints associated with the boundary conditions at ground surface.

We may summarize the discussion of the above two cases as follows:

(a) If $p_w(t) = p_a\big|_{\rm atm}$, then,

- If $R(t) \geq I(t)$, use (6.3.69).
- If $R(t) < I(t)$, use (6.3.68) with $I = R(t)$.

(b) If $p_w(t) < p_a\big|_{\rm atm}$, use (6.3.68) with $I = R(t)$.

Let us comment about the case of evaporation or evapotranspiration, produced by solar radiation reaching ground surface. Also in this case, the boundary conditions must be based on the equality of water and water vapor fluxes across ground surface. In this case, as water leaves the soil in the form of vapor, the soil dries out. As the saturation at (i.e., just below) ground surface reaches the irreducible saturation level, S_{wr}, water effective permeability reduces to zero. In the isothermal flow models discussed here, it is usually assumed that when the soil at ground surface reaches S_{wr}, it can no longer transmit liquid water. The boundary condition has to be switched to one of no-flow at ground surface until the water saturation rises above S_{wr}. In more sophisticated, say, nonisothermal models, which are beyond the scope of this book, the water saturation can reduce to below the irreducible value. Obviously, as saturation drops to *below* the irreducible one, e.g., by evaporation and/or root uptake, the effective permeability to water vanishes. Under nonisothermal conditions, the situation may be more complicated, as a *drying front* may move up and down below ground surface.

G. Ponding

Finally, let us consider the possibility of ponding. This occurs when the liquid's pressure at ground surface satisfies the condition $p_w > p_a\big|_{\text{atm}}$. Suppose we allow ponding up to a maximum depth that can be specified as $p_w\big|_{\max}/\rho_w g$. Instead of condition (6.3.69) at $S_w = 1$, we treat $p_w|_{z=0}$ as an unknown, and introduce the condition

$$R(t) - I(t) \equiv R(t) - \frac{\mathsf{k}_z\big|_{\text{sat}}}{\mu_w}\left(\frac{\partial p_w}{\partial z} + \rho_w g\right) = \frac{1}{\rho_w g}\frac{\partial}{\partial t}\left(p_w\big|_{z=0} - p_a\big|_{\text{atm}}\right), \tag{6.3.70}$$

allowing $p_w\big|_{z=0}$ to rise up to $p_w\big|_{\max} + p_a\big|_{\text{atm}}$. We usually assume $p_a\big|_{\text{atm}} = 0$.

We do not allow p_w to rise above the specified maximum. If, as a result of $R(t)$, the water level tends to rise higher, it is set at the maximum value. As it drops and reaches zero, or atmospheric gas pressure, we switch to the condition (6.3.69), as long as $S_w = 1$.

We may also assume that ponding occurs *within* an external soil domain that may have different soil characteristics. Defining a specific yield of such a soil by S_y (similar to its definition for a phreatic aquifer), the boundary condition with ponding may be rewritten as

$$R(t) - I(t) \equiv R(t) - \frac{\mathsf{k}_z\big|_{\text{sat}}}{\mu_w}\left(\frac{\partial p_w}{\partial z} + \rho_w g\right) = S_y\frac{1}{\rho_w g}\frac{\partial}{\partial t}\left(p_w\big|_{z=0} - p_a\big|_{\text{atm}}\right). \tag{6.3.71}$$

6.3.3 Complete flow model

The discussion in Sec. 5.3, presented with respect to saturated flow, is valid also here. In Subs. 5.3.1, we started by understanding what are the requirements for constructing a *well-posed problem*, or a *well-posed model.* Needless to say that the entire discussion on this subject is valid also for modeling single or two phase flow in the unsaturated zone. Similarly, the presentation on the content of the conceptual model and on the complete model are also the same as in saturated flow models, except that if we consider air-water flow, we have two balance equations, appropriate constitutive relations for the two fluids, appropriate boundary conditions for the two equations, etc.

The discussion on primary variables in Sec. 7.9.4 is also valid for the case of two phases (air-water flow) considered here. We note that both air and water are single component phases. Thus, in our model of a non-deformable porous medium, under isothermal conditions, using Darcy's law to describe the velocity of the two phases, we have: NP $= 2$, NC $= 2$, and from (7.9.116) we obtain NF $= 2$. We may select any two variables, from the list S_w, S_a, p_w, p_a, ρ_w, ρ_a, $\mathbf{q}_w, \mathbf{q}_a$, provided that they are *independent* of each other. For example, we may select p_w and S_w and solve the mass balance equations (6.3.7), the first with boundary conditions on the flow of water, and the second for the flow of air. The other six equations are used to obtain the remaining six variables (at every time step). Depending on the numerical model and the code used for its solution, there is no need to write the two balance equations *explicitly* in terms of the two selected variables.

6.4 Methods of Solution

The flow equations presented in Subs. 6.3.1A, e.g., Richards' equation, (6.3.36), as well as the boundary conditions discussed in Subs. 6.3.2, are *nonlinear*, due to the dependence of the effective permeability on saturation. Another difficulty associated with unsaturated flow is that the two functional relationships, $\mathsf{K}_w = \mathsf{K}_w(S_w)$ and $p_c = p_c(S_w)$, appearing in the flow model, cannot be easily determined experimentally, and are subject to hysteresis (Subs. 6.1.7). These are fundamental features of multiphase flow models that make the solution of unsaturated flow problems very difficult.

Because of its nonlinearity, an analytical solution of unsaturated flow models is, generally, not feasible, even in the simple one-dimensional case. Nevertheless, there have been a number of attempts to develop analytical, semi-analytical, or approximate, solutions, such as those based on the series expansion technique of Phillip (1957a–e, 1958a, b; 1969), and the integral method of Parlange (1971, 1972). Although these and other methods have been applied to derive approximate analytical solutions for multiphase flow (Bear *et al.*, 1968; Philip, 1969; Braester, 1973; Parlange *et al.*, 1985; Broadbridge and White, 1988; Haverkamp *et al.*, 1990; Barry and Sander, 1991; Warrick *et al.*, 1991; Serrano, 1998; Si and Kachanoski, 2000; Lu *et al.*, 2007; Mollerup, 2007; Mollerup and Hansen, 2007; Nasseri *et al.*, 2008), with the present-day capabilities of numerical method, their usefulness has diminished.

In what follows, we shall briefly describe a few methodologies used for finding approximate semi-analytical solutions, as well as a few exact solutions, based on some special soil-water constitutive relations. The numerical solution technique is also briefly commented. A more extensive discussion of numerical solution is given in Chap. 8.

6.4.1 Analytical solutions

Let us examine the flow equation (6.3.35), expressed in terms of the suction head, rewritten in the form

$$\nabla \cdot [\mathsf{K}_w(\psi_w)\nabla\psi_w] - \frac{\partial \mathsf{K}_w}{\partial z} = -\frac{\partial \theta_w}{\partial t}. \tag{6.4.1}$$

This type of equation is known as the *Fokker-Planck equation.*

One attempt to find analytical solution of (6.4.1) is to *linearize* it by assuming the following moisture retention and effective hydraulic conductivity constitutive relations

$$\mathsf{K}_{rw} = \frac{\mathsf{K}_w}{\mathsf{K}_{\text{sat}}} = e^{-\alpha\psi_w}, \tag{6.4.2}$$

$$\theta^* = \frac{\theta_w - \theta_{wr}}{\theta_{\text{sat}} - \theta_{wr}} = e^{-\alpha\psi_w}, \tag{6.4.3}$$

where θ_{wr} and $\theta_{\text{sat}}(\equiv \phi)$ are, respectively, the irreducible and the saturation ($\equiv$ porosity) water content. We also note that (6.4.2) is identical to the Gardner (1958) model, (6.2.18). The above requirement is obviously rather restrictive. However, it allows the simplification of the governing equation (Srivastava and Yeh, 1991). By substituting (6.4.2) and (6.4.3) into (6.4.1), we obtain

$$\nabla^2 \mathsf{K}_w + \alpha\frac{\partial \mathsf{K}_w}{\partial z} = \frac{\alpha(\theta_{\text{sat}} - \theta_{wr})}{\mathsf{K}_{\text{sat}}}\frac{\partial \mathsf{K}_w}{\partial t}. \tag{6.4.4}$$

In a one-dimensional geometry, analytical solution can be found for homogeneous or layered homogeneous soil (Srivastava and Yeh, 1991). Once $\mathsf{K}_w(z,t)$ is solved for, the suction head is easily found through (6.4.2).

As an example, let us consider a case of steady flow in an inhomogeneous soil. Equation (6.3.35) then takes the form

$$\nabla\cdot[\mathsf{K}_{\text{sat}}(\mathbf{x})\,\mathsf{K}_{rw}(\mathbf{x},\psi_w)\nabla(\psi_w - z)] = 0, \tag{6.4.5}$$

where we have expressed the unsaturated hydraulic conductivity K as the product of a saturated hydraulic conductivity K_{sat}, which, for a heterogeneous medium, is a function of space, and a relative hydraulic conductivity K_{rw}, which is a function of both the suction head, ψ_w, and space (see Subs. 6.2.3). Similar to (6.4.2), we shall assume that the relative hydraulic conductivity can be expressed in terms of the Gardner (1958) model, as

$$\mathsf{K}_{rw} = e^{-\alpha(\mathbf{x})\psi_w}. \tag{6.4.6}$$

We note that here α, a soil index parameter related to pore size distribution, is a function of space. One way to simplify (6.4.5) is to apply Kirchhoff transform (Kirchhoff, 1894; Crank, 1956). In this transform, we assume that there exists a variable F, defined as

$$F(\mathbf{x}) = \int_{\psi_w(\mathbf{x})}^{\infty} \mathsf{K}_{rw}(\mathbf{x}, \psi')\, d\psi' = \frac{1}{\alpha(\mathbf{x})} e^{-\alpha(\mathbf{x})\psi_w(\mathbf{x})}. \tag{6.4.7}$$

We can see from the above that $\alpha F = \mathsf{K}_{rw}$. It can be shown that this new variable satisfies the linear PDE

$$\nabla \cdot [\mathsf{K}_{\text{sat}}(\mathbf{x})\, \nabla F(\mathbf{x})] + \frac{\partial}{\partial z} [\alpha(\mathbf{x}) \mathsf{K}_{\text{sat}}(\mathbf{x})\, F(\mathbf{x})] = 0. \tag{6.4.8}$$

For a homogeneous porous medium, this equation further reduces to

$$\nabla^2 F + \alpha \frac{\partial F}{\partial z} = 0, \tag{6.4.9}$$

known as the *Berger equation.* This equation is similar to the steady state form of (6.4.4). In order to solve (6.4.9), we need to transform also the boundary conditions. For example, for the suction head and the flux type boundary conditions, we have,

$$\psi_w = \Psi, \tag{6.4.10}$$
$$-\mathbf{q} \cdot \mathbf{n} = -\mathsf{K}_{\text{sat}}\, \mathsf{K}_{rw} \nabla(\psi - z) \cdot \mathbf{n} = Q. \tag{6.4.11}$$

These, in turn are, respectively, transformed into

$$F = \frac{1}{\alpha} e^{-\alpha \Psi}, \tag{6.4.12}$$
$$\mathsf{K}_{\text{sat}} \nabla F \cdot \mathbf{n} + \alpha\, n_z \mathsf{K}_{\text{sat}} F = Q, \tag{6.4.13}$$

where n_z is the z component of the outward unit normal vector, $\mathbf{n}$.

Next, we examine the moisture diffusivity equation, (6.3.37), which is a *nonlinear diffusion equation.* When the porous medium domain is homogeneous and unbounded in one of the spatial directions (such as a semi-infinite space), it is possible that the water content profile, $\theta(\mathbf{x}, t)$, becomes *self-similar.* By self-similar, we mean that the water content profile, rather than being a function of multiple, independent space and time variables, can be expressed as a function of a single, combined space-time variable. For example, we can assume that the water content profile, $\theta_w = \theta_w(x, t)$, can be expressed as a function of a single variable, $\eta = x/\sqrt{t}$, such that $\theta_w = \theta_w(\eta)$. If this is true, then the PDE (6.3.37) can be transformed into a (nonlinear) ordinary differential equation having the following form

$$\frac{d}{d\eta} \left[D_w(\theta_w) \frac{d\theta_w}{d\eta} \right] + \frac{\eta}{2} \frac{d\theta_w}{d\eta} = 0. \tag{6.4.14}$$

The above transformation is known as *Boltzmann transform.* This ordinary differential equation can be solved using some approximation method. For problems with radial symmetry, the Boltzmann transform can be applied with the variable $\eta = r/\sqrt{t}$ or $\eta = R/\sqrt{t}$, where r and R are the radial coordinate in cylindrical and spherical coordinates, respectively.

We note that two conditions must be satisfied for the Boltzmann transform to be valid, that is, θ_w is indeed a function of a single variable η: the initial condition $\theta_w(x,0)$, as well as the boundary condition $\theta_w(x_b,t)$, where x_b is the boundary location, must be functions of that single variable η only. For any problem with a non-zero, or non-constant, initial condition, the Boltzmann transform is, generally, not valid. This requirement has limited the application of this method.

When the gravity term is present, we refer to the full Richards' equation, (6.3.36), but in one-dimension,

$$\frac{\partial \theta_w}{\partial t} = \frac{\partial}{\partial z}\left[D_w(\theta_w)\frac{\partial \theta_w}{\partial z}\right] + \frac{d\mathsf{K}_w(\theta_w)}{d\theta_w}\frac{\partial \theta_w}{\partial z}. \tag{6.4.15}$$

Philip (1957a–e, 1958a, b, 1969) employed a power series in $t^{1/2}$,

$$z(\theta_w,t) = \phi_1(\theta_w)\,t^{1/2} + \phi_2(\theta_w)\,t + \phi_3(\theta_w)\,t^{3/2} + \ldots, \tag{6.4.16}$$

where $\phi_1 \gg \phi_2 \gg \phi_3 \gg \ldots$. Substituting (6.4.16) into (6.4.15), sorting terms of the same power of t, and integrating with respect to θ_w, a set of integrodifferential equations are obtained (Phillip, 1969):

$$\int_{\theta_i}^{\theta_w} \phi_1(\theta')\,d\theta' = -2\frac{D_w}{d\phi_1/d\theta_w}, \tag{6.4.17}$$

$$\int_{\theta_i}^{\theta_w} \phi_2(\theta')\,d\theta' = \frac{D_w(d\phi_2/d\theta_w)}{(d\phi_1/d\theta_w)^2} + (\mathsf{K}_w - \mathsf{K}_i), \tag{6.4.18}$$

$$\int_{\theta_i}^{\theta_w} \phi_3(\theta')\,d\theta' = \frac{2D}{3}\left[\frac{d\phi_3/d\theta_w}{(d\phi_1/d\theta_w)^2} - \frac{(d\phi_2/d\theta_w)^2}{(d\phi_1/d\theta_w)^3}\right], \tag{6.4.19}$$

and so on. In the above, θ_i is the initial uniform water content, and $\mathsf{K}_i = \mathsf{K}(\theta_i)$ is the initial hydraulic conductivity. These equations are solved by a finite difference approximation and numerical integration.

The solution methods discussed above have been widely used to solve unsaturated flow problems by analytical, semi-analytical, or by numerical methods, e.g., Gardner (1958), Knight and Philip (1974), Broadbridge and White (1988), Broadbridge *et al.* (1996), Yeh (1989), Ross and Bristow (1990), Srivastava and Yeh (1991), Tracy (1995, 2006), Tartakovsky *et al.* (1999), Williams (2000), Serrano (2004), and Ji *et al.* (2008).

Let us consider a three special cases in which an exact solution is possible.

A. Constant diffusivity

As the simplest case, we shall linearize the Richards' equation (6.4.15) by assuming that the diffusivity, $D_w(\theta_w)$, is a constant, and that the hydraulic conductivity K_w is a linear function of θ_w, i.e.,

$$D_w(\theta_w) = D_o, \qquad \frac{d\mathsf{K}_w(\theta_w)}{d\theta_w} = \frac{\mathsf{K}_w(\theta_1) - \mathsf{K}_w(\theta_o)}{\theta_1 - \theta_o} = k, \tag{6.4.20}$$

where θ_o and θ_1 may be considered as the initial and the final water contents of the considered transient problem. Equation (6.4.15) then becomes

$$\frac{\partial \theta_w}{\partial t} = D_o \nabla^2 \theta_w + k\frac{\partial \theta_w}{\partial z}. \tag{6.4.21}$$

This linear PDE can be solved by analytical, as well as numerical methods.

Assuming a one-dimensional, absorption problem, that is, neglecting the gravity effect, we obtain the following equation form (6.4.21)

$$\frac{\partial \theta_w}{\partial t} = D_o \frac{\partial^2 \theta_w}{\partial z^2}. \tag{6.4.22}$$

Consider the case of vertical infiltration into the ground under ponded (saturated) conditions at ground surface (Subs. 6.3.2C), which means that rainfall intensity exceeds maximum infiltration rate. This means,

$$\begin{aligned} \theta_w(z, 0) &= \theta_i, \quad z \le 0, \\ \theta_w(0, t) &= \theta_{\text{sat}}, \quad t > 0, \end{aligned} \tag{6.4.23}$$

where θ_i is the initial constant water content in the soil, and θ_{sat} ($\equiv$ porosity ϕ) is the saturation water content. The solution of (6.4.22) and (6.4.23) (Carslaw and Jaeger, 1959) is expressed in the normalized form,

$$\theta^* = \frac{\theta_w - \theta_i}{\theta_{\text{sat}} - \theta_i} = \text{erfc}\left[\frac{|z|}{2\sqrt{D_o t}}\right], \tag{6.4.24}$$

where erfc is the *complementary error function.*

We notice that the solution, presented in terms of the normalized water content, θ^*, is indeed self-similar, meaning that $\theta^* = \theta^*(\eta)$, with $\eta = z/\sqrt{t}$. Hence, this problem could also be solved by utilizing the Boltzmann transform introduced above, such that (6.4.14) is simplified to

$$D_o \frac{d^2\theta^*}{d\eta^2} + \frac{\eta}{2}\frac{d\theta^*}{d\eta} = 0, \tag{6.4.25}$$

the solution of which is $\theta^* = \text{erfc}(|\eta|/2\sqrt{D_o})$.

In Fig. 6.4.1, we plot this self-similar profile as $\theta^* = \text{erfc}(\eta^*)$, where $\eta^* = |z|/(2\sqrt{D_o t})$. We observe an *invasion front* type of profile for water content, from 100% saturation at ground surface, $\theta^*(0) = 1$, dropping down to a very small amount at $\eta^* = 2$, $\theta^*(2) = 0.00468$. Here we observe that $\eta^* = 0$ corresponds to $z = 0$, and $\eta^* = 2$ can be expressed as

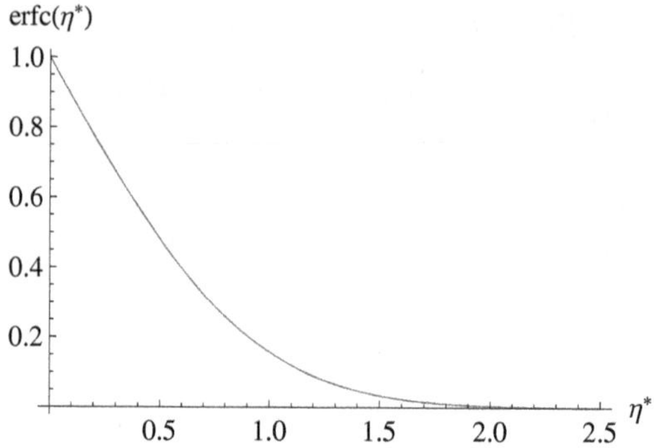

Figure 6.4.1: Complementary error function.

$$|z| = 4\sqrt{D_o t}. \tag{6.4.26}$$

Hence, this self-similar water content profile is growing in size (penetration distance, $|z|$) proportional to $\sqrt{t}$.

From the above solution, we can determine the flux entering the soil (*infiltration capacity*) as

$$I_c = - \left. q_z \right|_{z=0} = D_o \left. \frac{\partial \theta_w}{\partial z} \right|_{z=0} = (\theta_{\text{sat}} - \theta_i) \sqrt{\frac{D_o}{\pi}}\, t^{-1/2}. \tag{6.4.27}$$

Another option is to (approximately) keep the neglected gravity effect by utilizing (6.2.15) to find the flux. We obtain

$$I_c = \mathsf{K}_{\text{sat}} + (\theta_{\text{sat}} - \theta_i) \sqrt{\frac{D_o}{\pi}}\, t^{-1/2}. \tag{6.4.28}$$

This equation can be compared with the infiltration capacity equation (6.3.66) suggested by Richards.

However, (6.4.28) is not the correct solution that takes the effect of gravity into account. To consider the gravity effect, we can use the (linearized) equation (6.4.21), expressed in the one-dimensional form,

$$\frac{\partial \theta_w}{\partial t} = D_o \frac{\partial^2 \theta_w}{\partial z^2} + k \frac{\partial \theta_w}{\partial z}. \tag{6.4.29}$$

The solution of the this equation, subject to the same initial and boundary conditions, (6.4.23), is

$$\theta^* = \frac{1}{2} \left[\operatorname{erfc} \left(\frac{|z| - kt}{2\sqrt{D_o t}} \right) + \exp \left(\frac{k|z|}{D_o} \right) \operatorname{erfc} \left(\frac{|z| + kt}{2\sqrt{D_o t}} \right) \right]. \tag{6.4.30}$$

We notice that this solution does not fulfill the assumption of self-similarity.

B. Instantaneous source

Knight and Philip (1974), by assuming that the diffusivity takes on a certain form, obtained an exact solution of the one-dimensional nonlinear diffusion equation, (6.3.37),

$$\frac{\partial \theta_w}{\partial t} = \frac{\partial}{\partial x}\left[D_w(\theta_w)\frac{\partial \theta_w}{\partial x}\right], \tag{6.4.31}$$

subject to the instantaneous injection of uniformly distributed moisture over a domain of finite length. First, using the Kirchhoff transform,

$$F(\theta_w) = \int^{\theta_w} D_w(\theta')\, d\theta', \tag{6.4.32}$$

(6.4.31) can be expressed as

$$\frac{\partial F}{\partial t} = D_w \frac{\partial^2 F}{\partial x^2}. \tag{6.4.33}$$

In order to linearize the above equation, it is assumed that the diffusivity can be approximated by the following model:

$$D_w(\theta_w) = \frac{a}{(b-\theta_w)^2}, \tag{6.4.34}$$

where a and b are empirical constants. By defining a new variable

$$X(x,t) = \int_0^x \frac{1}{\sqrt{D_w(x',t)}}\, dx', \tag{6.4.35}$$

(6.4.33) can be transformed into

$$\frac{\partial F}{\partial t} = \frac{\partial^2 F}{\partial X^2} + \frac{2}{\sqrt{a}}\, q_x\big|_{x=0}\, \frac{\partial F}{\partial X}, \tag{6.4.36}$$

where

$$q_x = -\frac{\partial F}{\partial x} = -\frac{dF}{d\theta_w}\frac{\partial \theta_w}{\partial x} = -D_w\frac{\partial \theta_w}{\partial x} \tag{6.4.37}$$

is the specific discharge. Particularly, for problems with the boundary condition $q_x = 0$ at $x = 0$, equation (6.4.36) reduces to

$$\frac{\partial F}{\partial t} = \frac{\partial^2 F}{\partial X^2}, \tag{6.4.38}$$

which is a linear PDE.

Knight and Philip (1974) investigated the problem of an instantaneous injection of moisture at $t = 0$,

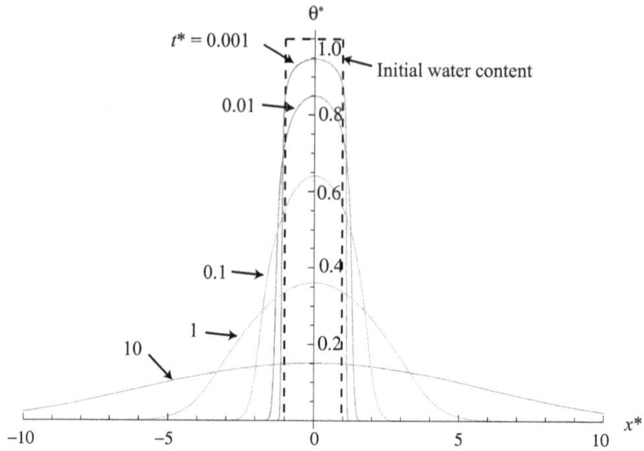

Figure 6.4.2: The spreading of an initial moisture content with time.

$$\theta_w = b, \quad \text{for} \quad |x| \le \frac{Q}{b-\theta_o}, \qquad \theta_w = \theta_o, \quad \text{for} \quad |x| > \frac{Q}{b-\theta_o}, \tag{6.4.39}$$

where θ_o is the background water content, and b is the empirical coefficient in (6.4.34). We note that within $|x| \le Q/(b-\theta_o)$, we have

$$(\theta_w - \theta_o) \times 2(Q/(b-\theta_o)) = (b-\theta_o) \times 2(Q/(b-\theta_o)) = 2Q.$$

Hence, $2Q$ is the volume of infiltrating moisture. Due to the symmetry of the condition about $x = 0$, we can conclude that the condition $q_x = 0$ at $x = 0$ exists; hence, (6.4.38) can be utilized for the solution of the problem. Thus, the exact solution of this problem is given by

$$x^* = 2\sqrt{\left[t^* \ln \frac{1-\theta^*}{\theta^* \sqrt{\pi t^*}}\right]} + \sqrt{\operatorname{erf}\left[\ln \frac{1-\theta^*}{\theta^* \sqrt{\pi t^*}}\right]}, \tag{6.4.40}$$

where

$$\theta^* = \frac{\theta-\theta_o}{b-\theta_o}, \quad x^* = \frac{b-\theta_o}{Q}x, \quad t^* = \frac{a}{Q^2}t. \tag{6.4.41}$$

In Fig. 6.4.2, the transient profile of the normalized moisture content, θ^*, is plotted versus the dimensionless distance x^*, at various dimensionless times, t^*. The solution (6.4.40) also shows that the water content at $x^* = 0$ decays with time as

$$\theta^*(0, t^*) = \frac{1}{1+\sqrt{\pi t^*}}. \tag{6.4.42}$$

C. One-dimensional infiltration

Consider the case of one-dimensional infiltration of water from ground surface, located at $z = L$, to the water table, at $z = 0$. The suction head at the water table is maintained at $\psi_w = 0$. Initially, there is a steady state downward flux of q_A in the soil column. At time $t = 0^+$, the downward flux at the soil surface is suddenly raised to q_B; and we seek the suction head profile along the soil column as a function of space and time, $\psi_w(z, t)$. Or, written explicitly,

$$q_z(z,0) = -q_A, \quad \psi(0,t) = 0, \quad q(L,t) = -q_B. \tag{6.4.43}$$

For the present case, we shall assume that the moisture retention and unsaturated hydraulic conductivity constitutive relations are given by the Gardner (1958) model, (6.4.2) and (6.4.3). The unsaturated flow equation is then given by (6.4.4), which, for one-dimensional flow takes the form,

$$\frac{\partial^2 \mathsf{K}_w}{\partial z^2} + \alpha \frac{\partial \mathsf{K}_w}{\partial z} = \frac{\alpha(\theta_{\text{sat}} - \theta_{wr})}{\mathsf{K}_{\text{sat}}} \frac{\partial \mathsf{K}_w}{\partial t}. \tag{6.4.44}$$

We also note that the specific flux in the z-direction is given by

$$q_z = \mathsf{K}_w \frac{\partial \psi_w}{\partial z} - \mathsf{K}_w = -\frac{1}{\alpha} \frac{\partial \mathsf{K}_w}{\partial z} - \mathsf{K}_w. \tag{6.4.45}$$

Let us introduce the following dimensionless variables

$$\begin{aligned} &z^* = \alpha z, \quad L^* = \alpha L, \quad t^* = \frac{\alpha \, \mathsf{K}_{\text{sat}} \, t}{\theta_{\text{sat}} - \theta_{wr}}, \\ &\mathsf{K}_{rw} = \frac{\mathsf{K}_w}{\mathsf{K}_{\text{sat}}}, \quad q_A^* = \frac{q_A}{\mathsf{K}_{\text{sat}}}, \quad q_B^* = \frac{q_B}{\mathsf{K}_{\text{sat}}}. \end{aligned} \tag{6.4.46}$$

Equation (6.4.44) then becomes

$$\frac{\partial^2 \mathsf{K}_{rw}}{\partial z^{*2}} + \frac{\partial \mathsf{K}_{rw}}{\partial z^*} = \frac{\partial \mathsf{K}_{rw}}{\partial t^*}. \tag{6.4.47}$$

The initial and boundary conditions, (6.4.43), are transformed into

$$\begin{aligned} &\mathsf{K}_{rw}(z^*, 0) = \mathsf{K}_o(z^*) = q_A^* - (q_A^* - 1)e^{-z^*}, \\ &\mathsf{K}_{rw}(0, t^*) = 1, \\ &\left[\frac{\partial \mathsf{K}_{rw}}{\partial z^*} + \mathsf{K}_{rw}\right]_{z^*=L^*} = q_B^*. \end{aligned} \tag{6.4.48}$$

The solution of the above problem, (6.4.47) and (6.4.48), is given by (Srivastava and Yeh, 1991)

$$\mathsf{K}_{rw} = q_B^* - (q_B^* - 1)e^{-z^*} - 4(q_B^* - q_A^*) \, e^{(L^*-z^*)/2} \, e^{-t^*/4}$$

$$\times \sum_{i=1}^{\infty} \frac{\sin(\lambda_i z^*) \sin(\lambda_i L^*) e^{-\lambda_i^2 t^*}}{1 + (L^*/2) + 2\lambda_i^2 L^*}, \tag{6.4.49}$$

where λ_i is the ith positive root of the following equation:

$$\tan(\lambda L^*) + 2\lambda = 0. \tag{6.4.50}$$

According to (6.4.2), the solution in terms of suction head is

$$\psi_w = -\frac{1}{\alpha} \ln \mathsf{K}_{rw}. \tag{6.4.51}$$

6.4.2 Numerical solutions

Numerical solutions of single-phase, unsaturated, or multiphase flows, in multiple spatial dimensions, can be accomplished by one of the standard numerical methods available for solving partial differential equations, such as the finite element, the finite difference, and the finite volume methods. These methods are discussed in some detail in Chap. 8, though not directly in the context of solving multiphase flow.

As the governing equations presented in Sec. 6.3, and in Subs. 6.5.3, are, generally, of the advection-diffusion type (the $\partial/\partial z$ term in the flow equations plays the role of advection), a formal numerical treatment can follow the standard numerical procedure (see, for example, Huyakorn and Pinder (1983)). However, compared to saturated flow and transport problems, the unsaturated and multiphase flow cases have a number of complications. First, the flow equations are nonlinear. In fact, they are highly nonlinear. An explicit time-stepping solution scheme generally leads to instability; hence, an iterative solution scheme is often needed. A second difficulty arises from the fact that multiphase flow is described by several coupled flow equations. Again, an iterative procedure is needed.

Saturated-unsaturated flow can be solved by using a number of commercial and public domain computer codes. In Chap. 8, a few of the codes, such as FEMWATER, FEFLOW, HYDRUS, and SUTRA, are reviewed.

6.5 Some Comments on Three Fluid Phases

The possibility that the void space is occupied by three rather than two fluid phases has already been presented and discussed in Subs. 1.1.5. There, in the discussion on subsurface contamination, we have introduced the third, *nonaqueous*, fluid phase, a NAPL (or DNAPL), as a contaminant. Indeed, in most cases, the NAPL is a contaminant, often a very toxic one. We have also mentioned that although the three phases maintain rather 'sharp' visible microscopic interfaces between them, they are actually miscible fluid phases. Thus, we have NAPL dissolving in water, gas dissolving in water, water and NAPL evaporating, etc. In this book, we do not intend to include an elaborate discussion on the flow of this third fluid phase (but we do discuss the transport of NAPL dissolved in water, and partitioning of the NAPL component between water and air in the void space). However, as this is an important subject, the readers should be exposed to some basic knowledge about it. Accordingly, the objective of this section is to give a brief presentation that shows how the two-phase flow and transport processes can be extended to three fluid phases. The interested reader can seek additional information in the literature.

6.5.1 Statics

We assume that the entire void space is occupied by three fluid phases: an aqueous phase (w), a NAPL (o), and a gaseous phase (a). Note that we have introduced here the term 'aqueous phase', as it may be composed of water as a chemical species, together with other dissolved chemical species. Similarly, both the NAPL and the gaseous phase may be composed of a number of chemical species. For the sake of simplicity, we shall, henceforth, use the word 'water' to represent the aqueous phase, the word 'oil' to represent the NAPL, and the word 'air' to represent the gaseous phase.

Let S_α, $\alpha = w, o, g$, denote the saturation of the three fluid phases that together occupy the void space, with

$$S_w + S_o + S_a = 1. \tag{6.5.1}$$

The discussion below is an extension of that presented in Subs. 6.1.4 on two fluid phases.

A. Capillary pressure

The concept of wettability, introduced for two fluid phases in Subs. 6.1.2, is applicable also to three fluid phases. In most cases, the oil ($\equiv$ NAPL) is the less wetting phase with respect to the solid than the aqueous phase, so that the latter phase is in the immediate contact with the solid. However, oil-wet, or mixed oil-water-wet soils (= solids) may be encountered, e.g., soils with high content of organic matter, or in cases where mineral surfaces exhibit natural organic coatings. This phenomenon has a strong influence on the behavior of fluid phases within the void space. In what follows, we shall consider the more common case of soils for which water is the wetting phase, oil is the *intermediate wetting phase*, and air (= gaseous phase) is the nonwetting phase.

A schematic representation of the three fluid phases in a typical cross-section of a 'pore' is shown in Fig. 6.5.1. These phases are separated from each other by two fluid-fluid (assumed sharp) interfaces: an air-oil interface, and a oil-water one. Actually, it is difficult to precisely define a 'pore' and the 'radius of a pore' in a porous medium. Intuitively, let us define a 'radius of a pore' (or effective radius) as the radius of the largest sphere that can be

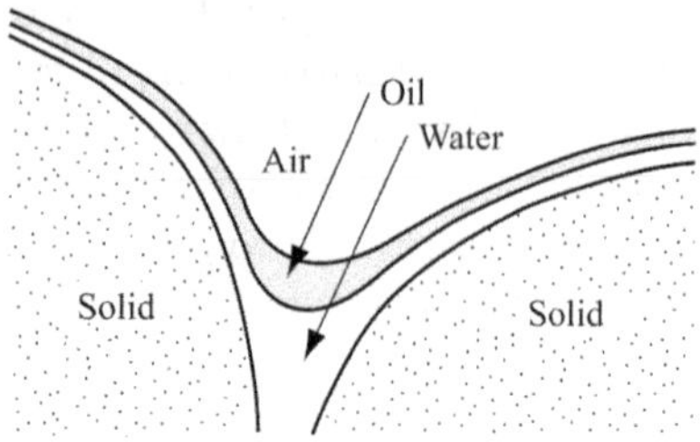

Figure 6.5.1: Schematic pore cross-section with three fluid phases.

placed in a considered portion of the void space, with the solid being tangent to the sphere at least at two points. Due to the assumed order of wettability of the three fluids, water occupies primarily pores with the smallest effective radii, air occupies pores with the largest radii, and oil occupies intermediate size pores. Accordingly, the mean radius of curvature of oil-water interfaces will always be smaller than that of air-oil interfaces.

The concept of capillary pressure, introduced earlier for two fluid phases, can be extended to three phases that occupy the void space. The interface curvature may be related to the respective capillary pressure by a generalization of Laplace's formula (6.1.15). In this case, it takes the form:

$$p_{cow} \equiv p_o - p_w = \frac{2}{r^*_{ow}}\gamma_{ow}, \qquad p_{cao} \equiv p_a - p_o = \frac{2}{r^*_{ao}}\gamma_{ao}, \tag{6.5.2}$$

where p_{cow} and p_{cao} are oil-water and air-oil capillary pressures, respectively, r^*_{ow} and r^*_{ao} are the average radii of air-water and air-oil interfaces, respectively, and γ_{ow} and γ_{ao} are the respective interfacial tensions. The effect of the respective contact angles, appearing as factors in $\cos\theta_{LG}$, can be included in (6.5.2), but are neglected in what follows. The average radii of curvature of each of the fluids are functions of the respective fluid saturations:

$$r^*_{ow} = r^*_{ow}(S_w), \quad r^*_{ao} = r^*_{ao}(S_\ell), \tag{6.5.3}$$

where $S_\ell = (S_w + S_o)$ is the total liquid saturation. We note that r^*_{ow} is a function of S_w only, since all pores with radii smaller than r^*_{ow} are assumed to be occupied by water only. However, all pores with radii smaller than r^*_{ao} are assumed to be occupied by *both* water and oil, having a combined saturation of S_ℓ $(= 1 - S_g)$. The main assumption here is that, with respect to air (which is the most nonwetting fluid), the two liquids behave as a *single* wetting fluid.

Since surface tension depends on temperature and concentration of dissolved matter, e.g., expressed by the mass fractions ω^γ_α, we could express the capillary pressure curves in the general forms:

$$p_a - p_o \equiv p_{cao} = p_{cao}(S_\ell, \gamma_{ao}(T, \omega^\gamma_o)), \tag{6.5.4}$$

$$p_o - p_w \equiv p_{cow} = p_{cow}(S_w, \gamma_{ow}(T, \omega_w^\gamma)), \tag{6.5.5}$$

in which the superscript γ represents all dissolved components.

On the basis of the discussion on the difference between capillary pressure curves and retention curves in two-phase flow, we can also introduce here the *retention curves*

$$p_a - p_o \equiv r_{cao} = r_{cao}(S_\ell, T, \omega_o^\gamma), \tag{6.5.6}$$

$$p_o - p_w \equiv r_{cow} = r_{cow}(S_w, T, \omega_w^\gamma). \tag{6.5.7}$$

By extending (6.5.1) and (6.5.3) to three fluid phases, we obtain

$$p_{cow}(S_w) = \frac{2}{r^*(S_w)}\gamma_{ow}, \qquad p_{cao}(S_\ell) = \frac{2}{r^*(S_\ell)}\gamma_{ao}. \tag{6.5.8}$$

Again, we may replace the surface tension by its product with the cosine of the contact angle.

This implies that, for a given pair of fluids, S_w is a function of p_{cow} only, and S_ℓ is a function of p_{cao} only. Based on our previous assumptions, for a given value of p_{cow}, the resulting saturation, $S_w^{\rm III}(p_{cow})$, in a three-fluid phase system at equilibrium should be identical, or almost identical, to the saturation $S_w^{\rm II}(p_{cow})$ for a two-phase, oil-water system, except for the influence of the $\cos\theta$'s. Here, the superscripts II and III denote two- and three-phase systems, respectively. Similarly, at a given value of p_{cow}, the saturation $S_\ell^{\rm III}(p_{cao})$ for a three phase system should be identical, or nearly identical, to the saturation $S_\ell^{\rm II}(p_{cao})$ for a two phase, air-oil system. Put it succinctly, we can write

$$S_w^{\rm III}(p_{cow}) = S_w^{\rm II}(p_{cow}), \qquad S_\ell^{\rm III}(p_{cao}) = S_\ell^{\rm II}(p_{cao}). \tag{6.5.9}$$

This protocol was first proposed on theoretical grounds by Leverett (1941), and verified experimentally by Lenhard *et al.* (1989).

The phenomenon of *hysteresis* in the relationship between capillary pressure and saturation, discussed in Subs. 6.1.7 for two-phase systems, occurs also in three-phase ones. Again, the reasons are nonwetting fluid entrapment, contact angle hysteresis, and the ink bottle effect.

B. Vertical equilibrium saturation distributions

Consider a sufficiently large spill of LNAPL, such that the percolating LNAPL will reach and accumulate on an underlying water table in the form of a floating lens that spreads out laterally (Fig. 6.5.2), primarily in the direction of the downward sloping water table. The figure does not show the plume of dissolved LNAPL that develops as a result of the dissolution of NAPL in water, and the movement of the latter.

The vertical distribution of LNAPL saturation, S_o, under the LNAPL source area will be different if the spill is of a small volume, such that the infiltrating LNAPL will become immobile when all the LNAPL is reduced to *residual LNAPL saturation*, S_{or}. A sufficiently larger spill will create the lens

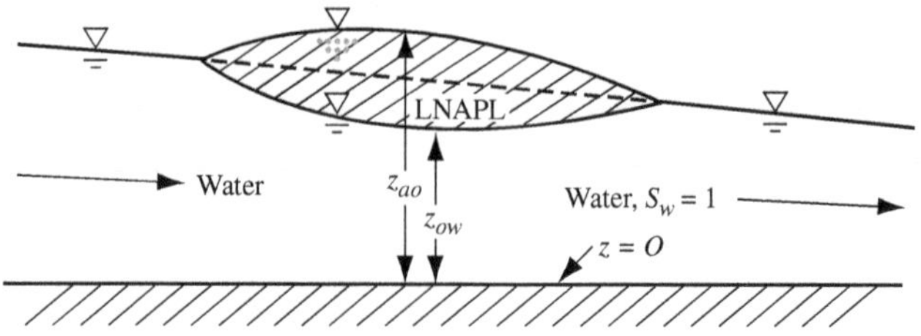

Figure 6.5.2: Schematic cross-section of an LNAPL lens above a sloping water table.

described above. Figure 6.5.3 shows the vertical distribution of LNAPL in the subsurface resulting from spills of increasing volumes ($\mathcal{U}_1 < \mathcal{U}_2 < \mathcal{U}_3, \ldots, \mathcal{U}_8$).

Because of the essentially horizontal movement of the lens, we may assume that at any instant, the LNAP in the subsurface is *hydrostatically distributed* along the vertical, and, hence, vertical flow (of all phases) is negligible. This is often called *vertical-equilibrium (VE-)hypothesis*, which is equivalent to stating that the *vertical pressure distribution within each phase is hydrostatic.* The vertical distribution *within* the LNAPL source area is shown schematically in Fig. 6.5.4. Outside the source area, a similar distribution occurs, except that the upper extent of the lens will be where the LNAPL saturation reaches the residual level. Above this point, no LNAPL will be present.

To determine the vertical distributions of the three fluids: air, water and LNAPL, under equilibrium conditions, assuming that fluid densities are constant, we define an *equivalent piezometric head* for each phase. Taking water density as the reference density for all fluids, i.e., $\rho_{\text{ref}} = \rho_w$, the equivalent piezometric heads, $h_{\text{ref},\alpha}$, are defined by

$$h_{\text{ref},w} = \pi_w + z, \qquad h_{\text{ref},o} = \pi_o + z\frac{\rho_o}{\rho_w}, \qquad h_{\text{ref},a} = \pi_a + z\frac{\rho_a}{\rho_w}, \tag{6.5.10}$$

where $\pi_\alpha = p_\alpha/\rho_w g$ denotes the equivalent pressure head in the α-phase (i.e., the height of an equivalent column of water that produces the pressure p_α), and z denotes the elevation above an arbitrary datum.

To facilitate the understanding of the distributions of the three fluids in subsurface, let us consider the two wells shown in Fig. 6.5.4: Well 1, which is screened along its entire length, and Well 2, which is screened only in the water-saturated zone. We note fluid-fluid interfaces in both wells. Well 1 has an air-LNAPL interface at the elevation z_{ao}, and an LNAPL-water interface at the elevation z_{ow}. Well 2 has only one air-water interface at the elevation z_{aw}. Pressure is continuous (i.e., no jumps) across these interfaces. Let us refer to the atmospheric pressure in the air at (i.e., immediately above) the air-LNAPL interface inside Well 1 as zero pressure ($p_a = 0$). This is, then, also the pressure in the LNAPL at that point. At the LNAPL-water interface, $p_o = p_w$ so that $p_{cow} = 0$ there. Note that $S_w = 1$ occurs at the

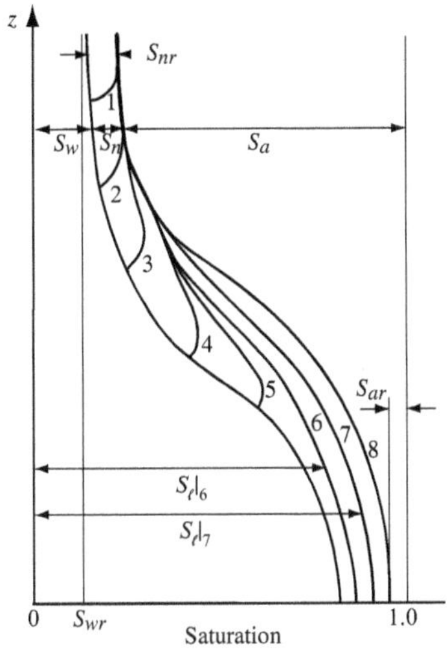

Figure 6.5.3: Equilibrium LNAPL distributions in a three-fluid system for various spill volumes.

LNAPL-water interface, where the LNAPL-water capillary pressure, p_{cow}, is zero, while the soil remains fully water saturated for some distance above this elevation. The reason is that LNAPL cannot enter pores until a certain capillary pressure is exceeded. Inside Well 2, the air-water interface, where the pressure is atmospheric, is at an elevation z_{aw}. We note that $S_\ell = 1$ occurs at the air-LNAPL interface, while air does not enter pores for some distance above this elevation. The thickness of the *capillary fringe* above the air-LNAPL interface is smaller than that for the LNAPL-water one, because the air-LNAPL capillary pressure increases with elevation in proportion to the LNAPL's specific gravity, ρ_o/ρ_w, while the LNAPL-water capillary pressure increases in proportion to $(1 - \rho_o/\rho_w)$.

Hydrostatic conditions require that $\partial h_{\text{ref},\alpha}/\partial z = 0$, i.e., the reference piezometric head, defined in (6.5.10), is a constant, independent of z, for $\alpha = a, o, w$. Since $\rho_a/\rho_w \simeq 0$, the pressure gradient in the air may be assumed to be negligible, or, equivalently, the pressure in the air may be taken as approximately constant, equal to $p_{\text{atm}} = 0$. As a consequence, we shall take the reference pressure head in the air, $p_{\text{atm}}/\rho_w g = \pi_a = 0$. For the LNAPL, the value of the constant for h_{ref} is determined by noting that at $z = z_{ao}$, $p_a - p_o = 0$, and hence $h_{\text{ref},o}\big|_{z=z_{ao}} = z_{ao}(\rho_o/\rho_w)$. Accordingly, within the LNAPL, $h_{\text{ref},o} = z_{ao}(\rho_o/\rho_w)$. For the water, the value of $h_{\text{ref},w}$ is determined by noting that at $z = z_{aw}$, $p_w = p_a = 0$. Hence, $h_{\text{ref},w} = z_{aw}$. Altogether,

$$\pi_w = z_{aw} - z, \qquad \pi_o = (z_{ao} - z)\frac{\rho_o}{\rho_w}, \tag{6.5.11}$$

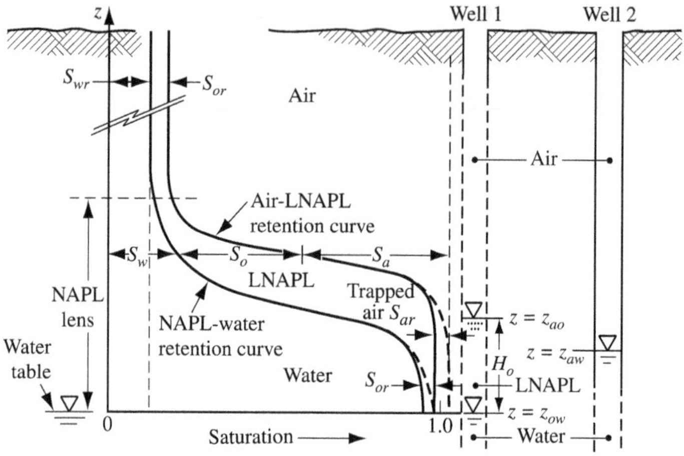

Figure 6.5.4: Equilibrium fluid distributions in a three-fluid system with LNAPL.

for the water and for the LNAPL, respectively.

Since $p_w = p_o$ at $z = z_{ow}$, we have

$$p\big|_{z=z_{ow}} = \rho_w g(z_{aw} - z_{ow}) = \rho_o g(z_{ao} - z_{ow}). \tag{6.5.12}$$

It follows that the various interface elevations are related to each other by

$$z_{aw} - z_{ow} = H_o \frac{\rho_o}{\rho_w}, \qquad H_o = z_{ao} - z_{ow}, \tag{6.5.13}$$

where H_o is the thickness of the layer of LNAPL inside Well 1, as long as the fluids in the well are in equilibrium with those in the soil. Stipulating any two of the three interface elevations, completely defines the three phase static vertical head distributions within the surrounding soil.

To determine the fluid saturation distributions, we recall that water saturation is controlled by the air-LNAPL capillary pressure, while total liquid saturation is controlled by the oil-water capillary pressure. Because of the different densities of the three fluids, we introduce here *equivalent capillary pressure heads*, defined, respectively, by

$$\pi_{ao} = \pi_a - \pi_o = \frac{p_{cao}}{\rho_w g}, \qquad \pi_{ow} = \pi_o - \pi_w = \frac{p_{cow}}{\rho_w g}. \tag{6.5.14}$$

In view of (6.5.11) and (6.5.13), we write

$$\pi_{ao} = (z - z_{ao})\frac{\rho_o}{\rho_w}, \qquad \pi_{ow} = (z - z_{ow})\left(1 - \frac{\rho_o}{\rho_w}\right). \tag{6.5.15}$$

The two equations in (6.5.15) express the relationships between the equivalent capillary pressure and the elevation in the soil for the LNAPL and for the water. On the other hand, following the discussion on capillary pressure presented above, since the three phase capillary pressure relationships, $\pi_{ao}(S_\ell)$ and $\pi_{ow}(S_w)$, between these equivalent pressure heads and the liquid saturation (i.e., combined water and LNAPL) and water, respectively, are known, we may readily compute the sought saturation distributions along the vertical, $S_\ell(z)$ and $S_w(z)$. For example, if hysteresis is disregarded, and the two-phase van Genuchten capillary pressure model, (6.1.23), is used, we obtain

$$S_w = (1 - S_{wr}) \left[1 + (A\beta_{ow}\pi_{ow})^B\right]^{-C} + S_{wr}, \tag{6.5.16}$$

$$S_\ell = (1 - S_{wr}) \left[1 + (A\beta_{ao}\pi_{ao})^B\right]^{-C} + S_{wr}, \tag{6.5.17}$$

where A, B and $C = 1 - 1/B$ are van Genuchten parameters for the soil, S_{wr} is the irreducible water saturation, and β_{ao} and β_{ow} are fluid-dependent scaling factors.

6.5.2 Motion equations for three fluids

Continuing to assume no momentum transfer across the microscopic interfaces between adjacent fluid phases, the motion equations for three fluid phases are similar to (6.2.3) and (6.2.4), except that an equation for the third phase is also required:

$$\mathbf{q}_{rw} = -\frac{\mathbf{k}_{w(on)}(S_w)}{\mu_w} \cdot (\nabla p_w + \rho_w g \nabla z), \tag{6.5.18}$$

$$\mathbf{q}_{ro} = -\frac{\mathbf{k}_{o(wn)}(S_w, S_n)}{\mu_o} \cdot (\nabla p_o + \rho_o g \nabla z), \tag{6.5.19}$$

$$\mathbf{q}_{rn} = -\frac{\mathbf{k}_{n(ow)}(S_n)}{\mu_n} \cdot (\nabla p_n + \rho_n g \nabla z). \tag{6.5.20}$$

In these equations, $\mathbf{k}_{w(on)}$ is the effective permeability to the wetting phase, in the presence of the intermediate and nonwetting phases. Similar definitions apply to $\mathbf{k}_{o(wn)}$ and $\mathbf{k}_{n(ow)}$.

Based on studies by Corey *et al.* (1956) and Snell (1962), although they actually studied only relative permeabilities of *isotropic* porous media, we assume that

$$\mathbf{k}_{w(on)} = \mathbf{k}_w(S_w), \qquad \mathbf{k}_{n(ow)} = \mathbf{k}_n(S_n), \qquad \mathbf{k}_{o(wn)} = \mathbf{k}_o(S_w, S_n).$$

Following are some key interpretations of these equations:

- The effective permeabilities to the wetting phase and to the nonwetting one *in a two-phase system*, are functions of their respective saturations

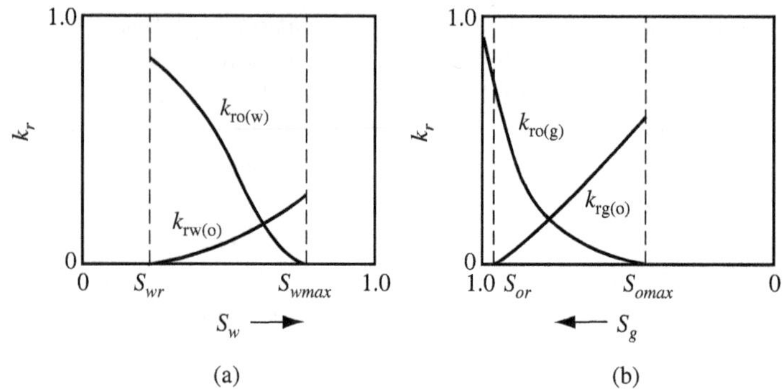

Figure 6.5.5: Two phase relative permeability curves: (a) NAPL-water, (o, w), and (b) Gas-NAPL, (n, o), in a three phase system.

only, i.e.,

$$\mathsf{k}_{w(n)} = \mathsf{k}_{w(n)}(S_w), \qquad \mathsf{k}_{n(w)} = \mathsf{k}_{n(w)}(S_n).$$

- In a *three-phase system* the effective permeabilities to the wetting and nonwetting phases are the same functions of their respective saturations as they are in a two-phase one, i.e.,

$$\mathsf{k}_{w(on)}(S_w) = \mathsf{k}_{w(o)}(S_w), \qquad \mathsf{k}_{n(ow)}(S_n) = \mathsf{k}_{n(o)}(S_n).$$

- The effective permeability to the intermediate wetting phase is a function of both the wetting and the nonwetting saturations, i.e.,

$$\mathsf{k}_{o(wn)} = \mathsf{k}_{o(wn)}(S_w, S_n).$$

It is rather difficult to obtain, experimentally, the effective permeabilities as functions of the various saturations (even for isotropic porous media). Practical approaches in petroleum reservoir engineering (where the three phases are hydrocarbon gas, liquid hydrocarbon, and aqueous solution, for the nonwetting, intermediate wetting, and wetting phases, respectively) are based on the *estimation* of three-phase effective (or relative) permeabilities. Two sets of two phase data are used: $\mathsf{k}_{o(w)} = \mathsf{k}_{o(w)}(S_w)$, which is the effective permeability to the o-phase in an o-w-system, and $\mathsf{k}_{o(n)}=\mathsf{k}_{o(n)}(S_n)$ in an o-n-system. The same approach is valid when the intermediate wetting phase is NAPL and the nonwetting phase is air. The underlying conceptual model, say, for a water-NAPL-air system, is that for the water, *both* the NAPL and the air may be considered as *more nonwetting* phases, while for the air, both the water and the NAPL are regarded as *more wetting* phases.

Figures 6.5.5a and b show relative permeability curves for three phases (w, o, n) in an isotropic porous medium. The point where $\mathsf{k}_{ro} = 0$ corresponds to $S_o = 1 - S_{w\ \max}$, where $S_{w\ \max}$ is the maximum value occurring in the NAPL-water system, rather than to the residual o-saturation, S_{or}, in a NAPL-water-air system. The latter saturation can be further reduced by increasing air saturation.

Stone (1970, 1973), Aziz and Settari (1979) Aleman and Slattery (1985) proposed methods and equations for determining three-phase relative permeabilities. As in the case of two-phase flow, hysteresis is also exhibited in three-phase flow.

6.5.3 Mass balance equation and complete model

The discussion in Sec. 6.3 can easily be extended to three fluid phases. Thus, (6.3.2) can be rewritten for the three phases: water (w), air (a) and the intermediate wetting fluid, NAPL (n), with $\mathbf{q}_\alpha (\equiv \theta_\alpha \mathbf{V}_\alpha)$ denoting the specific discharge of the α-phase, $f_{\alpha\to\beta}$ denoting the transfer of mass of an α-phase into the β-phase across their common (microscopic) interface, say by dissolution, and the symbol Γ'^α denoting a source of α-fluid (= added volume of α-phase per unit volume of porous medium, per unit time), other than through this interface (e.g., evaporation of water, dissolution of air in the water, dissolution of NAPL in water, volatilization of NAPL, etc.). In fact, to handle such transfers, we have to regard each phase as composed of a number of chemical species, or components, such that the transfer from one phase to the another is of individual components rather than of the phase as a whole. Because this subject will be discussed in detail in Chap. 7, let us assume here that there is no interphase mass transfer among the three considered phases. Accordingly, we can rewrite (6.3.3) and (6.3.4) for the mass balances of the three phases in the form

$$\frac{\partial}{\partial t}(\phi S_\alpha \rho_\alpha) = -\nabla\cdot(\rho_\alpha \mathbf{q}_\alpha) + \rho_\alpha \Gamma'^\alpha, \quad \alpha = w, n, a. \tag{6.5.21}$$

For the fluxes of the three fluid phases, following (6.5.20), we write:

$$\mathbf{q}_\alpha = -\frac{\mathbf{k}_\alpha}{\mu_\alpha}\cdot(\nabla p_\alpha + \rho_\alpha g \nabla z), \quad \alpha = w, n, a. \tag{6.5.22}$$

Following the discussion in Subs. 6.5.1, we now have two capillary pressure curves:

$$p_a - p_n \equiv p_{can} = p_{can}(S_\ell), \quad p_n - p_w \equiv p_{cnw} = p_{cnw}(S_w), \quad S_w + S_n = S_\ell, \tag{6.5.23}$$

where the sum of saturations is expressed by

$$S_w + S_n + S_a = 1. \tag{6.5.24}$$

In addition, we need the three constitutive relationships for the fluids' densities:

$$\rho_\alpha = \rho_\alpha(p_\alpha), \quad \alpha = w, n, a. \tag{6.5.25}$$

Altogether, we have 18 scalar equations for the 18 scalar variables:

$$p_\alpha, \; \mathbf{q}_\alpha, \; S_\alpha, \; \rho_\alpha, \; \alpha = w, n, a.$$

However, following the discussion in Subs. 7.9.4, we have only 3 primary variables for which we have to solve the three partial differential (mass balance) equations. Obviously, each partial differential equation requires appropriate initial and boundary conditions.

Chapter 7

MODELING CONTAMINANT TRANSPORT

The issue of contamination of water in the subsurface was introduced in Subs. 1.1.5. In that subsection, we have also listed a number of the more common sources of subsurface contamination.

Our objective in this chapter is to develop complete mathematical models that describe the transport in the subsurface of contaminants dissolved in the water that occupies the void space, or part of it. These dissolved species determine groundwater quality. We use here the term 'transport' as an abbreviation for 'movement, storage, and transformations', with the term 'transformations' indicating changes in concentrations of dissolved chemical species as consequences of chemical reactions, and interphase transfers, such as dissolution of the solid matrix, and precipitation. Such transformations, which appear in the mass balance equation for a chemical species as source or sink terms, are also discussed in this chapter.

Accordingly, we shall start this chapter by discussing the flux of a chemical species, by different modes, then construct the mass balance equation for such species, and finally discuss the source/sink terms that describe chemical reactions and interphase transfers.

Although we are using here the term *contaminant*, to emphasize that our primary interest is groundwater contamination, the discussion, the modeling, etc. in this chapter is applicable to any chemical species dissolved in the water (e.g., a tracer) that travels through the void space of a porous medium domain.

As in previous chapters, we shall consider macroscopic level models that describe and facilitate the prediction of the transport of one or more (possibly interacting) chemical species in a single- or multi-phase system. Such predictions are required in order to plan the management of an aquifer, or the cleanup of the subsurface. In principle, such models are obtained by first understanding the relevant phenomena that occur at the microscopic level, i.e., at points *within* the fluid phase, and modeling these phenomena at that level. The macroscopic models are then derived by averaging the microscopic ones. Both modeling levels are discussed in Sec. 1.3.

In principle, the structure and content of a contaminant transport model is the same as that of a flow model. The main difference is that in the flow

J. Bear, A.H.-D. Cheng, *Modeling Groundwater Flow and Contaminant Transport*, Theory and Applications of Transport in Porous Media 23,
DOI 10.1007/978-1-4020-6682-5_7, © Springer Science+Business Media B.V. 2010

model, the transported quantity is the mass of the fluid phase, while in a contaminant transport model, the transported quantity is the mass of the chemical species—the contaminant—carried by that phase. In the first case, the intensive quantity is the fluid mass density, while in the latter case, it is the concentration of the chemical species. Furthermore, we may have to consider, simultaneously, a number of interacting chemical species.

Note that although in this book we are interested primarily in groundwater in aquifers, especially with respect to groundwater quality and contaminant transport, we have also to consider the transport of contaminants through the unsaturated zone, as many chemical and biological activities that take place in this zone strongly affect the eventual contamination of groundwater in an underlying aquifer.

As explained in the previous chapter (see Sec. 6.5), in addition to water and air that together occupy the void space in the unsaturated zone (and water alone in the saturated zone), a third fluid phase—a Non-Aqueous Phase Liquid (NAPL)—may occupy part of the void space in both zones. In most cases, this NAPL (e.g., benzene and other hydrocarbons, and trichloroethylene) is also a contaminant. We have emphasized in the previous chapter that although a NAPL may be almost immiscible in water, its small, often very small, solubility is sufficient to render groundwater as being contaminated. In this chapter, we shall not treat the movement of NAPL as a separate, third phase, but focus only on contaminants that are dissolved in and carried by the water.

Figure 7.0.1 shows some typical cases of subsurface contamination:

(a) The migration of a contaminant that is leached from a landfill; the leachate travels through the vadose and then through the (saturated) aquifer, eventually draining to a river.
(b) An LNAPL (= Light NAPL) leaks from an underground storage tank and migrates through the vadose zone, eventually accumulating on an underlying water table.
(c) Different routes through which a DNAPL (= Dense NAPL) and an LNAPL can contaminate an aquifer.

In all these cases, dissolved contaminants are transported in the subsurface. Our objective in this section is to discuss modes of transport of such contaminants and the laws that governs their fluxes. Unless stated otherwise, the discussion will be at the macroscopic level. As throughout this book, the presentation is limited to isothermal conditions, although (man-made or naturally occurring) temperature changes may significantly affect the transport of solutes.

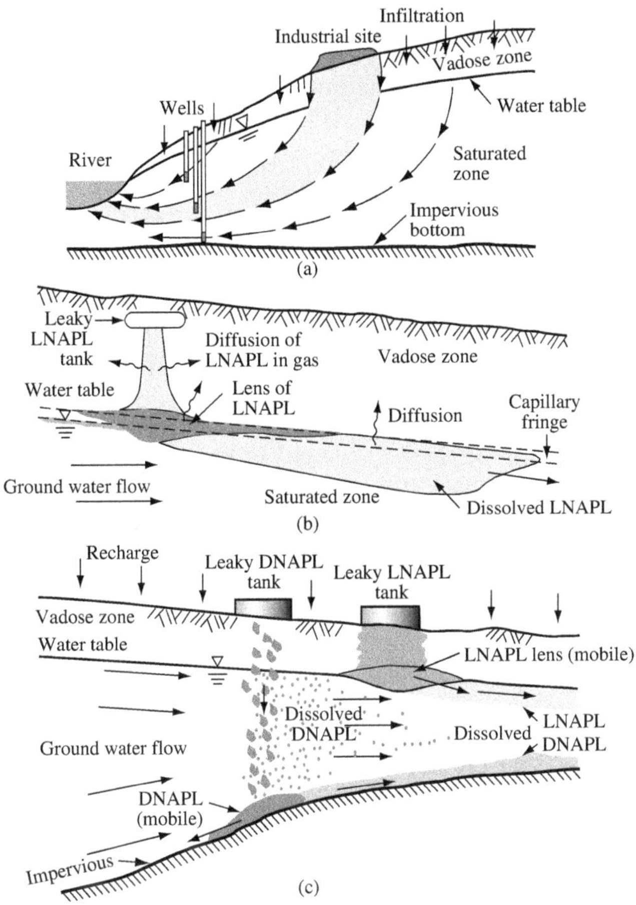

Figure 7.0.1: Examples of subsurface contamination.

7.1 Contaminant Fluxes

7.1.1 Measures of phase composition

Liquid and gas phases are comprised of many chemical species. This statement is also valid for the solid matrix. However, our main interest is in the aqueous phase that occupies the entire void space in the subsurface, or part of it. This phase is comprised primarily of water (H_2O), with minute quantities of various chemical species dissolved in it. Although chemical reactions that involve solid matrix minerals, e.g., dissolution and precipitation, may play a significant role in changing the structure and configuration of the solid matrix, unless otherwise specified, we shall simplify the discussion by assuming that the numerous minerals constituting the soil's solid matrix are represented by a single pseudo-species, referred to as 'solid'. On the other hand, each fluid phase that occupies the void space, or part of it, whether a liquid or a gas, may be composed of more than a single species of interest. It is, therefore, necessary to consider the composition of each individual phase. The quantity of a given chemical species within a phase may be expressed in a number of ways.

Concentration. The concentration of a component indicates the quantity of the latter in a unit volume of fluid phase. It can, thus, be measured in different ways, depending on the selected units for quantity and volume. As everywhere in this book, we shall use a subscript to denote a *phase* (e.g., α), and a superscript to denote a *component* (e.g., γ). To emphasize that we measure the concentration at a point in a porous medium domain, or in a sample taken at such point, we shall assume that the fluid's volume is $\mathcal{U}_{o\alpha}$, i.e., the volume of the α-phase within an REV.

Sometimes, the chemical species is referred to as a 'solvent' if it is the predominant species in a phase, or as a 'solute' if it constitutes only a small portion of a phase.

The common units for expressing the quantity of a species are the *gram*, abbreviated g (expressing the mass of the component), the *mole* (expressing the number of basic entities, such as molecules or ions), and the *equivalent* (expressing the number of equivalent weights). One mole of a substance contains as many atoms (or molecules, etc.) as there are atoms in exactly 12 g of carbon (^{12}C). This number is approximately 6.0221415×10^{23}, also known as the *Avogadro's number.*

In the SI system of units, the *kilogram* (= 1000 grams) is the standard unit for mass. The standard unit for volume in the metric system is the *liter*, defined as the volume of one kilogram of water at 20°C and pressure of one atmosphere. Other units of volume are the milliliter ($m\ell = 1/1000$ of a liter), equal to the cubic centimeter (cc, cm^3).

Mass concentration. The mass concentration, c^γ_α ($\equiv \rho^\gamma_\alpha$), expresses the mass of a γ-species, per unit volume of a fluid α-phase:

$$c^\gamma_\alpha = \frac{m^\gamma_\alpha}{\mathcal{U}_{o\alpha}}, \tag{7.1.1}$$

where m^γ_α denotes the mass of γ within an REV. This is the most commonly used measure for describing water quality. The common units are g/ℓ (= grams of γ per liter of fluid), or mg/ℓ (= milligrams of γ per liter of fluid).

Molar concentration. The molar concentration, or **molarity**, $[c^\gamma]$, expresses the number of γ-moles, N^γ_α, per unit volume of the α-phase:

$$[c^\gamma_\alpha] = \frac{N^\gamma_\alpha}{\mathcal{U}_{o\alpha}} = \frac{c^\gamma_\alpha}{M^\gamma}, \tag{7.1.2}$$

where M^γ is the molecular mass of the γ-species. In thermodynamic relationships, moles are the only measure of concentration. The common units are moles of γ per liter of α-phase (mol/ℓ), or mol/m^3 ($\equiv$ mmol/ℓ).

Molar fraction. The molar fraction, or mole fraction, n_α^γ, is useful when dealing with modeling transport of contaminants with chemical reactions (= *reactive transport*). It is defined as the ratio between the number of moles of γ and the total number of moles in the α-phase:

$$n_\alpha^\gamma = \frac{N_\alpha^\gamma}{N_\alpha}, \qquad N_\alpha = \sum_{(\gamma)} N_\alpha^\gamma, \quad \sum_{(\gamma)} n_\alpha^\gamma = 1. \tag{7.1.3}$$

Equivalent concentration, $c_\alpha^{\gamma eq}$, is another useful measure, defined as

$$c_\alpha^{\gamma eq} = \frac{N_\alpha^{\gamma eq}}{\mathcal{U}_{o\alpha}}, \tag{7.1.4}$$

where $N^{\gamma eq}$ denotes the number of *equivalents* of γ in α. It gives the quantity of γ that reacts with, or equal to the combining value of, a specified quantity of another substance with respect to a given reaction. For *redox* reactions (see any text on chemistry or geochemistry, e.g., Appelo and Postma (2005)), the mass of a substance associated with the loss or gain of one mole of electrons is commonly referred to as the 'equivalent weight' of that substance with respect to the reaction. In this case, $c_\alpha^{\gamma eq} = c_\alpha^\gamma i^\gamma / M_\alpha$, with i^γ denoting the *ionic charge* or *valence* of γ.

Other often encountered definitions of concentration are the *equivalents per liter* ($\equiv$ eq/ℓ), defined as the number of moles of a solute, multiplied by the *valence* of the solute species, per liter of solution, and *equivalents per million*, epm, defined as the number of moles of a solute, multiplied by the *valence* of the solute species, per 10^6 g of solution.

Mass fraction. The mass fraction, ω_α^γ, is the mass of a γ-species per unit mass of the α-phase:

$$\omega_\alpha^\gamma = \frac{m_\alpha^\gamma}{m_\alpha} = \frac{c_\alpha^\gamma}{c_\alpha}, \qquad \sum_{(\gamma)} \omega_\alpha^\gamma = 1. \tag{7.1.5}$$

This measure is applicable to a γ-species in solution in a fluid-phase, or as part of a solid phase. The unit ppm, 'parts per million', defines the number of grams of solute per million grams of solution. The mass fraction is widely used for aqueous contaminants in the saturated zone.

Electrical conductivity, EC, measures the ability of a solution to conduct electrical current. Although this is not a measure of concentration, it is included here because it is related to the ions that are present in the solution. The unit is the reciprocal of ohm-meters, or, in the SI system, siemens per meter (S/m). Often, the EC is measured in the reciprocal of milliohms or, microohms, known as millimhos ($\equiv$ mS), or micromhos ($\equiv$ μS), respectively.

7.1.2 Advective flux

We consider the transport of a γ-contaminant, actually, any dissolved chemical species, within a fluid phase that occupies the entire void space, or part of it, at a volumetric fraction θ. With $\mathbf{V}$ denoting the (intrinsic phase) average (mass-weighted) velocity of the phase, and c^γ denoting the (intrinsic phase) average concentration of the contaminant (expressed as mass of contaminant per unit phase volume), the *advective flux*, $\mathbf{J}^\gamma_{\text{adv}}$, of the considered component (= contaminant) is given by the product

$$\mathbf{J}^\gamma_{\text{adv}} = \theta \mathbf{V} c^\gamma. \tag{7.1.6}$$

This flux expresses the mass of the component passing through a unit area *of porous medium*, normal to $\mathbf{V}$, per unit time.

7.1.3 Diffusive flux

A. Definition of diffusive flux and Fick's law

A fluid phase is, usually, composed of a number of *chemical species* (Subs. 1.3.1), each made up of a large number of identical molecules (ions, atoms, etc.) that are in constant random motion (Brownian motion). Note that a solvent is one of the chemical species. At the microscopic level, each intensive quantity of a chemical species of a phase, e.g., concentration, may be regarded as a continuum. The behavior of this continuum is obtained by averaging the behavior of the individual molecules that comprise it. For example, each molecule has mass, momentum, and energy. The transport of these extensive quantities at the microscopic level is obtained by averaging their transport by the individual molecules. Let superscript γ denote a dissolved chemical species. We shall start by considering what happens at the microscopic level, without using any symbol to denote this fact.

The *total flux*, $\mathbf{j}^{tE}$, of an extensive quantity E, and $\mathbf{j}^{tE^\gamma}$ of the extensive quantity E^γ, of a γ-species, can be expressed as

$$\mathbf{j}^{tE} = e\mathbf{V}^E, \qquad \mathbf{j}^{tE^\gamma} = e^\gamma \mathbf{V}^{E^\gamma}, \tag{7.1.7}$$

with

$$\mathbf{j}^{tE} = \sum_{(\gamma)} \mathbf{j}^{tE^\gamma} = \sum_{(\gamma)} e^\gamma \mathbf{V}^{E^\gamma}, \tag{7.1.8}$$

where e (= quantity of E per unit volume of the phase) denotes the density of E, and $e = \sum_{(\gamma)} e^\gamma$. In these equations, the velocities, $\mathbf{V}^E$ of a particle of an E-continuum and $\mathbf{V}^{E^\gamma}$ of a particle of an E^γ-continuum of a γ-species, both of a phase, are defined as:

$$\mathbf{V}^E \equiv \left.\frac{\partial \mathbf{x}^E}{\partial t}\right|_{\boldsymbol{\xi}^E=\text{const.}}, \qquad \mathbf{V}^{E^\gamma} \equiv \left.\frac{\partial \mathbf{x}^{E^\gamma}}{\partial t}\right|_{\boldsymbol{\xi}^{E^\gamma}=\text{const.}}, \tag{7.1.9}$$

where $\mathbf{x}^E$ denotes the position vector of an E-particle with material coordinates $\boldsymbol{\xi}^E$. We recall that $\mathbf{V}(\equiv \mathbf{V}^m)$ denotes the mass weighted velocity, i.e., the velocity of the extensive quantity mass (i.e., $E \equiv m$).

Physically, $e\mathbf{V}^E$ represents the quantity of E passing through a unit area of the E-continuum, normal to the direction of $\mathbf{V}^E$, per unit time. This flux may be expressed as the sum of two fluxes:

$$e\mathbf{V}^E = e\mathbf{V} + e\,(\mathbf{V}^E - \mathbf{V}) \equiv e\mathbf{V} + \mathbf{j}^E. \tag{7.1.10}$$

We have, thus, decomposed the total flux of E, $\mathbf{j}^{tE}$, into two parts:

- an *advective E-flux*, $e\mathbf{V}$, carried by the *(mass-weighted) velocity* of the phase, $\mathbf{V}$, with respect to a *fixed* coordinate system, and
- a flux, $e\,(\mathbf{V}^E - \mathbf{V})$, relative to the advective one. This second flux, denoted by $\mathbf{j}^E$, is called the *diffusive flux of E* (with respect to the mass-weighted velocity):

$$\mathbf{j}^E = e\,(\mathbf{V}^E - \mathbf{V})\,. \tag{7.1.11}$$

We may now apply the above definitions to the particular case in which E^γ is the mass of a γ-species of a fluid phase, with $e^\gamma \equiv \rho^\gamma \equiv c^\gamma$, and with c^γ referred to as the *concentration* of the γ-species. For this case, the total mass flux is expressed by $c^\gamma\mathbf{V}^\gamma$. When decomposed into two parts, we obtain:

$$c^\gamma\mathbf{V}^\gamma = c^\gamma\mathbf{V} + c^\gamma(\mathbf{V}^\gamma - \mathbf{V}) = c^\gamma\mathbf{V} + \mathbf{j}^\gamma, \tag{7.1.12}$$

where

$$\sum_{\gamma=1}^{N} c^\gamma\mathbf{V}^\gamma \equiv \rho\mathbf{V}, \quad \text{and} \quad \mathbf{j}^\gamma = c^\gamma(\mathbf{V}^\gamma - \mathbf{V}), \tag{7.1.13}$$

is *the diffusive mass flux* of the γ-species, usually referred to as *molecular diffusion*. We note that for all (N) species within a phase,

$$\sum_{\gamma=1}^{N}\mathbf{j}^\gamma = \sum_{\gamma=1}^{N} c^\gamma(\mathbf{V}^\gamma - \mathbf{V}) = \sum_{\gamma=1}^{N} \rho(\mathbf{V} - \mathbf{V}) = 0. \tag{7.1.14}$$

We have thus decomposed the total mass flux of a γ-species into two parts:

- an *advective mass flux*, $c^\gamma\mathbf{V}$, carried by the mass-weighted velocity of the phase, with respect to a fixed coordinate system, and
- a *diffusive flux*, $c^\gamma(\mathbf{V}^\gamma - \mathbf{V})$, relative to the advective one.

Both fluxes are in terms of mass of chemical species per unit area *of fluid phase*.

Equations (7.1.12) and (7.1.13) give the total and the diffusive mass fluxes of a considered chemical species, respectively.

Still at the microscopic level, we consider a fluid containing only two species: γ and δ (= *binary system*). The mass flux of molecular diffusion of the γ-species, $\mathbf{j}^\gamma$, relative to the advective mass flux of the fluid phase, moving at the mass-weighted velocity, $\mathbf{V}$, is expressed by *Fick's law* of molecular diffusion, in the form:

$$\mathbf{j}^\gamma = c^\gamma(\mathbf{V}^\gamma - \mathbf{V}) = -\rho\mathcal{D}^{\gamma\delta}\nabla\omega^\gamma, \qquad \sum_{(\gamma)} \mathbf{j}^\gamma = 0, \tag{7.1.15}$$

$\omega^\gamma = \rho^\gamma/\rho$ denotes the mass fraction of γ, and the scalar $\mathcal{D}^{\gamma\delta}$ is the *coefficient of molecular diffusion* (dims. L^2/T) of the γ-component in a fluid phase that contains only two components, γ and δ.

The diffusive flux of the other component, δ, is given by $\mathbf{j}^\delta = -\rho\mathcal{D}^{\delta\gamma}\nabla\omega^\delta$. Note that the condition $\mathbf{j}^\gamma + \mathbf{j}^\delta = 0$ implies that for a binary system $\mathcal{D}^{\gamma\delta} = \mathcal{D}^{\delta\gamma}$. It is assumed that $\mathcal{D}^{\gamma\delta}$ is independent of c^γ. However, it is, in general, a function of pressure and temperature.

When $\nabla\rho = 0$, i.e., a *homogeneous fluid*, we may write Fick's law, (7.1.15), in terms of the concentration, c^γ, as

$$\mathbf{j}^\gamma \equiv c^\gamma(\mathbf{V}^\gamma - \mathbf{V}) = -\mathcal{D}^{\gamma\delta}\nabla c^\gamma. \tag{7.1.16}$$

Fick's law can also be written in terms of the gradient of molar concentration, $[c^\gamma]$.

Typical values of $\mathcal{D}^{\gamma\delta}$ at 25°C, for a solute in an aqueous phase, are in the range of 5–100$\times 10^{-6}$ cm^2/s. For example, for Ca^{2+}, $\mathcal{D}^{\gamma\delta} = 7.9 \times 10^{-6}$ cm^2/s; for K^+, $\mathcal{D}^{\gamma\delta} = 19.6 \times 10^{-6}$ cm^2/s; and for Cl^-, $\mathcal{D}^{\gamma\delta} = 20.3 \times 10^{-6}$ cm^2/s. Typical values for a dilute component in an air are: for water vapor, $\mathcal{D}^{\gamma\delta} = 2.2 \times 10^{-1}$ cm^2/s; and for TCE vapor, $\mathcal{D}^{\gamma\delta} = 7.8 \times 10^{-2}$ cm^2/s. The diffusivity of a broad range of compounds as a function of temperature and pressure are given by Poling *et al.* (2000). Fick's law, (7.1.15), also holds, as an approximation, for the diffusive flux of a γ-component in a multicomponent system, as long as the δ-component is the solvent component and all components, except δ and γ, are at dilute concentrations. Another case where Fick's law holds is when all components are at dilute concentrations, except for the δ-component. Then, $\mathbf{j}^\lambda = -\rho\mathcal{D}^{\lambda\delta}\nabla\omega^\lambda$ for $\lambda \neq \delta$, and

$$\mathbf{j}^\delta = -\sum_{\gamma(\neq\delta)} \mathbf{j}^\gamma = \sum_{\gamma(\neq\delta)} \rho\mathcal{D}^{\gamma\delta}\nabla\omega^\gamma. \tag{7.1.17}$$

This last equation follows from the necessary condition: $\sum_{(\gamma)} \mathbf{j}^\gamma = 0$.

B. Diffusion of ions and electroneutrality

The diffusion of an ion (considered as a γ-species) in an aqueous solution, away from any charged solid surface, is affected by the electrical field generated by all ions in the solution. In a dilute solution, this diffusive mass flux is given by

$$\mathbf{j}^\gamma = -\frac{\rho\mathcal{F}}{RT} z^\gamma \mathcal{D}^\gamma \omega^\gamma \nabla\varphi_e - \rho\mathcal{D}^\gamma\nabla\omega^\gamma, \tag{7.1.18}$$

where φ_e denotes the potential of the electrical field, and z^γ is the electrical charge of the ion. Here, $\mathcal{F}$ is *Faraday's constant*, defined as the charge of one mole of singly-charged ions (= 9.65×10^4 Coulombs/mole). Equation (7.1.18) is derived from the *Nernst-Planck equations* (Probstein, 1994).

It is observed experimentally that in (non-organic) electrolytic solutions, the condition of *electroneutrality* holds: the net charge at any given point in a solution, away from charged surfaces, is essentially zero. That is,

$$\sum_{(\lambda)} z^\lambda n^\lambda = 0. \tag{7.1.19}$$

Electroneutrality requires that the diffusion fluxes satisfy the condition

$$\sum_{(\lambda)} z^\lambda \mathbf{j}^\lambda / M^\lambda = 0. \tag{7.1.20}$$

The electrical field between the ions, which is proportional to the gradient, $-\nabla\varphi_e$, counteracts the tendency of molecular diffusion to disturb charge neutrality. Therefore, by substituting (7.1.18) into the condition (7.1.20) and solving for $-\nabla\varphi_e$, we obtain

$$-\nabla\varphi_e = \frac{RT}{\mathcal{F}} \frac{\sum_{(\lambda)} z^\lambda \mathcal{D}^\lambda \nabla\omega^\lambda / M^\lambda}{\sum_{(\lambda)} (z^\lambda)^2 \mathcal{D}^\lambda \omega^\lambda / M^\lambda}. \tag{7.1.21}$$

Substituting this expression into (7.1.18) gives

$$\mathbf{j}^\gamma = z^\gamma \rho \mathcal{D}^\gamma \omega^\gamma \frac{\sum_{(\lambda)} z^\lambda \mathcal{D}^\lambda \nabla\omega^\lambda / M^\lambda}{\sum_{(\lambda)} (z^\lambda)^2 \mathcal{D}^\lambda \omega^\lambda / M^\lambda} - \rho \mathcal{D}^\gamma \nabla\omega^\gamma. \tag{7.1.22}$$

This expression is the diffusive mass flux of an ionic species in an electrically neutral dilute solution; it is used for modeling the transport of multiple ionic species (Lichtner, 1995).

C. Macroscopic Fick's law

Our next step is to average the microscopic level advective and diffusive flux expressions presented above in order to obtain their macroscopic counterparts. This goal can be achieved by volume averaging over an REV (Sec. 1.3.3), or by various methods of 'homogenization' (Subs. 1.3.4).

In the passage from (7.1.16) to its macroscopic counterpart, the configuration of the solid-fluid interface, and conditions on it, affect the transformation of the (local) concentration gradient into a gradient of the average concentration (which is the state variable at the macroscopic level). Bear and Bachmat (1990), who used volume (REV) averaging, presented 'averaging rules', which should be employed in order to average mathematical models written at the microscopic level. We recall that taking an average involves integration (e.g.,

(1.3.2)), which takes into account the configuration and the conditions on the domain over which integration is performed. Following Bear and Bachmat (1990), and noting the various underlying assumptions, we average (7.1.15) for a fluid of constant density and constant coefficient of molecular diffusion, obtaining an expression for the macroscopic form of Fick's law of molecular diffusion. It expresses the macroscopic diffusive flux, $\mathbf{J}^\gamma$, of a γ-species within a fluid phase that occupies the entire void space, or part of it, in the form:

$$\mathbf{J}^\gamma = -\mathcal{D}^\gamma \mathsf{T}^*(\theta) \cdot \nabla c^\gamma = -\boldsymbol{\mathcal{D}}^{*\gamma}(\theta) \cdot \nabla c^\gamma, \tag{7.1.23}$$

where $c^\gamma (\equiv \overline{c^\gamma}^\alpha)$ is now the concentration of the γ-species at the macroscopic level, θ denotes the volumetric fraction of the considered phase, and $\boldsymbol{\mathcal{D}}^{*\gamma}$ $(= \boldsymbol{\mathcal{D}}^{*\gamma}(\theta)) = \mathcal{D}^\gamma \mathsf{T}^*(\theta)$, a second rank symmetric tensor, is the coefficient of molecular diffusion within a phase *in a porous medium*. Note that $\mathbf{J}^\gamma$ denotes the flux of γ *per unit area of the fluid within a porous medium cross-section.* For brevity, we have dropped the superscript δ in $\mathcal{D}^{\gamma\delta}$. For saturated flow, we replace θ by the porosity, ϕ.

The symbol T^*, a second rank symmetric tensor, represents the *tortuosity* of the porous medium (e.g., Bear and Bachmat, 1990). In an isotropic porous medium, the components of the tortuosity tensor, T^*_{ij}, may be represented as $\mathsf{T}^*\delta_{ij}$, in which $\mathsf{T}^*(< 1)$ is a scalar tortuosity, and δ_{ij} is the Kronecker delta, with $\delta_{ij} = 1$ for $i = j$, and $\delta_{ij} = 0$ for $i \neq j$.

In indicial notation, (7.1.23) takes the form:

$$J_i^\gamma = -\mathcal{D}^\gamma \mathsf{T}^*_{ij}(\theta) \frac{\partial c^\gamma}{\partial x_j}, = -\mathcal{D}^{*\gamma}_{ij}(\theta) \frac{\partial c^\gamma}{\partial x_j}. \tag{7.1.24}$$

Note that *Einstein summation convention*, introduced in Subs. 4.1.4, is used in (4.1.4), as in all indicial notation equations in this book.

For a fluid of variable density, the macroscopic diffusive flux is

$$\mathbf{J}^\gamma = -\rho \boldsymbol{\mathcal{D}}^{*\gamma}(\theta) \cdot \nabla \omega^\gamma, \tag{7.1.25}$$

where all variables and the coefficient are at the macroscopic level. Solute diffusion is also affected by the fact that the fluid's viscosity near the solid surfaces is higher than that in the interior of the fluid phase (Olsen and Kemper, 1968). We could include this affect as a factor affecting $\boldsymbol{\mathcal{D}}^{*\gamma}$.

The tortuosity of a phase is a macroscopic geometrical coefficient that expresses the effects of the microscopic surface that bounds that phase on the diffusive flux. In fact, Bear and Bachmat (1990) show that the same tortuosity appears when considering the diffusive flux of any extensive quantity, say heat, that is confined to the fluid-occupied domain. As such, it depends on the configuration of the phase within the void space. Hence, in systems with multiple phases, each of the tortuosity components is a function of the saturations. Some authors relate the tortuosity in an isotropic porous medium to the volumetric fraction of the phase. For example, Millington (1959) gave the equation

$$\mathsf{T}^*(\theta) = \frac{\theta^{7/3}}{\phi^2}. \tag{7.1.26}$$

Note that when used to express the flux *per unit area of porous medium*, the tortuosity given by (7.1.26) takes the form $\theta\mathsf{T}^*(\theta) = \theta^{10/3}/\phi^2$. Because the value of the tortuosity falls in the range zero to one, the value of the diffusivity in a fluid that occupies the entire void space, or part of it, is smaller than the corresponding value in an open fluid body.

D. Surface diffusion

Surface diffusion is the phenomenon of the net motion on the solid's surface of adsorbed atoms (often referred to as 'adatoms'), or molecules, ions or clusters of adatoms. When integrated over the entire surface area within a unit volume of porous medium, it is expressed as a macroscopic flux analogous to the diffusive flux that results from the random motion of the ions or atoms of a chemical species within the fluid that occupies the void space, or part of it. In fact, surface diffusion will also occur within the films that coat the solid surface in voids occupied by air in the unsaturated zone. Surface diffusion could be significant in fine grained porous media, e.g., clay, due to their huge specific surface.

An important example is the compacted clay or *bentonite* that serves as a barrier to radionuclides in a geological repository of high-level nuclear waste. Due to its huge specific surface, the barrier effect is achieved by adsorbing the escaped radionuclides; nevertheless, surface diffusion may play an important role since very long time periods are considered. Jahnke and Radke (1987) visualized the total (macroscopic) flux of a γ-ion as composed of two parts: J^γ_{diff}, which is the usual Fickian diffusion, and $J^\gamma_{\text{s.diff}}$, which describes surface diffusion, with

$$J^\gamma_{\text{diff.total}} = J^\gamma_{\text{diff}} + J^\gamma_{\text{s.diff}}, \tag{7.1.27}$$

in which both fluxes are described as Fickian expressions, i.e., both fluxes are proportional to the gradients in the respective concentrations. The model will then involve two variables: concentration in the fluid, and concentration adsorbed on the solid. Jahnke and Radke (1987) developed an expression for an equivalent diffusive flux that combines the two phenomena under the assumption of an equilibrium ion-exchange isotherm.

The subject of surface diffusion will not be further considered in this book.

7.1.4 Hydrodynamic dispersion

We shall start by considering the transport of a contaminant (actually, any solute) in a single fluid phase that occupies the entire void space. Later, we shall extend the discussion to multiple phases.

Consider the flow of a fluid phase (f), say, water, through a porous medium domain. At some initial time, let a portion of this fluid phase be labeled by

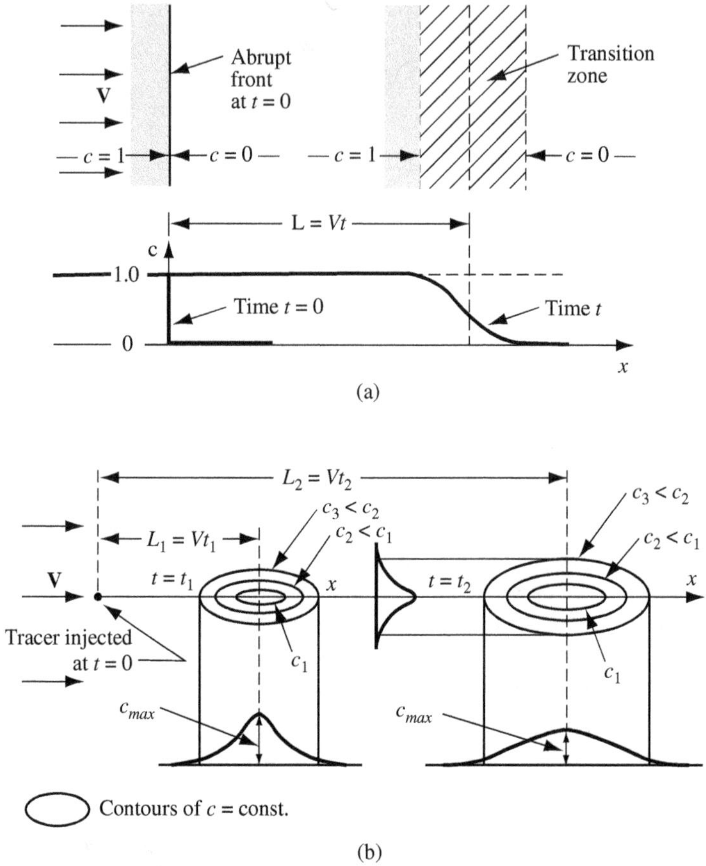

Figure 7.1.1: Longitudinal and transversal spreading of a solute. (a) Longitudinal spreading of an initially sharp front, (b) Spreading of a solute slug injected at a point.

some identifiable solute, a tracer. Let us start by conducting two *conceptual* field experiments.

Figure 7.1.1a shows saturated, (macroscopic) two-dimensional uniform flow, at an average velocity $\mathbf{V}$ ($\equiv \overline{\mathbf{V}}^f$), in the x-direction, in a porous medium domain, say, an aquifer. At an initial time, $t = 0$, an (assumed) abrupt straight line front divides the domain into two parts: one, $x < 0$, occupied by water with a dissolved solute at the concentration $c = 1$, and the other, $x \geq 0$, occupied by water at the solute concentration of $c = 0$. Using Darcy's law to calculate $\mathbf{V}$ ($= -(\mathsf{K}/\phi) \cdot \nabla h$), we may obtain the position of the (assumed) abrupt front, at $x = L$, at any later time, t, through the expression $L = \mathbf{V}t$. *On the basis of Darcy's law alone*, the two kinds of fluids would continue to occupy distinct subdomains, separated by an abrupt moving front.

However, in our 'experiment', by measuring concentrations at a number of observation wells scattered in the domain, we observe that no such front exists. Instead, we observe the development of a *transition zone* across which the solute concentration varies from $c = 1$ to $c = 0$. Experience shows that as flow continues, the width of the transition zone increases. This spreading of the solute-labeled fluid, and the evolution of a transition zone, instead of a sharp front, cannot be explained by the average movement of the fluid at a velocity calculated by Darcy's law.

As a second (conceptual) experiment, consider the injection of a small quantity of tracer at a point $x = 0$, $y = 0$, at some initial time, $t = 0$, into a tracer-free (macroscopic) two-dimensional uniform flow in a porous medium domain. Making use of the (averaged) velocity as calculated by Darcy's law, we should expect the tracer-labeled fluid slug to move as a volume of fixed shape, reaching the point $x = \mathbf{V}t$ at time t. Again, field observations (shown in Fig. 7.1.1b) reveal a completely different picture. We observe a spreading of the solute, not only in the direction of the (averaged) uniform flow, but also *normal to it*. The area occupied by the solute-labeled fluid, which has the shape of an ellipse in the two-dimensional flow domain considered here, will continue to grow, both *longitudinally*, i.e., in the direction of the uniform flow, and *transversely*, i.e., normal to it. The concentration peak appears to move at the averaged velocity. Curves of equal concentration have the shape of confocal ellipses. Again, this spreading cannot be explained by considering the averaged flow alone, especially noting the spreading perpendicular to the direction of the uniform averaged flow, and the ever-growing subdomain occupied by solute-labeled fluid.

The spreading phenomenon in a porous medium domain as described above is called *hydrodynamic dispersion*. It is an unsteady, *irreversible process* (in the sense that the initial tracer distribution cannot be obtained by reversing the direction of the uniform flow) in which the mass of a tracer continuously 'mixes' with the non-labeled portion of the moving fluid.

The phenomenon of dispersion may be demonstrated also by a simple laboratory experiment. Consider steady flow of water at a constant discharge, Q, in a column of homogeneous porous material. At $t = 0$, tracer-marked water (e.g., water with NaCl at a concentration that is sufficiently low so that the effect of density variations on the flow pattern is negligible), at $c = c_1$, starts to displace the indigenous unmarked water ($c = c_o$) in the column . Let the tracer concentration, $c = c(t)$, be measured in the effluent leaving the column and presented in a graphical form, called a *breakthrough curve*, as a relationship between the relative tracer concentration, ϵ ($\equiv (c(t) - c_o)/(c_1 - c_o)$) and time. In the absence of dispersion, the breakthrough curve would take the form of the dashed line shown in Fig. 7.1.2, where $\mathcal{U}_{\text{column}}$ is the pore volume in the column. This would be indicative of the movement of a persistent sharp front between the labeled and unlabeled fluids. In reality, due to hydrodynamic dispersion, the breakthrough curve will take the form of the S-shaped curve shown as a solid line in the figure.

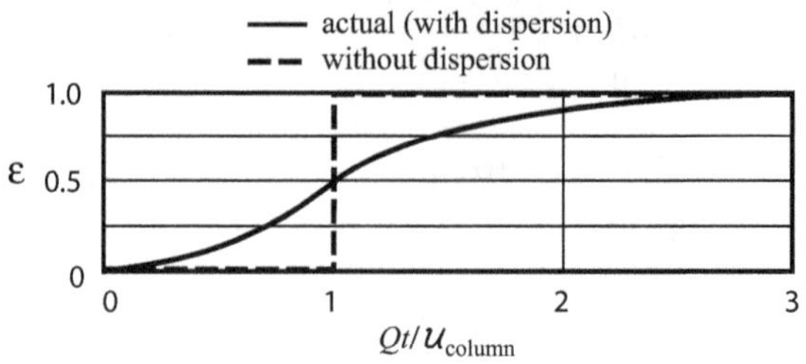

Figure 7.1.2: Breakthrough curve in one-dimensional flow in a column of homogeneous porous material; $\varepsilon \equiv (c(t) - c_o)/(c_1 - c_o)$

As stated above, we cannot explain all the above observations on the basis of the average flow velocity. We must refer to what happens at the *microscopic level*, viz., *inside* the REV. There, we observe (Fig. 7.1.3) velocity variations in both magnitude and direction across any pore cross-section (even when the averaged flow is uniform), and between flow paths. We recall that even in a straight circular capillary tube, we have a parabolic distribution of fluid velocity (see any text on fluid mechanics), with zero velocity at the (stationary) solid surface, and a maximum velocity at the center of the tube. The void space may be visualized as an assembly of interconnected tubes, with varying diameters. The maximum velocity itself varies according to the size and shape of the pores. Because of the shape of the interconnected pore space, the (microscopic) streamlines deviate from the mean direction of flow (Fig. 7.1.3a and b). Altogether, we note that the velocity at the microscopic level varies in magnitude and direction from point to point within the fluid present in the void space. As a consequence, any initial cloud of closely-spaced tracer particles will spread out, with each fluid particle traveling along its own microscopic streamline. Therefore, the shape of the initial cloud will gradually change, and so will the fluid volume occupied by it. This phenomenon is referred to as *mechanical dispersion*, where the term 'mechanical' is used to remind us that this part of the spreading is due to fluid mechanical phenomena, and 'dispersion' is just another word for 'spreading.'

As flow continues, the tracer particles, which originate from any small subdomain in the fluid within the void space, will occupy an ever growing volume of the flow domain. The two basic factors that produce mechanical dispersion are, therefore, *flow* and the *presence of a pore system* through which the flow takes place.

Although this spreading is in both the longitudinal direction, viz., that of the (local) average flow, and in the direction transverse to the latter, it is primarily in the former direction. Very little spreading in the direction perpendicular to the average flow is produced by velocity variations alone. Also, such velocity variations alone cannot explain the ever-growing width

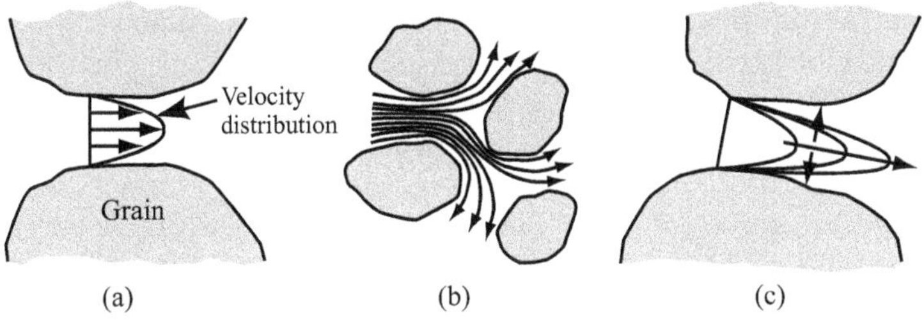

Figure 7.1.3: Dispersion due to mechanical spreading (a,b), and molecular diffusion (c).

(normal to the direction of flow) of a plume of tracer-labeled fluid particles originating at a point source.

In order to explain the observed spreading, especially transverse to the flow direction, we must refer to an additional phenomenon that takes place in the void space, viz., *molecular diffusion*, discussed in Subs. 7.1.3.

As shown in Subs. 7.1.3, molecular diffusion of a component in a fluid, caused by the random motion of the molecules (Brownian motion), produces an additional flux of the component's particles (at the microscopic level) from regions of higher concentrations to those of lower ones. This flux is relative to the advective one, produced by the velocity of the fluid phase. This means, for example, that as component particles spread *along* each microscopic stream-tube, as a result of mechanical dispersion, a concentration gradient is produced, which, in turn, produces an additional flux of the component by the mechanism of molecular diffusion. The latter phenomenon tends to equalize the concentration along every microscopic stream-tube. At the same time, a concentration gradient is also produced between adjacent stream-tubes, causing *lateral molecular diffusion, across streamlines* (Fig. 7.1.3c), which tends to equalize the concentration *across* pores. It is this phenomenon, combined with the randomness of the streamlines, that explains the observed ever-growing extent of transverse dispersion.

Thus, the deviations in solute concentration within a fluid phase, from the concentrations obtained by assuming advection only (at the average velocity), are due to two *simultaneous* phenomena: variations in the microscopic velocity of the phase, with respect to the averaged velocity, and molecular diffusion. In this way, molecular diffusion contributes to the dispersive flux. This contribution is *in addition* to the diffusive flux *at the macroscopic level*, as described by (the averaged) Fick's law. The latter is the only flux that takes place when the averaged velocity is zero. It may thus be concluded, that even when the macroscopic effect of diffusion is relatively small, it is only the combination of microscopic velocity variations and molecular diffusion that produces mechanical dispersion.

Molecular diffusion makes dispersion, even in purely laminar flow, an *irreversible* phenomenon. Irreversibility is exhibited, for example, by the growing width of a transition zone around an initially sharp tracer front in uniform flow, as the direction of the flow is reversed. A second source of irreversibility arises from the procedure used for averaging microscopic velocities, for example, averaging over an REV. There is an inevitable continual and growing loss of information at the microscopic level as one solves the governing equations at the macroscopic level forward in time. This fact exhibits itself in the smearing of an initially sharp macroscopic concentration front. This loss of information is irretrievable and, hence, the solutions to the equations are also not reversible. Thus, it is possible for macroscopic theories to be irreversible even without considering molecular diffusion. Whereas the first type of irreversibility, i.e., that caused by molecular diffusion, predicts the irreversibility of physical observable values (i.e., actual concentrations), the second type of irreversibility is manifested in the mathematical equations describing the dispersion phenomena.

We refer to the flux that causes mechanical dispersion (of a component) as *dispersive flux*. It is a *macroscopic* flux that takes into account the effect of the variations in the microscopic velocity in the vicinity of a point. We note that the decomposition of the average of the total (local) advective flux into an advective flux at the average velocity and a dispersive flux, is a consequence of the averaging process that we have chosen to employ.

We use the term *hydrodynamic dispersion* to denote the spreading (at the macroscopic level) that results from *both* mechanical dispersion and molecular diffusion. Actually, the separation between the two processes is rather artificial, as they are inseparable. An exception is in the absence of motion, when only molecular diffusion takes place. Because molecular diffusion is a relatively slow process, its overall effect on dispersion is more significant at low velocities.

In general, variations in tracer concentration cause changes in the fluid's density and viscosity. These, in turn, affect the flow regime (i.e., velocity distribution) that depends on these properties. We use the term *ideal tracer* when the concentration of the latter does not affect the fluid's density and viscosity. At relatively low concentrations, the ideal tracer approximation is sufficient for most practical purposes. However, in certain cases, for example, in the case of seawater intrusion into a region of freshwater, the density may vary appreciably, and the ideal tracer approximation should not be used; we have to take into account the effect of concentration changes on the fluid's density and viscosity.

7.1.5 Dispersive flux

From the discussion in previous subsection, it follows that the dispersive flux, as a macroscopic flux of a component relative to the flux carried by the average velocity ($\equiv$ advective flux), is a consequence of the fact that both the velocity and the concentration vary from point to point within a fluid phase that occupies the entire void space, or part of it. Let us relate this flux to measurable quantities such as (average) velocity and (average) concentration. For the sake of simplicity, we shall make use of the volume averaging approach.

The advective flux of a component (per unit area of fluid) at a (microscopic) point, $\mathbf{x}'$, within a fluid phase (f) that occupies part of the void space within an REV centered at a point $\mathbf{x}$, is given by $c\mathbf{V}$. The intrinsic phase average of this flux is $\overline{c\mathbf{V}}^f$. In order to express this flux in terms of the average values, $\overline{c}^f$ and $\overline{\mathbf{V}}^f$, the velocity, $\mathbf{V}(\mathbf{x}',t;\mathbf{x})$, and the component concentration, $c(\mathbf{x}',t;\mathbf{x})$, are decomposed into two parts: an intrinsic phase average value and a deviation from that value, in the form:

$$\begin{aligned} \mathbf{V}(\mathbf{x}',t;\mathbf{x}) &= \overline{\mathbf{V}}^f(\mathbf{x},t) + \mathring{\mathbf{V}}(\mathbf{x}',t;\mathbf{x}), \\ c(\mathbf{x}',t;\mathbf{x}) &= \overline{c}^f(\mathbf{x},t) + \mathring{c}(\mathbf{x}',t;\mathbf{x}). \end{aligned} \tag{7.1.28}$$

Because an average value is constant over the REV, we have $\overline{\overline{(..)}^f}^f = \overline{(..)}^f$. As a consequence,

$$\overline{\mathring{\mathbf{V}}}^f = 0, \qquad \text{and} \qquad \overline{\mathring{c}}^f = 0. \tag{7.1.29}$$

To obtain the average flux (still per unit area of fluid), we write:

$$\overline{c\mathbf{V}}^f = \overline{(\overline{c}^f + \mathring{c})(\overline{\mathbf{V}}^f + \mathring{\mathbf{V}})}^f = \overline{\overline{c}^f\overline{\mathbf{V}}^f}^f + \overline{\overline{c}^f\mathring{\mathbf{V}}}^f + \overline{\mathring{c}\overline{\mathbf{V}}^f}^f + \overline{\mathring{c}\mathring{\mathbf{V}}}^f. \tag{7.1.30}$$

Because the average of the deviations vanishes, the second and third terms on the right-hand side of (7.1.30) vanish, leaving the relationship:

$$\overline{c\mathbf{V}}^f = \overline{c}^f\overline{\mathbf{V}}^f + \overline{\mathring{c}\mathring{\mathbf{V}}}^f. \tag{7.1.31}$$

From this equation, it follows that the average (= macroscopic) flux of a component at a point in a porous medium domain (= centroid of an REV) is equal to the sum of two macroscopic fluxes:

- an advective flux, $\overline{c}^f\overline{\mathbf{V}}^f$, expressing the mass of the component carried by the fluid at the latter's average velocity, $\overline{\mathbf{V}}^f$. The (microscopic) mass averaged velocity, $\mathbf{V}$ was introduced in Subs. 7.1.3, and
- a flux, $\mathbf{J}^{*\gamma} \equiv \overline{\mathring{c}\mathring{\mathbf{V}}}^f$, that results from the variation of c and $\mathbf{V}$ within the REV for which the point, $\mathbf{x}$, serves as a centroid. Recalling the discussion in the previous subsection, this is the flux that produces the mechanical spreading (≡ the dispersion) of the component. We refer to it as the *dispersive flux*. It is a macroscopic flux caused by the variations in the microscopic fluid velocity and in fluid concentration in the vicinity of a (macroscopic) point in a porous medium domain. This flux is introduced

to circumvent the lack of information concerning the detailed velocity variations at the microscopic level. This lack of information is the consequence of our ignorance as to the detailed geometry of the surfaces that bound the fluid phase.

7.1.6 Dispersion coefficient and dispersivity

Our next objective is to express the dispersive flux in terms of averaged (and *measurable*) quantities, such as averaged velocity and averaged concentration. Investigations over a period of about three decades, starting around the mid-50's (e.g., de Josselin de Jong, 1958; Saffman, 1959; Bear, 1961a; Scheidegger, 1961; Bear, 1972; and Bear and Bachmat, 1990, p. 401), have led to the conclusion that the dispersive flux of a component (per unit area of fluid) in a porous medium can be expressed as a *Fickian-type* law (i.e., a law that resembles Fick's (linear) law of molecular diffusion) in the form:

$$\mathbf{J}^* \equiv \overline{\mathring{c}\mathring{\mathbf{V}}}^f = -\mathsf{D} \cdot \nabla \overline{c}^f, \tag{7.1.32}$$

or, in indicial notation:

$$J_i^* \equiv \overline{\mathring{c}\mathring{V}_i}^f = -\mathsf{D}_{ij} \frac{\partial \overline{c}^f}{\partial x_j}, \tag{7.1.33}$$

where the D_{ij}'s (dims. L^2/T) are components of a coefficient, D, called the *coefficient of mechanical (or advective) dispersion*, or the *dispersion coefficient.* This coefficient is a second rank tensor that relates the flux vector $\mathbf{J}^*$ to the driving force vector $-\nabla\overline{c}^f$. Equation (7.1.33) is valid for the general case of an anisotropic porous medium. The dispersion coefficient is characterized by:

- The D_{ij}-matrix is *non-negative definite* (or positive definite). This is a consequence of thermodynamics: the *rate of entropy production*, $\dot{\mathcal{S}}$, is related to the thermodynamic driving force, $\mathbf{X}$, and the thermodynamic flux, $\mathbf{Y}$, (referred to by De Groot and Mazur (1962) as *conjugated flux and force*, respectively) by $\dot{\mathcal{S}} = Y_i X_i$. Here, the driving force $\mathbf{X}$ is proportional to the negative concentration gradient, $-\nabla\overline{c}^f$. In this case, the rate of entropy production can be expressed by

$$\dot{\mathcal{S}} = \chi\left(-\mathsf{D}_{ij}\frac{\partial \overline{c}^f}{\partial x_j}\right) \times \chi\left(-\frac{\partial \overline{c}^f}{\partial x_i}\right) \geq 0, \quad \text{or} \quad \chi^2\mathsf{D}_{ij}\frac{\partial \overline{c}^f}{\partial x_j}\frac{\partial \overline{c}^f}{\partial x_i} \geq 0, \tag{7.1.34}$$

  in which, $\mathbf{Y} = \chi\mathbf{J}^* = -\chi\mathsf{D}\cdot\nabla\overline{c}^f$ and $\mathbf{X} = -\chi\nabla\overline{c}^f$. In the above, χ is a parameter that depends on the extensive quantity considered; for each such quantity, it transforms the flux and the driving force, in the form of a gradient of an appropriate scalar considered (here $\nabla\overline{c}^f$), into conjugated thermodynamic flux and force (De Groot and Mazur, 1962).
- The D_{ij}-matrix is *symmetric*, i.e.,

$$\mathsf{D}_{ij} = \mathsf{D}_{ji}. \tag{7.1.35}$$

This is a consequence of the conjugated force and flux relation (De Groot and Mazur, 1962), i.e., they satisfy

$$\frac{\partial Y_i}{\partial X_j} = \frac{\partial Y_j}{\partial X_i}. \tag{7.1.36}$$

Because we have circumvented the need to know the details (of velocity and concentration) at the microscopic level by 'escaping' to the macroscopic level, we are left with the need to determine a set of coefficients, in this case, D_{ij}. This situation is always the case whenever we try to overcome the lack of information about details at the microscopic level by moving to the macroscopic level.

It is interesting to note that although Darcy's law, (4.1.27), and the Fick's type law that governs the dispersive flux, (7.1.33), look similar, there is a basic difference between the coefficients K_{ij} and D_{ij}: the former is a function of the microscopic geometry of the void space (and of fluid properties), while the latter depends also on the macroscopic velocity field.

Several authors (e.g., Nikolaevskii, 1959; Bear, 1961a; Scheidegger, 1961; Bear and Bachmat, 1967, 1990) have derived the following expression for the components D_{ij}:

$$\mathsf{D}_{ij} = \mathsf{a}_{ijk\ell} \frac{\overline{V_k}^f \overline{V_\ell}^f}{\overline{V}^f} f(\mathrm{Pe}, r), \tag{7.1.37}$$

where $\overline{V}^f \left(\equiv |\overline{\mathbf{V}}^f|\right)$ is the magnitude of the average velocity, r represents the ratio between characteristic lengths, in the direction of the flow and normal to it, within a pore, and Pe is a *Peclet number* defined by:

$$\mathrm{Pe} \equiv \frac{\overline{V}^f \Delta_f}{\mathcal{D}_f}, \tag{7.1.38}$$

which expresses the ratio between the rates of transport of the considered component, respectively, by advection and by diffusion (see a detailed discussion on dimensionless numbers in Subs. 7.7). In this definition, Δ_f is the *hydraulic radius* of the fluid occupied portion of the void space, serving as a characteristic length of the void space, and $\mathcal{D}_f$ denotes the coefficient of molecular diffusion in the fluid phase. Bear and Bachmat (1990) suggested an expression for $f(\mathrm{Pe}, r)$. However, as is common in practice, we shall assume $f(\mathrm{Pe}, r) \approx 1$, so that the coefficient of dispersion is expressed in the form

$$\mathsf{D}_{ij} = \mathsf{a}_{ijk\ell} \frac{V_k V_\ell}{V}, \tag{7.1.39}$$

in which $V_k \equiv \overline{V_k}^f$. Henceforth, for simplicity, we shall continue to drop the notation for intrinsic phase averaging.

The coefficients $\mathsf{a}_{ijk\ell}$ (dims. L) appearing in (7.1.39) are components of a fourth rank tensor, $\mathbf{a}$, called the *dispersivity* of the porous medium. It expresses the effect, on the flow, of the microscopic configuration of the interface between the considered fluid phase and all other phases within the REV. In a saturated system, this interface is that between the fluid and the solid. When a fluid occupies only part of the void space, each of the dispersivity components, $\mathsf{a}_{ijk\ell}$, is a function of the volumetric fraction of the fluid.

In a three-dimensional space, the dispersivity tensor, $\mathsf{a}_{ijk\ell}$, has $3^4 = 81$ components. However, because of various symmetry considerations, the number of independent coefficients is smaller. Specifically,

(a) From the expression for the rate of entropy production, $\dot{\mathcal{S}}$, and following the discussion leading to (7.1.34), we have

$$\dot{\mathcal{S}} = \chi\left(-\mathsf{D}_{ij}\frac{\partial \overline{c}^f}{\partial x_i}\right) \times \chi\left(-\frac{\partial \overline{c}^f}{\partial x_j}\right) = \chi^2 \mathsf{a}_{ijk\ell}\frac{\partial \overline{c}^f}{\partial x_i}\frac{\partial \overline{c}^f}{\partial x_j}\frac{V_k V_l}{V} \geq 0. \quad (7.1.40)$$

It follows that $\mathsf{a}_{ijk\ell}$ is positive definite. This means that all *principal minors* of $\mathsf{a}_{ijk\ell}$ are positive.

(b) The values of the $\mathsf{a}_{ijk\ell}$ are invariant under the permutation of indices,

$$\mathsf{a}_{ijk\ell} = \mathsf{a}_{ij\ell k}, \quad \mathsf{a}_{ijk\ell} = \mathsf{a}_{jik\ell}. \quad (7.1.41)$$

Hence, only 36 of the 81 components are *independent* of each other. It is interesting to note that the 36 components are constrained by $2^6 - 1 = 63$ constraints. As the material has more symmetry properties, the number of independent coefficients decreases, until, when the material is isotropic, this number is reduced to two (Bear *et al.*, 2009).

A. Isotropic porous medium

In an *isotropic porous medium*, it has been demonstrated (Bear and Bachmat, 1990; see also Sirotine and Chaskolskaya, 1984, p. 651–2) that the 36 independent components reduce to *two*. This can be shown by considering fourth rank tensors that satisfy the relationships (7.1.41) and are invariant under the action of full rotational (orthogonal) symmetry. The two coefficients are designated as a_L and a_T, and are called the *longitudinal* and the *transverse dispersivities* of the porous medium, respectively. The parameter a_L is a length that characterizes the microscopic configuration of the phase within the REV. Thus, for a phase that completely fills the void space in a porous medium, a_L should be of the order of magnitude of the size of a typical pore. Furthermore, by the positive definiteness of $\mathsf{a}_{ijk\ell}$, it follows that

$$\mathsf{a}_L \geq 0, \quad \mathsf{a}_T \geq 0. \quad (7.1.42)$$

De Josselin de Jong (1958) and laboratory column experiments (e.g., Bear, 1961b) have shown that a_T is 8 to 24 times smaller than a_L.

In terms of a_L and a_T, components of the dispersivity tensor for an *isotropic porous medium* can be expressed in the form:

$$\mathsf{a}_{ikj\ell} = \mathsf{a}_T\delta_{ij}\delta_{k\ell} + \frac{\mathsf{a}_L - \mathsf{a}_T}{2}(\delta_{ik}\delta_{j\ell} + \delta_{i\ell}\delta_{jk}), \tag{7.1.43}$$

where δ_{ij} is the Kronecker delta. The coefficient of dispersion, with $f(\mathrm{Pe}, r) \simeq 1$, can then be expressed as

$$\mathsf{D}_{ij} = \left[\mathsf{a}_T\delta_{ij} + (\mathsf{a}_L - \mathsf{a}_T)\frac{V_iV_j}{V^2}\right]V, \quad V = |\mathbf{V}|, \tag{7.1.44}$$

in which V_i denotes the ith component of the average velocity vector $\mathbf{V}$.

In Cartesian coordinates, with V_x, V_y, and V_z denoting average velocity components in the x, y, and z directions, respectively, we obtain from (7.1.44):

$$\begin{aligned}
\mathsf{D}_{xx} &= \left[\mathsf{a}_T + (\mathsf{a}_L - \mathsf{a}_T)\frac{V_x^2}{V^2}\right]V = \frac{1}{V}\left(\mathsf{a}_LV_x^2 + \mathsf{a}_TV_y^2 + \mathsf{a}_TV_z^2\right),\\
\mathsf{D}_{yy} &= \left[\mathsf{a}_T + (\mathsf{a}_L - \mathsf{a}_T)\frac{V_y^2}{V^2}\right]V = \frac{1}{V}\left(\mathsf{a}_TV_x^2 + \mathsf{a}_LV_y^2 + \mathsf{a}_TV_z^2\right),\\
\mathsf{D}_{zz} &= \left[\mathsf{a}_T + (\mathsf{a}_L - \mathsf{a}_T)\frac{V_z^2}{V^2}\right]V = \frac{1}{V}\left(\mathsf{a}_TV_x^2 + \mathsf{a}_TV_y^2 + \mathsf{a}_LV_z^2\right),\\
\mathsf{D}_{xy} &= \left[(\mathsf{a}_L - \mathsf{a}_T)\frac{V_xV_y}{V^2}\right]V = \mathsf{D}_{yx},\\
\mathsf{D}_{xz} &= \left[(\mathsf{a}_L - \mathsf{a}_T)\frac{V_xV_z}{V^2}\right]V = \mathsf{D}_{zx},\\
\mathsf{D}_{yz} &= \left[(\mathsf{a}_L - \mathsf{a}_T)\frac{V_yV_z}{V^2}\right]V = \mathsf{D}_{zy}.
\end{aligned} \tag{7.1.45}$$

Like any second rank tensor, $\mathbf{D}$ also has *three principal directions.* Using these principal directions as Cartesian coordinate axes, x_1, x_2, x_3, we may write $\mathbf{D}$, in the matrix form:

$$\mathbf{D} = \begin{bmatrix} \mathsf{D}_{x_1x_1} & 0 & 0 \\ 0 & \mathsf{D}_{x_2x_2} & 0 \\ 0 & 0 & \mathsf{D}_{x_3x_3} \end{bmatrix}. \tag{7.1.46}$$

In the special case of uniform flow, say $V_x = V$, $V_y = V_z = 0$, equation (7.1.45) reduces to $\mathsf{D}_{xx} = \mathsf{a}_LV$, $\mathsf{D}_{yy} = \mathsf{a}_TV$, $\mathsf{D}_{zz} = \mathsf{a}_TV$, $\mathsf{D}_{xy} = \mathsf{D}_{xz} = \mathsf{D}_{yz} = 0$; or, in matrix form:

$$\mathbf{D} = \begin{bmatrix} \mathsf{a}_L & 0 & 0 \\ 0 & \mathsf{a}_T & 0 \\ 0 & 0 & \mathsf{a}_T \end{bmatrix}V = \begin{bmatrix} \mathsf{D}_L & 0 & 0 \\ 0 & \mathsf{D}_T & 0 \\ 0 & 0 & \mathsf{D}_T \end{bmatrix}, \tag{7.1.47}$$

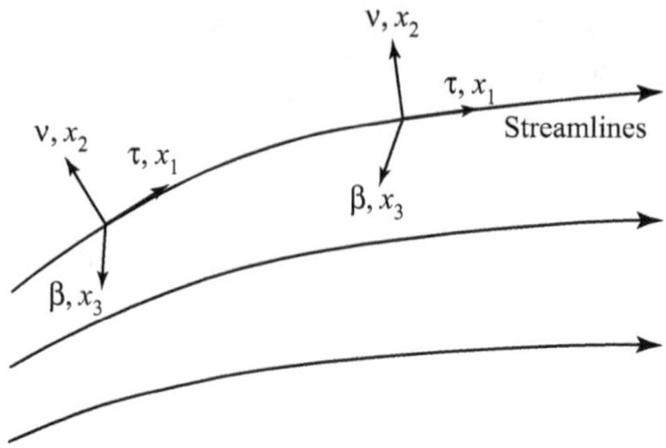

Figure 7.1.4: Principal directions of dispersion coefficient in an isotropic porous medium.

where D_L and D_T are, respectively, the longitudinal and transversal dispersion coefficients of an isotropic porous medium.

We have already mentioned that the tensor $\mathbf{D}$ (and its principal directions) depends also on the (macroscopic) velocity field. Specifically, if we consider a point on a macroscopic (instantaneous) streamline in a flow domain, we may construct at that point:

- a unit vector, $\boldsymbol{\tau}$, in the direction of the *tangent* to the streamline, (i.e., in the direction of the flow),
- a unit vector, $\boldsymbol{\nu}$, called the *principal normal* to the streamline (defined by $\kappa\boldsymbol{\nu} = d\boldsymbol{\tau}/ds$, where s is the distance measured along the streamline, and κ is the curvature of the streamline at the point), and
- a unit vector, $\boldsymbol{\beta}$ $(= \boldsymbol{\tau} \times \boldsymbol{\nu})$, normal to both $\boldsymbol{\tau}$ and $\boldsymbol{\nu}$ (Fig. 7.1.4).

In an *isotropic porous medium*, the principal directions of the tensor $\mathbf{D}$ *coincide with the directions of these three unit vectors.* As such, as the velocity varies, these directions may vary from point to point and in time.

If, *locally*, we select $\boldsymbol{\tau}$, $\boldsymbol{\nu}$ and $\boldsymbol{\beta}$, as *basis vectors* of the coordinate system, x_1, x_2, x_3, then $\mathbf{D}$ takes the form (7.1.46). In such a case, $\mathsf{D}_{x_1x_1}$ is called *coefficient of longitudinal dispersion*, while $\mathsf{D}_{x_2x_2}$ and $\mathsf{D}_{x_3x_3}$ are called *coefficients of transverse dispersion.*

B. Transverse isotropy (axial symmetry)

In an *anisotropic porous medium*, the number of independent dispersivity coefficients is larger, depending on the kind of symmetry exhibited by the anisotropic medium. As an example, we may consider a porous medium with *transverse isotropy*, i.e., a porous medium with one axis of rotational symmetry. This means that in any plane perpendicular to that axis, the material is isotropic, i.e., it does not exhibit property changes with direction, while medium properties in the direction parallel to this axis are different (see Sec. 2.5).

For an axially symmetrical porous medium, with the vector $\mathbf{e}$ (components e_i) indicating the axis of symmetry, there exist *six* independent $\mathsf{a}_{ijk\ell}$-coefficients. The dispersivity components can then be expressed by (Bear *et al.*, 2009)

$$\mathsf{a}_{ijk\ell} = \mathsf{a}_1\delta_{ij}\delta_{k\ell} + \frac{\mathsf{a}_2}{2}(\delta_{ik}\delta_{j\ell} + \delta_{i\ell}\delta_{jk}) + \mathsf{a}_3 e_i e_j \delta_{k\ell} + \mathsf{a}_4 e_k e_\ell \delta_{ij} + \frac{\mathsf{a}_5}{2}(e_i e_k \delta_{j\ell} + e_j e_k \delta_{i\ell} + e_i e_\ell \delta_{jk} + e_j e_\ell \delta_{ik}) + \mathsf{a}_6 e_i e_j e_k e_\ell, \quad (7.1.48)$$

with a_1 through a_6 indicating the six independent dispersivity coefficients. Fel and Bear (2009) determined the constraints that the six a_i's have to satisfy as a consequence of the positive definiteness of the entropy production, expressed by (7.1.34). We note that by dropping terms associated with e_i in (7.1.48), we obtain the isotropic case, described by (7.1.43), i.e., with $\mathsf{a}_1 = \mathsf{a}_T$, and $\mathsf{a}_2 = \mathsf{a}_L - \mathsf{a}_T$, and $\mathsf{a}_3 = \mathsf{a}_4 = \mathsf{a}_5 = \mathsf{a}_6 = 0$. These six dispersivity coefficients, a_1–a_6, *are properties of the porous medium only*, meaning that they are independent of the flow taking place in the porous medium, and the chosen coordinate system.

Based on (7.1.39), the corresponding expression for the dispersion coefficients D_{ij} is

$$\mathsf{D}_{ij} = \left[\mathsf{a}_1\delta_{ij} + \mathsf{a}_2\frac{V_iV_j}{V^2} + \mathsf{a}_3 e_i e_j + \mathsf{a}_4\delta_{ij}\frac{(V_k e_k)^2}{V^2} + \mathsf{a}_5\frac{V_k e_k}{V}\frac{V_i e_j + V_j e_i}{V^2} + \mathsf{a}_6 e_i e_j \frac{(V_k e_k)^2}{V^2}\right] V. \quad (7.1.49)$$

We note that the dispersion coefficient, which is used for determining the dispersive flux by means of (7.1.32), depends not only on the porous medium (through the dispersivity coefficients), but also on the velocity vector. As velocity may vary in space and time, so does the dispersion coefficients.

In order to model solute transport in a transversely isotropic porous medium under general flow conditions, we need to determine the *six* independent dispersivities, or dispersivity coefficients. These a_i coefficients can be determined by conducting tracer tests in the field, and comparing tracer concentrations within a plume with available analytical or numerical solutions. Generally, this parameter determination is conducted as an inverse solution procedure in which the optimal solution is obtained by minimizing the sum of square errors between the observed and theoretically predicted concentrations (Sec. 11.3).

As the expression for dispersive flux, (7.1.32), and the solute transport equation (7.2.23) introduced in Sub. 7.2.2, involve six dispersion coefficients D_{ij} (six, instead of nine, because of the symmetry $\mathsf{D}_{ij} = \mathsf{D}_{ji}$), the first step in a parameter estimation procedure is to determine these six components of the dispersion coefficient. Given a transversely isotropic aquifer with known axis of symmetry (i.e, known vector components e_i), and known flow (i.e., known V_i), in principle, it is possible to determine the dispersion coefficients from information on observed concentrations during a controlled experiment.

Once the six dispersion components D_{ij} have been determined for a given location through such experiments, the next step is to solve the following linear system of equations, based on (7.1.49), for the six coefficients, a_1–a_6,

$$[\mathsf{EV}]\begin{Bmatrix}\mathsf{a}_1\\ \mathsf{a}_2\\ \mathsf{a}_3\\ \mathsf{a}_4\\ \mathsf{a}_5\\ \mathsf{a}_6\end{Bmatrix}=\begin{Bmatrix}\mathsf{D}_{11}\\ \mathsf{D}_{22}\\ \mathsf{D}_{33}\\ \mathsf{D}_{12}\\ \mathsf{D}_{13}\\ \mathsf{D}_{23}\end{Bmatrix}, \tag{7.1.50}$$

in which $[\mathsf{EV}]$ is a 6×6 matrix containing the expressions of e_i and V_i, as defined in (7.1.49). However, a matrix analysis shows that the matrix $[\mathsf{EV}]$ *is of rank four*, meaning that of the six equations defined in (7.1.50) (or (7.1.49)), only four are *linearly independent*, and the other two are linearly dependent on the rest. This means that (7.1.50) can yield at most four a_i values.

While the above conclusion has been proven for the general case of any orientation of coordinate system, it is of interest to examine some special cases. In the case of a horizontally layered material, selecting the z-axis to coincide with the axis of material symmetry, i.e., $e_3 = 1$ and $e_1 = e_2 = 0$, we express (7.1.49) as

$$\begin{aligned}
\mathsf{D}_{xx} &= \left(\mathsf{a}_1 + \mathsf{a}_2\frac{V_x^2}{V^2} + \mathsf{a}_4\frac{V_z^2}{V^2}\right)V,\\
\mathsf{D}_{yy} &= \left(\mathsf{a}_1 + \mathsf{a}_2\frac{V_y^2}{V^2} + \mathsf{a}_4\frac{V_z^2}{V^2}\right)V,\\
\mathsf{D}_{zz} &= \left[\mathsf{a}_1 + \mathsf{a}_3 + (\mathsf{a}_2 + \mathsf{a}_4 + 2\mathsf{a}_5 + \mathsf{a}_6)\frac{V_z^2}{V^2}\right]V,\\
\mathsf{D}_{xy} = \mathsf{D}_{yx} &= \mathsf{a}_2\frac{V_xV_y}{V^2}\,V,\\
\mathsf{D}_{xz} = \mathsf{D}_{zx} &= (\mathsf{a}_2 + \mathsf{a}_5)\frac{V_xV_z}{V^2}\,V,\\
\mathsf{D}_{yz} = \mathsf{D}_{zy} &= (\mathsf{a}_2 + \mathsf{a}_5)\frac{V_yV_z}{V^2}\,V.
\end{aligned} \tag{7.1.51}$$

By eliminating the factor $(\mathsf{a}_2 + \mathsf{a}_5)$ between the fifth and sixth equations in (7.1.51), we can clearly see that D_{xz} and D_{yz} are related to each other. Further analysis shows that the matrix is of rank four, and the following constraints must be satisfied in order for the system of equations to have a solution:

$$\mathsf{D}_{yz} = \frac{V_y}{V_x}\,\mathsf{D}_{xz},$$
$$\mathsf{D}_{xx} = \mathsf{D}_{yy} + \frac{V_x^2 - V_y^2}{V_x V_y}\,\mathsf{D}_{xy}. \tag{7.1.52}$$

Hence, *only four of the six dispersion coefficients are independent.*

Often, it is convenient to use a *local coordinate system* that coincides with the flow direction at the considered location. For example, in the case of uniform flow in the direction of the x_1-axis (see Fig. 7.1.4), such that $V_1 = V$ and $V_2 = V_3 = 0$, we can show the following interdependence among dispersion coefficients:

$$\mathsf{D}_{xy} = \frac{e_y}{e_z}\,\mathsf{D}_{xz},$$
$$\mathsf{D}_{yy} = \mathsf{D}_{zz} + \frac{e_y^2 - e_z^2}{e_y e_z}\,\mathsf{D}_{yz}. \tag{7.1.53}$$

This is the same conclusion as the above, that is, only four of the dispersion coefficients are independent.

The important conclusions of the above analysis are:

(a) In a single field experiment, in which the flow conditions remain unchanged, it is possible to determine only four dispersion coefficients at any one location, due to the required interdependency given either by (7.1.52) or by (7.1.53).
(b) Given these four independent dispersion coefficients, it is not possible to resolve the six dispersivity coefficients, a_1–a_6, from (7.1.50), due to the rank deficiency of the matrix.
(c) However, as demonstrated below (see also Fel and Bear, 2009), it is possible to determine the six dispersivity coefficients if two experiments are conducted.
(d) In a forward modeling problem, in which values of six dispersion coefficients are required as input, one needs to check the consistency of the assigned dispersion values. These values need to be either determined from (7.1.49), based on the six dispersivity coefficients, or satisfy the relations as shown in (7.1.52) or (7.1.53).

Next, let us consider two special flow cases in the layered medium considered above. In the following discussion, we shall choose the z-axis to coincide with the material axis of symmetry, i.e., $e_3 = 1$ and $e_1 = e_2 = 0$.

In the first case, we consider uniform flow *normal* to the layers, that is, in the z-direction, such that $V_3 = V$ and $V_1 = V_2 = 0$. Using this condition in (7.1.49), we obtain

$$\mathbf{D}^V = \begin{bmatrix} \mathsf{a}_{TH}^V & 0 & 0 \\ 0 & \mathsf{a}_{TH}^V & 0 \\ 0 & 0 & \mathsf{a}_{LV}^V \end{bmatrix} V, \qquad \begin{aligned} \mathsf{a}_{TH}^V &= \mathsf{a}_1 + \mathsf{a}_4, \\ \mathsf{a}_{LV}^V &= \mathsf{a}_1 + \mathsf{a}_2 + \mathsf{a}_3 + \mathsf{a}_4 + 2\mathsf{a}_5 + \mathsf{a}_6, \end{aligned} \tag{7.1.54}$$

where the superscript $(.)^V$ is used to emphasize that the flow direction is vertical, a_{TH}^V is the transverse dispersivity in the horizontal direction (only one value because of the isotropy in the horizontal plane), and a_{LV}^V is the longitudinal dispersivity in the vertical direction. Altogether, to describe dispersion in a layered horizontal porous medium, when the flow is uniform and normal to the layers, we need only one longitudinal and one transversal dispersivities.

As a second case, we consider uniform flow *parallel* to the layers, say, in the $+x$-axis direction, such that $V_1 = V$ and $V_2 = V_3 = 0$. Equation (7.1.49) becomes:

$$\mathbf{D}^H = \begin{bmatrix} \mathsf{a}_{LH}^H & 0 & 0 \\ 0 & \mathsf{a}_{TH}^H & 0 \\ 0 & 0 & \mathsf{a}_{TV}^H \end{bmatrix} V, \quad \begin{aligned} \mathsf{a}_{LH}^H &= \mathsf{a}_1 + \mathsf{a}_2, \\ \mathsf{a}_{TH}^H &= \mathsf{a}_1, \\ \mathsf{a}_{TV}^H &= \mathsf{a}_1 + \mathsf{a}_3, \end{aligned} \tag{7.1.55}$$

where a_{TH}^H and a_{TV}^H are, respectively, the transverse dispersivities in the horizontal and in the vertical directions, and a_{LH}^H is the longitudinal dispersivity in the horizontal direction. Thus, to describe dispersion in a layered horizontal porous medium, when flow is uniform and parallel to the layers, we need one longitudinal and two transversal dispersivities.

As observed in the cases discussed above, under uniform flow conditions, we can only determine two, three, or four independent dispersion coefficients in a single experiment, depending on whether the flow is perpendicular, parallel, or at an angle, to the material symmetry axis. This implies that *at least two flow tests in different flow directions are needed, and one of the two directions must be inclined with respect to the direction of the material symmetry axis.* For example, if we conduct a horizontal flow test, and obtain result as in (7.1.55), we can determine three dispersivity coefficients, a_1, a_2 and a_3. For the second test, the flow should be neither in the vertical, nor in the horizontal, direction, as there will not be sufficient information to determine the remaining three coefficients. The flow of the second test must be in an inclined direction with the horizontal plane and the vertical axis, which will provide four additional equations. The remaining three coefficients can then be determined under overdeterminancy condition. Similar statement was presented by Fel and Bear (2009) for the special case of flow in the horizontal direction, and making a 45° angle with the axis of symmetry.

In the above, we have assumed that the direction of the axis of symmetry is known *a priori*, i.e., we know the three values: e_1, e_2 and e_3 that appear in (7.1.49). If this direction is not known, we have to use the experimental data to solve the inverse problem also for two of these three components of $\mathbf{e}$ (because $e_1^2 + e_2^2 + e_3^2 = 1$), for a total of 8 unknown values. In this case, two flow tests in two different inclined directions (with respect to the materials axis of symmetry) are sufficient for the determination of these 8 unknowns.

C. Anisotropy with tetragonal symmetry

As an example of such porous medium material, we may consider one that is made up of orderly packed solid boxes $a \times a \times c$, with equal spacing between the boxes in all directions (or cubes with 3 different spaces). For this case, the 36 independent $\mathsf{a}_{ijk\ell}$-components can be expressed by 7 independent, parameters, which are subject to certain constraints (Bear *et al.*, 2009). It is interesting to note that this case is not identical to the case with axial symmetry (such as a stratified aquifer), considered above. Here, we also need information on the directions in which the boxes, $a \times a \times c$, are positioned in space, e.g., in the form of two of the three e_i's. This case is analyzed in detail by Bear *et al.* (2009).

D. Anisotropy with orthorhombic symmetry

An example is a porous medium material made up of orderly packed solid boxes $a \times b \times c$ with equal spacing between the boxes in all directions (or cubes with three different spaces). For this case, the 36 independent dispersivity components can be expressed by *twelve* independent parameters. We also need information on the directions in which the boxes, $a \times b \times c$, are oriented in space (and this, as indicated earlier, requires information on two e_i's).

It is possible to analyze three special cases of flow, each one with uniform flow parallel to one of the three axes. To describe dispersion in each of these three cases we need only three coefficients: a longitudinal dispersivity and two transversal ones.

In each of the material symmetry cases discussed above, the number of independent coefficients is accompanied by a number of constraints that these coefficients have to satisfy. The information concerning the number of independent coefficients and the constraints among them (Bear *et al.*, 2009) is important when experiments are conducted aimed at determining the values of these coefficients for a specific porous medium, by using an inverse method.

Similar to the discussion presented with respect to the experimental procedure for determining the dispersivity coefficients in the case of transverse isotropy, here also, a number of independent experiments will be required. Also, in practice, because of the inaccuracy and uncertainty in the measured values during experiments, (say of concentrations and piezometric heads) an optimization procedure that minimizes the overall error may be called for.

E. Other models for dispersion in anisotropic domains

Some authors, on the basis of field observations, have suggested that for flow parallel to the horizontal stratification in a stratified (= layered) aquifer, transverse dispersion is much smaller in the vertical direction than in the horizontal one, i.e., $\mathsf{a}^H_{TH} \gg \mathsf{a}^H_{TV}$ in (7.1.55) (Robson, 1974, 1978; Garabedian *et al.*, 1991; Gelhar *et al.*, 1992). Based on the above observation, Burnett and Frind (1987) (see also Jensen *et al.*, 1993; Zheng and Bennett, 1995) suggested a 'working model' for transversely isotropic porous medium, in which the dispersion tensor is defined by *three* dispersivities only (rather than six, see Subs. 7.1.6B): a longitudinal dispersivity, a_L, and two transversal dispersivities, a horizontal one, a_{TH}, and a vertical one, a_{TV}. The components of the dispersion tensor in three dimensions, with the z-axis as the axis of material symmetry, are presented as:

$$\begin{aligned}
\mathsf{D}_{xx} &= \frac{1}{V}\left(\mathsf{a}_L V_x^2 + \mathsf{a}_{TH} V_y^2 + \mathsf{a}_{TV} V_z^2\right),\\
\mathsf{D}_{yy} &= \frac{1}{V}\left(\mathsf{a}_{TH} V_x^2 + \mathsf{a}_L V_y^2 + \mathsf{a}_{TV} V_z^2\right),\\
\mathsf{D}_{zz} &= \frac{1}{V}\left(\mathsf{a}_{TV} V_x^2 + \mathsf{a}_{TV} V_y^2 + \mathsf{a}_L V_z^2\right),\\
\mathsf{D}_{xy} = \mathsf{D}_{yx} &= \frac{1}{V}\left(\mathsf{a}_L - \mathsf{a}_{TH}\right) V_x V_y,\\
\mathsf{D}_{xz} = \mathsf{D}_{zx} &= \frac{1}{V}\left(\mathsf{a}_L - \mathsf{a}_{TV}\right) V_x V_z,\\
\mathsf{D}_{yz} = \mathsf{D}_{zy} &= \frac{1}{V}\left(\mathsf{a}_L - \mathsf{a}_{TV}\right) V_y V_z.
\end{aligned} \tag{7.1.56}$$

These expressions can be compared with those for the isotropic case, (7.1.45). Burnett and Frind (1987) further *assumed* that $\mathsf{a}_{TH} \gg \mathsf{a}_{TV}$. Most solute transport codes (Sec. 8.8), such as MT3D (Zheng, 1990), MOC3D (Konikow *et al.*, 1996), and PHAST (Parkhurst *et al.*, 2004), use this formulation. The relations presented in (7.1.56), however, are not consistent with (7.1.49). In fact, Lichtner *et al.* (2002, 2008) have demonstrated that (7.1.56) does not conform with tensor transformation rules, suggesting that it is not an acceptable model.

Based on a turbulence model investigated by Batchelor (1959), using a method introduced by Robertson (1940), Poreh (1965) suggested a model that is based on four dispersivity coefficients,

$$\mathsf{D}_{ij} = \left[\alpha_1\delta_{ij} + \alpha_2\frac{V_iV_j}{V^2} + \alpha_3 e_i e_j + \frac{\alpha_4}{2}\frac{e_iV_j + e_jV_i}{V}\right]V. \tag{7.1.57}$$

Comparing the above with (7.1.49), we observe that Poreh's (1965) model is equivalent to setting $V_k e_k/V$ to zero for terms associated with a_4 and a_6, and to 1 for the term associated with a_5. Here we notice that

$$\frac{V_k e_k}{V} = \frac{\mathbf{V}\cdot\mathbf{e}}{V} = \cos\theta, \tag{7.1.58}$$

where θ is the angle between the material axis and the flow direction. A reason for the absence of such terms might be that the *turbulence* model lacks the *material anisotropy* aspect, because fluid as a material is *isotropic*; while in a porous medium we have the additional effect of material anisotropy, represented by the $\cos\theta$ term.

Lichtner *et al.* (2002) examined the Poreh (1965) model and discussed the need for introducing $\cos\theta$ as a factor in the constitutive model. As a result, a three parameter model (called a 'four parameter model' in Lichtner *et al.* (2002)) was proposed. By selecting z as the material axis of symmetry

($e_3 = 1$, $e_1 = e_2 = 0$), the components of the dispersion coefficient in Lichtner *et al.* (2002) model are

$$
\begin{aligned}
\mathsf{D}_{xx} &= \left[\mathsf{a}_L \frac{V_x^2}{V^2} + \mathsf{a}_{TH} \frac{V_y^2}{V_x^2 + V_y^2} + \mathsf{a}_T \frac{V_x^2 V_z^2}{V^2 (V_x^2 + V_y^2)}\right] V, \\
\mathsf{D}_{yy} &= \left[\mathsf{a}_L \frac{V_y^2}{V^2} + \mathsf{a}_{TH} \frac{V_x^2}{V_x^2 + V_y^2} + \mathsf{a}_T \frac{V_y^2 V_z^2}{V^2 (V_x^2 + V_y^2)}\right] V, \\
\mathsf{D}_{zz} &= \left[\mathsf{a}_L \frac{V_z^2}{V^2} + \mathsf{a}_T \frac{V_x^2 + V_y^2}{V^2}\right] V, \\
\mathsf{D}_{xy} = \mathsf{D}_{yx} &= \left[\mathsf{a}_L \frac{V_x V_y}{V^2} + \mathsf{a}_{TH} \frac{V_x V_y}{V_x^2 + V_y^2} + \mathsf{a}_T \frac{V_x V_y V_z^2}{V^2 (V_x^2 + V_y^2)}\right] V, \\
\mathsf{D}_{xz} = \mathsf{D}_{zx} &= \left[(\mathsf{a}_L - \mathsf{a}_T) \frac{V_x V_z}{V^2}\right] V, \\
\mathsf{D}_{yz} = \mathsf{D}_{zy} &= \left[(\mathsf{a}_L - \mathsf{a}_T) \frac{V_y V_z}{V^2}\right] V,
\end{aligned} \tag{7.1.59}
$$

where

$$
\begin{aligned}
\mathsf{a}_L &= \alpha_1 + \alpha_2 - \alpha_3 \cos^2\theta, \\
\mathsf{a}_T &= \alpha_1 + \alpha_3 (1 - \cos^2\theta), \\
\mathsf{a}_{TH} &= \alpha_1,
\end{aligned} \tag{7.1.60}
$$

with α_1, α_2, and α_3 as the three material coefficients (which are different from those defined in (7.1.57)). The relation (7.1.59) also shows that at any location in the flow field, only three of the dispersivity coefficients can be independent.

For vertical flow ($\theta = 0$), (7.1.59) reduces to

$$
\mathsf{D}_{ij}^{V} = \begin{bmatrix} \alpha_1 & 0 & 0 \\ 0 & \alpha_1 & 0 \\ 0 & 0 & \alpha_1 + \alpha_2 - \alpha_3 \end{bmatrix} V = \begin{bmatrix} \mathsf{a}_{TH}^{V} & 0 & 0 \\ 0 & \mathsf{a}_{TH}^{V} & 0 \\ 0 & 0 & \mathsf{a}_{LV}^{V} \end{bmatrix} V, \tag{7.1.61}
$$

and for horizontal flow ($\theta = 90°$),

$$
\mathsf{D}_{ij}^{H} = \begin{bmatrix} \alpha_1 + \alpha_2 & 0 & 0 \\ 0 & \alpha_1 & 0 \\ 0 & 0 & \alpha_1 + \alpha_3 \end{bmatrix} V = \begin{bmatrix} \mathsf{a}_{LH}^{H} & 0 & 0 \\ 0 & \mathsf{a}_{TH}^{H} & 0 \\ 0 & 0 & \mathsf{a}_{TV}^{H} \end{bmatrix} V. \tag{7.1.62}
$$

In the above, we observe

$$
\begin{aligned}
\mathsf{a}_{LV}^{V} &= \alpha_1 + \alpha_2 + \alpha_3, \quad \mathsf{a}_{LH}^{H} = \alpha_1 + \alpha_2, \\
\mathsf{a}_{TV}^{H} &= \alpha_1 + \alpha_3, \quad \mathsf{a}_{TH}^{H} = \mathsf{a}_{TH}^{V} = \alpha_1.
\end{aligned} \tag{7.1.63}
$$

We note that the dispersivities, a^V_{LV}, a^H_{LH}, etc., defined above, are not the same as those in (7.1.54) and (7.1.55). From (7.1.61) and (7.1.62), we note the important consequence that $D^H_{TH} = D^V_{TH}$, i.e., the horizontal transverse dispersion coefficient of the horizontal flow is equal to the transverse dispersion coefficient of the vertical flow. We also note that $D^H_{TV} = D^V_{LV} - D^H_{LH} + D^H_{TH}$. In other words, if we conduct a horizontal and a vertical uniform flow tests, the dispersion coefficients obtained are related to each other.

F. Additional comments on dispersion

Before leaving the discussion on the flux of a dissolved chemical species, let us mention a number of phenomena, which may further affect the spreading and travel time of solutes in porous media:

Multiphase flow In multiphase flow, a dispersivity is associated with each fluid phase. Thus, each of the dispersivity components, e.g., the longitudinal and transversal dispersivities, depends on the phase configuration within the void space. Hence, each of these components is a function of the phase saturation. However, very little information on these functions is available to date. Probably because of this reason, the dependence on saturation is usually overlooked in practice.

Non-Fickian dispersion model In recent years, several researchers (e.g., Berkowitz *et al.*, 2000, 2002; Berkowitz and Scher, 2001) have demonstrated that even in a relatively homogeneous porous medium, the dispersive flux cannot be expressed as a Fickian-type law.

Consider a solute slug injected into an aquifer with uniform flow. After waiting some initial short period, so that the use of the continuum approach will be justified, the tracer spreads out such that, in a Fickian model, contours of constant solute concentration can be described as confocal ellipsoids (in 3-D), indicating a binormal distribution. Such an experiment is shown, conceptually, in Fig. 7.1.1b. The *size* of the contaminant cloud, estimated by its longitudinal and transverse standard deviations, σ_L, and σ_T, respectively, grows with the square root of time

$$\sigma_L = \sqrt{2D_L t} = \sqrt{2D_L L/V}, \quad \text{and} \quad \sigma_T = \sqrt{2D_T t} = \sqrt{2D_T L/V}, \tag{7.1.64}$$

where $L(= Vt)$ is the distance that the center of mass has traveled during time t, at the mean velocity V.

It is well known from the dispersion phenomena observed in pipe flow (Taylor, 1953, 1954), natural streams (Liu and Cheng, 1980), and groundwater flow, that at small times, the dispersion coefficient is not a constant; rather it grows with time and reaches a constant only after a sufficiently long time (or large distance traveled). Also, the concentration cloud is skewed toward the source, and becomes Gaussian only after a certain distance traveled. Other models based on laboratory observation and random walk models (Berkowitz *et al.*, 2000, 2002; Berkowitz and Scher, 2001) have indicated that the center of mass of the cloud does not travel with the flow velocity, and $L \sim t^\beta$, $\sigma \sim t^\beta$, where $0 < \beta < 1$. It is of interest to observe that for a Fickian model, $L \sim t$, $\sigma \sim t^{1/2}$, and $L/\sigma \sim t^{1/2}$; while for this non-Fickian model, $L/\sigma \sim$ constant.

Ion exclusion Because of the electrical charge on certain solid surfaces, a chemical species that is an ion may be repelled from the solid wall of the void space, where the water velocity is small (recalling that we assume that water is adsorbed to the solid wall), and move mainly in the regions of higher velocity inside the void space. The average velocity of the water carrying and dispersing the species is, thus, higher than for non-ionic contaminants. As a consequence, the advective flux of the contaminant will be higher, and so will the coefficient of dispersion, which is proportional to the average velocity. This phenomenon has also been called *charge exclusion* (Gvirtzman and Magaritz, 1989; Gvirtzman and Gorelick, 1991).

Size exclusion Some molecules, or ions, are so large that their travel is restricted to the larger pores. As a consequence, they are carried (by advection) at a higher average water velocity. The higher average velocity also results in a higher coefficient of dispersion. This phenomenon is more prevalent in fine-grained soils and for large molecules, like organic macromolecules.

Although this section deals with a dissolved component, we would like also to point out that the magnitude of exclusion is particularly important when considering the transport of microorganisms and of colloidal particles that may carry contaminants, because of their relatively large size.

7.1.7 Total flux

We may now combine the three modes of transport of a chemical species—advection, dispersion, and diffusion, and write the total macroscopic flux (per unit area of a fluid f-phase), $\mathbf{J}^\gamma_{\text{total}}$, in the form:

$$\begin{aligned}\mathbf{J}^\gamma_{\text{total}} &= \overline{c}^f \overline{\mathbf{V}}^f + \mathbf{J}^\gamma + \mathbf{J}^{*\gamma} \\ &= \overline{c}^f \overline{\mathbf{V}}^f - \mathsf{D}^\gamma_h \cdot \nabla \overline{c}^f,\end{aligned} \tag{7.1.65}$$

where

$$\mathsf{D}^\gamma_h \equiv \mathsf{D} + \mathcal{D}^{*\gamma} \tag{7.1.66}$$

denotes the *coefficient of hydrodynamic dispersion* of the chemical species. Note that the average in $\overline{c}^f$ is an intrinsic phase averages. Henceforth, we shall omit the symbol that denotes the intrinsic phase average.

7.1.8 Field-scale heterogeneity

In Subs. 7.1.4, the phenomenon of solute dispersion was shown to be a consequence of the heterogeneity of the porous medium *at the microscopic scale*, i.e., due to the presence of a solid matrix and a void space within the REV. A grain or pore diameter, or the hydraulic radius of the pore space, was suggested as the *scale* of this heterogeneity. This heterogeneity produces the velocity variations that take place inside the void space. The dispersive flux, a macroscopic level concept obtained by averaging over an REV, was introduced as a means for circumventing the need to know the details of the velocity distribution and of other transport features at the microscopic level.

As emphasized in Subs. 1.3.4, subsurface domains, which are the porous medium domains of interest in this book, are highly heterogeneous with respect to their macroscopic coefficients, e.g., porosity and permeability. We have introduced the term 'megascopic level', obtained by smoothing out variations at the macroscopic level, and introduced the concept of 'scale of heterogeneity', indicating that at the macroscopic level, variations, say in permeability, may occur at different scales. In fact, this multiple scale heterogeneity is a dominant factor in the subsurface. Because pressure propagates very fast, the effect of this inherent heterogeneity is less noticeable when considering fluid flow. However, its effect on the transfer of the mass of a dissolved chemical species is significant.

In principle, it should be possible to solve a transport problem at the macroscopic level in any heterogeneous domain in which the spatial variations of the permeability and the other relevant coefficients are known. Indeed, in small scale field problems, e.g., in the vicinity of an injection well, or for a small distance downstream of a pollution source, the formation properties (porosity, permeability, dispersivity) may be known (or estimated), and the problem of predicting the concentration distribution of the injected solute, or of the advancing pollution plume, can be solved by making use of the (macroscopic level deterministic) model described in this chapter. However, usually, especially if we are interested in a pollution plume that advances a large distance, sometimes measured in kilometers, we face a situation similar to that which is encountered at the microscopic level, viz., that the detailed information about the spatial variation of the relevant parameters is not known, due to the heterogeneity inherent in such domains. The way we overcome the lack of information about the heterogeneity at the microscopic level (resulting from pore scale heterogeneity) is to use homogenization, or averaging over an REV, as discussed in Sec. 1.3. One may visualize this averaging as a *smoothing operation*. As a consequence of the averaging process, the phenomenon of dispersion was introduced. The same averaging, or smoothing approach, may also be applied to heterogeneities that are encountered at the macroscopic level, to obtain a continuum at the *megascopic level*. Such an averaging volume was referred to as the *representative macroscopic volume* (RMV) (Subs. 1.3.4B). As indicated in (1.3.14), the characteristic size of this volume, ℓ^*, is constrained by

$$d^* \ll \ell^* \ll L,$$

where d^* is a length characterizing the macroscopic heterogeneity that we wish to smooth out, and L is a length characterizing the porous medium domain. Similar to what happens during microscopic-to-macroscopic smoothing, here, the information about the heterogeneity at the macroscopic level appears at the megascopic level in the form of various coefficients.

Denoting the volume of an RMV by $\overline{\mathcal{U}}_o$, and the *macroscopic value* of e by $\overline{e}$, we define the *megascopic value* of e by:

$$\overline{\overline{e}}(\mathbf{x},t) \equiv \frac{1}{\overline{\mathcal{U}}_o} \int_{\overline{\mathcal{U}}_o} \overline{e}(\mathbf{x}',t;\mathbf{x})\, d\overline{\mathcal{U}}_o(\mathbf{x}'), \tag{7.1.67}$$

where $\mathbf{x}$ and $\mathbf{x}'$ denote the centroid of the RMV and of a point (of the porous medium regarded as a continuum) inside it, respectively. With this definition, we may now derive the total flux of a γ-component at the megascopic level, by averaging (7.1.65) over an RMV. For saturated flow, we obtain:

$$\begin{aligned}\overline{\overline{\theta \mathbf{J}^{t\gamma}}} &= \overline{\overline{\overline{c}^f \mathbf{q}}} + \overline{\overline{\phi\,(\mathbf{J}^{\gamma} + \mathbf{J}^{*\gamma})}} \\ &= \overline{\overline{\overline{c}^f}}\,\overline{\overline{\mathbf{q}}} + \overline{\overline{\hat{\overline{c}}^f\,\hat{\mathbf{q}}}} + \overline{\overline{\phi\,(\mathbf{J}^{\gamma} + \mathbf{J}^{*\gamma})}} \\ &\approx \overline{\overline{\overline{c}^f}}\,\overline{\overline{\mathbf{q}}} + \overline{\overline{\hat{\overline{c}}^f\,\hat{\mathbf{q}}}},\end{aligned} \tag{7.1.68}$$

where a double bar over a macroscopic value indicates a megascopic value obtained by averaging over an RMV, with $\overline{c} = \theta\overline{c}^f$, and $\hat{(..)}$, defined by:

$$\hat{\overline{(..)}}^f = \overline{(..)}^f - \overline{\overline{\overline{(..)}^f}},$$

is the deviation of a macroscopic value at any point within an RMV, from its average over the RMV. We note that the flux on the left-hand side of (7.1.68) (and hence all other terms) is per unit area of porous medium.

As could have been expected, the megascopic total flux contains two *new* additional dispersive fluxes, which result from the variability of the relevant macroscopic quantities. One is $\overline{\overline{\hat{\overline{c}}^f\,\hat{\mathbf{q}}}}$, which will be referred to as the *macrodispersive flux* of the chemical species. The other is the average over the RMV of the sum of the dispersive and diffusive fluxes at the macroscopic level. Note that on the last line of (7.1.68) we have neglected the second dispersive flux as being much smaller than the first.

Altogether, the total flux is again the sum of an advective flux and a dispersive one. There is no analogy here to the diffusive flux, as we have neglected it. At very low velocities, we may not neglect the average of the macroscopic diffusive flux.

We have to express the dispersive flux at the megascopic level in terms of megascopic quantities, in the same manner as is done for describing transport at the macroscopic level. We usually *assume* that a Fickian-type dispersion law, e.g., (7.1.32), is also valid for describing the macrodispersive flux. A *macrodispersivity*, a_{ijkm}, can be defined in the same way as the dispersivity was defined earlier in (7.1.39). Bear (1979), while developing the vertically integrated mass balance equation for a component of a phase, suggested for the general case of an anisotropic porous medium, an expression for macrodispersivity in the form:

$$\mathsf{a}_{ijkm} = \frac{\overline{\hat{\mathsf{K}}_{in}\hat{\mathsf{K}}_{j\ell}}}{\overline{\mathsf{K}_{kn}}\;\overline{\mathsf{K}_{m\ell}}}\,\widetilde{L}, \tag{7.1.69}$$

where K_{ij} denotes the ij-th component of the hydraulic conductivity tensor, and $\widetilde{L}$ is a length that characterizes the inhomogeneity of the aquifer, resulting from stratification. It is a fourth rank tensor, which is analogous to the dispersivity at the macroscopic level (i.e., with a_L and a_T, etc.). In an isotropic porous medium, the macrodispersivity reduces to a scalar. Gelhar (1976) and Gelhar *et al.* (1979) analyzed the dependence of macrodispersion on permeability variations. For horizontal flow in an isotropic confined aquifer, they suggested that

$$\mathsf{a}_L = \frac{1}{3}\frac{L_1^2\sigma^2_{\ln \mathsf{k}}}{\mathsf{a}_T}, \tag{7.1.70}$$

in which L_1 is a correlation distance (= distance along which permeabilities are still correlated), $\sigma_{\ln \mathsf{k}}$ is the standard deviation of $\ln \mathsf{k}$, and a_T is the transverse dispersivity.

Altogether, we may summarize this topic by suggesting that dispersion and macrodispersion are analogous phenomena, in that both are consequences of velocity variations that are due to heterogeneity, but at different scales. Dispersion arises from velocity variations *within* the void space (i.e., at the microscopic level), caused by the presence of the solid surfaces. Macrodispersion is produced by macroscopic velocity variations, caused by variations in the permeability and porosity. In both cases, the flux is the sum of an advective flux and a (hydrodynamic) dispersive one, written at the respective levels. The structure of the coefficient of dispersion is the same in both cases, and so is the relationship between the coefficient of dispersion, the dispersivity, and the average velocity. In practice, we use exactly the same mathematical model (except that in the case of field scale, we usually neglect the flux due to molecular diffusion), but select the magnitude of the dispersivity according to the scale of heterogeneity.

In laboratory column experiments, the porous medium is more or less homogeneous, say with respect to permeability and porosity. The scale of heterogeneity is that of the size of a grain or a pore. Indeed, the magnitude of longitudinal dispersivity found in numerous column experiments is approximately equal to a pore- or grain-size. However, under field conditions, the scale of heterogeneity, due to variability in permeability and porosity, is much larger. In fact *this scale grows with the size of the domain.* Gelhar *et al.* (1992) compiled a large number of field experiments and presented the observed longitudinal dispersivities, a_L, as a function of the travel distance, L_s, as shown in Fig. 7.1.5. It is clear that macrodispersivity is proportional to the size of

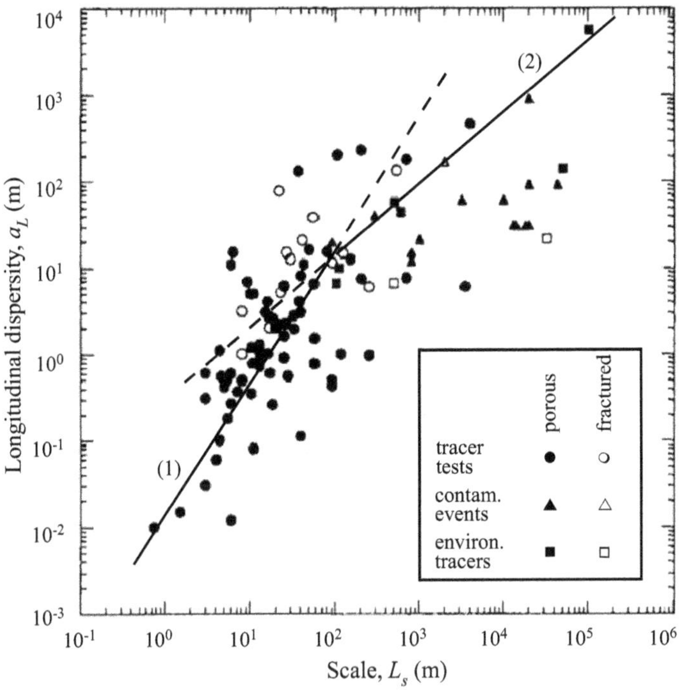

Figure 7.1.5: Longitudinal dispersivity versus plume travel distance for various types of observations and media (Gelhar et al., 1992). Line marked as (1): equation (7.1.71); and (2): equation (7.1.72).

the field, although the data shows a wide range of scatter. Lallemand-Barrés and Peaudecerf (1978) analyzed published values of dispersivity and showed that, on the average, the dispersivity increases with the distance (between a few meters and 10 km), between the source and the point of observation. As a 'rule of thumb', they concluded that the dispersivity can be approximated as 1/10 of the distance traveled. This is often referred to as a 'scale effect'.

Based on the argument of self-similar (fractal) hierarchy of logarithmic hydraulic conductivity, Neuman (1990) suggested a universal scaling law and presented the following equations based on the least square fit of the data:

$$\mathsf{a}_L = 0.017\, L_s^{1.5}; \quad L_s \leq 100 \text{ m}; \tag{7.1.71}$$

$$\mathsf{a}_L = 0.32\, L_s^{0.83}; \quad L_s > 100 \text{ m}. \tag{7.1.72}$$

These two empirical formulas are plotted in Fig. 7.1.5. Gelhar *et al.* (1992, 1993), however, cautioned the use of these power laws by pointing out the large scatter in data (2–3 orders of magnitude) in Fig. 7.1.5.

Gelhar and Axness (1983) and Dagan (1984) (see also (7.1.70)) showed that the longitudinal dispersivity is also proportional to the product of the variance of the logarithm of the hydraulic conductivity, and the correlation length scale, i.e.,

$$\mathsf{a}_L \sim L_1 \sigma^2_{\ln \mathsf{k}}. \tag{7.1.73}$$

This can explain the range of scatter observed in Fig. 7.1.5.

In practice, often, the first estimate of the longitudinal dispersivity, prior to actual calibration, is taken as 1/10 of the size of the *domain of interest.* Thus, for example, for a domain of interest which is 10 m long, we estimate $\mathsf{a}_L = 1$ m. For a domain that is hundreds of meters in size, we estimate a_L in the range of tens of meters. The horizontal transverse dispersivity is estimated as approximately equal to about 1/10 of the longitudinal one. The vertical transverse dispersivity, in a layered horizontal aquifer, is 1–2 orders of magnitude smaller than the horizontal one (Gelhar *et al.*, 1992). Obviously, these are merely orders of magnitude and may be used as initial or preliminary estimates only. In each particular case, the actual value should be determined by some model calibration procedure.

Apart from the different magnitude of the dispersivity to be employed, the expressions for advective and dispersive fluxes presented in Subs. 7.1.2 and 7.1.5 may be assumed to remain valid when modeling field conditions.

7.2 Balance Equation for Single Species

As in the case of fluid flow, the flux equation, (7.1.65), contains two variables: the total flux and the concentration. This means that we need one more equation; this is the mass balance equation for the γ-species. We have not counted the velocity as a variable as we can obtain it by writing and solving the relevant flow model.

We have already introduced the concept of the balance of an extensive quantity in Chap. 5, where we considered the balance of fluid mass. In this section, we are interested in the balance of the mass of a chemical species, or a component dissolved in a fluid phase. We shall introduce this topic through a very simple model. In spite of its simplicity, this model may provide useful insight in practice, especially during the initial stage of an investigation, by indicating whether or not a more sophisticated model is required.

7.2.1 Single cell model

As explained in Subs. 5.1.1, a balance is written for an extensive quantity within *a specified spatial domain*, for a *specified period* of time. Let us denote the time interval by Δt, and the volume of the porous medium domain for which the balance is written by $\mathcal{U}_o$. A balance of this kind describes the integrated behavior within the domain, usually referred to as a *cell*, or *compartment*, during the balance period. Sometimes, we consider a number of

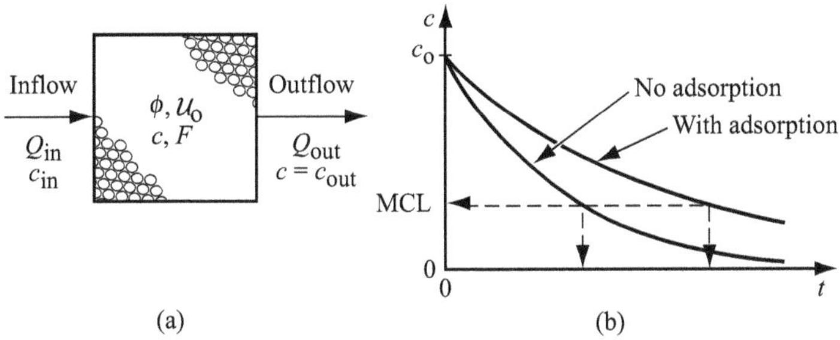

Figure 7.2.1: (a) A single cell model. (b) $c = c(t)$ in the cell.

adjacent cells, in a one-, two-, or three-dimensional configuration. To obtain the integrated behavior in the cell *at a point in time*, we divide both sides of the balance equation by Δt, and let Δt shrink to zero. When both the finite volume around a point in space and the time interval around a point in time are shrunk to the limiting value of zero, we obtain the description of the behavior at a point in space and in time in the form of a partial differential equation. This is of interest whenever state variables and fluid- and solid-phase properties vary spatially within a domain.

We start by considering a cell (Fig. 7.2.1) of finite volume and a finite time interval. In writing a balance for a cell, it is assumed that all fluid and porous medium properties (e.g., porosity, ϕ, and partitioning coefficient, K_d) and all state variables, (e.g., fluid density, component concentration, and fluid pressure) are uniform within $\mathcal{U}_o$. This assumption means that the quantity the considered chemical species present in the cell is continuously mixed to form a uniform mixture.

Let us express the verbal statement of balance presented in (5.1.1) in a mathematical form for the mass of a γ-species (concentration c) within a fluid phase that occupies the void space at a saturation S. We assume that the solid matrix is nondeformable, i.e., $\partial\phi/\partial t = 0$, and that the fluid is incompressible, i.e., $\partial\rho/\partial p = 0$.

The quantity of the γ-species within the cell is expressed by $\phi S\mathcal{U}_o c^\gamma$, so that the balance of γ within $\mathcal{U}_o$, during a time interval Δt, takes the form:

$$(\phi S\mathcal{U}_o c^\gamma)\big|_{t+\Delta t} - (\phi S\mathcal{U}_o c^\gamma)\big|_t = \Delta t\,(Q_{\text{in}}c^\gamma_{\text{in}} - Q_{\text{out}}c^\gamma + f^\gamma\mathcal{U}_o + \phi S\mathcal{U}_o\rho\Gamma^\gamma)\,. \tag{7.2.1}$$

Here, c^γ_{in} is the γ-concentration in the incoming fluid. The total rates of fluid inflow and outflow are given by Q_{in} and Q_{out}, respectively. The symbol f^γ denotes the *rate* (= mass per unit time) at which the γ-species moves from the solid and from all other fluid phases into the considered fluid phase across their common microscopic boundaries, per unit volume of porous medium, and Γ^γ denotes the *rate* of production of γ within the fluid phase (e.g., by chemical reactions), per unit mass of fluid (of mass density ρ).

In order to write the balance for a point in time, we divide (7.2.1) by Δt and let $\Delta t \to 0$, obtaining

$$\lim_{\Delta t \to 0} \frac{(\phi S \mathcal{U}_o c^\gamma)\big|_{t+\Delta t} - (\phi S \mathcal{U}_o c^\gamma)\big|_t}{\Delta t} = Q_{\text{in}} c^\gamma_{\text{in}} - Q_{\text{out}} c^\gamma + f^\gamma \mathcal{U}_o + \phi S \mathcal{U}_o \rho \Gamma^\gamma, \tag{7.2.2}$$

or, according to the definition of a derivative

$$\frac{d}{dt}(\phi S \mathcal{U}_o c^\gamma) = Q_{\text{in}} c^\gamma_{\text{in}} - Q_{\text{out}} c^\gamma + f^\gamma \mathcal{U}_o + \phi S \mathcal{U}_o \rho \Gamma^\gamma. \tag{7.2.3}$$

Dividing this equation by $\mathcal{U}_o$, we obtain a balance *per unit volume of porous medium*. Recalling that we have assumed uniformity within the cell, the balance equation that describes the behavior within the cell becomes

$$\frac{d}{dt}(\phi S c^\gamma) = \frac{1}{\mathcal{U}_o}\left(Q_{\text{in}} c^\gamma_{\text{in}} - Q_{\text{out}} c^\gamma\right) + f^\gamma + \phi S \rho \Gamma^\gamma. \tag{7.2.4}$$

In a similar way, we can derive a mass balance equation for a γ-species that adsorbs on the solid. We obtain

$$\frac{d}{dt}(\rho_b F^\gamma) = -f^\gamma + \rho_b \Gamma^\gamma_s, \tag{7.2.5}$$

where F^γ denotes the mass of γ *per unit mass of solid*, $\rho_b (\equiv (1-\phi)\rho_s)$ denotes the *bulk density of the solid*, and Γ^γ_s denotes the *rate of production of* γ adsorbed on the solid, per unit mass of solid.

We shall consider the following four cases:

CASE A. Let the fluid phase be a liquid that completely saturates the porous medium domain. Suppose that $Q_{\text{in}} = Q_{\text{out}}$, and that the γ-species does not adsorb, decay, or undergo any chemical transformation. Then, (7.2.4) reduces to

$$\frac{dc^\gamma}{dt} = \frac{Q_{\text{in}}}{\phi \mathcal{U}_o}(c^\gamma_{\text{in}} - c^\gamma). \tag{7.2.6}$$

For the conditions, $c^\gamma = c^\gamma_o$ at $t = 0$ and $c^\gamma_{\text{in}} = 0$ for $t \geq 0$, the solution of this equation is

$$c^\gamma(t) = c^\gamma_o \exp\left[-\left(\frac{Q_{\text{in}}}{\phi \mathcal{U}_o}\right) t\right]. \tag{7.2.7}$$

The lower curve in Fig. 7.2.1b presents (7.2.7) in graphical form. The product $\phi \mathcal{U}_o$ that expresses the volume of void space within $\mathcal{U}_o$ is usually referred to as 'one pore volume'. The quotient $\phi \mathcal{U}_o / Q_{\text{in}}$ gives the time required to flush the fluid through the cell once, assuming complete flushing. We refer to this time interval as *residence time*, because it is the average time that a liquid particle stays in the cell while undergoing continuous, but incomplete flushing. Because of the continuous mixing that we have assumed to take place in the cell, its concentration, as indicated by (7.2.7), is gradually reduced. It may require the flushing of a number of pore volumes in order to reduce the initial concentration in the cell to a desired level, e.g., down to the MCL (maximum contaminant level), in the case of a contaminant.

CASE B. We shall continue under the same assumptions as in Case A, except that the γ-species may adsorb onto the solid under equilibrium conditions, obeying the linear isotherm, $F^\gamma = K_d c^\gamma$. A detailed discussion on adsorption under equilibrium conditions is presented in Subs. 7.3.3. The mass of γ in the liquid, per unit volume of porous medium, is given by ϕc^γ, while, e.g. according to the linear isotherm (7.3.66), the mass of γ adsorbed on the solid, per unit volume of porous medium, is given by $(1-\phi)\rho_s F^\gamma = \rho_b F^\gamma = \rho_b K_d c^\gamma$. The total mass of γ per unit volume of porous medium, $\rho^\gamma_{\rm pm}$, is, therefore,

$$\rho^\gamma_{\rm pm} = \phi c^\gamma + \rho_b K_d\, c^\gamma = \phi\left(1 + \frac{\rho_b K_d}{\phi}\right) c^\gamma = \phi R_d\, c^\gamma, \tag{7.2.8}$$

where

$$R_d \equiv 1 + \frac{\rho_b K_d}{\phi} \geq 1 \tag{7.2.9}$$

is a coefficient that expresses the partitioning of γ within a unit volume of porous medium: for one unit of mass of γ in the liquid, we have $\rho_b K_d/\phi$ mass units on the solid. For a nonlinear isotherm, say, $F = K_d\,(c^\gamma)^n$, the coefficient R_d takes the form:

$$R_d = R_d(c^\gamma) = 1 + \frac{\rho_b K_d\,(c^\gamma)^{n-1}}{\phi}. \tag{7.2.10}$$

For the linear isotherm, the balance equation is

$$\frac{dc^\gamma}{dt} = -\left(\frac{Q_{\rm in}}{R_d \phi \mathcal{U}_o}\right) c^\gamma, \tag{7.2.11}$$

with its solution

$$c^\gamma(t) = c^\gamma_o \exp\left[-\left(\frac{Q_{\rm in}}{R_d \phi \mathcal{U}_o}\right) t\right]. \tag{7.2.12}$$

Note that (1) the same mass balance equation can be obtained by summing (7.2.4) and (7.2.5), since the terms expressing the exchange between the fluid and the solid cancel each other, and (2) the solutions (7.2.7) and (7.2.12) are similar, except for the role played by R_d.

The upper curve in Fig. 7.2.1b presents (7.2.12) in graphical form. Since $R_d > 1$ (under the assumption of $K_d > 0$), we note that the time required for reducing the concentration in the cell to a desired level is longer with adsorption than without it. The flushing of the γ-contaminant is slower than that of the host liquid. We say that the movement of the contaminant is *retarded*, relative to the liquid. The coefficient R_d is, therefore, referred to as *retardation coefficient* (or *factor*). This coefficient is further discussed in Subs. 7.4.2. We may now define an *effective residence time* equal to the quotient $R_d\phi\mathcal{U}_o/Q_{\text{in}}$; it indicates the average time that the γ-contaminant stays in the cell that undergoes continuous, but incomplete flushing.

CASE C. The conditions are similar to those of Case B, except that the contaminant undergoes a first order (e.g., radioactive) decay described by

$$\rho\Gamma^{\gamma} = -\lambda c^{\gamma}, \quad \Gamma^{\gamma}_{s} = -\lambda F^{\gamma}, \tag{7.2.13}$$

following from (7.3.53). The balance equation is

$$\frac{dc^{\gamma}}{dt} = -\left(\frac{Q_{\text{in}}}{R_d\phi\mathcal{U}_o} + \lambda\right)c^{\gamma}, \tag{7.2.14}$$

with its solution

$$c^{\gamma}(t) = c^{\gamma}_{o}\exp\left[-\left(\frac{Q_{\text{in}}}{R_d\phi\mathcal{U}_o} + \lambda\right)t\right]. \tag{7.2.15}$$

Note that if the liquid occupies only part of the void space, at saturation S_ℓ, assumed to remain constant during the flushing, we must replace ϕ by ϕS_ℓ in (7.2.8) through (7.2.15). The effective residence time is, then, $S_\ell\phi\mathcal{U}_oR_d/Q_{\text{in}}$.

CASE D. Here, the void space is occupied by a liquid water phase at a constant saturation S_w and a gas phase at saturation S_g. The gas phase pressure is assumed to stay at approximately a constant value so that its density does not change appreciably. The γ-contaminant is a volatile one, partitioned between the liquid, the solid, and the gas (at constant pressure). The partitioning of a volatile species between a liquid and a gas, through a common interface, under equilibrium conditions, is assumed to obey Henry's law (presented and discussed in Subs. 7.3.5), written here in the form

$$c^{\gamma}_{g} = \mathcal{H}c^{\gamma}_{w}. \tag{7.2.16}$$

Inflow and outflow are only of the gaseous phase (as in 'Vapor Extraction'; Subs. 7.10.4). The evaporation and condensation of water is not considered.

The mass of γ per unit volume of porous medium in the liquid (w), in the gas (g), and on the solid, are now

$$\phi S_w c^{\gamma}_{w}, \quad \phi S_g c^{\gamma}_{g}, \quad \text{and} \quad \rho_b F^{\gamma},$$

respectively. Let us choose c_g as the unknown variable of the problem. We wish to predict future c^{γ}_{g}-concentrations. Obviously, once $c^{\gamma}_{g}(t)$ is known, we can calculate $c^{\gamma}_{\ell}(t)$, and $F^{\gamma}(t)$.

Thus, the total mass of γ per unit volume of porous medium can be written as

$$\phi S_g c_g^\gamma + \phi S_w c_w^\gamma + \rho_b F^\gamma = S_g\phi\left(1+\frac{S_w}{S_g\mathcal{H}}+\frac{\rho_b K_d}{S_g\phi\mathcal{H}}\right)c_g^\gamma = S_g\phi R_v c_g^\gamma$$
$$= S_w\phi\left(\frac{S_g\mathcal{H}}{S_w}+R_d\right)c_w^\gamma = S_g\phi\left(1+\frac{S_w R_d}{S_g\mathcal{H}}\right)c_g^\gamma, \tag{7.2.17}$$

where

$$R_d \equiv 1+\frac{\rho_b K_d}{S_w\phi}, \quad \text{and} \quad R_v \equiv 1+\frac{S_w}{S_g\mathcal{H}}+\frac{\rho_b K_d}{S_g\phi\mathcal{H}}. \tag{7.2.18}$$

Here, R_v is another kind of retardation coefficient, this time for the concentration of a volatile species in a gas phase. Without volatilization, $\mathcal{H}=0$, $c_g^\gamma = 0$, and the total mass of the species per unit volume of porous medium is expressed by $S_w\phi R_d c_w^\gamma$. The solution, under the same conditions on $c_g^\gamma(t)$ as those for $c^\gamma(t)$ in Case B above, is

$$c_g^\gamma(t) = c_{go}^\gamma \exp\left[-\left(\frac{Q_{\text{in}}}{R_v\phi\mathcal{U}_o}\right)t\right], \tag{7.2.19}$$

where we recall that the inflow of the gas has been assumed to be equal to the outflow (that removes the mass of γ from the cell).

The single cell models described in this subsection may be employed to obtain preliminary estimates of clean-up times required in order to reduce concentrations to below permissible levels. In the following subsection, we shall extend our analysis to balance equations that describe transport in porous medium domains visualized as continua that involve spatial variations in material properties, fluxes, and state variables.

7.2.2 Fundamental balance equation

Our objective here is to develop the differential balance equation for the mass of a chemical species in a fluid phase that fully or partly occupies the void space. We shall follow the same methodology as used for developing the balance equation for the mass of a fluid phase in both saturated and unsaturated flow.

We consider the case of a γ-species (e.g., a contaminant) in a fluid α-phase (liquid, or gas) that occupies the entire void space, or part of it, at a fluid content θ_α $(=\phi S_\alpha)$ that is allowed to vary in space. Since we are considering here only a single fluid phase, and only a single chemical species, the subscript α and the superscript γ will be omitted wherever possible.

The starting point may be the microscopic balance equation (5.1.1) for any extensive quantity E, in which E is replaced by 'mass of a γ-species', and the resulting differential E-balance equation (5.1.4), at the microscopic level, in which e is replaced by the concentration $c \equiv c^\gamma$.

The macroscopic differential mass balance equation for a component in a fluid phase may be obtained either by writing the microscopic balance equation (5.1.4) for the mass of the component, and averaging it, or directly from the averaged equation (5.1.5). We shall follow the second approach, making the following substitutions in (5.1.5): replace $\overline{e}^{\alpha}$ by c^{γ} (= average concentration of the γ-component in the fluid-phase), replace $\overline{\mathbf{V}}^{\alpha}$ by $\mathbf{V}$ (= average (mass-weighted) fluid velocity), replace $\overline{\mathring{e}\mathring{\mathbf{V}}}^{\alpha}$ by $\mathbf{J}^*$ (= dispersive flux of the γ-component), replace $\overline{\mathbf{j}^E}^{\alpha}$ by $\mathbf{J}^{\gamma}$ (= macroscopic diffusive flux of γ), and, finally, replace $\overline{\rho\Gamma^E}^{\alpha}$ by $\rho\Gamma^{\gamma}$ (= source of mass of γ per unit volume of the fluid phase). We obtain

$$\frac{\partial(\theta c^{\gamma})}{\partial t} = -\nabla\cdot\theta\,(c^{\gamma}\mathbf{V} + \mathbf{J}^* + \mathbf{J}^{\gamma}) - f^{\gamma}_{\alpha\to\beta} + \theta\rho\Gamma^{\gamma}, \tag{7.2.20}$$

where $f^{\gamma}_{\alpha\to\beta}$ replaces the surface integral term in (5.1.5). It denotes the rate of transfer of γ from the interior of the α-phase to all other (β) phases across their common (microscopic) interfaces lying *within the* REV. These interfaces include both fluid-fluid and fluid-solid portions. In the case of saturated flow, it is comprised only of the latter, and we have $(\mathbf{V}-\mathbf{u})\cdot\mathbf{n}\equiv 0$, where $\mathbf{V}$ is the fluid velocity, $\mathbf{u}$ is the velocity at the interface, and $\mathbf{n}$ is the unit normal vector on the interface.

The various terms in (7.2.20) are:

θc^{γ}	the mass of γ per unit volume of porous medium;
$\partial(\theta c^{\gamma})/\partial t$	the rate of increase of θc;
$\theta\,(c^{\gamma}\mathbf{V} + \mathbf{J}^* + \mathbf{J}^{\gamma})$	$(= \theta\mathbf{J}^{t\gamma})$ the total flux of γ, (per unit area of porous medium) by advection, dispersion and diffusion;
$-\nabla\cdot\theta\mathbf{J}^{t\gamma}$	the excess of inflow over outflow of γ, per unit volume of porous medium, per unit time;
$f^{\gamma}_{\alpha\to\beta}$	the net rate of transfer of γ from the α-phase to all β-phases, per unit volume of porous medium;
$\theta\rho\Gamma^{\gamma}$	the rate of net production of γ, per unit volume of porous medium.

For a single fluid that occupies the entire void space, we replace θ by ϕ.

The macroscopic mass balance equation for a chemical species in a fluid phase, (7.2.20), may include one or more *source terms* (with a negative source referred to as a *sink*), each expressing the rate at which mass of that species is added to the phase by a particular process, per unit volume of porous medium. The relevant processes may include, among many others, chemical reactions among various species, adsorption, ion exchange, mineral precipitation, dissolution, interphase transfer, decay and growth phenomena, and biotransformation. Sources and sinks may also take the form of injection into and withdrawal from the aquifer domain of the considered fluid phase that contains the considered chemical species.

The interphase mass transfer, expressed by $f^{\gamma}_{\alpha\to\beta}$, may be due to a number of processes: adsorption (from the liquid phase to the solid), evaporation or volatilization (i.e., a liquid-gas transfer), dissolution (i.e., solid-liquid transfer), and liquid-liquid transfer. These processes may occur simultaneously. If required, the term $f^{\gamma}_{\alpha\rightarrow\beta}$ in the balance equation may be replaced by $\sum_{(j)} f^{\gamma}_{\alpha\rightarrow\beta,j}$, where j indicates the jth process.

Although (7.2.20) may be regarded as a single equation in three unknowns: $c\,(\equiv c^{\gamma})$, $f^{\gamma}_{\alpha\rightarrow\beta}$, and Γ^{γ}, in practice, we have to write a separate balance equation for every γ-species in the system, and for every phase within the system. In this way, the list of variables includes all the concentrations c^{γ}_{α}, the volumetric fractions θ_{α}, the exchange terms, $f^{\gamma}_{\alpha\rightarrow\beta}$, and the source terms, Γ^{γ}_{α}. In general, each of the $f^{\gamma}_{\alpha\rightarrow\beta}$-terms may depend on both c^{γ}_{α} and c^{γ}_{β}. Obviously, we also need the values of $\mathbf{V}_{\alpha}$, which we obtain from a model that describes the movement of the fluid phases present in the system.

For a single γ-component in a single phase, we have to solve (7.2.20) for $c^{\gamma} = c^{\gamma}(\mathbf{x}, t)$ within a specified porous medium domain. We need expressions for $f^{\gamma}_{\alpha\rightarrow\beta}$ in terms of c^{γ} $(\equiv c^{\gamma}_{\alpha})$, or in terms of c^{γ}_{α} and c^{γ}_{β}.

The nature of $f^{\gamma}_{\alpha\rightarrow\beta}$ depends on the process that causes the interphase transfer. It also depends on whether we assume that equilibrium conditions prevail, or not. We usually assume that no sink or source of the mass of the considered component exists on the microscopic interphase boundary. Therefore, the condition of *no-jump* in the normal flux of the component across the boundary prevails at every point on the latter, and $f^{\gamma}_{\alpha\rightarrow\beta} = -f^{\gamma}_{\beta\rightarrow\alpha}$.

As a result of the passage of chemical species across (microscopic) interphase boundaries, the total quantity of each species, say within an REV, or a unit volume of porous medium, is *redistributed*, or *repartitioned* between the adjacent phases. The driving force is the tendency of phases and species to reach a state of equilibrium. Sometimes, equilibrium is achieved very quickly (compared to the other transport mechanisms), and we consider partitioning under conditions of equilibrium. However, in many cases, the processes are relatively slow, and we have to consider the system as being continuously under nonequilibrium conditions. A *kinetic approach* is required (see 7.3.5C–E). For practical purposes (actually, for the sake of convenience, or because of lack of knowledge and data), in many cases, equilibrium conditions are *assumed* to prevail, at least as a sufficiently good approximation.

The knowledge of the laws that govern chemical and biological reactions, as well as the partitioning of species between adjacent phases, is required in order to handle the source terms that appear in the mass balance equation of a solute. In principle, a chemical species can reach the microscopic interphase boundary from the interior of a phase by two modes of transport: by advection and by diffusion. Hence, strength of the source ($\equiv$ net rate of transfer) of a γ-species, in an α-phase, is expressed as:

$$f^{\gamma}_{\alpha\rightarrow\beta} = -\frac{1}{\mathcal{U}_o}\int_{\mathcal{S}_{\alpha\beta}} \left[c^{\gamma}_{\alpha}\left(\mathbf{V}_{\alpha} - \mathbf{u}\right) + \mathbf{j}^{\gamma}_{\alpha}\right]\cdot \mathbf{n}_{\alpha}\, dS, \tag{7.2.21}$$

where β denotes all the other phases within the REV, of volume $\mathcal{U}_o$, $\mathcal{S}_{\alpha\beta}$ denotes the total (possibly moving) α-β-surface within the REV, $\mathbf{n}_{\alpha}$ denotes the outward unit normal vector on this surface, and $\mathbf{u}$ denotes the velocity of points on the interphase boundary. The minus sign stems from the fact that $\mathbf{n}_\alpha$ is directed outward from the domain occupied by the α-phase. Here we have assumed that no sources or sinks exist on the boundary, i.e., on the α-β-surface within the REV.

When $\mathcal{S}_{\alpha\beta}$ is a material surface with respect to the α-phase, viz., $\mathbf{V}_\alpha = \mathbf{u}$, the above expression reduces to

$$f^\gamma_{\alpha\to\beta} = -\frac{1}{\mathcal{U}_o}\int_{\mathcal{S}_{\alpha\beta}} \mathbf{j}^\gamma_\alpha \cdot \mathbf{n}_\alpha \, dS. \tag{7.2.22}$$

This means that the chemical species can reach and cross interphase boundaries *only by diffusion.*

The last term on the right-hand side of (7.2.20) expresses a source (= rate of production) that is a consequence of processes that occur *inside* the phase. Examples are: chemical reactions that consume or produce the considered component, and radioactive (and other) decay phenomena. As will be explained below, we include here also the source (or sink) due to pumping (or injection) of the considered component from (or into) an aquifer. The various source terms have to be expressed in terms of the concentrations of the chemical species involved in the processes.

Equation (7.2.20) is written in vector notation, and, as such, it is independent of the selected coordinate system. In may be interesting to see how this equation is written, for example, in a *cartesian coordinate* system, x, y, z. Omitting, for the sake of simplicity the source terms, and neglecting the flux due to molecular diffusion, (7.2.20) is written as:

$$\frac{\partial(\theta c^\gamma)}{\partial t} = -\frac{\partial}{\partial x_i}\theta\left\{c^\gamma V_i - \frac{\partial}{\partial x_j}\left(\mathsf{D}_{ij}\frac{\partial c^\gamma}{\partial x_j}\right)\right\}, \tag{7.2.23}$$

in which the vector V_i denotes the average velocity and the coefficient of dispersion, D_{ij}, is expressed by (7.1.39). Let us consider a few examples.

Uniform flow in an isotropic aquifer The mass balance equation (7.2.23) for uniform flow in the $+x$-direction in a homogeneous isotropic aquifer, i.e., $\theta = \phi =$ const., $\mathsf{K}_{xx} = \mathsf{K}_{yy} = \mathsf{K}_{zz} =$ const., $V_x = V$, $V_y = V_z = 0$, with the coefficient D_{ij} expressed by (7.1.47), is

$$\frac{\partial c^\gamma}{\partial t} = a_L V\frac{\partial^2 c^\gamma}{\partial x^2} + a_T V\left(\frac{\partial^2 c^\gamma}{\partial y^2} + \frac{\partial^2 c^\gamma}{\partial z^2}\right) - V\frac{\partial c^\gamma}{\partial x}. \tag{7.2.24}$$

Uniform flow in a horizontal layered aquifer (transverse isotropy) The mass balance equation (7.2.23) for uniform flow in the $+x$-direction in a homogeneous aquifer with transverse isotropy, i.e., $\mathsf{K}_{xx} = \mathsf{K}_{yy} \neq \mathsf{K}_{zz}$, $\theta = \phi =$ const., $V_x = V$, $V_y = V_z = 0$, with the coefficient D_{ij} expressed by (7.1.55), is

$$\frac{\partial c^\gamma}{\partial t} = \mathsf{a}^{H}_{LH} V \frac{\partial^2 c^\gamma}{\partial x^2} + \mathsf{a}^{H}_{TH} V \frac{\partial^2 c^\gamma}{\partial y^2} + \mathsf{a}^{H}_{TV} V \frac{\partial^2 c^\gamma}{\partial z^2} - V \frac{\partial c^\gamma}{\partial x}. \tag{7.2.25}$$

The third term will vanish if we assume $c^\gamma = c^\gamma(x, y)$.

When the uniform flow is in the vertical z-direction, i.e., $V_x = V_x = 0$, $V_z = V$, the equation takes the form

$$\frac{\partial c^\gamma}{\partial t} = \mathsf{a}^{V}_{TH} V \left(\frac{\partial^2 c^\gamma}{\partial x^2} + \frac{\partial^2 c^\gamma}{\partial y^2} \right) + \mathsf{a}^{V}_{LV} V \frac{\partial^2 c^\gamma}{\partial z^2} - V \frac{\partial c^\gamma}{\partial z}. \tag{7.2.26}$$

Sources and sinks of a chemical species within a phase, at (macroscopic) points within a porous medium domain, may be divided into two groups:

- **Sources and sinks of chemical species that result from phenomena occurring at (microscopic) points *within* the phase.** Radioactive decay of a species may serve as an example. Another example is when a species disappears from a phase (or is created within it) as a result of a chemical reaction in which the considered species participates as a reactant (or appears as a product). We have denoted the strength of such a (macroscopic) source by $\theta \rho \Gamma^\gamma$, where Γ^γ denotes the rate of generation of the mass of the γ-species per unit mass of the fluid phase, and ρ denotes the mass density of the latter. Chemical reactions that occur solely within a phase and away from the influence of interfacial forces are referred to as 'homogeneous reactions.'
- **Sources and sinks of chemical species that result from the transfer of the species into a considered phase (or out of it for a sink) across the (microscopic) interphase boundaries.** Adsorption of a species of an aqueous liquid on the solid matrix of an aquifer, may serve as an example. Ion exchange is a second example. Additional examples are precipitation of a species, dissolution of the solid phase, evaporation of a volatile species dissolved in a liquid phase into a gaseous phase that occupies part of the void space, condensation of a species present in the gaseous phase, and dissolution of a species present in one liquid phase to another. We have denoted such a source by $f^\gamma_{\alpha \to \beta}$. In all these examples, a chemical species is transferred from one phase (α) to an adjacent one (β), across their common (microscopic) boundary, $S_{\alpha\beta}$. The strength of the source is equal to the rate of transfer of the mass of the species, per unit volume of porous medium. Sometimes, the strength of the source is expressed per unit area of the interphase boundary, and then translated into a strength per unit volume through multiplication by the *specific surface* area of that boundary. Often, transfers that occur across interphase boundaries are referred to as 'inhomogeneous', or 'heterogeneous' reactions.

We may distinguish between two modes of reaction:

- **Under equilibrium conditions.** By summing the mass balance equations for γ for all phases present in the system, we obtain a single mass balance equation for γ *in the porous medium as a whole*. No transfer term appears in this equation, since $\sum_{(\alpha)} f^{\gamma}_{\alpha \to \beta} = 0$, where β denotes all phases, including the solid. However, in this single balance equation we have all the c^{γ}_{α}'s as variables, and additional equations are required. We shall see later how various equilibrium (partitioning) relationships among these variables are introduced in order to enable a solution.

- **Under nonequilibrium conditions**. For this case, nothing will be gained by summing up the phase equations, unless we wish to write a mass balance for the total mass. Usually, we leave the balance equations for the individual phases, and introduce expressions for the rates of interphase mass transfer.

An important consequence of the above discussion is that one of the first decisions that have to be made in constructing the conceptual model of a problem that involves chemical reactions is whether or not equilibrium may be assumed to prevail. Contaminant transport with chemical reactions is discussed in Subs. 7.9.2.

Whenever necessary, we shall make a distinction between chemical *species* and *chemical components*, as defined in Subs. 1.3.1. Also, we shall use the terms 'water', 'aqueous phase' or 'aqueous liquid' interchangeably; however, we shall usually use 'water' in the context of non-reacting species, while 'aqueous liquid' or 'aqueous phase' will be used when considering reacting species, as then 'water' (H_2O) is regarded as a chemical species. As an example for distinguishing between 'water' as a (liquid) phase and 'water' as a species (and between 'air' as a (gaseous) phase and 'air' as a species (overlooking the fact that air is composed of a number of gaseous species)), consider the following example.

Example: Mass balance for a liquid-gas system, with (net) evaporation and (net) air dissolution

We consider unsaturated flow, i.e., a liquid phase (ℓ, S_ℓ, ρ_ℓ), primarily water (w), and a gaseous phase (g, S_g, ρ_g), primarily air (a). Air (a) dissolves in water ($f^a_{g\to\ell}$), and liquid water (w) evaporates ($f^w_{\ell\to g}$).

The mass balance for the species water, m^w_{pm}:

$$\frac{\partial}{\partial t}\phi\left(\omega^w_\ell \rho_\ell S_\ell + \omega^w_g \rho_g S_g\right) = -\nabla\cdot\phi\left(S_\ell\rho_\ell \mathbf{V}_\ell c^w_\ell + S_g\rho_g\mathbf{V}_\ell c^w_g\right)$$
$$-\nabla\cdot\phi\left(S_\ell \mathbf{J}^w_\ell + S_g\mathbf{J}^w_g\right) + S_\ell\phi\rho_\ell\Gamma^w_\ell, \qquad (7.2.27)$$

where water evaporation and air dissolution do not appear, since $f^w_{\ell\to g}\equiv 0$, $f^a_{g\to\ell}\equiv 0$, no water is produced by chemical reactions in the gas, but it is possible that water will be produced by such reaction in the liquid phase at a rate Γ^w_ℓ per unit mass of water.

The mass balance for the species air, m^a_{pm}:

$$\frac{\partial}{\partial t}\phi\left(\omega_\ell^a\rho_\ell S_\ell+\omega_g^a\rho_g S_g\right)=-\nabla\cdot\phi\left(S_\ell\rho_\ell\mathbf{V}_\ell c_\ell^a+S_g\rho_g\mathbf{V}_\ell c_g^a\right)$$
$$-\nabla\cdot\phi\left(S_g\mathbf{J}_g^a+S_\ell\mathbf{J}_\ell^a\right)+S_\ell\phi\rho_g\Gamma_\ell^a, \tag{7.2.28}$$

and it is possible that gas (air) will be produced by reactions in the liquid phase at a rate Γ_ℓ^a per unit mass of liquid.

The specific discharge of each fluid phase in (7.2.27) and (7.2.28) is expressed by Darcy's law, e.g.,

$$q_\alpha\equiv S_\alpha\phi\mathbf{V}_\alpha=-\frac{\mathsf{k}_\alpha(\theta_\alpha)}{\mu_\alpha}\left(p_\alpha+\frac{p_\alpha}{\rho_\alpha g}\right),\quad \alpha=\ell,g. \tag{7.2.29}$$

The sum of dispersive and diffusive fluxes (per unit area of porous medium) of a γ-species (a,w) in an α-phase (g,ℓ), per unit area of porous medium, in (7.2.27) and (7.2.28) is expressed as:

$$S_\alpha\phi\mathbf{J}_\alpha^\gamma=-S_\alpha\phi\left[\mathcal{D}^\gamma\mathsf{T}_\alpha^*(\theta_\alpha)+\mathsf{D}_\alpha(\theta_\alpha)\right]\rho_\alpha\cdot\nabla\omega_\alpha^\gamma, \tag{7.2.30}$$

where $c_\alpha^\gamma\equiv\rho_\alpha\omega_\alpha^\gamma$, $\boldsymbol{\mathcal{D}}_\alpha^{*\gamma}$ $(=\boldsymbol{\mathcal{D}}^{*\gamma}(\theta_\alpha))=\mathcal{D}^\gamma\mathsf{T}^*(\theta_\alpha)$, a second rank symmetric tensor, is the sum of the coefficient of dispersion and the coefficient of molecular diffusion within an α-phase *in a porous medium.* The mass fractions, ω_ℓ^w and ω_g^w, are expressed by *Henry's law,* e.g., in the form (7.2.16), and by the *psychrometric law* (that expresses vapor mass fraction as a function of temperature and capillary pressure), respectively.

In the next section, we shall elaborate on various source terms, due to chemical reactions, both of the Γ-type and of the $f_{\alpha\to\beta}^\gamma$-type, that appear in the mass balance equation (7.2.20). We shall also show how these sources are incorporated in the mass balance equation for a considered species. Since sources are due to various kinds homogeneous and inhomogeneous chemical reactions, we shall start by presenting a brief introduction to such phenomena. However, first, in the next subsection, we shall discuss pumping and artificial recharge through wells that also represent point sinks and sources of the Γ-type that have to be incorporated in the mass balance equations.

7.2.3 Pumping and injection

We often refer to pumping and injection wells in an aquifer also as 'sinks and sources', respectively. In this case an injection well, through which a fluid that contains the γ species is injected, acts as a source of γ. A pumping well removes fluid that contains γ at the concentration prevailing at the location. These sources and sinks are represented in the Γ-type source appearing in the macroscopic mass balance equation (7.2.20), although they do not appear as sources in the microscopic solute transport model; there they appear in the form of boundary conditions.

In a three-dimensional domain, such sources and sinks may take the form of points or of line segments approximating the screened segments of wells.

When the screen length of a pumping well is much smaller than the aquifer's thickness, the well is often approximated as a point sink located at its center. Sources and sinks in the form of points in a three-dimensional flow domain were discussed in Subs. 5.1.4.

When considering relatively small domains, the circumference of a well may be considered as a boundary, with the interior of the well being excluded from the domain. Pumping and injection rates will then be specified as flux conditions (for both fluids and solutes) on such a boundary (see discussion on boundary condition on wells in Subs. 5.2.3G).

Consider a fluid phase (liquid, or gas) containing a component at a *known* concentration, c_R, that is added to a porous medium domain as a *distributed external fluid source* at a rate $R_{\text{ext}} = R_{\text{ext}}(\mathbf{x}, t)$ (= volume of fluid phase added per unit volume of porous medium per unit time). Then, the source term appearing in (7.2.20) is expressed by

$$\theta \rho \Gamma^{\gamma} = R_{\text{ext}}(\mathbf{x}, t) c_R^{\gamma}. \tag{7.2.31}$$

When sources exist at N isolated points, $\mathbf{x}^{(m)}$, $m = 1, 2, ...N$, the source term in (7.2.20) takes the form:

$$\theta \rho \Gamma^{\gamma} = \sum_{(m)} R_{\text{ext}}^{(m)}(\mathbf{x}^{(m)}, t) \delta(\mathbf{x} - \mathbf{x}^{(m)}) c_R^{\gamma(m)}, \tag{7.2.32}$$

where $R_{\text{ext}}^{(m)}$ represents the rate of injection (in terms of volume added per unit time), at some point $\mathbf{x}^{(m)}$ and time t, of fluid at the *known* concentration, $c_R^{\gamma(m)}$, and $\delta(\mathbf{x} - \mathbf{x}^{(m)})$ denotes the *Dirac delta function*, defined formally by (5.1.77). For a distributed sink, and for a collection of point sinks (e.g., pumping wells), the corresponding expressions are

$$\theta \rho \Gamma^{\gamma} = -P_{\text{ext}}(\mathbf{x}, t) c^{\gamma}(\mathbf{x}, t), \tag{7.2.33}$$

and

$$\theta \rho \Gamma^{\gamma} = -\sum_{(m)} P_{\text{ext}}^{(m)}(\mathbf{x}^{(m)}, t) \delta(\mathbf{x} - \mathbf{x}^{(m)}) c^{\gamma}(\mathbf{x}, t), \tag{7.2.34}$$

respectively, where the withdrawn fluid is at the *unknown concentration*, $c^{\gamma}(\mathbf{x}, t)$, and the terms P_{ext} and $P_{\text{ext}}^{(m)}$ denote the magnitude of the fluid discharge rates (= volume per unit time) of the respective sinks.

7.3 Sources and Sinks

The objective of this section is to briefly introduce and discuss chemical processes that constitute sources (or sinks) of a γ-species in a porous medium domain. These processes appear as either Γ^{γ} or $f_{\alpha \to \beta}^{\gamma}$ terms in the mass balance equation, (7.2.20).

The discussion here should be considered merely as very a brief introduction to some of the essentials and to the employed terminology; in order to seriously consider chemistry and microbiology of contamination, much more knowledge and experience is required. Books on chemistry and microbiology should be consulted (e.g., Weber, 1972; Stumm and Morgan, 1995; Paul and Clark, 1996; Hiemenz and Rajagopalan, 1997; Sawyer *et al.*, 2002; Schwarzenbach *et al.*, 2002; Sposito, 2004; Appelo and Postma, 2005). As emphasized at the very beginning of this book, the incorporation of the chemical and biological aspects in a model calls for close cooperation between experts in transport in porous media, numerical methods, geohydrology, chemistry, and microbiology.

7.3.1 Conditions for chemical equilibrium

We are interested in chemical reactions that occur as solutes are being transported by advection, dispersion and diffusion within the fluid that occupies the void space of a porous medium domain. Given the time required for a reaction to reach equilibrium (as obtained from batch experiments, i.e., at the microscopic level), we wish to discuss the conditions under which the assumption of equilibrium within an REV—usually referred to as *Local Equilibrium Assumption*, LEA—is justified. We shall start by considering homogeneous reactions (Subs. 7.2.2).

The use of dimensionless numbers as a tool for comparing terms in a balance equation, which, actually, means comparing processes in a transport problem, will be presented in detail in Sec. 7.7. There, in the example of modeling contaminant transport, we shall introduce dimensionless numbers that are relevant to the discussion here: the Strouhal (St), the Peclet (Pe) and the Damköhler (Dm^I, Dm^{II}) numbers. They are defined as:

$$
\begin{aligned}
\mathrm{St} &\equiv \frac{L_c}{V_c t_c} = \frac{t_{c,\mathrm{adv}}}{t_c},\\
\mathrm{Pe} &\equiv \frac{L_c V_c}{\mathcal{D}} = \frac{t_{c,\mathrm{diff}}}{t_{c,\mathrm{adv}}},\\
\mathrm{Dm}^I &\equiv \frac{L_c/V_c}{t_{c,\mathrm{react}}} = \frac{t_{c,\mathrm{adv}}}{t_{c,\mathrm{react}}},\\
\mathrm{Dm}^{II} &\equiv \frac{L_c^2/\mathcal{D}}{t_{c,\mathrm{react}}} = \frac{t_{c,\mathrm{diff}}}{t_{c,\mathrm{react}}},
\end{aligned}
\tag{7.3.1}
$$

where t_c is a characteristic time, such as the duration over which the problem is being modeled. In determining these numbers, the domain of interest is the REV. Hence, the characteristic length, L_c, is the size of the REV, the characteristic fluid velocity, V_c, is the maximum fluid velocity within the REV, and $\mathcal{D}$ is the coefficient of diffusion at the microscopic level.

The characteristic time of reaction, $t_{c,\mathrm{react}}$, is determined by a batch experiment. It is defined as the time at which the concentration of an important species in the reaction, $c(t)$, decays to the extent that the value of some factor, w, becomes close to unity, $w \sim 1$. This factor is defined by

$$w\Delta c_c = \left|c(t_{c,\text{react}}) - c(0)\right|, \tag{7.3.2}$$

in which Δc_c is a characteristic concentration change.

We also need the dimensionless diffusion time, $t^\star_{\text{diff}}$, and the dimensionless reaction time, $t^\star_{\text{react}}$, defined by

$$\begin{aligned} t^\star_{\text{diff}} &\equiv \frac{L_c^2/\mathcal{D}}{t_c} = \frac{t_{c,\text{diff}}}{t_c} = \text{St} \cdot \text{Pe}, \\ t^\star_{\text{react}} &\equiv \frac{t_{c,\text{react}}}{t_c} = \frac{\text{St}}{\text{Dm}^I}. \end{aligned} \tag{7.3.3}$$

In order to ensure chemical equilibrium, we require that the concentrations be approximately uniform within the REV. This means that the characteristic time for diffusion must be much smaller than that for advection. This means that

$$\text{Pe} \ll 1. \tag{7.3.4}$$

This condition implies not only that diffusive fluxes, due to concentration gradients, are small, but also that reaction rates (which are functions of concentration) are approximately uniform throughout the REV.

Next, we require that the characteristic time for diffusion must be much smaller than the characteristic time of the problem, so that

$$t^\star_{\text{diff}} \ll 1, \quad \text{or, equivalently,} \quad \text{St} \ll \text{Pe}. \tag{7.3.5}$$

We also require that the characteristic time for the reaction be much smaller than that for advection and diffusion, i.e.,

$$\text{Dm}^I \gg 1, \quad \text{Dm}^{II} \gg 1. \tag{7.3.6}$$

Finally, we require that the characteristic time for the reaction be much smaller than the characteristic time of the problem:

$$t^\star_{\text{react}} \ll 1, \quad \text{or, equivalently,} \quad \text{St} \ll \text{Dm}^I. \tag{7.3.7}$$

Altogether, conditions (7.3.4) through (7.3.7) ensure chemical equilibrium of a homogeneous reaction within an REV.

Next, we analyze the conditions for equilibrium of heterogeneous reactions. This type of reactions is often studied in the laboratory by using batch experiments in which, for example, two fluid phases, usually a gas and a liquid, are in contact at a flat interface, with both phases kept well-mixed so that concentrations are uniform in their respective domains. Let $\mathcal{S}_{\alpha,\beta}$ denote the interface between an α-phase and a β-phase. We may use (7.3.2) to define a characteristic time of the reaction, $t_{c,\text{react}}$. However, in this case, the characteristic time also depends on the involved volumes of the phases relative to the area of the interface. In fact the characteristic time is proportional to the volume-to-surface ratio $\mathcal{U}_\alpha/S_{\alpha,\beta}$ $(= \ell_\alpha)$, where ℓ_α is a characteristic length of the phase. Thus, $\kappa_\alpha \equiv \ell_\alpha/t_{c,\text{react}}$ is a quantity that is independent of the size of the experimental system and is, therefore, a true characteristic of the reaction at the interface.

Let us now consider a porous medium REV containing the two phases. In this case, $\ell_\alpha = \mathcal{U}_{o,\alpha}/S_{\alpha,\beta}$ where the volume and surface area now refer to those of the α-phase in the REV. The characteristic length L_c for the α-phase is equal to ℓ_α. The relevant Peclet number for the α-phase, and the relevant Damköhler and Strouhal numbers (see also Sec. 7.7) are:

$$\begin{aligned}
(\text{Pe})_\alpha &= \frac{(V_\alpha)_c \ell_\alpha}{\mathcal{D}_\alpha}, \\
(\text{Dm}^I)_\alpha &= \frac{\ell_\alpha/(V_\alpha)_c}{t_{c,\text{react}}} = \frac{1/(V_\alpha)_c}{1/\kappa_\alpha}, \\
(\text{St})_\alpha &= \frac{\ell_\alpha}{(V_c)_\alpha t_c}, \\
(\text{Dm}^{II})_\alpha &= \frac{\ell^2_\alpha/\mathcal{D}_\alpha}{t_{c,\text{react}}} = \frac{\ell_\alpha/\mathcal{D}_\alpha}{1/\kappa_\alpha}.
\end{aligned} \tag{7.3.8}$$

Similar expressions apply also to the dimensionless numbers of the β-phase. Therefore, the conditions for equilibrium of a heterogeneous reaction, (7.3.4) through (7.3.7), must hold for both phases.

7.3.2 Equilibrium chemical reactions

Chemical and biological reactions often play a key role in the fate of contaminants in the subsurface. For example, the migration of a contaminant from its source may be significantly *attenuated* by degradation caused by such reactions. Other examples are the immobilization of a contaminant by chemical precipitation onto the solid, and the transformation of a contaminant, radioactively, chemically, or biologically, into another chemical compound that may, or may not, be harmful.

The term 'natural attenuation' is used when such phenomena occur in the subsurface without human intervention. Often, natural attenuation reduces the concentration of a contaminant to permissible levels within an acceptable time interval, so that human intervention is not required.

Many remediation techniques rely on reactive processes. The effectiveness of *soil vapor extraction*, for example, depends on the partitioning of volatile contaminants between the aqueous and gaseous phases. Some methods utilize chemical or biological reactions to degrade a contaminant into harmless products by introducing appropriate reactants or nutrients into the subsurface, either through an injection well or by placement into a permeable trench, also known as a 'Permeable Reactive Barrier (PRB)' (Subs. 7.10.6). Other methods convert the contaminant into products that have a low mobility.

A. Rate of chemical reactions

We start by focusing on reactions that occur in a closed batch system. Later, we shall consider reactions in a porous medium during transport.

A chemical reaction is described by an equation, e.g.,

$$\begin{aligned} &\mathrm{A} \rightarrow \mathrm{C}, \qquad \mathrm{A} \rightarrow \mathrm{B}+\mathrm{C}, \qquad \mathrm{A}+\mathrm{B} \rightarrow \mathrm{AB}, \\ &\mathrm{A}+\mathrm{A} \rightarrow \mathrm{A}_2, \qquad a\mathrm{A}+b\mathrm{B} \rightarrow c\mathrm{C}+d\mathrm{D}, \end{aligned} \tag{7.3.9}$$

in which A, B, AB, C, and D are species or compounds and a, b, c, d are numbers. In each equation, the species on the left-hand side are referred to as *reactants*, while those on the right-hand side are called *products*. The first two equations describe *unimolecular reactions*. The subsequent two involving *bimolecular reactions*, describe *association*. The last equation describes a general case with two reactants and two products. In all reactions, the involved species may be atoms, molecules, free radicals, or ions.

Stoichiometry is the term used for the balancing of equations such as those presented above, making sure that the same number of each kind of atom appears on both sides of each equation. This balance enables the calculation of the amounts of each involved compound. Each such balance equation is referred to as *stoichiometric equations.*

Still in a closed batch, i.e., no product or reactant is removed, when reactions are *reversible*, we replace the symbol $\rightarrow$ by $\rightleftharpoons$. The equation then represents a state of *equilibrium.*

For a homogeneous reaction, i.e., one that occurs *within* a fluid phase, the *reaction rate* expresses the decrease in the concentration of a reactant, or the increase in that of a product, per unit time. The reaction rate per unit volume of fluid phase is defined as *specific reaction rate.* We shall often use these terms interchangeably, as long as no confusion may result.

Consider a homogeneous reaction described by

$$\mathrm{A}+\mathrm{B} \rightarrow \mathrm{C}. \tag{7.3.10}$$

The *reaction rate*, R_r, is defined as

$$R_r = -\frac{d[\mathrm{A}]}{dt} = -\frac{d[\mathrm{B}]}{dt} = \frac{d[\mathrm{C}]}{dt}, \tag{7.3.11}$$

in which $[\mathrm{A}](\equiv [c^\gamma_\alpha] \equiv c^\mathrm{A}/M^\mathrm{A})$ represents the *molar concentration* of A, with M^A denoting the molar mass of A. Thus the *reaction rate expresses the number of moles (produced or disappearing) per unit volume of solution per unit time.* We note that we have defined a *single* rate for the entire reaction.

The derivatives in the above equation have their usual meaning, e.g.,

$$\frac{dc^{\mathrm{A}}}{dt} = \lim_{\Delta t \to 0} \frac{c^{\mathrm{A}}|_{t+\Delta t} - c^{\mathrm{A}}|_t}{\Delta t}. \tag{7.3.12}$$

For a more complicated chemical reaction, e.g., expressed by the stoichiometric equation

$$\mathrm{A} + 2\mathrm{B} \rightarrow 3\mathrm{C} + \mathrm{D}, \tag{7.3.13}$$

the *reaction rate*, R_r, is given by

$$R_r = -\frac{d[\mathrm{A}]}{dt} = -\frac{1}{2}\frac{d[\mathrm{B}]}{dt} = \frac{1}{3}\frac{d[\mathrm{C}]}{dt} = \frac{d[\mathrm{D}]}{dt}. \tag{7.3.14}$$

In general terms, a stoichiometric equation can be written as

$$\chi^{\mathrm{A}}\mathrm{A} + \chi^{\mathrm{B}}\mathrm{B} + \ldots \rightarrow \chi^{\mathrm{P}}\mathrm{P} + \chi^{\mathrm{Q}}\mathrm{Q} + \ldots, \tag{7.3.15}$$

with the reaction rate expressed as

$$R_r = -\frac{1}{\chi^{\mathrm{A}}}\frac{d[\mathrm{A}]}{dt} = -\frac{1}{\chi^{\mathrm{B}}}\frac{d[\mathrm{B}]}{dt} = \ldots = \frac{1}{\chi^{\mathrm{P}}}\frac{d[\mathrm{P}]}{dt} = \frac{1}{\chi^{\mathrm{Q}}}\frac{d[\mathrm{Q}]}{dt} = \ldots, \tag{7.3.16}$$

in which the χ^γ ($\chi^\gamma > 0$), for γ = A, B, P, Q,..., are the *stoichiometric coefficients.* They describe the relative number of moles of each reactant and those of each product that participate in the considered reaction. Note that the stoichiometric equation imposes constraints on the rates of production (and disappearance) of the species involved in the reaction.

A stoichiometric equation can also be written in the compact form:

$$\sum_{(\gamma)} \nu^\gamma \mathcal{M}^\gamma \rightarrow 0, \tag{7.3.17}$$

or, for a reversible reaction:

$$\sum_{(\gamma)} \nu^\gamma \mathcal{M}^\gamma \rightleftharpoons 0, \tag{7.3.18}$$

in which $\mathcal{M}^\gamma$ denotes the chemical symbol of the respective γ-species, and ν^γ denotes its corresponding *stoichiometric coefficient.* Note that the signs in the above forms are opposite to those normally used when writing actual reactions. Following standard convention, $\nu^\gamma < 0$ for a reactant and $\nu^\gamma > 0$ for a product. Then, the reaction rate, R_r, is given by

$$R_r = \frac{1}{\nu^\gamma}\frac{d[\gamma]}{dt}. \tag{7.3.19}$$

The reaction rate is measured in moles per second (in a given volume of solution), or in moles per liter per second.

For the reaction expressed by (7.3.15), the reaction rate often takes the form (see Subs. 7.3.2B and C):

$$R_r = k[\mathrm{A}]^{\lambda^{\mathrm{A}}}[\mathrm{B}]^{\lambda^{\mathrm{B}}}\ldots, \tag{7.3.20}$$

in which the product is only over the reactant species and the λ^γ's are powers, which, in general, are not necessarily equal to the χ^γ's. Equation (7.3.20) is an example of a *rate law*. The coefficient k is called the *rate constant* of the reaction. The rate law expresses the reaction rate, R_r, as a function of the concentrations of all reactants present in the solution. For a given reaction, the rate law must be determined *experimentally*. The reaction expressed by (7.3.20) is said to be λ^{A}-order in A, λ^{B}-order in B, etc. The total order of the rate law is the sum of these exponents.

For example, for the reaction A + B $\rightarrow$ C, we may have

$$R_r \equiv -\frac{d[\mathrm{A}]}{dt} = k[\mathrm{A}][\mathrm{B}]. \tag{7.3.21}$$

Note that in this example, the reaction has a second-order rate law.

The reversible reaction (7.3.18) actually consists of a forward and a reverse reaction. For example, the reversible reaction,

$$\mathrm{A} + 2\mathrm{B} \underset{k_{\mathrm{rev}}}{\overset{k_{\mathrm{for}}}{\rightleftharpoons}} \mathrm{C}, \tag{7.3.22}$$

is the result of the two reactions:

$$\mathrm{A} + 2\mathrm{B} \overset{k_{\mathrm{for}}}{\longrightarrow} \mathrm{C}, \quad \text{and} \quad \mathrm{A} + 2\mathrm{B} \overset{k_{\mathrm{rev}}}{\longleftarrow} \mathrm{C}. \tag{7.3.23}$$

The rate of a reversible reaction is the sum of the reaction rates of the forward and reverse reactions. The forward reaction rate may be given by

$$R_{r,\mathrm{for}} \left(= -\frac{d[\mathrm{A}]}{dt} = -\frac{1}{2}\frac{d[\mathrm{B}]}{dt} \right) = k_{\mathrm{for}}[\mathrm{A}][\mathrm{B}], \tag{7.3.24}$$

and the reverse reaction rate by

$$R_{r,\mathrm{rev}} \left(= -\frac{d[\mathrm{C}]}{dt} \right) = -k_{\mathrm{rev}}[\mathrm{C}]. \tag{7.3.25}$$

The resulting reaction rate is

$$R_r = R_{r,\mathrm{for}} - R_{r,\mathrm{rev}} = k_{\mathrm{for}}[\mathrm{A}][\mathrm{B}] - k_{\mathrm{rev}}[\mathrm{C}]. \tag{7.3.26}$$

For ideal, or dilute, solutions, where the solvent does not participate in the reaction, the rate law for an elementary reversible reaction can be shown to be (de Groot and Mazur, 1962)

$$R_r = k_{\mathrm{for}} \prod_{(\gamma,\ \nu^\gamma<0)} [\mathrm{X}^\gamma_{\mathrm{react}}]^{-\nu^\gamma} - k_{\mathrm{rev}} \prod_{(\gamma,\ \nu^\gamma>0)} [\mathrm{X}^\gamma_{\mathrm{prod}}]^{\nu^\gamma}, \tag{7.3.27}$$

where we have followed the standard sign convention for ν mentioned earlier, and $[X^\gamma]$ denotes the concentration (in this case, in terms of molar fraction) of the (reactant or product) γ-species.

Consider the reaction $A \rightleftharpoons B$, in which both forward (f) and reverse (r) reactions are first-order (see below), but with different constants:

$$A \xrightarrow{k_f} B; \qquad R_{rf} = -\left.\frac{d[A]}{dt}\right|_1 = k_f[A] = \left.\frac{d[B]}{dt}\right|_1, \tag{7.3.28}$$

$$A \xleftarrow{k_r} B; \qquad R_{rr} = -\left.\frac{d[B]}{dt}\right|_2 = k_r[B] = \left.\frac{d[A]}{dt}\right|_2. \tag{7.3.29}$$

In this equation, we note how, simultaneously, A is depleted by the forward reaction, at a rate $k_f[A]$, and produced by the reverse reaction, at a rate $k_r[B]$. The net rate of production of A is

$$\frac{d[A]}{dt} \equiv \left.\frac{d[A]}{dt}\right|_1 + \left.\frac{d[A]}{dt}\right|_2 = -k_f[A] + k_r[B]. \tag{7.3.30}$$

From

$$\frac{d[B]}{dt} \equiv \left.\frac{d[B]}{dt}\right|_1 + \left.\frac{d[B]}{dt}\right|_2 = -\frac{d[A]}{dt}, \tag{7.3.31}$$

it follows that

$$\frac{\mathrm{d}}{\mathrm{d}t}\Big([A]+[B]\Big) = 0, \quad \text{or,} \quad [A]+[B] = [A]\big|_{t=0} + [B]\big|_{t=0}. \tag{7.3.32}$$

Thus, if $[B]\big|_{t=0} = 0$, we may integrate (7.3.30) to obtain

$$[A] = \left(\frac{k_r + k_f\, e^{-(k_f+k_r)t}}{k_f + k_r}\right)[A]\big|_{t=0}. \tag{7.3.33}$$

As $t \to \infty$, the concentrations reach their equilibrium values:

$$[A] = \frac{k_r}{k_f + k_r}[A]\big|_{t=0}, \qquad [B] = \frac{k_f}{k_f + k_r}[A]\big|_{t=0}. \tag{7.3.34}$$

Thus, at equilibrium,

$$k_f[A] = k_r[B], \quad \text{and} \quad K_{\text{eq}} = \frac{k_f}{k_r}, \tag{7.3.35}$$

with K_{eq} referred to as the *(thermodynamic) equilibrium constant* of the considered reaction. If K_{eq} is known for a reaction, and one of the rate constants is also known, the other one can be determined by this relationship.

For the reaction

$$A + B \underset{k_r}{\overset{k_f}{\rightleftharpoons}} C + D, \tag{7.3.36}$$

which is second order in both directions, suppose that

$$\text{Forward reaction}: \quad \left.\frac{d[\mathrm{A}]}{dt}\right|_f = -k_f[\mathrm{A}][\mathrm{B}], \tag{7.3.37}$$

$$\text{Reverse reaction}: \quad \left.\frac{d[\mathrm{A}]}{dt}\right|_r = k_r[\mathrm{C}][\mathrm{D}], \tag{7.3.38}$$

$$\text{Net gain in A}: \quad \frac{d[\mathrm{A}]}{dt} = -k_f[\mathrm{A}][\mathrm{B}] + k_r[\mathrm{C}][\mathrm{D}]. \tag{7.3.39}$$

At equilibrium, the net gain in A vanishes, so that

$$K_{\mathrm{eq}} = \frac{[\mathrm{C}][\mathrm{D}]}{[\mathrm{A}][\mathrm{B}]} = \frac{k_f}{k_r}. \tag{7.3.40}$$

This equation is a special case of the *law of mass action* to be discussed later. The coefficient K_{eq} $(= k_f/k_r)$ is the *equilibrium constant* defined above.

For the reaction described by the general stoichiometric equation

$$a\mathrm{A} + b\mathrm{B} + \ldots \rightleftharpoons p\mathrm{P} + q\mathrm{Q} + \ldots. \tag{7.3.41}$$

we obtain the general form of the *law of mass action*:

$$K_{\mathrm{eq}} = \frac{\{\mathrm{P}\}^p\{\mathrm{Q}\}^q \ldots}{\{\mathrm{A}\}^a\{\mathrm{B}\}^b \ldots}, \tag{7.3.42}$$

in which the bracketed quantities $\{\cdot\}$ denote *activities.*

The activity $\{A\}$ of a species A, is related to its *molar concentration*, m^{A}, which is the number of moles of the latter per kg mass of solvent (usually, in the context of aqueous solutions), by

$$\{\mathrm{A}\} = \gamma^{\mathrm{A}}\mathrm{m}^{\mathrm{A}}, \tag{7.3.43}$$

where γ^{A} is the *activity coefficient* of A. For dilute solutions, $\gamma^{\mathrm{A}} \approx 1$ and $\{\mathrm{A}\}$ $\approx m^{\mathrm{A}}$.

The activity, $\{\mathrm{A}\}$, is often related to the molality, $\hat{m}^{\mathrm{A}}$ (= concentration defined as moles of a species per kilogram of solvent), by a factor known as the *activity coefficient* (dimensionless), γ^{A}, defined by

$$\gamma^{\mathrm{A}} \equiv \frac{\{\mathrm{A}\}}{\hat{m}^{\mathrm{A}}}, \qquad \gamma^{\mathrm{A}} > 0. \tag{7.3.44}$$

Note that γ^{A} depends on the standard state selected for the species, and that $\gamma^{\mathrm{A}} \to 1$ as $\hat{m}^{\mathrm{A}} \to 0$. This definition of the activity coefficient is the standard one used in geochemistry.

For an ionic aqueous species, the activity coefficient, as defined by (7.3.44), is given by various empirical formulas, e.g. by Helgeson (1969), in the form:

$$\log \gamma^{\mathrm{A}} = -\frac{\mathcal{A}\left(z^{\mathrm{A}}\right)^2 \sqrt{I}}{1 + \mathcal{B} r^{\gamma} \sqrt{I}} + \mathcal{C} I, \tag{7.3.45}$$

where z^{A} is the charge on the A-species and r^{γ} denotes the effective diameter of the hydrated ion (in cm). The coefficients $\mathcal{A}$, $\mathcal{B}$, and $\mathcal{C}$ are temperature-dependent constants that are independent of γ. The symbol I denotes the *ionic strength* of the solution, defined by

$$I \equiv \frac{1}{2} \sum_{(\gamma)} \hat{m}^{\gamma} \left(z^{\gamma}\right)^2, \tag{7.3.46}$$

where γ denotes the γth ionic species. This equation gives good agreement with experimental data for ionic strengths up to around 1 molal solution. At higher ionic strengths, more complicated expressions are required (e.g., Pitzer, 1979).

When a species participates in several chemical reactions that cause its concentration within a fluid phase to increase (or decrease), we express the strength of the source (= rate of production) of that species in the (microscopic) balance equation of the latter, say, in (7.2.20), by

$$\rho_{\alpha} \Gamma_{\alpha}^{\gamma} = \sum_{(j)} \left.\frac{dc^{\gamma}}{dt}\right|_{j\text{th hom chem react}} = M^{\gamma} \sum_{(j)} \nu_j^{\gamma} R_{r,j}, \tag{7.3.47}$$

in which $R_{r,j}$ is the reaction rate of the jth homogeneous chemical reaction in the fluid α-phase and ν_j^{γ} is the stoichiometric coefficient of the γ-species in the jth reaction. This rate of production is in addition to the rates of production resulting from other sources.

Although we have referred above to the characteristic time of the reaction described by a given stoichiometric equation, often, the actual reaction goes through a number of intermediate steps that do not appear explicitly in the stoichiometric equation and in the corresponding rate law. However, when such an intermediate step is much slower than the one explicitly referred to, it dictates the *rate-determining*, or *rate-limiting step* of the entire reaction.

B. Half-life for first order reactions

Consider the reaction

$$\mathrm{A} \xrightarrow{k} \mathrm{B}. \tag{7.3.48}$$

The *first-order rate law* for the consumption of a reactant A, is expressed in the form:

$$R_r \equiv -\frac{d[\mathrm{A}]}{dt} = k[\mathrm{A}], \tag{7.3.49}$$

in which k is referred to as the *first-order rate constant* (dims. T^{-1}).

An example of a first-order reaction is the radioactive decay of tritiated water (HTO) into ordinary water, where T stands for *tritium* ($\equiv \mathrm{H}^3$):

$$\text{HTO} \xrightarrow{k} \text{H}_2\text{O}, \tag{7.3.50}$$

for which the rate constant is $k = 1.78 \times 10^{-9}$ sec^{-1}.

By integrating (7.3.49) from [A]=$[\text{A}]_o$ at $t = 0$, to any time, t, we obtain

$$[\text{A}](t) = \text{A}_o\, e^{-kt}, \tag{7.3.51}$$

often referred to as the *integrated rate law.* A plot of $\ln([\text{A}](t)/[\text{A}]_o)$ versus time, will yield a straight line with a slope $-k$. A larger k indicates a faster rate of disappearance, or decay, of the A-species. From (7.3.49), we may define the *half-life*, $t_{1/2}$, of the A-species in the considered reaction, i.e., the time in which its concentration will be reduced by a factor 2:

$$t_{1/2} = \frac{\ln 2}{k} = \frac{0.693}{k}. \tag{7.3.52}$$

In a first-order reaction, e.g., (7.3.49), the half-life of the reactant is independent of the concentration. The half-life, $t_{1/2}$, may be considered as a characteristic reaction time.

Radioactive and certain other decay phenomena, A $\rightarrow$ products (sometimes referred to as *unimolecular reactions*), may be expressed as the first-order rate law:

$$\frac{dN}{dt} = -\lambda N, \tag{7.3.53}$$

in which λ is the *first order rate constant* for the radioactive decay, and N is the number of atoms of the radioactive material. We can also use molar concentrations instead of N. Integrating the above expression from $N = N_o$ at $t = 0$, to any t, gives

$$N(t) = N_o\, e^{-\lambda t}. \tag{7.3.54}$$

The half-life is defined by (7.3.52), in which k is replaced by λ. In principle, no equilibrium can be reached until the radioactive material has completely disappeared.

C. Half-life for higher order reactions

The total order of the rate law is determined empirically. Consider, for example, the case

$$\text{A} + \text{B} \rightarrow \text{C}, \tag{7.3.55}$$

with the second-order rate law (but first-order in the reactants A and B),

$$R_r \equiv -\frac{d[\text{A}]}{dt} = k''[\text{A}][\text{B}], \tag{7.3.56}$$

in which k'' (dims. M^{-1}T^{-1}) is a *second-order rate constant.* In this case, integration is not possible, until we know the stoichiometry of the reaction. Formally, we could write

$$t_{1/2} = \frac{\ln 2}{k''[\mathrm{B}]}, \tag{7.3.57}$$

with [B]=[B_o], and k''[B] referred to as a *pseudo-first-order rate constant.* However, since [B] is a function of time, we cannot determine the half-life, $t_{1/2}$, unless [B] $\approx$ constant.

D. Temperature dependence of reaction rate

It is found empirically that the rate of many chemical reactions increases with a rise in temperature, following the relationship

$$k = Ae^{-E_a/RT}, \tag{7.3.58}$$

known as the *Arrhenius equation* (see discussion in Lasaga and Kirkpatrick, 1981). Here, A is a *pre-exponential factor*, and E_a is like an *activation energy* of the reaction that expresses the minimum energy required for reactants to form products. The rate coefficient k increases as the temperature increases.

E. Reactions at the microscopic and macroscopic levels

In Subs. 1.3, we have introduced the concept of a *microscopic representative elementary volume* (abbreviated μREV) at every point within a fluid phase (unless the phase domain is too small). The behavior of the phase as a continuum is obtained by averaging the behavior of the individual molecules over the μREV. This means that the μREV has a valid thermodynamic state given by pressure, temperature, and species concentration. The concept of a μREV is now extended to include chemical reactions.

In order to interact, molecules of chemical species must collide at the proper 'orientation', and have the required amount of energy to perform the interaction, that is, to break and make chemical bonds. Another factor is the chemical structure of the interacting species. An interacting species may have to follow a rather intricate path before it can interact with other species. This is especially true in the case of large organic molecules. It is clear that the net rate of production (or consumption) of each species participating in a reaction, as indicated by the number of molecules (and ions) reacting per unit time, is a statistical quantity. In order to describe a reaction at the microscopic level, its rate must be statistically averaged over the μREV. We, therefore, require that the size of the μREV be much greater than the mean distance traveled by a molecule before reacting with another molecule.

An important reason for defining averaged reaction rates is that one can often write a *rate law* that relates a microscopic reaction rate to microscopic conditions within the μREV. For homogeneous reactions, the rate law depends on the thermodynamic state within the μREV. For heterogeneous reactions, the rate law usually depends on the thermodynamic state of the phases on both sides of a microscopic interface between phases and on the physical and chemical conditions on the interface.

When constructing transport models at the macroscopic level, we need values of state variables, coefficients, etc., at the macroscopic level. *Macroscopic* rate of reaction of a species at a point in a porous medium domain is defined as the volume-averaged microscopic rate over all points in the REV that is centered at the considered point. For homogeneous reactions, it is usually assumed that the deviations in the thermodynamic state within the REV are sufficiently small such that *the same form of the microscopic rate law can be used at the macroscopic level.* Developing rate laws for heterogeneous reactions can be more involved because the spatial distribution of the microscopic conditions and properties of the interface must be related to macroscopic variables and parameters.

F. First order reaction in a porous medium

When we consider a γ-species in a fluid α-phase (concentration c_α^γ) that participates in a first order reaction within a porous medium domain, e.g., undergoes radioactive decay, the sink term, expressing the rate of disappearance of the species in a macroscopic mass balance equation, say, (7.2.20), is given by

$$\theta_\alpha \rho_\alpha \Gamma_\alpha^\gamma = -\theta_\alpha \lambda c_\alpha^\gamma. \tag{7.3.59}$$

G. Decay and other degradation phenomena in porous media

These types of sources have already been presented in Subs. 7.2.1. Let us summarize them as follows.

For radioactive decay, the source term takes the form:

$$\theta \rho \Gamma^\gamma = -\theta \lambda c^\gamma. \tag{7.3.60}$$

The minus sign indicates that we have here a sink. For an adsorbed component (Case B in Subs. 7.2.1, and detailed discussion on adsorption in Subs. 7.3.3) that undergoes radioactive decay, the source term takes the form:

$$\Gamma_s^\gamma = -\lambda F^\gamma, \tag{7.3.61}$$

in which λ is the coefficient of radioactive decay.

If we assume that adsorption follows a linear isotherm, the last expression can be replaced by

$$\Gamma_s^\gamma = -\lambda K_d c^\gamma. \tag{7.3.62}$$

For any other decay or degradation of a considered component, the source term takes the form:

$$\theta \rho \Gamma^\gamma = -\theta k_f c^\gamma, \tag{7.3.63}$$

where k_f represents the *degradation rate constant* for the component in the fluid phase. For an adsorbed component, we have

$$\Gamma_s^\gamma = -k_s F^\gamma, \tag{7.3.64}$$

where k_s represents the *degradation rate constant* for the adsorbate.

7.3.3 Equilibrium adsorption

A. The phenomenon of adsorption

Adsorption, or *sorption*, , the opposite of *desorption*, is the phenomenon in which part of the mass of a chemical species present in a liquid (*adsorbate*) that occupies the void space, or part of it, accumulates on the solid matrix (*adsorbent*) at the liquid-solid interface. Adsorption is caused by the attraction of the species to the surface of a solid, or by reactions of one or more species present in the liquid with the solid. Although we are interested here, primarily, in adsorption of a species onto a solid in contact with a liquid phase, adsorption may also take place between a solid and a gaseous phase.

A simple treatment of adsorption uses the concept of an *adsorption isotherm*, discussed in the next subsection. This concept assumes that *equilibrium conditions* prevail, and that under these conditions, the amount of adsorbed species on a solid, say, within an REV, is solely a function of the concentration of the species in the liquid. This assumption is valid as long as the concentration of all other dissolved species affecting adsorption do not change appreciably in time. In general, however, this condition does not hold and a more complicated analysis is required, involving the analysis of reactions at the solid surface. Examples of such reactions are ion exchange and surface complexation.

One possible primary driving force for adsorption is the *lyophobic* ('solute disliking') nature of a solute relative to that of the solvent. Another possible reason is the high affinity of the solute for the solid (Weber, 1972). The affinity of a dissolved species to the solid surface is due to *physical causes*, such as electrical attraction, *van der Waals attraction*, i.e., intermolecular forces of attraction between molecules of the solid and those of the adsorbed species, and *chemisorption*, i.e., chemical interaction between the solid and the adsorbed species. However, the most significant factor is the degree of solubility of the dissolved species. Other factors that may affect the adsorption and desorption of a chemical species are the physical, electrical, and chemical characteristics of the species and of the solid's surface, temperature, and the presence of other species in the liquid phase (e.g., through the pH that results from their presence in the liquid phase).

In some theories (e.g., Weber, 1972), the solid is always assumed to be covered by a thin fluid boundary layer, or film, that has properties and composition different from those of the bulk fluid. When considering 'equilibrium', we mean equilibrium between the adsorbed species and the concentration of the species in that film. Then, since, for our modeling purposes, we need a macroscopic description of adsorption, we make certain assumptions, e.g., that because of diffusion and the short distances involved, the average concentration, say in an REV, is the same as that close to the solid, so that we can express the isotherm in terms of average the concentration.

Because advective fluid flow within this boundary layer is negligible, to reach the solid surface, the adsorbate has first to pass from the bulk solution through this layer by *molecular diffusion*. Then, after passing through the boundary layer, the adsorbate can interact with the solid. The desorbed species can return to the bulk solution in a similar way.

In certain porous media, the solid matrix itself is porous, but with much smaller pores. A porous medium of this kind is often referred to as a 'double porosity medium' (Subs. 7.3.6H), because the porous medium and the solid each has its own porosity. The terms *micropores* and *macropores* are often used. Usually, the permeability of the porous solid is much smaller than that of the porous medium as a whole. As a consequence, we often *assume* that the liquid occupying the void space of the porous solid is (practically) immobile (see Subs. 6.1.4). Under such conditions, an adsorbate must first diffuse *into* the (liquid saturated) solid matrix, then diffuse within the small pores that constitute the pore space of the latter, and finally adsorb on its (very large) surface area.

In many cases, when considering the rate of adsorption, or the characteristic time involved, the rate determining (or rate limiting) step is not the chemical interaction with the solid, but the diffusion through the film and (in the case of a porous matrix) through the tiny pores within the solid matrix.

B. Adsorption isotherms

In saturated flow, when adsorption of a dissolved chemical species takes place, its total mass, say within every REV of the porous medium, is *partitioned* between the solution and the solid matrix. Any increase in the quantity of the species in the liquid is associated with an appropriate increase in its quantity on the solid, and vice versa. In *desorption*, the quantity of the species on the solid decreases; this is associated with an appropriate decrease of the quantity in solution. In unsaturated flow, we have to consider partitioning of the total mass of the species in the REV, among the solid, the liquid and the gaseous phases. An *adsorption isotherm* is a function that relates the quantity of a species adsorbed on the solid to its quantity in the liquid phase that occupies the void space (or part of it), *at a fixed temperature*, under conditions of (chemical) equilibrium between the two quantities.

Let the symbol F denote the mass of a species (= adsorbate) adsorbed on the solid (= adsorbent), per unit mass of the latter. The concentration F may be measured in kg/kg, or in moles/kg, while the concentration in the liquid, c, is measured in kg/ℓ, or in moles/ℓ. Although it would seem more natural to define the quantity of the species on the solid per *unit surface area of the solid*, the reference to 'unit mass of solid', stems from the way this quantity is measured in the laboratory.

Different adsorbate-adsorbent pairs have different isotherms, stemming from the different mechanisms involved. The isotherm for a given adsorbate-adsorbent pair can be obtained by performing a *batch adsorption experiment.* A fixed amount of porous medium (soil) is mixed in separate containers with an aqueous solution at different concentrations, and the change in the latter, resulting from adsorption, is recorded as time elapses, until the systems reach equilibrium. By performing a mass balance within each container, the adsorbed quantity is computed, yielding a point on the isotherm. If the time for reaching equilibrium is too long, equilibrium may not be assumed.

Following are examples of the more commonly encountered isotherms:

(i) Freundlich (1907) suggested the *nonlinear isotherm*

$$F = bc^m, \tag{7.3.65}$$

where b and m are constant coefficients (functions of temperature), and c denotes the concentration of the adsorbate in the solution. The case $m < 1$ means that as F increases, it becomes more difficult for additional quantities of the adsorbate to be adsorbed. The opposite situation is described by $m > 1$.

(ii) For $m = 1$, and replacing the symbol b by the more commonly used symbol K_d, the relationship (7.3.65) reduces to the *linear isotherm*:

$$F = K_d\, c. \tag{7.3.66}$$

The coefficient K_d, which expresses the affinity of the species for the solid, relative to that for the liquid (usually for an aqueous phase), is called the *distribution coefficient*, or *partitioning coefficient.* From (7.3.66), it follows that K_d ($\equiv F/c$) gives, *at every instant*, the mass of the species on the solid, per unit mass of the latter, per unit concentration of the species in the liquid phase. It describes the partitioning of the total amount of the species between the solid surface and the liquid phase, say, in a unit volume of porous medium. We note that K_d has the dimensions of mass per unit volume, and should be described by the corresponding units (e.g., kg/ℓ).

Sometimes, K_d for the adsorption process differs from that for the desorption one. This implies that the process is not completely reversible, often due to surface catalysis. Another observation is that, often, especially in chemisorption, there exists a limit to the adsorptive capacity of a solid surface. This requires a modification of the isotherm (7.3.66).

In unsaturated flow, as the larger pores are occupied by air, part of the solid's surface is less readily accessible to pore water. This may make K_d in (7.3.66) a function of the saturation. On the other hand, we recall that water is the *wetting liquid*, and, as such, it is everywhere adjacent to the solid surface, albeit at some places as a very thin film, with diffusion of chemical species through it. Hence, we may conclude that (7.3.66) is valid also in unsaturated flow, unless the moisture content is very low, a situation that, under certain conditions, may occur close to ground surface.

(iii) Langmuir (1915, 1918) suggested the nonlinear equilibrium isotherm:

$$F = \frac{k_3 c}{1 + k_4 c}, \quad k_3, k_4 = \text{constant coefficients.} \tag{7.3.67}$$

Note that $F \to \text{const.}$ as $c \to \infty$.

(iv) Lindstrom *et al.* (1971) and van Genuchten (1974), present the nonlinear isotherm:

$$F = k_5\, c \exp(-2k_6 F), \quad k_5, k_6 = \text{constant coefficients.} \tag{7.3.68}$$

Some of the above isotherms can be expressed as $F = K_d(c)\, c$, where $K_d \equiv F/c$. In such cases, it can be shown that for any γ-species, we have

$$\frac{m^\gamma\big|_{\text{in the fluid}}}{m^\gamma\big|_{\text{in the porous medium}}} = \frac{c\,\mathcal{U}_w}{c\,\mathcal{U}_w + F\, m_s} = \frac{1}{1 + \frac{\rho_b K_d(c)}{\phi}}, \tag{7.3.69}$$

in which ϕ is the porosity, and ρ_b denotes the bulk density of the soil.

Altogether, an adsorption isotherm tells us what portion of a species will be adsorbed when the solid and aqueous phases are in equilibrium, i.e., when the net rate of mass transfer of the species between the aqueous liquid and the solid is zero. Of course, groundwater systems are never strictly at equilibrium, with advective and diffusive fluxes of species always occurring within the liquid (= water). Nevertheless, as will be explained in Subs. 7.3.6, in many cases, the time characterizing the adsorption reaction may be sufficiently fast, relative to the times characterizing advection and diffusion in the liquid phase within the void space, so that equilibrium may be assumed to prevail. However, we may also encounter cases where equilibrium is not a valid assumption, and a kinetic approach will be required. Nonequilibrium phase partitioning will be discussed in Subs. 7.3.6.

C. Adsorption of hydrophobic organic species

Nonionic and nonpolar organic molecules are *hydrophobic* (= 'water hating'), i.e, they have a low affinity for water, which is a highly polar substance. Hydrophobic substances 'prefer' to associate with other nonpolar substances like themselves. This feature of organic compounds explains their relatively low solubility in water, the degree of hydrophobicity being inversely proportional to their solubility. Organic matter is often hydrophobic; it may occur in the form of grains, globules, or films that coat soil grains.

A commonly used measure of hydrophobicity of an organic γ-species, i.e., of the degree to which it will preferentially dissolve in water, rather than in an organic solvent, is its coefficient of partitioning between *octanol* and water,

$$K^\gamma_{ow} = \frac{c^\gamma_o}{c^\gamma_w}, \tag{7.3.70}$$

where c^γ_o and c^γ_w represent the γ-concentrations in octanol (o) and in water, respectively. The octanol is used here as a reference liquid phase for assessing the organic liquid-water partitioning behavior of an organic solute. A larger value of K_{ow}^{γ} indicates a higher tendency to dissolve in an organic liquid, rather than in water, and thus becoming less mobile in the subsurface.

Adsorption of hydrophobic species is controlled, to a large extent, by the presence in the soil of organic matter, as a solid phase, which exhibits a greater attraction to hydrophobic substances than for water. For most soils, the concentration of organic matter is in the range 0.5–1.0%. Since adsorption is a surface phenomenon, it is usually more important how the organic matter is distributed and made available for adsorption. Thus, soils with very low bulk organic matter can exhibit significant retardation for organic compounds. Additionally, in soils that are very low in organic matter, adsorption of organic compounds onto the clay fraction of the soil can become dominant. When the soil organic carbon content exceeds a few tenths of a percent, even moderately hydrophobic substances will be adsorbed onto it (McCarty *et al.*, 1981). The linear partitioning coefficient, K_d, may, then, be estimated by

$$K_d = K_{oc} f_{oc}, \tag{7.3.71}$$

where K_{oc} is an organic carbon normalized partitioning coefficient, e.g., mass of adsorbed species per unit mass of organic soil per unit concentration of species in solution (dims. L^3M^{-1}), and f_{oc} is the mass fraction of organic carbon in the soil (dims. $M\,M^{-1}$), i.e., mass of organic carbon per unit mass of soil. Since the value of K_{oc} is primarily a measure of the hydrophobic tendency of the adsorbing species, it may be anticipated that it should have a high correlation with the octanol-water partitioning coefficient. Indeed, this is the case. A number of studies have led to various empirical formulas for estimating K_{oc}. Most of these formulas are of the form:

$$\log K_{oc} = A \log K_{ow} + B, \tag{7.3.72}$$

where A and B are empirical coefficients, which depend on the composition of the organic matter and on the experimental conditions. Lyman *et al.* (1982) suggested to use $A = 0.937$ and $B = 0.006$, when K_{oc} is expressed in cm^3g^{-1}. Similarly, correlations have been shown to exist between K_{oc} and water solubility, S_w^{γ}, of the form:

$$\log K_{oc} = A' \log S_w^{\gamma} + B', \tag{7.3.73}$$

where A' and B' are empirical coefficients. For example, Means *et al.* (1980) proposed $A' = -0.686$, and $B' = 4.273$, when S_w^{γ} is in mg per liter.

7.3.4 Ion exchange

In ion exchange reactions, ions that are held by electrostatic forces to a charged functional group on the surface of the solid matrix are exchanged for ions of a similar charge present in the aqueous solution. This exchange will continue until equilibrium is reached for all ions present in the system. A process of this kind takes place, for example, in synthetic ion exchange resins. In the soil, ion exchange processes occur primarily on clay minerals (in connection with cation exchange), but also on oxides/oxihydroxides (in connection with both cation and anion exchange). The explanation is based on the observation that oxides/oxihydroxides can be positively charged (anion exchange) or negatively charged (cation exchange), depending on their point of zero charge and the pH of the water, when they are not charged, they are not available for ion-exchange. It is of interest to note that, as charge forces act over larger distances, compared to hydrophobic adsorption, ion-exchange is an extremely fast process.

Some ion exchange theories envision the presence of a thin liquid boundary layer that covers the solid. The time characterizing ion exchange depends on the relative times of (1) transport of the ions from the bulk solution to the boundary layer, (2) diffusion through the layer, (3) (in the case of a porous solid matrix) diffusion into the solid, (4) exchange of ions on the solid, (5) their diffusion through the layer, and (6) transport back to the bulk solution. The limiting rate is often dictated by the various diffusive steps, rather than by the actual exchange process (Weber, 1972).

Without going into details, which can be found in texts that deal with clay minerals (e.g., Grim, 1968), the structure of most clay minerals can be described as composed of layers of aluminum silicates, each layer being made up of sheets of SiO_4 tetrahedral units and $Al(OH)_xO_{6-x}$ octahedral ones, where the two kinds of units share some oxides with each other. For example, *kaolinite*, $Al_2Si_2O_5(OH)_4$, is composed of tetrahedral SiO_4 and octahedral $Al(OH)_4O_2$ sheets. The Al^{3+} ion in the octahedral unit may be substituted by such ions as Mg^{2+}, Fe^{2+}, Mg^{2+}, and Mn^{2+}. This isomorphic substitution creates an excess negative charge on the sheets, such that the clay surface can now attract other positive ions from the solution.

Reversible ion exchange for a univalent component may take the form:

$$X^+ + AS \rightleftharpoons A^+ + XS, \tag{7.3.74}$$

in which X^+ is a dissolved cationic species, XS is the species in the adsorbed state on the solid, denoted by S, and A^+ is the cation, initially on the solid, remaining in the liquid phase. Upon reaching a state of equilibrium, we obtain, according to the *law of mass action* (Subs. 7.3.2A; and, e.g., Chang and Cruickshank, 2003),

$$K^X_{eqA} = \frac{\{XS\}\{A^+\}}{\{AS\}\{X^+\}}, \tag{7.3.75}$$

in which K^X_{eqA} is the *equilibrium coefficient* for the exchange reaction, and the bracketed terms represent thermodynamic concentrations, or activities (Subs. 7.3.2A).

At a low concentration of a considered species, X^+, relative to that of an exchangeable cation, A^+, and when the exchange between the two does not cause a significant change in the activity ratio, $\{A^+\} / \{AS\}$, the partitioning coefficient, K_d, may be obtained from

$$K_d = \frac{K^{X}_{eqA}}{\{A^+\} / \{AS\}} = \frac{\{XS\}}{\{X^+\}}. \tag{7.3.76}$$

For a dilute solution, the activities are approximately equal to the respective molalities, and we have

$$K_{eq} = \frac{(\hat{m}^{P})^p (\hat{m}^{Q})^q \ldots}{(\hat{m}^{A})^a (\hat{m}^{B})^b \ldots}, \tag{7.3.77}$$

where the concentrations are in molals. Also, for such a solution, the activities are proportional to the molar concentrations, so that

$$K'_{eq} = \frac{[P]^p [Q]^q \ldots}{[A]^a [B]^b \ldots}, \tag{7.3.78}$$

(i.e., the mass action law introduced earlier), where K'_{eq} is an equilibrium constant that is proportional to K_{eq}, and includes factors involving the densities of the phases.

7.3.5 Volatilization and dissolution

Volatilization and dissolution are examples of *heterogeneous reactions*, which, as defined earlier, are reactions that take place between species that exist in different (adjacent) phases. An example is the dissolution of CO_2 gas (denoted as $CO_2(g)$) in water to form carbonic acid, H_2CO_3. This reaction is described by the stoichiometric equation

$$CO_2(g) + H_2O \rightleftharpoons H_2CO_3. \tag{7.3.79}$$

Note that in the above equation, and in what follows, the symbol for a species A in an α-phase is denoted either as A_α or $A(\alpha)$, with $\alpha = g$ for gas, and $\alpha = aq$ for the aqueous phase.

A special case is the dissolution of an *ionic solid.* When such a solid is placed in water, it dissolves and equilibrium is established between the ions in the (saturated) solution and the (excess) solid phase. For example, in the case of solid *silver chloride* (AgCl) in contact with a saturated solution of the sodium chloride (adjacent to the solid-liquid interface), equilibrium is reached, with

$$AgCl(s) \rightleftharpoons Ag^+ + Cl^-. \tag{7.3.80}$$

The mass action law leads to

$$\frac{[\mathrm{Ag}^+][\mathrm{Cl}^-]}{[\mathrm{AgCl}(s)]} = K. \tag{7.3.81}$$

In the above expression, the concentration of the pure solid is constant and fixed, no matter how much solid is in contact with the liquid. Hence, when the saturated solution is in equilibrium with excess solid, we have

$$[\mathrm{Ag}^+][\mathrm{Cl}^-] = K[\mathrm{AgCl}(s)] = K_{\mathrm{sp}}, \tag{7.3.82}$$

where the constant K_{sp} is called the *solubility product* of this solution. The value of this coefficient can be determined only experimentally. A larger value means higher solubility of the solid.

Another example is the dissolution of *calcium sulfate*:

$$\mathrm{CaSO}_4 \rightleftharpoons \mathrm{Ca}^{2+} + \mathrm{SO}_4^{2-}. \tag{7.3.83}$$

Then, with $K_{\mathrm{sp,\,gypsum}} = 10^{-4.60}$ at 25°C, we have:

$$K_{\mathrm{sp,\,gypsum}} = [\mathrm{Ca}^{2+}][\mathrm{SO}_4^{2-}] = 10^{-4.60}. \tag{7.3.84}$$

The case of solid matrix dissolution is discussed in Subs. 7.3.6F.

A special case of a heterogeneous reaction is the transformation of a single species in a phase into another single species in another phase that is present on the other side of a common interface. Such reaction is often referred to as a *binary heterogeneous reactions*. An example is the dissolution of nitrogen gas in water,

$$\mathrm{N}_2(g) \rightleftharpoons \mathrm{N}_2(aq). \tag{7.3.85}$$

Another example is *evaporation*, i.e., the transformation of liquid water into water vapor:

$$\mathrm{H}_2\mathrm{O}(aq) \rightleftharpoons \mathrm{H}_2\mathrm{O}(g). \tag{7.3.86}$$

Other examples of binary heterogeneous reactions are when a gaseous species dissolves into a liquid, when a component in a liquid phase volatilizes, and when a solid species dissolves into a single-species aqueous phase. In all of these cases, the reaction involves the transformation of a chemical species, say $\mathcal{C}$, residing in one phase, say α, into the same component residing in another phase, say, β. This reaction can be represented in the symbolic form

$$\mathcal{C}(\alpha) \rightleftharpoons \mathcal{C}(\beta). \tag{7.3.87}$$

The condition for chemical equilibrium in this reaction is

$$\mu^{\mathcal{C}(\alpha)} = \mu^{\mathcal{C}(\beta)}, \tag{7.3.88}$$

i.e., two phases, whose only interaction is by dissolution or volatilization of their components, are in equilibrium with each other only if the *chemical potentials* of these components are equal.

Let us briefly discuss the concept of chemical potential introduced by Willard Gibbs (1928, see Denbigh, 1981, p. 78). Gibbs defined the chemical potential in the following way:

> "If to any homogeneous mass we suppose an infinitesimal quantity of any substance to be added, the mass remaining homogeneous and its entropy and volume remaining unchanged, the increase of the energy of the mass divided by the quantity of the substance added is the *chemical potential* for the substance in the mass considered."

Thus, we may define the chemical potential, μ^γ, of a species γ of a phase, as

$$\mu^\gamma \equiv \left(\frac{\partial G}{\partial N^\gamma}\right)\bigg|_{p,T,N^\delta;\delta\neq\gamma}, \tag{7.3.89}$$

where $G = G(p, T, N^1, N^2, \ldots, N^k)$ denotes the *Gibbs free energy* (see any textbook on Thermodynamics), N^γ denotes the number of moles of γ in the phase, and p and T denote the pressure and temperature, respectively.

The chemical potential is analogous to temperature and pressure. While a temperature difference determines the rate and direction of heat movement from one body to another, and a pressure difference determines the motion of mass, the difference in chemical potential produces the movement of a within a phase, or from one phase to another. It also determines the direction of chemical reactions. A system is in chemical equilibrium when the Gibbs free energy is at a minimum level.

From the definition of G, it follows that the chemical potential of a chemical species expresses the increase in the capacity of the latter to do work (other than work of expansion), per unit amount of added species, at constant temperature and pressure.

Finally, two phases, α and β, are said to be in *chemical equilibrium* with respect to a γ-species, if

$$\mu^\gamma_\alpha = \mu^\gamma_\beta. \tag{7.3.90}$$

In general, a system is said to be in *thermodynamic equilibrium* if the thermodynamic potentials (= chemical potentials and temperature) are uniform within each homogeneous region of the system and do not change with time. We often distinguish between thermodynamic equilibrium, thermal equilibrium (when the temperature is uniform), chemical equilibrium (when the chemical potential is uniform), and mechanical equilibrium (when the pressure is uniform). (See any textbook on Thermodynamics for these potentials.) Bear and Nitao (1995) discuss these definitions for a porous medium domain.

The *chemical potential* of a species γ can also be expressed as

$$\mu^\gamma = \mu^{o\gamma} + RT \ln\{\gamma\}, \tag{7.3.91}$$

in which $\{\gamma\}$ denotes the *activity* of the γ-species, and $\mu^{o\gamma}$ is the value of the chemical potential at some standard state for that species.

Inversely, we can use (7.3.91) as the a for the activity of γ,

$$\{\gamma\} \equiv \exp\left\{\frac{\mu^{\gamma} - \mu^{o\gamma}}{RT}\right\}. \tag{7.3.92}$$

In this context, the standard state of a γ-species refers to a fixed specified composition of the system at some pressure and temperature of interest, usually the same as those of the system under consideration. Therefore, the chemical potential at the standard state, $\mu^{o\gamma}$, is independent of composition, but may be considered as a function of pressure and temperature. The activity and the standard potential are linked by the chosen standard state. The standard state of an aqueous solute species is usually taken as that existing in an aqueous solution at infinite dilution, without any other solute species.

With the above definition of chemical potential, consider the general binary heterogeneous reaction in which a species A in an α-phase transforms into a species B in a β-phase,

$$\mathrm{A}(\alpha) \rightleftharpoons \mathrm{B}(\beta). \tag{7.3.93}$$

The condition of chemical equilibrium, (7.3.88) or (7.3.90), for this reaction takes the form

$$\mu^{\mathrm{A}(\alpha)} = \mu^{\mathrm{B}(\beta)}. \tag{7.3.94}$$

By applying (7.3.94) to species A and B, we obtain

$$K_{\mathrm{eq}} = \frac{\{\mathrm{B}\}}{\{\mathrm{A}\}}, \tag{7.3.95}$$

where K_{eq} is the *equilibrium constant*, defined as

$$K_{\mathrm{eq}} = K_{\mathrm{eq}}(p,T) \equiv \exp\left\{-\frac{(\mu^{o\mathrm{B}} - \mu^{o\mathrm{A}})}{RT}\right\}. \tag{7.3.96}$$

Let p^{B} denote the *vapor pressure* of species B. To understand the meaning of vapor pressure, we consider a closed container partly filled by a liquid. The other part is filled only by the vapor of that liquid. Under equilibrium conditions, the rate of evaporation and that of condensation are equal. Under such conditions, the vapor is said to be *saturated*, and the pressure in the gaseous phase (= vapor) is called 'vapor pressure'. The saturated vapor pressure is a function of the temperature. If the container is open to the air, then the vapor pressure is seen as a *partial pressure* (of the air), along with the other constituents of the air.

If a gas is close to being an ideal gas, then

$$K_{\mathrm{eq}}(p,T) = \frac{p^{\mathrm{B}}}{\{\mathrm{A}\}}. \tag{7.3.97}$$

In most cases, we may assume that K_{eq} is primarily a function of T only, i.e.,

$$K_{\text{eq}}(T) = \frac{p^{\text{B}}}{\{\text{A}\}}. \tag{7.3.98}$$

For a gas that is close to an ideal one, we have $p^{\text{B}} \approx n^{\text{B}} p_g$, so that this equation implies that

$$K_{\text{eq}}(T)/p_g = n^{\text{B}}/\{\text{A}\}. \tag{7.3.99}$$

If an α-phase is a dilute solution, then $\{A\} \approx \hat{m}^{\text{A}} \approx n^{\text{A}}\hat{\rho}_\alpha$, where n^{A} is the mole fraction and $\hat{\rho}_\alpha$ is the molar density of the α-phase. From (7.3.98), we then have

$$\hat{K}_{\text{eq}}(T) = p^{\text{B}}/n^{\text{A}}, \tag{7.3.100}$$

where we define $\hat{K}_{\text{eq}}(T) \equiv K_{\text{eq}}(T)\hat{\rho}_\alpha$. Also, from (7.3.99), we have

$$\hat{K}_{\text{eq}}(T)/p_g = n^{\text{B}}/n^{\text{A}}. \tag{7.3.101}$$

Equation (7.3.100) is another form of *Henry's law*, (7.2.16), used particularly when applied to a species B that, at the pressure and temperature of interest, exists only in its pure state as a gas. It states that, *at equilibrium*, the mole fraction of a dissolved gas species is proportional to its vapor

If a species A in an α-phase can exist at temperature T, as its mole fraction in the phase tends to one, $n^{\text{A}} \to 1$, then it follows from (7.3.100) that $\hat{K}_{\text{eq}}(T) = p^{\text{A}}_{\text{sat}}(T)$. Here, $p^{\text{A}}_{\text{sat}}(T)$ is the *saturated vapor pressure,* which is defined as the vapor pressure in the presence of a non-gaseous phase consisting only of species A. Therefore, in this case, we have

$$p^{\text{A}}_{\text{sat}}(T) = p^{\text{B}}/n^{\text{A}}, \tag{7.3.102}$$

and

$$p^{\text{A}}_{\text{sat}}(T)/p_g = n^{\text{B}}/n^{\text{A}}. \tag{7.3.103}$$

The first equation, (7.3.102), states that the vapor pressure of a volatile species is equal to the saturated vapor pressure times the mole fraction of the species dissolved in the liquid solution that is in contact with its vapor. This relationship is called *Raoult's law.*

Consider a contaminant, say a VOC (= volatile organic compound), in the unsaturated domain, in which the void space is occupied by two fluid phases: a gas, g, and a liquid, ℓ, say, an aqueous phase, at saturations S_g and S_ℓ, respectively. Let the considered contaminant be denoted by the superscript v. The VOC is present as (1) a *vapor* in the gaseous phase at the concentration c^v_g, (2) a *solute* (= dissolved component) in the aqueous liquid, at concentration c^v_ℓ. No 'free product' (i.e., no liquid VOC) is present in the void space.

From (7.3.94) and (7.3.91), we can develop another form of *Henry's law* (e.g., Chang and Cruickshank, 2003)

$$\mathcal{H} \equiv \mathcal{H}^v_{g,\ell} = \frac{c^v_g}{c^v_\ell}, \tag{7.3.104}$$

where we recall that c^v denotes mass concentration (= mass of the solute per volume of gaseous or liquid phase). Altogether, Henry's law expresses fluid to fluid partitioning, or *volatilization*, of a chemical species between a liquid *under conditions of equilibrium*, in which the species appears as a solute, and a gas, where the species appears as a vapor.

7.3.6 Nonequilibrium reactions

Under practical circumstances, the rates of many reactions are fast relative to the time required for liquids and dissolved chemical species to travel through an REV, so that *local (chemical) equilibrium* (or approximately so, in terms of macroscopic variables) may be assumed to prevail. Under such conditions, the system's behavior may be described, with sufficient accuracy, by using thermodynamic relationships. However, in general, it should be recognized that reaction rates are finite and that these relationships do not always hold. This is the topic of the next subsection.

A. Elements of Chemical kinetics

Under *nonequilibrium* conditions, we have to introduce appropriate expressions for the rate of production, or disappearance, of every species in the system that is involved in a chemical reaction. As explained in Subs. 7.3.2A, these rates are, in turn, related to the rates of the reactions. Let us first consider a liquid phase in a well-mixed isolated system that is closed to the transfer of mass from the outside, and is kept at constant temperature and pressure. A well-mixed beaker of liquid may serve as an example. Only homogeneous reactions are considered. In such a closed system, the microscopic balance equation of a γ-species in the liquid phase (density ρ) is given by

$$\frac{dc^\gamma}{dt} = \rho\Gamma^\gamma. \tag{7.3.105}$$

From (7.3.47), it follows that the source term, $\rho\Gamma^\gamma$, may be expressed as

$$\rho\Gamma^\gamma = \sum_{(j)} \left.\frac{dc^\gamma}{dt}\right|_{j\text{th hom chem react}} = M^\gamma \sum_{(j)} \nu_j^\gamma R_{r,j}. \tag{7.3.106}$$

Therefore, the following system of ordinary differential equations

$$\frac{dc^\gamma}{dt} = M^\gamma \sum_{(j)} \nu_j^\gamma R_{r,j}, \tag{7.3.107}$$

must be solved for the c^γ's, where the $R_{r,j}$'s must be known functions of the concentrations. The initial concentrations must also be specified. Often, the reaction rates, $R_{r,j}$'s, are functions of concentration, expressed in some molar unit such as moles per unit volume of liquid. In the latter case, it is more convenient to write the system of differential equations in the form:

$$\frac{d\,[\gamma]}{dt} = \sum_{(j)} \nu_j^{\gamma} R_{r,j}. \tag{7.3.108}$$

As an example, consider the following simple reaction:

$$\mathrm{A} \xrightarrow{k} \mathrm{B}, \tag{7.3.109}$$

with species A obeying a first-order reaction, $R_r^{\mathrm{A}} = -kc^{\mathrm{A}}$, where k denotes a first-order rate constant. We then have

$$\begin{aligned} \frac{dc^{\mathrm{A}}}{dt} &= M^{\mathrm{A}} R_r^{\mathrm{A}} = \nu^{\mathrm{A}} M^{\mathrm{A}} R_r = -M^{\mathrm{A}} k c^{\mathrm{A}}, \\ \frac{dc^{\mathrm{B}}}{dt} &= M^{\mathrm{B}} R_r^{\mathrm{B}} = \nu^{\mathrm{B}} M^{\mathrm{B}} R_r = M^{\mathrm{B}} k c^{\mathrm{A}}, \end{aligned} \tag{7.3.110}$$

where $\nu^{\mathrm{A}} = -1$ and $\nu^{\mathrm{B}} = 1$.

When the reaction is reversible, we have the following forward and reverse reactions:

$$\mathrm{A} \xrightarrow{k_{\mathrm{for}}} \mathrm{B}, \qquad \mathrm{B} \xrightarrow{k_{\mathrm{rev}}} \mathrm{A}. \tag{7.3.111}$$

Let the forward reaction be first-order and the reverse reaction be second-order, i.e.,

$$R_{r,\mathrm{for}} = k_{\mathrm{for}}\, c^{\mathrm{A}}, \qquad R_{r,\mathrm{rev}} = k_{\mathrm{rev}} (c^{\mathrm{B}})^2. \tag{7.3.112}$$

Then, the resulting system of balance equations is

$$\begin{aligned} \frac{dc^{\mathrm{A}}}{dt} &= M^{\mathrm{A}} (\nu_{\mathrm{for}}^{\mathrm{A}} R_{r,\mathrm{for}} + \nu_{\mathrm{rev}}^{\mathrm{A}} R_{r,\mathrm{rev}}) = M^{\mathrm{A}} (-k_{\mathrm{for}}\, c^{\mathrm{A}} + k_{\mathrm{rev}} (c^{\mathrm{B}})^2), \\ \frac{dc^{\mathrm{B}}}{dt} &= M^{\mathrm{B}} (\nu_{\mathrm{for}}^{\mathrm{B}} R_{r,\mathrm{for}} + \nu_{\mathrm{rev}}^{\mathrm{B}} R_{r,\mathrm{rev}}) = M^{\mathrm{B}} (k_{\mathrm{for}}\, c^{\mathrm{A}} - k_{\mathrm{rev}} (c^{\mathrm{B}})^2). \end{aligned} \tag{7.3.113}$$

B. Chemical kinetics at the macroscopic level

We need expressions for kinetic chemical reactions in porous media, i.e., at the macroscopic level. From (7.3.47), it follows that the macroscopic source term of a γ-species in a liquid α-phase is given by

$$\theta_\alpha \overline{\rho_\alpha \Gamma^\gamma}^{\alpha} = \theta_\alpha M^\gamma \sum_{(j)} \nu_j^\gamma \overline{R_{r,j,\mathrm{hom}}}^{\alpha}, \tag{7.3.114}$$

where $\overline{R_{r,j,\mathrm{hom}}}^{\alpha}$ denotes the volume-averaged rate of the jth homogeneous reaction. This term serves as a source due to homogeneous reactions in the mass balance equation of the γ-species. We shall assume that macroscopic rate laws for all homogeneous reactions in the α-phase are known functions of the macroscopic concentrations $\overline{c^\lambda}^\alpha$ in that phase, with λ denoting all species in the α-phase that participate in the reaction.

C. Interphase exchange at the macroscopic level

Next we consider sources due to heterogeneous reactions, i.e., due to interphase transfers. Let $\mathcal{S}_{\alpha\beta}$ denote the interface between an α and a β-phase. Consider a γ-species in the α-phase that is involved in a heterogeneous reaction, which is occurring either within or along one of the sides of $\mathcal{S}_{\alpha\beta}$. A flux of γ-species passing through the α-face of $\mathcal{S}_{\alpha\beta}$ will result from the reaction. The corresponding specific flux, in units of moles per unit time per unit interfacial area, will be denoted by R^γ_{heter}. If the reaction occurs almost instantaneously, compared to the other relevant time scales for transport of the γ-species, then the flux of the species will be equal to its reaction rate per unit interfacial surface area. In what follows, we shall assume that this equivalence holds.

The rate of the reaction per unit interfacial surface area, R_{heter}, is given by

$$R_{\text{heter}} = \frac{R^\gamma_{\text{heter}}}{\nu^\gamma}. \tag{7.3.115}$$

The total mass flux R^γ_{heter} of a γ-species due to all heterogeneous reactions from $\mathcal{S}_{\alpha\beta}$ is given by

$$R^\gamma_{\text{heter},\alpha\beta} = M^\gamma \sum_{(j)} \nu^\gamma_j R_{\text{heter},j}, \tag{7.3.116}$$

where $R_{\text{heter},j}$ is the reaction rate of the jth heterogeneous reaction on $\mathcal{S}_{\alpha\beta}$ that involves species in the α-phase.

The term $f^\gamma_{\alpha\rightarrow\beta}$ appearing in (7.2.21) for a γ-species is the rate of mass per unit volume of porous medium that leaves the α-phase and reacts at the interface with another phase, β. It is, therefore, the total rate at which a γ-species reacts in all *heterogeneous* reactions between the α and β-phases. Thus, we have

$$f^\gamma_{\alpha\rightarrow\beta} = \frac{1}{\mathcal{U}_o} \sum_{(j)} \int_{\mathcal{S}_{\alpha\beta}} R^\gamma_{\text{heter},j}\, dS = (\Sigma_{\alpha\beta}) M^\gamma \sum_{(j)} \nu^\gamma_j \overline{R_{\text{heter},j}}^{\alpha\beta}, \tag{7.3.117}$$

where $\overline{R_{\text{heter},j}}^{\alpha\beta}$ is the reaction rate of the jth heterogeneous chemical reaction, surface-averaged over the portion of the interface, within the REV, of specific surface $(\Sigma_{\alpha\beta})$, and ν^γ_j is the stoichiometric coefficient of the γ-species in the reaction. Note that $f^\gamma_{\alpha\rightarrow\beta}$ is also the exchange between phases due to advection and diffusion, defined by (7.2.21). Usually, $f^\gamma_{\alpha\rightarrow\beta}$ will be expressed as some function of macroscopic quantities, which must be assigned to (7.3.117), resulting in an equation that must be satisfied, in addition to the relevant balance equations.

In order to understand the nature of heterogeneous liquid-liquid or liquid-solid reactions, and the expressions that describe such reactions (actually, interphase transfers), it is convenient to visualize two thin film regions that exist in the α and β-phases next to and on both sides of the $\mathcal{S}_{\alpha\beta}$-interface. The term 'film' is used here to denote a thin layer that is a subregion of the phase itself, but is not a separate entity, as, for example, is a film of water on a solid. We may then associate with each point $\mathbf{x}'$ on the interface, a microscopic concentration, $c^{\star\lambda}(\mathbf{x}')$, for any λ-species, defined as the concentration, c^{λ}, in the film region of the α-phase immediately next to $\mathbf{x}'$. The macroscopic quantity $\widetilde{c^{\star\lambda}}^{\alpha\beta}$, defines the average of $c^{\star\lambda}$ over $\mathcal{S}_{\alpha\beta}$. When considering heterogeneous reactions whose rates depend only on conditions right next to the interface, i.e., in the film layer, it is reasonable to assume that the reaction rate of the heterogeneous reactions can be described by

$$\widetilde{R_{rj',\mathrm{het}}}^{\alpha\beta} = F_{rj',\mathrm{het}}(\widetilde{c^{\star\lambda}}^{\alpha\beta}), \tag{7.3.118}$$

where λ stands for all species in the α and β-phases that are involved in the reaction; and $F_{rj',\mathrm{het}}$ represents a functional relationship.

Next, we assume that the $f^{\gamma}_{\alpha\rightarrow\beta}$'s have the following known functional dependence:

$$f^{\gamma}_{\alpha\rightarrow\beta} = f^{\gamma}_{\alpha\rightarrow\beta}(\overline{c^{\lambda}}^{\alpha}, \widetilde{c^{\star\lambda}}^{\alpha\beta}, \Sigma_{\alpha\beta}, \theta_{\alpha}, \theta_{\beta}). \tag{7.3.119}$$

D. Nonequilibrium fluid-to-fluid mass transfer

Let us consider the case of mass transfer between two fluid phases, resulting from a heterogeneous reaction across their common interface. The dissolution of a hydrocarbon species in water, or the volatilization of a species dissolved in a NAPL into a vapor in the gas phase, may serve as examples. Partitioning *at the interface* is assumed to occur almost instantaneously; this means that chemical equilibrium is always assumed there. For the sake of simplicity, let us assume that no reactions occur *within* the fluid phases. However, nonequilibrium conditions in the phases may exist in the form of concentration gradients and movement of species. Interphase mass transfer rates are, therefore, controlled by diffusive and advective transport of species from within each fluid phase to the interface.

Suppose we consider the transformation between a species in a gas phase (g) and a species in a liquid phase (ℓ), resulting in the exchange of a component between the two phases, primarily through diffusion. In the macroscopic balance equation of a γ-species in an α-phase ($\alpha = g, \ell$), the term $f^{\gamma}_{\alpha\rightarrow\beta}$ defined by (7.2.21), expresses the exchange. However, we need to specify the actual functional dependence indicated by (7.3.119). We also require that the fluxes at the interface satisfy the equality

$$f^{\gamma}_{\alpha\to\beta} = -f^{\lambda}_{\beta\to\alpha}. \tag{7.3.120}$$

The superscript λ in the above equation denotes the species as the relevant component in the β-phase.

As stated earlier, heterogeneous reactions result in mass transfer between adjacent phases across their common interface. When such a reaction is at equilibrium, a relationship exists between the concentrations on both sides of the interface. It has the functional form:

$$g_{\alpha\beta}(\overparen{c^{\star\gamma}}^{\alpha\beta}, \overparen{c^{\star\lambda}}^{\alpha\beta}) = 0. \tag{7.3.121}$$

An example is Raoult's law (e.g., Chang and Cruickshank, 2003) which can be expressed in the form:

$$\overparen{c^{\star\gamma}}^{\alpha\beta} = \mathcal{K}^{\gamma\lambda}\overparen{c^{\star\lambda}}^{\alpha\beta}. \tag{7.3.122}$$

As mentioned earlier, two 'films' are associated with the gas-liquid interface: a 'gas film' on the gas side, and a 'liquid film' on the liquid side. The transfer of a chemical component from the gaseous phase (in which it exists as a component) to be dissolved in the liquid phase, and vice versa, may be visualized as made up of several steps:

- transport of the component from the bulk gaseous phase to the surface of the 'gas film',
- diffusion through the gas film,
- diffusion through the 'liquid film',
- transport into the bulk liquid.

A characteristic time is associated with each step. The characteristic time of the entire process is determined by the slowest process. This will be the 'rate limiting step' for the entire transfer process. Usually, diffusion through the liquid phase is the 'rate limiting step'. There would be an additional step if the actual reaction would take place on the interface itself. However, here we have assumed that the heterogeneous reaction is an equilibrium one.

In Subs. 7.7, we shall discuss the conditions that justify the assumption of equilibrium for a heterogeneous reaction in an REV. The conditions were expressed in terms of the Strouhal, Peclet, and Damköhler numbers, using a characteristic length ℓ_α that is equal to the local volume-to-surface-area ratio of the phase. If conditions (7.3.6) and (7.3.7), based on these numbers, are satisfied, then the reaction at the interface is fast enough so that it may be considered to be in equilibrium. However, it is still possible that conditions (7.3.4) or (7.3.5) do not hold, because advection might dominate over diffusion, or diffusion could take place slowly. In such a case, we need expressions for the rate of interphase mass transfer. Usually such expressions are derived empirically.

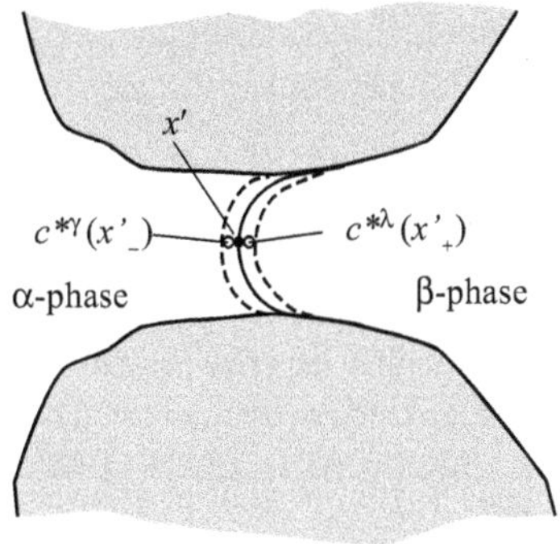

Figure 7.3.1: Diffusion at an interface (not drawn to scale).

The mechanism that drives the transfer (in an effort to bring the system closer to equilibrium) is the difference in concentration across the interface (rigorously, the difference in chemical potentials), visualized as a thin film. Therefore, the rate of transfer, $f^{\gamma}_{\alpha\to\beta}$, of the mass of a γ-species from an α-phase to an adjacent β-phase across the interface $\mathcal{S}_{\alpha\beta}$ is very often assumed to be proportional to the difference in concentration of that component between the two phases. According to Fick's law, this difference acts as a driving force for the transfer of mass. In other words, the rate of mass transfer is commonly described by a first-order mass transfer relationship of the form (Fig. 7.3.1):

$$f^{\gamma}_{\alpha\to\beta} = \alpha^{\star\gamma}_{\alpha}\big(\widetilde{c^{\star\gamma}}^{\alpha\beta} - \overline{c^{\gamma}}^{\alpha}\big). \tag{7.3.123}$$

Similarly, for the transfer from the β-phase, we have

$$f^{\lambda}_{\beta\to\alpha} = \alpha^{\star\lambda}_{\beta}\big(\widetilde{c^{\star\lambda}}^{\alpha\beta} - \overline{c^{\lambda}}^{\beta}\big). \tag{7.3.124}$$

Here, $\alpha^{\star\gamma}_{\alpha}$ and $\alpha^{\star\lambda}_{\beta}$ are *mass transfer coefficients*, which, in general, must be determined experimentally for the specific conditions of interest.

Next, we derive an empirical relation which provides an estimate of the mass transfer coefficient. Suppose that diffusion dominates over advection, that is, (7.3.4) holds for both phases. Then, the diffusive flux of the γ-species, $\mathbf{j}^{\gamma}\cdot\mathbf{n}_{\alpha}$, through the interface is approximated by

$$\mathbf{j}^{\gamma}\cdot\mathbf{n}_{\alpha} = -\mathcal{D}^{\gamma}\big(\widetilde{c^{\star\gamma}}^{\alpha\beta} - \overline{c^{\gamma}}^{\alpha}\big)/\Delta_{\alpha}, \tag{7.3.125}$$

where $\mathcal{D}^{\gamma}$ is the coefficient of molecular diffusion of the γ-species in the α-phase, and Δ_{α} is the length characterizing the mean size of the phase (e.g., $\Delta_{\alpha} = \ell_{\alpha}$). One may think of this length as representing the thickness of the assumed 'film' (or 'films' of the two liquid phases) representing the interface, plus the typical distance which the component has to diffuse within the bulk of each fluid. Substituting (7.3.125) into (7.2.22), we obtain

$$f^{\gamma}_{\alpha\to\beta} = (\Sigma_{\alpha\beta})\mathcal{D}^{\gamma}\big(\widetilde{c^{\star\gamma}}^{\alpha\beta} - \overline{c^{\gamma}}^{\alpha}\big)/\Delta_{\alpha}. \tag{7.3.126}$$

Thus, the mass transfer coefficient in this case, given by $\alpha^{\star\gamma}_{\alpha} = (\Sigma_{\alpha\beta})\mathcal{D}^{\gamma}_{\alpha}/\Delta_{\alpha}$, is seen to increase with the specific interfacial area of the phase and with the molecular diffusivity of the considered species. Also, note that the transfer coefficient depends on the volumetric fraction of the phase. Similarly, for the transfer in the β-phase, we have

$$f^{\lambda}_{\beta\to\alpha} = (\Sigma_{\alpha\beta})\mathcal{D}^{\lambda}\big(\widetilde{c^{\star\lambda}}^{\alpha\beta} - \overline{c^{\lambda}}^{\beta}\big)/\Delta_{\beta}. \tag{7.3.127}$$

By substituting (7.3.126) and (7.3.127) into (7.3.120), and using (7.3.122), it is possible to eliminate $\widetilde{c^{\star\gamma}}^{\alpha\beta}$ and $\widetilde{c^{\star\lambda}}^{\alpha\beta}$ from these equations. The resulting expression for the transfer terms is

$$f^{\gamma}_{\alpha\to\beta} = -f^{\lambda}_{\beta\to\alpha} = \frac{\Sigma_{\alpha\beta}}{\Delta_{\alpha}/\mathcal{D}^{\gamma} + \mathcal{K}^{\gamma\lambda}\Delta_{\beta}/\mathcal{D}^{\lambda}}\left(\mathcal{K}^{\gamma\lambda}\overline{c^{\lambda}}^{\beta} - \overline{c^{\gamma}}^{\alpha}\right). \tag{7.3.128}$$

The approximation used for (7.3.125) essentially assumes that the concentration within a phase varies, more or less, linearly with the distance from the interface. Such a situation may be expected to hold for a *diffusion-dominated system* under quasi-steady conditions. However, if the system is dominated by advection, i.e., if (7.3.4) does not hold, then the concentration profile, and, hence, the resulting mass transfer coefficients, will depend also on the magnitude of the mean fluid velocity.

Although we have used the concentration, c^{γ}, in the above formal derivations, we could have used the mass fraction ω^{γ} $(= c^{\gamma}/\rho_{\alpha})$ instead. The latter *must* be used when there are significant density variations in the fluid phase, which is often the case for a gas phase. The reason is that the diffusive flux used to obtain (7.3.125), as an approximation, is proportional to the mass fraction and not to the concentration.

E. Fluid-to-solid mass transfer

Next, we consider adsorption/desorption, i.e., the transfer of a component from a fluid to the surface of a solid (and, in a reversible case, also back to the fluid). Let γ and δ denote the component dissolved in the fluid phase and the component in its adsorbed state, respectively. As before, we visualize a thin film of fluid at the interface between the fluid and the solid. The mean concentration of the film is denoted by $\widetilde{c^{\star\gamma}}^{\alpha s}$. The surface of the solid is visualized as being populated by possible discrete adsorption sites for the species. We use the term *mole fraction*, n^{δ}, a macroscopic intensive quantity, for the ratio of the number of moles of occupied sites to the number of moles of all possible sites. Under nonequilibrium conditions, the transfer of a component (at the macroscopic level) from the fluid to the surface of the solid (and, in a reversible case, also back to the fluid) varies with time, according to some macroscopic reaction rate law that is a function of $\widetilde{c^{\star\gamma}}^{\alpha s}$, and n^δ.

Let $\hat{\rho}_s^\delta$ denote the number of adsorbed moles per unit surface area of the solid, and M^δ denote the molecular mass of the adsorbed species (= molecular mass of the component). For a first-order reaction rate law, the macroscopic mass balance equation for the adsorbed species is given by

$$M^\delta \hat{\rho}_s^\delta (\Sigma_{\alpha s}) \frac{dn^\delta}{dt} = k_1 \widetilde{c^{\star\gamma}}^{\alpha s} - k_2 n^\delta, \tag{7.3.129}$$

where k_1 and k_2 are *first-order rate constants*. The quantity F^δ, defined as the mass of adsorbed species per unit mass of solid, is often used instead of n^δ, with $\rho_b F^\delta = M^\delta \hat{\rho}_s^\delta (\Sigma_{\alpha s} n^\delta)$. To be consistent with the discussion in previous subsections, we drop the superscript δ, and write

$$f^\gamma_{\alpha\to s} \equiv -\rho_b \frac{dF}{dt} = k_r F - k_f \widetilde{c^{\star\gamma}}^{\alpha s}, \tag{7.3.130}$$

where the coefficients k_f (dims. T^{-1}) and k_r (dims. $\mathrm{ML}^{-3}\mathrm{T}^{-1}$) are related to k_1 and k_2. We note that at equilibrium, $dF/dt = 0$. This leads to $F = (k_f/k_r)\overline{c^\gamma}^\alpha$, i.e., $K_d = (k_f/k_r)$.

One simple model that has been suggested is to visualize a thin film on the interface between the solid surface and the bulk liquid. Molecular diffusion takes place through this film, with the driving force proportional to the difference in concentration across it, or to some function of both concentrations. The characteristic time of this diffusion may be much longer than that of the reactions on the solid surface. The rate of transfer is then expressed by an equation similar to (7.3.123), i.e.,

$$f^\gamma_{\alpha\to s} = \alpha^{\star\gamma} \left(\widetilde{c^{\star\gamma}}^{\alpha s} - \overline{c^\gamma}^\alpha \right), \tag{7.3.131}$$

where $\alpha^{\star\gamma}$ is a *mass transfer coefficient*. Here, $\widetilde{c^{\star\gamma}}^{\alpha s}$ is the average concentration in the film; it can be eliminated by equating the right-hand sides of (7.3.130) and (7.3.131), resulting in the expression:

$$f^\gamma_{\alpha\to s} = \frac{\alpha^{\star\gamma}}{\alpha^{\star\gamma} + k_f} (k_r F - k_f \overline{c^\gamma}^\alpha). \tag{7.3.132}$$

Some other examples of empirical equations for the rate of nonequilibrium adsorption are:

$$f^\gamma_{\alpha\to s} = \alpha^{\star\gamma} \left(k_{20} \overline{c^\gamma}^\alpha + k_{21} - F \right),$$

$$f^{\gamma}_{\alpha\to s} = \alpha^{\star\gamma}\left(\frac{k_{22}\overline{c^{\gamma}}^{\alpha}}{1+k_{23}\overline{c^{\gamma}}^{\alpha}} - F\right),$$

$$f^{\gamma}_{\alpha\to s} = \alpha^{\star\gamma}\left(k_{24}\left(\overline{c^{\gamma}}^{\alpha}\right)^{k_{25}} - F\right), \tag{7.3.133}$$

in which the various k's are experimentally determined coefficients. Note that many of these relationships have the form:

$$f^{\gamma}_{\alpha\to s} = \alpha^{\star\gamma}\left(F_{\text{equil}}(\overline{c^{\gamma}}^{\alpha}) - F\right), \tag{7.3.134}$$

in which $F_{\text{equil}}\left(\overline{c^{\gamma}}^{\alpha}\right)$ denotes the value of F when the solid is in equilibrium with the fluid at concentration $\overline{c^{\gamma}}^{\alpha}$.

Various authors (e.g., Selim *et al.*, 1976; Rao *et al.*, 1979; Rao and Davidson, 1980; Jury, 1983; van Genuchten and Dalton, 1986; Parker and Valocchi, 1986) suggested *two-site equilibrium-kinetic* models in which the assumption is made that the total number of sorption sites on the solid surface is made up of two parts. On one part, a fraction p of the total surface area, adsorption is assumed to be *instantaneous*, so that equilibrium is always assumed to prevail. On the other, a fraction $1-p$ of the total area, kinetic conditions prevail, so that adsorption is assumed to be time dependent. Together, we have

$$F = F_{\text{eq}} + F_{\text{kin}}, \tag{7.3.135}$$

where F_{eq} and F_{kin} represent the mass of adsorbate on the first and second type sites, respectively, per unit mass of solid. We shall assume that for the first type of sites, the linear equilibrium isotherm is applicable, viz.,

$$F_{\text{eq}} = pK_d\,\overline{c^{\gamma}}^{\alpha}. \tag{7.3.136}$$

For sites of the second type—the kinetic, nonequilibrium sites—we use the linear reversible rate equation:

$$\rho_b \left.\frac{dF_{\text{kin}}}{dt}\right|_{\text{ads}} = k_{13}[(1-p)K_d\,\overline{c^{\gamma}}^{\alpha} - F_{\text{kin}}], \tag{7.3.137}$$

in which k_{13} is a *first-order rate coefficient.* An example of a balance equation in which this model is employed is given in Subs. 7.3.2B.

Some authors, e.g., Brusseau *et al.* (1992), considered *two-site kinetic sorption models.* In such models, the entire number of available sites on the solid are, again, divided into two parts. However, in both parts the sorption process is under non-equilibrium conditions, but with different coefficients. One may also consider *two-site equilibrium sorption models.*

F. Solid matrix dissolution

The solid matrix itself (e.g., carbonate rocks) may dissolve in water (see Subs. 7.3.5). It is also possible that the solid matrix comprising the aquifer is composed of a number of minerals, each with its own dissolution properties.

Thus, it is possible that the dissolution of a specific mineral takes place over only part of the (fluid-solid) interface, while the precipitation of that mineral may occur over the entire surface. This is another example of fluid-to-solid mass transfer.

We shall demonstrate this case by using a rather simplified model of saturated flow and a solid matrix comprised of a single dissolvable mineral. It is usually assumed that next to the solid surface there exists (always) a very thin layer of water saturated by the dissolved mineral, say, at concentration c_{sat}. It is also assumed that the force that drives the flux of the dissolved mineral from this layer to the bulk fluid in the void space, at concentration c, is expressed as

$$f_{s \to f} = \alpha_{sf} \Sigma_{sf} (c_{\text{sat}} - c), \tag{7.3.138}$$

where α_{sf} is a transfer coefficient, and Σ_{sf}, a function of the porosity ϕ, is the specific surface area of the solid. The mass balance for the mineral species in solution is:

$$\frac{\partial \phi c}{\partial t} = -\nabla \cdot \phi (c\mathbf{V} - \mathbf{D}_h \cdot \nabla c) + f_{s \to f}, \tag{7.3.139}$$

in which the velocity is described by the appropriate flow model.

As a result of dissolution, the solid's mass decreases. With ρ_s denoting the solid's density, the mass balance for the solid phase takes the form:

$$\rho_s \frac{\partial \phi_s}{\partial t} = -f_{s \to f}, \qquad \phi_s = 1 - \phi. \tag{7.3.140}$$

Altogether, we have two variables: c and ϕ to solve for, and two mass balance equations: for the dissolved species and for the mass of the solid. The flow in the void space is described by the usual flow model, except that here we may want to take into account the change in porosity as dissolution progresses (and, hence, also of permeability and specific surface area).

A similar model can be applied to the case where the mineral is exposed only over part of the solid-fluid interface.

G. Intragranular adsorption and exchange

Sometimes, the solid comprising the solid matrix is porous. The porous medium has, then, a primary pore space outside of the solid matrix and a secondary pore space, with much smaller pores, inside it. The term 'double porosity medium' is often employed to describe such a material. An example is a porous material whose solid matrix consists of porous grains that were formed from the consolidation of smaller granules. A fractured porous rock, comprised of blocks of porous rock surrounded by fractures, may serve as another example. The fractures may be conceptualized as comprising the primary porosity, while the void space in the porous rock blocks represent the secondary pore space. In the present chapter, our interest in double porosity models stems from the fact that a significant portion of the material dissolved in the fluid can be adsorbed onto the solid surfaces that are inside the porous solid matrix, because of its extremely large specific surface area. This large surface area increases the exchange of dissolved material between the solid and the fluid. These attributes are also important in non-geologic settings. As an example, consider a container of *activated carbon* pellets that is often used for filtering gases. The pellets are porous, with very small pores. Large amounts of unwanted vapors can be adsorbed onto the internal surface of the porous carbon pellets. In many chemical plants, beds of porous pellets, made of catalytic material, are used to accelerate a desired reaction (Satterfield, 1991).

H. Double porosity porous medium

Double porosity porous medium domain is a term used to describe a porous medium in which the solid matrix is porous. In such a medium, the void space is composed of two parts: the (intragranular) void space *inside* the grains (or solid matrix), and the (intergranular) void space between the grains. The pores in the former are much smaller than those in the latter. A porous medium of this kind is, often, modeled as composed of two *overlapping continua*: one made of the void space with large pores, and the other made only of the void space with the smaller ones. An example is a porous material whose solid matrix consists of porous grains that were formed from the consolidation of smaller granules. A fractured porous rock, comprised of blocks of porous rock surrounded by fractures, may serve as another example. Thus, a (macroscopic) point in space belongs, simultaneously, to both continua. Mass of fluid phases and contaminants may be transported within each of the two continua, while being exchanged between them. When modeling the mass transport of a fluid phase, each continuum is assumed to have its own fluid pressure, temperature, concentration, and velocity fields. A difference in pressure (at a macroscopic point) between the two continua, will cause fluid mass to be transferred from high to low pressure. The rate of mass transfer may be expressed as a transfer coefficient times the pressure difference (e.g., Barenblatt *et al.*, 1960).

Because of the very small size of the pores within the porous solid, the permeability of that continuum is often assumed to be negligible, as compared to that of the larger pores, and the fluid occupying this very low permeability continuum is, usually, assumed to be immobile. However, contaminants may *diffuse* through the immobile fluid. The process in which a species present in the fluid occupying the void space enters a porous solid by diffusion (and is adsorbed onto its large internal surface area) is referred to as *absorption* (see discussion on HK-LK models in Subs. 7.5.2).

Let us consider a γ-species that is present in both the large pores (filled with mobile water and denoted by m) and the smaller pore (filled with immobile water and denoted by im). The macroscopic mass balance equation for a γ-species in the large pores is

$$\frac{\partial \phi_m \overline{c^\gamma}^m}{\partial t} = -\nabla \cdot \phi_m \left(\overline{c^\gamma}^m \overline{\mathbf{V}}^m + \overline{\overset{\circ}{c}{}^\gamma \overset{\circ}{\mathbf{V}}}^m + \overline{\mathbf{j}^\gamma}^m\right) - f^\gamma_{im\to m}, \tag{7.3.141}$$

where $f^\gamma_{im\to m}$ expresses the transfer of γ from the immobile to the mobile water. The mass balance equation for the species in the smaller pore space, including the amount adsorbed on the inner solid surfaces, is

$$\frac{d}{dt}\left(\phi_{im}\overline{c^\gamma}^{im} + \rho_{b,im}K^\gamma_d\overline{c^\gamma}^{im}\right) = f^\gamma_{im\to m}, \tag{7.3.142}$$

where linear equilibrium adsorption is assumed. In the model considered here, we also assume that diffusion in the immobile water is neglected, except for the diffusion to and from the $\mathcal{S}_{m,im}$ interface. Note that ϕ_{im} is not the porosity of the matrix, but the ratio of the volume of void space *in the matrix* to the total volume of the porous medium, so that $\phi = \phi_m + \phi_{im}$.

Similar to the expression in (7.2.22), the transfer source term is given by

$$f^\gamma_{im\to m} = -\frac{1}{\mathcal{U}_o}\int_{\mathcal{S}_{im,m}} \mathbf{j}^\gamma_{im}\cdot\mathbf{n}_{im}\,dS, \tag{7.3.143}$$

where $\mathcal{S}_{im,m}$ denotes the interface between the mobile and immobile water. The mobile water is assumed to be well-mixed within the REV at each considered point, so that the mean concentration on $\mathcal{S}_{im,m}$ is equal to that in the mobile water. We then make the following approximation:

$$\mathbf{j}^\gamma_{im}\cdot\mathbf{n}_{im} = -\mathcal{D}^\gamma(\overline{c^\gamma}^m - \overline{c^\gamma}^{im})/\Delta_{im}, \tag{7.3.144}$$

where Δ_{im} is the mean distance over which diffusion occurs from inside the matrix to the interface $\mathcal{S}_{im,m}$. It is related to the mean radius of the matrix grains or blocks, ℓ_{im}, by the tortuosity, τ_{im}, of the pore space in the matrix:

$$\ell_{im} = \tau_{im}\Delta_{im}. \tag{7.3.145}$$

By substituting (7.3.144) into (7.3.143), we obtain

$$f^\gamma_{im\to m} = \tau_{im}\phi_{im}(\Sigma')\mathcal{D}^\gamma(\overline{c^\gamma}^m - \overline{c^\gamma}^{im})/\ell_{im}, \tag{7.3.146}$$

where we assumed that

$$\Sigma_{im,m} = \phi_{im}\Sigma', \tag{7.3.147}$$

in which Σ' is the specific area of the interface surrounding the porous solid grains, or porous blocks, within an REV. Equations (7.3.141) and (7.3.142), together with (7.3.146), form a set of equations from which one can solve for the unknown variables, $\overline{c^\gamma}^m$ and $\overline{c^\gamma}^{im}$.

Equation (7.3.144) is valid only if the (microscopic) concentration profile in the void space of the matrix is approximately linear with respect to some path length within the pore space, as we proceed from the boundary between the immobile and mobile water to some point in the interior of the pore space of the matrix. The linear approximation, in turn, is valid under quasi-steady conditions and in the absence of chemical reactions. The characteristic time, $t_{c,\text{diff}}$, required for reaching quasi-steady conditions by diffusion, is given by

$$t_{c,\text{diff}} = \Delta_{im}^2/\mathcal{D}^\gamma. \tag{7.3.148}$$

Therefore, the above model is accurate only for resolving (macroscopic) concentration changes that occur over times that are much greater than $t_{c,\text{diff}}$. If the grains or blocks that comprise the matrix are very large, then this characteristic time will be so large that predictions based on this model will be erroneous. If we envision a sharp concentration front in the mobile water traveling at some velocity V, it follows that there exists a zone of width $L_c = t_{c,\text{diff}}V$ in which the matrix is undergoing transition to quasi-steady conditions. This zone is not modeled accurately because of the assumption of a linear concentration profile in the matrix. Therefore, only predictions of spatial gradients in the (macroscopic) concentration over distances that are much greater than L_c are valid. If the matrix is very large, then this distance can be so large that, again, the results will be meaningless.

7.3.7 Biotransformations

This is an important subject. However, within the scope of this book, only a brief, partial review is presented.

A. What is biotransformation

Biodegradation and *biotransformation* are terms used for processes in which a chemical species in solution, or adsorbed on the solid, is transformed by *biochemical reactions* into other products. The latter may be more toxic than the reactants, or less harmful. In recent years, the subject of natural and artificially-enhanced biotransformations has attracted much attention as more and more efforts have been made to develop *in situ* biotransformation-based techniques for the removal of contaminants from the subsurface, or to transform them into less harmful products. The term *bioremediation* is often used for remediation techniques based on biotransformations.

On the other hand, '*natural attenuation*' is a term used for describing the combined physical, chemical and biological processes (e.g., dispersion, adsorption, volatilization, chemical reactions, and biotransformation) that reduce the mass (and also the mobility and toxicity) of a pollutant as it travels in the subsurface to permissible levels, without human intervention.

Under natural conditions, microorganisms of various kinds (such as bacteria, fungi, yeasts, protozoa) are always present in the soil, either as bodies suspended in the aqueous phase, or as a thin layer, usually referred to as *biofilm*, on the solid surface. This subsection introduces the biochemical reactions that are based on the activity of these subsurface microorganisms. Biochemical reactions serve as the basis for bioremediation techniques.

The goal of *biodegradation* of organic contaminants, e.g., NAPLs, is to convert them to *biomass* (= mass of microorganisms), water, inorganic salts, CO_2, and other harmless gases. The transformation of hydrocarbons to CO_2 and H_2O by indigenous microorganisms may serve as an example. For inorganic contaminants, the goals are to transform them into products that are less toxic, and/or have lower mobility, so that they do not travel very far, or to make them more mobile, so that they can be easily removed.

The actual process of biotransformation is the result of a chemical reaction, or a set of reactions, in which a contaminant interacts with organic compounds produced by microorganisms. In many cases, these compounds are *enzymes*, which are organic catalysts, usually, *proteins*. A *catalyst* is a compound that is used to increase the rate of a reaction, without being consumed in the process and without changing the equilibrium of the reaction. In the case considered here, the enzyme accelerates the conversion of a contaminant into harmless products.

In order for biodegradation of a contaminant at a site to proceed, the following conditions must be met:

- Appropriate microorganisms must be present in a sufficient quantity in the formation. Otherwise, it may be necessary to artificially introduce special microorganisms that have been grown in a laboratory.
- The contaminant must be accessible to the microorganisms. If the reactive bonds of the contaminant molecules are attached to the solid surface, then the contaminant cannot be degraded by the microorganisms. The contaminant may not be able to react because it cannot pass through the outer membrane of the microorganism's cell in order to come into contact with the enzyme inside the cell.
- The physical, chemical, and thermal environment must be conducive to the survival of the microorganisms.
- An important consideration is the pH of the aqueous solution.
- There must be an adequate supply of *nutrients* for the microorganisms to obtain the energy that enables them to live, grow, and produce the chemicals needed for the reactions that transform the contaminant. A substance which supplies such nutrients is called a *substrate*.

Respiration is the process in which cells obtain the energy required for performing cell functions. An *electron acceptor* is required for respiration to occur. *Aerobic* microorganisms dissolve oxygen while serving as an electron acceptor for respiration. *Anaerobic* microorganisms use other electron acceptors, such as NO_3^- or SO_4^{-2}, instead of oxygen. Carbon (C) is another nutrient that is needed for the respiration and biomass building. Certain chemicals, such as Nitrogen (N), Phosphorus (P), and Sulfur (S), are also necessary nutrients for cell functioning and growth.

Actually, cell respiration is a chain of oxidation reactions, involving organic molecules, whose final products are water and carbon dioxide. Electrons are transferred at each stage in the chain until they reach the final oxidant, or *terminal electron acceptor*, at the end of the chain. Usually, an enzyme, produced by the cell, is required at each stage in the chain. The importance of respiration stems from (1) the need for an appropriate terminal electron acceptor for the survival of the desired microorganisms, and (2) the role that respiration plays in the transformation of contaminants. The electron donor at the beginning of the respiration chain is, often (but not always), the contaminant itself, as it is being oxidized. In some rare cases, the contaminant may serve as the terminal electron acceptor. Whether the contaminant is *oxidized* or *reduced*, a specific enzyme produced by the cell is required for the desired transformation to take place. Enzymes, therefore, have a crucial role in many biotransformations. Although we have referred to reactions as if they proceed inside a single type of organism, in many cases, several types of organisms are involved in the same reaction chain, with the product from reactions within one type of organism sequentially feeding into reactions within other types.

As mentioned earlier, *aerobic* microorganisms use oxygen as the terminal electron acceptor for respiration. Biodegradation in the presence of oxygen is called *aerobic biodegradation*. Microorganisms that do not use oxygen as the terminal electron acceptor are called *anaerobes*, and biodegradation under such conditions is called *anaerobic*.

B. Example: Degradation of petroleum hydrocarbons

Subsurface contamination by spilled petroleum hydrocarbons (e.g., gasoline, diesel, and aviation fuel) and their derivatives (such as alcohols, ketones, and esters) has been successfully bioremediated at many sites (Atlas and Bartha, 1992; Venosa *et al.*, 1996; Salanitro *et al.*, 1997; Dojka *et al.*, 1998; Macnaughton *et al.*, 1999; Gieg *et al.*, 1999; Banat *et al.*, 2000). Because petroleum hydrocarbons can serve as energy sources, they are readily oxidized by microorganisms (under the action of enzymes), as part of their respiration process. They also serve as a source of carbon for cell growth. A *terminal electron acceptor* is still required for this degradation process to occur. The most efficient terminal electron acceptor is oxygen. The following reaction may serve as an example:

$$2C_6H_6 + 15O_2 \rightarrow 12CO_2 + 6H_2O. \tag{7.3.149}$$

The availability of electron acceptors in sufficient quantity may act as a *limiting factor* in the biodegradation of petroleum hydrocarbons. Although oxygen may be initially available in the vadose zone, it often becomes nearly depleted before all the hydrocarbon has been consumed. In *bioventing*, a technology used for removing contaminants from the vadose zone, air is injected at a relatively low rate into the vadose zone in order to supply the necessary oxygen (e.g., Hoeppel *et al.*, 1991; Dupont, 1993; Moller, 1996; Lee *et al.*, 2006; Rayner *et al.*, 2007).

In another technology, called *biosparging*, oxygen is introduced by injecting air below the water table. Another technique is to introduce a chemical, such as hydrogen peroxide (H_2O_2), which dissociates into water and oxygen through the reaction

$$2H_2O_2 \rightarrow 2H_2O + O_2. \tag{7.3.150}$$

Hydrogen peroxide has the added benefit of directly oxidizing organic compounds. However, at high concentrations, it can inhibit the activity of microorganisms. Also, the released oxygen gas in the form of trapped bubbles, can reduce the permeability to the aqueous phase.

In the absence of oxygen, or when it has been depleted, *anaerobic conditions* prevail. *Nitrate* (NO_3^-), or other oxidized forms of nitrogen, if present, may serve as electron acceptors for some bacteria. The resulting reaction is

$$C_6H_6 + 6H^+ + 6NO_3^- \rightarrow 6CO_2 + 3N_2 + 6H_2O. \tag{7.3.151}$$

Following the depletion of *nitrate*, *oxidized ferric iron* (Fe^{3+}) is consumed by the bacteria as an electron acceptor. For the biodegradation of hydrocarbons, next in the sequence of priority as electron acceptors, are sulfates and methane (e.g., Rifai *et al.*, 1995). In general, base3d on microbiological considerations, the sequence of preferred electron acceptors under anaerobic conditions appears to be:

- NO_3^- (the process is called *denitrification*),
- Fe^{3+} (iron reduction),
- Mn^{4+} (manganese reduction),
- SO_4^{2-} (sulfate reduction),
- CO_2 (methanogenesis, produces methane (CH_4)).

Besides electron acceptors, other nutrients, such as N, P, and S, can act as limiting factors for biodegradation, especially in large spills, where a significant amount of biological growth must be sustained.

C. Rate of biodegradation

Usually, the biochemical reactions that degrade or transform a contaminant, take place *only* within the aqueous phase, whether outside or inside a microorganism. If the transformation occurs within a microorganism, the contaminant has to pass through the microorganism's outer cell membrane before it can react with an enzyme. In other cases, the transformation may be due to a reaction occurring with an enzyme on the outer surface of the microorganism. In some cases, when the contaminant is a growth substrate, microorganisms may be attracted to higher aqueous concentrations that may exist near the adsorbed contaminant. Or, if the active microorganisms are attached to the solid, reactions will be faster near the solid than in the rest of the aqueous phase. The contaminant may also react with chemicals released into the aqueous phase by the microorganisms. In all cases, the reaction is at the microscopic level; its rate will depend on the local value of the microscopic aqueous contaminant concentration. However, as always, we have to employ some averaging technique in order to derive the required macroscopic level model.

Practically, it is difficult to experimentally distinguish whether the contaminant that is being degraded is adsorbed on the solid surface, or it is desorbed from the surface and is then degrading in the aqueous phase.

For modeling purpose, we consider the time-variation in the concentration, $c(\mathbf{x}, t)$, of a considered biodegradable contaminant. In the mass balance equation, which is part of the mathematical model, biodegradation appears as a sink term that expresses the rate at which the considered species is removed from the aqueous solution by bio-reactions. For example, in the presence of both adsorption and biodegradation, the macroscopic mass balance equation for a biodegrading γ-species contaminant may be rewritten as a special case of (7.4.8), in the form:

$$\frac{\partial}{\partial t}\left(\theta + \rho_b K_d^\gamma\right) c^\gamma = -\nabla \cdot \theta \left(c^\gamma \mathbf{V} + \mathbf{J}^{*\gamma} + \mathbf{J}_{\text{diff}}^\gamma\right) + \left.\frac{dc_{\text{pm}}^\gamma}{dt}\right|_{\text{bioreact}}, \qquad (7.3.152)$$

with c_{pm}^γ denoting the total mass of the γ-species per unit volume of porous medium, whether the species is in suspension in the fluid, adsorbed on the solid, attached to the solid, or just accumulated next to the solid. The last term of the above equation expresses the rate of growth (source) of the γ-species. The negative of the last term on the right-hand side expresses the rate of decrease in the mass of the γ-species (as a consequence of consuming the contaminant during the process of biodegradation), per unit volume of porous medium.

D. Michaelis-Menten kinetics

As explained earlier, in most of bioremediation processes, the contaminant is degraded or transformed through a reaction that requires the presence of enzymes that are produced by microorganisms. Therefore, let us focus first on the rate law for a single substrate that reacts in the presence of an enzyme to form a single product. The rate law for such a reaction was first introduced by Michaelis and Menten (1913). An important assumption underlying their rate law is that *the substrate is available in sufficient quantity*, so that the quantity of the enzyme, not of the substrate, is limiting the reaction (Bailey and Ollis, 1986).

The considered reaction takes place in two stages. In the first stage, the substrate (S) and the enzyme (E) form a complex (ES). This reaction can be represented as

$$\text{S} + \text{E} \rightleftharpoons \text{ES}. \qquad (7.3.153)$$

This reaction is assumed to be reversible and in equilibrium. The mass action law implies that

$$\frac{1}{K_m} \equiv K_{\text{eq}} = \frac{[\text{ES}]}{[\text{S}]\,[\text{E}]}, \qquad (7.3.154)$$

where $1/K_m$ is equal to the equilibrium constant of the reaction. Note that, although, like all chemical reactions, the biochemical reactions considered in this subsection occur at the microscopic level, we assume that they are valid also at the macroscopic level, i.e., concentrations are averaged over an REV.

In the second stage, the complex ES forms a product (P), while the enzyme, E, remains unaltered through the catalytic reaction

$$\mathrm{ES} \rightarrow \mathrm{P} + \mathrm{E}. \tag{7.3.155}$$

This reaction is assumed to be irreversible and to proceed according to a first-order rate law given by

$$\frac{d[\mathrm{P}]}{dt} = \lambda_m [\mathrm{ES}], \tag{7.3.156}$$

where λ_m is a rate coefficient. Let $[\mathrm{E}]_{\mathrm{tot}}$ denote the concentration of the total amount of enzyme, so that $[\mathrm{E}]_{\mathrm{tot}} \equiv [\mathrm{E}] + [\mathrm{ES}]$. By using (7.3.154), we obtain $[\mathrm{E}]_{\mathrm{tot}} = [\mathrm{E}]\,(1 + [\mathrm{S}]/K_m)$. Therefore, $[\mathrm{E}] = [\mathrm{E}]_{\mathrm{tot}}/(1 + [\mathrm{S}]/K_m)$. By substituting this expression into (7.3.154), and solving for [ES], we obtain $[\mathrm{ES}] = [\mathrm{E}]_{\mathrm{tot}}[S]/(K_m + [S])$. Then, from (7.3.156) and $d[\mathrm{S}]/dt = -d[\mathrm{P}]/dt$, we obtain the rate law for the free substrate in the form:

$$\frac{d[\mathrm{S}]}{dt} = -\lambda_m [\mathrm{E}]_{\mathrm{tot}} \frac{[\mathrm{S}]}{K_m + [\mathrm{S}]} \qquad \left(= -\frac{\lambda_{\max}[\mathrm{S}]}{K_m + [\mathrm{S}]} \right), \tag{7.3.157}$$

where $\lambda_{\max}(\equiv \lambda_m [\mathrm{E}]_{\mathrm{tot}})$ is the maximum reaction rate. The constant K_m is seen to be equal to the concentration of the substrate at which the rate is equal to one-half of the maximum rate.

For computing the source term, the microscopic balance equation of the substrate requires an expression for $d[\mathrm{S}]_{\mathrm{total}}/dt$, where the total substrate concentration is given by $[\mathrm{S}]_{\mathrm{total}} = [\mathrm{S}] + [\mathrm{ES}] + [\mathrm{P}]$. Under the conditions

$$[\mathrm{P}] \ll [\mathrm{S}], \qquad [\mathrm{ES}] \ll [\mathrm{S}], \tag{7.3.158}$$

we obtain $d[\mathrm{S}]_{\mathrm{total}}/dt = d[\mathrm{S}]/dt$, which allows (7.3.157) to serve as the source term in the microscopic balance equation of the substrate S. Note that a factor equal to the molecular mass of the substrate must be present when the balance equation is expressed in terms of mass.

In some situations, there exists an *inhibitor* species (I) that competes with the substrate for the enzyme by forming a complex (EI) according to the reaction

$$\mathrm{E} + \mathrm{I} \rightleftharpoons \mathrm{EI}. \tag{7.3.159}$$

If the reaction is in equilibrium, then the mass action law gives

$$\frac{1}{K_I} = \frac{[\mathrm{EI}]}{[\mathrm{I}]\,[\mathrm{E}]}. \tag{7.3.160}$$

The total enzyme concentration is given, in this case, by $[\mathrm{E}]_{\mathrm{tot}} \equiv [\mathrm{E}] + [\mathrm{ES}] + [\mathrm{EI}] = [\mathrm{E}](1 + [\mathrm{S}]/K_m + [\mathrm{I}]/K_I)$. Using (7.3.154) and (7.3.156), and applying the same type of algebra manipulations that were used above, we obtain

$$\frac{d[\mathrm{S}]}{dt} = -\lambda_m [\mathrm{E}]_{\mathrm{tot}} \frac{[\mathrm{S}]/K_m}{1 + [\mathrm{S}]/K_m + [\mathrm{I}]/K_I}. \tag{7.3.161}$$

This is the *Michaelis-Menten rate law in the presence of an inhibitor.* If the substrate and the inhibitor interact, then the numerator will also contain terms of the type $[\mathrm{S}][\mathrm{I}_i]/K_{SI_i}$.

The rate laws presented here can be easily written in terms of mass concentrations instead of molar concentrations. For example, $c^{\mathrm{S}} = M^{\mathrm{S}}[\mathrm{S}]$, where M^{S} is the molecular mass of the substrate. Thus, (7.3.157) becomes

$$\frac{dc^{\mathrm{S}}}{dt} = -\lambda'_m c^{\mathrm{Etot}} \frac{c^{\mathrm{S}}}{K'_m + c^{\mathrm{S}}}, \tag{7.3.162}$$

where $c^{E_{\mathrm{tot}}}$ is the mass concentration of the total amount of enzyme, both in solution and combined with substrate.

At the macroscopic level, the Michaelis-Menten rate law for the source term on the right-hand side of (7.3.152) becomes

$$\left.\frac{dc^{\mathrm{S}}_{\mathrm{pm}}}{dt}\right|_{\mathrm{bioreact}} = -\lambda'_m c^{\mathrm{Etot}}_{\mathrm{pm}} \frac{c^{\mathrm{S}}}{K'_m + c^{\mathrm{S}}} \quad \left(= -\frac{\lambda'_{\max} c^{\mathrm{S}}}{K'_m + c^{\mathrm{S}}}\right), \tag{7.3.163}$$

where it is assumed that (7.3.158) holds. This expression is the desired source term to be used in the macroscopic balance equation of the substrate. In this equation, $c^{\mathrm{S}}_{\mathrm{pm}}$ is the substrate mass per unit volume of porous medium, $c^{\mathrm{Etot}}_{\mathrm{pm}}$ is the total enzyme mass per unit volume of porous medium, and c^{S} is the macroscopic concentration of the substrate material dissolved in the aqueous phase (as mass of substrate per unit volume of aqueous phase).

E. Monod kinetics

We derive the Monod rate law for a porous medium as a macroscopic law. Whereas Michaelis-Menten kinetics is used to describe the degradation of a contaminant, *Monod kinetics* describes also the growth of microorganisms. Since this growth requires enzymatic reactions of substrates, it is, perhaps, not surprising that the mathematical form of the rate law for the Monod kinetics is identical to that of Michaelis-Menten kinetics. As mentioned earlier, whenever possible, it is often convenient not to distinguish between the degradation rate of a contaminant on the solid phase and in the aqueous phase. Instead, the total rate, $\frac{dc_{\mathrm{pm}}}{dt}|_{\mathrm{bioreact}}$, is used in the macroscopic balance equation (7.3.152).

The critical assumption underlying Monod kinetics is that the rate of growth of microorganisms is limited by the amount of substrate, which, in our case, is the contaminant.

A second assumption is that this growth rate has the following form:

$$\left.\frac{dc_{pm}^{org}}{dt}\right|_{bioreact} = \chi_{max} c_{pm}^{org} \frac{c}{c + K_s}, \tag{7.3.164}$$

in which c_{pm}^{org} is the mass of microorganisms per unit mass of porous medium, c denotes the aqueous concentration of the substrate, χ_{max} is the maximum specific growth rate (= maximum attainable growth rate per unit mass of microorganisms), and K_s is concentration at which the specific growth rate is equal to one-half of its maximum attainable value.

Finally, we assume that the rate of microorganism growth is proportional to the rate at which the substrate in the aqueous phase is being consumed,

$$\left.\frac{dc_{pm}^{org}}{dt}\right|_{bioreact} = -Y \left.\frac{dc_{pm}}{dt}\right|_{bioreact}, \tag{7.3.165}$$

where Y is a constant called the *yield.*

From (7.3.164) and (7.3.165), we have

$$\left.\frac{dc_{pm}}{dt}\right|_{bioreact} = -\frac{\chi_{max}}{Y} c_{pm}^{org} \frac{c}{c + K_s}. \tag{7.3.166}$$

This equation is the *Monod rate law* for a contaminant in a porous medium, assuming that the aqueous phase is well-mixed (within each REV). As noted earlier, the mathematical form of this rate law is identical to that of the Michaelis-Menten rate law.

At sufficiently high concentrations, the contaminant, as a substrate, can sometimes become toxic to the microorganisms and inhibit their growth to the extent that their growth rate will begin to decrease. A rate equation that has been used to describe this situation is

$$\left.\frac{dc_{pm}^{org}}{dt}\right|_{bioreact} = \lambda c_{pm}^{org} \frac{c}{c^2/K_i + c + K_s}, \tag{7.3.167}$$

for constants $\lambda > 0$, $K_i > 0$. Note that the rate decreases to zero when the concentration becomes large, instead of tending to a constant value as does the rate expressed by the Monod law. The constant K_i is called the *substrate inhibition constant.* For example, Bouwer and McCarthy (1984) present the Monod rate equation for the case of anaerobic degradation of a hydrocarbon (HC).

If microorganisms can exist in both a mobile suspended state and an immobile state (for example, they may be attached or have a strong attraction to the solid), then a balance equation is required for each of these two states. We also need then an expression for the rate of detachment of immobile microorganisms that are attached to the solid, e.g., a first-order law in the form

$$f_{im \to mob} = k_1 c^{im-org} - k_2 c^{mob-org}, \tag{7.3.168}$$

where the coefficients k_1 and k_2 depend on the macroscopic flow velocity, the characteristics of the microorganisms, and those of the porous medium. In practice, in order to account for the apparent upper limit in the mass of microorganisms that can become immobile on the solid surfaces, these coefficients may depend on the concentration. When the 'partitioning' between the suspended and the attached microorganisms can be assumed to be in equilibrium, then only one balance equation is needed, instead of the two balance equations that were described above. That balance equation is then for the total amount of microorganisms (= immobile + mobile). The equilibrium relationship between the concentrations of the mobile and the immobile microorganisms must be known.

If the biodegradation reaction consumes oxygen that is in limited supply (rate-limiting), then (7.3.164) is usually multiplied by an appropriate factor that involves the concentration of the dissolved oxygen gas, c^{O_2}, to give

$$\left.\frac{dc_{\text{pm}}^{\text{org}}}{dt}\right|_{\text{bioreact}} = \chi_{\max} c_{\text{pm}}^{\text{org}} \frac{c}{c + K_s} \frac{c^{O_2}}{c^{O_2} + K_s^{O_2}}. \tag{7.3.169}$$

This rate law is an example of *dual Monod kinetics*. Note that the corresponding degradation rate for the contaminant follows from (7.3.165). The same kind of factor can be used for electron acceptors other than oxygen.

Raising the temperature of a system usually increases the rate of biodegradation of a contaminant, until some temperature level is reached at which the biological activity begins to decline. Below this level, a temperature dependent factor, based on the Arrhenius equation, (7.3.58), has been used to modify the microorganism and substrate reaction rates.

As a rule of thumb, Van't Hoff's Rule (Atkins and De Paula, 2006) can be applied to get an approximate value for the reaction rate at a different temperatures, when Arrhenius parameters (Subs. 7.3.2D) are unavailable. This rule implies that for every rise in temperature of 10°C, the rate of (bio)chemical reactions doubles. This rule is applicable only to the temperature range where the specific microorganism can survive.

7.4 Complete Mathematical Model with Sources

In this section, our objective is to insert specific expressions for the source terms that appear in (7.2.20).

7.4.1 Balance equations with sources

In the case of adsorption, the term $f^{\gamma}_{\alpha\rightarrow\beta}$, that expresses the rate of interphase exchange in (7.2.20), takes the form:

$$(f^{\gamma}_{\alpha\rightarrow\beta} \equiv) f^{\gamma}_{\alpha\rightarrow s} = \left.\frac{d(\rho_b F)}{dt}\right|_{\text{ads}}, \tag{7.4.1}$$

where ρ_b $(= (1-\phi)\rho_s)$ denotes the *bulk density of the solid matrix.*

According to the methodology proposed in Subs. 7.2.2, to eliminate the rate of interphase exchange, we sum up the balance equations for the component in all the phases present in the system. We need, therefore, the macroscopic balance equation for the γ-component *on the solid phase.* We obtain this equation from (5.1.5), in which the α subscript is replaced by s, θ is replaced by $\theta_s(\equiv 1-\phi)$, and $\theta\overline{e}^\alpha$ is replaced by $\theta_s\rho_s F(\equiv \rho_b F)$. We assume that no flux of γ, whether advective, dispersive, or diffusive, takes place within the solid phase and/or on its surface, i.e., $\overline{\mathbf{V}}^\alpha \equiv \mathbf{V}_s = 0$, $\mathbf{J}^*_s = 0$, $\mathbf{J}^\gamma_s = 0$. The resulting mass balance for the component on the solid surface, takes the form:

$$\frac{\partial(\rho_b F)}{\partial t} = -f^\gamma_{s\to\alpha} + \rho_b\Gamma^\gamma_s, \tag{7.4.2}$$

to be compared with (7.2.5).

By summing up the two balance equations for the considered component: (7.2.20) for the component in the fluid, and (7.4.2) for the component on the solid, we obtain the *mass balance equation for the component in the porous medium as a whole*, in the form:

$$\frac{\partial(\theta c + \rho_b F)}{\partial t} = -\nabla\cdot\theta(c\mathbf{V} - \mathsf{D}\cdot\nabla c - \boldsymbol{\mathcal{D}}^*\cdot\nabla c) + \theta\rho\Gamma^\gamma + \rho_b\Gamma^\gamma_s. \tag{7.4.3}$$

In indicial notation, (7.4.3) takes the form:

$$\frac{\partial(\theta c + \rho_b F)}{\partial t} = -\frac{\partial}{\partial x_i}\theta\left(cV_i - \mathsf{D}_{ij}\frac{\partial c}{\partial x_j} - \mathcal{D}^*_{ij}\frac{\partial c}{\partial x_j}\right) + \theta\rho\Gamma^\gamma + \rho_b\Gamma^\gamma_s. \tag{7.4.4}$$

We have, thus, eliminated the terms that represent the rate of transfer of the considered component from the fluid phase to the solid and vice versa. Note that (7.4.3) is valid only for a constant density fluid. Otherwise, the diffusive flux for a binary system, $-\boldsymbol{\mathcal{D}}^*\cdot\nabla c$, must be replaced by $-\rho\boldsymbol{\mathcal{D}}^*\cdot\nabla\omega$.

Let us use this opportunity to demonstrate how we include in (7.4.3) sources that (1) are due to pumping from and injection into an aquifer (Subs. 7.2.3), (2) are represented by $\theta\rho\Gamma^\gamma_\alpha$ for the component in the fluid, and $\rho_b\Gamma^\gamma_s$ for the component on the solid, say, first order decay, with different rate constants for the component in the fluid and on the solid, and (3) are due to radioactive sources (Subs. 7.3.2). With such sources, (7.4.3) takes the form:

$$\begin{aligned}\frac{\partial(\theta c + \rho_b F)}{\partial t} = &-\nabla\cdot\theta(c\mathbf{V} - \mathsf{D}\cdot\nabla c - \boldsymbol{\mathcal{D}}^*\cdot\nabla c)\\ &-\rho_b(k_s+\lambda)F - \theta(k_f+\lambda)c\\ &+\Sigma_{(m)}R^{(m)}_{\text{ext}}(\mathbf{x}^{(m)},t)\delta(\mathbf{x}-\mathbf{x}^{(m)})c^{(m)}_R(\mathbf{x}^{(m)},t)\\ &-\Sigma_{(r)}P^{(r)}_{\text{ext}}(\mathbf{x}^{(r)},t)\delta(\mathbf{x}-\mathbf{x}^{(r)})c(\mathbf{x},t).\end{aligned} \tag{7.4.5}$$

In this equation, the source terms for injection and pumping are obtained from (7.2.31) through (7.2.34). The terms that describe radioactive decay within the fluid phase and on the solid are obtained from (7.3.59) and (7.3.62), respectively, while those due to degradation within the fluid phase and on the solid, are obtained from (7.3.63) and (7.3.64), respectively.

We recall that the mass balance equation for water is (5.1.7). In this equation, let us (1) use the Dirac delta function, as defined by (5.1.77), to express point sinks and sources due to pumping and injection wells, and (2) assume that the fluid is of constant density and that the solid matrix is incompressible (i.e., $\partial\phi\rho/\partial t = 0$). Under these conditions, (5.1.7) reduces to

$$\nabla\cdot\mathbf{q} = \sum_{(m)} R^{(m)}_{\text{ext}}(\mathbf{x}^{(m)}, t)\delta(\mathbf{x} - \mathbf{x}^{(m)}) - \sum_{(m)} P^{(m)}_{\text{ext}}(\mathbf{x}^{(m)}, t)\delta(\mathbf{x} - \mathbf{x}^{(m)}). \quad (7.4.6)$$

We multiply both sides by c and subtract it from (7.4.5) to obtain:

$$\begin{aligned}\theta\frac{\partial c}{\partial t} &= -\theta\mathbf{V}\cdot\nabla c + \nabla\cdot(\theta\mathbf{D}_h\cdot\nabla c) - \frac{\partial(\rho_b F)}{\partial t}\\ &\quad -\rho_b(k_s+\lambda)F - \theta(k_f+\lambda)c\\ &\quad +\sum_{(m)} R^{(m)}_{\text{ext}}(\mathbf{x}^{(m)}, t)\delta(\mathbf{x} - \mathbf{x}^{(m)})\left[c^{(m)}_R(\mathbf{x}^{(m)}, t) - c(\mathbf{x}, t)\right]. \quad (7.4.7)\end{aligned}$$

As a general methodology, we summed the balance equations in order to eliminate the unknown interphase transfer rates. However, the result is that we have fewer balance equations. Following the discussion in Sec. 7.3.3, let us assume that the adsorption process is sufficiently fast, so that *chemical equilibrium* may be assumed to prevail. This assumption will enable us to make use of the thermodynamic relationships that express the partitioning of the component between the solid and the liquid phases. These are the isotherms considered in Subs. 7.3.3.

Equation (7.4.3) contains the two variables: $c(\mathbf{x}, t)$ and $F(\mathbf{x}, t)$. Hence, we now make use of an appropriate isotherm that expresses the relationship between c and F. For example, if the adsorption of the considered species on the considered solid calls for the use of the linear Freundlich isotherm (7.3.66), we obtain from (7.4.3) the mass balance equation for the component in the porous medium as a whole, in the form:

$$\begin{aligned}\frac{\partial}{\partial t}(\theta + \rho_b K_d)\,c \equiv \frac{\partial}{\partial t}\theta R_d c &= -\nabla\cdot\theta(c\mathbf{V} - \mathbf{D}\cdot\nabla c - \boldsymbol{\mathcal{D}}^*\cdot\nabla c)\\ &\quad +\theta\rho\Gamma^\gamma + \rho_b\Gamma^\gamma_s, \quad (7.4.8)\end{aligned}$$

where R_d $(\equiv 1 + \rho_b K_d/\theta)$ is the *retardation coefficient* discussed in Case B and Case C in Subs. 7.2.1 and in detail in Subs. 7.4.2 below. This equation involves only the single variable $c(\mathbf{x}, t)$, expressing the concentration of the component in the liquid phase. In all the above mass balance equations, for saturated flow, we replace θ by ϕ.

Once c and F are known, we can determine the desired (instantaneous) rate of transfer from the fluid to the solid phase, by using the relationship:

$$f^{\gamma}_{\alpha\to s} = \frac{\partial(\rho_b F)}{\partial t} - \rho_b \Gamma^{\gamma}_{s}. \tag{7.4.9}$$

Actually, (7.4.3) and (7.4.8) contain two additional fluid variables: $\theta(\mathbf{x},t)$ and $\mathbf{V}(\mathbf{x},t)$. For a deformable porous medium, $\theta_s(=1-\phi)$ is yet another variable. Hence, the complete model describing the $c(\mathbf{x},t)$-distribution, will contain additional equations which describe fluid flow and solid matrix deformation.

When equilibrium cannot be assumed, we have to express the rate of transfer *in both balance equations* in terms of an appropriate rate transfer expression. In principle, this expression will have the form: $f^{\gamma}_{\alpha\to s} = f(c,F)$. For example, by using (7.3.132), we obtain $f^{\gamma}_{s\to\alpha} = -k'_f c + k'_r F$, and the model is written in the form:

$$\frac{\partial(\theta c)}{\partial t} = -\nabla\cdot\theta\,(c\mathbf{V}+\mathbf{J}^*+\mathbf{J}^{\gamma}) + f^{\gamma}_{s\to\alpha} + \theta\rho\Gamma^{\gamma}, \tag{7.4.10}$$

$$\frac{\partial(\rho_b F)}{\partial t} = -f^{\gamma}_{s\to\alpha} + \rho_b\Gamma^{\gamma}_{s}. \tag{7.4.11}$$

The two equations have to be solved simultaneously for c, F, and $f^{\gamma}_{s\to\alpha}$.

Before leaving this subject, let us consider the case, often referred to as a 'two-site' adsorption model. As explained in Subs. 7.3.6E, in such a model, the assumption is made that the total number of sorption sites on the solid surface is made up of two parts: on one, a fraction p of the total, adsorption is assumed to occur *instantaneously*, so that equilibrium is *assumed* to prevail always, while on the other, a fraction $1-p$, kinetic (or nonequilibrium) conditions prevail. The corresponding concentrations on the solid are F_{eq} and F_{kin}, satisfying (7.3.135).

Using the isotherm (7.3.136), where p is replaced by p^*, to describe equilibrium adsorption, the balance equation (7.4.3) takes the form:

$$\frac{\partial}{\partial t}\left(\theta + p^*\rho_b K_d\right)c = -\nabla\cdot\theta(c\mathbf{V} - \mathbf{D}\cdot\nabla c - \boldsymbol{\mathcal{D}}^*\cdot\nabla c) - \left.\frac{d\rho_b F_{\text{kin}}}{dt}\right|_{\text{ads}} + \theta\rho\Gamma^{\gamma} + \rho_b\Gamma^{\gamma}_{s}, \tag{7.4.12}$$

in which $(d\rho_b F_{\text{kin}}/dt)\big|_{\text{ads}}$ is expressed by an appropriate rate transfer expression, say, (7.3.137), rewritten here for convenience in the form:

$$\left.\frac{d\rho_b F_{\text{kin}}}{dt}\right|_{\text{ads}} = \rho_b k_{13}\left[(1-p^*)K_d c - F_{\text{kin}}\right]. \tag{7.4.13}$$

The mass balance equation for the adsorbed component can be rewritten as

$$\frac{\partial}{\partial t}\left(p^* \rho_b K_d c + \rho_b F_{\text{kin}}\right) = \left.\frac{d\rho_b F_{\text{kin}}}{dt}\right|_{\text{ads}} + \rho_b \Gamma_s^\gamma. \tag{7.4.14}$$

Equations (7.4.12), (7.4.14), and (7.4.13) have now to be solved simultaneously for c, F_{kin}, and $d\rho_b F_{\text{kin}}/dt|_{\text{ads}}$.

7.4.2 Retardation

The retardation factor, R_d, has already been mentioned in connection with the single-cell model presented in Subs. 7.2.1. Let us elaborate on the interpretation of this coefficient.

Consider the case of a liquid that occupies the entire void space, i.e., $\theta = \phi$. We shall assume that (a) no external sources or sinks exists, (b) $\rho_s =$ constant, and $\partial\phi/\partial t = 0$, i.e., a nondeformable solid matrix, (c) no degradation or decay phenomena takes place, and (d) the considered component adsorbs to the solid under conditions of equilibrium, following a linear isotherm, with $K_d > 0$ and $\partial K_d/\partial t = 0$. Then, (7.4.8) reduces to the form:

$$R_d \phi \frac{\partial c}{\partial t} = -\nabla \cdot \phi(c\mathbf{V} - \mathbf{D} \cdot \nabla c - \boldsymbol{\mathcal{D}}^* \cdot \nabla c), \tag{7.4.15}$$

where

$$R_d \equiv 1 + \frac{\rho_b K_d}{\phi} \quad (> 1) \tag{7.4.16}$$

is the *retardation factor* defined in (7.2.9).

The explanation for the significance of R_d, as presented in the discussion on the cell-model, was only applicable to a well-mixed system. In order to see its significance in the case of advective transport of a component, and to understand its interpretation as a 'retardation factor', let us further simplify (7.4.15). We assume that the porous medium is homogeneous, i.e., $\nabla R_d = 0$. Then, (7.4.15) may be rewritten as

$$\phi \frac{\partial c}{\partial t} = -\nabla \cdot \phi \left(c\frac{\mathbf{V}}{R_d} - \frac{\mathbf{D}_h}{R_d} \cdot \nabla c \right), \tag{7.4.17}$$

where $\mathbf{D}_h$ $(= \mathbf{D} + \boldsymbol{\mathcal{D}}^*)$ denotes the coefficient of hydrodynamic dispersion.

For comparison, we shall also refer to the case without adsorption, i.e., $K_d = 0$, for which the balance equation (7.4.17) takes the form:

$$\phi \frac{\partial c}{\partial t} = -\nabla \cdot \phi(c\mathbf{V} - \mathbf{D}_h \cdot \nabla c). \tag{7.4.18}$$

We note that (7.4.17) and (7.4.18) are similar, except that in the former equation, the average fluid velocity is replaced by $\mathbf{V}/R_d$, and the coefficient of hydrodynamic dispersion is replaced by $\mathbf{D}_h/R_d$. We recall that (except when velocities are very low) the major component comprising $\mathbf{D}_h$ is the coefficient of mechanical dispersion, $\mathbf{D}$, which is proportional to $\mathbf{V}$. Thus, under the

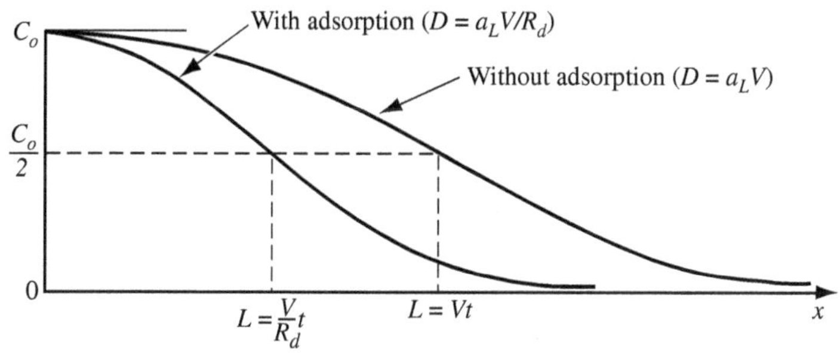

Figure 7.4.1: Effect of retardation.

assumption of equilibrium adsorption, described by a linear isotherm, the effect of adsorption is to *retard* the advance of the component (as part of it is adsorbed onto the solid). Instead of advancing with the fluid, moving at a velocity $\mathbf{V}$, the mean movement of the contaminant is at the reduced, or retarded velocity, $\mathbf{V}/R_d$. At the same time, spreading occurs as if the coefficient of mechanical dispersion, D, is also reduced by the factor R_d, along with the coefficient of molecular diffusion in a porous medium, $\mathcal{D}^*$.

In the absence of equilibrium between F and c, the form (7.4.17) is not valid. However, as long as equilibrium is assumed, even when described by a nonlinear adsorption isotherm, the phenomenon of retardation still exists. Although, in order to explain the phenomenon of retardation, we have reduced (7.4.3) to the simpler form (7.4.17), this phenomenon exists also in the more general case where a variety of source terms may exist.

Figure 7.4.1 shows the effect of retardation in the example of a semi-infinite column. The curves were obtained by an analytical solution. We note that in the case with adsorption, the point $c = 0.5$ advances at a speed V/R_d, and that the curve is steeper, indicating a smaller coefficient of hydrodynamic dispersion.

In unsaturated flow, with a contaminant that is partitioned between the liquid phase and the solid according to the linear isotherm (7.3.66), the retardation factor takes the form:

$$R_d(\theta) \equiv 1 + \frac{\rho_b K_d}{\theta}, \tag{7.4.19}$$

which now depends on $\theta(\mathbf{x}, t)$. In employing the isotherm (7.3.66), with the same K_d as in saturated flow, we have assumed that, due to the thin film of water (= the wetting-phase) which covers the solid in the air filled portion of the void space, the entire surface area of the solid provides sites for adsorption. Otherwise, K_d would be a function of the volumetric fraction, i.e., $K_d = K_d(\theta)$.

7.4.3 Initial condition and boundary conditions

The solution of any of the mass balance equations presented above for the concentration of a contaminant, e.g., (7.4.8) with $\theta \to \phi$, requires initial and boundary conditions.

In this subsection, we shall continue to consider only the case of a constant density fluid. The case of modeling flow and transport of a variable density fluid will be discussed in Subs. 9.3.1 of Chap. 3 that deals with seawater intrusion.

A. Initial conditions

Initial conditions state the spatial distribution of the state variable, here the concentration, c, of the chemical species, at some initial time, usually taken as $t = 0$:

$$c(\mathbf{x}, 0) = f_1(\mathbf{x}), \tag{7.4.20}$$

where $f_1(\mathbf{x})$ is a known function.

B. General boundary conditions

The general boundary conditions discussed in Subs. 5.2.2B are valid also here, with the intensive property $\overline{e}^\alpha$ replaced by the (macroscopic) concentration c. Briefly, in a solute transport problem, there are two conditions that have to be satisfied on every boundary segment:

- Continuity in the (macroscopic) concentration across the boundary,

$$[\![\; c \;]\!]_{1,2} = 0, \tag{7.4.21}$$

  where the symbol $[\![\; (..) \;]\!]$ denotes the jump in (..), and 1 and 2 denote the two sides of the boundary.
- In the absence of sources and sinks on the boundary, which is the usual case, there exists continuity of the (normal component of the) flux of the mass of the chemical species, across the boundary,

$$[\![\; \phi \left[c(\mathbf{V} - \mathbf{u}) - \mathbf{D}_h \cdot \nabla c \right] \;]\!]_{1,2} \cdot \mathbf{n} = 0, \tag{7.4.22}$$

  in which the total flux is defined by (7.1.65), and we have taken into account that the boundary may be moving at a speed $\mathbf{u}$. If sources are present, the jump is equal to the strength of the sources.

As in the case of fluid mass flux, discussed in Subs. 5.2.2B, since we have assumed that the boundary is material with respect to the solid matrix, viz., $(\mathbf{V}_s - \mathbf{u})\big|_1 = (\mathbf{V}_s - \mathbf{u})\big|_2 = 0$, we may rewrite (7.4.22) in the form:

$$[\![\; (c\mathbf{q}_r - \theta \mathbf{D}_h \cdot \nabla c) \;]\!]_{1,2} \cdot \mathbf{n} = 0, \tag{7.4.23}$$

where $\mathbf{q}_r$ ($\equiv \theta(\mathbf{V}_f - \mathbf{V}_s)$) (= specific discharge relative to the solid) is expressed by Darcy's law.

For (7.4.23) to become a condition for c on a boundary segment, the information on what happens on the external side of the latter (in this case, the total flux relative to the boundary) *must be known* as a function of space and time.

Note that the kind of PDE that expresses the mass balance is such that only *one* boundary condition—value of c or of the flux—has to be specified on every boundary segment. The value of the other condition can be obtained form the solution.

We recall that to transform the no-jump conditions presented above into a boundary condition, we must know the value of the concentration or the flux on the external side of the boundary.

C. Particular cases

When necessary, subscripts 1 and 2 will be used to denote the internal and external sides of a boundary surface, respectively. The latter is described by $F(\mathbf{x}, t) = 0$, with a normal unit vector given by $\mathbf{n} = \nabla F/|\nabla F|$.

Boundary of prescribed concentration. When the values of $c(\mathbf{x}, t)$ are imposed as a known function, $f_2(\mathbf{x}, t)$, at all points of a boundary segment, $\mathcal{B}$, due to phenomena that take place on the external side of the domain, independent of what happens within the latter, we employ the boundary condition

$$c(\mathbf{x}, t) = f_2(\mathbf{x}, t) \qquad \text{on} \quad \mathcal{B}, \tag{7.4.24}$$

where $c(\mathbf{x}, t)$ denotes the concentration on the internal side of the boundary and $f_2(\mathbf{x}, t)$ is the known c on the external side. This is a *first type*, or *Dirichlet* boundary condition.

Boundary of prescribed flux. When phenomena occurring in the external domain impose a *known* total flux of the considered component normal to a boundary segment, $\mathcal{B}$, say, $f_3(\mathbf{x}, t)$, at all points of the boundary, regardless of what happens within the domain itself, the condition obtained from (7.4.23) is

$$(c\mathbf{q}_r - \phi \mathbf{D}_h \cdot \nabla c) \cdot \mathbf{n} = f_3(\mathbf{x}, t) \qquad \text{on} \quad \mathcal{B}. \tag{7.4.25}$$

Since both c and ∇c appear in (7.4.25), this is a *Robin*, or *third type boundary condition*. When $\mathbf{q}_r = 0$, equation (7.4.25) reduces to a *Neumann*, or *second type boundary condition.*

A boundary of special interest is the *impervious boundary*. For such a boundary, with $f_3(\mathbf{x}, t) = 0$, and $\mathbf{q}_r \cdot \mathbf{n} = 0$, equation (7.4.25) reduces to

$$(\mathbf{D}_h \cdot \nabla c) \cdot \mathbf{n} = 0 \qquad \text{on} \quad \mathcal{B}. \tag{7.4.26}$$

This is a particular case of the *Neumann* boundary condition.

For an impervious boundary surface that coincides with the vertical xz-plane, with $V_y = 0$, $V_x, V_z \neq 0$, $n_y = 1$, $n_x, n_y = 0$, $\mathsf{D}_{hyx} = \mathsf{D}_{hyz} = 0$, $\mathsf{D}_{hyy} = \mathsf{a}_T V + \mathcal{D}^*$, the condition (7.4.26), of zero total flux normal to an impervious boundary, reduces to

$$(\mathsf{a}_T V + \mathcal{D}^*) \frac{\partial c}{\partial y} = 0, \quad \text{or} \quad \frac{\partial c}{\partial y} = 0. \tag{7.4.27}$$

Boundary between two porous media. Along such a boundary, discontinuities may exist in all solid matrix characteristics. Here, neither the concentration nor the flux is *a priori* known on the boundary, as each side serves as an external side to the other one.

Hence, two conditions must be satisfied on such a boundary. The first is that of no-jump in component concentration, expressed in the form:

$$c\big|_1(\mathbf{x}, t) = c\big|_2(\mathbf{x}, t) \qquad \text{on} \quad \mathcal{B}. \tag{7.4.28}$$

The second condition is that of continuity of the normal component of the total flux of the considered component,

$$(c\mathbf{q}_r - \phi \mathbf{D}_h \cdot \nabla c)\big|_1 \cdot \mathbf{n} = (c\mathbf{q}_r - \phi \mathbf{D}_h \cdot \nabla c)\big|_2 \cdot \mathbf{n}. \tag{7.4.29}$$

Because $\mathbf{q}_r|_1 \cdot \mathbf{n} = \mathbf{q}_r|_2 \cdot \mathbf{n}$, and $c|_1 = c|_2$, the above equation reduces to

$$(\phi \mathbf{D}_h \cdot \nabla c)\big|_1 \cdot \mathbf{n} = (\phi \mathbf{D}_h \cdot \nabla c)\big|_2 \cdot \mathbf{n}. \tag{7.4.30}$$

To understand the reason for requiring *two* conditions on such a boundary, rather than one, we note that the partial differential (balance) equations cannot be solved for domains with discontinuities in the coefficients. To overcome this difficulty, we divide the problem domain along the surfaces of discontinuity into subdomains in each of which no discontinuity exists. We write a complete model for each of these subdomain. Such a model requires that a boundary condition be specified along the surface of discontinuity (which now serves as a segment of the boundary of both subdomains). We need one condition for each side, for a total of two conditions. Because each of these conditions involves the variables for the two adjacent subdomains, the two models are coupled, and have to be solved simultaneously.

Boundary with a body of liquid. We consider the boundary between a porous medium domain (denoted by subscript pm) and a body of liquid (denoted by subscript ℓb), assumed to be a 'well-mixed' domain, that is, at a *known* uniform concentration of the considered component. A large lake and a river may serve as examples (Fig. 7.4.2).

We assume that the saturated zone is in good hydraulic contact with the body of liquid, which contains the chemical species at a known uniform concentration, c_o. The boundary is assumed to be stationary. The condition

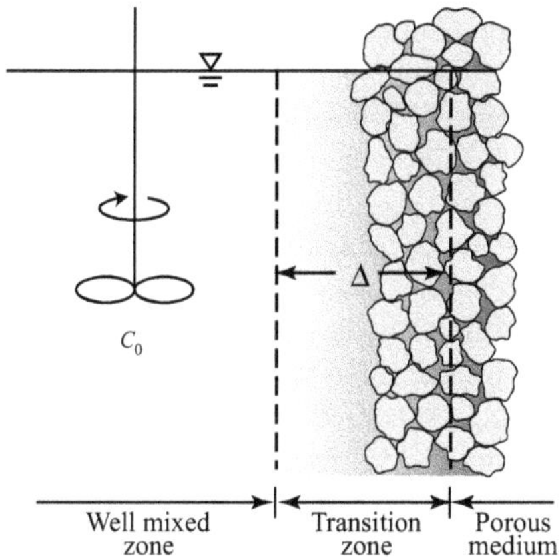

Figure 7.4.2: Boundary with a body of liquid.

of no-jump in the normal component of the total flux of the considered species takes the form

$$(c_o\mathbf{V})\big|_{\ell b}\cdot\mathbf{n}-(c\mathbf{q}_r+\phi\mathbf{J}^\gamma+\phi\mathbf{J}^{*\gamma})\big|_{\text{pm}}\cdot\mathbf{n}=0, \tag{7.4.31}$$

where $\phi\big|_{\ell b}=1$. This equation expresses the continuity in the mass flux across the boundary, of the component in the water, by advection, diffusion and dispersion. Since the liquid body is assumed to be at a uniform concentration, only advection takes place in it. The use of $\mathbf{q}_r$ stems from the assumption that the boundary is material with respect to the solid matrix. When $\mathbf{V}_s=0$, we have $\mathbf{q}_r\equiv\mathbf{q}$.

Expressing the dispersive and diffusive fluxes in terms of ∇c, we may rewrite (7.4.31) in the form:

$$(c_o\mathbf{V})\big|_{\ell b}\cdot\mathbf{n}-(c\mathbf{q}_r-\phi\mathbf{D}_h^\gamma\cdot\nabla c)\big|_{\text{pm}}\cdot\mathbf{n}=0. \tag{7.4.32}$$

Consequently, when $\mathbf{V}\big|_{\ell b}\cdot\mathbf{n}=\mathbf{q}_r\cdot\mathbf{n}=0$, i.e., no advection takes place across the boundary, the dispersive flux, $\mathbf{J}^{*\gamma}$, vanishes, and (7.4.32) reduces to

$$(\mathcal{D}^{*\gamma}\cdot\nabla c)\big|_{\text{pm}}\cdot\mathbf{n}=0. \tag{7.4.33}$$

This implies that no transport of mass by molecular diffusion takes place across such a boundary, even when $c\big|_{\text{pm}}\neq c_o$. This conclusion is unacceptable. Under the physical conditions of this example, we should expect transport by molecular diffusion of the component to take place between the porous medium domain and the adjacent liquid body, as this remains the only possible mode of transport.

The error in the conclusion expressed by (7.4.33) stems from the assumption that a 'well-mixed' zone exists on the external side of the boundary. This assumption, in the absence of advection, combined with the sharp boundary approximation, must yield no mass flux by diffusion across it. In order to reinstate the diffusive-dispersive flux, which takes place in reality, we introduce the concept of a *transition zone*, *boundary layer*, or *buffer zone*, at the boundary (Fig. 7.4.2). We may associate the width of this transition zone, Δ, with the magnitude of an REV, assuming that the abrupt boundary passes through its midpoint. Instead of the boundary between the body of liquid and the porous medium, we now consider the boundary between the latter and the transition zone. Assuming that the sum of dispersive and diffusive fluxes through the transition zone is proportional to the average concentration gradient, and that the latter is proportional to the concentration difference $c_o - c$, we express the condition of continuity of flux at the boundary by

$$c_o \mathbf{V}\big|_{\ell b} \cdot \mathbf{n} + \alpha^\star (c_o - c) = (c\mathbf{q}_r - \phi \mathbf{D}_h^\gamma \cdot \nabla c)\big|_{\text{pm}} \cdot \mathbf{n}, \tag{7.4.34}$$

where $\alpha^\star$ is a *transfer coefficient*, such that $\alpha^\star(c_o - c)$ represents the sum of diffusive and dispersive fluxes through the transition zone.

Since $\mathbf{V}\big|_{\ell b} \cdot \mathbf{n} = \mathbf{q}_r \cdot \mathbf{n}$, equation (7.4.34) reduces to

$$(c_o - c\big|_{\text{pm}})(\mathbf{q}_r \cdot \mathbf{n} + \alpha^\star) = -\phi \mathbf{D}_h^\gamma \cdot \nabla c\big|_{\text{pm}} \cdot \mathbf{n}, \tag{7.4.35}$$

which now serves as the boundary condition.

In the absence of advection, or when $|\mathbf{q}_r \cdot \mathbf{n}| \ll \alpha^\star$, equation (7.4.35) reduces to

$$\alpha^\star (c_o - c\big|_{\text{pm}}) = -\phi \boldsymbol{D}^{*\gamma} \cdot \nabla c\big|_{\text{pm}} \cdot \mathbf{n}. \tag{7.4.36}$$

We note that if we accept (7.4.35), then $c_o\big|_{\ell b} \neq c\big|_{\text{pm}}$ on the boundary, i.e., a jump in concentration takes place on the boundary. This is a consequence of introducing the transition zone and the 'well-mixed zone' approximation.

When $|\mathbf{q}_r \cdot \mathbf{n}| \gg \alpha^\star$, equation (7.4.35) reduces to

$$(c_o - c\big|_{\text{pm}})\mathbf{q}_r \cdot \mathbf{n} = -\phi \mathbf{D}_h^\gamma \cdot \nabla c\big|_{\text{pm}} \cdot \mathbf{n}, \tag{7.4.37}$$

which is a boundary condition of the third type, identical to (7.4.32), yet is based on different reasoning.

Phreatic surface. The phreatic surface serves as the upper boundary of the saturated domain. At the same time, it serves as the lower boundary of the unsaturated one.

The condition for liquid mass transport at a phreatic surface is presented in Subs. 5.2.3E. Here we consider the condition for the transport of a chemical species dissolved in the water. The boundary condition is derived from the requirement of no-jump in the component's flux normal to the (possibly moving) phreatic surface.

As has already been stated in Subs. 5.2.3E, the main difficulty associated with the phreatic surface as a boundary stem from the fact that its shape and position are not known *a priori*. In fact, in many flow problems, they are the very objective of modeling. In principle, the shape of the phreatic surface can be described by the equation $F(\mathbf{x}, t) = 0$. However, since the pressure is atmospheric everywhere on this surface, or $h(\mathbf{x}, t) = z$, i.e., the piezometric head is equal to the elevation, the equation that describes the shape of the phreatic surface may be written also (in cartesian coordinates) in the form

$$F = F(\mathbf{x}, t) = h(x, y, z, t) - z = 0. \tag{7.4.38}$$

Equations (5.2.1), (5.2.2), and (5.2.3) are applicable.

As is usually done in groundwater hydrology, we neglect the details of the movement of water through the unsaturated zone, and consider only some mean value of natural replenishment, infiltrating at ground surface, percolating through the unsaturated zone, and reaching the phreatic surface at a rate $\mathbf{N}$. Let the concentration of the considered component in the infiltrating water, as it approaches the phreatic surface, be denoted by c_N. We assume that (here and elsewhere in this chapter), in spite of concentration differences, the mass density of the water remains constant.

In the unsaturated zone, just above the phreatic surface, the moisture content is assumed to be equal to the residual moisture content, $\theta = \theta_{wr}$, which is a known constant. The component's total flux, relative to the moving phreatic surface, is given by $c_N\,(\mathbf{N} - \theta_{wr}\mathbf{u})\cdot\mathbf{n}$. The total flux in the saturated zone, relative to the moving phreatic surface, is expressed by $\phi\,[c(\mathbf{V} - \mathbf{u}) - \mathbf{D}_h \cdot \nabla c]$. Thus, the no-jump condition can be expressed as:

$$\{\phi\,[c(\mathbf{V} - \mathbf{u}) - \mathbf{D}_h \cdot \nabla c]\} \cdot \mathbf{n} = c_N(\mathbf{N} - \theta_{wr}\mathbf{u}) \cdot \mathbf{n}. \tag{7.4.39}$$

When combined with the flow boundary condition (5.2.28), we obtain

$$(c - c_N)\left(\mathbf{N} \cdot \nabla F + \theta_{wr}\frac{\partial F}{\partial t}\right) - \phi(\mathbf{D}_h \cdot \nabla c)\big|_{\text{sat}} \cdot \nabla F = 0. \tag{7.4.40}$$

We may insert $F = h(x, y, z, t) - z$ in the last equation. This is a third type boundary condition for c. We note that $[\![\, c \,]\!]_{\text{sat, unsat}} \equiv c - c_N \neq 0$. Thus, the unsaturated zone just above the phreatic surface acts as a 'well-mixed zone' in the sense discussed earlier. However, we have simplified the expression for the flow through the transition zone by neglecting the dispersive-diffusive flux through it.

Seepage face. The seepage face is discussed in Subs. 5.2.3F. Here, the water leaving the porous medium domain through the seepage face carries the dissolved chemical species.

Because there is no porous medium on the external side of this boundary, the condition of continuity of flux of a component takes the form

$$\left(\phi c\mathbf{V} - \phi\mathbf{D}_h \cdot \nabla c\right)\Big|_{\text{pm}} \cdot \mathbf{n} = (c\mathbf{V})\Big|_{\text{env}} \cdot \mathbf{n}, \tag{7.4.41}$$

where symbols '*pm*' and '*env*' denote the porous medium domain and its external environment, respectively, and we have assumed a stationary seepage face, $\mathbf{u} \equiv 0$. With

$$c\Big|_{\text{pm}} = c\Big|_{\text{env}}, \qquad (\phi\mathbf{V})\Big|_{\text{pm}} \cdot \mathbf{n} = \mathbf{V}\Big|_{\text{env}} \cdot \mathbf{n}, \tag{7.4.42}$$

i.e., assuming neither volatilization, nor precipitation, (7.4.41) reduces to the boundary condition

$$(\mathbf{D}_h \cdot \nabla c)\Big|_{\text{pm}} \cdot \mathbf{n} = 0. \tag{7.4.43}$$

This is a boundary condition of the second type.

Artificial boundary. For a numerical solution involving an unbounded domain, or practically so (e.g., a domain which is much larger than the domain of interest), it is often necessary to truncate the modeled domain to a finite size, as the computer cannot handle a discrete system with an infinite number of unknowns (or it is uneconomical to handle a very large number of nodes). The truncation of the domain requires the introduction of an *artificial boundary*, on which the boundary condition is unknown; a certain approximation is called for.

When such an artificial boundary is introduced, the most prudent thing to do is to make sure that it is sufficiently far away from the region of interest, or region of major activities, that is, the region where significant concentration changes occur. A simple and often used condition on such a boundary is to fix and maintain the concentration on it equal to the initial concentration there (in some computer codes this is referred to as to 'clamp' the condition). Another option is to assume that the dispersive flux normal to such a boundary is negligible (i.e., transport normal to the boundary is by advection only), i.e., $\nabla c \cdot \mathbf{n} = 0$ on this boundary.

As simulation time increases, the zone of major activities may expand (e.g., as a contaminant plume advances), and the selected location of the artificial boundary may no longer be appropriate; it may have to be moved farther away. A trial-and-error approach may be required in order to determine the appropriate location of such a boundary or the condition to be used on it.

In a *Lagrangian method* (Secs. 7.6 and 8.6), individual particles are tracked for their locations in relation to cells fixed in space. In that case, we simply allow the particles to exit a boundary cell, transported by the dominant velocity in the cell, without additional treatment. In other words, the dispersion, which is handled by an *Eulerian scheme* (and thus requires a boundary condition), will not be executed.

7.4.4 Complete model for single component

So far, we have been discussing models of transport of a single component in the water (= aqueous phase), with or without adsorption, volatilization, and chemical reactions. Cases of multiple components with chemical reactions, together with their complete model, are discussed in Sec. 7.9.

We have repeatedly emphasized that fluid velocities and saturations have to be known, as they are required as input to the contaminant transport model. This information may be obtained by solving the flow model separately. In some situations, it is convenient to state and solve the flow and the contaminant transport problems simultaneously as a single model. This is the only option whenever the density is concentration-dependent.

The mathematical model of a single, or multiphase flow problem, combined with the transport of a single chemical component, possibly with first order decay, adsorption, and volatilization, consists of the following parts:

(a) A mathematical description of the configuration of the surface that bounds the porous medium problem domain.
(b) A list of the dependent variables. These are the concentrations, c_α^γ, of the γ-component within all α-fluid phases present in the system. In the case of adsorption, F^γ is included in the list of state variables. For the flow model, depending on the number of fluid phases that are in motion, we may add such variables as piezometric heads, pressures, saturations, etc., following the discussion on the complete flow model in Subs. 5.3.3.
(c) Flux equations for the mass of the fluid phases. Darcy's law (for both saturated and unsaturated flow) is usually employed.
(d) For unsaturated flow, soil-water retention and unsaturated hydraulic conductivity relations.
(e) Partial differential ($\equiv$ mass balance) equations for the relevant fluid phases.
(f) Mass and momentum balance equations for the solid, when the latter is deformable.
(g) Partial differential equations that describe mass balances of the considered component within all fluid phases present in the system and on the solid. These balance equations may contain source terms that correspond to decay, adsorption, and volatilization of the component.
(h) Dispersive, and diffusive flux equations for the mass of the considered component.
(i) Constitutive equations for the fluid phases, for the solid (in the case of a deformable solid), and for the component. These include also thermodynamic relationships that describe the partitioning of the component between adjacent phases under equilibrium conditions, or transfer functions for nonequilibrium conditions.
(j) Expressions for the various external sources and sinks for the mass of the fluid phases and the considered component.
(k) Statement of initial conditions for each of the relevant state variables.

(l) Statement of boundary conditions for each of the relevant extensive quantities—mass of fluid phases, and of the considered component.
(m) Numerical values, or functional relationships for all the coefficients that appear in the various balance equations and constitutive relations included in the model

Based on the above standard content of a model, we usually end up with a large number of variables that describe the state of the system. To obtain a closed set of equations, within the framework of a *well-posed problem*, we need an equal number of equations. However, following the discussion on *primary variables* (or *degrees of freedom*) of a problem, presented in Subs. 7.9.4, the next step is to determine the number of *primary variables* (or *degrees of freedom*) of the problem. The number of partial differential equations of balance that has to be solved is then equal to the number of the primary variables. All other variables are obtained from the known values of the selected primary variables, using the remaining equations.

Actually, when we assume that adsorption and volatilization take place *under equilibrium conditions*, a single mass balance equation can be written for the component *in the porous medium as a whole*. The single variable is, then, the concentration of the component within every fluid phase in the porous medium. The boundary condition for such an equation is written for the concentration or total mass flux of the component in the porous medium as a whole. Similarly, initial conditions are written for the concentration or total mass of the component in all the phases present in the system.

The statement of the transport model for a particular site must also include information on all relevant porous medium coefficients, such as porosity, permeability and dispersivity and their spatial distributions. Information is also required concerning the coefficients that appear in the constitutive relations, e.g., decay and growth coefficients, partitioning coefficients and reaction rate coefficients.

7.4.5 Some analytical solutions

Due to the irregularity of boundaries and the heterogeneity of aquifers, analytical solutions of cases of practical interest are not feasible; numerical techniques (see Chap. 8) must be employed to obtain solutions of contaminant transport models in real aquifer domains. Only for a very limited number of rather simple, mostly one-dimensional problems, can analytical solutions be derived. The objective of this subsection is to present some of these solutions in order to gain insight into the produced pattern of contaminant transport. In all cases, we shall assume that the density and viscosity of the solution remain unchanged, and that the solid matrix is homogeneous, isotropic, and nondeformable. Only transport by advection-dispersion-diffusion with and without adsorption ($\equiv$ first order decay) in saturated flow will be considered.

A. Solute transport in an infinite homogeneous column

Consider one-dimensional flow along the x-axis. The specific discharge, q, satisfies the continuity equation $\partial q/\partial x = 0$, which leads to $q = q(t)$, i.e., q may vary in time, but is constant at any given time. The partial differential (mass balance) equation that governs the solute distribution, say, (7.4.18), reduces for this 1-d case, with the porosity $\phi = \text{const.}$, to

$$\frac{\partial c}{\partial t} = \mathsf{D}_h \frac{\partial^2 c}{\partial x^2} - \frac{q}{\phi}\frac{\partial c}{\partial x}, \qquad -\infty < x < +\infty, \tag{7.4.44}$$

in which $\mathsf{D}_h = \mathsf{a}_L|q|/\phi + \mathcal{D}$ denotes the coefficient of hydrodynamic dispersion.

The initial and boundary conditions are:

$$\begin{array}{llll} \text{for } t=0, & -\infty < x < 0, & c = c_1, & \\ & 0 \le x < +\infty, & c = c_o\,(< c_1); & \\ \text{for } t>0, & x = -\infty, & c = c_1, & \\ & x = +\infty, & c = c_o. & \end{array} \tag{7.4.45}$$

Bear (1960) solves this problem by applying the Laplace transform to (7.4.44). The solution is

$$\varepsilon(x,t) \equiv \frac{c(x,t)-c_o}{c_1-c_o} = \frac{1}{2}\text{erfc}\left\{\frac{x - \int_0^t [q(t)/\phi]\,dt}{2\left[\int_0^t (\mathsf{a}_L|q|/\phi + \mathcal{D})\,dt\right]^{1/2}}\right\}, \tag{7.4.46}$$

where $\text{erf}\,x = (2/\sqrt{\pi})\int_o^x e^{-\alpha^2}\,d\alpha$, and $\text{erfc}\,x = 1 - \text{erf}\,x = (2/\sqrt{\pi})\int_x^\infty e^{-\alpha^2}\,d\alpha$. For $q = \phi V = \text{const.}$, and neglecting molecular diffusion, i.e., $\mathcal{D} \ll \mathsf{a}_L V$, the above solution reduces to

$$\varepsilon(x,t) \equiv \frac{c(x,t)-c_o}{c_1-c_o} = \frac{1}{2}\text{erfc}\left[\frac{x - Vt}{2\,(\mathsf{D}t)^{1/2}}\right] = \frac{1}{2}\text{erfc}\left[\frac{x-Vt}{\sqrt{2}\,\sigma}\right], \tag{7.4.47}$$

where $\mathsf{D} = \mathsf{a}_L|V|$, $\sigma^2 = 2\mathsf{D}t = 2\mathsf{a}_L|V|t = 2\mathsf{a}_L L$ is the variance of the distribution, and $L = Vt$ is the (average) distance traveled by the water during t. Figure 7.4.3 shows (7.4.47) in graphical form. From this figure, it follows that the point $\varepsilon = 0.5$ travels with the mean flow, and that the variance of the concentration distribution is proportional to the length traveled, L. The corresponding conclusions from (7.4.46), where $q = q(t)$, are that the point $\varepsilon = 0.5$ travels, again, with the mean flow, i.e., with the velocity V, and that σ^2 is *proportional to the total path traveled.*

Figure 7.4.4 presents the experimental result of a fluctuating flow, $q = q(t)$, moving up and down an infinite column. Figure 7.4.4c shows that σ^2 of the concentration increases linearly with the cumulative distance traveled, regardless of the flow direction.

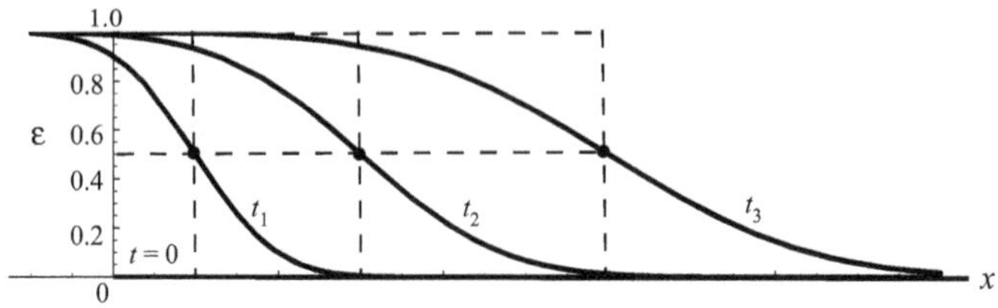

Figure 7.4.3: Concentration distribution in flow in an infinite column. $(0 < t_1 < t_2 < t_3 \ldots)$

B. Solute transport in an infinite homogeneous column with adsorption

Let us assume q = constant. For this case, with adsorption described by a linear equilibrium isotherm, e.g., $F = K_d c$, we have to solve the partial differential equation (7.4.15), rewritten as

$$\frac{\partial c}{\partial t} = \frac{\mathsf{D}_h}{R_d}\frac{\partial^2 c}{\partial x^2} - \frac{q}{R_d\phi}\frac{\partial c}{\partial x}, \quad -\infty < x < +\infty, \tag{7.4.48}$$

where $R_d = 1 + \rho_b K_d/\phi$ denotes the retardation factor. Note that (7.4.48) can be obtained from (7.4.44) by replacing q/ϕ by $q/R_d\phi$ and D_h by D_h/R_d. Hence, for the initial and boundary conditions given by (7.4.45), the solution is

$$\varepsilon(x,t) \equiv \frac{c(x,t) - c_o}{c_1 - c_o} = \frac{1}{2}\text{erfc}\left[\frac{R_d x - Vt}{2\,(R_d \mathsf{D}_h t)^{1/2}}\right], \tag{7.4.49}$$

obtained form (7.4.47) by the above stated substitution.

C. Injection of a solute slug into an infinite homogeneous column

Initially, the entire column is at the solute concentration $c = 0$. A very small volume is injected at $t = 0$ during a very short time into the column at $x = 0$. The velocity of the liquid in the column is $V = q/\phi$ = constant. Because of hydrodynamic dispersion, as the slug moves with the water in the $+x$ direction, it spreads out, occupying an ever increasing portion of the column. The solute concentration along the column is governed by (7.4.44).

For an observer moving with the average velocity, the governing equation becomes

$$\frac{\partial c}{\partial t} = \mathsf{D}_h\frac{\partial^2 c}{\partial x'^2}, \quad x' = x - Vt. \tag{7.4.50}$$

The initial condition takes the form

$$c(x,0) = \frac{M}{\phi}\delta(x), \quad M = \int_{-\infty}^{+\infty} \phi\, c(\xi,t)\, d\xi, \tag{7.4.51}$$

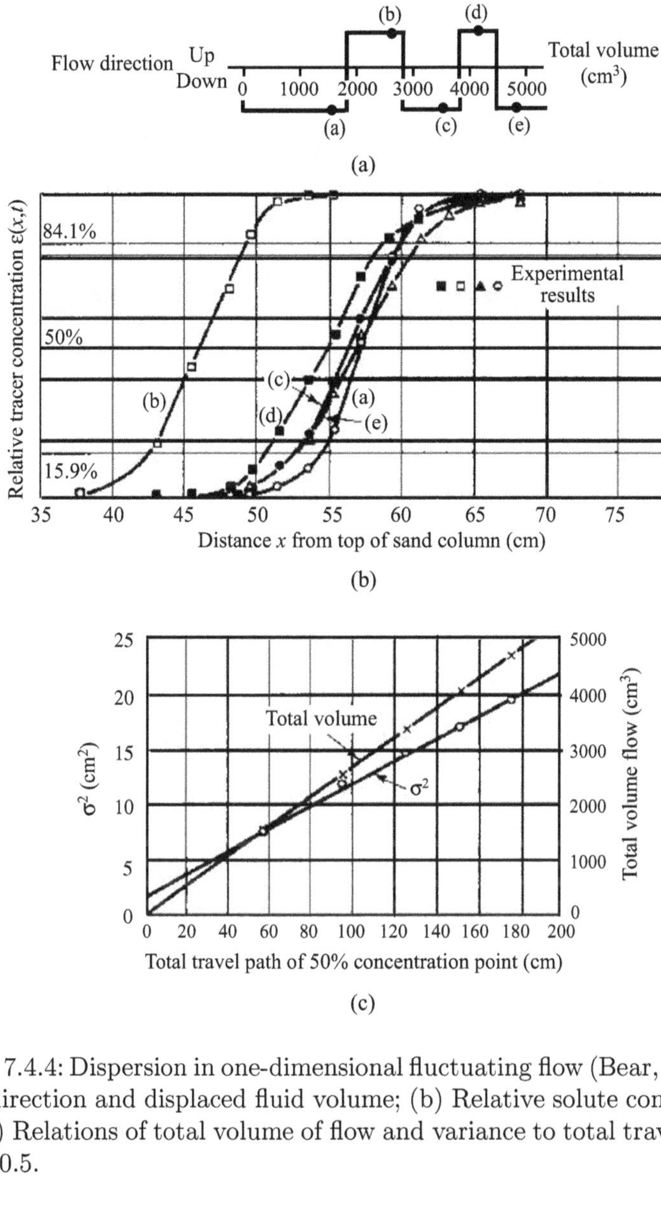

Figure 7.4.4: Dispersion in one-dimensional fluctuating flow (Bear, 1961b): (a) Flow direction and displaced fluid volume; (b) Relative solute concentration; and (c) Relations of total volume of flow and variance to total travel distance of $\epsilon = 0.5$.

in which M denotes the mass of solute in the slug, and $\delta(x)$ is the *Dirac delta function*, defined by (5.1.77). For the one-dimensional case considered here, it takes the form

$$\delta(x) = \begin{cases} \lim_{a\to 0} 1/a, & -a/2 < x < a/2, \quad a > 0, \\ 0, & \text{elsewhere}, \end{cases} \tag{7.4.52}$$

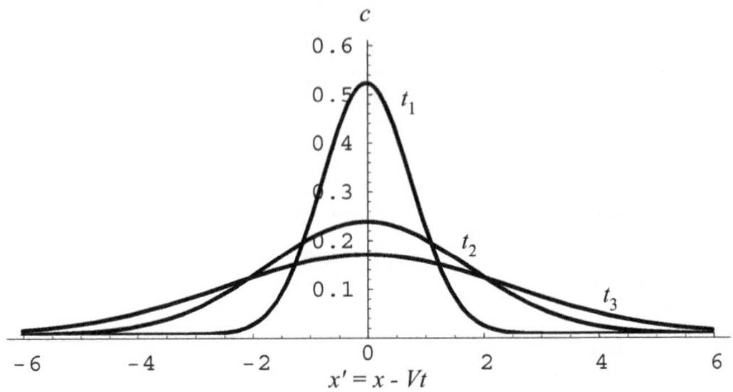

Figure 7.4.5: Spreading of a slug concentration. ($t_1 < t_2 < t_3$)

The boundary conditions specified for $c(x', t)$ are

$$\lim_{|x'|\to\infty} c(x', t) = 0. \tag{7.4.53}$$

The solution is (Crank, 1956)

$$c(x,t) = \frac{M/\phi}{(4\pi \mathsf{D}_h t)^{1/2}} \exp\left(-\frac{x'^2}{4\mathsf{D}_h t}\right) = \frac{M/\phi}{(4\pi \mathsf{D}_h t)^{1/2}} \exp\left[-\frac{(x-Vt)^2}{4\mathsf{D}_h t}\right]. \tag{7.4.54}$$

Figure 7.4.5 shows solute concentration resulting from an instantaneous slug.

D. Solute transport in an infinite two-dimensional domain with a point source

Wilson and Miller (1978) developed an analytical solution for the mass balance equation describing solute transport with first order decay in uniform flow (constant velocity V in the $+x$-direction) in an infinite two-dimensional domain:

$$\frac{\partial c}{\partial t} = \mathsf{D}_L \frac{\partial^2 c}{\partial x^2} + \mathsf{D}_T \frac{\partial^2 c}{\partial y^2} - V\frac{\partial c}{\partial x} - \lambda c, \tag{7.4.55}$$

with a continuous point source at $(0,0)$. Their solution is

$$c(x,y,t) = \frac{c_o Q}{4\pi\phi(\mathsf{D}_L \mathsf{D}_T)^{1/2}} \exp\left(\frac{x}{B}\right) W\left(u, \frac{r}{B}\right), \tag{7.4.56}$$

where $c_o Q$ is the injection rate at the point-source, ϕ is the porosity, and

$$B = \frac{2\mathsf{D}_L}{V}, \quad d = 1 + \frac{2B\lambda}{V},$$

$$u = \frac{r^2}{4d\mathsf{D}_L t}, \qquad r = \sqrt{\left(x^2 + y^2 \frac{\mathsf{D}_L}{\mathsf{D}_T}\right) d}\,. \tag{7.4.57}$$

In the above, the function $W(u, r/B)$ is the *Hantush-Jacob leaky aquifer well function* (Hantush and Jacob, 1955), which has been tabulated by Hantush (1956, 1964). (For a computer program for its evaluation, see Cheng (2000).) For easier evaluation, Wilson and Miller (1978) provided the following approximate solution of (7.4.56),

$$c(x,y,t) = \frac{c_o Q}{4\sqrt{\pi}\phi(Vr\mathsf{D}_T)^{1/2}} \exp\left(\frac{x-r}{B}\right) \operatorname{erfc}\left(\frac{2u - r/B}{2u^{1/2}}\right). \tag{7.4.58}$$

For $r/B > 10$, the error of the above solution is less than 1%.

E. Solute transport in a semi-infinite three-dimensional domain with an areal source

Solute transport in a three-dimensional homogeneous isotropic aquifer, with uniform flow at a velocity V in the $+x$-direction is sometimes described by the mass balance equation

$$\frac{\partial c}{\partial t} = \mathsf{D}_L \frac{\partial^2 c}{\partial x^2} + \mathsf{D}_T \frac{\partial^2 c}{\partial y^2} + \mathsf{D}_T \frac{\partial^2 c}{\partial z^2} - V \frac{\partial c}{\partial x} - \lambda c. \tag{7.4.59}$$

Domenico (1987) derived a semi-analytical solution for the case of a semi-infinite aquifer ($x \geq 0$) with a symmetrical contaminant source in the form of an area of size $Y \times Z$, normal to the $+x$-axis, centered at the origin $(0,0,0)$. The solution takes the form of the product of the concentration distribution along the central line and a lateral dispersion factor, $\mathcal{F}$,

$$c(x,y,z,t) = \mathcal{F} \cdot \omega(x,y,z). \tag{7.4.60}$$

$$\mathcal{F} = \frac{c_o}{2} \exp\left(\frac{Vx}{2\mathsf{D}_L}\right) \left[\exp(-\beta x) \operatorname{erfc} \gamma^- + \exp(\beta x) \operatorname{erfc} \gamma^+\right], \tag{7.4.61}$$

where

$$\beta = \left(\frac{V^2}{4\mathsf{D}_L^2} + \frac{\lambda}{\mathsf{D}_L}\right)^{1/2}, \tag{7.4.62}$$

$$\gamma^- = \frac{x - (V + 4\lambda\mathsf{D}_L)^{1/2} t}{2(\mathsf{D}_L t)^{1/2}}, \qquad \gamma^+ = \frac{x + (V + 4\lambda\mathsf{D}_L)^{1/2} t}{2(\mathsf{D}_L t)^{1/2}}, \tag{7.4.63}$$

and

$$\begin{aligned} \omega(x,y,z) = \frac{1}{4} &\left\{ \operatorname{erf}\left[\frac{y + Y/2}{2(\mathsf{a}_T x)^{1/2}}\right] - \operatorname{erf}\left[\frac{y - Y/2}{2(\mathsf{a}_T x)^{1/2}}\right] \right\} \\ &\left\{ \operatorname{erf}\left[\frac{z + Z/2}{2(\mathsf{a}_T x)^{1/2}}\right] - \operatorname{erf}\left[\frac{z - Z/2}{2(\mathsf{a}_T x)^{1/2}}\right] \right\}. \end{aligned} \tag{7.4.64}$$

In the above, a_T is the transversal dispersivity, and ω, independent of time and species, is a function of the transversal dispersivity and of the spatial coordinates, with $\omega = 1$ on the central line and $\omega = 0$ as $y, z \to \infty$.

F. Continuous injection of a radioactive solute into an infinite homogeneous column

For this case, the solute mass balance equation takes the form

$$\frac{\partial c}{\partial t} = \mathsf{D}_h \frac{\partial^2 c}{\partial x^2} - \frac{q}{\phi}\frac{\partial c}{\partial x} - \lambda c, \tag{7.4.65}$$

where the radioactive decay is described by (7.3.53), in which we replace N by c. The *elementary solution* of an instantaneous injection is

$$c(x,t,t') = \frac{dM}{[4\pi \mathsf{D}_h(t-t')]^{1/2}} \exp\left\{-\frac{[x - V(t-t')]^2}{4\mathsf{D}_h(t-t')} - \lambda(t-t')\right\}, \tag{7.4.66}$$

in which $dM = c_o V dt'$ is the mass of the solute, at concentration $c = c_o$, injected during the time interval dt' at $x = 0$ and $t = t'$. For a continuous injection at a constant rate, we obtain by integration

$$c(x,t) = \frac{c_o V}{(4\pi \mathsf{D}_h)^{1/2}} \exp\left(\frac{Vx}{2\mathsf{D}_h}\right) \int_0^t \frac{1}{\sqrt{\tau}} \exp\left(-\frac{a}{\tau} - b\tau\right) d\tau, \tag{7.4.67}$$

where $a = x^2/4\mathsf{D}_h$, $b = V^2/4\mathsf{D}_h + \lambda$.

As $t \to \infty$, we obtain

$$c(x,\infty) = \frac{c_o}{(1 + 4\lambda \mathsf{D}_h/V^2)^{1/2}} \exp\left\{\frac{Vx}{2\mathsf{D}_h}\left[1 - \left(1 + 4\lambda \mathsf{D}_h/V^2\right)^{1/2}\right]\right\}. \tag{7.4.68}$$

For $x = 0$, (7.4.67) can be integrated to yield

$$c(0,t) = \frac{c_o}{(1 + 4\lambda \mathsf{D}_h/V^2)^{1/2}} \operatorname{erf}\left(\frac{V^2 t}{4\mathsf{D}_h} + \lambda t\right), \tag{7.4.69}$$

which shows that $c(0,t) \neq c_o$. As $t \to \infty$, we obtain

$$c(0,\infty) = \frac{c_o}{(1 + 4\lambda \mathsf{D}_h/V^2)^{1/2}} \neq c_o. \tag{7.4.70}$$

G. Movement of a radioactive solute in a semi-infinite homogeneous column, $x \geq 0$

The flow along the column is at a constant specific discharge q. Initially, the concentration along the column, $x > 0$, is $c = 0$. For the boundary conditions

$$x = 0, \quad c = c_o, \quad \text{or} \quad q(c - c_o) = \phi \mathsf{D}_h \partial c/\partial x, \tag{7.4.71}$$

Ogata and Banks (1961) presented a solution to this problem (see also Bear, 1979, p. 268). They also presented the following approximate solution, which is valid when x/a_L is sufficiently large, a condition usually satisfied in practice:

$$c(x,t) = \frac{c_o}{2} \operatorname{erfc}\left[\frac{x - Vt}{2(\mathsf{D}_h t)^{1/2}}\right], \tag{7.4.72}$$

which is the same as (7.4.47), developed for an infinite column.

The above equation can also serve as a basis for the laboratory determination of the coefficient of hydrodynamic dispersion, D_h, and from it, neglecting molecular diffusion, of the longitudinal dispersivity, $\mathsf{a}_L = \mathsf{D}_h/V$. A typical experiment is one in which a column is initially filled with water at a certain solute (= tracer) concentration. During the experiment, the water in the column is displaced by water having another known constant concentration, introduced at one end of the column, say at $x = 0$, from $t = 0$ onward. The concentration of the effluent at $x = L$, $t \geq 0$ is recorded. Although the column is of a finite length, L, the solution (7.4.72) derived for a semi-infinite column is usually employed to describe the concentration of the effluent at $x = L$, as it is basically the same as (7.4.72).

Thus, the relationship (7.4.72), written for the column outlet at $x = L$, takes the form

$$c(L,t) = \frac{c_o}{2} \operatorname{erfc}\left[\frac{L - Vt}{2(\mathsf{D}_h t)^{1/2}}\right] = \frac{c_o}{2} \operatorname{erfc}\left[\frac{1 - \frac{\mathcal{U}}{\mathcal{U}_\mathrm{P}}}{2\left(\frac{\mathsf{D}_h}{LV}\right)^{1/2}\left(\frac{\mathcal{U}}{\mathcal{U}_\mathrm{P}}\right)^{1/2}}\right], \tag{7.4.73}$$

in which we have made use of $\mathcal{U} = \phi VtA$ as the effluent volume, and $\mathcal{U}_\mathrm{P} = \phi AL$ as one pore volume contained in the column, with A denoting the column's cross-sectional area.

During the experiment, c is recorded as a function of $\mathcal{U}$. Figure 7.4.6 shows results of a typical experiment. The coefficient D_h is derived from the slope of the curve at the time corresponding to one pore volume (i.e., $\mathcal{U} = \mathcal{U}_\mathrm{P}$), given by

$$\left.\frac{d(c/c_o)}{d(\mathcal{U}/\mathcal{U}_\mathrm{P})}\right|_{\mathcal{U}/\mathcal{U}_\mathrm{P}=1} = \frac{1}{2\sqrt{\pi}}\left(\frac{LV}{\mathsf{D}_h}\right)^{1/2}, \quad V = \frac{QL}{\mathcal{U}_\mathrm{P}}, \tag{7.4.74}$$

where $Q = \phi VA$ is the column discharge. Another possibility is to obtain D_h from the same curve, using

$$\mathsf{D}_h = \frac{1}{2}\sigma^2 LV = \frac{1}{2}\left(\frac{\mathcal{U}_{0.841} - \mathcal{U}_{0.159}}{2\mathcal{U}_\mathrm{P}}\right)^2 LV, \tag{7.4.75}$$

based on the observation that (7.4.73) describes a *normal distribution*, with mean at $\mathcal{U}/\mathcal{U}_\mathrm{P} = 1$, and standard deviation $\sigma = (2\mathsf{D}_h/LV)^{1/2}(\mathcal{U}/\mathcal{U}_\mathrm{P})^{1/2}$.

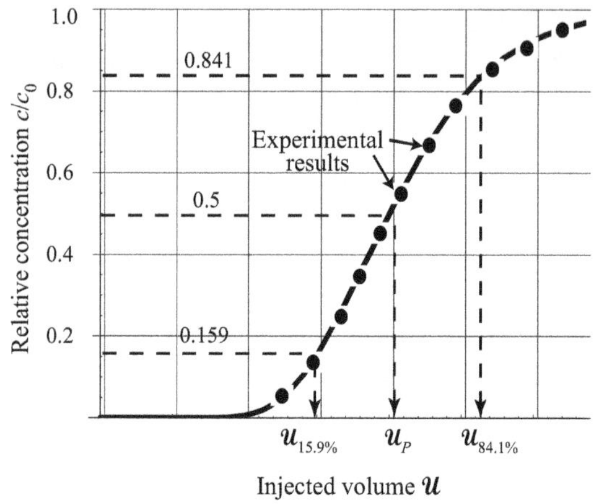

Figure 7.4.6: Computing D_h from a column displacement experiment.

Actually, Fig. 7.4.6 is not exactly a normal distribution with respect to $\mathcal{U}(t)$; it is only an approximation.

Useful analytical solutions for one-dimensional cases were also presented by Bastian and Lapidus (1956) and Gershon and Nir (1969); see discussion of these solutions in Bear (1979).

Clement (2001), extending the work of Sun *et al.* (1999a, b), presented a generalized method for solving the multispecies transport equations, coupled with a first-order reaction network. The method is flexible for solving one-, two-, or three-dimensional advection-dispersion equations that are coupled with a set of first-order reactions. A major limitation is that the method cannot be used for solving multispecies transport equations with different retardation factors.

Assuming uniform flow in the $+x$-direction, and an isotropic porous medium, the three-dimensional mass balance equation for predicting the fate and transport of n chemical γ-species coupled by a set of first-order reactions (e.g., radioactive decay), can be written as

$$R_d^\gamma \frac{\partial c^\gamma}{\partial t} - \mathsf{D}_L \frac{\partial^2 c^\gamma}{\partial x^2} - \mathsf{D}_T \frac{\partial^2 c^\gamma}{\partial y^2} - \mathsf{D}_T \frac{\partial^2 c^\gamma}{\partial z^2} + V \frac{\partial c^\gamma}{\partial x}$$
$$= \sum_{\delta=1}^{\gamma-1} y^{\gamma,\delta} \lambda^\delta c^\delta - \lambda^\gamma c^\gamma + \sum_{\delta=\gamma+1}^{n} y^{\gamma,\delta} \lambda^\delta c^\delta, \quad \gamma = 1, 2, \ldots, n, \tag{7.4.76}$$

where c^γ (dims. ML^{-3}) denotes the concentration of the γ-species, R_d^γ is the retardation coefficient of the γ-species, $y^{\gamma,\delta}$ (dims. MM^{-1}) is the effective yield factor, which describes the mass of a γ-species produced from a δ-species, λ^γ (dims. T^{-1}) is the first-order destruction rate constant of the γ-species, V (dims. LT^{-1}) denotes the fluid's velocity, $\mathsf{D}_L(\equiv \mathsf{a}_L V)$, and $\mathsf{D}_T(\equiv \mathsf{a}_T V)$ (dims. $\mathrm{L}^2\mathrm{T}^{-1}$) are the longitudinal and transverse dispersion coefficients, respectively. Note that the above equation is applicable when the degradation is limited only to the liquid phase.

Srinivasan and Clement (2008a, b) presented an analytical solution for a case involving n sequentially decaying contaminants simultaneously subjected to advection, dispersion and linear equilibrium adsorption processes in a semi-infinite domain ($0 < x < \infty$). The general governing equation for this transport problem is expressed as

$$R_d^\gamma \frac{\partial c^\gamma(x,t)}{\partial t} - \mathsf{D}_L \frac{\partial^2 c^\gamma(x,t)}{\partial x^2} + V\frac{\partial c^\gamma(x,t)}{\partial x} = \begin{cases} y^\gamma \lambda^{\gamma-1} c^{\gamma-1}(x,t) - \lambda^\gamma c^{\gamma-1}(x,t), & \gamma = 2,3,\dots,n, \\ -\lambda^\gamma c^\gamma(x,t), & \gamma = 1, \end{cases} \tag{7.4.77}$$

in which y^γ is the effective yield factor that describes the mass of a γ-species produced from the $(\gamma-1)$-species. Equation (7.4.77) is solved for a generic exponentially distributed initial condition given by

$$c^\gamma(x,0) = c_o^\gamma e^{-\kappa^\gamma x}, \quad 0 < x < \infty, \quad \gamma = 1,2,\dots,n, \tag{7.4.78}$$

where c_o^γ is the initial source concentration of the γ-species at the origin (dims. ML^{-3}) and κ^γ is the first-order decay parameter of the initial distribution of the γ-species (dims. L^{-1}).

The boundary condition at $+\infty$ is given by

$$\frac{\partial c^\gamma(\infty,t)}{\partial x} = 0, \quad t > 0, \quad \gamma = 1,2,\dots,n. \tag{7.4.79}$$

For the case of the Dirichlet boundary condition, the source term, at $x = 0$, is described by

$$c^\gamma(0,t) = \begin{cases} \sum_{\gamma_1=1}^{\gamma} B_\gamma^{\gamma_1} e^{-\lambda^{\gamma_1} t}, & t < 0 < t_o, \\ 0, & t > t_o, \end{cases} \quad \gamma = 1,2,\dots,n, \tag{7.4.80}$$

where $B_\gamma^{\gamma_1}$ (dims. ML^{-3}) is the source boundary concentration of the γ_1-specie that contributes to species γ, and λ^{γ_1} (dims. T^{-1}) is the first-order decay rate constant of the corresponding $B_\gamma^{\gamma_1}$ term.

For the case of the Robin (also called third type) boundary condition, the source term, at $x = 0$, is described by

$$-\mathsf{D}_L \left.\frac{\partial c^\gamma(x,t)}{\partial x}\right|_{x=0} + Vc^\gamma(0,t) = \begin{cases} \sum_{\gamma_1=1}^{\gamma} B_\gamma^{\gamma_1} e^{-\lambda^{\gamma_1} t}, & t < 0 < t_o, \\ 0, & t > t_o, \end{cases} \quad \gamma = 1,2,\dots,n. \tag{7.4.81}$$

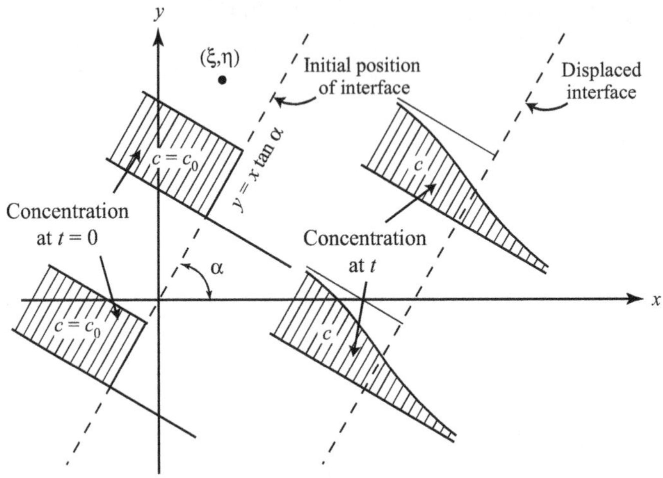

Figure 7.4.7: The transition zone at a moving interface in uniform planar flow.

In (7.4.77), it is assumed that decay occurs only in the liquid phase. When degradation occurs also on the solid surface (i.e., the adsorbed species also undergoes decay), the above model holds, with λ^γ replaced by $R_d^\lambda \lambda^\gamma$. Solutions of (7.4.78), subject to the boundary conditions (7.4.79)–(7.4.81), are given in Srinivasan and Clement (2008a).

Srinivasan and Clement (2008b) discussed special cases, implementation and tests.

H. Dispersion of an initially sharp front in uniform planar flow

We assume uniform flow in the $+x$ direction. Initially, an abrupt sharp front along $y = x \tan\alpha$ separates the two zones, with concentrations $c = c_o$ and $c = 0$ (Fig. 7.4.7). The governing equation for this case is

$$\frac{\partial c}{\partial t} = \mathsf{D}_L \frac{\partial^2 c}{\partial x^2} + \mathsf{D}_T \frac{\partial^2 c}{\partial y^2} - \frac{q}{\phi}\frac{\partial c}{\partial x}, \tag{7.4.82}$$

where $\mathsf{D}_L = \mathsf{a}_L q/\phi + \mathcal{D}$, $\mathsf{D}_T = \mathsf{a}_T q/\phi + \mathcal{D}$, with a_L and a_T denoting the longitudinal and transversal dispersivities, respectively.

The sought distribution is obtained by integrating the elementary solution for a point source, similar to (7.4.54), over the region occupied by tracer labeled water. With M denoting the tracer mass injected at a point (ξ, η) at $t = 0$, the concentration areal distribution at any later time is given by

$$c(x, y, t) = \frac{M/\phi}{4\pi(\mathsf{D}_L \mathsf{D}_T)^{1/2} t} \exp\left[-\frac{(x - \xi - Vt)^2}{4\mathsf{D}_L t} - \frac{(y - \eta)^2}{4\mathsf{D}_T t}\right]. \tag{7.4.83}$$

The mass conservation requires that

$$\frac{M}{\phi} = \int_{-\infty}^{+\infty} \int_{-\infty}^{+\infty} c(x', y', t)\, dx'dy'. \tag{7.4.84}$$

Curves of $c =$ const., described by (7.4.83), have the shape of ellipses centered at $(\xi + Vt, \eta)$.

To obtain the areal concentration distribution resulting from a moving front, we integrate the effect of an infinite number of small point sources, each with $M = c_o \phi \, d\xi \, d\eta$, initially located in the semi-infinite space behind the front. The result is

$$c(x, y, t) = \frac{c_o}{2} \text{erfc} \left\{ \frac{(x - Vt) \sin \alpha - y \cos \alpha}{[4(\mathsf{D}_L \sin^2 \alpha + \mathsf{D}_T \cos^2 \alpha) t]^{1/2}} \right\}, \tag{7.4.85}$$

which describes a normal distribution perpendicular to the displaced front (Fig. 7.4.7). When the initial front is parallel to the x-axis, $\alpha = 0$, and initially $c = 0$ for $y \leq 0$, $c = c_o$ for $y > 0$, (7.4.85) reduces to

$$c(y, t) = \frac{c_o}{2} \text{erfc} \left[\frac{y}{(4\mathsf{D}_T t)^{1/2}} \right], \tag{7.4.86}$$

which is independent of x and D_L (and also of V!).

Verruijt (1971) studied the case of a steady state distribution in the semi-infinite plane $x > 0$. The concentration is fixed on the x-axis as

$$\begin{array}{lll} x = 0, & 0 < y < \infty, & c = 0, \\ x = 0, & -\infty < y < 0, & c = c_o. \end{array} \tag{7.4.87}$$

For sufficiently large values of x, he obtained

$$c(y, t) = \frac{c_o}{2} \text{erfc} \left[\frac{y}{2(\mathsf{a}_T x)^{1/2}} \right]. \tag{7.4.88}$$

For large values of x, (7.4.88) describes a normal distribution in the y-direction, with the width of the transition zone being proportional to $x^{1/2}$. Lines of constant concentration have the form of parabolas, at least for large x. It is of interest to note that from (7.4.88) it follows that $c(y, t)$ is independent of the velocity $V(= q/\phi)$.

I. Continuous injection at the origin into a uniform steady flow in an infinite plane

As in the case of continuous injection presented in Subs. 7.4.5F, the resulting distribution is obtained by integrating the elementary instantaneous injection solution of (7.4.82). The effect of an instantaneous slug of tracer of mass $dM = c_o Q \, dt$ is

$$dc(x,y,t) = \frac{dM}{2\pi[2\mathsf{D}_L t]^{1/2}[2\mathsf{D}_T t]^{1/2}} \exp\left[-\frac{(x-Vt)^2}{4\mathsf{D}_L t} - \frac{y^2}{4\mathsf{D}_T t}\right]. \qquad (7.4.89)$$

For a continuous injection, we obtain

$$\frac{c(x,y,t)}{c_o} = \frac{Q}{4\pi[\mathsf{D}_L\mathsf{D}_T]^{1/2}} \int_0^t \frac{1}{t-\tau} \exp\left[-\frac{1}{t-\tau}\left(\frac{x^2}{4\mathsf{D}_L} + \frac{y^2}{4\mathsf{D}_T}\right) + \frac{2xV}{4\mathsf{D}_L} - \frac{(t-\tau)V^2}{4\mathsf{D}_L}\right] d\tau. \qquad (7.4.90)$$

7.5 Immobile Water and Double Porosity Models

Up to this point, we have considered the transport of contaminants through the void space. The entire fluid phase was mobile and the solid matrix was impermeable to the fluid. However, under certain conditions, a portion of the fluid phase may be regarded as immobile or at least moving at a much slower rate as compared to that of the remaining fluid. In the unsaturated zone, capillarity may create pendular domains with practically immobile water. Another example is dead-end pores in saturated flow. In zones of very low permeability, e.g., clay lenses, the fluid is practically immobile. Yet another possibility is when the solid matrix itself is porous, but with very small pores, hence with very low permeability. In all these cases, the fluid in the low permeability domains is *practically immobile*, although the entire fluid occupied domain is a single fluid continuum at the microscopic level. In these instances, we may conceptualize the porous medium domain as occupied by two overlapping fluid continua—one is mobile (and the flow in it is governed by Darcy's law), and the other immobile. Exchanges of fluid mass as well as of chemical spices takes place between the two continua. Two such cases are discussed below.

7.5.1 Immobile water

In the discussions on the conceptual model of water and air distribution in the unsaturated zone and on water and air flow (Chap. 6), we have introduced the concept of *irreducible saturation*, S_{wr}, of a wetting phase as the saturation below which the phase is immobile. Similarly, the air phase (= nonwetting fluid) is assumed to be immobile below the residual saturation, S_{ar}. Flow of a fluid phase occurs when the saturation is above the irreducible or residual saturation. However, depending on the pore sizes and shapes, e.g., the presence of *dead-end* pores, a certain portion of the water subdomain may be discontinuous, *immobile*, or stagnant, or practically so, even at saturations above the residual value. For example, water in pores with narrow throats may be practically stagnant in the unsaturated zone in the soil. Similarly, water is immobile when present in the form of isolated pendular rings, even at relatively high saturations. The fraction of the void space that contains an *immobile wetting liquid* is not constant, but is rather some function of the saturation, reaching its maximum when $S_w = S_{wr}$. However, it does not necessarily vanish at full saturation, as even then, a wetting liquid phase in part of the void space may be *practically* immobile.

Similarly, in three-phase flow, parts of all three phases may be discontinuous or stagnant. Again, the immobile fraction of the saturation of a phase is not a constant; it depends on the saturation of the phase.

Immobile water may occur also in clay bodies imbedded in a sandy soil and in porous blocks (with fine pores) in a fractured rock formation. If the suction, or pressure, in the immobile portion equilibrates rapidly with that in the mobile portions, then the usual approach presented in Chap. 4 may be used for the flow equations. Otherwise, in the mathematical model, we regard the fluid phase to consist of *mobile and immobile portions.*

Some authors describe the exchange relationship between the mobile, $S_{\alpha m}$, and immobile, $S_{\alpha im}$, parts of the saturation, S_α, of an α-phase, in the form of a rate equation

$$\left.\frac{dS_{\alpha m}}{dt}\right|_{m\to im} = C_{m,im}(\Psi_{\alpha m} - \Psi_{\alpha im}), \tag{7.5.1}$$

where $C_{m,im}$ is a coefficient that may depend on the saturation. It has to be determined experimentally. The symbol Ψ_α denotes the capillary potential in the α-phase, defined by (6.1.11).

It is of interest to note that once we accept the notion (or *model*) that only part of a phase is actually in motion, and that this part is a function of the saturation (and is not identical to the 'irreducible saturation' of the phase, $S_{\alpha r}$), then all the coefficients that represent properties of the phase occupied portion of the void space, hitherto regarded as functions of the saturation, should actually be regarded as functions only of the mobile part of the saturation.

Thus, in the continuum approach, we may declare, as part of the conceptual model, that a liquid phase, say water, may be visualized as *two overlapping continua*: one is of mobile water, while the other is of immobile water (noting that in the thermodynamic sense, our system consists only of a single liquid phase). We shall refer to these two phases as *apparent phases.* Mass balance equations can then be written separately for the mobile and the immobile portions of each participating phase.

The immobile portion of a phase is, usually, in direct contact with the mobile portion, thus enabling the transfer of components from one to the other by the mechanism of molecular diffusion. Even when part of the immobile portion of a wetting fluid phase is in the form of isolated pendular rings, transfer of components between them may take place by molecular diffusion through the wetting fluid film that covers the solid surface in the larger pores, occupied by the nonwetting phase. Similarly, isolated air bubbles may contain vapor of a volatile component present in the liquid.

In what follows, we consider an example of transport of a contaminant component in the unsaturated zone in the soil. The water (= wetting fluid) is visualized as made up of two apparent fluid phases: mobile and immobile water. We denote the concentrations of the considered component in the two phases by c_m and c_{im}, respectively.

The conceptual model is comprised of the following assumptions:

[**A.1**] We regard both the mobile water and the immobile water as two continua, each filling up the *entire* domain. The water contents of these two continua, θ_{im} and θ_m, satisfy $\theta_{im} + \theta_m = \theta_w$.

[**A.2**] A component is transported in the mobile water by advection, dispersion, and diffusion, while in the immobile phase, it can be transported only by diffusion.

[**A.3**] Adsorption takes place on the *entire* surface of the solid matrix, as this surface is everywhere in contact with water. However, we have to refer separately to the contact with each of the two 'apparent phases'. We use p^* (< 1) and $1 - p^*$ to denote the areal fractions of the total area of the solid matrix, or the fractions of the total number of adsorption sites, that are in contact with the mobile and immobile (apparent) water phases, respectively. We assume that the same fractions represent also the corresponding fractions of the solid mass that interacts with the two water phases. Obviously, this last assumption may be a very poor one for certain solid matrix configurations.

[**A.4**] Equilibrium adsorption is described by the linear isotherms:

$$F_m = p^* K_d c_m, \qquad \text{and} \quad F_{im} = (1 - p^*) K_d c_{im}, \tag{7.5.2}$$

in which F_m and F_{im} denote the concentrations of the component on the two portions of the solid's surface.

[**A.5**] The component undergoes decay, with λ denoting the rate constant. The component can be exchanged (by diffusion) between the two water continua. The rate at which this exchange takes place, say, from the mobile ('apparent') phase to the immobile one (in terms of mass of component per unit volume of porous medium per unit time) is proportional to the difference between their respective concentrations. This exchange means that the two 'apparent phases' are not in chemical equilibrium. Using the symbol $\alpha^{\star\gamma}$ to denote the coefficient of proportionality, this rate can be expressed as $\alpha^{\star\gamma}(c_m - c_{im})$. The *transfer coefficient*, $\alpha^{\star\gamma}$, is proportional to the molecular diffusivity of the component in the water, to the total area of the surface of contact between the two fluid phases, and inversely proportional to some length characterizing the distance between the two phases within an REV. In principle, this coefficient need not be a constant. However, it is often approximated as such, rather than making it dependent on the saturations of the two phases. For a porous medium model comprised of spherical immobile water zones, Parker and Valocchi (1986) suggested

$$\alpha^{\star\gamma} = 15\,\mathcal{D}\theta_w \frac{1 - \theta_m/\theta_w}{r_{im}^2}, \tag{7.5.3}$$

where r_{im} is the radius of the spheres.

[**A.6**] The water density is constant, independent of changes in component concentration.

[**A.7**] The solid matrix is rigid and stationary.

For this conceptual model, the balance equation for a component in the mobile phase takes the form:

$$\frac{\partial \theta_m c_m}{\partial t} = -\nabla \cdot [c_m \mathbf{q} - \theta_m \mathbf{D}_{hm}(\theta_m) \cdot \nabla c_m] - f^{\gamma}_{m\to s} - f^{*}_{w,m\to im} c_m$$
$$+ f^{*}_{w,im\to m} c_{im} + \alpha^{\star\gamma}(c_{im} - c_m) - \theta_m \lambda c_m, \tag{7.5.4}$$

where $\mathbf{q}$ denotes the specific discharge of the mobile phase, and $f^{\gamma}_{m\to s}$ denotes the rate at which mass of the component leaves the mobile phase, to be adsorbed onto the solid (per unit volume of porous medium). Here, $f^{*}_{w,m\to im}$ denotes the (positive) rate of transfer of water phase volume from the mobile to the immobile water (per unit volume of porous medium), and similarly, $f^{*}_{w,im\to m}$ is the corresponding (positive) rate from the immobile to the mobile water. However,

$$f^{*}_{w,m\to im} \equiv \begin{cases} f_{w,m\to im}, & \text{if} \quad f_{w,m\to im} > 0, \\ 0, & \text{if} \quad f_{w,m\to im} \le 0, \end{cases} \tag{7.5.5}$$

$$f^{*}_{w,im\to m} \equiv \begin{cases} f_{w,im\to m}, & \text{if} \quad f_{w,im\to m} > 0, \\ 0, & \text{if} \quad f_{w,im\to m} \le 0, \end{cases} \tag{7.5.6}$$

where $f_{w,im\to m} = -f_{w,m\to im}$.

The mass balance equation for the component in the immobile phase is

$$\frac{\partial \theta_{im} c_{im}}{\partial t} = \nabla \cdot [\theta_{im} \boldsymbol{\mathcal{D}}^{*}_{im}(\theta_{im}) \cdot \nabla c_{im}] - f^{\gamma}_{im\to s} - f^{*}_{w,im\to m} c_{im}$$
$$+ f^{*}_{w,m\to im} c_m + \alpha^{\star\gamma}(c_m - c_{im}) - \theta_{im} \lambda c_{im}. \tag{7.5.7}$$

The mass balance equation for the component on the portion of the solid surface that is in contact with the mobile water phase, is

$$\frac{\partial \rho_b F_m}{\partial t} = f^{\gamma}_{m\to s} - \rho_b \lambda F_m. \tag{7.5.8}$$

For the portion of the solid surface that is in contact with the immobile water phase, we write

$$\frac{\partial \rho_b F_{im}}{\partial t} = f^{\gamma}_{im\to s} - \rho_b \lambda F_{im}. \tag{7.5.9}$$

In order to eliminate the terms that represent transfers due to adsorption, we sum up the balance equations for the component in the mobile phase and its adsorbed part on the solid, obtaining

$$\frac{\partial \theta_m R_{d\,m} c_m}{\partial t} = -\nabla \cdot [c_m \mathbf{q} - \theta_m \mathbf{D}_{hm}(\theta_m) \cdot \nabla c_m] - f^*_{w,m\to im} c_m + f^*_{w,im\to m} c_{im} + \alpha^{\star\gamma}(c_{im} - c_m) - \theta_m R_{d\,m} \lambda c_m, \qquad (7.5.10)$$

where

$$R_{d\,m} \equiv 1 + \frac{\rho_b p^* K_d}{\theta_m} \qquad (7.5.11)$$

is the *retardation factor* for the mobile water.

For the component in the immobile water, under the same conditions, we obtain

$$\frac{\partial \theta_{im} R_{d\,im} c_{im}}{\partial t} = \nabla \cdot [\theta_{im} \boldsymbol{\mathcal{D}}^*_{im}(\theta_{im}) \cdot \nabla c_{im}] + f^*_{w,m\to im} c_m - f^*_{w,im\to m} c_{im} + \alpha^{\star\gamma}(c_m - c_{im}) - \theta_{im} R_{d\,im} \lambda c_{im}, \qquad (7.5.12)$$

where

$$R_{d\,im} \equiv 1 + \frac{\rho_b (1 - p^*) K_d}{\theta_{im}}. \qquad (7.5.13)$$

If we neglect diffusion in the immobile phase, the first term on the right-hand side of (7.5.12) should be deleted.

To complete the contaminant transport model, we need to solve also the following flow model:

$$\frac{\partial \theta_m}{\partial t} = -\nabla \cdot \mathbf{q} + f_{w,im\to m}, \qquad \frac{\partial \theta_{im}}{\partial t} = f_{w,m\to im}, \qquad (7.5.14)$$

where Darcy's law,

$$\mathbf{q} = -\mathsf{K}_m(\theta_m) \nabla \left[\Psi_m(\theta_m) + z\right], \qquad (7.5.15)$$

is used for the flux of the mobile water. Thus, the two coupled equations in (7.5.14) can be solved to yield θ_m and θ_{im}. The exchange term, $f_{w,m\to im}$, in these equations may be computed by (7.5.1), as

$$f_{w,m\to im} = C_{m,im} \left[\Psi_m(\theta_m) - \Psi_{im}(\theta_{im})\right]. \qquad (7.5.16)$$

7.5.2 Double porosity medium

Finally, consider the saturated flow of water carrying a γ-component in a porous medium domain in which the solid matrix (not necessarily in the form of grains) or part of it, is porous (see, also, Subs. 7.3.6F). The total void space is made up of two parts: (1) the void space within the porous solid matrix, where individual pores are, usually, very small, and (2) the remainder of the void space, with much larger pores. Accordingly, the permeability to water flow inside the porous solid matrix is much smaller than that of the remaining void space. Clay aggregates may serve as an example of such a porous solid matrix. Such a domain is often referred to as a 'double porosity' porous medium.

A *double porosity porous medium* domain, say an aquifer, may also consist of a very low permeability (LK) portion in the form of isolated lenses that are embedded in a high permeability (HK) porous material. An example is an aquifer made of sand and gravel in which lenses of clay are embedded. Another example is an aquifer composed of an interconnected network of HK-channels embedded in a LK-material, such as a network of sand and gravel channels embedded in a clay or silty-clay matrix.

A double porosity problem domain can be regarded as a single inhomogeneous domain as long as its size is of the order of magnitude of the *scale of heterogeneity*, namely, the length of a low permeability lens, or the distance between lenses. However, if the problem domain is much larger than this scale of heterogeneity, such that it includes many such lenses, the averaging approach may be employed to 'homogenize' the heterogeneous domain. The transformation of such a heterogeneous domain into a single continuum, or into two overlapping continua: one of high permeability (HK), and the other of very low permeability (LK), by averaging over an RMV, is introduced in Subs. 7.3.6H.

The transport of a component in a double porosity domain can be described by a conceptual model that envisions mobile water in the larger pores, say between clay aggregates, or in the HK-material, and immobile water in the porous solid matrix, or in the LK-lenses. Obviously, following the concept of a continuum, the LK aggregates, or lenses, should be in sufficient number, and well distributed within the HK-domain, to justify regarding them as a continuum. The considered component may be exchanged *by molecular diffusion* between the two 'apparent water phases': the mobile water in the larger pores, or HK-subdomain, and the immobile water in the small pores, or LK one. In this way, the very low permeability continuum plays the role of distributed sinks for the considered component traveling in the high permeability continuum. Adsorption may take place on the relatively large surface area inside the LK-continuum. The component reaches these sites by diffusion through the void space inside the LK-continuum. If the LK-zone is made up of disjoint subdomains, as in the case of isolated aggregates, or lenses, there is no transport of the component in it, whether by advection, dispersion, or diffusion. In practice, these LK-zones, regarded as a continuum, play an important role in problems of contaminant transport. In the first stages of contamination, they are invaded by the contaminant. This is a relatively slow process, as the contaminant is transported by diffusion. The contaminant is stored in the LK-zones, mainly in the form of an adsorbate. At a later stage, when the concentration of the contaminant in the HK-material is reduced (say, by remediation activities), the lenses serve as a source for the contaminant for prolonged periods, releasing the contaminant stored in them into the water in the HK-zone.

The model that describes the transport of a contaminant in such a two-media domain is very similar to the one discusses for the mobile-immobile water case. With LK (l) and HK (h) denoting the low and high permeability continua, respectively, let us first summarize the conceptual model:

[**A.1**] The HK- and the LK-subdomains are considered as continua, each occupying the entire aquifer domain. Both subdomains are assumed to be interconnected, although one may consider the possibility that one of these subdomains is not interconnected.

[**A.2**] Both the LK- and the HK-materials are rigid, homogeneous, and isotropic.

[**A.3**] A single contaminant is present in the aquifer.

[**A.4**] In the HK-continuum, the contaminant may be transported by advection, dispersion and diffusion. However, we shall assume that transport by diffusion is negligible with respect to the other modes of transport.

[**A.5**] In the LK-continuum, transport by all three modes is negligible, as distances traveled in this continuum are much smaller than those in the HK-one. That is, there is no flux of contaminant on a length scale much larger than an REV so that no significant fluxes occur between any two macroscopic points in the continuum. However, we shall leave diffusion as the only mode of exchange between the two continua, across their common interface. Thus, diffusion can occur within the LK-continuum on the length scale of an REV, so that the contaminant can travel from within the LK-continuum to the LK-HK interface. This exchange will be expressed as a an averaged rate of transfer (per unit volume of porous medium) at every point within the considered porous medium domain.

[**A.6**] The considered contaminant may adsorb on the solid matrix in both subdomains. Equilibrium conditions prevail and a *linear isotherm* will be used. However, the specific surface area of the LK-HK interface is much smaller than that within the LK material, and may be neglected.

We now translate this conceptual model into a mathematical one. The core of the model consists of the following balance equations:

Mass balance for γ in the HK-subdomain:

$$\frac{\partial}{\partial t}\left[n_h\left(\phi_h+\rho_{bh}K_{dh}\right)c_h\right]=-\nabla\cdot\left[n_h\phi_h\left(c_h\mathbf{V}-\mathbf{D}_h\cdot\nabla c_h\right)\right]-f_{h\rightarrow l}. \quad (7.5.17)$$

Mass balance for γ in the LK-subdomain:

$$\frac{\partial}{\partial t}\left[n_l\left(\phi_l+\rho_{bl}K_{dl}\right)c_l\right]=-f_{l\rightarrow h}, \quad (7.5.18)$$

where

n_β Volumetric fraction of β-continuum within the aquifer, $\beta = h, l$.
ϕ_β Volumetric fraction of void space (= porosity) in the β-continuum.
$\rho_{b\beta}$ Bulk density of a solid matrix in the β-material ($= \rho_s(1-\phi_\beta)$).
c_β Component concentration in the water within the β-continuum.
$\mathbf{V}$ Velocity of water in the HK-continuum.
$f_{\beta\to\delta}$ Rate of transfer of γ from the β-continuum to the δ-one, across their common interface per unit volume of total aquifer.
$K_{d\beta}$ Partitioning coefficient in the β-continuum.
$\mathbf{D}_h$ Coefficient of dispersion in the HK-material.

The volumetric fractions obey the relationship:

$$n_h + n_l = 1. \tag{7.5.19}$$

The two mass balance equations may also be rewritten in the form:

$$n_h \phi_h R_{dh} \frac{\partial c_h}{\partial t} = -\nabla \cdot n_h \phi_h \left(c_h \mathbf{V} - \mathbf{D}_h \cdot \nabla c_h \right) - f_{h\to l}, \tag{7.5.20}$$

and

$$n_\ell \phi_\ell R_{dl} \frac{\partial c_l}{\partial t} = -f_{l\to h}, \tag{7.5.21}$$

where

$$R_{d\beta} \equiv 1 + \frac{\rho_{ba} K_{d\beta}}{n_\beta \phi_\beta} \tag{7.5.22}$$

denotes the retardation factor for the β-continuum, $\beta = h, l$.

We have here two balance equations in the two variables c_a, $a = h, l$ (assuming that the velocity, $\mathbf{V}$, in the HK-material can be determined from an appropriate flow model). However, we need the value of the rate of exchange, $f_{h\to l}$, appearing in these equations.

We now focus on the rate of exchange of a component between the LK- and the HK-materials. Very often, in a double-continua transport model, as considered here, the rate of transfer is expressed as

$$f_{h\to l} = \alpha^* \left(c_h - c_l \right), \tag{7.5.23}$$

in which c_h and c_l denote the average values of the concentrations in the HK- and LK-domains, respectively, and α^* is a transfer coefficient. Inherent in the use of such an expression are the assumptions that the transfer coefficient, α^*, is a constant, and that its value can be evaluated, like any other coefficient, by model calibration. It can be estimated as the ratio $\Sigma_{hl}\mathcal{D}^*/\Delta_{hl}$, in which Σ_{hl} is the specific area of the HK-LK interface, $\mathcal{D}^*$ is the diffusivity in the LK-porous medium, and Δ_{hl} is a length that characterizes the distance between the HK-LK interface and the interior of the LK-material (visualized as a continuum).

7.6 Eulerian-Lagrangian Formulation

So far, each of the balance equations presented in this section has been written as an *Eulerian* formulation, i.e., from the point of view of what happens *at a fixed (macroscopic) point in space.* However, a balance equation may be written also as a *Lagrangian formulation*, viz., from the point of view of an observer that moves with the considered (in this case, contaminant) quantity, as it travels within the domain.

Let us rewrite (7.4.7) in the form:

$$\begin{aligned}\theta\frac{\mathrm{D}c}{\mathrm{D}t} &\equiv \theta\left(\frac{\partial c}{\partial t} + \mathbf{V}\cdot\nabla c\right)\\ &= \nabla\cdot(\theta\mathbf{D}_h\cdot\nabla c) - \frac{\partial(\rho_b F)}{\partial t} - \rho_b(k_s+\lambda)F - \theta(k_f+\lambda)c\\ &\quad + \sum_{(m)} R_{\text{ext}}^{(m)}(\mathbf{x}^{(m)},t)\delta(\mathbf{x}-\mathbf{x}^{(m)})\left[c_R^{(m)}(\mathbf{x}^{(m)},t) - c(\mathbf{x},t)\right], \qquad (7.6.1)\end{aligned}$$

where $\mathrm{D}c/\mathrm{D}t$ denotes the *material derivative* of c. The material derivative appearing on the left-hand side of (7.6.1) is a *Lagrangian expression.* It expresses the rate of change of a property, here c, of a fluid particle *of a fixed identity* as it travels at the fluid's velocity, $\mathbf{V}$. In the absence of dispersion, adsorption, decay, and sources, the right-hand side of (7.6.1) vanishes, reducing the balance equation to

$$\frac{\mathrm{D}c}{\mathrm{D}t} \equiv \frac{\partial c}{\partial t} + \mathbf{V}\cdot\nabla c = 0. \qquad (7.6.2)$$

This may be interpreted as a statement that, *under the assumptions specified above*, the concentration of a fluid particle does not change as that particle travels. This is the Lagrangian formulation of the mass balance equation for a considered component in the fluid.

Equation (7.6.1), in which the right-hand side does not vanish, is referred to as a *Eulerian-Lagrangian formulation* of the component's mass balance equation. The Eulerian-Lagrangian formulation, (7.6.1), serves as a basis for a number of numerical codes. The main advantage stems from the observation that we have eliminated the advective flux—a term that produces *numerical dispersion* when solving the model by numerical methods that are based on finite difference and finite element techniques. This difficulty prompted the development of numerical techniques that are based on the Eulerian-Lagrangian formulation.

In the absence of adsorption, decay, and sources, (7.6.1) reduces to

$$\theta\frac{\mathrm{D}c}{\mathrm{D}t} = \nabla\cdot\theta\left(\mathbf{D}_h\cdot\nabla c\right). \qquad (7.6.3)$$

If we further assume that $|c\mathbf{V}| \gg |\mathbf{D}_h \cdot \nabla c|$, i.e., that advection dominates, the component's mass balance equation, (7.6.3), further reduces to (7.6.2), which serves as a basis for the 'advection only' models discussed in Subs. 7.8.1. The above condition for advection dominance can be expressed by a *Peclet number* (Sec. 7.7):

$$\text{Pe} \equiv \frac{L_c V_c}{\mathsf{D}_{hc}} \gg 1, \tag{7.6.4}$$

where L_c is a length characterizing changes in concentration within the domain, V_c is a characteristic velocity, and D_{hc} is a characteristic coefficient of hydrodynamic dispersion. Thus, the (macroscopic) Peclet number defined above represents the ratio between the advective and dispersive fluxes. At the megascopic level (the so called 'field level'), we should have used the coefficient of mechanical dispersion, D, rather than D_h, representing hydrodynamic dispersion, as diffusion is negligible at the megascopic level, except for very low velocities, so that $\mathsf{D}_h \approx \mathsf{D} = \mathsf{a}_L V$, and the condition becomes Pe $= L_c/\mathsf{a}_L \gg 1$.

If a relatively narrow zone of transition exists, across which concentration varies significantly, we may have also to define a Peclet number that compares the width of the transition zone with the transversal dispersivity. A detailed discussion on the Peclet number is presented in Sec. 7.7.

There should not be any difficulty to extend the above discussion to a case with adsorption, decay, and sources.

7.7 Evaluating Dominance of Effects

Before leaving the discussion on modeling contaminant transport, let us focus attention on the various terms that appear in the mass balance equation for a considered chemical species. Throughout this book, we have been emphasizing that each term in the balance equation expresses a contribution resulting from some *physical or chemical phenomenon*: rate of accumulation (per unit volume) due to advective flux, or due to dispersive flux, or to the total flux, or rate of disappearance due to some sink activity, such as chemical reaction, decay, or adsorption, etc. Obviously, not all these contributions are of the same *order of magnitude*. It may, therefore, be of interest, once a complete model has been developed, and before attempting to solve it, to analyze the order of magnitude of the various terms (which express corresponding phenomena) in order to investigate the possibility of deleting terms that express *non-dominant effects*, thus simplifying the remaining model, without affecting its solution significantly. The level of dominance of a term (representing a physical phenomenon) is determined by comparing it with other terms that appear in the same balance equation. Obviously, the magnitude of each term is to be estimated for the range of conditions that prevail in a considered problem.

We shall demonstrate the approach through two examples, following the methodology presented by Bear and Bachmat (1990, 1992). In all cases, it is assumed that the macroscopic model is a valid description of the considered phenomena.

Example 1: Consider the microscopic balance equation for a chemical species that undergoes chemical reactions,

$$\frac{\partial c}{\partial t} = -\frac{\partial}{\partial x_i}\left(cV_i - \mathcal{D}_{ij}\frac{\partial c}{\partial x_j}\right) + \left.\frac{dc}{dt}\right|_{\text{chem react}}. \tag{7.7.1}$$

The last term on the right-hand side expresses the rate (per unit fluid phase volume) at which the considered component is produced by the chemical reactions. For example, when the reactions are decay following a first order kinetics, in which the component disappears, we would write

$$\left.\frac{dc}{dt}\right|_{\text{chem react}} = -k_f c. \tag{7.7.2}$$

For radioactive decay, we usually use the same equation, with k_f replaced by the decay constant λ.

- **Step 1**: For every dependent variable, independent variable, and coefficient that appears in the considered equation, we introduce a *dimensionless variable* (denoted by an asterisk) that represents the ratio between the considered (dimensional) quantity and a corresponding *characteristic quantity* denoted by subscript c, of the same dimension. In the example considered here,

$$\frac{\partial c}{\partial t} = \left(\frac{\partial c}{\partial t}\right)^* \frac{(\Delta c)_c}{(\Delta t)_c} \equiv \frac{\partial c^*}{\partial t^*}\frac{(\Delta c)_c}{t_c}, \tag{7.7.3}$$

$$\frac{\partial cV_i}{\partial x_i} \equiv c\frac{\partial V_i}{\partial x_i} + V_i\frac{\partial c}{\partial x_i} = c^*\frac{\partial V_i^*}{\partial x_i^*}\frac{(\Delta V)_c c_c}{L_c^{(V)}} + V_i^*\frac{\partial c^*}{\partial x_i^*}\frac{V_c(\Delta c)_c}{L_c^{(c)}}, \tag{7.7.4}$$

$$\frac{\partial}{\partial x_i}\left(\mathcal{D}_{ij}\frac{\partial c}{\partial x_j}\right) = \frac{\partial}{\partial x_i^*}\left(\mathcal{D}_{ij}^*\frac{\partial c^*}{\partial x_j^*}\right)\frac{\mathcal{D}_c(\Delta c)_c}{L_c^{(c)2}}, \tag{7.7.5}$$

$$\lambda c = \lambda^* c^* \lambda_c c_c, \quad x_i = x_i^* L_c^{(\cdot)}, \tag{7.7.6}$$

where $t_c \equiv (\Delta t)_c$, and $L_c^{(c)}$ and $L_c^{(V)}$ are lengths characterizing spatial changes in c and $\mathbf{V}$, respectively, and $\mathcal{D}_c$ is a characteristic coefficient of dispersion. We may relate these lengths to the gradients in the respective quantities, e.g.,

$$\frac{(\Delta V)_c}{L_c^{(V)}} = \max\left|\frac{dV}{dx}\right|, \qquad \text{viz.,} \quad L_c^{(V)} = \frac{|V|_{\max}}{|dV/dx|_{\max}}, \tag{7.7.7}$$

or we can use some characteristic length of the domain. Usually, we assume a single common characteristic length, associated with all the domain's dimensions. We shall do so here, denoting it by L_c. This is not essential, as we could proceed with different characteristic lengths. A similar discussion applies to the characteristic time, which may take on different values, depending on the particular transported quantity and mode of transport. Here, t_c denotes the characteristic time for a change in concentration at a point in the domain. We note that the characteristic *rate* of a process is inversely proportional to its characteristic time.

Note that, at first, we made a distinction between c_c and $(\Delta c)_c$, but later, we made them equal. The same is true for t_c and $(\Delta t)_c$. We have a characteristic time for the total change in c, by all processes, but there may be a characteristic time for individual processes, and they need not be equal.

The maximum velocity within a considered domain may be taken as the characteristic velocity.

- **Step 2**: We insert these relationships into the considered balance equation. Here, with $t_c = (\Delta t)_c$, $L_c \equiv L_c^{(c)} \equiv L_c^{(V)}$, we obtain:

$$\frac{\partial c^*}{\partial t^*}\frac{(\Delta c)_c}{(\Delta t)_c} = -c^*\frac{\partial V_i^*}{\partial x_i^*}c_c\frac{(\Delta V)_c}{L_c} - V_i^*\frac{\partial c^*}{\partial x_i^*}\frac{V_c(\Delta c)_c}{L_c} + \frac{\partial}{\partial x_i^*}\left(\mathcal{D}_{ij}^*\frac{\partial c^*}{\partial x_j^*}\right)\frac{\mathcal{D}_c(\Delta c)_c}{L_c^2} - \lambda^* c^* \lambda_c c_c. \tag{7.7.8}$$

- **Step 3**: We rewrite the original balance equation in one of the three forms:

$$\mathrm{St}\,\frac{\partial c^*}{\partial t^*} = -\frac{\partial}{\partial x_i^*}\left(c^*V_i^* - \frac{1}{\mathrm{Pe}}\mathcal{D}_{ij}^*\frac{\partial c^*}{\partial x_j^*}\right) - \mathrm{Dm}^{I}\lambda^* c^*, \tag{7.7.9}$$

$$\frac{1}{\mathrm{Fo}}\frac{\partial c^*}{\partial t^*} = -\frac{\partial}{\partial x_i^*}\left(\mathrm{Pe}\,c^*V_i^* - \mathcal{D}_{ij}^*\frac{\partial c^*}{\partial x_j^*}\right) - \mathrm{Dm}^{II}\lambda^* c^*, \tag{7.7.10}$$

$$\frac{\mathrm{St}}{\mathrm{Pe}}\frac{\partial c^*}{\partial t^*} = -\frac{1}{\mathrm{Pe}}\frac{\partial}{\partial x_i^*}\left(c^*V_i^* - \frac{1}{\mathrm{Pe}}\mathcal{D}_{ij}^*\frac{\partial c^*}{\partial x_j^*}\right) - \mathrm{Dm}^{III}\lambda^* c^*, \tag{7.7.11}$$

in which, with c_c denoting $(\Delta c)_c$, V_c denoting $(\Delta V)_c$, and $t_c \equiv t_{c,\mathrm{accum}}$,

$$\mathrm{St} \equiv \frac{L_c}{V_c t_c} = \frac{L_c/V_c}{t_c} = \frac{t_{c,\mathrm{adv}}}{t_c} = \text{Strouhal number},$$
$$\mathrm{Pe} \equiv \frac{L_c V_c}{\mathcal{D}_c} = \frac{L_c^2/\mathcal{D}_c}{L_c/V_c} = \frac{t_{c,\mathrm{diff}}}{t_{\mathrm{adv}}} = \text{Peclet number},$$
$$\mathrm{Fo} \equiv \frac{t_c}{L_c^2/\mathcal{D}} = \frac{t_c}{t_{c,\mathrm{diff}}} = \frac{1}{\mathrm{St\ Pe}} = \text{Fourier number},$$
$$\mathrm{Dm}^{I} \equiv \frac{L_c/V_c}{1/\lambda} = \frac{t_{c,\mathrm{adv}}}{t_{c,\mathrm{react}}} = \text{1st kind Damköhler number},$$
$$\mathrm{Dm}^{II} \equiv \frac{L_c^2/\mathcal{D}}{1/\lambda} = \frac{t_{c,\mathrm{diff}}}{t_{c,\mathrm{react}}} = \mathrm{Pe\ Dm}^{I} = \text{2nd kind Damköhler number},$$

$$\mathrm{Dm}^{III} \equiv \frac{\mathrm{Dm}^{I}}{\mathrm{Pe}} = \frac{\mathrm{Dm}^{II}}{\mathrm{Pe}^2} = \frac{\lambda \mathcal{D}}{V_c^2}$$
$$= \frac{t_{c,\mathrm{adv}}}{t_{c,\mathrm{diff}}} \frac{t_{c,\mathrm{adv}}}{t_{c,\mathrm{react}}} = \text{3rd kind Damköhler number.} \quad (7.7.12)$$

Note that the characteristic times of advection, diffusion, and chemical reactions, are, respectively:

$$t_{c,\mathrm{adv}} \equiv \frac{L_c}{V_c}, \qquad t_{c,\mathrm{diff}} \equiv \frac{L_c^2}{\mathcal{D}_c}, \qquad t_{c,\mathrm{react}} \equiv \frac{1}{\lambda}. \quad (7.7.13)$$

Since, if the reference (or characteristic) values are properly selected (see discussion in Bear and Bachmat, 1990, 1992), the asterisk'ed terms are *of order one*, the dominance of a term is determined by *the magnitude of the dimensionless numbers that appear in that term.* We note that in (7.7.9), the rate of accumulation by advection (= inverse of characteristic time of advection) is used as a reference. In (7.7.10), the rate of accumulation by diffusion (= inverse of characteristic time of diffusion) is used as a reference time.

Lichtner (1993) pointed out that no matter how small the Damköhler number becomes, the system will always approach local equilibrium away from an advancing front provided time increases indefinitely, i.e. asymptotically.

To better understand these statements, and the interpretation of these numbers, let us take another step.

- **Step 4**: Let us take the ratio between any two terms in (7.7.1) which we wish to compare. For simplicity, consider a one-dimensional case. For example,

$$\frac{|\partial c/\partial t|}{|\partial cV/\partial x|} = \mathrm{St}\frac{\partial c^*/\partial t^*}{\partial c^* V^*/\partial x^*}. \quad (7.7.14)$$

Since every term with an asterisk is of order one, the Strouhal number, St, indicates *the ratio between two characteristic time intervals: that required for a significant change in concentration to spread out throughout the considered domain by advection, and that* ($t_c \equiv t_{c,\mathrm{accum}}$) *required for local changes in concentration to take place.*

Example 2: Let us compare advective and diffusive fluxes:

$$\frac{|cV|}{|\mathcal{D}\partial c/\partial x|} = \mathrm{Pe}\frac{c^* V^*}{\mathcal{D}^* \partial c^*/\partial x^*}. \quad (7.7.15)$$

Thus, the *Peclet number gives the ratio between the advective and diffusive fluxes.* For Pe $\gg 1$, advection dominates over diffusion. For Pe $\ll 1$, diffusion dominates.

One can easily interpret the Peclet number also as a ratio between two time scales: one ($t_{c,\mathrm{diff}} = L_c^2/\mathcal{D}$) that is required for spreading by diffusion, the other ($t_{c,\text{adv}} = L_c/V_c$) for spreading by advection. When the former time scale is smaller than the latter, diffusion dominates over advection.

Example 3: Let us compare the rate of production of a source (here, a sink due to radioactive decay) with the rate of accumulation (or spreading out) by advection. This is expressed by the *Damköhler number* of the first kind defined above. Again, this number may be interpreted as the ratio between two characteristic times: that of advection, and that of production ($= 1/\lambda$). It is also possible to define another (second kind) Damköhler number, by replacing the time required for spreading by advection, by the time required for spreading by diffusion, or by comparing the source term in the balance equation ($|\lambda c|$), with that expressing accumulation by diffusion ($|\nabla \cdot \mathcal{D}\nabla c|$). We obtain

$$\text{Dm}^{II} = \frac{L_c^2/\mathcal{D}}{1/\lambda} = \text{Pe Dm}^{I}. \tag{7.7.16}$$

Note that Dm^{II} is independent of the characteristic velocity, and that Dm^{III} is not a ratio between two characteristic times. We refer to all three as 'Damköhler numbers', as they involve the characteristic time of reaction.

By examining whether a dimensionless number is much smaller or much larger than unity, we may learn the relative significance of various transport processes. Following are some examples:

- When $\text{Dm}^{I} \ll 1$, advection dominates over the source term. In other words, the time required for transport by advection is much smaller than that required for production or removal by the source. Conversely, when $\text{Dm}^{I} \gg 1$, the source term dominates over advection.
- When $\text{Dm}^{II} \ll 1$, the spreading of the contaminant by diffusion dominates over the source term. In other words, the time required for spreading by diffusion is much smaller than that required for production, or removal by the source.

From (7.7.9) through (7.7.11), it follows that:

- When $\text{Pe} \ll 1$, transport by diffusion dominates over that by advection. If also:
  - $\text{Dm}^{II} \gg 1$, the reaction is referred to as a *fast reaction.*
  - $\text{Dm}^{II} \ll 1$, the reaction is referred to as a *slow reaction.*
- When $\text{Pe} \gg 1$, transport by advection dominates over that by diffusion. If also:
  - $\text{Dm}^{I} \gg 1$, the reaction is referred to as a *fast reaction.*
  - $\text{Dm}^{I} \ll 1$, the reaction is referred to as a *slow reaction.*
- When $\text{Pe} \gg 1$, and also $\text{Dm}^{II} \gg 1$, we have a process dominated by advection and reaction. If also
  - $\text{Dm}^{I} \gg 1$, the reaction is referred to as a (relatively) *fast reaction.*

- $\mathrm{Dm}^I \ll 1$, the reaction is referred to as a (relatively) *slow reaction.*

- When $\mathrm{Pe} \gg 1$, and $\mathrm{Dm}^I \ll 1$, the situation implies $\mathrm{Dm}^{III} \ll 1$; the process is dominated by advection and reaction.
- When $\mathrm{Pe} \ll 1$, and $\mathrm{Dm}^I \ll 1$, the situation implies $\mathrm{Dm}^{III} \gg 1$; the process is dominated by diffusion and reaction.

Example 4: The above discussion, related to the microscopic level, could also be repeated for the *macroscopic* balance equation. For example, for the case of (equilibrium) adsorption described by a linear isotherm, with constant K_d, and for a constant porosity, this equation takes the form:

$$R_d \frac{\partial c}{\partial t} = -\nabla \cdot (c\mathbf{V} - \mathbf{D} \cdot \nabla c) - k_f c. \tag{7.7.17}$$

By applying the methodology presented above, we may define similar dimensionless numbers, this time at the macroscopic level, and use them to identify non-dominant effects. In this case, $\mathcal{D}$ will be replaced by $\mathbf{D}_h$ (or by $\mathbf{D}$ when $\mathsf{a}_L V \gg \mathcal{D}^*$, with $\mathcal{D}^*$ denoting the coefficient of molecular diffusion in a porous medium). In the presence of adsorption, $R_d \neq 1$, we replace $\mathbf{V}$ in all the definitions of the dimensionless numbers, by the *retarded velocity*, $\mathbf{V}/R_d$, recalling that $\mathbf{D}$ is also replaced by $\mathbf{D}/R_d$. In all dimensionless numbers, L_c denotes a characteristic length of the domain of interest. We also use the *retarded velocity*, V/R_d, in the definition of the Strouhal number, St, discussed below..

Once a balance equation has been written in a dimensionless form, the dimensionless numbers may be employed to determine the dominance of processes that are involved. For example, they may be used to determine whether a reaction is fast or slow, with respect to other transport processes, and, hence, whether the assumption of chemical equilibrium is permitted.

It may be of interest to note that the Reynolds number, Re, mentioned in the discussion on the range of validity of Darcy's law (Subs. 4.3.1), is obtained by the same methodology as the ratio between inertial forces and viscous ones in the macroscopic momentum balance equation.

Let us discuss two useful dimensionless numbers (Bear and Bachmat, 1990, 1992). One is the *Fourier number*, $\mathrm{Fo}^{(E)}$ (see also Sec. 7.7):

$$\mathrm{Fo}^{(E)} = \frac{|\nabla \cdot \mathbf{D}^{(E)} \cdot \nabla e|_c}{|\partial e/\partial t|_c} = \frac{t_c^{(E)}}{L_c^{(E)}/\mathsf{D}_c^{(E)}}, \tag{7.7.18}$$

which expresses the ratio between the time interval required for the introduction of changes in the density of an extensive quantity (E) into a system, say, in the vicinity of a point, and that required for these changes to spread throughout the system by the dispersion-diffusion process.

The *Strouhal number*, St, was presented earlier. Let us introduce the symbol $\mathrm{St}^{(E)}$ and define it as

$$\mathrm{St}^{(E)} = \frac{|\partial e/\partial t|_c}{|\nabla \cdot e\mathbf{V}|_c} = \frac{L_c^{(E)}/V_c}{t_c^{(E)}}. \tag{7.7.19}$$

It expresses the reciprocal of the ratio between the time during which significant changes in the density of an extensive quantity take place, and that required for these changes to spread out throughout the system by advection.

In the next section, we shall discuss the case in which the appropriate Peclet number is sufficiently large such that advection dominates. The mass balance for a γ-component, say, (7.4.15) (noting all the assumptions underlying it), takes the form:

7.8 Transport Without Dispersion

7.8.1 Transport by advection only

So far in this chapter, we have been considering the movement of contaminants by the three modes of motion: advection, dispersion and diffusion, as discussed in Sec. 7.1. In particular, in Subs. 7.1.5, we were led to the conclusion that the dispersive flux must be taken into account as a mode of motion, in addition to advection and molecular diffusion, which also appear at the microscopic level of description. In spite of this conclusion, in many cases of practical interest, sometimes (but certainly not always) as a first approximation, we may neglect both dispersion and molecular diffusion as fluxes that are much smaller than the advective one, i.e.,

$$|c\mathbf{V}| \gg |\mathbf{D}_h \cdot \nabla c|, \tag{7.8.1}$$

or, in indicial notation:

$$|cV_i| \gg \left| \mathsf{a}_{ijk\ell} \frac{V_k V_\ell}{V} \frac{\partial c}{\partial x_j} + \mathcal{D}^*_{ij} \frac{\partial c}{\partial x_j} \right|, \tag{7.8.2}$$

in which we recall that $\mathcal{D}^*_{ij}$ denotes the coefficient of molecular diffusion in a porous medium.

According to the methodology presented in Sec. 7.7, we compare terms that appear in a model, i.e., in the balance equation of an extensive quantity, and delete terms that are much smaller than other ones. As exemplified in (7.8.1), we compare fluxes, or other terms, recalling that each term in a balance equation represents an added quantity per unit volume and unit time, due to some process.

Let us apply this methodology to the case considered here. In order to compare the terms appearing in (7.8.2), we introduce the following dimensionless quantities, indicated by an asterisk (*) (except in $\mathcal{D}^*_{ij}$, where the asterisk symbol is already used for a dimensional quantity):

$$(\mathcal{D}^*_{ij})^* = \frac{\mathcal{D}^*_{ij}}{\mathcal{D}^*_c}, \qquad \mathsf{a}^*_{ijk\ell} = \frac{\mathsf{a}_{ijk\ell}}{\mathsf{a}_c}, \qquad c^* = \frac{(\Delta c)}{(\Delta c)_c},$$
$$\left(\frac{\partial c}{\partial x_j}\right)^* = \frac{\partial c/\partial x_j}{|\partial c/\partial x_j|_c} = \frac{\partial c/\partial x_j}{(\Delta c)_c/L^{(c)}_c}, \qquad V^*_i = \frac{V_i}{V_c}, \tag{7.8.3}$$

where the subscript c denotes characteristic values, and the superscript identifies the quantity associated with the symbol. For example, $L^{(c)}_c$ denotes a characteristic length for the spatial variation of c, i.e., a length over which c undergoes a characteristic change, Δc. The characteristic values are selected such that:

$$V_c = \left|V_i\right|_{\max}, \quad \left|\frac{\partial c}{\partial x_j}\right|_c = \left|\frac{\partial c}{\partial x_j}\right|_{\max}, \quad \mathcal{D}^*_c = \text{Max}|\mathcal{D}^*_{ij}|, \quad \text{for all} \quad i,j, \tag{7.8.4}$$

$$(\Delta c)_c = (\Delta c)\big|_{\max}, \quad L^{(c)}_c = \frac{(\Delta c)\big|_{\max}}{\left|\partial c/\partial x_j\right|_{\max}}, \tag{7.8.5}$$

and a_c is the largest value of the dispersivity components. For example, in an isotropic porous medium, we may select $\mathsf{a}_c = \mathsf{a}_L$.

With these dimensionless quantities, let us start by comparing the diffusive and dispersive flux, i.e.,

$$\left|\mathsf{D}_{ij}\frac{\partial c}{\partial x_j}\right| \equiv \left|\mathsf{a}_{ijk\ell}\frac{V_k V_\ell}{V}\frac{\partial c}{\partial x_j}\right| \gg \left|\mathcal{D}^*_{ij}\frac{\partial c}{\partial x_j}\right|. \tag{7.8.6}$$

Inserting the dimensionless quantities, we obtain

$$\frac{\mathsf{D}_c}{L^{(c)}_c}\left|\mathsf{D}^*_{ij}\left(\frac{\partial c}{\partial x_j}\right)^*\right| \equiv \frac{\mathsf{a}_c V_c}{L^{(c)}_c}\left|\mathsf{a}^*_{ijk\ell}\frac{V^*_k V^*_\ell}{V^*}\left(\frac{\partial c}{\partial x_j}\right)^*\right| \gg \frac{\mathcal{D}^*_c}{L^{(c)}_c}\left|\mathcal{D}^*_{ij}\left(\frac{\partial c}{\partial x_j}\right)^*\right|. \tag{7.8.7}$$

Since the criteria for selecting the characteristic values are that all dimensionless quantities appearing in the comparison of terms should be of the order of magnitude of unity, the above condition reduces to

$$\text{Pe}^{(\mathsf{a},\mathcal{D}^*)} \equiv \frac{\mathsf{a}_c V_c}{\mathcal{D}^*_c} \gg 1, \tag{7.8.8}$$

where $\text{Pe}^{(\mathsf{a},\mathcal{D}^*)}$ is a Peclet number associated with the diffusive and dispersive fluxes. Thus, when this Peclet number is much larger than one, these fluxes may be neglected.

Employing the same procedure to compare the advective flux with the dispersive one, we obtain:

$$|cV_i| \gg \left|\mathsf{D}_{ij}\frac{\partial c}{\partial x_j}\right|, \qquad c_c V_c|c^* V^*_i| \gg \mathsf{D}_c\left(\frac{\partial c}{\partial x_j}\right)_c\left|\mathsf{D}^*_{ij}\left(\frac{\partial c}{\partial x_j}\right)^*\right|, \tag{7.8.9}$$

$$\frac{c_c}{(\partial c/\partial x_j)_c} = \frac{c|_{\max}}{(\partial c/\partial x_j)|_{\max}} = L_c^{(c)}, \qquad Pe^{(L^{(c)},D)} \equiv \frac{V_c L_c^{(c)}}{\mathsf{D}_c} \gg 1. \quad (7.8.10)$$

This means that advective flux dominates whenever this Peclet number, which expresses the ratio between advective and dispersive fluxes, is much larger than unity. Note that one may define all kinds of Peclet numbers, depending on the selected length scale, and on the reference to the coefficients of diffusion, or dispersion.

We can now return to the case in which the appropriate Peclet number is sufficiently large such that advection dominates. The mass balance for a γ-component, say, (7.4.15) (noting all the assumptions underlying it), takes the form:

$$R_d \phi \frac{\partial c}{\partial t} = -\nabla \cdot \phi c \mathbf{V}. \quad (7.8.11)$$

There should be no difficulty in writing analogous equations for other cases, e.g., unsaturated flow, or when sources are present. We shall, therefore, continue the discussion by referring to (7.8.11).

When combined with the mass balance equation, which for the case of steady, incompressible, saturated flow in a nondeformable porous medium reduces to

$$\nabla \cdot \mathbf{V} = 0. \quad (7.8.12)$$

Equation (7.8.11) can be written in the form:

$$\frac{\mathrm{D}_R c}{\mathrm{D}t} = 0, \quad (7.8.13)$$

where $\mathrm{D}_R c/\mathrm{D}t \, (\equiv \partial c/\partial t + (V/R_d(\theta)) \cdot \nabla c)$ denotes the material derivative from the point of view of a particle of a contaminant that travels at the (retarded) velocity $\mathbf{V}/R_d(\theta)$. For saturated flow, we replace θ by ϕ. We may interpret (7.8.13) as stating that the concentration of a *fixed* particle of a contaminant, say, described by its concentration, c, remains unchanged as the particle is displaced at the retarded velocity, $\mathbf{V}/R_d$ (without adsorption, $R_d = 1$), with $\mathbf{V}$ denoting the fluid's velocity.

To get a better insight of the transport by advection only, let us rewrite (7.8.13) for the case $R_d = 1$ in a two dimensional field of flow in the xy-plane, in the form:

$$\frac{\mathrm{D}c}{\mathrm{D}t} \equiv \frac{\partial c}{\partial t} + V_x \frac{\partial c}{\partial x} + V_y \frac{\partial c}{\partial y} = 0. \quad (7.8.14)$$

This is a *linear hyperbolic partial differential equation* in the three independent variables, x, y, and t.

The solution of this equation, $c = c(x, y, t)$, can, at least in principle, be represented by lines of constant concentration of moving particles, referred to as *characteristic curves*. The variation of concentration of a *fixed* particle that moves along a characteristic curve vanishes, i.e.,

$$dc \equiv \frac{\partial c}{\partial t} dt + \frac{\partial c}{\partial x} dx + \frac{\partial c}{\partial y} dy = 0. \tag{7.8.15}$$

From the last two equations, we may conclude that the direction of the characteristic curve is defined by the relationship:

$$1 : V_x : V_y = dt : dx : dy.$$

Hence,

$$dx = V_x \, dt, \qquad dy = V_y dt. \tag{7.8.16}$$

This means that the direction of a characteristic curve at a point coincides with that of the velocity, or tangent to the *streamline* at that point. Thus, in the model of *advection only*, the concentration of a fluid particle remains constant as it moves along a streamline at the velocities it encounters along the latter. In a case with adsorption, the above discussion remains valid, with the velocity, $\mathbf{V}$ replaced by the retarded velocity, $\mathbf{V}/R_d$, and the material derivative replaced by $\mathrm{D}_R(..)/\mathrm{D}t$.

A particular velocity field of interest arises when pumping and injection wells are superimposed on the natural flow in an aquifer. We shall discuss such cases in the following subsection.

7.8.2 Velocity field

The velocity of a particle at a point at a certain instant can be obtained by first solving for the piezometric head, $h(\mathbf{x}, t)$, and then using Darcy's law and the porosity, or the volumetric fraction of the fluid, in order to determine the fluid's velocity.

Of special interest is the case of steady two-dimensional horizontal flow in a homogeneous aquifer (xy-plane) in which a number of pumping and injection wells are operating. Each well may have its own schedule of operation, $Q(t)$. We shall use the symbol Q to denote the rate of both pumping and injection, with a negative rate to indicate the latter. The flow field is assumed to be *quasi-steady*; that is, any change in Q is assumed to be sufficiently slow such that transient effects may be neglected. The velocity at time t of a *fluid* particle, p, located at point (x_p, y_p), which is at a distance $r_{p,m}$ from the mth well, located at point (x_m, y_m), is given by

$$V_p(x_p, y_p, t) = -\frac{Q_m(t)}{2\pi r_{p,m} \phi B}, \tag{7.8.17}$$

where $Q_m(t)$ denotes the well's rate of discharge, B denotes the aquifer's thickness, and the negative sign indicates that the velocity points toward the well for the case of pumping. The problem is a linear one and superposition is permitted. Hence, together, all the wells produce at the considered point, a velocity which is the *vector sum* of the velocities induced by the individual wells that are active at that instant. With $\mathbf{V}_o$ denoting the undisturbed velocity of flow in the aquifer, viz., as produced by the regional natural groundwater gradient (subject to boundary conditions, natural replenishment, etc.), not associated with the wells, the position $(x_p|_{t+\Delta t}, y_p|_{t+\Delta t})$ of the particle at time $t + \Delta t$, is given by:

$$
\begin{aligned}
x_p\big|_{t+\Delta t} &= x_p\big|_t + \left[V_{ox} - \sum_{(m)} \frac{Q_m(t)}{2\pi\phi B} \frac{x_p\big|_t - x_m}{r_{p,m}^2} \right] \Delta t, \\
y_p\big|_{t+\Delta t} &= y_p\big|_t + \left[V_{oy} - \sum_{(m)} \frac{Q_m(t)}{2\pi\phi B} \frac{y_p\big|_t - y_m}{r_{p,m}^2} \right] \Delta t, \\
r_{p,m} &= \sqrt{\left(x_p\big|_t - x_m\right)^2 + \left(y_p\big|_t - y_m\right)^2}.
\end{aligned} \tag{7.8.18}
$$

We recall that in the presence of adsorption, the movement of contaminant particles is determined by the retarded velocity, equal to the fluid's one divided by R_d.

A *front*, or *isochrone* (≡ equal time), is the *locus* of all points that at some earlier instant of time had a common property. For example, the property may be a specific concentration. Another example of interest is the cluster composed of all the particles that left an injection well at a specified time.

By applying (7.8.18) repeatedly to all points on a front (in practice, to a selected number of such points (= particles) spaced along the front), for a given time increment, and then repeating the procedure for successive time increments, we can follow the movement of a front, say an iso-concentration curve.

Because the velocity is infinite at a point source/sink, it is convenient to start the process of displacing particles introduced through an injection well at some finite small radius from the latter. Similarly, it is convenient to assume that any particle reaching some small distance from a pumping well is captured by the latter. Figure 7.8.1 shows the superposition of velocities and frontal movement.

By locating contaminant particles on the rim of some initial plume of contaminated water in an aquifer, and employing the technique outlined above, the movement of the plume can be followed. A case commonly encountered in practice, is when pumping is employed as part of a 'pump and treat' technique to remove contaminants from an aquifer, often injecting the treated water back into the aquifer. By tracing the advancing plume, the optimal location and discharge rates of pumping and injection wells can be determined, and cleanup time can be estimated.

The advantage of this approach, when applicable, is that one need not solve for the distribution of piezometric heads, $h = h(x, y, t)$, in the entire field at the end of every time interval in order to determine the velocity. Obviously, the effect of dispersion is not accounted for in this approach.

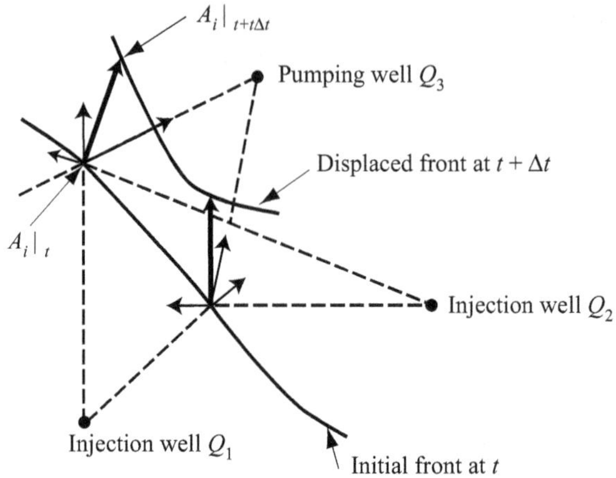

Figure 7.8.1: Superposition of velocities at an advancing front.

7.8.3 Travel time

From the above discussion, it follows that the incremental distance, Δs, traveled by a contaminant particle at the velocity V_s, during a time interval, Δt, can be determined from

$$\Delta s = V_s \, \Delta t. \tag{7.8.19}$$

In two-dimensional flow in the xy-plane, when the particle is at a point (x, y), the components in the x and y directions of its incremental displacement, Δs, are defined, respectively, by:

$$\Delta x = V_x(x, y, t)\Delta t, \qquad \Delta y = V_y(x, y, t)\Delta t. \tag{7.8.20}$$

We note that the velocity is taken here at the beginning of the time interval, Δt, and is assumed to remain unchanged during that interval. We may regard (7.8.20) as an integration of (7.8.16). Such an integration, equivalent to a *forward finite difference*, is called *Euler's method.*

If ds denotes a length increment along a streamline, the time required for a contaminant particle to be displaced a distance $s(t)$ along the streamline, in the presence of adsorption described by a linear isotherm, is given by

$$t = \int_0^{s(t)} \frac{ds}{V_s(s)/R_d} = -\frac{\phi R_d}{\mathsf{K}} \int_0^{s(t)} \frac{ds}{\partial h/\partial s}. \tag{7.8.21}$$

For the sake of simplicity, we may use Darcy's law for saturated flow in an isotropic porous medium to determine the component of the fluid's velocity, V_s, along the streamline.

When the piezometric head is derived numerically, we replace (7.8.21) by

$$t = \sum_{(i)} (\Delta t)_i = \frac{\phi R_d}{\mathsf{K}} \sum_{(i)} \frac{(\Delta s)_i^2}{(\Delta h)_i}, \qquad (7.8.22)$$

in which $(\Delta s)_i$ is the displacement along the streamline during $(\Delta t)_i$, and $(\Delta h)_i$ is the drop in piezometric head along $(\Delta s)_i$.

In unsteady flow, $h = h(\mathbf{x}, t)$ and the velocity (magnitude and direction) may vary in time. We cannot define streamlines, but we may obtain *flow lines*, which may be used to follow a particle as it is being displaced, recalculating its velocity as it varies in time and space.

Altogether, the movement of contaminant particles may be determined in two ways. We may be able to develop an analytic expression for the streamlines. For example, Bear (1979, p. 370) presents analytical expressions for the streamlines in the case of a single pumping or injection well in uniform flow, and for a 'doublet' composed of a pair of pumping and injection wells in uniform flow, with an angle between the direction of the uniform flow and that of the line connecting the two wells. Then, when the streamlines are known, we can follow the displacement of contaminant particles along them. When an analytic expression cannot be developed, which is always the case in unsteady flow, we can obtain the displacement by moving the particle one incremental displacement at a time, making use of the relationship $\Delta s = V_s \Delta t$ for each such increment. Since the velocity, V_s, is determined at the beginning of Δt, at the current location of the particle, an error is introduced due to variations in velocity during Δt. The reason is that the displacement is along the tangent to the streamline, rather than along the streamline. The deviation is reduced by reducing Δt, and using higher order numerical integration methods, such as the *trapezoidal rule*, or *Runge-Kutta methods.* This approach serves as a basis for a large number of computer programs used for tracking particles as they move in a flow field (Delay *et al.*, 1994; USEPA, 2005; Tompson, 1993).

Bear and Jacobs (1965) and Bear (1979, p. 282) studied the shape of an advancing front separating a body of (say, tracer labeled) water injected into an aquifer through a well from the indigenous, uniformly flowing, water in the latter. They obtained their solution by tracking injected particles along streamlines originating at the well.

7.9 Multiple Components and Reactive Transport

Only a few examples, namely, a radioactive decay chain, and relatively simple cases of chemical reactions, both with adsorption, are included in this section. Precipitation/dissolution reactions are fundamentally different from adsorption reactions in that an isotherm approach does not exist (Lichtner and Carey, 2006).

7.9.1 Radionuclide decay chain

We consider the case of a single liquid phase (e.g., an aqueous phase) that occupies part of the void space at a liquid volumetric fraction, θ. Let a chain of species that undergo decay due to radioactive disintegration,

$$\mathrm{A}^1 \to \mathrm{A}^2 \to \mathrm{A}^3 \to \ldots \to \mathrm{A}^N, \tag{7.9.1}$$

be present as solutes in the liquid.

When determined with respect to a unit volume of the phase, the rate of decay is related to the current concentration. For example, assuming this rate to be linearly proportional to the current mass, we have for the concentration, c^γ, of a γ-species:

$$\begin{aligned} \left.\frac{dc^1}{dt}\right|_{\mathrm{rad}} &= -\lambda^1 c^1, \\ \left.\frac{dc^\gamma}{dt}\right|_{\mathrm{rad}} &= -\lambda^\gamma c^\gamma + \lambda^{\gamma-1} c^{\gamma-1}, \qquad \gamma = 2, 3, \ldots, N, \end{aligned} \tag{7.9.2}$$

where λ^γ is the decay rate constant. We notice that the rate of change of species $\gamma(= 2, 3, \ldots)$, is equal to the sum of its loss through decay (negative), which is proportional to the current species concentration c^γ, and its gain through production (positive), which is proportional to the concentration of the preceding species in the chain, $c^{\gamma-1}$. Given an initial set of species concentrations c_o^γ, the set of ordinary differential equations (7.9.2) can be solved (Bateman, 1910). It is obvious that the concentration of the first species is

$$c^1 = c_o^1 e^{-\lambda^1 t}, \tag{7.9.3}$$

and of species 2

$$c^2 = c_o^2 e^{-\lambda^2 t} + \frac{c_o^1 \lambda^1}{\lambda^2 - \lambda^1}\left(e^{-\lambda^1 t} - e^{-\lambda^2 t}\right). \tag{7.9.4}$$

Expressions for the subsequent species are lengthier and can be found in Sadana (1991, p. 74). In the above, c^1, $c^2, \ldots, c^N$ are each a function of space and time, and often $\lambda^N = 0$, i.e., a stable species.

Bateman (1910) provided a solution of first-order chain reactions in batch reactors. To derive analytical solutions for radionuclide transport with such first-order chain reactions, Sun *et al.* (1999a) developed a linear transform. Equation (7.9.1) can be written in matrix form as

$$\frac{d\mathbf{c}}{dt} = \mathbf{A}\,\mathbf{c} = \begin{bmatrix} -\lambda^1 & 0 & \cdots & 0 & 0 & \cdots & 0 & 0 \\ \lambda^1 & -\lambda^2 & \cdots & 0 & 0 & \cdots & 0 & 0 \\ \vdots & \vdots & \vdots & \vdots & \vdots & \vdots & \vdots & \vdots \\ 0 & 0 & \cdots & \lambda^{i-1} & -\lambda^i & \cdots & 0 & 0 \\ \vdots & \vdots & \vdots & \vdots & \vdots & \vdots & \vdots & \vdots \\ 0 & 0 & \cdots & 0 & 0 & \cdots & \lambda^{N-1} & -\lambda^N \end{bmatrix} \mathbf{c}. \tag{7.9.5}$$

The reaction matrix $\mathbf{A}$ can be diagonalized,

$$\mathbf{A} = \mathbf{S}\,\mathbf{\Lambda}\,\mathbf{S}^{-}, \tag{7.9.6}$$

where

$$\mathbf{\Lambda} = \mathrm{diag}(\mathbf{A}), \tag{7.9.7}$$

and $\mathbf{S}$ and $\mathbf{S}^{-}$ are transform matrices

$$S(i,j) = \begin{cases} (-1)^{i-j} \prod_{l=j}^{i-1} \frac{\lambda^l}{\lambda^j - \lambda^{l+1}}, & j < i \\ 1 & j = i \\ 0 & j > i \end{cases},$$

$$S^{-}(i,j) = \begin{cases} \prod_{l=j}^{i-1} \frac{\lambda^l}{\lambda^l - \lambda^i}, & j < i \\ 1 & j = i \\ 0 & j > i \end{cases}. \tag{7.9.8}$$

Multiplying by $\mathbf{S}^{-}$, (7.9.5) is transformed into the following set of *uncoupled* equations:

$$\frac{d\mathbf{a}}{dt} = \frac{d(\mathbf{S}^{-}\,\mathbf{c})}{dt} = \mathbf{\Lambda}\,\mathbf{S}^{-}\,\mathbf{c} = \begin{bmatrix} -\lambda^1 & 0 & \cdots & 0 & \cdots & 0 \\ 0 & -\lambda^2 & \cdots & 0 & \cdots & 0 \\ \vdots & \vdots & \vdots & \vdots & \vdots & \vdots \\ 0 & 0 & \cdots & -\lambda^i & \cdots & 0 \\ \vdots & \vdots & \vdots & \vdots & \vdots & \vdots \\ 0 & 0 & \cdots & 0 & \cdots & -\lambda^N \end{bmatrix} \mathbf{a}. \tag{7.9.9}$$

In the transformed domain, $\mathbf{a} = \mathbf{S}^{-}\,\mathbf{c}$. The ODE's are independent, and the solution is:

$$a^i = a_0^i e^{-\lambda^i t}. \tag{7.9.10}$$

Finally, the solution in the $\mathbf{c}$-domain is

$$\mathbf{c} = \mathbf{S}\,\mathbf{a}. \tag{7.9.11}$$

Sun *et al.* (1999) proved that Bateman equation (1910) is a special case of the transform for first-order chain reaction in a batch reactor. Sun *et al.*'s (1999) transform has been used recently for deriving analytical solutions of radionuclide transport with first-order chain reactions.

Let us assume that each of the radioactive species can also be adsorbed onto the solid matrix, that equilibrium conditions prevail, and that a linear isotherm is applicable, say (7.3.66), with $K_d^1, K_d^2, \ldots, K_d^N$ denoting the respective distribution coefficients. Obviously, the coefficient is equal to zero for a species that does not adsorb. The net increase in the mass of a species results from the reduction due to radioactive decay, simultaneously with an increase due to the decay of the species that precedes it in the chain. Accordingly, the source term for a γ-species in the liquid phase takes the form:

$$\rho\Gamma^\gamma = -\lambda^\gamma c^\gamma + \lambda^{\gamma-1}c^{\gamma-1}. \tag{7.9.12}$$

For the γth species on the solid, the source term becomes:

$$\Gamma_s^\gamma = -K_d^\lambda \lambda^\gamma c^\gamma + K_d^{\gamma-1}\lambda^{\gamma-1}c^{\gamma-1}, \tag{7.9.13}$$

with $\gamma = 1, 2, \ldots, N$, and $c^o = 0$. We have assumed the same rates of decay for a species in the liquid and on the solid. They may be different.

Inserting these expressions into the mass balance equation (7.4.8) for the γth species, leads to

$$\begin{aligned} \frac{\partial(\theta R_d^\gamma c^\gamma)}{\partial t} = &- \nabla \cdot (c^\gamma \mathbf{q} - \theta \mathbf{D}_h^\gamma \cdot \nabla c^\gamma) \\ &- \theta(\lambda^\gamma R_d^\gamma c^\gamma - \lambda^{\gamma-1} R_d^{\gamma-1} c^{\gamma-1}), \end{aligned} \tag{7.9.14}$$

where

$$R_d^\gamma \equiv 1 + \frac{\rho_b K_d^\gamma}{\theta} \tag{7.9.15}$$

is the *retardation factor* for the γth species. In the case of a single liquid that occupies the entire void space, we replace the moisture content, θ, by the porosity, ϕ.

Equations in (7.9.14), with $\gamma = 1, 2, \ldots, N$, represent N equations to be solved simultaneously for the N concentrations, c^γ.

7.9.2 Chemically reacting species

In a problem of subsurface contamination, we are interested in predicting the future concentrations of one or more specific species. However, these species may interact chemically with other species present in the considered phase or phases, such that we often have to consider *all* the interacting species, simultaneously. Hence, let us start by considering the case of interacting species within a phase that occupies the entire void space or part of it.

A. Rate of reaction

Let NS denote the number of γ-species, $\gamma = 1, 2, ..., \mathrm{NS}$, that participate in chemical reactions within a liquid phase. We shall use the symbols c^γ and $[c^\gamma]$ $(= c^\gamma/M^\gamma)$ to denote the mass and molar concentrations, respectively, of a γ-species, with M^γ denoting the molar mass of γ. Together, the NS chemical species participate in NR (<NS) *independent* chemical reactions. Each reaction is described by a *stoichiometric equation*, which can symbolically be represented in the form of (7.3.17). Usually, an individual γ-species participates only in NJ^γ (<NR) of these reactions. However, if we use $\nu_j^\gamma = 0$ for the *stoichiometric coefficient* of a γ-species that does not participate in the jth reaction, we may write the NR stoichiometric equations in the form:

$$\sum_{(\gamma)} \nu_j^\gamma \mathcal{M}^\gamma = 0, \qquad \begin{cases} j = 1, 2, ..., \mathrm{NR}, \\ \gamma = 1, 2, ..., \mathrm{NS}, \end{cases} \qquad \begin{cases} \nu_j^\gamma > 0, \text{ for products,} \\ \nu_j^\gamma < 0, \text{ for reactants,} \\ \nu_j^\gamma = 0, \text{ for species that do} \\ \qquad\qquad \text{not participate,} \end{cases} \tag{7.9.16}$$

in which the $\mathcal{M}^\gamma$ and ν^γ denote the chemical symbol of a γ-species and its corresponding stoichiometric coefficient.

It may not always be obvious which reactions, within a given set of reactions, are really independent of each other. Certain mathematical techniques may have to be employed in order to determine the number of independent equations that describe these reactions. Here, we shall assume that we already know which are the NR independent reactions.

In general, the rate of a homogeneous reaction can be expressed by a *rate law* of the form:

$$R_{r,j} = f_j' \left([\gamma_1], [\gamma_2], \ldots, [\gamma_{NS}], p, T\right) \equiv f_j([c^\gamma], p, T), \tag{7.9.17}$$

i.e., a function of pressure, p, temperature, T, and the molar concentrations (or concentrations), of all the species that participate in that reaction.

B. Mass balance equation

We may start from the mass balance equation (7.2.20). When homogeneous chemical reactions are the only source of γ, the rate of production of γ per unit volume of porous medium, due to such reactions, is expressed as $\theta\rho\Gamma^\gamma$. We may write this rate in the form (Subs. 7.3.2):

$$\begin{aligned} \theta\rho\Gamma^\gamma &= \theta \sum_{(j)} \left.\frac{dc^\gamma}{dt}\right|_{j\text{th hom react}} = \theta \sum_{(j)} M^\gamma \left.\frac{d[c^\gamma]}{dt}\right|_{j\text{th hom react}} \\ &= \theta M^\gamma \sum_{(j)} \nu_j^\gamma R_{r,j} = M^\gamma \sum_{(j)} \nu_j^\gamma R_{\mathrm{pm},\, j}, \end{aligned} \tag{7.9.18}$$

where $R_{\mathrm{pm},\, j} (\equiv \theta R_{r,j})$ denotes the rate of the jth reaction, expressed in terms of the number of reacting moles per unit volume of porous medium.

Heterogeneous reactions give rise to the source term $f_{\alpha\to\beta}^\gamma$ in (7.2.20). It can be expressed as

$$f^{\gamma}_{\alpha\to\beta} = M^{\gamma}\sum_{(j)}\nu^{\gamma}_{j}R_{r,j}\Sigma_{\alpha\beta} = M^{\gamma}\sum_{(j)}\nu^{\gamma}_{j}R_{\text{pm},\,j}, \tag{7.9.19}$$

where $R_{\text{pm},\,j}(\equiv R_{r,j}\Sigma_{\alpha\beta})$ is the reaction rate per unit volume of porous medium, and $\Sigma_{\alpha\beta}$ is the surface area of the $\alpha\beta$-interface, per unit volume of porous medium. Thus, the sum of all source terms in the balance equation is given, symbolically, by:

$$\theta\rho\Gamma^{\gamma} + f^{\gamma}_{\beta\to\alpha} = M^{\gamma}\sum_{j=1}^{NR'}\nu^{\gamma}_{j}R_{\text{pm},\,j}, \tag{7.9.20}$$

where NR$'$ refers to both homogeneous and heterogeneous reactions. In this equation we have assumed that for the heterogeneous reactions, the flux of the γ-species is from the β- to the α-phase and the corresponding ν^{γ}_{j} coefficients are positive.

Let us introduce the *balance operator* symbol, $\mathcal{B}(..)$, defined as

$$\mathcal{B}(c) \equiv \frac{\partial}{\partial t}\theta c + \nabla\cdot\theta\left(c\mathbf{V} - \mathbf{D}_h\cdot\nabla c\right). \tag{7.9.21}$$

It expresses the net rate of accumulation of the γ-species, per unit volume of porous medium. When we consider a model that involves a number of fluid phases, we shall use the symbol $\mathcal{B}_{\alpha}(c^{\gamma}_{\alpha})$, in which $\mathbf{V}_{\alpha}$ is the velocity of the α-phase in the mass balance equation of the γ-species in the α-phase.

When adsorption is the sole heterogeneous reaction, we may use the symbol $\mathcal{B}_{\text{ads}}(c^{\gamma}_{\text{ads}})$, defined by

$$\mathcal{B}_{\text{ads}}(c^{\gamma}_{\text{ads}}) \equiv \frac{\partial c^{\gamma}_{\text{ads}}}{\partial t}\Sigma_{s\ell} \equiv \frac{\partial(\rho_b F^{\gamma})}{\partial t}, \qquad \rho_b F^{\gamma} \equiv c^{\gamma}_{\text{ads}}\Sigma_{s\ell}, \tag{7.9.22}$$

to express the mass of an adsorbed γ-species added by adsorption to the surface of the solid phase, per unit volume of porous medium. The symbol c^{γ}_{ads} denotes the concentration of the γ-adsorbate, in mass of γ per unit surface area of the solid, and $\Sigma_{s\ell}$ denotes the surface area of the solid, per unit volume of porous medium.

We note that from the linearity of the $\mathcal{B}$ operator, it follows that:

$$\mathcal{B}([c^{\gamma}]) = \frac{1}{M^{\gamma}}\mathcal{B}(c^{\gamma}), \tag{7.9.23}$$

where M^{γ} is the molecular mass of the species. Therefore, $\mathcal{B}([c^{\gamma}])$ is, formally, equal to the net rate of accumulation of the γ-species, in number of moles per unit volume of porous medium.

In general, because $\mathbf{D}_h$ depends on the coefficient of molecular diffusion, the operator $\mathcal{B}$ depends on the considered component. However, in what follows, we shall first assume that $\mathcal{B}$ is independent of γ. After that, we shall consider the more general case of a γ-dependent $\mathcal{B}$, denoting it as $\mathcal{B}^{\gamma}$.

For the NS interacting species that are present in the fluid phase, we have NS mass balance equations of the form of (7.2.20), with the source term given by (7.9.20). In terms of the $\mathcal{B}(c^\gamma)$, we can express the balance equations for all γ-species present in the pore space in the compact form:

$$\mathcal{B}(c^\gamma) = M^\gamma \sum_{j=1}^{NR'} \nu_j^\gamma R_{\mathrm{pm},\,j}, \quad \gamma = 1, 2, ..., \mathrm{NS}, \tag{7.9.24}$$

where $R_{\mathrm{pm},\,j}$ is defined in (7.9.18), and NR$'$ refers to the total number of homogeneous and heterogeneous reactions. Or, in terms of the molar concentration, in the form:

$$\mathcal{B}([c^\gamma]) = \sum_{j=1}^{NR'} \nu_j^\gamma R_{\mathrm{pm},\,j}, \quad \gamma = 1, 2, ..., \mathrm{NS}. \tag{7.9.25}$$

In each of the above sets of NS mass balance equations, we have NS molar concentration variables, say $[c^\gamma]$. If all reaction rates, $R_{\mathrm{pm},\,j}$, are assumed to be known functions of the concentrations, then we have NS equations in NS unknowns. However, often, some or all of the reaction rates are not known. On the other hand, because of the mass action law, not all species' concentrations are independent of each other, and some of the balance equations become redundant.

C. Primary and secondary species

Consider a set of (homogeneous) equilibrium reactions, written in the notation of (7.3.18) as

$$\sum_{(\gamma)} \nu_r^\gamma \mathcal{M}^\gamma \rightleftharpoons 0, \qquad r = 1, \ldots, \mathrm{NR}, \tag{7.9.26}$$

where $\mathcal{M}^\gamma$ ($\gamma = 1, \ldots, \mathrm{NS}$) denotes the species in the reactions, with NS equal to the number species and NR equal to the number of reactions. Note that if a species doesn't take part in a reaction, we just set $\nu_r^\gamma = 0$. We assume that the reactions are independent, i.e., no reaction can be written in terms of a combination of other reactions. This condition is equivalent to saying that the rows of the matrix of stoichiometric coefficients,

$$\begin{bmatrix} \nu_1^1 & \nu_1^2 & \nu_1^3 & \ldots & \nu_1^{NS} \\ \nu_2^1 & \nu_2^2 & \nu_2^3 & \ldots & \nu_2^{NS} \\ \vdots & \vdots & \vdots & \ddots & \vdots \\ \nu_{NR}^1 & \nu_{NR}^2 & \nu_{NR}^3 & \ldots & \nu_{NR}^{NS} \end{bmatrix}, \tag{7.9.27}$$

are, in the terminology of linear algebra, *linearly independent*. It is, then, always possible to rewrite the given set of reactions in the form:

$$\mathrm{Q}^r \rightleftharpoons \sum_{\gamma=1}^{NC} \lambda_r^\gamma \mathrm{P}^\gamma, \quad r = 1,\dots,\mathrm{NR}, \quad \gamma = 1,\dots,\mathrm{NC}, \quad \lambda_r^\gamma = -\sum_{r=1}^{NR} \nu_r^\gamma [\boldsymbol{\nu}_r]^{-1}, \tag{7.9.28}$$

where the P^γ ($\gamma = 1,\dots,\mathrm{NC}$)'s are the set of NC *primary, or basis, species*, also referred to as *components* (see below), the Q^r's ($r = 1,\dots,\mathrm{NR}$) are the set of *secondary species*, and λ_r^γ is another kind of stoichiometric coefficients (see the references given by Lichtner (1996)). Note that a distinct secondary species is associated with each reaction, while all other species in the reaction are primary. This equivalent representation of the set of reactions is called a *canonical form*. For a given set of reactions, this form is not necessarily unique. It is constructed by first identifying a (non-unique) set of NC = NS − NR primary species from which all other NR species can be expressed through the appropriate reactions.

The canonical form is convenient for cataloging and storing properties of reactions in a database, with the common ionic forms of each element often used as the primary species (e.g., the common ionic forms of iron are Fe^{2+} and Fe^{3+}). All species participating in a reaction can be written in terms of their primary forms; their properties can be stored together with the stoichiometric coefficients for the canonical reaction associated with each of the secondary species. A table of values for the equilibrium constant, K_r, of the reactions can also be stored for different temperatures.

An important advantage of the canonical form is that the law of mass action for each reaction takes the form:

$$\{\mathrm{Q}^r\} = \frac{1}{K_r} \prod_{\gamma=1}^{NC} \{\mathrm{P}^\gamma\}^{\lambda_r^\gamma}, \tag{7.9.29}$$

so that the activity (denoted as {..}) of each secondary species is expressed directly in terms of those of the primary species.

In Subs. 1.3.1, we defined *components* as the smallest set of species required to completely define the chemical composition of a phase under equilibrium conditions. Thus, the primary species may be considered as the components of such a system. Note that the set of components is not unique.

The chemical analysis of an aqueous solution is often reported in terms of the number of moles of the various elements in the system. It is possible to translate the number of moles of each element into the corresponding number of moles of each primary species. As illustrated in the following example, this translation can be easily done by associating a unique primary species with each element.

Example of the canonical form

Consider the set of reactions:

$$\mathrm{OH}^- + \mathrm{H}^+ \rightleftharpoons \mathrm{H_2O},$$

$$\begin{aligned} CO_3^{2-} + H^+ &\rightleftharpoons HCO_3^-, \\ CO_2(aq) + H_2O &\rightleftharpoons HCO_3^- + H^+, \\ HSiO_3^- + H^+ &\rightleftharpoons SiO_2(aq) + H_2O. \end{aligned} \tag{7.9.30}$$

Selecting H^+, H_2O, HCO_3^-, $SiO_2(aq)$ as the set of primary species, the canonical form of the reactions is

$$\begin{aligned} OH^- &\rightleftharpoons -H^+ + H_2O, \\ CO_3^{2-} &\rightleftharpoons -H^+ + HCO_3^-, \\ CO_2(aq) &\rightleftharpoons H^+ - H_2O + HCO_3^-, \\ HSiO_3^- &\rightleftharpoons -H^+ + H_2O + SiO_2(aq). \end{aligned} \tag{7.9.31}$$

Suppose an experimental analysis of a solution yields the number of moles of the elements: H, O, C, and Si. To convert to the amount of each primary species, we may associate each mole of H^+ with one mole of H, each mole of H_2O with one mole of O, each mole of HCO_3^- with a mole of C, and each mole of $SiO_2(aq)$ with one mole of Si.

An alternative canonical form is obtained by choosing OH^-, H_2O, CO_3^{2-}, $HSiO_3^-$ as the set of primary species. We then obtain:

$$\begin{aligned} H^+ &\rightleftharpoons -OH^- + H_2O, \\ HCO_3^- &\rightleftharpoons -OH^- + H_2O + CO_3^{2-}, \\ CO_2(aq) &\rightleftharpoons -2OH^- + H_2O + CO_3^{2-}, \\ SiO_2(aq) &\rightleftharpoons HSiO_3^- - OH^-. \end{aligned} \tag{7.9.32}$$

We may associate each mole of OH^- with a mole of O, each mole of H_2O with two moles of H, each mole of CO_3^{2-} with a mole of C, and each mole of $HSiO_3^-$ with a mole of Si.

D. Speciation

The procedure for determining the concentrations of all *species* in a given chemical system under equilibrium, given the total amount of all relevant *components* is called *speciation.* Here, we consider speciation within a single-phase solution. However, taking into account also heterogeneous (inter-phase) reactions, we may consider also speciation in a multiphase system.

Let n^γ_{total} denote the molar fraction of the γ component (= basis species) contained within all species in a solution (i.e., number of γ-moles per mole of solution). Then, the molar balance equations for the components are

$$n^\gamma_{\text{total}} = n^\gamma + \sum_{(r)} \lambda^\gamma_r n'^r, \quad \gamma = 1, \dots, \text{NC}, \tag{7.9.33}$$

where n^γ is the mole fraction of the γ-component, subscript r runs over all reactions, the n'^r's are the mole fractions of the secondary species associated with the rth reaction, and the λ_r^γ's are the stoichiometric coefficients for the γ-components in the rth reaction. Note that this yields a system of NC equations, where NC is the number of components.

The mole fractions can be written in terms of molar concentrations by the relationships

$$n^\sigma = \hat{m}^\sigma / \sum_{(\delta)} \hat{m}^\delta, \tag{7.9.34}$$

where the sum is taken over all species. Therefore, (7.9.33) can be viewed as written in terms of molar concentrations, $\hat{m}^\delta$, thus constituting the set of unknown variables that needs to be determined.

We also need the system of NR equations given by (7.9.29). It is assumed that the activity of each species is a known function of the $\hat{m}^\delta$'s, using, for example, (7.3.44) and (7.3.45), so that (7.9.29) is a system of equations with $\hat{m}^\delta$ as the unknown variables. Therefore, (7.9.29) and (7.9.33) together form a system of NC + NR = NS equations in the NS unknowns, $\hat{m}^\delta$. This system is nonlinear, and, therefore, it must, usually, be solved by numerical means.

Because, in many cases, the concentrations of certain species can vary by many orders of magnitude, it is, usually, preferable to use the logarithm of concentrations as the unknown variables, in order to avoid an ill-conditioned system of equations (Wolery, 1983). In some cases, the set of primary species may have to be changed in order to obtain a suitable set.

PHREEQC (Appelo and Postma, 2005; Parkhurst and Appelo, 1999) is probably the most commonly used public-domain computer program designed to perform a wide variety of low temperature aqueous geochemical speciation calculations for natural waters of the kind described above. It is based on the ion-association aqueous model described above. It can: (1) perform speciation and saturation-index calculations in a batch-reactor. It can also solve a one-dimensional flow and transport model involving reversible reactions, which include aqueous, mineral, gas, solid-solution, surface-complexation, and ion-exchange equilibria, and irreversible reactions, such as specified mole transfers of reactants, kinetically controlled reactions, mixing of solutions, and temperature changes. In fact, PHREEQC can also perform *inverse modeling*, which finds sets of mineral and gas mole transfers that account for differences in composition between waters, within specified compositional uncertainty limits. PHREEQC version 2, can also perform: kinetically controlled reactions, solid-solution equilibria, fixed-volume gas-phase equilibria, variation of the number of exchange or surface sites in proportion to a mineral or kinetic reactant, diffusion or dispersion in 1-D transport, 1-D transport coupled with diffusion into stagnant zones, and isotope mole balance in inverse modeling.

E. Equilibrium reactions

We start from the case in which all reactions are in equilibrium, i.e., NR = NR_{eq}. In Subs. 7.3.2, we saw that any set of independent equilibrium reactions can be transformed into the *canonical* form:

$$\mathrm{Q}^i \rightleftharpoons \sum_{j=1}^{NC} \nu_i^{\mathrm{P}^j} \mathrm{P}^j, \qquad i = 1, \ldots, \mathrm{NR}\,(= \mathrm{NR}_{\mathrm{eq}}), \tag{7.9.35}$$

in which P^j ($j = 1, \ldots, \mathrm{NC}$) is the set of *primary species* (or *components*), Q^i ($i = 1, \ldots, \mathrm{NR}_{\mathrm{eq}}$) is the set of *secondary species*, and $\nu_i^{\mathrm{P}^j}$ denotes the stoichiometric coefficient of the ith canonical reaction associated with the primary species P^j. Here, $\mathrm{NC} = \mathrm{NS} - \mathrm{NR}_{\mathrm{eq}}$ is the number of *components.* When the reactions are written in a canonical form, then the system of balance equations (7.9.25), becomes

$$\mathcal{B}([c^{\mathrm{Q}^i}]) = -R_{\mathrm{pm}\,i}, \qquad i = 1, 2, \ldots, \mathrm{NR}_{\mathrm{eq}}, \tag{7.9.36}$$

$$\mathcal{B}([c^{\mathrm{P}^j}]) = \sum_{i=1}^{NR_{\mathrm{eq}}} \nu_i^{\mathrm{P}^j} R_{\mathrm{pm}\,i}, \qquad j = 1, 2, \ldots, \mathrm{NC}. \tag{7.9.37}$$

Substituting the reaction rates, $R_{\mathrm{pm}\,i}$, from (7.9.36) into (7.9.37), gives

$$\mathcal{B}([c^{\mathrm{P}^j}]) = -\sum_{i=1}^{NR_{\mathrm{eq}}} \nu_i^{\mathrm{P}^j} \mathcal{B}([c^{\mathrm{Q}^i}]), \qquad j = 1, 2, \ldots, \mathrm{NC}. \tag{7.9.38}$$

From the linearity of the balance operator, we then obtain

$$\mathcal{B}([c^{*\mathrm{P}^j}]) = 0, \qquad j = 1, 2, \ldots, \mathrm{NC}, \tag{7.9.39}$$

where the total concentration (in units of molar concentration) of a primary species, P^j, is defined as

$$[c^{*\mathrm{P}^j}] \equiv [c^{\mathrm{P}^j}] + \sum_{i=1}^{NR_{\mathrm{eq}}} \nu_i^{\mathrm{P}^j} [c^{\mathrm{Q}^i}], \qquad j = 1, 2, \ldots, \mathrm{NC}. \tag{7.9.40}$$

Note that the total molar concentration $[c^{*\mathrm{P}^j}]$ can be positive or negative, depending on the sign of the stoichiometric coefficients $\nu_i^{\mathrm{P}^j}$, and the relative magnitudes of the primary and secondary species concentrations (Lichtner, 1985).

Any given set of reactions written in canonical form can lead to a unique decomposition of every species in terms of the primary species. For example, if CO_3^{2-}, and H^+ are primary species, then one mole of the secondary species H_2CO_3 contains one mole of CO_3^{2-}, and two moles of H^+. In this way, it is possible to consider the total number of moles of a primary species as it exists within all species present in a system. This number, per unit volume of the phase, is defined as the *total concentration of the primary species.* It is common to associate a *chemical element* with a primary species, where the *element* is a constituent of the primary species. If there are N moles of an element per mole of primary species, and the element does not occur in any other primary species, then the total concentration of that element is equal to N times the total concentration of the primary species.

The set of NC balance equations (7.9.39) must be solved in terms of the total concentrations, $[c^{*\mathrm{P}^j}]$. Note that the equations are decoupled, so that they may be solved individually. Boundary conditions and initial conditions involving the primary species must be expressed in terms of the total concentration, using the mass action law (7.9.29).

For a dilute solution, the mass action law is given by (7.3.78), which is rewritten here as:

$$[c^{\mathrm{Q}^i}] = \frac{1}{K'_{\mathrm{eq}\,i}} \prod_{j=1}^{NR_{\mathrm{eq}}} ([c^{\mathrm{P}^j}])^{\nu^{\mathrm{P}^j}}. \tag{7.9.41}$$

Then, substituting this expression into (7.9.40), gives:

$$[c^{*\mathrm{P}^j}] \equiv [c^{\mathrm{P}^j}] + \sum_{i=1}^{NR_{\mathrm{eq}}} \left(\frac{\nu_i^{\mathrm{P}^j}}{K'_{\mathrm{eq}\,i}} \prod_{j=1}^{NR_{\mathrm{eq}}} ([c^{\mathrm{P}^j}])^{\nu^{\mathrm{P}^j}} \right), \quad j = 1, 2, \ldots, \mathrm{NC}, \tag{7.9.42}$$

which is a function solely of the primary species concentrations. Thus, in this case, the total concentrations are easily computed from the primary species concentrations. However, in order to obtain the primary species concentrations from the total concentrations, a nonlinear equation must, in general, be solved numerically. Once the primary species are found, then (7.9.41) can be used to obtain the concentrations of the secondary species.

When the dilute solution assumption is not valid, the law of mass action to be used is (7.9.29), which is rewritten here as

$$\{\mathrm{Q}^i\} = \frac{1}{K_{\mathrm{eq}\,i}} \prod_{j=1}^{NR_{\mathrm{eq}}} \{\mathrm{P}^j\}^{\nu^{\mathrm{P}j}}, \tag{7.9.43}$$

where the $\{..\}$'s denote the activities of the species, each of which is usually some nonlinear function of the other species concentrations. Thus, to convert from primary concentrations to total concentrations, as required for the initial and boundary conditions, we need to first find the concentrations of the secondary species in terms the primary species by solving the nonlinear system of equations presented in (7.9.43), and then substitute these concentrations into (7.9.39).

F. Nonequilibrium reactions

We now consider the case where some of the reactions are *kinetic*, i.e, not in equilibrium (say, $\mathrm{NR_{eq}} < \mathrm{NR}$). Let $\mathrm{NR_{ne}}$ $(= \mathrm{NR} - \mathrm{NR_{eq}})$ represent the number of nonequilibrium reactions. As in the case of equilibrium, we select the primary and secondary species on the basis of the canonical form of the equilibrium reactions. As before, the stoichiometric coefficients are denoted as $\lambda_i^{\mathrm{P}^j}$. The nonequilibrium reactions do not have to be expressed in any special form; they include the usual stoichiometric coefficients, ν_k^γ, for the γ-species participating in the kth nonequilibrium reaction ($k = 1, 2, \ldots, \mathrm{NR_{ne}}$). However, it is often convenient to write them also in canonical forms. If there are no equilibrium reactions, then there are no secondary species, and all species are primary. The resulting balance equations are:

$$\mathcal{B}^{\mathrm{Q}^i}([c^{\mathrm{Q}^i}]) = -R^{\mathrm{eq}}_{\mathrm{pm}\,i} + \sum_{k=1}^{NR_{\mathrm{ne}}} \nu_k^{\mathrm{Q}^i} R^{\mathrm{ne}}_{\mathrm{pm}\,k}, \qquad i = 1, 2, \ldots, \mathrm{NR_{eq}}, \tag{7.9.44}$$

$$\mathcal{B}^{\mathrm{P}^j}([c^{\mathrm{P}^j}]) = \sum_{i=1}^{NR_{\mathrm{eq}}} \nu_i^{\mathrm{P}^j} R^{\mathrm{eq}}_{\mathrm{pm}\,i} + \sum_{k=1}^{NR_{\mathrm{ne}}} \nu_k^{\mathrm{P}^j} R^{\mathrm{ne}}_{\mathrm{pm}\,k}, \quad j = 1, 2, \ldots, \mathrm{NC}, \tag{7.9.45}$$

where the number of *components* is given by $\mathrm{NC} \equiv \mathrm{NS} - \mathrm{NR_{eq}}$. Note that now we have allowed the balance operator to be different for different γ-species, so that the coefficient $\mathbf{D}_h$ can depend on the relevant species, say, because of molecular diffusion. Also, some of the species may now be immobile on the solid phase, or they may have different advective velocities, such as in the case of colloids, or in the presence of ion exclusion phenomena. We have also made a distinction between the equilibrium reaction rates, $R^{\mathrm{eq}}_{\mathrm{pm}\,i}$, and the nonequilibrium reaction rates, R^{ne}_{ki}. We assume that the R^{ne}_{ki}'s are known functions of the species concentrations.

By substituting the equilibrium reaction rates appearing in (7.9.44) into (7.9.45), we obtain the system of balance equations:

$$\mathcal{B}^{\mathrm{P}^j}([c^{\mathrm{P}^j}]) + \sum_{i=1}^{NR_{\mathrm{eq}}} \nu_i^{\mathrm{P}^j} \mathcal{B}^{\mathrm{Q}^i}([c^{\mathrm{Q}^i}]) = \sum_{k=1}^{NR_{\mathrm{ne}}} a_k^{\mathrm{P}^j} R^{\mathrm{ne}}_{\mathrm{pm}\,k}, \quad j = 1, 2, \ldots, \mathrm{NC}, \tag{7.9.46}$$

where

$$a_k^{\mathrm{P}^j} \equiv \nu_k^{\mathrm{P}^j} + \sum_{i=1}^{NR_{\mathrm{eq}}} \lambda_i^{\mathrm{P}^j} \nu_k^{\mathrm{Q}^i}. \tag{7.9.47}$$

This set of NC partial differential equations, combined with the $\mathrm{NR_{eq}}$ algebraic equations (7.9.43), gives a total of NS equations in the NS unknown species concentrations, $[c^\gamma]$. This type of coupled equations is called a system of *algebraic-differential equations* (ADE) (although the term is most often used in the context of ordinary, not partial, differential equations).

We may express (7.9.46) by using the total concentrations as the unknowns, as long as the $\mathcal{B}^\gamma$-operators are independent of γ. However, this may not be advantageous whenever the nonequilibrium rates, R^{ne}_{ki}, are functions of the species concentrations, instead of the total concentrations. In some cases, it may be better to simply regard the species concentrations as the unknown variables. If the solution is dilute, the mass action law, (7.9.41), can be used to eliminate the secondary species concentrations, reducing (7.9.46) to a smaller set of NC balance equations, with NC primary species concentrations, $\hat{c}^{\mathrm{P}^j}$,

The system of equations described above can be quite nonlinear and the involved concentrations may vary over many orders of magnitude, causing the system to be numerically ill-conditioned. To overcome this difficulty, the logarithm of the concentrations are often used as the unknown variables instead of the concentrations themselves, similar to what is done in dealing with speciation (Subs. 7.9.2D). Also, whenever the concentration of a primary species becomes very small, the set of primary variables may have to be changed.

We emphasize that (7.9.46) may include equilibrium solid-fluid reactions through the corresponding mass action law (7.9.43), and nonequilibrium reactions through appropriate mass balance equations and reaction rates. We also recall that we have referred to the exchange of a species between two adjacent phases as a 'heterogenous reaction'; it may occur under equilibrium or nonequilibrium conditions. It is, thus, included in this analysis. In fact, the method of summing component balance equations to eliminate exchange terms, which we have presented earlier for heterogeneous reactions, is a special case of the more general procedure presented here. Let us demonstrate the above procedure through a few examples.

Example 1. Consider a case of saturated flow, with two chemical species, A and B (NS $=2$), which participate in a single fast chemical reversible reaction A $\rightleftharpoons$ B (NR $=1$). We assume that the system is continuously in equilibrium (NR$_{\mathrm{eq}}=1$). The reaction is already written in its canonical form:

$$\mathrm{A} \rightleftharpoons \mathrm{B}, \qquad \lambda^{\mathrm{B}}=1, \tag{7.9.48}$$

where A is a secondary species and B is the primary species. The two mass balance equations are:

$$\begin{aligned} \mathcal{B}([c^{\mathrm{A}}]) &= -R_{\mathrm{pm}}, \\ \mathcal{B}([c^{\mathrm{B}}]) &= \lambda^{\mathrm{B}} R_{\mathrm{pm}} = R_{\mathrm{pm}}, \quad \text{as} \quad \lambda^{\mathrm{B}}=1, \end{aligned} \tag{7.9.49}$$

leading to the balance equation

$$\mathcal{B}([c^{\mathrm{A}}]+[c^{\mathrm{B}}]) \equiv \mathcal{B}([c^{*\mathrm{B}}])=0. \tag{7.9.50}$$

The component with total concentration $[c^{*\mathrm{B}}]$ is a conservative one. No sources appear in its balance equation.

We now solve the last equation for $[c^{*\mathrm{B}}]=[c^{*\mathrm{B}}](\mathbf{x},t)$ within the considered domain, subject to appropriate initial and boundary conditions on $[c^{*\mathrm{B}}]$. Since we have assumed equilibrium conditions, we use the mass action law for a dilute solution,

$$K'_{\mathrm{eq}}=\frac{[c^{\mathrm{B}}]}{[c^{\mathrm{A}}]}, \tag{7.9.51}$$

to solve *algebraically* for $[c^{\mathrm{A}}]$ and $[c^{\mathrm{B}}]$:

$$[c^{\mathrm{A}}] = \frac{1}{1+K'_{\mathrm{eq}}}[c^{*\mathrm{B}}], \qquad [c^{\mathrm{B}}] = \frac{K'_{\mathrm{eq}}}{1+K'_{\mathrm{eq}}}[c^{*\mathrm{B}}]. \tag{7.9.52}$$

We note that in this problem, we have two chemical species, but only one chemical degree of freedom (for which a PDE has to be solved).

Example 2. We return to the first example of A $\rightleftharpoons$ B, but now the chemical reaction is slow, i.e., the system is under nonequilibrium conditions. For this case, $R_{\mathrm{pm}} = \theta f'([c^{\mathrm{A}}],[c^{\mathrm{B}}])$. For example, $f'([c^{\mathrm{A}}],[c^{\mathrm{B}}]) = k[c^{\mathrm{A}}][c^{\mathrm{B}}]$.

Again, with $\nu^{\mathrm{A}} = -1, \nu^{\mathrm{B}} = 1$, the two balance equations are:

$$\begin{aligned} \mathcal{B}([c^{\mathrm{A}}]) &= \nu^{\mathrm{A}} R_{\mathrm{pm}} = -\theta f'([c^{\mathrm{A}}],[c^{\mathrm{B}}]), \\ \mathcal{B}([c^{\mathrm{B}}]) &= \nu^{\mathrm{B}} R_{\mathrm{pm}} = \theta f'([c^{\mathrm{A}}],[c^{\mathrm{B}}]). \end{aligned} \tag{7.9.53}$$

By eliminating $f'([c^{\mathrm{A}}],[c^{\mathrm{B}}])$, we obtain the PDE:

$$\mathcal{B}\left([c^{\mathrm{A}}]+[c^{\mathrm{B}}]\right) \equiv \mathcal{B}([c^{*\mathrm{B}}]) = 0. \tag{7.9.54}$$

We solve this *homogeneous* PDE, subject to appropriate initial and boundary conditions. We obtain the solution in the form of $[c^*] = [c^{*\mathrm{B}}](\mathbf{x},t)$. We then have to solve another, this time *inhomogeneous* PDE,

$$\mathcal{B}\left([c^{\mathrm{B}}]\right) = \theta f'\left([c^{*\mathrm{B}}]-[c^{\mathrm{B}}],[c^{\mathrm{B}}]\right), \tag{7.9.55}$$

for $[c^{\mathrm{B}}]$. Finally, we compute $[c^{\mathrm{A}}](\mathbf{x},t)$ by using $[c^{\mathrm{A}}] = [c^{*\mathrm{B}}]-[c^{\mathrm{B}}]$.

In Example 1, the number of chemical degrees of freedom was one. Here it is two, since we cannot use the law of mass action. Therefore, we need to solve for two unknowns, $[c^{*\mathrm{B}}]$ and $[c^{\mathrm{B}}]$.

Example 3. As a second example of equilibrium reactions, we consider the case of the *carbonate* system. This system contains NS $=7$ chemical species:

$$H_2O,\ OH^-,\ H^+,\ H_2CO_3,\ HCO_3^-,\ CO_3^{2-},\ CO_2(aq),$$

which participate in the following $\mathrm{NR}_{\mathrm{eq}} = 4$ equilibrium chemical reactions:

$$\begin{array}{lll} H_2O & \rightleftharpoons H^+ & + OH^-, \\ H_2CO_3 & \rightleftharpoons HCO_3^- & + H^+, \\ HCO_3^- & \rightleftharpoons CO_3^{2-} & + H^+, \\ CO_2(aq) + H_2O & \rightleftharpoons H_2CO_3. & \end{array} \tag{7.9.56}$$

Thus, the number of components is NC $=$ NS - $\mathrm{NR}_{\mathrm{eq}} = 3$.

The carbonate system is important, because, in natural systems, it includes some of the important reactions affecting the pH ($\equiv -\log a^{\mathrm{H}^+}$), which has a major effect on other aqueous and mineral reactions. Note that in the above example, we have ignored carbonate complexes involving cations such as $MgCO_3$, $CaCO_3$, $CaHCO_3^+$, and $MgHCO_3^+$. Also note that $CO_2(aq)$ and H_2CO_3 are usually treated as equivalent species.

For the sake of illustration, let us select the species

$$\mathrm{P}^1{=}\mathrm{H}^+,\ \mathrm{P}^2{=}\mathrm{H_2CO_3},\ \mathrm{P}^3{=}\mathrm{H_2O},$$

as the primary species, and

$$\mathrm{Q}^1{=}\mathrm{OH}^-,\ \mathrm{Q}^2{=}\mathrm{HCO_3^-},\ \mathrm{Q}^3{=}\mathrm{CO_3^{2-}},\ \mathrm{Q}^4{=}\mathrm{CO_2(aq)},$$

as secondary species.

The corresponding reactions, written in canonical form, are given by:

$$\begin{aligned}
R_1:&\quad \mathrm{OH^-} &&\rightleftharpoons \mathrm{H_2O}\ \text{-}\ \mathrm{H^+},\\
R_2:&\quad \mathrm{HCO_3^-} &&\rightleftharpoons \mathrm{H_2CO_3}\ \text{-}\ \mathrm{H^+},\\
R_3:&\quad \mathrm{CO_3^{2-}} &&\rightleftharpoons \mathrm{HCO_3^-}\ \text{-}\ \mathrm{H^+} = \mathrm{H_2CO_3}\ \text{-}\ 2\ \mathrm{H^+},\\
R_4:&\quad \mathrm{CO_2(aq)} &&\rightleftharpoons \mathrm{H_2CO_3}\ \text{-}\ \mathrm{H_2O},
\end{aligned} \tag{7.9.57}$$

with the corresponding stoichiometric coefficients given by:

$$\left[\lambda_i^{\mathrm{Q}^j}\right] = \begin{bmatrix} -1 & 0 & 1\\ -1 & 1 & 0\\ -2 & 1 & 0\\ 0 & 1 & -1 \end{bmatrix}. \tag{7.9.58}$$

Note how in the above equation, every primary species reaction has coefficients that multiply every secondary species, Q_j; the coefficients λ^P of the primary species are all equal to 1 by the definition of the canonical formulation.

The species H_2CO_3 differs from the species CO_2 by a single water (H_2O) molecule and, therefore, these species are considered equivalent since they only differ by hydration and one of them can be omitted (usually H_2CO_3). In this context, it would be useful to point out that all species present in an aqueous solution are hydrated, i.e., surrounded by some number of loosely bound water molecules

The mass balance equations for the four secondary species are:

$$\mathcal{B}([c^{\mathrm{Q}^1}]) = \mathcal{B}([c^{\mathrm{OH}^-}]) = -R_{\mathrm{pm}\,1}, \tag{7.9.59}$$

$$\mathcal{B}([c^{\mathrm{Q}^2}]) = \mathcal{B}([c^{\mathrm{HCO}_3^-}]) = -R_{\mathrm{pm}\,2}, \tag{7.9.60}$$

$$\mathcal{B}([c^{\mathrm{Q}^3}]) = \mathcal{B}([c^{\mathrm{CO}_3^{2-}}]) = -R_{\mathrm{pm}\,3}, \tag{7.9.61}$$

$$\mathcal{B}([c^{\mathrm{Q}^4}]) = \mathcal{B}([c^{\mathrm{CO}_2(aq)}]) = -R_{\mathrm{pm}\,4}. \tag{7.9.62}$$

The mass balance equations for the three primary species are given by:

$$\begin{bmatrix} \mathcal{B}([c^{\mathrm{P}^1}])\\ \mathcal{B}([c^{\mathrm{P}^2}])\\ \mathcal{B}([c^{\mathrm{P}^3}]) \end{bmatrix} = \begin{bmatrix} -1 & -1 & -2 & 0\\ 0 & 1 & 1 & 1\\ 1 & 0 & 0 & -1 \end{bmatrix} \begin{bmatrix} R_{\mathrm{pm}\,1}\\ R_{\mathrm{pm}\,2}\\ R_{\mathrm{pm}\,3}\\ R_{\mathrm{pm}\,4} \end{bmatrix}, \tag{7.9.63}$$

where the middle matrix in the above equation is the transpose of the one in (7.9.58). Thus, rewriting (7.9.63) as individual equations, we obtain:

$$\mathcal{B}([c^{\mathrm{P}^1}]) \equiv \mathcal{B}([c^{\mathrm{H}^+}]) \quad = -R_{\mathrm{pm}\,1} - R_{\mathrm{pm}\,2} - 2R_{\mathrm{pm}\,3}, \tag{7.9.64}$$

$$\mathcal{B}([c^{\mathrm{P}^2}]) \equiv \mathcal{B}([c^{\mathrm{H_2CO_3}}]) = R_{\mathrm{pm}\,2} + R_{\mathrm{pm}\,3} + R_{\mathrm{pm}\,4}, \tag{7.9.65}$$

$$\mathcal{B}([c^{\mathrm{P}^3}]) \equiv \mathcal{B}([c^{\mathrm{H_2O}}]) \quad = R_{\mathrm{pm}\,1} - R_{\mathrm{pm}\,4}. \tag{7.9.66}$$

Substituting (7.9.59) through (7.9.62) into the right-hand sides of the above equations, and from the definition of total concentrations, we obtain:

$$\begin{aligned}
\mathcal{B}([c^{*\mathrm{H}^+}]) = 0, \quad & [c^{*\mathrm{H}^+}] \equiv [c^{\mathrm{H}^+}] - [c^{\mathrm{Q}^1}] - [c^{\mathrm{Q}^2}] - 2[c^{\mathrm{Q}^3}],\\
\mathcal{B}([c^{*\mathrm{H_2CO_3}}]) = 0, \quad & [c^{*\mathrm{H_2CO_3}}] \equiv [c^{\mathrm{H_2CO_3}}] + [c^{\mathrm{Q}^2}] + [c^{\mathrm{Q}^3}] + [c^{\mathrm{Q}^4}],\\
\mathcal{B}([c^{*\mathrm{H_2O}}]) = 0, \quad & [c^{*\mathrm{H_2O}}] \equiv [c^{\mathrm{H_2O}}] + [c^{\mathrm{Q}^1}] - [c^{\mathrm{Q}^4}].
\end{aligned} \tag{7.9.67}$$

There is a one-to-one correspondence between the number of moles of carbon and the number of moles of the primary species H_2CO_3. Thus,

$$[c^{*\mathrm{H_2CO_3}}] \equiv [c^{\mathrm{H_2CO_3}}] + [c^{\mathrm{HCO_3^-}}] + [c^{\mathrm{CO_3^{2-}}}] + [c^{\mathrm{CO_2(aq)}}] \tag{7.9.68}$$

is the total number of moles of carbon per unit volume of aqueous phase. The negative of the total concentration $[c^{\mathrm{H}^+}]$ is the *total alkalinity* of the system. It is defined as the equivalent amount of a base that is titratable with a strong acid.

The results of laboratory analyses of groundwater samples are often reported in terms of the total concentrations of chemical elements. Hence, in the model that represents the groundwater transport problem, the initial conditions will likely be presented also in terms of these units.

When the initial conditions are given in terms of the concentrations of the primary species, instead of total concentrations, we need to determine also the secondary species concentrations, in order to determine also the total concentrations. When the solution is dilute, we may use the following law of mass action for the considered reactions:

$$[c^{\mathrm{OH}^-}] = \frac{[c^{\mathrm{H_2O}}]}{[c^{\mathrm{H}^+}]K'_{\mathrm{eq1}}}, \qquad [c^{\mathrm{HCO_3^-}}] = \frac{[c^{\mathrm{H_2CO_3}}]}{[c^{\mathrm{H}^+}]K'_{\mathrm{eq2}}}, \tag{7.9.69}$$

$$[c^{\mathrm{CO_3^{2-}}}] = \frac{[c^{\mathrm{H_2CO_3}}]}{([c^{\mathrm{H}^+}])^2 K'_{\mathrm{eq3}}}, \quad [c^{\mathrm{CO_2(aq)}}] = \frac{[c^{\mathrm{H_2CO_3}}]}{[c^{\mathrm{H_2O}}]K'_{\mathrm{eq4}}}, \tag{7.9.70}$$

in which the K'_{eq}s are known equilibrium coefficients.

Often, the quantity of water (H_2O) involved in the reactions is negligible, so that the balance equation for water is not needed. Then, the number of components is reduced from three to two, NC=2, and the remaining two balance equations are:

$$\mathcal{B}([c^{*H^+}]) = 0, \quad \text{and} \quad \mathcal{B}([c^{*H_2CO_3}]) = 0. \tag{7.9.71}$$

Example 4. Here, we wish to consider an example with cation exchange (Subs. 7.3.4) and two phases: an aqueous solution and a solid (Kinzelbach, 1992). The example is the same as Example 3 above, except for the cation exchange. The species in the aqueous phase are mobile, while those on the solid are immobile. The species in solution participate in the same chemical reactions as in Example 3, where we have assumed that the system is continuously under conditions of equilibrium. In addition to the species in Example 3, we have the following cations in the aqueous solution:

$$Ca^{2+},\ Mg^{2+},\ Na^+,\ K^+,$$

and their counterparts adsorbed on the solid surface:

$$Ca^{2+}(s),\ Mg^{2+}(s),\ Na^+(s),\ K^+(s).$$

Altogether, we have NS = 15 chemical species.

The ion exchange reactions are:

$$\begin{aligned} Mg^{2+}(s) + Ca^{2+} &\rightleftharpoons Mg^{2+} + Ca^{2+}(s),\\ 2Na^+(s) + Ca^{2+} &\rightleftharpoons 2Na^+ + Ca^{2+}(s),\\ 2K^+(s) + Ca^{2+} &\rightleftharpoons 2K^+ + Ca^{2+}(s), \end{aligned} \tag{7.9.72}$$

The entire set of reactions, written in canonical form, is

$$\begin{aligned} R_1:&\quad OH^- \rightleftharpoons H_2O - H^+,\\ R_2:&\quad HCO_3^- \rightleftharpoons H_2CO_3 - H^+,\\ R_3:&\quad CO_3^{2-} \rightleftharpoons H_2CO_3 - 2\,H^+,\\ R_4:&\quad CO_2(aq) \rightleftharpoons H_2CO_3 - H_2O,\\ R_5:&\quad Mg^{2+}(s) \rightleftharpoons Mg^{2+} + Ca^{2+}(s) - Ca^{2+},\\ R_6:&\quad Na^+(s) \rightleftharpoons Na^+ + \tfrac{1}{2}Ca^{2+}(s) - \tfrac{1}{2}Ca^{2+},\\ R_7:&\quad K^+(s) \rightleftharpoons K^+ + \tfrac{1}{2}Ca^{2+}(s) - \tfrac{1}{2}Ca^{2+}, \end{aligned} \tag{7.9.73}$$

where we selected the following primary species:

$$P^1{=}H^+,\ P^2{=}H_2CO_3,\ P^3{=}H_2O,$$
$$P^4{=}Mg^{2+},\ P^5{=}Na^+,\ P^6{=}K^+.\ P^7{=}Ca^{2+},\ P^8{=}Ca^{2+}(s).$$

The secondary species are:

$$Q^1{=}OH^-,\ Q^2{=}HCO_3^-,\ Q^3{=}CO_3^{2-},\ Q^4{=}CO_2(aq),$$
$$Q^5{=}Mg^{2+}(s),\ Q^6{=}Na^+(s),\ Q^7{=}K^+(s).$$

The resulting stoichiometric matrix for the primary species, obtained from the canonical form of the reactions, is

$$\left[\lambda_i^{Q^j}\right] = \begin{bmatrix} -1 & 0 & 1 & 0 & 0 & 0 & 0 & 0 \\ -1 & 1 & 0 & 0 & 0 & 0 & 0 & 0 \\ -2 & 1 & 0 & 0 & 0 & 0 & 0 & 0 \\ 0 & 1 & -1 & 0 & 0 & 0 & 0 & 0 \\ 0 & 0 & 0 & 1 & 0 & 0 & -1 & 1 \\ 0 & 0 & 0 & 0 & 1 & 0 & -\frac{1}{2} & \frac{1}{2} \\ 0 & 0 & 0 & 0 & 0 & 1 & -\frac{1}{2} & \frac{1}{2} \end{bmatrix}. \tag{7.9.74}$$

The balance equations for the secondary species, which have not already been given in Example 3, are:

$$\begin{aligned} \mathcal{B}_{\text{ads}}([c^{Q^5}]) &\equiv \mathcal{B}_{\text{ads}}([c^{\text{Mg}^{2+}(s)}]) &= -R_{\text{pm}\,5} = f_{\ell\to s}^{\text{Mg}^{2+}}, \\ \mathcal{B}_{\text{ads}}([c^{Q^6}]) &\equiv \mathcal{B}_{\text{ads}}([c^{\text{Na}^{+}(s)}]) &= -R_{\text{pm}\,6} = f_{\ell\to s}^{\text{Na}^{+}}, \\ \mathcal{B}_{\text{ads}}([c^{Q^7}]) &\equiv \mathcal{B}_{\text{ads}}([c^{\text{K}^{+}(s)}]) &= -R_{\text{pm}\,7} = f_{\ell\to s}^{\text{K}^{+}}. \end{aligned} \tag{7.9.75}$$

In the above equations, the molar concentration, $[c]$, of an adsorbed cation species is defined as the number of moles attached to the solid per unit surface area of the solid; the mass balance equation operator is defined by (7.9.22).

By taking the transpose of the stoichiometric matrix, we obtain the balance equations for the primary species, in the form:

$$\begin{aligned} \mathcal{B}_\ell([c^{P^4}]) \equiv \quad & \mathcal{B}_\ell([c^{\text{Mg}^{2+}}]) = R_{\text{pm}\,5} = -f_{\ell\to s}^{\text{Mg}^{+2}}. \\ \mathcal{B}_\ell([c^{P^5}]) \equiv \quad & \mathcal{B}_\ell([c^{\text{Na}^{+}}]) = R_{\text{pm}\,6} = -f_{\ell\to s}^{\text{Na}^{+}}, \\ \mathcal{B}_\ell([c^{P^6}]) \equiv \quad & \mathcal{B}_\ell([c^{\text{K}^{+}}]) = R_{\text{pm}\,7} = -f_{\ell\to s}^{\text{K}^{+}}, \\ \mathcal{B}_\ell([c^{P^7}]) \equiv \quad & \mathcal{B}_\ell([c^{\text{Ca}^{2+}}]) = R_{\text{pm}\,5} + \tfrac{1}{2}R_{\text{pm}\,6} + \tfrac{1}{2}R_{\text{pm}\,7}, \\ & = -f_{\ell\to s}^{\text{Mg}^{2+}} - \tfrac{1}{2}f_{\ell\to s}^{\text{Na}^{+}} - \tfrac{1}{2}f_{\ell\to s}^{\text{K}^{+}}, \\ \mathcal{B}_{\text{ads}}([c^{P^8}]) \equiv \mathcal{B}_{\text{ads}}&([c^{\text{Ca}^{2+}(s)}]) = -R_{\text{pm}\,5} - \tfrac{1}{2}R_{\text{pm}\,6} - \tfrac{1}{2}R_{\text{pm}\,7}, \\ & = +f_{\ell\to s}^{\text{Mg}^{2+}} + \tfrac{1}{2}f_{\ell\to s}^{\text{Na}^{+}} + \tfrac{1}{2}f_{\ell\to s}^{\text{K}^{+}}. \end{aligned} \tag{7.9.76}$$

As before, we eliminate the balance equations for the secondary species to obtain the balance equations for the primary species:

$$\begin{aligned} \mathcal{B}_\ell([c^{\text{Mg}^{2+}}]) + \mathcal{B}_{\text{ads}}([c^{\text{Mg}^{2+}(s)}]) &= 0, \\ \mathcal{B}_\ell([c^{\text{Na}^{+}}]) + \mathcal{B}_{\text{ads}}([c^{\text{Na}^{+}(s)}]) &= 0, \\ \mathcal{B}_\ell([c^{\text{K}^{+}}]) + \mathcal{B}_{\text{ads}}([c^{\text{K}^{+}(s)}]) &= 0, \\ \mathcal{B}_\ell([c^{\text{Ca}^{2+}}]) + \mathcal{B}_{\text{ads}}([c^{\text{Ca}^{2+}(s)}]) &= 0, \\ \mathcal{B}_{\text{ads}}(c^{*\text{Ca}^{2+}(s)}) &= 0. \end{aligned} \tag{7.9.77}$$

In the above equations, the concentration

$$[c^{*\text{Ca}^{2+}(s)}] \equiv [c^{\text{Ca}^{2+}(s)}] + [c^{\text{Mg}^{+}(s)}] + \tfrac{1}{2}[c^{\text{Na}^{+}(s)}] + \tfrac{1}{2}[c^{\text{K}^{+}(s)}] \tag{7.9.78}$$

may be interpreted as the total amount of cations on the solid surface, in equivalent moles of a *doubly-charged cation* per unit volume of porous medium. This concentration is equal to twice the *cation exchange capacity* measured in equivalents, using a solution with a *singly-charged cation.*

We have four balance equations, in addition to the ones in Example 3, for a total of eight balance equations. There are eleven unknowns:

$$[c^{*\mathrm{H}^+}],\ [c^{*\mathrm{H_2CO_3}}],\ [c^{*\mathrm{H_2O}}],$$

$$[c^{\mathrm{Ca}^{2+}}],\ [c^{\mathrm{Mg}^{2+}}],\ [c^{\mathrm{Na}^+}],\ [c^{\mathrm{K}^+}],\ [c^{\mathrm{Ca}^{2+}(s)}],\ [c^{\mathrm{Mg}^{2+}(s)}],\ [c^{\mathrm{Na}^+(s)}],\ [c^{\mathrm{K}^+(s)}].$$

The last three unknown concentrations can be expressed in terms of the other ones, using the three equilibrium conditions for the cation exchange reactions:

$$[c^{\mathrm{Mg}^{+2}(s)}] = \frac{[c^{\mathrm{Mg}^{+2}}][c^{\mathrm{Ca}^{+2}(s)}]}{K'_{\mathrm{Ca/Mg}}[c^{\mathrm{Ca}^{+2}}]}, \qquad \left([c^{\mathrm{Na}^+(s)}]\right)^2 = \frac{\left([c^{\mathrm{Na}^+}]\right)^2[c^{\mathrm{Ca}^{+2}(s)}]}{K'_{\mathrm{Ca/Na}}[c^{\mathrm{Ca}^{+2}}]}, \tag{7.9.79}$$

$$\left([c^{\mathrm{K}^+(s)}]\right)^2 = \frac{\left([c^{\mathrm{K}^+}]\right)^2[c^{\mathrm{Ca}^{+2}(s)}]}{K'_{\mathrm{Ca/K}}[c^{\mathrm{Ca}^{+2}}]}, \tag{7.9.80}$$

where $K'_{\mathrm{Ca/Mg}}$, $K'_{\mathrm{Ca/Na}}$, and $K'_{\mathrm{Ca/K}}$ are known selectivity coefficients (see any chemistry textbook, e.g., Schwarzenbach *et al.* (2002), and Sparks (2003)) written here for molar concentrations.

By inserting these expressions into the balance equations, we obtain a set of eight equations in eight unknowns. The above equilibrium relationships assume a dilute solution. Otherwise, the concentrations need to be replaced by their respective activities, and a system of nonlinear equations for the mass action laws must be solved, together with the balance equations.

Example 5. This example is similar to Example 3, except that we introduce an additional species, CO_2(g) that can participate in a non-equilibrium heterogeneous reaction— its dissolution in the aqueous phase. We also have a mineral called *calcite*, $CaCO_3$(s), which can dissolve in the aqueous phase as a non-equilibrium reaction. Precipitation of the mineral is also possible. This chemical system is called the *calcium carbonate* system.

So far, we have considered the solid matrix as consisting of a single inert solid phase. In fact, for reacting minerals, the solid matrix may be regarded as comprised of several phases. In this example, the solid matrix consists partly of the calcite mineral (subscript *calcite*) and partly of a non-reactive solid (subscript *inert*). The volumetric fractions of these phases will be denoted by $\theta_{\mathrm{calcite}}$ and θ_{inert}, respectively. It is assumed that θ_{inert} is known. Note that the porosity in this model varies with time as it is related to $\theta_{\mathrm{calcite}}$ through:

$$1 - \phi = \theta_{\mathrm{calcite}} + \theta_{\mathrm{inert}}, \qquad \Delta\phi = -\Delta\theta_{\mathrm{calcite}}. \tag{7.9.81}$$

We select the following species as primary:

$$P^1{=}H^+,\ P^2{=}H_2CO_3,\ P^3{=}H_2O,\ P^4{=}CO_2(g),\ P^5{=}CaCO_3(s),$$

and:

$$Q^1{=}OH^-,\ Q^2{=}HCO_3^-,\ Q^3{=}CO_3^{2-},\ Q^4{=}CO_2(aq),$$

as secondary. The corresponding reactions, with the equilibrium reactions presented in canonical form, are:

$$\begin{array}{lrcl}
R_1^{eq}: & OH^- & \rightleftharpoons & H_2O - H^+, \\
R_2^{eq}: & HCO_3^- & \rightleftharpoons & H_2CO_3 - H^+, \\
R_3^{eq}: & CO_3^{2-} & \rightleftharpoons & H_2CO_3 - 2\,H^+, \\
R_4^{eq}: & CO_2(aq) & \rightleftharpoons & H_2CO_3 - H_2O, \\
R_1^{ne}: & CO_2(g) & \rightleftharpoons & CO_2(aq), \\
R_2^{ne}: & CaCO_3(s) & \rightleftharpoons & Ca^{2+} + CO_3^{2-}.
\end{array} \tag{7.9.82}$$

The non-equilibrium reactions do not have to be presented in canonical form, but they must be consistent with whatever rate law is used.

For non-equilibrium reaction rates, we use the rate law defined by (7.3.128):

$$R_{\mathrm{pm}\,1}^{ne} = \Sigma_{\ell,g}\alpha^{\star CO_2}\left\{[c^{CO_2(g)}] - [c^{CO_2(aq)}]/K_{eq}^{'CO_2}\right\}. \tag{7.9.83}$$

Using the rate law proposed by Steefel and Lasaga (1994), we obtain:

$$R_{\mathrm{pm}\,2}^{ne} = \Sigma_{\ell,\mathrm{calcite}}k_{\mathrm{calcite}}\left\{1 - \frac{[c^{Ca^{+2}}][c^{CO_3^{2-}}]}{K_{eq}^{'\mathrm{calcite}}}\right\}, \tag{7.9.84}$$

where $K_{eq}^{'CO_2}$ and $K_{eq}^{'\mathrm{calcite}}$ are the equilibrium coefficients for the reactions, $\alpha^{\star CO_2}$ is a transfer coefficient, and k_{calcite} is a rate constant. The specific interfacial areas, $\Sigma_{\ell,g}$ and $\Sigma_{\ell,\mathrm{calcite}}$, are functions of the fluid and solid volumetric fractions.

The balance equations for the secondary species are:

$$\begin{aligned}
\mathcal{B}_\ell([c^{Q^1}]) &\equiv \mathcal{B}_\ell([c^{OH^-}]) &&= -R_{\mathrm{pm}\,1}^{eq}, \\
\mathcal{B}_\ell([c^{Q^2}]) &\equiv \mathcal{B}_\ell([c^{HCO_3^-}]) &&= -R_{\mathrm{pm}\,2}^{eq}, \\
\mathcal{B}_\ell([c^{Q^3}]) &\equiv \mathcal{B}_\ell([c^{CO_3^{2-}}]) &&= -R_{\mathrm{pm}\,3}^{eq} + R_{\mathrm{pm}\,2}^{ne}, \\
\mathcal{B}_\ell([c^{Q^4}]) &\equiv \mathcal{B}_\ell([c^{CO_2(aq)}]) &&= -R_{\mathrm{pm}\,4}^{eq} + R_{\mathrm{pm}\,1}^{ne}.
\end{aligned} \tag{7.9.85}$$

The balance equations for the primary species are:

$$\begin{aligned}
\mathcal{B}_\ell([c^{P^1}]) &\equiv \mathcal{B}_\ell([c^{H^+}]) &&= -R_{\mathrm{pm}\,1}^{eq} - R_{\mathrm{pm}\,2}^{eq} - 2R_{\mathrm{pm}\,3}^{eq}, \\
\mathcal{B}_\ell([c^{P^2}]) &\equiv \mathcal{B}_\ell([c^{H_2CO_3}]) &&= R_{\mathrm{pm}\,2}^{eq} + R_{\mathrm{pm}\,3}^{eq} + R_{\mathrm{pm}\,4}^{eq}, \\
\mathcal{B}_\ell([c^{P^3}]) &\equiv \mathcal{B}_\ell([c^{H_2O}]) &&= R_{\mathrm{pm}\,1}^{eq} - R_{\mathrm{pm}\,4}^{eq}, \\
\mathcal{B}_g([c^{P^4}]) &\equiv \mathcal{B}_g([c^{CO_2(g)}]) &&= -R_{\mathrm{pm}\,1}^{ne}, \\
\mathcal{B}_{\mathrm{calcite}}([c^{P^5}]) &\equiv \mathcal{B}_{\mathrm{calcite}}([c^{CaCO_3(s)}]) &&= -R_{\mathrm{pm}\,2}^{ne}.
\end{aligned} \tag{7.9.86}$$

The molar concentration of the precipitating species (in this case, calcite) is defined as the number of moles per unit volume of porous medium, with the balance operator for calcite taking the form:

$$\mathcal{B}_{\text{calcite}}([c^{\text{CaCO}_3(s)}]) \equiv \frac{\partial [c^{\text{CaCO}_3(s)}]}{\partial t}. \tag{7.9.87}$$

We solve for the equilibrium reaction rates of the secondary species in the balance equations, and then substitute these rates into the balance equations of the primary species to obtain the following final set of balance equations for each component:

$$\begin{aligned}
\mathcal{B}_\ell([c^{*\text{P}^1}]) &\equiv \mathcal{B}_\ell([c^{*\text{H}^+}]) &&= -2R^{\text{ne}}_{\text{pm}\,2},\\
\mathcal{B}_\ell([c^{*\text{P}^2}]) &\equiv \mathcal{B}_\ell([c^{*\text{H}_2\text{CO}_3}]) &&= R^{\text{ne}}_{\text{pm}\,1} + R^{\text{ne}}_{\text{pm}\,2},\\
\mathcal{B}_\ell([c^{*\text{P}^3}]) &\equiv \mathcal{B}_\ell([c^{*\text{H}_2\text{O}}]) &&= -R^{\text{ne}}_{\text{pm}\,1},\\
\mathcal{B}_g([c^{\text{P}^4}]) &\equiv \mathcal{B}_g([c^{\text{CO}_2(g)}]) &&= -R^{\text{ne}}_{\text{pm}\,1},\\
\mathcal{B}_s([c^{\text{P}^5}]) &\equiv \mathcal{B}_s([c^{\text{CaCO}_3(s)}]) &&= -R^{\text{ne}}_{\text{pm}\,2}.
\end{aligned} \tag{7.9.88}$$

Altogether, we have here NS $=$ 9 species concentrations, NC $=$ 5 balance equations and $\text{NR}_{\text{eq}} = 4$ mass action laws, for the equilibrium reactions, i.e., a total of NC $+$ NR_{eq} equations. If the solution is dilute, then the mass action laws can be used to eliminate the secondary species concentrations from the balance equations, and we have NC $=$ 5 balance equations for the NC $=$ 5 primary variables. In some formulations, however, the *total concentration* is solved for as an independent variable. This has the advantage that the accumulation and flux terms in the mass balance equations are linear in the total concentrations, which is useful if no heterogeneous reactions are considered. Also fundamentally the mass balance equation is formulated in terms of total concentrations.

From (7.9.81), we obtain the change in porosity in the form $-\Delta\theta_{\text{calcite}}$ $(= -\Delta([c^{\text{CaCO}_3(s)}]M^{\text{CaCO}_3(s)}/\rho_{\text{calcite}}))$. When the changes in porosity, relative to the initial porosity, is significant, the flow field will be affected (by changes in void space, permeability and capillary pressure) and the flow field based on the unaltered porosity cannot be used in the transport equations; the transport and flow equations must be solved simultaneously as a single coupled set of equations.

7.9.3 Three multicomponent phases

We consider the flow of three multicomponent fluid phases under isothermal conditions that, together, occupy the entire void space. Each phase is made up of a number of components. Some components move from one phase to an adjacent one by such mechanisms as dissolution, volatilization, and condensation. Components may also adsorb to the solid surface, but, apart from these heterogeneous reactions, they do not interact with each other chemically

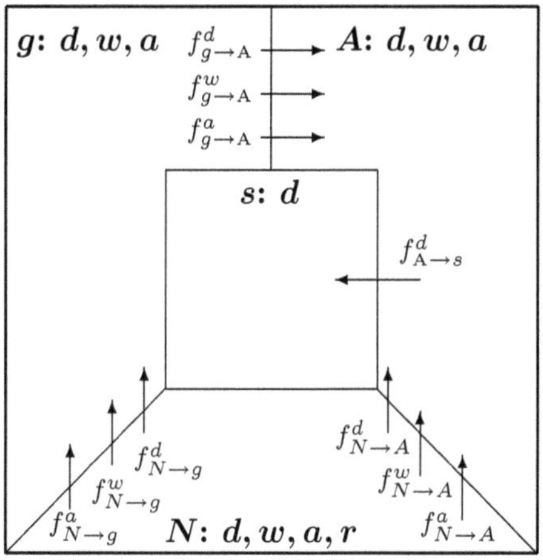

Figure 7.9.1: Schematic diagram for transport in three phase flow.

within a fluid phase. The term *compositional model*, originating in reservoir engineering, is often used for such a model.

The considered solid and fluid phases are:

- **A nonaqueous phase** (e.g., oil), N, containing a volatile component, d, that can dissolve in the aqueous phase and volatilize into the gaseous phase, dissolved water, w, and the rest of the phase, r, regarded as a non-volatile component. We shall assume that the d-component can evaporate from both the aqueous and the nonaqueous phases, to become a component in the gaseous one.
- **An aqueous phase** (e.g., water), A, that contains 'pure water', w, as a component, the d-component as a solute, and dissolved gas (air), a.
- **A gas** (e.g., air), g, composed of two components: 'dry air', a, the volatile d-component, and the vapor w.
- **A rigid stationary solid**, s, that does not dissolve, and on which the d-component can be adsorbed, but only from the A-phase.

The three phases, their components, and the interphase transfer rates, are shown, schematically, in Fig. 7.9.1. No other transfers will be considered.

To simplify the presentation, we make use of the *balance operator*, $\mathcal{B}$, defined by (7.9.21):

$$\mathcal{B}^\gamma_\alpha(\omega^\gamma_\alpha) \equiv \frac{\partial}{\partial t}(\theta_\alpha \omega^\gamma_\alpha \rho_\alpha) + \nabla \cdot \theta_\alpha \left[\omega^\gamma_\alpha \rho_\alpha \mathbf{V}_\alpha - \rho_\alpha \boldsymbol{\mathcal{D}}^{*\gamma}_\alpha \cdot \nabla \omega^\gamma_\alpha - \mathbf{D}_\alpha \cdot \nabla \omega^\gamma_\alpha \rho_\alpha\right]. \tag{7.9.89}$$

Note that the *mass fraction*, ω^γ_α, is used rather than the usual mass concentrations, c^γ_α. One reason is that the density function of a fluid phase is of the form $\rho_\alpha = \rho_\alpha(p, T, \omega^\gamma_\alpha)$. If the mass concentration is used, we are left with the circular result, $\rho_\alpha = \rho_\alpha(p, T, c_\alpha^\gamma/\rho_\alpha)$. The molecular diffusive flux term in the above definition may be replaced by (7.1.17) in order to account for the interaction of multiple components.

We shall assume that no external sources and sinks of phases and components, no decay and growth, and no chemical interactions exist. The only sources and sinks of components in a considered phase are due to adsorption, dissolution, and volatilization.

The mass balance equations for the components in the A-phase are:

$$\mathcal{B}_{\rm A}^w(\omega_{\rm A}^w) = f_{N\to{\rm A}}^w + f_{g\to{\rm A}}^w, \quad \mathcal{B}_{\rm A}^a(\omega_{\rm A}^a) = f_{N\to{\rm A}}^a + f_{g\to{\rm A}}^a, \tag{7.9.90}$$

$$\mathcal{B}_{\rm A}^d(\omega_{\rm A}^d) = f_{N\to{\rm A}}^d + f_{g\to{\rm A}}^d + f_{s\to{\rm A}}^d. \tag{7.9.91}$$

The mass balance equations for the components in the g-phase are:

$$\mathcal{B}_g^w(\omega_g^w) = f_{N\to g}^w - f_{g\to{\rm A}}^w, \quad \mathcal{B}_g^a(\omega_g^a) = f_{N\to g}^a - f_{g\to{\rm A}}^a, \tag{7.9.92}$$

$$\mathcal{B}_g^d(\omega_g^d) = f_{N\to g}^d - f_{g\to{\rm A}}^d, \tag{7.9.93}$$

and, for the N-phase, they are:

$$\mathcal{B}_N^w(\omega_N^w) = -f_{N\to{\rm A}}^w - f_{N\to g}^w, \quad \mathcal{B}_N^a(\omega_N^a) = -f_{N\to{\rm A}}^a - f_{N\to g}^a, \tag{7.9.94}$$

$$\mathcal{B}_N^d(\omega_N^d) = -f_{N\to{\rm A}}^d - f_{N\to g}^d. \tag{7.9.95}$$

The mass balance equation for the adsorbed d-component on the solid is:

$$\mathcal{B}_{\rm ads}^d(c_{\rm ads}^d) = -f_{s\to{\rm A}}^d, \qquad c_{\rm ads}^d \equiv (1-\phi)\rho_s F^d, \tag{7.9.96}$$

where the balance operator is defined by:

$$\mathcal{B}_{\rm ads}^d(c_{\rm ads}^d) \equiv \frac{\partial c_{\rm ads}^d}{\partial t}. \tag{7.9.97}$$

In order to eliminate the interphase transfers, we sum up the mass balance equations for the d-component in all phases, i.e.,

$$\mathcal{B}_{\rm A}^d(\omega_{\rm A}^d) + \mathcal{B}_N^d(\omega_N^d) + \mathcal{B}_g^d(\omega_g^d) + \mathcal{B}_{\rm ads}^d(c_{\rm ads}^d) = 0. \tag{7.9.98}$$

We then obtain:

$$\begin{aligned}
&\frac{\partial}{\partial t}\left[\theta_{\rm A}\omega_{\rm A}^d\rho_{\rm A} + \theta_N\omega_N^d\rho_N + \theta_g\omega_g^d\rho_g + (1-\phi)\rho_s F^d\right] \\
&\quad +\nabla\cdot\left(\theta_{\rm A}\omega_{\rm A}^d\rho_{\rm A}\mathbf{V}_{\rm A} + \theta_N\omega_N^d\rho_N\mathbf{V}_N + \theta_g\omega_g^d\rho_g\mathbf{V}_g\right) \\
&\quad -\nabla\cdot\left(\theta_{\rm A}\rho_{\rm A}\boldsymbol{\mathcal{D}}_{\rm A}^{*d}\cdot\nabla\omega_{\rm A}^d + \theta_N\rho_N\boldsymbol{\mathcal{D}}_N^{*d}\cdot\nabla\omega_N^d + \theta_g\rho_g\boldsymbol{\mathcal{D}}_g^{*d}\cdot\nabla\omega_g^d\right) \\
&\quad -\nabla\cdot\left(\theta_{\rm A}\mathbf{D}_{\rm A}\cdot\nabla\omega_{\rm A}^d\rho_{\rm A} + \theta_N\mathbf{D}_N\cdot\nabla\omega_N^d\rho_N + \theta_g\mathbf{D}_g\cdot\nabla\omega_g^d\rho_g\right) = 0.
\end{aligned} \tag{7.9.99}$$

In a similar way, we obtain the mass balance equation for the water component:

$$\frac{\partial}{\partial t}\left[\theta_{A}\omega_{A}^{w}\rho_{A}+\theta_{N}\omega_{N}^{w}\rho_{N}+\theta_{g}\omega_{g}^{w}\rho_{g}\right]$$
$$+\nabla\cdot\left(\theta_{A}\omega_{A}^{w}\rho_{A}\mathbf{V}_{A}+\theta_{N}\omega_{N}^{w}\rho_{N}\mathbf{V}_{N}+\theta_{g}\omega_{g}^{w}\rho_{g}\mathbf{V}_{g}\right)$$
$$-\nabla\cdot\left(\theta_{A}\rho_{A}\boldsymbol{\mathcal{D}}_{A}^{*w}\cdot\nabla\omega_{A}^{w}+\theta_{N}\rho_{N}\boldsymbol{\mathcal{D}}_{N}^{*w}\cdot\nabla\omega_{N}^{w}+\theta_{g}\rho_{g}\boldsymbol{\mathcal{D}}_{g}^{*w}\cdot\nabla\omega_{g}^{w}\right)$$
$$-\nabla\cdot\left(\theta_{A}\mathbf{D}_{A}\cdot\nabla\omega_{A}^{w}\rho_{A}+\theta_{N}\mathbf{D}_{N}\cdot\nabla\omega_{N}^{w}\rho_{N}+\theta_{g}\mathbf{D}_{g}\cdot\nabla\omega_{g}^{w}\rho_{g}\right)=0, \tag{7.9.100}$$

and for the air component:

$$\frac{\partial}{\partial t}\left[\theta_{A}\omega_{A}^{a}\rho_{A}+\theta_{N}\omega_{N}^{a}\rho_{N}+\theta_{g}\omega_{g}^{a}\rho_{g}\right]$$
$$+\nabla\cdot\left(\theta_{A}\omega_{A}^{a}\rho_{A}\mathbf{V}_{A}+\theta_{N}\omega_{N}^{a}\rho_{N}\mathbf{V}_{N}+\theta_{g}\omega_{g}^{a}\rho_{g}\mathbf{V}_{g}\right)$$
$$-\nabla\cdot\left(\theta_{A}\rho_{A}\boldsymbol{\mathcal{D}}_{A}^{*a}\cdot\nabla\omega_{A}^{a}+\theta_{N}\rho_{N}\boldsymbol{\mathcal{D}}_{N}^{*a}\cdot\nabla\omega_{N}^{a}+\theta_{g}\rho_{g}\boldsymbol{\mathcal{D}}_{g}^{*a}\cdot\nabla\omega_{g}^{a}\right)$$
$$-\nabla\cdot\left(\theta_{A}\mathbf{D}_{A}\cdot\nabla\omega_{A}^{a}\rho_{A}+\theta_{N}\mathbf{D}_{N}\cdot\nabla\omega_{N}^{a}\rho_{N}+\theta_{g}\mathbf{D}_{g}\cdot\nabla\omega_{g}^{a}\rho_{g}\right)=0. \tag{7.9.101}$$

We note that the mass balance equation for any γ-component has the following form:

$$\frac{\partial}{\partial t}\left[\sum_{\alpha=A,N,g}\theta_{\alpha}\omega_{\alpha}^{\gamma}\rho_{\alpha}+(1-\phi)\rho_{s}F_{s}^{\gamma}\right]+\nabla\cdot\left[\sum_{\alpha=A,N,g}\theta_{\alpha}\omega_{\alpha}^{\gamma}\rho_{\alpha}\mathbf{V}_{\alpha}\right]$$
$$-\nabla\cdot\left[\sum_{\alpha=A,N,g}\theta_{\alpha}\rho_{\alpha}\boldsymbol{\mathcal{D}}_{\alpha}^{*\gamma}\cdot\nabla\omega_{\alpha}^{\gamma}\right]-\nabla\cdot\left[\sum_{\alpha=A,N,g}\theta_{\alpha}\mathbf{D}_{\alpha}\cdot\nabla\omega_{\alpha}^{\gamma}\rho_{\alpha}\right]=0,$$
$$\gamma=d,w,a;\quad F_{s}^{w}=F_{s}^{a}=0. \tag{7.9.102}$$

In this equation, the advective fluxes, $\theta_A\mathbf{V}_A$, $\theta_N\mathbf{V}_N$, and $\theta_g\mathbf{V}_g$, are given by the motion equation, which, neglecting momentum exchange between adjacent fluid phases, is:

$$\theta_{\alpha}\mathbf{V}_{\alpha}=-\frac{\mathbf{k}_{\alpha}}{\mu_{\alpha}}\cdot\left(\nabla p_{\alpha}+\rho_{\alpha}\nabla z\right),\quad \alpha=A,N,g. \tag{7.9.103}$$

We assume that in these balance equations, the following variables:

$$\phi,\ \rho_s,\ \rho_A,\ \rho_g,\ \rho_N,\ \mu_A,\ \mu_g,\ \mu_N,$$

$$\mathbf{k}_A,\ \mathbf{k}_g,\ \mathbf{k}_N,\ \boldsymbol{\mathcal{D}}_A^{*d},\ \boldsymbol{\mathcal{D}}_g^{*d},\ \boldsymbol{\mathcal{D}}_N^{*d},\ \mathbf{D}_A,\ \mathbf{D}_g,\ \mathbf{D}_N,$$

are either known constants or functions of the thermodynamic state, as defined by the pressure, temperature, and mass fractions in the appropriate phase. The remaining variables, or unknowns, are:

$$p_{\mathrm{A}},\ p_g,\ p_N,\ \theta_{\mathrm{A}},\ \theta_g,\ \theta_N,$$

$$c_s^d,\ \omega_{\mathrm{A}}^d,\ \omega_g^d,\ \omega_N^d,\ \omega_{\mathrm{A}}^w,\ \omega_g^w,\ \omega_N^w,\ \omega_{\mathrm{A}}^a,\ \omega_g^a,\ \omega_N^a.$$

The following constraints, on the volumetric fractions:

$$\theta_{\mathrm{A}} + \theta_g + \theta_N = \phi, \tag{7.9.104}$$

and on the phase pressures through the capillary pressure relations:

$$p_g - p_N = p_{cgN}(\theta_g), \qquad p_N - p_{\mathrm{A}} = p_{cN\mathrm{A}}(\theta_{\mathrm{A}}), \tag{7.9.105}$$

allow us to eliminate some of the unknown variables. For example, we can eliminate θ_N in (7.9.104), by using the equation $\theta_N = \phi - \theta_{\mathrm{A}} - \theta_g$, leaving θ_{A} and θ_g. Using (7.9.105), we may eliminate two of the phase pressures, say p_N and p_{A}, leaving only p_g. Thus, after eliminating the above variables, we are left with the following thirteen unknown variables:

$$p_g,\ \theta_{\mathrm{A}},\ \theta_g,\ c_s^d,\ \omega_{\mathrm{A}}^d,\ \omega_g^d,\ \omega_N^d,\ \omega_{\mathrm{A}}^w,\ \omega_g^w,\ \omega_N^w.\ \omega_{\mathrm{A}}^a,\ \omega_g^a,\ \omega_N^a.$$

To further reduce the number of unknown variables, we now assume equilibrium conditions among the phases, and use appropriate thermodynamic relationships to relate the concentrations of the components in adjacent phases to each other. For example, we may use the linear isotherm for the adsorption on the solid:

$$K_d^d = \frac{F^d}{\omega_{\mathrm{A}}^d \rho_{\mathrm{A}}}, \tag{7.9.106}$$

to eliminate the variable c_s^d, leaving twelve unknowns in our list. For the fluid components, when the solutions are dilute, we may use Henry's law:

$$\mathcal{H}_{g,\mathrm{A}}^d = \frac{\omega_g^\gamma}{\omega_{\mathrm{A}}^\gamma}, \qquad \mathcal{H}_{g,N}^\gamma = \frac{\omega_g^\gamma}{\omega_N^\gamma}, \qquad \gamma = d, w, a, \tag{7.9.107}$$

in which $\mathcal{H}_{g,\mathrm{A}}^d$ and $\mathcal{H}_{g,N}^\gamma$ are *Henry's law coefficients*, appropriately converted to ratios of mass fractions. Other forms of Henry's law, e.g., in terms of the mole fraction, m_α^γ, may have to be used. These six relationships allow us to express six of the mass fractions in terms of the others, leaving six unknowns in our list. However, the mass fractions must also satisfy the following constraints:

$$\omega_{\mathrm{A}}^d + \omega_{\mathrm{A}}^w + \omega_{\mathrm{A}}^a = 1, \quad \omega_g^d + \omega_g^w + \omega_g^a = 1, \quad \omega_N^d + \omega_N^w + \omega_N^a = 1. \tag{7.9.108}$$

In order to satisfy these constraints, we select for each component a mass fraction in some particular phase, which will be its corresponding 'basis phase'. For example, suppose we select the gas phase to be the basis phase for all components (although, in general, the basis phase does not have to be the same for all components), so that the 'basis mass fractions' are ω_g^d, ω_g^w, and ω_g^a. We then use Henry's law to express the other mass fractions in terms of these, and substitute the results into (7.9.108) to obtain the following system of three linear equations:

$$\omega_g^d + \omega_g^w + \omega_g^a = 1, \tag{7.9.109}$$

$$\mathcal{H}_{g,\mathrm{A}}^d \omega_g^d + \mathcal{H}_{g,\mathrm{A}}^w \omega_g^w + \mathcal{H}_{g,\mathrm{A}}^a \omega_g^a = 1, \tag{7.9.110}$$

$$\mathcal{H}_{g,N}^d \omega_g^d + \mathcal{H}_{g,N}^w \omega_g^w + \mathcal{H}_{g,N}^a \omega_g^a = 1, \tag{7.9.111}$$

which may be solved for ω_g^d, ω_g^w, and ω_g^a; these may, therefore, be considered as functions of pressure and temperature.

Thus, we are left with the following three unknowns, or *primary variables*:

$$p_g, \quad \theta_\mathrm{A}, \quad \theta_g,$$

to be determined by solving the three balance equations in (7.9.102).

A situation may arise in an investigated domain, in which initially we have only two phases, say, air and water, in part of the domain, and three phases—air, water, NAPL—in the remaining part, with the boundary between the two domains behaving like a moving front. Similarly, as time evolves, a phase, say NAPL, may disappear from part of the investigated domain. In both cases, we have to switch form a two-phase system to a three-phase one, or vice versa. Usually, certain features have to be included in the computer program in order to take care of these options.

Obviously, a complete model also requires initial conditions and boundary conditions for the selected mass balance equations.

7.9.4 Primary variables

A rather large number of variables may, sometimes, be required to describe the complete behavior of a system involving many chemical species, or a number of multicomponent fluid phases (sometimes, under nonisothermal conditions), in a possibly deformable solid matrix. However, on the basis of balance and thermodynamic relationships, this number can be significantly reduced, thus simplifying the task of solving the mathematical model in order to predict the future behavior of the system. The *number of degrees of freedom* is the smallest number of independent variables needed to fully define a system's present and future behavior. We shall refer to these variables as *primary variables*. Values of all other state variables of the system can be obtained from the primary ones through the use of constitutive relationships and definitions. A case that requires special attention is when all phases and components within a system are *at equilibrium*, or when the rate of transformation of the system from one state to another is sufficiently slow so that it can be assumed to be continuously close to equilibrium. Under such conditions, the number of primary variables can further be reduced.

Gibbs phase rule (see any text on thermodynamics, e.g., Denbigh, 1981) states that the state of a system composed of NP phases and NC non-reacting components, under conditions of equilibrium, is fully defined by NF state variables, with NF determined by the relationship

$$\mathrm{NF} = \mathrm{NC} - \mathrm{NP} + 2. \tag{7.9.112}$$

For example, in the case of a single fluid phase composed of a single component, say H_2O, NF $= 2$. This means that the state of the system at equilibrium is fully defined by two independent variables, say the pressure, p, and the temperature, T. We could, however, select also T and the phase density, ρ, with the constitutive relation, $\rho = \rho(p, T)$, as long as this relationship can, at least in theory, be solved for p as a function of ρ and T. As a second example, consider two fluid phases: a liquid, consisting of a single component water, and a gas, composed only of water vapor. This means that we have two phases and one chemical component (see definition of phases and components in Sec. 1.3), and by Gibbs' phase rule, we have one degree of freedom. Suppose we select T as the independent variable. Since we have here liquid water and water vapor in equilibrium, the gas pressure is determined as a function of T by $p_g = p_{\text{sat}}(T)$, where p_{sat} is the saturated vapor pressure of water at which the system can exist at any given temperature. Once we know p and T, we can determine the densities of the phases, or the value of any other thermodynamic property.

Let us generalize Gibbs law by considering a model that describes the behavior at the *macroscopic* level of a system composed of multiple multicomponent fluid phases within a possibly deformable porous medium domain. We recall that at such level, the behavior *at a point* means the averaged behavior within an REV centered at the considered point.

Based on balance considerations and on thermodynamic relationships, Bear and Nitao (1995) showed that when conditions of thermodynamic equilibria prevail (or are assumed to prevail as a good approximation) among all phases and components present within a deformable porous medium under nonisothermal conditions, the number of degrees of freedom, NF, in a problem of heat and mass transport, involving NP fluid phases and NC components, is given by the relationship

$$\mathrm{NF} = \mathrm{NC} + \mathrm{NP} + 4. \tag{7.9.113}$$

Under conditions of nonequilibrium between the phases, this rule becomes

$$\mathrm{NF} = \mathrm{NC} \times \mathrm{NP} + 2 \times \mathrm{NP} + \mathrm{NC} + 4. \tag{7.9.114}$$

In both cases, when Darcy's law is used to determine the velocities of the fluid phases, NF is reduced by NP. When the solid matrix is nondeformable, NF is reduced by 3, leading to the relationship

$$\mathrm{NF} = \mathrm{NC} + 1. \tag{7.9.115}$$

Furthermore, if the system is isothermal, then

$$\mathrm{NF} = \mathrm{NC}. \tag{7.9.116}$$

These rules are, thus, extensions of the well known *Gibbs phase rule* to transport phenomena in porous media. The number of degrees of freedom for reactive transport problems was also discussed by Saaltink *et al.* (1998), and by Molins *et al.* (2004).

The above discussion is applicable to NC *non-reacting* components. For a system with chemical reactions, let NS be the number of reacting species and $\mathrm{NR_{eq}}$ be the number of equilibrium reactions. Then, by expressing the reactions in the form a canonical set of equations, and making use of the *law of mass action*, there are $\mathrm{NC} = \mathrm{NS} - \mathrm{NR_{eq}}$ components, or independent species concentrations. Thus, the number of degree of freedom, NF, in the above equations for a non-reacting system still applies to a reacting system, as long as we use $\mathrm{NC} = \mathrm{NS} - \mathrm{NR_{eq}}$.

Once we have determined the number of degrees of freedom of a given problem, and select the most convenient variables to be declared as primary ones, we identify the (same number of) balance (partial differential) equations which have to be solved in order to determine the values of these variables within the problem's spatial and temporal domains. All other variables are, subsequently, determined by using the remaining equations—constitutive relations and definitions.

7.9.5 Methods of solution for reactive transport models

In principle, a reactive transport model is composed of

(a) A set of equations that describe the flow of the participating phase(s). The set includes the mass balance equation and Darcy's law for each participating phase. By solving this set, we obtain the future velocity distributions within each phase, and, in the case of multiple phases, also the saturations. These equations are presented in Chaps. 5 and 6.
(b) A set of equations that describe how the future concentrations of the chemical species present in the phase(s) vary as a result of flow (advection, dispersion), diffusion, and chemical reactions within the participating phase(s), and transfers across interphase boundaries (including the solid phase). These equations are discussed in Subs. 7.4 and 7.9.2B.
(c) Equations that describe the variations in concentration of the participating chemical species as a result of (homogeneous and heterogeneous) chemical reactions. These are discussed in Subs. 7.9.2A.

The equations in (a) and (b) are coupled when the density of the fluid phases is concentration-dependent. Otherwise, the set (a) can be solved to yield the future velocity distribution of the considered phase(s). Density variable flow is discussed in Subs. 9.3.1. Considered separately, the set of equation in (c) is underlying the process of speciation presented in Subs. 7.9.2D.

Under non-isothermal conditions, the *reactive transport model* becomes even more complicated, as we have to add the energy balance equation (and the equation describing energy flux) to the above set of equations. As elsewhere in this book, we shall limit our discussion to isothermal conditions.

All the components of a reactive transport model have already been presented and discussed in different forms in earlier sections of this chapter (e.g., Subs. 7.9.2D–F. Often, the number of unknown variables is quite large and many of the model (algebraic and partial differential) equations are coupled and nonlinear. Obviously, an analytic solution is not possible, except for very special simple cases. A numerical solution and an appropriate computer code are required. Several authors, e.g., Lichtner (1985), Yeh and Tripathi (1989, 1991), Steefel and MacQuarrie (1996), Chilakapati *et al.* (1998), Saaltink *et al.* (1998, 2000), Hammond *et al.* (2005), Fang *et al.* (2006), Kräutle and Knabner (2005, 2007), and Zhang *et al.* (2007), have presented numerical techniques and computer codes for solving reactive transport models. This last reference also lists a large number of additional relevant publications.

Among the codes developed and presented in the literature in recent years, we may mention RETRASO (= REactive TRAnport of SOlues) (Saaltink *et al.*, 2004), PHAST (Program for Simulating Ground-Water Flow, Solute Transport, and Multicomponent Geochemical Reactions) (Parkhurst *et al.*, 2004), and TOUGH (Transport of Unsaturated Groundwater and Heat) (Pruess *et al.*, 1999); see Sec. 8.8.

Actually, a number of examples of relatively simple reactive transport models and their solution were presented in Subs. 7.9.2F.

Because the reactive transport model involves a large number of variables and, hence, of equations, and because of the nonlinear nature of the many of the model equations, special numerical approaches have been proposed. Basically, the various methods can be divided into two groups (Saaltink *et al.*, 2004):

- The *Operator splitting*, or the *two-step approach*. The latter includes the *Sequential Iteration Approach* (SIA), and the *Sequential Non-Iteration Approach* (SNIA). In these techniques, the transport (i.e, mass balance) equations (listed as (b) above) and the equations that describe the chemical reactions (listed as (c) above) are solved separately. In the transport part, we freeze changes in concentration that result from chemical (homogeneous) reactions, and compute concentration changes, for the next time step, due only to advection, dispersion, diffusion and interphase exchange. The second step involves "freezing" the flow and considering speciation only, based on the assumption of equilibrium, say within each (grid) cell, regarded as a batch operation. In many cases, the speciation code PHREEQC is used. In the SIA version, we iterate between the two steps, in an effort to achieve convergence.

- The single-step, or the *Direct Substitution Approach* (DSA), or the *Global Implicit Approaches* (GIA). In this group of approaches, the chemical equations are substituted into the transport equations, and the latter are then solved simultaneously, employing the Newton-Raphson technique. The obvious disadvantage is the need to solve many equations simultaneously, resulting in high computational costs. Saaltink *et al.* (2000) present a comparison between the DSA and SIA and SNIA techniques. For example, they claim that the latter two techniques, generally, show slow convergence, are less robust and more stiff.

Many computer codes have been developed, most of them based on the SIA approach. Saaltink *et al.* (2004) list a large number of such codes. Some codes are applicable to unsaturated (i.e., two-phase) case. For example, Mayer *et al.* (2002) developed a code, MIN3P, which solved a model of reactive transport in saturated or unsaturated flow without assuming equilibrium between the minerals and the water.

As an example, let us consider the multispecies multi-reaction case presented by Saaltink *et al.* (1998, 2000) as an example of the SIA and the DSA approaches. In this example, we consider NS chemical species, which participate in NR reactions. The example is based on the assumption that the chemical reactions are sufficiently fast, so that the *Local Equilibrium Assumption* (LEA; Subs. 7.3.1) can be made. Under this assumption, the law of mass action that relates concentrations of products to those of reactants is valid. This law was already presented as (7.3.40) or (7.3.42), and is rewritten here in the form:

$$\boldsymbol{\mathcal{N}}_e \log\{\boldsymbol{\gamma}\} = \log \mathbf{K}_{\text{eq}}, \quad \text{or} \quad \nu_{ij} \log\{\gamma^j\} = \log K_{eq,i}, \tag{7.9.117}$$

where $\boldsymbol{\mathcal{N}}$ is an NR×NS matrix (components ν_{ij}) that contains the stoichiometric coefficients of the NR reactions (see (7.9.27)), $\{\boldsymbol{\gamma}\}$ is a vector (components $\{\gamma^j\}$) that denotes the activity of all NS chemical species, and $\mathbf{K}_{\text{eq}}$ is a vector (components $K_{eq,i}$) of the NR equilibrium constants.

For an aqueous species, the vector of activities, $\{\boldsymbol{\gamma}\}$, and that of molar concentrations, $[\mathbf{c}]$, are related to each other by the activity coefficient, $\boldsymbol{\gamma}$ (see (7.3.43)):

$$\log\{\boldsymbol{\gamma}\} = \log \boldsymbol{\gamma}([\mathbf{c}]) + \log[\mathbf{c}], \tag{7.9.118}$$

in which the activity coefficient of each j-species, γ^j, is a function of all aqueous concentrations. The Debye-Hückel equation is an example of such function. For dilute aqueous solutions, the activity coefficient is approximated as unity. Although we shall continue to develop the model in terms of activities, the model could have been developed also in terms of concentrations.

Following the discussion in Subs. 7.9.4, with NS species and NR chemical equilibrium equations, the number of degrees of freedom in the case of equilibrium reactions considered here is NC = NS − NR activities of chemical *components*, each being a function of the vector of concentrations, $[\mathbf{c}]$. These can, at least in principle, be determined from the (stoichiometric) chemical equations. Let us use the symbol $\{\gamma\}_{\rm P}$ and $\{\gamma\}_{\rm Q}$ to denote the activities that correspond to the NC primary species, or components, and to the NR secondary species, respectively,

$$\{\gamma\} = \begin{pmatrix} \{\gamma\}_{\rm P} \\ \{\gamma\}_{\rm Q} \end{pmatrix}. \tag{7.9.119}$$

Then, (7.9.117) can be rewritten as

$$\boldsymbol{\mathcal{N}}_{\rm P} \log\{\gamma\}_{\rm P} + \boldsymbol{\mathcal{N}}_{\rm Q} \log\{\gamma\}_{\rm Q} = \log \mathbf{K}_{\rm eq}, \tag{7.9.120}$$

in which we have divided $\boldsymbol{\mathcal{N}}$ into one part, $\{\gamma\}_{\rm P}$, that refers to the primary species and a second part that refers to the secondary ones, $\{\gamma\}_{\rm Q}$. We shall use the notation

$$\boldsymbol{\mathcal{N}} = (\{\gamma\}_{\rm P} \,|\, \{\gamma\}_{\rm Q}). \tag{7.9.121}$$

Different sets of primary and secondary species may be selected, but they must be selected such that $\{\gamma\}_{\rm Q}$ is a full ranked matrix. Saaltink *et al.* (1998), following Steefel and Yabusaki (1995), discussed constraints on the selection of the sets of primary and secondary species. Their conclusion was that difficulties can be avoided by selecting a set of primary variables that does not include the lowest activities. Saaltink *et al.* (1998) also showed that when the system involves species with constant activities, the number of primary species can further be reduced. This happens when the reactions involve mineral or gas phases. For example, the activity is unity for a pure mineral, unity for water and partial pressure for gas at low pressure. Thus, the number of primary species can be reduced to NS – NR – NT, where NT is the number of constant activity species.

So far, we have considered only equilibrium reactions (for which the mass action law is applicable). Let us assume that some of the reactions are sufficiently slow so that we have to take into account the *reaction rate*, $R_{r,j}$, introduced in Subs. 7.3.2A and 7.3.6. In this symbol, the j indicates the number of the considered reaction and $R_{r,j}$ is defined as the amount of reactant evolving to products, per unit time, in the jth chemical reaction. In principle, the reaction rate depends on the concentrations of all species present in the considered solution,

$$R_{r,j} = R_{r,j}([\mathbf{c}]).$$

At this point, we have to consider the mass balance equations for the NS species. Using the *balance operator* defined by (7.9.21), we can write the mass balance equation—reactive transport equation—in the form (7.9.25), rewritten here as

$$\mathcal{B}([c^\gamma]) = \sum_{j=1}^{NR'} \nu_j^\gamma R_{{\rm pm},j}, \quad \gamma = 1, 2, ..., {\rm NS}. \tag{7.9.122}$$

We can also make use of (7.9.44) and (7.9.45).

Instead, let us follow Saaltink *et al.* (2000). They divide all chemical reactions into three groups: aqueous complexations, sorption, and precipitation dissolution, described below by (7.9.124), (7.9.125) and (7.9.120), respectively. They write the overall reactive transport equation for all species present in single phase flow, assuming that ϕ is a constant, in the form:

$$\frac{\partial \mathbf{u}_a}{\partial t} + \frac{\partial \mathbf{u}_d}{\partial t} + \frac{\partial \mathbf{u}_m}{\partial t} = \mathcal{L}(\mathbf{u}_a) + \mathbf{U}\boldsymbol{\mathcal{N}}'_k \mathbf{R}_{r,k}([\mathbf{c}_a]), \tag{7.9.123}$$

$$\log[\mathbf{c}_2] = \boldsymbol{\mathcal{N}}_a (\log[\mathbf{c}_1] + \log \gamma_1) - \log \boldsymbol{\gamma}_2 + \log \mathbf{K}_{eq,a}([\mathbf{c}_a]), \tag{7.9.124}$$

$$\log[\mathbf{c}_d] = \boldsymbol{\mathcal{N}}_d (\log[\mathbf{c}_1] + \log \boldsymbol{\gamma}_1) - \log \gamma_d + \log \mathbf{K}_{eq,d}, \tag{7.9.125}$$

$$0 = \boldsymbol{\mathcal{N}}_m (\log[\mathbf{c}_1] + \log \boldsymbol{\gamma}_1) - \log \gamma_d + \log \mathbf{K}_m, \tag{7.9.126}$$

$$[\mathbf{c}_a] = \begin{pmatrix} [\mathbf{c}_1] \\ [\mathbf{c}_2] \end{pmatrix}, \quad \mathbf{u}_a = \mathbf{U}_a[\mathbf{c}_a], \quad \mathbf{u}_d = \mathbf{U}_d[\mathbf{c}_d], \quad \mathbf{u}_m = \mathbf{U}_m[\mathbf{c}_m]. \tag{7.9.127}$$

In the above equations, $\mathcal{L}(c) \equiv -(1/\phi)\nabla\cdot(\mathbf{q}c - \mathbf{D}_h \cdot \nabla c) + m$ is a linear operator (to be compared with the operator $\mathcal{B}$), with m denoting a source term (injection), the vectors $[\mathbf{c}_a]$, $[\mathbf{c}_d]$ and $[\mathbf{c}_m]$ contain, respectively, concentrations of aqueous, adsorbed and precipitated species, with $[\mathbf{c}_a]$ divided into the vectors $[\mathbf{c}_1]$ and $[\mathbf{c}_2]$ denoting the concentrations of primary and secondary species, respectively, obeying (7.9.124). The vector γ, which has the same subscripts as $[\mathbf{c}]$, refers to the activity coefficients. The matrices $\boldsymbol{\mathcal{N}}_a$, $\boldsymbol{\mathcal{N}}_d$, and $\boldsymbol{\mathcal{N}}_m$ contain the stoichiometric constants of all the equilibrium aqueous, adsorbed and precipitated reactions, respectively, while $\boldsymbol{\mathcal{N}}'_k$ of all kinetic chemical reactions. The vectors $\mathbf{u}_a$, $\mathbf{u}_d$ and $\mathbf{u}_m$, contain the total concentrations of a chemical species in, respectively, aqueous, adsorbed and precipitated forms. The vector $\mathbf{R}_{r,k}$ contains the rates of the kinetic reactions; these rates can normally be expressed as functions of all concentrations. The last three equations in (7.9.127) relate the concentrations of the species with the total concentrations, with $\mathbf{U}_a$, $\mathbf{U}_d$ and $\mathbf{U}_m$ referred to as 'component matrices'.

In the SIA approach, the first step is to solve the transport equation (7.9.123), with the total aqueous concentrations of every chemical component (vector $\mathbf{u}_a$) as unknowns. The concentrations of adsorbed species, minerals and kinetic reactions as source-sink term (vector $\mathbf{f}$) are treated by the previous iteration:

$$\frac{\partial \mathbf{u}'_a}{\partial t} = \mathcal{L}(\mathbf{u}'_a) + \mathbf{f}^{n-1}, \tag{7.9.128}$$

where n denotes the iteration number.

In the second step, the source-sink terms are updated. To achieve this goal, Saaltink *et al.* (2000) calculate the concentrations of the components (i.e., primary species), $\mathbf{c}_1$, and the minerals in equilibrium, c_m, from the total aqueous concentrations, $\mathbf{u}_a$, by means of the chemical equations:

$$\mathbf{u}_a^i + \mathbf{u}_d^{i-1} + \mathbf{u}_m^{i+1} = \mathbf{U}_a \begin{pmatrix} [\mathbf{c}_1^i] \\ [\mathbf{c}_2] \left([\mathbf{c}_1^i]\right) \end{pmatrix} + \mathbf{U}_d[\mathbf{c}_d] \left([\mathbf{c}_1^i]\right) + \mathbf{U}_m[\mathbf{c}_m^i], \quad (7.9.129)$$

$$0 = \mathcal{N}_m \left(\log[\mathbf{c}_1^i] + \log \gamma_1\right) + \log \mathbf{K}_m. \quad (7.9.130)$$

Note that (7.9.124) and (7.9.125) were inserted into the second and third equations of (7.9.127), and, because these equations are nonlinear, the Newton-Raphson scheme is applied for its solution. In a numerical scheme, this is done for every node separately. From $[\mathbf{c}_1]$ and $[\mathbf{c}_m]$, the new source-sink term is calculated as:

$$\mathbf{f}^i = \mathbf{U}_d \frac{\partial[\mathbf{c}_d]([\mathbf{c}_1^i])}{\partial t} + \mathbf{U}_m \frac{\partial[\mathbf{c}_m]}{\partial t} + \mathbf{U}\mathcal{N}_k^i \mathbf{R}_{r,k} \left([\mathbf{c}_1^i], [\mathbf{c}_2]([\mathbf{c}_1^i])\right). \quad (7.9.131)$$

The process is repeated until convergence is reached.

Saaltink *et al.* (2000) discuss also on the DSA approach to the solution of this problem.

7.10 Remediation Techniques

The issue of subsurface contamination and a review of contamination sources were introduced in Sec. 1.1.5. Obviously, the best strategy is to make every effort to prevent subsurface contamination. Unfortunately, too often, especially in industrialized countries, contamination of the subsurface does occur, whether within the framework of planned activities, or due to accidents. When these happen, remediation activities are called for, as part of aquifer management (Chap. 11).

The objective of this section is to briefly describe a number of techniques commonly employed for aquifer remediation. They are brought here because, whenever remediation of a contaminated sight is being considered, and the most appropriate technique is being sought, the implementation of that technique must first be examined by constructing its model and running it for different scenarios. The outcomes of the various runs are introduced as inputs to the management models.

7.10.1 General considerations

The remediation strategy for a given contaminated site depends on the local hydrogeological conditions, on the kind of source, and on the intended use of groundwater—for drinking, for other domestic uses, for industry, for agriculture, or for recreation, nature, and environmental protection. The general objective of remediation is to protect *both* human health and the environment. The driving force for control and remediation activities, from their inception through implementation, are laws and regulations which have been established in most countries to ensure clean and safe water for water supply, and a clean and safe environment.

Because it is obvious that it is practically impossible to restore a contaminated aquifer to pristine conditions, clean-up goals for drinking water are dictated by laws and regulations, such as (in the USA) the Resource Conservation and Recovery Act (RCRA), the Comprehensive Environmental Response, Compensation, and Liability Act (CERCLA), and the National Environmental Policy Act (NEPA). These set up drinking water standards, such as Maximum Contaminant Levels (MCLs). Standards are also set up for groundwater used for other purposes. Similar laws and regulations exist in Europe, where standards have been set up by the European Community (e.g., the Drinking Water Directive (DWD)) and in other countries around the world. More detail about environmental law, regulation, and policy can be found in Rechtschaffen and Gauna (2002), Percival *et al.* (2006), and Fiorino (2006).

An obvious first step in any remediation strategy is either the containment or the removal of the source itself. This is relatively easy when the source is above ground surface, or at a relatively shallow depth. In the latter case, the contaminated soil is removed and hauled to a treatment or a disposal facility. We often refer to such source as a *primary source.*

Removal is impractical when the contaminant is a NAPL, spilled at ground surface in a quantity that spreads out in the subsurface, mainly downward, eventually becoming immobile as isolated NAPL ganglia (at residual NAPL saturation) down to a large depth (Fig. 7.0.1). The NAPL in the subsurface acts as *secondary source*, releasing contaminants (by dissolution) into the water that percolates through the vadose zone, eventually reaching an underlying aquifer. Preventing infiltration by covering the ground surface above this secondary source by an impervious natural (e.g., compacted clay) or man made (e.g., plastic lining) material, may reduce the effect of such source. In what follows, we shall present methods that can handle such a source.

When a sufficiently large volume of DNAPL is spilled above a phreatic aquifer (Fig. 7.0.1), some NAPL may eventually reach groundwater *below* the water table and become a source by dissolution in the flowing groundwater for prolonged periods of time. The intensity with which the dissolved chemicals are released from the liquid hydrocarbon over time is referred to as *source strength*, Γ, (dims. M/T) (Testa and Winegardner, 2000). Source strength can be expressed as $\Gamma = K_m A$, where K_m is the mass exchange coefficient (dims. M/L^2T), and A (dims. L^2) is the contact area. or interface, across which the mass exchange occurs. Quantification of A is very difficult, reflecting the complexity of DNAPL distribution in the pore space. Estimates of K_m for certain products, such as gasoline and tar oil ($1.0\ mg/m^2/s$), fuel oil, diesel and kerosene ($0.01\ mg/m^2/s$), have been attempted by the USGS (Hult, 1984).

In general, four measures may be undertaken in connection with subsurface contamination:

- **Prevention**. Activities (at ground surface) that prevent the creation of pollution sources.

- **Abatement**. Activities that are undertaken, once a source has been created, to prevent the contaminants from reaching the subsurface, or an underlying aquifer.
- **Clean-up**, or **remediation**. Activities that are undertaken, once the subsurface has been contaminated, to remove contaminants from the subsurface.
- **Restoration**. Efforts that are made to return the subsurface to its original, unpolluted, conditions, or at least to conditions that are acceptable according to prevailing regulations. The term 'remediation' is often employed as a synonym for clean-up and/or restoration.

An optimal remediation strategy will often include a combination of a number of strategies, each aimed at achieving a specific goal. There is no need to emphasize that, like any management, or decision making process, associated with the subsurface—the unsaturated zone, an aquifer, or both—investigations leading to decisions on remediation, once the presence of a contaminant in the subsurface has been established, should include:

- Gathering information on the source extent, above and below ground surface.
- Gathering information on the chemical/biological nature of the contaminants and the extent of the contaminated subsurface zone. This includes information on the possibility of *natural attenuation* of the contaminants, their interaction with the solid matrix and/or with groundwater along their pathway. This activity is often referred to as *monitored natural attenuation.*
- Geohydrological investigations that should lead to a complete conceptual geohydrological model (structure, properties, replenishment, sources and sinks, boundaries and boundary conditions, etc.) of the relevant subsurface domain.
- Determining the goals (and alternative goals, if relevant) for remediation, based on prevailing regulations.

With the above information, various remediation techniques can be explored. The feasibility of application of every technique to the considered site should be investigated by means of simulation models of flow and contaminant transport. These models serve as tools for evaluating potential sources that might have produced a discovered plume, and as tools for evaluating the fate and transformation of contaminants in the subsurface under various considered clean-up alternatives (including the one of 'do nothing'). Models are also used for predicting the future spatial distribution of contaminants in the considered domain, once a proposed clean-up program has been implemented, as a basis for designing the monitoring and follow-up network of observations and sampling. Model results can also be used for the evaluation of costs, clean-up duration and level, compliance with regulations, etc., such that the optimal method can be selected on the basis of agreed criteria. The ultimate goal of remediation is the removal of contaminants from the contaminated domain to the extent that the concentration of whatever remains in the aquifer is acceptable, e.g., below *maximum contaminant level* (MCL).

Review of various remediation techniques can be found in Nyer (1992), Avogadro and Ragaini (1993), Ward *et al.* (1997), Meyers (1998), Wise *et al.* (2000a, 2000b), Suthersan and Payne (2004), and Houlihan and Berman (2006).

Although this section is devoted to remediation, we wish to emphasize that **prevention is always preferable** to allowing contamination, followed (sometimes many years later) by (often very costly) remediation.

Some examples of commonly used remediation techniques are briefly described below. Two techniques, *air sparging* and *soil venting* were mentioned in Subs. 7.3.7B.

7.10.2 Caps and cutoff walls

The term *control* is sometimes used to describe the activities that are aimed at preventing, or at least reducing the possibility and effects of groundwater contamination.

One example of a control measure takes the form of an impervious clay blanket (or plastic sheets lining) placed under a landfill to act as a barrier that prevents landfill leachate from infiltrating into the subsurface (Lo *et al.*, 1997; Rowe, 2005). Instead, the blanket diverts the leachate into a collecting system that conveys it to a treatment plant. Sometimes, when a high water table is present under the landfill, under the assumption that sooner or later leachate will leak through the blanket to the subsurface, a drainage system is placed under the landfill area to intercept the contaminated groundwater.

A clay blanket or cap can also be placed on top of an area of contaminated (mostly by heavy metals) sediment on the bottom of an open body of water, such as a river, a lake or the sea, to prevent the release of toxic substances from the sediment to the open water (Palermo, 1998; Mohan *et al.*, 2000).

Another type of control measure is *physical containment* of contaminated groundwater. Typically, this involves the construction of a *low permeability barrier* (e.g., a slurry wall, or a grout curtain) in the subsurface, designed to surround the contaminated subsurface domain (Starr and Cherry, 1994; Devlin and Parker, 1996; Philip, 2001). The ground surface bounded by the barrier is often covered with an impermeable cap which prevents further contamination through ground surface, forestalls any water level rise by infiltration in the area surrounded by the barrier, and prevents toxic gases from escaping into the atmosphere.

A *slurry wall* is a relatively narrow (usually, 0.6–1.0 m wide) trench dug down to some depth below the water table. The trench is then filled with low permeability material, e.g., clay, a mixture of soil and bentonite, or soil and cement. The typical desired permeability of such a mixture is about 10^{-8} cm/sec. In this way, a barrier to the movement of water immediately below the water table is created (as well as a barrier to any lateral movement in the vadose zone). The objective of this barrier, which usually takes the form of a closed loop, is to capture, or *contain* contaminants that float on the water table (LNAPL), and those that dissolve in the water, but are present in the layer of water immediately below the water table. In most cases, the slurry wall is not completely impervious, but its permeability is much lower than that of the undisturbed soil. The immediate advantage of the barrier is that it isolates a contaminated groundwater body from its surrounding. This is especially true if the slurry filled trenched is keyed into an underlying aquiclude (in which case it is referred to as a *cutoff wall*). Then the contaminated groundwater body is completely isolated and trapped. It can then be pumped and treated. If such aquiclude does not exist, the isolation is incomplete. A 'free product' (LNAPL) floating on the water table cannot escape, as it is stopped by the slurry wall, but dissolved components will continue to be transported downstream by the water, albeit, sometimes slowed down by the low velocity zone created by the barrier when it reaches some depth below the water table. In spite of this conclusion, a barrier may achieve the goal of containing a contaminant within property boundaries, at least temporarily

An issue that has to be addressed when designing a barrier, or cutoff wall, is the chemical effects of the contaminants on the materials of which the barrier is constructed. In fact, the chemical reactions of contaminants with the barrier wall materials can be used as a strategy for contaminant containment and removal (Subs. 7.10.6).

7.10.3 Pump-and-treat

'*Pump-and-treat*' is, probably, the most common technology used for remediating a contaminated aquifer, once the (chemical) nature and the extent of the subsurface plume have been identified (Mackay and Cherry, 1989; Keely, 1989; Mercer *et al.*, 1990). Cohen *et al.* (1998) presented design guidelines for pump-and-treat systems, with many references. Basically, this technique involves the removal of the contaminant from the aquifer with water pumped through specially installed wells, located (areal distribution and depth of screen) in some optimal manner, e.g., so as to maximize the removal of contaminant mass with a minimum volume of total pumped water. Obviously, determining the location of the wells/screens is a multiple objective management problem: we wish to minimize time of clean-up, say until reaching MCL, to maximize contaminant mass removed, to minimize costs (or to achieve maximum clean-up within a specified budget), etc.

The pumped water is sent to a *treatment facility*, where it is treated according to the chemical nature of the contaminant (often by *air stripping* or *liquid-phase granular activated charcoal*) to a level that allows the treated water to be discharged to the drainage system, or to a nearby stream. Under certain conditions, depending on prevailing regulations, it is permitted

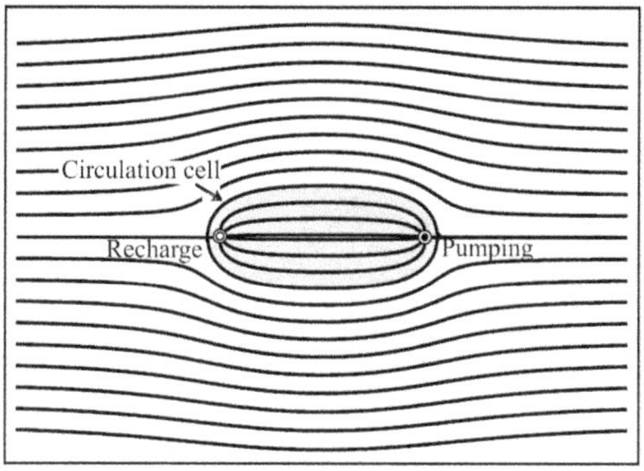

Figure 7.10.1: A circulation cell in a uniform flow (flow from left to right).

to re-inject the treated water back into the aquifer, using artificial recharge technologies (Sec. 3.4). This 'pump-treat-inject' technique is a more efficient option as it saves water (Bear and Sun, 1998; Chang *et al.*, 2007). Furthermore, the combined flow pattern, created by both pumping and injection, can be designed to be much more efficient from the point of view of optimizing the entire pump and treat operation. At the same time, the placement of wells should be such as to avoid the creation of stagnation zones, or zones of very low velocity, from which the contaminant cannot be removed.

A *circulation cell* is a special case of the *pump-treat-inject* technique. In this case, we have a pumping well and a recharging ($\equiv$ injecting) one of equal strength, located along a streamline, with the recharging well placed upstream of the pumping one (see Fig. 7.10.1). The resulting pattern is flow from the recharging well to the pumping one, within a closed elliptically-shaped, or, in 3-D flow, ellipsoidally-shaped, domain within the aquifer. Ideally, the aquifer domain occupied by the contaminant to be removed should be enclosed within the water divide that delineates the circulation cell, thus isolating the contaminant plume from the rest of the aquifer. The recharge-pumping operation acts like a push-pull one, and contaminated water is gradually pumped out of the aquifer, treated to remove the contaminants, and re-injected. In this technique the volume of water that is pumped (= injected) is minimal. Sometimes, various additives may be added to the injected water. e.g., nutrients, in order to enhance bio-transformations, and surfactants to affect surface tension (in multiphase flow). In addition, surfactants will increase the solubility of contaminants (say, from DNAPL ganglia within the saturated zone), thus enhancing the efficiency of the pump-and-treat technology (Fountain *et al.*, 1991, 1996).

One efficient strategy for determining the location and rate of discharge of the pumping wells, is to install them in such a way that they create (individually or through their combined effect) a *capture zone* that will force the interception of the plume of contaminated groundwater (again, meaning the

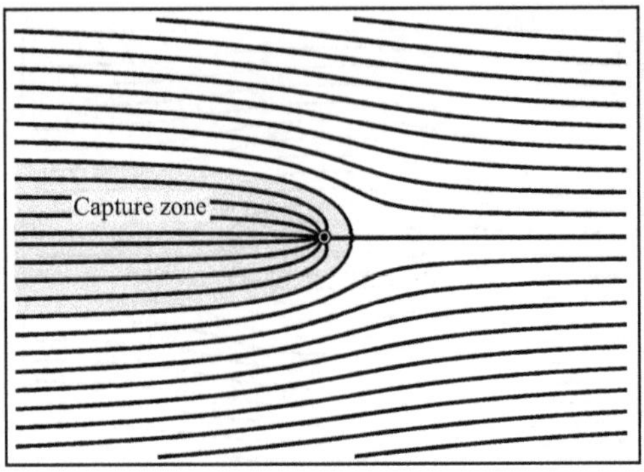

Figure 7.10.2: An interception well in uniform flow (flow from left to right).

minimization of total pumped water per unit mass of removed contaminant). Fig. 7.10.2 shows an interception well in uniform flow in an aquifer.

Obviously, the entire pump-and-treat operation should be accompanied by continuous monitoring (taking and analyzing water samples) of the evolution of the contaminated zone. The data continuously obtained from the monitoring system enhances the efficient management of the clean-up operations. As the clean-up operation evolves, pumping and injection wells may have to be relocated, existing wells may have to be turned off, new wells may have to be added (Hoffman, 1993; Bear and Sun, 1998).

Methods for evaluating pump and treat system performance, and tools for creating optimal efficient pumping schemes are discussed in Gorelick *et al.* (1993), Marryott *et al.* (1993), McKinney and Lin (1995), Huang and Mayer (1997), Fetter (1999), and Aly and Peralta (1999).

Source areas often contain contaminants in the form of isolated ganglia of a NAPL phase residing in fine-grained (almost) water saturated sediments. Because of their low permeability, it is practically impossible to remove such NAPL ganglia, or water with dissolved NAPL, from such formations by the flow produced by pumping wells. Instead, dissolved contaminants escape from such low permeability formations by the mechanism of molecular diffusion (discussed in Sec. 7.5) into the flow pattern in the higher permeability aquifer material. This is a rather slow process, so that the clean-up of such source area by the pump-and-treat technique is practically impossible.

Since it is difficult to force flow through low permeability formations, the pump and treat strategy is often used for the *hydraulic containment* of the source area, or *plume interception.* Extraction wells are placed down-gradient of the source area in such a pattern that the entire source area falls within the capture zone of the extraction wells. Such a strategy intercepts the dissolved contaminant, preventing it from migrating downstream, away from the source area. Furthermore, the interception wells can be placed upstream of pumping wells, protecting the latter from contamination by intercepting the advancing plume.

Another option of the *hydraulic containment* technique is the creation, by pumping, of a 'crater' in the water table, such that contaminated water cannot escape from it, except by pumping. This kind of containment can be consider a variation on the pump-and-treat technique. The pumped water can be sent for treatment. It is also possible to re-inject the treated water, creating a flow pattern that contains the source area. Such operation can continue until the entire source is removed.

7.10.4 Soil vapor extraction

Soil Vapor Extraction (SVE), also known as '*soil venting*', is a most effective *in situ* clean-up technology that reduces concentrations of volatile constituents in the water, the gaseous phase, and that adsorbed to soil in the unsaturated (vadose) zone (Hinchee *et al.*, 1991; Wright *et al.*, 1997; Brusturean *et al.*, 2007; Machackova *et al.*, 2008). It is a technology suitable particularly for cleanup of *source areas* that are contaminated by VOC's (Volatile Organic Compounds). Dissolved volatile components will partition between the water and the gaseous phase, in an attempt to reach equilibrium. In this technology, fully or partially screened wells are installed within the vadose zone, designed to pump (vacuum) air containing contaminants in the form of vapor. The volatile contaminant constituents volatilize and the vapors are drawn toward the extraction wells. At the same time, fresh air is drawn into the subsurface through ground surface. It is also possible to enhance the operation by injecting fresh air into the formation at appropriate locations through specially installed injection wells (see Fig. 7.10.3). In this way, subsurface volatile contaminants are gradually removed. The extracted air is released to the atmosphere after being treated, often by activated carbon absorption, or combustion technologies, to avoid air pollution.

Another necessary condition for the vapor extraction technique to be cost-effective is that the soil's permeability to gas, at the prevailing gas saturation, needs to be sufficiently high, so that a significant flux of gas may be produced by the planned level of vacuum imposed at the extraction well.

In areas of high groundwater levels, simultaneous pumping of water may be required in order to offset the effect of *up-welling* induced by the vacuum.

In addition, lowering the water table by pumping, increases the volume of unsaturated zone, and brings more (contaminated) water in contact with air, thus enhancing volatilization.

7.10.5 Air sparging

When a volatile contaminant, e.g., a petroleum hydrocarbon, or a VOC, such as a chlorinated solvent, after percolating through the vadose zone, reaches an underlying phreatic aquifer and accumulates on the water table in the form

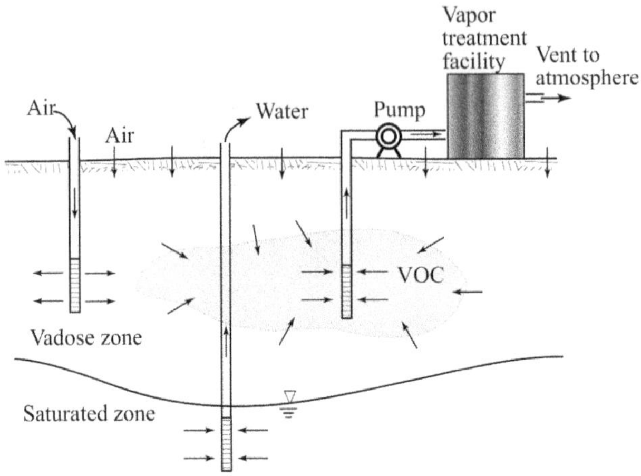

Figure 7.10.3: Soil vapor extraction together with water pumping and air injection.

of a lens, because its density is lower than that of water, it can be removed (at least, partly) by *air sparging.* The same technique is also applicable when the volatile contaminant is carried by the water as a dissolved component.

In this technique, air is injected into the saturated zone through a specially designed well, with its screen located below the water table (see Fig. 7.10.4). The injected air spreads out in the contaminated formation, both vertically and horizontally, creating a two-phase (air-water), or three-phase (air-NAPL-water), zone in the form of a cone around the well, thus bringing (mobile) air into contact either directly with the VOC or NAPL, or with water containing dissolved VOC or NAPL. The volatile contaminant will partition between the aqueous phase (water) or the NAPL and the gaseous one (air), and the contaminant vapor will be carried and removed from the formation by vacuuming the air. The latter is achieved by installing and operating a soil vapor extraction (SVE) system (discussed above). The removed vapor-loaded air is delivered to an air-stripping facility.

As contaminant vapor is being removed from the void space, partitioning will continuously take place between the adsorbed contaminant and the aqueous solution. The latter, in turn, tries continuously to reach equilibrium with the gaseous phase. In this way, the contaminant is removed from both the aqueous phase and the (surface of the) solid one. Air-sparging removes also adsorbed contaminants from the saturated zone by creating air-water interfaces, thus enabling solid-aqueous phase, and aqueous phase-gaseous phase partitioning. The same partitioning phenomenon takes place also in the unsaturated zone.

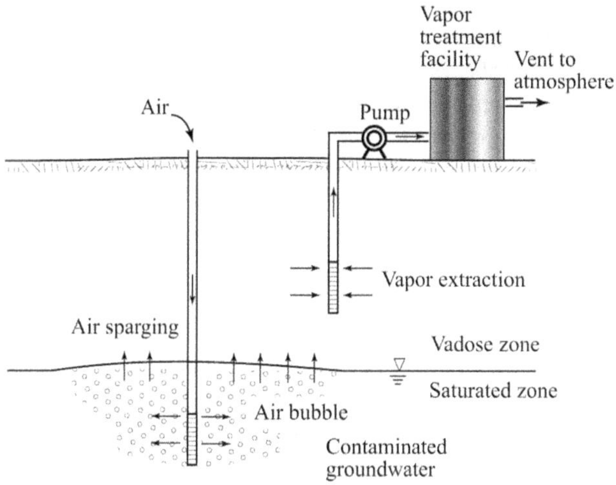

Figure 7.10.4: Air sparging together with soil vapor extraction.

An important by-product of air-sparging is the enhancement of biodegradation (of degradable organic compounds) in the saturated zone (as well as in the unsaturated one) by increasing the availability of dissolved oxygen.

Since the effectiveness of the air-sparging technique depends on the creation of air-water, or air-NAPL (microscopic) interfaces, it is important to ensure that the surface area of these interfaces be as large as possible. This is done by designing the screen and the injection well to create small stream-tubes of air (to maximize surface area). Tiny bubbles have a larger specific surface area, but to enable flow, they need to coalesce to thicker stream-tubes. However, if the air will move along a small number of thicker preferential pathways, the air-water surface area will be relatively small, thus reducing the efficiency of this technique. The creation of such pathways depends also on the homogeneity or heterogeneity of the porous medium domain surrounding the injection well. The presence of horizontal impervious lenses may adversely affect the air's flow pattern, creating zones of (almost) immobile air.

The rate at which air is circulated will determine its ability to cause volatilization of VOCs and their removal from the formation. This rate, in turn, depends on the permeability of the formation. Too low a permeability will rule out the use of air sparging as a remediation technique. The injection of air into a saturated zone of a sufficiently high permeability, will create an unsaturated, air-water zone in the vicinity of the injecting well. As long as the air is at funicular saturations (Fig. 6.1.5), effective permeability to air exists, and the air will flow and spread out around the well, both horizontally and vertically under the produced pressure gradient. In this way, the effectiveness of the system increases, as air can move through a larger volume of contaminated soil. Soil anisotropy, with horizontal permeability larger than the vertical one, will produce more lateral spreading.

Although increasing the rate of air injection (by increasing the pressure at the injection well) seems to increase the efficiency of the operation, when the rate becomes too high, the air saturation in the vicinity of the injection well (i.e., within the cone of influence) will rise, by displacing the water in the void space, This, in turn, will reduce the air-water specific interface area, say, per unit volume of porous medium, thus reducing the efficiency of the operation. The optimal rate of air injection, as well as spacing of wells, etc., may be determined by modeling the air sparging operation. Such a model may be made to simulate a single injection well and a single vapor extraction well, as well as a number of wells of both kinds distributed so as to optimize the removal of the contaminant.

More information on this technique can be found, for example, in Marley *et al.* (1992), Hinchee and Ong (1992), Johnson *et al.* (1993), Rayner *et al.* (2007), Mohamed *et al.* (2007), Klinchuch *et al.* (2007), and Aivalioti and Gidarakos (2008).

7.10.6 Permeable reactive barrier

A *permeable reactive barrier* (PRB) consists of an immobile (permeable) porous material emplaced in the subsurface to extract (e.g., by adsorption) or degrade (e.g., by chemical reactions) contaminants dissolved in groundwater passing through it (Morrison and Spangler, 1993; Day *et al.*, 1999; Benner *et al.*, 1999; Puls *et al.*, 1999; Phillips *et al.*, 2000; Scherer *et al.*, 2000; Blowes *et al.*, 2000; Czurda and Haus, 2002). Usually, the PRB is created by digging a trench of sufficient depth and length ahead of and perpendicular to an advancing plume, and filling it up to the water table by a physical, biological, or chemical reagent. The contaminants carried by the water are treated as they pass through this barrier. Sometimes, nutrients to promote the growth of indigenous microbes, or colonies of non-native microbes, tailored to a particular contaminant, are added. The barrier can also be constructed by injecting these reacting materials into the subsurface. Examples of PRBs are the placement of walls of elemental iron filings in the subsurface to chemically reduce concentrations of contaminants (Gavaskar *et al.*, 1998), such as chlorinated hydrocarbons (reductive dehalogenation), or uranium (reduction and immobilization). The transport of the dissolved contaminants to and through the PRB can be enhanced by pumping downstream of the barrier.

The advantages of PRBs are that once they are emplaced, ground surface can be restored to its previous beneficial use, and the system itself requires little if any maintenance. The disadvantages include the dependence on the slow movement of groundwater to deliver the contaminants to the reactant, and the finite life of the reactant. However, because, usually, no pumping is required, this technology may be more cost-effective than pump-and-treat.

Additional remediation techniques include:

- *Bioremediation* (or *biotransformation*, or *biodegradation*; Mercer *et al.*, 1990). Here microorganisms are introduced into the subsurface in order to degrade NAPL in-situ (or, nutrients are introduced to enhance the growth of such microorganisms).
- The use of *surfactants* to increase the solubility of hydrophobic compounds (e.g., Abdul and Ang, 1994; Mansell *et al.*, 1996; Fountain, 1997).
- Electro-kinetic enhanced remediation technique, in which a low intensity electric current between a pair (or pairs) of electrodes implanted in the subsurface, is used to mobilize a NAPL contaminant. Due to the electric potential difference between the electrodes, electro-osmosis (pore water flow induced by applied electric field) takes place within the contaminated domain (Rohrs *et al.*, 2002; Suer and Lifvergren, 2003; Niqui-Arroyo *et al.*, 2006; Yeung, 2006; De La Rosa-Perez *et al.*, 2007; Korolev *et al.*, 2008).

Information on additional remediation methods can be found in the literature, e.g., bioremediation (Atlas and Bartha, 1992; Pardieck *et al.*, 1992), thermal treatment by electrical heating (Iben *et al.*, 1996), or by steam injection (Tse *et al.*, 2001), natural bioattenuation (Salanitro, 1993), bioventing (Dupont, 1993).

Chapter 8

NUMERICAL MODELS AND COMPUTER CODES

As stated in the Preface, and emphasized repeatedly, the objective of this book is to present and discuss the underlying fundamentals, as well as the actual construction, of groundwater flow and solute transport models. Such models can predict the future behavior, e.g., in the form of water levels and solute concentrations, in specified subsurface domains. The relevant domains of interest here are aquifers and the unsaturated zone. So far, we have been focussing only on conceptual and mathematical models. We have repeatedly emphasized, and we shall do so again in Chap. 11, that optimal management decisions should not be made unless we use models to predict the consequences of implementing the proposed decision alternatives. By analyzing these consequences, or forecasts, we can make sure that constraints are not violated, and that the optimal decision alternative is, indeed, selected. Such forecasts can be made by solving the mathematical models that *simulate* the behavior in the domain of interest, in response to the implementation of various proposed alternative decisions. Unfortunately, although analytical solutions are preferable, they are seldom possible for problems of practical interest, because of the irregular boundaries of the problem domain, the heterogeneity of the domain, with respect to its physical parameters, and, sometimes, the nonlinearity of the equations. Instead, computer-based numerical methods are used in practice for solving (or 'running') these models.

The use of numerical techniques and computer codes was introduced as **Step 4** in the modeling process described in Subs. 1.2.2. The objective of this chapter is to provide some basic information on such techniques and on the use of computer programs to solve (or 'run', or 'simulate') flow and solute transport problems in practice. A number of numerical techniques are reviewed in this chapter, but no attempt is made to present a thorough, or critical, review. In each case, an example is used to demonstrate how model equations—of either a flow or a solute transport problem—are treated by the considered numerical technique. For computer codes, we have presented and discussed a selection of codes that we consider to be the more commonly used by hydrogeologist.

Nowadays, computer-aided numerical solutions are the major (perhaps, the only) tool for solving problems in practice. Sometimes, the term 'com-

J. Bear, A.H.-D. Cheng, *Modeling Groundwater Flow and Contaminant Transport*, Theory and Applications of Transport in Porous Media 23,
DOI 10.1007/978-1-4020-6682-5_8,

puter simulation' is employed. The rapid progress in computer technologies has made computers faster, with larger storage capacities, with parallel computing capabilities, etc. It is possible nowadays to solve larger and more complex problems faster, cheaper, and more accurately.

What is a 'numerical solution'? While an analytical solution seeks to determine the spatial and temporal distribution of the problem's state variables, e.g., $h = h(\mathbf{x}, t)$ and $c = c(\mathbf{x}, t)$, as continuous functions of space and time, a numerical solution provides information on these variables only at a selected set of points in space and time. Information on what happens at all other points of interest is obtained by interpolation. Actually, nowadays, there are also numerical methods that are based on tracing the movement of a set of water and solute 'particles' through space and time; we shall mention them in Sec. 8.6.

In this way, the problem is transformed from one described by a *mathematical model*, written in terms of a small number of variables, which are *continuous* functions of space and time, e.g., $h(x, y, z, t)$ and $c(x, y, z, t)$, to one described by a *numerical model*, written in terms of many *discrete* values of these variables, defined at *specified* points in space and time, e.g., h_j^n for h at a point in space marked as j and a time level marked as n. The continuous function $h(x, y, z, t)$ is, thus, *approximated* by using values of h at a set of specified points in time and space, combined with a certain interpolation procedure. The small number of *partial differential equations* (PDEs) that contain the continuous variables is replaced by a large number (often, a very large number) of linear *algebraic* equations that contain the discrete variables. Although the computer is incapable of having a routine procedure to solve a PDE in a domain with arbitrary geometry, it can solve the set of simultaneous linear equations through repetitive steps rather rapidly. The set of instructions, or commands that tell the computer how to solve these equations is called a 'computer program' or 'computer code'. Once these discrete values are solved for, we can obtain information on what happens at *every* point in the space and time domains of interest by appropriate interpolation.

It is not our intention to cite examples of numerical solutions of mathematical models, as numerous such examples can readily be found in the literature. Nor will we elaborate on specific computer codes (with the exception of one or two widely used codes), as they are many, each designed for a specific problem, or class of problems. Instead, we shall focus on some principles and basic ideas that underlie the construction of numerical models and the use of computer programs for their solution. Many codes, covering most of the problems encountered in practice, are available nowadays, either commercially, or in the public domain. Sometimes, an available code has to be modified, or extended to cover a particular problem. A summary on computer codes is given, for example, by Holzbecher and Sorek (2005). Summaries on computer codes for density dependent solute transport are given by Sorek and Pinder (1999), and Langevin *et al.* (2004). For more details on numerical techniques, the reader should consult any of the numerous textbooks available on the subject, focussing on numerical methods for the solution of boundary value problems.

The issues of code verification, the importance of using only properly verified codes, etc., are discussed in Subs. 1.2.2.

8.1 Finite Difference Methods

As mentioned earlier, the first step in most numerical methods is to replace the mathematical model, composed of partial differential equations, accompanied by initial and boundary conditions, written in terms of *continuous* state variables like $h(\mathbf{x}, t)$ and $c(\mathbf{x}, t)$, by a numerical model, written in terms of *discrete* variables, such as $h_j^n \equiv h(\mathbf{x}_j, t_n)$, which are the corresponding state variables at grid points $\mathbf{x}_j$ and times t_n.

We start by introducing a few definitions. Let h_{exac} denote the *exact solution* of the PDE and h_{FD} its *finite difference* approximation. Let h_{num} denote the numerical solution of the equations formed by the finite difference approximation. We refer to the difference $|h_{\text{exac}} - h_{\text{FD}}|$ as the *truncation error*, due to the truncation of the Taylor series expansion in the finite difference formulation (discussed below). The difference $|h_{\text{FD}} - h_{\text{num}}|$ is referred to as the *numerical roundoff error*, at a considered point j and at all time steps, due to the inability of the computer to carry numbers to an infinite number of digits. The total error is made up of the sum of the truncation error and the roundoff error. The condition for the *convergence* of a solution is that $|h_{\text{exac}} - h_{\text{FD}}| \to 0$ everywhere in the solution domain as the 'grid size', Δx, Δy, etc., is made smaller, approaching zero. The condition for *stability* is that given an error in the solution process, either from truncation, or from roundoff error, that error will not grow exponentially from one time step to the next.

With these definitions, our objective is to determine h_{num} such that over the entire space and time domains of interest, the condition $|h_{\text{exac}} - h_{\text{num}}| \leq \varepsilon$ is satisfied, where ε is an *a-priori* specified error criterion. This is normally accomplished by solving the problem with a few refinements of the mesh and of the time step, observing the difference between the successive solutions.

The first step in a typical implementation of the Finite Difference Method (abbreviated FDM) is to draw an orthogonal grid across the modeled domain. Figure 8.1.1a demonstrates a grid for a two-dimensional planar domain. The grid is obtained by dividing the axes into segments, and drawing lines parallel to the axes. The segments on the axes may be equal (uniform grid) or different (variable grid). In general, lines are made closer in areas where we wish to obtain more detailed information on the behavior of the state variable. Figure 8.1.1b shows such a variable grid.

By replacing the derivatives that appear in the PDE by approximate expressions written in terms of variable values at grid points, shown as dots in Fig. 8.1.1, we obtain the corresponding finite difference equations of the problem. Recalling that the PDE actually expresses a balance of an extensive

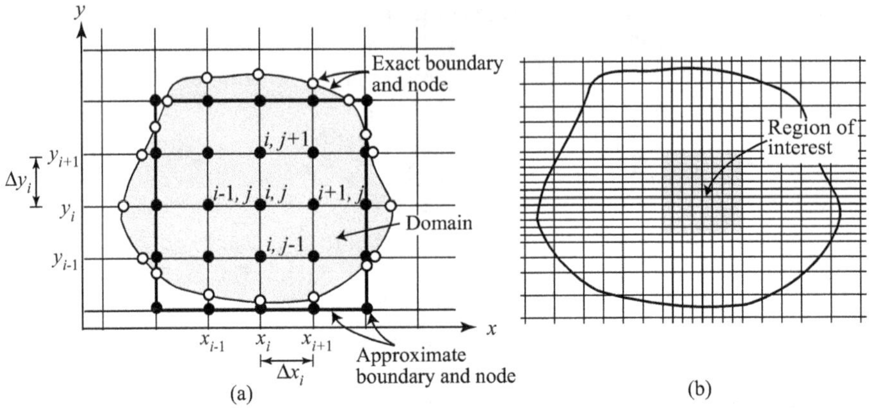

Figure 8.1.1: A grid for the grid-centered FDM: (a) Equal spacing, (b) Variable spacing.

quantity, e.g., mass, a second, physically-based, approach is to consider the balance of the considered extensive quantity in an element of area $\Delta x_i \times \Delta y_j$ (in a 2-D domain), called a 'cell'. Actually, this is the first step in developing any balance equation, using the *control box approach* (Fig. 5.1.1), but without going to the limit by letting the dimensions of the control box go to zero. In this case, the nodes at which the discrete values of the considered variable are represented are placed at the centers of the cells. We shall present this second approach in Subs. 8.1.3.

8.1.1 Laplace equation

Let us demonstrate the finite difference method by applying it to the PDE known as the *Laplace equation*:

$$\frac{\partial^2 h}{\partial x^2} + \frac{\partial^2 h}{\partial y^2} = 0. \tag{8.1.1}$$

It describes steady flow in a two-dimensional homogeneous isotropic confined aquifer; see (5.4.55).

Let $h = h(x, y)$ be a sufficiently smooth function so that it can be expanded into a *Taylor series* about x in the positive direction:

$$h(x + \Delta x, y) = h(x, y) + (\Delta x)\frac{\partial h}{\partial x}\bigg|_{x,y} + \frac{(\Delta x)^2}{2}\frac{\partial^2 h}{\partial x^2}\bigg|_{x,y} + \frac{(\Delta x)^3}{3!}\frac{\partial^3 h}{\partial x^3}\bigg|_{x,y} + \dots. \tag{8.1.2}$$

The implication here is that Δx is a small quantity, and each successive term in the above equation is of smaller and smaller magnitude, and, therefore, can be dropped in an approximation. By keeping only the first two terms on

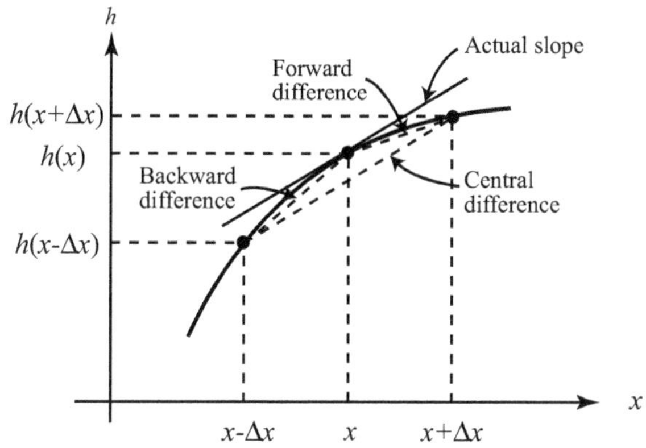

Figure 8.1.2: Geometrical interpretation of the forward, backward and central difference approximations of the derivative of a function.

the right hand side, the above equation can be rearranged as

$$\left.\frac{\partial h}{\partial x}\right|_{x,y} = \frac{h(x+\Delta x, y) - h(x,y)}{\Delta x} + \mathrm{O}(\Delta x), \tag{8.1.3}$$

where the term $\mathrm{O}(\Delta x)$ indicates that the *truncated* terms (meaning the *error* in this approximation) are 'of the order Δx'. In other words, the error, ε, is bounded by $\varepsilon < C\Delta x$, where C is a finite, positive constant, independent of Δx. Hence, as we make Δx smaller and smaller, the error will go to zero at the same rate as Δx goes to zero. The above formula is called the *forward difference approximation.* Figure 8.1.2 gives a graphical illustration of this approximation of the first derivative.

In a similar way, we may expand $h(x)$ into a Taylor series about x in the negative direction:

$$h(x-\Delta x, y) = h(x,y) - (\Delta x)\left.\frac{\partial h}{\partial x}\right|_{x,y} + \frac{(\Delta x)^2}{2}\left.\frac{\partial^2 h}{\partial x^2}\right|_{x,y} - \frac{(\Delta x)^3}{3!}\left.\frac{\partial^3 h}{\partial x^3}\right|_{x,y} + \ldots, \tag{8.1.4}$$

from which we obtain the *backward difference approximation* of the first derivative, in the form

$$\left.\frac{\partial h}{\partial x}\right|_{x,y} = \frac{h(x,y) - h(x-\Delta x, y)}{\Delta x} + \mathrm{O}(\Delta x). \tag{8.1.5}$$

By subtracting (8.1.4) from (8.1.2), we obtain the *central difference approximation* of the first derivative,

$$\left.\frac{\partial h}{\partial x}\right|_{x,y} = \frac{h(x+\Delta x,y)-h(x-\Delta x,y)}{2\Delta x} + \mathrm{O}\left((\Delta x)^2\right). \tag{8.1.6}$$

We note that this is a better approximation than either (8.1.3) or (8.1.5), as the truncation error, $\mathrm{O}\left((\Delta x)^2\right)$, is of a higher order in Δx. Figure 8.1.2 gives a graphical interpretation of the three approximations. Clearly, the central difference approximation gives a better representation of the true slope at x. Our interest here, however, lies in the second derivative. To obtain an approximation for the second derivative, we add (8.1.4) to (8.1.2), obtaining

$$\left.\frac{\partial^2 h}{\partial x^2}\right|_{x,y} = \frac{h(x+\Delta x,y)-2h(x,y)+h(x-\Delta x,y)}{(\Delta x)^2} + \mathrm{O}\left((\Delta x)^2\right). \tag{8.1.7}$$

Or, we can refer to the grid system shown in Fig. 8.1.1a, such that $h_{i,j}$ stands for $h(x_i, y_j)$, and assume constant spacing $\Delta x_{i-1} = \Delta x_i = \Delta x$, to obtain the finite difference approximation

$$\left.\frac{\partial^2 h}{\partial x^2}\right|_{i,j} \approx \frac{h_{i+1,j}-2h_{i,j}+h_{i-1,j}}{(\Delta x)^2}. \tag{8.1.8}$$

Similarly, we can develop a formula that represents an approximation of the second derivative, $\partial^2 h/\partial y^2$.

With these finite difference formulas, we can replace (8.1.1) at a grid point (i, j) (see Fig. 8.1.1a) by its approximation

$$\frac{h_{i+1,j}-2h_{i,j}+h_{i-1,j}}{(\Delta x)^2} + \frac{h_{i,j+1}-2h_{i,j}+h_{i,j-1}}{(\Delta y)^2} = 0, \tag{8.1.9}$$

where we have assumed equal spacing in both the x and y directions. The truncation error for the above expression is $\varepsilon = \mathrm{O}\left((\Delta x)^2\right) + \mathrm{O}\left((\Delta y)^2\right)$. For the special case $\Delta x = \Delta y$, we obtain from (8.1.9)

$$h_{i,j} = \frac{1}{4}\left(h_{i+1,j}+h_{i-1,j}+h_{i,j-1}+h_{i,j+1}\right), \tag{8.1.10}$$

i.e., the value of the variable at a grid point is equal to the average of its four neighboring points.

The formula for variable spacing is somewhat lengthier and can be found in any book on the finite difference method, e.g., Smith (1986) and Huyakorn and Pinder (1983).

We note that the finite difference equation, either (8.1.9) or (8.1.10), is a *linear algebraic equation.* At each node in the solution domain, where the value of h is not known *a-priori*, we can generate an equation like (8.1.9). For n such internal nodes, we have n unknown h-values, and n such equations to solve for them; hence, this is a closed set of equations. Although, in principle, such a linear system can be solved by a matrix equation solver, using an algorithm such as *Gauss elimination* (e.g., Press *et al.*, 1992), in practice,

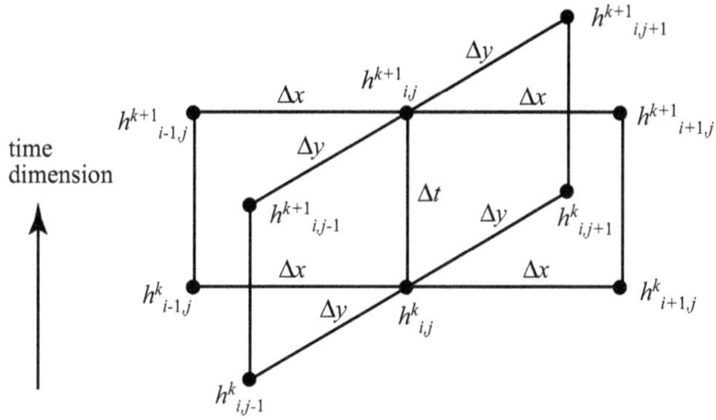

Figure 8.1.3: Grid system for the approximation in time.

such system is often solved by some iterative method (see any book on the finite difference approach, e.g., Smith, 1986).

8.1.2 Diffusion equation

The Laplace equation contains only space derivatives. As a second example, let us demonstrate the FDM by applying it to an equation that contains both space and time derivatives. This is the *diffusion-type equation* that is often encountered in groundwater flow

$$\frac{S}{\mathsf{T}}\frac{\partial h}{\partial t} = \frac{\partial^2 h}{\partial x^2} + \frac{\partial^2 h}{\partial y^2}, \tag{8.1.11}$$

in which S and T are constant over the entire domain. It describes transient flow in a two-dimensional homogeneous isotropic confined aquifer; see (5.4.53). In the FDM, the following difference formula is used for the time derivative:

$$\left.\frac{\partial h}{\partial t}\right|_{i,j,k} = \frac{h_{i,j}^{k+1} - h_{i,j}^{k}}{\Delta t}, \tag{8.1.12}$$

where the superscript k is used to denote a grid point in the time domain (Fig. 8.1.3), such that $h_{i,j}^k \equiv h(x_i, y_j, t^k)$, etc., and $\Delta t \equiv t^{k+1} - t^k$. Depending on the time reference, k, $k+1$, or $k+\frac{1}{2}$, selected for the spatial derivatives as the grid point in the time domain, this formula represents a forward, a backward, or a central difference, respectively. These approximation schemes also corresponds to the explicit, implicit, and Crank-Nicolson (1947) schemes to be discussed below.

For the spatial derivatives on the right hand side of (8.1.11), we can use the central difference approximation, (8.1.9). However, we may ponder on the question as to which time level should we assign the spatial derivatives?

Generally, we use a weighted average of the two time levels, k and $k+1$, such that (8.1.11) is approximated as

$$\frac{S}{\mathsf{T}}\frac{h_{i,j}^{k+1}-h_{i,j}^{k}}{\Delta t} = (1-\theta)\left[\frac{h_{i-1,j}^{k}-2h_{i,j}^{k}+h_{i+1,j}^{k}}{(\Delta x)^2}+\frac{h_{i,j-1}^{k}-2h_{i,j}^{k}+h_{i,j+1}^{k}}{(\Delta y)^2}\right] + \theta\left[\frac{h_{i-1,j}^{k+1}-2h_{i,j}^{k+1}+h_{i+1,j}^{k+1}}{(\Delta x)^2}+\frac{h_{i,j-1}^{k+1}-2h_{i,j}^{k+1}+h_{i,j+1}^{k+1}}{(\Delta y)^2}\right], \quad (8.1.13)$$

where $0 \leq \theta \leq 1$ is a weighting factor. In the above equation, we assume that all values of h are known at the current time level, k, as this is the nature of a well-posed problem described by the diffusion equation, and we attempt to find the values of h at time level $k+1$. In other words, values like $h_{i,j}^k$, $h_{i-1,j}^k$, etc., are known, while $h_{i,j}^{k+1}$, $h_{i-1,j}^{k+1}$, etc., are sought for.

Let us discuss three different methods for solving such equations.

Explicit scheme ($\theta = 0$). This scheme completely ignores the spatial derivative terms at time level $k+1$, so that (8.1.13) reduces to

$$h_{i,j}^{k+1} = h_{i,j}^{k} + \frac{\mathsf{T}\Delta t}{S}\left[\frac{h_{i-1,j}^{k}-2h_{i,j}^{k}+h_{i+1,j}^{k}}{(\Delta x)^2}+\frac{h_{i,j-1}^{k}-2h_{i,j}^{k}+h_{i,j+1}^{k}}{(\Delta y)^2}\right]. \quad (8.1.14)$$

The error associated with this approximation is $\varepsilon = \mathrm{O}\left((\Delta x)^2\right)+\mathrm{O}\left((\Delta y)^2\right)+\mathrm{O}(\Delta t)$. We note that since all values on the right hand side of (8.1.14) are known, we can obtain $h_{i,j}^{k+1}$ at each (i,j)-node *explicitly*, without the need for a simultaneous solution of a (linear) system of equations.

This scheme is very easy to implement. However, it has a serious drawback: the scheme is only *conditionally stable*, that is, its stability can be guaranteed only if the following condition is satisfied

$$\frac{\mathsf{T}}{S}\left[\frac{\Delta t}{(\Delta x)^2+(\Delta y)^2}\right] \leq \frac{1}{2}. \quad (8.1.15)$$

As discussed at the beginning of this subsection, in a stable scheme, any error generated by the approximation is dampened as the solution progresses in time (although new errors are constantly being generated at each time step). In an unstable scheme, any error will grow exponentially, such that after a few time steps, the numerical solution has no relevance to the true solution. The implication of (8.1.15) is that for a given grid system, the size of the time step, Δt, is restricted. In order to use a larger time increment, i.e., to reach the large-time solution sooner, the grid size, Δx and Δy, must be correspondingly reduced; this means increased computational costs.

Fully implicit scheme ($\theta = 1$). This scheme uses time level $k+1$ to represent the spatial derivative. Equation (8.1.13) then becomes

$$\frac{S}{\mathsf{T}}\frac{h_{i,j}^{k+1}-h_{i,j}^{k}}{\Delta t}=\left[\frac{h_{i-1,j}^{k+1}-2h_{i,j}^{k+1}+h_{i+1,j}^{k+1}}{(\Delta x)^2}+\frac{h_{i,j-1}^{k+1}-2h_{i,j}^{k+1}+h_{i,j+1}^{k+1}}{(\Delta y)^2}\right]. \tag{8.1.16}$$

The error is the same as in the explicit scheme, i.e., $\varepsilon = \mathrm{O}\left((\Delta x)^2\right) + \mathrm{O}\left((\Delta y)^2\right) + \mathrm{O}(\Delta t)$. The difference is that these equations are *coupled*, when written for individual nodes, and cannot be solved individually. These linear algebraic equations form a system that has to be solved simultaneously by either elimination or iteration procedures. However, a significant feature of the implicit scheme is that it is *unconditionally stable*. Hence, selecting any grid dimensions, there is no restriction on the size of the time steps used. This means that rather large time steps can be used to reach a larger time sooner. Unfortunately, as indicated by the error estimate above, the error will still grow with the size of the selected time step.

Crank-Nicolson scheme ($\theta = 1/2$). In this case, following (8.1.13), the spatial derivative terms are taken as the average at the time levels k and $k+1$. This scheme is also *implicit*, i.e., a simultaneous solution is needed. In fact, we may view this scheme as being centered at the time level $k+\frac{1}{2}$, with k and $k+1$ representing 'before' and 'after' steps, respectively. Hence, both the left and right hand sides are equivalent to a central difference approximation in time (compare with the central difference formula in space, (8.1.6)). It can be shown (Smith, 1986) that the error associated with this scheme is $\varepsilon = \mathrm{O}\left((\Delta x)^4\right) + \mathrm{O}\left((\Delta y)^4\right) + \mathrm{O}\left((\Delta t)^2\right)$, which is superior to the other two schemes. The Crank-Nicolson scheme is also unconditionally stable; hence, it is widely used for the solution of diffusion-type equations.

8.1.3 Cell-centered approach

As mentioned earlier, a second approach for the development of a finite difference scheme is to use a *cell-centered*, rather than a *grid-centered*, approximation. Particularly, the cell-centered approach is adopted in the widely used computer code, MODFLOW, developed by the U.S. Geological Survey (see Sec. 8.8). The cell-centered approach may be considered as a special case of the *Finite Volume Method* (Loudyi, 2005) discussed in Sec. 8.2.

The cell-centered approach is based on the mass (or any other relevant extensive quantity) balance of the cell. We shall illustrate this approach by using the flow equation for an inhomogeneous confined aquifer, (5.4.53). Let us consider the grid system shown in Fig. 8.1.4, with rectangular cells identified by the column and row numbers. For cell (i,j), the piezometric head is represented at the center of the cell, as $h_{i,j}$, and the discharges are represented at the mid-points of the four boundary segments; they are denoted by $(Q_x)_{i-\frac{1}{2},j}$, etc. For this cell, the mass (or volume, for incompressible fluid) balance requires that the excess of inflow over outflow be equal to the added

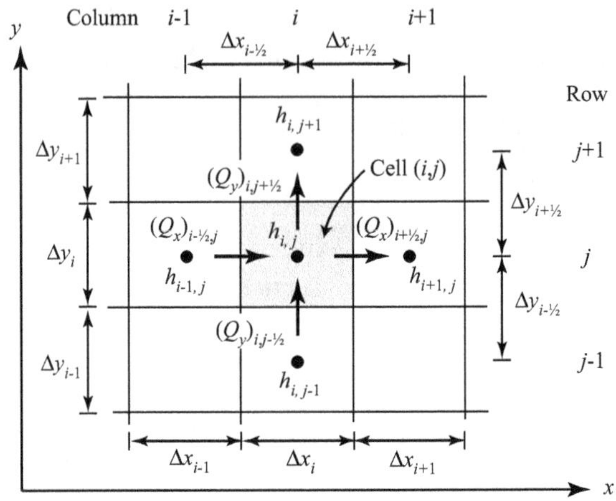

Figure 8.1.4: Cell-centered finite difference approach.

storage in the cell (see Sec. 5.1):

$$S_{i,j}\Delta x_i \Delta y_i \frac{h_{i,j}^{k+1} - h_{i,j}^{k}}{\Delta t} = (Q_x)_{i-\frac{1}{2},j} - (Q_x)_{i+\frac{1}{2},j}$$
$$+(Q_y)_{i,j-\frac{1}{2}} - (Q_y)_{i,j+\frac{1}{2}} + N_{i,j}\Delta x_i \Delta y_i - P_{i,j}\Delta x_i \Delta y_i, \qquad (8.1.17)$$

where $S_{i,j}$ is the storage coefficient in cell (i,j), and $N_{i,j}$ and $P_{i,j}$ are, respectively, the recharge and the pumping rates within that cell. Particularly, if a pumping or a recharge well exists within a cell, with discharge Q_w (negative for recharge), then $P_{i,j} = Q_w/\Delta x_i \Delta y_i$, i.e., effectively spreading the pumping over the entire cell area.

Similar to the grid-centered approach, we may raise the question: at which time step should the discharge be evaluated? This leads to the same considerations as in the explicit, implicit, and Crank-Nicolson schemes, discussed above. In the implicit scheme (employed in the MODFLOW computer code), we express the discharge by

$$(Q_x)_{i-\frac{1}{2},j} = -\mathsf{T}_{i-\frac{1}{2},j}\Delta y_j \frac{h_{i,j}^{k+1} - h_{i-1,j}^{k+1}}{\Delta x_{i-\frac{1}{2}}}$$
$$(Q_x)_{i+\frac{1}{2},j} = -\mathsf{T}_{i+\frac{1}{2},j}\Delta y_j \frac{h_{i+1,j}^{k+1} - h_{i,j}^{k+1}}{\Delta x_{i+\frac{1}{2}}}, \qquad (8.1.18)$$

with a similar expression for the Q_y terms. In the above equation, we note that:

$$\Delta x_{i-\frac{1}{2}} = \frac{\Delta x_{i-1} + \Delta x_i}{2}, \quad \Delta x_{i+\frac{1}{2}} = \frac{\Delta x_i + \Delta x_{i+1}}{2}. \qquad (8.1.19)$$

The transmissivity can be approximated as either a simple average

$$\mathsf{T}_{i-\frac{1}{2},j} = \frac{\mathsf{T}_{i-1,j} + \mathsf{T}_{i,j}}{2}, \quad \text{etc.,} \tag{8.1.20}$$

or as a *harmonic mean*

$$\mathsf{T}_{i-\frac{1}{2},j} = \frac{\Delta x_{i-1} + \Delta x_i}{(\Delta x_{i-1}/\mathsf{T}_{i-1,j}) + (\Delta x_i/\mathsf{T}_{i,j})}, \quad \text{etc.} \tag{8.1.21}$$

Combining (8.1.17) and (8.1.18), we obtain a linear equation in terms of the (cell-centered) nodal values of h, similar to (8.1.16). An iterative procedure is typically used for the solution of the simultaneous system of equations.

8.1.4 Boundary and boundary conditions

As illustrated in Figs. 8.1.1 and 8.1.4, the cells formed by the FDM grid have either a rectangular (2-D) or a cubical (3-D) shape. It is usually difficult to use these shapes to fit an irregular geometry of the boundary of a considered domain. In Fig. 8.1.1, we demonstrate two options for fitting the geometry: either we create nodes located on the boundary (marked as white dots) so as to represent the boundary more accurately, or we use the regular grid system (marked by black dots) to obtain a rough approximation of the boundary. In the former case, special finite difference formula and extra bookkeeping efforts are needed in order to keep track of the boundary nodes. A strength of the FDM, in contrast to other numerical methods, such as the Finite Volume and Finite Element methods, is that the nodes are arranged in a regular pattern. In this way, they can be referred to by their row and column numbers (or array indices in 3-D cases), making their relations with neighboring nodes easily identifiable, without too much bookkeeping. To retain this advantage, the latter approach of using a regular grid to approximate the geometry is often adopted.

Figure 8.1.5 gives an example of how a solution domain is approximated by the cell-centered FDM. All cells with (cell-centered) nodes falling within the domain are considered domain cells. As in many codes, e.g., in MODFLOW, all domain cells are designated as *variable-head cells*, meaning that the head values are unknown in such cells, and have to be solved for. The finite difference equation (8.1.17) is applicable to each variable-head cell. Cells with nodes falling *immediately outside* the boundary, or *on* the boundary, are referred to as *boundary cells.*

We consider two types of boundary conditions: Type 1: a boundary of prescribed head (Dirichlet condition), and Type 2: an impermeable boundary (Neumann condition). Accordingly, we assign two types of boundary cells: *constant-head cells* and *no-flow cells.* A constant-head cell means that the head value is prescribed on the cell (node), although its value may change with

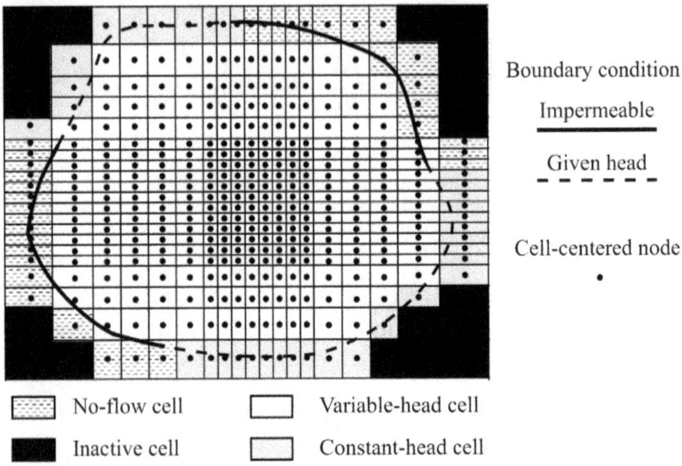

Figure 8.1.5: The approximation of a solution domain and the designation of boundary conditions by the cell-centered finite difference approach.

time, depending on the given boundary condition. The no-flow cell means that no discharge is allowed to cross its boundary.

We may use Fig. 8.1.4 to illustrate the implementation of these boundary conditions. We assume that the center cell, (i, j), is a variable-head cell, while the cell to the left, $(i-1, j)$, is a constant-head cell. The discharge term $(Q_x)_{i-\frac{1}{2},j}$ in (8.1.17) is expressed by the first equation in (8.1.18), with $h^{k+1}_{i-1,j}$ replaced by the known boundary head value. On the other hand, if the left cell is a no-flow cell, we simply set $(Q_x)_{i-\frac{1}{2},j} = 0$ in (8.1.17). The finite difference equation (8.1.17) is not applied on these boundary cells, as there is no unknown value of the variable there. Finally, cells outside the boundary cells are called *inactive cells*, as no action is required on them.

Other boundary conditions can be modeled in a similar way. For example, a *constant-flow cell* can be assigned on a boundary with prescribed (non-zero) normal specific discharge, q_n (Subs. 5.2.3B).

In this case, the discharge $(Q_x)_{i-\frac{1}{2},j}$ in (8.1.17), contributed by the constant flow cell, $(i-1, j)$, to the adjacent variable-head cell, (i, j), can be assigned the value

$$(Q_x)_{i-\frac{1}{2},j} = q_x \Delta y_j B_{i-\frac{1}{2},j}, \tag{8.1.22}$$

where q_x is the x-component of q_n, and B is the aquifer's thickness.

When the aquifer is adjacent to a river with a semi-permeable bed (see Subs. 5.2.3C), the flux contribution to and from the river is proportional to the head difference between the river and the adjacent cell. This leads to the third type (Robin) boundary condition,

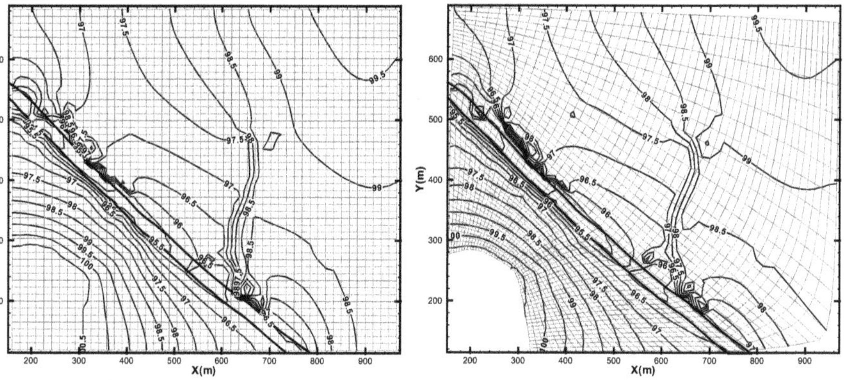

(a) (b)

Figure 8.2.1: Comparison of (a) An FDM Cartesian grid (50 × 50 mesh), and (b) An FVM non-orthogonal grid (50 × 50 mesh) (Loudyi *et al.*, 2007). (Figure courtesy of D. Loudyi and R.A. Falconer.)

$$(Q_x)_{i-\frac{1}{2},j} = \frac{h_R - h_{i,j}}{c_r} \Delta y_j B_{i-\frac{1}{2},j}, \tag{8.1.23}$$

where h_R is the head in the river, and c_r is the resistance (= reciprocal of the *leakance*) of the semipervious layer. Other boundary conditions (Subs. 5.2.3) can be modeled in a similar way.

8.2 Finite Volume Methods

The *Finite Volume Method* (FVM), which was originated in computational fluid dynamics (CFD) to address some of the difficulties of the FDM, has become one of the most popular methods used in CFD (Leveque, 2002; Versteeg and Malalasekera, 1995). A deficiency of the FDM, as described in the preceding section, is that it relies on a regular, Cartesian mesh. As a result, it can create relatively large errors around the boundary of a considered domain. The FVM allows the distortion of a rectilinear mesh to conform to the geometry of a problem's boundary. The resulting 'cells', in a two-dimensional domain, are simple convex *quadrilaterals*. Figure 8.2.1 illustrates both a Cartesian FDM-grid, and a distorted FVM-grid (Loudyi *et al.*, 2007). The distorted grid not only conforms better to the problem domain, but can also be made denser in regions where the solution is expected to have a steeper variation, thus leading there to a more accurate solution. This distorted Cartesian grid is a 'structured' grid, as it is possible to keep track of the cells and grid nodes by their row and column numbers. It is, however, still somewhat limited in its ability to conform to a complex geometry of a problem boundary. A more conforming 'unstructured' cell system can be developed for the FVM by using triangular shaped elements, similar to those employed in the FEM (Sec. 8.3).

A possible option is to use polygonal elements formed by the Thiessen network (e.g., Bear, 1979, p. 447). The trade-off is that, like for the FEM, a more complex mathematical formulation, a more tedious preparation of input data, and more bookkeeping, are needed. Altogether, this is not the preferred approach for FVM. Hence, in this section, the discussion is limited to *structured quadrilateral elements.*

Another advantage of the FVM is that it is based on the integration of the principle of conservation of extensive quantities, such as mass, heat, and momentum. Unlike the node-centered FDM, where the conservation principle is satisfied only if the grid size approaches zero, the FVM conserves the extensive quantity within each local cell. The cell-centered FDM has a similar property. In fact, the FDM can be considered as a special case of the FVM, using a Cartesian grid. Other advantages of the FVM include the ability to deal with discontinuities, such as a shock front (i.e., a stationary or moving front of solution discontinuity), and with discontinuities in material properties. The (grid-centered) FDM relies on the existence of derivatives and cannot easily satisfy the conservation principles and discontinuity conditions.

As an example of an extensive quantity, let us consider the mass of a dissolved species. The problem of transport of a single dissolved species without adsorption in a saturated porous medium, as described by (7.4.3), is repeated here as

$$\phi\frac{\partial c}{\partial t} = -\nabla\cdot\phi(c\mathbf{V} - \mathbf{D}_h\cdot\nabla c) + \phi\,\rho\,\Gamma^{\gamma}. \tag{8.2.1}$$

In this equation, we assume that the flow velocity, $\mathbf{V}$, is already known from the solution of the corresponding flow equation, and the solute concentration, c, is sought from the solution of (8.2.1), subject to appropriate initial and boundary condition. We can now average the above equation by integrating it over the finite volume, Ω_i, of cell i,

$$\int_{\Omega_i}\phi\frac{\partial c}{\partial t}\,d\mathbf{x} = -\int_{\Omega_i}\nabla\cdot(\phi c\mathbf{V})\,d\mathbf{x} + \int_{\Omega_i}\nabla\cdot(\phi\,\mathbf{D}_h\cdot\nabla c)\,d\mathbf{x} + \int_{\Omega_i}\phi\,\rho\,\Gamma^{\gamma}\,d\mathbf{x}, \tag{8.2.2}$$

in which $d\mathbf{x} \equiv dx\,dy\,dz$. By applying the *divergence (Gauss) theorem*, we can transform the first two terms on the right hand side into surface integrals:

$$\int_{\Omega_i}\phi\frac{\partial c}{\partial t}\,d\mathbf{x} = -\int_{\partial\Omega_i}\phi\,c\,\mathbf{V}\cdot\mathbf{n}\,d\mathbf{x} + \int_{\partial\Omega_i}\phi\,(\mathbf{D}_h\cdot\nabla c)\cdot\mathbf{n}\,d\mathbf{x} + \int_{\Omega_i}\phi\,\rho\,\Gamma^{\gamma}\,d\mathbf{x}, \tag{8.2.3}$$

where $\partial\Omega_i$ is the surface area of the boundary of Ω_i.

Equation (8.2.3), which is the integral form of the mass balance equation (8.2.1), is the basis for the FVM approximation. While in the FDM, the differential equation is approximately satisfied *at a point*, in the FVM, the integral equation is approximately satisfied in a *finite volume cell* of finite

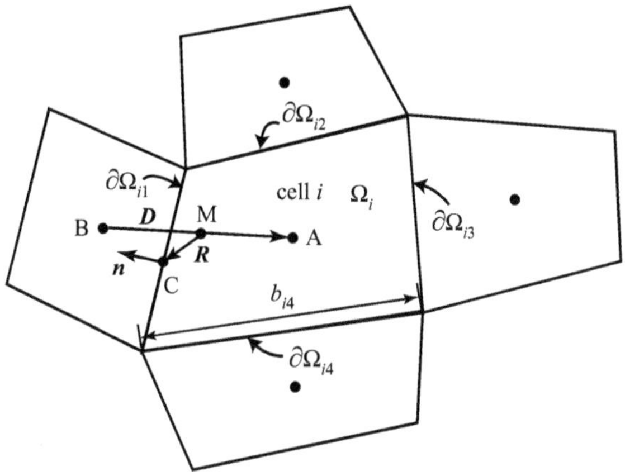

Figure 8.2.2: Schematic view of a FVM quadrilateral cell system.

size. As in all numerical methods, (8.2.3) needs to be transformed into a discrete form. Similar to other numerical methods, there are different versions of FVM. For example, the FVM can be cell-centered, edge-centered, or vertex-centered (Lewis *et al.*, 2006). The grid system can be a *structured Cartesian grid* (Fig. 8.2.1a), or a *structured, non-Cartesian, quadrilateral grid* (Fig. 8.2.1b), or a *totally unstructured, arbitrary polygonal grid.* In fact, the cell-centered, Cartesian, uniform grid FVM is the same as the cell-centered FDM, introduced in Subs. 8.1.3; hence, there is no need to discuss this special, but widely used, case here. Instead, we shall demonstrate below how to approximate (8.2.3), using a structured quadrilateral grid.

Consider the two-dimensional finite volume cell with index i, together with its neighboring cells, shown in Fig. 8.2.2. Note that although a quadrilateral cell with four sides is shown here, most of the discussion below is not limited to this number of cell sides.

First, we rewrite the left hand side of (8.2.3) in the form

$$\int_{\Omega_i} \phi \frac{\partial c}{\partial t}\, d\mathbf{x} = |\Omega_i| \phi_i \frac{\partial c_i}{\partial t} = |\Omega_i| \phi_i \frac{c_i^{k+1} - c_i^k}{\Delta t}, \tag{8.2.4}$$

where ϕ_i is the porosity, assumed to be constant over the cell, and c_i is the average concentration over the ith cell, assigned to the node located at the cell's (geometric) center, superscripts k and $k+1$ denote time steps, and $|\Omega_i|$ is the area of the cell. The two surface integral terms in (8.2.3), representing the advective and dispersive fluxes, can be summed over the surfaces of the finite volume cell, $\partial\Omega_{ij}$, $j = 1, \ldots, n$, and approximated as

$$\sum_{j=1}^{n} \int_{\partial\Omega_{ij}} \phi\, c\, \mathbf{V} \cdot \mathbf{n}\, d\mathbf{x} = \sum_{j=1}^{n} b_{ij}\, \phi_{ij}\, c_{ij}\, \mathbf{V}_{ij} \cdot \mathbf{n}_{ij}, \tag{8.2.5}$$

$$\sum_{j=1}^{n} \int_{\partial\Omega_{ij}} \phi\,(\mathbf{D}_h \cdot \nabla c) \cdot \mathbf{n}\, d\mathbf{x} = \sum_{j=1}^{n} b_{ij}\, \phi_{ij}\,(\mathbf{D}_{h(ij)} \cdot (\nabla c)_{ij}) \cdot \mathbf{n}_{ij}, \qquad (8.2.6)$$

where c_{ij}, $\mathbf{V}_{ij}$, etc., are the surface averaged values of c and $\mathbf{V}$, respectively, assigned to the mid-point of the edge, $\partial\Omega_{ij}$, and b_{ij} is the length of $\partial\Omega_{ij}$ (Fig. 8.2.2).

We note that in the cell-centered scheme, the discrete, unknown concentration values are assigned to the cell center, denoted as points A and B in Fig. 8.2.2. However, in (8.2.5) and (8.2.6), values of concentrations and concentration gradients, c_{ij} and $(\nabla c)_{ij}$, are referred to the center of the boundary segment, e.g., point C in Fig. 8.2.2. It is necessary to assign these edge concentration values to the center node. Although there are many different ways to approximate the concentration and concentration gradients at node C (Karimian and Straatman, 2006), only one technique will be presented here as an illustration. According to Demirdžić *et. al* (1992), the following approximation

$$c_C = \frac{1}{2}(c_A + c_B) + \frac{1}{2}[(\nabla c)_A + (\nabla c)_B] \cdot \mathbf{R}, \qquad (8.2.7)$$

is of second order accuracy (equivalent to a central difference scheme). The vector $\mathbf{R}$ is marked in Fig. 8.2.2; the point M is shown as the mid-point of the segment AB. Through the above equation, the concentration at the boundary node C is decomposed into a linear combination of concentrations and concentration gradients at the cell-center nodes, A and B. The following approximation formula relates the concentration gradient at node C, to the values at cell-center nodes:

$$(\nabla c)_C = \frac{c_A - c_B}{\mathbf{D} \cdot \mathbf{n}_C}\,\mathbf{n}_C + \left\{ \frac{(\nabla c)_A + (\nabla c)_B}{2} - \frac{[(\nabla c)_A + (\nabla c)_B] \cdot \mathbf{D}}{2\mathbf{D} \cdot \mathbf{n}_C}\,\mathbf{n}_C \right\}. \qquad (8.2.8)$$

This expression is second order accurate (Demirdžić and Muzafarija, 1995).

Finally, it is necessary to approximate the concentration gradients at A and B, $(\nabla c)_A$ and $(\nabla c)_B$, by the nodal concentration values. One way to construct this relation is through the approximation:

$$(\nabla c)_A \cdot \mathbf{D} = c_B - c_A. \qquad (8.2.9)$$

We note that two unknowns, $(\partial c/\partial x)_C$ and $(\partial c/\partial y)_C$, to be solved for, appear in the single equation (8.2.8). This relation, however, can also be constructed with respect to each of the neighboring cells. Hence, based on either triangular, quadrilateral, or other polygonal cells, we always have more equations than required for solving for the two unknowns. These two components of the gradient are, therefore, solved for by the *least square method*, to obtain the *optimal solution*. Through back substitution, we note that (8.2.7) and

(8.2.8), and then (8.2.5) and (8.2.6), and finally (8.2.3), are expressed in terms of concentration values at the cell-centered nodes. We also note that the approximation algorithm described above is not unique; there exist other approximation schemes that assign velocity, concentration, etc., to cell centers, edge centers, or vertices. A comparison of the accuracy of the different approximation schemes can be found in Loudyi (2005).

There remains one more issue to be addressed—at which time level should all these variables be represented? As with the FDM, we can select time level k for an *explicit scheme*, level $k+1$ for an *implicit scheme*, or $k+\frac{1}{2}$ for a *Crank-Nicolson scheme*. In the actual implementation of FVM, however, a mixed scheme is typically used for the different terms. Particularly, since the gradient needs to be calculated by the least square method, it is, typically, modeled by the explicit scheme; otherwise, iterations will be required within any given time step.

8.3 Finite Element Methods

The origin of the *Finite Element Method* (FEM) may be traced to Courant (1943). He used the works of Ritz (1908) and Galerkin (1915), which are based on the *variational integral representation* (virtual work) of a PDE and on the minimization of such functional to obtain approximate solutions to elliptic PDEs. In Courant's approximate solution, the domain was subdivided into a number of triangular subdomains (elements), and the solution was represented, piecewise, in each subdomain as a summation of simple *basis functions*, such as polynomials. The full development of the method, however, had to wait until the wide availability of electronic computers. Driven by the needs of the aerospace industry, Clough (1960) and co-workers applied the method to the solution of problems in structural mechanics, and coined the term *Finite Element Method* (FEM).

Because the FEM uses triangular and various irregularly shaped elements, the subdivision of a considered domain can be *unstructured*, meaning that the nodes forming the elements can be arbitrarily scattered in the domain in sparse or concentrated patterns to form elements of various sizes. This flexibility is useful not only for irregularly shaped domains, but also for concentrating elements in regions where larger variations exist in the considered variable, and where a better accuracy is required. The rectilinear grid based FDM has some difficulty to conform to these requirements.

There exist a few different theoretical formulations for the FEM. In structural mechanics, the development of the FEM is, usually, based on the physical consideration of the energy principle. The latter takes the form of the virtual work principle, or the principle of minimum total potential energy. This formulation has its root in the Ritz method mentioned above. For *field problems*, such as problems governed by the Laplace or the diffusion equations, a different finite element formulation, known as the *weighted residual method* (Zienkiewicz, 1971), particularly the version called *Galerkin method*, is often used. Only the weighted residual formulation is discussed here. A more complete presentation of the FEM, as applied to porous medium flow and transport, can be found in Pinder and Gray (1977), and Huyakorn and Pinder (1983).

8.3.1 Weighted residual methods

Let us start by presenting the general concept of the *weighted residual method.* Consider a function of space, $u(\mathbf{x})$, which can be a scalar, or a vector, satisfying a (system of) PDEs. We can, symbolically, express the system of equations in the following operator form:

$$\mathcal{L}\left(u(\mathbf{x})\right) = f(\mathbf{x}), \quad \mathbf{x} \in \Omega, \tag{8.3.1}$$

where $\mathcal{L}$ is a partial differential operator, f is the inhomogeneous right hand side, and Ω is the solution domain, bounded by the surface $\partial\Omega$. The above system is subject to a set of *essential* and *natural* boundary conditions to form a well-posed system,

$$\begin{aligned} \mathcal{S}(u(\mathbf{x})) &= g_1(\mathbf{x}), \quad \mathbf{x} \in \partial\Omega_S, \\ \mathcal{N}(u(\mathbf{x})) &= g_2(\mathbf{x}), \quad \mathbf{x} \in \partial\Omega_N, \end{aligned} \tag{8.3.2}$$

where $\mathcal{S}$ and $\mathcal{N}$ are the partial differential operators corresponding to these conditions, and $\partial\Omega_S$ and $\partial\Omega_N$ denote the boundary segments that correspond to these conditions, with $\partial\Omega_S \cup \partial\Omega_N = \partial\Omega$, $\partial\Omega_S \cap \partial\Omega_N = \emptyset$, and $\partial\Omega_S \neq \emptyset$. By 'essential boundary condition' we mean that values are given to the function u itself (as in the Dirichlet condition), while by 'natural boundary condition' we mean that boundary values are given in the form of derivatives of the function (e.g., as in the Neumann condition).

One basic approach to finding an approximate solution is to express it as a series expansion; for example,

$$u(\mathbf{x}) \approx \hat{u}(\mathbf{x}) = \sum_{i=1}^{n} a_i N_i(\mathbf{x}), \quad \mathbf{x} \in \Omega, \tag{8.3.3}$$

where $\hat{u}$ denotes the approximate solution, N_i are some given basic functions, known as the *basis functions*, and a_i are a set of unknown coefficients to be determined. Examples of basis functions include polynomials (e.g., Lagrangian, or Chebyshev polynomials), sine and cosine functions (as in the Fourier series), exponential function, etc. It is desirable to select these basis functions from a *complete set* of functions, which can represent any functions of a given class.

We note that the approximation function on the right hand side of (8.3.3) is defined for the entire domain Ω; hence, it is a *global* interpolation function. Depending on the selection of the basis function, global interpolation is typically highly continuous, meaning that it can be differentiated many times.

The *finite element* idea, on the other hand, is based on *local* interpolation functions that are piecewise continuous. The finite element formulation will be discussed in Subs. 8.3.2.

To ensure that (8.3.3) is indeed an approximate solution of the given system, we have to show that the governing equation (8.3.1), as well as the boundary condition (8.3.2), are satisfied at least in an approximate sense. Our goal is to express these requirements in terms of a set of *linear algebraic equations* defined in terms of the unknown coefficients a_i, and to solve this linear system for these coefficients. In essence, this general procedure defines most numerical methods. One specific procedure for achieving this goal is the *weighted residual method* described below.

Since the approximate solution, $\hat{u}$, needs to satisfy the governing equation, we can substitute it on the left hand side of (8.3.1), and find its difference with respect to the exact expression:

$$\mathcal{L}(\hat{u}(\mathbf{x})) - \mathcal{L}(u(\mathbf{x})) = \mathcal{L}(\hat{u}(\mathbf{x})) - f(\mathbf{x}) = R(\mathbf{x}), \tag{8.3.4}$$

where R is called the (error) *residual.* Similar expressions can be defined for the boundary conditions. The residual is not the solution error itself, which is given by $\hat{u} - u$; instead, it is an indication of error. Particularly, if we can make the residual equal to zero everywhere, i.e., $R = 0$ for all $\mathbf{x} \in \Omega$ and $\partial\Omega$, then we have the exact solution. Since this is not possible for the approximate solution, we seek to minimize the residual in a certain mathematical sense. The *method of weighted residual* seeks to minimize the residual as a *weighted average* over the domain, i.e.,

$$\int_\Omega R(\mathbf{x})\, w_i(\mathbf{x})\, d\mathbf{x} = \int_\Omega [\mathcal{L}(\hat{u}(\mathbf{x})) - f(\mathbf{x})]\, w_i(\mathbf{x})\, d\mathbf{x} = 0, \tag{8.3.5}$$

for a finite set of *weighting functions*, w_i, $i = 1, \ldots, n$. Since it is also necessary to satisfy the boundary conditions, (8.3.2), the following boundary residuals also need to be minimized:

$$\int_{\partial\Omega_S} [\mathcal{S}(\hat{u}(\mathbf{x})) - g_1(\mathbf{x})]\, w_i(\mathbf{x})\, d\mathbf{x} + \int_{\partial\Omega_N} [\mathcal{N}(\hat{u}(\mathbf{x})) - g_2(\mathbf{x})]\, w_i(\mathbf{x})\, d\mathbf{x} = 0. \tag{8.3.6}$$

Here, the weighting functions, w_i, need not be the same as those in (8.3.5), and the weights for the two types of boundary conditions may be different.

As pointed out by Brebbia and Dominguez (1977), different choices of weighting function lead to different numerical methods. For example, the finite difference method can be considered as a method that uses the *Dirac delta function* as the weighting function, while also enforcing the boundary conditions to be exact. The boundary element method (Brebbia, 1978) (Sec. 8.4) uses the *fundamental solution* as the weighting function. The Galerkin method, on the other hand, uses the same basis function of the approximate solution as the weighting function.

To gain a basic understanding of the weighted residual method, let us present three different versions: the collocation, the subdomain, and the Galerkin methods.

A. Collocation method

If, in the weighted residual formula, (8.3.5), we use the Dirac delta function as the weighting function, i.e., $w_i(\mathbf{x}) = \delta(\mathbf{x} - \mathbf{x}_i)$, we are no longer integrating over a region; instead, we pick up a discrete value at $\mathbf{x}_i$. Equation (8.3.5) becomes

$$\mathcal{L}(\hat{u}(\mathbf{x}))\big|_{\mathbf{x}=\mathbf{x}_i} = f(\mathbf{x}_i), \quad \mathbf{x}_i \in \Omega. \tag{8.3.7}$$

For example, if the approximate solution $\hat{u}$, as given by (8.3.3), is used, its substitution into (8.3.7) yields

$$\mathcal{L}\left\{\sum_{j=1}^{n} a_j N_j(\mathbf{x})\right\}_{\mathbf{x}=\mathbf{x}_i} = \sum_{j=1}^{n} a_j \mathcal{L}\left\{N_j(\mathbf{x})\right\}_{\mathbf{x}=\mathbf{x}_i} = f(\mathbf{x}_i). \tag{8.3.8}$$

In the above equation, we notice that the PDE operator, $\mathcal{L}$, operates only on the basis functions N_j, which are some simple algebraic expressions, such as polynomials, to interpolate the solution. For a given PDE operator (governing equation), and a given choice of basis functions (of which there are many possible choices), the differentiations contained in the expression $\mathcal{L}\{N_j(\mathbf{x})\}$ can be analytically carried out. With the substitution of $\mathbf{x} = \mathbf{x}_i$, the expression becomes a known constant. Hence, (8.3.8) is a *linear algebraic equation* with the coefficients a_j as unknowns. Enforcing (8.3.8) at a point $\mathbf{x}_i \in \Omega$ ensures that the governing equation is exactly satisfied at that location. To have the governing equation approximately satisfied over the entire solution domain, we select a finite set of points, $\mathbf{x}_i$, $i = 1, \ldots, m_1$, distributed all over the domain, and enforce the governing equation (8.3.8) on them. As a result, we obtain a set of linear algebraic equations.

To satisfy the boundary conditions, it is also necessary to enforce (8.3.6) on a set of boundary points, i.e.,

$$\begin{aligned} \mathcal{S}(\hat{u}(\mathbf{x}_i)) &= g_1(\mathbf{x}_i), \quad \mathbf{x}_i \in \partial\Omega_S, \\ \mathcal{N}(\hat{u}(\mathbf{x}_i)) &= g_2(\mathbf{x}_i), \quad \mathbf{x}_i \in \partial\Omega_N. \end{aligned} \tag{8.3.9}$$

Following the same procedure that transforms (8.3.7) into (8.3.8), we also obtain a set of linear algebraic equations with the coefficients a_j as unknowns. If we select m_1 points to enforce the governing equation, and m_2 points for the boundary conditions, and let $m = m_1 + m_2 = n$, where n is the number of terms in the series approximation of $\hat{u}$, we obtain a system of linear algebraic equations that contain n unknowns (a_j, $j = 1, \ldots, n$), and n equations. The system can be solved for the unknown coefficients a_j. With a_j given, the approximate solution $\hat{u}$, as defined in (8.3.3), is found.

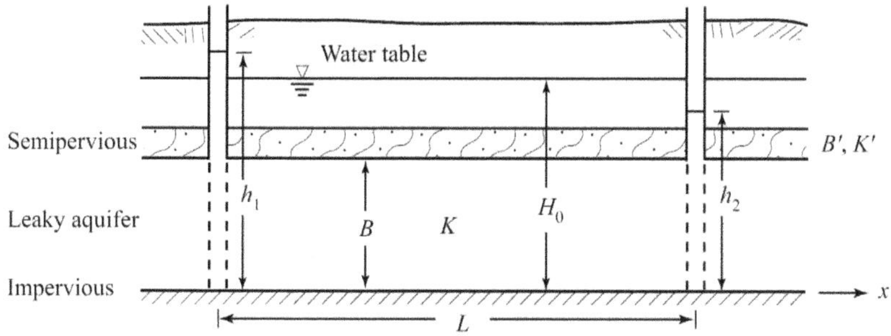

Figure 8.3.1: Flow in a leaky confined aquifer.

As an illustration, let us consider the problem of the leaky confined aquifer shown in Fig. 8.3.1. Assuming a constant head, H_o, in the upper, phreatic aquifer, and an impervious base at the bottom, the governing equation for a steady, one-dimensional flow is given by the simplified from of (5.4.58):

$$\frac{d^2h}{dx^2} + \frac{H_o - h}{\lambda} = 0, \tag{8.3.10}$$

where $\lambda = \mathsf{K}BB'/\mathsf{K}'$ is the leakage factor, K and K' are, respectively, the hydraulic conductivity of the leaky aquifer and of the semipervious layer, and B and B' are the thicknesses of the layers. Given the boundary conditions

$$h(0) = h_1 \text{ and } h(L) = h_2, \tag{8.3.11}$$

the above problem can be solved to yield the exact solution

$$h = H_o + C_1 e^{x/\sqrt{\lambda}} + C_2 e^{-x/\sqrt{\lambda}}, \tag{8.3.12}$$

where

$$C_1 = \frac{e^{L/\sqrt{\lambda}}(H_o - h_2) - (H_o - h_1)}{1 - e^{2L/\sqrt{\lambda}}},$$
$$C_2 = \frac{e^{2L/\sqrt{\lambda}}(H_o - h_1) - e^{L/\sqrt{\lambda}}(H_o - h_2)}{1 - e^{2L/\sqrt{\lambda}}}. \tag{8.3.13}$$

For $\mathsf{K} = 100$ m/day, $\mathsf{K}' = 1$ m/day, $B = 10$ m, $B' = 20$ m, $H_o = 36$ m, $h_1 = 32$ m, $h_2 = 24$ m, and $L = 1{,}000$ m, the head distribution between $x = 0$ and L is plotted as the solid line in Fig. 8.3.2.

To find an approximate solution, we assume that it can be represented by the polynomial

$$h \approx \hat{h} = a_0 + a_1 x + a_2 x^2 + a_3 x^3 + a_4 x^4. \tag{8.3.14}$$

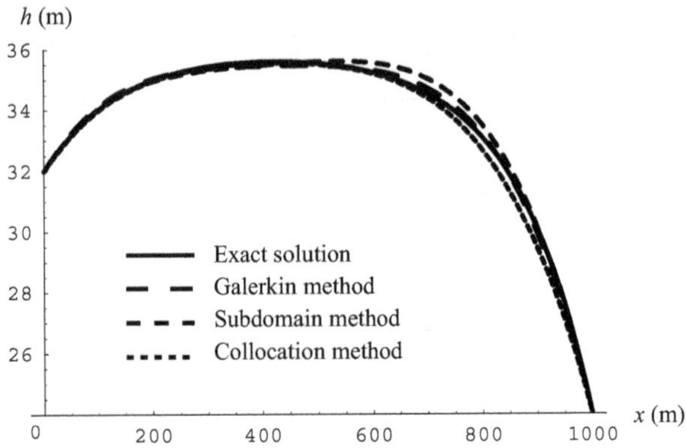

Figure 8.3.2: Comparison of exact and approximate solutions: Exact solution (8.3.12); collocation method (8.3.18); subdomain method (8.3.21); and Galerkin method (8.3.31).

It is obvious that if we include more terms on he right hand side of (8.3.14), the solution will be better represented. However, the price of such improved accuracy is an increase in computational efforts.

The approximate solution needs to satisfy the two boundary conditions in (8.3.11); hence, we obtain the two equations

$$\begin{aligned} 32 &= a_0, \\ 24 &= a_0 + 10^3\, a_1 + 10^6\, a_2 + 10^9\, a_3 + 10^{12}\, a_4. \end{aligned} \tag{8.3.15}$$

Inside the domain, the error of this approximation can be found by substituting the approximate solution (8.3.14) into the governing equation (8.3.10), to obtain the residual R, defined in (8.3.4),

$$R = 0.00005(36 - a_0 - a_1 x - a_2 x^2 - a_3 x^3 - a_4 x^4) + 2a_2 + 6a_3 x + 12a_4 x^2. \tag{8.3.16}$$

Rather than making the residual vanish everywhere, we select three locations, $x = 250$, 500 and 750 m, to enforce the zero residual condition there, resulting in the equations:

$$\begin{aligned} 0 &= 0.0018 - 0.00005\, a_0 - 0.0125\, a_1 - 1.125\, a_2 + 719\, a_3 + 5.55 \times 10^5\, a_4, \\ 0 &= 0.0018 - 0.00005\, a_0 - 0.025\, a_1 - 10.5\, a_2 - 3250\, a_3 - 1.25 \times 10^5\, a_4, \\ 0 &= 0.0018 - 0.00005\, a_0 - 0.0375\, a_1 - 26.1\, a_2 - 1.66 \times 10^5 a_3 \\ &\quad -9.07 \times 10^6 a_4. \end{aligned} \tag{8.3.17}$$

The above system, (8.3.15) and (8.3.17), can be solved for a_0–a_4; the approximate solution is

$$\hat{h} = 32 + 0.0265\,x - 7.87 \times 10^{-5} x^2 + 1.14 \times 10^{-7} x^3 - 7.02 \times 10^{-11} x^4. \quad (8.3.18)$$

For comparison, the exact solution is also plotted (as the long dashed lines) on Fig. 8.3.2.

B. Subdomain method

In the subdomain method, we use the weighting function

$$\begin{aligned} w_i(\mathbf{x}) &= 1, \quad \text{if } \mathbf{x} \in \Omega_i, \\ &= 0, \quad \text{if } \mathbf{x} \notin \Omega_i, \end{aligned} \qquad (8.3.19)$$

where $\Omega_i (\in \Omega)$ is a subdomain. In other words, rather than enforcing the governing equation on a set of selected points, as in the collocation method, we require that the averaged residual vanishes over a set of subdomains. It is obvious that the collection of subdomains should cover the entire domain, $\sum_{i=1}^{n} \Omega_i = \Omega$. Similar considerations should be applied to the boundary conditions. In what follows, to illustrate the method. we shall use the same one-dimensional leaky aquifer problem presented in the preceding subsection.

We assume that the approximate solution, $\hat{h}$, is given by the same polynomial presented in (8.3.14). By requiring that the two boundary conditions, at $x = 0$ and $x = L$, be satisfied, we arrive at the same two equations as (8.3.15). To satisfy the governing equation over a subdomain $x_1 \le x \le x_2$, we integrate the residual (8.3.16) over that subdomain

$$\begin{aligned} \int_0^L w(x) R(x)\, dx &= \int_{x_1}^{x_2} R(x)\, dx \\ &= \big(0.0018\,x - 0.00005\,a_0 x + 2\,a_2 x - 0.000025\,a_1 x^2 + 3\,a_3 x^2 \\ &\quad - 0.0000167\,a_2 x^3 + 4\,a_4 x^3 - 0.0000125\,a_3 x^4 \\ &\quad - 0.00001\,a_4 x^5\big)\Big|_{x_1}^{x_2}. \end{aligned} \qquad (8.3.20)$$

We may select three subdomains, $(0, L/3)$, $(L/3, 2L/3)$, and $(2L/3, L)$, to enforce the zero weighted residual condition, and obtain from (8.3.20) three linear equations in terms of a_0–a_4. Together with (8.3.15), we can solve for the coefficients to obtain the approximate solution

$$\hat{h} = 32 + 0.0286\,x - 9.59 \times 10^{-5} x^2 + 1.53 \times 10^{-7} x^3 - 9.39 \times 10^{-11} x^4. \quad (8.3.21)$$

This solution is plotted as the short dashed lines in Fig. 8.3.2.

C. Galerkin method

The Galerkin method uses the basis functions $N_i(\mathbf{x})$ as the weighting function for the approximation (see (8.3.3)). In other words, (8.3.5) becomes

$$\int_{\Omega} [\mathcal{L}(\hat{u}(\mathbf{x})) - f(\mathbf{x})]\, w_i(\mathbf{x})\, d\mathbf{x} =$$
$$\int_{\Omega} \left[\mathcal{L}\left(\sum_{j=1}^{n} a_j N_j(\mathbf{x}) \right) - f(\mathbf{x}) \right] N_i(\mathbf{x})\, d\mathbf{x} = 0. \tag{8.3.22}$$

Again, we shall use the leaky aquifer example to illustrate the method.

In the Galerkin method, there is no particular way to enforce boundary conditions; hence, the selected basis functions must automatically satisfy the boundary conditions. To facilitate the selection of basis functions, we transform the boundary value problem defined in (8.3.10) and (8.3.11) into one with null boundary conditions, by defining a new variable

$$\phi = h - \frac{h_2 - h_1}{L}\, x - h_1. \tag{8.3.23}$$

Substituting ϕ into (8.3.10) and (8.3.11), we obtain a new equation,

$$\frac{d^2\phi}{dx^2} + \frac{H_o - \phi}{\lambda} - \frac{\frac{h_2 - h_1}{L}\, x + h_1}{\lambda} = 0, \tag{8.3.24}$$

with

$$\phi(0) = 0 \text{ and } \phi(L) = 0. \tag{8.3.25}$$

We approximate the solution by

$$\hat{\phi} = x(x - L)(a_0 + a_1 x + a_2 x^2), \tag{8.3.26}$$

noting that this is the same 4th degree polynomial as (8.3.14), but here it satisfies the null boundary conditions in (8.3.25).

Substituting (8.3.26) into (8.3.24), we find the residual to be

$$\begin{aligned} R(x) = {} & 0.0002 + 2\, a_0 - 2000\, a_1 + 0.0000004\, x + 0.05\, a_0 x + 6\, a_1 x, \\ & -6000\, a_2 x - 0.00005\, a_0 x^2 + 0.05\, a_1 x^2 + 12\, a_2 x^2 - 0.00005\, a_1 x^3 \\ & +0.05\, a_2 x^3 - 0.00005\, a_2 x^4. \end{aligned} \tag{8.3.27}$$

To minimize this residual, we identify the three basis functions, associated with the coefficients a_0, a_1 and a_2:

$$N_1 = x(x - L), \quad N_2 = x^2(x - L), \quad N_3 = x^3(x - L). \tag{8.3.28}$$

Minimizing against N_1, we obtain

$$\begin{aligned} \int_0^L R(x) N_1(x)\, dx = {} & -6.67 \times 10^4 \\ & -2 \times 10^9\, a_0 - 1 \times 10^{12}\, a_1 - 5.76 \times 10^{14}\, a_2 = 0. \end{aligned} \tag{8.3.29}$$

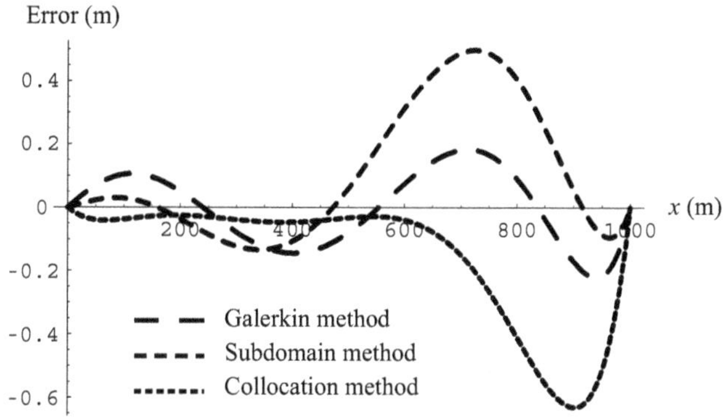

Figure 8.3.3: Comparison of errors in the three approximate solutions: collocation method (8.3.18); subdomain method (8.3.21); and Galerkin method (8.3.31).

Similarly, we can obtain two additional linear equations, using N_2 and N_3 as weights. Solving these equations, we obtain

$$a_0 = -3.77 \times 10^{-5}, \quad a_1 = 6.18 \times 10^{-8}, \quad a_2 = -9.22 \times 10^{-11}. \tag{8.3.30}$$

Finally, after restoring the original variable, we obtain

$$\hat{h} = 32 + 0.0297\,x - 0.0000995\,x^2 + 1.54 \times 10^{-7}\,x^3 - 9.22 \times 10^{-11}\,x^4. \tag{8.3.31}$$

This approximate solution is plotted in Fig. 8.3.2 as the long dashed line for comparison with the other solutions. We notice that all three approximate solutions, (8.3.18), (8.3.21), and (8.3.31), are 4th degree polynomials; they differ only by the coefficients.

To gain a clearer picture of the accuracy of the approximate solutions, as obtained by each of the methods, we present in Fig. 8.3.3 the solution error as the difference between the approximate and the exact solution. We observe that the choice of weighting function makes a difference in the error. Particularly, the use of Dirac delta function (collocation method) minimizes the error on a set of selected nodes, but not everywhere, which results in the largest *maximum* error in the domain. On the other hand, the subdomain and the Galerkin method distribute the error over the domain by minimizing the integrated error, which is weighted differently for these two methods. Particularly, we observe that Galerkin method is the most effective in spreading the error such that the maximum error in the domain is the smallest. The lesson learned here seems to suggest that using weights that require integration is the best strategy in terms of minimizing the maximum error, although extra effort is needed in performing the integration. However, we shall demonstrate in the following sections that the high accuracy of the weighted residual meth-

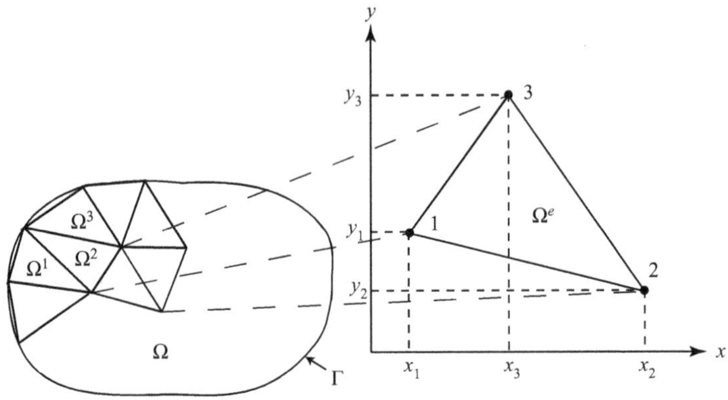

Figure 8.3.4: Dividing the domain into triangular elements.

ods is destroyed, when we start to subdivide the domain into *elements* for the purpose of integration, and to use low degree polynomials to interpolate the solution within each element. In fact, this is the path followed by the finite element method.

8.3.2 Galerkin finite element methods

As mentioned earlier in this section, there are, generally, two formulations for the FEM: one based on the variational principle, and the other based on the method of weighted residuals (Zienkiewicz, 1977). Only the latter, particularly the more popular *Galerkin method*, is discussed in this book.

A. Finite element discretization

The *finite element method* is based on subdividing the solution domain into a large number of *elements*. Figure 8.3.4 shows an example of a two-dimensional domain subdivided into *triangular elements*. Rather than define a single, continuous, global function for the approximate solution, such as (8.3.3), we represent the solution by using a number of local interpolation functions, each defined within a given element. We then patch them together to form a *piecewise continuous* function. Within each element, the solution is typically represented by a small number of discrete values, and a low degree polynomial is used for the interpolation.

For the simplest representation, we select the triangular element shown as an inset in Fig. 8.3.4. At the three vertices of the element, the discrete values of head are given as h_1, h_2 and h_3. Inside the element, the head is interpolated, using the linear function

$$\hat{h} = a + bx + cy, \quad (x, y) \in \Omega^e, \tag{8.3.32}$$

where Ω^e is the subdomain defined by the e-th element. Since this solution must satisfy the condition that $\hat{h} = h_1$, h_2 and h_3, at $(x, y) = (x_1, y_1)$, (x_2, y_2) and (x_3, y_3), respectively, we have sufficient conditions to uniquely determine the three coefficients a, b and c, in terms of these coordinates and the discrete head values. For the final expression, it is customary to sort and collect the nodal values, h_1, h_2 and h_3, and express them in the form

$$\hat{h}(x,y) = \lfloor N_1 \; N_2 \; N_3 \rfloor \begin{Bmatrix} h_1 \\ h_2 \\ h_3 \end{Bmatrix} = N_1 h_1 + N_2 h_2 + N_3 h_3, \tag{8.3.33}$$

where N_1, N_2 and N_3 are called, in the finite element terminology, *shape functions*. Each of them is a linear function

$$N_i = a_i + b_i x + c_i y, \quad i = 1, 2, 3, \tag{8.3.34}$$

with

$$\begin{aligned} a_1 &= (x_2 y_3 - x_3 y_2)/2A, \; b_1 = (y_2 - y_3)/2A, \; c_1 = (x_3 - x_2)/2A, \\ a_2 &= (x_3 y_1 - x_1 y_3)/2A, \; b_2 = (y_3 - y_1)/2A, \; c_2 = (x_1 - x_3)/2A, \\ a_3 &= (x_1 y_2 - x_2 y_1)/2A, \; b_3 = (y_1 - y_2)/2A, \; c_3 = (x_2 - x_1)/2A, \end{aligned} \tag{8.3.35}$$

where

$$A = x_1 y_2 + x_2 y_3 + x_3 y_1 - x_1 y_3 - x_2 y_1 - x_3 y_2, \tag{8.3.36}$$

is the area of the element. It can easily be shown that at the vertex (x_1, y_1), we have $N_1 = 1$, $N_2 = N_3 = 0$, with similar behavior at the other two vertices. Hence, $\hat{h}$ takes the value of h_1, h_2 and h_3 at the respective nodes; thus, it is a linear function in the form (8.3.32).

When these elements are patched together, they share sides and vertices. To ensure *compatibility*, it is necessary that whenever several elements share the same node, the same nodal value will be used for all elements. This will ensure the continuity of the function at the nodes, as well as on the edges of the element. However, it does not guarantee the continuity of the derivatives. In fact, the latter are *discontinuous* upon crossing from one element to the other. By patching the Ω^e elements, $e = 1, \ldots, n_e$, together, the approximate function becomes

$$\hat{h} = \sum_{e=1}^{n_e} \sum_{i=1}^{3} N_i h_i^e. \tag{8.3.37}$$

Or, after sorting the terms for the nodal values, h_j, $j = 1, \ldots, m$, where j is the node's global index, and m is the total number of nodes in the finite element domain, (8.3.37) can be expressed as

$$\hat{h} = \sum_{i=1}^{m} N_i h_i = \mathbf{N}\,\mathbf{h}, \tag{8.3.38}$$

where N_i is the combined shape function, incorporating all the contributions to the j-th node. Note that we have written this equation also in matrix form.

The above kind of *local* interpolation may be compared with the *global* interpolation used in Subs. 8.3.1A, B and C, in the forms of the collocation, the subdomain, and the Galerkin methods. We notice that the polynomial interpolation given in (8.3.14) is defined for the entire domain, and that only a single 'element' exists. The accuracy of the solution is improved by increasing the polynomial's degree. Such functions are highly continuous (i.e., having high order derivatives), as opposed to the local interpolation, for example (8.3.37), which is not continuous, even in the first derivative. Methods based on global interpolation are more accurate than those based on local interpolation. This issue will be revisited in Subs. 8.5.

B. Weak formulation for flow problems

In Subs. 8.3.1, we have shown that the Galerkin method, which uses the basis function ($\equiv$ shape function) as the weighting function, gives a better solution accuracy than other methods; hence, the Galerkin weighted residual scheme is most often adopted in the finite element discretization presented in the preceding subsection.

As a demonstration, let us apply the Galerkin FEM to the governing equation of a leaky confined aquifer in a two-dimensional domain, (5.4.57), expressed in the form

$$\mathcal{L}(h) \equiv \frac{\partial}{\partial x}\left(\mathsf{T}\frac{\partial h}{\partial x}\right) + \frac{\partial}{\partial y}\left(\mathsf{T}\frac{\partial h}{\partial y}\right) - S\frac{\partial h}{\partial t} + \frac{H_o - h}{c_r} = 0, \tag{8.3.39}$$

where H_o is the steady piezometric head in the upper unconfined aquifer, and c_r is the resistance of the confining semi-pervious layer. The boundary and initial conditions are

$$h(\mathbf{x},t) = g(\mathbf{x},t), \quad \mathbf{x} \in \partial\Omega_S; \qquad -\mathsf{T}\frac{\partial h}{\partial n} = q(\mathbf{x},t), \quad \mathbf{x} \in \partial\Omega_N;$$
$$h(\mathbf{x},0) = f(\mathbf{x}), \quad \mathbf{x} \in \Omega. \tag{8.3.40}$$

Based on the weighted residual method, (8.3.5), we aim at minimizing the residuals of both the governing equation and the boundary conditions in a certain mathematical sense. For the Galerkin method, we use the shape functions, N_i, as the weights, and require that the following weighted residuals vanish, i.e.,

$$\int_\Omega N_i \left[\frac{\partial}{\partial x}\left(\mathsf{T}\frac{\partial h}{\partial x}\right) + \frac{\partial}{\partial y}\left(\mathsf{T}\frac{\partial h}{\partial y}\right) - S\frac{\partial h}{\partial t} + \frac{H_o - h}{c_r}\right] d\mathbf{x}$$
$$- \int_{\partial\Omega_N} N_i \left[\mathsf{T}\frac{\partial h}{\partial n} + q\right] d\mathbf{x} = 0. \tag{8.3.41}$$

We note that in the above equation, the residual of the essential (Dirichlet) boundary condition is not included, because it is automatically satisfied by the adoption of the approximation function (8.3.37) that uses the nodal head values for interpolation. Equation (8.3.41) is called a *strong formulation*, because, when it is satisfied for an arbitrary weighting function at all points of Ω, the original partial differential equation, (8.3.40), must be satisfied at all points. We also notice that in order to satisfy the differential equation, the function h must be at least twice differentiable.

The differentiability requirement poses a problem for the piecewise linear approximation (8.3.37), as once the shape functions are differentiated twice, they become zero and (8.3.41) vanishes. Although this problem can be solved by using higher degree polynomials, it is generally undesirable to do so in FEM, as this makes the element more complicated. Hence, rather than increase the degree of the polynomial approximation, we reduce the formulation's differentiation requirement by introducing the *weak formulation*. This is accomplished by making use of Green's theorem

$$\int_{\Omega} N_i \nabla \cdot (\mathsf{T}\nabla h)\, d\mathbf{x} = -\int_{\Omega} \nabla N_i \cdot (\mathsf{T}\nabla h)\, d\mathbf{x} + \int_{\partial\Omega} N_i \mathsf{T}\frac{\partial h}{\partial n}\, d\mathbf{x}. \tag{8.3.42}$$

Substituting (8.3.42) into (8.3.41), we obtain

$$\int_{\Omega} \nabla N_i \cdot (\mathsf{T}\nabla h)\, d\mathbf{x} + \int_{\Omega} N_i \left[S\frac{\partial h}{\partial t} - \frac{H_o - h}{c_r} \right] d\mathbf{x} + \int_{\partial\Omega_N} N_i\, q\, d\mathbf{x} - \int_{\partial\Omega_S} N_i \mathsf{T}\frac{\partial h}{\partial n}\, d\mathbf{x} = 0. \tag{8.3.43}$$

Furthermore, we notice that the last integral on the left hand side can be dropped because the shape functions N_i, selected as weighting functions, are associated with the discrete unknown h-values, and they vanish at any boundary, $\partial\Omega_S$, with essential boundary condition. Hence,

$$\int_{\Omega} \nabla N_i \cdot (\mathsf{T}\nabla h)\, d\mathbf{x} + \int_{\Omega} N_i \left[S\frac{\partial h}{\partial t} - \frac{H_o - h}{c_r} \right] d\mathbf{x} + \int_{\partial\Omega_N} N_i\, q\, d\mathbf{x} = 0. \tag{8.3.44}$$

Equation (8.3.44) is then the *weak formulation* of (8.3.39) and (8.3.40). In this form, h is differentiated only once; hence, h can be represented by the linear shape function $\mathbf{N}$. As noted before, the linear shape function vanishes in the *strong* form, (8.3.41), when it is differentiated twice; hence, it cannot be used in the strong formulation. Another reason for using the weak formulation is that, as demonstrated below, when the same shape function, $\mathbf{N}$, is used to approximate h, the solution matrix representing the linear system is *symmetric*, which means a significant reduction in computational effort.

The next step in the finite element formulation is to substitute the piecewise approximation, (8.3.37) or (8.3.38), into (8.3.43), obtaining

$$\left[\int_{\Omega} SN_i\mathbf{N}\,d\mathbf{x}\right]\frac{\partial\mathbf{h}}{\partial t}+\left[\int_{\Omega}\nabla N_i\cdot(\mathbf{T}\nabla\mathbf{N})\,d\mathbf{x}+\int_{\Omega}\frac{1}{c_r}N_i\mathbf{N}\,d\mathbf{x}\right]\mathbf{h}$$
$$-\int_{\Omega}N_i\frac{H_o}{c_r}\,d\mathbf{x}+\int_{\partial\Omega_N}N_i\,q\,d\mathbf{x}=0, \qquad (8.3.45)$$

in which we have used $\mathbf{h}$ to denote a column vector of the discrete nodal values of h, $\lfloor h_1, h_2, \ldots \rfloor^T$. In (8.3.45), we note that the discrete piezometric head values, $\mathbf{h}$, are not functions of space, and can, therefore, be moved out of the integration. Finally, when we execute (8.3.45) for $i = 1, \ldots, m$, where m is the number of nodes and the head is an unknown, we obtain the matrix equation

$$\mathsf{C}\frac{\partial\mathbf{h}}{\partial t}+\mathsf{A}\,\mathbf{h}+\mathbf{F}=0, \qquad (8.3.46)$$

where C and A are $m \times m$ matrices, and $\mathbf{F}$ is a $m \times 1$ matrix, with components

$$\begin{aligned} C_{ij} &= \int_{\Omega} SN_iN_j\,d\mathbf{x} \\ A_{ij} &= \int_{\Omega}\nabla N_i\cdot(\mathbf{T}\nabla N_j)\,d\mathbf{x}+\int_{\Omega}\frac{1}{c_r}N_i\,N_j\,d\mathbf{x} \\ F_i &= -\int_{\Omega}N_i\frac{H_o}{c_r}\,d\mathbf{x}+\int_{\partial\Omega_N}N_i\,q\,d\mathbf{x}. \end{aligned} \qquad (8.3.47)$$

Thus, (8.3.46) is a system of first-order ordinary differential equations, with time derivative terms: dh_1/dt, dh_2/dt, etc. This system can be solved by utilizing a *finite difference* scheme for its time discretization.

As discussed in Subs. 8.1.2, the variable $\mathbf{h}$ can be expressed at two time levels, k and $k+1$, which are Δt apart, as $\mathbf{h}^k$ and $\mathbf{h}^{k+1}$. The time derivative term in (8.3.47) is then expressed by the first-order Taylor series expansion

$$\frac{\partial\mathbf{h}}{\partial t}=\frac{\mathbf{h}^{k+1}-\mathbf{h}^k}{\Delta t}. \qquad (8.3.48)$$

The rest of the terms in (8.3.47) can be represented as a weighted average between the two time levels. For example, we can express $\mathbf{h}$ as

$$\mathbf{h}=(1-\theta)\mathbf{h}^k+\theta\mathbf{h}^{k+1}, \qquad (8.3.49)$$

where θ is a weighting factor for the time discretization.

With these representations, equation (8.3.46) becomes

$$\left(\frac{\mathsf{C}}{\Delta t}+\theta\mathsf{A}\right)\mathbf{h}^{k+1}+\left[-\frac{\mathsf{C}}{\Delta t}+(1-\theta)\mathsf{A}\right]\mathbf{h}^k+(1-\theta)\mathbf{F}^k+\theta\mathbf{F}^{k+1}=0. \qquad (8.3.50)$$

We note that in (8.3.50), all quantities at time level k are known, and we seek the solution at time level $k+1$. Similar to the finite difference method, discussed in Subs. 8.1.2, by setting $\theta = 0$, 1 and 1/2, we obtain the explicit, implicit, and Crank-Nicolson schemes, respectively. Or, by setting $\theta = 2/3$ or $1/3$, we obtain a Galerkin type weighting process (Zienkiewicz, 1977). Particularly, for the explicit scheme, (8.3.50) becomes

$$\frac{\mathsf{C}}{\Delta t}\mathbf{h}^{k+1} = \left[\frac{\mathsf{C}}{\Delta t} - \mathsf{A}\right]\mathbf{h}^{k} - \mathbf{F}^{k}. \tag{8.3.51}$$

As the matrix C, defined in (8.3.47), is in a *diagonal* form, the above system of equations can be solved recursively, without the decomposition of the matrix. Similar to the FDM, the explicit scheme is only conditionally stable; hence, only small time steps can be used. The implicit, Crank-Nicolson, and Galerkin schemes are unconditionally stable.

C. Weak formulation for transport problems

As an example of a solute transport problem, we refer to the advective-dispersive transport equation (8.2.1). To find the approximate solution of this equation, we seek to minimize the following weighted residual expression, with respect to a set of trial (basis) functions:

$$\int_{\Omega} N_i\left[\nabla\cdot\phi(c\mathbf{V} - \mathsf{D}_h\cdot\nabla c) + \phi\frac{\partial c}{\partial t} - \phi\,\rho\,\Gamma^{\gamma}\right]\,d\mathbf{x} = 0, \tag{8.3.52}$$

where the N_i's, $i = 1,\ldots,m$, are shape functions that interpolate the concentration, c, within the finite element domain, similar to that defined in (8.3.38). Since the shape functions are used as weighting functions, this is a Galerkin formulation. Applying the divergence theorem to the dispersive transport part of (8.3.52), we obtain

$$\begin{aligned}
&\int_{\Omega}[\nabla N_i\cdot(\phi\mathsf{D}_h\cdot\nabla N_j) + N_i(\phi\mathbf{V}\cdot\nabla N_j)]\,c_j\,d\mathbf{x}\\
&+\int_{\Omega}N_i\left[\phi\frac{\partial c_j}{\partial t}N_j - \phi\rho\Gamma^{\gamma}\right]\,d\mathbf{x}\\
&-\int_{\partial\Omega}[N_i(\phi\mathsf{D}_h\cdot\mathbf{n})\cdot\nabla N_j]\,c_j\,d\mathbf{x} = 0,
\end{aligned} \tag{8.3.53}$$

where we have replaced the concentration by its discrete representation $c = \sum_{j=1}^{m} N_j c_j$. The above equation forms the basis for the finite element discretization.

D. Stabilized finite element for advection-dominated transport

It is well known that it is more difficult to solve the advection-dispersion equation (8.2.1) numerically when the transport is dominated by advection, i.e., it is characterized by a large Peclet number (see Subs. 7.7). Consider the special case of *advection-only* flow. In such a case, an initially sharp front will be transported unchanged at the advective velocity. In a numerical solution, however, we typically observe that the concentration oscillates near the front; the numerical solution can alternately overshoot and undershoot the true solution from one node to the next. This phenomenon exists for all numerical methods, FDM, FEM, and FVM. Let us use a finite difference example to illustrate the cause of this effect.

Consider the transport of solute by advection only in a one-dimensional flow of an incompressible fluid (i.e., $\partial V/\partial x = 0$), in a homogeneous domain ($\phi = \text{const.}$), such that (8.2.1) reduces to

$$\frac{\partial c}{\partial t} + V\frac{\partial c}{\partial x} = 0. \tag{8.3.54}$$

We can write a finite difference approximation of (8.3.54), using the central difference formula for the spatial derivative, and the explicit time stepping scheme. The concentration c_i^{k+1}, at node i and time-step $k+1$, takes the form

$$c_i^{k+1} = c_i^k - \frac{V\Delta t}{2\Delta x}\left(c_{i+1}^k - c_{i-1}^k\right). \tag{8.3.55}$$

Let us assume that the sharp front of a step function is located at node i at time-step k, such that $c_{i-1}^k = c_i^k = c_o$ and $c_{i+1}^k = 0$. We can clearly see from (8.3.55) that at time-step $k+1$, the concentration is $c_i^{k+1} > c_o$, that is, the concentration at node i at the next time step becomes greater than the constant concentration c_o, while it should be equal. For node $i+1$, if we make the velocity $V = \Delta x/\Delta t$, then the front should travel to node $i+1$ at time step $k+1$, with concentration $c_{i+1}^{k+1} = c_o$. However, from the finite difference formula, we calculate $c_{i+1}^{k+1} = c_o/2 < c_o$. Hence, the interpolation based on central difference, which uses information from points both ahead of and behind the front, causes the concentration near the front to overshoot and undershoot, thus developing *oscillations*.

One way to avoid this oscillation error is to use backward differences in space, that is, to use only information from behind the front (i.e., only in the upstream direction), not ahead of it, leading to

$$c_i^{k+1} = c_i^k - \frac{V\Delta t}{\Delta x}\left(c_i^k - c_{i-1}^k\right). \tag{8.3.56}$$

This type of *upstream*, or *upwind* approximation has been widely used in all numerical methods, e.g., Leveque (2008), and Ozisik (1994), for finite difference, Zhu (1991), Lazarov *et al.* (1996), and Leveque (2002), for finite volume, and Therrien and Sudicky (1996), for control volume finite element implementation. In this subsection, we shall discuss the finite element techniques for reducing the oscillation error.

Similar to the FDM and the FVM, the upwind method has been developed also for the FEM (Heinrich *et al.*, 1977; Huyakorn and Nilkuha, 1979; Huyakorn and Pinder, 1983). Referring to the advection-dispersion transport equation (8.2.1), and comparing it with the Galerkin formulation (8.3.52), the *upstream weighted residual formulation* can be expressed as

$$\int_{\Omega} W_i \left[\nabla \cdot \phi(c\mathbf{V} - \mathbf{D}_h \cdot \nabla c)\right] \, d\mathbf{x} + \int_{\Omega} N_i \left[\phi \frac{\partial c}{\partial t} - \phi \, \rho \, \Gamma^{\gamma}\right] \, d\mathbf{x} = 0, \quad (8.3.57)$$

where the W_i's are the *upstream weighting functions*, which are biased toward the upstream direction. Similar to (8.3.53), the weak formulation then becomes

$$\begin{aligned} &\int_{\Omega} \left[\nabla W_i \cdot (\phi \mathbf{D}_h \cdot \nabla N_j) + W_i(\phi \mathbf{V} \cdot \nabla N_j)\right] c_j \, d\mathbf{x} \\ &+ \int_{\Omega} N_i \left[\phi \frac{\partial c_j}{\partial t} N_j - \phi \rho \Gamma^{\gamma}\right] \, d\mathbf{x} \\ &- \int_{\partial \Omega} \left[W_i(\phi \mathbf{D}_h \cdot \mathbf{n}) \cdot \nabla N_j\right] c_j \, d\mathbf{x} = 0. \qquad (8.3.58) \end{aligned}$$

This type of FEM formulation, which uses a weight that is not a shape function, causing a non-symmetric matrix, is called a *Petrov-Galerkin formulation.*

Although the upwind scheme can smooth out the oscillations, it is a lower order approximation, compared to the central difference based one. This lower order approximation will, in turn, introduce an error called *numerical dispersion.* For an illustration, we can continue with the finite difference analysis presented through (8.3.54)–(8.3.56). Let us examine the error generated by using the backward difference scheme. From the Taylor series expansion, we obtain

$$V \frac{\partial c}{\partial x} = V \frac{c(x) - c(x - \Delta x)}{\Delta x} + V \frac{\Delta x}{2} \frac{\partial^2 c}{\partial x^2} + \dots \qquad (8.3.59)$$

The backward difference contained in (8.3.56) refers only to the first term on the right hand side. By comparing the (truncated) second term on the right hand side of (8.3.59) with the dispersive transport terms in (8.2.1), we conclude that the error caused by the truncation is equivalent to introducing the effect of dispersion, with a dispersion coefficient equal to $V\Delta x/2$. This *numerical dispersion* has the effect of smearing out the sharp front, as if real dispersion exists.

One way to overcome this error is to combine the central difference approach with the upwind difference one, such that the *artificial dispersion* will compensate for the numerical dispersion in the upwind scheme. This method needs to be carefully constructed such that it does not generate a *cross-wind dispersion*, that is, dispersion in the direction perpendicular to the flow, as some schemes do. These considerations led to the development of the *streamline-upwind Petrov-Galerkin (SUPG) method* (Hughes and Brooks, 1979; Brooks and Hughes, 1982; Hughes *et al.*, 1989), in which an artificial dispersion operator is constructed in the *streamline* direction only. In this way, the cross-wind dispersion is avoided. This is done by modifying the

Galerkin formulation presented as (8.3.53) in the following way:

$$\int_{\Omega} \nabla N_i \cdot (\phi \mathbf{D}_h \cdot \nabla c) + N_i(\phi \mathbf{V} \cdot \nabla c) + \phi \frac{\partial c}{\partial t} - \phi \rho \Gamma^{\gamma} \, d\mathbf{x}$$
$$- \int_{\partial\Omega} N_i(\phi \mathbf{D}_h \cdot \mathbf{n}) \cdot \nabla c \, d\mathbf{x}$$
$$+ \sum_{e=1}^{n_e} \int_{\Omega^e} \tau_{\mathrm{SUPG}} \mathbf{V} \cdot \nabla N_i \left[\nabla \cdot \phi(c\mathbf{V} - \mathbf{D}_h \cdot \nabla c) + \phi \frac{\partial c}{\partial t} - \phi \, \rho \, \Gamma^{\gamma} \right] d\mathbf{x} = 0, \tag{8.3.60}$$

where n_e is the number of elements, and τ_{SUPG} is a stabilization factor that depends on the *local Peclet number*, the element size, and the flow (streamline) direction. For example, we can choose τ_{SUPG} as (Brooks and Hughes, 1982; Galeao *et al.*, 2004)

$$\tau_{\mathrm{SUPG}} = \frac{\delta}{2|\mathbf{V}|} \xi(\mathrm{Pe}), \tag{8.3.61}$$

where δ is the element size, Pe is the local Peclet number, defined as

$$\mathrm{Pe} = \frac{|\mathbf{V}|h}{2\mathrm{D}_h}, \tag{8.3.62}$$

and ξ is a dimensionless function, for example (Brooks and Hughes, 1982),

$$\xi(\mathrm{Pe}) = \coth(\mathrm{Pe}) - \frac{1}{\mathrm{Pe}}. \tag{8.3.63}$$

The SUPG method has been applied to solve groundwater solute transport problems by Gordon *et al.* (2000, 2001), with reactive transport by Couto and Malta (2008), and in partially saturated flow problems by Kees *et al.* (2008).

Further developments of the stabilized finite element methods can be found in Masud and Hughes (2002), Brezzi *et al.* (2005), and Hughes *et al.* (2006), for porous medium flow problems, in Hughes (1995), Franca *et al.* (1992), Franca and Frey (1992), Brezzi *et al.* (1992), and Malta and Loula (1998), for advection-diffusion problems, and in Tezduyar *et al.* (1992a, b), for moving interface problems.

8.3.3 Meshless finite element methods

One disadvantage of the conventional finite element methods is its need to assemble the element *connectivity data*. Referring to Fig. 8.3.4, we observe that each *finite element* is defined by a number of nodes, which are defined by a global numbering system. It is necessary to provide, as computer program input, the nodes that define every element. We also observe that each node is shared by a number of elements. All this information in required in order to properly assign the contribution of the weighted integration of the Galerkin scheme to the correct discrete nodal variables. The creation of such information can be quite cumbersome. According to the finite element community, "*... even with powerful mesh generators, three-dimensional meshing is still an extremely burdensome task and that the conversion of solid models to finite element data is time-consuming and often introduces numerous ambiguities ...*" (Belytschko *et al.*, 1994).

Since the mid-1990s, the finite element community has been engaged in an intensive effort to develop *meshless* or *element-free* methods that eliminate the need of creating element connectivity data as input. In addition to simplifying input data preparation, meshless methods are also known for their easier 'remeshing' for moving boundary or interface problems. A number of meshless finite element methods have emerged. These include diffuse element method (Nayroles *et al.*, 1992), element-free Galerkin method (Belytschko *et al.*, 1994; Sakurai and Kawahara, 2004; Modaressi and Aubert, 1998), partition of unity method (Melenk and Babuška, 1996), h-p clouds method (Duarte and Oden, 1996), local Petrov-Galerkin method (Atluri and Zhu, 1998; Atluri and Shen, 2002; Atluri, 2004), and reproducing kernel particle method (Liu *et al.*, 1995; Li and Liu, 2007). (See also Sec. 8.5 for the meshless radial basis function collocation method.)

8.3.4 Control volume finite element methods

Control volume finite element methods (CVFEM) are hybrid methods that combine the advantages of the finite element and the finite volume methods. Generally, these methods follow two lines of development (Martinez, 2006), one type is developed for convection-diffusion problems (Baliga and Patankar, 1980; Ramadhyani and Patankar, 1985; Schneider and Raw, 1985), and another type for multiphase flow (Forsyth, 1991; Letniowski and Forsyth, 1991; Forsyth *et al.*, 1995; Therrien and Sudicky, 1996).

For the discretization, the basic approach follows that of the finite volume method. Particularly, the unstructured, cell-centered, arbitrary polygonal grid (Sec. 8.2) can be used, and elements are formed by *Delauney triangulation* (Joe, 1986). The conservation laws are applied to the control volumes of the FVM, such that the local conservation of the properties is satisfied (while the Galerkin finite element only satisfies global conservation). To improve upon the low accuracy of the FVM approximation, higher order shape functions and a similar time-marching Galerkin finite element procedure can be applied (Therrien and Sudicky, 1996). Directional upwind schemes are implemented to minimize numerical dispersion of advection dominant problem. The resulting formulation is, typically, solved by iterative techniques, such as the Newton-Raphson method.

8.4 Boundary Element Methods

The numerical methods presented so far, the FDM, FVM, and FEM, require domain discretization, meaning that discrete values of the unknowns are assigned to points distributed over the solution domain. However, domain discretization is not always necessary. There exists a class of numerical methods, referred to as *boundary methods*, in which the discrete unknowns need to be distributed only over the solution *boundary*. The advantage of the boundary methods is that the size of the solution system (the number of unknowns) is much smaller, thus, both the computer storage requirement and the computational effort are significantly reduced. Another advantage of the boundary methods is in the preparation of the solution mesh. The reduction in spatial dimensions in the formulation of a boundary method (recalling that a three-dimensional domain has a two-dimensional boundary) can make the mesh preparation easier.

The most critical advantage of the boundary methods, however, lies in their flexibility in dealing with moving boundary problems, e.g., a moving free surface or a moving interface between two fluids. In domain methods, whenever the boundary location is moved, the interior mesh needs to be adjusted, as elements and nodes can be crossed. The logistics of such an operation can be a daunting job. Boundary methods do not have interior mesh, hence such a problem does not exist.

The major disadvantage of boundary methods is that they cannot be applied to all types of governing equations. The boundary methods require information on some solution of the governing equation, such as a general solution, or a fundamental solution (free-space Green's function). In general, these solutions are available only for linear governing equations with constant coefficients (i.e., homogeneous materials), or piecewise constant coefficients (by dividing the domain into homogeneous subdomains). Thus, altogether, the application of the boundary methods is limited to special problems.

There exist a number of boundary methods (Cheng and Cheng, 2005); among the most known are the Boundary Element Method (BEM) (Brebbia, 1978; Dominguez and Brebbia, 1989; Liggett and Liu, 1983), the Method of Fundamental Solutions (MFS) (Mathon and Johnston, 1977; Fairweather and Karageorghis, 1998), the Trefftz Method (TM) (Herrera, 1984; Li *et al.*, 2007, 2008), and the Analytic Element Method (AEM) (Strack, 1989, 2003; Haitjema, 1995; Bakker and Strack, 2003). Here, we shall discuss only the Boundary Element Method (BEM).

Let us consider a general two-dimensional, linear, second order partial differential operator of the form

$$\mathcal{L}(u) \equiv A\frac{\partial^2 u}{\partial x^2} + 2B\frac{\partial^2 u}{\partial x \partial y} + C\frac{\partial^2 u}{\partial y^2} + D\frac{\partial u}{\partial x} + E\frac{\partial u}{\partial y} + Fu, \tag{8.4.1}$$

where $A, B, \ldots, F$ are coefficients that may be functions of x, y, but not of u. Our goal is to solve the problem

$$\mathcal{L}(u) = f(\mathbf{x}), \tag{8.4.2}$$

subject to a set of boundary conditions.

Applying the generalized Green's theorem to the above equation (Greenberg, 1971), we obtain the following reciprocal integral relation:

$$\int_{\Omega} [u\mathcal{L}^*(v) - v\mathcal{L}(u)] \; d\mathbf{x} = \int_{\partial\Omega} [u\mathcal{B}_n^*(v) - v\mathcal{B}_n(u)] \; d\mathbf{x}, \tag{8.4.3}$$

where Ω is the solution domain, $\partial\Omega$ denotes its boundary, $\mathcal{L}^*$ is the adjoint operator of $\mathcal{L}$,

$$\mathcal{L}^*(v) = \frac{\partial^2 Av}{\partial x^2} + 2\frac{\partial^2 Bv}{\partial x \partial y} + \frac{\partial^2 Cv}{\partial y^2} - \frac{\partial Dv}{\partial x} - \frac{\partial Ev}{\partial y} + Fv, \tag{8.4.4}$$

and $\mathcal{B}_n$ and $\mathcal{B}_n^*$ are generalized normal derivatives,

$$\mathcal{B}_n(u) = A\frac{\partial u}{\partial x}n_x + 2B\frac{\partial u}{\partial y}n_x + C\frac{\partial u}{\partial y}n_y + Dun_x \tag{8.4.5}$$

$$\mathcal{B}_n^*(v) = \frac{\partial Av}{\partial x}n_x + 2\frac{\partial Bv}{\partial x}n_y + \frac{\partial Cv}{\partial y}n_y - Evn_y, \tag{8.4.6}$$

in which n_x and n_y are components of the boundary's outward normal vector. We note that (8.4.3) is obtained by performing *integration by parts* twice on the term $\int_{\Omega} v\mathcal{L}(u) \; d\mathbf{x}$, thus relieving u from the operator, and creating a new operator for v, together with two boundary integrals. The operators $\mathcal{L}^*$, $\mathcal{B}_n$, and $\mathcal{B}_n^*$ are the result of these operations.

The Laplacian operator, $\mathcal{L} = \nabla^2$, can serve as an example. In this case, the operator is *self adjoint*, meaning that we also have $\mathcal{L}^* = \nabla^2$. Equation (8.3.2) becomes

$$\int_{\Omega} \left(u\nabla^2 v - v\nabla^2 u\right) \; d\mathbf{x} = \int_{\partial\Omega} \left(u\frac{\partial v}{\partial n} - v\frac{\partial u}{\partial n}\right) \; d\mathbf{x}. \tag{8.4.7}$$

We identify the above equation as the well-known *Green's second identity* (Morse and Feshbach, 1953; Greenberg, 1998). Another example for this operator is $\mathcal{L} = \nabla \cdot (K(x,y)\nabla)$, i.e., the operator for Darcy's flow in an inhomogeneous domain. In this case, the reciprocity relation is given by (Cheng, 1984, 1987)

$$\int_{\Omega} [u\nabla \cdot (K\nabla v) - v\nabla \cdot (K\nabla u)] \; d\mathbf{x} = \int_{\partial\Omega} \left(uK\frac{\partial v}{\partial n} - vK\frac{\partial u}{\partial n}\right) \; d\mathbf{x}. \tag{8.4.8}$$

We should also remark that although the time variable is not modeled in the above operators, the time dimension can be handled by a time stepping scheme, similar to what is done in the FDM and FEM. Another way to handle the time dependence is to utilize the Laplace transform to eliminate the time variable (Liggett and Liu, 1979) and then, numerically, invert the result back to the time domain (Davies and Martin, 1979; Cheng, *et al.*, 1994).

The key to the boundary element formulation is the availability of a *fundamental solution*, also known as the *free space Green's function*. The fundamental solution of a partial differential operator, $\mathcal{L}^*$, is one that satisfies

$$\mathcal{L}^* \left(G(\mathbf{x}-\mathbf{x}')\right) = \delta(\mathbf{x}-\mathbf{x}'), \tag{8.4.9}$$

without the need to fulfill boundary conditions (hence the term 'free space'). In the above equation, $\delta(\mathbf{x}-\mathbf{x}')$ is the Dirac delta function with its singularity located at $\mathbf{x}'$. It has the following properties

$$\begin{aligned} \delta(\mathbf{x}-\mathbf{x}') &= 0, && \text{for} \quad \mathbf{x} \neq \mathbf{x}', \\ \delta(\mathbf{x}-\mathbf{x}') &\to \infty, && \text{for} \quad \mathbf{x} = \mathbf{x}', \\ \int_\Omega \delta(\mathbf{x}-\mathbf{x}')\, d\mathbf{x} &= 1, && \text{for} \quad \mathbf{x}' \in \Omega, \\ \int_\Omega \delta(\mathbf{x}-\mathbf{x}')\, d\mathbf{x} &= 0, && \text{for} \quad \mathbf{x}' \notin \Omega, \\ \int_\Omega f(\mathbf{x})\delta(\mathbf{x}-\mathbf{x}')\, d\mathbf{x} &= f(\mathbf{x}'), && \text{for} \quad \mathbf{x}' \in \Omega. \end{aligned} \tag{8.4.10}$$

An example of a fundamental solution for the Laplace equation is

$$\nabla^2 G(\mathbf{x}-\mathbf{x}') = \delta(\mathbf{x}-\mathbf{x}'). \tag{8.4.11}$$

The solution is given by

$$\begin{aligned} G &= \frac{\ln r}{2\pi} \qquad \text{(2-D)}, \\ &= -\frac{1}{4\pi r} \qquad \text{(3-D)}, \end{aligned} \tag{8.4.12}$$

where $r = \|\mathbf{x}-\mathbf{x}'\|$ is the Euclidean norm.

To obtain the boundary integral equation, we substitute (8.4.2) and (8.4.11) (G for v) into (8.4.3), and find

$$\begin{aligned} \alpha(\mathbf{x}')u(\mathbf{x}') = &\int_{\partial\Omega} \left[u(\mathbf{x})B_n^* \left(G(\mathbf{x},\mathbf{x}')\right) - G(\mathbf{x},\mathbf{x}')B_n\left(u(\mathbf{x})\right)\right]\, d\mathbf{x} \\ &+ \int_\Omega G(\mathbf{x},\mathbf{x}')f(\mathbf{x})\, d\mathbf{x}, \end{aligned} \tag{8.4.13}$$

where α is a constant generated by the delta function, and $\alpha = 1$ for $\mathbf{x} \in \Omega$, $\alpha = 0$ for $\mathbf{x} \notin \Omega$, and $\alpha = 1/2$ for $\mathbf{x} \in \partial\Omega$. We notice that all volume integrals vanish in the above equation, except for the last term on the right hand side. This integral contains the integration of a known function, and there is no representation of discrete unknowns. In the case of the Laplace equation, $\nabla^2 u = 0$, we further obtain from (8.4.7) or (8.4.13)

$$\alpha(\mathbf{x}')u(\mathbf{x}') = \int_{\partial\Omega} \left[u(\mathbf{x}) \frac{\partial G(\mathbf{x}, \mathbf{x}')}{\partial n(\mathbf{x})} - G(\mathbf{x}, \mathbf{x}') \frac{\partial u(\mathbf{x})}{\partial n(\mathbf{x})} \right] d\mathbf{x}. \tag{8.4.14}$$

Equations (8.4.13) and (8.4.14) are *boundary integral equations*, ready to be discretized.

The procedure for approximately representing a continuous variable, using a set of piecewise continuous 'elements', is similar to the FEM, except that only *boundary elements* are needed, thus reducing the number of spatial dimension. For a two-dimensional problem, these elements are straight lines, or curved segments. In a three-dimensional problem domain, a triangular mesh may be formed. The discrete unknowns are interpolated within an element, using *shape functions*, as in (8.3.37). They are then assembled into a global form, $\hat{u} = \mathbf{N}\,\mathbf{u}$, as in (8.3.38), where $\mathbf{u} = \lfloor u_1, u_2, \ldots, u_n \rfloor^T$ is a column matrix containing the n discrete unknowns used for interpolation of the continuous variable u.

To form a linear solution system, we may select node i on the boundary, $i = 1, \ldots, n$, one at a time, to place the delta function there; hence, $\mathbf{x}' = \mathbf{x}_i$, $u = u_i$, and (8.4.14) becomes

$$\frac{1}{2} u_i = \int_{\partial\Omega} \left[\mathbf{N}(\mathbf{x})\mathbf{u} \frac{\partial G(\mathbf{x}, \mathbf{x}_i)}{\partial n(\mathbf{x})} - G(\mathbf{x}, \mathbf{x}_i)\mathbf{N}(\mathbf{x})\mathbf{u_n} \right] d\mathbf{x}, \tag{8.4.15}$$

where $\mathbf{u_n}$ represents the column vector of the discrete values of $\partial u/\partial n$. We can apply (8.4.15) to each discrete node on the boundary, $i = 1, \ldots, m$, and obtain the matrix equation

$$\mathsf{H}\mathbf{u} - \mathsf{G}\mathbf{u_n} = 0, \tag{8.4.16}$$

where $\mathbf{u}$ and $\mathbf{u_n}$ are column matrices, and H and G are $m \times m$ matrices, with elements given by

$$\begin{aligned} H_{ij} &= \int_{\partial\Omega} N_j(\mathbf{x}) \frac{\partial G(\mathbf{x}, \mathbf{x}_i)}{\partial n(\mathbf{x})} \, d\mathbf{x} - \frac{\delta_{ij}}{2}, \\ G_{ij} &= \int_{\partial\Omega} N_j(\mathbf{x}) \, G(\mathbf{x}, \mathbf{x}_i) \, d\mathbf{x}. \end{aligned} \tag{8.4.17}$$

Equation (8.4.16) is a system of n linear equations, in terms of n discrete values of u, and n discrete value of $\partial u/\partial n$. However, in order to solve it, only n unknowns may be admitted. We realize that on the boundary of a well-posed boundary value problem, either the Dirichlet condition (given u values), or the Neumann condition (given values of $\partial u/\partial n$) is specified. Hence, only half of the $2n$ discrete values of u and $\partial u/\partial n$ are unknowns, while the other half can be evaluated and summed to become the right hand side of a linear system (the column matrix $\mathbf{b}$ below). The remaining unknowns can be assembled into a column matrix, $\mathbf{y}$, and the system of equations, (8.4.16), can then be reassembled and condensed into the new matrix form

$$\mathbf{A}\,\mathbf{y} = \mathbf{b}, \tag{8.4.18}$$

where $\mathbf{A}$ is of size $n \times n$, and $\mathbf{y}$ and $\mathbf{b}$ are of size $n \times 1$. This linear system can be solved for $\mathbf{y}$, i.e., the missing part of the boundary data, $\mathbf{u}$ or $\mathbf{u_n}$. Once the missing boundary data are available, the right hand side of (8.4.14) is completely defined. Equation (8.4.14) then allows us to place $\mathbf{x}'$ at any point of interest and evaluate the u value there.

The boundary element method, due to its flexibility in adjusting the mesh to accommodate a moving boundary, has been applied to locating the steady and unsteady free surface in porous medium flow problems (Liggett, 1977) and to the moving interface between saltwater and freshwater in a coastal aquifer (Liu, *et al.*, 1981; Taigbenu *et al.*, 1984). The BEM has also been applied to flow in a heterogeneous porous medium for a certain class of permeability distributions (Cheng, 1984, 1987). For multilayered aquifers, an integrodifferential equation formulation (Herrera, 1970), combined with he Laplace transform, has allowed BEM to entirely eliminate the discretization of the aquitards, such that only aquifer boundaries need to be modeled (Cheng and Morohunfola, 1993a, b). Other applications of the BEM include the simulation of stochastic groundwater flow (Cheng and Lafe, 1991; Cheng *et al.*, 1993), and stochastic saltwater intrusion (Naji *et al.*, 1999; Amaziane *et al.*, 2005). Because the covariance of the piezometric head is modeled, there exist n^2 unknowns for a system with n discrete nodes. A boundary formulation significantly reduces the number of nodes in the discretization, making the size of the solution system tolerable.

8.5 Radial Basis Function Collocation Methods

To enable the local interpolation of the sought solution function within a small surrounding region, the numerical methods presented so far, FDM, FEM, and BEM, always involve a *mesh*, or an assemblage of *elements*, that facilitate the interpolation. In a complicated solution geometry, or in a solution domain where a local refinement is needed in order to represent certain refined features, the effort involved in constructing such mesh and its *connectivity data*, that is, how each node is associated with other nodes in an interpolation scheme, and how each element shares the common nodes with other elements, is not trivial. Despite the availability of automatic mesh generation software, human intervention is often needed. Hence, mesh generation is typically the most labor intensive part of the numerical modeling effort. This difficulty has caused the FEM community to develop *meshless* or *element free* FEMs (Belytschko, 1994; Melenk and Babuška, 1996; Duarte and Oden, 1996; Atluri and Zhu, 1998). However, as pointed out by Atluri and Zhu (1998) and Idelsohn and Onate (2006), not all the so-claimed meshless methods are truly meshless; some may still involve an unseen, 'shadow', mesh.

The collocation method briefly discussed in Subs. 8.3.1A is meshless. As demonstrated, the collocation method minimizes the residual at a set of discrete points that bear no relation to each other; hence, no connectivity data is needed. This is a highly desirable feature. However, in the weighted residual examples illustrated in Subs. 8.3.1, it appears that the methods that distribute the error by performing a weighted integration over the domain, like the Galerkin method, are more accurate. Thus, the point collocation method is not the most accurate one. On the other hand, in the FEM implementation of the Galerkin method, the domain is subdivided into elements, and low degree polynomials are used for interpolation; this destroys the high accuracy of the Galerkin method, as, along the edges of every element, there exists a discontinuity and the governing equation is not satisfied. Hence, for a problem represented by a continuous field, the point collocation is *much more accurate* than the FEM.

The mathematical formulation of the collocation method is rather simple. We start from a boundary value problem in terms of spatial variables only, as defined in (8.3.1) and (8.3.2):

$$\mathcal{L}\{u(\mathbf{x})\} = f(\mathbf{x}), \quad \mathbf{x} \in \Omega, \tag{8.5.1}$$

subject to the essential and natural boundary conditions

$$\begin{aligned} \mathcal{S}\{u(\mathbf{x})\} &= g(\mathbf{x}), \quad \mathbf{x} \in \partial\Omega_S, \\ \mathcal{N}\{u(\mathbf{x})\} &= h(\mathbf{x}), \quad \mathbf{x} \in \partial\Omega_N, \end{aligned} \tag{8.5.2}$$

where $\partial\Omega_S$ and $\partial\Omega_N$ are the parts of the boundary with essential and natural boundary conditions, respectively. In addition, we introduce the interior condition: on a number of interior nodes, $\mathbf{x}_i$, the function values are *specified*, i.e.,

$$u(\mathbf{x}_i) = \bar{u}_i; \quad i = 1, \ldots, m; \quad \mathbf{x}_i \in \Omega. \tag{8.5.3}$$

As discussed in Subs. 5.3.1 and 8.3.1, in a well-posed boundary value problem, i.e., one that guarantees the existence and uniqueness of a solution, we must have $\partial\Omega_S \cup \partial\Omega_N = \partial\Omega$, $\partial\Omega_S \cap \partial\Omega_N = \emptyset$, and $\partial\Omega_S \neq \emptyset$. In other words, on every part of the boundary, at least one type of boundary condition, either of the essential or of the natural type, must be specified, but not both. Also, the boundary conditions cannot be only of the natural type, or else the solution will be non-unique (i.e., determined only to within an additive constant). Another condition that must be satisfied is that no interior condition can be prescribed, that is, $m = 0$. If any of the above conditions is violated, $\partial\Omega_S \cup \partial\Omega_N \neq \partial\Omega$, or $\partial\Omega_S \cap \partial\Omega_N \neq \emptyset$, or $\partial\Omega_S = \emptyset$, or $m \neq 0$, we are faced with a different type of problem—an *ill-posed problem.*

In modeling groundwater flow, we are sometimes faced with an ill-posed problem. For example, although a mathematical solution requires boundary conditions on all parts of the boundary, in reality, we may not know the conditions on certain parts of the boundary. At the same time, we may have a number of field observation of piezometric head *within* the domain. These values cannot be admitted in a well-posed problem; nevertheless, these are *true values*, and we would like to make use of this information. Another situation is that on part of a boundary, we may have both the Dirichlet and the Neumann boundary conditions, say, from observations (this is also known as the *Cauchy condition*). This is exactly what happens in geo-prospecting: while we do not have access to an underground boundary to obtain the boundary condition, we measure two boundary conditions on the accessible portion of the boundary, i.e., at ground surface. Using this information, we attempt to describe the unknown field below. We shall demonstrate how the collocation method can solve well-posed as well as ill-posed problems.

To find an approximate solution of (8.5.1)–(8.5.3), we express u by the summation of basis functions:

$$u(\mathbf{x}) \approx \hat{u}(\mathbf{x}) = \sum_{j=1}^{n} a_j N_j(\mathbf{x}), \quad \mathbf{x} \in \Omega, \tag{8.5.4}$$

where the N_j's are basis functions, and the a_j's are coefficients to be determined. Examples of basis functions include Chebyshev polynomials (Boyd, 2001), Fourier series, fundamental solutions (method of fundamental solutions) (Fairweather and Karageorghis, 1998), and general solutions (Trefftz method) (Herrera, 1984; Li, *et al.*, 2007). In this section, we shall discuss a method based on *Radial Basis Functions* (Kansa, 1990a, 1990b; Cheng *et al.*, 2003).

A radial basis function (RBF) is a function that depends only on the radial distance, r, between two points, $\mathbf{x}$ and $\mathbf{x}_j$, such that $N_j(\mathbf{x}) = N(\|\mathbf{x} - \mathbf{x}_j\|)$, where $\|\cdot\|$ represents the Euclidean norm. Commonly used RBFs include the conical function, r^{2k-1}, the polyharmonic splines, $r^{2k}\ln r$, and the multiquadric function, $(r^2 + C^2)^{k-\frac{1}{2}}$, where $k = 1, 2, 3, \ldots$. Particularly, Hardy's (1971) multiquadric function has been shown to be the one that is most efficient for interpolating a continuous function (Franke, 1982). The first attempt to apply RBF for the collocation solution of partial differential equation was made by Kansa (1990a, 1990b). It has been shown (Madych, 1992; Cheng, *et al.*, 2003; Huang, *et al.*, 2007) that the solution error has an *exponential* convergence with respect to the size of the grid, as well as the constant C in the multiquadric expression.

To solve a boundary value problem, we need to make sure that (8.5.4) approximately satisfies the governing equation (8.5.1). This is done by substituting (8.5.4) into (8.5.1), and enforcing the equation on a set of discrete points,

$$\mathcal{L}\{\hat{u}(\mathbf{x})\} = \mathcal{L}\left\{\sum_{j=1}^{n} a_j\, N(\|\mathbf{x}-\mathbf{x}_j\|)\right\} = \sum_{j=1}^{n} a_j\, \mathcal{L}\{N(\|\mathbf{x}-\mathbf{x}_j\|)\} = f(\mathbf{x}),$$

$$\text{for} \quad \mathbf{x}=\mathbf{x}_i, \quad \mathbf{x}_i \in \Omega, \quad i = 1,\ldots,n_1. \tag{8.5.5}$$

It is also necessary to satisfy the boundary conditions, (8.5.2); hence,

$$\sum_{j=1}^{n} a_j\, \mathcal{S}\{N(\|\mathbf{x}-\mathbf{x}_j\|)\} = g(\mathbf{x}),$$

$$\text{for} \quad \mathbf{x}=\mathbf{x}_i, \quad \mathbf{x}_i \in \partial\Omega_S, \quad i = n_1+1,\ldots,n_2, \tag{8.5.6}$$

$$\sum_{j=1}^{n} a_j\, \mathcal{N}\{N(\|\mathbf{x}-\mathbf{x}_j\|)\} = h(\mathbf{x}),$$

$$\text{for} \quad \mathbf{x}=\mathbf{x}_i, \quad \mathbf{x}_i \in \partial\Omega_N, \quad i = n_2+1,\ldots,n_3. \tag{8.5.7}$$

If there exists any interior condition (for an ill-posed problem), we can also collocate for them, following (8.5.3), to obtain:

$$\sum_{j=1}^{n} a_j\, N(\|\mathbf{x}-\mathbf{x}_j\|) = \bar{u}(\mathbf{x})$$

$$\text{for} \quad \mathbf{x}=\mathbf{x}_i, \quad \mathbf{x}_i \in \Omega, \quad i = n_3+1,\ldots,n_4. \tag{8.5.8}$$

These linear equations can be assemble into a matrix equation

$$\begin{bmatrix} \mathbf{G} \\ \mathbf{S} \\ \mathbf{N} \\ \mathbf{H} \end{bmatrix} \{\mathbf{a}\} = \begin{Bmatrix} \mathbf{f} \\ \mathbf{g} \\ \mathbf{h} \\ \bar{\mathbf{u}} \end{Bmatrix}, \tag{8.5.9}$$

where elements of the sub-matrices are

$$\begin{aligned} G_{ij} &= \mathcal{L}\{N(\|\mathbf{x}-\mathbf{x}_j\|)\}\big|_{\mathbf{x}=\mathbf{x}_i} \\ S_{ij} &= \mathcal{S}\{N(\|\mathbf{x}-\mathbf{x}_j\|)\}\big|_{\mathbf{x}=\mathbf{x}_i} \\ N_{ij} &= \mathcal{N}\{N(\|\mathbf{x}-\mathbf{x}_j\|)\}\big|_{\mathbf{x}=\mathbf{x}_i} \\ H_{ij} &= N(\|\mathbf{x}-\mathbf{x}_j\|)\big|_{\mathbf{x}=\mathbf{x}_i}, \end{aligned} \tag{8.5.10}$$

and the elements on the right hand side of (8.5.9) are

$$f_i = f(\mathbf{x}_i), \quad g_i = g(\mathbf{x}_i), \quad h_i = h(\mathbf{x}_i), \quad \bar{u}_i = \bar{u}(\mathbf{x}_i). \tag{8.5.11}$$

If we select n in (8.5.4) to be $n = n_1 + n_2 + n_3 + n_4$, equation (8.5.9) is a $n \times n$ linear system, which can be solved for a_j, $j = 1, \ldots, n$. The approximate solution (8.5.4) is, therefore, fully defined as a continuous function.

We notice that in the above procedure, we do not distinguish between well-posed and ill-posed problems. The only difference is that in an ill-posed problem, where a boundary condition is missing on part of the boundary, we do not select these points for collocation. When both types of boundary conditions are available, we collocate for both types, using points on the boundary that do not coincide with each other. Collocation at interior points is conducted whenever necessary. Hence, the well- and ill-posed problems are solved in the same way. This direct solution procedure may be compared with other types of numerical methods. In the FEM, FDM, etc., ill-posed problems cannot be directly solved. These problems can only be solved as well-posed ones by *assuming* the missing conditions and discarding the redundant conditions. Iterative solution procedures, typically utilizing optimization schemes, are needed.

As an example of the above general derivation, consider the equation describing steady flow in a confined aquifer (see (5.4.53)),

$$\mathcal{L}\{h\} \equiv \frac{\partial}{\partial x}\left(\mathsf{T}\frac{\partial h}{\partial x}\right) + \frac{\partial}{\partial y}\left(\mathsf{T}\frac{\partial h}{\partial y}\right) = 0. \tag{8.5.12}$$

We can approximate the piezometric head, h, by the *inverse multiquadric* function

$$\hat{h} = \sum_{j=1}^{n} a_j \, N_j = \sum_{j=1}^{n} a_j \, \frac{1}{\sqrt{r_j^2 + C^2}}, \tag{8.5.13}$$

where $r_j = \sqrt{(x - x_j)^2 + (y - y_j)^2}$, and $\mathbf{x}_j$, $j = 1, \ldots, n$, are a set of scattered points in the domain. To satisfy the governing equation, we substitute (8.5.13) into (8.5.12), obtaining

$$\sum_{j=1}^{n} a_j \left[\frac{\partial \mathsf{T}}{\partial x} \frac{x_j - x}{(r_j^2 + C^2)^{3/2}} + \frac{\partial \mathsf{T}}{\partial y} \frac{y_j - y}{(r_j^2 + C^2)^{3/2}} + \mathsf{T} \frac{r_j^2 - 2C^2}{(r_j^2 + C^2)^{5/2}} \right] = 0. \tag{8.5.14}$$

By enforcing this equation on a set of points $\mathbf{x} = \mathbf{x}_i$, $i = 1, \ldots, n_1$, we obtain the coefficients for (8.5.9) as

$$\begin{aligned} G_{ij} = \; & \frac{1}{[(x_i - x_j)^2 + (y_i - y_j)^2 + C^2]^{3/2}} \left[\frac{\partial \mathsf{T}}{\partial x}(x_j - x_i) + \frac{\partial \mathsf{T}}{\partial y}(y_j - y_i) \right. \\ & \left. + \mathsf{T} \frac{(x_i - x_j)^2 + (y_i - y_j)^2 - 2C^2}{(x_i - x_j)^2 + (y_i - y_j)^2 + C^2} \right]. \end{aligned} \tag{8.5.15}$$

Similarly, if we are given a Neumann condition $-\mathsf{T}(\partial h / \partial n) = q$, we find

$$N_{ij} = -\mathsf{T}\frac{n_x(x_j - x_i) + n_y(y_j - y_i)}{[(x_i - x_j)^2 + (y_i - y_j)^2 + C^2]^{3/2}}, \qquad (8.5.16)$$

where n_x and n_y are the directional cosines of the boundary's outward normal, **n**. For the Dirichlet condition, $h = \bar{h}$, we find

$$S_{ij} = \frac{1}{[(x_i - x_j)^2 + (y_i - y_j)^2 + C^2]^{1/2}}. \qquad (8.5.17)$$

Hence, the solution of boundary value problem involves the selection of a set of interior and boundary nodes on which we enforce the governing equation, or boundary condition. The matrix coefficients are calculated according to the above formulas, (8.5.15)–(8.5.17). No construction of elements is required.

In the above, we have discussed the advantages of the RBF collocation method. For example, it uses unstructured nodes without the need to define elements, has high adaptability for moving boundary problems, solutions are highly continuous and highly accurate, can improve accuracy without refining mesh, is easy to formulate, and solves a certain type of inverse problems efficiently. The method, however, also has deficiencies. One observed deficiency is that the solution matrix is fully populated, and non-symmetric. When the matrix becomes large in size, it is inefficient to solve, and is often *ill-conditioned*, that is, the matrix *condition number* is large. Ill-conditioned matrix is generally unstable, that is, its solution error may be unbounded. (See, e.g., Press *et al.* (2007) or any book on numerical analysis for a discussion on matrix condition number and solution stability.)

A number of remedies have been developed to overcome these difficulties. For example, rather than using direct interpolation, as given by (8.5.4)–(8.5.7), *Hermite interpolation* can be used, resulting in a symmetric matrix solution system (Fasshauer, 1999; Chen, 2002; La Rocca *et al.*, 2005, 2008). Also, there are a number of techniques developed for matrix preconditioning aimed at reducing the condition number of the solution system (Beatson *et al.*, 1999; Kansa and Hon, 2000). One reason that the solution matrix of the RBF collocation method is fully populated is that the interpolation functions used are *globally* defined, for example (8.5.13). These global functions are compared to the *locally* defined, *compactly supported*, interpolation functions (meaning that the interpolation function becomes zero outside a certain range, such as an element) of FEM. This non-compact situation of the RBF can be improved by using the *compactly supported radial basis functions* (Wendland, 1998; Wong *et al.*, 2002), and the multilevel scheme (Chen *et al.*, 2002). In fact, as demonstrated by Beatson and Light (1993) and Ling (2005), by using *quasi-interpolation*, a matrix system and its solution is not even required, as the interpolation coefficient can be directly evaluated without a linear system, thus eliminating the ill-conditioning problem of matrix solution. The quasi-interpolation method, however, is limited to problems in rectangular (2-D), or cuboidal (3-D) domains.

8.6 Eulerian-Lagrangian Methods

As discussed in Subs. 8.3.2D, a common source of errors in the numerical solution of the transport equation, e.g., (8.2.1), is in the approximation of the advection term in the numerical model. This error manifests itself in the form of *numerical dispersion.* In a transport problem that is dominated by advection, this 'apparent dispersion' tends to artificially smooth out any sharp concentration front. This difficulty is present in most of numerical methods based on the *Eulerian formulation* (Sec. 7.4). Special numerical techniques are needed in order to overcome this difficulty. This section is devoted to the discussion of techniques known as *Eulerian-Lagrangian methods,* which are most successful in overcoming the above mentioned difficulty.

In an Eulerian approach, in order to describe the concentration of a chemical species dissolved in water (as an example of an intensive quantity), we adopt an observation frame that is fixed in space, known as the *Eulerian frame.* We denoted the concentration c at a fixed point in space, (x, y, z), and at a time t, as $c(x, y, z, t)$. To obtain the mass balance *at a point* (x, y, z) in the considered domain, we draw a small box (= control volume) around that (fixed) point (see Fig. 5.1.1), and determine the species mass balance for the box during a small time interval around t. The PDE that describes the transport of the dissolved species *at a point* is obtained by letting the dimensions of the box shrink to zero around the point (= center of the box), and letting the time interval shrink to zero around the considered point in time. The resulting PDE contains partial derivatives, such as $\partial c/\partial t$, $\partial c/\partial x$, $\partial^2 c/\partial x^2$, etc. In presenting these partial derivatives, we have made use of the definitions

$$\frac{\partial c}{\partial t} = \left.\frac{\partial c(x, y, z, t)}{\partial t}\right|_{x,y,z=\text{const.}}, \qquad \frac{\partial c}{\partial x} = \left.\frac{\partial c(x, y, z, t)}{\partial x}\right|_{y,z,t=\text{const.}}, \qquad \text{etc.} \tag{8.6.1}$$

All this has been discussed in detail in Secs. 7.1 through 7.4.

In the numerical methods discussed so far, we have approximated the above partial derivatives by some discrete expressions, using a system of grid points or cells (for FDM), elements (FEM), or volumes (FVM), laid upon a system of fixed coordinates (e.g., Cartesian). All these methods are based on the Eulerian approach.

In the *Lagrangian approach,* we examine the mass balance from the point of view of an observer moving with the velocity $\mathbf{V}$ of a set of particles in the flow domain. We shall describe this approach in more detail in the following subsections.

8.6.1 Lagrangian method

Let us focus on a given *fluid particle,* and denote its position in a certain coordinate system, such as Cartesian, as $\boldsymbol{\xi} = (\xi_1, \xi_2, \xi_3)$. As the particle is in

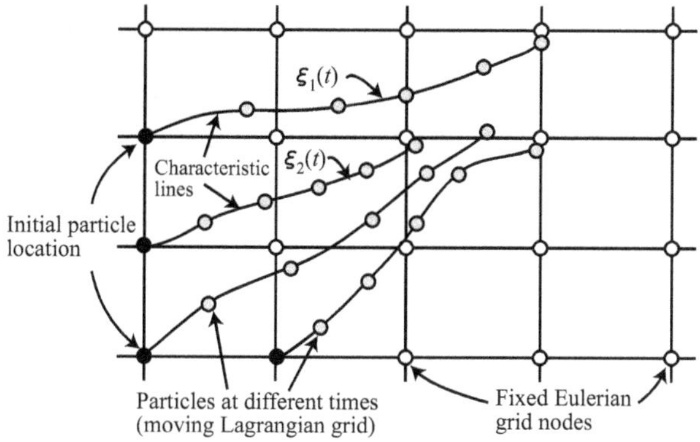

Figure 8.6.1: An Eulerian fixed and a Lagrangian moving coordinate systems.

a flow field, its position changes with time, such that $\boldsymbol{\xi} = \boldsymbol{\xi}(t)$, with its initial position given by $\boldsymbol{\xi}(0)$.

Starting from the Cartesian grid system shown in Fig. 8.6.1, with grid points marked as white circles, we consider it as an *Eulerian coordinate system*, because the grid point positions are *fixed in time*.

At a given time, say $t = 0$, we may select a set of particles located on the Cartesian grid points. A few of these particles are shown in Fig. 8.6.1 as dark circles. As time progresses, these particles will move with the prevailing flow, with their successive new locations at discrete times marked as gray circles. At any instant of time, these particles form a *deformed* grid, called a *Lagrangian coordinate system*. The time trajectories of these particles, shown as solid lines in Fig. 8.6.1, are known as *pathlines* of the flow field. In the Lagrangian approach, these pathlines are also called *characteristic lines*.

Using mathematical representation, we can denote the locations of these particles as $\boldsymbol{\xi}_i(t)$, $i = 1, \ldots, m$, where m denotes the particle number in some numbering system. The initial positions of these particles, $\boldsymbol{\xi}_i(0)$, fall on the *Eulerian grid* points.

In a solute transport model, each of these particles is interpreted as a concentrated fluid mass, representing the lumped mass of a surrounding initial volume. These particles may also be regarded as carriers of certain intensive quantities, such as solute mass concentration, temperature, etc. While the fluid mass is conserved on these particles (due to mass conservation), the intensive quantities that they carry, generally, are not. Particularly, in a model of solute transport in a porous medium, the solute mass can be transported by mechanisms other than *advection*, that is, fluid flow; for example, it can be transported by diffusion and hydrodynamic dispersion, as discussed in Secs. 7.1 through 7.4. Hence, the solute mass is not conserved in the mass of the carrier, and can be passed from one particle to the next, based on the prevailing transport mechanism.

We can express the concentration, or any other intensive quantity that is being transported, as a function of the Lagrangian coordinates, and time, in the form $c = c(\boldsymbol{\xi}(t), t)$, to be compared with the Eulerian expression $c = c(\mathbf{x}, t)$. We can now refer to another kind of time derivative, known as the *material derivative* (see (7.6.1)), $\mathrm{D}c/\mathrm{D}t$, which represents the time rate of change of concentration following the (moving) mass of the carrier fluid (particles). In other words, we identify a particle by its initial location and follow its subsequent trajectory $\boldsymbol{\xi}(t)$, such that to this observer, $\boldsymbol{\xi}(t) = \text{const.}$, when observing the time rate change of concentration carried by that particle.

By invoking the chain rule, we can relate the material derivative to the local derivatives by the relationship

$$\begin{aligned}\left.\frac{\mathrm{D}\,c(\boldsymbol{\xi}(t), t)}{\mathrm{D}t}\right|_{\boldsymbol{\xi}=\text{const.}} &= \left.\frac{\partial c(\mathbf{x}, t)}{\partial t}\right|_{\mathbf{x}=\text{const.}} + \mathbf{V}\cdot\nabla c(\mathbf{x}, t)\\ &= \frac{\partial c}{\partial t} + \nabla\cdot(\mathbf{V}c),\end{aligned} \tag{8.6.2}$$

where $\mathbf{V}$ ($\equiv d\boldsymbol{\xi}(t)/dt$) is the velocity at which the fluid particles are transported, and $\partial c/\partial t$, $\partial c/\partial x$ etc. (as contained in ∇c), are the *local derivatives*, defined in (8.6.1), which are different from the *material derivative*. In writing the second line of the above equation, we have used the condition $\nabla\cdot\mathbf{V} = 0$, which is based on the assumption of constant fluid density and porosity (cf. (5.1.8)). If we examine the terms on the right hand side of (8.6.2), we realize that they account for two transport mechanisms: the time rate of change of concentration at a point, and the net advective flux.

The Eulerian-Lagrangian formulation for solute transport has been briefly discussed in Sec. 7.6. Here we shall consider the solute balance equation (7.4.18), describing solute transport in a homogeneous domain (ϕ, a_L, a_T = const.), without sources,

$$\frac{\partial c}{\partial t} + \nabla\cdot(\mathbf{V}c - \mathbf{D}\cdot\nabla c) = 0, \tag{8.6.3}$$

in which $\mathbf{D}$ is the coefficient of dispersion (a tensor) defined by (7.1.37). Based on the material derivative defined in (8.6.2), the above equation can be rewritten in the form

$$\frac{\mathrm{D}c}{\mathrm{D}t} = \nabla\cdot(\mathbf{D}\cdot\nabla c)\,. \tag{8.6.4}$$

We observe that the left hand side in (8.6.4) is a Lagrangian expression, while the right hand side remains Eulerian.

Let us consider the special case of transport by *advection only*. This means,

$$\frac{\mathrm{D}c}{\mathrm{D}t} = 0. \tag{8.6.5}$$

In this case, if we take the Lagrangian approach, by selecting a set of initial fluid particles, and assigning to each of them a concentration based on the initial concentration in the field, then, as the particles are transported by the flow, the concentration they carry remains unchanged at any time, and any new location. Hence, at any given time, the concentration at a point, such as the points of an Eulerian grid, or the center of a cell, can be obtained by some interpolation scheme, using the concentration represented by a set of nearby particles, or particles that fall within a cell.

The concept of an invariant concentration along a pathline can be used to select a set of 'particles' that do not necessarily represent a fixed amount of fluid mass. For example, we can deploy an arbitrary number of particles along solute concentration contour lines just for the purpose of tracking the movement of these lines. In such case, the time history of the contour lines gives a convenient way to visualize the (advective) transport process. This technique can be used to assess whether pollutants released from a contaminated site (represented by a group of particles) will reach a pumping well (see, for example, Cheng (2000), Sec. 5.5; or the computer code MODPATH, Pollock (1994)). Or, it can be used to track the movement of 'sharp' interfaces between two immiscible fluids, such as between freshwater and saltwater in coastal aquifers (Subs. 9.2.1), or between water and oil in *water flooding* operation in petroleum reservoirs. In this way, the movement of the sharp change in property is modeled exactly, without smearing (i.e., without numerical dispersion), that is present in the Eulerian approach. It should be emphasized, however, that in this model (of a sharp interface), we have neglected the dispersion of concentration caused by advection, which is hardly true for porous medium flow.

In order to track the particles in a Lagrangian approach, a set of *characteristic lines* (Fig. 8.6.1) needs to be constructed (O'Neill, 1981; Varoglu and Finn, 1982). Given a set of particles initially located at $\boldsymbol{\xi}_i(0)$, $i = 1, \ldots, m$, we can trace their subsequent locations by the time integration,

$$\boldsymbol{\xi}_i(t_{k+1}) = \boldsymbol{\xi}_i(t_k) + \int_{t_k}^{t_{k+1}} \mathbf{V}(\boldsymbol{\xi}_i(t), t)\, dt, \tag{8.6.6}$$

where t_k and $t_{k+1}(= t_k + \Delta t)$ are two different times, and $\mathbf{V}$ is the particle's velocity.

Other than for the simple case, in which the flow field is given by an analytical solution, this velocity needs to be computed using a numerical method, from an *Eulerian flow equation* (Subs. 5.1.4). For unsteady, nonuniform flow, this velocity (we may refer to it as the 'Eulerian velocity') is a function of space and time. With the linear, spatial interpolation of piezometric head used in the selected numerical method, the calculated velocity is approximated as a constant within an element, thus suffering a jump when a particle crosses to an adjacent element. And, in the time dimension, the temporal change in velocity is often crudely approximated using a time stepping scheme. These issues need to be carefully addressed in a particle tracking scheme in order to minimize the error in tracking particle locations. An accurate technique for tracking particles—an *adaptive pathline-based tracking algorithm*—is presented by Bensabat *et al.* (2000).

8.6.2 Method of characteristics

In Subs. 8.6.1 we discussed the advantage of treating the left hand side of (8.6.4) by a Lagrangian numerical technique. We also commented that the right hand side of (8.6.4), the dispersive transport term, is in an Eulerian form; hence is not suitable for a Lagrangian treatment. To overcome the difficulty, a number of numerical methods have combined the two approaches by handling the advection term by a Lagrangian method, and the diffusion-dispersion and reaction terms by an Eulerian one. These methods are generally referred to as *Eulerian-Lagrangian methods.* In this subsection, we shall discuss a class of methods known as *methods of characteristics* (Pinder and Cooper, 1970; Konikow and Bredehoeft, 1978; Neuman, 1981, 1984; Molz, 1986), which has been adopted, for example, in the popular computer codes MOC (Konikow and Bredehoeft, 1978) and MT3D (Zheng, 1990) (see Sec. 8.8).

One method that handles advective and non-advective transport separately uses *operator splitting* (see, e.g., Sun *et al.* (2008) for a review of the technique). Consider, for example, equation (8.6.4). We can split the concentration into two parts:

$$c = c_L + c_r, \tag{8.6.7}$$

where c_L is the part that satisfies the Lagrangian operator,

$$\frac{\mathrm{D}c_L}{\mathrm{D}t} = 0, \tag{8.6.8}$$

and c_r is a residual concentration, satisfying

$$\frac{\mathrm{D}c_r}{\mathrm{D}t} - \nabla \cdot [\mathbf{D} \cdot \nabla(c_L + c_r)] = 0. \tag{8.6.9}$$

It is easy to show that the original governing equation, (8.6.4), is recovered by summing (8.6.8) and (8.6.9).

In the numerical implementation, a Lagrangian (moving) grid is used to solve (8.6.8) by time stepping from t_k to t_{k+1}. The information obtained on c_L is then transferred to (8.6.9) as a source term. Equation (8.6.9) is then solved on an Eulerian (fixed) grid, using any standard numerical method, e.g., FDM, or FEM. This is the approach taken in the computer code MOC (see Sec. 8.8). This technique, however, requires the deployment of a large number of particles. It is not a straightforward process when dealing with complex boundary conditions and nonlinearities. In fact, the need for han-

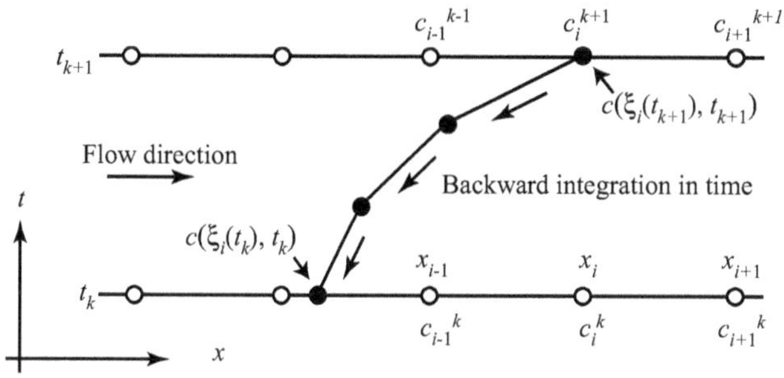

Figure 8.6.2: A backward particle tracking scheme.

dling numerous particles, and the highly deformed grid, makes this technique rather cumbersome and time consuming.

In a different, one-step, strategy, a *backward particle tracking* technique is employed (Molz, 1986). In this technique, particles are deployed at time t_{k+1} at locations coinciding with the Eulerian node locations. The corresponding particle locations at t_k are found by back tracking:

$$\boldsymbol{\xi}_i(t_k) = \boldsymbol{\xi}_i(t_{k+1}) - \int_{t_k}^{t_{k+1}} \mathbf{V}(\boldsymbol{\xi}_i(t), t)\, dt. \tag{8.6.10}$$

We note that in order to allow the use of larger time steps, and to ensure accurate particle locations, the integration is, typically, divided into smaller time increments, making use of high order integration techniques, such as Runge-Kutta and predictor-corrector methods (Press *et al.*, 2007). Figure 8.6.2 illustrates the integration process for a one-dimensional problem.

With the one-step backward scheme, we can express the Lagrangian time derivative term in (8.6.4) as $\mathrm{D}c/\mathrm{D}t = \left[c(\boldsymbol{\xi}_i(t_{k+1}), t_{k+1}) - c(\boldsymbol{\xi}_i(t_k), t_k)\right]/\Delta t$. Here we notice that in the one step process, we have selected Lagrangian particle location $\boldsymbol{\xi}_i(t_{k+1})$, on which the concentration $c(\boldsymbol{\xi}_i(t_{k+1}), t_{k+1})$ is represented, to coincide with the Eulerian node x_i, on which we observe the concentration c_i^{k+1} (Fig. 8.6.2). The value $c(\boldsymbol{\xi}_i(t_k), t_k)$, can be obtained by interpolation, using the known concentration at the Eulerian coordinates at time t_k. To obtain the dispersion term in (8.6.4), we approximate it by a central finite difference, and use an implicit scheme (Sec. 8.1). Finally, we can express (8.6.4), in its one-dimensional form, by its finite difference approximation:

$$\frac{c_i^{k+1} - c(\xi_i(t_k), t_k)}{\Delta t} - D\frac{c_{i+1}^{k+1} - 2c_i^{k+1} + c_{i-1}^{k+1}}{(\Delta x)^2} = 0. \tag{8.6.11}$$

These equations allow us to solve for the concentration at time t_{k+1} on an Eulerian grid, from the known concentration at t_k. The backward particle tracking is renewed at every time step, such that the forward particle locations are always on a regular Eulerian grid. As there is no continuously deforming Lagrangian grid, the backward particle tracking scheme is less cumbersome as compared to the forward scheme. However, it has been found that the backward scheme still introduces a certain numerical dispersion error (Yeh and Tripathi, 1987). To combine the advantages and minimize the shortcomings of these two schemes, Neuman (1981, 1984) mixed their use: near a steep advection front, a forward tracking algorithm is adopted, while backward tracking algorithm is employed away from the front.

There exist a variety of other numerical methods aimed at improving the accuracy and efficiency of solving advection-dispersion transport equations. Among these, we can mention the Modified Method of Characteristics (Douglas and Russell, 1982; Ewing *et al.*, 1984; Chiang *et al.*, 1989), the Eulerian-Lagrangian Localized Adjoint Methods (Celia *et al.*, 1990; Herrera *et al.*, 1993; Binning and Celia, 1996, 2002; Wang *et al.*, 1995, 1999; Russell and Celia, 2002), the Petrov-Galerkin Finite Element Method (Christie *et al.*, 1976; Barrett and Morton, 1984; Liu *et al.*, 1988; Miller and Rabideau, 1993), the Streamline Diffusion Finite Element Method (Brooks and Hughes, 1982; Hughes and Mallet, 1986; Johnson *et al.*, 1987, 1990; Hansbo and Szepessy, 1990), and the Discontinuous Galerkin Method (Riviere *et al.*, 2000; Aizinger *et al.*, 2000; Sun and Wheeler, 2005; Wang *et al.*, 2007). For further review and discussion, see for example Al-Lawatia *et al.* (1999), and Ewing and Wang (2001).

8.6.3 Random walk method

The *random walk method* (Prickett *et al.*, 1981; Kock and Prickett, 1989) takes a different approach from the Eulerian-Lagrangian method. Rather than deploying a number of fluid particles that carry with them the solute concentration, the random walk method distributes *solute mass particles.* The concentration of solute at a point is interpreted by taking a small volume and counting the number of solute particles located in it. By this definition, we recognize that the obtained concentration profile can be 'lumpy', that is, not smooth, unless a very large number of particles are used. Nevertheless, the method is simple to program, as it does not involve the cumbersome handling of two coordinate systems of an Eulerian-Lagrangian method, and there is no partial differential equation to solve (except for the flow part of the problem, which can be handled by any Eulerian numerical method).

In the random walk method, a number of solute particles are deployed in the flow domain, where the solute concentration exists. This can be an advantage, as unlike the Eulerian-Lagrangian method, in which fluid particles are needed everywhere in the solution domain, in a random walk model, solute particles of needed only where significant solute concentration exists, which may mean smaller number of particles needed.

In the random walk method, solute particles are transported by two separate processes, first by advection, and then by dispersion. The advection part is carried out in a similar Lagrangian process as described in Subs. 8.6.1; that is, for a given time increment, each (conserved) solute mass is carried by the prevailing flow to a new location. At this new location, we then consider the effect of dispersion.

The dispersion is considered as a random walk process. To understand such process, we consider a one-dimensional transport of a packet of solute mass, modeled as a concentrated mass (Dirac delta function), by a constant velocity flow, in an homogeneous medium. According to the analytical solution (7.4.54) given in Subs. 7.4.5 C, the concentration as function of space and time is

$$\frac{c(x,t)}{M/\phi} = \frac{1}{(4\pi \mathsf{D}_h t)^{1/2}} \exp\left[-\frac{(x - Vt)^2}{4\mathsf{D}_h t}\right]. \tag{8.6.12}$$

We notice that the concentration profile is a *normal distribution* (or *Gaussian distribution*), centered at $x = Vt$ (see Fig. 7.4.5). We can further express (8.6.12) into the following form

$$\frac{c(x,t)}{M/\phi} = \frac{1}{\sqrt{2\pi}\,\sigma} \exp\left[-\frac{(x - \overline{x})^2}{2\sigma^2}\right], \tag{8.6.13}$$

where $\overline{x} = Vt$ and $\sigma = \sqrt{2\mathsf{D}_h t}$. We may view the right hand side of (8.6.13) as a (Gaussian) *probability density function* with mean $\overline{x}$ and standard deviation σ. Hence, for a particle being transported by the advective velocity V for a time t, rather than landing on the exact location $x = Vt$, it has a chance of arriving ahead or behind that location, with a probability given by the right hand side of (8.6.13).

In actual implementation, consider a three-dimensional flow. With a given time increment Δt, a solute particle is first transported by the prevailing advective velocity $\mathbf{V}$, along the pathline, to a new position. To model dispersion in the longitudinal direction, a random number generator is used to generate an incremental distance $\Delta\ell$ (positive or negative) following a Gaussian distribution with zero mean and standard deviation $\sigma = \sqrt{2\mathsf{a}_L V \Delta t}$, where a_L is the longitudinal dispersivity. (Refer to, for example, Press *et al* (2007), for random number generator that gives a Gaussian distribution.) The incremental distance is then used to move the particle forward or backward in the pathline direction, to account for the effect of dispersion. To avoid too large a displacement from being created, $\Delta\ell$ is typically restricted to within $\pm 6\sigma$. The same procedure is then applied to the two perpendicular directions of the pathline, to simulate the effect of transverse dispersion. In this case, the standard deviation of the Gaussian distribution is given by $\sigma = \sqrt{2\mathsf{a}_T V \Delta t}$, where a_T is the transverse dispersivity. More detail about this method can be found in Prickett *et al.* (1981).

It is obvious from the above description, for the random walk method, a large number of particles are needed in order to obtain smooth concentration and accurate result. This, in fact, can be accomplished by two ways, either by using a large number of particles in a single simulation, or by conduction many simulations and averaging the results. Another difficulty associated with the method is the need to use very small time steps when the domain is heterogenous, with high contrasts in hydraulic conductivities and dispersivities (Anderson *et al.*, 1993).

8.6.4 Modified Eulerian-Lagrangian method

Before leaving this subject, let us address one more issue involved in the numerical modeling of advective-dispersive transport—the apparent advection caused by field heterogeneity. So far in this section, to simplify the illustration, we have considered a solute transport model that assumes an incompressible fluid, and a homogeneous field (which means constant dispersivity and other field coefficients). In what follows, we shall discuss a *modified Eulerian-Lagrangian scheme* that takes into consideration the effects of field inhomogeneity. The example below will also involve two fluid phases.

Consider the solute balance equation (7.4.3),

$$\frac{\partial(\theta c)}{\partial t} = -\nabla \cdot \theta(c\mathbf{V} - \mathbf{D}_h \cdot \nabla c) + \theta\rho\Gamma + f. \tag{8.6.14}$$

where c and θ are the concentration of the considered species and the volume fraction of the fluid phase, $\mathbf{D}_h$ is the coefficient of hydrodynamic dispersion, Γ is the source term in the fluid phase, and f represents the net rate of gain of the considered species entering the considered phase through its interphase boundaries. By expanding and dividing by θc, we obtain

$$\begin{aligned}\frac{1}{c}\frac{\partial c}{\partial t} + \frac{1}{\theta}\frac{\partial \theta}{\partial t} &= -\nabla \cdot \mathbf{V} - \frac{1}{\theta}\mathbf{V} \cdot \nabla\theta - \frac{1}{c}\mathbf{V} \cdot \nabla c \\ &+ \frac{1}{c}\mathbf{D}_h \cdot \nabla(\nabla c) + \frac{1}{\theta c}\nabla \cdot (\theta\,\mathbf{D}_h) \cdot \nabla c + \frac{1}{c}\rho\Gamma + \frac{1}{\theta c}f,\end{aligned} \tag{8.6.15}$$

in which the second and third terms on the right hand side represent advective transport. We also note that the fifth term has also the appearance of an effective transport at a velocity $(1/\theta)\nabla \cdot (\theta\,\mathbf{D}_h)$. Hence, to take full advantage of the Lagrangian method, Bear *et al.*, (1997) and Sorek *et al.*, (1999) proposed a *modified Eulerian-Lagrangian formulation*, which utilizes a modified velocity. In this formulation, (8.6.15) is rewritten as

$$\frac{1}{c}\frac{\mathrm{D}_{\tilde{\mathrm{v}}}\,c}{\mathrm{D}t} + \frac{1}{\theta}\frac{\mathrm{D}_{\mathrm{V}}\,\theta}{\mathrm{D}t} = -\nabla \cdot \mathbf{V} + \frac{1}{c}\mathbf{D}_h \cdot \nabla(\nabla c) + \frac{1}{c}\rho\Gamma + \frac{1}{\theta c}f, \tag{8.6.16}$$

where we have used the subscripts v and $\tilde{\mathrm{v}}$ in the material derivatives to identify the velocity at which the quantity is being transported; particularly, we notice that the concentration is transported at the *modified velocity*

$$\bar{\mathbf{V}} = \mathbf{V} - \frac{1}{\theta}\nabla \cdot (\theta\, \mathbf{D}_h). \tag{8.6.17}$$

The modified velocity consists of two parts, the actual fluid velocity, $\mathbf{V}$, and an *apparent velocity*, $-\frac{1}{\theta}\nabla \cdot (\theta\, \mathbf{D}_h)$, associated with the heterogeneous characteristics of the medium and the flow field. The first term gives an advective transport in the direction of decreasing piezometric head gradient, while the second term has an apparent advection effect in the direction of decreasing hydrodynamic dispersion. It has been demonstrated that the modified Eulerian-Lagrangian numerical scheme gives highly accurate result in advection-dominated flow (Sorek *et al.*, 1999).

8.7 Matrix Solution

For most numerical method, the discretization of the governing equations leads to a matrix solution system

$$\mathbf{A} \cdot \mathbf{x} = \mathbf{b}, \tag{8.7.1}$$

where $\mathbf{A}$ is an $N \times N$ matrix, which can be *fully populated*, *banded*, or *sparsely populated*, $\mathbf{x}$ and $\mathbf{b}$ are $N \times 1$ column matrices, and N is the size of the matrix, which is equal to the number of discrete unknowns modeled in the numerical solution. In the above, $\mathbf{A}$ and $\mathbf{b}$ contain known quantities, and $\mathbf{x}$ contains unknowns to be solved for; these, are the discrete values of solution represented at a set of nodes, or the coefficients of a series approximating the solution.

In principle, the solution of (8.7.1) is

$$\mathbf{x} = \mathbf{A}^{-1} \cdot \mathbf{b}, \tag{8.7.2}$$

where $(\ )^{-1}$ indicates matrix inverse. In reality, however, the matrix is never inverted; instead, some other procedure, such as the *Gauss elimination*, or LU-decomposition, is used (Press *et al.*, 2007). For these solution methods, the required computer CPU time is generally proportional to N^3, where N is the number of unknowns of the discrete solution system; hence, the computation time increases very fast with respect to the refinement of the solution mesh. Consider, for example, a problem solved to a certain unsatisfactory accuracy with a relatively coarse mesh; and, for better accuracy, it is desirable to refine the mesh. If the spacing between the nodes is halved, then for a three-dimensional problem, this will produce a discrete system size that is $2^3 = 8$ times the original one. According to the $O(N^3)$ rule mentioned above, the computational effort grows by $8^3 = 512$ folds. Assuming that the employed numerical method uses the central difference formula of the FDM, or the quadratic element of the FEM, the error is, generally, reduced to $(1/2)^2 = 1/4$ of the previous one, a minimal improvement. We can clearly see the inefficiency of the numerical solution, and the challenge associated with achieving increasing accuracy.

Luckily, in most numerical methods, the resulting matrix $\mathbf{A}$ is either banded, blocked, or is sparsely populated, meaning that most of the matrix elements are null in value, and no operation is needed on them. Special matrix solution techniques exist to take advantage of this fact.

There exist a substantial number of computer subroutines that can handle matrices that have recognizable banded and/or blocked patterns, such as tridiagonal,, banded diagonal, block diagonal, block triangular, etc. A good example is the IMSL (1997) subroutine library. If the user does not have sufficient knowledge about the banded/blocked structure of the matrix, the matrix can be *analyzed.* The NAG (2006) subroutine library has such capability. Generally, the computational effort for solving banded and blocked systems is proportional to $O(M \times N^2)$, where M is associated with the band or the block size, which is a significant improvement over N^3 for large N-values.

8.7.1 Conjugate gradient method

When the matrix is sparse, and particularly when there is no recognizable banded or blocked pattern, an iterative search technique based on *optimization* (see Sec. 11.2) is often much more efficient.

A widely used iterative matrix solution technique, used for solving groundwater transport problems, is the *conjugate gradient method.* In this method, we set up an *objective function* Z and seek to minimize it with respect to a set of *decision variables* $\mathbf{x} = (x_1, x_2, \ldots, x_N)$ (see terminology in Subs. 11.2.1):

$$\text{Minimize } Z(\mathbf{x}) = \frac{1}{2}\mathbf{x} \cdot \mathbf{A} \cdot \mathbf{x} - \mathbf{b} \cdot \mathbf{x}. \tag{8.7.3}$$

We observe that Z is minimized if $\partial Z/\partial x_1 = \partial Z/\partial x_2 = \ldots = 0$, or

$$\nabla Z = \mathbf{A} \cdot \mathbf{x} - \mathbf{b} = 0, \tag{8.7.4}$$

and that this equation is exactly (8.7.1). Hence, solving the linear system (8.7.1) is equivalent to solving the optimization (minimization) problem of (8.7.3) or (8.7.4).

To start the solution, we need an initial guess of the solution vector, $\mathbf{x}_0$, to evaluate the function Z. Consider the objective function Z that takes the form of a surface with a valley (Fig. 11.2.6). We seek an optimal path that will lead us to the bottom of the valley in as few steps as possible, using only available information around the trial solution (see also Subs. 11.2.4B and C). Let us denote the direction of descent by the vector $\mathbf{p}_k$, with $k = 0, 1, 2, \ldots$ indicating successive steps. In addition to the direction, we need a step size, denoted as α_k. to move in that direction. Hence, in the next trial, we seek a better $\mathbf{x}$ value by determining

$$\mathbf{x}_{k+1} = \mathbf{x}_k + \alpha_k \mathbf{p}_k. \tag{8.7.5}$$

The process is continued until the minimum is reached.

Different methods, such as the *method of steepest descent*, and the *conjugate gradient method*, can be used to find the descending directions and step sizes. We shall describe below only the conjugate gradient method, considered to be the more efficient method.

There exist a few variations of conjugate gradient method. If the matrix A is *symmetric* and *positive definite*, we can utilize the *ordinary conjugate gradient method*. In this method, we start by making an initial guess, $\mathbf{x}_0$, substituting it for the true solution, $\mathbf{x}$. Next we use the new value evaluate the *residual*

$$\mathbf{r}_0 = \mathbf{b} - \mathsf{A} \cdot \mathbf{x}_0. \tag{8.7.6}$$

From this residual, we select the first search direction

$$\mathbf{p}_0 = \mathbf{r}_0. \tag{8.7.7}$$

We then repeat the following steps for $k = 0, 1, 2, ...,$

$$\alpha_k = \frac{\mathbf{r}_k \cdot \mathbf{r}_k}{\mathbf{p}_k \cdot \mathsf{A} \cdot \mathbf{p}_k}, \tag{8.7.8}$$

$$\mathbf{x}_{k+1} = \mathbf{x}_k + \alpha_k \mathbf{p}_k, \tag{8.7.9}$$

$$\mathbf{r}_{k+1} = \mathbf{r}_k - \alpha_k \, \mathsf{A} \cdot \mathbf{p}_k, \tag{8.7.10}$$

$$\beta_k = \frac{\mathbf{r}_{k+1} \cdot \mathbf{r}_{k+1}}{\mathbf{r}_k \cdot \mathbf{r}_k}, \tag{8.7.11}$$

$$\mathbf{p}_{k+1} = \mathbf{r}_{k+1} + \beta_k \, \mathbf{p}_k. \tag{8.7.12}$$

This procedure continues until the residual is smaller than a certain desired error tolerance.

8.7.2 Preconditioning

The conjugate gradient method, as other iterative methods, works most effectively when the matrix A is sparse, and well conditioned, that is, its condition number is small. The *condition number* of a matrix is defined as

$$\kappa(\mathsf{A}) = \|\mathsf{A}^{-1}\| \cdot \|\mathsf{A}\|, \tag{8.7.13}$$

where $\|..\|$ indicates the *matrix norm* (see any textbook on matrix analysis, e.g., Golub and van Loan, 1996; Jennings and McKeown, 1992). When the condition number of A is large, the solution of the system (8.7.1) becomes unstable, that is, small errors introduced to the right hand side vector $\mathbf{b}$ (such as roundoff error due to computer's limited precision) can cause large errors in the evaluation of $\mathbf{x}$; hence, it is desirable to *precondition* the matrix A, i.e., to reduce its condition number, prior to the application of the conjugate gradient method.

Preconditioning means pre-multiplying the matrix system (8.7.1) by a preconditioning matrix $\mathbf{M}^{-1}$,

$$(\mathbf{M}^{-1} \cdot \mathbf{A}) \cdot \mathrm{x} = \mathbf{M}^{-1} \cdot \mathrm{b}, \tag{8.7.14}$$

such that

$$\overline{\mathbf{A}} \cdot \mathrm{x} = \overline{\mathrm{b}}, \tag{8.7.15}$$

in which the new matrix, $\overline{\mathbf{A}} = \mathbf{M}^{-1} \cdot \mathbf{A}$, has a smaller condition number. There exist many choices of the preconditioning matrix. In fact, the perfect preconditioning matrix is $\mathbf{M}^{-1} = \mathbf{A}^{-1}$, or $\mathbf{M} = \mathbf{A}$, such that $\overline{\mathbf{A}} = \mathbf{I}$, i.e., the identity matrix, which has the condition number 1. In this case, the system (8.7.15) can be solved in one step, without iterations. However, finding $\mathbf{A}^{-1}$ is equivalent to solving the system by the *method of elimination*, and there is no advantage in applying the iterative method. Hence, our goal is to find preconditioning matrices that approximate $\mathbf{A}^{-1}$.

The simplest preconditioning matrix is the *Jacobi preconditioner*, given by

$$M_{ij} = \begin{cases} A_{ii}, & \text{for } i = j, \\ 0, & \text{otherwise}, \end{cases} \tag{8.7.16}$$

where M_{ij} and A_{ij} are the matrix elements of $\mathbf{M}$ and $\mathbf{A}$, respectively. In other words, we take only the diagonal terms of $\mathbf{A}$, such that its inverse is

$$M_{ij}^{-1} = \begin{cases} 1/A_{ii}, & \text{for } i = j, \\ 0, & \text{otherwise}. \end{cases} \tag{8.7.17}$$

Other, more sophisticated, preconditioning methods include the *incomplete LU conjugate gradient* (Chin *et al.*, 1992; Chow and Saad, 1997) and the *incomplete Choleski conjugate gradient* (Kershaw, 1978; Kuiper, 1981; Ajiz and Jennings, 1984) method. In the *LU decomposition*, a matrix can be factored into the product of a *lower triangular matrix* $\mathbf{L}$ (all elements above the diagonal are null elements), and an *upper triangular matrix* $\mathbf{U}$ (all elements below the diagonal are null), such that

$$\mathbf{A} = \mathbf{L} \cdot \mathbf{U}. \tag{8.7.18}$$

Or, in the *Choleski decomposition*, a positive definite symmetric matrix can be factored into

$$\mathbf{A} = \mathbf{L} \cdot \mathbf{L}^T, \tag{8.7.19}$$

where the upper triangular matrix is simply the transpose of the lower triangular matrix (Golub and van Loan, 1996; Jennings and McKeown, 1992; Press *et al.*, 2007). These triangular matrix pairs are then used as preconditioner to pre- and post-multiply the matrix $\mathbf{A}$.

As an example, consider the incomplete Choleski conjugate gradient method. We may precondition the system (8.7.1) in the form

$$[\mathbf{L}^{-1} \cdot \mathbf{A} \cdot (\mathbf{L}^T)^{-1}] \cdot (\mathbf{L}^T \cdot \mathbf{x}) = \mathbf{L}^{-1} \cdot \mathbf{b}, \tag{8.7.20}$$

such that it becomes

$$\overline{\mathbf{A}} \cdot \overline{\mathbf{x}} = \overline{\mathbf{b}}. \tag{8.7.21}$$

Once we apply the conjugate gradient method to solve for $\overline{\mathbf{x}}$, we can recover $\mathbf{x}$ from

$$\mathbf{x} = (\mathbf{L}^T)^{-1} \cdot \overline{\mathbf{x}}. \tag{8.7.22}$$

In the above procedure, we recall that $\mathbf{A}$ is a sparse matrix. However, in the complete LU decomposition, or complete Choleski decomposition, new nonzero matrix elements, or "fill-ins", can be generated, thus defeating the purpose of having a sparse matrix. To circumvent this difficulty, the decomposition needs to be *incomplete*, that is, whenever a nonzero element is generated at a zero element location of $\mathbf{A}$, it should be deleted and stored in a cancelation matrix, such that the (incomplete) decomposition takes the form

$$\mathbf{A} = \mathbf{L} \cdot \mathbf{L}^T + \mathbf{C}, \tag{8.7.23}$$

in which $\mathbf{L}$ has the same sparse structure as $\mathbf{A}$, and $\mathbf{C}$ is a cancelation matrix. Further details of these preconditioned conjugate gradient methods can be found in the literature (Jennings and McKeown, 1992; Hackbusch, 1993; Golub and van Loan, 1996; Saad, 2003).

8.8 Computer Codes

Once a numerical model has been constructed, and a numerical method has been chosen to enable the (approximate) solution of the PDEs included in a mathematical model that represents a given physical problem, a computer code (or program) is used to execute the large number of repetitive calculation steps involved in the solution of the numerical model. As already mentioned in the preamble of this chapter, we use the term 'computer code' for the set of instructions, or commands, that define the set of tasks to be carried out by a computer. This section reviews some of the more widely used computer codes used for solving flow and transport problems.

Computer codes have been developed by individual researchers, by governmental agencies, or by commercial entities. In this section, we focus mainly on *public domain codes*, i.e., codes that are accessible to the public for free. Among the most notable sources for such codes are government agencies, such as the U.S. Geological Survey (USGS), the U.S. Environmental Protection Agency (USEPA), the U.S. Army Corps of Engineers (USACE), the various U.S. Department of Agriculture (USDA) laboratories, and the National Laboratories. Here we should mention that many of the public domain codes have commercial versions, which often provide more user-friendly preprocessors and postprocessors.

As there exist a large number of codes, with overlapping capabilities, the selection here is, generally, based on the popularity of their adoption, and on a broad spectrum coverage. Codes that are largely based on analytical solutions, or are limited to lower spatial dimensions, have not been included. There is no attempt to recommend one code over another, as each code may serve a different purpose, and there is no code that is without deficiency. The reader should also be aware of the dynamic nature of code development; newer versions of existing codes, with new capabilities, are continuously being developed and announced.

One last word of caution: whenever selecting a code, it is of utmost importance to check the code for its documented verification; or else the code may not be appropriate for the consider case (see **Step 5** in Subs. 1.2.2).

With these comments in mind, following is a brief review of some of the more widely used computer codes.

MODFLOW (Modular Finite-Difference Ground-Water Flow Model). This is a computer program that simulates three-dimensional groundwater flow by using a cell-centered FDM (Subs. 8.1.3). It was first released by the U.S. Geological Survey in 1984 as a public domain computer code (McDonald and Harbaugh, 1984). It went through several revisions, MODFLOW-88, MODFLOW-96, MODFLOW-2000 (Harbaugh *et al.*, 2000; Hill *et al.*, 2000), and the current version is MODFLOW-2005 (Harbaugh, 2005). MODFLOW is one of the most widely used computer codes for groundwater flow simulation, partly because it is a public-domain, open-source code. Also, it has been developed with a *modular* structure, meaning that the source code is easier to understand, and users can develop their own subroutines to link with the main program. Particularly, in MODFLOW-2000 and in later versions, users can add multiple non-groundwater flow equations, such as the contaminant transport equation, to enhance its modeling capability. The USGS has also developed several major extensions to MODFLOW. For example, **MODFLOWP** (Hill, 1992) is an inverse modeling program, based on nonlinear regression, with the purpose of calibrating for aquifer parameters, using observed groundwater heads. MODFLOWP has been incorporated into MODFLOW-2000 and its later version. **MODPATH** (Pollock, 1994) is a particle tracing postprocessing computer code for MODFLOW. The program allows the user to deploy a group of particles and to trace their pathlines, either forward or backward in time, based on the assumption of advective transport only.

MOC3D (Three-Dimensional Method-of-Characteristics Ground-Water Flow and Transport Model). This USGS computer code was originally released as **MOC**, a two-dimensional flow and transport code (Konikow and Bredehoeft, 1978). It uses a cell-centered finite difference scheme for the flow equation, and the method of characteristics for solute transport. In the solute transport equation, the advective term is handled by the Lagrangian particle tracking technique (Sec. 8.6). Hydrodynamic dispersion and fluid source terms are solved by using a two-step explicit finite difference procedure. The newer version, MOC3D (Konikow *et al.*, 1996), solves three-dimensional problems, utilizing MODFLOW as its flow solver. It is distributed as a module of the MODFLOW-96 and later versions. The latest version of MOC3D (Goode, 1999) adds capabilities to simulate aging of groundwater, double porosity exchange, and simple reactions. For modeling saltwater intrusion, which involves variable density effect, the two-dimensional MOC code was modified to **MOCDENSE** (A two-constituent solute transport model for groundwater with variable density) by Sanford and Konikow (1985), and the three-dimensional code to **MOCDENSE3D** by Oude Essink (2001, 2003).

MT3DMS (Modular 3-D Multi-Species Transport Model for Simulation of Advection, Dispersion, and Chemical Reactions of Contaminants in Groundwater Systems). The code was first released as **MT3D** by the U.S. Environmental Protection Agency (Zheng, 1990). MT3D interfaces directly with MODFLOW for the flow solution. For solute transport, it utilizes an Eulerian-Lagrangian scheme, based on a *forward tracking* method of characteristics (MOC) (Konikow and Bredehoeft, 1978) (Sec. 8.6), on a modified method of characteristics (MMOC), as well as on a combination of these two methods (HMOC). A major drawback of MOC is that a large number of particles needs to be deployed, particularly in three-dimensional cases. It heavily taxes the computer storage and CPU time, as otherwise the solution will lack the desirable accuracy. MMOC, on the other hand, approximates the advection term by tracking the nodal points of a fixed grid *backward* in time, and utilizes an interpolation scheme (Cheng *et al.*, 1984). MMOC requires the maintenance of much fewer particles than MOC and is, therefore, more efficient. However, it introduces numerical dispersion (Sec. 8.6) on a sharp concentration front, thus losing part of the advantage of the MOC formulation. The hybrid MOC/MMOC technique (Neuman, 1984) attempts to combine the advantages of the two techniques, based on an automatic adaptation of the transport process. When sharp concentration fronts are present, the advection term is solved by the forward-tracking MOC technique through the use of moving particles dynamically distributed around each front. Away from such fronts, the advection term is solved by the MMOC technique, with nodal points tracked backward in time. When a front dissipates due to dispersion and chemical reactions, the forward tracking stops automatically and the corresponding particles are removed. MT3D was later extended and released by the U.S. Army Corps of Engineers, Engineering and Research Development Center (Zheng and Wang, 1999), as MT3DMS, where the 'MS' stands for *multispecies* transport. Its most current release is version 5 (Zheng, 2006).

RT3D (Multi-Species Reactive Flow and Transport Simulation Software). RT3D is a code supported by the Pacific Northwest National Laboratory (PNNL) (Clement, 1997). It simulates three-dimensional, multispecies, reactive transport of chemical compounds in groundwater. It utilizes MODFLOW and MT3D as the flow and transport codes, and adds reaction kinetic modules, such as instantaneous reaction, first-order, rate-limited, BTEX (benzene, toluene, ethylbenzene, and xylene) degradation, rate-limited sorption, double Monod model, sequential first-order decay, and aerobic/anaerobic chlorinated ethene dechlorination (Clement, 1997; Clement *et al.*, 1998, 2000).

HST3D (Heat and Solute Transport in Three-Dimensional Groundwater Systems). HST3D is a USGS finite difference code for three-dimensional advective-dispersive heat and mass transport (Kipp, 1987, 1997). It provides a transient simulation of non-isothermal flow in a saturated aquifer, with transport of a single solute, linear adsorption, and linear decay.

FEMWATER (Three-Dimensional Finite Element Model of Water Flow Through Saturated-Unsaturated Media). The original code was developed in two parts: a **3DFEMWATER** code for three-dimensional density-dependent flow through variably saturated porous media (Yeh, 1987), and a **3DLEWASTE** code for three-dimensional Eulerian-Lagrangian model of waste transport (Yeh, 1990). It has been combined into a single coupled flow and transport code, FEMWATER (Lin *et al.*, 1996). The code is available from U.S. EPA.

Random-Walk (Random-Walk Solute Transport Model for Selected Groundwater Quality Evaluations). Random-Walk is an open source Fortran code that simulates solute transport with advection, dispersion, and chemical reactions, in one or two spatial dimensions, developed by Illinois State Water Survey (Prickett *et al.*, 1981). The solution for groundwater flow is based on a finite difference formulation. The solute transport portion of the code is based on a particle-in-a-cell technique for the advection mechanisms, and a random-walk technique for the dispersion effects.

GMS (Groundwater Modeling System). GMS is a software pre-processor, post-processor, and graphic user interface (GUI) implementation of a number of public domain groundwater modeling computer codes, which include FEMWATER, MODFLOW, MODPATH, MT3DMS, RT3D, UTCHEM, and PEST. It was developed by the U.S. Army Engineer Research and Development Center for use by governmental agencies. It is available as a commercial software.

SWAT (Soil and Water Assessment Tool). SWAT (Arnold and Fohrer, 2005; Neitsch *et al.*, 2005) is a basin scale, continuous time model designed to predict the impact of management on water, sediments, and agricultural chemical yields in ungaged watersheds. Major model components include weather, hydrology, soil temperature and properties, plant growth, nutrients, pesticides, bacteria and pathogens, and land management. In SWAT, a watershed is divided into multiple sub-watersheds, which are then further subdivided into hydrologic response units that consist of homogeneous land use, management, and soil characteristics. SWAT is a public domain code supported by the U.S.D.A. Agricultural Research Service at the Grassland, Soil and Water Research Laboratory.

HYDRUS (Movement of Water, Heat, and Multiple Solutes in Variably Saturated Media). HYDRUS-1D and -2D (Simunek *et al.*, 1999) are public domain software packages developed for the U.S. Salinity Laboratory. HYDRUS-3D is a commercial software. The program is a FE model that solves Richards' equation, (6.3.36, for saturated-unsaturated water flow and Fickian-based advection-dispersion equations for heat and solute transport. The flow equation incorporates a sink term to account for water uptake by plant roots. The solute transport equations consider advective-dispersive transport in the liquid phase, and diffusion in the gaseous phase. The equations also include provisions for nonlinear and/or nonequilibrium reactions between the solid and liquid phases, linear equilibrium reactions between the liquid and gaseous phases, zero-order production, and first-order degradation reactions. This software is widely used in agriculture applications.

FEFLOW (Finite Element Subsurface Flow and Transport Simulation System). FEFLOW is a commercial software package based on the finite element method. It simulates three-dimensional, density-dependent, saturated-unsaturated flow, chemical mass transport, and solid and fluid heat transport in porous media.

PEST (Model-independent parameter estimation). PEST was developed by Doherty (2005) as a model-independent nonlinear parameter estimation computer code. It has undergone a number of modifications and the current version is offered as a freeware. 'Model-independent' means that PEST communicates with the application program through its input and output files, without the need to access and modify the source code of the application program. It has been used for parameter estimation for groundwater flow and transport models (Doherty, 2003; Moore and Doherty, 2006). It has also been incorporated into a number of groundwater flow and transport computer programs, such as MODFLOW, MT3D, and GMS for parameter estimation and calibration purposes. Some of these implementations are available only as commercial packages.

SUTRA (Model for 2D or 3D Saturated-Unsaturated, Variable-Density Ground-Water Flow, with Solute or Energy Transport). SUTRA is a Galerkin-based finite element code that solves groundwater flow and transport problems under saturated and unsaturated conditions. Particularly, it models variable density, and, hence, has been widely used for simulating saltwater intrusion (Chap. 9). SUTRA is a USGS code and was initially released in 1984 as a two-dimensional code (Voss, 1984, 1999). Its three-dimensional version was released in 2003 (Voss and Provost, 2002).

SEAWAT (Computer Program for Simulation of Three-Dimensional Variable-Density Ground-Water Flow). SEAWAT is a USGS code that combines MODFLOW and MT3DMS into a single computer program for the purpose of simulating saltwater intrusion (Langevin *et al.*, 2003). It is a finite difference, Eulerian-Lagrangian code, in contrast to SUTRA, which is a finite element code.

CODESA-3D (Coupled Variable Density and Saturation 3D Model). This is a finite element code, developed by the Center for Advanced Studies, Research and Development in Sardinia (CRS4), Italy. It solves the convective-dispersive, variable density transport equation in saturated and variably saturated porous media (Gambolati *et al.*, 1999). Its functions are similar to those of SUTRA.

SHARP (A Quasi-Three-Dimensional, Numerical FDM that Simulates Freshwater and Saltwater Flow Separated by a Sharp Interface in a Layered Coastal Aquifer Systems). SHARP is a quasi-three-dimensional (layered aquifers, with Dupuit assumption), implicit finite difference model that simulates saltwater-freshwater movement in coastal aquifers (Essaid, 1990a, 1990b, 1999). The saltwater and freshwater are separated by a sharp interface. It is a USGS code.

ParFlow (Modeling Surface and Subsurface Flow on High Performance Computers). ParFlow is a parallel, three-dimensional, variably saturated groundwater flow code, developed by Lawrence Livermore National Laboratory (LLNL), that is especially suitable for large-scale, high-resolution problems (Ashby and Falgout, 1996; Tompson *et al.*, 1999). ParFlow has been extended to coupled surface-subsurface flow to enable the simulation of hill slope runoff and channel routing in an integrated fashion. The components implemented into ParFlow enable large scale, high resolution watershed simulations (Maxwell and Miller, 2005; Kollet and Maxwell, 2006).

TOUGH (Transport of Unsaturated Groundwater and Heat). TOUGH (Pruess *et al.*, 1999) is a multi-dimensional numerical code for simulating the coupled transport of water, vapor, non-condensible gas, and heat in porous and fractured media. TOUGH uses an integral finite difference method for space discretization, and first-order fully implicit time difference. The code, developed by the Lawrence Berkeley National Laboratory (LBNL), is primarily designed for geothermal reservoir studies and high-level nuclear waste isolation. It provides options for specifying injection or withdrawal of heat and fluids, double-porosity, dual-permeability, multiple interacting media, different fluid mixtures (water, vapor, tracer, CO_2, air, and hydrogen), Klinkenberg effects (Subs. 4.2.1) and binary diffusion in the gas phase, capillary and phase adsorption effects for the liquid phase, heat transport by means of conduction (with thermal conductivity dependent on water saturation), convection, and binary diffusion.

NUFT (Nonisothermal, Unsaturated Flow and Transport with Chemistry). NUFT (Nitao, 1998) is a Lawrence Livermore National Laboratory (LLNL) code designed to simulate coupled fluid movement (multiple liquids and gas), heat transport (including thermal radiation), and chemical reactions in saturated or unsaturated porous media. Chemical interactions that modify the physical properties of the porous media are also considered. Applications of the code have primarily addressed simulations of the long-term evolution of rock in the vicinity of deep geological high level nuclear waste repositories, thermal perturbation of sedimentary basins, and mineral and chemical evolution associated with subsurface sequestration of CO_2. Finite element and finite difference solution options, together with internal structured, external unstructured, and multigrid meshes, are also available.

STOMP (Subsurface Transport Over Multiple Phases). STOMP (Nichols *et al.*, 1997; White and Oostrom, 2000, 2006; Ward *et al.*, 2005) is a general purpose computer code, developed by Pacific Northwest National Laboratory, for simulating multidimensional subsurface flow and transport. The simulator's modeling capabilities address a variety of subsurface environments, including nonisothermal conditions, fractured media, multiple-phase systems, nonwetting fluid entrapment, soil freezing conditions, nonaqueous phase liquids, first-order chemical reactions, radioactive decay, solute transport, dense brines, nonequilibrium dissolution, and surfactant-enhanced dissolution and mobilization of organics.

SLAEM/MLAEM (Single/Multi-Layer Analytic Element Method). These are commercial codes based on Strack's (1989) *analytic element method.* The method involves the superposition of analytic functions, each representing a particular geohydrological feature in a two-dimensional, steady-state flow in an infinite aquifer (using the Dupuit assumption). Analytic elements have been developed to model uniform flow, rainfall infiltration, rivers, creeks, lakes or polders, wells, cracks, slurry walls, and inhomogeneities in aquifer properties. MLAEM models multilayer aquifers using leakage areas to connect them. SLAEM is the single-layer version of MLAEM.

WHPA (Wellhead Protection Area). WHPA is a semi-analytical groundwater flow simulation code by EPA, used for delineating the capture zone in a wellhead protection area by particle tracking (Blandford and Huyakorn, 1991). It is applicable to two-dimensional, steady groundwater flow in a homogeneous confined, unconfined, or leaky aquifers.

BIOPLUME. BIOPLUME is a two-dimensional computer code for simulating contaminant transport of a single and multiple hydrocarbons with oxygen-limited and reactant-limited bioreactions (Rafai *et al.*, 1998), developed by EPA. Its transport code is based on the USGS MOC.

NAPL Simulator. The NAPL Simulator is an EPA three-dimensional computer code based on Hermite collocation finite element discretization (Guarnaccia *et al.*, 1997). It simulates NAPL contamination in three interrelated zones: a vadose zone, a capillary zone, and a water-table aquifer zone. Three mobile phases are accommodated: water, NAPL, and gas. The 3-phase k-S-P sub-model accommodates capillary and fluid entrapment hysteresis. NAPL dissolution and volatilization are accounted for through rate-limited mass transfer sub-models.

UTCHEM (University of Texas Chemical Compositional Simulator). Originally a three-dimensional finite difference model for multiphase flow, multicomponent transport and chemical flooding, this code has been modified to become a general purpose NAPL simulator (University of Texas, 2000). Physical, chemical and biological process models, important in describing the fate and transport of NAPLs in contaminated aquifers, have been incorporated into this simulator. These include multiple organic NAPL phases, the dissolution and/or mobilization of NAPLs by non-dilute remedial fluids, chemical and microbiological transformations, and changes in fluid properties as a site is being remediated. The model allows for non-equilibrium interphase mass transfer, sorption, geochemical reactions, and temperature dependence of pertinent chemical and physical properties. It can simulate the flow and transport of remedial fluids of variable density, temperature and viscosity, including surfactants, co-solvents, and other enhancement agents. The biodegradation model includes inhibition, sequential use of electron acceptors, and co-metabolism. It can be used for modeling a very general class of bioremediation processes. The code is a public domain code maintained by the University of Texas, Center for Petroleum & Geosystems Engineering.

BIOMOC (A Multispecies Solute-Transport Model with Biodegradation). BIOMOC is a USGS two-dimensional code based on MOC. It simulates the transport and biotransformation of multiple reacting solutes (Essaid and Bekins, 1997). A number of biological transformation processes, including single, multiple, and minimum Monod kinetics and competitive, noncompetitive, and Haldane inhibition, are also included.

PHREEQC (A Computer Program for Speciation, Batch-Reaction, One-Dimensional Transport, and Inverse Geochemical Calculations). PHREEQC is a USGS computer program designed to perform a wide variety of low-temperature aqueous geochemical speciation calculations (Parkhurst and Appelo, 1999). The PHREEQC code differs from other codes reviewed in this section in that PHREEQC does not model flow and transport; rather, it models speciation (Subs. 7.9.2D). Given a set of species present in an aqueous phase, PHREEQC has all the chemical relationships that can predict what chemical processes will occur, and what will be the resulting set of species and their concentrations, assuming chemical equilibrium. PHREEQC is based on an ion-association aqueous model and has capabilities for (1) speciation and saturation-index calculations; (2) batch-reaction and one-dimensional transport calculations, involving reversible reactions, which include aqueous, mineral, gas, solid-solution, surface-complexation, and ion-exchange equilibria, and irreversible reactions, which include specified mole transfers of reactants, kinetically controlled reactions, mixing of solutions, and temperature changes; and (3) inverse modeling, which finds sets of mineral and gas mole transfers that account for differences in composition between waters, within specified compositional uncertainty limits. PHREEQC is often used in conjunction with a flow and transport code (see PHAST below).

PHAST (Program for Simulating Ground-Water Flow, Solute Transport, and Multicomponent Geochemical Reactions). PHAST is a USGS code that simulates multicomponent, reactive solute transport in three-dimensional saturated groundwater flow (Parkhurst *et al.*, 2004). The flow and transport calculations are based on a specialized version of HST3D that is restricted to constant fluid density and constant temperature. The geochemical reactions are simulated with the geochemical model PHREEQC.

Chapter 9

SEAWATER INTRUSION

In many parts of the world, coastal aquifers constitute an important source of water. Often, coastal areas are also heavily populated, a fact that makes the demand for freshwater even more acute. Due to the proximity and contact with the sea, the planning and management of such aquifers requires special attention associated with the danger of seawater (or saltwater) intrusion. In fact, this phenomenon constitutes one of the major constraints in the management of groundwater in costal aquifers. As seawater intrusion progresses, the part of the aquifer close to the sea becomes saline, and pumping wells that operate close to the coast have to be abandoned. Also, the area above the intruding seawater wedge, which remains fresh by natural replenishment, is lost as a source of freshwater.

The above remarks emphasize the special features of coastal aquifers that affect management decisions. The objective of this chapter is to present and discuss models that describe seawater intrusion. Particularly, we shall discuss a model that couples flow and solute transport under variable density conditions. Certain issues related to coastal aquifer management will also be presented.

9.1 Occurrence and Exploration

9.1.1 Occurrence of seawater intrusion

In general, a hydraulic gradient toward the sea exists in a coastal aquifer, with the sea serving as the recipient for the excess of freshwater, i.e., for the difference between aquifer replenishment, both natural and artificial, and the quantity of water pumped from the aquifer. Because of the presence of the sea, seawater occupies the void space of the aquifer formation beneath the sea. This seawater zone in the aquifer extends also to some distance landward of the coast. As a consequence, a zone of transition exists in the aquifer, across which the water in the aquifer varies from the lighter fresh aquifer water flowing to the sea to the heavier seawater in the aquifer. For reasons to be discussed later in this chapter, let us refer to this transition zone as the 'interface', or 'interface zone', between freshwater and seawater.

J. Bear, A.H.-D. Cheng, *Modeling Groundwater Flow and Contaminant Transport*, Theory and Applications of Transport in Porous Media 23,
DOI 10.1007/978-1-4020-6682-5_9,

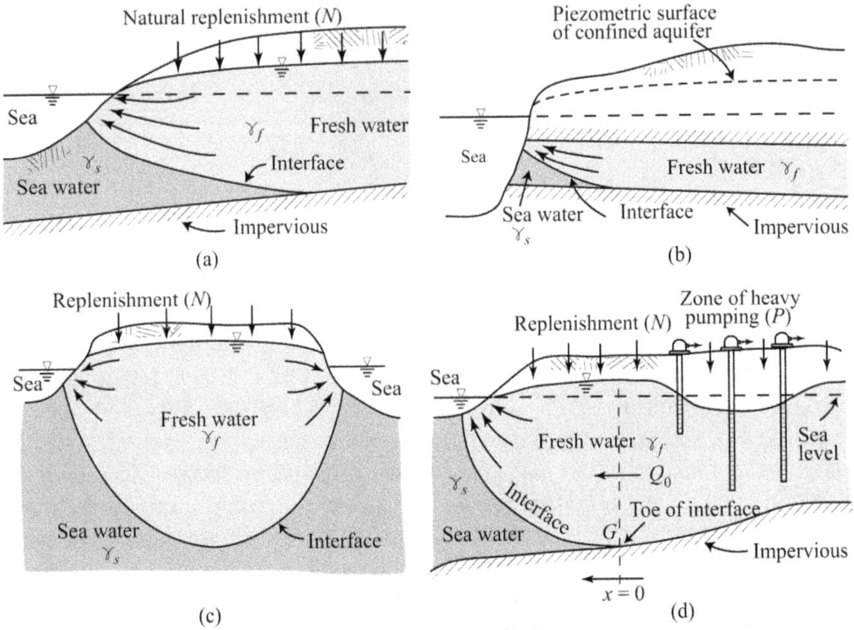

Figure 9.1.1: Typical vertical cross sections of seawater intrusion in coastal aquifers. (a) Unconfined aquifer with replenishment; (b) Confined aquifer; (c) Freshwater lens on an island; (d) Unconfined aquifer with pumping.

Figure 9.1.1 presents some typical cross sections with interfaces in coastal aquifers under natural conditions (a, b, c) and with pumping (d). Like all figures that describe aquifers, these are also highly distorted figures, not drawn to scale. In this figure, the surface indicated as an 'interface' represents a transition zone. The detailed shape of the transition zone depends, among other factors, to be discussed below, on whether this zone is advancing inland or retreating. Figure 9.1.2 shows two schematic cases of coastal aquifer cross-sections with transition zones. In all cases, the aquifer domain occupied by seawater has the form of a wedge.

When we start pumping from a coastal aquifer, or when the rate of existing pumping is increased, the freshwater discharge to the sea is reduced, water levels (or piezometric heads in a confined aquifer) drop close to the sea, and the transition zone rises. The entire seawater and transition zone wedge advances landward, until a new equilibrium is reached. Wells that operate within the wedge zone pump saline water and have to be abandoned. When pumping takes place in a well located above the transition zone, the saltwater upcones towards the well. Unless the well is at a sufficient distance above this

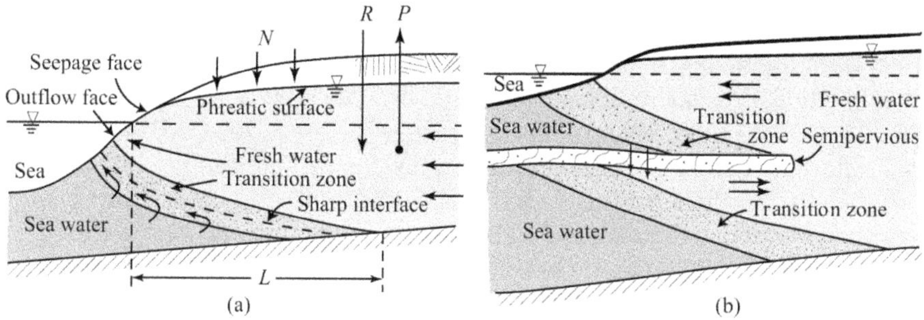

Figure 9.1.2: Typical cross-sections of transition zones in coastal aquifers.

zone and/or the rate of pumping is sufficiently small, the well will eventually pump saline water.

Since the famous works of Badon-Ghyben (1888) and Herzberg (1901), and the less known work of Du Commun (1828, see Konikow and Reilly, 1999), extensive research has been carried out, leading to the understanding of the mechanisms that govern seawater intrusion. The dominant factors are the flow regime in the aquifer above the intruding seawater wedge, the variable density, and hydrodynamic dispersion. Reviews of the phenomenon of seawater intrusion, and of the research that has been carried out on this subject, both theoretical work, and field and laboratory investigations, may be found in many books and publications, and will not be repeated here (e.g., Bear, 1972, 1979; Bear and Verruijt, 1987; Reilly and Goodman, 1985, Bear *et al.*, 1999; Cheng and Ouazar, 2004; Bear, 2005b; Goswami and Clement, 2007).

Seawater and freshwater are often referred to as 'miscible liquids', although, actually, they constitute a single liquid phase—water—with different concentrations of total dissolved solid (TDS), such as salt. For the sake of simplicity, we shall continue to refer to them as two liquids—freshwater, and seawater. The passage from the portion of the aquifer that is occupied by the former to that occupied by the latter, takes the form of a transition zone, rather than a sharp interface. Under certain circumstances, depending on the extent of seawater intrusion, and on certain aquifer properties, this transition zone, which is, primarily, a result of hydrodynamic dispersion (Chap. 7) of the dissolved matter, may be rather wide. Under other conditions, it may be rather narrow, relative to the aquifer's thickness, and the passage from the zone occupied by freshwater to that occupied by seawater may be approximated as a *sharp interface*. Often, the term 'interface' is used for the iso-density surface that is midway between freshwater and seawater. In this chapter, the term 'interface' will be used either for a 'sharp interface', or, interchangeably for the 'transition zone'.

Information on seawater intrusion into coastal aquifers can be found in the books by Bear *et al.* (1999), and by Cheng and Ouazar (2004), as well as in many articles. For example, on the management problem (van Dam, 1999;

Maimone *et al.*, 2004), on modeling with the sharp interface approximation (Bear, 1999; Essaid, 1999), and on analytical solutions (Cheng and Ouazar, 1999), on modeling as a variable density flow and transport problem (Bear, 1999). Numerical solutions are presented by Voss (1999), by Gambolati *et al.* (1999), by Sorek *et al.*, (1999), by Langevin *et al.* (2004). Discussions on specific numerical codes are presented by Sorek and Pinder (1999). Case studies on specific locations can also by found, e.g., on seawater intrusion in Egypt (Sherrif, 1999), Israel (Melloul and Zeitoun, 1999), the Netherlands (Oude Essink, 1999, 2004; Stakelbeek, 1999), Italy (Barrocu *et al.*, 2004), Mexico (Marin *et al.*, 2004), Hawaii (Voss, 1999), Florida (Swarzenski and Kindinger, 2004), California (Johnson and Whitaker, 2004), as well as in other locations in the U.S. (Konikow and Reilly, 1999).

9.1.2 Exploration of saltwater intrusion

In the following, we shall discuss methods for detecting the presence of seawater in coastal aquifers.

A. Water sampling

There exist many well-established methods for detecting the presence of *water* in geological formations; they can be roughly categorized into *geological methods* and *geophysical methods* (Todd, 1980). The detection of *saltwater* or *seawater*, however, is more difficult. As an example, consider the geological method that consists of drilling an observation well at a location of interest. Unless the screen through which the water that enters the borehole is short, we cannot tell at which depth this kind of water (or saltwater) is present in the formation. When the screen is long (or consists of multiple short screens) water enters the borehole from various depths and is then mixed in the borehole.

In aquifers where a thick saltwater-freshwater transition zone exists, it is of interest to monitor the continuous change in salt concentration across the transition zone. Monitoring such information requires taking samples of *unmixed* water at different depths. This can be achieved by the use of *packers* that block seal sections of the well, selectively, such that each section has its own screen and water samples taken from different sections are not mixed with each other. Obviously, such operation is tedious and the data obtained is limited to the well's location and to the elevation of the screened portions of the well at that location. In order to cover a large area, a large number of wells need to be drilled, and the associated costs are usually high. The data obtained, however, is the most direct and, probably, the most accurate. Data obtained by this method is often used to *calibrate* data obtained by indirect methods, such as geophysical methods.

B. Geophysical investigations

Geophysical methods make use of physical properties, such as the velocity of seismic waves, electrical conductivity, electromagnetic permeability, thermal conductivity, etc., of the geomaterials, such as the soil, and the water occupying the void space, with the latter varying with the concentration of dissolved matter. The typical advantages of the geophysical methods are:

- Measurements are taken at ground surface, although some techniques also utilize existing, or specially drilled borehole. Sometimes, measurements can be conducted from airborne by aircrafts. Usually, this means hence less time consumed and lower costs.
- Measurements cover a large surface area, thus making the methods more suitable for field scale investigations.

The disadvantage of the surface based geophysical methods, however, is that they are less accurate, and often require calibration from direct measurements. The optimal solution to this problem is to conduct an *integrated geophysical survey* that combines results from geophysical techniques with data obtained from direct water sampling.

Several geophysical methods that can be used for detecting the presence of saline water in geological formations are briefly discussed below. A more extensive presentation can be found in Stewart (1999).

DC (direct current) resistivity. This electrical method, which is one of the earliest geophysical methods (Swartz, 1937, 1939), is one of the most widely used methods for the detection of saltwater underground in coastal environments (Fretwell and Stewart, 1981; Stewart *et al.*, 1983; Hagemeyer and Stewart, 1990; Nowroozi *et al.*, 1999; Leroux and Dahlin, 2006). The method is based on the increase in electrical conductivity with increasing pore water salinity. The principal advantages of the method are its simplicity and the relatively low cost of the required equipment.

In the DC method, an electrical current is introduced into the ground through *current electrodes* driven into the soil. The resulting electrical potential (voltage) is measured between two *potential electrodes*. The measured resistance represents the integrated resistivity over the electrically heterogeneous soil. In the more advanced equipment, multi-conductor cables connect a large number of electrodes along a profile. A receiver then cycles sequentially through the electrodes, producing a series of closely-spaced soundings along the profile. A computer program then performs 2-D or 3-D resistivity inversions on the field data. The inversion, however, is not unique, and is dependent on the assumed 'model', e.g., the number of soil layers, and their thickness. This technique may not be robust enough to accurately determine a continuously varying salt concentration in a transition zone. In the case of a sharp interface, however, there is a distinct change in electrical conductivity across such interface; hence, the interface location can, generally, be determined quite accurately (Stewart, 1990).

Frequency domain electromagnetic methods. The frequency domain electromagnetic method (FDEM) involves the generation of an electromagnetic field, which induces current in the soil, which, in turn, causes the subsurface to create a magnetic field. By measuring this secondary magnetic field, various subsurface properties and features can be deduced (Stewart, 1999). The currents in the soil, termed *eddy currents*, are induced by time-varying magnetic fields produced by a frequency-controlled AC in a transmitter coil. The transmitted electromagnetic field is called the *primary field*. The induced eddy currents in the soil produce a secondary field, which is usually 90° out of phase with the primary one. The ratio of the out-of-phase component of the secondary field to the in-phase component of the primary one is an indication of the terrain's conductivity. The method has been effectively applied to detect the saltwater-freshwater interface (Kauahikaua, 1987; Anthony, 1992; Martinez *et al.*, 1995). The depth of investigation by the FDEM is primarily a function of the frequency of the primary field, with lower frequencies having greater penetration.

Airborne electromagnetic methods. The frequency domain electromagnetic method creates eddy currents through electromagnetic induction, so that, actually, no contact with the ground is required (Stewart, 1999). This means that a frequency EM system can be flown by fixed wing aircrafts or by helicopters. The typical airborne system uses several receiver-transmitter coil pairs at varying frequencies. These coil pairs are placed in a 'bird', which is towed behind the aircraft at elevations of 25–50 m above ground surface (Fraser, 1972, 1978, 1979). The depth of investigation is determined by the transmitter's frequencies. Common frequencies range from 200 to 56,000 Hz, yielding penetration depths from tens of meters to less than one meter, respectively. The output, as an apparent resistivity map for each frequency, is produced by an inverse method (Sengpiel, 1988; Huang and Fraser, 1996; Sengpiel and Siemon, 2000). Interpretation is normally qualitative, and, typically, needs to be calibrated with ground surface data. The airborne method is generally applied to surveys of large areas; it has been successfully applied to the detection of saltwater intrusion in freshwater aquifers, and to the exploration of freshwater lenses in saltwater environments (Sengpiel, 1983; Fitterman and Deszcz-Pan, 1998; Meng *et al.*, 2006; Steuer *et al.*, 2008).

Time domain electromagnetic methods. The time domain (transient) electromagnetic method (TDEM) employs a transmitter that drives an AC through a square loop of insulated electrical cable laid on the ground. The current consists of equal periods of time-on and time-off, with base frequencies that range from 3 to 75 Hz, producing an electromagnetic field. Similar to the FDEM, the electrical current generates a primary, time-varying electromagnetic field, which in turn creates a secondary electromagnetic field.

TDEM soundings can be used to detect saltwater at depths of 5 m to several hundred meters below land surface. The TDEM method has several significant advantages over DC soundings, notably depths of investigation up to twice the transmitter coil dimension, and the ability to sound through a conductive, near-surface unit, such as a clay confining layer. TDEM equipment, however, is more expensive and complicated to use than DC equipment, and the interpretation of TDEM data requires sophisticated interpretation software (Stewart, 1999). TDEM has been successfully applied to many field investigations (e.g., Fitterman and Stewart, 1986; Hoekstra and Blohm, 1990; Kontar and Ozorovich, 2006; Nielsen *et al.*, 2007; Duque *et al.*, 2008)

Other geophysical investigation techniques that have been employed for the detection of saline water in the subsurface include the *ground penetrating radar* (Soldal *et al.*, 1994; Tronicke *et al.*, 1999); the *loop-loop electromagnetic method* (Mitsuhata *et al.*, 2006), the *very low frequency (VLF) electromagnetic method* (McNeill, 1990; McKenzie, 1990), and various borehole geophysical methods; see Stewart (1999) for a review. For an overview of the general principles and practices of geophysical methods, see Telford *et al.* (1990), Ward (1990), and Beck (1991).

C. Geochemical investigations

As a part of exploration of freshwater contaminated by intruding saline water, it is important to identify the origin of the latter. In coastal aquifers, seawater encroachment inland is the most common reason for the increase in salinity; however, other sources or processes can contribute to groundwater salinity. Custodio (1997) lists a number of salinity sources that can contaminate freshwater supply, which are not directly related to seawater encroachment. These include entrapped fossil seawater, sea-spray accumulation, evaporite rock dissolution, displacement of old saline groundwater from underlying or adjacent aquifers or aquitards through natural, or man-imposed advection or by thermal convection, leaking aquitards through fault systems, and anthropogenic pollution from various sources, including sewage effluents, industrial effluents, mine water, road deicing salts, effluents from water softening or de-ionization plants, and agriculture return flows.

In general, seawater has a uniform chemistry due to the long residence time of the major constituents. Its main features are (Jones *et al.*, 1999): predominance of Cl^- and Na^+, with a molar ratio of 0.86, an excess of Cl^- over the alkali ions (Na and K), and Mg greatly in excess of Ca^{2+} (Mg/Ca $= 4.5$–5.2). In contrast, continental fresh groundwater are characterized by a highly variable chemical composition, although the predominant anions are HCO_3^-, SO_4^{2-} and Cl^-. If not anthropogenetically polluted, the fundamental cations are Ca^{2+} and Mg^{2+} and, to a lesser extent, the alkali ions, Na^+ and K^+. In most cases Ca^{2+} predominates over Mg^{2+}. Seawater solutes are specifically characterized by $Mg > SO_4 + HCO_3$, whereas meteoric waters (dilute or saline), even if dominated by re-solution of marine salts, reflect $Na > Cl$. In contrast, sedimentary basin fluids can carry significant Ca and perhaps K excess over $SO_4 + HCO_3$, due to diagenetic carbonate or silicate reactions.

Geochemists, generally, use the following criteria to define the signature and to distinguish the sources of salinization (Jones *et al.*, 1999):

Salinity: In coastal aquifers, a time series of steadily increasing chloride concentrations can indicate the early evolution of a salinity breakthrough from seawater, due to the over-exploitation of groundwater and reduction of piezometric head.

Cl/Br ratios: The Cl/Br ratio can be used as a reliable tracer, as both Cl and Br, usually, behave conservatively, except in the presence of very high amounts of organic matter. Seawater (Cl/Br weight ratio = 297) is distinguished from relics of evaporated seawater (hypersaline brine Cl/Br $<$ 297), evaporite-dissolution products (over 1000), and anthropogenic sources like sewage effluents (Cl/Br ratios up to 800), or agriculture-return flows (low Cl/Br ratios).

Na/Cl ratios: Na/Cl ratios of saltwater intrusion are usually lower than the marine values (i.e., < 0.86, molar ratio). On the other hand, high (> 1) Na/Cl ratios, typically, characterize anthropogenic sources, like domestic waste waters. Thus, low Na/Cl ratios, combined with other geochemical parameters, can foretell the arrival of saltwater intrusion, even at relatively low chloride concentrations, during early stages of salinization.

Ca/Mg, Ca/($HCO_3 + SO_4$) ratios: One of the most conspicuous features of saltwater intrusion is the enrichment of Ca over its concentration in seawater. High Ca/Mg and Ca/($HCO_3 + SO_4$) ratios (> 1) are further indicators of the intrusion of seawater.

O and H isotopes: The stable O and H isotopes can be used to describe the mixing process between saline water and freshwater. Fresh groundwater is generally depleted in both ^{18}O and ^{2}H (deuterium) relative to seawater. Mixing of fresh and seawater should result in a linear correlation. Different sources with high salinity (e.g., agriculture return flows, sewage effluents) would result in different slopes due to evaporation processes that would change the isotopic composition.

Boron isotopes: One of the processes that modify the chemistry of seawater intrusion is the adsorption of potassium, boron and lithium onto clay minerals in the host aquifer. These elements are relatively depleted in saline water associated with seawater intrusion. Hence, the boron isotopic composition of groundwater can be used as a tool to discern the salinization sources, in particular to distinguish seawater from anthropogenic contamination such as domestic waste water.

As a conclusion of this section, we observe that many of the techniques discussed above for the exploration of an environment intruded by saline water, are of a qualitative nature. Hence, in island or coastal regions, an exploration program usually requires the conjunctive use of two or more complementary geophysical, geological, and geochemical methods. Using several methods can increase the confidence level of the interpretation of observed data. The collected data can be used to validate the numerical model constructed for the simulation of seawater intrusion for management purposes.

9.2 Sharp Interface Models

Although, as emphasized above, in reality, a transition zone occurs, the *sharp interface approximation* is still, sometimes, used, primarily, to enable relatively easy solutions, both analytical and numerical, of certain seawater intrusion problems. However, nowadays, with the availability of new improved numerical techniques, including methods for coping with nonlinearities that are inherent in the transition zone model, and with fast and large memory computers (even microcomputers), numerical solutions of three-dimensional models that take the transition zone into account should not pose special difficulties. Indeed, a number of models and computer codes that consider seawater intrusion as a solute transport problem have already been developed, e.g., MOC3D (Konikow *et al.*, 1996); SUTRA (Voss and Provost, 2002). Our objective in presenting this subsection is not only to present a more complete and comprehensive description of the subject, but also to provide modeler with another tool that may be useful under certain circumstances, as when the transition zone is narrow relative to the thickness of the freshwater zone overlying it.

9.2.1 Sharp interface

The concept underlying the sharp interface model is that the freshwater and the seawater are envisioned as two immiscible liquids (but without the capillary pressure that exists between immiscible fluids). Let us denote them by subscripts s and f, respectively. The two liquids have different densities and viscosities, with each of them occupying a distinct portion of the flow domain. Figure 9.2.1 shows an example of a domain that is occupied by two liquids, each occupying a distinct subdomain: the s-liquid in subdomain R_s, and the f-liquid in subdomain R_f. The two adjacent regions are separated from each other by a sharp, possibly moving, interface. Sources and sinks of liquid (i.e., artificial recharge and pumping wells) may exist in both subdomains. Our objective is to model the flow in the two subregions, as well as the shape and displacement of the interface.

Following (4.1.4), we define a piezometric head for each of the two subregions:

$$h_\alpha = z + \frac{p}{\rho_\alpha g}, \quad \text{in} \quad R_\alpha, \quad \alpha = f, s, \tag{9.2.1}$$

where p denotes pressure, ρ denotes the liquid's density, z denotes elevation above some datum level, and g denotes gravity acceleration.

Based on the discussion of a complete, well-posed mathematical model in Sec. 5.3, we can state the considered problem in the following way: determine h_f in R_f, and h_s in R_s, such that the following (mass balance) equations are

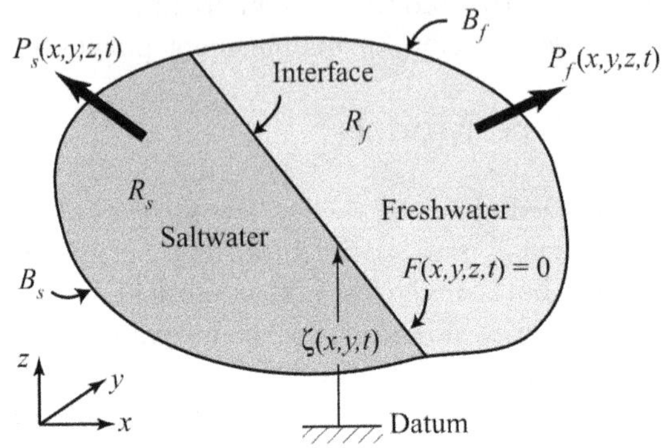

Figure 9.2.1: An interface between zones of different immiscible liquids.

satisfied:

$$S_{of}\frac{\partial h_f}{\partial t} = \nabla \cdot (\mathbf{K}_f \cdot \nabla h_f) - P_f(x, y, z, t), \qquad \text{in} \quad R_f, \tag{9.2.2}$$

$$S_{os}\frac{\partial h_s}{\partial t} = \nabla \cdot (\mathbf{K}_s \cdot \nabla h_s) - P_s(x, y, z, t), \qquad \text{in} \quad R_s, \tag{9.2.3}$$

in which $P_\alpha = P_\alpha(x, y, z, t)$ represents sinks of the α-phase, $\mathbf{K}_\alpha(x, y, z) = \mathbf{k}(x, y, z)\, g\rho_\alpha/\mu_\alpha$ is the hydraulic conductivity, $\mathbf{k}$ is the permeability (both $\mathbf{K}$ and $\mathbf{k}$ are tensors), μ_α is the dynamic viscosity, and $S_{o\alpha}(x, y, z)$ is the specific storativity of the α-phase (see (5.1.75)). We also note that the fluxes are defined as $\mathbf{q}_f = -\mathbf{K}_f \cdot \nabla h_f$, and $\mathbf{q}_s = -\mathbf{K}_s \cdot \nabla h_s$. In writing Darcy's law for the two liquids, we have overlooked the possibility that the solid matrix is not stationary; otherwise, $\mathbf{q}$ should be replaced by $\mathbf{q}_r$.

In addition, we have to specify initial conditions at $t = 0$: for h_f in R_f and h_s in R_s. Boundary conditions have to be specified: for h_f on B_f and for h_s on B_s; they are the usual ones encountered in flow of a single liquid. However, the boundary conditions on the interface separating the two liquids requires special attention.

Similar to the case of a phreatic surface (Sec. 5.2.3), which is, actually, an interface between two fluids, air and water, the location and shape of the interface considered here can also be expressed in the form $F = F(x, y, z, t) = 0$, with F describing an *a priori* unknown (until the problem is solved) surface. Denoting the elevation, z, of points on the interface by $\zeta = \zeta(x, y, t)$ (Fig. 9.2.1), the relationship for F becomes

$$z = \zeta(x, y, t), \quad \text{or} \quad F(x, y, z, t) \equiv z - \zeta(x, y, t) = 0. \tag{9.2.4}$$

The pressure at a point $p(x, y, z, t)$ on the interface is the same when the latter is approached from both sides. Hence, from the definitions of h_f and h_s in (9.2.1), we obtain

$$\rho_f(h_f - \zeta) = \rho_s(h_s - \zeta), \tag{9.2.5}$$

or

$$\zeta = h_s \frac{\rho_s}{\rho_s - \rho_f} - h_f \frac{\rho_s}{\rho_s - \rho_f} = h_s(1+\delta) - h_f\delta, \tag{9.2.6}$$

where h_f and h_s take on their respective values on the interface, and

$$\delta = \frac{\rho_f}{\rho_s - \rho_f}. \tag{9.2.7}$$

If we can solve (9.2.2) for $h_f = h_f(x, y, z, t)$ and (9.2.3) for $h_s = h_s(x, y, z, t)$, in their respective domains, (9.2.6) becomes the sought *equation for the shape of the (possibly moving) interface*. We can rewrite it in the form:

$$F \equiv z - h_s(1+\delta) + h_f\,\delta = 0, \tag{9.2.8}$$

where, as this relationship is valid only for points z on the interface, i.e., $z \equiv \zeta$, we recall that $h_f = h_f(x, y, \zeta, t)$, and $h_s = h_s(x, y, \zeta, t)$.

Once we have the location of the boundary ($\equiv$ the interface), the boundary conditions on it—one for each side—are obtained from the fact that the interface is a *material surface* with respect to the mass of each of the liquids; no liquid mass crosses it. The two conditions are

$$(\mathbf{V}_\alpha - \mathbf{u}) \cdot \mathbf{n} = 0, \quad \alpha = f, s, \tag{9.2.9}$$

in which $\mathbf{V}_\alpha$ are the velocity of the respective fluids, $\mathbf{u}$ is the speed of displacement of the interface F, and $\mathbf{n}$ denotes the outward unit vector on F, with

$$\frac{\mathrm{D}F}{\mathrm{D}t} \equiv \frac{\partial F}{\partial t} + \mathbf{u} \cdot \nabla F, \quad \mathbf{n} = \frac{\nabla F}{|\nabla F|}, \quad \mathbf{u} \cdot \nabla F = -\frac{\partial F}{\partial t}. \tag{9.2.10}$$

In addition, the interface is also a material surface with respect to the solid, and hence,

$$(\mathbf{V}_{\text{solid}} - \mathbf{u}) \cdot \mathbf{n} = 0. \tag{9.2.11}$$

From the above two equations, it follows that on the interface, which serves as a common boundary to both subdomains, we have

$$(\mathbf{q}_{r\alpha} - \phi\,\mathbf{u}) \cdot \mathbf{n} = 0, \quad \alpha = f, s. \tag{9.2.12}$$

Making use of Darcy's law and (9.2.10), we obtain the two conditions on the interface, for the R_f and R_s subdomains, respectively, in the form

$$\phi\,\delta \frac{\partial h_f}{\partial t} - \phi(1+\delta)\frac{\partial h_s}{\partial t} = \mathsf{K}_f \cdot [\nabla z - (1+\delta)\nabla h_s + \delta \nabla h_f] \cdot \nabla h_f, \tag{9.2.13}$$

$$\phi\,\delta \frac{\partial h_f}{\partial t} - \phi(1+\delta)\frac{\partial h_s}{\partial t} = \mathsf{K}_s \cdot [\nabla z - (1+\delta)\nabla h_s + \delta \nabla h_s] \cdot \nabla h_s. \tag{9.2.14}$$

In principle, we can find solutions for h_f and h_s in their respective domains by solving the governing equations (9.2.2) and (9.2.3), together with the boundary (interface) conditions, (9.2.12)–(9.2.14). Unfortunately, the interface conditions, (9.2.13) and (9.2.14), are nonlinear coupled partial differential equations in the variables h_f and h_s. It is practically impossible to directly solve the coupled system in order to determine the shape and position of the interface. Instead, numerical methods can be employed to approximately find the interface location.

As a simple example, consider the case of a coastal aquifer, and let the two liquids and the porous medium be assumed incompressible, meaning that the specific storativity is zero. Then, (9.2.2) and (9.2.3) reduce to

$$\nabla \cdot (\mathbf{K}_f \cdot \nabla h_f) = P_f(x, y, z, t), \qquad \text{in} \quad R_f, \tag{9.2.15}$$

$$\nabla \cdot (\mathbf{K}_s \cdot \nabla h_s) = P_s(x, y, z, t), \qquad \text{in} \quad R_s. \tag{9.2.16}$$

In a transient interface problem, the initial interface is, usually, given at a non-equilibrium location, and our goal is to predict its future movement. At any given time, at which we know the interface shape and location, we can solve (9.2.15) and (9.2.16) simultaneously, as fixed domain problems in the respective freshwater and saltwater domains. To have a well-posed problem, in addition to the boundary conditions, two interface conditions are needed on the interface. The interface conditions are given by the continuity of pressure, expressed as (9.2.8), and the continuity of flux normal to the interface, given by

$$(\mathbf{K}_f \cdot \nabla h_f) \cdot \mathbf{n} = (\mathbf{K}_s \cdot \nabla h_s) \cdot \mathbf{n}, \tag{9.2.17}$$

where $\mathbf{n}$ is the unit vector normal of the interface. Once the problem is solved, we obtain, as a part of the solution, the normal flux on the interface, $\mathbf{q}_\alpha \cdot \mathbf{n}$. Then, according to (9.2.9), we can obtain the interface velocity $\mathbf{u}$ at that instant as

$$\mathbf{u} \cdot \mathbf{n} = \mathbf{V}_\alpha \cdot \mathbf{n} = -\frac{1}{\phi} (\mathbf{K}_\alpha \cdot \nabla h_\alpha) \cdot \mathbf{n}. \tag{9.2.18}$$

In a numerical solution, we can multiply a small time increment by the normal component of interface displacement velocity, $\mathbf{u} \cdot \mathbf{n}$, in order to obtain, approximately, the interface normal displacement in that time increment; the interface can then be moved by that increment to a new location. The new shape and position of the interface defines a new problem at the next time step. The process continues and the interface is continuously displaced, until an equilibrium, stationary, location is reached. This technique was used, for example, by Liu *et al.* (1982) and Taigbenu *et al.* (1984) to simulate the transient interface of saltwater intrusion.

In Subs. 9.2.5, we shall present some approximate models and solutions that are based on the Dupuit assumption of essentially horizontal flow, thus eliminating the boundary conditions on the sharp interface.

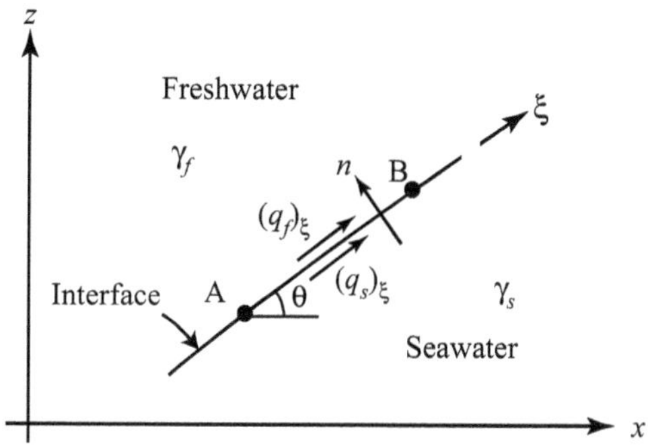

Figure 9.2.2: Interface at equilibrium between two immiscible fluids.

It is of interest to determine the slope at a point on a stationary interface. Figure 9.2.2 shows a segment AB along an (assumed sharp) stationary interface, in two-dimensional flow in a vertical xz-plane. The components of the specific discharge tangential to the interface in the two liquid regions are given by

$$\begin{aligned}(q_f)_\xi &= -\frac{\mathsf{k}g\rho_f}{\mu_f}\frac{\partial h_f}{\partial \xi} = -\frac{\mathsf{k}}{\mu_f}\left(\frac{\partial p}{\partial \xi} + g\rho_f\frac{\partial z}{\partial \xi}\right),\\ (q_s)_\xi &= -\frac{\mathsf{k}g\rho_s}{\mu_s}\frac{\partial h_s}{\partial \xi} = -\frac{\mathsf{k}}{\mu_s}\left(\frac{\partial p}{\partial \xi} + g\rho_s\frac{\partial z}{\partial \xi}\right).\end{aligned} \tag{9.2.19}$$

By eliminating $\partial p/\partial\xi$ from both equations, we obtain

$$\sin\theta \equiv \frac{\partial z}{\partial \xi} = \frac{(q_f)_\xi\,\mu_f - (q_s)_\xi\,\mu_s}{\mathsf{k}g(\rho_s - \rho_f)}, \tag{9.2.20}$$

where θ is the angle that the interface makes with the $+x$ direction. For stationary saltwater, $\mathbf{q}_s = 0$; as q_f increases, the angle θ also increases. As a coast is approached, the shape of a stationary interface follows from this conclusion: q_f increases as the coast is approached (because the total discharge moves through a decreasing cross-section), and, hence, θ also increases, and the interface approaches the vertical direction.

9.2.2 Ghyben-Herzberg approximation

Beginning with Du Commun (1828), Badon-Ghyben (1888) and Herzberg (1901), investigation on the interface in coastal aquifers have aimed at determining the relationships between its shape and position, and the various hydrological components of the groundwater balance near the coast.

Consider the interface shown in Fig. 9.2.3. The U-tube superimposed on this figure is intended to demonstrate the conceptual model proposed by Du

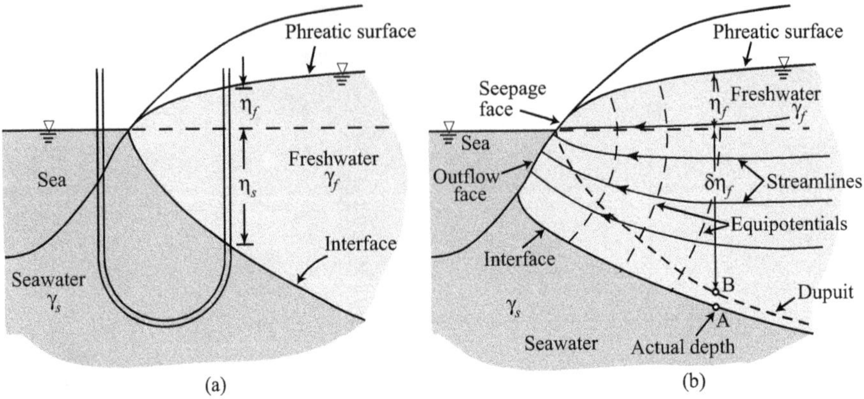

Figure 9.2.3: (a) The Ghyben-Herzberg interface model. (b) Actual flow pattern.

Commun (1828), Badon-Ghyben (1888) and Herzberg (1901). Essentially, they assumed that a static equilibrium exists under steady state conditions, with stationary seawater and a hydrostatic pressure distribution in the seaward flowing freshwater zone. This means that the flow is (essentially) horizontal and equipotentials (= surfaces of equal piezometric head) are vertical lines, or surfaces. This, in fact, is identical to the Dupuit assumption (Sec. 4.5). With the notation of Fig. 9.2.3a, we have

$$\eta_s = \delta\,\eta_f, \tag{9.2.21}$$

where $\delta = \rho_f/(\rho_s - \rho_f)$. The above relation is called the *Ghyben-Herzberg approximation.* It means that at any distance from the sea, the depth of an assumed stationary interface below sea level is δ times the height of the freshwater table above it. For example, for $\rho_s = 1.025$ g/cm^3, $\rho_f = 1.000$ g/cm^3, we obtain $\delta = 40$; i.e., at any distance from the sea, the depth of a *stationary interface* below sea level is 40 times the height of the freshwater table above it. Obviously, as the sea is approached, the assumption of essentially horizontal flow is no longer valid. Moreover, this assumption does not provide for an outflow surface through which freshwater can drain to the sea.

Figure 9.2.3b shows the actual flow conditions near the sea. We note the difference between the (actual) point A and point B, determined from the equipotential which is the depth predicted by the Ghyben-Herzberg approximation formula (9.2.21). The discrepancy stems from the difference between the actual shape of the equipotentials and the vertical ones assumed by the Dupuit approximation. In a confined aquifer, η_s in (9.2.21) is the depth of a point on the interface below sea level, while $\eta_f \equiv h_f$ is the freshwater piezometric head. Bear and Dagan (1964a) suggested that for steady flow, (9.2.21) gives an error that is less than 5%, provided $\pi \mathsf{K} B/Q\,\delta > 8$, where Q is the freshwater discharge to the sea, and B is the (constant) thickness of a confined aquifer. In the case of a phreatic aquifer, the Ghyben-Herzberg approximation also overlooks the presence of the seepage face shown in Fig. 9.2.3b. Note that a stationary interface always terminates on the sea bottom at some distance from the coast, and that at that point (M), it terminates on the sea bottom as a curve that is *tangent to the vertical.*

Note that by inserting $h_s = \text{const.} \equiv 0$, into (9.2.6), we obtain $\zeta = -\delta h_f$, which, with $\zeta = -\eta_s$ (see Figs. 9.2.1 and 9.2.3a), gives $\eta_s = \delta h_f$. This is similar to the Ghyben-Herzberg relationship, $\eta_s = \delta \eta_f$. This means that when the freshwater pressure is not hydrostatic (i.e., without applying the Dupuit assumption), rather than interpreting h_f as the free surface location η_f, as in Fig. 9.2.3b, we can interpret it as the actual freshwater head at the interface (point A in Fig. 9.2.3b). Then, the similar Ghyben-Herzberg relationship holds. This condition has been used by Glover (1959) and Detournay and Strack (1988) in an analytical solution, and by Naji *et al.* (1998b) in a numerical solution, for locating the saltwater-freshwater interface.

9.2.3 Upconing

In a costal aquifer, whenever a pumping well operates above a freshwater-saltwater interface, it creates drop of head (or a cone of depression in the case of a phreatic aquifer) above the well and a drop of freshwater head below the well. According to the Ghyben-Herzberg approximation (9.2.21), this will cause a local rise of interface. We use the term *interface upconing*, or just *upconing*, whenever the rising interface is such that everywhere above it, the flow of freshwater is towards the well. In this subsection, we consider the upconing under the *sharp interface approximation*. In Subs. 9.3.2B, we shall discuss the more realistic case of upconing with a *transition zone*. In both cases, 'upconing' describes the shape of the interface (or transition zone) that takes the form of a local rise in interface elevations, as a consequence of the drop in piezometric head caused by the pumping. Under certain conditions, the rising interface may reach the pumping well. In the case of a sharp interface, an equilibrium stationary upconed interface is possible up to a certain critical pumping rate (Fig. 9.2.4).

When the pumping rate is increased from one (say, steady) rate to a higher (steady) one, yet, below some critical value, a new equilibrium, with a higher upconed (sharp) interface, is established, following a transition period. At the critical pumping rate, the interface is very unstable and any small increase in pumping rate will, quickly, bring the interface, and with it saltwater, into the pumping well. The fast rising upconed interface will reach the pumping well in a cusp-like form (Fig. 9.2.4). When pumping stops, the upconed interface undergoes a gradual decay towards the initial steady state interface. Since, at the same time, seaward flow of freshwater takes place above the interface, the decaying upconed interface mound is also displaced seaward.

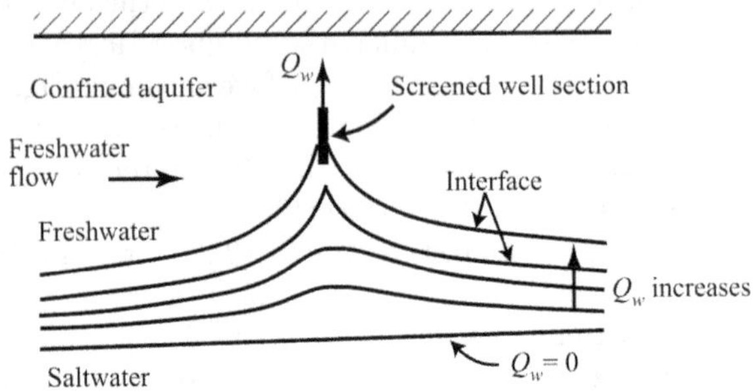

Figure 9.2.4: Interface upconing below a pumping well above a sloping interface.

Because the phenomenon of upconing is a consequence of the drop in piezometric head, it is sensitive both to the distance of the pumping well above the interface and to its pumping rate. Figures 9.2.5a, b, and c show, qualitatively, some of the features of the freshwater flow pattern, and the upconing that occurs when a pumping well operates above the interface in a three-dimensional aquifer flow domain. The figures show a vertical cross-section normal to the coast. Figure 9.2.5d demonstrates also a two-dimensional flow situation, with the well replaced as a gallery parallel to the coast.

Figure 9.2.5a shows a well that is either sufficiently far above the interface, or is pumping at a sufficiently low rate. Under such conditions, the reduction of freshwater head just above the interface is small. We observe that although the interface rises toward the well, there is no upconing. That is, the interface everywhere rises toward the sea, and a *peak* does not exist. This implies that the freshwater flow just above the interface is always in a seaward direction, and there is no reverse flow toward the well. We note how the pumping creates a *capture zone*, bounded by a *water divide.* We also observe a *stagnation point* between the well and the sea. The situation is the basis for the 'coastal collector technique', in the form of an array of shallow wells above the interface, along a line parallel to the coast, mentioned in Sec. 9.4.

Figure 9.2.5b shows a situation in which the well is either sufficiently near the interface, or is pumping at a sufficiently high rate causing a significant reduction in freshwater head above interface. As a consequence, a local upconing, similar to that demonstrated in Fig. 9.2.4, will develop. As suggested by (9.2.20), the flow above an interface that rises landward must also be in a landward direction. Figure 9.2.5b also shows the (three-dimensional) capture zone of the pumping well. We notice the existence of a *stagnation line*, which is a part of the water divide. Figure 9.2.5c shows a similar situation, but for a phreatic aquifer. Recall that the figure shows a vertical cross-section through the well normal to the sea. the flow pattern will be different in other cross-sections.

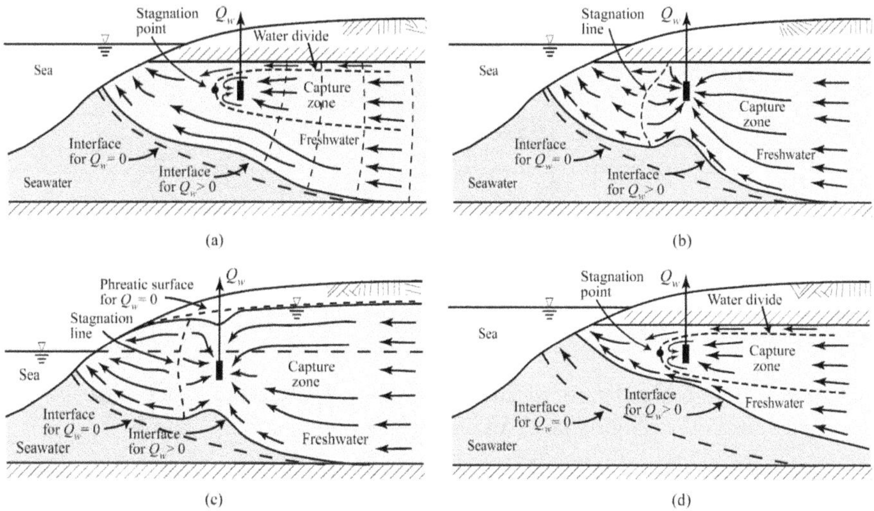

Figure 9.2.5: Upconing and freshwater flow in the vicinity of a well pumping above the interface. (a), (b), and (c): Three-dimensional flow. (d) Two dimensional flow (with a gallery well).

Figure 9.2.5d demonstrates a two-dimensional flow situation, i.e., a gallery well parallel to the coast, rather than a vertical well with a screened section. The flow pattern produced by a gallery can also be approximately achieved by an array of closely spaced wells along a line parallel to coast, such that their *radii of influence* overlaps. We note in Fig. 9.2.5d that as long as the pumping rate (per unit coastline length) is smaller than the freshwater outflow rate, the two-dimensional capture zone cannot encroach on the interface, as a freshwater outflow path must be provided to allow the unpumped portion of freshwater to reach the sea. In the case of three-dimensional flow (Figs. 9.2.5a–c), a freshwater outflow path is not required at a given cross-section, as freshwater can flow around the capture zone. The important conclusion is that, *under a two-dimensional flow situation, as along as there is a freshwater outflow just above the interface, the latter's profile must always be rising toward the sea.* In other words, a local upconing with an interface peak cannot develop, and no salinization of the gallery will occur (at least under the assumption of a sharp interface). This means that a large part of the freshwater flowing to the sea can be intercepted and collected, without the interface rising to the gallery. (In Subs. 9.3.2B, we shall discuss the more realistic case of upconing with a *transition zone*.) This type of *skimming wells* (see also Sec. 9.4) that distribute the pumping rate over a line or over an area to avoid concentrated local upconing beneath a point well has been practiced in Israel (Schmorak and Mercado, 1969), Palestinian territories (Aliewi *et al.*, 2001), and Pakistan (Saeed and Ashraf, 2005; Saeed, 2008).

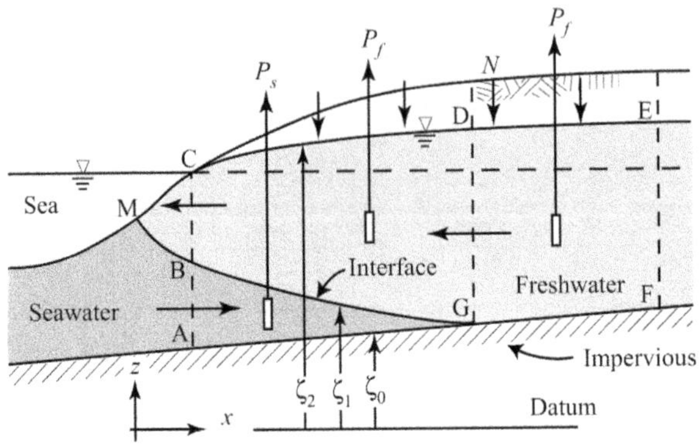

Figure 9.2.6: Nomenclature for integration over the thickness of the freshwater and the seawater regions, and for boundary conditions.

9.2.4 *Essentially horizontal flow model*

We continue to assume the existence of a sharp interface, but we introduce the *hydraulic approach*, based on the (Dupuit) assumption of essentially horizontal flow. The flow equations are obtained by integrating the 3-D flow model presented in Subs. 9.2.1, separately for each region, over the vertical thickness of the respective region. Figure 9.2.6 shows a phreatic aquifer with three regions: a freshwater region, a freshwater region above the interface, and a seawater region below the interface.

For the freshwater region above the interface, we integrate (9.2.2) from the interface at $\zeta_1(x,y,t)$ to the phreatic surface with accretion, at $\zeta_2(x,y,t)$. For the saltwater region, we integrate (9.2.3) from the aquifer's (impervious) bottom at $\zeta_o(x,y)$ to the interface, at $\zeta_1(x,y,t)$. The integration is based on the Leibnitz rule (Bear, 1979; Bear, 1999; Bear and Bachmat, 1990), which takes into account the conditions on the (possibly moving) boundaries of integration (see discussion in Sec. 5.4.1). By integrating (9.2.2) and (9.2.3), we obtain for the freshwater region above the interface and for the seawater region below it, respectively,

$$\int_{\zeta_1}^{\zeta_2}\left(\nabla\cdot\mathbf{q}_f+S_{of}\frac{\partial h_f}{\partial t}+P_f\right)dz=\nabla'\cdot B_f\,\widetilde{\mathbf{q}'_f}+\mathbf{q}_f\big|_{F_2}\cdot\nabla F_2-\mathbf{q}_f\big|_{F_1}\cdot\nabla F_1$$
$$+S_{of}\left(B_f\frac{\partial\widetilde{h_f}}{\partial t}+\widetilde{h_f}\frac{\partial B_f}{\partial t}+h_f\big|_{F_2}\frac{\partial F_2}{\partial t}-h_f\big|_{F_1}\frac{\partial F_1}{\partial t}\right)+B_f\widetilde{P_f}=0,\quad(9.2.22)$$

$$\int_{\zeta_o}^{\zeta_1}\left(\nabla\cdot\mathbf{q}_s+S_{os}\frac{\partial h_s}{\partial t}+P_s\right)dz=\nabla'\cdot B_s\,\widetilde{\mathbf{q}'_s}+\mathbf{q}_s\big|_{F_1}\cdot\nabla F_1-\mathbf{q}_s\big|_{F_o}\cdot\nabla F_o$$
$$+S_{os}\left(B_s\frac{\partial\widetilde{h_s}}{\partial t}+\widetilde{h_s}\frac{\partial B_s}{\partial t}+h_s\big|_{F_1}\frac{\partial F_1}{\partial t}-h_s\big|_{F_o}\frac{\partial F_o}{\partial t}\right)+B_s\widetilde{P_s}=0,\quad(9.2.23)$$

in which $F_2 = z - \zeta(x,y,t)$, $F_1 = z - \zeta_1(x,y,t)$, $F_o = z - \zeta_o(x,y,t)$, ∇' denotes the gradient operator in the xy-plane, the prime in $\mathbf{q}'$ indicates the specific discharge vector in the xy-plane, $B_f = \zeta_2 - \zeta_1$, $B_s = \zeta_1 - \zeta_o$, and the tilde $\widetilde{(..)}$ symbol indicates the average over the relevant vertical length. With $h_f\big|_{F_2} \simeq h_f\big|_{F_1} = \widetilde{h_f}$, $h_s\big|_{F_o} \simeq h_s\big|_{F_1} = \widetilde{h_s}$, which expresses the assumption of essentially horizontal flow in both domains, we obtain for the freshwater and saltwater domains, respectively,

$$\nabla' \cdot B_f\widetilde{\mathbf{q}'_f} + \mathbf{q}_f\big|_{F_2} \cdot \nabla F_2 - \mathbf{q}_f\big|_{F_1} \cdot \nabla F_1 + S_{of}B_f\frac{\partial \widetilde{h_f}}{\partial t} + B_f\widetilde{P_f} = 0, \quad (9.2.24)$$

$$\nabla' \cdot B_s\widetilde{\mathbf{q}'_s} + \mathbf{q}_s\big|_{F_1} \cdot \nabla F_1 - \mathbf{q}_s\big|_{F_o} \cdot \nabla F_o + S_{os}B_s\frac{\partial \widetilde{h_s}}{\partial t} + B_s\widetilde{P_s} = 0. \quad (9.2.25)$$

We now introduce the conditions on the top and bottom bounding surfaces. On the impervious bottom, $F_o(x,y,z) = 0$, the condition is

$$\mathbf{q}_s\big|_{F_o} \cdot \nabla F_o = 0. \quad (9.2.26)$$

On the interface, $F_1(x,y,z,t) = 0$, the condition is obtained from (9.2.9), i.e.,

$$\mathbf{q}_f\big|_{F_1} \cdot \nabla F_1 \equiv -\phi\,\delta\frac{\partial \widetilde{h_f}}{\partial t} + \phi(1+\delta)\frac{\partial \widetilde{h_s}}{\partial t} = \phi\,\mathbf{u} \cdot \nabla F_1 = -\phi\frac{\partial F_1}{\partial t}, \quad (9.2.27)$$

in which $h_f|_{F_1} \simeq \widetilde{h_f}$, and $h_s|_{F_1} \simeq \widetilde{h_s}$. The magnitude of the error introduced by these assumptions depends on the deviation of the actual flow from the assumed horizontal one in the two domains.

On the phreatic surface, $F_2(x,y,z,t) = 0$, with the accretion rate $\mathbf{N} = -N\nabla z$, we obtain

$$F_2(x,y,z,t) \equiv z - \zeta_2(x,y,t) = z - h_f\big|_{F_2} \simeq z - \widetilde{h_f} = 0, \quad (9.2.28)$$

$$\mathbf{q}_f\big|_{F_2} \cdot \nabla F_2 = \mathbf{N} \cdot \nabla F_2 - \phi_{\text{eff}}\frac{\partial F_2}{\partial t} = -N + \phi_{\text{eff}}\frac{\partial \widetilde{h_f}}{\partial t}. \quad (9.2.29)$$

With these conditions, we obtain the flow equations for the two domains,

$$-\nabla' \cdot B_f\,\widetilde{\mathbf{q}'_f} + N - B_f\widetilde{P_f} = (\phi_{\text{eff}} + S_{of}B_f + \phi\,\delta)\,\frac{\partial \widetilde{h_f}}{\partial t} - \phi(1+\delta)\frac{\partial \widetilde{h_s}}{\partial t}, \quad (9.2.30)$$

$$-\nabla' \cdot B_s\,\widetilde{\mathbf{q}'_s} - B_s\widetilde{P_s} = [S_{os}B_s + \phi(1+\delta)]\,\frac{\partial \widetilde{h_s}}{\partial t} - \phi\,\delta\frac{\partial \widetilde{h_f}}{\partial t}, \quad (9.2.31)$$

in which we, usually, assume that

$$B_fS_{of} \ll \phi_{\text{eff}}, \quad B_sS_{os} \ll \phi.$$

Or, in terms of interface and phreatic surface elevations, and with $\phi_{\text{eff}} \approx \phi$,

$$-\nabla' \cdot \mathbf{Q}'_f + N - B_f \widetilde{P_f} = \phi \frac{\partial(\zeta_2 - \zeta_1)}{\partial t}, \quad \mathbf{Q}'_f \equiv B_f \widetilde{\mathbf{q}'_f} = -B_f \widetilde{\mathsf{K}'_f} \cdot \nabla \widetilde{h_f}, \quad (9.2.32)$$

$$-\nabla' \cdot \mathbf{Q}'_s - B_s \widetilde{P_s} = \phi \frac{\partial \zeta_1}{\partial t}, \quad \mathbf{Q}'_s \equiv B_s \widetilde{\mathbf{q}'_s} = -B_s \widetilde{\mathsf{K}'_s} \cdot \nabla \widetilde{h_s}, \quad (9.2.33)$$

$$\zeta_2 = \widetilde{h_f}, \quad \zeta_1 = (1+\delta)\widetilde{h_s} - \delta \widetilde{h_f}. \quad (9.2.34)$$

For a confined aquifer, we delete N in (9.2.32), and note that $\partial \zeta_2 / \partial t \equiv 0$.

Since we have assumed 'essentially horizontal flow', the flow domain, for which we have to solve (9.2.24) and (9.2.25) for $\widetilde{h_f}$ and $\widetilde{h_s}$, is bounded by vertical surfaces that pass: (a) through the toe of the interface (point G in Fig. 9.2.6), and (b) through the coast. To complete the delineation of the considered aquifer domain, we add type (c) vertical surfaces . Appropriate boundary conditions have to be specified on all these boundaries. The conditions on type (c) surfaces are the common ones (of specified flux, or specified head), and need not be elaborated here. Along the boundary of type (b), the considered freshwater domain becomes an aquifer without the seawater wedge, i.e., $\zeta_1 \equiv \zeta_o$, $F_1 \equiv F_o$. In the case of a phreatic aquifer, (9.2.24) reduces to (Bear, 1979)

$$-\nabla' \cdot B_f \widetilde{\mathbf{q}'_f} + N - B_f \widetilde{P_f} = (\phi_{\text{eff}} + S_{of} B_f) \frac{\partial \widetilde{h_f}}{\partial t}, \quad (9.2.35)$$

in which we usually neglect the effect of elastic storativity, as it is much smaller than the specific yield, denoted here as ϕ_{eff}. As the interface advances or retreats, this boundary, between the two aquifer freshwater domains, is also moving. On the common boundary (DG), we have to maintain the same piezometric head and the same normal flux. On the surface that passes through the coastline, we have two parts: the freshwater part, CB, and the seawater one, BA. The boundary between these two parts (point B) is not fixed, but moves as the interface advances or retreats. For the freshwater portion, we assume (Bear, 1979) that the aquifer subdomain BCM acts as a resistance to the flow, so that across it the head on BC is reduced to that dictated by the sea level. This condition is expressed as

$$Q_{fo} = \frac{\widetilde{h_f}\big|_{BC}}{\text{Resist.}} = -\mathsf{K}_f \overline{BC} \left.\frac{\partial \widetilde{h_f}}{\partial x}\right|_{\overline{BC}}, \quad (9.2.36)$$

or

$$\frac{\widetilde{h_f}}{\alpha} + \overline{BC} \frac{\partial \widetilde{h_f}}{\partial x} = 0, \quad \text{on} \quad \overline{BC}, \quad (9.2.37)$$

where α is a coefficient. The above equation is a third type boundary condition. A similar condition, but with a different coefficient, can be written for the seawater portion of this boundary (AB).

Note that at no point in the development of the flow equations presented above, say (9.2.30) and (9.2.31), and the corresponding boundary conditions, we made use of the Ghyben-Herzberg approximation (9.2.21), which corresponds to steady flow with a stationary interface, and $h_s = \text{const.}$, $h_f = \widetilde{h_f}$. Instead, we have used the relationship (9.2.6), based only on the no-jump in pressure as the interface is approached from both sides.

A number of examples of analytical solutions of the above equations in simple cases is presented in the next subsection.

9.2.5 Some analytical solutions for stationary interface

We consider a number of simple examples to demonstrate the shape of the (assumed) sharp interface and the relationship that exists between the extent of seawater intrusion and the flow of freshwater to the sea. This relationship underlies all aquifer management decisions that aim at a sustainable yield of a coastal aquifer. In all examples, the aquifer's bottom is horizontal, the flow is assumed to be everywhere perpendicular to the coastline, with a stationary interface, i.e., steady flow. This means that freshwater is moving, but seawater is stationary. We also assume that the geometry of the freshwater flow domain above the interface (= seawater wedge of length L) is such that the *Dupuit assumption* of essentially horizontal flow is valid (and, equivalently, so is the Ghyben-Herzberg approximation).

A. Interface in a confined aquifer

Figure 9.2.7a shows a cross-section of a confined aquifer of constant thickness $B(\ll L)$. Let the interface toe (point G) be located at the origin $x = 0$. The seaward freshwater flow at this point is Q_{fo}; it is the difference between the total inflow to the aquifer through its right-side boundary and the pumping from the aquifer in the coastal aquifer strip to the right of point G. As suggested above, we assume that the Dupuit assumption of essentially horizontal flow is valid for the freshwater domain above the interface.

For this (one-dimensional) case, the freshwater flow equation reduces to

$$\frac{dQ_f}{dx} \equiv \frac{d}{dx}\left[-\mathsf{K}_f \eta(x)\frac{dh_f(x)}{dx}\right] = 0, \tag{9.2.38}$$

or

$$-\mathsf{K}_f\,\eta(x)\frac{dh_f(x)}{dx} = \text{const.} = Q\Big|_{x=0} = Q_{fo}, \tag{9.2.39}$$

where η is the thickness of freshwater above the saltwater wedge (see Fig. 9.2.7a). Based on the Ghyben-Herzberg approximation, we can express the freshwater head as

$$h_f = \frac{\eta + b}{\delta}. \tag{9.2.40}$$

Using the above expression in (9.2.39), we can easily obtain

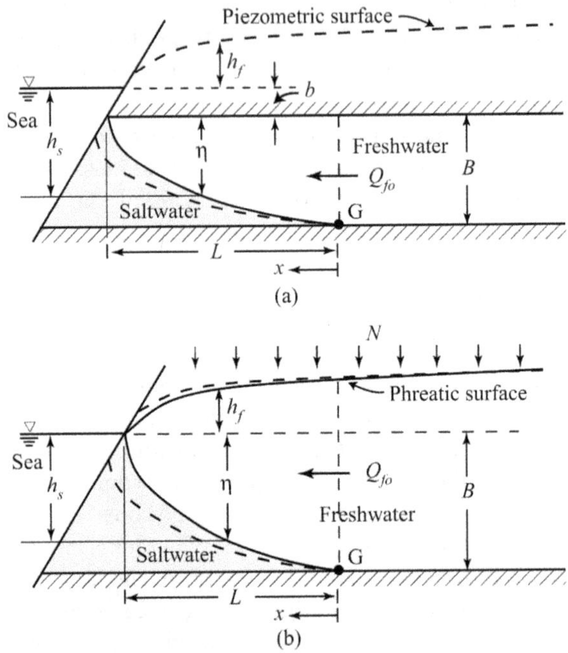

Figure 9.2.7: Stationary interface derived by the Dupuit (Ghyben-Herzberg) approximation.

$$\eta^2 = \frac{2\delta Q_{fo}}{\mathsf{K}_f}(L - x). \tag{9.2.41}$$

In the above, we have used the boundary condition $\eta = 0$ at $x = L$. Equation (9.2.41) shows that the interface has the shape of a parabola. Using the condition $\eta = B$ at $x = 0$, we obtain from (9.2.41)

$$B^2 = \frac{2\delta Q_{fo}}{\mathsf{K}_f} L. \tag{9.2.42}$$

This equation clearly shows the relationship between the length of the seawater wedge, L, and the discharge of freshwater to the sea, Q_{fo}. As Q_{fo} increases, L decreases. This means that *within the framework of coastal aquifer management, the extent of seawater intrusion, expressed by L, is a decision variable*; it is controlled by controlling Q_{fo}, or, alternatively, by controlling the recharge and/or pumping in the coastal aquifer strip.

B. Interface in a phreatic aquifer

This case is shown in Fig. 9.2.7b. The aquifer is uniformly replenished, say, from precipitation, at the rate N. Again, assuming essentially horizontal flow,

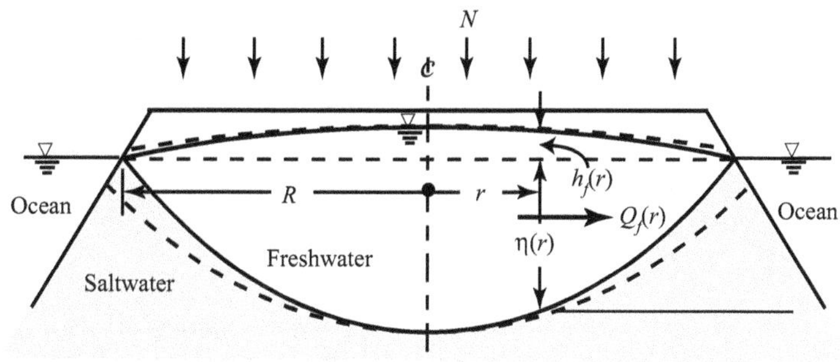

Figure 9.2.8: Interface in an oceanic island.

the flow equation takes the form

$$-\frac{dQ_f}{dx} + N \equiv -\frac{d}{dx}\left[-\mathsf{K}_f(h_f(x) + \eta(x))\frac{dh_f}{dx}\right] + N = 0. \tag{9.2.43}$$

Substituting $h_f = \eta/\delta$ into the above and integrating, noting that $Q_f = Q_{fo}$ at $x = 0$, we obtain

$$\eta^2 - B^2 = -\frac{\delta^2}{\mathsf{K}_f(1+\delta)}\,(2Q_{fo} + Nx)\,x. \tag{9.2.44}$$

When the recharge is zero, $N = 0$, the interface shape is parabolic. Using the boundary condition, $\eta = 0$ at $x = L$, in (9.2.44), we obtain,

$$B^2 = \frac{\delta^2}{\mathsf{K}_f(1+\delta)}\,(2Q_{fo} + NL)\,L. \tag{9.2.45}$$

In the above, we observe a relationship among L, Q_{fo}, and N. Hence, the intrusion distance, L, can be controlled by the outflow rate, Q_{fo}, as well as by the rate of artificial recharge, N.

C. Interface in an oceanic island

We consider the stationary interface in a circular oceanic island of radius R, with accretion, N (Fig. 9.2.8). As in the previous examples, we assume that the freshwater flow above the (assumed) sharp interface is essentially horizontal, so that the Dupuit assumption is applicable (and so is the Ghyben-Herzberg approximation). The freshwater flow is described by

$$-\frac{dQ_f}{dr} + 2\pi r N = 0, \tag{9.2.46}$$

where

$$Q_f = -2\pi r \mathsf{K}(h_f + \eta)\frac{dh_f}{dr}. \tag{9.2.47}$$

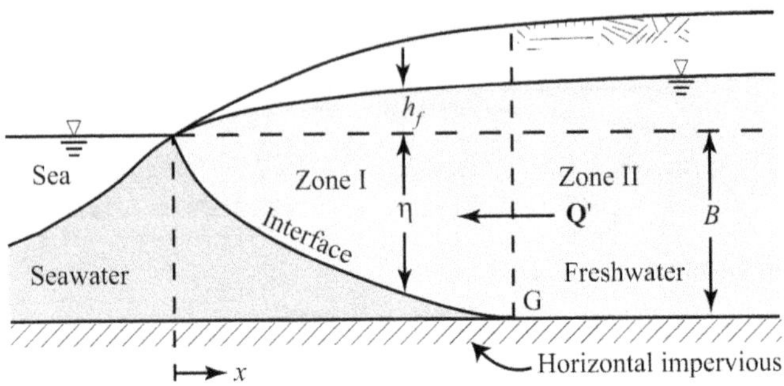

Figure 9.2.9: Nomenclature for Strack's potential for a phreatic aquifer.

Again, Ghyben-Herzberg approximation gives $h_f = \eta/\delta$. By integration from η at r to $\eta = 0$ at $r = R$, we obtain the elliptic shape of the interface, described by

$$\eta^2 = \frac{N\delta^2(R^2 - r^2)}{2\mathsf{K}(1+\delta)}. \tag{9.2.48}$$

D. Strack's solution

Whether a coastal aquifer is phreatic or confined, we can distinguish in it two freshwater zones: a freshwater only zone, and a freshwater-saltwater zone, separated by a (possibly moving) interface. We shall denote these zones as zone II and I, respectively (see Fig. 9.2.9). Assuming essentially horizontal flow in both zones, i.e., the Dupuit assumption is valid, it is easy to write the model describing the flow in each of these zones, with the piezometric head in each zone as the state variable to be solved for. On the common boundary, we require the continuity of both the piezometric head and the flux.

Strack (1976) developed a model that describes the freshwater flow in both zones, using a *harmonic potential* (i.e., one that satisfies the Laplace equation) as the *single* variable of state for both zones. This potential is continuous across the boundary between the two zones. Figure 9.2.9 shows the case of a phreatic aquifer. Strack (1976) presents also the case of a confined aquifer.

For the flow in zone I (phreatic flow above the interface), Strack's potential, φ, is defined as:

$$\varphi = \frac{1}{2}(h_f + \eta)\,h_f = \frac{1}{2}(1+\delta)\,h_f^2, \tag{9.2.49}$$

$$\mathbf{Q}' = -\mathsf{K}\,(h_f + \eta)\,\nabla' h_f = -\mathsf{K}\nabla\varphi, \tag{9.2.50}$$

where symbols are explained in Fig. 9.2.9, and $\mathbf{Q}'$ is the freshwater discharge per unit aquifer width, through the entire thickness of the aquifer.

For the flow in zone II (freshwater only zone), Strack introduces the potential

$$\varphi = \frac{1}{2}\left[(h_f + B)^2 - \frac{1+\delta}{\delta}B^2\right], \tag{9.2.51}$$

$$\mathbf{Q}' = -\mathsf{K}(h_f + B)\nabla' h_f = -\mathsf{K}\nabla\varphi. \tag{9.2.52}$$

At point G (= interface 'toe'), $B = h_f\delta$, and the two values of φ, defined by (9.2.49) and (9.2.51), become identical,

$$\varphi = \varphi_{\text{toe}} = \frac{(1+\delta)B^2}{2\delta^2}. \tag{9.2.53}$$

Also the flows at this point, when approached from both sides, become identical. Since, for the steady flow considered here, $\nabla \cdot \mathbf{Q}' = 0$ in each zone, we obtain

$$\frac{\partial^2 \varphi}{\partial x^2} + \frac{\partial^2 \varphi}{\partial y^2} = 0, \tag{9.2.54}$$

which means that Strack's potentials, φ, are harmonic functions in the respective zones. In fact, as the potentials and their derivatives are continuous across the zone's boundary, they become a single function that satisfies the Laplace equation in the combined zone (zone I and II). The saltwater intrusion problem can be solved as one with a single potential in a single zone. Once the potential φ is solved for, the location of the toe is defined by the equipotential line whose value is given by (9.2.53).

Strack (1976) demonstrates the application of Strack's potential by considering the case of a well located at a distance x_w from a coast that has the shape of a straight line (Fig. 9.2.10). The well is pumping at a rate $Q_w =$ const., which is superimposed on a uniform freshwater flow at a specific discharge Q'_o (= discharge per unit width through the entire thickness of the aquifer) perpendicular to the coast. The solution of this problem can be obtained by superposition, and by the method of images. The derived potential is then (Strack, 1976)

$$\varphi = \frac{Q'_o}{\mathsf{K}}x + \frac{Q_w}{4\pi\mathsf{K}}\ln\frac{(x-x_w)^2 + y^2}{(x+x_w)^2 + y^2}. \tag{9.2.55}$$

It is easy to verify that along the coastline, $x = 0$, the potential φ is zero, and that $Q' = Q'_o$ as $x \to \infty$.

The equation describing the toe of the interface in the xy-plane can now be obtained by setting φ in (9.2.55) to the value of (9.2.53):

$$\frac{(1+\delta)B^2}{2\delta^2} = \frac{Q'_o}{\mathsf{K}}x + \frac{Q_w}{4\pi\mathsf{K}}\ln\frac{(x-x_w)^2 + y^2}{(x+x_w)^2 + y^2}. \tag{9.2.56}$$

This equation defines the trajectory of the toe location. In Fig. 9.2.10, several such toe locations, corresponding to different pumping rates, are plotted. First, we notice that when there is no pumping, $Q_w = 0$, the toe location is a straight line parallel to the coast, at a distance x_o, as determined from

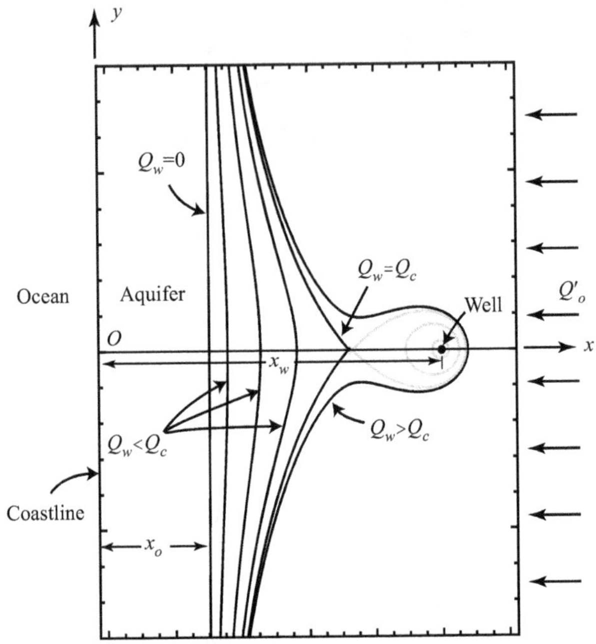

Figure 9.2.10: Locations of the toe of the saltwater wedge corresponding to different pumping rates of a single well near the coast.

(9.2.56)

$$x_o = \frac{(1+\delta)B^2\mathsf{K}}{2\delta^2 Q'_o}. \tag{9.2.57}$$

As Q_w increases, the toe of the interface advances inland toward the well. The figure shows several profiles of the advancing interface, marked as $Q_w < Q_c$. As learned from the solution for a pumping well, surrounding such well there exists a *cone of depression* where the potential is low, and can even be negative (i.e., below sea level). A low potential value means that saltwater can exist below the freshwater. Hence, (9.2.56) has a second possible solution for the toe, delineated by the grey curves in Fig. 9.2.10, corresponding to different pumping rates. Such regions, however, are protected from seawater intrusion because there exists a higher potential region between the well and the coast that serves as a barrier to seawater intrusion. This situation can be noted on the cross-sectional view along the axis $y = 0$, marked as interface 1 and phreatic surface 1 in Fig. 9.2.11. The peak of the groundwater mound, marked as point B_1, is located at the distance x_s to coast:

$$x_s = x_w\sqrt{1 - \frac{Q_w}{\pi Q'_o x_w}}. \tag{9.2.58}$$

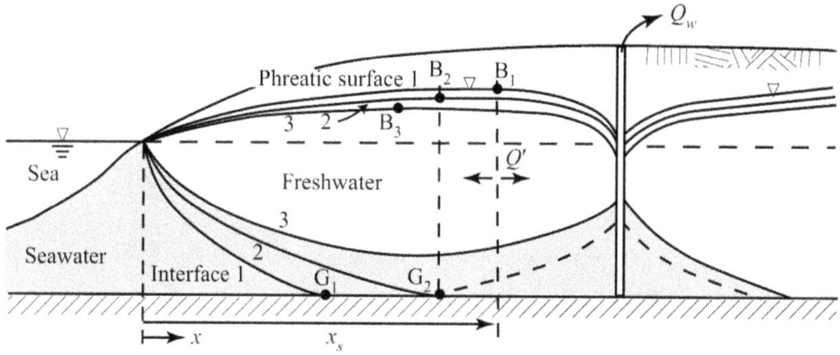

Figure 9.2.11: Vertical cross-sectional view of the interface along $y = 0$.

This point is a *stagnation point*, or *groundwater divide*: left of B_1, the flow is to the sea, while to the right of B_1, the flow is to the well.

As the pumping rate continues to increase, a critical situation is reached, at which that the toe location forms a wedge and touches the low potential zone surrounding the well. This situation is marked by the curve corresponding to $Q_w = Q_c$ in Fig. 9.2.10, and interface 2 in Fig. 9.2.11. At this critical situation, the toe location along the $y = 0$ axis, x_G, coincides with the stagnation distance, x_s (Fig. 9.2.11). Hence, from (9.2.56) with $y = 0$, and (9.2.58), we obtain

$$\frac{(1+\delta)B^2\mathsf{K}}{2\delta^2 Q'_o x_w} = \sqrt{1 - \frac{Q_c^*}{\pi}} + \frac{Q_c^*}{2\pi} \ln\left(\frac{1 - \sqrt{1 - \frac{Q_c^*}{\pi}}}{1 + \sqrt{1 - \frac{Q_c^*}{\pi}}}\right), \tag{9.2.59}$$

where

$$Q_c^* = \frac{Q_c}{Q'_o x_w}. \tag{9.2.60}$$

By finding the root of the algebraic equation (9.2.59), we can solve for the critical pumping rate, Q_c. When the pumping rate exceeds this critical value, $Q_w > Q_c$, these two low potential zones become connected, and the saltwater invades a large region surrounding the well, as shown in Figs. 9.2.10 and 9.2.11. Figure 9.2.12 shows a three-dimensional perspective of the interface before and after the invasion of the well by seawater.

E. Upconing

Bear and Dagan (1964b) and Yih (1964) presented analytical solutions of upconing for steady, *two-dimensional* (vertical plane) domain (i.e., well $\equiv$ horizontal drain, or gallery well), in the absence of seaward flow, by the *hodograph method*. These, and additional steady state approximate models for the estimation of upconing, are presented in Bear (1979). Dagan and Bear

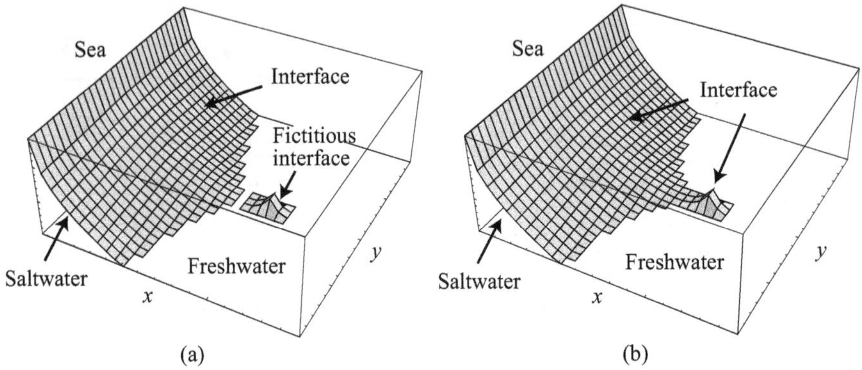

Figure 9.2.12: Saltwater-freshwater interface due to uniform flow and a pumping well near the coast. (a) $Q_w < Q_c$; (b) $Q_w > Q_c$.

(1968) presented a solution for time dependent interface upconing, using the *method of small perturbations.* This approach is limited to relatively small interface rises, e.g., less than 1/3 the distance from the initial interface to the well (sink). The solution is also summarized in Bear (1979).

9.2.6 Multilayered aquifers

Often, the coastal aquifer is made up of a number of sub-aquifers separated by impervious, and/or semi-pervious layers that extend from some distance seaward of the coast up to some distance landward of it. When these sub-aquifers are open to the sea, they enable freshwater discharge to the sea. A seawater wedge may encroach each of the sub-aquifers that are hydraulically connected to the sea, the extent of the intruding wedge depending on the rate of freshwater flow to the sea in that sub-aquifer, on the aquifer's hydraulic conductivity, and on the geometry of the sub-aquifer. Figures 9.1.2b and 9.3.1 show examples of a two-layered aquifer separated by a semipervious layer, showing transition zones (Sec. 9.3) between fresh and saltwater.

9.3 Transition Zone Modeling

As already mentioned in Sec. 9.1, the coastal aquifer domain, occupied by seawater, and the inland aquifer domain, occupied by freshwater, are separated by a *transition zone.* This is a consequence of the fact that the two (so-called 'miscible') liquids are actually a *single* liquid—water—with different concentrations of dissolved salts.

The width of the transition zone is dictated by three phenomena:

- Advection of water—fresh and mixed—towards the sea (or, under certain conditions, landward).

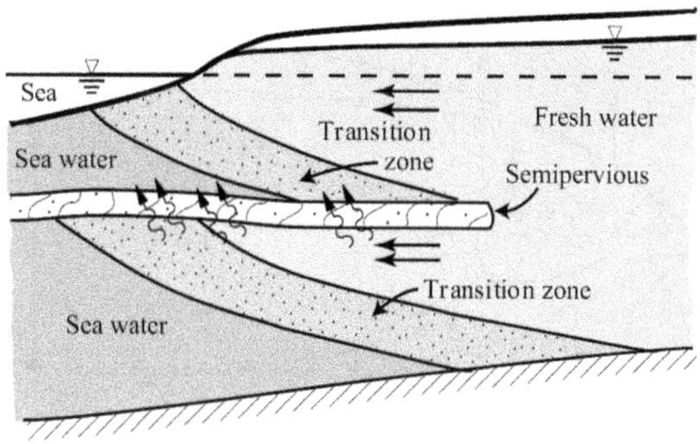

Figure 9.3.1: Transitions zone in a multilayered coastal aquifer with a semipervious layer.

- Recirculation of seawater and mixed water.
- Hydrodynamic dispersion (dispersion and molecular diffusion).

Hydrodynamic dispersion is discussed in Subs. 7.1.4, and need not be repeated here. Suffice it to summarize that a transition zone exists; across it, the salinity of the water varies from that of freshwater to that of seawater. The width of this zone grows as it is being displaced in response to changes in the flow regime and in the discharge of water to the sea. The transition zone is also fed by a flux of salt from the seawater zone.

9.3.1 Variable density model

Figure 9.3.2 shows a phreatic coastal aquifer with a transition zone between seawater and freshwater. The considered flow domain is ABCDEMFA. Basically, the mathematical model describing seawater intrusion in a coastal aquifer consists of:

(a) Mass balance equation for the water (= salt solution).
(b) Flux equation (e.g., Darcy's law) for the water.
(c) Mass balance equation for the dissolved salts.
(d) Flux equation for the dissolved salts.

The first two equations, often combined into a single *flow equation* for the water, are discussed in Chaps. 4 and 5. The last two equations, which can be combined to form a single mass balance equation for the dissolved salts, are discussed in Chap. 7. However, in these three chapters, we focussed our attention on the case in which the liquid's density remains unchanged (or approximately so). Here, we intend to construct a complete variable density flow and solute transport model. The solution of such model will provide the shape and position of the transition zone between freshwater and seawater in a coastal aquifer.

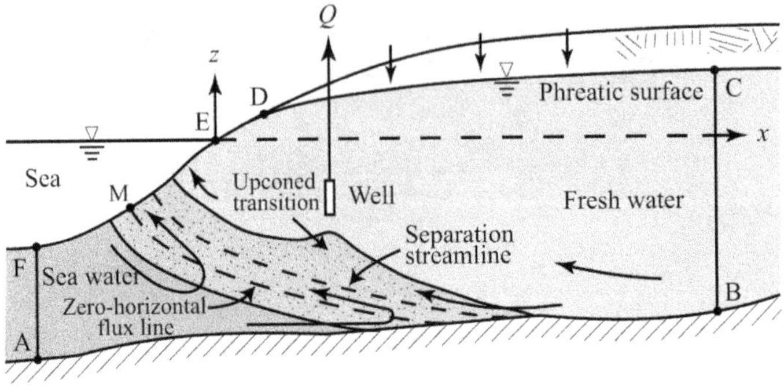

Figure 9.3.2: The transition zone with upconing in a simplified vertical cross section of a coastal aquifer.

In addition to the four equations listed above, the complete model requires appropriate initial and boundary conditions.

Although, in principle, the flow part of the variable density model can be written in terms of pressure, p, as a state variable, a more (numerically) efficient model is obtained by expressing the flow model in terms of a *reference piezometric head.*

Accordingly, following Bear and Zhou (2006), we introduce a reference piezometric head, $h'(x, y, z, t)$, associated with a reference water density. We often select the density of freshwater, ρ_f as a reference density, so that the reference head is defined as

$$h' = z + \frac{p}{\rho_f g}, \tag{9.3.1}$$

in which z and g denote the vertical coordinate and gravity acceleration.

A. Density and viscosity

To enable the use of the variable density mathematical model, beyond the case of seawater intrusion, discussed here, we consider TDS (= Total Dissolved Solids) in the range where the *volume of water may vary with concentration.* To facilitate the discussion, we introduce a *normalized salt mass fraction*, C, defined by

$$C = \frac{\omega - \omega_f}{\omega_s - \omega_f}, \quad 0 \leq C \leq 1.0, \tag{9.3.2}$$

in which ω, ω_s and ω_f denote the salt (or TDS) mass fraction in water, in seawater, and in freshwater, respectively, with the latter serving as a reference mass fraction. Note that $c \equiv \omega\rho$, with c denoting the TDS concentration (= mass of TDS per unit volume of fluid).

The constitutive equation that expresses the relationship, $\rho = \rho(p, c)$, between the fluid density, ρ, the pressure, p, and the salt mass fraction, ω (=

mass of dissolved salt per unit mass of fluid), under isothermal conditions, is

$$\rho = \rho_o \exp\left[\beta'_p(p - p_o) + \beta'_\omega(\omega - \omega_o)\right], \tag{9.3.3}$$

where ρ_o, p_o, and ω_o are reference values of density, pressure and salt mass fraction, respectively, β'_p $(= (1/\rho)\partial\rho/\partial p)$ is the coefficient of water compressibility (at constant salt concentration), and β'_ω $(= (1/\rho)\partial\rho/\partial\omega)$ is a coefficient that introduces the effect of change in salt mass fraction on the fluid's density (at constant pressure). Usually, we select $\omega_o = \omega_f$, i.e., equal to the mass fraction of freshwater. The reference pressure is such that for freshwater at $p = p_o$ and $\omega = \omega_o$, the density is $\rho_o = \rho_f$, i.e., the density of freshwater is selected as a reference density.

The linearized approximation of (9.3.3) is

$$\rho = \rho_o\left[1 + \beta''_p(p - p_o) + \beta''_\omega(\omega - \omega_o)\right], \tag{9.3.4}$$

where $\beta''_p = (1/\rho_o)\partial\rho/\partial p$, and $\beta''_\omega = (1/\rho_o)\partial\rho/\partial\omega$. In what follows, we shall assume that for the range of pressures considered here, $\beta''_\omega|\Delta\omega| \gg \beta''_p|\Delta p|$, so that we may employ the approximation:

$$\rho = \rho_f(1 + \beta_c C), \quad \beta_c \equiv \beta''_\omega(\omega_s - \omega_f), \tag{9.3.5}$$

where β_c may be referred to as a *density difference factor* (dimensionless). In spite of the above assumption, we do take into account the effect of pressure in the expression for the specific storativity appearing in the mass balance equation (5.1.54).

To obtain the effect of concentration on the fluid's dynamic viscosity, we use the constitutive relationship for dynamic viscosity in the form (Lever and Jackson, 1985)

$$\mu = \mu_{fo}\,\mu_r = \mu_{fo}\left(1 + 1.85\,\omega - 4.1\,\omega^2 + 44.50\,\omega^3\right), \tag{9.3.6}$$

in which the viscosity, μ_{fo}, corresponds to $\omega = 0$, and $\mu_r = \mu/\mu_{fo}$ is the relative viscosity.

B. Darcy's law

The specific discharge relative to the solid, $\mathbf{q}_r$, is expressed by Darcy's law (4.2.5), rewritten here for convenience in the form

$$\mathbf{q}_r = \phi(\mathbf{V} - \mathbf{V}_s) = -\frac{\mathbf{k}}{\mu}\cdot(\nabla p + \rho g \nabla z), \tag{9.3.7}$$

in which $\mathbf{V}$ and $\mathbf{V}_s$ denote the velocity vectors of the water and of the solid matrix, respectively, ϕ denotes the porosity, the second rank tensor, $\mathbf{k}$, denotes the permeability, and $\mathbf{q}_r$ denotes the specific discharge vector, relative to the solid matrix. We usually assume, also here, that $\mathbf{V}_s \approx 0$, and, therefore,

$\mathbf{q}_r \approx \mathbf{q}$. In terms of h' and C, Darcy's law can be rewritten in the form

$$\mathbf{q} = -\frac{\mathbf{K}_f}{\mu_r}\left(\nabla h' + \beta_c C \nabla z\right), \quad \mathbf{K}_f = \frac{\rho_f g \mathbf{k}}{\mu_f}, \tag{9.3.8}$$

with $\mathbf{K}_f$ denoting the hydraulic conductivity with respect to the reference density and viscosity (of the freshwater).

C. Mass balance equation for the water

The basic mass balance equation is (5.1.7). Let us rewrite this equation (a) without neglecting the *dispersive flux of the total fluid mass*, $\phi\,\mathbf{J}^{*\rho}$, and (b) replacing the source term $\rho\Gamma^m$ by $\rho_R Q_R - \rho Q_P$. We obtain

$$\frac{\partial}{\partial t}(\phi\rho) = -\nabla\cdot(\rho\mathbf{q} + \phi\,\mathbf{J}^{*\rho}) + \rho_R Q_R - \rho Q_P, \tag{9.3.9}$$

in which

$$\phi\,\mathbf{J}^{*\rho} \equiv -\phi\,\mathbf{D}\cdot\nabla\rho = -\phi\rho_f\beta_c\mathbf{D}\cdot\nabla C, \tag{9.3.10}$$

is the *dispersive flux of the total mass*, with $\mathbf{D}$, a second rank symmetric tensor, denoting the coefficient of hydrodynamic dispersion (dims. L^2/T), Q_R and Q_P denoting, symbolically, the rates of injection of water into and withdrawal of water from the aquifer (dims. $1/T$), and ρ_R denoting the density of the injected water. When water is withdrawn and injected through (point) wells, the source terms in (9.3.9) may, symbolically, be written as

$$\rho_R Q_R = \sum_m \rho_{Rm} Q_{Rm}(t)\,\delta(\mathbf{x}-\mathbf{x}_m), \tag{9.3.11}$$

$$\rho Q_P = \sum_m \rho_m Q_{Pm}(t)\,\delta(\mathbf{x}-\mathbf{x}_m), \tag{9.3.12}$$

in which Q_{Rm} and ρ_{Rm} are the injection rate, and the density of the water injected through a well at point $\mathbf{x}_m$, respectively, Q_{Pm} and ρ_m are the withdrawal rate and density of the water withdrawn through a well at point $\mathbf{x}_m$, and δ denotes the *Dirac delta function.*

A few comments are given below on the fact that (9.3.9) includes a term that expresses the dispersive flux of the total fluid mass. In the case of a fluid of variable density, significant density gradients may develop. For example, it is possible that a rather narrow transition zone will develop, with a relatively large density gradient across it, such that with flow that is more or less normal to such a gradient, significant lateral dispersion may take place. Bear and Bachmat (1990, p. 290) presented and discussed a method, using appropriate Peclet numbers (that define the ratio between advective and dispersive fluxes), for examining the conditions under which the dispersive flux of the total mass may be neglected as being much smaller than the advective one. Note that there is no diffusive flux of the total dissolved mass. In the case of seawater intrusion, we may encounter a rather narrow transition zone, with flow parallel to it. The appropriate Peclet number may be less than or not much larger than one, so that we cannot conclude, *a priori*, that advection dominates over dispersion.

In view of the relationship $\rho = \rho(p, C)$, we make use of (9.3.8) and (9.3.10) to modify (9.3.9), rewriting it in terms of the reference piezometric head, h', and the normalized mass fraction, C, in the form:

$$S_o \frac{\partial h'}{\partial t} + \phi\beta_c \frac{\rho}{\rho_f}\frac{\partial C}{\partial t} = \nabla \cdot \left[\frac{(1+\beta_c C)}{\mu_r}\mathbf{K}_f \cdot (\nabla h' + \beta_c C \nabla z) + \phi\beta_c \mathbf{D} \cdot \nabla C\right] + \frac{\rho_R}{\rho_f} Q_R - (1+\beta_c C) Q_P, \qquad (9.3.13)$$

in which S_o, defined by (5.1.75), denotes the aquifer's specific storativity.

D. Mass balance equation for the dissolved salt

For water of variable density, the general mass balance equation for the total dissolved solids (TDS) in the water can be written in the form

$$\frac{\partial \phi\rho\omega}{\partial t} = -\nabla \cdot (\rho\omega\mathbf{q} - \phi\rho\mathbf{D}_h \cdot \nabla\omega) + \rho_R\omega_R Q_R - \rho\omega Q_P, \qquad (9.3.14)$$

in which $\mathbf{D}_h$ denotes the coefficient of hydrodynamic dispersion (a second rank symmetric tensor), $\mathbf{D}_h = \mathbf{D} + \boldsymbol{\mathcal{D}}^*$, and $\boldsymbol{\mathcal{D}}^*$ denotes the coefficient of molecular diffusion in a porous medium. These coefficients are defined and discussed in Subs. 7.1.3 through 7.1.6.

The salt transport equation, (9.3.14) can be written in terms of the normalized mass fraction, C, with $\omega_f = 0$, in the form (Bear, 1999; Zhou *et al.*, 2005)

$$\frac{\partial \phi\rho C}{\partial t} = -\nabla(\rho C\mathbf{q} - \phi\rho\mathbf{D}_h \cdot \nabla C) + \rho_R C_R Q_R - \rho C Q_P. \qquad (9.3.15)$$

Recalling that the flux $\mathbf{q}$ is related to h' and C by (9.3.8), and noting the dependence of ρ on c (hence on C) and on p (hence on h'), the description of seawater intrusion as a variable density problem, i.e., as a transition zone one, is obtained by the simultaneous solution of the two equations: (9.3.13) and (9.3.15), for the two primary variables h' and C.

D. Initial and boundary conditions

Figure 9.3.2 shows the boundaries of the flow domain, ABCDEMFA, in a seawater intrusion (= variable density flow and salt transport) problem:

The initial conditions are

$$h' = h_{fo}(x, y, z, 0), \quad C = C_o(x, y, z, 0), \qquad (9.3.16)$$

where h_{fo} and C_o are known distributions.

For the flow equation, (9.3.13), *neglecting the dispersive flux of the total fluid mass*, the following boundary conditions are assigned on the various segments comprising the boundary of the flow domain:

- On AB (impervious bottom):

$$q_n = \mathbf{q}\cdot\mathbf{n} = 0. \tag{9.3.17}$$

- On BC (lateral boundary on land side)

$$h' = h_{fp}(x,y,z,t), \quad \text{or} \quad q_n = q_{np}(x,y,z,t). \tag{9.3.18}$$

- On CD (phreatic surface)

$$\rho q_n = -\rho_N N n_z + \frac{(\phi\rho - \theta_{wo}\rho_N)}{|\nabla F|}\frac{\partial h'}{\partial t}. \tag{9.3.19}$$

- On DE (seepage face)

$$h' = \zeta_{DE}. \tag{9.3.20}$$

- On EF (sea bottom)

$$h' = \beta_c h_{\text{sea}}. \tag{9.3.21}$$

- On FA (sea side boundary)

$$h' = \beta_c H_{FA}. \tag{9.3.22}$$

In these conditions, h_{fp} is the prescribed reference head on a Dirichlet-type boundary, q_{np} is the prescribed fluid flux on a second-type boundary, $\mathbf{n}(=\nabla F/|\nabla F|)$ is the outward normal unit vector on a boundary represented as $F = F(x,y,z,t)$, with n_x, n_y, n_z denoting components of $\mathbf{n}$ in three dimensions, $F(x,y,z,t) \equiv h' - z = 0$ represents the shape of the phreatic surface, $\mathbf{N}\cdot\mathbf{n} \equiv -Nn_z$ denotes the rate of replenishment, ρ_N is the density of the replenishment, θ_{wo} is the irreducible water content assumed to prevail in the unsaturated zone, above the phreatic surface, ζ_{DE} is the elevation of a point on the seepage face, above sea level, and H_{sea} and H_{FA} are the depth of seawater at a point on the sea bottom and on the sea-side lateral boundaries, respectively. If we wish to take into account the dispersive flux of the fluid's mass, we replace $\rho\mathbf{q}$ by $\rho\mathbf{q} - \phi\rho\beta_c\mathbf{D}\cdot\nabla C$.

The boundary conditions for the salt balance equation, (9.3.15), are:

- On AB and DE

$$q_{n,\text{disp}} \equiv \phi\,\mathbf{J}_{\text{disp}}\cdot\mathbf{n} = \phi\,\mathbf{D}_h\cdot\nabla C = 0. \tag{9.3.23}$$

- On BC

$$C = 0. \tag{9.3.24}$$

- On CD

$$q_{n,\text{disp}} = -\left(\frac{\rho_N}{\rho}C - C_N\right)\left(Nn_z + \frac{\theta_{wo}}{|\nabla F|}\frac{\partial h'}{\partial t}\right), \tag{9.3.25}$$

where $q_{n,\text{disp}}$ is the hydrodynamic dispersive flux normal to a 2nd- or 3rd-type boundary, and C_N is the normalized mass fraction of the replenishment water.

- On FA

$$C = 1.0. \tag{9.3.26}$$

- On EF, the sea bottom boundary, we may have either only landward flow from the sea, in which case the condition there is the same as on FA, or we may have both an inflow portion, MF, and an outflow portion, EM, separated by a point with zero-normal fluid flux.
- On MF, the inflow portion, we usually employ either a third-type condition

$$q_{n,\text{disp}} = -\mathbf{D}_h \cdot \nabla C = (1.0 - C)q_n, \tag{9.3.27}$$

or a Dirichlet condition $C = 1.0$, as (9.3.26).
- On EM, the outflow portion, the assumption is usually made that the fluid's concentrations are identical (i.e., continuous) on both sides of this boundary. Since the fluid fluxes are also identical, the condition becomes a second-type condition,

$$q_{n,\text{disp}} \equiv -\mathbf{D}_h \cdot \nabla C = 0. \tag{9.3.28}$$

The (moving) point M, between the inflow and outflow portions of the boundary, is *a priori* unknown. In a numerical solution, during each iteration, we check whether flow along the sea bottom is directed inward or outward, and then assign the appropriate boundary condition accordingly.

We note an *inconsistency* in the salt mass fraction specified on the sea bottom segment EM, between the flow boundary condition (based on the assumption of hydrostatic pressure of seawater on the sea side of EM) and the transport boundary condition (based on the assumption of equality of mass fraction ($C < 1$)). As a consequence, a large error in the salt mass balance may occur in a numerical simulation, when such an inconsistency occurs in the specified flow and transport boundary conditions. To overcome this inconsistency, we may assume the presence of a buffer zone on the sea bottom, which contains outflowing water. We then take the density of this water into account when determining the head condition on this boundary. Iterations may be required in a numerical solution. The thickness of this buffer zone is a calibration parameter.

Altogether, the variable density flow and transport model presented above requires the simultaneous solution of two nonlinear equations. Obviously, no analytical solution is possible. A numerical solution is needed, using an appropriate computer program. A number of such programs have been developed, and reviewed in the literature (see Bear *et al.*, 1999, and Cheng and

Ouazar, 2004), such as FEFLOW (Diersch, 1988), MOCDENSE (Sanford and Konikow, 1985), SUTRA (Voss, 1984; Voss and Souza, 1987), SEAWAT (Langevin *et al.*, 2003, 2004), DSTRAM (Huyakorn *et al.*, 1987), CODESA-3D (Galeati *et al.*, 1992; Putti and Paniconi, 1995), and SWIFT (Ward *et al.*, 1984; Ma *et al.*, 1997).

Some codes applicable to seawater intrusion problems are briefly reviewed in Sec. 8.8. However, in this chapter, we shall mention two examples solved by using a computer code called FEAS (a Finite Element Aquifer Simulator) which has been successfully used for a number of practical modeling projects in the coastal aquifer in Israel (Zhou, 1999; Bensabat *et al.*, 2000; Zhou *et al.*, 2001; Zhou *et al.*, 2005).

9.3.2 Examples

Following are two examples of solutions of seawater intrusion problems as variable density flow and transport ones, using the FEAS code. The detailed discussions, the numerical model, and the computer program are discussed in Bear and Zhou (2006).

A. A typical case of seawater intrusion

This example involves the simulation of a seawater wedge under natural equilibrium conditions, and seawater intrusion under pumping conditions in a phreatic aquifer. The aquifer strip is 11 km long (AB in Fig. 9.3.2), and 1 km wide (perpendicular to the cross-sectional view). The aquifer bottom is of a constant slope of 0.8%, varying from 20 m below sea level at the land-side boundary (B in Fig. 9.3.2), to 100 m at the coast (below E), and further to 108 m at A. The sea bottom, EF, has a slope of 5%. Under natural equilibrium conditions, a freshwater flux of 0.015 m/d enters the aquifer at its land-side boundary, BC, and the rate of natural replenishment is 0.1 m/yr. The aquifer is homogeneous and isotropic, with a hydraulic conductivity of 30 m/d and a porosity of 0.2. The longitudinal dispersivity is 10 m and the transversal dispersivity is 1 m. Three wells, located along the centerline of the domain's width at distances of 5,000, 7,000, and 9,000 m from the coastline, pump, when active, a total of one third of the total natural replenishment.

The aquifer bottom (AB) is assumed to be impervious to both flow and solute transport. The vertical, sea-side boundary (AF) and the sea bottom (FE) are assumed to be at hydrostatic pressure imposed by seawater. For transport, the mass fraction of freshwater is specified at the land-side boundary (BC), and the mass fraction of seawater is specified at the vertical, sea-side boundary (AF). On the sea bottom (FE), a mixed-type boundary condition is specified. This type of condition is automatically switched from the first-type condition (9.3.26) to the second-type condition of zero dispersive flux, (9.3.28), depending on the direction of the normal groundwater flux.

To obtain the equilibrium situation without pumping, the model is run with the initial condition of an arbitrary phreatic surface, and with freshwater

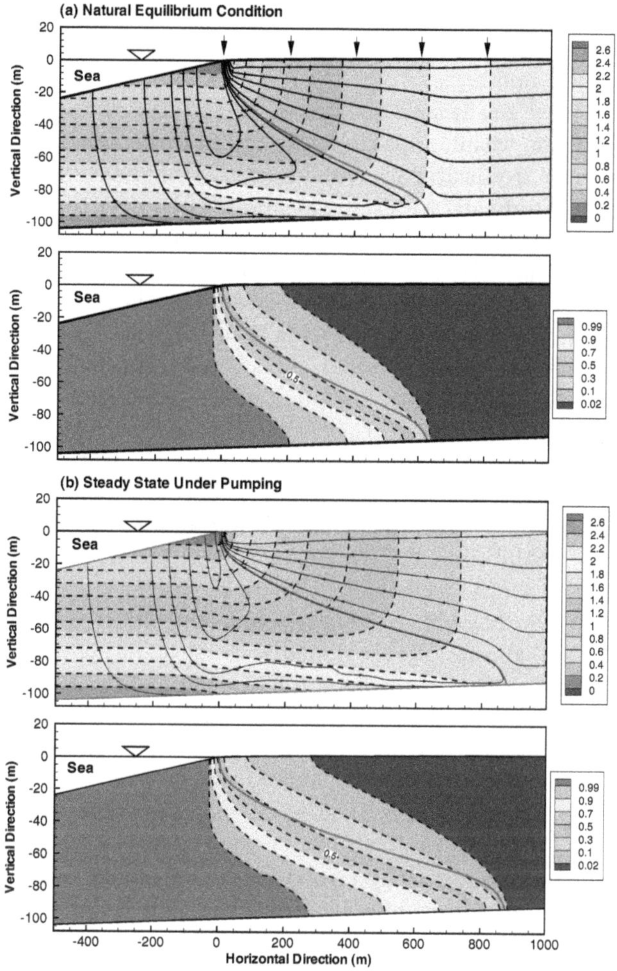

Figure 9.3.3: Numerical solution for the transition zone in a vertical cross-section of a coastal aquifer (Bear and Zhou, 2006). (a) Steady state (no pumping). (b) With pumping. The upper diagram shows the flow pattern with equipotential lines (solid lines) and streamlines (dashed lines). The lower diagram shows the contour lines of salt mass fraction. (Figure courtesy of Q. Zhou.)

concentration on the land side and the seawater concentration on the sea side, separated by a vertical sharp interface. This model is run until a steady state seawater wedge (= transition zone) is established. This steady state, shown in Fig. 9.3.3(a) can be used as initial conditions for any other runs with a non-stationary interface produced by pumping. The figure shows the flow pattern (upper diagram) and the salt mass fraction (lower diagram) at the cross section along the centerline. Note that these figures show only part of the investigated aquifer.

With the total freshwater discharged to the sea, the seawater wedge is formed in a region within approximately 610 m from the coast. The freshwater zone receives freshwater from the land-side boundary and from natural replenishment, producing a seaward freshwater flow over the seawater wedge. Within the seawater wedge, a *recirculating* seawater flow develops. Seawater fills the wedge primarily from the sea bottom, and is swept away back to the sea within the transition zone. As demonstrated by the streamlines, the freshwater and seawater flow originating from their respective zones, are separated (sharply) by a 'separation streamline', which is the envelope of all streamlines originating at the sea. The starting point of this separation streamline at the aquifer bottom is referred to as the 'toe' of the interface. This term is usually used in sharp interface models. At the toe of the separation streamline, the contrast in the horizontal hydraulic gradient between the seaward freshwater flow and the landward seawater flow is balanced, and the entire system is in a condition of equilibrium.

Unlike the 'sharp interface approximation' discussed earlier, in the more realistic model considered here, a rather wide transition zone occurs between the seawater and the freshwater domains, because of hydrodynamic dispersion. Across this zone, the salt mass fraction varies from that of seawater to that of freshwater. The shape and width of the transition zone depend on both the flow field and on (the velocity-dependent) dispersion. We note that the 50% salt mass fraction contour is not close to the separation streamline, which may be considered as equivalent to the *sharp interface* in the case of no dispersion.

The equilibrium condition described above is then taken as the initial conditions for the case with pumping. Figure 9.3.3(b) shows the steady state flow pattern (upper diagram) and salt mass fraction (lower diagram) for the case with pumping. As shown in the figure, by pumping freshwater far away from the coast, the elevations of the phreatic surface are lowered. The lowered freshwater hydraulic head in the seawater wedge disturbs the natural equilibrium conditions, inducing further seawater movement into the coastal aquifer. The seawater intrusion starts from the toe of the separation streamline at the aquifer bottom, where the reference head on the sea side of the seawater wedge is higher than on the inland side, lowered by pumping. The landward gradient of reference head leads to seawater intrusion. Once the seawater has advances landward, the seawater that invades the expanded wedge moves upward along the advancing separation streamline. The upward seawater movement results in a higher salt mass fraction in the overlying zone, above the already invaded zone. Meanwhile, dispersion also induces an increase in the salt mass fraction in this zone. Both advection and dispersion contribute to the widening of the intruding seawater wedge in the vertical direction. The higher salt mass fraction and wider transition zone gradually increases the reference head in the expanded intruding wedge. Meanwhile, the intruding wedge also changes the reference head of the freshwater zone in contact with it by reducing the cross-sectional area for freshwater flow to the sea, and by reducing the reference head immediately inland of the seawater wedge. As a result, a further landward movement of seawater occurs. The speed of seawater intrusion is fast immediately after the initiation of the disturbance caused by pumping. With the intruding seawater wedge becoming wider and extending farther landward, the landward gradient of the reference head on the sea side of the wedge becomes smaller, and seawater intrusion is slowed down. A new equilibrium is finally reached by the contrast in the reduced landward freshwater flow and the recirculating seawater flow.

As shown in Fig. 9.3.3(a) and (b), the seawater wedge changes from 610 m to 860 m from the coast by pumping a third of the total freshwater discharge. The water table at the land-side boundary is lowered from 10.0 m above the sea level to 6.4 m.

After shutting off pumping, the seawater wedge slowly retreats. However, the recovery process is relatively slow. In the case considered here, after 13 years, the transition zone is only approximately halfway between the initial equilibrium state under pumping conditions and that under natural equilibrium conditions without pumping.

B. Upconing

This example involves the simulation of regional seawater intrusion in a confined aquifer, combined with a local upconing of the transition zone. The aquifer strip is 4,000 m long perpendicular to the coast, with 1,000 m of it extending from the coast to the sea. It is 1,000 m wide parallel to the coast, and 200 m thick. The aquifer top and bottom are horizontal. The sea bottom has a slope of 5%. An inflow flux of 0.05 m/d takes place through the vertical land-side boundary, whereas the vertical sea-side boundary and the sea bottom are assumed to be at the hydrostatic pressure of seawater. For transport, the salt mass fraction is 0 and 1 at the land-side and sea-side boundary, respectively. Like in example A above, a mixed-type boundary condition is specified at the sea bottom. A hydraulic conductivity of 30 m/d and a porosity of 0.2 are used for this confined aquifer. The longitudinal and transversal dispersivities are 10 m and 1 m, respectively.

Figure 9.3.4 shows the result of the simulation, using FEAS (Bear and Zhou, 2006). Only a part of the total simulated domain is shown. In Fig. 9.3.4(b), the initial (without pumping) steady state seawater wedge is shown as the dashed line marking the 0.5 mass fraction. We observe a wide transition zone between the dashed lines marking the 0.1 and 0.9 mass fraction, and a toe located at about 1,080 m.

A pumping well is installed above the transition zone at $x = 706$ m, along the centerline of the width, and is screened from -24 m to -40 m below sea level. The well start to pump at a constant rate of 3,000 m^3/d. After 8.9 years of simulation, the resultant flow and concentration pattern, at the

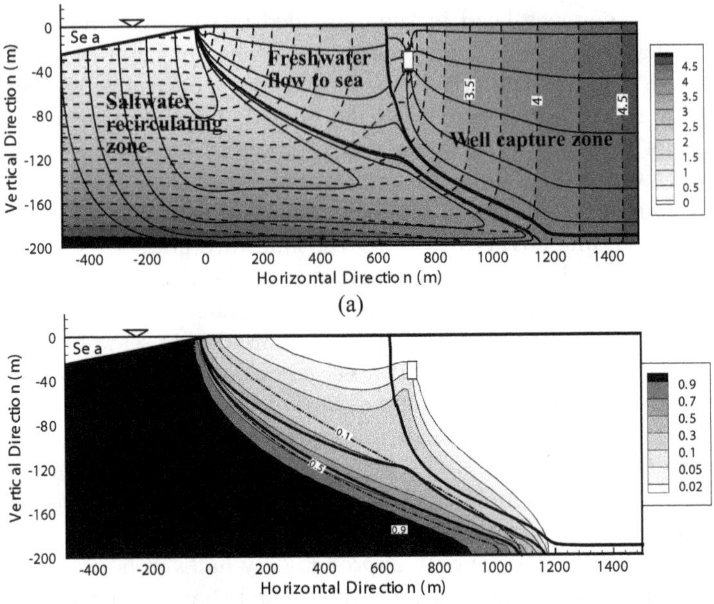

Figure 9.3.4: Transient seawater upconing in a confined aquifer at the end of 8.9 years (Bear and Zhou, 2006). (a) Dashed lines: contours of reference hydraulic head; solid lines: streamlines; dark solid lines: the streamlines separating the well capture zone, the freshwater outflow zone, and the saltwater recirculating zone. (b) Mass fraction contours (solid lines), in comparison with the initial steady-state transition zone (dashed lines). (Figure courtesy of Q. Zhou.)

vertical cross-section where the well is located, is shown as the contour lines in Fig. 9.3.4(a) and (b). A local seawater upconing can be clearly observed. In Fig. 9.3.4(a), the dashed lines are equipotential lines and solid lines are streamlines. By observing the streamline pattern, we can identify two separating streamlines, marked as thick solid lines, separating the flow domain into three zones. On the land side, we obverse a well *capture zone*; all streamlines within this zone end up in the pumping well. On the sea side, we find a saltwater recirculating zone; streamlines in this zone originate in the sea and flow back to the sea. In between, there is a freshwater outflow zone; freshwater flows underneath the capture zone and is drained to the sea.

Figure 9.3.4(b) shows the contour lines of constant mass fraction corresponding to the pumping case. These contour lines can be compared to the dashed lines, corresponding to the initial, unpumped case. For the upconing, we observe that the mixed water of low mass fractions upcones toward the pumping well more significantly than the rise of the higher mass fraction contours. During this stage, in the vicinity of the well, only mixed water is affected by pumping, whereas mixed water in the area close to the aquifer bottom and that close to the coast does not change significantly. This is because pumping above the transition zone lowers the hydraulic heads on its landside, but this does not change the balance between freshwater and seawater at the toe of the separation streamline. The enhanced seawater flow from the sea brings more seawater to the upconed seawater zone. This gradually increases the hydraulic head below the transition zone.

More details can be found in Zhou *et al.* (2005), and Bear and Zhou (2006).

9.4 Management of Coastal Aquifer

In principle, the approach and methodology for the management of a coastal aquifer are not different from those for any other aquifer (Chap. 11), whether as a stand-alone aquifer, or as an element of a water resources system. Like for the management of any aquifer system, (1) a policy should exist that dictates management goals, as well as the objective function and the technical and non-technical constraints that govern the selection of the optimal management scheme, and (2) a calibrated hydrogeological (flow and solute transport) model of the considered aquifer is required that describes flow and solute/contaminant transport phenomena that take place in it. This model will provide the response of the managed aquifer to any proposed management scheme, thus enabling the decision maker to evaluate and compare proposed management alternatives, in terms of the objective function, and to eliminate those schemes that violate (externally) imposed constraints. One may visualize the water mass balance equation and the solute mass balance equations, together with their appropriate boundary conditions as *constraints* that have to be satisfied continuously. An example of how these two models—the flow and transport model and the management model—are incorporated into a single optimization model is given by Bear (1979, p. 495), albeit not for a coastal aquifer.

Unfortunately, a *practical* management model that incorporates the hydrogeological flow and transport model(s) that describe seawater intrusion, as a constraint that has to be continuously satisfied, is still unavailable and trial and error approaches, as well as approaches based on comparison of alternative management schemes are employed. In the latter approach, a number of alternative management schemes are examined, e.g., with different volumes of annual withdrawal of water and different areal distribution of pumping and artificial recharge, and the models are used in order to forecast future water levels, aquifer salinity, salinity of pumped water, and other parameters of interest. Alternatives that violate imposed constraints are discarded. The alternative that maximizes or minimizes a selected objective function is selected as the optimal management scheme.

Like for any aquifer, calibration (see Subs. 11.3.2) will provide the natural replenishment, whether from precipitation, as inflow through aquifer boundaries, say, from adjacent aquifers, or as infiltration from streams and lakes. Knowledge of the natural replenishment is essential for determining the sustainable yield of the aquifer. In principle, unless imported water is added by artificial recharge techniques, in the long run, the sustainable yield cannot exceed the natural replenishment. In fact, it must be smaller than the replenishment, such that the outflow, i.e., the difference between replenishment and abstraction, flushes out to the sea salts and contaminants, which, otherwise, would accumulate in the aquifer.

The major difference between an inland aquifer and a coastal one stems from the latter's interaction with the sea. The presence of the sea constitutes a continuous threat to the water quality, e.g., salinity, of groundwater in the vicinity of the sea through the mechanism of seawater intrusion.

In earlier sections, we have presented two kinds of models that describe seawater intrusion:

- Models based on the assumption that the two 'kinds' of water, freshwater and seawater, are separated by a sharp interface.
- Models that regard the two 'kinds' of water as a single liquid phase, with different concentrations of total dissolved salts (TDS), and with a transition zone that separates the domain occupied only by freshwater and the one occupied only by seawater.

Nowadays, we have a number of good (and practical) computer codes that enables the solution of transition zone models (see Sec. 8.8), and, unless there exist special circumstances that justify the use of a sharp interface model, transition zone models should be preferred.

The sustainable management of a coastal aquifer is based on understanding the strong relationship that exists between the discharge of freshwater from the aquifer to the sea and the extent of seawater intrusion, associated with water levels near the coast. Good management seeks to determine the optimal sustainable yield of the aquifer such that the interface (i.e., the transition zone) will not advance beyond a desired distance. The objective is to avoid salinization of existing and future wells, and not 'preventing seawater intrusion'.

A coastal aquifer may be in one of following two situations:

- **Virgin, or initial developing conditions**. Pumping and seawater intrusion take place, but no wells have been salinized (yet). We wish to continue to operate the aquifer and increase water extraction (say, by increasing pumping in existing wells, by moving wells to new locations, and/or by adding new wells) up to its sustainable yield.
- **Excessive seawater intrusion conditions**. Withdrawal has already exceeded sustainable yield, and significant seawater intrusion takes place. Some wells have already been salinized. We wish to restore the aquifer to its sustainable yield conditions.

In the first case, withdrawal can be increased up to the estimated sustainable yield. In fact, for a number of years, the withdrawal may even be allowed to exceed the ultimate sustainable yield (obviously, if consumers are available), in order to make beneficial use of the fact that the interface is advancing landward. We often refer to this excess pumping as 'pumping the one-time-reserve'. In the second case, withdrawal has to be reduced. To speed up restoration, withdrawal may at first, and for some years, be reduced to below sustainable yield, and/or increasing artificial recharge. We have to take into account two important factors characterizing the restoration process:

- *Changes in the position of the interface are slow*, so that the restoration to sustainable yield, with an optimal position of the interface, may take many years.
- Furthermore, since, in reality, no sharp interface exists, as the transition zone retreats seaward, it tends to widen.
- It may take a very long period, e.g., tens of years, to really flush the saline water and reduce water salinity near the coast to desirable levels.

Since the extent of seawater intrusion depends on the rate at which aquifer freshwater is allowed to drain to the sea, by maintaining appropriate water table elevations at a certain distance from the sea, often called a 'freshwater mound', we can arrest the intruding seawater wedge at a desired (optimal) distance from the sea. Beyond the 'freshwater mound', we can manage the aquifer with water levels fluctuating above and even below sea level. The 'mound' can be maintained by controlling pumping and/or by artificial recharge through infiltration basins, or through an array of wells along a line parallel to the coast. Artificial recharge can be implemented with imported water, and even with treated sewage (obviously, taking into account that part of the water, if not intercepted, will also flow landward).

Various 'engineering' techniques have also been proposed from time to time for controlling, and even preventing seawater intrusion. Among them, we can mention a 'water table trough' parallel to the coast, produced by pumping through an array of wells in saltwater zone, and an impervious or semipervious barrier constructed by various geotechnical techniques, such as (Johnson and Whitaker, 2004) slurry walls, deep soil mixing, and jet grouting.

Once optimal pumping has been determined for a segment of a coastal aquifer, the rate of freshwater drained to the sea can be further reduced by installing a 'coastal collector', also known as skimming wells (Wang, 1965; Reilly and Goodman, 1987; Aliewi *et al.*, 2001; Asghar *et al.*, 2002; Saeed and Ashraf, 2005; Rao *et al.*, 2006; Saeed, 2008). This is an array of low capacity wells that are installed along a line parallel to the coast, at a rather small distance from it, with screens that are sufficiently high above the interface zone. These wells intercept part of the freshwater that would otherwise flow to the sea (see Subs. 9.2.3 and Fig. 9.2.5a). If these wells are sufficiently close to each other, then they created a two-dimensional flow pattern that is similar to a gallery, as shown in Fig. 9.2.5d.

As mentioned in the preamble to this subsection, many coastal aquifers are contaminated from ground surface by point and distributed sources. Furthermore, groundwater in coastal aquifers, especially close to the coast, is, to some extent, saline, because of airborne salts originating in the sea. Altogether, aquifer water draining to the sea prevent the accumulation of salts and other contaminants in the aquifer. The management of the aquifer should take this aspect (of flushing of contaminants) into account.

To summarize, the management of a coastal aquifer requires a calibrated model of the hydrogeology of the aquifer, preferably a three-dimensional variable density flow and transport model. The selected management scheme should be based on an objective function and constraints dictated by a prevailing water policy, established within the framework of some system that is higher in the hierarchy, e.g., the regional or national one. The basic feature of management of a coastal aquifer is the determination of a sustainable yield that will arrest the seawater wedge, or transition zone, at a desired distance from the sea by allowing a certain discharge of freshwater to be drained to the sea.

Chapter 10

MODELING UNDER UNCERTAINTY

Up to this point, all the modeling efforts presented in this book have been based on the *deterministic approach*. This approach assumes (1) a full knowledge and understanding of the physical/chemical processes that occur within an investigated domain, (2) a mathematical model that correctly represents these processes and can be used to predict future responses to imposed excitations, (3) the availability of information (measured or derived) on all parameters and coefficients that appear in the mathematical model, and, on the domain's geometry, as well as on the initial and boundary conditions. This situation is, however, far from that encountered in practice. In the real world, we need to address the following questions:

Model uncertainty. Is the selected conceptual model appropriate for representing, albeit to an acceptable degree of approximation, the processes that take place within the problem domain? For example, in view of the geological structure of the aquifer, was our decision to model it as a two-dimensional domain justified? Or, is the domain really homogeneous, as assumed? Or, is it correct to assume vertical flow only in the unsaturated zone?

Parameter uncertainty. Unlike processes in man-manufactured systems, there always exist numerous unknown factors that affect phenomena and processes in nature, e.g., geological and hydrological processes. This is particularly true when we are interested in the quantitative aspects of such phenomena. Most of the uncertainty associated with modeling may be attributed to the enormous heterogeneity of subsurface domains. This is, certainly, true for large aquifers, but also, and very much so, for relatively small domains in the unsaturated zone. This heterogeneity in lithology, soil types, etc., say, as manifested by heterogeneity in permeability and porosity values, may cover a range of many orders of magnitude within a domain. Not recognizing its existence, or not attempting to account for its effects in the modeling process, may lead to large errors of prediction. Some heterogeneity may be identified from logs of boreholes, or by geophysical methods. In most cases, however, the data are too scarce and too sparse to yield a deterministic description of the domain's various properties. Hence, it is essential to address the uncertainty involved in such modeling conditions. This is particularly so when the

J. Bear, A.H.-D. Cheng, *Modeling Groundwater Flow and Contaminant Transport*, Theory and Applications of Transport in Porous Media 23,
DOI 10.1007/978-1-4020-6682-5_10, © Springer Science+Business Media B.V. 2010

distances between monitoring locations are much larger than the distance at which the relevant properties are still *correlated*. A similar situation exists when the coefficients are unknown, and have to be determined by an inverse (= parameter estimation) method (Sec. 11.3), especially when the locations at which the piezometric heads are observed are too sparse and cannot provide the details of the field's heterogeneity.

Boundary uncertainty. Are the selected domain's boundaries and the associated boundary conditions appropriate? Usually, the exact location of certain underground features, e.g., a fault, or a water divide, that serve as boundaries are uncertain. When conditions on boundaries are not available, the boundary location is taken at a large distance from the domain of interest, in hope that the influence of the conditions assumed to prevail on them on the predictions within the domain of interest will be small. Nevertheless, this is a potential source of errors.

Initial conditions uncertainty. The initial conditions within the modeled domain may not be known. Sometimes, they are determined by running a simulation based on the *pre-development* conditions until a steady state is reached; the latter is then used as initial condition for the period of interest. Or, initial conditions may be generated from a groundwater contour map from an early day survey.

In general, we may classify the uncertainties associated with modeling into two types. The first is *aleatoric uncertainty* (or *intrinsic uncertainty*), associated with the inherent unpredictability of nature (Ross *et al.*, 2009). Most hydrological events, such as precipitation and streamflow, cannot be predicted, or, at least, cannot be predicted beyond a certain time into the future, due not only to their *complexity*, but also to the inherent *instability* of the physical processes. When such uncertain hydrological data are used as part of the input to a deterministic model, the output will also be uncertain.

The second type is *epistemic uncertainty* (or *information uncertainty*). Here, we assume that we know how to describe the system's behavior, making use of a set of non-random parameters and coefficients. However, we lack the complete knowledge of the values of these parameters and their spatial distribution, thus making our prediction uncertain. In this case, the degree of uncertainty can be reduced by spending more time and efforts (and the associated costs) on gathering more information, to be used in the calibration process. The decision to gather such information is a managerial one, based on a cost vs. benefit analysis.

Despite their importance, uncertainty issues are often insufficiently addressed in groundwater modeling. Typically, general groundwater textbooks do not cover these issues, although a number of books dedicated to uncertainty and stochastic modeling are available (e.g., Dagan, 1989; Dagan and Neuman, 1997; Gelhar, 1993; Zhang, 2002; Rubin, 2003). Hence, this chapter is dedicated to the introduction of some fundamental issues of modeling groundwater flow and transport under conditions of uncertainty, and to the various tools used for their analysis.

Ideally, for a stochastic analysis technique to be successful, it needs to have the following characteristics:

- The resultant methodology should be computationally tractable.
- The stochastic analysis should indicate how to reduce uncertainty by collecting and using more measurements and observed data to be used in the calibration process.
- It should account for different types of errors, including errors in the data used for model development, errors in estimated parameters, and errors in the conceptual model.
- It should be generally applicable, and not be limited to rarely occurring conditions and assumptions.
- It should provide prediction of uncertainty, and the measures of uncertainty should be understandable to nonspecialists, such as policy makers.

We shall begin this chapter by defining the various statistical measures of a stochastic process and their use as tools in uncertainty analysis. Next, we shall discuss the Monte Carlo technique, which uses a deterministic model to analyze a large number of samples in the probabilistic space. To provide the multiple realizations of the modeled field required in the Monte Carlo simulation, the random field generation technique is presented. The kriging method is introduced for generating the complete information on the parameters of a modeled field, when the values of these parameters are known only at a number of sampled locations. This method provides the best estimate of missing data. Other techniques for stochastic analysis, such as the solution of stochastic differential equations and the perturbation technique, are introduced as tools for solving problems with uncertain boundary conditions or uncertain parameters.

10.1 Stochastic Processes

10.1.1 Random process

Briefly, a *random process* is a process for which we cannot predict the outcome of an *experiment* (a trial of the process), prior to performing it. This is contrary to a *deterministic process*, whose outcome can be controlled, hence, predicted. The result of a random process, or experiment, can be quantitative, such as integers or real numbers, or qualitative, such as 'head' or 'tail' of a coin. To render the result amenable to a mathematical analysis, the qualitative outcomes are often mapped onto numbers, e.g., 0 and 1, respectively, for head and tail.

We may argue that all physical phenomena, such as tossing a coin or throwing a dice, are controlled by physical laws, such as Newton's law of motion, and, therefore, are predictable. In reality, however, physical processes can be too complex, or too unstable, such that the predicted outcome can deviate significantly from the actual outcome. In other words, the tolerance range for the outcomes, subject to any change of initial condition or in the environment, is extremely small. As the sources of these disturbances cannot be eliminated, their influence may grow, sometimes exponentially, throughout the process, leading to unpredictable outcomes. This is indeed the situation for throwing a dice, and for many hydrological and weather related processes.

When the outcome of a random process is expressed in numbers, we call it a *random variable*, which can be *discrete*, such as the integers representing the outcomes of the coin and dice example, or *continuous*, such as the positive real numbers. The collection of all the possible outcomes of an experiment is called a (statistical) *population*, or *sample space*. For example, the set of numbers $\{1, 2, 3, 4, 5, 6\}$ is the population of the outcomes of throwing a dice. A *sample* is a selection from the population, or an execution of an experiment that produces a (random) outcome.

For example, we can express the discharge of a stream at a given cross-section as a function of time, $Q = Q(t)$. However, the discharge at a fixed future time, say $t = t_1$, as $Q(t_1)$, cannot be controlled by any action at the present. Hence, this discharge is unpredictable. The quantity $Q(t_1)$ is, thus, a random variable. As $Q(t_1)$ can be any value on the positive real number axis, its population consists of points along the positive real axis, $[0, \infty]$. It is also a continuous random variable.

Often, our interest lies in the prediction of a continuous record, rather than a single number. For example, we may be interested in the discharge at a specified location as a function of time, $Q = Q(t)$. Then $Q(t)$, being a *function* and being random, is a *random function.*

A *stochastic process* is *a random process evolving in time* (Lawler, 2006). It can be made of a set of indexed variables, such as $Q_1 = Q(t_1)$, $Q_2 = Q(t_2), \ldots$, or a continuous (random) function, $Q(t)$. A stochastic process need not to be a function of time; it can be a function of a spatial variable, say x. Furthermore, it can be a function of multiple spatial variables, e.g., (x, y, z) in a multi-dimensional space. In the subsections to follow, we shall describe the modeling of hydrogeological parameters, such as the hydraulic conductivity K, as a multi-dimensional stochastic process, $\mathsf{K}(x, y, z)$.

10.1.2 Quantifying uncertainty as stochastic process

Uncertainty in groundwater modeling is caused, mostly, by the heterogeneity in aquifer properties, primarily, permeability and porosity. In most practical cases, there are never enough data to describe the spatial distributions of these properties in sufficient detail, and interpolation is used to fill in missing data. Hence, the interpolated field is characterized by a high degree of uncertainty. For example, if we examine the variation of a certain aquifer parameter, say K, as a function of x, within a spatial domain, we may be faced with profiles like those shown as Fig. 10.1.1a and b. Not knowing the

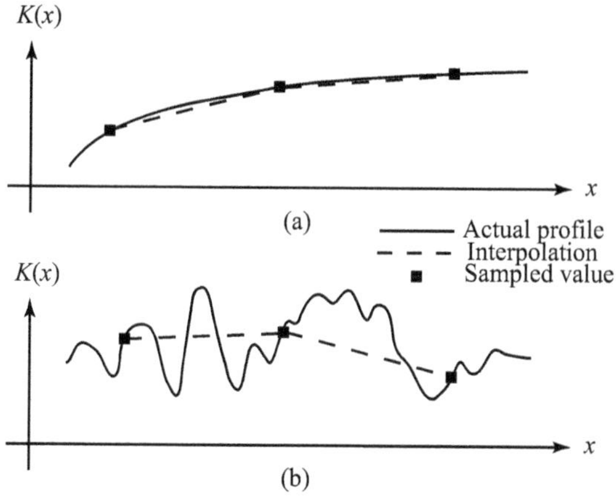

Figure 10.1.1: Actual and interpolated parameter profiles, based on sparsely sampled data: (a) Smooth profile (predictable); (b) Highly heterogeneous, random-like profile (unpredictable).

actual variation, and (as is usually the case) constrained by an operational budget, we have sampled three locations for the K-values, To approximate K at the unsampled locations, we may use interpolation, shown as the dash lines in Fig. 10.1.1. We notice in Fig. 10.1.1a that the values at the unsampled locations are quite predictable, with only small errors. On the other hand, the profile in Fig. 10.1.1b cannot be predicted from the small number of available sampled points. For this case, if we observe the parameter at one point, our ability to tell the parameter value at a distance away deteriorates very fast, as the distance increases. Hence, as far as our knowledge about this parameter is concerned, it behaves as an unpredictable event, i.e., as a random event. Aquifer properties are more likely to be represented by the profile shown in Fig. 10.1.1b. This random profile is similar to the recorded data of other random processes in nature, e.g., streamflow, turbulent air flow, and seismic waves. Hence, similar mathematical techniques can be used for its analysis.

One way to quantify uncertainty, whether intrinsic or informational, associated with a phenomenon, or a process, is to envision it as a *random process*, also called a *stochastic process*. We shall use Fig. 10.1.2 to illustrate this concept.

Consider streamflow discharge at a given cross-section as a *random function* of time, $Q(t)$, where t denotes (future) time. As we do not know what will the actual value of $Q(t)$ be, this function can be any of the profiles presented as $Q_1(t)$, $Q_2(t)$, $Q_3(t)$, etc., in Fig. 10.1.2. Each function may equally likely be the true $Q(t)$, and is a *sample* in the probability sampling space. Such sample is also called a *realization*. All possible samples, or realizations,

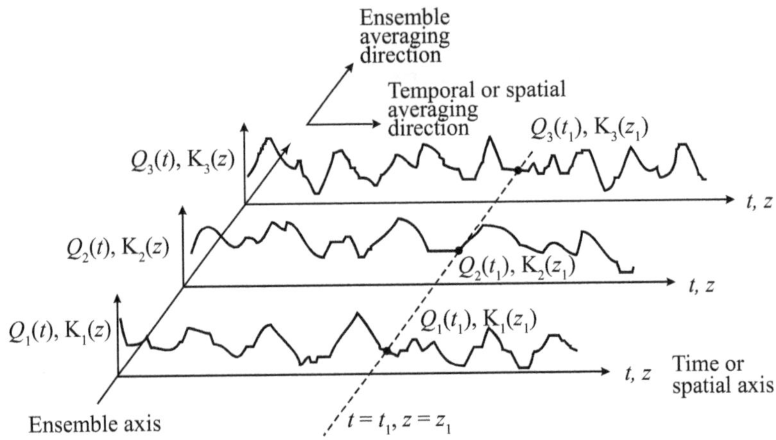

Figure 10.1.2: Three samples from the ensemble of the random function $Q(t)$ or $\mathsf{K}(z)$.

and there are infinite many of them, form the *population* of the probability space. This space, shown in Fig. 10.1.2, is also called the *ensemble space.*

Let us now consider the vertical hydraulic conductivity profile, $\mathsf{K}(z)$, at a given location, (x, y). This profile can look like those presented in Fig. 10.1.2. Unlike the case of streamflow discharge, in which we consider a future event, which is unpredictable due to the random nature of the process, the hydraulic conductivity *does exist*, irrespective of whether the process that created it was random or deterministic. Nevertheless, the profile is regarded as unpredictable due to our lack of information about it. Hence, the hydraulic conductivity profile is a random (stochastic) process just as streamflow.

A stochastic process can be a continuous random function, or a sequence of *indexed*, discrete data. For example, we can express the function $\mathsf{K}(z)$ as a sequence of discrete values, $\mathsf{K}_1 = \mathsf{K}(z_1)$, $\mathsf{K}_2 = \mathsf{K}(z_2)$, $\mathsf{K}_3 = \mathsf{K}(z_3)$, etc. A common and important characteristic of stochastic processes is that the sequence of data is *correlated* to each other. In other words, each datum in the sequence depends to some extent on the one preceding it, with a certain probability that their values are close to each other. For, example, if the preceding datum is in the large value range, then there is a high probability that the ensuing datum will take on a large value too, though not necessarily so.

Statistical analysis can be applied to the realizations of the data sequence, or a random function, in order to extract useful information about the process, such as expectation, standard deviation, covariance, correlation coefficient, correlation length, reliability, confidence limit, etc. Such quantities, computed from the realizations, are often referred to as 'empirical', because the available realizations are, typically, a small sample of the entire population space. In what follows, we shall present a brief introduction to these concepts.

10.1.3 Ensemble statistics

We shall define a few statistical concepts and quantities that are essential to the understanding of stochastic analysis.

Ensemble. The term 'ensemble', or 'statistical ensemble' originated in statistical mechanics and thermodynamics. There, it defines an idealization consisting of a large number (sometimes infinitely many) of 'mental copies' of a system, considered all at once, each of which representing a possible state (= realization) that the real system *might be* in. On this ensemble, we can perform a statistical analysis and find empirical *statistical measures*, such as *mean* and *standard deviation.*

Ensemble average of a random variable. Figure 10.1.2 shows a few realizations of hydraulic conductivity profiles. The spatial distribution of this coefficient may be regarded as a random function of depth, or a stochastic process, and the figure shows a number of realizations of such process.

Focusing our attention on the value at a fixed depth, say z_1, then $\mathsf{K}(z_1)$ is a random variable. There exist many samples (realizations) of the random variable, $\{\mathsf{K}_1(z_1), \mathsf{K}_2(z_1), \ldots, \mathsf{K}_n(z_1)\}$ (see Fig. 10.1.2). If each of these values has an equal probability of occurring, or having actually occurred, we can find the *mean* (*expectation*) of the random variable $\mathsf{K}(z_1)$ as the *ensemble average*

$$E\left[\mathsf{K}(z_1)\right] \equiv \overline{\mathsf{K}(z_1)} = \frac{1}{n}\sum_{i=1}^{n}\mathsf{K}_i(z_1), \tag{10.1.1}$$

where n is the number of samples.

Ensemble average of a random function. We are, however, more interested in the *random function* $\mathsf{K}(z)$, than in the random variable $\mathsf{K}(z_1)$. Thus, the mean of a random function can be formally defined as

$$E\left[\mathsf{K}(z)\right] \equiv \overline{\mathsf{K}}(z) = \frac{1}{n}\sum_{i=1}^{n}\mathsf{K}_i(z). \tag{10.1.2}$$

As these functions are likely to be obtained from measurements and, thus, presented in discrete rather than analytical form, the average can be viewed as taken one z-value at a time: $z_1, z_2, \ldots$, leading to a similar operation as described by (10.1.1). We emphasize that the mean of a function of a spatial variable is, generally, a function of the same spatial variable, as indicated by the argument of $\overline{\mathsf{K}}(z)$.

Statistical homogeneity. If, after ensemble averaging, the mean turns out not to be a function of space, but $\overline{\mathsf{K}(z)}$ = constant, then the process is referred to as *homogeneous*, or *statistically homogeneous*, as long as there

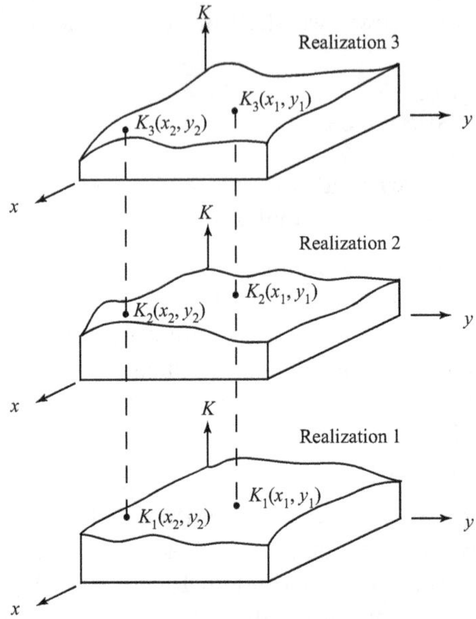

Figure 10.1.3: Ensemble averaging of a two-dimensional random function.

is no danger of confusion with the usage of the same term in the deterministic sense. In the deterministic case, which may be considered as the case with a single realization of the entire population, 'homogeneous' means $\mathsf{K}(z) = \mathsf{K} =$ constant. It is important to understand that in the statistically homogeneous case, each realization of the hydraulic conductivity profile is, generally, *heterogeneous*. The terminology used above for the random spatial functions can be compared to the random time functions: when the mean is not a function of time, the random process is said to have a *stationary* (corresponding to homogeneous) mean.

Ensemble average of a multivariate function. The ensemble average applies to *multivariate functions*, e.g., $\mathsf{K}(x, y, z)$, as well as to *univariate* ones, e.g., $\mathsf{K}(z)$. For example, the mean defined in (10.1.2) can be extended to

$$E\left[\mathsf{K}(x, y, z)\right] \equiv \overline{\mathsf{K}}(x, y, z) = \frac{1}{n}\sum_{i=1}^{n} \mathsf{K}_i(x, y, z), \tag{10.1.3}$$

i.e., taken with respect to the many realizations of the function.

Figure 10.1.3 illustrates the process of taking an ensemble average of a two-dimensional random function, $\mathsf{K}(x, y)$. As it is difficult to take the average of a function, it is often taken one discrete point, e.g., (x_1, y_1), (x_2, y_2), etc., at a time. This is illustrated on the figure.

Variance. The *variance* is defined as

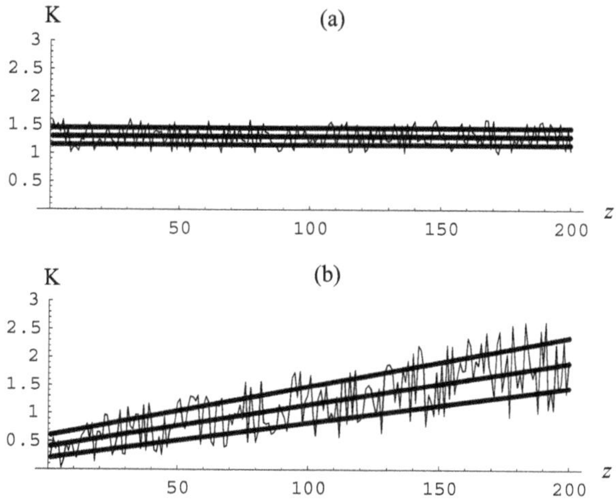

Figure 10.1.4: One-dimensional spatial random functions with mean ± standard deviation: (a) With homogeneous mean and homogeneous standard deviation; (b) Inhomogeneous mean and inhomogeneous standard deviation.

$$\sigma_K^2(\mathbf{x}) \equiv \overline{[\mathsf{K}(\mathbf{x}) - \overline{\mathsf{K}}(\mathbf{x})]^2} = \overline{\mathsf{K}^2(\mathbf{x})} - \overline{\mathsf{K}(\mathbf{x})}^2, \tag{10.1.4}$$

in which the vector $\mathbf{x}$ denotes the point (x, y, z), the overbar denotes an ensemble averaging operation, like the one defined in (10.1.3), and the subscript K indicates that this variance is of the variable K. We note that, generally, the variance is also a function of space. For the special case of $\sigma_K^2(\mathbf{x}) = \sigma_K^2 =$ cosntant, the variance is homogeneous (i.e., stationary).

Standard deviation. The square root of the variance, denoted as $\sigma_K(\mathbf{x})$, is called *standard deviation.* It is a measure of the average magnitude of the deviations from the mean (= average). Figure 10.1.4 shows two one-dimensional spatial random functions with their mean and plus/minus standard deviation envelopes.

If we assume that the probability distribution of a random variable is Gaussian (see any book on statistics), then these two envelopes enclose about 68% of all possible values. We notice that the random function in the upper diagram has a homogeneous (= stationary) mean that remains constant with distance, while the lower function has an inhomogeneous (= nonstationary) mean that varies with distance. We also observe that the upper diagram has a homogeneous standard deviation, while the lower one has an inhomogeneous standard deviation.

Covariance and correlation. Another important statistical measure is the *covariance* between two variables. For example, the hydraulic conductivity, K, at one location and the storativity, S, at another one may be correlated to each other. The covariance is defined as

$$\begin{aligned} c_{KS}(\mathbf{x}_1, \mathbf{x}_2) &\equiv \overline{[\mathsf{K}(\mathbf{x}_1) - \overline{\mathsf{K}}(\mathbf{x}_1)][S(\mathbf{x}_2) - \overline{S}(\mathbf{x}_2)]} \\ &= \overline{\mathsf{K}(\mathbf{x}_1)S(\mathbf{x}_2)} - \overline{\mathsf{K}(\mathbf{x}_1)}\;\overline{S(\mathbf{x}_2)}, \end{aligned} \tag{10.1.5}$$

where $\mathbf{x}_1$ and $\mathbf{x}_2$ indicate two different locations.

Let us consider the trend of the two quantities, $\mathsf{K}(\mathbf{x}_1)$ and $S(\mathbf{x}_2)$, where $\mathbf{x}_1$ and $\mathbf{x}_2$ are any pair of points. When the observed $\mathsf{K}(\mathbf{x}_1)$-value is larger than its mean, is there, in the same realization, a similar or opposite trend in $S(\mathbf{x}_2)$? If the trends of these two quantities are frequently consistent (or oppositely consistent), then the covariance takes a large positive (or large negative) value. If there is no consistent trend between the two observed quantities, i.e., when one quantity is above average, the other has an equal chance of being above or below the average, then the summation of terms under the ensemble average operation tends to cancel out, and the covariance is small or zero. A similar quantity, known as the *correlation*, is defined as

$$r_{KS}(\mathbf{x}_1, \mathbf{x}_2) = \overline{\mathsf{K}(\mathbf{x}_1)S(\mathbf{x}_2)}. \tag{10.1.6}$$

Autocovariance and autocorrelation. Given a spatial function, it is also of interest to find the correlation between the *same* quantity as observed at two different locations, $\mathbf{x}_1$ and $\mathbf{x}_2$. To achieve this goal, we use (10.1.5), and define

$$c_{KK}(\mathbf{x}_1, \mathbf{x}_2) = \overline{\mathsf{K}(\mathbf{x}_1)\mathsf{K}(\mathbf{x}_2)} - \overline{\mathsf{K}(\mathbf{x}_1)}\;\overline{\mathsf{K}(\mathbf{x}_2)}, \tag{10.1.7}$$

as *autocovariance* of $\mathsf{K}(\mathbf{x})$. Similarly, we can define an *autocorrelation* as r_{KK}.

We notice that in a inhomogeneous case, the covariance and autocovariance are functions of the relevant pair of points in space. Even in the discrete case, the number of such values is huge. For example, on a three-dimensional $n \times n \times n$ grid, the number of discrete autocovariances is roughly $n^6/2$. It is obviously impossible to process the available data in order to generate such information. Hence, for most practical applications, the covariance is always considered to be homogeneous.

For a homogeneous model, we assume that the covariance is not a function of a pair of locations; instead, we assume that it is a function of the distance vector $\boldsymbol{\chi}(= \mathbf{x}_2 - \mathbf{x}_1)$, separating these two locations, . Expressing the two points as $\mathbf{x}$ and $\mathbf{x} + \boldsymbol{\chi}$, we can write

$$c_{KK}(\mathbf{x}, \mathbf{x} + \boldsymbol{\chi}) \equiv \overline{\mathsf{K}(\mathbf{x})\mathsf{K}(\mathbf{x} + \boldsymbol{\chi})} - \overline{\mathsf{K}(\mathbf{x})}\;\overline{\mathsf{K}(\mathbf{x} + \boldsymbol{\chi})} = c_{KK}(\boldsymbol{\chi}). \tag{10.1.8}$$

We note that $c_{KK}(\boldsymbol{\chi})$ is not a function of the location $\mathbf{x}$, but is still a function of the distance vector, $\boldsymbol{\chi}$, between the two points. This means that the covariance is not only a function of the *distance* separating the two points, but also of the vector's *direction*. This implies that the covariance field may be *statistically anisotropic*. For a *statistically isotropic* field, the covariance function, $c_{KK}(\boldsymbol{\chi})$ further simplifies to $c_{KK}(\chi)$, where $\chi = |\boldsymbol{\chi}|$. Here, the covariance is a function of the single (scalar) variable χ, i.e., the distance separating the two points. It is easy to construct this parameter, even with a limited number of available spatial data points. In Subs. 10.2.1, within the framework of the *kriging* technique, we shall discuss the empirical construction of the covariance function.

When the covariance is determined for a single location, $\mathbf{x}$, i.e., $\chi = 0$, we use the notation $c_{KS}(\mathbf{x}, \mathbf{x}) \equiv \sigma_{KS}(\mathbf{x})$. Furthermore, if the two quantities are the same quantity, the covariance definition becomes identical to that of the variance, i.e., $\sigma_{KK}(\mathbf{x}) \equiv \sigma_K^2(\mathbf{x})$.

Weakly homogeneous (stationary) process. A random process is said to be *strongly homogeneous* (or *strongly stationary*) if *all* its statistical moments are homogeneous. In practice, however, due to our limitation in observing and recording a sufficient quantity of random data, we are often limited to the evaluation of only the first two moments. i.e., the mean and the standard deviation. If homogeneous conditions exist for these two moments, they become independent of location, and the autocovariance is a function of the separating distance only. The random process is then referred to as *weakly homogeneous* (or *weakly stationary*).

Correlation coefficient. The covariance is a dimensional quantity; hence, it provides an absolute rather than a relative magnitude of the variation. For a measure that shows a relative magnitude, a *cross correlation coefficient* can be defined by normalizing the covariance:

$$R_{KS}(\mathbf{x}_1, \mathbf{x}_2) = \frac{c_{KS}(\mathbf{x}_1, \mathbf{x}_2)}{\sqrt{\sigma_K^2(\mathbf{x}_1)\, \sigma_S^2(\mathbf{x}_2)}}. \qquad (10.1.9)$$

When the two related quantities are identical, except that they are taken at different locations, we introduce the *autocorrelation coefficient*, $R_{KK}(\mathbf{x}_1, \mathbf{x}_2)$. For a (statistically) homogeneous and isotropic case, it reduces to $R_{KK}(\chi)$. We notice that for a zero separation distance, i.e., $\chi = 0$, the autocorrelation coefficient is $R_{KK}(0) = 1$. This means that the normalized correlation between a quantity and itself, at the same location, is unity. Also, the following bounds exist: $-1 \leq R_{KK}(\chi) \leq 1$.

Correlation length. For most natural phenomena, the autocovariance approaches zero as the separating distance between the two locations, at which the quantity is measured, becomes larger. The distance for which $R_{KK}(\chi)$ drops to an insignificant magnitude is referred to as the *correlation length scale* (or, for a function of time, *correlation time scale*). More rigorously, these scales can be defined in terms of an *integral scale* (Dagan, 1989),

$$I = \int_0^\infty R_{KK}(\chi)\, d\chi. \qquad (10.1.10)$$

Figure 10.1.5 shows two autocorrelation coefficients, one with a small, and the other with a large correlation length. When the correlation length is larger, we anticipate the data profile to be smoother, and vice versa.

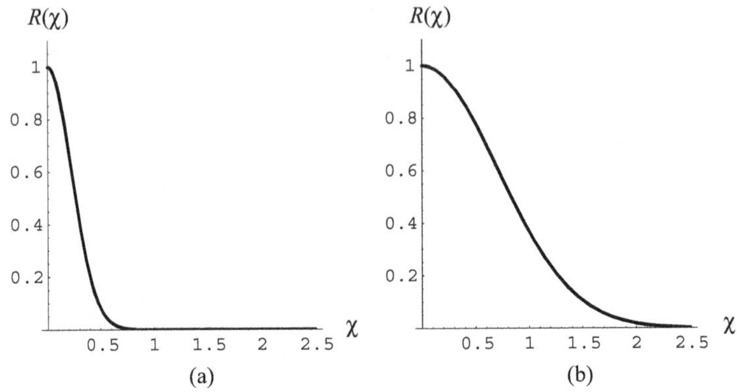

Figure 10.1.5: Correlation coefficient: (a) With a small correlation length ($I = 0.28$), (b) With a large correlation length ($I = 0.89$).

Figure 10.1.4, demonstrates the effect of different mean and standard deviation on the hydraulic conductivity profiles. Figure 10.1.6, on the other hand, compares two hydraulic conductivity profiles with the same mean and standard deviation, but with different correlation lengths. The profile with the smaller correlation length is rougher, while that with the larger correlation length is smoother.

10.1.4 Spatial (or temporal) statistics

The *spatial statistics* (or *temporal statistics* for a function of time) differs from the *ensemble statistics*, presented in the preceding subsection, in a most important way—the former is performed on a single realization.

Spatial mean. Given a single realization of a random function in time, or in a one-dimensional space (see Fig. 10.1.2), the *spatial mean* (or *temporal mean* for a time function), $\langle \mathsf{K} \rangle(z)$, of a random variable $\mathsf{K}(z)$, is defined by

$$\langle \mathsf{K} \rangle(z) = \frac{1}{L} \int_{z-L/2}^{z+L/2} \mathsf{K}(z')\, dz', \tag{10.1.11}$$

where the brackets $\langle\ \rangle$ are used to denote the *spatial average*, and L is a sufficiently large length, such that the average becomes independent of the selection of L, yet small enough, such that the average may still vary with distance in the inhomogeneous case (see Fig. 10.1.4b). We note that the criterion for the selection of the averaging 'cell'—not too small such that the result becomes dependent on the size of the cell, and not too large in order not to lose the large scale trend—bears some resemblance to the criterion for selecting the size of the REV, discussed in Subs. 1.3.3.

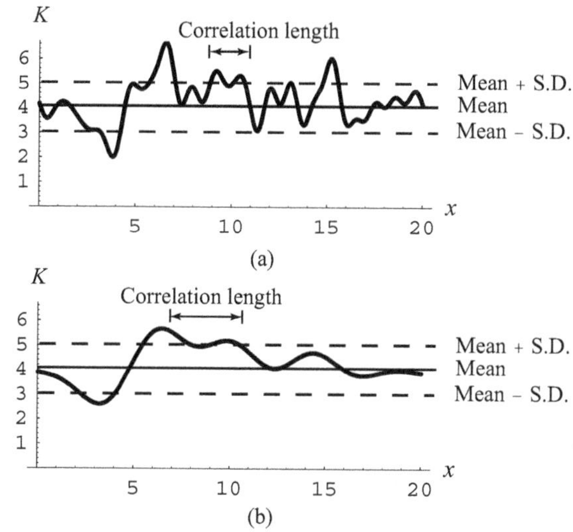

Figure 10.1.6: Two hydraulic conductivity logs with the same homogeneous mean and standard deviation, but different correlation lengths.

For a random function in a three-dimensional space, the above definition is modified to a volume, rather than line, averaging,

$$\langle \mathsf{K}\rangle(\mathbf{x}) = \frac{1}{L^3}\int_{\mathcal{U}_L(\mathbf{x})} \mathsf{K}(\mathbf{x}')\, d\mathcal{U}(\mathbf{x}'). \tag{10.1.12}$$

In the above equation, $\mathcal{U}_L(\mathbf{x})$ represents a cube with a side L, centered at $\mathbf{x}$. Similarly, we can apply the volume averaging to the second order statistics.

Spatial covariance and correlation. A *spatial autocorrelation* is defined as

$$\begin{aligned}\Psi_{KK}(\mathbf{x}, \mathbf{x}+\chi) &= \langle \mathsf{K}(\mathbf{x})\mathsf{K}(\mathbf{x}+\chi)\rangle \\ &= \frac{1}{L^3}\int_{\mathcal{U}_L(\mathbf{x})} \mathsf{K}(\mathbf{x}')\,\mathsf{K}(\mathbf{x}'+\chi)\, d\mathcal{U}(\mathbf{x}'). \end{aligned}\tag{10.1.13}$$

A *spatial autocovariance* is defined by

$$\psi_{KK}(\mathbf{x}, \mathbf{x}+\chi) = \langle [\mathsf{K}(\mathbf{x}) - \langle \mathsf{K}\rangle(\mathbf{x})]\,[\mathsf{K}(\mathbf{x}+\chi) - \langle \mathsf{K}\rangle(\mathbf{x}+\chi)]\rangle\,. \tag{10.1.14}$$

The variance is, simply, $\psi_{KK}(\mathbf{x}, \mathbf{x})$.

If a process is homogeneous, then the mean is not a function of location, $\langle \mathsf{K}\rangle(\mathbf{x}) =$ constant. Similarly, a homogeneous, isotropic autocovariance is a function of the separating distance only, i.e., $\psi_{KK}(\mathbf{x}, \mathbf{x}+\chi) = \psi_{KK}(\chi)$.

Altogether, in this subsection, we have introduced various statistical measures that will be used below in methods employed for coping with uncertainty. We have also shown that for every statistical measure defined in the ensemble sense, there is a corresponding one in the spatial sense. An important consequence of this observation will be discussed in the next subsection.

10.1.5 Ergodicity hypothesis

In Subs. 10.1.3 and 10.1.4, we introduced two types of averaging processes: the ensemble averaging, and the spatial (temporal) averaging, to produce statistical measures. In this subsection, we discuss which of these two analyses is relevant to analysis of uncertainty in groundwater flow transport.

In an uncertainty analysis, we seek to describe events, or to predict future ones, in terms of a *probability*. For example, given a location in an aquifer, we may ask 'what is the probability that the hydraulic conductivity there falls within a given range of values?' Or, given the probability information regarding the input parameters needed in a modeling effort, we may ask 'what is the probability that the predicted water levels, or solute concentrations, will be below or above a certain specified level?' To answer these and similar questions, ensemble averaging is needed, since probability is defined as the percentage of samples in the ensemble that fits a certain criterion.

To perform ensemble averaging, we need knowledge about the samples. If information is available about the *entire* population, then the averaging is straightforward, and the statistical measures can be exactly calculated. However, this is not the case for hydrological events. For example, our interest may lie in the probability of stream flow discharge at a fixed location, at any time of the year, expressed as $Q(t)$, where t is time from January 1 to December 31. In this case, each sample, or realization, is the flow record of a full year. It is obvious that it is not possible to have the record for all the years, and, particularly, it is not possible to have the record for future years. However, it is important that we have a record for *some* years. These historical records can be used as the basis for an ensemble analysis to obtain *empirical statistics.*

The situation is different for predicting the hydraulic conductivity. For example, we are given a groundwater flow domain in which the hydraulic conductivity is measured at a number of locations distributed over the domain. If we select an unmeasured location, and wish to know the hydraulic conductivity at that location, in the probabilistic sense, how do we find its statistics, and what is the ensemble space from which we should find its statistics? As we are given a single physical domain, and the domain is likely to be *unique*, that is, no other similar domain exists that can be claimed to be the outcome of the same process, a physical ensemble space does not exist. However, this single physical realization, that is, the given groundwater domain, is *unknown*. Hence, through *thought construction*, many *imaginary* realizations are considered to exist, each of which is a *possible* representation of the considered groundwater domain. Together, these realizations form an ensemble population that allows an ensemble statistics to be defined.

The difficulty inherent in this type of analysis lies in the fact that, as such ensemble population does not physically exist, we cannot collect a set of samples to perform an ensemble statistical analysis. This difficulty can be overcome if we accept an important *assumption* in statistical analysis—the ergodicity hypothesis, typically assumed for natural processes. The *ergodicity hypothesis* is a concept borrowed from statistical mechanics (Khinchin, 1949). It may be stated as follows:

> *for a stationary random process, a large number of observations made on a single system, at N arbitrary instants of time, have the same statistical properties as observing N arbitrarily chosen systems, from an ensemble of similar systems* (McQuarrie, 2000)

Or, put simply, with the ergodicity hypothesis, we can assume that the average of a process over time is equal to the average of that process over its statistical ensemble.

The consequence of the above hypothesis is that, rather than performing ensemble statistics (see Subs. 10.1.3) on a set of samples (which are not available), we can perform spatial (or temporal) statistics (see Subs. 10.1.4) based on data points taken from the same sample at different spatial locations (or times), as the statistical properties are the same. We should caution, however, that such hypothesis is possible only for a homogeneous (stationary) random process; in other words, the statistical moments (mean, covariance, etc.), must not be functions of space (or time).

With this assumption, we can now state that *in a homogeneous process, the ensemble mean and autocorrelation are equal to their spatial counterparts*, i.e.,

$$\overline{\mathsf{K}} \equiv \langle \mathsf{K} \rangle = \text{constant}, \quad \text{and} \quad c_{KK}(\chi) = \psi_{KK}(\chi). \tag{10.1.15}$$

We note that *an ergodic process must be homogeneous*, but that a homogeneous process need not necessarily be ergodic. The implication of (10.1.15) is that, in the absence of available realizations, we can use the considered aquifer as a single realization and conduct the spatial analysis as described in Subs. 10.1.4. The statistics obtained: mean, standard deviation, covariance, etc., may then be considered as ensemble statistics.

Although (10.1.15) is the definition of ergodicity, strictly speaking, this relation is not the one actually used. In an investigated field, when only sparsely distributed data points exist, it is difficult to perform volume (or areal, in a two-dimensional domain) averaging; hence, an ensemble averaging is often performed. For example, if the data consists of n measured hydraulic conductivity values, $\{\mathsf{K}(\mathbf{x}_1), \mathsf{K}(\mathbf{x}_2), \ldots, \mathsf{K}(\mathbf{x}_n)\}$, we can estimate the mean as

$$\overline{\mathsf{K}} = \frac{1}{n}\sum_{i=1}^{n} \mathsf{K}(\mathbf{x}_i). \tag{10.1.16}$$

Similarly, we can find the covariance as an ensemble average based on (10.1.8), using all combinations of pairs of points from the available field data. This is permitted, in view of the homogeneity assumption—data can be taken from any location in the field, as the statistical result should be independent of location. In this way, effectively, the ergodicity assumption can be restated as follows:

> *Under the assumption of homogeneous first and second statistical moments, the spatially distributed data from a single available realization can be used as an ensemble for statistical analysis.*

In the next section, we shall demonstrate the use of such empirically obtained statistical moments for the creation of multiple realizations of random parameter fields within the framework of the Monte Carlo simulations technique.

10.2 Tools for Uncertainty Analysis

In this section, we discuss the various tools used when facing information uncertainty in groundwater modeling. Some simple tools give rudimentary information that allows the manager to understand trends and impacts, while others, more elaborate tools, provide statistical measures that can be used for optimizing operations, and for assessing the risk to management goals.

We start from the well-known geostatistical tool called *kriging*. In hydrogeology, parameter information is typically known only at a small number of discrete sampling, or monitoring points. However, to model the system, parameter values are needed for the entire field. How can we use the available data in order to reasonably estimate missing values by interpolation? The kriging technique, discussed below, provides the *best estimate* (in the statistical sense) for the missing values.

Let us focus, first, on the concept of *sensitivity analysis* (Subs. 10.2.2). Faced with the uncertain values of model parameters, the modeler may want to know how sensitive is the envisaged model prediction to the various uncertain values of these parameters. If the predicted outcome is sensitive to a ceratin parameter, then it will be highly beneficial to invest resources in order to obtain more accurate information concerning that parameter, rather than waste the limited resources on the less sensitive ones.

Another most versatile modeling tool for analyzing a system under conditions of parameter uncertainty is the *Monte Carlo Simulation*. In this technique, we construct a large number of realizations of the considered domain, say with respect to a property like hydraulic conductivity. Each realization is investigated, yielding a forecast. The collective behavior of all forecasts is then analyzed, providing the probabilistic information needed for making management decisions under conditions of uncertainty.

As stated above, the Monte Carlo technique requires the generating of a large number of *realizations* of the parameter field. This is done by a *random field generation* algorithms. Following is an introduction to this subject.

10.2.1 Kriging

Kriging is a geostatistical technique that interpolates between any data values in a random field. The technique was pioneered by Matheron (1963, 1973), based on the original work of Krige (1951). Originally, kriging was introduced to maximize the probability of locating mineral deposits. Nowadays, it is widely used in many other fields, such as hydrogeology, environmental science, and remote sensing, as an interpolation tool for spatial data.

Consider a two- or three-dimensional groundwater domain for which we need to know the spatial distribution of a hydrogeological property like hydraulic conductivity, or transmissivity. Usually, we obtain such information from pumping tests conducted at a (rather small number of) selected sites. On the other hand, for the purpose of groundwater modeling, we need information on these parameters everywhere, or at the grid points of a numerical model. Usually, this grid is rather dense and certainly does not coincide with the points of data measurements. How do we estimate the values of these parameters at the numerical grid points? Furthermore, knowing that the aquifer is (usually) highly heterogeneous, any estimate that we make will contain errors, at least at the unmeasured points. What is, then, our estimate of the range of such errors, e.g., expressed as standard deviation? These questions can be answered by making use of the theory of *geostatistics* (Journel and Huijbregts, 1978; Kitanidis, 1997).

Consider an aquifer domain for which a set of spatially distributed hydraulic conductivity values, $\mathsf{K}(\mathbf{x}_i)$, is known at locations $\mathbf{x}_i$, $i = 1, \ldots, n$. We are interested in finding the mean hydraulic conductivity, in the probabilistic sense, at an unobserved point, $\mathbf{x}$. Recall (Subs. 10.1.5) that, so far, an ensemble sampling space, from which this mean can be taken, does not exist; hence, the mean has to be taken as a spatial average following the ergodic assumption (Subs. 10.1.5). This is accomplished by

$$\overline{\mathsf{K}}(\mathbf{x}) = \sum_{i=1}^{n} \lambda_i \mathsf{K}(\mathbf{x}_i), \quad \sum_{i=1}^{n} \lambda_i = 1, \tag{10.2.1}$$

where λ_i is the *weight* applied to an observation point i.

Field observations have indicated that hydraulic conductivity is more likely to be *log-normally* than *normally* distributed (Gelhar, 1993). Hence, it is common to perform the statistical analysis on the logarithm of the hydraulic conductivity, i.e.,

$$\overline{Y}(\mathbf{x}) = \sum_{i=1}^{n} \lambda_i Y(\mathbf{x}_i), \quad Y \equiv \log \mathsf{K}. \tag{10.2.2}$$

There are many options for selecting the weights, λ_i. For example, if we consider the field to be statistically homogeneous, and anticipate that the mean will be a constant for the entire field, then the weight is simply taken as $\lambda_i = 1/n$. On the other hand, if the random field has a large scale trend, i.e., $\overline{Y}$ is a function of space, then the averaging should take into account only nearby points. This is implemented by removing points at a distance larger than a certain specified radius, r_o, such that $\lambda_i = 0$ if $r_i \equiv |\mathbf{x} - \mathbf{x}_i| > r_o$. Within this radius, it is still possible to assign different weights to different points. For example, we may make the weights *inversely* proportional to the distance, or to the square of the distance, i.e.,

$$\lambda_i = \frac{1/r_i^2}{\sum_{j=1}^{m}(1/r_j^2)}, \tag{10.2.3}$$

where m is the number of points enclosed within the circle, or sphere, of radius r_o. In this weighting scheme, we take into consideration that data that are nearer to the considered point have a stronger correlation with what happens at the considered point, and those that are farther away have a weaker correlation, and may, therefore, be disregarded. The above procedure, however, is not rigorous in the statistical sense. To have a more rational estimate, we shall now introduce the *kriging* idea, which is based on a spatial statistical analysis.

To perform a geostatistical analysis, we have first to apply a certain statistical model to the random parameter field. We assume that the first statistical moment, that is, the mean of the considered property, Y, is *inhomogeneous*, such that $\overline{Y} = \overline{Y}(\mathbf{x})$. We also assume that the second moment is *homogeneous* (i.e., stationary), such that the (auto)covariance is not a function of the spatial coordinate, $\mathbf{x}$, but is a function of the distance, r, between the two points. These assumptions lead to $c_{YY} = c_{YY}(r)$ (see (10.1.8)). To perform the geostatistical analysis, we need to estimate these statistical quantities (mean and covariance) from the observations at the limited number of points.

In geostatistics, instead of using the covariance, $c_{YY}(r)$, it is customary to use the quantity called *semivariogram*, γ. This quantity is related to the covariance by

$$\gamma(r) = \frac{1}{2}\overline{[Y(\mathbf{x}+\mathbf{r}) - Y(\mathbf{x})]^2} = c_{YY}(0) - c_{YY}(r), \tag{10.2.4}$$

where $c_{YY}(0)(\equiv \sigma_Y^2)$ is the (homogeneous) variance, which is a constant. We note that at $r = 0$, $\gamma(0) = 0$.

The construction of the semivariogram is based on available data. For example, given data measured at points $\mathbf{x}_i$, we sort their separating distances $r_{ij} = |\mathbf{x}_i - \mathbf{x}_j|$ into different ranges, Δr, $2\Delta r, \ldots$. For the data that fall within the same range, we can use (10.2.4) to evaluate $\gamma(\Delta r)$, $\gamma(2\Delta r)$, ... These data can then be plotted as the semivariogram shown in Fig. 10.2.1. From the fitted curve, it follows that at a large distance, say, $r > r_o$, the curve levels off, meaning that the data cease to be correlated, i.e., the covariance $c_{YY}(r) \to 0$, as $r \to \infty$. The asymptotic value marked as a_1 is the estimated variance, $c_{YY}(0) = \sigma_Y^2$. Although, theoretically, the intercept should be $\gamma(0) = 0$ at

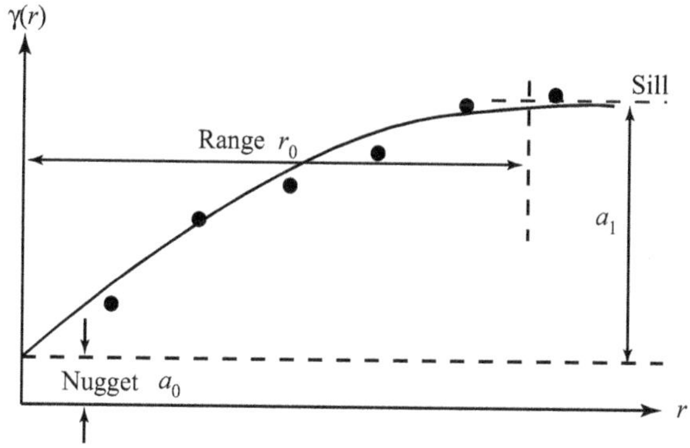

Figure 10.2.1: Empirical semivariogram constructed from data.

$r = 0$, it is typically not zero. This is attributed to a spatially uncorrelated data noise (measurement error), known as the *nugget effect.*

Based on the shape of the curve, it is often fitted to one of several known models, such as the *spherical model,*

$$\gamma(r) = \begin{cases} a_o + a_1 \left[\frac{3}{2} \frac{r}{r_o} - \frac{1}{2} \left(\frac{r}{r_o} \right)^3 \right], & \text{for } 0 < r < r_o, \\ a_o + a_1, & \text{for } r \geq r_o. \end{cases} \tag{10.2.5}$$

Other models, such as the *exponential model,* the *Gaussian model,* the *hole effect model,* can be found the literature on geostatistics (e.g., Deutsch and Journel, 1998).

With the constructed semivariogram, we are ready to estimate the property Y at any *unsampled point,* $\mathbf{x}$, by one of several kriging methods. A common assumption in all methods is that the *residual,* defined as the difference between a value and its mean, $Y(\mathbf{x}) - \overline{Y}(\mathbf{x})$, can be estimated by a weighted sum of residuals at the surrounding points (all points within the range r_o, defined in the semivariogram shown in Fig. 10.2.1), as

$$Y^*(\mathbf{x}) - \overline{Y}(\mathbf{x}) = \sum_{i=1}^{n(\mathbf{x})} \lambda_i(\mathbf{x})[Y(\mathbf{x}_i) - \overline{Y}(\mathbf{x}_i)], \tag{10.2.6}$$

where $\mathbf{x}_i$ indicates a *sampled point,* and the asterisk superscript is used to denote an *estimated* value. The weights λ_i in (10.2.6), which are selected differently at different locations, have still to be determined. The estimate, Y^*, is said to be *'unbiased'*, if the mean of the residual of the estimate satisfies

$$\overline{Y^*(\mathbf{x}) - Y(\mathbf{x})} = 0. \tag{10.2.7}$$

This is evident by taking the mean of (10.2.6). As we are free to choose the weights, we usually choose λ_i as the set that minimizes the variance of the above residual, i.e.,

$$\sigma^2_{Y^*}(\mathbf{x}) = \overline{[Y^*(\mathbf{x}) - Y(\mathbf{x})]^2} = \overline{[(Y^*(\mathbf{x}) - \overline{Y}(\mathbf{x})) - (Y(\mathbf{x}) - \overline{Y}(\mathbf{x}))]^2}. \quad (10.2.8)$$

This can be accomplished by differentiating the above expression with respect to the λ_i's, one at a time, and setting the derivatives to zero. First, we substitute (10.2.6) into (10.2.8) to obtain

$$\begin{aligned}\sigma^2_{Y^*}(\mathbf{x}) &= \overline{[\textstyle\sum_{i=1}^{n(\mathbf{x})} \lambda_i(\mathbf{x})(Y(\mathbf{x}_i) - \overline{Y}(\mathbf{x}_i)) - (Y(\mathbf{x}) - \overline{Y}(\mathbf{x}))]^2} \\ &= \textstyle\sum_{i=1}^{n(\mathbf{x})} \sum_{j=1}^{n(\mathbf{x})} \lambda_i(\mathbf{x})\lambda_j(\mathbf{x}) c_{YY}(\mathbf{x}_i - \mathbf{x}_j) + c_{YY}(0) \\ &\quad -2\textstyle\sum_{i=1}^{n(\mathbf{x})} \lambda_i(\mathbf{x}) c_{YY}(\mathbf{x}_i - \mathbf{x}),\end{aligned} \quad (10.2.9)$$

where $c_{ab}(\mathbf{x}_1, \mathbf{x}_2)$ is the covariance (see (10.1.5)), $c_{aa}(\mathbf{x}_1, \mathbf{x}_2)$ is the autocovariance, and we have used the stationarity assumption, $c_{aa}(\mathbf{x}_1, \mathbf{x}_2) = c_{aa}(\mathbf{x}_1 - \mathbf{x}_2)$ (see (10.1.8)). We also note the relation $c_{aa}(\mathbf{x}_1, \mathbf{x}_1) = c_{aa}(0) = \sigma^2_a$.

In order to use (10.2.9) for the determination of the λ_i's, which is required for the estimate of the hydraulic conductivity, we need prior knowledge of the mean values, $\overline{Y}(\mathbf{x}_i)$, $i = 1, \dots, n(\mathbf{x})$ and $\overline{Y}(\mathbf{x})$, and of the covariances, $c_{YY}(\mathbf{x}_i - \mathbf{x}_j)$ and $c_{YY}(\mathbf{x}_i - \mathbf{x})$, $i, j = 1, \dots, n(\mathbf{x})$. We have already decided that the covariances will be provided by the empirically constructed variogram, e.g., (10.2.5) and (10.2.4) (Fig. 10.2.1). As for the mean, different assumptions lead to different kriging methods. We shall discuss below the three most popular kriging methods: the *simple*, the *ordinary*, and the *universal* kriging.

A. Simple kriging ('sk')

In this method, we assume that the field is statistically homogeneous, that is, without a spatial trend. This means that the mean is a *constant*, $\overline{Y}(\mathbf{x}) = \overline{Y}(\mathbf{x}_i) = \overline{Y}$. For example, we can estimate this quantity as a simple average of all the data points in the field,

$$\overline{Y} = \frac{1}{n} \sum_{i=1}^{n} Y(\mathbf{x}_i). \quad (10.2.10)$$

Using this value in (10.2.6), we can estimate the logarithm of the hydraulic conductivity at an unsampled point as

$$Y^*_{\text{sk}}(\mathbf{x}) = \overline{Y} + \sum_{i=1}^{n(\mathbf{x})} \lambda_i^{\text{sk}}(\mathbf{x})[Y(\mathbf{x}_i) - \overline{Y}], \quad Y = \log K, \quad (10.2.11)$$

in which the subscript and superscript 'sk' stand for 'simple kriging'. To determine the weights, we minimize the variance (10.2.9) by differentiating it with respect to λ_i^{sk}, $i = 1, \ldots, n(\mathbf{x})$, and setting the derivatives equal to zero. This leads to the following linear system of equations:

$$\sum_{j=1}^{n(\mathbf{x})} \lambda_j^{\text{sk}}(\mathbf{x})\, c_{YY}(\mathbf{x}_i - \mathbf{x}_j) = c_{YY}(\mathbf{x}_i - \mathbf{x}); \quad i = 1, \ldots, n(\mathbf{x}). \tag{10.2.12}$$

As the values of the covariances are known, (10.2.12) can be solved for the λ_i^{sk}-values. Then using (10.2.11), we can calculate the estimated Y on the system's grid points. This, in turn, can be used to plot contours of constant Y-values. The advantage of the kriging method, compared to any other deterministic interpolation scheme, is that not only does it yield the *best* estimate, based on a least square fit, but it also provides information about the size of the estimation error. This is done by substituting (10.2.12) into (10.2.9), thus yielding a formula for the variance of the estimation error, in the form

$$\sigma_{\text{sk}}^2(\mathbf{x}) = c_{YY}(0) - \sum_{i=1}^{n(\mathbf{x})} \lambda_i^{\text{sk}}(\mathbf{x})\, c_{YY}(\mathbf{x}_i - \mathbf{x}). \tag{10.2.13}$$

This error can also be presented as a contour map.

B. Ordinary kriging ('ok')

Rather than assuming that the mean is constant over the entire domain, we recognize that it may have a spatial trend; hence, we express it as a function of space, $\overline{Y}(\mathbf{x})$. Since, typically, we do not have sufficient data to determine such a trend, we seek to eliminate the need for such information. First, we assume that the mean in the neighborhood of $\mathbf{x}$, represented by the group of points $\mathbf{x}_i$, $i = 1, \ldots, n(\mathbf{x})$, can be approximated by the mean at $\mathbf{x}$; that is, $\overline{Y}(\mathbf{x}_i) = \overline{Y}(\mathbf{x})$. Using this condition in (10.2.6), we can rearrange it to become

$$Y_{\text{ok}}^*(\mathbf{x}) = \sum_{i=1}^{n(\mathbf{x})} \lambda_i^{\text{ok}}(\mathbf{x})\, Y(\mathbf{x}_i) + \left[1 - \sum_{i=1}^{n(\mathbf{x})} \lambda_i^{\text{ok}}(\mathbf{x})\right] \overline{Y}(\mathbf{x}). \tag{10.2.14}$$

We shall require that the weights sum to unity, i.e.,

$$\sum_{i=1}^{n(\mathbf{x})} \lambda_i^{\text{ok}}(\mathbf{x}) = 1, \tag{10.2.15}$$

and (10.2.14) reduces to

$$Y_{\text{ok}}^*(\mathbf{x}) = \sum_{i=1}^{n(\mathbf{x})} \lambda_i^{\text{ok}}(\mathbf{x})\, Y(\mathbf{x}_i). \tag{10.2.16}$$

By taking the mean of both sides of the above equation, together with the approximation $\overline{Y}(\mathbf{x}_i) = \overline{Y}(\mathbf{x})$ adopted above, we observe that the condition of unbiased estimate, (10.2.7), is satisfied. Equation (10.2.16) is then the formula for estimating an unsampled value based on ordinary kriging. We notice that the knowledge of the mean is not required in the above formula.

We still need to determine the weights in (10.2.16). Since we already have one constraint equation for λ_i^{ok}, in the form of (10.2.15), the minimization procedure described above for simple kriging will produce an overdetermined linear system. This problem is then solved by introducing a *Lagrangian multiplier*, $\mu^{\text{ok}}(\mathbf{x})$, and a new functional to be minimized:

$$L = \sigma_{Y^*}^2(\mathbf{x}) + 2\mu^{\text{ok}}(\mathbf{x}) \left[1 - \sum_{i=1}^{n(\mathbf{x})} \lambda_i^{\text{ok}}(\mathbf{x}) \right], \tag{10.2.17}$$

where $\sigma_{Y^*}^2$ is given by (10.2.9). The functional L is minimized with respect to both λ_i^{ok} and μ^{ok}. The particular form of (10.2.17) ensures that the required minimization condition based on

$$\frac{\partial L}{\partial \mu^{\text{ok}}} = 2 \left[1 - \sum_{i=1}^{n(\mathbf{x})} \lambda_i^{\text{ok}}(\mathbf{x}) \right] \equiv 0, \tag{10.2.18}$$

is identically zero, and does not produce an extra equation.

Differentiating (10.2.17) with respect to λ_i^{ok}, produces the following set of equations:

$$\sum_{j=1}^{n(\mathbf{x})} \lambda_j^{\text{ok}}(\mathbf{x}) c_{YY}(\mathbf{x}_i - \mathbf{x}_j) + \mu^{\text{ok}}(\mathbf{x}) = c_{YY}(\mathbf{x}_i - \mathbf{x}); \quad i = 1, \ldots, n(\mathbf{x}). \tag{10.2.19}$$

Together with (10.2.15), the system of $n(\mathbf{x}) + 1$ equations can be used to solve for the $n(\mathbf{x}) + 1$ unknowns, λ_i^{ok}'s and μ^{ok}'s. We can then obtain the estimated field, $Y_{\text{ok}}^*(\mathbf{x})$, from (10.2.16); while the variance of estimation error can be obtained from

$$\sigma_{\text{ok}}^2(\mathbf{x}) = c_{YY}(0) - \sum_{i=1}^{n(\mathbf{x})} \lambda_i^{\text{ok}}(\mathbf{x})\, c_{YY}(\mathbf{x}_i - \mathbf{x}) - \mu^{\text{ok}}(\mathbf{x}). \tag{10.2.20}$$

C. Universal kriging ('kt')

This approach is also known as the *kriging with a trend* ('kt') model. In this method, it is assumed that the global trend can be fitted to the function

$$\overline{Y(\mathbf{x})} = \sum_{k=1}^{m} a_k\, f_k(\mathbf{x}), \tag{10.2.21}$$

where the f_k's are *basis functions*, such as 1, x, y, x^2, y^2,... for polynomial basis functions, and the a_k's are unknown coefficients; the latter will be eliminated in the formulation process. Similar to ordinary kriging, we need to introduce m Lagrangian multipliers, $\mu_k^{\text{kt}}(\mathbf{x})$, $k = 1, \ldots, m$. The universal kriging estimator is given by

$$Y_{\text{kt}}^*(\mathbf{x}) = \sum_{i=1}^{n(\mathbf{x})} \lambda_i^{\text{kt}}(\mathbf{x})\, Y(\mathbf{x}_i). \tag{10.2.22}$$

The weights and Lagrangian multipliers are solved from the system

$$\sum_{j=1}^{n(\mathbf{x})} \lambda_j^{\text{kt}}(\mathbf{x})\, c_{YY}(\mathbf{x}_i - \mathbf{x}_j) + \sum_{k=1}^{m} \mu_k^{\text{kt}}(\mathbf{x}) f_k(\mathbf{x}_i) = c_{YY}(\mathbf{x}_i - \mathbf{x}), \qquad i = 1, \ldots, n(\mathbf{x}), \tag{10.2.23}$$

and

$$\sum_{j=1}^{n(\mathbf{x})} \lambda_j^{\text{kt}}(\mathbf{x})\, f_k(\mathbf{x}_j) = f_k(\mathbf{x}), \quad k = 1, \ldots, m. \tag{10.2.24}$$

D. Cokriging ('ck')

So far, we have discussed the three simplest and most commonly used kriging methods. A number of important kriging issues remain unaddressed. One important issue that has not been addressed is kriging of correlated data, or *cokriging*. Sometimes two sets of different, but correlated, physical data are observed, e.g., the hydraulic conductivity and the piezometric head. In such case, we may use the observed piezometric head to assist in the estimate of the logarithm of the hydraulic conductivity.

Similar to (10.2.6), we can write

$$Y_{\text{ck}}^*(\mathbf{x}) = \overline{Y}(\mathbf{x}) + \sum_{i=1}^{n(\mathbf{x})} \lambda_i^{\text{ck}}(\mathbf{x})[Y(\mathbf{x}_i) - \overline{Y}(\mathbf{x}_i)] + \sum_{i=1}^{m(\mathbf{x})} \mu_i^{\text{ck}}(\mathbf{x})[h(\mathbf{x}_i') - \overline{h}(\mathbf{x}_i')], \tag{10.2.25}$$

where the $h(\mathbf{x}_i')$'s are observed head values at a set of nearby sampling points, $\mathbf{x}_i'$, $i = 1, \ldots, m(\mathbf{x})$, which may be different from the sampling points, $\mathbf{x}_i$, of the hydraulic conductivity, and the μ_i's are weights applied to the observed head samples. To perform cokriging, it is necessary to obtain not only the covariance function of the hydraulic conductivity, $c_{YY}(r)$, but also that of the piezometric head, $c_{hh}(r)$, as well as the cross covariance $c_{Yh}(r)$. In practice, these covariances, particularly the cross covariance, are difficult to come by. Hence, it is rather difficult to implement cokriging. In Subs. 11.3.2B, however, we shall discuss the use of an inverse solution procedure that assists in the identification of the covariances, as well as the spatial trend of the mean.

With this information available, we can derive the weights from the following set of equations:

$$\sum_{j=1}^{n(\mathbf{x})} \lambda_j^{\mathrm{ck}}(\mathbf{x})\, c_{YY}(\mathbf{x}_i - \mathbf{x}_j) + \sum_{k=1}^{m(\mathbf{x})} \mu_k^{\mathrm{ck}}(\mathbf{x}) c_{Yh}(\mathbf{x}_i - \mathbf{x}'_k) = c_{YY}(\mathbf{x}_i - \mathbf{x}),$$
$$i = 1, \ldots, n(\mathbf{x}), \qquad (10.2.26)$$

$$\sum_{j=1}^{n(\mathbf{x})} \lambda_j^{\mathrm{ck}}(\mathbf{x})\, c_{Yh}(\mathbf{x}_i - \mathbf{x}'_j) + \sum_{k=1}^{m(\mathbf{x})} \mu_k^{\mathrm{ck}}(\mathbf{x}) c_{hh}(\mathbf{x}'_i - \mathbf{x}'_k) = c_{hY}(\mathbf{x}'_i - \mathbf{x}),$$
$$i = 1, \ldots, m(\mathbf{x}). \qquad (10.2.27)$$

Additional details on other kriging issues can be found in textbooks on geostatistics, e.g., Deutsch and Journel (1998), Goovaerts (1997), and Stein,(1999). Particularly, we should mention that Deutsch and Journel (1998) present computer source codes in Fortran for a number of kriging methods. A review on kriging is given by de Marsily (1986); see also Ezzedine (2005), and Sun and Sun (2005).

Before leaving the discussion on kriging, let us recapitulate a few key points. In modeling flow and transport in groundwater domains, we are faced (1) with highly heterogenous fields, and (2) with measurements that are taken only at a relatively small number of locations. Since, usually, fields are highly heterogeneous, a large uncertainty is involved in determining (by some inverse method) the values of parameters at unsampled points. This makes the estimated spatial distribution of permeability, effectively, a random process.

Given these underlying conditions, our objective is to make the 'best' (in the probabilistic sense) estimation of the parameters of interest at unsampled points. When we focus our attention on such a point, there is a high probability that the value there is similar to that at nearby sampled points (where the probability of the measured value is 100%). At the same time, there is a small probability that it is also similar to the value at a far away point. It is, thus, reasonable to estimate the value at the considered point as a weighted average of the values at some nearby points, with weights, $\lambda_1, \lambda_2, \ldots$. The question is, 'what is a rational way of choosing these weights?' A good starting point is to use weights based on the inverse of the distance, or the inverse of the squared distance. However, this is somewhat arbitrary. There are cases in which sampling points are clustered together, and it seems more reasonable to count them as a single point in the weighted average. Methods that limit the number of points per quadrant also address this issue of uneven distribution of sampling points, but, again, the algorithm is arbitrary. Kriging methods, on the other hand, are based on the theory of random field analysis. The weights are determined with the objective of minimizing the square estimation error. The problem of data redundancy, due to clustering of sampling points, is taken into consideration by detecting the closeness of these data points; the calculated weights automatically adjust for it. Finally, and most important, kriging gives an error map that shows how the error increases as we move away from sampled points. In regions with too few sampled data, larger errors should be anticipated. If the error is not acceptable, a decision can be made to seek additional data in these selected regions.

So far, the discussion in this subsection has focused on using kriging in order to provide the best estimate of domain coefficients. Our interest has been in the use of interpolated parameter fields for groundwater flow and transport modeling. However, there are a few issues involved in using kriged results as input to groundwater models in the event of information uncertainty. For example, it is well known that in a highly heterogeneous field, kriged results tend to smooth out the heterogeneity (e.g., Deutsch and Journel, 1998). Hence, generally, the resulting parameter field does not resemble the true field, in terms of the smaller scale heterogeneity features. If this smoothed parameter map is used in modeling, certain effects associated with heterogeneity, such as macrodispersion, may not be accurately modeled.

Another issue is that although, when used in groundwater modeling, kriging provides the best estimate of the parameter field and the variance of estimation error, this information does not translate into the error in the predicted results, e.g., heads or solute concentrations. In other words, despite the fact that the best estimates are used as input, we may not claim that the output is also the best estimate. In this sense, kriging is just a convenient interpolation tool for generating the full field information needed for numerical models. Often kriging is used as a selection of interpolation tools in a software for plotting model output in the form of contour lines. Again, in this case, kriging is just a convenient interpolation tool and cannot always claim a statistical advantage over other interpolation methods. For a correct representation of a random field employed for producing statistical predictions of the model results, methods like *random field generation* (Subs. 10.2.4) and *Monte Carlo simulation* (Subs. 10.2.3) are needed. These will be discussed in subsequent sections.

10.2.2 Sensitivity analysis

Sensitivity analysis is the study of how a system's response varies with changes in the various factors on which such response depends (Hill and Tiedeman, 2007). A system is said to be sensitive to a factor if a small change in the latter causes a large change in the system's response. For the purpose of groundwater modeling, the system of interest is a groundwater aquifer, a vadose zone, or a groundwater remediation site, the behavior of which depends on the various parameters that appear in the model. In the case of an aquifer system, the response may be the future piezometric heads or solute concentrations at certain locations and at certain times, or it may be the extent of land subsidence at a certain time. Model parameters include the geometry of the domain's boundary, the initial piezometric head distribution, boundary conditions, forcing functions, physical coefficients, etc. Very often, due to the lack of sufficient data, the information concerning these factors may not be reliably available, say from a model calibration process (Step 7 in Subs. 1.2.2). In other cases, these factors, e.g., boundary conditions, or natural replenishment, may be associated with future events that cannot be reliably predicted. These uncertainties limit our confidence in the response or output of the model. As described in Step 9 in Subs. 1.2.2, good modeling practice requires that the modeler provides an evaluation of the confidence in the model, possibly assessing the uncertainties associated with the modeling process and with model predictions. Under such circumstances, a sensitivity analysis may assist decision makers to assess the impact of these uncertainties on the model's forecasts, which, in turn, affect their decisions. The results of a sensitivity analysis may also guide future data collection activities aimed at reducing modeling uncertainties. Input information, e.g., model coefficients, to which model results are more sensitive, would require narrowing the range of uncertainty; this, in turn, would require more investment in data acquisition and improved characterization of the modeled site.

While there exist various types of model factors that call for a sensitivity analysis, in this section we focus on parameter sensitivity, namely, on a measure of the system's sensitivity to changes in the values of its physical coefficients. For an aquifer system, the physical coefficients are transmissivity, storativity, hydraulic conductivity, aquifer thickness, etc. If an input parameter involves an error of a certain magnitude, it is of interest to find out the magnitude of error in the model's output, i.e., in the model's prediction. If the estimated error is too large to be acceptable, sensitivity analysis may be used to determine the degree of refinement needed in the input data in order to achieve an acceptable accuracy in the output.

Sensitivity analysis is closely related to *parameter estimation*, or to the *inverse problem*, introduced as Step 7, in Subs. 1.2.2 and in Sec. 11.3. Usually, it is difficult to determine the values of physical parameters of an investigated domain by direct measurements. Instead, indirect means, such as observing aquifer response to changes produced by pumping activities, are employed. Mathematically, an *inverse problem* is solved. If the response that we choose to observe is sensitive to a certain parameter, then that parameter stands a good chance to be accurate. On the other hand, if the response is insensitive to the variation of a certain parameter, then the mathematical problem can becomes ill-posed in the presence of data noise associated with the measured responses.

Generally, we can express a quantity of interest as a *performance function*, E. The drawdown in an aquifer, or the concentration of a pollutant at a certain location and time, may serve as examples. The performance function depends on a number of parameters, p_1, p_2, ..., i.e., $E = E(p_1, p_2, \ldots)$. These parameters may vary, either because we have control over them, e.g., through controlling the pumping rate, or because their values are uncertain, such as

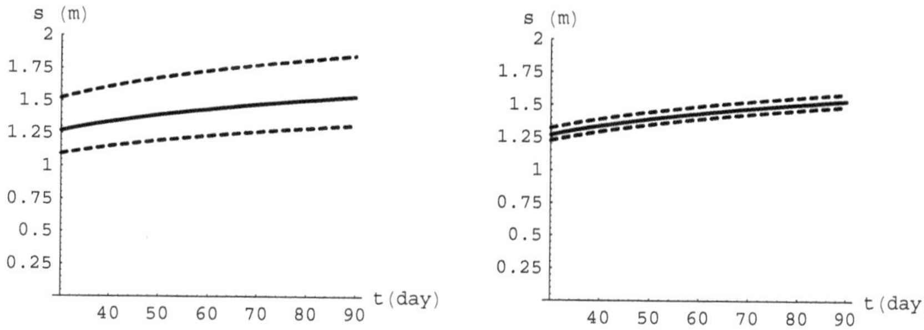

Figure 10.2.2: Sensitivity of drawdown at 1 km from a pumping well due to (a) ±20% change in transmissivity; and (b) ±20% change in storativity.

the uncertainty associated with aquifer parameters—hydraulic conductivity, storativity, dispersivity, etc. We are interested in the answer to the following question: 'If the value of the parameter, p_i, is changed by a certain amount, say, Δp_i, what will be the corresponding change, ΔE, in the performance function?' We may express the answer as a sensitivity coefficient, $\Delta E/\Delta p_i$, or, more precisely, by $\partial E/\partial p_i$.

As an example, consider, a pumping test (see Subs. 11.3.1) conducted in a well in a confined aquifer. During a short time interval, drawdown is caused within a certain radius of influence. The pumping test analysis is based on the solution for drawdown, $s = s(r,t)$, known as Theis solution (e.g., Bear, 1979)

$$s(r,t) = \frac{Q}{4\pi \mathsf{T}} W(u), \qquad u = \frac{r^2 S}{4\mathsf{T}t}, \tag{10.2.28}$$

where Q is the pumping rate, r is the radial distance from the well, t is time, and $W(u)$ is the *well function* of u. Let the analysis lead to the aquifer parameter values $\mathsf{T} = 300$ m^2/day and $S = 0.0001$. These parameters are to be used for predicting drawdown at a larger distance and longer time. Because of the limited extent of the sampled aquifer, we suspect that the estimates of both parameters may be wrong by ±20%. Figure 10.2.2 presents the drawdown at a distance of 1 km from the well, caused by pumping 900 m^3/day for the period from 1 to 3 months: the solid line represents the predicted drawdown, while the dashed lines present the drawdown corresponding to the range of ±20% parameter values. We clearly observe that the drawdown is sensitive to transmissivity changes, but is less sensitive to changes in the storativity. In fact, at the end of two months, the ±20% transmissivity changes causes −14% and +20% changes in drawdown, while the same percentage changes in storativity, cause only −3.0% and +3.7% changes in drawdown. Thus, we may conclude that a certain change in the value of transmissivity causes a

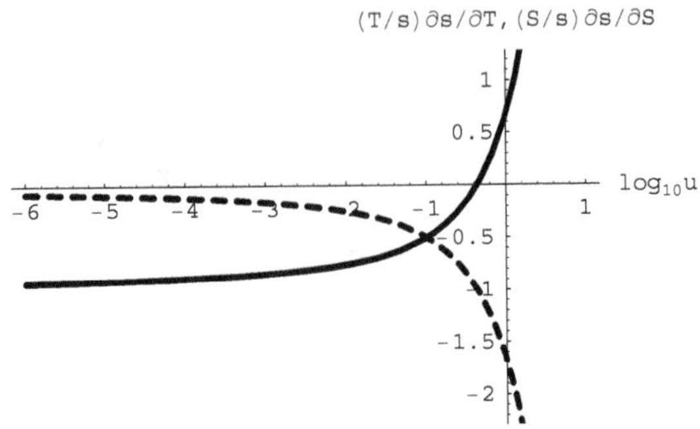

Figure 10.2.3: Normalized sensitivity coefficient for a pumping well in a confined aquifer. Solid line: C_T; dash line: C_S.

proportional (but opposite) change in the drawdown, while for storativity, the effect is, roughly, only 15% (also opposite). This type of information can be useful to a manager in evaluating model predictions.

We can refine the above concept by making the changes in parameter values infinitesimally small. This leads to *sensitivity coefficients* defined as the derivatives $\partial s/\partial \mathsf{T}$ and $\partial s/\partial S$. These quantities, however, are dimensional and do not give the correct notion of the proportionality between the considered changes. Hence, it is better to use the *normalized sensitivity coefficients*, defined as

$$C_T = \frac{\mathsf{T}}{s}\frac{\partial s}{\partial \mathsf{T}}, \qquad C_S = \frac{S}{s}\frac{\partial s}{\partial S}. \tag{10.2.29}$$

These quantities are functions of distance and time, as well as of T and S (but they are not functions of the well's discharge, Q). We also note that the solution (10.2.28) is written in terms of the dimensionless variable u that combines distance and time. Hence, it is possible to express the result in terms of u, rather than, separately, in terms of r and t (Cheng, 2000). Figure 10.2.3 presents C_T and C_S as functions of u in a semi-log scale. These curves are universal, for all values of r, t, T and S.

To interpret Fig. 10.2.3, we note that $u \sim r^2$ and $u \sim t^{-1}$. Hence, the right side of the abscissa indicates a large distance, or small time, while the left side shows small distance, or long time. We also observe that the two curves intersect at $u = 0.102$. For any $u < 0.102$, the drawdown is more sensitive to changes in transmissivity than in storativity. Particularly, as $u \to 0$, corresponding to a small distance from the well, or a long time after pumping, $C_T \to -1$ and $C_S \to 0$. Thus, a certain percentage change in transmissivity causes an equal, but opposite, percentage change in drawdown. As for storativity, we note that the sensitivity diminishes with time. We also note that the right portions of the curves go to positive and negative infinity, respectively, indicating infinite sensitivity. However, realizing that this part corresponds to $t \to 0$ or $r \to \infty$, when the magnitude of drawdown is practically zero, these behaviors are of no practical value.

We should remark that the above example is based on an analytical solution; hence, the sensitivity coefficients could be obtained analytically. However, in general, the solution of a problem is obtained by a numerical method, using a set of prescribed parameters. The performance function is expressed in terms of discrete values at the nodes, $\mathbf{E} = \{E_1, E_2, \ldots\}$. We also note that the parameter vector, $\mathbf{p} = \{p_1, p_2, \ldots\}$, does not necessarily represent different parameters. In a numerical solution, $\mathbf{p}$ could refer to the same parameter, e.g., the hydraulic conductivity, but in different zones, $\mathbf{p} = \{\mathsf{K}_1, \mathsf{K}_2, \ldots\}$. Hence, we may express these sensitivity coefficients in the matrix form

$$\frac{\partial \mathbf{E}}{\partial \mathbf{p}} = \begin{bmatrix} \frac{\partial E_1}{\partial p_1} & \frac{\partial E_2}{\partial p_1} & \cdots & \frac{\partial E_n}{\partial p_1} \\ \frac{\partial E_1}{\partial p_2} & \frac{\partial E_2}{\partial p_2} & \cdots & \frac{\partial E_n}{\partial p_2} \\ \vdots & \vdots & \ddots & \vdots \\ \frac{\partial E_1}{\partial p_m} & \frac{\partial E_2}{\partial p_m} & \cdots & \frac{\partial E_n}{\partial p_m} \end{bmatrix}, \tag{10.2.30}$$

where we assume that there are m discrete parameters, and n discrete E-values. This matrix is called the *sensitivity matrix*, because it expresses the sensitivity of the solutions to changes in the individual parameters. This matrix also plays an important role in the perturbation solution of uncertainty problems (Subs. 10.3.2), and in inverse (parameter estimation) problems (Sec. 11.3).

In a numerical solution, the derivatives in (10.2.30) cannot be obtained analytically, and need to be evaluated numerically. This is done by solving the problem, using a set of selected parameters, say $\mathbf{p}^o = \{p_1^o, p_2^o, \ldots\}$, to obtain $\mathbf{E}^o = \{E_1^o, E_2^o, \ldots\}$. We then perturb each parameter value by a small amount, $\Delta\mathbf{p} = \{\Delta p_1, \Delta p_2, \ldots\}$, one at a time. In other words, we first perturb p_1^o, keeping all other parameters unchanged. A new problem is solved with the set of parameters $\{p_1^o + \Delta p_1, p_2^o, p_3^o, \ldots\}$, to obtain a new solution $\mathbf{E}^{(1)}$. We then find the first row of the sensitivity matrix as $\Delta\mathbf{E}^{(1)}/\Delta p_1$, where $\Delta\mathbf{E}^{(1)} = \mathbf{E}^{(1)} - \mathbf{E}^o$. This process continues by solving the problem m times, each time with one parameter perturbed. The obtained sensitivity matrix is valid only around that set of parameters, namely, $\mathbf{p}^o$. For the sensitivity of the solution around a very different set of parameters, the analysis needs to be performed again.

The simple example presented in this section shows how the sensitivity coefficient can be used to assess the impact of uncertain data on a model's prediction. In a more complex situation, we may be faced with different and multiple kinds of uncertainty, requiring an assessment of their overall impact. In principle, we can, independently, analyze the sensitivity with respect to each such factor. In the *worst case scenario*, the resultant range of variation is the summation of all the individual changes. Such a scenario, however, is highly unlikely to occur, as it requires that all factors work simultaneously in the same unfavorable direction and to their maximum extent. In this case, the simple sensitivity analysis no longer provides a realistic assessment, and a full uncertainty analysis that utilizes the probabilistic information is needed, using techniques such as the Monte Carlo simulation described in the next subsection.

10.2.3 Monte Carlo simulation

The Monte Carlo method assumes that the modeled phenomena can be represented by a *deterministic* mathematical model, with known coefficient values. For example, groundwater flow in an isotropic confined aquifer is described by the (deterministic) governing equation (5.4.37), repeated here for convenience, as

$$S(x,y)\frac{\partial h(x,y,t)}{\partial t} = \nabla \cdot [\mathbf{T}(x,y) \cdot \nabla h(x,y,t)] + R(x,y,t). \qquad (10.2.31)$$

Given information on the geometry of the domain's boundary, the initial and boundary conditions, the recharge rate, $R(x,y,t)$, and full knowledge of the transmissivity, $\mathbf{T}(x,y)$, and storativity, $S(x,y)$, equation (10.2.31) can be solved to yield a unique prediction (solution) of future head values, $h = h(x,y,t)$. However, uncertainty is associated with the fact that we are uncertain about part of the input information. For example, since transmissivity and storativity are measured at wide apart locations, their values cannot be smoothly interpolated to yield reliable information throughout the field.

The Monte Carlo method deals with the uncertainty described above as a probability issue. Instead of attempting to obtain the uncertain, or missing information needed as input, it uses the available (reliable, measured) data to produce the statistical characteristics (e.g., mean, standard deviation, covariance) of the coefficients associated with the considered flow domain (say, the transmissivity), and then uses them to create a large number of *realizations*, each of which is a possible manifestation of the unknown reality. All these realizations follow the same statistical characteristics as those of the actual measured information. Each of these (equally likely to occur) realizations, e.g., in the form of maps of $\mathsf{T}(x,y)$, $S(x,y)$, and $R(x,y,t)$, is used as input to the deterministic flow and transport model, producing the model's prediction as output. A large number of simulations is conducted in this way, each making use (as input) of one of the realizations of the spatial distribution of the model parameters. Each of the produced outputs contains detailed information on the distributions of the sought variables, say, space and time distributions of water levels and contaminant concentrations. Each of these spatial and temporal functions, for example the concentration $c_i(x,y,z,t)$, where subscript i indicates the simulation number, is a sample taken from the ensemble space. The ensemble statistics described in Subs. 10.1.3 can then be applied to this information. In this way, instead of a single deterministic prediction, obtained by solving the given mathematical model with known parameters, we obtain many solutions, one for each realization of the parameter field. From them, we obtain the statistical characterization of the solution. For example, for a specified location and a specified point in time, we obtain the mean concentration, $\overline{c}(x, y, z, t)$, its standard deviation $\sigma_c(x, y, z, t)$, as well as its correlation length and correlation time. The correlation length tells us the persistence of a value in the spatial direction (for example, if one value is observed at a certain location, how far away may we anticipate that value to stay about the same) and in time (how long into the future that value will not change significantly). In fact, the Monte Carlo method produces much more than merely statistical information. The large number of (artificially produced) samples allows us to construct a histogram of any output prediction, such as concentration at a fixed location and fixed time, and gives the probability distribution of that quantity.

By applying a probabilistic (or statistical) analysis to these many 'equally likely to occur' outcomes, we can provide quantitative, albeit probabilistic, answers to questions like, 'what is the probability that the water level will be below a certain value at this location and time, in response to certain anticipated precipitation and imposed pumping'. Such information can, in turn, be used to answer management questions, such as 'what is the probability that the aquifer will no longer be able to sustain the current production rate within a specified future period?', 'what is the probability that a given site will be invaded by a contaminant plume at concentrations that exceed dictated limits?', or 'what is the *risk* to public health?' With such risk information, mitigation measures can be undertaken. We have used here the term 'risk' to mean 'the probability of certain undesirable event happening', e.g., a certain concentration limit being exceeded. In management, this means the failure of a designed system. Hence, another definition is that *risk is the probability of failure of a designed system.* The *reliability* of a system is then the complementary part of risk (i.e., reliability $= 1 -$ risk). When applied to groundwater flow and transport, the Monte Carlo method, as described above, can produce certain risk factors, such as the risk of supplying pumped water with certain carcinogen exceeding a specified MCL (maximum contaminant level). This information can be used as part of the input to a complex system, such as the risk to public health. The evaluation of the probabilistic risk of a complex system, containing many independent or correlated risk factors, is known as *risk analysis*, or *safety assessment.*

The Monte Carlo method is a versatile technique that is not limited only to modeling under conditions of parameter uncertainty, as described above. There are many other uncertainties associated with the modeling effort. For example, natural replenishment, which is a random phenomenon resulting from precipitation, is usually introduced as input to groundwater models. By analyzing its historical records, the statistical characteristics of precipitation at monitoring points can be obtained and then used to derive synthetic sequences of future precipitation, with different probabilities of occurrence.

The relationship between precipitation and replenishment can, for example, be obtained from the discussion in Subs. 1.1.6. A similar statistical analysis can be conducted for river stages, which often serve as boundary conditions in groundwater model. Another typical source of uncertainty is associated with the location of domain boundaries, or with boundaries of stratigraphic units of different hydraulic properties.

Although the Monte Carlo procedure seems simple and straightforward, it is actually not; let us discuss some issues associated with it.

First, to generate the many realizations of the spatial distribution of transmissivity and storativity, we need to determine their probability distributions, i.e., their *probability density functions* (pdf's). In most fields, however, the amount of measured data is insufficient for such purpose. Fortunately, based on a few studies in which a large quantity of data was indeed available for a given site, or could be compiled from different similar sites, it was concluded that the hydraulic conductivity is *log-normally* distributed (e.g., Freeze, 1975; Hoeksema and Kitanidis, 1985a; Gelhar 1986, 1993). The same conclusion may be extended to transmissivity and storativity. With this assumption, the full determination of the pdf's is reduced to determining only two statistical moments: the mean and the standard deviation of the required distributions.

To determine these statistical moments of the pdf, a large number of samples in the ensemble space is needed. Unfortunately, such ensemble space does not exist in groundwater modeling, as each investigated hydrogeological field is unique. As discussed in Subs. 10.1.5, this obstacle can be overcome by making the assumption of *ergodicity*. The ergodicity hypothesis allows us to conduct the spatial statistical analysis by using data collected at different locations of the same field, and using the results as the sought ensemble statistics. However, often, the amount of spatial data is insufficient for producing a statistical model with a spatial trend. In other words, we may not be able to reliably obtain a mean that varies from location to location. Most likely, we can only obtain a statistically homogeneous mean, meaning that a constant mean is obtained for the entire field.

For the second moment, e.g., the covariance, the amount of data is most definitely insufficient for determining its spatial trend. Hence, the covariance is almost always assumed to be statistically homogeneous; in other words, we assume that the variance and the correlation length are everywhere the same. Due to the lack of sufficient data for the construction of a detailed *empirical statistical model*, the simulated results are often questionable

The generation of random realizations, required in the Monte Carlo simulation, calls for the generation of a sequence of random numbers. Although such sequence can be obtained, for example, by throwing a dice repeatedly, we use a computerized *pseudo random number generator*. This is based on a mathematical algorithm, programmed for a computer, that can generate a seemingly random sequence of numbers, say, between 0 and 1, with a certain precision (single, or double precision). Actually, the process is only *pseudo-random*, because, given the same 'seed number', the same sequence of numbers will be generated every time. Using an algebraic transformation, meaning replacing one variable by another, defined by a functional relation, this sequence of random numbers, can be mapped onto a Gaussian (normal) probability distribution. With another transformation, the normal probability distribution can be mapped onto a log-normal probability distribution (Press *et al.*, 1992).

Armed with a random number generator and a probability distribution, we can now generate the random fields (of parameters, such as storativity and transmissivity) needed as input for the Monte Carlo simulations of the considered mathematical model. We divide the modeled inhomogeneous domain into a number of small cells, each assumed to be homogeneous. Selecting one cell to start from, we can 'randomly' assign to it a parameter value, say, the value of transmissivity. We then move to the next cell and assign to it another random transmissivity value. We continue this process, until the entire transmissivity field is defined. To obtain these transmissivity values, we start by inserting a 'seed number' (between zero and one) into a random number generator. The random number generator then produces a sequence of pseudo-random (i.e., almost random) numbers, uniformly distributed between zero and one. Based on the assumed pdf (for example, log-normal distribution), and a provided mean and standard deviation, these numbers are then mapped (= transformed) into the sought transmissivity values. Note that by conducting a statistical analysis on these random transmissivity values, we should return to the same mean and standard deviation originally used for their generation. These transmissivity values can then be assigned, cell by cell, as described above. This process, however, generates an *incoherent* random parameter field that does not exhibit a spatial correlation, as a real field should. In other words, the value selected for one cell bears no correlation with those assigned to neighboring cells, and the resulting field is rougher than it should be. In fact, the smaller the cell, the rougher is the field. In reality, however, if the value in one cell is higher than average, there is a high probability that the values at neighboring cells will also be higher than the average. Thus, the values at neighboring cells should be *conditionally* generated, on the basis of known information concerning the covariance. The correct generation of spatially correlated random parameter field is discussed in Subs. 10.2.4.

Furthermore, once the transmissivity has been generated for all cells, we need to generate the spatial distribution of storativity. If we allow the random number generator to arbitrarily select a storativity value for individual cells, that value may not be physically feasible, as it has been observed that a large transmissivity is often associated with a low storativity (Freeze, 1975; Dagan, 1979). The conclusion is that the selection of transmissivity and storativity must be based on a *joint probability density function* for these two parameters. This function can be obtained by compiling the storativity and transmissivity data pairs on a histogram that sorts these values into ranges.

The above discussion (in which we have used transmissivity and storativity just as examples of two porous medium parameters) indicates that the generation of a multidimensional, multivariate, and spatially correlated random field of the relevant model parameters, to be used as input in a Monte Carlo simulation, is a complicated process; special algorithms are required for its generation (e.g., Mantoglou and Wilson, 1982; Mantoglou and Gelhar, 1987; Tompson *et al.*, 1989; Tompson and Gelhar, 1990; Robin *et al.*, 1993; Bellin and Rubin, 1996). Obviously, if sufficient attention is not paid, and a wrong representation is made of the random parameter fields, a wrong output will be produced. The generation of random parameter fields is further discussed in the next subsection.

10.2.4 Generation of random field

Generally, there are two types of random parameter fields that can be generated: *conditional* and *unconditional*. The conditional (or *constrained*) random parameter field must satisfy the requirement that its values at sampled points should be exactly equal to those actually measured, or observed, there. These measured values are *true* values (barring measurement and other errors); hence, the generated realizations should conform to this constraint. In an unconditional (i.e., *unconstrained*) random parameter field generation, the observed values are ignored. These types, and the techniques used to generate them, are discussed below.

A. Generation of an unconditional one-dimensional random field

There exist a number of algorithms for the generation of an unconditional random parameter field, e.g., the spectral method, the autoregressive moving average method, and the turning bands method, to mention just a few. In view of the declared scope of this chapter, we shall demonstrate below, in some detail, only the *spectral method*, as applied to the generation of a one-dimensional random parameter field. The purpose is to demonstrate the concepts developed so far, in particular the concepts of ensemble and spatial averages, homogeneous and inhomogeneous mean and standard deviation, spatial covariance, and correlation length. The generation of a multidimensional random parameter field is mathematically more challenging, but can generally be constructed by using a summation of a number of unidirectional processes. We shall also mention briefly a very simple case, based on the *turning bands method*. More information on other, more advanced, algorithms can be found in books on geostatistics (e.g., Journel and Huijbregts, 1978; Christakos, 1992).

Figure 10.2.4 compiles 23 core samples of sea bottom sediments taken at a shallow water site (Badiey *et al.*, 1994, 1996). The left diagram plots the shear modulus (on a logarithmic scale) versus depth, as inferred from the number of blow counts needed to drive a pile a fixed distance. The right diagram gives the porosity profile interpreted by using an empirical formula.

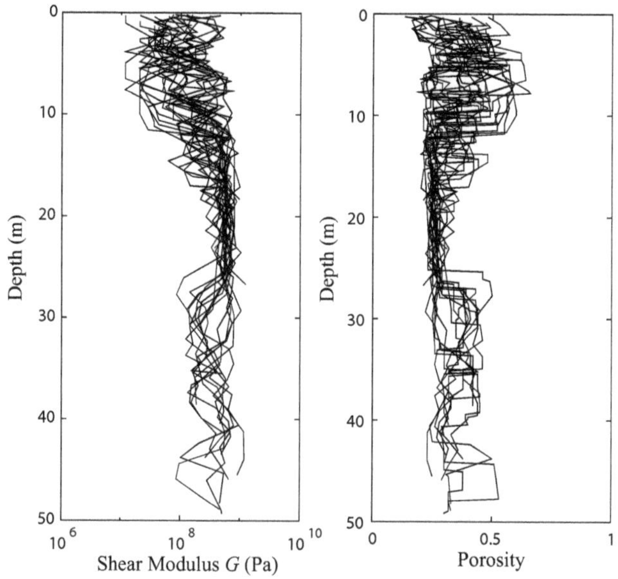

Figure 10.2.4: Shear modulus and porosity versus depth for 23 cores at the New Jersey Atlantic Generating Station site (Badiey *et al.*, 1996).

Our goal is to analyze the spatial variability of the site and to generate a large number of synthetic core profiles that exhibit the same geostatistical features, to be used as input in Monte Carlo simulations.

We start by observing in Fig. 10.2.4 that the soil beneath the seabed is highly heterogeneous in the depth direction. In the horizontal direction, we find that cores are spaced too far apart to be correlated, and that the few available cores do not permit reliable interpolation in the horizontal direction. In other words, the small amount of data does not allow us to construct a reliable three-dimensional spatial distribution of soil properties. Hence, these data will be used only for analyzing the spatial variability with depth.

Accordingly, we assume that the available cores constitute 23 samples that form an *ensemble* space. These data offer us the opportunity to determine a spatially dependent ensemble average. Based on (10.1.2) and (10.1.4), we can obtain both the mean and the standard deviation as functions of depth, z. Particularly, we assume that the shear modulus is log-normally distributed (i.e., $Y(z) = \log G(z)$); hence, the average is taken with respect to $Y(z)$. The result is presented in Fig. 10.2.5 as the mean profile plus and minus the standard deviation envelopes. We can clearly observe that the soil beneath the seabed has a very strong trend in the depth direction, indicating that its mean properties are certainly not homogeneous. Therefore, we need to express the mean logarithmic shear modulus and the porosity, ϕ, as $\overline{Y}(z)$ and $\overline{\phi}(z)$, respectively. We also observe that not only the mean, but also

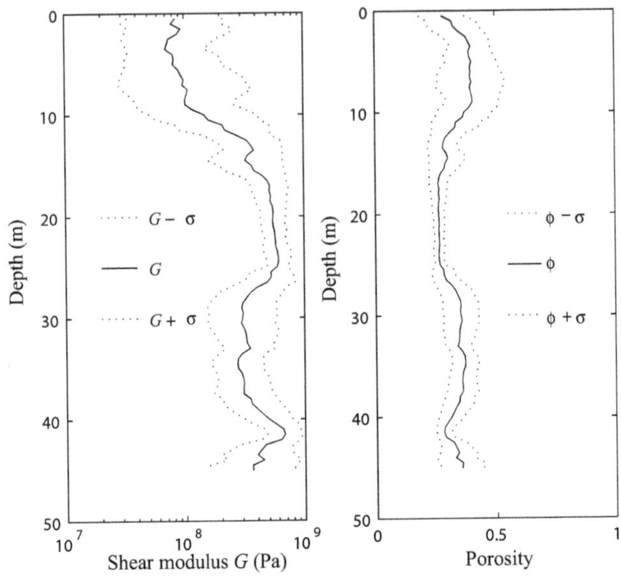

Figure 10.2.5: Mean and plus/minus standard deviation envelopes for shear modulus and porosity (Badiey *et al.*, 1996).

the standard deviation varies with depth; hence, the standard deviations are expressed as $\sigma_Y(z)$ and $\sigma_\phi(z)$.

Next, we need to analyze the covariance (autocovariance), as given by (10.1.5) and (10.1.8). As discussed earlier, with respect to the variance, it is clear that the covariance is also depth-dependent. However, both are based on the observation that the profiles seem to have the same correlation length in that direction. In the absence of sufficient data for a more detailed analysis, we assume that the covariance has certain universal features. For example, we assume that the statistically inhomogeneous covariance can be expressed as a homogeneous correlation coefficient, modulated by an inhomogeneous variance, $\sigma_Y^2(z)$, in the form of an envelope function (Priestley, 1965, 1967),

$$c_{YY}(z, \chi) = \sigma_Y^2(z) R_Y(\chi), \tag{10.2.32}$$

where χ is the distance between any two points, and R_Y is the autocorrelation coefficient defined in (10.1.9). This assumption allows us to construct a single correlation coefficient function, similar to the semivariogram idea used in kriging. This empirically constructed relation is shown in Fig. 10.2.6. It is a fairly typical correlation profile. The integral scale, a measure of the correlation length, defined as the integral of the positive part of the correlation coefficient (see (10.1.10)), is 1.5 m.

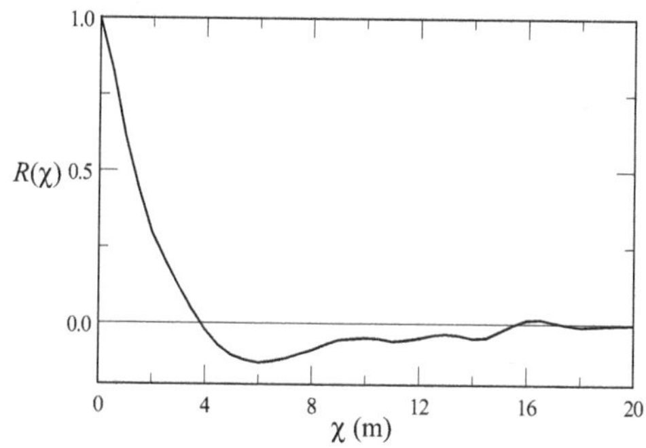

Figure 10.2.6: Correlation coefficient obtained by performing ensemble statistics on core data.

To prepare for the application of the spectral technique for the generation of a random field, we need the *power spectral density function*, defined as the Wiener-Khinchine transformation of the covariance (Bendat and Piersol, 2000). Applying this transformation to the correlation coefficient

$$S_Y^o(\omega) = \frac{1}{\pi}\int_{-\infty}^{\infty} R_Y(\chi)e^{-i\omega\chi}\,d\chi, \tag{10.2.33}$$

where ω is a spectral frequency, we can express the power spectral density as

$$S_Y(z,\omega) = \sigma_Y^2(z)S_Y^o(\omega). \tag{10.2.34}$$

The function $S_Y^o(\omega)$ is plotted in Fig. 10.2.7.

Without going into further mathematical details, we shall generate multiple realizations of the one-dimensional, nonstationary profiles as the inverse of a summation of Fourier components of discrete frequencies, each shifted by a random phase (Shinozuka and Jan, 1972),

$$Y(z) = \overline{Y}(z) + 2\,\sigma_Y(z)\sum_{n=0}^{N-1}\sqrt{S_Y^o(\omega_n)\,\Delta\omega}\,\cos(n\,\Delta\omega\,z + \theta_n), \tag{10.2.35}$$

where θ_n is a random phase angle, uniformly distributed over $[0, 2\pi]$, $\Delta\omega$ is a frequency increment, determined from the maximum profile length to be generated, Z_o, as $\Delta\omega = 2\pi/Z_o$, and N is the number of terms in the series, with $N = \omega_u/\Delta\omega$, and ω_u the cutoff frequency beyond which the power spectral density becomes negligible and may be truncated. Here, we have assumed that Y is normally distributed. Using the actual data of $\overline{Y}$, $\sigma_Y(z)$, and $S_Y^o(\omega)$ in Figs. 10.2.5 and 10.2.7, we can now generate any number of realizations of the shear modulus profile.

Figure 10.2.8 shows a collection of these simulated shear modulus profiles, to be compared with the real data presented in Fig. 10.2.4. We note that the

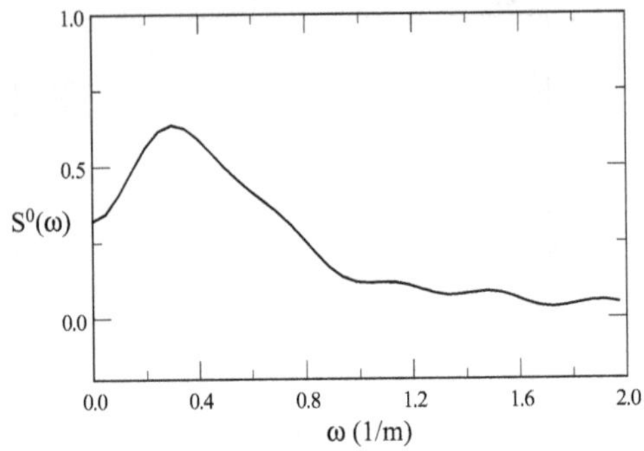

Figure 10.2.7: Power spectral density function.

real data are truncated at various depths, due to the irregularity of the field sampling, while the artificially generated ones cover the full depth of 45 m. The visual comparison of these two sets of profiles, in terms of the spatial trend of the mean, the scattering of the profiles, as well as the correlation length, appears to be satisfactory.

B. Generation of a multidimensional unconditional random field

So far, we have presented the generation algorithm only for a one-dimensional random parameter field. We shall now show that multidimensional random parameter fields can be generated as an extension of the unidirectional process. This is based on the *space transformation* theory, which states that points in an n-dimensional space ($n = 2, 3$) can always be mapped onto coordinates of a lower dimensional object, such as a set of lines or surfaces (Christakos, 1992). This theory allows us to treat the multidimensional random parameter field in a lower dimensional space. Typically, this is much easier. If the lower dimensional random field is second order homogeneous (weakly stationary), then the space transformation preserves the second order statistical structure, e.g., the correlation length, in the higher dimensional random field.

Following is a brief description of the *turning bands* method (Matheron, 1973; Journal, 1974; Mantoglou and Wilson, 1982) for the generation of two-dimensional, Gaussian, second order homogeneous random parameter fields.

Figure 10.2.9, shows a two-dimensional field. We select the origin, O, as a starting point for lines in the directions $\boldsymbol{\theta}_i$, $i = 1, \ldots, N$, where N is the number of lines, typically around 16. Along each line, we apply a unidirectional algorithm, e.g., the spectral method introduced above, to generate realizations of statistically homogeneous random profiles $Y_{1,i}(\theta_i)$, where the subscript 1 indicates the one-dimensional nature of the profile. We then lay the spatial grid, shown as dots in Fig. 10.2.9, at which the random parameter values will be generated. The parameter value at a grid point, represented by

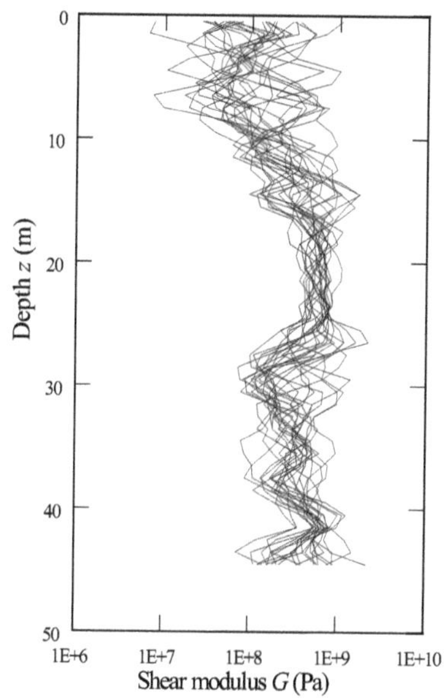

Figure 10.2.8: Simulated shear modulus profiles.

the vector **s**, can be expressed as the summation of the values on the N lines at the projection points, $\mathbf{s} \cdot \boldsymbol{\theta}_i$, i.e.,

$$Y_2(\mathbf{s}) = \frac{1}{\sqrt{N}} \sum_{i=1}^{N} Y_{1,i}(\mathbf{s} \cdot \boldsymbol{\theta}_i), \tag{10.2.36}$$

where Y_2 is the two-dimensional random parameter field. More can be found in Mantoglou and Wilson (1982), Tompson *et al.* (1989), and Christakos (1992).

C. Generation of a conditional random field

The techniques presented so far in this subsection generate only unconditional random fields, meaning that the fields are not constrained and cannot match the observed values at the sampled points. As the observed values are considered true values, it is desirable that the generated random fields be *conditional* to these observed values.

One technique to generate conditional random fields is to use kriging in conjunction with the unconditional random field generation algorithms. For example, we can create a conditional random parameter field based on the following formula (Journal and Huijbregts, 1978; Delhomme, 1979):

$$Y_c(\mathbf{x}) = Y^*(\mathbf{x}) + [Y_u(\mathbf{x}) - Y_u^*(\mathbf{x})], \tag{10.2.37}$$

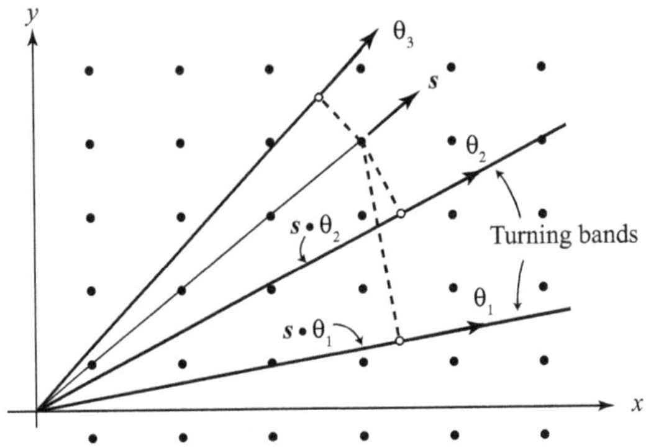

Figure 10.2.9: Turning bands algorithm for the generation of a two-dimensional random field.

in which $Y_c(\mathbf{x})$ represents the *conditionally* generated logarithm of the hydraulic conductivity at $\mathbf{x}$, and $Y_u(\mathbf{x})$ is the value generated by some *unconditional* algorithm. We also note that Y^* and Y_u^* are kriged values, based on data from the same set of locations. The difference between them is that Y^* is a kriged value based on the true (observed) parameter values, while Y_u^* is based on one realization of the unconditionally generated random field.

Since kriging is an exact interpolator, meaning that the exact data values are returned on the observed locations, it can easily be seen that at the sampled points, $\mathbf{x}_i$, the quantity in the square brackets in (10.2.37) vanishes. Hence, we obtain $Y_c(\mathbf{x}_i) = Y^*(\mathbf{x}_i) = Y(\mathbf{x}_i)$, in which Y stands for the true value, and the required condition is satisfied. Thus, (10.2.37) takes each unconditionally random field and turns it into a conditional one.

10.3 Examples of Uncertainty Problems

As mathematical models of groundwater flow and transport are often formulated as partial differential equations, one strategy to solve uncertainty problems of groundwater is to formulate and solve them as *stochastic differential equations* (SDEs). There are generally two types of SDEs. In one type, an SDE is a differential equation in which one or more of its parameters is described by a stochastic (random) process; thus, the resulting solution itself becomes a random process.

For example, consider the steady state groundwater flow equation (5.1.76),

$$\nabla\left(\mathbf{K}\cdot\nabla h\right) = P - R, \tag{10.3.1}$$

where P and R are pumping and recharge rates, respectively. This PDE is subject to the following boundary conditions,

$$h = h_o \quad \text{on } \Gamma_D; \qquad q_n = q_o \quad \text{on } \Gamma_N, \tag{10.3.2}$$

where Γ_D and Γ_N are the respective parts of the boundary. In (10.3.1) and (10.3.2), we can consider either the physical parameter, i.e. the hydraulic conductivity $\mathbf{K}$, and/or the forcing terms, i.e. the recharge R and the pumping P, and/or the prescribe head and flux boundary conditions, as random variables. If any of the above-mentioned quantities is random, then, obviously, the solution of (10.3.1) and (10.3.2), in terms of the piezometric head h, is random. Equation (10.3.1) hence is a stochastic differential equation.

One way to solve the above SDE is by Monte Carlo simulations, as described in Subs. 10.2.3. In this technique, we generate a large number of realizations of the random input variables, such as the random hydraulic conductivity field, the random recharge, and the random boundary conditions. We then attempt to solve the PDE, one realization at a time. Since, in a given realization, the parameters are all single-valued, the PDE is solved as a *deterministic* boundary value problem.

A difficulty involved in implementing the Monte Carlo method is the large amount of CPU time required for the simulations. A numerical solution of a (deterministic) three-dimensional heterogeneous regional scale aquifer, may require significant computing resources. In the Monte Carlo technique, the same problem is solved hundreds to a few thousand times, and this number may even increase exponentially when there exist, simultaneously, a number of correlated or uncorrelated uncertain variables. Nevertheless, as mentioned in Subs. 10.2.3, the Monte Carlo technique, due to its versatility, is still a powerful tool for handling parameter uncertainty in modeling flow and transport problems (Freeze, 1975; Smith and Freeze, 1979).

In this section, we focus on the second type of stochastic differential equation, in which we formulate a (partial) differential equation that directly describes the probability distribution function, or the statistical measures, of a stochastic process. One of the most known equation of this kind is the Fokker-Planck equation of statistical physics. In this equation, the time evolution of a probability distribution function is solved as if it were a deterministic equation.

This type of SDE, however, does not arise easily, and needs to be formulated on the basis of the problem on hand. Generally, the formulation, and its solution, is very difficult. This is particularly so if the random quantity involved is the coefficients of the PDE (such as random hydraulic conductivity), and if the governing equation is nonlinear. Typically, an approximation is needed in order to formulate these equations. Often, the approximations are crude, such that they may not have mathematical convergence properties, and their applications are limited to certain ranges of input random parameters (typically requires small variation from mean).

One such approximation technique is the *perturbation method.* By separating the different order terms in the perturbed equations, (approximate) differential equations are obtained in terms of the mean, standard deviation, etc., of the required solution. Particularly, in the first order approximation, the equation for the mean is often the same as the original deterministic equation. This fact often provides a certain meaning and justification to the averaging based deterministic solution, such as the megascopic scale averaging of heterogeneity (Subs. 1.3.4) and homogenization (Subs. 4.2.3).

The advantage of these *non-Monte-Carlo-simulation-based methods* is that they require the solution of the differential equation system only *once* (or once per approximation order); this should be compared to the hundreds or thousands of times required of the Monte Carlo simulations. Another advantage is that in certain simple cases, analytical solutions are possible. Analytical solutions provide explicit relations that show how the predicted outcomes are affected by the individual parameters. Monte Carlo simulations, and in fact all numerical solutions, give numerical results that correspond to a set of input values, without the general parametric insight.

There exist a range of techniques to formulate and solve stochastic differential (as well as integral) equations. These include the exact stochastic differential and integral equations (Cheng and Lafe, 1991; Cheng *et al.*, 1993), the perturbation method (Dagan, 1982a, 1985; Graham and McLaughlin, 1989a, 1989b), the Neumann expansion method (Zeitoun and Braester, 1991; Orr and Neuman, 1994; Serrano, 1995), the spectral analysis method (Bakr *et al.*, 1978; Gutjahr *et al.*, 1978; Gelhar and Axness, 1983; Dagan, 1988), Green's function method (Dagan, 1982a; Neuman and Orr, 1993; Rubin and Dagan, 1988, 1989), the stochastic finite element method (Sagar, 1978; Dettinger and Wilson, 1981; Guadagnini and Neuman, 1999a, 1999b), and the stochastic boundary element method (Lafe and Cheng, 1993; El Harrouni *et al.*, 1997; Naji *et al.*, 1999).

In the following subsections, we shall discuss a few of these solution techniques.

10.3.1 Random boundary conditions

Consider steady flow in an unconfined, homogeneous aquifer, subject to certain boundary conditions and to natural replenishment. The governing equation that predicts the piezometric head ($\equiv$ water table elevation), h, is

$$\nabla^2_{\mathbf{x}}\, \varphi(\mathbf{x}) = -f(\mathbf{x}), \qquad (10.3.3)$$

where $\varphi(\equiv h^2)$ is a potential (see Sec. 4.5 and Subs. 5.4.3), $f = 2N/\mathsf{K}$ is a dimensionless natural replenishment factor, N is the rate of natural replenishment from precipitation, K is the (constant) hydraulic conductivity, $\mathbf{x} = (x_1, x_2)$ is the Cartesian coordinate vector, $\nabla^2_{\mathbf{x}}$ $(= \partial^2/\partial x_1^2 + \partial^2/\partial x_2^2)$ denotes the two-dimensional Laplacian operator, and the subscript $\mathbf{x}$ is used to indicate the coordinate system for the Laplacian operator.

Let (10.3.3) be subject to two types of boundary conditions: a Dirichlet type and a Neumann type, one on each segment of the boundary. Thus,

$$\begin{aligned} \varphi &= h_o^2 && \text{on } \mathbf{x} \in \Gamma_\varphi, \\ \partial\varphi/\partial n &= -2Q'_o/\mathsf{K} && \text{on } \mathbf{x} \in \Gamma_Q, \end{aligned} \tag{10.3.4}$$

where Γ_φ and Γ_Q denote the boundary segments with Dirichlet and Neumann conditions, respectively, and Q'_o is the discharge per unit width of the aquifer's boundary (see (4.4.2)). For a well-posed problem, we also need to know the recharge rate, f. Let h_o, Q'_o, and the forcing function, f, be random functions of space. These random inputs to the solution system render the sought solution, $\varphi = \varphi(\mathbf{x})$, also a random function. Hence, the system represented by (10.3.3) and (10.3.4) is a *stochastic model.*

To extract from this system of equations useful information in terms of statistical moments, rather than employing the Monte Carlo simulation method (i.e., solving it one realization at a time), let us directly apply the ensemble average defined in (10.1.2) to the governing equation and boundary conditions. This leads to the following system, written in terms of the *mean* $\overline{\varphi}$,

$$\nabla^2_{\mathbf{x}}\,\overline{\varphi}(\mathbf{x}) = -\overline{f}(\mathbf{x}), \tag{10.3.5}$$

$$\begin{aligned} \overline{\varphi} &= \overline{h_o^2}, && \text{on } \mathbf{x} \in \Gamma_\varphi, \\ \partial\overline{\varphi}/\partial n &= -2\overline{Q'_o}/\mathsf{K}, && \text{on } \mathbf{x} \in \Gamma_Q. \end{aligned} \tag{10.3.6}$$

We notice that this system is *linear*, meaning that superposition is allowed. Hence, the ensemble averaging operator, which involves summation followed by division, can be applied to all the above equations without affecting their form. This means that the system (10.3.5) and (10.3.6) is identical in form to (10.3.3) and (10.3.4). The implication of the above observation is that we can solve the stochastic system by using the same solution tools as the deterministic system, whether analytical or numerical. Another implication is that if the *mean* of the boundary condition and forcing function (right hand side of (10.3.3)) are used as input, the deterministic solution is, in fact, the *mean response* of the system. The same conclusion cannot be made for the case of parameter uncertainty discussed in Subs. 10.3.2.

The next quantity of interest is the covariance, defined as

$$c_{ab}(\mathbf{x},\boldsymbol{\xi}) = \overline{[a(\mathbf{x}) - \overline{a}(\mathbf{x})][b(\boldsymbol{\xi}) - \overline{b}(\boldsymbol{\xi})]} = \overline{a'(\mathbf{x})b'(\boldsymbol{\xi})}, \tag{10.3.7}$$

(see (10.1.5)), where a and b stand for two different quantities, e.g., head and discharge, or precipitation rate and river stage, that we wish to correlate, and $\mathbf{x} \equiv (x_1, x_2)$ and $\boldsymbol{\xi} \equiv (\xi_1, \xi_2)$, are two locations in the considered domain. Although different symbols are used to distinguish between the two locations, $\mathbf{x}$ and $\boldsymbol{\xi}$ are points in the *same* Cartesian coordinate system. We also note that the prime denotes a deviation from the mean, e.g., $a' = a - \overline{a}$ and $b' = b - \overline{b}$. The covariance is the most general form of the second statistical moment. It correlates two different quantities at two different locations. The correlation of the *same* quantity at two different locations is expressed by the *autocovariance*, $c_{aa}(\mathbf{x},\boldsymbol{\xi})$. To complete the picture, the variance expresses the deviation form the mean of a single variable at the same location. The variance is defined as $c_{aa}(\mathbf{x},\mathbf{x}) = \sigma_a^2(\mathbf{x})$.

To derive the differential equation for the covariance, we subtract (10.3.5) from (10.3.3), obtaining

$$\nabla_{\mathbf{x}}^2 \varphi'(\mathbf{x}) = -f'(\mathbf{x}). \tag{10.3.8}$$

This equation governs the perturbed quantities (from the mean). The same treatment can be applied to the boundary conditions.

Next, we write an equation identical to (10.3.8), but at a different location, say, $\boldsymbol{\xi}$. We can multiply these two equations by each other: left side by left side, right side by right side, and take the ensemble average of the resultant equation. This produces the stochastic differential equation for the covariance

$$\nabla_{\mathbf{x}}^2 \nabla_{\boldsymbol{\xi}}^2 c_{\varphi\varphi}(\mathbf{x},\boldsymbol{\xi}) = c_{ff}(\mathbf{x},\boldsymbol{\xi}). \tag{10.3.9}$$

In the above equation, the product of the Laplacian operator can be expanded to yield

$$\nabla_{\mathbf{x}}^2 \nabla_{\boldsymbol{\xi}}^2 \equiv \frac{\partial^4}{\partial x_1^2 \partial \xi_1^2} + \frac{\partial^4}{\partial x_1^2 \partial \xi_2^2} + \frac{\partial^4}{\partial x_2^2 \partial \xi_1^2} + \frac{\partial^4}{\partial x_2^2 \partial \xi_2^2}. \tag{10.3.10}$$

We notice that the covariances, hence also the governing equation, are defined in a *four-dimensional space*, $(x_1, x_2; \xi_1, \xi_2)$. The following boundary conditions are needed

$$\begin{array}{ll} c_{\varphi\varphi}(\mathbf{x},\boldsymbol{\xi}) & \text{on } \mathbf{x},\boldsymbol{\xi} \in \Gamma_\varphi, \\ c_{Q'Q'}(\mathbf{x},\boldsymbol{\xi}) & \text{on } \mathbf{x},\boldsymbol{\xi} \in \Gamma_Q, \\ c_{\varphi Q'}(\mathbf{x},\boldsymbol{\xi}) & \text{on } \mathbf{x} \in \Gamma_\varphi \text{ and } \boldsymbol{\xi} \in \Gamma_Q, \\ c_{Q'\varphi}(\mathbf{x},\boldsymbol{\xi}) & \text{on } \mathbf{x} \in \Gamma_Q \text{ and } \boldsymbol{\xi} \in \Gamma_\varphi. \end{array} \tag{10.3.11}$$

Together with information on aquifer recharge, and on the autocovariance, c_{ff}, and its covariance with the aquifer's boundary conditions, $c_{f\varphi}$ and $c_{fQ'}$, the system composed of (10.3.10) and (10.3.11) constitutes a well-posed boundary value problem that can be solved by utilizing available numerical techniques. For example, Cheng and Lafe (1991) used the boundary element method. The discussion of such numerical techniques is beyond the scope of the present book. Instead, in what follows, we shall present a one-dimensional example for which an analytical solution can be derived. Through such example, we hope to gain not only an understanding of the stochastic differential equation solution system, but also of the physical meaning behind the obtained solution.

Consider a phreatic aquifer in the form of an elongated strip flanked on its two opposite sides by rivers with random stages: H_1 and H_2. The aquifer's replenishment, $N(x)$, is a result of random precipitation, (see Fig. 10.3.1). The governing equation (10.3.3), simplified to one dimension, becomes

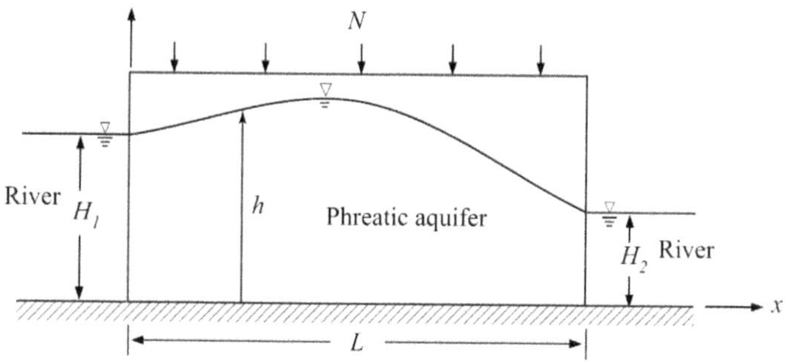

Figure 10.3.1: A phreatic aquifer flanked by two rivers with surface replenishment.

$$\frac{d^2\varphi(x)}{dx^2} = -f(x), \qquad (10.3.12)$$

where $f = 2N/\mathsf{K}$. The boundary conditions, expressed in terms of the present solution variable, are

$$\varphi(0) = \varphi_1 = H_1^2, \quad \varphi(L) = \varphi_2 = H_2^2. \qquad (10.3.13)$$

The solution to such problem is

$$\varphi(x) = -\int_0^x \int_0^{x''} f(x')\, dx'\, dx'' + \left[\varphi_2 - \varphi_1 + \int_0^L \int_0^{x''} f(x')\, dx'\, dx''\right] \frac{x}{L} + \varphi_1. \qquad (10.3.14)$$

Given the spatial distribution of the recharge function, $f(x)$, the above expression can be integrated. Although (10.3.14) looks like an explicit solution, $\varphi(x)$ is, in fact, a random function, because the input to the system, φ_1, φ_2 and f, are random. The solution (10.3.14) is just one realization of many, and, therefore, is of no value.

In a stochastic analysis, our goal is to obtain a solution expressed in terms of the statistical moments, such as the mean and standard deviation of the water table elevation, of the flux, etc. In order to find solutions in terms of these moments, the boundary conditions and recharge need to be processed in order to provide the relevant statistical measures. For example, f is processed into $\overline{f}$ and c_{ff}, φ_1 into $\overline{\varphi_1}$ and $\sigma^2_{\varphi_1}$, etc. With these boundary conditions, we can obtain a solution by taking the ensemble average of (10.3.14), yielding

$$\overline{\varphi}(x) = -\int_0^x \int_0^{x''} \overline{f}(x')\, dx'\, dx''$$

$$+\left[\overline{\varphi}_2 - \overline{\varphi}_1 + \int_0^L \int_0^{x''} \overline{f}(x')\, dx'\, dx''\right] \frac{x}{L} + \overline{\varphi}_1. \quad (10.3.15)$$

We note that (10.3.15) has the same form as (10.3.14), except that mean boundary conditions and mean recharge are used.

Although the solution for the mean is (almost) trivial, our interest lies in the second moment, i.e., the variance of the solution. For example, we may ask: 'If we know that the stage in the left river can vary by a certain magnitude, with respect to the mean, characterized by the standard deviation, and the right river stage by another magnitude, what will be the range of variations in the water table elevation in the middle of the aquifer?', or 'What will happen if the left river stage and right river stage are correlated?'. They can be positively correlated, meaning that if one river stage is high, then the other tends also to be high, because they may be influenced by the same weather pattern. Or, they may be negatively correlated, that is, if one is high, then the other is low, as when water is diverted from one river to the other. What will be the range of water table uncertainty in such cases? What is the effect of spatial correlation of the replenishment on aquifer response? Replenishment patterns may have a large correlation length, such that when replenishment is high at one location, it is also high at all locations. Or, it can be uncorrelated, such that it varies from one location to the next. In addition, the recharge rate and the river stage may be correlated, as they are both the consequence of precipitation, one through infiltration, and the other through runoff. All these questions can be quantitatively answered from the solution for the covariance presented below.

To obtain the solution for the covariance, we have first to find the solution for the deviations from the mean, $\varphi' = \varphi - \overline{\varphi}$. This is obtained by subtracting (10.3.15) from (10.3.14),

$$\varphi'(x) = -\int_0^x \int_0^{x''} f'(x')\, dx'\, dx'' + \left[\varphi_2' - \varphi_1' + \int_0^L \int_0^{x''} f'(x')\, dx'\, dx''\right] \frac{x}{L} + \varphi_1', \quad (10.3.16)$$

where f', φ_1', and φ_2' are deviations. Recall that the covariance is the ensemble average of the product of two fluctuating quantities, observed at two locations, x and ξ, say, $c_{ab}(x,\xi) = \overline{a'(x)b'(\xi)}$. We can write an equation identical to (10.3.16), but located at ξ, multiply the two equations by each other, and take the ensemble average of the result. This operation produces the rather lengthy, but general solution for the autocovariance of φ,

$$c_{\varphi\varphi}(x,\xi) = \left(1 - \frac{x}{L}\right)\left(1 - \frac{\xi}{L}\right)\sigma^2_{\varphi_1} + \frac{x\xi}{L^2}\sigma^2_{\varphi_2} + \left(\frac{x}{L} + \frac{\xi}{L} - \frac{2x\xi}{L^2}\right)\sigma_{\varphi_1\varphi_2}$$

$$
\begin{aligned}
&+\frac{x\xi}{L^2}\int_0^L\int_0^L\int_0^{\xi''}\int_0^{x''} c_{ff}(x',\xi')\,dx'\,d\xi'\,dx''\,d\xi'' \\
&-\frac{x}{L}\int_0^{\xi}\int_0^L\int_0^{\xi''}\int_0^{x''} c_{ff}(x',\xi')\,dx'\,d\xi'\,dx''\,d\xi'' \\
&-\frac{\xi}{L}\int_0^L\int_0^x\int_0^{\xi''}\int_0^{x''} c_{ff}(x',\xi')\,dx'\,d\xi'\,dx''\,d\xi'' \\
&+\int_0^{\xi}\int_0^x\int_0^{\xi''}\int_0^{x''} c_{ff}(x',\xi')\,dx'\,d\xi'\,dx''\,d\xi'' \\
&-\left(1-\frac{\xi}{L}\right)\int_0^x\int_0^{x''} c_{f\varphi_1}(x')\,dx'\,dx'' - \frac{\xi}{L}\int_0^x\int_0^{x''} c_{f\varphi_2}(x')\,dx'\,dx'' \\
&-\left(1-\frac{x}{L}\right)\int_0^{\xi}\int_0^{\xi''} c_{f\varphi_1}(\xi')\,d\xi'\,d\xi'' - \frac{x}{L}\int_0^{\xi}\int_0^{\xi''} c_{f\varphi_2}(\xi')\,d\xi'\,d\xi'' \\
&+\left(1-\frac{\xi}{L}\right)\frac{x}{L}\int_0^L\int_0^{x''} c_{f\varphi_1}(x')\,dx'\,dx'' + \frac{2x\xi}{L^2}\int_0^L\int_0^{x''} c_{f\varphi_2}(x')\,dx'\,dx'' \\
&+\left(1-\frac{x}{L}\right)\frac{\xi}{L}\int_0^L\int_0^{\xi''} c_{f\varphi_1}(\xi')\,d\xi'\,d\xi'',
\end{aligned}
\tag{10.3.17}
$$

where we note the definitions $\sigma_{\varphi_1}^2 \equiv \overline{\varphi_1'\varphi_1'}$, $\sigma_{\varphi_1\varphi_2} = \overline{\varphi_1'\varphi_2'}$, $c_{f\varphi_1}(x) = \overline{\varphi_1' f(x)}$, etc. The above solution can be simplified for some special cases.

First, consider the case of random river stages, but deterministic recharge. In this case, (10.3.17) drastically simplifies to

$$
c_{\varphi\varphi}(x,\xi) = \left(1-\frac{x}{L}\right)\left(1-\frac{\xi}{L}\right)\sigma_{\varphi_1}^2 + \frac{x\xi}{L^2}\sigma_{\varphi_2}^2 + \left(\frac{x}{L}+\frac{\xi}{L}-\frac{2x\xi}{L^2}\right)\sigma_{\varphi_1\varphi_2}.
\tag{10.3.18}
$$

Particularly, the variance becomes

$$
\sigma_{\varphi}^2(x) = \left(1-\frac{x}{L}\right)^2\sigma_{\varphi_1}^2 + \frac{x^2}{L^2}\sigma_{\varphi_2}^2 + 2\left(\frac{x}{L}-\frac{x^2}{L^2}\right)\sigma_{\varphi_1\varphi_2}.
\tag{10.3.19}
$$

In the above equation, we observe that σ_{φ}^2 changes from $\sigma_{\varphi}^2(0) = \sigma_{\varphi_1}^2$ on the left side of the aquifer to $\sigma_{\varphi}^2(L) = \sigma_{\varphi_2}^2$ on the right side, as dictated by the boundary conditions. It then varies quadratically between these two values. To gain a further understanding of this solution, let us be more specific.

Let us assume that the variance of river fluctuation on the right side is twice that on the left side, i.e., $\sigma_{\varphi_2}^2 = 2\sigma_{\varphi_1}^2$. We can normalize the variance in (10.3.19) by $\sigma_{\varphi_1}^2$ and rewrite it in the form

$$
\begin{aligned}
\frac{\sigma_{\varphi}^2(x)}{\sigma_{\varphi_1}^2} &= \left(1-\frac{x}{L}\right)^2 + \frac{x^2}{L^2}\frac{\sigma_{\varphi_2}^2}{\sigma_{\varphi_1}^2} + 2\frac{x}{L}\left(1-\frac{x}{L}\right)\frac{\sigma_{\varphi_2}}{\sigma_{\varphi_1}}R_{\varphi\varphi}(0,L) \\
&= 1 - 2\frac{x}{L} + 3\frac{x^2}{L^2} + 2\sqrt{2}\,\frac{x}{L}\left(1-\frac{x}{L}\right)R_{\varphi\varphi}(0,L),
\end{aligned}
\tag{10.3.20}
$$

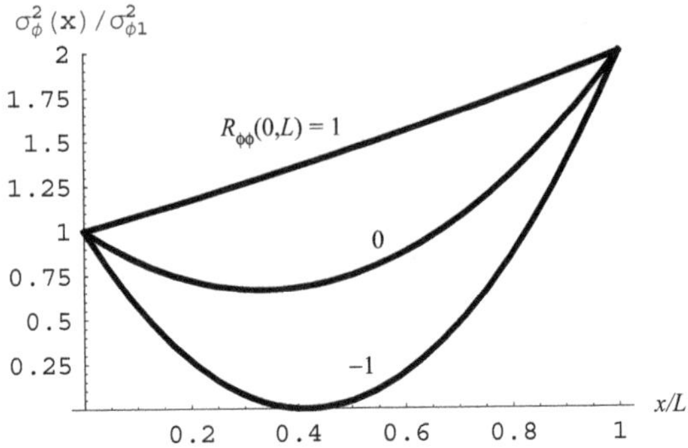

Figure 10.3.2: Head variance for the two river problem—random river stages.

where

$$R_{\varphi\varphi}(0, L) = \frac{c_{\varphi\varphi}(0, L)}{\sigma_{\varphi_1}\,\sigma_{\varphi_2}} = \frac{\sigma_{\varphi_1\varphi_2}}{\sigma_{\varphi_1}\,\sigma_{\varphi_2}}, \tag{10.3.21}$$

is the autocorrelation coefficient between the left and right river banks.

Next, let the left and right river banks be *uncorrelated*, i.e., $R_{\varphi\varphi}(0, L) = 0$. In this case, (10.3.20) is plotted as the middle curve in Fig. 10.3.2. In this figure, we make the interesting observation that the variance of the water table elevation, which characterizes the magnitude of uncertainty, is smaller in the aquifer's central portion than on either the left or the right portions. In fact, this observation should not be surprising—if the two river stages are not correlated, there is a higher chance that their effect will compensate each other. On the other hand, if the two river stages are *fully correlated*, $R_{\varphi\varphi}(0, L) = 1$, i.e., if one river stage is high, the other one will also be high, with full certainty, then the head variance is shown as the upper curve in Fig. 10.3.2. We observe that the head variance is bounded between the two boundary values. Finally, if the two stages are *fully*, but *negatively*, correlated, $R_{\varphi\varphi}(0, L) = -1$, and their effects always offset each other, and the variance is low in the middle of aquifer.

Following similar derivation, we can obtain the variance for the discharge per unit aquifer width, in the form

$$\sigma^2_{Q'} = \frac{\mathsf{K}^2}{4L^2}\left(\sigma^2_{\varphi_1} + \sigma^2_{\varphi_2} - 2\sigma_{\varphi_1\varphi_2}\right). \tag{10.3.22}$$

We note that, due to the restriction of the current model to the steady state assumption (i.e., large correlation time), the discharge variance is not a function of space. The continuity requirement dictates that the discharge be constant of space.

Similarly, we can normalize the above equation to the form

$$\frac{4L^2\,\sigma^2_{Q'}}{\mathsf{K}^2\,\sigma^2_{\varphi 1}} = 1 + \frac{\sigma^2_{\varphi 2}}{\sigma^2_{\varphi 1}} - 2\,\frac{\sigma_{\varphi 2}}{\sigma_{\varphi 1}}\,R_{\varphi\varphi}(0,L). \qquad (10.3.23)$$

By substituting values of $\sigma^2_{\varphi 2} = 2\sigma^2_{\varphi 1}$, and $R_{\varphi\varphi}(0,L) = -1, 0, 1$, we may readily conclude that the discharge variance is smallest when the two river stages are fully correlated, and largest if they are negatively so.

Next we examine the effect of random natural replenishment, particularly its correlation length, on the variance of water table elevations. In this case, we also assume that the river stages are deterministic, such that (10.3.17) reduces to

$$\begin{aligned} c_{\varphi\varphi}(x,\xi) = {} & \frac{x\xi}{L^2}\int_0^L\int_0^L\int_0^{\xi''}\int_0^{x''} c_{ff}(x',\xi')\,dx'\,d\xi'\,dx''\,d\xi'' \\ & -\frac{x}{L}\int_0^{\xi}\int_0^L\int_0^{\xi''}\int_0^{x''} c_{ff}(x',\xi')\,dx'\,d\xi'\,dx''\,d\xi'' \\ & -\frac{\xi}{L}\int_0^L\int_0^x\int_0^{\xi''}\int_0^{x''} c_{ff}(x',\xi')\,dx'\,d\xi'\,dx''\,d\xi'' \\ & +\int_0^{\xi}\int_0^x\int_0^{\xi''}\int_0^{x''} c_{ff}(x',\xi')\,dx'\,d\xi'\,dx''\,d\xi''. \end{aligned} \qquad (10.3.24)$$

The autocovariance function, $c_{ff}(x,\xi)$, can be constructed from the spatial distribution of precipitation, based on historical records, and the above expression can be integrated. However, as discussed in Subs. 10.2.1, due to the lack of dense spatial data, the autocovariance (or equivalently, the semivariogram) is, typically, assumed to be statistically homogeneous. In other words, rather than being a function of any two locations, x and ξ, the covariance $c_{ff}(x,\xi)$ is a function of the separating distance, $\chi(= |x-\xi|)$, only, and $c_{ff}(x,\xi) = c_{ff}(\chi)$. We also note that $\sigma_f^2 = c_{ff}(0)$ is constant everywhere.

A typical model of the autocovariance function, used for fitting observed data, takes the form

$$c_{ff}(\chi) = \sigma_f^2\,e^{-\chi^2/\ell^2}, \qquad (10.3.25)$$

where ℓ is a correlation length related to the integral scale I (see (10.1.10)) by

$$\ell = \frac{2}{\sqrt{\pi}}\,I. \qquad (10.3.26)$$

Despite the simplicity of the function in (10.3.25), the integration in (10.3.24) cannot be carried out analytically. Let us examine two limiting cases.

First, consider the fully correlated case, that is, the correlation length $\ell \to \infty$. Equation (10.3.25) becomes $c_{ff}(\chi) = \sigma_f^2$, and (10.3.24) can be easily integrated to yield

$$c_{\varphi\varphi}(x,\xi) = \frac{\sigma_f^2}{4}\,x\xi(L-x)(L-\xi). \qquad (10.3.27)$$

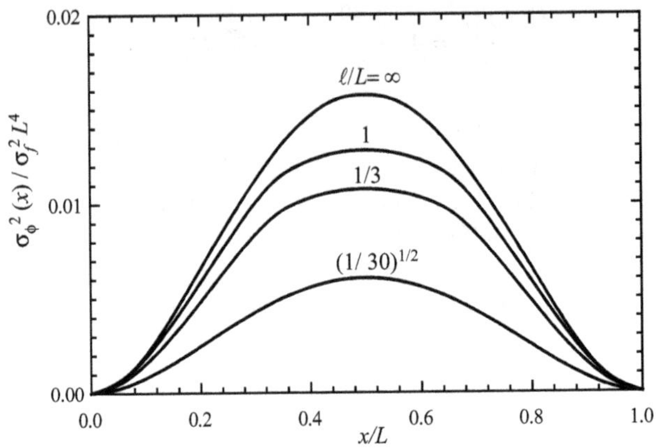

Figure 10.3.3: Head variance for the two river problem—random recharge.

The variance is given by

$$\sigma_\varphi^2(x) = \frac{\sigma_f^2}{4} x^2 (L-x)^2. \tag{10.3.28}$$

We plot this head variance, normalized by the recharge variance, as the top curve in Fig. 10.3.3. We observe that the variance is largest in the middle of the aquifer, and drops to zero, as constrained by the deterministic river stage conditions at the banks. On this figure, we also plot the head variance, obtained numerically (Cheng and Lafe, 1991), for various ℓ/L values. We notice that as the correlation length decreases, the recharge distribution fluctuates more in space, and the variance in water table elevations decreases.

The above stochastic analysis is based on the steady state groundwater flow equation (10.3.3). The transient analysis is presented by Cheng *et al.* (1993) and Cheng (2000).

Through the above examples, we have demonstrated, albeit only for rather simplified cases, that for aquifer flow problems with deterministic (i.e., fully known) aquifer parameters, but random boundary conditions and natural replenishment, a direct formulation and solution for the statistical moments, such as the mean and the covariances, is possible. Generally, of course, a numerical method is needed for such solution.

10.3.2 Uncertain parameters

The random boundary condition and recharge problems, discussed in the previous subsection, are *linear* ones. In other words, if we solve the problem with one realization, that is, one set of boundary and recharge conditions, and then solve it for another realization, the sum of these two solutions is also a solution that satisfies the sum of the governing equations and boundary conditions. This superposition property allows the ensemble average to be directly applied to the differential equation to produce the stochastic differential equations for the mean and the covariances.

The general stochastic groundwater flow and transport problems, however, are nonlinear. Consider, for example, the problem of *random parameters*, i.e., the hydraulic conductivity, or the dispersion coefficient, is a random spatial function. In such cases, the solution of a problem with a spatially varying hydraulic conductivity cannot be added to another solution with a different spatially varying hydraulic conductivity, as the sum of the two solutions *does not* satisfy the sum of these two governing equations with different hydraulic conductivities. As a consequence, the ensemble averaging cannot be directly applied to the governing PDE. Hence, we must seek a different approach, which, typically, will involves approximation. One such approximation method is the *perturbation method*, briefly introduced below.

A. Perturbation method

The perturbation method is a well-known technique for solving nonlinear, deterministic problems (Nayfeh, 2000). In this method, we express the parameters and the solution as a sum of terms with descending order of magnitude. For example, we may express the solution, h, which depends on the parameter K, in the form

$$
\begin{aligned}
h &= h_o + \epsilon h_1 + \epsilon^2 h_2 + \ldots, \\
\mathsf{K} &= \mathsf{K}_o + \epsilon \mathsf{K}_1 + \epsilon^2 \mathsf{K}_2 + \ldots,
\end{aligned}
\tag{10.3.29}
$$

where $h_o, h_1, h_2, \ldots$ and $\mathsf{K}_o, \mathsf{K}_1, \mathsf{K}_2, \ldots$ are perturbation quantities of different orders, and ϵ ($\ll 1$) is a small parameter. The above expressions are then substituted into the original (nonlinear) governing equation. The latter is expanded and sorted into a polynomial in terms of ϵ. Terms of the same polynomial degree are considered to be the same order of magnitude. When terms of the same order of magnitudes are collected and set to zero, we obtain a set of *linear* equations, each representing the contribution to the solution in a certain order of magnitude. These linear equations are solved and the final solution is the sum of the different order of magnitude terms.

For a problem described by a stochastic model, the perturbation method is applied in a different way. We start by considering multiple realizations in an ensemble space. For example, consider a groundwater problem with uncertain hydraulic conductivity, K. The hydraulic conductivity and the piezometric head, h, are random functions. From their ensemble spaces, we can find their (ensemble) mean, $\overline{\mathsf{K}}$ and $\overline{h}$. For each realization, the perturbations, or deviations, are defined as $\mathsf{K}' = \mathsf{K} - \overline{\mathsf{K}}$ and $h' = h - \overline{h}$. We note that by this definition, the ensemble mean of the perturbations is zero, $\overline{h'} \equiv 0$ and $\overline{\mathsf{K}'} \equiv 0$. Our goal is to approximately separate the solution into two parts: one that can be written and solved in terms of the mean, and one that is written and solved in terms of the perturbed quantities. However, for the separation to be reasonably accurate, similar to the deterministic perturbation technique described above, it is necessary that the perturbed quantities, K' and h', be *one order of magnitude smaller than the mean.*

From our experience, we know that in a highly heterogeneous formation, the hydraulic conductivity can vary over several orders of magnitude. Hence, its deviation from the mean is certainly not small! In the perturbation solution method, we often use $Y = \ln \mathsf{K}$, instead of K, as the parameter to be perturbed. Although the range of variation of Y is much smaller, the *small perturbation* assumption is still not satisfied. The perturbation method has been used as one of the major tools for analyzing stochastic models; reasonably good results (as confirmed by comparing with Monte Carlo simulations) have been obtained.

Let us consider a groundwater problem in which the goal is to predict the piezometric head, $h(\mathbf{x}, t)$, and/or the contaminant concentration, $c(\mathbf{x}, t)$, when the physical parameters appearing in the model are uncertain. Because of this uncertainty, we regard these variables as random functions that depend on a range of input parameters (data), such as the aquifer's parameters, the recharge, the boundary conditions, and the geometry of the domain. Some of the input parameters are known (certain), while others are not (i.e., they are uncertain). Let us denote the *uncertain input parameters* as $\boldsymbol{\zeta} \equiv (\zeta_1, \zeta_2, \ldots, \zeta_n)$. These parameters may cover many situations. As examples of such parameters, we may mention the constant, but uncertain, hydraulic conductivity of a homogeneous domain, or the discrete transmissivity or storativity values prescribed for each zone of a heterogeneous aquifer that is subdivided into a number of zones (or elements in the FE numerical solution). Other examples are the recharge rate in a zone, the value on a boundary segment, or, in the case of an uncertain boundary, the uncertain Cartesian coordinates (x, y) of a boundary node. Hence, in general, h is not only a function of space and time, but also a function of all the certain and uncertain input information. Since our focus here is on uncertainty, we shall show, explicitly, the dependency only on the uncertain parameters, and write $h = h(\boldsymbol{\zeta}) = h(\zeta_1, \zeta_2, \ldots, \zeta_n)$. In order to solve the uncertainty problem, we need to know the statistical moments of these uncertain input parameters, e.g., the mean, $\overline{\zeta_1}$, $\overline{\zeta_2}$, ..., the variance, $\sigma^2_{\zeta_1}$, $\sigma^2_{\zeta_2}$, ..., and the covariance, $\sigma_{\zeta_1\zeta_2}$, $\sigma_{\zeta_1\zeta_3}$, ... Typically, the statistical input stops at the second moment, as higher moments are either of no interest as a part of the prediction, or they cannot be reliably obtained from the limited amount of data.

Next, we consider a single realization of the uncertainty problem, with a given set of input parameters, $\zeta_1, \zeta_2, \ldots$. Each of these parameters can be expressed in terms of a perturbation from its mean, $\zeta_1' = \zeta_1 - \overline{\zeta_1}$, $\zeta_2' = \zeta_2 - \overline{\zeta_2}$, ... In the perturbation method, it is essential to assume that the perturbation is much smaller than the mean, i.e., $\zeta_1' \ll \overline{\zeta_1}$, $\zeta_2' \ll \overline{\zeta_2}$, ... With this assumption, we can write the Taylor series expansion of the function, $h(\boldsymbol{\zeta})$, with respect to these (small) perturbation parameters, around their respective mean values. For example,

$$h(\boldsymbol{\zeta}) = h(\overline{\boldsymbol{\zeta}}) + \sum_{i=1}^{n} \frac{\partial h(\overline{\boldsymbol{\zeta}})}{\partial \overline{\zeta_i}} \zeta_i' + \frac{1}{2} \sum_{i=1}^{n} \sum_{j=1}^{n} \frac{\partial^2 h(\overline{\boldsymbol{\zeta}})}{\partial \overline{\zeta_i} \partial \overline{\zeta_j}} \zeta_i' \zeta_j' + O(\zeta'^3), \tag{10.3.30}$$

where the higher order terms may, generally, be discarded.

In the above development, we have assumed that h is an analytical expression such that we can take its derivatives. In a general problem, however, h is obtained by using a numerical method, so that it is expressed in terms of discrete values at the nodes, $\mathbf{h} = \{h_1, h_2, \ldots\}$. We also note that the parameter vector $\boldsymbol{\zeta}$ can represent the different values of the same parameter in different zones, e.g., $\boldsymbol{\zeta} = \{\mathsf{K}_1, \mathsf{K}_2, \ldots\}$, where K is the hydraulic conductivity. Hence, the first derivative in (10.3.30) is generalized to $\partial \mathbf{h}/\partial \overline{\boldsymbol{\zeta}}$, which is similar to the sensitivity matrix (10.2.30). It can be approximately obtained by following the procedure described in Subs. 10.2.2.

We can obtain stochastic solutions of different orders. In the *first order approximation*, we keep only the first two terms on the right hand side of (10.3.30). Knowing that (10.3.30) represents a realization, we can take the ensemble mean of this equation, obtaining

$$\overline{h(\boldsymbol{\zeta})} \approx h(\overline{\boldsymbol{\zeta}}). \tag{10.3.31}$$

In performing the above operation, we have made use of the fact that the mean of the perturbations of a quantity is zero; here, $\overline{\zeta'} = 0$. Equation (10.3.31) is a simple and powerful result. It states that in order to obtain the mean of a solution, $\overline{h(\boldsymbol{\zeta})}$, that is accurate to first order (in terms of the Taylor series expansion), all we need to do is to solve a *deterministic* problem, using the estimated mean values of parameters, $\overline{\boldsymbol{\zeta}}$, as input data. This statement is similar to the one made in the preceding section for the random boundary condition problems (see (10.3.5) and (10.3.6)). However, the difference is that (10.3.5) and (10.3.6) are exact mathematical statements, while the statement in (10.3.31) is an *approximation.*

To obtain a more accurate estimate, we retain the third term in (10.3.30), and derive a *second order approximation.* After taking the ensemble mean, we obtain

$$\overline{h(\boldsymbol{\zeta})} \approx h(\overline{\boldsymbol{\zeta}}) + \frac{1}{2} \sum_{i=1}^{n} \sum_{j=1}^{n} \frac{\partial^2 h(\overline{\boldsymbol{\zeta}})}{\partial \overline{\zeta_i} \partial \overline{\zeta_j}} \sigma_{\zeta_i \zeta_j}, \tag{10.3.32}$$

where $\sigma_{\zeta_i \zeta_j} = \overline{\zeta_i' \zeta_j'}$ is the covariance between the properties ζ_i and ζ_j, with $i = j$, and $\sigma_{\zeta_i \zeta_i} = \sigma^2_{\zeta_i}$ expresses the variance. We note that the correction term for the mean is based on the covariances of the input parameters. The latter are readily available as a part of the input information.

To obtain the variance of the predicted piezometric head, we first subtract (10.3.32) from (10.3.30), obtaining

$$h'(\boldsymbol{\zeta}) = h(\boldsymbol{\zeta}) - h(\overline{\boldsymbol{\zeta}}) \approx \sum_{i=1}^{n} \frac{\partial h(\overline{\boldsymbol{\zeta}})}{\partial \overline{\zeta_i}} \zeta_i'. \tag{10.3.33}$$

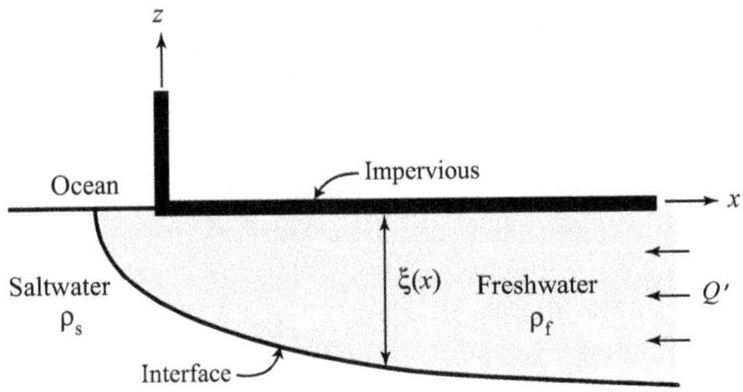

Figure 10.3.4: Glover's (1959) solution for seawater-freshwater interface.

In the above equation, we did not retain the higher order terms because they will produce the third and fourth moments that are not modeled in the second order approximation. We continue by taking the self product of (10.3.33), and then the ensemble mean. This yields the variance in the form

$$\sigma_h^2 \approx \sum_{i=1}^{n} \sum_{j=1}^{n} \frac{\partial h(\overline{\boldsymbol{\zeta}})}{\partial \overline{\zeta_i}} \frac{\partial h(\overline{\boldsymbol{\zeta}})}{\partial \overline{\zeta_j}} \sigma_{\zeta_i \zeta_j}. \tag{10.3.34}$$

It is thus the sum of all the input covariances, multiplied by the sensitivity coefficients. Or, since h is, generally, a vector, we can find the covariance between two such values, say h_1 and h_2, in the form $\sigma_{h_1 h_2}$.

Sometimes, we are interested in correlating two different quantities, such as head and discharge, or head and concentration. Denoting the second quantity as $g = g(\boldsymbol{\eta})$, where $\boldsymbol{\eta} = (\eta_1, \eta_2, \ldots)$ are the same or a different sets of random parameters, we can express the covariance between h and g in the form

$$\sigma_{hg} \approx \sum_{i=1}^{n} \sum_{j=1}^{m} \frac{\partial h(\overline{\boldsymbol{\zeta}})}{\partial \overline{\zeta_i}} \frac{\partial g(\overline{\boldsymbol{\eta}})}{\partial \overline{\eta_j}} \sigma_{\zeta_i \eta_j}. \tag{10.3.35}$$

B. An analytical example

To gain some physical understanding of the formulas developed above, let us examine a simple problem, for which an analytical solution can be obtained. Consider a two-dimensional seawater intrusion problem (Chap. 9). Figure 10.3.4 shows a vertical cross-section normal to the coast of a homogeneous (K = const.) confined coastal aquifer of infinite depth, with a steady state seawater-freshwater interface. A uniform groundwater discharge (per unit length of shoreline) of Q' enters the aquifer from the landside. Adopting the Ghyben-Herzberg approximation (Subs. 9.2.2), the seawater head, h_s, in the seawater region is considered to be constant and steady. The constant seawater and freshwater densities are denoted by ρ_s, ρ_f, respectively. The parabolic interface elevation, expressed by $z = \xi(x)$, takes the form of the Glover solution (Glover, 1959; Cheng and Ouazar, 1999),

$$\xi(x) = \sqrt{\frac{2\delta Q' x}{\mathsf{K}} + \frac{\delta^2 Q'^2}{\mathsf{K}^2}}, \tag{10.3.36}$$

where $\delta = \rho_f/(\rho_s - \rho_f)$, as introduced in (9.2.6).

With this data, we pose the following question: if the input parameters, e.g., K and Q', appearing in (10.3.36), are uncertain, and we know only their statistical measures, $\overline{\mathsf{K}}$, $\overline{Q'}$, σ_K^2, $\sigma_{Q'}^2$, and $\sigma_{KQ'}$, what is the mean shape of the interface, and what is its standard deviation?

To answer this question, we turn to the perturbation equations, (10.3.32) and (10.3.34), for mean and variance, respectively. In order to derive a shorter mathematical expression as the final result, we express (10.3.36) in a different way, by solving for x as a function of ξ,

$$x(\xi) = \frac{\mathsf{K}}{2\delta Q'}\xi^2 - \frac{\delta Q'}{2\mathsf{K}}. \tag{10.3.37}$$

Thus, rather than finding the mean depth of the interface, $\overline{\xi}$, at a given x, we seek the mean location, $\overline{x}$, of a specified interface depth, ξ. The results are consistent within the same perturbation approximation order. As observed in (10.3.32) and (10.3.34), to obtain the mean and variance, $\overline{x}(\xi)$ and $\sigma_x^2(\xi)$, respectively, we perform the differentiations, $\partial x/\partial\mathsf{K}$, $\partial x/\partial Q'$, etc. Since (10.3.37) is an analytic expression, we readily obtain

$$\begin{aligned}\overline{x} &= \frac{\overline{\mathsf{K}}}{2\delta\overline{Q'}}\xi^2 - \frac{\delta\overline{Q'}}{2\overline{\mathsf{K}}} \\ &\quad + \frac{\delta\overline{Q'}}{2\overline{\mathsf{K}}}\left[\frac{\overline{\mathsf{K}}^2}{\delta^2\overline{Q'}^2}\frac{\sigma_{Q'}^2}{\overline{Q'}^2}\xi^2 - \frac{\sigma_K^2}{\overline{\mathsf{K}}^2} + \left(1 - \frac{\overline{\mathsf{K}}^2\xi^2}{\delta^2\overline{Q'}^2}\right)\frac{\sigma_{KQ'}}{\overline{Q'}\,\overline{\mathsf{K}}}\right], \end{aligned}\tag{10.3.38}$$

$$\sigma_x^2 = \frac{1}{4}\left(\frac{\overline{\mathsf{K}}}{\delta\overline{Q'}}\xi^2 + \frac{\delta\overline{Q'}}{\overline{\mathsf{K}}}\right)^2\left(\frac{\sigma_{Q'}^2}{\overline{Q'}^2} + \frac{\sigma_K^2}{\overline{\mathsf{K}}^2} - 2\frac{\sigma_{KQ'}}{\overline{Q'}\,\overline{\mathsf{K}}}\right). \tag{10.3.39}$$

To obtain a better understanding of the above solution, let us consider the following numerical example: $\delta = 40$, $\overline{\mathsf{K}} = 69$ m/day and $\overline{Q'} = 3.9$ m^2/day. We also assume that the *coefficients of variation* (= the standard deviation divided by the mean) of both K and Q' are 20%. Hence, $\sigma_K^2/\overline{\mathsf{K}}^2 = \sigma_{Q'}^2/\overline{Q'}^2 = 0.04$. Finally, we assume that the outflow rate and the hydraulic conductivity are uncorrelated, and, hence, $\sigma_{KQ'} = 0$. With these data, we can calculate the mean from (10.3.38) and the variance from (10.3.39). The resulting mean interface location, together with the (plus/minus) standard deviation envelopes, are plotted in Fig. 10.3.5.

In this figure, we note that the range of uncertainty in the interface location is rather large, and that this range increases as we move inland. This happens

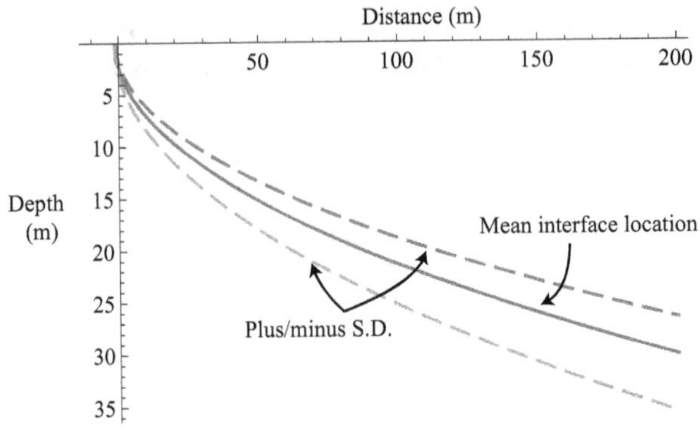

Figure 10.3.5: The mean seawater-freshwater interface, with ± standard deviation envelopes.

despite the fact that we have assumed only 20% for the coefficient of variation in the input parameters. In reality, these input uncertainties are likely to be larger. These predictions, of course, are inherent in the underlying physics. In fact, in Chap. 9 we learn that the seawater-freshwater interface is sensitive to the various physical conditions, as enhanced by the small density difference between seawater and freshwater.

Under such conditions, how do we cope with a degree of uncertainty that is too large to be acceptable for management purposes? The situation can be significantly improved if we make a few actual observations of the interface location. Once we know these (several) 'true' locations, the uncertainty problem is transformed into a *conditional probability* problem (Dagan and Zeitoun, 1988). In other words, at the observation points, the uncertainty (standard deviation) envelopes will be 'pinned' to zero width, due to the 'certainty' at these points. Then the uncertainty cannot grow in the unconstrained fashion shown in Fig. 10.3.5.

Additional analytical solutions of uncertainty problems can be found in the literature. For example, Cheng and Ouazar (1995) discuss the uncertain groundwater drawdown predicted by the Theis solution, resulting from uncertainty in the transmissivity and the storativity. Naji *et al.* (1998a) discuss the uncertain location of the interface in a coastal aquifer with pumping wells.

Let us recapitulate. In the above example, we have demonstrated the perturbation procedure for the solution of a stochastic model. First, we set up a deterministic model, whether analytical or numerical. Then, we acquire the statistical measures of the input data, mean and standard deviation, considered to be uncertain. The effect of parameter uncertainty (standard deviation) on the output uncertainty is enhanced (or dampened) by the *sensitivity* of the output with respect to that parameter, with sensitivity defined as the partial derivative given in (10.3.34). Depending on the case, the partial derivatives are obtained analytically or numerically. We need to find the approximate sensitivity of the solution, by perturbing the uncertain input data by a small amount and observe the changes in the output. In a numerical solution, the uncertain input parameters may be the many discrete values of the same physical parameter at different locations. Finally, the mean and the standard deviation of the predicted solution are assembled, based on (10.3.32) and (10.3.34).

Chapter 11

OPTIMIZATION, INVERSE, AND MANAGEMENT TOOLS

The management of a groundwater system—an aquifer, or a system of aquifers—alone, or conjunctively with surface water sources, aims at achieving certain goals through a set of decisions concerning the development and/or operation of the system. Typically, the same goal or goals can be achieved by different *management alternatives*, each involving a specific set of decisions. Management means the selection and implementation of the best, or *optimal* set of decisions.

To define what constitutes an optimal decision for a given system, a quantitative measure of achievement is needed. Usually, the latter is expressed in terms of certain economic outcomes, although other measures of achievement are also possible. Mathematically, this measure of achievement is expressed in the form of a (scalar) *objective function* associated with the behavior of the considered system. Determining the value of the objective function, requires information on the future behavior of the considered system, in response to the implementation of these decisions *prior* to implementing them. This information can be obtained from the solution of models that are believed to represent the physical reality of the system. In fact, most of this book is devoted to the construction of such models.

In general, there are three types of groundwater management problems: (a) problems associated with water quantity, such as water supply, water allocation, extraction operations, and conjunctive utilization of surface water and groundwater, (b) problems associated with water quality, such as controlling the quality of supplied water, protection of groundwater quality, and remediation of contaminated aquifers, and (c) problems associated with water policy, management directives, and regulations. Only the first two types will be discussed in this chapter. The discussion will focus on the fundamentals of selecting the optimal alternative set of decisions for aquifer management, i.e., one that maximizes or minimizes a specified objective function, without violating constraints imposed on the system. The simulation-optimization approach to the problem will be presented. Management problems belonging to the third type, must address such issues as the socio-economic aspects of allocating water resources to different economic sectors, monetary and so-

J. Bear, A.H.-D. Cheng, *Modeling Groundwater Flow and Contaminant Transport*, Theory and Applications of Transport in Porous Media 23,
DOI 10.1007/978-1-4020-6682-5_11,

cial costs, the impact on the environment, the effects on ecosystems, and the effects on the health of the population.

Altogether, the main objective of this chapter is to present a brief introduction to groundwater resources management, based on mathematical optimization techniques that are commonly used in practice. We shall also discuss some parameter identification and inverse problems, an important step in the groundwater modeling process (Step 7 in Subs. 1.2.2), which are typically solved by employing optimization techniques.

11.1 Groundwater Management

Not all decision alternatives are permissible, or *feasible*, as some may be forbidden by specified technical, economic, legal, and societal *constraints*. A management alternative will be considered unfeasible if the system's response to its implementation violates at least one specified constraint. In the sections to follow, we shall discuss some procedures for searching and selecting the optimal solution for certain types of problems.

Let us begin by presenting some typical examples of optimization terms and concepts that pertain to groundwater management. (See Sec. 11.2 for the definition of terms.)

Examples of state variables: (a) Water levels; (b) Concentrations of specified chemical species; (c) Extent of land subsidence; (d) Extent of seawater intrusion.

Examples of decision variables: (a) Number and locations of new (pumping and/or artificial recharge) wells; (b) Pumping schedule and rates of existing or new wells; (c) Extent of seawater intrusion (as this value is closely associated with the rate of freshwater discharge to the sea).

Examples of objective functions: (a) Total net benefits from operating the system during a specified period of time, or present worth of total net benefits, if timing of future costs and benefits is taken into account, and we wish to maximize this value; (b) Total cost of clean-up operations aimed at removing contaminants from an aquifer, and we wish to minimize this cost; (c) Total cost of a unit volume of water supplied to consumers, and we wish to minimize this cost; (d) Total consumption of energy required for operating the system, and we wish to minimize this quantity; (e) Sum of absolute values of the difference between certain desired water levels and actual ones (or sum of squares of differences) at all points of a grid covering the aquifer, and we wish to minimize this sum; (f) Total mass of a contaminant removed by clean-up operations, and we wish to maximize this amount; (g) Expected land subsidence, and we wish to minimize this quantity.

Generally, the value of an objective function is associated with certain relevant physical processes. For example, the objective function 'cost of pumping', which we wish to minimize, may depend on the total pumped volume of water, as well as on the water table elevations at the pumping wells during pumping (as these elevations affect the cost of energy required for raising the water to ground surface). Water table elevations, in turn, depend on aquifer properties, such as transmissivity and storativity, and on the number of wells, on their location, and on certain well characteristics.

Examples of constraints: (a) Water levels at specified locations should not rise above specified maximum elevations, or should not drop below specified minimum ones; (b) Spring discharge should not drop below a specified minimum; (c) Base flow in a stream fed by groundwater should not drop below a specified minimum; (d) Concentrations of certain chemicals in solution in the water pumped at specified locations should not exceed specified threshold values; (e) Land subsidence should not exceed specified values; (f) Total pumped water volume should satisfy at least the demand for water in a given region; (g) Pumping (and/or artificial recharge) rates should not exceed the installed pumping (and/or artificial recharge) capacities; (h) The residence time for water injected into an aquifer, before being pumped, should exceed a certain minimum period; (i) The length of the intruding seawater wedge in a coastal aquifer should not exceed a specified value.

Examples of management decisions: (a) What should be the total volume of water to be pumped from a considered aquifer, say annually. (b) Should the annual withdrawal from the aquifer be at a fixed rate, or may it vary from year to year? (c) What should be the spatial and temporal distributions of pumping and artificial recharge? (d) What should the quality of the pumped water be? (e) How should we allocate water of different qualities to different users, e.g., domestic, agricultural, and industrial ones, in order to achieve certain economic goals? (f) How can we balance the conjunctive use of groundwater and surface water? (g) How can groundwater availability be increased by making use of artificial recharge techniques, say, by injecting treated wastewater, or by injecting intercepted flash floods? (h) What should be the water levels maintained in streams and lakes that are hydraulically connected to the aquifer? (i) What should be the capacity of new installations for pumping and/or artificial recharge, their location, and the time schedule for their construction? (j) Where should we locate pumping wells for the removal of contaminants, within the framework of clean-up operations?

As stated earlier, in all cases, a specified goal can be achieved by implementing alternative sets of decisions. In order to reach the optimal set of decisions, we have to calculate the actual value of the objective function for every such considered set (= management alternative). In view of this statement of the groundwater management problem, it is obvious that forecasting the response of an aquifer system to a suggested management alternative is an intrinsic part of the procedure for determining the optimal management scheme. We must know the future values of relevant state variables that will occur in the aquifer as a result of implementing a proposed set of decisions (a) in order to examine whether they violate specified constraints, and (b) in order to examine the value of the resulting level of achieving the objective function. This book is dedicated to the tool that enables us to make the required forecasts, namely, the *model* of the investigated system. The definition of a model and the process of modeling were discussed in Sec. 1.2.

Sometimes, our knowledge of the modeled physical system is *uncertain*. For example, we may not know the system's physical parameters to a sufficient degree of accuracy, due insufficient measured data. In some cases, uncertainty may be associated with the prediction of future events, e.g., natural replenishment. Then the mathematical optimization model changes from *deterministic* to *stochastic* (Chap. 10). In fact, in practice, conditions of uncertainty are more the norm than an exception. In a *stochastic management problem*, the input parameters are given as statistical measures, e.g., as the mean and variance of these parameters. As a consequence, the optimal solution will depend on the desired reliability of prediction. In other words, the more we want to guarantee the success of the management scheme, of course, without violating the imposed constraints, the more conservative measures need to be taken, leading to results which are farther from being optimal.

11.2 Optimization

11.2.1 Optimization problem

In mathematics and in systems analysis, the term *optimization*, refers to the procedure in which one seeks to determine the *best* solution to a problem by assigning values to a set of problem variables, referred to as *decision variables*, or *design variables*, such that a certain function, called the *objective function*, which depends on these variables, will attain its maximum or its minimum value. The terms 'decision variable', and 'objective function' were defined earlier. While seeking the optimal values of the decision variables, we have to make sure that the constraints imposed on these values, or on the system's state variables, are not violated. In general, there are many solutions that satisfy the imposed constrains. Each such solution is referred to as a *feasible solution*. The optimal set of decision variables is obtained by systematically searching throughout the *decision variable space*, with the objective of finding the feasible solution (= set of decision variables) that constitutes the optimal solution, i.e., leads to either the maximum or the minimum value of the objective function, or sufficiently close to these values. We notice that, as optimization is a mathematical procedure, the objective must be quantifiable. This may involve assigning numerical values to qualitative descriptions.

An optimization problem can be *constrained* or *unconstrained*. In an unconstrained problem, all solutions defined in the *decision variables space* are *feasible solutions* (although only one of these feasible solutions is the optimal one). In a constrained problem, certain ranges of the decision variables produce *infeasible solutions*; these regions must be excluded from the search for optimal solution. The infeasible solution regions are delineated by the imposed constraints. The latter can be applied to the decision variables themselves, e.g., the pumping rate at a well may be limited to a maximum value, determined by the capacity of the equipment, and to a minimum value set by the feasible economical operating condition. In general, the constraints can be algebraic equalities or inequalities, involving linear or nonlinear combinations of the decision variables. The constraints may also depend, implicitly, on the decision variables through another system that has to be solved. For example, we may want to maximize the pumping in a region, but may also want to prevent the resulting land subsidence from endangering a historical building. In such a case, a separate model—the land subsidence model—should be added as a constraint.

With the above introduction, we can now formulate the mathematical statement of optimization. We shall use Z (a real-value) to denote the objective function. The value of Z is assumed to depend on a set of n decision variables, x_1, x_2, ..., $x_n(\equiv \mathbf{x})$. We shall assume the existence of a set of m functions f_j, $j = 1, \ldots, m$, that express m constraints, each of which may depend on location, time, and many other parameters. The jth-constraint is expressed by making f_j either greater than, equal to, or less than a certain constraining value, b_j. Our goal is to determine the set of decision variables, $\mathbf{x}$, so as to maximize or minimize the objective function, Z, without violating the m constraints. Let us put this statement into the following concise mathematical formulation:

Determine the values of the decision variables, x_i, $i = 1, 2, \ldots, n$, so as to

$$\begin{aligned}&(\text{Maximize, Minimize})\; Z(\mathbf{x}), \text{with respect to } \mathbf{x},\\ &\text{subject to:}\quad f_j(\mathbf{x})\;(>, =, <)\; b_j,\; j = 1, \ldots, m.\end{aligned} \tag{11.2.1}$$

The process of finding the solution to this problem is called *optimization*. Note that the decision variables may be limited to a set of discrete values.

It is important to note that when an optimization problem is associated with a groundwater system, we have to include 'satisfying the flow model' as one of the constraints. This means that the mass flow model, comprising the balance equation that describes the flow in the groundwater system, say in terms of water table elevations, together with the initial and boundary conditions, constitutes a set of constraints that have to be satisfied. If groundwater quality aspects are also involved, then satisfying the appropriate transport model, say in terms of concentrations of the participating chemical species, also constitutes a constraint. This comment is valid for both steady state and transient flow and transport problems.

As mentioned above, the process of optimization involves conducting a search in the multidimensional space that contains all the feasible values of the decision variables. Generally, this space is continuous and contains an infinite number of possible values. Although, in the solution process, the search space is discretized, its multidimensional nature makes a brute force search difficult, if not impossible. Hence, a well guided and efficient search algorithm is essential to the success of optimization.

There exist a large number of different optimization techniques; only a few of them will be discussed in this chapter.

Linear programming is an easy and efficient technique for finding the optimal solution for the class of problems in which the objective function is *linearly* dependent on the decision variables, and all the constraints are also *linear* equalities, or inequalities. This technique is discussed in Subs. 11.2.2. If either the objective function, or the constraints, or both, are nonlinear functions of the decision variables, linear programming is no longer applicable, and *nonlinear programming* techniques are required. Nonlinear constraints, however, are rather difficult to handle. Often, problems with such constraints are converted to ones *without* constraints ($\equiv$ *unconstrained optimization problem*) by utilizing a *penalty* method. These unconstrained problems can then be solved by the *gradient search* method described in Subs. 11.2.4. The gradient method requires the objective function to be *twice differentiable* with respect to the decision variables, in order to facilitate the identification of search direction.

Many optimization problems have objective functions and search spaces that are discontinuous, discrete, multiply connected, or nonconvex, with each such space having a large number of local optima. It is difficult to handle such cases by the traditional techniques, such as the *gradient search* and *nonlinear programming* ones. In recent years, a new generation of optimization techniques, generally known as *metaheuristics*, e.g., *genetic algorithm* and *simulated annealing*, have emerged for dealing with these problems; some of them are discussed in Subs. 11.2.5.

11.2.2 Linear programming

Linear programming (LP) is a technique that solves optimization problems with linear objective functions and linear equality and inequality constraints. We can set up the mathematical formulation of such a problem in the following *standard form:*

> Given an objective function Z, which is linearly dependent on a set of n decision variables, $\{x_1, x_2, \ldots, x_n\}$ ($\equiv \mathbf{x}$), determine the values of $x_1, x_2, \ldots, x_n$ so as to *maximize* Z,
>
> $$\text{Maximize } Z(\mathbf{x}) = \sum_{i=1}^{n} c_i x_i, \qquad (11.2.2)$$
>
> where the c_i's are constant coefficients, subject to the $m = (m_1 + m_2 + m_3)$ constraints,
>
> $$\sum_{i=1}^{n} a_{ij} x_i \leq b_j, \quad j = 1, \ldots, m_1, \qquad (11.2.3)$$

$$\sum_{i=1}^{n} a_{ij}x_i \geq b_j, \quad j = m_1 + 1, \ldots, m_1 + m_2, \tag{11.2.4}$$

$$\sum_{i=1}^{i} a_{ij}x_i = b_j, \quad j = m_1 + m_2 + 1, \ldots, m_1 + m_2 + m_3, \tag{11.2.5}$$

where, following common convention, the b_j's are set to be non-negative, and the a_{ij}'s can be positive, negative, or zero. The coefficients, a_{ij}, b_i, c_i must all be known constants.

In the above statement, following the usual convention, we have defined LP-problem as a maximization one. We could equally have defined it as a minimization problem. Particularly, we note that any minimization problem, say Minimize Z, can be easily transformed into a maximization problem, if we define $Z' = -Z$, and then use Maximize Z'.

Also, following convention, we require the decision variables to satisfy the *primary constraints*, i.e., that the x_i's be real, non-negative values,

$$x_i \geq 0, \quad i = 1, \ldots, n. \tag{11.2.6}$$

Similar to the above, if a decision variable, x_i, is negative, we can define another one, $x'_i(= -x_i)$, which is positive. Or, if $x_i \geq -c$, where c is a positive constant, we can shift it by that amount, using a new variable, $x'_i = x_i + c$, such that $x'_i \geq 0$.

Any vector $\mathbf{x}$ that satisfies all the constraints in (11.2.6)–(11.2.5) is a *feasible solution.* The feasible solution that yields the maximum (optimum) value of the objective function in (11.2.2) is called the *optimal solution.* For management purposes, we may consider each feasible solution a *decision.* Our objective is to select the particular decision that maximizes the objective function, as prescribed by the decision maker.

Once an optimization problem is cast into the standard linear programming form, (11.2.2)–(11.2.5), the common algebraic procedure for solving it is the *simplex method*, developed by Dantzig in the late 40's (Dantzig, 1963). This procedure is well suited for solution by a digital computer. In what follows, we shall use a simple example to explain the concept and the critical steps leading to an LP solution.

A. A graphical view of LP

When an LP optimization problem involves no more than three decision variables, it is possible to *visualize* the solution through a graphical presentation. We shall use this fact to gain some further understanding of the nature of the LP solution of an optimization problem.

Consider the relatively simple case of groundwater pumping costs presented by Bear (1979, p. 501). Figure 11.2.1a gives the aerial view of a phreatic aquifer bounded on three sides by impermeable formations, and on one side by a river of constant stage. A cross-sectional view is shown in Fig. 11.2.1b.

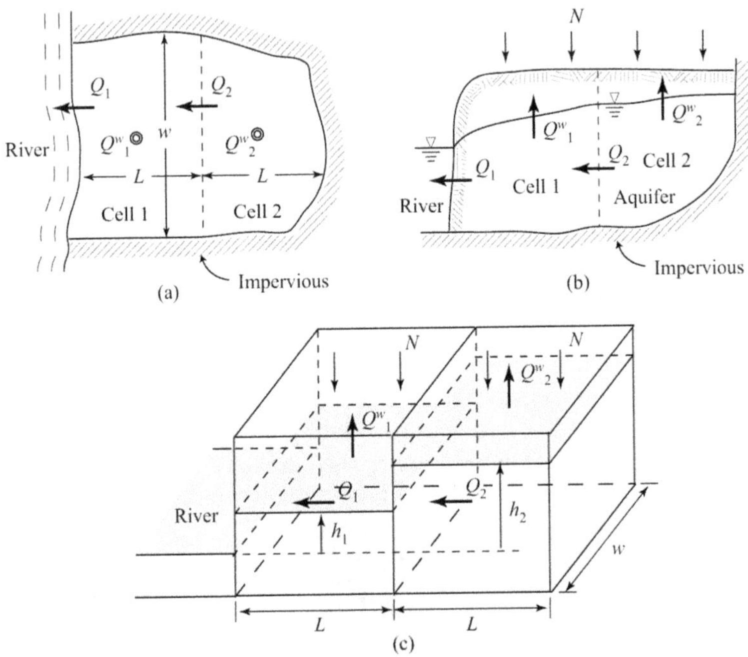

Figure 11.2.1: A two-cell aquifer model with pumping wells: (a) Aerial view; (b) Side view; (c) A two-cell aquifer model.

The aquifer is replenished at the rate N. Water is pumped by two wells, with discharge rates Q_1^w and Q_2^w in zones (cells) 1 and zone 2, respectively. The decision maker's goals and constraints are stated as follows:

(a) The combined pumping rates from these two wells must be sufficient to meet at least the minimum demand, D, i.e., $Q_1^w + Q_2^w \geq D$.
(b) The costs of pumping per unit discharge, are C_1 and $C_2 (\neq C_1)$, for zones 1 and 2, respectively.
(c) The decision maker's goal is to minimize the total cost, $C_1 Q_1^w + C_2 Q_2^w$, while still meeting the demand.
(d) The average piezometric head in zone 1 and zone 2, h_1 and h_2, should not fall below some prescribed values, h_1^o and h_2^o, respectively.

Question: Assuming a steady state regime, what are the discharge rates for these two wells, such that the operation costs, $Z = Z(Q_1^w, Q_2^w)$ will be minimal?

Before solving this problem, we must realize that a feasible solution may or may not exist with the above goal and constraints. However, if a feasible solution does exist, we wish to find the optimal one, i.e., the one that minimizes the operation costs, Z. Thus, our optimization goal is to determine Q_1^w and Q_2^w, so as to

$$\text{Minimize } Z(Q_1^w, Q_2^w) = C_1 Q_1^w + C_2 Q_2^w, \tag{11.2.7}$$

subject to the constraints:

$$Q_1^w + Q_2^w \geq D, \tag{11.2.8}$$
$$h_1 \geq h_1^o, \tag{11.2.9}$$
$$h_2 \geq h_2^o. \tag{11.2.10}$$

Furthermore, there exist two obvious constraints that have to be satisfied:

$$Q_1^w \geq 0, \tag{11.2.11}$$
$$Q_2^w \geq 0. \tag{11.2.12}$$

In addition, the flow model, which consists of appropriate balance equations for the cells, must be satisfied. As stated above, this is an additional constraint. In fact, this constraint expresses certain relationships between piezometric heads and pumping rates. For simplicity, the flow model is described by the Finite Volume Method (FVM) (Sec. 8.2). Figure 11.2.1 shows the flow domain as represented by the two cells. The transmissivity, T, is assumed to be constant in both cells. We express the mass balance equations (constant water density) for the two cells (Fig. 11.2.1c) in the form

$$NwL - Q_1^w + \mathsf{T}w\frac{h_2 - h_1}{L} - \mathsf{T}w\frac{2h_1}{L} = 0, \tag{11.2.13}$$
$$NwL - Q_2^w - \mathsf{T}w\frac{h_2 - h_1}{L} = 0. \tag{11.2.14}$$

In these equations, we have used the approximations $Q_1 = \mathsf{T}wh_1/(L/2)$ and $Q_2 = \mathsf{T}w(h_2 - h_1)/L$, in accordance with the FVM. We can easily solve the above equations for h_1 and h_2, as functions of Q_1^w and Q_2^w. Substituting these expressions into (11.2.9) and (11.2.10), we can now summarize the entire system in the standard form of an LP problem:

Determine Q_1^w and Q_2^w, such that the objective function, $Z(Q_1^w, Q_2^w)$, satisfies:

$$\text{Minimize } Z(Q_1^w, Q_2^w) = C_1 Q_1^w + C_2 Q_2^w, \tag{11.2.15}$$

subject to the constraints:

$$Q_1^w + Q_2^w \geq D, \tag{11.2.16}$$
$$-\frac{L}{2\mathsf{T}w}Q_1^w - \frac{L}{2\mathsf{T}w}Q_2^w \geq h_1^o - \frac{L^2N}{\mathsf{T}}, \tag{11.2.17}$$
$$-\frac{L}{2\mathsf{T}w}Q_1^w - \frac{3L}{2\mathsf{T}w}Q_2^w \geq h_2^o - \frac{2L^2N}{\mathsf{T}}. \tag{11.2.18}$$
$$Q_1^w \geq 0, \tag{11.2.19}$$
$$Q_2^w \geq 0. \tag{11.2.20}$$

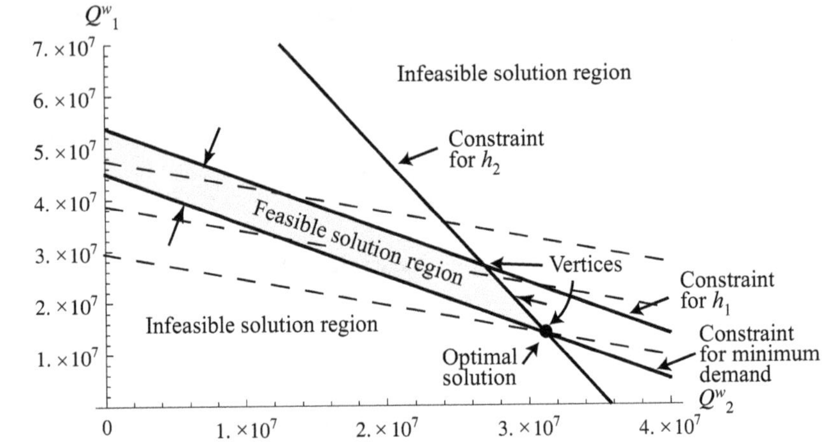

Figure 11.2.2: Graphical solution of a LP problem with two decision variables.

All the above expressions are *linear* with respect to the decision variables, Q_1^w and Q_2^w.

Since the above equations contain only two variables, it is possible to examine the feasible solution domain, as defined by the above constraints, in a two-dimensional plot. First, let us introduce specific parameter values. We assume that $D = 45\,\mathrm{Mm}^3/\mathrm{yr}$ ($1\ \mathrm{Mm}^3 = 10^6\,\mathrm{m}^3$), $C_1 = 0.2\,\mathrm{MU/m}^3$ (MU $\equiv$ monetary unit), $C_2 = 0.1\,\mathrm{MU/m}^3$, $w = 10\,\mathrm{km}$, $L = 10\,\mathrm{km}$, $\mathrm{T} = 10{,}000\,\mathrm{m}^2/\mathrm{d}$, $N = 0.360\,\mathrm{m/yr}$, $h_1^o = 2.5\,\mathrm{m}$, and $h_2^o = 5$ m. Substituting the above data into (11.2.16)–(11.2.18), we obtain

$$Q_1^w + Q_2^w \geq 4.5 \times 10^7, \tag{11.2.21}$$

$$-1.37 \times 10^{-7} Q_1^w - 1.37 \times 10^{-7} Q_2^w \geq -7.36, \tag{11.2.22}$$

$$-1.37 \times 10^{-7} Q_1^w - 4.11 \times 10^{-7} Q_2^w \geq -14.73, \tag{11.2.23}$$

where all terms are in m^3/yr. If we set the above equations into the format of equalities, they will appear as straight lines in the Q_2^w–Q_1^w plane. This is shown in Fig. 11.2.2. We then select the half planes that correspond to the inequality signs, as indicted by the arrows. The intersection of all these half planes, including the ones defined by (11.2.19) and (11.2.20), is then the *feasible solution domain*, shown as the shaded convex polygonal region in Fig. 11.2.2.

A feasible solution domain may or may not exist, depending on the problem setup. If a problem is not properly constrained, there would be no intersection, a feasible domain does not exist, and the optimization problem has no solution. Or, if a feasible domain does exist, but it is not bounded on all sides, there is a direction that the objective function may continue to grow indefinitely; again, a bounded optimal solution does not exist.

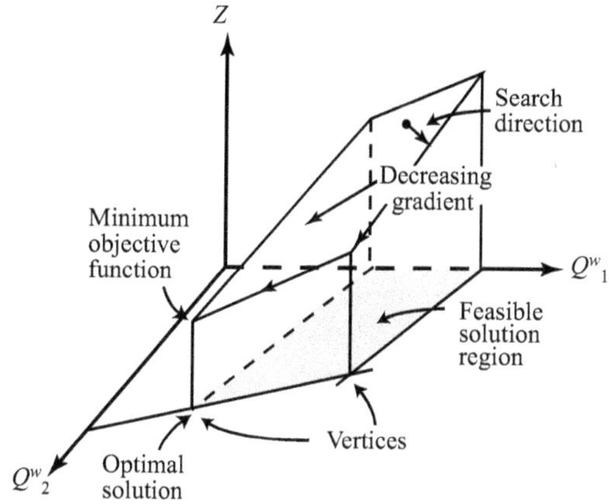

Figure 11.2.3: A three-dimensional view of pumping cost (Z) versus pumping rates Q_1^w and Q_2^w, and the corresponding feasible solution domain.

As our goal is to find the optimal solution inside the feasible domain, we may ask the following question: 'Can the optimal solution occur at a point *in the interior* of the feasible domain?' The answer is 'no'. This can be easily explained as follows. As the objective function is a linear function of the decision variables, it has a constant, non-zero (vector) gradient everywhere in the decision variable space. This means that at any interior feasible solution point, we can always find a direction in which the objective function will increase (or decrease in a minimization problem). We can continue to move in that direction to improve the objective function, until we hit a boundary (straight line), which corresponds to one of the constraints. On this boundary line, we can still find a direction along which we can move to improve the objective function, with the exception that the objective function is a constant on that line; in that case, the solution is non-unique. Excluding the latter case, we continue to move along the straight boundary line until we reach an intersection of two such lines, which is a vertex of the polygon. This vertex may or may not be our optimal solution, but, at least, we can conclude the following: *In an LP problem, if an optimal solution exists and is unique, it can only be located at one of the vertices of the feasible solution domain.*

Figure 11.2.3 shows how the considered pumping cost problem, with two decision variables, can be extended to a three-dimension format. This is done by plotting the objective function Z ($\equiv$ cost) as the vertical axis. We observe that if we start from an interior point on the inclined plane, we can find a decreasing direction to reach an edge. From the edge, we can move to a vertex, and continue on to the next vertex with a smaller value of the objective function. For this problem, the vertex at which the minimum cost solution is located is identified as the intersection of the two constraints: (11.2.22) and (11.2.23). Solving these two equalities for the coordinates of the vertex, we obtain the optimal pumping rate as $Q_1^w = 13.7\,\mathrm{Mm}^3/\mathrm{yr}$, and $Q_2^w = 31.2\,\mathrm{Mm}^3/\mathrm{yr}$. The minimum cost is then calculated as $5.87 \times 10^6\,\mathrm{MU/yr}$.

The above example involves two decision variables. In general, however, an optimization problem may involve a large number of decision variables. In such a case, the decision variable space corresponding to Fig. 11.2.2 becomes n-dimensional. As we have observed in the two-dimensional space, the feasible solution domain is a *polygon.* In the case of three decision variables, and a three-dimensional feasible domain, it is a *polyhedron,* while for n dimensions, it is an n-dimensional *polytope.* The feasible solution domain is bounded by constraints which are $(n-1)$-dimensional geometric objects—*lines* in two-dimensional space, *planes* in three-dimensional space, and *hyperplanes* in higher dimensions.

When we start from an interior point, we can move in the direction of decreasing value of the objective function to reach an $(n-1)$-dimensional hyperplane. From there, we continue to the intersection of two hyperplanes of $(n-2)$ dimensions. This process continues until we reach a zero-dimensional object—a point (vertex), where n hyperplanes intersect simultaneously. We may then search around for the next vertex that corresponds to a smaller value of the objective function. Although this process sounds simple, it is not so in practice. In the next subsection, we shall present a general solution method, known as the *simplex method.*

B. The simplex method

As demonstrated in the previous subsection, the optimal solution of an LP problem must be located on one of the vertices of the feasible domain. This important conclusion significantly reduces the effort of finding the optimal solution. Rather than searching everywhere in the feasible domain, all we have to do is to check the value of the objective function on *all vertices* of that domain, and then select the one at which the objective function reaches a maximum or a minimum. Although this process sounds easy, it is not all that simple in a general problem that may involve many decision variables and constraints. Consider, for example, a problems with n decision variables and m inequality constraints. Since the decision variables must satisfy *non-negativity constraints*, the total number of constrains is really $m+n$. If we select any n of these constraints, to be set as equalities, and solve the resulting equations, simultaneously, we shall obtain the coordinates of the vertex where all these equations intersect. At most, we should have to examine $C(n+m, n)$ vertices, where $C(\cdot,\cdot)$ is the binomial coefficient. For example, if we have an optimization problem that contains 30 decision variables and 20 additional constraints, which is a rather modest optimization problem in practice, we have to calculate the location of $C(50, 30) = 4.7 \times 10^{13}$ vertices. Thus, using brute force is certainly not a feasible solution technique; we need a smart search algorithm.

The simplex method is one such smart algorithm. In it, we starts from one of the vertices, e.g., the origin. We then choose an adjacent vertex such that the objective function value increases, or, at least, does not decrease. Usually, there exist many such neighboring vertices, and a *pivot rule* is specified in order to determine which vertex is to be selected. Various pivot rules exist, but we shall not elaborate on them. This process continues until no such vertex can be found. This means that we have solved the optimization problem.

Without dwelling on the details of the simplex algorithm, let us outline the steps needed in order to set up an LP problem as an input to a simplex computer program (e.g., Press *et al.*, 2007).

The first step is to express it in the standard mathematical format of (11.2.2) through (11.2.5). For the pumping cost problem, represented by (11.2.7) through (11.2.12), we have to start by transforming (11.2.7) from a problem of minimizing the cost to one of maximizing the *negative* cost, by reversing the sign of the objective function.

Next, we recall that in order to express the constraint equations in terms of the decision variables only, we were required, as a first step, to solve the linear system, (11.2.13) and (11.2.14), that represents the flow model. In fact, in the simple example considered here, we solved these equations *analytically* in order to arrive at (11.2.17) and (11.2.18). In practice, however, this is not feasible, as this linear system represents the discrete solution of the flow model as obtained by a numerical method, because no analytical solution is possible. One way to overcome this difficulty is to incorporate the linear system of equations that represent the flow model in the LP system of equations. This can be accomplished by expanding the definition of decision variables to include variables of state. For the simple two-cell model presented above, this means that we add h_1 and h_2 as decision variables, so that $Z = Z(Q_1^w, Q_2^w, h_1, h_2)$. While doing so, we are allowed to keep (11.2.9) and (11.2.10) in their original form, and include the two equalities, (11.2.13) and (11.2.14), as part of the set of constraints in the LP set of equations.

Another option, stemming from the linearity of the considered problem, is to start by solving the linear system in terms of *influence functions*, or in terms of an *influence matrix*. For example, let us assume that in the optimization problem, we have n_1 discrete head values, $\mathbf{h}(\equiv h_1, h_2, \ldots)$ and n_2 pumping rates, $\mathbf{Q^w}(\equiv Q_1^w, Q_2^w, \ldots)$. Since the problem is set up such that the head values are *linearly dependent* on the pumping rates, we can condense all these relations into the matrix form:

$$\{\mathbf{h}\} = [\mathbf{A}]\{\mathbf{Q^w}\}, \tag{11.2.24}$$

where $[\mathbf{A}]$ is a $n_1 \times n_2$ coefficient matrix. We notice that if we set $Q_1^w = 1$, and $Q_2^w - Q_3^w = \cdots = 0$ in (11.2.24), we obtain a sequence of equations: $h_1 = A_{11}$, $h_2 = A_{21}$, ..., i.e., the first column of the matrix $\mathbf{A}$. This means that, using a numerical technique, we can construct the influence matrix by solving it n_2 times, each time setting one of the decision variables, Q_i^w, to unity, and making all other Q_i^w-values equal to zero. In this way, the *influence matrix* is assembled, column by column. Equation (11.2.24) is then used to eliminate the **h**-variable in the constraint equations.

There is one additional step that needs to be carried out in order to construct the *restricted normal form* of the problem statement, as required in the simplex method. In this form, all additional constraints must be *equalities.* This goal is achieved by adding to the problem certain additional fictitious decision variables, S_i, called *slack variables,* one for each inequality, noting that constraints that are already equalities do not need them. For example, the constraint (11.2.9), $h_1 \geq h_1^o$, is now expressed as $h_1 - S_1 = h_1^o$ by subtracting an unknown S_1 on the left hand side. As these S_i's are regarded as decision variables, it is necessary that $S_i \geq 0$. For constraint equations with a '$\leq$' sign, instead of a '$\geq$' sign, we have to *add* a slack variable to the left hand side. The values of the S_i's will be solved for together with all other decision variables. At the end, when we find them as a part of the solution, we just discard them.

To summarize the above description, let us rewrite the LP problem (11.2.7)–(11.2.12) in the following form:

$$\text{Maximize } Z(Q_1^w, Q_2^w, h_1, h_2, S_1, S_2, S_3) = -0.2\, Q_1^w - 0.1\, Q_2^w, \qquad (11.2.25)$$

subject to the primary constraints:

$$Q_1^w \geq 0,\ Q_2^w \geq 0,\ h_1 \geq 0,\ h_2 \geq 0,\ S_1 \geq 0,\ S_2 \geq 0,\ S_3 \geq 0, \qquad (11.2.26)$$

and to the additional constraints:

$$\begin{aligned} Q_1^w + Q_2^w - S_1 - 4.5 \times 10^7 &= 0, & (11.2.27)\\ h_1 - S_2 - 2.5 &= 0, & (11.2.28)\\ h_2 - S_3 - 5 &= 0, & (11.2.29)\\ -1.01 \times 10^7 h_1 + 3.65 \times 10^6 h_2 - Q_1^w + 3.65 \times 10^7 &= 0, & (11.2.30)\\ 3.65 \times 10^6 h_1 - 3.65 \times 10^6 h_2 - Q_2^w + 3.65 \times 10^7 &= 0. & (11.2.31) \end{aligned}$$

With the above equations, we can now prepare a matrix, called *tableau,* to be used as input in the LP computer program. Table 11.2.1 shows this tableau for the problem considered here. The values of the coefficients can easily be understood if we compare this table with (11.2.25)–(11.2.31). With the above input, the simplex computer program will, more or less, take over and provide the optimal solution of the considered system.

The simplex method is one of the most widely used computer algorithms for solving a linear optimization problem. Since it is available from many commercial and public domain sources, we shall not attempt to offer any recommendation, other than to point out one such source, vis., the computer program listing in the book *Numerical Recipes* (Press *et al.*, 2007).

Let us summarize the simplex solution steps:

(a) Identify the decision variables and define the objective function.

const	Q_1^w	Q_2^w	h_1	h_2	S_1	S_2	S_3
0	−0.2	−0.1	0	0	0	0	0
-4.5×10^7	1	1	0	0	−1	0	0
−2.5	0	0	1	0	0	−1	0
−5	0	0	0	1	0	0	−1
3.65×10^7	−1	0	-1.01×10^7	3.65×10^6	0	0	0
3.65×10^7	0	−1	3.65×10^6	-3.65×10^6	0	0	0

Table 11.2.1: Tableau for the pumping cost optimization problem.

(b) State the constraint equations. Remember that the flow model, or the flow and transport model, is also a constraint that must be satisfied.

(c) Examine the constraint equations for variables. If there exist non-decision variables, they need to be either incorporated as decision variables, or eliminated. Two options are possible:

a. Solve the linear system representing the flow model to produce influence functions, and then use them to eliminate the non-decision variables in the constraint equations, or

b. Keep the linear solution system as constraint equations without solving it, but expand the list of decision variables in the objective function, such that all variables in the constraint equations are decision variables.

(d) Construct the matrix known as *tableau*, and it becomes the input of the simplex computer program.

C. Flow model as a constraint

Let us examine how the simplex optimization procedure can be combined with a flow model as a constraint, particularly its numerical equivalent. Figure 11.2.4 shows the map of an aquifer bounded on three sides by impermeable boundaries and on one side by a river at a constant stage. Five wells ($n_w = 5$) operate in the aquifer. The decision maker's goal is to maximize the total pumping rate,

$$\text{Maximize } Z(Q_1^w, \dots, Q_{n_w}^w) = \sum_{k=1}^{n_w} Q_k^w, \tag{11.2.32}$$

subject to the constraints:

(a) The piezometric head in the aquifer should not drop below certain minimum levels,

$$h_i \geq h_i^{\min}, \quad i = 1, \dots, n_h, \tag{11.2.33}$$

where n_h is the number of nodes in the aquifer's numerical model grid.

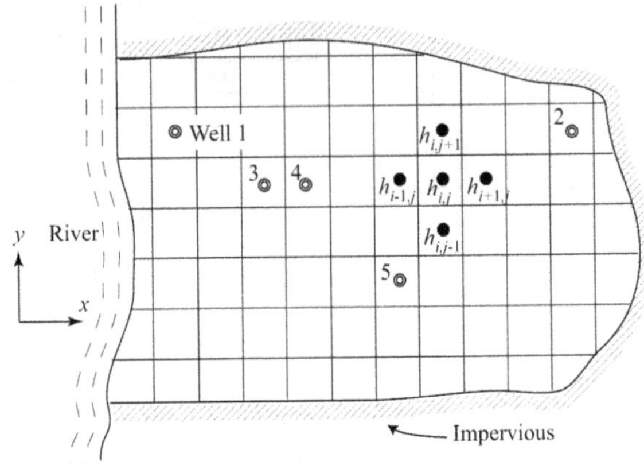

Figure 11.2.4: Map view of an aquifer with pumping wells.

(b) The pumping rates at the individual wells should not exceed the maximum capacity of the respective pumps,

$$Q_k^w \leq Q_k^{w,\max}, \quad k = 1, \ldots, n_w. \tag{11.2.34}$$

(c) In addition, the flow model itself constitutes a constraint. It takes the form of a set of finite difference equations that express mass balance equations, initial and boundary conditions.

Prior to solving the optimization problem, the decision maker, on the basis of his knowledge of the physical system, may already have some idea about the desired distribution of pumping. For example, he may decide to pump as much as possible in well 1 without any significant impact on the drawdown, due to the fact that this well is rather close to the river; the pumping in this well is constrained only by its maximum pumping capacity. For well 2, as it is near an impermeable fault, the opposite is true. Wells 3 and 4 are near to each other, and may interfere with each other's drawdown; hence, these wells, probably, cannot pump at their maximum capacity and will be restricted by the head constraint. This is important *a-priori* qualitative information. However, in order to obtain the quantitative information required for optimal operation, the decision maker still needs an optimization tool.

Examining (11.2.32)–(11.2.34), we identify the pumping rates, Q_k^w, as decision variables. However, the constraints in (11.2.33) are not expressed in terms of these decision variables; hence, we need to expand the list of decision variables so as to include in it also the discrete piezometric head values at the nodes. Thus, the objective function is modified to the form $Z(Q_1^w, \ldots, Q_{n_w}^w; h_1, \ldots, h_{n_h})$.

In addition, the flow model, must also be satisfied. This is done by first constructing the mathematical flow model and then rewriting it by introducing a discretization scheme as required for its numerical solution.

Let us assume that the considered aquifer is confined, inhomogeneous, and anisotropic, with principal axes in the x and y directions. We also assume that the flow is nearly horizontal so that the vertically integrated flow approximation is applicable. Furthermore, we are interested in long term, steady state operation conditions. With the above assumptions, the mathematical model describing the flow in this aquifer is

$$\frac{\partial}{\partial x}\left(\mathsf{T}_{xx}\frac{\partial h}{\partial x}\right)+\frac{\partial}{\partial y}\left(\mathsf{T}_{yy}\frac{\partial h}{\partial y}\right)=\sum_{k=1}^{n_w} Q_k^w \delta(x-x_k, y-y_k), \qquad (11.2.35)$$

where T_{xx} and T_{yy} are the principal transmissivities, $\delta(\cdot,\cdot)$ is the Dirac delta function, and (x_k, y_k) are well locations. Obviously, we have to add appropriate boundary conditions.

For the numerical model, we use, as an example, the cell-centered finite difference scheme (Subs. 8.1.3), illustrated in Fig. 11.2.4. Equation (11.2.35) is then replaced by the finite difference equation,

$$\begin{aligned}&-\mathsf{T}_{i-\frac{1}{2},j}\,\Delta y\frac{h_{i,j}-h_{i-1,j}}{\Delta x}+\mathsf{T}_{i+\frac{1}{2},j}\,\Delta y\frac{h_{i+1,j}-h_{i,j}}{\Delta x}\\&-\mathsf{T}_{i,j-\frac{1}{2}}\,\Delta x\frac{h_{i,j}-h_{i,j-1}}{\Delta y}+\mathsf{T}_{i,j+\frac{1}{2}}\,\Delta x\frac{h_{i,j+1}-h_{i,j}}{\Delta y}=Q_{i,j}^w. \qquad (11.2.36)\end{aligned}$$

Applying this equation to all active cells, assuming Dirichlet and Neumann conditions on boundary segments, and leaving the pumping rates as unknowns, we obtain the following matrix equation:

$$[\mathbf{A}]\,\{\mathbf{h}\}-\{\mathbf{Q^w}\}=\{\mathbf{b}\}, \qquad (11.2.37)$$

where $\{\mathbf{h}\}$ is a column matrix containing the unknown head values, $h_{i,j}$, numbered as $\lfloor h_1, h_2, \ldots, h_{n_h}\rfloor^T$, with $\lfloor..\rfloor^T$ denoting the transpose of the matrix, $\{\mathbf{b}\}$ is the 'right hand side' obtained by substituting the boundary values into (11.2.36), and the unknown pumping rates are included in $\{\mathbf{Q^w}\} = \lfloor 0, \ldots, Q_1^w, \ldots, Q_2^w, \ldots, 0\rfloor^T$. This matrix equation, which expresses the flow model in its discretized form, serves as a constraint in the optimization model.

With the above information, we are ready to construct the tableau for the problem as input to the simplex computer program. Table 11.2.2 represents such a construction. Note that we have omitted the slack variables, which are associated with the inequality signs. We have, however, included the inequality signs in the table. Once the type of the inequality is supplied as input to the program, the latter can, usually, automatically, assign values to the slack variables. This table represents the complete input needed for the solution by the simplex program.

h_1	h_2	$\cdots$	h_{n_h}	Q_1^w	Q_2^w	Q_3^w	Q_4^w	Q_5^w	sign	RHS
a_{11}	a_{12}	$\cdots$	a_{1n_h}	0	0	0	0	0	$=$	b_1
a_{21}	a_{22}	$\cdots$	a_{2n_h}	0	0	0	0	0	$=$	b_2
$\vdots$	$\vdots$	$\ddots$	$\vdots$	$\vdots$	$\vdots$	$\vdots$	$\vdots$	$\vdots$	$\vdots$	$\vdots$
$\vdots$	$\vdots$	$\ddots$	$\vdots$	-1	0	0	0	0	$=$	$\vdots$
$\vdots$	$\vdots$	$\ddots$	$\vdots$	$\vdots$	$\vdots$	$\vdots$	$\vdots$	$\vdots$	$\vdots$	$\vdots$
$\vdots$	$\vdots$	$\ddots$	$\vdots$	0	-1	0	0	0	$=$	$\vdots$
$\vdots$	$\vdots$	$\ddots$	$\vdots$	$\vdots$	$\vdots$	$\vdots$	$\vdots$	$\vdots$	$\vdots$	$\vdots$
$\vdots$	$\vdots$	$\ddots$	$\vdots$	0	0	0	0	-1	$=$	$\vdots$
$\vdots$	$\vdots$	$\ddots$	$\vdots$	$\vdots$	$\vdots$	$\vdots$	$\vdots$	$\vdots$	$\vdots$	$\vdots$
a_{n_h1}	a_{n_h2}	$\cdots$	$a_{n_hn_h}$	0	0	0	0	0	$=$	b_{n_h}
1	0	$\cdots$	0	0	0	0	0	0	$\geq$	$h_1^{\min}$
0	1	$\cdots$	0	0	0	0	0	0	$\geq$	$h_2^{\min}$
$\vdots$	$\vdots$	$\ddots$	$\vdots$	$\vdots$	$\vdots$	$\vdots$	$\vdots$	$\vdots$	$\vdots$	$\vdots$
0	0	$\cdots$	1	0	0	0	0	0	$\geq$	$h_{n_h}^{\min}$
0	0	$\cdots$	0	1	0	0	0	0	$\leq$	$Q_1^{\max}$
0	0	$\cdots$	0	0	1	0	0	0	$\leq$	$Q_2^{\max}$
$\vdots$	$\vdots$	$\ddots$	$\vdots$	$\vdots$	$\vdots$	$\vdots$	$\vdots$	$\vdots$	$\vdots$	$\vdots$
0	0	$\cdots$	0	0	0	0	0	1	$\leq$	$Q_5^{\max}$

Table 11.2.2: Tableau for the maximum pumping rate problem, using the finite difference method to represent the flow model.

11.2.3 Nonlinear problems and unconstrained optimization

As discussed in the previous section, linear programming optimization requires that both the objective function and the constraints be *linear* functions of the decision variables. If one of the above conditions is violated, the problem becomes *nonlinear*.

Figure 11.2.5a illustrates a case of a linear objective function, with two design variables, subject to nonlinear constraints. The contour lines describing the objective function are shown as the parallel dashed lines. As discussed in Subs. 11.2.2, and observed in Fig. 11.2.5a, the optimal (maximal or minimal) value of the objective function must be located on the *boundary* of the feasible solution region. However, unlike the linear constraint case, which further requires the optimal objective function value to be located on a vertex, in the nonlinear constraint case, the optimal value can be located at any point on the (curved) boundary (or, in higher dimensions, a curved plane or hyperplane).

Furthermore, if the objective function is *nonlinear*, then, as illustrated in Fig. 11.2.5b, the optimal solution may exist in the *interior* of that domain,

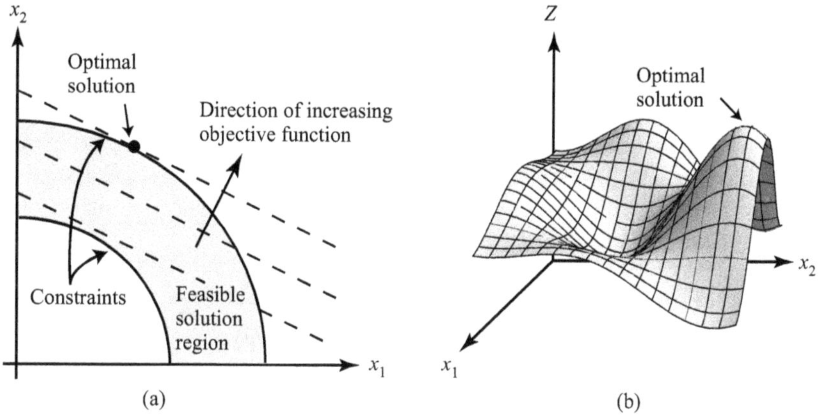

Figure 11.2.5: Nonlinear optimization problem. (a) Linear objective function and nonlinear constraints; (b) Nonlinear objective function.

and need not be on one of its bounding curves (or surfaces). This means that in order to find the optimal solution of a nonlinear optimization problem, we need to search *everywhere* in the feasible solution domain, rather than inspect only vertices on its boundary. As, generally, the search space may be vast (especially, for a large number of decision variables), an intelligent search algorithm is required.

Generally, there are two types of search strategies for the optimal solution of nonlinear problems: *constrained* and *unconstrained*. The constrained search strategy seeks to optimize the original *constrained optimization problem*. These search methods are, generally, referred to as *nonlinear programming* (NLP) methods. They include, among others, *quadratic programming*, *separable convex programming*, and *geometric programming*, each suitable for a specific type of problems (Avriel, 2003).

The unconstrained search strategy, on the other hand, seeks to modify the original constrained optimization problem by transforming it into an *unconstrained* one. Once the unconstrained problem has been formulated, a class of *gradient search methods* can be applied to its solution. Generally speaking, the unconstrained gradient search methods are more general, efficient, and versatile than the constrained nonlinear programming ones; hence, we shall limit the discussion below only to the unconstrained gradient search methods.

Before starting the discussion on the gradient search methods, we must address the issue of constraints. As all optimization problems have constraints, how do we cope with them in an unconstrained search technique? The way to get around the presence of constraints is to treat them as part of the objective function, through assigning *penalties* to the latter. This means that rather than setting barriers that remove the infeasible solution domain from the search space, we open up the entire domain for search, but impose a heavy *penalty* in subdomains of infeasible solutions in order to 'discourage' the search from moving in these directions. The penalty is usually imposed by creating steep and unfavorable gradients in the objective function in those domains.

As an illustration, consider the (linear) pumping cost minimization problem defined by (11.2.7)–(11.2.12). The constraints (11.2.8)–(11.2.10) can be incorporated into the objective function by modifying the definition of the objective function in (11.2.7) to the form

$$\begin{aligned}\text{Minimize } & Z'\left(Q_1^w, Q_2^w\right) = C_1 Q_1^w + C_2 Q_2^w \\ & + \mathrm{H}\left(D - Q_1^w - Q_2^w\right) C_Q (Q_1^w + Q_2^w - D) \\ & + \mathrm{H}\left(h_1^o - h_1\right) C_h (h_1 - h_1^o) + \mathrm{H}(h_2^o - h_2)\, C_h (h_2 - h_2^o). \quad (11.2.38)\end{aligned}$$

In the above equation, $\mathrm{H}(x)$ is the Heaviside unit step function, which is equal to 1 for $x \geq 0$, and to 0 for $x < 0$, and $(C_Q, C_h) \gg (C_1, C_2)$ are, respectively, the penalty cost factors (with proper units chosen) for the well discharge and piezometric head constraints. Equation (11.2.38) states that as long as we stay within the constraints, (11.2.8)–(11.2.10), the arguments of the Heaviside function are negative, and all the terms added to the objective function become zero; hence, the original problem is unaltered. On the other hand, if any of the constraints is violated, there is a large cost, linearly proportional to the extent of violation, added to the total cost.

We notice that in the penalty procedure, the primary constraints, (11.2.11) and (11.2.12), are not included in (11.2.38), because these are automatically satisfied if we define the range of the search space to be $Q_1^w \geq 0$ and $Q_2^w \geq 0$. We also note that the flow equations, (11.2.13) and (11.2.14), which form a set of equality constraints, are not included in the objective function as penalty, because they are not directly treated as constraints. The flow equations are independently solved, once a set of the design variable values (in this case, Q_1^w and Q_2^w) is given. In a general multidimensional problem, this means solving a finite element or finite difference system for which all input values are given. Once the piezometric head values are solved for, they are used to evaluate the penalty part of the objective function in (11.2.38). Unlike the linear (and nonlinear) programming methods (Subs. 11.2.2C), which incorporate flow equations as constraints, and in that process it becomes necessary to expand the list of design variables to include the piezometric head, unconstrained optimization uses only the original set of design variables (i.e., well discharges) in the objective function; this has the effect of reducing the size of the design variable search space.

With the above modification of the objective function, we have turned the constrained minimization problem for the objective function Z to an unconstrained one in terms of Z'. Any of the unconstrained search algorithms can then be used for solving this modified problem.

11.2.4 Gradient search method

In general, we can classify the gradient search methods into three categories, depending on the required type of information (Sun, 1994; Press *et al.*, 2007):

(a) *Search methods*: (Non-gradient based) search methods that require the evaluation of an objective function only. Example: the *Powell method.*
(b) *Gradient methods*: Methods hat require knowledge of the objective function, as well as information on its gradients (derivatives). Examples: the *steepest descent* and the *conjugate gradient methods.*
(c) *Second order methods*: Further to the above, the second derivatives, known as the *Hessian matrix*, are needed. Examples: the *quasi-Newton* and the *BFGS methods.*

The pros and cons of the above search strategies can be summarized as follows. It is obvious that in an unconstrained optimization problem, the value of the objective function, corresponding to a specified set of decision variables, is the easiest to evaluate. To evaluate its gradient, perturbations are made on the decision variables, and finite difference formula is used to approximate the derivatives. In a problem with n decision variables, n such derivatives have to be evaluated. It is even more complex and costly to construct the second derivatives at a point. In fact, the second derivative, or the *Hessian matrix*, is often approximated by using the stored gradient information of previous iteration steps. The reason for using higher order methods is that they converge much faster and fewer iterations are needed in order to find the final, optimal solution.

A. Bracketing method

This method is based on the idea of *bracketing* in finding the *root* of an equation. In root finding, given a continuous, one-dimensional function, $Z(x)$, our goal is to find the root x_r, such that $Z(x_r) = 0$. We start from an initial pair of 'brackets', $[x_1, x_2]$, that delimit a relatively large segment, on which we hope that a root exists. By evaluating the function at these bracket points, we check the result: if we obtain $Z(x_1)Z(x_2) < 0$ (that is, the functions are of opposite sign), we are guaranteed a root in the range between these brackets. In the simplest method, we can narrow down this range by bisecting the interval. Choosing $x_3 = (x_1 + x_2)/2$, we can check the signs of $Z(x_1)Z(x_3)$ and $Z(x_2)Z(x_3)$. By discarding the interval in which the root does not exist, we have halved the interval. This process continues until the interval is small enough for the desirable accuracy.

The search for the minimum of a function is similar to the search for the zero value. The process, however, begins with three points, say $x_1 > x_2 > x_3$. If we find that $Z(x_2) < Z(x_1)$ and $Z(x_2) < Z(x_3)$, a minimum must exist between x_1 and x_3. We can then bisect the intervals $[x_1, x_2]$ and $[x_2, x_3]$ and check the above condition for the existence of a minimum. The interval without a minimum is discarded. This process continues until a prescribed

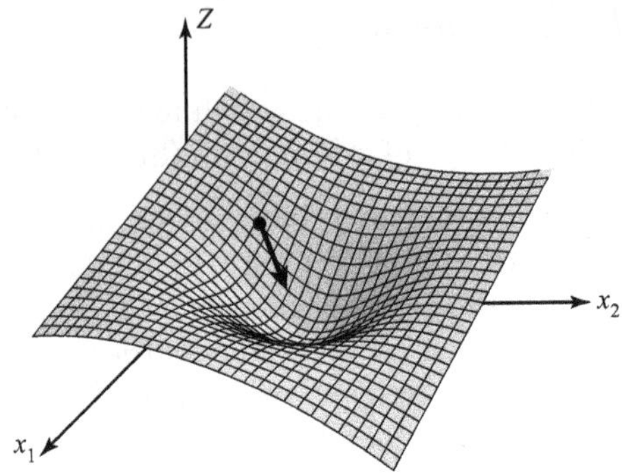

Figure 11.2.6: Method of steepest descent.

precision is reached. This method can be extended to multidimensional functions, using methods such as the Powell method (Press *et al.*, 2007), in which the bracketing is conducted in successive one-dimensional directions. In other words, we search for the minimum by varying one variable at a time. The process is iterated until the change in the objective function becomes sufficiently small. Obviously, this is not an efficient algorithm.

B. Method of steepest descent

The *method of steepest descent* is a gradient search method in which the evaluation of local gradients (first derivatives) is needed. To illustrate the concept, we shall use an example with two decision variable. Figure 11.2.6 shows an objective function in a two-dimensional search space, $Z(x_1, x_2)$. We start with a pair of decision variables and evaluate the objective function. This point is shown on the objective function surface. Our goal is to descend to the bottom of the surface by taking some finite size steps. The question is: 'what is the best direction to place our next 'trial''?

We observe that there are many downhill directions that lead to a decrease in the objective function. However, there is only one direction that is the *steepest*; it is given by the negative gradient

$$-\nabla Z = -\frac{\partial Z}{\partial x_1}\mathbf{i} - \frac{\partial Z}{\partial x_2}\mathbf{j}. \tag{11.2.39}$$

Hence, we need to evaluate the two partial derivatives at the current location. These partial derivatives can point to the direction of steepest descent. In addition to the direction, we also need a step size in an effort of moving downward. A certain empirical formula is used to determine the proper step size that can guide the solution smoothly downhill to the bottom of the 'valley'.

C. Conjugate gradient method

Despite its simplicity, the method of steepest descent does not work well for highly distorted gradient fields, e.g., for a 'valley' that is long and narrow. In such a case, the best direction is that which connects the current point to the bottom of the valley. However, this direction generally does not coincide with the local direction of steepest descent. By 'conjugating' the current gradient and the gradients of the previous steps, we can improve the search for that direction. This improved method is called the *conjugate gradient method.* In Subs. 8.7.1 we have demonstrated the steps in a conjugate gradient method for solving a linear equation system by minimizing an objective function. Details of this and other variations of the conjugate gradient method can be found in books on numerical methods and optimization (e.g., Press *et al.*, 2007; Sun, 1994).

D. BFGS method

We notice that the information based on a gradient suggests only a direction, but not the size of the step to be taken. The latter is determined empirically, through experimenting. To obtain a suggestion of step size, we need information on the higher derivatives; this leads to *second order methods.* As these are the more efficient and widely used method, let us present them, particularly the BFGS method (Broyden, 1970; Fletcher, 1970; Goldfarb, 1970; Shanno, 1970) in more details.

Assume that in the process of successive trials, in search of the optimum, we are in the ith trial location, expressed by the decision variable vector, $\mathbf{x}^i$. Let the true solution be located at $\mathbf{x}$, which is a small distance $\Delta\mathbf{x}^i$ away. We can approximate the objective function at $\mathbf{x}$ by a multivariate Taylor series expansion, including up to the second order terms:

$$Z(\mathbf{x}) \approx Z(\mathbf{x}^i) + \left(\Delta\mathbf{x}^i\right)^T \cdot \nabla Z(\mathbf{x}^i) + \frac{1}{2}\left(\Delta\mathbf{x}^i\right)^T \cdot \mathsf{H}(\mathbf{x}^i) \cdot \Delta\mathbf{x}^i, \qquad (11.2.40)$$

where $\mathbf{x}$, $\mathbf{x}^i$, and $\Delta\mathbf{x}^i$ are $n \times 1$ column matrices, the notation $(.)^T$ indicates matrix transpose, and H is the *Hessian matrix*, defined as

$$\mathsf{H} = \begin{bmatrix} \frac{\partial^2 Z}{\partial x_1^2} & \frac{\partial^2 Z}{\partial x_1 \partial x_2} & \cdots & \frac{\partial^2 Z}{\partial x_1 \partial x_n} \\ \frac{\partial^2 Z}{\partial x_2 \partial x_1} & \frac{\partial^2 Z}{\partial x_2^2} & \cdots & \frac{\partial^2 Z}{\partial x_2 \partial x_n} \\ \vdots & \vdots & \ddots & \vdots \\ \frac{\partial^2 Z}{\partial x_n \partial x_1} & \frac{\partial^2 Z}{\partial x_n \partial x_2} & \cdots & \frac{\partial^2 Z}{\partial x_n^2} \end{bmatrix}. \qquad (11.2.41)$$

Next, we take the gradient of (11.2.40) with respect to $\mathbf{x}$, obtaining

$$\nabla Z(\mathbf{x}) \approx \nabla Z(\mathbf{x}^i) + \mathsf{H}(\mathbf{x}^i) \cdot \Delta\mathbf{x}^i. \qquad (11.2.42)$$

In this equation, we note that $\mathbf{x}^i$ is regarded as a constant when differentiating with respect to $\mathbf{x}$, while $\Delta\mathbf{x}^i = \mathbf{x} - \mathbf{x}^i$ is not a constant. For Z to be a minimum, we require that at $\mathbf{x}$, we have

$$\nabla Z(\mathbf{x}) = 0. \tag{11.2.43}$$

Using the above condition, we can solve (11.2.42) for $\Delta\mathbf{x}^i$, obtaining

$$\Delta\mathbf{x}^i = -\mathbf{H}^{-1}(\mathbf{x}^i) \cdot \nabla Z(\mathbf{x}^i), \tag{11.2.44}$$

where the superscript in $(.)^{-1}$ denotes matrix inverse. Hence, $\Delta\mathbf{x}^i$ is the increment for the next trial. In other words, the iterative formula,

$$\begin{aligned} \mathbf{x}^{i+1} &= \mathbf{x}^i + \Delta\mathbf{x}^i \\ &= \mathbf{x}^i - \mathbf{H}^{-1}(\mathbf{x}^i) \cdot \nabla Z(\mathbf{x}^i), \end{aligned} \tag{11.2.45}$$

can be used to update the vector of decision variables, until convergence is achieved. At that point, the minimum solution has been achieved. This method is called *Newton's method.*

Although the Newton method converges very fast near the minimum, it has several drawbacks. One is that it requires second derivatives of the objective function. When these are not available analytically, evaluating the derivatives numerically at every iteration step can be inaccurate and time consuming. The major difficulty, however, is that when the trial solution, $\mathbf{x}^i$, is still far from the minimum, the Hessian matrix may not be *positive definite.* In other words, the search sequence may not converge. Alternative methods, known as *quasi-Newton methods*, are often used instead.

In the quasi-Newton method, the inverse of the Hessian matrix, $\mathbf{H}^{-1}$, is replaced by a symmetric positive definite matrix $\mathbf{G}^i$, which evolves with iteration steps. Eventually, it is expected that it will approximate $\mathbf{H}^{-1}$ near the true minimum. Hence, the equation

$$\mathbf{x}^{i+1} = \mathbf{x}^i - \mathbf{G}^i \cdot \nabla Z(\mathbf{x}^i) \tag{11.2.46}$$

is used to update the selection of decision variables. Different selections of $\mathbf{G}^i$ lead to different quasi-Newton methods. Particularly, the BFGS method (which is a quasi-Newton method) uses

$$\mathbf{G}^1 = \mathbf{I}, \tag{11.2.47}$$

$$\begin{aligned} \mathbf{G}^{i+1} &= \mathbf{G}^i + \left[1 + \frac{(\Delta\mathbf{g}^i)^T \cdot \mathbf{G}^i \cdot \Delta\mathbf{g}^i}{(\Delta\mathbf{x}^i)^T \cdot \Delta\mathbf{g}^i}\right] \frac{\Delta\mathbf{x}^i \cdot (\Delta\mathbf{x}^i)^T}{(\Delta\mathbf{x}^i)^T \cdot \Delta\mathbf{g}^i} \\ &\quad - \frac{\Delta\mathbf{x}^i \cdot (\Delta\mathbf{x}^i)^T \cdot \mathbf{G}^i + \mathbf{G}^i \cdot \Delta\mathbf{g}^i \cdot (\Delta\mathbf{x}^i)^T}{(\Delta\mathbf{x}^i)^T \cdot \Delta\mathbf{g}^i}, \end{aligned} \tag{11.2.48}$$

where $\mathbf{I}$ is the identity matrix, and

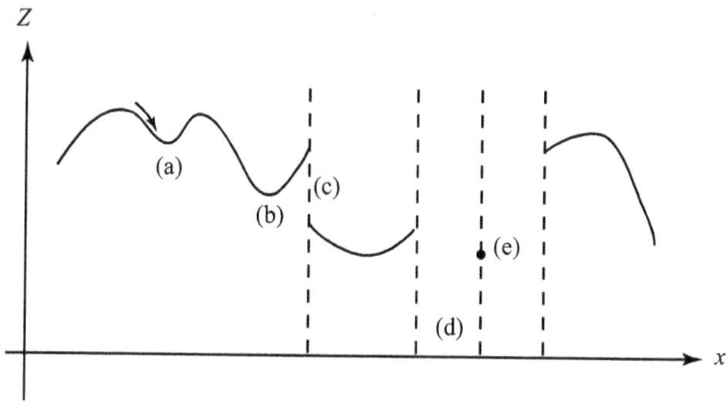

Figure 11.2.7: A schematic objective function showing the difficulty in finding the global minimum in optimization: (a) A local minimum that can trap a gradient search; (b) A possible global minimum that can be found only if we start the search nearby; (c) A discontinuity with gradients on two sides, pointing in different directions; (d) A discontinuous search space; and (e) a discrete search space.

$$\Delta \mathbf{g}^i = \nabla Z(\mathbf{x}^{i+1}) - \nabla Z(\mathbf{x}^i). \tag{11.2.49}$$

This completes the definition of the iteration steps in the BFGS scheme.

We note that although the BFGS method is classified as a second order method, it actually avoids the evaluation of the second derivative by approximating the Hessian matrix, using first derivatives. Hence, it has the efficiency of a Newton method, yet without the deficiency of the cumbersome evaluation of second derivatives.

We shall not go into further details of the BFGS method, nor of other gradient methods for optimization. Subroutines for a number of the search methods mentioned above are presented by Press *et al.* (2007).

In the optimization methods presented so far, we have made a number of assumptions concerning the objective function and the search space. We have assumed that the objective function is continuous, and that it has up to second derivatives. We have also assumed that there exists only one global minimum or maximum, i.e. there is no *local minimum* or *local maximum* that can 'trap' a gradient search. Finally, for the feasible solution domain, we have assumed that it is *simply connected* and *convex*; a domain is said to be convex if the straight line that connects any two points within it will not step outside it. If any of the above conditions is violated, the methods presented above, such as the LP, NLP, and gradients methods, generally, are not applicable. A schematic illustration of some of the problematic conditions mentioned above, are shown in Fig. 11.2.7, using a case with a single decision variable.

Let us continue by focusing on a problem with multiple minima. As the gradient search relies on the *local* gradient to *descend* to the bottom of a 'valley' (see Fig. 11.2.6), it always settles to the valley that the local gradient points to. It has no way of feeling the existence of any other valleys beyond this close-by one, although other 'valleys' may have lower minima. To identify the possible existence of other 'valleys', the search algorithm must be able to *climb uphill* or to *jump* to another location that is some distance away, and to start a new exploration track that is beyond the local trap.

The search for a *global minimum* in the presence of many *local minima* is a haphazard business; generally, it depends on an insightful initial guess that is close to the true solution. One way to overcome this difficulty is to solve the problem, starting from a large number of different initial trials. Sometimes a stochastic technique is employed to randomly select these initial trials points (Ramarao *et al.*, 1995; Jang and Choe, 2002). However, when the number of local minima is large, sometimes extremely large, even the multiple initial trials technique may not be feasible. We need some principle to guide the selection of initial trials. Along these directions, a number of heuristic optimization methods, particularly those based on biological evolution principles, have been developed. Some of these methods are reviewed below.

11.2.5 Genetic algorithm and simulated annealing

Since the 1980s, a class of non-traditional optimization techniques has been developed in order to deal with the difficult situations mentioned above, in which the traditional, and even the heuristic, optimization techniques fail. These assorted, sometimes curious, techniques are generally referred to as *metaheuristics* (meaning *beyond heuristic*) (Osman and Laporte, 1996; Glover and Kochenberger, 2003). Examples of these techniques, as applied to groundwater optimization problems, include *simulated annealing* (Dougherty and Marryott, 1991; Marryott *et al.*, 1993; Meyer *et al.*, 1994) and *genetic algorithm* (Ritzel *et al.*, 1994; McKinney and Lin, 1994; Cieniawski *et al.*, 1995). These techniques borrow some seemingly irrelevant physical or biological phenomena, and draw an analogy between them and the optimization problem on hand. Some of these techniques are still controversial. However, when the traditional methods fail, these techniques can deliver good and, sometimes, even very good results.

A. Simulated annealing

Simulated annealing (SA) is a probabilistic search algorithm for the global optimum of a given function in a large search space (Kirkpatrick *et al.*, 1983; Kirkpatrick, 1984; Cerny, 1985). It mimics the physical phenomenon of *annealing* in metallurgy. It is well known that when a molten metal cools rapidly and solidifies, the atoms tend to adjust their position locally, so that they fall into some local minimum potential energy well. Such metal is disorderly in terms of its atomic structure and is of inferior quality with respect to material strength. On the other hand, if the metal is cooled slowly, such that each atom has sufficient energy to climb out of the local energy well to find some new location that will be better from the *global* minimum energy point of view, then a global crystalline structure can be formed, and the resulting metal is stronger. The lesson to be learned here is that in optimization, we should resist the temptation to let the solution follow a rapid path that leads to a minimum, because it is likely to be a local minimum. Instead, we should allow the solution to have sufficient *kinetic energy* such that it will not be trapped by the local potential energy wells. By analogy to this physical process, the SA algorithm starts from some initial trials of feasible solutions. In each successive iteration step, the current solution is replaced by a random 'nearby' solution, chosen with a probability that depends not only on the difference between the corresponding objective function values, but also on a global parameter Θ, called the 'temperature', the value of which is gradually decreased during the process. When (initially), Θ is large, the search direction changes almost randomly, not guided by the objective function gradient direction. As Θ decreases, an indication that we are getting closer to the 'equilibrium state', the search directions become less random, and are mostly 'downhill'. We shall not present any further details on the Simulated Annealing method, as abundance of references can be found in the literature (e.g., Van Laarhoven and Aarts, 1987; Davis, 1987; Aarts and Korst, 1989).

B. Genetic algorithm

Genetic algorithm (GA) is an optimization technique that mimics the *natural selection* (or, *survival of the fittest*) process of biological evolution (Holland, 1975; Goldberg, 1989; Michalewicz, 1992). Borrowing the terminology of evolutionary biology, we may consider the *discretized* feasible solution space as a *population*, or a collection of many *individuals*, each representing a feasible solution. Each individual is characterized by a parameter called *fitness to survive*, measured by the corresponding value of the objective function. In a maximization problem, a solution with a larger objective function value means an individual with a higher value of fitness to survive. The size of the population is generally huge, so that it is not feasible to check the fitness of *all* individuals in order to see which one has the highest fitness (i.e., finding the optimal solution).

In the GA method, initially, a small number of individuals (feasible solutions) are randomly selected to form a *family*. The fitness to survive of each family member is checked by evaluating the corresponding value of its objective function. A probabilistic *selection* rule is applied to select members to be included in a *breeding program*, as explained below. The more fit members have a higher chance to be selected. However, for the purpose of maintaining diversity, less fit members are also selected, but with a smaller probability. Next, the selected members go through certain *reproduction processes*, such as *crossover* and *mutation*, to create 'offsprings'. A new family, with the same number of members, is obtained. All fittest individuals discovered so far will be stored. This process continues, generation after generation, until, after a while, there is little improvement in the fittest individuals, or a prescribed maximum number of generations is reached. The program is then terminated.

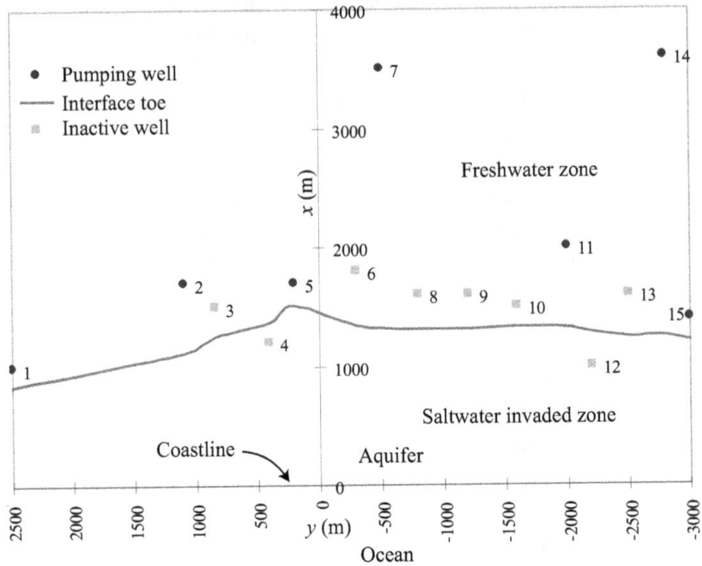

Figure 11.2.8: Optimum pumping pattern in a saltwater invaded coastal aquifer.

Although there is no guarantee that the true optimal solution has been found, the process, generally, finds a *near optimal* solution better than most other methods.

Let us use an example (Cheng *et al.*, 2000) to illustrate some of the GA optimization steps described above.

Consider the coastal aquifer shown as plane view in Fig. 11.2.8. The coast is along the line $x = 0$. The aquifer is unconfined, with an average thickness $d =$ 15 m, a hydraulic conductivity $\mathsf{K} = 40$ m/day, and freshwater and saltwater densities of $\rho_f = 1.000$ g/cm^3 and $\rho_s = 1.025$ g/cm^3, respectively. Inflow from the landside boundary is uniform at a rate (per unit aquifer width) of $Q' = 40$ m^2/day toward the sea, and there is no natural replenishment ($N = 0$). The figure shows the locations of 15 pumping wells. The wells' data is presented in Table 11.2.3.

The management question is: 'What is the maximum total pumping rate that can be sustained, without pumping saltwater?' A well is considered destroyed, and has to be shut-off, when the advancing saltwater front (i.e., the toe of saltwater wedge, see Fig. 9.2.6) reaches its bottom. We realize that although there are 15 wells, an optimal pumping pattern may not require all wells to be active. Indeed, it may be beneficial to shut down some of the near-coast wells and allow saltwater to invade that region, while allowing inland wells to pump more. In addition to the saltwater intrusion constraint, the

Well no.	x^w (m)	y^w (m)	$Q^{\max}$ (m^3/d)	$Q^{\min}$ (m^3/d)	Case 1 Q^w (m^3/d)	Case 2 Q^w (m^3/d)	Case 2 x^{toe} (m)
1	1000	2500	600	150	255	201	836
2	1700	1100	1300	150	402	351	1117
3	1500	850	1100	150	0	0	1257
4	1200	400	800	150	0	0	1372
5	1700	200	1300	150	0	150	1514
6	1800	−300	1400	150	158	0	1344
7	3500	−500	1500	150	728	1497	1323
8	1600	−800	1200	150	150	0	1311
9	1600	−1200	1200	150	0	0	1315
10	1500	−1600	1100	150	0	0	1332
11	2000	−2000	1500	150	232	155	1319
12	1000	−2200	600	150	0	0	1287
13	1600	−2500	1200	150	0	0	1241
14	3600	−2800	1500	150	1500	1387	1251
15	1400	−3000	1000	150	185	150	1213
Total					3610	3891	

Table 11.2.3: Pumping well locations, constraints, and maximum pumping rate solution.

pumping rate of each well is limited to a specified range, $Q_i^{\min} \leq Q_i^w \leq Q_i^{\max}$, where the subscript indicates the well's number. The constraining values are shown in Table 11.2.3.

We are now ready to formulate the mathematical statement of the optimization problem:

$$\text{Maximize } Z(Q_1^w, \ldots, Q_{15}^w) \left(\equiv \sum_{i=1}^{15} Q_i^w \right), \tag{11.2.50}$$

with respect to the decision variables Q_i^w, and subject to the constraints

$$Q_i^{\min} \leq Q_i^w \leq Q_i^{\max}, \quad \text{or} \quad Q_i^w = 0, \tag{11.2.51}$$

$$x_i^{\text{toe}} < x_i^w \quad \text{if} \quad Q_i^w \neq 0, \tag{11.2.52}$$

for $i = 1, \ldots, 15$, where x_i^w is the x-coordinate of well i, and x_i^{toe} is the x-coordinate of the saltwater front advancing towards well i.

The optimization problem as defined above is not amenable to treatment by conventional optimization techniques for the following reasons. First, the search space is not continuous, as it consists of the options $\{0, [Q_i^{\min}, Q_i^{\max}]\}$ for each well. In other words, there is a gap in the feasible pumping rate between the two values, 0 and $Q_i^{\min}$. Another difficulty arises when the location of the interface toe approaches the pumping well. For each well, there exists a critical pumping rate such that when the well is pumped at that rate, the saltwater front will stay arrested at a certain distance in front of the well, and not right at the well (Strack, 1976; Cheng and Ouazar, 1999). When that critical rate is slightly exceeded, the next equilibrium position of the saltwater front is some distance *behind* the well. In other words, there is no equilibrium position for the saltwater front within a certain range from the well, so that the solution of x^{toe} is discontinuous with respect to a continuously increasing pumping rate. Yet another difficulty is the existence of multiple local maxima of the objective function.

Before selecting an appropriate solution procedure, it may be interesting to have an idea about the size of the search space—how many possible pumping rate combinations are there? If we assume that the pumping rate of each of the 15 wells is represented by 20 discrete values, a very crude precision, there are $20^{15} \approx 10^{19}$ possible combinations! Finding the optimal solution, or even a near optimal solution, among such a huge number of possible combinations, without using gradient information, is practically impossible.

All these considerations make the LP, NLP, and gradient search methods infeasible; hence, the GA method is employed for the solution of this optimization problem.

The first step is to define the search space i.e., the decision variable space, as given by (11.2.51). Depending on the desirable precision, the pumping rates are represented by a number of discrete values, $\{0, Q_i^{\min}, Q_i^3, Q_i^4, \ldots, Q_i^{\max}\}$. A feasible pumping program is selected from these discrete values.

The next step is to represent the discrete feasible solutions by some code, e.g., a *binary code*. For example, if we allow the pumping rate of well 1 to be represented by $2^8 = 256$ discrete values, then each pumping rate can be assigned a binary code of 8 bits length. For example, [1 0 1 1 1 0 0 0] or [1 1 1 0 0 0 1 0]. The same procedure is repeated for wells 2 through 15. We can then string the codes together to form a single binary code of 120 bit length, [1 0 1 1 1 0 0 0 1 1 ... 1 1 0]. If we randomly generate a binary code of such length by choosing 0 or 1 for each bit, it will represent a feasible pumping program. Analogous to a genetic code, such string is called a *chromosome*.

We need to define the objective function of the problem, or the *fitness of individuals to survive*. As we intend to conduct an unconstrained search, we need to incorporate constraints into the objective function as penalties. We note that of the two constraints, (11.2.51) and (11.2.52), only the second one, (11.2.52), needs to be incorporated. The first constraint is automatically satisfied in the process of defining the search space. Hence, we modify the original optimization statement involving the objective function Z, as defined by (11.2.50), by adding a penalty, i.e.,

$$\text{Maximize } Z'(Q_1^w, \ldots, Q_{15}^w) = \sum_{i=1}^{15} Q_i^w - r\,\mathrm{H}(Q_i^w)\,\mathrm{H}(x_i^{\text{toe}} - x_i^w)\left(\frac{x_i^{\text{toe}}}{x_i^w} - 1\right)^2, \tag{11.2.53}$$

where H is the Heaviside unit step function, and r is an empirical penalty factor. We note that the creation of a penalty scheme is somewhat *ad hoc* and empirical; there is no fixed rule on its selection. In (11.2.53), we observe that when both conditions $x_i^{\text{toe}} > x_i^w$ (the well location is invaded) and $Q_i^w > 0$ (the well is a pumping one) are true, a penalty that is proportional to the square of the distance invaded by seawater beyond the well is imposed to reduce the objective function value, thus making the selection of this pumping rate less desirable.

To begin the search process, we use a random number generator to generate a small 'family' of individuals (feasible solutions), typically containing 8 members. For each of these individual, we need to evaluate its fitness to survive ($\equiv$ objective function). This is done by solving a set of saltwater intrusion flow equations (see Secs. 9.2 and 9.3), subject to the proper boundary conditions, with the design variables (the individual well pumping rates) given as part of the data input. For example, we may employ the Strack model as described in Subs. 9.2.5D for solving the saltwater intrusion problem. We can, of course, use another model, including one that involves a transition zone, rather than a sharp interface. Once the problem is solved, either analytically or numerically, we can find the saltwater-freshwater interface location, and, subsequently, to determine whether and which individual wells are invaded by seawater. This information is then used to evaluate the objective function (11.2.53). This step of solving the relevant flow problem is usually the most (computer) time consuming part, as we need to solve a saltwater intrusion model in order to determine whether wells are being invaded by seawater.

As explained in Subs. 11.2.3, we view the step of solving the flow model as satisfying the requirement that the equations included in the flow model must be satisfied as a constraint. We also observe that this step is independent of the unconstrained optimization process.

The next step is to subject the selected 'family of individuals' to a number of GA optimization operations, such as *selection*, *crossover*, and *mutation*, each mimicking some biological process (Holland, 1975; Goldberg, 1989; Michalewicz, 1992):

Selection Some members of the family are selected for the breeding program, based on their fitness. Probabilistic rules, such as the *roulette wheel selection*, or the *tournament selection*, are applied to allow the chance for the less fit individuals to reproduce.

Crossover The selected individuals are allowed to mate. The children are produced by taking segments of each parent's chromosome and combine them, using *one-point crossover*, *two-point crossover*, or *cut and splice* rule.

Mutation Each bit in the chromosome is allowed a small probability to flip (mutate), thus creating very different new individuals.

Once a new family has been created, the process is repeated until termination. Or, put into the format of a pseudo-code:

(1) Choose initial population.
(2) Evaluate the fitness of each individual in the population.
(3) Select best-ranking individuals to reproduce by a *selection* rule.
(4) Breed a new generation through *crossover* and *mutation*.

(5) Evaluate the fitness of each offspring.
(6) Store the best (so far) individual.
(7) Check for the termination criterion.

   a. If not satisfied, go to step 3.
   b. If satisfied, terminate.

Table 11.2.3 shows two near-optimal solutions from two different GA runs (Cheng *et al.*, 2000), as Case 1 and Case 2, which are seeded differently. Case 2 is selected as the (near) optimal solution, with the shown toe location, x^{toe}. The saltwater front is plotted in Fig. 11.2.8, in relation to the pumping well locations. We note that in the optimal solution, only 7 wells are operating, the remaining 8 are shut down. Of the wells that are not pumped, some have been invaded by saltwater. It is also of interest to note that from a different simulation of Case 1, a somewhat inferior solution was produced (about 7% less production), although this may still be an acceptable solution. However, the pumping patterns for the two cases are different—Case 1 has 8 pumping wells, but only 6 of them operate also in Case 2. This is a consequence of the multiplicity of local maxima, indicating the need for a techniques like the GA to sort them out.

11.2.6 Chance constrained optimization

In Chap. 10, we have discussed the uncertainties associated with groundwater modeling, and the option of stochastic modeling. We have also discussed the meaning of a deterministic solution, obtained by using mean properties as input. Briefly, the deterministic solution, based on mean properties as input, can be roughly viewed as the *mean response of the groundwater system.* In groundwater optimization, however, the use of such deterministic solution as the basic simulation tool may be quite erroneous. Realizing that the *mean response* means that there is a 50% chance for success and 50% chance for failure, and that the process of optimization will push the design right to the limit, without leaving any slack, the optimal result is, generally, unacceptable from the point of view of redundancy considerations. For example, in the coastal aquifer pumping problem discussed in the preceding subsection, if the mean aquifer properties would be used in the model, and the optimization would push the saltwater front to just beyond the critical distance in front of the wells, then in real life, due to uncertainty, there is a 50% probability that some of the pumping wells will be invaded by seawater. Hence, it is dangerous to overlook the issue of uncertainty in the process of optimization.

Continuing to use the saltwater intrusion problem as an example, let us examine the (deterministic) constraint (11.2.52), $x_i^{\text{toe}} < x_i^w$. In the presence of uncertainty, obeying this constraint cannot be guaranteed; we can address this issue only in probabilistic terms. Hence, let us modify the deterministic constraint, turning it into a probabilistic one:

$$\text{Prob}\left(x_i^{\text{toe}} < x_i^w\right) > \mathcal{R}, \tag{11.2.54}$$

where 'Prob' stands for 'the probability that', and $\mathcal{R}$ is a reliability measure, set by the manager. For example, if the reliability is set to be $\mathcal{R} = 90\%$, then there is a 10% chance that the constraint will be violated. The next step is to conduct the optimization procedure under this *probabilistic constraint.*

Optimization under probabilistic constraints is a rather difficult task. Often, an approximation is sought by using an equivalent deterministic constraint. One popular method that allows the conversion from a probabilistic constraint to a deterministic optimization is the *chance constrained programming*, pioneered by Charnes and Cooper (1959, 1963). Although the theoretical background of this method may be difficult, requiring a few qualifying assumptions, the concept is quite simple.

Consider, for example, the constraint (11.2.52), which states that the saltwater toe should not invade the active wells, $x_i^{\text{toe}} < x_i^w$. If the aquifer parameters, e.g., hydraulic conductivity and outflow rate, are uncertain, we can conduct a stochastic solution of saltwater intrusion with a given set of pumping rates, using one of the solution techniques described in Chap. 10. As a result, we obtain the mean toe location as $\overline{x}_i^{\text{toe}}$, and its standard deviation as $\sigma_{x_i^{\text{toe}}}$. As stated above, if we modify the deterministic constraint, using the mean toe location, $\overline{x}_i^{\text{toe}} < x_i^w$, then we are at 50% risk. On the other hand, if we add a *'cushion'* in front of the toe location, say, one standard deviation, $\overline{x}_i^{\text{toe}} + \sigma_{x_i^{\text{toe}}} < x_i^w$, we are safer; only about 32% at risk (68% reliable), assuming a Gaussian probabilistic distribution. If we wish to be safer, we can add two standard deviations and be at 95% reliability. In fact, we can prescribe a desirable reliability, $\mathcal{R}$, and then, based on the value of standard cumulative normal probability distribution, F, obtain the needed (deterministic) constraint:

$$\bar{x}_i^{\text{toe}} + F^{-1}(\mathcal{R})\,\sigma_{x_i^{\text{toe}}} < x_i^w, \tag{11.2.55}$$

where F^{-1} is the inverse of F.

The next step is to examine the effect of uncertainty on pumping. Adding a 'cushion' between the predicted mean toe location and the pumping well means that the mean toe location should be guided to stay farther away from the well. To satisfy this requirement, we are forced to pump less. The size of the 'cushion' is determined by two factors: the size of the standard deviation, and the desirable reliability. The higher the desired reliability, the larger should the F^{-1}-value be. The reliability level depends on many considerations, and the decision maker may have some control over of them. For example, it may be possible to reduce the size of standard deviation associated with the uncertainty in aquifer parameter by assigning more resources to data collection. This leads to a tradeoff between cost of data collection and the benefit of being able to pump more water.

In order to perform unconstrained optimization, the constraint (11.2.55) needs to be converted into a penalty term in the objective function. Hence, the *chance constrained* optimization statement, corresponding to the *deterministic* optimization statement, (11.2.53), becomes:

$$\text{Maximize } Z''(Q_1^w, \ldots, Q_{15}^w) = \sum_{i=1}^{15} Q_i^w - r\,\mathrm{H}(Q_i^w)\times$$

$$\mathrm{H}\Big(\overline{x}_i^{\mathrm{toe}} + F^{-1}(\mathcal{R})\,\sigma_{x_i^{\mathrm{toe}}} - x_i^w\Big)\left[\frac{\overline{x}_i^{\mathrm{toe}} + F^{-1}(\mathcal{R})\,\sigma_{x_i^{\mathrm{toe}}}}{x_i^w} - 1\right]^2. \quad (11.2.56)$$

11.2.7 Multiobjective optimization

Up to this point, we have discussed only optimization problems involving a *single* objective, expressed by a single objective function. A decision maker, however, is often faced with *multiple*, sometimes *conflicting*, goals and objectives. The reasons for multiple objectives stem from the competition among various groups of interest and consumers, including nature as one, for the limited quantity and quality of available water, from the competition between immediate needs and benefits versus long term sustainability of the sources, or from the possible conflict between local, regional and national interests, to mention but a few. In real life, multiobjective optimization is the rule rather than an exception. The considered objectives may be expressed in terms of different measures, e.g., monetary value for operational cost, meters for piezometric head drawdown, or parts per million for contamination. In this subsection, we shall briefly discuss the issues and basic methodologies for solving multiobjective optimization problems.

A. Non-inferior solution

Mathematically, a *multiobjective* optimization problem, can be formulated as:

Determine the values of the decision variables, x_i, $i = 1, 2, \ldots, n$, such that

$$\text{(Maximize, Minimize)}\ \{Z_1(\mathbf{x}),\ Z_2(\mathbf{x}),\ \ldots,\ Z_p(\mathbf{x})\},\ \text{with respect to } \mathbf{x},$$
$$\text{subject to:}\quad f_j(\mathbf{x})\ (>,=,<)\ b_j,\ j = 1, \ldots, m, \qquad (11.2.57)$$

where Z_1, Z_2, $\ldots Z_p$ are different objective functions.

An *optimal solution* of the above problem is a decision variable vector, $\mathbf{x}(\equiv \{x_1, x_2, \ldots, x_n\})$, that gives the maximum (or minimum) values of all objective functions, $Z_1(\mathbf{x})$, $Z_2(\mathbf{x})$, $\ldots$, $Z_p(\mathbf{x})$, within the feasible solution domain (as set by the constraints). For most practical problems, however, such solution does not exist. As mentioned above, objective functions are often in conflict with each other; hence it is not possible to *simultaneously* attain the maximum value for *all* objective functions.

In view of this conflicting situation, leading to the non-existence of solution, our goal in optimization must to be modified. The goal of a multiobjective optimization is then to *seek a set of decision variables such that it is not possible to further improve upon the achievement of one objective without making at least one other objective suffer.* This solution may be called a *non-inferior solution*, or a *Pareto solution*, named after the Italian sociologist and economist Vilfredo Pareto (1848-1923), who used the concept in studying economic efficiency and income distribution. Such a solution, when it exists, is often nonunique. The collection of such non-inferior solutions is a *Pareto set*, which forms a hypersurface, called *Pareto front*, in the decision variable or the objective function space. On the Pareto front, we can find the *trade-off* between the different objectives; that is, how many units of one objective function the decision maker is willing to trade for one unit of another objective function, such that the decision maker has no preference with respect to these two objective functions. A simple example will be used to illustrate the above points.

Consider the following multiobjective optimization problem that involves two decision variables and two objective functions (Haimes *et al.*, 1975):

$$\begin{aligned}&\text{Minimize } Z_1(x_1, x_2) = x_1\\&\text{Minimize } Z_2(x_1, x_2) = 10 - x_1 - x_2\\&\text{subject to:}\quad 0 \le x_1 \le 5; \quad 0 \le x_2 \le 5.\end{aligned} \tag{11.2.58}$$

In Fig. 11.2.9a, we present the solution space for the two decision variables, (x_1, x_2), as a two-dimensional plot. The feasible solution region, as bounded by the constraints in (11.2.58), is shown as the shaded area. As each pair of decision variable values allows us to evaluate the objective functions (Z_1, Z_2) in (11.2.58), we can map the feasible solution region onto the objective function space as the shaded region in Fig. 11.2.9b. Within this feasible solution region, our goal is to find either the optimal solution, or the non-inferior solution, of the given optimization problem.

We start by selecting a point somewhere inside of feasible solution region, marked as D in Fig. 11.2.9; it corresponds to D' in the objective function space. Since our goal is to minimize the objective functions, it is easy to see that there are directions that we can move in (indicated by the arrows) such that we improve (minimize) one objective function, either Z_1 or Z_2, without hurting the other. For example, we can move in the direction parallel to the Z_2 axis in Fig. 11.2.9b, thus reducing Z_2 until reaching the point E'. Hence, we conclude that D cannot be a non-inferior solution. In fact, similar to the discussion in Subs. 11.2.2 for linear programming, it can be proven that *if all the objective functions and all the constraints are linear functions of the decision variables, the non-inferior solution can be located only on the boundary of the feasible solution domain.* For the present simple case, this is easily observed on Fig. 11.2.9. The above statement can be compared to the single objective functions case, in which the optimal solution is located on a *vertex* of the feasible solution domain. We should also emphasize that the above statements (of having the solution always located on boundary or on

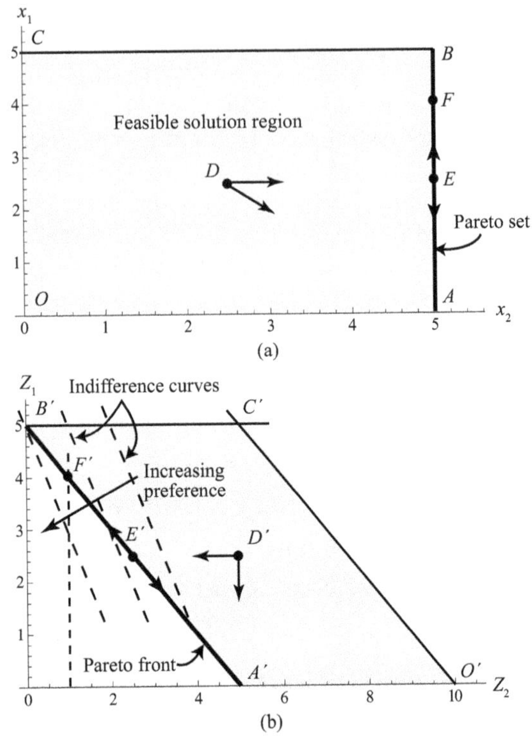

Figure 11.2.9: Multiobjective optimization problem: (a) The decision variable space; (b) The objective function space.

a vertex) is *not true*, if any of the objective functions or of the constraints is nonlinear.

In addition to E, we can identify by visual inspection that all points located on the boundary $\overline{AB}$ on Fig. 11.2.9a, which are mapped onto $\overline{A'B'}$ on Fig. 11.2.9b, fulfill the non-inferior solution criterion. Hence, $\overline{AB}$ defines a *Pareto set*, and $\overline{A'B'}$ is a *Pareto front*. If we select a point belonging to the Pareto set, marked by E and E', respectively in Fig. 11.2.9a and b, we observe that we can move along the boundary $\overline{AB}$ seeking to minimize Z_1 or Z_2. However, it is clear that if we move in one direction to improves one objective function, the other will deteriorate, indicating a conflict. In fact, this is exactly the definition of a non-inferior solution—*in any direction that we move, at least one objective function will suffer*. Altogether, there is no *optimal solution* in this example, and we are given a multiplicity of solutions as the Pareto set.

Since the non-inferior solution set is on a boundary (a line in the present linear, two-objective case, and a hypersurface in a nonlinear, multiobjective case), we can express the surface as a function, $F(Z_1,\ Z_2, \ldots,\ Z_p) = 0$. This function allows us, in principle, to solve for any one objective function as an implicit function of the rest, i.e.,, $Z_i = F_i(Z_1, \ldots, Z_{i-1}, Z_{i+1}, \ldots, Z_p)$. These functions, known as *tradeoff functions*, tell us how one objective function depends on the rest. We can define the *tradeoff rate functions* between the ith and jth objective function as

$$T_{ij}(\mathbf{x}) = \frac{\partial F_i(\mathbf{x})}{\partial Z_j(\mathbf{x})}. \qquad (11.2.59)$$

These functions have the properties:

$$T_{ii} = 1, \qquad T_{ij} = \frac{1}{T_{ji}}. \qquad (11.2.60)$$

For example, in the present case, the tradeoff function is $Z_1 = F_1(Z_2) = -Z_2 + 5$, and the tradeoff rate is $T_{12} = \partial F_1/\partial Z_2 = -1$. We can see that the tradeoff ratio between Z_1 and Z_2 is that one part of improvement in Z_1 causes one part of deterioration in Z_2.

B. Preferred solution

The situation of multiplicity of solutions is unsuitable for decision making. To overcome this difficulty, narrowing the range of action alternatives, the decision maker needs to seek and impose additional conditions. Ideally, the decision maker should aim at a single, *optimal* solution. However, in view of the subjectivity in the decision process, this solution is often called a *preferred solution*, rather than an optimal solution.

Haimes *et al.* (1975) lists three schools of thought concerning decision making under conditions of multiple objectives. The first suggests that one should not bother to narrow down the non-inferior solution set in order to reach a preferred solution. Instead, the decision maker should have the freedom to choose *any* of the inferior solutions that is found to be appropriate. The second school suggests that all the non-inferior solutions are possible options. However, the decision maker should attempt to quantify the *utility*, that is, the societal benefits or costs associated with each of the considered objectives, and use this information to justify the selection of a preferred solution in a less subjective way. According to the third school, it is argued that the combined utility of the considered objectives can be determined beforehand, so that the multiple objectives can be reduced to a *single* objective of maximizing the utility. In the following, we shall briefly discuss some of the above ideas, following Haimes *et al.* (1975).

Utility function approach. In this approach, we assume that the minimization of each objective function, Z_i, has a certain benefit to society, measured in terms of a certain utility, which is to be maximized. Hence we can define a *utility function* $U = U(Z_1, Z_2, \ldots, Z_p)$. This function is also known as *social preference function*, or *social welfare function*. The optimization problem then becomes:

$$\text{Maximize } U(Z_1, Z_2, \ldots, Z_p), \quad \text{subject to all constraints.} \tag{11.2.61}$$

The difficulty in this method lies in the determination of the utility function. The utility function is, generally, nonlinear, and coupled, that is, the utility generated by achieving one objective may also depend on another. For example, when the pumped water volume is ample to meet the water demand, the public will be more concerned about long term detrimental effects, such as the lowering of piezometric heads, and give more utility gain to minimizing the drawdown. On the other hand, if the pumped volume is dangerously near the minimum demand, the public may care less about these *good conscience* impacts. In other words, the (negative) utility of lowering the piezometric head may be evaluated differently when the other objectives have high or low values. In a simple approach, we can assume that such interference among objectives does not exist, and the total utility is the sum of the individual utilities

$$U(\mathbf{Z}) = U_1(Z_1) + U_2(Z_2) + \cdots + U_p(Z_p). \tag{11.2.62}$$

Indifference function approach. In this approach, the decision maker is asked for his preference of objectives. For example, at a certain feasible solution point, is he willing to trade a units of objective 1 for b units of objective 2, and vice versa? If the answer is yes, it means that to this decision maker, these items are of equal value, and this decision maker is *indifferent* to this trade-off. It is our goal to determine these *iso-preference* surfaces, that is, surfaces on which the decision maker or the public is socially indifferent to one decision versus another.

Take, for example, the case of the two objective functions discussed above. The decision maker may be willing to trade two units of Z_1 for one unit of Z_2. If this position is the same for the entire feasible solution space, we may plot the *indifference curves* shown as dashed lines on Fig. 11.2.9. Any feasible solution that falls on the same curve makes no difference to the this decision maker, but solutions on different curves mean different preferences.

Another way to quantify a decision maker's preference structure is through a *value function*, $v(\mathbf{Z})$. The value function can be linear or nonlinear, with empirical coefficients. The decision maker's attitude to possible tradeoffs among objectives may be used to determine these coefficients. In the above case, if we assume that v is a linear function of Z_1 and Z_2, the decision maker's attitude to tradeoffs determines the value function $v(\mathbf{Z}) = Z_1 + 2Z_2$. The indifference curves are the contour curves of the value function. As our problem is a minimization problem, the increasing preference is in the direction of decreasing value function, marked by the arrow in Fig. 11.2.9b. An optimal solution is then identified to be the point B' in Fig. 11.2.9b (or point B in Fig. 11.2.9a).

Lexicographic approach. In the lexicographic approach, we assume that the decision maker can rank the objectives in terms of their importance, say $\{Z_1, Z_2, \ldots, Z_p\}$, with Z_1 being the most important. The higher ranked objectives take precedence over all the lower ranked ones. The multiobjective optimization is conducted as a sequence of single objective optimizations, starting form Z_1,

$$\begin{aligned}&\text{Minimize } Z_1(\mathbf{x}), \text{ with respect to } \mathbf{x},\\ &\text{subject to:} \quad f_j(\mathbf{x})\,(>,=,<)\,b_j,\; j=1,\ldots,m.\end{aligned} \tag{11.2.63}$$

Suppose we find all the possible solutions of the problem as stated above, and call this set $\mathbf{y}_1$. We then conduct the sequential optimization as

$$\begin{aligned}&\text{Minimize } Z_k(\mathbf{x}), \text{ with respect to } \mathbf{x},\\ &\text{subject to:} \quad \mathbf{x}\in\mathbf{y}_k;\; k=2,\ldots,p.\end{aligned} \tag{11.2.64}$$

This process continues until a single solution is reached. This is, then, the optimal solution. Any remainder, unused objectives are simply discarded at this point.

As an example, we again examine the of two objective function case defined in (11.2.58). Let us assume that the objective function Z_2 is more important than Z_1. As a first step, we shall ignore Z_1 by dropping the first equation in (11.2.58), and conduct the following single objective optimization:

$$\begin{aligned}&\text{Minimize } Z_2(x_1,x_2) = 10 - x_1 - x_2,\\ &\text{subject to:} \quad 0\le x_1\le 5; \quad 0\le x_2\le 5.\end{aligned} \tag{11.2.65}$$

The solution of the above statement is simply $x_1 = 5$ and $x_2 = 5$, or point B in Fig. 11.2.9a. As this is a single solution, no more optimization can be performed; hence, it is the optimal solution. The objective Z_1 plays no role, and is discarded.

Conversely, if, for this decision maker, Z_1 is the more important objective function, then the following optimization problem should be executed first

$$\begin{aligned}&\text{Minimize } Z_1(x_1,x_2) = x_1,\\ &\text{subject to:} \quad 0\le x_1\le 5; \quad 0\le x_2\le 5.\end{aligned} \tag{11.2.66}$$

The solution of the above problem gives the solution of $x_1 = 0$ and $0 \le x_2 \le 5$. In this case, apparently, there is room for further optimization. By further minimizing Z_2 as in (11.2.65), with the new constraint $x_1 = 0$, we obtain the optimal solution of $x_1 = 0$ and $x_2 = 5$, or point A in Fig. 11.2.9a.

Parametric approach. We may assume that the relative importance of the objective functions, $Z_1, Z_2, \ldots, Z_p$, can be determined, such that different *weights*, $w_1, w_2, \ldots, w_p$, can be assigned to the different objectives. This is similar to the case of the linear utility function discussed above. Once the (single) composite objective function is defined, we can perform the single objective function optimization

$$\text{Minimize } Z(\mathbf{x}) = \sum_{i=1}^{p} w_i Z_i(\mathbf{x}), \quad \text{subject to constraints.} \tag{11.2.67}$$

For example, for the two objective function case examined above, the decision maker may decide that the utility of objective 2 is twice as important as that of objective 1; hence, the problem of optimal decision reduces to

$$\begin{aligned} &\text{Minimize } Z(\mathbf{x}) = Z_1(\mathbf{x}) + 2Z_2(\mathbf{x}) = 20 - x_1 - 2x_2, \\ &\text{subject to:} \quad 0 \le x_1 \le 5; \quad 0 \le x_2 \le 5. \end{aligned} \tag{11.2.68}$$

We notice that the new objective function, Z, is just like the value function v defined in the indifference function approach. Minimizing this objective function, we obtain the optimal solution $x_1 = 5$ and $x_2 = 5$.

The ε-constraint approach. The ε-constraint approach selects $p - 1$ objectives from among the p ones, say $\{Z_1, Z_2, \ldots, Z_{p-1}\}$, and sets their maximum allowable level to $\{\varepsilon_1, \varepsilon_2, \ldots, \varepsilon_{p-1}\}$. These conditions then enter the original multiobjective optimization problem as additional constraints; hence, we may modify (11.2.57) to the following statement:

$$\begin{aligned} &\text{Minimize } Z_p(\mathbf{x}), \text{ with respect to } \mathbf{x}, \\ &\text{subject to:} \quad f_j(\mathbf{x}) \; (>, =, <) \; b_j, \; j = 1, \ldots, m, \\ &\text{and } Z_k(\mathbf{x}) \le \varepsilon_k, \; k = 1, \ldots, p-1. \end{aligned} \tag{11.2.69}$$

This approach can be made equivalent to the utility function approach by declaring that the objective functions to be minimized, $\{Z_1, Z_2, \ldots, Z_{p-1}\}$, have constant utility as long as they do not exceed certain levels; beyond these levels, these objectives become harmful and infinitely undesirable to the public. This can be expressed as

$$U_k(Z_k) = \begin{cases} \text{constant}; & Z_k \le \varepsilon_k, \\ -\infty; & Z_k > \varepsilon_k, \end{cases} \quad k = 1, \ldots, p-1. \tag{11.2.70}$$

Or, we can follow the lexicographic approach, with Z_1 regarded as the most important objective, and Z_p as the least important one. We begin by minimizing Z_1, subject to $Z_1 \le \varepsilon_1$, and the original set of constraints, to obtain the solution set $\mathbf{y}_1$. The process continues until we reach the least important objective Z_p, and obtain the solution $\mathbf{y}_p$.

Take the simple example (11.2.58) as an illustration. If the decision maker decides that objective 2 should not be allowed to exceed the value 1, that is, $Z_2 \le 1$, then the optimization problem is modified to

$$\text{Minimize } Z_1(x_1, x_2) = x_1$$

$$\text{subject to: } 10 - x_1 - x_2 \leq 1;\ 0 \leq x_1 \leq 5;\ 0 \leq x_2 \leq 5. \quad (11.2.71)$$

We clearly identify the optimal solution as points F and F' on Fig. 11.2.9.

C. Pairwise comparison weight determination

In the parametric and the utility approaches discussed above, as in many other multiobjective decision making processes, we are often faced with the need to assign relative *weights*, or *utility* values, to many different objective functions. For example, we may assign the weights $\mathbf{w} = \{w_1, w_2, \ldots, w_p\}$ to objective functions $1, 2, \ldots, p$, respectively, as a measure of importance of the objective functions in the decision making process. Generally, but not always, the weights are normalized, such that $w_1 + w_2 + \ldots + w_p = 1$.

In reality, however, it is difficult for a decision maker to assign weights to too many different objectives, or even to rank them, particularly if the objectives are *non-commensurate*, and their number is large. Psychologists have long discovered that it is difficult for people to compare and rank several objects simultaneously. It has been claimed that human short term memory cannot store and compare more than seven (Miller, 1956), or four (Cowan, 2000), chunks of information. For example, when one is asked whether he or she prefers apples to oranges, oranges to pears, and pears to apples, a contradictory, circular ranking may result. According to Arrow's (1951) *impossibility theorem*, this is particularly so if multiple decision makers are involved, each representing a different interest. For example, three managers are asked to rank A, B, and C; and they decide, respectively, that $A > B > C$, $B > C > A$ and $C > A > B$. Since two of them agree that $A > B$, we may accept their ruling, based on majority. However, two of them also agree that $B > C$ and $C > A$. Thus, the decisions based on majority can produce contradictions. When a single manager is involved, the manager can be asked to resolve the conflict himself. When multiple managers are involved, this may not be possible.

These difficulties have resulted in the design of a number of weighting, and collaborative weighting, methods, such as the *entropy method*, the *direct evaluation (ranking) method*, and the *pairwise comparison method* (Pomerol, 2000). Only the last type of methods will be discussed here.

For the purpose of distributing the 100% weight among objectives, the pairwise comparison method, first introduced by Thurstone (1927), breaks the simultaneous comparison of multiple objectives into a sequence of comparisons between any two objectives, and then attempts to resolve the individual weights.

For example, we can compare the relative importance between one unit of objective i and one unit of objective j, and decide that the former is 3 times more important than the latter. This implies a weight ratio $w_i/w_j = 3$, although we know neither w_i, nor w_j. We then assign this ratio to a matrix element $a_{ij}(= w_i/w_j) = 3$. From such definition, we easily deduce the relation $a_{ji} = 1/a_{ij}$, and $a_{ii} = 1$. When all the *pairwise comparisons* of the p

objectives are complete, we can assemble a $p \times p$ square *comparison matrix*, $\mathbf{A}$, with elements a_{ij}. However, as these weights w_i are not known objectively *a priori*, it is unlikely that a decision maker will assign the comparison elements *consistently* when comparing multiple pairs of objectives. For example, if any three elements, a_{ij}, a_{ik}, and a_{jk}, are picked from the matrix, likely the relation $a_{kj} = a_{ij}/a_{ik}$ will *not* be satisfied. Hence, once the task of constructing the $\mathbf{A}$-matrix is complete, it is important to check its *consistency*. In fact, if the number of objectives is large, and comparisons are made in a random order, the chance of 100% consistency by a decision maker is nearly impossible.

As a part of the *analytic hierarchy process* for planning and priority setting, Saaty (1980) proposed the *eigenvalue method* to check the consistency of the comparison matrix, and to extract the weight vector from it. We notice that if a comparison matrix is consistent, it takes the form

$$\mathbf{A} = \begin{bmatrix} w_1/w_1 & w_1/w_2 & \cdots & w_1/w_p \\ w_2/w_1 & w_2/w_2 & \cdots & w_2/w_p \\ \vdots & \vdots & \ddots & \vdots \\ w_p/w_1 & w_p/w_2 & \cdots & w_p/w_p \end{bmatrix}. \tag{11.2.72}$$

It is easy to show, by simple multiplication, that the following relation exists

$$\mathbf{A} \cdot \mathbf{w} = p\mathbf{w}, \tag{11.2.73}$$

where p is the size of the matrix (number of objectives), and $\mathbf{w}$ is the weight vector. From the above relationship, we observe that $\mathbf{w}$ is an *eigenvector*, and p is an *eigenvalue*; in fact, p is the largest and the only non-null eigenvalue of $\mathbf{A}$. Saaty (1997, 1980) proposed to use this relation to check the consistency of a comparison matrix filled out by a decision maker, using the pairwise comparison procedure. We shall use an example to illustrate Saaty's approach.

A decision maker is asked to compare, pairwise, five different objectives, e.g., volume of pumped water, energy cost, drawdown at certain critical locations, water quality due to the threat of saltwater intrusion, etc., and comes up with the comparison matrix

$$\mathbf{A} = \begin{bmatrix} 1 & 1/3 & 7 & 5 & 6 \\ 3 & 1 & 8 & 4 & 5 \\ 1/7 & 1/8 & 1 & 1/4 & 1/3 \\ 1/5 & 1/4 & 4 & 1 & 2 \\ 1/6 & 1/5 & 3 & 1/2 & 1 \end{bmatrix}. \tag{11.2.74}$$

Upon checking its eigenvalues, we find that the largest one is $\lambda_{\max} = 5.323$, which is different from the anticipated value of $p = 5$, for a consistent matrix. We can define a *coefficient of inconsistency* as

p	3	4	5	6	7	8	9
CRI	0.58	0.90	1.12	1.24	1.32	1.41	1.45

Table 11.2.4: Coefficient of random inconsistency versus matrix rank.

$$\text{CI} = \frac{\lambda_{\max} - p}{p - 1}, \tag{11.2.75}$$

and obtain CI = 0.081 for the present case. This value is to be compared with the *coefficient of random inconsistency*, CRI, which is obtained by calculating CI for randomly filled reciprocal matrices. Saaty (1980) computed these values, and Table 11.2.4 presents a few such values for different p values (matrix size). (See Bozóki and Rapcsak (2008) for further discussion of CRI.) We then define the *inconsistency ratio* as

$$\text{IR} = \frac{\text{CI}}{\text{CRI}}. \tag{11.2.76}$$

For the present case, we find IR = 0.072. According to Saaty (1980), if IR $< 10\%$, then the inconsistency is acceptable; hence, we conclude that the comparison matrix (11.2.74) is reasonable and can be accepted. On the other hand, if the inconsistency condition is not acceptable, i.e., IR $> 10\%$, the decision maker needs to repeat the pairwise comparison, and revise the matrix, re-subjecting the new matrix to the consistency test.

Once the comparison matrix is accepted, the weight vector comes from the eigenvector (corresponding to the largest eigenvalue) of (11.2.74), which is found to be $\mathbf{w} = \{4.46, 6.58, 0.50, 1.51, 1.00\}$. Since, in the optimization process, only the relative weight matters, not the absolute weight, it is customary to normalize the weights such that $\sum_{i=1}^{p} w_i = 1$; hence, the weight vector becomes $\mathbf{w} = \{0.317, 0.468, 0.036, 0.108, 0.071\}$.

Another way to obtain weights from a comparison matrix is to use the row (or column) geometric mean (Boender *et al.*, 1989; Koczkodaj and Orlowski, 1999), defined as

$$w_i = \frac{r_i}{\sum_{j=1}^{p} r_j}, \quad i = 1, 2, \ldots, p, \tag{11.2.77}$$

where

$$r_i = \left(\prod_{j=1}^{p} a_{ij} \right)^{1/p}. \tag{11.2.78}$$

Applying this formula to the matrix (11.2.74), we obtain the weight vector $\mathbf{w} = \{0.315, 0.463, 0.037, 0.112, 0.074\}$, which is close to the one obtained by the eigenvector method.

In Subs. 11.2.7E, we shall discuss the use of these weights in a multi-criteria decision process to select a preferred alternative.

D. Collaborative weighting method

As discussed in the above subsection, the decision making process often involves a collaborative effort of multiple managers. As each manager may have a different opinion, it is necessary to find a way to combine the inputs from the various managers. One way to do so is to utilize the least squares regression to minimize the deviation of opinions.

In a collaborative effort, each manager fills in a comparison matrix $\mathbf{A}$, defined in (11.2.72). If these matrices are compiled together, they form a three-dimensional matrix, with elements a_{ijk}, where the index k indicates the input provided by the kth manager. In fact, as different managers may have different technical background, they may or may not feel comfortable to provide input on *all* pair comparison elements. In such case, the manager may leave certain elements unfilled, or null.

De Graan (1980) and Lootsma (1982) proposed the *least squares logarithmic regression method* (see also Limayem and Yannou, 2004), which seeks to minimize the following least square error between the (logarithm of the) consistent matrix and the inconsistent (manager filled) matrices,

$$\sum_{i=1}^{p}\sum_{j=1}^{p}\sum_{k=1}^{d}\alpha_{ijk}\left[\log a_{ijk}-\log\frac{w_i}{w_j}\right]^2, \tag{11.2.79}$$

where d is the number of decision makers, and α_{ijk} is a number equal to 1 when the decision maker k chooses to express an opinion, and 0 otherwise. In the latter case, a_{ijk} is set to an arbitrary non-zero, positive number to avoid the ill-defined $\log 0$ value, which will prevent algebraic operation.

The minimization of (11.2.79) leads to the following set of linear algebraic equations

$$\theta_i \sum_{j=1,j\neq i}^{p} d_{ij} - \sum_{j=1,j\neq i}^{p-1} d_{ij}\theta_j = \sum_{j=1,j\neq i}^{p}\sum_{k=1}^{d}\alpha_{ijk}\log a_{ijk}, \quad i=1,2,\ldots,p-1, \tag{11.2.80}$$

and

$$\theta_p = 0, \tag{11.2.81}$$

where

$$d_{ij}=\sum_{k=1}^{d}\alpha_{ijk}, \quad i,j=1,2,\ldots,p, \text{ and } i\neq j. \tag{11.2.82}$$

In the above equations, we identify d_{ij} as the number of decision makers who expressed an opinion on the pair comparison a_{ij}. Equations (11.2.80) and (11.2.81) are a set of p equations that can be used to solve for the p unknowns, $\theta_1, \theta_2, \ldots, \theta_p$. Once these are known, we obtain the weights as

$$w_i = e^{\theta_i}, \quad i=1,2,\ldots,p. \tag{11.2.83}$$

We note that in the above equation, w_p is selected to be 1, according to (11.2.81). This set of weights is then normalized such that $\sum_{i=1}^{p} w_i = 1$.

We note that the collaborative procedure described above should work if there is only one decision maker. This gives $d = 1$, and $d_{ij} = 1$ (all comparison matrix elements must be filled). We can assemble the following matrix system, based on (11.2.80) and (11.2.81),

$$\begin{bmatrix} p-1 & -1 & -1 & \cdots & -1 & 0 \\ -1 & p-1 & -1 & \cdots & -1 & 0 \\ -1 & -1 & p-1 & \cdots & -1 & 0 \\ \vdots & \vdots & \vdots & \ddots & \vdots & \vdots \\ -1 & -1 & -1 & \cdots & p-1 & 0 \\ 0 & 0 & 0 & \cdots & 0 & 1 \end{bmatrix} \begin{Bmatrix} \theta_1 \\ \theta_2 \\ \theta_3 \\ \vdots \\ \theta_{p-1} \\ \theta_p \end{Bmatrix} = \begin{Bmatrix} \log \prod_{j=1}^{p} a_{1j} \\ \log \prod_{j=1}^{p} a_{2j} \\ \log \prod_{j=1}^{p} a_{3j} \\ \vdots \\ \log \prod_{j=1}^{p} a_{(p-1)j} \\ 0 \end{Bmatrix}. \tag{11.2.84}$$

If we use the comparison matrix (11.2.74) as an example, we can solve for $\theta_1, \theta_2, \ldots, \theta_5$, from which we obtain $w_1, w_2, \ldots, w_5$. Once normalized, we obtain the weight vector $\mathbf{w} = \{0.315, 0.463, 0.037, 0.112, 0.074\}$, which is exactly the same as that obtained from the row geometric mean method, given as (11.2.77) and (11.2.78). This means that the de Graan (1980) method is an extension of the row geometric mean method to the multiple decision makers environment.

E. Analytic hierarchy process

Up to this point, we have been discussing single- or multi-objective optimization in a continuous decision variable space, subject to constraints, in order to obtain an optimal, or a preferred solution, using mathematical programming, or other search techniques. In this subsection, we shall discuss a different type of decision process, namely, one in which the decision maker is required to select a preferred alternative from amongst a given set of alternatives, on the basis of qualitative criteria. This process is (mathematically) simpler than the single- and multi-objective optimization techniques described in the sections above. We shall focus on the *analytic hierarchy process* proposed by Saaty (1977, 1980, 1994).

Consider a water manager who is faced with the following three alternatives in his planning for future water supply of a community:

- Alternative 1: Use surface water only.
- Alternative 2: Use groundwater only.
- Alternative 3: Conjunctive use of 50% surface water and 50% groundwater.

Although both surface water and groundwater supplies involve uncertainties, and a probabilistic model is called for, for the sake of simplicity, we shall consider only a deterministic model here.

Intensity of Importance	Definition
1	Equal importance
3	Moderate importance
5	Strong importance
7	Very strong or demonstrated importance
9	Extreme importance
2,4,6,8	For compromised between the above values

Table 11.2.5: A scale of numbers used to assign numerical values to judgments by comparing two elements for their relative importance.

	Public benefit	Company benefit	Priorities
Public benefit	1	5	0.833
Company benefit	1/5	1	0.167

Table 11.2.6: Comparison matrix for benefits

To select a preferred alternative, the manager needs to consider the benefits and costs associated with them. For benefits the manager finds that the following two criteria are important:

(1) public benefits,
(2) company benefits.

To compare the relative importance of these criteria, the manager follows Saaty's (1980, 1994) suggestion to use the scales shown in Table 11.2.5. For example, using these scales, the manager decides that the public interest is of *strong importance* as compared to company benefits; he then enters the number 5 in row 1, column 2, position of the comparison matrix shown in Table 11.2.6. The rest of the matrix is filled by recalling from Subs. 11.2.7C that the diagonal elements must be unity, and the reciprocal relation $a_{ij} = 1/a_{ji}$ must be satisfied.

For this simple matrix, we can easily see that it is *consistent* and the coefficient of inconsistency, as defined in (11.2.75), is CI $= 0$. Following the procedure presented earlier, we find the weights w_i, given as the normalized eigenvector of the matrix, to be $\{0.833, 0.167\}$. These values are entered as the last column in Table 11.2.6; they are referred to as *priorities*. In this way, we have quantified the priorities of the public and company benefits.

Following a hierarchial structure set up by the manager, each of the considered criteria may contain sub-criteria. For example, the public benefit criterion may contain three sub-criteria

(1) water quantity
(2) water quality
(3) recreation (e.g., public use of a lake)

The priorities of these sub-criteria can also be established, for example, in the form shown in Table 11.2.7. We observe in this table that the manager

	Water quantity	Water quality	Recreation	Weights	Priorities
Water quantity	1	2	9	0.589	0.491
Water quality	1/2	1	8	0.357	0.297
Recreation	1/9	1/8	1	0.054	0.045

Table 11.2.7: Comparison matrix for sub-criteria of public benefits.

Water Quantity	Alt. 1	Alt. 2	Alt. 3	Priorities
Alt. 1	1	2	3	0.540
Alt. 2	1/2	1	2	0.297
Alt. 3	1/3	1/2	1	0.163

Table 11.2.8: Comparison matrix for the decision alternatives concerning the *water quantity* sub-criterion.

has judged that water quantity is slightly more important than water quality (between *equal importance* and *moderate importance*, with a numeric value of 2), and that recreation is of the lowest importance.

The comparison matrix in Table 11.2.7 has a CI = 0.018 and an inconsistency ratio IR = 3.2%; hence, it is acceptable. The calculated weights are shown in the next to last column. As these are sub-criteria of a main criterion (public benefits), these weights are multiplied by the weight of the parent criterion, 0.833, to obtain the individual priorities of the sub-criteria shown in the last column of Table 11.2.7. The priorities of the benefit criteria, as obtained in Tables 11.2.6 and 11.2.7, are entered as the second column in Table 11.2.9.

Next, we need to separately examine the importance of *each* criterion and sub-criterion related to the three considered alternatives. This relation is presented in Table 11.2.8 for the 'water quantity' sub-criterion.

To construct the table, the manager asks a question like 'will there be sufficient water quantity?' and compares the effect of the three alternatives, pair by pair. The manager may decide that surface water is more abundant, and has a better chance to supply the needed quantity; in Table 11.2.8, the manager then assigns a 3:1 importance for surface water versus groundwater, and a 2:1 importance for surface water versus conjunctive use. Similarly, a 2:1 importance ratio is assigned for conjunctive use versus groundwater only. The calculated priorities for the alternatives, as shown in the last column, are entered into Table 11.2.9 as the row marked 'water quantity'.

Questions of the above type need to be asked for each of the other criteria—water quality, recreation, and company benefits—leading to tables similar to Table 11.2.8. We shall omit the details of such tables, and present only the resulting priorities in the respective rows of Table 11.2.9.

The compiled Table 11.2.9 can be utilized to 'score' the alternatives. We observe in the column marked alternative 1 (surface water) that it has a 0.540 score in the water quantity criterion, which, in turn, has a weight of 0.491. Hence, we can form a composite score by multiplying each score

Benefits	Priorities			
	Criteria	Alt. 1	Alt. 2	Alt. 3
Water quantity	0.491	0.540	0.297	0.163
Water quality	0.297	0.127	0.238	0.625
Recreation	0.045	0.474	0.474	0.053
Company benefits	0.167	0.105	0.258	0.637
Benefit Synthesis		0.345	0.281	0.374
Cost Synthesis		0.372	0.346	0.282
Benefit/Cost Ratio		0.927	0.812	1.326

Table 11.2.9: Benefit/cost analysis for multi-criteria optimization.

by the corresponding weight of its criterion, and then sum. For example, for alternative 1, we obtain $0.491 \times 0.540 + 0.297 \times 0.137 + 0.045 \times 0.474 + 0.167 \times 0.105 = 0.345$. This and other composite scores are entered in the row marked 'benefit synthesis' in Table 11.2.9. Comparing the three composite scores, 0.345, 0.281, and 0.374, respectively for alternatives 1, 2, and 3, we conclude that if we address benefit concerns only, the decision should favor alternative 3, i.e., groundwater.

The same analysis needs to be performed for the costs. Omitting the details, we present only the final cost synthesis as the next to the last row in Table 11.2.9. A higher weight in cost indicates a less favorable choice. In the last row, we divide the benefit by the cost, to obtain the benefit/cost ratio. Based on the ratios, the third alternative, using groundwater only for water supply, is the preferred solution to the water supply problem.

In the above presentation, we have demonstrated the basic concept by using simple cases. Real-life multiobjective optimization problems are much more complicated and difficult to solve. More extensive discussion of multi-objective optimization can be found in the books by Haimes *et al.* (1975), Haimes (1977), Chankong and Haimes (1983), Saaty (1980), and Pomerol (2000).

11.3 Inverse Problem

In Chaps. 1–9, we have focused our attention mainly on the *forward* (or *direct*, or *forecasting*) problems. We have constructed *mathematical models* that describe flow and transport, assuming that for a considered site, the values of the *parameters* that appear in these models are known. We suggested that our task is to solve the mathematical models by analytical or by numerical means in order to *forecast* the system's behavior in the future. However, in constructing a model for a specific site, the question will arise: 'how do we obtain the values of the various model parameters?' The answer to this question was already provided as Step 7 of the Modeling Process presented in Sec. 1.2.2—they are obtained by solving an *inverse problem*, or a *parameter estimation problem*, or *identification problem*, or *calibration problem.*

Unlike the forward problem, the inverse problem does not have a unique solution, and hence, we seek the optimal solution, employing techniques that have been discussed earlier in this chapter. This explains why we have included the discussion on the inverse problem in this chapter.

As an illustration, let us consider the mathematical model (5.4.53), describing flow in a confined aquifer. Rewritten for an isotropic aquifer, it takes the form:

$$S\frac{\partial h}{\partial t} = \nabla' \cdot (\mathsf{T}\nabla' h) - P(x, y, t) + R(x, y, t). \tag{11.3.1}$$

In a forecasting problem, the storativity S, transmissivity T, the pumping and recharge rates, P and R, together with a set of initial and boundary conditions, are known, and we seek the solution of (11.3.1) for the unknown response $h(x, y, t)$.

The *inverse problem*, on the other hand, aims at determining the unknown parameters that appear in a mathematical model, making use of observed data on the system's state variables. For example, with known values of P and R, we may observe data on h at various locations for a period of time, and seek a solution of (11.3.1) for the values of $S(x, y)$ and $\mathsf{T}(x, y)$. Obviously, in practice, in almost all cases, the data collected will not be sufficient for determining S and T as continuous spatial functions; hence, we attempt to determine discrete values of these parameters at the nodes of a numerical solution system, or within certain homogeneous zones. In other cases, we may know the piezometric head and aquifer parameters, and we seek to determine the unknown recharge or pumping that took place in the past. Or, we may rely on the head information to help us determine the location of an unknown boundary, such as an impermeable fault. In rare occasions, we may wish to determine the proper aquifer model to be used, e.g., confined, leaky, or multilayered; this is referred to as *system identification*. In the following, we shall focus only on the problem of parameter estimation.

In real-world applications, solving the inverse problem needs to precede the solution of the forward problem, because the system and its parameters must be known before any forecast can be made. In general, however, inverse problems are more difficult to solve. Unlike forward problems, which are mathematically *well-posed*, inverse problems are generally *ill-posed*, meaning that the solution is *nonunique*, and/or *unstable*. In addition, we always have difficulties resulting from insufficient observed data and from inaccuracy in these data.

It is important emphasize again that the knowledge of head alone, no matter how complete, is not sufficient to provide a unique determination of the hydraulic conductivity or the transmissivity. This is evident in the one-dimensional steady flow problem shown in Fig. 4.1.2. As long as the porous medium filling up the column is homogeneous, the shape of the piezometric line depends only on the boundary conditions, independent of the value of K. In other words, even with information on the complete head distribution along the column, the hydraulic conductivity cannot be uniquely determined without some auxiliary data, such as the flux through the column, or an independent measurement of the hydraulic conductivity, at least at one point.

The same conclusion remains valid also for two- and three-dimensional flow domains, except that, as is well-known from differential equation theory (e.g., Garabedian, 1998), to solve such problem, i.e., to obtain the complete distribution of $\mathsf{K}(x, y, z)$, the value of K must be known at least at one point on every streamline in the flow domain (or, if the flow domain is divided into stream-tubes, one such value at some point along every stream-tube). Solving the parameter estimation problem subject to such information is referred to as a *conditional* parameter estimation.

Next, we consider the effect of data errors. Observed data always contain errors. To what extent do they affect the parameter estimation procedure? In Subs. 10.2.2, we discussed model sensitivity. In a forward problem, the predicted data, such as piezometric head, may be sensitive to one parameter and less sensitive to another. In order for these parameters to be reliably determined, the observed data needs to be sensitive to the estimated parameters. The main conclusion from the above discussion is that the availability of good quality data is essential for a reliable parameter estimation.

In reality, however, data, especially good quality data, in sufficient quantity is rarely available. In a regional scale problem, it is very labor intensive and expensive to survey the entire field and to produce the required piezometric head contour maps. In most cases, groundwater level surveys are conducted infrequently by different regional and national water and geological authorities, and these data need to be patched together. In many cases, the data are collected at different times and in parts of the region that do or do not overlap, due to budget, manpower, and jurisdiction constraints. While the consistency and resolution of the spatial data can be questionable, the time resolution is even worse. Very few tests are conducted with a dense enough time resolution. Certain parameters, such as storativity, require transient data for their determination.

Although the above discussion deals with flow models, the same principles are applicable to contaminant transport models. For such models, the estimations required are the (field-scale) dispersivity coefficient, porosity, and hydraulic conductivity. In this case, piezometric head, and concentration data are needed, and the calibration of such models is a much more difficult task. To obtain concentration data, artificial tracer tests are sometimes used in field experiments (Sudicky, 1986; LeBlanc *et al.*, 1991; Garabedian *et al.*, 1991; Hess *et al.*, 1992; Gelhar *et al.*, 1992). Such tests, however, are very expensive to conduct.

Despite the difficulties, as discussed in Subs. 1.2.2, parameter estimation is an essential step in constructing any groundwater model. In what follows, we shall discuss examples of parameter estimation on a local scale (pumping tests), and on a regional scale.

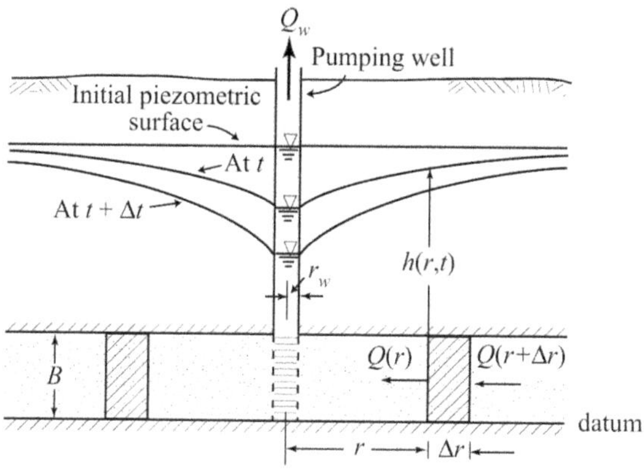

Figure 11.3.1: Unsteady flow to a well in a confined aquifer.

11.3.1 Pumping test

A pumping test is a controlled field experiment aimed at determining the basic aquifer parameters in the vicinity of a pumping well. For example, for a confined aquifer, the sought parameters are the transmissivity, T, and the storativity, S; for a leaky aquifer, in addition to T and S, we have to determine the leakage factor, λ. Sometimes we may also seek information on aquifer boundaries, or on the quality of well completion.

The analysis of a pumping test, as a tool for determining aquifer coefficients, is based on the theory usually referred to as *hydraulics of wells.* Many articles, books and chapters in books exist on this subject and need not be repeated here (e.g., Bear, 1972, 1979; Cheng, 2000; Walton, 1990; Vuković and Soro, 1992; Batu, 1998; Renard, 2005). Here, we focus only on the analysis of a pumping test, assuming that the reader is familiar with the underlying theory.

There exist many kinds of pumping tests. Here, we shall focus only on the pumping test in a confined aquifer. This test is based on a number of underlying assumptions: (1) the aquifer is confined, homogeneous, isotropic, and of infinite areal extent; (2) the well's radius is infinitesimally small (which is practically true for distances $r \gg r_w$); (3) the well's discharge is constant; and (4) water is instantaneously removed from elastic storage as the piezometric head declines.

Figure 11.3.1 shows a well in an infinite homogeneous aquifer, with two *cones of depression*, at t and $t + \Delta t$. By writing the water balance for the control volume between r and $r + \Delta r$, dividing this balance by $\Delta r \Delta t$, and passing to the limit as $\Delta r \to 0$ and $\Delta t \to 0$, we obtain the balance equation in the form of the partial differential equation,

$$S\frac{\partial h}{\partial t} = \mathsf{T}\left(\frac{\partial^2 h}{\partial r^2} + \frac{1}{r}\frac{\partial h}{\partial r}\right). \tag{11.3.2}$$

which is the same as (5.4.53), written in radial coordinates for a homogeneous isotropic aquifer, with $P, R = 0$. The boundary and initial conditions are

$$\begin{array}{lll} h(r,0) = h_o, \quad r_w \leq r \leq \infty, & \text{(initially uniform head)}, & \\ h(\infty,t) = h_o, \quad t \geq 0, & \text{(no influence at infinity)}, & (11.3.3) \\ \lim_{r=r_w\to 0} 2\pi r \mathsf{T}\frac{\partial h}{\partial r} = Q_w,\ t > 0, & (Q_w \text{ is the well's discharge}). & \end{array}$$

Equation (11.3.2) may also be written in terms of the drawdown, $s(r,t)$,

$$S\frac{\partial s}{\partial t} = \mathsf{T}\left(\frac{\partial^2 s}{\partial r^2} + \frac{1}{r}\frac{\partial s}{\partial r}\right), \quad s(r,t) = h_o - h(r,t), \tag{11.3.4}$$

with boundary conditions

$$s(r,0) = 0, \quad s(\infty,t) = 0, \quad \lim_{r=r_w\to 0} 2\pi r\mathsf{T}\frac{\partial s}{\partial r} = -Q_w. \tag{11.3.5}$$

For $Q_w = Q_w(t)$, the solution of (11.3.4) and (11.3.5) is (Muskat, 1937; Carslaw and Jaeger, 1959)

$$s = \frac{1}{4\pi\mathsf{T}}\int_0^t \frac{Q_w(\tau)}{\tau - t}\exp\left\{-\frac{r^2 S}{4\mathsf{T}(t-\tau)}\right\} d\tau. \tag{11.3.6}$$

For $Q_w(t) = Q_w = \text{const.}$, (11.3.6) reduces to

$$s = \frac{Q_w}{4\pi\mathsf{T}}\int_u^\infty \frac{e^{-x}}{x}\, dx = -\frac{Q_w}{4\pi\mathsf{T}}\mathrm{Ei}(-u). \tag{11.3.7}$$

This is also the solution given by Theis (1935)

$$s(r,t) = h_o - h(r,t) = \frac{Q_w}{4\pi\mathsf{T}}W(u), \tag{11.3.8}$$

where

$$W(u) \equiv -\mathrm{Ei}(-u) = \int_u^\infty \frac{e^{-x}}{x}\, dx \tag{11.3.9}$$

is the *well function* of $u(= Sr^2/4\mathsf{T}t)$ for a confined aquifer (Jacob, 1940). The function Ei is the *exponential integral* (e.g., Abramowitz and Stegun, 1972). Figure 11.3.2 gives values of $W(u)$. Theis (1935) obtained (11.3.8) by analogy to heat flow.

The well function can be expressed in the form of the series

$$W(u) = -0.5772 - \ln u + u - \frac{u^2}{2\times 2!} + \frac{u^3}{3\times 3!} - \frac{u^4}{4\times 4!} + \ldots \tag{11.3.10}$$

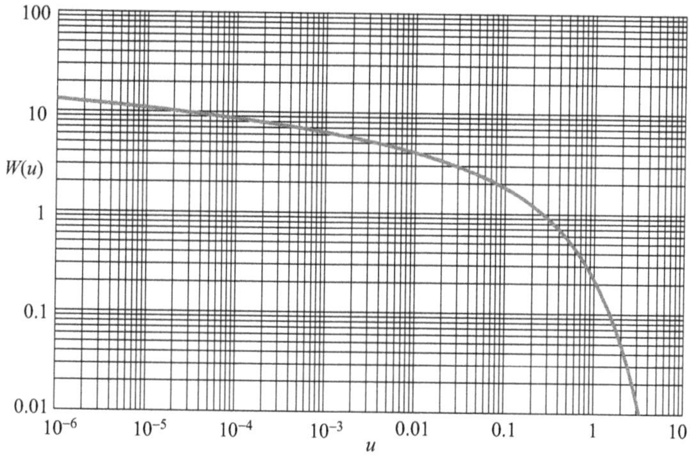

Figure 11.3.2: Type-curve for a confined aquifer.

For small values of u, say, $u < 0.01$ (i.e., for a large time at a given distance), this series may be approximated by its first two terms (Cooper and Jacob, 1946; Jacob, 1950)

$$s(r,t) \simeq \frac{Q_w}{4\pi\mathsf{T}}\left(-0.5772 - \ln\frac{r^2S}{4\mathsf{T}t}\right) = \frac{Q_w}{4\pi\mathsf{T}}\ln\frac{2.25\mathsf{T}t}{r^2S}. \tag{11.3.11}$$

With this approximation, plotting $s = s(\ln t)$, $s = s(\ln r)$, and $s = s(\ln(r^2/t))$, gives straight lines (Fig. 11.3.3).

During a pumping test, a well is pumped at a (usually) constant rate Q_w, and the drawdown produced by this pumping is observed in the pumping well itself and/or in observation well(s) in its vicinity. Let the recorded drawdown data be denoted as $s = \{\hat{s}_1, \hat{s}_2, \ldots\}$, corresponding to times $t = \{\hat{t}_1, \hat{t}_2, \ldots\}$. The traditional technique for determining the aquifer parameters is a graphical solution procedure known as the *type-curve* technique. It is based on the type-curve, $W(u)$, presented above. Actually, type-curves exist also for other types of aquifers, for other types of boundary conditions that prevail during the test, and for other underlying assumptions (e.g., the Hantush-Jacob (Hantush and Jacob, 1955) and the Hantush-Neuman (Hantush, 1960; Neuman, 1968) type-curves, etc.; see, for example, Bear (1979, Chap. 11), or Cheng (2000, Chaps. 3 and 4)). As an example, we shall focus here on the type-curve $W(u)$.

A. Theis type-curve and graphical solution

The type-curve pumping test analysis technique, or the Theis method, is based on a comparison between the type-curve and the test data of drawdown vs. time, both plotted on a transparent log-log paper as s versus r^2/t (see

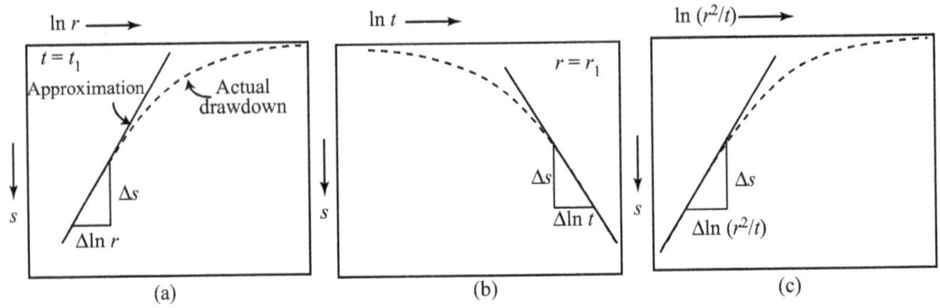

Figure 11.3.3: Straight line approximations of the well function for a confined aquifer.

Fig. 11.3.4), where r is the distance from the observation well to the pumping well.

The analysis is based on the observation that (11.3.8), rewritten as

$$s(r,t) = \frac{Q_w}{4\pi \mathsf{T}} W(u), \qquad \frac{r^2}{t} = \frac{4\mathsf{T}}{S} u, \tag{11.3.12}$$

can be expressed as

$$\log s = \log \frac{Q_w}{4\pi \mathsf{T}} + \log W(u), \tag{11.3.13}$$

$$\log \frac{r^2}{t} = \log \frac{4\mathsf{T}}{S} + \log u. \tag{11.3.14}$$

Since Q_w, T and S are constants, the relationship between s and r^2/t, and between $W(u)$ and u, are similar in form, when both are plotted on a logarithmic paper, except for a certain displacement of the curves with respect to each other, caused by the first terms on the right hand side of (11.3.13) and (11.3.14). This means that a plot of $W(u)$ vs. u (called type-curve) on a logarithmic paper is similar in form to that of s vs. r^2/t (called data curve), plotted on the same paper. The similarity between the two curves, as expressed by (11.3.13) and (11.3.14), indicates that if we plot the two curves on separate sheets of paper and superimpose the data curve on the segment of the type-curve corresponding to the data curve, we shall find that the s and $W(u)$ scales are displaced with respect to each other by an amount, $a = \log Q_w/4\pi\mathsf{T}$, while the $\log(r^2/t)$ and the s scales are displaced by the amount $b = \log 4\mathsf{T}/S$ (see Fig. 11.3.4). Reading the two displacements on the two sheets, the transmissivity and storativity values can be obtained from

$$\mathsf{T} = \frac{Q_w}{4\pi 10^a}, \qquad S = \frac{4\mathsf{T}}{10^b}. \tag{11.3.15}$$

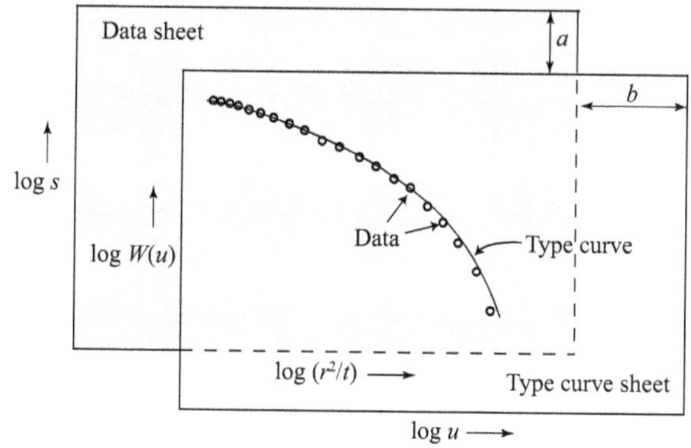

Figure 11.3.4: Schematic type-curve solution procedure: the type-curve is translated horizontally by b and vertically by a to find a visual match of the data.

The type-curve solution procedure described above, which requires a visual matching of curves, is rather *subjective*, as two hydrogeologists may come up with different estimates of aquifer parameters from what they consider as the *best visual match*. To achieve consistency of the solution, and to facilitate the use of automated (or computerized) procedures, a *mathematical* solution is required. This goal can be accomplished through the theory of *optimization*, and particularly, the procedure of *least squares*. We shall discuss these techniques below.

B. Cooper-Jacob solution and linear least square

The Cooper-Jacob solution (Cooper and Jacob, 1946) seeks to find the transmissivity of a confined aquifer from large time pumping test data. The solution of drawdown from a single well pumping at a constant discharge, Q_w, is presented as (11.3.12). Cooper and Jacob (1946) noted that when u is sufficiently small, say, $u < 0.01$, the well function $W(u)$ can be approximated as

$$W(u) \approx -\gamma - \ln u = \ln \frac{0.561}{u}, \tag{11.3.16}$$

where $\gamma = 0.5772157\ldots$ is the Euler constant. Under such conditions, the drawdown can be expressed in one of the following forms:

$$s = (2.3\,Q_w/2\pi\mathsf{T})[\tfrac{1}{2}\log(2.25\mathsf{T}t/S) - \log r], \tag{11.3.17}$$

$$s = (2.3\,Q_w/4\pi\mathsf{T})[\log t - \log(r^2S/2.25\mathsf{T})], \tag{11.3.18}$$

$$s = (2.3\,Q_w/4\pi\mathsf{T})[\log(2.25\mathsf{T}/S) - \log r^2/t]. \tag{11.3.19}$$

Plotting the test results as s vs. $\log r$ (at a given t), on a semi-log paper, (a rather unusual situation, as we rarely have many observation wells), or s

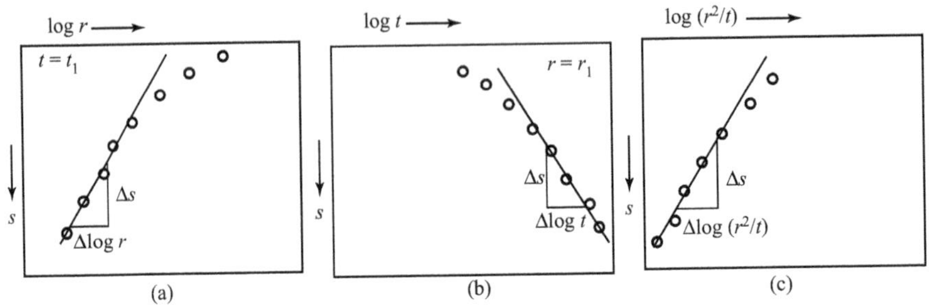

Figure 11.3.5: Straight line approximations of drawdown during a pumping test.

vs. $\log t$ (for a given r) (i.e., a single observation well), or s vs. $\log(r^2/t)$ (for an *interference test*, with a number of observation wells), the straight lines shown, respectively, on Fig. 11.3.5 are obtained. Actually, the data lie along straight lines only for sufficiently large t, or sufficiently small r. Figure 11.3.5 shows how the value of T can be derived from the slope of the straight line.

Since the solution is approximated as a straight line (Fig. 11.3.5), we need, in principle, only two data points, say, $\hat{s}_1$ and $\hat{s}_2$ taken at $\hat{t}_1$ and $\hat{t}_2$ in an observation well located at a distance r. We can then determine the slope m of the straight line and its intercept b, from

$$m = \frac{\hat{s}_1 - \hat{s}_2}{\log(\hat{t}_1/\hat{t}_2)}, \qquad b = \hat{s}_1 - m \log \hat{t}_1. \tag{11.3.20}$$

The transmissivity T and storativity S can then be obtained from

$$\mathsf{T} = \frac{2.30\, Q_w}{4\pi m}, \qquad S = \frac{2.25\, \mathsf{T}}{r^2 10^{b/m}}. \tag{11.3.21}$$

The above parameter estimation example is an *exact* inverse procedure. However, despite the simple result, such a procedure is, actually, undesirable because the drawdown measurements may contain errors. The errors will be transferred, and even enhanced, to the estimated parameters, particularly if the observed response (drawdown) is *insensitive* to the estimated parameters (Subs. 10.2.2).

To obtain more reliable aquifer parameters, it is better to use *all* the drawdown data recorded during the pumping test (barring some *outliers* that should be excluded from the data), to obtain the straight line that 'best' fits all the available data.

Thus, we may modify the definition of an inverse problem: *finding the set of parameters that minimizes the* discrepancy *between the observed data and that predicted on the basis of the estimated parameters.* Particularly, if the *discrepancy* is defined as the *weighted sum of the squared residuals*, we have

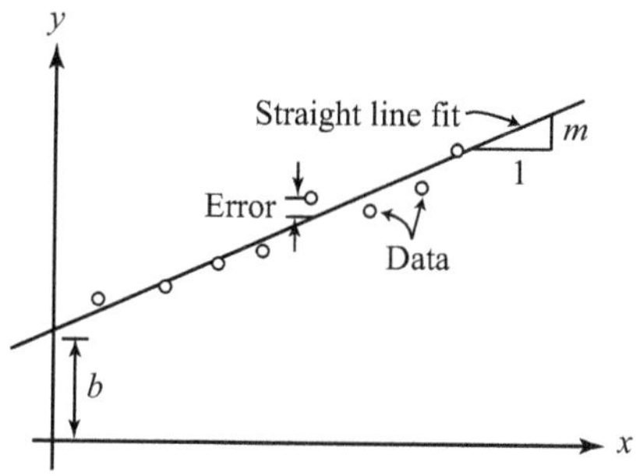

Figure 11.3.6: Linear least square fit.

the so-called *least square method.* Here, *residual* means the difference between the observed data and the prediction based on the estimated parameters.

Let us discuss the basic idea underlying the *linear least square method.* The *nonlinear least square method* will be discussed separately in the ensuing subsection.

Linear Least Square Method. In this method, we are given a set of (x, y) data points, as $\{(\hat{x}_1, \hat{y}_1), (\hat{x}_2, \hat{y}_2), \ldots, (\hat{x}_n, \hat{y}_n)\}$, which can be plotted as points in the xy-plane. We assume that this set of points is approximately represented by a straight line, with a certain amount of scatter caused by data errors (see Fig. 11.3.6). Our goal is to find the straight line

$$y(x) = mx + b, \tag{11.3.22}$$

that will best fit the data. Here, the criterion for best fit is that the (weighted) sum of the squared difference between the data and the straight line,

$$E = \sum_{i=1}^{n} w_i[\hat{y}_i - y(\hat{x}_i)]^2 = \sum_{i=1}^{n} w_i[\hat{y}_i - (m\hat{x}_i + b)]^2, \tag{11.3.23}$$

will be minimized with respect to the two free constants in (11.3.22), m and b. The use of squared differences ('errors') is to avoid the situation that positive and negative errors cancel each other.

In (11.3.23), we have introduced weights, w_i, for the individual residuals. The weights may allow the user of the least square method to exercise some *subjective* judgement, e.g., that outlier data points should be excluded (by assigning zero weight), or that a certain group of data should be considered more relevant than other data (e.g., large time data versus small time data). In a least square procedure that does not assign preference to any data point, we assign $w_i = 1$ to all weights.

The best fit can be achieved by separately differentiating (11.3.23) with respect to m and to b, and setting these equations to zero,

$$\frac{\partial E}{\partial m} = \sum_{i=1}^{n} -2w_i\hat{x}_i[\hat{y}_i - (m\hat{x}_i + b)] = 0, \tag{11.3.24}$$

$$\frac{\partial E}{\partial b} = \sum_{i=1}^{n} -2w_i[\hat{y}_i - (m\hat{x}_i + b)] = 0. \tag{11.3.25}$$

By solving these equations, we obtain the well-known least square formulas that determine the slope and intercept of the straight line that best fits the given data points:

$$m = \frac{w\Xi_{xy} - \Xi_x\Xi_y}{w\Xi_{xx} - (\Xi_x)^2}, \qquad b = \frac{\Xi_{xx}\Xi_y - \Xi_x\Xi_{xy}}{w\Xi_{xx} - (\Xi_x)^2}, \tag{11.3.26}$$

where

$$\Xi_x = \sum_{i=1}^{n} \hat{x}_i, \quad \Xi_y = \sum_{i=1}^{n} \hat{y}_i, \quad \Xi_{xx} = \sum_{i=1}^{n} \hat{x}_i^2, \quad \Xi_{xy} = \sum_{i=1}^{n} \hat{x}_i\hat{y}_i, \quad w = \sum_{i=1}^{n} w_i. \tag{11.3.27}$$

The above linear least square procedure can be applied to the long time pumping test data, using $(\log \hat{t}, \hat{s})$ as the $(\hat{x}, \hat{y})$ data pairs. Once the slope and intercept are determined from (11.3.27), the transmissivity and storativity can be evaluated from (11.3.21).

We note that, actually, the criterion for 'long time' refers to the dimensionless parameter u, rather than to the time itself. As (11.3.16) is an approximation for small u, we may define 'long time' as $u < 0.01$, such that the error in (11.3.16) is less than 5%. In practice, as u depends on the parameters to be estimated, S and T, u values are not known *a-priori.* The selection of data that satisfy the above criterion is normally conducted by visual inspection of the straight line behavior in the semi-log drawdown versus time plot.

C. Theis solution and nonlinear least square

In most cases, depending on the values the transmissivity and storativity of the tested aquifer, the duration of a pumping test is too short for the Cooper-Jacob inverse procedure to be applicable. Then, the Theis solution, (11.3.8), must be used for predicting drawdown. Obviously, the linear least square procedure, used to fit the test data to a straight line, is not applicable, and a *nonlinear least square* procedure must be used in order to fit the observed data to the type-curve solution.

Nonlinear Least Square. Again, we are given a set of (x, y) data, as $\{(\hat{x}_1, \hat{y}_1), (\hat{x}_2, \hat{y}_2), \ldots, (\hat{x}_n, \hat{y}_n)\}$, which can be plotted as points in the xy plane. Our goal is to adjust the parameters $\mathbf{p}$ of the solution *curve*, $y = y(x, \mathbf{p})$, representing the *model*, or *theory*, such that the data points will be best fitted by the type-curve solution. The criterion for the *best fit* is the minimization of the *sum of the square errors* between the model (Theis solution) and the measured data,

$$E(\mathbf{p}) = \sum_{i=1}^{n} [\hat{y}_i - y(\hat{x}_i, \mathbf{p})]^2 . \tag{11.3.28}$$

Note that in the above equation, we have taken the value $\hat{y}_i$ (e.g., the drawdown $\hat{s}_i$) observed at $\hat{x}_i$ (e.g., at pumping time $\hat{t}_i$), and found its difference with respect to the *predicted* value y evaluated at $x = \hat{x}_i$, $y(\hat{x}_i, \mathbf{p})$ (such as drawdown $s(\hat{t}_i, \mathsf{T}, S)$ predicted by the Theis equation (11.3.8)).

Our goal is to determine the *unique* set of parameters, p_i, $i = 1, 2, \ldots, m$, that will *minimize* $E(\mathbf{p})$, as defined in (11.3.28). This can be accomplished by differentiating E with respect to $\mathbf{p}(= \{p_1, p_2, \ldots, p_m\})$, and setting the resulting derivatives to zero:

$$\frac{\partial E}{\partial p_j} = 0; \qquad j = 1, \ldots, m. \tag{11.3.29}$$

By substituting the square error in (11.3.28) into the above equation, we obtain

$$\sum_{i=1}^{n} [\hat{y}_i - y(\hat{x}_i, \mathbf{p})] \frac{\partial y(\hat{x}_i, \mathbf{p})}{\partial p_j} = 0; \quad j = 1, \ldots, m, \tag{11.3.30}$$

in which $\hat{x}_i$ and $\hat{y}_i$ are known (observed) data, while the p_i's are the sought (unknowns) parameters. Equation (11.3.30) contains m equations, in the m unknowns, p_j, $j = 1, \ldots, m$. In principle, such system can be solved for the p_j-parameters. However, the equations in (11.3.29) are *nonlinear*. Unlike a linear system of equations, which can be easily solved, there is no general algorithm that can be used for the solution of a nonlinear system. Its solution depends on the actual functions involved. In what follows, we shall explore the solution of such nonlinear system, when the function y in (11.3.30) is the Theis solution for s, as given in (11.3.8) and (11.3.9).

In a pumping test, we are given the following information: the constant pumping rate, Q_w, the distance of the observation wells from the pumping well, r, and the discrete drawdowns, $\{\hat{s}_1, \hat{s}_2, \ldots\}$ recorded at $\{\hat{t}_1, \hat{t}_2, \ldots\}$. These data pairs $(\hat{t}_i, \hat{s}_i)$ correspond to the $(\hat{x}_i, \hat{y}_i)$ data referred to in the nonlinear least square procedure. Hence, the square error to be minimized, (11.3.28), can be expressed as

$$E(\mathsf{T}, S) = \sum_{i=1}^{n} \left[\hat{s}_i - \frac{Q_w}{4\pi \mathsf{T}} W\left(\frac{r^2 S}{4\mathsf{T}\hat{t}_i} \right) \right]^2 . \tag{11.3.31}$$

By differentiating (11.3.31) with respect to T and S, and setting the resultant derivatives to zero, we obtain the following two equations

$$\sum_{i=1}^{n} \left[\hat{s}_i - \frac{Q_w}{4\pi \mathsf{T}} W\left(\frac{r^2 S}{4\mathsf{T} t_i} \right) \right] \left[\exp\left(-\frac{r^2 S}{4\mathsf{T} t_i} \right) - W\left(\frac{r^2 S}{4\mathsf{T} t_i} \right) \right] = 0, \tag{11.3.32}$$

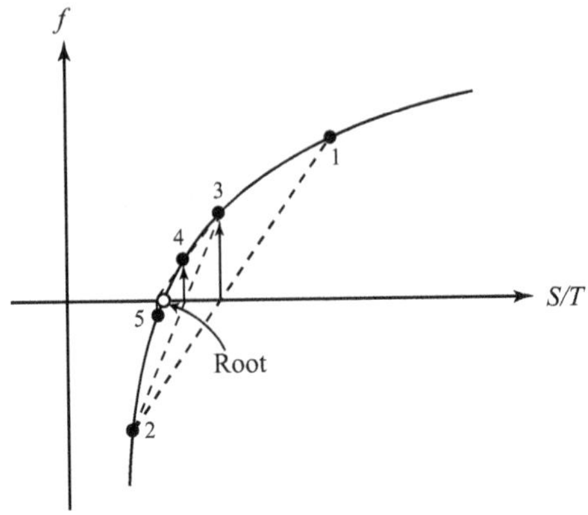

Figure 11.3.7: Illustration of finding the root of a single variable function, using the secant method. Successive trials are indicated as 1, 2, 3,....

$$\sum_{i=1}^{n}\left[\hat{s}_i - \frac{Q_w}{4\pi \mathsf{T}} W\left(\frac{r^2S}{4\mathsf{T}t_i}\right)\right]\exp\left(-\frac{r^2S}{4\mathsf{T}t_i}\right) = 0. \tag{11.3.33}$$

These two equations can be used to solve for the two parameters T and S. To make the solution procedure simpler, we rewrite these two equations in the form

$$\frac{Q_w}{4\pi \mathsf{T}} = \sum_{i=1}^{n}\hat{s}_i \exp\left(-\frac{r^2S}{4\mathsf{T}t_i}\right)\Big/\sum_{i=1}^{n} W\left(\frac{r^2S}{4\mathsf{T}t_i}\right)\exp\left(-\frac{r^2S}{4\mathsf{T}t_i}\right). \tag{11.3.34}$$

Equation (11.3.34) is then used to eliminate the factor $Q_w/4\pi\mathsf{T}$ in (11.3.33), to obtain the *single* equation (Cheng, 2000),

$$\sum_{i=1}^{n}\left\{\hat{s}_i - \left[\sum_{i=1}^{n}\hat{s}_i \exp\left(-\frac{r^2S}{4\mathsf{T}t_i}\right)\Big/\sum_{i=1}^{n} W\left(\frac{r^2S}{4\mathsf{T}t_i}\right)\exp\left(-\frac{r^2S}{4\mathsf{T}t_i}\right)\right]\times W\left(\frac{r^2S}{4\mathsf{T}t_i}\right)\right\}\exp\left(-\frac{r^2S}{4\mathsf{T}t_i}\right) = 0, \tag{11.3.35}$$

which is written in terms of the single unknown, S/T. The left hand side of (11.3.35) is essentially a function of the single variable, $f(x)$. Our goal is to find its *root*, x, such that $f(x) = 0$, with x representing S/T. We shall not elaborate here on this step. Instead, we shall use Fig. 11.3.7 to illustrate a procedure known as the *secant method* (Press *et al.*, 2007) for root finding.

Once the root for S/T is found from (11.3.35), we can use it to evaluate the right hand side of (11.3.34), which leads to the solution of T. Together with the root, the value of S is then determined. We note that the resultant parameters are the pair that minimizes the square error between the observed and the calculated values (Theis equation). A Fortran program for this least square based parameter determination for the Theis solution (confined aquifer) and for the Hantush-Jacob solution (leaky aquifer), can be found in Cheng (2000).

11.3.2 Regional scale parameter estimation

Parameter estimation on a regional scale is a much more challenging problem than local parameter estimation, because more often than not, there is not sufficient data to support a reliable solution of the inversion problem. The heterogeneity of the field and the uncertainty caused by the lack of information make a stochastic inversion an attractive solution. In what follows, we shall discuss both deterministic and stochastic parameter estimation techniques.

A. Deterministic parameter estimation

To solve an inverse problem, we first need to perform *parameterization* of the missing information. For example, the hydraulic conductivity is generally a two-dimensional or three-dimensional spatial function. As we cannot solve an inverse problem to obtain a function, as the latter has an infinite number of degrees of freedom, we need to parameterize the function. We may approximate the function by some low degree of freedom interpolation, $\mathsf{K}(x, y, z) = \sum_{i=1}^{m} \alpha_i \phi_i(x, y, z)$, where the α_i's are constant parameters to be determined in the inverse problem, and the ϕ_i's are given *basis functions*, e.g., polynomials. Or, we may divide the aquifer domain into zones, either based on our knowledge of the region's hydrogeostratigraphy, or arbitrarily, based on a selected computational grid. The hydraulic conductivity is then expressed as a set of discrete values in the selected subregions. On other occasions, the missing information may be the location and time of release of a contaminant source. Or, we may wish to search for a geometric shape, such as the location and size of a breach area between two aquifers, or the unknown location of a fault. In such case, the parameters may be the origin and radius of a circle that approximates the breached area, or a set of Cartesian coordinates (x_i, y_i) that defines line segments that approximate a sought boundary. We can denote these parameters as the vector $\boldsymbol{\theta}$.

A common approach for solving an inverse problem is to solve it as an optimization (minimization) one. The first step is to define the discrepancy (error) between the observed data, $\mathbf{u}^{\rm obs}$, and the predicted (by the model) data, $\mathbf{u}^{\rm model}$. One way to define this difference is to express it as the sum of the *distance* between observed and model-predicted data,

$$E = \sum_{i=1}^{n} |u_i^{\rm obs} - u_i^{\rm model}|, \qquad u_i^{\rm model} = u_i^{\rm model}(\boldsymbol{\theta}). \tag{11.3.36}$$

Or, more commonly, we define the discrepancy as the sum of the square errors,

$$E = \sum_{i=1}^{n} (u_i^{\text{obs}} - u_i^{\text{model}})^2. \tag{11.3.37}$$

We have here an optimization problem: the objective function, which we wish to minimize with respect to the parameters $\boldsymbol{\theta}$, is the error defined in (11.3.36) or (11.3.37), while the decision variables are the parameters $\boldsymbol{\theta}$. We can, thus, express the inverse problem in the form

$$\text{Minimize } Z(\boldsymbol{\theta}) = \sum_{i=1}^{n} (u_i^{\text{obs}} - u_i^{\text{model}})^2. \tag{11.3.38}$$

Sometimes, observations are made on two state variables, e.g., piezometric head and concentration (e.g., in the case of a contaminant transport model). In such cases, we will have multiple measures for error, and the optimization problem becomes one with two objectives.

Once the inverse problem is cast into an optimization problem, we have many different optimization solution methods at our disposal. Linear and nonlinear programming, the method of steepest descent, higher order gradient search, the genetic algorithm, and other methods may be employed. These methods were already discussed in Sec. 11.2.

Despite the above statement that an inverse problem is nothing but an optimization problem, some cautionary remarks are appropriate. First, generally, inverse problems are not well-posed and, thus, may not have unique solutions. In fact, we can make two statements: the inverse problem is not unique, because there is no unique selection of the objective functions that we wish to minimize; and it is also not unique because it depends on the amount, as well as on the spatial and temporal distribution of the measured data used for the calibration. In an inhomogeneous case, it depends also on how we have divided the heterogeneous region into homogeneous subregions, thus determining the number of parameters to be estimated. Inverse problems can also be unstable, because of the lack of sensitivity of the data to the parameters to be determined, and because of the lack of a sufficient amount of measured data. Hence, the success of solving a parameter estimation problem depends not only on the methodology selected for performing the inverse/optimization procedure, but also on the amount, quality, and type of available data.

B. Estimating the parameters of a geostatistical model

As indicated in the preceding subsection, the solution of a parameter estimation problem, especially in the (usual) case of a heterogeneous domain, is non-unique and unstable. In practice, a more reliable estimate is obtained if we have more actually observed data. For example, a much more reliable solution of the inverse model can be obtained if, in addition to observed data on state variables, we have a good amount of information on the model coefficients themselves.

Suppose we wish to solve the inverse problem for a heterogeneous aquifer, i.e. to determine the spatial distribution of its transmissivity and storativity. In addition to data on the piezometric head, we have the results for T and S obtained from pumping tests conducted at a number of locations across the aquifer. Once we have this set of sampled parameter values, kriging can be utilized to obtain the estimated values of these parameters across the entire field (Subs. 10.2.1). However, since the values the piezometric head depend on those of T and S, we can now use the observed data of piezometric head in order to improve the estimates of T and S, obtained by the kriging method.

The spatial statistical structure of the piezometric head contains information on the statistical structure of the coefficients T and S. As discussed in Subs. 10.2.1, (1) kriging itself does not provide the large scale *trend* (*drift*) of the parameter field, and (2) the kriging model requires as input information on the covariance (semivariogram), which is typically constructed by some empirical means. The availability of head observations may allow us to estimate the trend, as well as the covariance, in an inverse procedure.

Consider an aquifer in which we have observed the transmissivity T, expressed in the form, $Y = \log \mathsf{T}$, at m locations, $\mathbf{Y} = \{Y_1, Y_2, \ldots, Y_m\}$. We also have observed values of the piezometric head at n locations, $\mathbf{h} = \{h_1, h_2, \ldots, h_n\}$. For the purpose of modeling the heterogeneous transmissivity as a random field, we assume that the inhomogeneous (nonstationary) mean of the log-transmissivity (trend) has the following simple linear structure

$$\overline{Y}(\mathbf{x}) = \beta_1 + \beta_2 x + \beta_3 y. \tag{11.3.39}$$

We also assume that the covariance is homogeneous (stationary) and is of the following form (Hoeksema and Kitanidis, 1985b)

$$c_{YY}(\mathbf{x}_i, \mathbf{x}_j) = \psi_1 \delta_{ij} + \psi_2 e^{-r_{ij}/\psi_3}, \tag{11.3.40}$$

where $r_{ij} = |\mathbf{x}_i, \mathbf{x}_j|$ is the distance between the two locations, $\mathbf{x}_i$ and $\mathbf{x}_j$, and δ_{ij} is the Kronecker delta. The covariance is homogeneous because it depends only on the distance between any two points, and not on their location. Our goal is to determine the *drift parameters*, $\boldsymbol{\beta} = \{\beta 1, \beta_2, \beta_3\}$, and the *covariance parameters*, $\boldsymbol{\psi} = \{\psi_1, \psi_2, \psi_3\}$, by utilizing the inverse procedure.

The criterion for their determination can be based on the minimization of the *maximum likelihood estimate* given by (Sun, 1994)

$$\begin{aligned} \text{Minimize } Z(\boldsymbol{\psi}, \boldsymbol{\beta}) &= \log |\mathbf{C}_D(\boldsymbol{\psi})| \\ &+ [\mathbf{Z}_D - \mathbf{M}_D(\boldsymbol{\beta})]^T \mathbf{C}_D^{-1}(\boldsymbol{\psi}) [\mathbf{Z}_D - \mathbf{M}_D(\boldsymbol{\beta})], \end{aligned} \tag{11.3.41}$$

where $\mathbf{Z}_D$ is an $(m+n) \times 1$ vector, consisting of $\mathbf{Y}_D$, an $m \times 1$ vector of the logarithm of the hydraulic conductivity, and $\mathbf{h}_D$, an $n \times 1$ vector of head values, at the sampling points; $\mathbf{M}_D(\boldsymbol{\beta})$ is an $(m+n) \times 1$ vector consisting of the mean of the logarithmic hydraulic conductivity, evaluated at the data point, $\overline{\mathbf{Y}}_D(\boldsymbol{\beta})$, using the model given in (11.3.39), and the mean of head at data points, $\overline{\mathbf{h}}_D$. $\mathbf{C}_D$ is an $(m+n) \times (m+n)$ covariance matrix, consisting of covariances of the hydraulic conductivity and the head at observation points, given as

$$\mathbf{C}_D(\psi) = \begin{bmatrix} \mathbf{C}_{D,YY} & \mathbf{C}_{D,Yh} \\ \mathbf{C}_{D,Yh}^T & \mathbf{C}_{D,hh} \end{bmatrix}. \tag{11.3.42}$$

With the covariance expressed as (10.1.5), the sub-matrices are

$$\mathbf{C}_{D,YY} = \overline{\mathbf{Y}'_D \, \mathbf{Y}'^T_D}, \tag{11.3.43}$$

$$\mathbf{C}_{D,Yh} = \overline{\mathbf{Y}'_D \, \mathbf{h}'^T_D}, \tag{11.3.44}$$

$$\mathbf{C}_{D,hh} = \overline{\mathbf{h}'_D \, \mathbf{h}'^T_D}, \tag{11.3.45}$$

where $\mathbf{C}_{D,YY}$, $\mathbf{C}_{D,Yh}$ and $\mathbf{C}_{D,hh}$ are, respectively, $m \times m$, $m \times n$, and $n \times n$ matrices, and $\mathbf{Y}'_D = \mathbf{Y}_D - \overline{\mathbf{Y}}_D$ and $\mathbf{h}'_D = \mathbf{h}_D - \overline{\mathbf{h}}_D$ are the perturbations from the mean.

In (11.3.42), $\mathbf{C}_{D,YY}(\psi)$ can be evaluated from the model, (11.3.40), with assumed ψ values. The other covariances, involving head values, are evaluated by a numerical solution of the stochastic groundwater flow model. In stochastic models, such as those discussed in Chap. 10, we can construct a sensitivity matrix, $\mathbf{D}$, such that the perturbations of head, $\mathbf{h}'$, at the N grid nodes of the numerical solution, can be correlated to the perturbations of the hydraulic conductivity at the same nodes, $\mathbf{Y}'$, i.e.,

$$\mathbf{h}' = \mathbf{D}\mathbf{Y}', \tag{11.3.46}$$

where $\mathbf{h}'$, $\mathbf{D}$, and $\mathbf{Y}'$ are, respectively, $N \times 1$, $N \times N$ and $N \times 1$ matrices.

From the numerical solution, we can interpolate to find the head perturbations at the data points through an $n \times N$ weighting matrix, $\mathbf{W}_D$, such that

$$\mathbf{h}'_D = \mathbf{W}_D \, \mathbf{h}' = \mathbf{W}_D \, \mathbf{D} \, \mathbf{Y}'. \tag{11.3.47}$$

Substituting the above equation into the covariance matrices in (11.3.44) and (11.3.45), leads to

$$\mathbf{C}_{D,Yh}^T = (\mathbf{W}_D \, \mathbf{D}) \, \overline{\mathbf{Y}' \, \mathbf{Y}'^T_D}, \tag{11.3.48}$$

$$\mathbf{C}_{D,hh} = (\mathbf{W}_D \, \mathbf{D}) \, \overline{\mathbf{Y}' \, \mathbf{Y}'^T} (\mathbf{W}_D \, \mathbf{D})^T. \tag{11.3.49}$$

With all the above quantities defined, the objective function (11.3.41) can be evaluated with a given set of ψ- and $\boldsymbol{\beta}$-values, and minimized with respect to these decision variables, using some optimization procedure, such as the gradient search technique.

Once these trend and covariance parameters have been found, the hydraulic conductivity for the entire field can be estimated, using the cokriging formula, (10.2.25), which uses both the hydraulic conductivity and head data, rather than the hydraulic conductivity data alone.

References

Aarts, E.H.L. and Korst, J. *Simulated annealing and Boltzmann machines. A stochastic approach to combinatorial optimization.* Wiley, Chichester/New York, 1989.

Abbott, M.B., Bathurst, J.C., Cunge, J.A., O'Connell, P.E., and Rasmussen, J. An introduction to the European Hydrologic System—Système Hydrologique Européen, "SHE" 1: History and philosophy of a physically based distributed modeling system. *J. Hydrology*, **87**:45–59, 1986a.

Abbott, M.B., Bathurst, J.C., Cunge, J.A., O'Connell, P.E., and Rasmussen, J. An introduction to the European Hydrologic System—Système Hydrologique Européen, "SHE" 2: Structure of a physically based distributed modeling system. *J. Hydrology*, **87**:61–77, 1986b.

Abdul, A.S. and Ang, C.C. In-situ surfactant washing of polychlorinated-biphenyls and oils from a contaminated field site—Phase-II, pilot study. *Ground Water*, **32**:727–734, 1994.

Abramowitz, M. and Stegun, I.A. *Handbook of Mathematical Functions.* Dover, 1972.

Adamson, A.W. *Physical Chemistry of Surfaces*, 4th ed. Wiley, New York, 664 p., 1982.

Aiken, G.R. and Kuniansky, E.l. (Eds.) *U.S. Geological Survey Artificial Recharge Workshop Proceedings.* Sacramento, California, U.S. Geological Survey Open-File Report 02-89, 88 p., 2002.

Aitchison, G.D. and Donald, I.B. Effective stresses in unsaturated soils. *Proc. 2nd Australian-New Zealand Conf. Soil Mech. Foundation Eng.*. Inst. Engrs., 1956.

Aivalioti, M.V. and Gidarakos, E.L. In-well air sparging efficiency in remediating the aquifer of a petroleum refinery site. *J. Environ. Eng. Sci.*, **7**:71–82, 2008.

Aizinger, V., Dawson, C., Cockburn, B., and Castillo, P. The local discontinuous Galerkin method for contaminant transport. *Adv. Water Res.*, **24**:73–87, 2000.

Ajiz, M.A. and Jennings, A. A robust incomplete Choleski-conjugate gradient algorithm. *Int. J. Numer. Methods Eng.*, **20**:949–966, 1984.

J. Bear, A.H.-D. Cheng, *Modeling Groundwater Flow and Contaminant Transport*, Theory and Applications of Transport in Porous Media 23,
DOI 10.1007/978-1-4020-6682-5,

Aleman, M.A. and Slattery, J.C. A linear stability analysis for immiscible porous media contamination by organic compounds, 2. Numerical simulation. *Water Resour. Res.*, **21**:19–26, 1985.

Aliewi, A.S., Mackay, R., Jayyousi, A., Nasereddin, K., Mushtaha, A., and Yaqubi, A. Numerical simulation of the movement of saltwater under skimming and scavenger pumping in the Pleistocene aquifer of Gaza and Jericho areas, Palestine. *Transp. Porous Media*, **43**:195–212, 2001.

Allaire, G. Homogenization of the Navier-Stokes equations in open sets perforated with tiny holes. 1. Abstract framework, a volume distribution of holes. *Arch. Rat. Mech. Analys.*, **113**:209–259, 1991.

Allaire, G. One-phase Newtonian flow. Chap. 3. In: *Homogenization and Porous Media*, (ed.) U. Hornung, Springer, 45–76, 1997a.

Allaire, G. Mathematical approaches and methods, Appendix A. In: *Homogenization and Porous Media*, (ed.) U. Hornung, Springer, 226–250, 1997b.

Al-Lawatia, M., Sharpley, R.C., and Wang, H. Second-order characteristic methods for advection-diffusion equations and comparison to other schemes. *Adv. Water Res.*, **22**:741–768, 1999.

Allen, R.G. Evaporation modeling: Potential. In: *Encyclopedia of Hydrological Sciences*, Vol. 1, Art. 41, (eds.) J.J. McDonnell and M.G. Anderson, 603–613, 2005.

Allison, G.B., Gee, G.W., Tyler, S.W. Vadose-zone techniques for estimating groundwater recharge in arid and semiarid regions. *Soil Sci. Soc. Am. J.*, **58**:6–14, 1994.

Aly, A.H. and Peralta, R.C. Optimal design of aquifer cleanup systems under uncertainty using a neural network and a genetic algorithm. *Water Resour. Res.*, **35**:2523–2532, 1999.

Amaziane, B., Naji, A., Ouazar, D., and Cheng, A.H.-D. Chance-constrained optimization of pumping in coastal aquifers by stochastic boundary element method and genetic algorithm. *Computers, Materials & Continua*, **2**:85–96, 2005.

American Society of Civil Engineers. *Hydrology Handbook*, 2nd ed. ASCE Manuals and Reports on Engineering Practices No. 28, 1996.

American Water Works Association. *AWWA Standards: Water Wells.* AWWA-A100-97, Denver, Colorado, 57 p., 1997.

Anderson, M.P., Ward, D.S., Lappala, E.G., and Prickett, T.A. Computer models for subsurface water. Chap. 22 In: *Handbook of Hydrology*, (ed.) D.R. Maidment, McGraw-Hill, 22.1–22.34, 1993.

Anderson, M.P. and Woessner, W.W. The role of the postaudit in model validation. *Adv. Water Res.*, **15**:167–173, 1992.

Anderson, W.G. Wettability literature survey—Part 4: Effects of wettability on capillary pressure. *J. Petrol. Technol.*, **39**:1283–1300, 1987.

Anthony, S.S. Electromagnetic methods for mapping fresh-water lenses on Micronesian atoll islands. *J. Hydrology*, **137**:99–111, 1992.

Appelo, C.A.J. and Postma, D. *Geochemistry, Groundwater and Pollution*, 2nd ed. CRC Press, 652 p., 2005.

Aris, R. *Vectors, Tensors, and the Basic Equations of Fluid Mechanics.* Prentice-Hall, Englewood Cliffs, NJ, 286 p., 1962.

Arnold, J.G. and Fohrer, N. SWAT2000: Current capabilities and research opportunities in applied watershed modeling. *Hydrol. Process*, **19**:563–572, 2005.

Arrow, K.J. *Social Choice and Individual Values.* Wiley, New York, 1951.

Arya, L.M., Leij, F.J., Shouse, P.J., and van Genuchten, M.Th. Relationship between the hydraulic conductivity function and the particle-size distribution. *Soil Sci. Soc. Am. J.*, **63**:1063–1070, 1999a.

Arya, L.M., Leij, F.J., van Genuchten, M.Th., and Shouse, P.J. Scaling parameter to predict the soil water characteristic from particle-size distribution data. *Soil Sci. Soc. Am. J.*, **63**:510–519, 1999b.

Arya, L.M. and Paris, J.F. Physicoempirical model to predict the soil moisture characteristic from particle size distribution and bulk density data. *Soil Sci. Soc. Am. J.*, **45**:1023–1030, 1981.

Asghar, M.N., Prathapar, S.A., and Shafique, M.S. Extracting relatively-fresh groundwater from aquifers underlain by salty groundwater. *Agr. Water Manage.*, **52**:119–137, 2002.

Ashby, S.F. and Falgout, R.D. A parallel multigrid preconditioned conjugate gradient algorithm for groundwater flow simulations. *Nucl. Sci. Eng.*, **124**:145–159, 1996.

Atkins, P. and De Paula, J. *Physical Chemistry.* W.H. Freeman, 2006.

Atlas, R.M. and Bartha, R. Hydrocarbon biodegradation and oil-spill bioremediation. *Adv. Microb. Ecol.*, **12**:287–338, 1992.

Atluri, S.N. *The Meshless Method (MLPG) for Domain and BIE Discretizations.* Tech Science Press, 680 p., 2004.

Atluri, S.N. and Shen, S. *The Meshless Local Petrov-Galerkin (MLPG) Method.* Tech Science Press, 440 p., 2002.

Atluri, S.N. and Zhu, T. A new meshless local Petrov-Galerkin (MLPG) approach in computational mechanics. *Comput. Mech.*, **22**:117–127, 1998.

Auriault, J.-L., Geindreau, C. and Boutin, C. Filtration law in porous media with poor separation of scales. *Transp. Porous Media*, **60**:89–108, 2005.

Auriault, J.-L., Lebaigue, O., and Bonnet, G. Dynamics of two immiscible fluids flowing through deformable porous-media. *Transp. Porous Media*, **4**:105–128, 1989.

Avogadro, A. and Ragaini, R.C. *Technologies for Environmental Cleanup: Soil and Groundwater.* Springer, 466 p., 1993.

Avriel, M. *Nonlinear Programming: Analysis and Methods.* Dover, 2003.

Avroam, D.G. and Payatakes, A.C. Flow regimes and relative permeabilities during steady-state two-phase flow in porous media. *J. Fluid Mech.*, **293**:207–236, 1995.

Aziz, K. and Settari, A. *Petroleum Reservoir Simulation.* Appl. Sci. Publ., London, 476 p., 1979.

Babuška, I. Homogenization approach in engineering. In: *Lecture Notes in Economics and Mathematical Systems*, (eds.) M. Beckmann and H.P. Kunzi, Springer-Verlag, 137–153, 1975.

Babuška, I. Solution of interface by homogenization. I, II, III. *SIAM J. Math. Anal.*, **7**:603–634, 635–645, 1976, **8**:923–937, 1977.

Badiey, M., Cheng, A.H.-D., and Jaya, I. Deterministic and stochastic analyses of acoustic plane wave reflection from inhomogeneous porous seafloor. *J. Acoust. Soc. Am.*, **99**:903–913, 1996.

Badiey, M., Jaya, I., and Cheng, A.H.-D. A shallow water acoustic/geoacoustic experiment near the New Jersey Atlantic Generating Station site. *J. Acoust. Soc. Am.*, **96**:3593–3604, 1994.

Badon-Ghyben, W. Nota in verband met de voorgenomen putboring nabij Amsterdam (Notes on the probable results of well drilling near Amsterdam). *Tijdschrift van het Koninklijk Instituut van Ingenieurs*, Hague, 1888/9, 8–22, 1888.

Bailey, J.E. and Ollis, D.F. *Biochemical Engineering Fundamentals.* McGraw-Hill, 928 p., 1986.

Bakhvalov, N. and Panasenko, G. *Homogeneisation: Averaging Processes in Periodic Media.* Kluwer, Dordrecht, 1989.

Bakker, M. and Strack, O.D.L. Analytic elements for multiaquifer flow. *J. Hydrology*, **271**:119–129, 2003.

Bakr, A.A., Gelhar, L.W., Gutjahr, A.L., and MacMillan, J.R. Stochastic analysis of spatial variability in subsurface flows. 1. Comparison of one-dimensional and three-dimensional flows. *Water Resour. Res.*, **14**:263–271, 1978.

Baliga, B.R. and Patankar, S.V. A new finite-element formulation for convection-diffusion problems. *Numer. Heat Tr. A-Appl.*, **3**:393–409, 1980.

Banat, I.M., Makkar, R.S., and Cameotra, S.S. Potential commercial applications of microbial surfactants. *Appl. Microbiol. Biot.*, **53**:495–508, 2000.

Barenblatt, G.I., Zheltov, I.P., and Kochina, I.N. Basic concepts in the theory of seepage of homogeneous liquids in fissured rocks. *J. Appl. Math. Mech. (P.M.M.)*, **24**:852–864, 1960.

Barends, F.B.J., Brouwer, F.J.J., and Schroeder, F.H. (Eds.) *Land Subsidence: Proc. 5th Int. Symp. on Land Subsidence.* Hague, Netherlands, IHAS, **234**, 1995.

Barrett, J.W. and Morton, K.W. Approximate symmetrization and Petrov-Galerkin methods for diffusion-convection problems. *Comp. Meth. Appl. Mech. Eng.*, **45**:97–122, 1984.

Barrocu, G., Sciabica, M.G., and Muscas, L. Geographical information systems and modeling of saltwater intrusion in the Capoterra Alluvial Plain (Sardinia, Italy). Chap. 9. In: *Coastal Aquifer Management-Monitoring, Modeling, and Case Studies*, (eds.) A.H.-D. Cheng and D. Ouazar, Lewis Publ., 183–206, 2004.

Barry, D.A. and Sander, G.C. Exact-solutions for water infiltration with an arbitrary surface flux or nonlinear solute adsorption. *Water Resour. Res.*, **27**:2667–2680, 1991.

Bastian, W.C. and Lapidus, L. Longitudinal diffusion in ion exchange and chromatographic column, finite column. *J. Phys. Chem.*, **60**:816–817, 1956.

Batchelor, G.K. *The Theory of Homogeneous Turbulence.* Cambridge Univ. Press, 1959.

Bateman, H. The solution of a system of differential equations occurring in the theory of radioactive transformation. *Proc. Cambridge Phil. Soc.*, **15**:423–427, 1910.

Batu, V. *Aquifer Hydraulics: A Comprehensive Guide to Hydrogeologic Data Analysis.* Wiley-Interscience, 752 p., 1998.

Bear, J. On the tensor form of dispersion. *J. Geophys. Res.*, **66**:1185–1197, 1961a.

Bear, J. Some experiments on dispersion. *J. Geophys. Res.*, **66**:2455–2467, 1961b.

Bear, J. *Dynamics of Fluids in Porous Media.* American Elsevier, 764 p., 1972 (also published by Dover, 1988; translated into Chinese).

Bear, J. *Hydraulics of Groundwater.* McGraw-Hill, New York, 569 p., 1979. (Also as Dover edition, 2007.)

Bear, J. Conceptual and mathematical modeling. Chap. 5. In: *Seawater Intrusion into Coastal Aquifers—Concepts, Methods and Practices.* (eds.) J. Bear, A.H.-D. Cheng, S. Sorek, D. Ouazar & I. Herrera, Kluwer, 127–161, 1999.

Bear, J. Modeling Solute Transport Phenomena. Vol. 4, Pt. 13, Art. 152 in *Encyclopedia of Hydrological Sciences*, (eds.) J.J. McDonnell and M.G. Anderson, Wiley, 2341–2354, 2005a.

Bear, J. Sea water intrusion into coastal aquifers. Vol. 4, Pt. 13, Art. 157 in *Encyclopedia of Hydrological Sciences*, (eds.) J.J. McDonnell and M.G. Anderson, Wiley, 2431–2442, 2005b.

Bear, J. and Bachmat, Y. A generalized theory on hydrodynamic dispersion in porous media. *I.A.S.H. Symp. Artificial Recharge and Management of Aquifers*, **72**:7–16, 1967.

Bear, J. and Bachmat, Y. Macroscopic modeling of transport phenomena in porous media, 2. Applications to mass, momentum, and energy transport. *Transp. Porous Media*, **1**:241–269, 1986.

Bear, J. and Bachmat, Y. *Introduction to Modeling Phenomena of Transport in Porous Media.* Kluwer, Dordrecht, 553 p., 1990.

Bear, J. and Bachmat Y. Deletion of nondominant effects in modeling transport in porous media. *Transp. Porous Media*, **7**:15–38, 1992.

Bear, J., Braester, C., and Menier, P.C. Effective and relative permeabilities of anisotropic porous-media. *Transp. Porous Media*, **2**:301–316, 1987.

Bear, J., Cheng, A.H.-D., Sorek, S., Ouazar, D, and Herrera, I. (Eds.) *Seawater Intrusion into Coastal Aquifers—Concepts, Methods and Practices.* Kluwer, 625 p., 1999.

Bear, J. and Corapcioglu, M.Y. Mathematical-model for regional land subsidence due to pumping. 1. Integrated aquifer subsidence equations for vertical displacement only. *Water Resour. Res.*, **17**:937–946, 1981a.

Bear, J. and Corapcioglu, M.Y. Mathematical-model for regional land subsidence due to pumping. 2. Integrated aquifer subsidence equations for vertical and horizontal displacements. *Water Resour. Res.*, **17**:947–958, 1981b.

Bear, J., Corapcioglu, M.Y., and Bulkarishna, J. Modeling of centrifugal filtration in unsaturated deformable porous medium. *Adv. Water Res.*, **7**:150–167, 1984.

Bear, J. and Dagan, G. Moving interfaces in coastal aquifers. *J. Hyd. Div., ASCE*, **90**:193–216, 1964a.

Bear, J. and Dagan, G. Some exact solutions of interface problems by means of the hodograph method. *J. Geophys. Res.*, **69**:1563–1572, 1964b.

Bear, J., Fel, L., and Zimmels, Y. Effects of material symmetry on coefficients of transport in anisotropic porous media. Submitted for publication, 2009.

Bear, J., Nichols, E., Ziagos, J., and Kulshrestha, A. *Effect of contaminant diffusion into and out of low-permeability zones*, Lawrence Livermore National Laboratory, Rep. UCRL-JD-115626, 1994.

Bear, J. and Nitao, J.J. On equilibrium and primary variables in transport in porous media. *Transp. Porous Media*, **18**:151–184, 1995.

Bear, J., Sorek, S., and Borisov, V. On the Eulerian-Lagrangian formulation of balance equation in porous media. *Numer. Meth. Part. Diff. Eqs.*, **13**:505–530, 1997.

Bear, J. and Sun, Y. Optimization of pump-treat-inject (PTI) design for the remediation of a contaminated aquifer: multistage design with chance constraints. *J. Contam. Hydrol.*, **29**:223–242, 1998.

Bear, J. and Verruijt, A. *Modeling Groundwater Flow and Pollution.* D. Reidel Publ. Co., Dordrecht, the Netherlands, 414 p., 1987.

Bear, J., Zaslavsky, D., and Irmay, S. *Physical Principles of Percolation and Seepage*, UNESCO, 465 p., 1968.

Bear, J. and Zhou, Q. Sea water intrusion into coastal aquifers. Chap. 12. In: *The Handbook of Groundwater Engineering*, 2nd ed., (ed.) J.W. Delleur, CRC Press, 2006.

Beatson, R.K., Cherrie, J.B., and Mouat, C.T. Fast fitting of radial basis functions: Methods based on preconditioned GMRES iteration. *Adv. Comput. Math.*, **11**:253–270, 1999.

Beatson, R.K. and Light, W.A. Quasi-interpolation by thin-plate splines on a square. *Constr. Approx.*, **9**:407–433, 1993.

Beavers, G.S. and Joseph, D.D. Boundary conditions at a naturally permeable wall. *J. Fluid Mech.*, **30**:197–207, 1967.

Beck, A.E. *Physical Principles of Exploration Methods.* Wuerz Publ., Winnipeg, 292 p., 1991.

Bellin, A. and Rubin, Y. HYDRO_GEN: A spatially distributed random field generator for correlated properties. *Stoch. Hydrol. Hydraul.*, **10**:253–278, 1996.

Belytschko, T., Lu, Y.Y. and Gu, L. Element-free Galerkin methods. *Int. J. Numer. Methods Eng.*, **37**:229–256, 1994.

Bendat, J.S. and Piersol, A.G. *Random Data: Analysis & Measurement Procedures*, 3rd ed. Wiley-Interscience, 594 p., 2000.

Benner, S.G., Blowes, D.W., Gould, W.D., Herbert, R.B., and Ptacek, C.J. Geochemistry of a permeable reactive barrier for metals and acid mine drainage. *Environ. Sci. Technol.*, **33**:2793–2799, 1999.

Bensabat, J., Zhou, Q. and Bear, J. An adaptive pathline-based particle tracking algorithm for the Eulerian-Lagrangian method. *Adv. Water Res.*, **23**:383–397, 2000.

Bensoussan, A., Lions, J.L., and Papanicolaou, G. *Asymptotic Analysis of Periodic Structures.* North-Holland, Amsterdam, 1978.

Bentsen, R.G. and Manai, A.A. On the conventional cocurrent and counter-current modeling of two-phase flow. *Transp. Porous Media*, **11**:243–262, 1993.

Berkowitz, B., Klafter, J., Metzler, R., and Scher, H. Physical pictures of transport in heterogeneous media: Advection-dispersion, random-walk and fractional derivative formulations. *Water Resour. Res.*, **38**, art. no. 1191, 2002.

Berkowitz, B. and Scher, H. The role of probabilistic approaches to transport theory in heterogeneous media. *Transp. Porous Media*, **42**:241–263, 2001.

Berkowitz, B., Scher, H., and Silliman, S.E. Anomalous transport in laboratory-scale, heterogeneous porous media. *Water Resour. Res.*, **36**:149–158, Correction: **36**:1371, 2000.

Beven, K., Calver, A., and Morris, E. *The Institute of Hydrology Distributed Model.* U.K. Institute of Hydrology Report No. 98, 1987.

Binning, P. and Celia, M.A. A finite volume Eulerian-Lagrangian localized adjoint method for solution of the contaminant transport equations in two-dimensional multiphase flow systems. *Water Resour. Res.*, **32**:103–114, 1996.

Binning, P. and Celia, M.A. A forward particle tracking Eulerian-Lagrangian localized adjoint method for solution of the contaminant transport equation in three dimensions. *Adv. Water Res.*, **25**:147–157, 2002.

Biot, M.A. General theory of three-dimensional consolidation. J . Appl. Phys., **12**:155–164, 1941.

Blandford, T.N. and Huyakorn, P.S. *WHPA: A Modular Semi-Analytical Model for The Delineation of Wellhead Protection Areas, Version 2.0.* U.S. Environmental Protection Agency, 1991.

Blaney, H.F. and Criddle, W.D. Determining water requirements on irrigated areas from climatological and irrigation data. USDA Soil Cons. Serv. SDS-TP-96, 1950.

Blowes, D.W., Ptacek, C.J., Benner, S.G., McRae, C.W.T., Bennett, T.A., and Puls, R.W. Treatment of inorganic contaminants using permeable reactive barriers. *J. Contam. Hydrol.*, **45**:123–137, 2000.

Boender, C.G.E., Degraan, J.G., and Lootsma, F.A. Multi-criteria decision-analysis with fuzzy pairwise comparisons. *Fuzzy Sets and Systems*, **29**:133–143, 1989.

Boonstra, J. Well hydraulics and aquifer tests. Chap. 8. In: *The Handbook of Groundwater Engineering*, (ed.) J.W. Delleur, CRC Press and Springer, 1998.

Bouwer, H. Artificial recharge of groundwater: systems, design, and management. In: *Hydraulic Design Handbook.* (ed.) L.W. Mays, McGraw-Hill, New York, 24.1–24.44, 1999.

Bouwer, E.J. and McCarthy, P.L. Modeling of trace organics biotransformation in the subsurface. *Ground Water*, **22**:433–440, 1984.

Bowen, R.M. Incompressible porous-media models by use of the theory of mixtures. *Int. J. Eng. Sci.*, **18**:1129–1148, 1980.

Bowen, R.M. Porous media model formulations by the theory of mixtures. In: *Fundamentals of Transport Phenomena in Porous Media*, (eds.) J. Bear and M.Y. Corapcioglu, Martinus Nijhoff Publ., The Netherlands, 63–120, 1984.

Boyd, J.P. *Chebyshev and Fourier spectral methods*, 2nd ed. Dover, 2001.

Bozóki, S. and T. Rapcsak On Saaty's and Koczkodaj's inconsistencies of pairwise comparison matrices. *J. Global Optimization*, **42**:157–175, 2008.

Braester, C. Moisture variation at soil surface and advance of wetting front during infiltration at constant flux. *Water Resour. Res.*, **9**:687–694, 1973.

Bras, R.L. *An Introduction to Hydrologic Science*. Addison-Wesley, Reading, MA, 643 p., 1990.

Brebbia, C.A. *The Boundary Element Method for Engineers*. Pentech Press/ Halstead Press, London/New York, 1978.

Brebbia, C.A. and Dominguez, J. Boundary element methods for potential problems. *Appl. Math. Modelling*, **1**:372–378, 1977.

Brezzi, F., Bristeau, M.O., Franca, L.P., Mallet, M., and Roge, G. A relationship between stabilized finite-element methods and the Galerkin method with bubble functions. *Comput. Meth. Appl. Mech. Eng.*, **96**:117–129, 1992.

Brezzi, F., Hughes, T.J.R., Marini, L.D., and Masud, A. Mixed discontinuous Galerkin methods for Darcy flow. *J. Sci. Comput.*, **22**:119–145, 2005.

Brinkman, H.C. Calculations of the flow of heterogeneous mixture through porous media. *Appl. Sci. Res.*, **2**:81–86, 1948.

Broadbridge, P., Edwards, M.P., and Kearton, J.E. Closed-form solutions for unsaturated flow under variable flux boundary conditions. *Adv. Water Res.*, **19**:207–213, 1996.

Broadbridge, P. and White, I. Constant rate rainfall infiltration—A versatile nonlinear model. 1. Analytical solution. *Water Resour. Res.*, **24**:145–154, 1988.

Brooks, A.N. and Hughes, T.J.R. Streamline upwind Petrov-Galerkin formulations for convection dominated flows with particular emphasis on the incompressible Navier-Stokes equations. *Comp. Meth. Appl. Mech. Eng.*, **32**:199–259, 1982.

Brooks, R.H. and Corey, A.T. Hydraulic properties of porous media. Colorado State Univ., Hydrology Papers no. 3, Fort Collins, Colorado, 27 p., 1964.

Brooks, R.H. and Corey, A.T. Properties of porous media affecting fluid flow. *J. Irrig. Drain. Div., ASCE*, **92**:61–87, 1966.

Broyden, C.G. Convergence of single-rank quasi-Newton methods. *Math. Comput.*, **24**:365–382, 1970.

Brusseau, M.L., Jessup, R.E., and Rao, P.S.C. Modeling solute transport influenced by multiprocess nonequilibrium and transformation reactions. *Water Resour. Res.*, **28**:175–182, 1992.

Brusturean, G.A., Todinca, T., Perju, D., Carre, J., and Rusnac, C. Soil vapor extraction of synthetic gasoline mixture: Experimental observations and model predictions. *Revista De Chimie*, **58**:1268–1273, 2007.

Brutsaert, W. Probability laws for pore size distribution. *Soil Sci.*, **101**:85–192, 1966.

Brutsaert, W. The adaptability of an exact solution to horizontal infiltration. *Water Resour. Res.*, **4**:785–789, 1968.

Brutsaert, W. *Evaporation into the Atmosphere: Theory, History, and Applications.* D. Reidel Publishing Company, Dordrecht, Holland, 1982.

Buckingham, E. Studies on the movement of soil moisture. *Bull. No. 38. Bureau of Soils, USDA*, Washington, DC, 1907.

Burnett, R.D. and Frind, E.O. Simulation of contaminant transport in 3 dimensions. 2. Dimensionality effects. *Water Resour. Res.*, **23**:695–705, 1987.

Campbell, G.S. Simple method for determining unsaturated conductivity from moisture retention data. *Soil Sci.*, **117**:311–314, 1974.

Carman, P.C. Fluid flow through granular bed. *Trans. Inst. Chem. Engrs.* (London), **15**: 150–156, 1937.

Carsel, R.F. and Parrish, R.S. Developing joint probability-distributions of soil-water retention characteristics. *Water Resour. Res.*, **24**:755–769, 1988.

Carslaw, H.S. and Jaeger, J.C. *Conduction of Heat in Solids.* Oxford Univ. Press, 1959.

Celia, M.A., Reeves, P.C., and Ferrand, L.A. Recent advances in pore scale models for multiphase flow in porous-media. *Rev. Geophys.*, **33**:1049–1057, 1995.

Celia, M.A., Russell, T.F., Herrera, I., and Ewing, R.E. An Eulerian-Lagrangian localized adjoint method for the advection-diffusion equation. *Adv. Water Res.*, **13**:187–206, 1990.

Cerny, V. Thermodynamical approach to the traveling salesman problem—an efficient simulation algorithm. *J. Optimiz. Theory App.*, **45**:41–51, 1985.

Chang, L.C., Chu, H.J., and Hsiao, C.T. Optimal planning of a dynamic pump-treat-inject groundwater remediation system. *J. Hydrology*, **342**: 295–304, 2007.

Chang, R. and Cruickshank, B. *Chemistry*, 8th ed. McGraw-Hill, 1120 p., 2003.

Chankong, V. and Haimes, Y.Y. *Multiobjective Decision Making: Theory and Methodology.* North-Holland, 406 p., 1983.

Charnes, A. and Cooper, W.W. Chance-constrained programming. *Mgmt. Sci.*, **6**:73–79, 1959.

Charnes, A. and Cooper, W.W. Deterministic equivalents for optimizing and satisfying under chance constraints. *Oper. Res.*, **11**:18–39, 1963.

Chen, C.S., Ganesh, M., Golberg, M.A., and Cheng, A.H.-D. Multilevel compact radial functions based computational schemes for some elliptic problems. *Compt. Math. Applic.*, **43**:359–378, 2002.

Chen, F., Mitchell, K., Xue, Y., Pan, H., Koren, V., Duan, Q.Y., Ek, M., and Betts, A. Modeling of land-surface evaporation by four schemes and comparison with FIFE observations. *J. Geophys. Res.*, **101**:7251–7268, 1996.

Chen, W. Symmetric boundary knot method. *Eng. Anal. Bound. Elem.*, **26**:489–494, 2002.

Chen, Z.X., Lyons, S.L. and Qin, G. Derivation of the Forchheimer law via homogenization. *Transp. Porous Media*, **44**:325–335, 2001.

Cheng, A.H.-D. Darcy's flow with variable permeability—a boundary integral solution. *Water Resour. Res.*, **20**:980–984, 1984.

Cheng, A.H.-D. Heterogeneities in flows through porous media by the boundary element method. Chap. 6. In: *Topics in Boundary Element Research*, **4**: *Applications in Geomechanics*, (ed.) C.A. Brebbia, Springer-Verlag, 129–144, 1987.

Cheng, A.H.-D. *Multilayered Aquifer Systems—Fundamentals and Applications.* Marcel Dekker, New York, 384 p., 2000.

Cheng, A.H.-D., Abousleiman, Y., Ruan, F. and Lafe, O.E. Boundary element solution for stochastic groundwater flow: Temporal weakly stationary problems. *Water Resour. Res.*, **29**:2893–2908, 1993.

Cheng, A.H.-D. and Cheng, D.T. Heritage and early history of the boundary element method. *Eng. Anal. Bound. Elem.*, **29**:268–302, 2005.

Cheng, A.H.-D., Golberg, M.A., Kansa, E.J., and Zammito, G. Exponential convergence and *h-c* multiquadric collocation method for partial differential equations. *Numer. Meth. Part. Diff. Eqs.*, **19**:571–594, 2003.

Cheng, A.H.-D. and Lafe, O.E. Boundary element solution for stochastic groundwater flow: Random boundary condition and recharge. *Water Resour. Res.*, **27**:231–242, 1991.

Cheng, A.H.-D. and Morohunfola, O.K. Multilayered leaky aquifer systems: I. Pumping well solution. *Water Resour. Res.*, **29**:2787–2800, 1993a.

Cheng, A.H.-D. and Morohunfola, O.K. Multilayered leaky aquifer systems: II. Boundary element solution. *Water Resour. Res.*, **29**:2801–2811, 1993b.

Cheng, A.H.-D. and Ouazar, D. Theis solution under aquifer parameter uncertainty. *Ground Water*, **33**:11–15, 1995.

Cheng, A.H.-D. and Ouazar, D. Analytical solutions. Chap. 6. In: *Seawater Intrusion in Coastal Aquifers—Concepts, Methods, and Practices*, (eds.) J. Bear, A.H.-D. Cheng, S. Sorek, D. Ouazar & I. Herrera, Kluwer, 163–191, 1999.

Cheng, A.H.-D. and Ouazar, D. (Eds.) *Coastal Aquifer Management-Monitoring, Modeling, and Case Studies.* Lewis Publ., 280 p., 2003.

Cheng, A.H.-D., Sidauruk, P., and Abousleiman, Y. Approximate inversion of the Laplace transform. *Mathematica J.*, **4**:76–82, 1994.

Cheng, R.T., V. Casulli, and Milford, S.N. Eulerian-Lagrangian solution of the convection-dispersion equation in natural coordinates. *Water Resour. Res.*, **20**:944–952,1984.

Chiang, C.Y., Wheeler, M.F., and Bedient, P.B. A modified method of characteristics technique and mixed finite-elements method for simulation of groundwater solute transport. *Water Resour. Res.*, **25**:1541–1549, 1989.

Chilakapati, A., Ginn, T., and Szecsody, J. An analysis of complex reactions networks in groundwater modeling. *Water Resour. Res.*, **34**:1767–1780, 1998.

Childs, E.C. The transport of water through heavy clay soils, I. *J. Agr. Sci.*, **26**:114–127, 1936.

Childs, E.C. *An Introduction to the Physical Basis of Soil Water Phenomena.* Wiley, New York, 1969.

Childs, E.C. and Collis-George, N. The permeability of porous materials. *Proc. Roy. Soc. London, ser. A.*, **201**:392–405, 1950.

Chin, P., Dazevedo, E.F., Forsyth, P.A., and Tang, W.P. Preconditioned conjugate-gradient methods for the incompressible Navier-Stokes equations. *Int. J. Numer. Methods Fluids*, **15**:273–295, 1992.

Chow, E. and Saad, Y. Experimental study of ILU preconditioners for indefinite matrices. *J. Computat. Appl. Math.*, **86**:387–414, 1997.

Chow, V.T. Sequential generation of hydrological information. In: *Handbook of Applied Hydrology.* (ed.) V.T. Chow, Chap. 8, pt. IV, 8.91–8.97, McGraw-Hill, New York, 1964.

Christie, I., Griffiths, D.F., Mitchell, A.R., and Zienkiewicz, O.C. Finite-element methods for 2nd order differential equations with significant 1st derivatives. *Int. J. Numer. Methods Eng.*, **10**:1389–1396, 1976.

Cieniawski, S.E., Eheart, J.W., and Ranjithan, S. Using genetic algorithms to solve a multiobjective groundwater monitoring problem. *Water Resour. Res.*, **31**:399–409, 1995.

Cioranescu, D. and Donato, P. *An Introduction to Homogenization.* Oxford Univ. Press, 1999.

Clement, T.P. *RT3D: A Modular Computer Code for Simulating Reactive Multispecies Transport in 3-Dimensional Groundwater Systems.* PNNL-11720, Pacific Northwest National Laboratory, Richland, Washington, 1997.

Clement, T.P. Generalized solution to multispecies transport equations coupled with a first-order reaction network. *Water Resour. Res.*, **37**:157–163, 2001.

Clement, T.P., Johnson, C.D., Sun, Y.W., Klecka, G.M., and Bartlett, C. Natural attenuation of chlorinated solvent compounds: Model development and field-scale application. *J. Contam. Hydrol.*, **42**:113–140, 2000.

Clement, T.P., Sun., Y.W., Hooker, B.S., and Petersen, J.N. Modeling multi-species reactive transport in groundwater aquifers. *Ground Water Monit. Rem.*, **18**:79–92, 1998.

Clough, R.W. The finite element method in plane stress analysis, ASCE Struct. Div. Proc. 2nd Conf. Electronic Computation, 345–378, 1960.

Cohen, R.M., Mercer, J.W., and Greenwald, R.M. *EPA Groundwater Issue, Design Guidelines for Conventional Pump-and-Treat Systems.* EPA 540/S-97/504, 1998.

Collins, R.E. *Flow of Fluids Through Porous Media.* Reinhold, New York, 270 p., 1961.

Conkling, H., *et al.* Ventura County Investigations. *California Div. Water Resour. Bull.* **6**, 244 p., 1934.

Cooper, H.H. and Jacob, C.E. A generalized graphical method for evaluating formation constants and summarizing well-field history. *Trans. Am. Geophys. Union,* **27**:526–534, 1946.

Corapcioglu, M.Y. and Brutsaert, W. Viscoelastic aquifer model applied to subsidence due to pumping. *Water Resour. Res.* **13**:597–604, 1977.

Corey, A.T. Measurement of water and air permeability in unsaturated soils. *Proc. Soil Sci. Soc. Am.*, **21**:7–10, 1957.

Corey, A.T., Rathjens, C.H., Henderson, J.H., and Wyllie, M.R.J. Three-phase relative permeability. *Trans. AIME,* **207**:349–351, 1956.

Courant, R. Variation methods for the solution of problems of equilibrium and vibration. *Bull. Amer. Math. Soc.*, **49**:1–43, 1943.

Courant, R. and Hilbert, D. *Methods of Mathematical Physics.* Wiley Interscience, New York, 1962.

Couto, P.R.L. and Malta, S.M.C. Interaction between sorption and biodegradation processes in the contaminant transport. *Ecological Modelling,* **214**:65–73, 2008.

Cowan, N. The magical number 4 in short-term memory: A reconsideration of mental storage capacity. *Behavioral and Brain Sciences,* **24**:87–185, 2000.

Crank, J. *Mathematics of Diffusion.* Oxford Univ. Press, 347 p., 1956.

Crank J. and Nicolson P. A practical method for numerical evaluation of solutions of partial differential equations of the heat conduction type. *Proc. Camb. Phil. Soc.*, **43**:50–64, 1947.

Crawford, N.H. and Linsley, R.K. *Digital Simulation in Hydrology: Stanford Watershed Model IV.* Stanford Univ., Dept. Civil Eng. Tech. Rep. 39, 1966.

Custodio, E. Studying, monitoring and controlling seawater intrusion in coastal aquifers. In: *Guidelines for Study, Monitoring and Control,* FAO Water Reports No. 11, 7–23, 1997.

Czurda, K.A. and Haus, R. Reactive barriers with fly ash zeolites for in situ groundwater remediation. *Appl. Clay Sci.*, **21**:13–20, 2002.

Dagan, G. Models of groundwater flow in statistically homogeneous porous formation. *Water Resour. Res.*, **15**:47–63, 1979.

Dagan, G. Stochastic modeling of groundwater-flow by unconditional and conditional probabilities. 1. Conditional simulation and the direct-problem. *Water Resour. Res.*, **18**:813–833, 1982a.

Dagan, G. Stochastic modeling of groundwater-flow by unconditional and conditional probabilities. 2. the solute transport. *Water Resour. Res.*, **18**:835–848, 1982b.

Dagan, G. Solute transport in heterogeneous porous formations. *J. Fluid Mech.*, **145**:151–177, 1984.

Dagan, G. Stochastic modeling of groundwater-flow by conditional and unconditional probabilities—the inverse problem. *Water Resour. Res.*, **21**:65–72, 1985.

Dagan, G. Time-dependent macrodispersion for solute transport in anisotropic heterogeneous aquifers. *Water Resour. Res.*, **24**:1491–1500, 1988.

Dagan, G. *Flow and Transport in Porous Formations.* Springer-Verlag, New York, 1989.

Dagan, G. and Bear, J. Solving the problem of interface upconing in a coastal aquifer by the method of small perturbations. *J. Hydraul. Res.*, **6**:15–44, 1968.

Dagan, G. and Neuman, S.P. (Eds). *Subsurface Flow and Transport: A Stochastic Approach.* Cambridge Univ. Press, Cambridge UK, 241 p., 1997.

Dagan, G. and Zeitoun, D.G. Seawater-freshwater interface in a stratified aquifer of random permeability distribution. *J. Contam. Hydrol.*, **29**:185–203, 1988.

Dalton, M.G., Huntsman, B.E., and Bradbury, K. Acquisition and interpretation of water-level data. In: *Practical Handbook of Ground-Water Monitoring*, Lewis Publ., Chelsea, Michigan, 367–395, 1991.

Dane, J.H. and Topp, G. (Eds.). *Methods of Soil Analysis Pt. 4: Physical Methods*, Soil Sci. Soc. Am., Madison, Wisconsin, 1689 p., 2002.

Dantzig, G.B.*Linear Programming and Extensions.* Princeton Univ. Press, Princeton, N.J., 627 p., 1963.

Darcy, H. *Les Fontaines Publiques de la Ville de Dijon.* Dalmont, Paris, 647 p., 1856.

Das, B.M. *Advanced Soil Mechanics.* Hemisphere Publ. Corp. New York, 511 p., 1983.

Davies, B. and Martin, B. Numerical inversion of Laplace transform: A survey and comparison of methods. *J. Comp. Phys.*, **33**:1–32, 1979.

Davis, L. *Genetic Algorithms and Simulated Annealing.* Morgan Kaufmann Publ., San Francisco, 216 p., 1987.

Davis, S.N. and de Wiest, R.J.M. *Hydrogeology.* Wiley, New York, 463 p., 1966.

Day, S.R., O'Hannesin, S.F., and Marsden, L. Geotechnical techniques for the construction of reactive barriers. *J. Hazardous Materials*, **67**:285–297, 1999.

De Boer, R. *Theory of Porous Media—Highlights in the Historical Development and Current State.* Springer-Verlag, Berlin, 2000.

De Graan, J.G. Extensions to the multiple criteria analysis of T. L. Saaty. *Report National Institute of Water Supply*, the Netherlands, 1980

De Groot, S.R. and Mazur, P. *Non-equilibrium Thermodynamics.* North-Holland Pub. Co., Amsterdam, The Netherlands, 510 p., 1962.

De Josselin de Jong, G. Longitudinal and transverse diffusion in granular deposits. *Trans. Am. Geophys. Union*, **39**:67–74, 1958.

De Josselin de Jong, G. Consolidatie in drie dimensies (in Dutch). L.G.M.-Mededelingen, **7**:57–73, 1963.

De La Rosa-Perez, D.A., Teutli-Leon, M.M.M., and Ramirez-Islas, M.E. Polluted soils electroremediation, a technical review for field application. *Revista Internacional De Contaminacion Ambiental*, **23**:129–138, 2007.

Delay, S., Bobek, C., Bill, B., and Bair, R. Wellhead protection software. U.S. Environmental Protection Agency, 1994.

Delesse, A. Pour déterminer la composition des roches. *Annales des Mines Paris 4*, **13**:379–388, 1848.

De Marsily, G. *Quantitative Hydrogeology*. Academic Press, 440 p., 1986.

Demirdžić, I., Lilek, Ž., and Perić, M. Fluid flow and heat transfer test problems for non-orthogonal grids: bench-mark solutions. *Int. J. Numer. Methods Fluids*, **15**:329–354, 1992.

Demirdžić, I. and Muzaferija, S. Numerical method for coupled fluid flow, heat transfer and stress analysis using unstructured moving meshes with cells of arbitrary topology. *Comput. Meth. Appl. Mech. Eng.* **125**:235–255, 1995.

Demond, A.H., Desai, F.N., and Hayes, K.F. Effect of cationic surfactants on organic liquid water capillary-pressure saturation relationships. *Water Resour. Res.*, **30**:333–342, 1994.

Denbigh, K.G. *The Principles of Chemical Equilibrium*, 4th ed. Cambridge Univ. Press, 1981.

Detay, M. *Water Wells: Implementation, Maintenance and Restoration*. Wiley, 1997.

Detournay, C. and Strack, O.D.L. A new approximate technique for the hodograph method in groundwater-flow and its application to coastal aquifers. *Water Resour. Res.*, **24**:1471–1481, 1988.

Detournay, E. and Cheng, A.H.D. Fundamentals of poroelasticity. Chap. 5. In: *Comprehensive Rock Engineering: Principles, Practice and Projects, Vol. II, Analysis and Design Method*, (ed.) C. Fairhurst, Pergamon Press, 113–171, 1993.

Dettinger, M.D. and Wilson, J.L. First order analysis of uncertainty in numerical models of groundwater flow, I. Mathematical development. *Water Resour. Res.*, **17**:149–161, 1981.

Deutsch, C.V. and Journel, A.G. *GSLIB, Geostatistical Software Library and User's Guide*, 2nd ed. Oxford Univ. Press, 369 p., 1998.

Devlin, J.F. and Parker, B.L. Optimum hydraulic conductivity to limit contaminant flux through cutoff walls. *Ground Water*, **34**:719–726, 1996.

Diersch, H.J. Finite element modeling of recirculating density driven saltwater intrusion processes in groundwater. *Adv. Water Res.*, **11**:25–43, 1988.

Dillon, P.J. (Ed.) *Management of Aquifer Recharge for Sustainability: Proc. 4th Int. Symp. Artificial Recharge of Groundwater*, Adelaide, South Australia, Balkema, Amsterdam, 567 p., 2002.

Doherty, J. Ground water model calibration using pilot points and regularization. *Ground Water*, **41**:170–177, 2003.

Doherty, J. *PEST: Software for Model-Independent Parameter Estimation, 5th ed.* Watermark Numerical Computing, Australia, 2005.

Dojka, M.A., Hugenholtz, P., Haack, S.K., and Pace, N.R. Microbial diversity in a hydrocarbon- and chlorinated-solvent-contaminated aquifer undergoing intrinsic bioremediation. *Appl. Environ. Microb.*, **64**:3869–3877, 1998.

Domenico, P.A. An analytical model for multidimensional transport of a decaying contaminant species. *J. Hydrology*, **91**:49–58, 1987.

Domenico, P.A. and Robbins, G.A. A dispersion scale effect in model calibrations and field tracer experiments. *J. Hydrology*, **70**:123–132, 1984.

Dominguez, J. and Brebbia, C.A. *Boundary Elements: An Introductory Course.* McGraw-Hill, 1989.

Dougherty, D.E. and Marryott, R.A. Optimal groundwater management, 1, Simulated annealing. *Water Resour. Res.*, **27**:2493–2508, 1991.

Douglas, J. and Russell, T.F. Numerical-methods for convection-dominated diffusion-problems based on combining the method of characteristics with finite-element or finite-difference procedures. *SIAM J. Numer. Anal.*, **9**:871–885, 1982.

Driscoll, F.G. *Groundwater and Wells*, 2nd ed. Reynolds Guyar Designs, 1986.

Duarte, C.A. and Oden, J.T. Hp clouds—an hp meshless method. *Numer. Meth. Part. Diff. Eqs.*, **12**:673–705, 1996.

Du Commun, J. On the cause of freshwater springs, fountains, etc. *Am. J. Sci. Arts*, **14**:174–175, 1828.

Dullien, F.A.L. *Porous Media*, 2nd ed. Academic Press, San Diego, 574 p., 1992.

Dullien, F.A.L. and Dong, M. Experimental determination of the flow transport coefficients. *Transp. Porous Media*, **25**:97–120, 1996.

Dupont, R.R. Fundamentals of bioventing applied to fuel contaminated sites. *Environmental Progress*, **12**:45–53, 1993.

Dupuit, J. *Études Théoriques et Pratiques sur les Mouvementdes des Eaux dans les Cannaux Découverts et à Travers les Terrains Perméables* , 2nd ed., Dunod, Paris, 304 p., 1863.

Duque, C., Calvache, M.L., Pedrera, A., Martn-Rosales, W., and López-Chicano, M. Combined time domain electromagnetic soundings and gravimetry to determine marine intrusion in a detrital coastal aquifer (Southern Spain). *J. Hydrology*, **349**:536–547, 2008.

Edelfsen, N.E. and Anderson, A.B. Thermodynamics of soil moisture. *Hilgardia*, **15**:31–298, 1943.

Ek, M. and Mahrt, L. OSU 1-D PBL model user's guide. Dept. Atmos. Sci., Oregon State Univ., Corvallis, Oregon, 1991.

Ekwurzel, B., Schlosser, P., Smethie, W.M., Plummer, L.N., Busenberg, E., Michel, R. L., Weppernig, R., and Stute, M. Dating of shallow groundwater—Comparison of the transient tracers H-3/He-3 chlorofluorocarbons, and KR-85. *Water Resour. Res.*, **30**:1693–1708, 1994.

El Harrouni, K., Ouazar, D., Wrobel, L.C., and Cheng, A.H.-D. Uncertainty analysis of groundwater flow with DRBEM. *Eng. Anal. Bound. Elem.*. **19**:217–221, 1997.

Ene, H.I. and Poliševski, D. *Thermal Flow in Porous Media.* D. Reidel Publ., Dordrecht, 194 p., 1987.

Ene, H.I. and Sanchez-Palencia, E. Equations et phénomènes de surface pour l'écoulement dan un modèle de milieu poreux. *J. Méc.*, **14**:73–108, 1975.

Environmental and Water Resources Institute, ASCE *Standard Guidelines for Artificial Recharge of Ground Water.* ASCE, Reston, VA, 2001.

Espinoza, R.D. Infiltration. Chap. 6. In: *The Handbook of Groundwater Engineering*, 2nd ed., (ed.) J.W. Delleur, CRC Press, 2006.

Essaid, H.I. *The Computer Model SHARP, a Quasi-Three-Dimensional Finite-Difference Model to Simulate Freshwater and Saltwater Flow in Layered Coastal Aquifer Systems.* U.S. Geological Survey Water-Resources Investigations Report 90-4130, 181 p., 1990a.

Essaid, H.I. A multilayered sharp interface model of coupled freshwater and saltwater flow in coastal systems—model development and application. *Water Resour. Res.*, **26**:1431–1454, 1990b.

Essaid, H.I. USGS SHARP model. Chap. 8. In: *Seawater Intrusion in Coastal Aquifers—Concepts, Methods, and Practices* (eds.) J. Bear, A.H.-D. Cheng, S. Sorek, D. Ouazar & I. Herrera, Kluwer, 213–247, 1999.

Essaid, H.I. and Bekins, B.A. *BIOMOC, A multispecies solute-transport model with biodegradation.* U.S. Geological Survey Water-Resources Investigations Report 97-4022, 68 p., 1997.

Ewing, R.E., Russell, T.F., and Wheeler, M.F. Convergence analysis of an approximation of miscible displacement in porous-media by mixed finite-elements and a modified method of characteristics. *Comp. Meth. Appl. Mech. Eng.*, **47**:73–92, 1984.

Ewing, R.E. and Wang, H. A summary of numerical methods for time-dependent advection-dominated partial differential equations. *J. Comput. Appl. Math.*, **128**:423–445, 2001.

Ezzedine, S.M. Stochastic modeling of flow and transport in porous and fractured media. In: *Encyclopedia of Hydrological Sciences*, Vol. 4, (ed.) M.G. Anderson, Wiley, 2367–2400, 2005.

Fair, G.M. and Hatch, L.P. Fundamental factors governing the streamline flow of water through sand. *J. Am. Water Works Assoc.*, **25**:1551–1556, 1933.

Fairweather, G. and Karageorghis, A. The method of fundamental solutions for elliptic boundary value problems. *Adv. Comput. Math.*, **9**:69–95, 1998.

Fang, Y., Yabusake, S.B., and Yeh, G.T. A general simulator for reaction-based biogeochemical processes. *Computat. Geosci.*, **32**:64–72, 2006.

Fasshauer, G.E. Solving differential equations with radial basis functions: Multilevel methods and smoothing. *Adv. Comput. Math.*, **11**:139–159, 1999.

Feddes, R.A., de Rooij, G.H., and van Dam, J.C. (Eds.) *Unsaturated-Zone Modeling: Progress, Challenges and Applications.* Springer, 364 p., 2007.

Fel, L.G. and Bear, J. Dispersion and dispersivity tensors in saturated porous media with uniaxial symmetry. arXiv:0094.3447v1, 2009

Fetter, C.W. *Contaminant Hydrogeology,* 2nd ed. Prentice-Hall, 500 p., 1999.

Fick, A. On liquid diffusion. *Philos. Mag. J. Sci.*, **10**:31–39, 1855.

Fiorino, D.J. *The New Environmental Regulation.* MIT Press, 304 p., 2006.

Fitterman, D.V. and Deszcz-Pan, M. Helicopter EM mapping of saltwater intrusion in Everglades National Park, Florida. *Exploration Geophysics,* **29**:240–243, 1998.

Fitterman, D.V. and Stewart, M.T. Transient electromagnetic sounding for groundwater. *Geophysics,* **51**:995–1005, 1986.

Fleming, G. *Deterministic Simulation in Hydrology.* American Elsevier, New York, 1974.

Fletcher, R. A new approach to variable metric algorithms. *Computer J.*, **13**:317–322, 1970.

Forchheimer, P. Wasserbewegung durch boden. *Z. Ver. Deutsch. Ing.*, **45**:1782–1788, 1901.

Forsyth, P.A. A control volume finite-element approach to NAPL groundwater contamination. *SIAM J. Sci. Stat. Comp.*. **12**:1029–1057, 1991.

Forsyth, P.A., Wu, Y.S., and Pruess, K. Robust numerical-methods for saturated-unsaturated flow with dry initial conditions in heterogeneous media. *Adv. Water Res.*, **18**:25–38, 1995.

Fountain, J.C. The role of field trials in development and feasibility assessment of surfactant-enhanced aquifer remediation. *Water Environ. Res.*, **69**:188–195, 1997.

Fountain, J.C., Klimek, A., Beikirch, M.G., and Middleton, T.M. The use of surfactants for in situ extraction of organic pollutants from a contaminated aquifer. *J. Hazardous Materials,* **28**:295–311, 1991.

Fountain, J.C., Starr, R.C., Middleton, T., Beikirch, M., Taylor, C., and Hodge, D. A controlled field test of surfactant-enhanced aquifer remediation. *Ground Water,* **34**:910–916, 1996.

Fourier, J.B.J. *Théorie Analytique de la Chaleur.* F. Didot, Paris, 1822.

Franca, L.P. and Frey, S.L. Stabilized finite-element methods. 2. The incompressible Navier-Stokes equations. *Comput. Meth. Appl. Mech. Eng.*, **99**:209–233, 1992.

Franca, L.P., Frey, S.L., and Hughes, T.J.R. Stabilized finite-element methods. 1. Application to the advective-diffusive model. *Comput. Meth. Appl. Mech. Eng.*, **95**:253–276, 1992.

Franke, O.L. and McClymonds, N.E. *Summary of the Hydrologic Situation on Long Island, New York, as a Guide to Water-Management Alternatives.* U.S. Geological Survey Professional Paper 627-F, 59 p., 1972.

Franke R. Scattered data interpolation: tests of some methods. *Math. Comput.*, **38**:181–200, 1982.

Fraser, D.C. New multicoil aerial electromagnetic prospecting system. *Geophysics,* **37**:518–537, 1972.

Fraser, D.C. Resistivity mapping with an airborne multicoil electromagnetic system. *Geophysics,* **43**:144–172, 1978.

Fraser, D.C. Multicoil-II airborne electromagnetic system. *Geophysics*, **44**:1367–1394, 1979.

Freeze, R.A. A stochastic-conceptual analysis of one-dimensional groundwater flow in nonuniform homogeneous media. *Water Resour. Res.*, **11**:725–741, 1975.

Fretwell, J.D. and Stewart, M. Resistivity study of a coastal karst terrain. *Ground Water*, **19**:219–223, 1981.

Freundlich, H. Uber die Adsorption in Losungen. *Z. Phys. Chem.*, **57**:385, 1907.

Friedman, S.P. and Seaton, N.A. On the transport properties of anisotropic networks of capillaries. *Water Resour. Res.*, **32**:339–347, 1996.

Galeao, A.C., Almeida, R.C., Malta, S.M.C., and Loula, A.E. Finite element analysis of convection dominated reaction-diffusion problems. *Appl. Numer. Math.*, **48**:205–222, 2004.

Galeati, G., Gambolati, G. and Neuman, S.P. Coupled and partially coupled Eulerian-Lagrangian model of freshwater-saltwater mixing. *Water Resour. Res.*, **28**:149–165, 1992.

Galerkin, B.G. Series solution of some problems of elastic equilibrium of rods and plates (in Russian). *Vestn. Inzh. Tech.*, **19**:897–908, 1915.

Gambolati, G. and Freeze, R.A. Mathematical simulation of the subsidence of Venice, 1, Theory. *Water Resour. Res.*, **9**:721–733, 1973.

Gambolati, G., Gatto, P., and Freeze, R.A. Mathematical simulation of the subsidence of Venice, 2, Result. *Water Resour. Res.*, **10**:563–577, 1974.

Gambolati, G., Putti, M. and Paniconi, C. Three-dimensional model of coupled density-dependent flow and miscible salt transport. Chap. 10. In: *Seawater Intrusion in Coastal Aquifers—Concepts, Methods, and Practices*, (eds.) J. Bear, A.H.-D. Cheng, S. Sorek, D. Ouazar & I. Herrera, Kluwer, 315-362, 1999.

Garabedian, P.R. *Partial Differential Equations*, 2nd ed. AMS/Chelsea Publ., 672 p., 1998.

Garabedian, S.P., LeBlanc, D.R., Gelhar, L.W., and Celia, M.A. Large-scale natural gradient tracer test in sand and gravel, Cape Code, Massachusetts, 2. Analysis of spatial moments for a nonreactive tracer. *Water Resour. Res.*, **27**:911–924, 1991.

Gardner, W.R. Some steady state solutions of the unsaturated moisture flow equation, with application to evaporation from a water table. *Soil Sci.*, **85**:228–232, 1958.

Gavaskar, A.R., Gupta, N., Sass, B.M., Janosy, R.J. and O'Sullivan, D. *Permeable Barriers for Groundwater Remediation.* Battelle Press, Columbus, OH, 176 p., 1998.

Gelhar, L.W. Stochastic analysis of flow in aquifers. *AWRA Symp. on Advances in Groundwater Hydrology*, Chicago, Ill, 1976.

Gelhar, L.W. Stochastic subsurface hydrology from theory to applications. *Water Resour. Res.*, **22**:135S–145S, 1986.

Gelhar, L.W. *Stochastic Subsurface Hydrology.* Prentice-Hall, Englewood Cliffs, NJ, 1993.

Gelhar, L.W. and Axness, C.L. Three dimensional stochastic analysis of macrodispersion in aquifers. *Water Resour. Res.*, **19**:161–180, 1983.

Gelhar, L.W., Gutjhar, A.L., and Naff, R.L. Stochastic analysis of macrodispersion in a stratified aquifer. *Water Resour. Res.*, **15**:1387–1397, 1979.

Gelhar, L.W., Welty, C., and Rehfekdt, K.R. A critical-review of data on field-scale dispersion in aquifers. *Water Resour. Res.*. **28**:1955–1974, 1992.

Gelhar, L.W., Welty, C., and Rehfekdt, K.R. Reply. *Water Resour. Res.*. **29**:1867–1869, 1993.

Gershon, N.D. and Nir, A. Effects of boundary conditions of models on tracer distribution in flow through porous mediums. *Water Resour. Res.*, **54**:830–839, 1969.

Gibbs, J.W. *The Scientific Papers of J. Willard Gibbs*, Longmans, Green and Co., 1906.

Gibbs, J.W. *Elementary Principles in Statistical Mechanics.* Longmans and Green, New York, 1928.

Gieg, L.M., Kolhatkar, R.V., McInerney, M.J., Tanner, R.S., Harris, S.H., Sublette, K.L., and Suflita, J.M. Intrinsic bioremediation of petroleum hydrocarbons in a gas condensate-contaminate aquifer. *Environ. Sci. Technol.*, **33**:2550–2560, 1999.

Glover, F. and Kochenberger, G.A. *Handbook of Metaheuristics.* Springer, 2003.

Glover, R.E. The pattern of fresh-water flow in a coastal aquifer. *J. Geophys. Res.*, **64**:457–59, 1959.

Goldberg, D.E. *Genetic Algorithms in Search, Optimisation and Machine Learning.* Addison-Wesley, 1989.

Goldfarb, D. A family of variable-metric methods derived by variational means. *Math. Comput.*, **24**:23–26, 1970.

Golub, G.H. and van Loan, C.F. *Matrix Computations*, 3rd ed. Johns Hopkins Press, 728 p., 1996.

Goode, D.J. *Age, Double Porosity, and Simple Reaction Modifications for the MOC3D Ground-Water Transport Model.* U.S. Geological Survey Water-Resources Investigations Report 99-4041, 34 p., 1999.

Goode, P. and Ramakrishnan, T.S. Momentum-transfer across fluid-fluid interfaces in porous-media—A network model. *AIChE J.*, **39**:1124–1134, 1993.

Goovaerts, P. *Geostatistics for Natural Resources Evaluation.* Oxford Univ. Press, New York, 483 p., 1997.

Gordon, E., Shamir, U., and Bensabat, J. Optimal management of a regional aquifer under salinization conditions. *Water Resour. Res.*, **36**:3193–3203, 2000.

Gordon, E., Shamir, U., and Bensabat, J. Optimal extraction of water from regional aquifer under salinization. *J. Water Resour. Plng. Mgmt., ASCE*, **127**:71–77, 2001.

Gorelick, S.M., Freeze, R.A., and Donohue, D. *Groundwater Contamination Optimal Capture and Containment.* CRC Press, 416 p., 1993.

Goswami, R.R. and Clement, T.P. Laboratory-scale investigation of saltwater intrusion dynamics. *Water Resour. Res.*, **43**:W04418, 2007.

Graham, W. and McLaughlin, D. Stochastic-analysis of nonstationary subsurface solute transport. 1. Unconditional moments. *Water Resour. Res.*, **25**:215–232, 1989a.

Graham, W. and McLaughlin, D. Stochastic-analysis of nonstationary subsurface solute transport. 2. Conditional moments. *Water Resour. Res.*, **25**:2331–2355, 1989b.

Gray, W.G. and O'Neill, K. On the general equations for flow in porous media and their reduction to Darcy's law. *Water Resour. Res.*, **12**:148–154, 1976.

Green, W.H. and Ampt, C.A. Studies on Soil Physics. 1: Flow of air and water through soils. *J. Agr. Sci.*, **4**:1–24, 1911.

Greenberg, M.D. *Application of Green's Functions in Science and Engineering.* Prentice-Hall, 1971.

Greenberg, M. *Advanced Engineering Mathematics*, 2nd ed. Prentice-Hall, 1998.

Grim, R.E. *Clay Mineralogy*, 2nd ed. McGraw-Hill, New York, 596 p., 1968.

Guadagnini, A. and Neuman, S.P. Nonlocal and localized analyses of conditional mean steady state flow in bounded, randomly nonuniform domains 1. Theory and computational approach. *Water Resour. Res.*, **35**:2999–3018, 1999a.

Guadagnini, A. and Neuman, S.P. Nonlocal and localized analyses of conditional mean steady state flow in bounded, randomly nonuniform domains 2. Computational examples. *Water Resour. Res.*, **35**:3019–3039, 1999b.

Guarnaccia, J., Pinder, G.F., and Fishman, M. *NAPL: Simulator Documentation.* U.S. Environmental Protection Agency, EPA/600/SR-97/102, 1997.

Gutjahr, A.L., Gelhar, L.W., Bakr, A.A. and MacMillan, JR. Stochastic analysis of spatial variability in subsurface flows. 2. Evaluation and application. *Water Resour. Res.*, **14**:953–959, 1978.

Gvirtzman, H. and Gorelick, S.M. Dispersion and advection in unsaturated porous-media enhanced by anion exclusion. *Nature*, **352**:793-795, 1991.

Gvirtzman, H. and Magaritz, M. Water and anion transport in the unsaturated zone traced by environmental tritium. In: *Inorganic Contaminants in the Vadose Zone*, (eds.) B. Bar-Yosef, N.J. Barrow and J. Goldshmidt. *Ecological Studies*, **74**, Springer-Verlag, Berlin, 190–198, 1989.

Hackbusch, W. *Iterative Solution of Large Sparse Systems of Equations.* Springer, 460 p., 1993.

Hagemeyer, T. and Stewart, M.T. Resistivity investigations of salt-water intrusion near a major sea-level canal. In: *Geotechnical and Environmental Geophysics*, v. II, Soc. Explor. Geophysicists, Inv. in Geophysics, n. 5, (ed.) S. Ward, 67–78, 1990.

Haimes, Y.Y. *Hierarchial Analysis of Water Resources Systems: Modeling and Optimization of Large-Scale Systems.* McGraw-Hill, 512 p., 1977.

Haimes, Y.Y., Hall, W.A., and Freedman, H.T. *Multiobjective Optimization in Water Resources Systems.* Elsevier, 200 p., 1975.

Haines, W.B. The hysteresis effect in capillary properties and the modes of moisture distribution associated therewith. *J. Agric. Sci.*, **20**:96–105, 1930.

Haitjema, H.M. *Analytic Element Modeling of Groundwater Flow.* Academic Press, 394 p., 1995.

Hammond, G.E., Valocchi, A.J., and Lichtner, P.C. Application of Jacobian-free Newton-Krylov with physics-based preconditioning to biogeochemical transport. *Adv. Water Res.*, **28**:359–376, 2005.

Hansbo, P. and Szepessy, A. A velocity pressure streamline diffusion finite-element method for the incompressible Navier-Stokes equations. *Comp. Meth. Appl. Mech. Eng.*, **84**:175–192, 1990.

Hantush, M.S. Analysis of data from pumping tests in leaky aquifers. *Trans. Am. Geophys. Union*, **37**:702–714, 1956.

Hantush, M.S. Modification of the theory of leaky aquifers. *J. Geophys. Res.*, **65**:3713–3726, 1960.

Hantush, M.S. Hydraulics of wells. In: *Advances in Hydroscience*, Vol. 1, (ed.) V.T. Chow, Academic Press, 281–442, 1964.

Hantush, M.S. and Jacob, C.E. Non-steady radial flow in an infinite leaky aquifer. *Trans. Am. Geophys. Union*, **36**:95–100, 1955.

Harbaugh, A.W. *MODFLOW-2005, the U.S. Geological Survey Modular Ground-Water model—the Ground-Water Flow Process.* U.S. Geological Survey Techniques and Methods 6-A16, 2005.

Harbaugh, A.W., Banta, E.R., Hill, M.C., and McDonald, M.G. *MODFLOW-2000, the U.S. Geological Survey Modular Ground-Water Model—User Guide to Modularization Concepts and the Ground-Water Flow Process*, U.S. Geological Survey Open-File Report 00-92, 121 p., 2000.

Hardy, R.L. Multiquadric equations of topography and other irregular surfaces. *J. Geophys. Res.*, **76**:1905–1915, 1971.

Hargreaves, G.H. Moisture availability and crop production. *Trans. Am. Soc. Agric. Eng.*, **18**:980–984, 1975.

Harlan, R.L., Kolm, K.E., Gutentag, E.D. *Water-Well Design and Construction.* Elsevier, 1989.

Harpez, Y. Artificial ground water recharge by means of wells in Israel. *J. Hyd. Div., ASCE*, **97**:1947–1964, 1971.

Harr, M.E. *Foundations of Theoretical Soil Mechanics.* McGraw-Hill, New York, 381 p., 1966.

Hassanizadeh, S.M. Derivation of basic equations of mass transport in porous media: Generalized Darcy's and Fick's laws. *Adv. Water Res.*, **9**:207–233, 1986.

Haverkamp, R., Parlange, J.Y., Starr, J.L., Schmitz, G., and Fuentes, C. Infiltration under ponded conditions. 3. A predictive equation based on physical parameters. *Soil Sci.*, **149**:292–300, 1990.

Healy, R.W. and Cook, P.G. Using groundwater levels to estimate recharge. *Hydrogeology J.*, **10**:91–109, 2002.

Heinrich, J.C., Huyakorn, P.S., Zienkiewicz, O.C., and Mitchell, A.R. An 'upwind' finite-element scheme for two-dimensional convective transport-equation. *Int. J. Numer. Methods Eng.*, **11**:131–143, 1977.

Helgeson, H.C. Thermodynamics of hydrothermal systems at elevated temperatures and pressures. *Am. J. Sci.*, **267**:729–804, 1969.

Herrera, I. Theory of multiple leaky aquifers. *Water Resour. Res.*, **6**:185–193, 1970.

Herrera, I. Trefftz method. In: *Topics in Boundary Element Research, 1, Basic Principles and Applications*, (ed.) C.A. Brebbia, Springer, Berlin, 225–253, 1984.

Herrera, I., Ewing, R.E., Celia, M.A., Russell, T.F. Eulerian-Lagrangian localized adjoint method: The theoretical framework. *Numer. Meth. Part. Diff. Eqs.*, **9**:431–457, 1993.

Herrera, I. and Figueroa, G.E. A correspondence principle for the theory of leaky aquifers. *Water Resour. Res.*, **5**:900–904, 1969.

Herzberg, A. Die Wasserversorgung einiger Nordseebder (The water supply of parts of the North Sea coast in Germany). *Z. Gasbeleucht. Wasserversorg.*, **44**:815–819, and **45**:842-844, 1901.

Hess, K.M., Wolf, S.H. and Celia, M.A. Large-scale natural gradient tracer test in sand and gravel, Cape Cod, Massachusetts. 3. Hydraulic conductivity variability and calculated macrodispersivities. *Water Resour. Res.*, **28**:2011–2027, 1992.

Hiemenz, P.C. and Rajagopalan, R. *Principles of Colloid and Surface Chemistry*, 3rd ed. Marcel Dekker, 650 p., 1997.

Hill, M.C. *A Computer Program (MODFLOWP) for Estimating Parameters of a Transient, Three-Dimensional, Ground-Water Flow Model Using Nonlinear Regression.* U.S. Geological Survey Open-File Report 91-484, 358 p., 1992.

Hill, M.C., Banta, E.R., Harbaugh, A.W., and Anderman, E.R. *MODFLOW-2000, the U.S. Geological Survey Modular Ground-Water Model—User Guide to the Observation, Sensitivity, and Parameter-Estimation Processes and Three Post-Processing Programs*, U.S. Geological Survey Open-File Report 00-184, 210 p., 2000.

Hill, M.C. and Tiedeman, C.R. *Effective Groundwater Model Calibration: With Analysis of Data, Sensitivities, Predictions, and Uncertainty.* Wiley-Interscience, 480 p., 2007.

Hillel, D. *Fundamentals of Soil Physics.* Academic Press, 413 p., 1980.

Hinchee, R.E., Downey, D.C., Dupont, R.R., Aggarwal, P.K., and Miller, R.N. Enhancing biodegradation of petroleum-hydrocarbons through soil-venting. *J. Hazardous Materials*, **27**:315–325, 1991.

Hinchee, R.E. and Ong, S.K. A rapid in situ respiration test for measuring aerobic biodegradation rate of hydrocarbons in soil. *J. Air & Waste Management Assoc.*, **42**:1305–1312, 1992.

Hoeksema, R.J. and Kitanidis, P.K. Analysis of spatial variability of properties of selected aquifers. *Water Resour. Res.*, **21**:563–572, 1985a.

Hoeksema, R.J. and Kitanidis, P.K. Comparison of Gaussian conditional mean and kriging estimation in the geostatistical approach to the inverse problem. *Water Resour. Res.*, **21**:825–836, 1985b.

Hoekstra, P. and Blohm, M.W. Case histories of time-domain electromagnetic soundings in environmental geophysics. In: *Geotechnical and Environmental Geophysics*, v. II, Soc. Explor. Geophysicists, Inv. in Geophysics, n. 5, (ed.) S. Ward, 1–16, 1990.

Hoeppel, R.E., Hinchee, R.E., and Arthur, M.F. Bioventing soils contaminated with petroleum-hydrocarbons. *J. Industrial Microbiology*, **8**:141–146, 1991.

Hoffman, F. Groundwater remediation using smart pump and treat. *Ground Water*, **31**:98–106, 1993.

Holland, J.H. *Adaptation in Natural and Artificial Systems.* Ann Arbor Science Press, Ann Arbor, Michigan, 1975.

Holzbecher, E. and Sorek, S. Numerical models of groundwater flow and transport. In: *Encyclopedia of Hydrological Sciences*, Vol. 4, Art. 155, (eds.) J.J. McDonnell and M.G. Anderson, 2401–2414, 2005.

Hornung, U. (Ed.) *Homogenization and Porous Media.* Springer, 1997.

Horton, R.E. An approach towards a physical interpretation of infiltration capacity. *Proc. Soil Sci. Soc. Am.*, **5**:399–417, 1940.

Houlihan, M.F. and Berman, M.H. Remediation of contaminated groundwater. Chap. 36. In: *The Handbook of Groundwater Engineering*, 2nd ed., (ed.) J.W. Delleur, CRC Press, 2007.

Huang, C.L. and Mayer, A.S. Pump-and-treat optimization using well locations and pumping rates as decision variables. *Water Resour. Res.*, **33**:1001–1012, 1997.

Huang, C.S., Lee, C.-F., and Cheng, A.H.-D. Error estimate, optimal shape factor, and high precision computation of multiquadric collocation method. *Eng. Anal. Bound. Elem.*, **31**:614–623, 2007.

Huang, H.P. and Fraser, D.C. The differential parameter method for multifrequency airborne resistivity mapping. *Geophysics*, **61**:100–109, 1996.

Hubbert, M.K. The theory of ground water motion. *J. Geol.*, **48**:785–944, 1940.

Hughes, T.J.R. Multiscale phenomena: Green's functions, the Dirichlet-to-Neumann formulation, subgrid scale models, bubbles and the origins of stabilized methods. *Comput. Meth. Appl. Mech. Eng.*, **127**:387–401, 1995.

Hughes, T.J.R. and Brooks, A.N. A multidimensional upwind scheme with no crosswind diffusion. In: *Finite Element Methods for Convection Dominated Flows*, AMD Vol. 34, (ed.) T.J.R. Hughes, ASME, 19–35, 1979.

Hughes, T.J.R., Franca, L.P., and Hulbert, G.M. A new finite-element formulation for computational fluid-dynamics. VIII. The Galerkin least-squares method for advective-diffusive equations. *Comput. Meth. Appl. Mech. Engng.*, **73**:173–189, 1989.

Hughes, T.J.R. and Mallet, M. A new finite-element formulation for computational fluid-dynamics. III. The generalized streamline operator for multidimensional advective-diffusive systems. *Comp. Meth. Appl. Mech. Eng.*, **58**:305–328, 1986.

Hughes, T.J.R., Masud, A., and Wan, J. A stabilized mixed discontinuous Galerkin method for Darcy flow. *Comput. Meth. Appl. Mech. Eng.*, **195**:3347–3381, 2006.

Hult, M.F. Ground-water contamination by crude oil at the Bemidji, Minnesota, research site; US Geological Survey Toxic Waste—ground-water contamination study. *USGS Water-Resources Investigations Report*, No. 84-4188, 1984.

Hunt, J.R., Sitar, N., and Udell, K.S. Non-aqueous phase liquid transport and cleanup, 1. Analysis of mechanisms. *Water Resour. Res.*, **24**:1247–1258, 1988.

Huyakorn, P.S., Andersen, P.F., Mercer, J.W. and White, H.O. Salt intrusion in aquifers: development and testing of a three dimensional finite element model. *Water Resour. Res.*, **23**:293–319, 1987.

Huyakorn, P.S. and Nilkuha, K. Solution of transient transport-equation using an upstream finite-element scheme. *Appl. Math. Modelling*, **3**:7–17, 1979.

Huyakorn, P.S. and Pinder, G.F. *Computational Methods in Subsurface Flow.* Academic Press, 473 p., 1983.

Hydrocomp International Inc. *Hydrocomp Simulation Programming Operations Manual.* Palo Alto, California, 1968.

Iben, I.E.T., Edelstein, W.A., Sheldon, R.B., Shapiro, A.P., Uzgiris, E.E., Scatena, C.R., Blaha, S.R., Silverstein, W.B., Brown, G.R., Stegemeier, G.L., and Vinegar, H.J. Thermal blanket for in-situ remediation of surficial contamination: A pilot test. *Environ. Sci. Technol.*, **30**:3144–3154, 1996.

Idelsohn, S.R. and Onate, E. To mesh or not to mesh. That is the question. *Comput. Methods Appl. Mech. Eng.*, **195**:4681–4696, 2006.

IMSL *IMSL Fortran Subroutines for Mathematical Applications, Math/Library Vol. 1 & 2*, 1997.

Irmay, S. On the hydraulic conductivity of unsaturated soil. *Trans. Am. Geophys. Union*, **35**:463–468, 1954.

Irmay, S. Solutions of the non-linear diffusion equation with a gravity term in hydrology. I.A.S.H. Symp. *Water in the Unsaturated Zone*, Wageningen, 1966.

Jacob, C.E. The flow of water in an elastic artesian aquifer. *Trans. Am. Geophys. Union*, **21**:574–586, 1940.

Jacob, C.E. Flow of groundwater. Chap. 5. In: *Engineering Hydraulics*, (ed.) H. Rouse, Wiley, 321–386, 1950.

Jager, W. and Mikelic, A. On the interface boundary condition of Beavers, Joseph, and Saffman. *SIAM J. Appl. Math.*, **60**:1111–1127, 2000.

Jahnke, F.M. and Radke, C.J. Electrolyte diffusion in compacted montmorillonite engineered barriers. Chap. 22. In: *Coupled Processes Associated with Nuclear Waste Repositories*, (ed.) C.F. Tsang, Academic Press, 287–297, 1987.

Jang, M. and Choe, J. Stochastic optimization for global minimization and geostatistical calibration. *J. Hydrology*, **266**:40–52, 2002.

Jennings, A. and McKeown, J.J. *Matrix Computation*, 2nd ed. Wiley, 1992.

Jensen, K.H., Bitsch, K., and Bjerg, P.L. Large-scale dispersion experiments in a sandy aquifer in Denmark—Observed tracer movements and numerical-analyses. *Water Resour. Res.*, **29**:673–696, 1993.

Ji, S.H., Park, Y.J., Sudicky, E.A., and Sykes, J.F. A generalized transformation approach for simulating steady-state variably-saturated subsurface flow. *Adv. Water Res.*, **31**:313–323, 2008.

Jikov, V.V., Kozlov, S.M., and Oleinik, O.A. *Homogenization of differential operators and integral functionals.* Springer-Verlag, Berlin, New York, 1994.

Joe, B. Delaunay triangular meshes in convex polygons. *SIAM J. Sci. Stat. Comp.*, **7**:514–539, 1986.

Johnson, A.I. (Ed.) *Land Subsidence: Proc. 4rd Int. Symp. on Land Subsidence*, UNESCO/IAHS, Houston, Texas, USA, IAHS, **200**, 1991

Johnson, A.I., Carbognin, L. and Ubertini, L. (Eds.) *Land Subsidence: Proc. 3rd Int. Symp. on Land Subsidence*, UNESCO/IAHS, Venice, IAHS, **151**, 1984.

Johnson, A.I. and Finlayson, D.J. (Eds.) *Artificial Recharge of Ground Water, Proc. 1st Int. Symp. Artificial Recharge of Ground Water*, Anaheim, California, ASCE Publications, New York, 644 p., 1988.

Johnson, A.I. and Pyne, R.D.G. (Eds.) *Artificial Recharge of Ground Water, II, Proc. 2nd Int. Symp. Artificial Recharge of Ground Water*, Orlando, Florida, ASCE Publications, Reston, VA, 938 p., 1994.

Johnson, C., Schatz, A.H., and Wahlbin, L.B. Crosswind smear and pointwise errors in streamline diffusion finite-element methods. *Math. Comput.*, **49**:25–38, 1987.

Johnson, C., Szepessy, A., and Hansbo, P. On the convergence of shock capturing streamline diffusion finite-element methods for hyperbolic conservation-laws. *Math. Comput.*, **54**:107–129, 1990.

Johnson, R.L., Johnson, P.C., McWhorter, D.B., Hinchee, R.E., and Goodman, I. An overview of in-situ air sparging. *Ground Water Monit. Rem.*, **13**:127–135, 1993.

Johnson, T.A. and Whitaker, R. Saltwater intrusion in the coastal aquifers of Los Angeles County, California. Chap. 2. In: *Coastal Aquifer Management-Monitoring, Modeling, and Case Studies*, (eds.) A.H.-D. Cheng and D. Ouazar, Lewis Publ., 29–48, 2004.

Jones, B.F., Vengosh, A., Rosenthal, E., and Yechieli, Y. Geochemical investigations. Chap. 3. In: *Seawater Intrusion into Coastal Aquifers—Concepts, Methods and Practices*, (eds.) J. Bear, A.H.-D. Cheng, S. Sorek, D. Ouazar & I. Herrera, Kluwer, 51–71, 1999.

Journel, A.G. Geostatistics for conditional simulation of ore bodies. *Economic Geology*, **69**:673–687, 1974.

Journel, A.G. and Huijbregts, C.J. *Mining Geostatistics.* Academic Press, New York, 600 p., 1978.

Jury, W.A. Chemical transport modeling: Current approaches and unresolved problems. In: *SSSA Special Publication*, **11**, SSSA and ASA, Madison Wisconsin, 49–64, 1983.

Kalaydjian, F. A macroscopic description of multiphase flow in porous media involving space-time evolution of fluid-fluid interface. *Transp. Porous Media*, **2**:537–552, 1987.

Kalaydjian, F. Origin and quantification of coupling between relative permeabilities for two-phase flows in porous media. *Transp. Porous Media*, **5**:215–229, 1990.

Kansa, E.J. Multiquadrics—a scattered data approximation scheme with applications to computational fluid-dynamics. 1. Surface approximations and partial derivative estimates. *Comput. Math. Applic.*, **19**:127–145, 1990a.

Kansa, E.J. Multiquadrics—a scattered data approximation scheme with applications to computational fluid-dynamics. 2. Solutions to parabolic, hyperbolic and elliptic partial-differential equations. *Comput. Math. Applic.*, **19**:147–161, 1990b.

Kansa, E.J. and Hon, Y.C. Circumventing the ill-conditioning problem with multiquadric radial basis functions: Applications to elliptic partial differential equations. *Comput. Math. Applic.*, **39**:123–137, 2000.

Karimian, S.A.M. and Straatman, A.G. Discretization and parallel performance of an unstructured finite volume Navier-Stokes equation solver. *Int. J. Numer. Methods Fluids*, **52**:591–615, 2006.

Kauahikaua, J. *Description of a Fresh-Water Lens at Laura Atoll, Majuro Atoll, Republic of the Marshall Islands, Using Electromagnetic Profiling*, U.S. Geol. Survey Open File Report 87-582, 32 p., 1987.

Keely, J.F. Performance evaluations of pump-and-treat remediations. *EPA Ground Water Issue*, EPA/540/4-89/005, Ada, OK, 1989.

Kees, C.E., Farthing, M.W., and Dawson, C.N. Locally conservative, stabilized finite element methods for variably saturated flow. *Comput. Meth. Appl. Mech. Eng.*, **197**:4610–4625, 2008.

Kershaw, D.S. Incomplete Cholesky-conjugate gradient method for iterative solution of systems of linear equations. *J. Computat. Phys.*, **26**:43–65, 1978.

Khinchin, A. *Mathematical Foundations of Statistical Mechanics.* Dover, 1949.

Kim, S. and Russel, W.B. Modeling of porous-media by renormalization of the Stokes equation. *J. Fluid Mech.*, **154**:269–286, 1985.

Kinzelbach, W. *Numerische Methoden zur Modellierung des Transports von Schadstoffen im Grundwasser*, 2nd ed. Oldenbourg Verlag, Mnchen, 313 p., 1992.

Kipp, K.L., Jr. *HST3D: A Computer Code for Simulation of Heat and Solute Transport in Three-Dimensional Ground-Water Flow Systems.* U.S. Geological Survey Water-Resource Investigation Report 86-4095, 1987.

Kipp, K.L., Jr. *Guide to the Revised Heat and Solute Transport Simulator: HST3D.* U.S. Geological Survey Water-Resource Investigation Report 97-4157, 149p., 1997.

Kirchhoff, G. *Vorlesungen über die Theorie der Wärme*, Barth, Leipzig, 1894.

Kirkham, D. and Powers, W.L. *Advanced Soil Physics.* Wiley-Interscience, New York, 534 p., 1972.

Kirkpatrick, S. Optimization by simulated annealing: Quantitative studies. *J. Stat. Phys.*, **34**:975–986, 1984.

Kirkpatrick, S., Gelatt, C.D., and Vecchi, M.P. Optimization by simulated annealing. *Science*, **220**(4598):671–680, 1983.

Kitanidis, P.K. *Introduction to Geostatistics: Applications in Hydrogeology.* Cambridge Univ. Press, 249 p., 1997.

Kleineidam, S., Rügner, H., Ligouis, B. and Grathwohl, P. Organic matter facies and equilibrium sorption of phenanthrene. *Environ. Sci. Technol.*, **33**:1637–1644, 1999.

Klinchuch, L.A., Goulding, N., James, S.R., and Gies, J. Deep air sparging-15 to 46 m beneath the water table. *Ground Water Monit. Rem.*, **27**:118–126, 2007.

Klinkenberg, L.J. The permeability of porous media to liquids and gases. *American Petroleum Institute, Drilling and Productions Practices*, 200–213, 1941.

Knight, J. and Philip, J.R. Exact solutions in nonlinear diffusion. *J. Eng. Math.*, **8**:219–227, 1974.

Knudsen, M.H.C. *The Kinetic Theory of gases.* Methuen, London, 1934; (3rd ed.), 64 p., 1950.

Kock, D.L. and Prickett, T.A. *User Manual for RD3D, a Three-Dimensional Mass Transport Random Walk Model Attachment of the USGS MODFLOW Three-Dimensional Flow Model.* Joint Eng. Tech. Assoc. & T.A. Prickett and Assoc., Sci. Publ. 3, Ellicott City, MD, 1989.

Koczkodaj, W.W. and Orlowski, M. Computing a consistent approximation to a generalized pairwise comparisons matrix. *Comput. Math. Applic.*, **37**:79–85, 1999.

Kohr, M. and Sekhar, G.P.R. Existence and uniqueness result for the problem of viscous flow in a granular material with a void. *Quart. Appl. Math.*, **65**:683–704, 2007.

Kollet, S.J. and Maxwell, R.M. Integrated surface-groundwater flow modeling: A free-surface overland flow boundary condition in a parallel groundwater flow model. *Adv. Water Res.*, **29**:945–958, 2006.

Konikow, L.F. and Bredehoeft, J.D. *Computer Model of Two-Dimensional Solute Transport and Dispersion in Ground Water*, Tech. Water-Resources Investigations, U.S. Geological Survey, Book 7, 1978.

Konikow, L.F., Goode, D.J., and Hornberger, G.Z. *A Three-Dimensional Method-of-Characteristics Solute-Transport Model (MOC3D).* U.S. Geological Survey Water-Resources Investigations Report 96-4267, 87 p., 1996.

Konikow, L.F. and Reilly, T.E. Seawater Intrusion in the United States. Chap. 13. In: *Seawater Intrusion into Coastal Aquifers—Concepts, Methods and Practices*, (eds.) J. Bear, A.H.-D. Cheng, S. Sorek, D. Ouazar & I. Herrera, Kluwer, 463–506, 1999.

Kontar, E.A. and Ozorovich, Y.R. Geo-electromagnetic survey of the fresh/salt water interface in the coastal southeastern Sicily. *Continental Shelf Research*, **26**:843–851, 2006.

Kool, J.B. and Parker, J.C.P. Development and evaluation of closed-form expressions for hysteretic soil hydraulic properties. *Water Resour. Res.*, **23**:105–114, 1987.

Kool, J.B., Parker, J.C.P., and Van Genuchten, M.Th. Parameter estimation for unsaturated flow and transport models–A review. *J. Hydrology*, **91**:255–293, 1987.

Koplik, J., Levine, H., and Zee, A. Viscosity renormalization in the Brinkman equation. *Phys. Fluids*, **26**:2864–2870, 1983.

Korolev, V.A., Romanyukha, O.V., and Abyzova, A.M. Electrokinetic remediation of oil-contaminated soils. *J. Environ. Sci. Health A—Toxic/Hazardous Substances & Environ. Eng.*, **43**:876–880, 2008.

Kräutle, S. and Knabner, P. A new numerical reduction scheme for fully coupled multicomponent transport-reaction problems in porous media. *Water Resour. Res.*, **41**:W09414, 2005.

Kräutle, S. and Knabner, P. A reduction scheme for coupled multicomponent transport-reaction problems in porous media: Generalization to problems with heterogeneous equilibrium reactions. *Water Resour. Res.*, **43**:W03429, 2007.

Krige, D.G. A statistical approach to some mine valuations and allied problems at the Witwatersrand. Master's thesis, University of Witwatersrand, 1951.

Krumbein, W.C. and Monk, G.D. Permeability as a function of the size parameters of unconsolidated sands. *Trans. Inst. Min. Met. Engrs.*, **15**:153–163, 1943.

Kubik, J. Elements of constitutive modelling of saturated porous materials. In: *Modelling Coupled Phenomena in Saturated Porous Materials*, (eds.) J. Kubik, M. Kaczmarek, and I. Murdoch, Institute of Fundamental Technological Research, Polish Academy of Sciences, Warsaw, Poland, 279–347, 2004.

Kuiper, L.K. A comparison of the incomplete Cholesky-conjugate gradient-method with the strongly implicit method as applied to the solution of two-dimensional groundwater-flow equations. *Water Resour. Res.*, **17**:1082–1086, 1981.

Ladyzhenskaya, O.A. *The mathematical theory of viscous incompressible flow.* Gordon and Breach, 1963.

Lafe, O. and Cheng, A.H.-D. Stochastic indirect boundary element method. Chap. 14. In: *Computational Stochastic Mechanics: Theory, Computational Methodology and Engineering Application*, (eds.) A.H.-D. Cheng and C.Y. Yang, CMP/Elsevier, 301–322, 1993.

Lall, U. and Sharma, A. A nearest neighbor bootstrap for resampling hydrologic time series. *Water Resour. Res.*, **32**:679–693, 1996.

Lallemand-Barrés, A. and Peaudecerf, P. Recherche de relationsentre les valeurs measurées de la dispersivité macroscopique d'un milieu aqiufère, ses autres caractéritsique et les conditions de mesure. *Bull. Bur. Rech. Géol. Min. Sêr*, **2**(Sec. III):277–284, 1978.

Land, C.S. Calculation of imbibition relative permeability for two- and three-phase flow from rock properties. *Trans. Amer. Instit. Mining Metal. and Petrol. Engineering*, **243**:149–156, 1968.

Langevin, C.D., Oude Essink, G.H.P., Panday, S., Bakker, M., Prommer, H., Swain, E.D., Jones, W., Beach, M. and Barcelo, M. MODFLOW-based tools for simulation of variable-density groundwater flow. Chap. 2. In: *Coastal Aquifer Management—Monitoring, Modeling, and Case Studies*, (eds.) A.H.-D. Cheng and D. Ouazar, Lewis Publ., 49–76, 2004.

Langevin, C.D., Shoemaker, W.B., and Guo, W. *MODFLOW-2000, the U.S. Geological Survey Modular Ground-Water Model-Documentation of the SEAWAT-2000 Version with the Variable-Density Flow Process (VDF) and the Integrated MT3DMS Transport Process (IMT)*. U.S. Geological Survey Open-File Report 03-426, 43 p., 2003.

Langmuir, I. Chemical reactions at low temperatures. *J. Amer. Chem. Soc.*, **37**:1139, 1915.

Langmuir, I. The adsorption of gases on plane surfaces of glass, mica and platinum. *J. Amer. Chem. Soc.*, **40**:1361–1403, 1918.

La Rocca, A. and Power, H. A double boundary collocation Hermitian approach for the solution of steady state convection-diffusion problems. *Comput. Math. Applic.*, **55**:1950–1960, 2008.

La Rocca, A., Rosales, A.H., and Power, H. Radial basis function Hermite collocation approach for the solution of time dependent convection-diffusion problems. *Eng. Anal. Bound. Elem.*, **29**:359–370, 2005.

Lasaga, A.C. and Kirpatrick, R.J. (Eds.) *Kinetics of Geochemical Processes*. Rev. in Mineralogy, Mineralogical Soc. of America, Book Crafters Inc., Chelsea, Michigan, 1981.

Lasseux, D., Quintard, M., and Whitaker, S. Determination of the permeability tensors for two-phase flow. *Transp. Porous Media*, **24**:107–137, 1996.

Lawler, G.F. *Introduction to Stochastic Processes*, 2nd ed. Chapman & Hall, 248 p., 2006.

Lazarov, R.D., Mishev, I.D., and Vassilevski, P.S. Finite volume methods for convection-diffusion problems. *SIAM J. Numer. Anal.*, **33**:31–55, 1996.

LeBlanc, D.R., Garabedian, S.P., Hess, K.M., Gelhar, L.W., Quadri, R.D., Stollenwerk, K.G., and Wood, W.W. Large-scale natural gradient tracer test in sand and gravel, Cape Cod, Massachusetts. 1. Experimental design and observed tracer movement. *Water Resour. Res.*, **27**:895–910, 1991.

Lee, T.H., Byun, I.G., Kim, Y.O., Hwang, I.S., and Park, T.J. Monitoring biodegradation of diesel fuel in bioventing processes using in situ respiration rate. *Water Sci. Technol.*, **53**:263–272, 2006.

Lehr, J., Hurlburt, S., Gallagher, B., and Voytek, J. *Design and Construction of Water Wells: A Guide for Engineers*. Van Nostrand Reinhold, 1988.

Lenhard, R.J., Parker, J.C., and Kaluarachchi, J.J. A model for hysteretic constitutive relations governing multiphase flow, 3. Refinement and numerical simulations. *Water Resour. Res.*, **25**:1727–1736, 1989.

Leroux, V. and Dahlin, T. Time-lapse resistivity investigations for imaging saltwater transport in glaciofluvial deposits. *Environmental Geology*, **49**:347–358, 2006.

Letniowski, F.W. and Forsyth, P.A. A control volume finite-element method for 3-dimensional NAPL groundwater contamination. *Int. J. Numer. Methods Fluids*, **13**:955–970, 1991.

Leveque, R.J. *Finite Volume Methods for Hyperbolic Problems*. Cambridge Univ. Press, 558 p., 2002.

Leveque, R.J. *Finite Difference Methods for Ordinary and Partial Differential Equations: Steady-State and Time-Dependent Problems*, SIAM, 350 p., 2008.

Lever, D.A. and Jackson, C.P. On the equations for the flow of a concentrated salt solution through a porous medium, Harwell Rep. AERE-R. 11765, HMSO, London, 1985.

Leverett, M.C. Capillary behaviour in porous media. *Trans. AIME*, **142**:341–358, 1941.

Lewis, R.W., Masters, I., and Rees, I. Coupled and uncoupled contaminant transport using advanced finite volume methods. *Comput. Mech.*, **37**:292–310, 2006.

Li, S. and Liu, W.K. *Meshfree Particle Methods*. Springer, 502 p., 2007

Li, Z.C., Lu, T.T., Hu, H.Y., and Cheng, A.H.-D. *Trefftz and Collocation Methods*. WIT Press, 2008.

Li, Z.C., Lu, T.T., Huang, H.T., and Cheng, A.H.-D. Trefftz, collocation, and other boundary methods—A comparison. *Numer. Meth. Part. Diff. Eqs.*, **23**:93–144, 2008.

Liang, Q. and Lohrenz, J. Dynamic method of measuring coupling coefficients of transport equations of two-phase flow in media flow. *Transp. Porous Media*, **15**:771–779, 1994.

Lichtner, P.C. Continuum model for simultaneous chemical reactions and mass transport in hydrothermal systems. *Gmchimica n Cosmochimica Acta*, **49**:779–800, 1985.

Lichtner, P.C. Scaling properties of kinetic mass transport equations. *Am. J. Sci.*, **293**, 257–296, 1993.

Lichtner, P.C. Principles and practice of reactive transport modelling. *Mat. Res. Soc. Symp. Proc.*, Kyoto, Japan, **353**:117-130, 1995.

Lichtner, P.C. Continuum formulation of multicomponent-multiphase reactive transport. In: *Reactive Transport in Porous Media*, (eds.) P.C. Lichtner, C.I. Steefel, and E.H. Oelkers, 1–81, 1996.

Lichtner, P.C. and Carey, J.W. Incorporating solid solutions in reactive transport equations using a kinetic discrete-composition approach. *Geochimica Et Cosmochimica Acta*, **70**:1356–1378, 2006.

Lichtner, P.C., Kelkar, S., and Robinson, B.A. New form of dispersion tensor for axisymmetric porous media with implementation in particle tracking. *Water Resour. Res.*, **38**:DOI 10.1029/200wr000100, 2002.

Lichtner, P.C., Kelkar, S., and Robinson, B.A. Critique of Burnett-Frind dispersion tensor for axisymmetric porous media. Los Alomos National Laboratory Report, LA-UR-08-04495, 2008.

Liggett, J.A. Location of free surface in porous media. *J. Hyd. Div., ASCE*, **103**:353–365, 1977.

Liggett, J.A. and Liu, P.L.-F. Unsteady flow in confined aquifers: A comparison of two boundary integral methods. *Water Resour. Res.*, **15**:861–866, 1979.

Liggett, J.A. and Liu, P.L.-F. *The Boundary Integral Equation Method for Porous Media Flow.* George Allen and Unwin, 1983.

Limayem, F. and Yannou, B. Generalization of the RCGM and LSLR pairwise comparison methods. *Comput. Math. Applic.*, **48**:539–548, 2004.

Lin, H.C., Richards, D.R., Yeh, G.T., Cheng, J.R., Chang, H.P., and Jones, N.L. *FEMWATER: A Three-Dimensional Finite Element Computer Model for Simulating Density Dependent Flow and Transport*, U.S. Army Engineer Waterways Experiment Station Technical Report, 129 p., 1996.

Lindstrom, F.T., Boersma, L., and Stockard, D. A theory on the mass transport of previously distributed chemicals in a water saturated sorbing porous medium. *Soil Sci.*, **112**:291–300, 1971.

Ling, L. Multivariate quasi-interpolation schemes for dimension-splitting multiquadric. *Appl. Math. Computat.*, **161**:195–209, 2005.

Lions, J.-L. *Some methods in the mathematical analysis of systems and their control*, Kexue Chubanshe Science Press, Beijing, 1981.

Liu, H. and Cheng, A.H.-D. Modified Fickian model for predicting dispersion. *J. Hyd. Div., ASCE*, **106**:1021–1040, 1980. (Also: Closure. **108**:152, 1982.)

Liu, P.L.-F., Cheng, A.H.-D., Liggett, J.A., and Lee, J.H. Boundary integral equation solutions to moving interface between two fluids in porous media. *Water Resour. Res.*, **17**:1445–1452, 1981.

Liu, W.K., Chang, H., Chen, J.S., and Belytschko, T. Arbitrary Lagrangian-Eulerian Petrov-Galerkin finite-elements for nonlinear continua. *Comp. Meth. Appl. Mech. Eng.*, **68**:259–310, 1988.

Liu, W.K., Jun, S., and Zhang, Y.F. Reproducing kernel particle methods. *Int. J. Numer. Methods Fluids*, **20**:1081–1106, 1995.

Lo, I.M.C., Mak, R.K.M., and Lee, S.C.H. Modified clays for waste containment and pollutant attenuation. *J. Env. Eng., ASCE*, **12**:25–32, 1997.

Lohman, S.W. *Ground-Water Hydraulics*, U.S. Geological Survey Professional Paper 708, 70 p., 1972.

Lootsma, F.A. Performance evaluation of nonlinear optimization methods via multi-criteria decision analysis and via linear model analysis. In: *Nonlinear Optimization*, Vol. 1, (ed.) M.J.D. Powell, Academic Press, 419–453, 1982.

Loudyi, D. A 2D finite volume model for groundwater flow simulations: Integrating non-orthogonal grid capability into MODFLOW. Ph.D. Dissertation, Cardiff University, UK, 2005.

Loudyi, D., Falconer, R.A. and Lin, B. Mathematical development and verification of a non-orthogonal finite volume model for groundwater flow applications. *Adv. Water Res.*, **30**:29–42, 2007.

Low, P.F. Viscosity of interlayer water in montmorillonite. *Soil Sci. Soc. Am. J.*, **40**:500–505, 1976.

Lu, Z.M., Zhang, D.X., and Robinson, B.A. Explicit analytical solutions for one-dimensional steady state flow in layered, heterogeneous unsaturated soils under random boundary conditions. *Water Resour. Res.*, **43**:W09413, 2007.

Luckner, L., van Genuchten, M. Th., and Nielsen, D.R. A consistent set of parametric models for the two-phase flow of immiscible fluids in the subsurface. *Water Resour. Res.*, **25**:2187–2193, 1989.

Lyman, W.J., Reehl, W.F. and Rosenblatt, D.H. (Eds.) Adsorption coefficients for soils and sediments. In: *Handbook of Chemical Property Estimation Methods*, McGraw-Hill, New York, 1982.

Ma, T.-S., Sophocleous, M., Yu, Y.-S., and Buddemeier, R.W. Modeling saltwater upconing in a freshwater aquifer in south-central Kansas. *J. Hydrology*, **201**:120–137, 1997.

Machackova, J., Wittlingerova, Z., Vlk, K., Zima, J., and Linka, A. Comparison of two methods for assessment of in situ jet-fuel remediation efficiency. *Water Air & Soil Pollution*, **187**:181–194, 2008.

Mackay, D.M. and Cherry, J.A. Groundwater contamination—pump-and-treat remediation, 2. *Environ. Sci. Technol.*, **23**:630–636, 1989.

Macnaughton, S.J., Stephen, J.R., Venosa, A.D., Davis, G.A., Chang, Y.J., and White, D.C. Microbial population changes during bioremediation of an experimental oil spill. *Appl. Environ. Microb.*, **65**:3566–3574, 1999.

Madych, W.R. Miscellaneous error bounds for multiquadric and related interpolators. *Comput. Math. Applic.*, **24**:121–138, 1992.

Maidment, D.R. (Ed.) *Handbook of Hydrology.* McGraw-Hill, 1993.

Maimone, M., Harley, B., Fitzgerald, R., Moe, H., Hossain, R., and Heywood, B. Coastal aquifer planning elements. Chap. 1. In: *Coastal Aquifer Management-Monitoring, Modeling, and Case Studies*, (eds.) A.H.-D. Cheng and D. Ouazar, Lewis Publ., 1–27, 2004.

Malta, S.M.C. and Loula, A.F.D. Numerical analysis of finite element methods for miscible displacements in porous media. *Numer. Meth. Part. Diff. Eqs.*, **14**:519–548, 1998.

Manabe, S. Climate and the ocean circulation, 1, The atmospheric circulation and the hydrology of the earth's surface. *Mon. Weather Rev.*, **97**:739–774, 1969.

Mansell, R.S., Rhue, R.D., Ouyang, Y., and Bloom, S.A. Microemulsion-mediated removal of residual gasoline from soil columns. *J. Soil Contamination*, **5**:309–327, 1996.

Mantoglou, A. and Gelhar, L.W. Stochastic modeling of large-scale transient unsaturated flow systems. *Water Resour. Res.*, **23**:37–46, 1987.

Mantoglou, A. and Wilson, J.L. The turning bands method for simulation of random-fields using line generation by a spectral method. *Water Resour. Res.*, **18**:1379–1394, 1982.

Marin, L.E., Perry, E.C., Essaid, H.I., and Steinich, B. Hydrogeological investigations and numerical simulation of groundwater flow in the karstic aquifer of northwestern Yucatan, Mexico. Chap. 12. In: *Coastal Aquifer Management-Monitoring, Modeling, and Case Studies*, (eds.) A.H.-D. Cheng and D. Ouazar, Lewis Publ., 257–277, 2004.

Marley, M.C., Hazebrouck, D.J., and Walsh, M.T. The application of in situ air sparging as an innovative soils and ground-water remediation technology. *Ground Water Monit. Rem.*, **12**:137–145, 1992.

Marryott, R.A., Dougherty, D.E., and, Stollar, R.L. Optimal groundwater-management, 2. Application of simulated annealing to a field-scale contamination site. *Water Resour. Res.*, **29**:847–860, 1993.

Marshall, T.J. and Holmes, J.W. *Soil Physics.* Cambridge Univ. Press, Cambridge, 1979.

Martinez, M.I., Troester, J.W., and Richards, R.T. Surface electromagnetic geophysical exploration of the groundwater resources of Isla de Mona, Puerto-Rico, a Caribbean carbonate island. *Carbonates and Evaporites*, **10**:184–192, 1995.

Martinez, M.J. Comparison of Galerkin and control volume finite element for advection-diffusion problems. *Int. J. Numer. Methods Fluids*, **50**:347–376, 2006.

Masud, A. and Hughes, T.J.R. A stabilized mixed finite element method for Darcy flow. *Comput. Meth. Appl. Mech. Engng.*, **191**:4341–4370, 2002.

Matalas, N.C. and Wallis, J.R. Generation of synthetic sequences. Chap. 3. In: *Systems Approach to Water Management*, (ed.) A.K. Biswas, 54–79, McGraw-Hill, New York, 1976.

Matheron, G. Principles of geostatistics. *Economic Geology*, **58**:1246–1266, 1963.

Matheron, G. The intrinsic random functions and their applications. *Adv. Appl. Probability*, **5**:439–468, 1973.

Mathon, R. and Johnston, R.L. The approximate solution of elliptic boundary-value problems by fundamental solutions. *SIAM J. Numer. Anal.*, **14**:638–650, 1977.

Mavis, F.T. and Tsui, T.P. Percolation and capillarity movements of water through sand prisms. *Studies in Eng. Bull.*, **18**, Univ. Iowa, Iowa City, 1939.

Maxwell, R.M. and Miller, N.L. Development of a coupled land surface and groundwater model. *J. Hydrometeorology*, **6**:233–247, 2005.

Mayer, K.U., Frind, E.O., and Blowes, D.W. Multicomponent reactive transport modeling in variably saturated porous media using a generalized formulation for kinetically controlled reactions. *Water Resour. Res.*, **38**:WR000862, 2002.

McCarty, P.L., Reinhard, M., and Rittman, B.E. Trace organics in groundwater. *Science and Technology*, **15**:40–51, 1981.

McCord, J.T., Stephens, D.B. and Wilson, J.L. Hysteresis and state-dependent anisotropy in modeling unsaturated hillslope hydrologic processes. *Water Resour. Res.*, **27**:1501–1518, 1991.

McCuen, R.H., Rawls, W.J., and Brakensiek, D.L. Statistical-analysis of the Brooks-Corey and the Green-Ampt parameters across soil texture. *Water Resour. Res.* **17**:1005–1013, 1981.

McDonald, M.G. and Harbaugh, A.W. *A modular Three-Dimensional Finite-Difference Ground-Water Flow Model*, U.S. Geological Survey Open-File Report 83-875, 528 p., 1984.

McKenzie, D. *Water-Resources Potential of the Freshwater Lens at Key West, Florida*, U.S. Geological Survey Water-Resources Investigation Report 90-4115, 24 p., 1990.

McKinney, D.C. and Lin, M.D. Genetic algorithm solution of groundwater-management models. *Water Resour. Res.*, **30**:1897–1906, 1994.

McKinney, D.C. and Lin, M.D. Approximate mixed-integer nonlinear programming methods for optimal aquifer remediation design. *Water Resour. Res.* **31**:731–740, 1995.

McNaughton, K.G. and Black. T.A. A study of evapotranspiration from a Douglas Fir forest using the energy balance approach. *Water Resour. Res.*, **9**:1579–1590, 1973.

McNeill, J.D. Use of electromagnetic methods for groundwater studies. In: *Geotechnical and Environmental Geophysics*, v. I, Soc. Explor. Geophysicists, Inv. in Geophysics, n. 5, (ed.) S. Ward, 191–218, 1990.

McQuarrie, D.A. *Statistical Mechanics*, 2nd ed. University Science Books, 641 p., 2000.

Meaner, O.E. (Ed.) *Hydrology*. Dover, New York, 712 p., 1942.

Means, J.C., Woods, S.G., Hassett, J.J., and Banwart, W.N. Sorption of polynuclear aromatic hydrocarbons by sediments and soils. *Envir. Sci. and Technol.*, **14**:1524–1528, 1980.

Mei, C.C. and Auriault, J.-L. Mechanics of heterogeneous porous media with several spatial scales. *Proc. Roy. Soc. Lond., A*, **426**, 391–423, 1989.

Melenk, J.M. and Babuška, I. The partition of unity finite element method: basic theory and applications. *Comp. Meth. Appl. Mech. Eng.*, **139**:289–314, 1996.

Melloul, A.J. and Zeitoun, D.G. A semi-empirical approach to intrusion monitoring in Israeli coastal aquifer. Chap. 16. In: *Seawater Intrusion into Coastal Aquifers—Concepts, Methods and Practices*, (eds.) J. Bear, A.H.-D. Cheng, S. Sorek, D. Ouazar & I. Herrera, Kluwer, 543–558, 1999.

Meng, Q.M., Hu, H., and Yu, Q.F. The application of an airborne electromagnetic system in groundwater resource and salinization studies in Jilin, China. *J. Environ. Eng. Geophys.*, **11**:103–109, 2006.

Mercer, J.W., Skipp, D.C., and Giffin, D. *Basics of Pump-and-Treat Ground-Water Remediation Technology*, EPA-600/8-90/003, 65 p., 1990.

Metcalf and Eddy, Inc., University of Florida, and Water Resources Engineers, Inc. *Storm Water Management Model*, Vol. 1, Final Report. Water Pollution Control Research Series 11024 DOC 07/71, U.S. Environmental Protection Agency, 1971.

Meyer, P.D., Valocchi, A.J., and Eheart, J.W. Monitoring network design to provide initial detection of groundwater contamination. *Water Resour. Res.*, **30**:2647–2659, 1994.

Meyers, R.A. *Encyclopedia of Environmental Analysis and Remediation.* Wiley, 5400 p., 1998.

Michaelis, L. and Menten, M.L. Die Kinetik der Invertinwerkung. *Biochemische Zeitschrift*, **49**:333–369, 1913. (For excerpted translation, see Mikuls Teich. *A Documentary History of Biochemistry, 1770-1940*, Fairleigh Dickinson University Press, Rutherford, NJ, 1992.)

Michalewicz, Z. *Genetic Algorithms + Data Structures = Evolution Programs.* Springer-Verlag, New York, 1992.

Mikelic, A. Homogenization theory and applications to filtration through porous media. In: *Filtration in Porous Media and Industrial Application*, (ed.) A. Fasano, Lecture Notes in Mathematics Vol. 1734, Springer-Verlag, Berlin, 127–214, 2000.

Miller, C.T. and Rabideau, A.J. Development of split-operator, Petrov-Galerkin methods to simulate transport and diffusion-problems. *Water Resour. Res.*, **29**:2227–2240, 1993.

Miller, G.A. The magical number seven, plus or minus two: Some limits on our capacity for processing information. *Psychological Review*, **63**:81–97, 1956.

Millington, R.J. Gas diffusion in porous media. *Science*, **130**:100–102, 1959.

Mishra, S., Parker, J.C., and Singhal, N. Estimation of soil hydraulic-properties and their uncertainty from particle-size distribution data. *J. Hydrology*, **108**:1–18, 1989.

Mitsuhata, Y., Uchida, T., Matsuo, K., Marui, A., and Kusunose, K. Various-scale electromagnetic investigations of high-salinity zones in a coastal plain. *Geophysics*, **71**:B167–B173, 2006.

Modaressi, H. and Aubert, P. Element-free Galerkin method for deforming multiphase porous media. *Int. J. Numer. Methods Eng.*, **42**:313–340, 1998.

Mohamed, A.M.I., El-menshawy, N., and Saif, A.M. Remediation of saturated soil contaminated with petroleum products using air sparging with thermal enhancement. *J. Environ. Manage.*, **83**:339–350, 2007.

Mohan, R.K., Brown, M.P., and Barnes, C.R. Design criteria and theoretical basis for capping contaminated marine sediments. *Appl. Ocean Res.*, **22**:85–93, 2000.

Molins S., Carrera, J., Ayora, C., and Saaltink, M.W.A. A formulation for decoupling components in reactive transport problems, *Water Resour. Res.*, **40**:w10301, 2004.

Moller, J., Winther, P., Lund, B., Kirkebjerg, K., and Westermann, P. Bioventing of diesel oil-contaminated soil: Comparison of degradation rates in soil based on actual oil concentration and on respirometric data. *J. Industrial Microbiology*, **16**:110–116, 1996.

Mollerup, M. Philip's infiltration equation for variable-head ponded infiltration. *J. Hydrology*, **347**:173–176, 2007.

Mollerup, M. and Hansen, S. Power series solution for falling head ponded infiltration with evaporation. *Water Resour. Res.*, **43**:W03425, 2007.

Molz, F.J., Widdowson, M.A., and Benefield, L.D. Simulation of microbial-growth dynamics coupled to nutrient and oxygen-transport in porous media. *Water Resour. Res.*, **22**:1207–1216, 1986.

Monteith, J.L. Evaporation and the environment. *Symp. Soc. Exploratory Biology*, **19**:205–234, 1965.

Moore, C. and Doherty, J. The cost of uniqueness in groundwater model calibration. *Adv. Water Res.*, **29**:605–623, 2006.

Morel-Seytoux, H.J. Two-phase flows in porous media. *Advances in Hydroscience*, **9**, (ed.) V.T. Chow, Academic Press, New York, 119–202, 1973.

Morrison, S.J. and Spangler, R.R. Chemical barriers for controlling groundwater contamination. *Environmental Progress*, **12**:175–181, 1993.

Morse, P.M. and Feshbach, H. *Methods of Theoretical Physics.* McGraw-Hill, 1953.

Mualem, Y. Modified approach to capillary hysteresis based on a similarity hypothesis. *Water Resour. Res.*, **9**:1324–1331, 1973.

Mualem, Y.A. A conceptual model of hysteresis. *Water. Resour. Res.*, **10**:514–520, 1974.

Mualem, Y.A. Hysteretical models for prediction of the hydraulic conductivity of unsaturated porous media. *Water Resour. Res.*, **12**:1248–1254, 1976.

Mualem, Y.A. Extension of the similarity hypothesis used for modeling the soil water characteristics. *Water Resour. Res.*, **13**:773–780, 1977.

Mualem, Y.A. Theory of universal hysteretical properties of unsaturated porous media. In: *Surface and Subsurface Hydrology, Proc. 3rd Int. Hydrology Symp.*, (ed.) H.J. Morel-Seytoux, 387–399. Water Resources Publ., Fort Collins, Colorado, 1979.

Mualem, Y.A. A modified dependent domain theory of hysteresis. *Soil Sci.*, **137**:283–291, 1984.

Munson, B.R., Young, D.F. and Okiishi, T.H. *Fundamentals of Fluid Mechanics*, 5th ed. McGraw-Hill, 2005.

Muskat, M. *The Flow of Heterogeneous Fluids through Porous Media.* McGraw-Hill, 763 p., 1937.

Naar, J. and Henderson, J.H. An imbibition model–Its application to flow behavior and the prediction of oil recovery. *Trans. Soc. Pet. Eng., AIME*, **222**:61–70, 1961.

NAG *NAG Library Manual*, 2006.

Naji, A., Cheng, A.H.-D., and Ouazar, D. Analytical stochastic solutions of saltwater/freshwater interface in coastal aquifers. *Stoch. Hydrol. Hydraul.*, **12**:413–430, 1998a.

Naji, A., Cheng, A.H.-D., and Ouazar, D. BEM solution of stochastic seawater intrusion problems. *Eng. Anal. Bound. Elem.*, **23**:529–537, 1999.

Naji, A., Ouazar, D., and Cheng, A.H.-D. Locating the saltwater-freshwater interface using nonlinear programming and h-adaptive BEM. *Eng. Anal. Bound. Elem.*, **21**:253–259, 1998b.

Narasimhan, T.N. Hydraulic characterization of aquifers, reservoir rocks, and soils: A history of ideas. *Water Resour. Res.*, **34**:33–46, 1998.

Narasimhan, T.N. Central ideas of Buckingham, 1907: A century later. *Vadose Zone J.*, **4**:434–441, 2005.

Nasseri, M., Shaghaghian, M.R., Daneshbod, Y., and Seyyedian, H. An analytic solution of water transport in unsaturated porous media. *J. Porous Media*, **11**:591–601, 2008.

Nayfeh, A.H. *Perturbation Methods.* Wiley, 437 p., 2000.

Nayroles, B., Touzot, G., and Villon, P. Generalizing the finite element method: diffuse approximation and diffuse elements. *Comput. Mech.*, **10**:307–318, 1992.

Neitsch, S.l., Arnold, J.G., Kiniry, J.R., and Williams, J.R., *Soil and Water Assessment Tool Theoretical Documentation*, Grassland, Soil and Water Research Laboratory, Agricultural Research Service, 2005.

Neuman, S.P. Transient flow of groundwater to wells in multiple-aquifer systems. Doctoral dissertation, Univ. California, Berkeley, 1968.

Neuman, S.P. A Eulerian-Lagrangian numerical scheme for the dispersion-convection equation using conjugate space-time grids. *J. Comput. Phys.*, **41**:270–294, 1981.

Neuman, S.P. Adaptive Eulerian-Lagrangian finite element method for advection-dispersion. *Int. J. Numer. Methods Eng.*, **20**:321–337, 1984.

Neuman, S.P. Universal scaling of hydraulic conductivities and dispersivities in geologic media. *Water Resour. Res.*, **26**:1749–1758, 1990.

Neuman, S.P. and Orr, S. Prediction of steady-state flow in nonuniform geologic media by conditional moments—exact nonlocal formalism, effective conductivities, and weak approximation. *Water Resour. Res.*, **29**:341–364, 1993.

Neuman, S.P. and Witherspoon, P.A. Theory of flow in confined two aquifer system. *Water Resour. Res.*, **5**:803–816, 1969a.

Neuman, S.P. and Witherspoon, P.A. Applicability of current theories of flow in leaky aquifers. *Water Resour. Res.*, **5**:817–829, 1969b.

Nichols, W.E., Aimo, N.J., Oostrom, M., and White, M.D. *STOMP Subsurface Transport Over Multiple Phases: Application Guide.* PNNL-11216 (UC-2010), Pacific Northwest National Laboratory, Richland, Washington, 1997.

Nield, D.A. Resolution of a paradox involving viscous dissipation Theory of universal hysteretical and nonlinear drag in a porous medium. *Transp. Porous Media*, **41**:349–357, 2000.

Nield, D.A. and Bejan, A. *Convection in Porous Media*, 3rd ed. Springer, 640 p., 2006.

Nielsen, L., Jørgensen, N.O., and Gelting, P. Mapping of the freshwater lens in a coastal aquifer on the Keta Barrier (Ghana) by transient electromagnetic soundings. *J. Appl. Geophys.*, **62**:1–15, 2007.

Nikolaevski, V.N. Convective diffusion in porous media, *J. Appl. Math. Mech. (P.M.M.)*, **23**:1042–1050, 1959.

Niqui-Arroyo, J.L., Bueno-Montes, M., Posada-Baquero, R., and Ortega-Calvo, J.J. Electrokinetic enhancement of phenanthrene biodegradation in creosote-polluted clay soil. *Environmental Pollution*, **142**:326–332, 2006.

Nitao, J.J. *Reference Manual for the NUFT Flow and Transport Code, Version 2.0.* Lawrence Livermore National Laboratory, UCRL-MA-130651, 1998.

Nitao, J.J. and Bear, J. Potentials and their role in transport in porous media. *Water Resour. Res.*, **32**:225–250, 1996.

Nowroozi, A.A., Horrocks, S.B., and Henderson, P. Saltwater intrusion into the freshwater aquifer in the eastern shore of Virginia: a reconnaissance electrical resistivity survey. *J. Appl. Geophys.*, **42**:1–22, 1999.

Nyer, E. *Practical Techniques for Groundwater and Soil Remediation.* CRC Press, 224 p., 1992.

Ochoa-Tapia, J.A. and Whitaker, S. Momentum-transfer at the boundary between a porous-medium and a homogeneous fluid. 2. Comparison with experiment. *Int. J. Heat & Mass Transfer*, **38**:2647–2655, 1995.

Odeh, A.S. Effect of viscosity ratio on relative permeability. *Trans. AIME*, **216**:346–352, and discussion by C. F. Wienaug, 352–353, 1959.

Ogata, A. and Banks, R.B. *A Solution of the Differential Equation of Longitudinal Dispersion in Porous Media*, U.S. Geological Survey, Professional Paper, 411-A, 1961.

Ohm, G.S. *Die Galvanische kette, mathematisch bearbeitet.* T.H. Riemann, Berlin 1827. (English translation: *The Galvanic Circuit Investigated Mathematically*, van Nostrand, New York, 269 p., 1891).

Olsen, S.R. and Kemper, W.D. Movement of Nutrients to plant roots. *Adv. Agron.*, **20**:91–151, 1968.

O'Neill, K. Highly efficient, oscillation free solution of the transport equation over long times and large spaces. *Water Resour. Res.*, **17**:1665–1675, 1981.

Orr, S. and Neuman, S.P. Operator and integrodifferential representations of conditional and unconditional stochastic subsurface flow. *Stoch. Hydrol. Hydraul.*, **8**:157–172, 1994.

Osman, I.H. and Laporte, G. Metaheuristics: A bibliography. *Annals of Operations Research*, **63**:513–623, 1996

Oude Essink, G.H.P. Impact of sea level rise in the Netherlands. Chap. 14. In: *Seawater Intrusion into Coastal Aquifers—Concepts, Methods and Practices*, (eds.) J. Bear, A.H.-D. Cheng, S. Sorek, D. Ouazar & I. Herrera, Kluwer, 507–530, 1999.

Oude Essink, G.H.P. Salt water intrusion in a three-dimensional groundwater system in the Netherlands: a numerical study. *Transp. Porous Media*, **43**:137–158, 2001.

Oude Essink, G.H.P. Modeling three-dimensional density dependent groundwater flow at the Island of Texel, the Netherlands. Chap. 4. In: *Coastal Aquifer Management—Monitoring, Modeling, and Case Studies*, (eds.) A.H.-D. Cheng and D. Ouazar, Lewis Publ., 77–94, 2004.

Ozisik, M.N. *Finite Difference Methods in Heat Transfer.* CRC Press, 432 p., 1994.

Palermo, M.R. Design considerations for in-situ capping of contaminated sediments. *Water Sci. Technol.*, **37**:315–321, 1998.

Pan, C.X., Hilpert, M., and Miller, C.T. Pore-scale modeling of saturated permeabilities in random sphere packings. *Phys. Rev. E*, **64**, DOI 066702, 2001.

Pan, H.-L. and Mahrt, L. Interaction between soil hydrology and boundary-layer development. *Boundary Layer Meteorol.*, **38**:185–202, 1987.

Pardieck, D.L., Bouwer, E.J., and Stone, A.T. Hydrogen-peroxide use to increase oxidant capacity for in situ bioremediation of contaminated soils and aquifers—A review. *J. Contam. Hydrol.*, **9**:221–242, 1992.

Parker, J.C. and Lenhard, R.J. A model for hysteretic constitutive relations governing multiphase flow, 1. Saturation-pressure relations. *Water Resour. Res.*, **23**:2187–2196, 1987.

Parker, J.C. and Valocchi, A.J. Constraints on the validity of equilibrium and first-order kinetic transport models in structured soils. *Water Resour. Res.*, **22**:399–407, 1986.

Parkhurst, D.L. and Appelo, C.A.J. *User's Guide to PHREEQC (Version 2)–A Computer Program for Speciation, Batch-Reaction, One-Dimensional Transport, and Inverse Geochemical Calculations.* U.S. Geological Survey Water-Resources Investigations Report 99-4259, 310 p., 1999.

Parkhurst, D.L., Kipp, K.L., Engesgaard, P., and Charlton, S.R. *PHASTA Program for Simulating Ground-Water Flow, Solute Transport, and Multicomponent Geochemical Reactions.* U.S. Geological Survey Techniques and Methods 6-A8, 2004.

Parlange, J.Y. Theory of water-movement in soils. 1. One-dimensional absorption. *Soil Sci.*, **111**:134–137, 1971.

Parlange, J.Y. Theory of water movement in soils. 8. One-dimensional infiltration with constant flux at surface. *Soil Sci.*, **114**:1–4, 1972.

Parlange, J.Y., Haverkamp, R., and Touma, J. Infiltration under ponded conditions. 1. Optimal analytical solution and comparison with experimental observations. *Soil Sci.*, **139**:305–311, 1985.

Parlange, M.B. and Hopmans, J.W. (Eds.) *Vadose Zone Hydrology: Cutting Across Disciplines.* Oxford Univ. Press. 480 p., 1999.

Paul, E.A. and Clark, F.E. *Soil Microbiology and Biochemistry*, 2nd ed. Academic Press, 340 p., 1996.

Pedras, M.H.J. and de Lemos, M.J.S. Macroscopic turbulence modeling for incompressible flow through undeformable porous media. *Int. J. Heat & Mass Transfer*, **44**:1081–1093, 2001.

Penman, H.L. Natural evaporation from open water, bare soil and grass. *Proc. Royal Soc. London, Ser. A*, **193**:120–145, 1948.

Percival, R.V., Miller, A.S., Schroeder, C.H., and Leape, J.P. *Environmental Regulation: Law, Science, And Policy*, 5th ed. Aspen Publ., 1202 p., 2006.

Peters, J.H. (Ed.) *Artificial Recharge of Groundwater, Proc. 3rd Int. Symp. Artificial Recharge of Groundwater*, Amsterdam, Balkema, Amsterdam, 492 p., 1998.

Philip, J.R. The theory of infiltration. 1. The infiltration equation and its solution. *Soil Sci.*, **83**:345–357, 1957a.

Philip, J.R. The theory of infiltration. 2. The profile at infinity. *Soil Sci.*, **83**:435–448, 1957b.

Philip, J.R. The theory of infiltration. 3. Moisture profile and relation to experiments. *Soil Sci.*, **84**:163–178, 1957c.

Philip, J.R. The theory of infiltration. 4. Sorptivity and algebraic infiltration equations. *Soil Sci.*, **84**:257–264, 1957d.

Philip, J.R. The theory of infiltration. 5. The influence of initial moisture content. *Soil Sci.*, **84**:329–339, 1957e.

Philip, J.R. The theory of infiltration. 6. Effect of water depth over soil. *Soil Sci.*, **85**:278–286, 1958a.

Philip, J.R. The theory of infiltration. 7. *Soil Sci.*, **85**:333–337, 1958b.

Philip, J.R. *Theory of Infiltration.* In: *Advances in Hydrosciences.* (ed.) V.T. Chow, 215–296, Academic Press, New York, 1969.

Philip, J.R. Flow through porous media. *Ann. Rev. Fluid Mech.*, **2**:177–204, 1970.

Philip, L.K. An investigation into contaminant transport processes through single-phase cement-bentonite slurry walls. *Engineering Geology*, **60**:209–221, 2001.

Phillips, D.H., Gu, B., Watson, D.B., Roh, Y., Liang, L., and Lee, S.Y. Performance evaluation of a zerovalent iron reactive barrier: Mineralogical characteristics. *Environ. Sci. Technol.*, **34**:4169–4176, 2000.

Pinder, G.F. and Cooper, H.H. A numerical technique for calculating the transient position of the saltwater front. *Water Resour. Res.*, **6**:875–882, 1970.

Pinder, G.F. and Gray, W.G. *Finite Element Simulation in Surface and Subsurface Hydrology.* Academic Press, New York, 295 p., 1977.

Pitzer, K.S. Theory: Ion interaction approach. In: *Activity Coefficients in Electrolyte Solutions*, Vol. I, (ed.) R.M. Pytkowicz, CRC Press, 157–208, 1979.

Plumb, O.A. and Whitaker, S. Diffusion, dispersion and adsorption in porous media: Small scale averaging and local volume averaging. In: *Dynamics of Fluids in Hierarchial Porous Media*, (ed.) J.H. Cushman, Academic, London, 1990.

Poland, J.F. *Guidebook to studies of land subsidence due to ground-water withdrawal*, UNESCO International Hydrological Programme, Paris, 331 p., 1984.

Poling, B.E., Prausnitz, J.M., and O'Connell, J.P. *Properties of Gases & Liquids*, 5th ed. McGraw-Hill, 2000.

Pollock, D.W. *User's Guide for MODPATH/MODPATH-PLOT, Version 3: A Particle Tracking Post-Processing Package for MODFLOW, the U.S. Geological Survey Finite-Difference Ground-Water Flow Model.* U.S. Geological Survey Open-File Report 94-464, 6 ch., 1994.

Polubarinova-Kochina, P.Ya. Theory of filtration of liquids in porous media. In: *Advances in Applied Mechanics*, (eds.) R. von Mises and Th. von Karman, **2**:153–225, Academic Press, New York, 1951.

Polubarinova-Kochina, P.Ya. *Theory of Ground Water Movement.* Princeton Univ. Press, Princeton, NJ, 1962.

Pomerol, J.-C. *Multicriterion Decision in Management.* Springer, 395 p., 2000.

Poreh, M. Dispersivity tensor in isotropic and axisymmetric mediums. *J. Geophys. Res.*, **70**:3909–3914, 1965.

Poulovassilis, A. The hysteresis of pore water: An application concept of independent domains. *Soil Sci.*, **97**:405–412, 1962.

Press, W.H., Teukolsky, S.A., Vetterling, W.T., and Flannery, B.P. *Numerical Recipes, the Art of Scientific Computing*, 3rd ed. Cambridge Univ. Press, 1256 p., 2007.

Prickett, T.A., Naymik, T.G., and Lonnquist, C.G. A random-walk solute transport model for selected groundwater quality evaluations. *Bull. Illinois State Water Survey*, **65**, 103 p., 1981.

Priestley, C.H.B. and Taylor, R.J. On the assessment of surface heat flux and evaporation using large scale parameters. *Monthly Weather Review*, **100**:81–92, 1972.

Priestley, M.B. Power spectral analysis of nonstationary random processes. *J. R. Stat. Soc. B*, **27**:204–236,1965.

Priestley, M.B. Evolutionary spectra and nonstationary processes. *J. Sound Vib.*, **6**:86–89, 1967.

Probstein, R.L. *Physicochemical Hydrodynamics: An Introduction*, 2nd ed. Wiley, 400 p., 1994.

Pruess, K., Oldenburg, C., and Moridis, G. *TOUGH2 User's Guide, ver. 2.0.* Lawrence Berkeley National Laboratory, 1999.

Puls, R.W., Paul, C.J., and Powell, R.M. The application of in situ permeable reactive (zero-valent iron) barrier technology for the remediation of chromate-contaminated groundwater: A field test. *Applied Geochemistry*, **14**:989–1000, 1999.

Putti, M. and Paniconi, C. Picard and Newton linearization for the coupled model of saltwater intrusion in aquifers. *Adv. Water Res.*, **18**:159–170, 1995.

Pyne, R.D.G. *Groundwater Recharge and Wells: A Guide to Aquifer Storage Recovery.* CRC Press, 400 p., 1995.

Raats, P.A.C. Analytical solutions of a simplified flow equation. *Trans. ASAE*, **19**:683–689, 1976.

Rafai, H.S., Newell, J., Gonzales, J.R., Dendrou, S., Dendrou, B., Kennedy, L., and Wilson, J.T. *BIOPLUME III: Natural Attenuation Decision Support System, User's Manual.* U.S. Environmental Protection Agency, EPA/600/R-98/010, 1998.

Ramadhyani, S. and Patankar, S.V. Solution of the convection-diffusion equation by a finite-element method using quadrilateral elements. *Numer. Heat Tr.*, **8**:595–612, 1985.

Ramarao, B. S., Lavenue, A.M., Demarsily, G., and Marietta, M.G. Pilot point methodology for automated calibration of an ensemble of conditionally simulated transmissivity fields. 1. Theory and computational experiments. *Water Resour. Res.*, **31**:475–493, 1995.

Rao, P.S.C. and Davidson, J.M. Estimation of pesticide retention and transformation parameters required in nonpoint source models. In: *Environmental Impact of Nonpoint Source Pollution*, (eds.) J.M. Overcash and J.M. Davidson, 23–67, Ann Arbor Science Publ. Inc., 1980.

Rao, P.S.C., Davidson, J.M., Jessup, R.E., and Selim, H.M. Evaluation of conceptual models for describing non-equilibrium adsorption-desorption of pesticides during steady-flow in soils. *Soil Sci. Soc. Am. J.*, **43**:22–28, 1979

Rao, S.V.N., Kumar, S., Shekhar, S., and Chakraborty, D. Optimal pumping from skimming wells. *J. Hydrol. Eng., ASCE*, **11**:464–471, 2006.

Rawls, W.J. and Brakensiek, D.L. Estimating soil-water retention from soil properties. *J. Irrig. Drain. Div., ASCE*, **108**:166–171, 1982.

Rayner, J.L., Snape, I., Walworth, J.L., Harvey, P.M., and Ferguson, S.H. Petroleum-hydrocarbon contamination and remediation by microbioventing at sub-Antarctic Macquarie Island. *Cold Regions Sci. Technol.*, **48**:139–153, 2007.

Rechtschaffen, C. and Gauna, E.P. *Environmental Justice: Law, Policy, and Regulation.* Carolina Academic Press, 467 p., 2002.

Reilly, T.E. and Goodman, A.S. Quantitative analysis of saltwater-freshwater relationships in ground-water systems—A historical perspective. *J. Hydrology*, **80**:125–160, 1985.

Reilly, T.E. and Goodman, A.S. Analysis of saltwater upconing beneath a pumping well. *J. Hydrology*, **89**:169–204, 1987.

Renard, P. Hydraulics of Wells and Well Testing. Vol. 4, Pt. 13, Art. 154, in *Encyclopedia of Hydrological Sciences*, (eds.) J.J. McDonnell and M.G. Anderson, Wiley, 2323–2340, 2005.

Renard, P. and de Marsily, G. Calculating equivalent permeability: a review. *Adv. Water Res.*, **20**:253–278, 1997.

Richards, L.A. Capillary conduction of liquids through porous mediums. *Physics*, **1**:318–333, 1931.

Richards, L.A. and Gardner, W. Tensiometers for measuring the capillary tension and soil water. *J. Am. Soc. Agron.*, **28**:352–358, 1936.

Rider, N.E. Water loss from various land surfaces. *Quart. J. Roy. Meteorological Soc.*, **83**:181–193, 1957.

Rifai, H., Newwell, C.J., Miller, R., Taffinder, S., and Rounsaville, M. Simulation of natural attenuation with multiple electron acceptors. *Bioremediation*, **3**:53–58, 1995.

Ritz, W. Über eine neue methode zur Lösung gewissen variations—Problems der mathematischen physik. *J. Reine Angew. Math.*, **135**:1–61, 1908.

Ritzel, B.J., Eheart, J.W., and Ranjithan, S. Using genetic algorithms to solve a multi-objective groundwater pollution containment-problem. *Water Resour. Res.*, **30**:1589–1603, 1994.

Riviere, B., Wheeler, M.F., and Banas, K. Part II. Discontinuous Galerkin method applied to a single phase flow in porous media. *Computat. Geosci.*, **4**:337–349, 2000.

Roberts, J. Numerical transpiration. In: *Encyclopedia of Hydrological Sciences*, Vol. 1, Art. 42, (eds.) J.J. McDonnell and M.G. Anderson, 615–625, 2005.

Robertson, H.P. The invariant theory of isotropic turbulence. *Proc. Cambridge Phil. Soc.*, **36**:209–223, 1940.

Robin, M.J.L., Gutjahr A.L., Sudicky, E.A. and Wilson, J.L. Cross-correlated random-field generation with the direct Fourier-transform method. *Water Resour. Res.*, **29**:2385–2397, 1993.

Robock, A., Vinnikov, K., Scholsser, C.A., Speranskaya, N.A., and Xue, Y.K. Use of midlatitude soil moisture and meteorological observations to validate soil moisture simulations with biosphere and bucket model. *J. Clim.*, **9**:15–35, 1995.

Robson, S.G. *Feasibility of digital water-quality modeling illustrated by application at Barstow, California.* U.S. Geological Survey Water-Resources Investigations Report 46-73, 66 p., 1974.

Robson, S.G. *Application of digital profile modeling techniques to groundwater solute transport at Barstow, California.* U.S. Geological Survey Water Supply Paper, 2050, 28 p., 1978.

Rodda, J.C. (Ed.) *Land Subsidence: Proc. 2nd Int. Symp. on Land Subsidence*, UNESCO/IAHS, Anaheim, CA, USA, IAHS, **121**, 1976.

Rohrs, J., Ludwig, G., and Rahner, D. Electrochemically induced reactions in soils-a new approach to the in-situ remediation of contaminated soils? Part 2: Remediation experiments with a natural soil containing highly chlorinated hydrocarbons. *Electrochimica Acta*, **47**:1405–1414, 2002.

Rose, W. Some problems connected with the use of classical descriptions of fluid/fluid displacement processes. In: *Fundamentals of Transport Phenomena*, (ed.) J. Bear, 229–240, 1972.

Rose, W. Measuring transport coefficients to describe coupled two-phase flows in porous media. *Transp. Porous Media*, **3**:163–171, 1988.

Rose, W. Coupling-coefficient for two-phase flow in porous space of simple geometry. *Transp. Porous Media*, **5**:97–102, 1990.

Rose, W. An upgraded viscous coupling measurement methodology. *Transp. Porous Media*, **28**:221–23, 1997.

Rose, W. and Rose, D. An upgraded porous medium coupled transport process algorithm. *Transp. Porous Media*, **59**:357–372, 2005.

Rosenzweig, R. and Shavit, U. The laminar flow field at the interface of a Sierpinski carpet configuration. *Water Resour. Res.*, **43**, w10402, 2007.

Ross, J.L., Ozbek, M.M. and Pinder, G.F. Aleatoric and epistemic uncertainty in groundwater flow and transport simulation. *Water Resour. Res.*, **45**, W00B15, 2009.

Ross, P.J. and Bristow, K.L. Simulating water-movement in layered and gradational soils using the Kirchhoff transform. *Soil Sci. Soc. Am. J.*, **54**:1519–1524, 1990.

Rowe, R.K. Long-term performance of contaminant barrier systems. *Geotechnique*, **55**:631–677, 2005.

Rubin, Y. *Applied Stochastic Hydrogeology.* Oxford Univ. Press, 2003.

Rubin, Y. and Dagan, G. Stochastic-analysis of boundaries effects on head spatial variability in heterogeneous aquifers. 1. Constant head boundary. *Water Resour. Res.*, **24**:1689–1697, 1988.

Rubin, Y. and Dagan, G. Stochastic-analysis of boundaries effects on head spatial variability in heterogeneous aquifers. 1. Impervious boundary. *Water Resour. Res.*, **25**:707–712, 1989.

Russell, T.F. and Celia, M.A. An overview of research on Eulerian-Lagrangian localized adjoint methods (ELLAM). *Adv. Water Res.*, **25**:1215–1231, 2002.

Russo, D. and Bouton, M. Statistical-analysis of spatial variability in unsaturated flow parameters. *Water Resour. Res.*, **28**:1911–1925, 1992.

Rutledge, A.T. *Model-Estimated Ground-Water Recharge and Hydrograph of Ground-Water Discharge to a Stream.* U.S. Geological Survey Water Resources Investigations Report 97-4253, 1997.

Rutledge, A.T. *Computer Programs for Describing the Recession of Ground-Water Discharge and for Estimating Mean Ground-Water Recharge and Discharge from Streamflow Records–Update.* U.S. Geological Survey Water Resources Investigations Report 98-4148, 1998.

Saad, Y. *Iterative Methods for Sparse Linear Systems*, 2nd ed. SIAM, 528 p., 2003.

Saaltink, M.W., Ayora, C., and Carrera, J. A mathematical formulation for reactive transport that eliminates mineral concentrations. *Water Resour. Res.*, **34**:1649–1656, 1998.

Saaltink, M.W., Batlle, F., Ayora, C., Carrera, J., and Olivella, S. RETRASO, a code for modeling reactive transport in saturated and unsaturated porous media. *Geologic Acta*, **2**:235–251, 2004.

Saaltink, M.W., Carrera, J., and Ayora, C. A comparison of two alternatives to simulate reactive transport in groundwater. *J. Geochemical Exploration*, **69–70**:97–101, 2000.

Saaty, T.L. A scaling method for priorities in hierarchical structures. *J. Mathematical Psychology*, **15**:234–281, 1977.

Saaty, T.L. *The Analytic Hierarchy Process—Planning, Priority Setting, Resource Allocation.* McGraw-Hill, 1980.

Saaty, T.L. How to make a decision—The analytic hierarchy process. *Interfaces*, **24**:19–43, 1994.

Sadana, A. *Biocatalysis, Fundamentals of Enzyme Deactivation Kinetics.* Prentice-Hall, Englewood Cliffs, New Jersey, 1991.

Saeed, M.M. *Skimming Wells: Current practices and Guidelines for Improving Design and Operation.* Higher Education Commission, Pakistan, 2008.

Saeed, M.M. and Ashraf, M. Feasible design and operational guidelines for skimming wells in the Indus basin, Pakistan. *Agr. Water Manage.*, **74**:165-188, 2005.

Saffman, P.G. A theory of dispersion in a porous medium. *J. Fluid Mech.*, **6**:321–349, 1959.

Sagar, B. Galerkin finite-element procedure for analyzing flow through random media. *Water Resour. Res.*, **14**:1035–1044, 1978.

Sakurai, H. and Kawahara, M. Three-dimensional groundwater flow analysis system using the element-free Galerkin method. *Int. J. Computat. Fluid Dyn.*, **18**:309–315, 2004.

Salanitro, J.P. The role of bioattenuation in the management of aromatic hydrocarbon plumes in aquifers. *Ground Water Monit. Rem.*, **13**:150–161, 1993.

Salanitro, J.P., Dorn, P.B., Huesemann, M.H., Moore, K.O., Rhodes, I.A., Jackson, L.M.R., Vipond, T. E., Western, M.M., and Wisniewski, H.L. Crude oil hydrocarbon bioremediation and soil ecotoxicity assessment. *Environ. Sci. Technol.*, **31**:1769–1776, 1997.

Salas, J.D. Analysis and modeling of hydrologic time series. Chap. 19. In: *Handbook of Hydrology*, (ed.) D.R. Maidment, McGraw-Hill, 1993.

Sanchez-Palencia, E. Comportement local et macroscopique d'un type de milieu physiques héterogénes. *Int. J. Engng. Sci.*, **12**:331-352, 1974.

Sanchez-Palencia, E. *Non-homogeneous Media and Vibration Theory.* Lecture Notes in Physics, **127**, Springer-Verlag, N.Y., 1980.

Sanford, W.E. and Konikow, L.F. *A Two-Constituent Solute Transport Model for Ground Water Having Variable Density.* U.S. Geological Survey Water Resources Investigation Report 85-4279, 89 p., 1985.

Santamarina, J.C., Klein, K.A., Wang, Y.H. and Prenke, E. Specific surface: determination and relevance. *Can. Geotech. J.*, **39**:233–241, 2002.

Satterfield, C.N. *Heterogeneous Catalysis in Industrial Practice*, 2nd ed. McGraw-Hill, 554 p., 1991.

Sawyer, C.N., McCarty, P.L. and Parkin, G.F. *Chemistry for Environmental Engineering and Science*, 5th ed. McGraw-Hill, 768 p., 2002.

Schaake, J.C., Koren, V.I., Duan, Q.Y., Mitchell, K., and Chen, F. Simple water balance model for estimating runoff at different spatial and temporal scales. *J. Geophys. Res. Atmosphere*, **101**(D3):7461–7475, 1996.

Schaap, M.G., Leij, F.J., and van Genuchten, M.Th. Neural network analysis for hierarchical prediction of soil hydraulic properties. *Soil Sci. Soc. Am. J.*, **62**:847–855, 1998.

Scheidegger, A.E. General theory of dispersion in porous media. *J. Geophys. Res.*, **66**:3273–3278, 1961.

Scherer, M.M., Richter, S., Valentine, R.L., and Alvarez, P.J.J. Chemistry and microbiology of permeable reactive barriers for in situ groundwater clean up. *Critical Rev. Environ. Sci. Technol.*, **30**:363–411, 2000.

Schmorak, S. and Mercado, A. Upconing of freshwater-saltwater interface below pumping wells, field study. *Water Resour. Res.*, **5**:1290–1311, 1969.

Schneider, F.N. and Owens, W.W. Sandstone and carbonate two- and three-phase relative permeability characteristics. *Soc. Pet. Eng. J.*, **10**:75–84, 1970.

Schneider, G.E. and Raw, M.J. A skewed, positive influence coefficient upwinding procedure for control-volume-based finite-element convection-diffusion computation. *Numer. Heat Tr.*, **9**:1–26, 1985.

Schowalter, T.T. Mechanics of secondary hydrocarbon migration and entrapment. *Water Sci. Technol.*, **23**:467–476, 1979.

Schwarzenbach, R.P., Gschwend, P.M., and Imboden, D.M. *Environmental Organic Chemistry*, 2nd ed. Wiley Interscience, 1328 p., 2002.

Scott, R.F. *Principles of Soil Mechanics*. Addison-Wesley, Reading, Mass., 550 p., 1963.

Selim, H.M., Mansell, R.S., and Zelazny, L.W. Modeling reactions and transport of potassium in soils. *Soil Sci.*, **122**:77–84, 1976.

Sengpiel, K.P. Resistivity depth mapping with airborne electromagnetic survey data. *Geophysics*, **48**:181–196, 1983.

Sengpiel, K.P. Approximate inversion of airborne em data from a multilayered ground. *Geophysical Prospecting*, **36**:446–459, 1988.

Sengpiel, K.P. and Siemon, B. Advanced inversion methods for airborne electromagnetic exploration. *Geophysics*, **65**:1983–1992, 2000.

Serrano, S.E. Forecasting scale-dependent dispersion from spills in heterogeneous aquifers. *J. Hydrology*, **169**:151–169, 1995.

Serrano, S.E. Analytical decomposition of the nonlinear unsaturated flow equation. *Water Resour. Res.*, **34**:397–407, 1998.

Serrano, S.E. Modeling infiltration with approximate solutions to Richard's equation. *J. Hydrol. Eng., ASCE*, **9**:421–432, 2004.

Shanno, D.F. Conditioning of quasi-Newton methods for function minimization. *Math. Comput.*, **24**:647–656, 1970.

Shavit, U., Rosenzweig, R., and Assouline, S. Free flow at the interface of porous surfaces: A generalization of the Taylor brush configuration. *Transp. Porous Media*, **54**:345–360, 2004.

Sherif, M. Nile Delta Aquifer in Egypt. Chap. 17. In: *Seawater Intrusion into Coastal Aquifers—Concepts, Methods and Practices*, (eds.) J. Bear, A.H.-D. Cheng, S. Sorek, D. Ouazar & I. Herrera, Kluwer, 559–590, 1999.

Shinozuka, M. and Jan, C.M. Digital simulation of random processes and its applications. *J. Sound Vib.*, **25**:111–128, 1972.

Shuttleworth, W.J. Evaporation. Chap. 4. In: *Handbook of Hydrology*, (ed.) D.R. Maidment, McGraw-Hill, New York, 4.1–4.53, 1993.

Shuttleworth, W.J. and Wallace, J.S. Evaporation from sparse crops-an energy combination theory. *Quart. J. Royal Meteorological Soc.*, **111**:839–855, 1985.

Si, B.C. and Kachanoski, R.G. Unified solution for infiltration and drainage with hysteresis: Theory and field test. *Soil Sci. Soc. Am. J.*, **64**:30–36, 2000.

Silin-Bekchurin, A.I. *Dynamics of Groundwater*. (in Russian) Moscow Izdat., Moscow Univ., 258 p., 1958.

Simmers, I. (Ed.) *Estimation of Natural Groundwater Recharge*, Nato Science Series, C 222. Reidel, Dordrecht, 1988.

Simunek, J., Sejna, M., and van Genuchten, M.Th. The HYDRUS-2D software package for simulating two-dimensional movement of water, heat, and multiple solutes in variably saturated media. Version 2.0, IGWMC-TPS-53, International Ground Water Modeling Center, Colorado School of Mines, Golden, Colorado, 251 p., 1999.

Singh, V.P. *Computer Models of Watershed Hydrology.* Water Resources Publ., 1995.

Sirotine, Y. and Chaskolskaya, M. *Fondaments de la physique des crystaux.* Edition Mir, 680 p., (Russian Ed., 1975) 1984.

Smith, G.D. *Numerical Solution of Partial Differential Equations: Finite Difference Methods*, 3rd ed. Oxford Univ. Press, 350 p., 1986.

Smith, L. and Freeze, R.A. Stochastic analysis of steady state groundwater flow in a bounded domain, 2. Two-dimensional simulations. *Water Resour. Res.*, **15**:1543–1559, 1979.

Snell, R.W. Three phase relative permeability and residual oil data. *J. Inst. Petrol.*, **12**:80–88, 1962.

Soldal, O., Mauring, E., Halvorsen, E., and Rye, N. Seawater intrusion and fresh groundwater hydraulics in fjord delta aquifers inferred from ground penetrating radar and resistivity profiles–Sunndalsøra and Esebotn, western Norway. *J. Appl. Geophys.*, **32**:305–319, 1994.

Sophocleous, M.A. Combining the soil water balance and water-level fluctuation methods to estimate natural groundwater recharge—practical aspects. *J. Hydrology*, **124**:229–241, 1991.

Sorek, S., Borisov, V., and Yakirevich, A. Modified Eulerian Lagrangian method for density dependent miscible transport. Chap. 11. In: *Seawater Intrusion in Coastal Aquifers—Concepts, Methods, and Practices*, (eds.) J. Bear, A.H.-D. Cheng, S. Sorek, D. Ouazar & I. Herrera, Kluwer, 363–398, 1999.

Sorek, S. and Pinder, G.F. Survey of computer codes and case histories. Chap. 12. In: *Seawater Intrusion in Coastal Aquifers—Concepts, Methods, and Practices*, (eds.) J. Bear, A.H.-D. Cheng, S. Sorek, D. Ouazar & I. Herrera, Kluwer, 399–461, 1999.

Sparks, D.L. *Environmental Soil Chemistry.* Academic Press, 352 p., 2003.

Sposito, G. *The Surface Chemistry of Natural Particles.* Oxford Univ. Press, New York, 256 p., 2004.

Srinivasan, V. and Clement, T.P. Analytical solutions for sequentially coupled one-dimensional reactive transport problems—Part I: Mathematical derivations. *Adv. Water Res.*, **31**:203–218, 2008a.

Srinivasan, V. and Clement, T.P. Analytical solutions for sequentially coupled one-dimensional reactive transport problems—Part II: Special cases, implementation and testing. *Adv. Water Res.*, **31**:219–232, 2008b.

Srivastava, R. and Yeh, T.C.J. Analytical solutions for one-dimensional, transient infiltration toward the water-table in homogeneous and layered soils. *Water Resour. Res.*, **27**:753–762, 1991.

Stakelbeek, A. Movement of brackish groundwater near a deep-well infiltration system in the Netherlands. Chap. 15. In: *Seawater Intrusion into Coastal Aquifers—Concepts, Methods and Practices*, (eds.) J. Bear, A.H.-D. Cheng, S. Sorek, D. Ouazar & I. Herrera, Kluwer, 531–541, 1999.

Stallman, R.W. Flow in the zone of aeration. In: *Advances in Hydroscience*, **4**, (ed.) V.T. Chow, Academic Press, New York, 151–195, 1967.

Starr, R.C. and Cherry, J.A. In-situ remediation of contaminated groundwater—The funnel-and-gate system." *Ground Water*, **32**:465–476, 1994.

Steefel, C.I. and MacQuarrie, K.T.B. Approaches to modeling of reactive transport in porous media. In: *Reactive Transport in Porous Media*, (eds.) P.C. Lichtner, C.I. Steefel, and E.H. Oelkers, 83–129, 1996.

Steefel, C.I. and Yabusaki, S.B. *OS3D/GIMRT, Software for Modeling Multicomponent-Multidimensional Reactive Transport, User Manual and Programmer's Guide.* Pac. Northwest Lab., Richland, Wash., 1995.

Stein, M.L. *Interpolation of Spatial Data: Some Theory for Kriging.* Springer, 247 p., 1999.

Stephens, D.B. *Vadose Zone Hydrology.* CRC Press, 347 p., 1996.

Stephens, D.B. and Heermann, S. Dependence of anisotropy on saturation in a stratified sand. *Water Resour. Res.*, **24**:770–778, 1988.

Steuer, A., Siemon, B., and Eberle, D. Airborne and ground-based electromagnetic investigations of the freshwater potential in the tsunami-hit area Sigli, northern Sumatra. *J. Environ. Eng. Geophys.*, **13**:39–50, 2008.

Stewart, M.T. Rapid reconnaissance mapping of fresh-water lenses on small oceanic islands. In: *Geotechnical and Environmental Geophysics*, v. II, Soc. Explor. Geophysicists, Inv. in Geophysics, n. 5, (ed.) S. Ward, 57–66, 1990.

Stewart, M.T. Geophysical investigations. Chap. 2, In; *Seawater Intrusion into Coastal Aquifers—Concepts, Methods and Practices*, (eds.) J. Bear, A.H.-D. Cheng, S. Sorek, D. Ouazar & I. Herrera, Kluwer, 9–50, 1999.

Stewart, M.T., Layton, M., and Lizanec, T. Application of resistivity surveys to regional hydrogeologic reconnaissance. *Ground Water*, **21**:42–48, 1983.

Stone, H.L. Probability model for estimating three-phase relative permeability. *Trans. Soc. Pet. Eng., AIME*, **249**:214–218, 1970.

Stone, H.L. Estimation of three-phase relative permeability and residual oil data. *J. Can. Petrol. Technol.*, **12**:53–61, 1973.

Strack, O.D.L. A single-potential solution for regional interface problems in coastal aquifers. *Water Resour. Res.*, **12**:1165–1174, 1976.

Strack, O.D.L. *Groundwater Mechanics.* Prentice-Hall, 732 p., 1989.

Strack, O.D.L. Theory and applications of the analytic element method. *Rev. Geophysics*, **41**, ISI:000183886400001, 2003.

Stumm, W. and Morgan, J.J. *Aquatic Chemistry, Chemical Equilibria and Rates in Natural Waters*, 3rd ed. Wiley Interscience, 1040 p., 1995.

Sudicky, E.A. A natural gradient experiment on solute transport in a sand aquifer—spatial variability of hydraulic conductivity and its role in the dispersion process. *Water Resour. Res.*, **22**:2069–2082, 1986.

Suer, P. and Lifvergren, T. Mercury-contaminated soil remediation by iodide and electroreclamation. *J. Env. Eng., ASCE*, **129**:441–446, 2003.

Sun, N.-Z. *Inverse Problems in Groundwater Modeling.* Kluwer, 1994.

Sun, N.-Z. and Sun, A.Y. Inverse methods for parameter estimation. In: *Encyclopedia of Hydrological Sciences*, Vol. 4, Article 162, (eds.) J.J. McDonnell and M.G. Anderson, Wiley, 2415–2430, 2005.

Sun, S.Y. and Wheeler, M.F. Discontinuous Galerkin methods for coupled flow and reactive transport problems. *Appl. Numer. Math.*, **52**:273–298, 2005.

Sun, Y., Buscheck, T.A., and Hao, Y. Modeling reactive transport using exact solutions for first-order reaction networks. *Transp. Porous Media*, **71**:217–231, 2008.

Sun, Y., Petersen, J.N., and Clement, T.P. Analytical solutions for multiple species reactive transport in multiple dimensions. *J. Contam. Hydrol.*, **35**:429–440, 1999a.

Sun, Y., Petersen, J.N., Clement, T.P., and Skeen, R.S. Development of analytical solutions for multispecies transport with serial and parallel reactions. *Water Resour. Res.*, **35**:185–190, 1999b.

Suthersan, S.S. and Payne, F.C. *In Situ Remediation Engineering.* CRC Press, 536 p., 2004.

Swartz, J.H. Resistivity studies of some salt-water boundaries in the Hawaiian Islands. *Trans. Am. Geophys. Union*, **18**:387–393, 1937

Swartz, J.H. Geophysical investigations in the Hawaiian Islands. *Trans. Am. Geophys. Union*, **20**:292–298, 1939.

Swartzendruber, D. The flow of water in unsaturated soils. Chap. 6. In: *Flow Through Porous Media*, (ed.) R.J.M. De Wiest, 215–292, Academic Press, New York, 1969.

Swarzenski, P.W. and Kindinger, J.L. Leaky coastal margins: Examples of enhanced coastal groundwater and surface-water exchange from Tampa Bay and Crescent Beach submarine spring, Florida, USA. Chap. 5. In: *Coastal Aquifer Management-Monitoring, Modeling, and Case Studies*, (eds.) A.H.-D. Cheng and D. Ouazar, Lewis Publ., 95–114, 2004.

Taigbenu, A.E., Liggett, J.A. and Cheng, A.H.-D. Boundary integral solution to seawater intrusion into coastal aquifers. *Water Resour. Res.*, **20**:1150–1158, 1984.

Tartakovsky, D.M., Neuman, S.P., and Lu, Z.M. Conditional stochastic averaging of steady state unsaturated flow by means of Kirchhoff transformation. *Water Resour. Res.*, **35**:731–745, 1999.

Taylor, G.I. Dispersion of soluble matter in solvent flowing slowly through a tube. *Proc. Royal Soc., London, Ser. A*, **219**:186–203, 1953.

Taylor, G.I. The dispersion of matter in turbulent flow through a pipe. *Proc. Royal Soc., London, Ser. A*, **223**:446–468, 1954.

Telford, W.M., Geldart, L.P. and Sheriff, R.E. *Applied Geophysics.* Cambridge Univ. Press, Cambridge, 770 p., 1990.

Terminology Committee of Commission I of the International Soil Science Society (ISSS) Terminology in Soil Physics. *ISSS Bull.*, **49**:914–920, 1976.

Terzaghi, K. Die berechnung der durchlassigkeitsziffer des tones aus dem verlauf der hydrodynamischen spannungserscheinungen. *Sitz. Akad. Wissen., Wien Math. Naturwiss. Kl., Abt. IIa*, **132**:105–124, 1923.

Terzaghi, K. Soil moisture and capillary phenomena in soils. Chap. 10-A. In: *Hydrology*, (ed.) O.E. Meaner, McGraw-Hill, New York, 331–363, 1942.

Terzaghi, K. *Theoretical Soil Mechanics.* Wiley, New York, 528 p., 1943.

Testa, S.M. and Winegardner, D.L. *Restoration of Contaminated Aquifers: Petroleum Hydrocarbons and Organic Compounds*, 2nd ed. CRC Press, 464 p., 2000.

Tezduyar, T.E., Behr, M., and Liou, J. A new strategy for finite-element computations involving moving boundaries and interfaces—the deforming-spatial-domain space-time procedure. 1. The concept and the preliminary numerical tests. *Comput. Meth. Appl. Mech. Eng.*, **94**:339–351, 1992a.

Tezduyar, T.E., Behr, M., Mittal, S., and Liou, J. A new strategy for finite-element computations involving moving boundaries and interfaces—the deforming-spatial-domain space-time procedure. 2. Computation of free-surface flows, two-liquid flows, and flows with drifting cylinders. *Comput. Meth. Appl. Mech. Eng.*, **94**:353–371, 1992b.

Theis, C.V. The relation between the lowering of the piezometric surface and the rate and duration of discharge of a well using ground-water storage. *Trans. Am. Geophys. Union*, **16**:519–524, 1935.

Therrien, R. and Sudicky, E.A. Three-dimensional analysis of variably-saturated flow and solute transport in discretely-fractured porous media. *J. Contam. Hydrol.*, **23**:1–44, 1996.

Thomas, H.A. Jr. and Fiering, M.B. Mathematical synthesis of streamflow sequences for the analysis of river basins by simulation. Chap. 12. In: *Design of Water-Resources Systems*, (eds.) R. Dorfman, H.A. Thomas, Jr., S.A. Marglin, and G. Fair, 459–493, Harvard Univ. Press, Cambridge, Mass., 1962.

Thornthwaite, C.W. An approach toward a rational classification of climate. *Geographical Review*, **38**:55–94, 1948.

Thornthwaite, C.W. and Hare, F.K. The loss of water to the air. *Meteorological Monographs*, **6**:163–180, 1965.

Thurstone, L.L. A law of comparative judgments. *Psychological Reviews*, **34**:273–286, 1927.

Tindall, J.A. and Kunkel, J.R. *Unsaturated Zone Hydrology for Scientists and Engineers*. Prentice-Hall, 642 p., 1999.

Tison, L.J. (Ed.) *Land Subsidence: Proc. 1st Int. Symp. on Land Subsidence*, Tokyo, Japan, IAHS, **89**, 1969.

Todd, D.K. *Annotated Bibliography on Artificial Recharge of Ground Water through 1954*. U.S. Geological Survey Water-Supply Paper 1477, 115 p., 1959.

Todd, D.K. *Groundwater Hydrology*, 2nd ed. Wiley, 535 p., 1980.

Tompson, A.F.B Numerical simulation of chemical migrations in physically and chemically heterogeneous porous media. *Water Resour. Res.*, **29**:3709–3726, 1993.

Tompson, A.F.B., Ababou, R. and Gelhar, L.W. Implementation of the 3-dimensional turning bands random field generator. *Water Resour. Res.*, **25**:2227–2243, 1989.

Tompson, A.F.B., Carle, S.F., Rosenberg, N.D., and Maxwell, R.M. Analysis of groundwater migration from artificial recharge in a large urban aquifer: A simulation perspective. *Water Resour. Res.*, **35**:2981–2998, 1999.

Tompson, A.F.B. and Gelhar, L.W. Numerical-simulation of solute transport in 3-dimensional, randomly heterogeneous porous-media. *Water Resour. Res.*, **26**:2541–2562, 1990.

Topp, G.C. Soil water hysteresis measured in a sandy loam compared with the hysteretic domain model. *Soil Sci. Soc. Am. Proc.*, **33**:645–651, 1969.

Topp, G.C. Soil water hysteresis in silt loam and clay loam soils. *Water Resour. Res.*, **7**:914–920, 1971.

Tracy, F.T. 1-D, 2-D, and 3-D analytical solutions of unsaturated flow in groundwater. *J. Hydrology*, **170**:199–214, 1995.

Tracy, F.T. Clean two- and three-dimensional analytical solutions of Richards' equation for testing numerical solvers. *Water Resour. Res.*, **42**, W08503, 2006.

Tronicke, J., Blindow, N., Gross, R., and Lange, M.A. Joint application of surface electrical resistivity- and GPR-measurements for groundwater exploration on the island of Spiekeroog–northern Germany. *J. Hydrology*, **223**:44–53, 1999.

Tse, K.K.C., Lo, S.L., and Wang, J.W.H. Pilot study of in-situ thermal treatment for the remediation of pentachlorophenol-contaminated aquifers. *Environ. Sci. Technol.*, **35**:4910–4915, 2001.

University of Texas. *UTCHEM-9.0 A Three-Dimensional Chemical Flood Simulator*, Vol. I: User's Guide, Vol. II: Technical Documentation, Reservoir Engineering Research Program Center for Petroleum and Geosystems Engineering, University of Texas at Austin, 2000.

U.S. Environmental Protection Agency. *Storm Water Management Model (SWMM)*, ver. 4.3, 1994.

U.S. Environmental Protection Agency. *Hydrological Simulation Program–FORTRAN (HSPF)*, ver. 11, 1997.

U.S. Environmental Protection Agency. *The Manual of Individual Water Supply Systems*. Fredonia Books, 168 p., 2001.

U.S. Environmental Protection Agency. *WhAEM2000, Wellhead Analytic Element Model*, ver. 3.2.0, 2005.

Van Dam, J.C. Exploitation, restoration and management. Chap. 4. In: *Seawater Intrusion into Coastal Aquifers—Concepts, Methods and Practices*, (eds.) J. Bear, A.H.-D. Cheng, S. Sorek, D. Ouazar & I. Herrera, Kluwer, 73–125, 1999.

Van Genuchten, M.Th. *Mass Transfer Studies of Sorbing Porous Media*. Ph.D Thesis, New Mexico State Univ., La Cruz, NM, 1974.

Van Genuchten, M.Th. Models for describing water and solute movement through soils with large pores. *Agronomy Abstracts, Amer. Soc. Agron.*, 1980.

Van Genuchten, M.Th. and Dalton, F.N. Models for simulating salt movement in aggregated field soils. *Geoderma*, **38**:165–183, 1986

Van Laarhoven, P.J.M. and Aarts, E.H.L. *Simulated Annealing: Theory and Applications*. Springer, 204 p., 1987.

Varoglu, E. and Finn, W.D.L. Utilization of the method of characteristics to solve accurately two-dimensional transport problems by finite elements. *Int. J. Numer. Methods Fluids*, **2**:173–184, 1982.

Vauclin, M., Haverkamp, R. and Vachaud, G. *Resolution Numerique d'une Equation de Diffusion Non-Lineare.* Presses Univer, 1979.

Venosa, A.D., Suidan, M.T., Wrenn, B.A., Strohmeier, K.L., Haines, J.R., Eberhart, B.L., King, D., and Holder, E. Bioremediation of an experimental oil spill on the shoreline of Delaware bay. *Environ. Sci. Technol.*, **30**:1764–1775, 1996.

Verhoff, A. and Campbell, C.L. Evaporation measurement. In: *Encyclopedia of Hydrological Sciences*, Vol. 1, Art. 40, (eds.) J.J. McDonnell and M.G. Anderson, 589–601, 2005.

Verruijt, A. Elastic storage in aquifers. In: *Flow Through Porous Media.* (ed.) R.J.M. De Wiest, Academic Press, New York, 331–376, 1969.

Verruijt, A. Steady dispersion across an interface of a porous medium. *J. Hydrology*, **14**:337–347, 1971.

Verruijt, A. *Computational Geomechanics.* Kluwer, 1995.

Versteeg, H.K. and Malalasekera, W. *An Introduction to Computational Fluid Dynamics.* Pearson/Prentice-Hall, 257 p., 1995

Voss, C.I. *SUTRA—Saturated Unsaturated Transport—A Finite-Element Simulation Model for Saturated-Unsaturated, Fluid-Density-Dependent Ground-Water Flow with Energy Transport or Chemically-Reactive Single-Species Solute Transport*, U.S. Geological Survey Water-Resources Investigation Report 84-4369, 409 p., 1984.

Voss, C.I. USGS SUTRA Code—History, practical use, and application in Hawaii. Chap. 9. In: *Seawater Intrusion in Coastal Aquifers—Concepts, Methods, and Practices*, (eds.) J. Bear, A.H.-D. Cheng, S. Sorek, D. Ouazar & I. Herrera, Kluwer, 249–313, 1999.

Voss, C.I. and Provost, A.M. *SUTRA, A Model for Saturated-Unsaturated Variable-Density Ground-Water Flow with Solute or Energy Transport.* U.S. Geological Survey Water-Resources Investigations Report 02-4231, 250 p., 2002.

Voss, C.I. and Souza, W.R. Variable density flow and solute transport simulation of regional aquifers containing a narrow freshwater-saltwater transition zone. *Water Resour. Res.*, **23**:1851–1866, 1987.

Vuković, M., and Soro, A. *Hydraulics of Water Wells, Theory and Application.* Water Resources Publ., 354 p., 1992.

Walton, W.C. *Groundwater Resources Evaluation.* McGraw-Hill, New York, 1970.

Walton, W.C. *Groundwater Pumping Tests.* CRC Press, 216 p., 1990.

Wang, F.C. Approximate theory for skimming well formulation in indus plain of West Pakistan. *J. Geophys. Res.*, **70**:5055–5063, 1965.

Wang, H., Ewing, R.E., Qin, G., Lyons, S.L., Al-Lawatia, M., and Man, S. A family of Eulerian-Lagrangian localized adjoint methods for multi-dimensional advection-reaction equations. *J. Computat. Phys.*, **152**:120–163, 1999.

Wang, H., Ewing, R.E., and Russell, T.F. Eulerian-Lagrangian localized adjoint methods for convection-diffusion equations and their convergence analysis. *IMA J. Numer. Anal.*, **15**:405–459, 1995.

Wang, K.X., Wang, H., and Al-Lawatia, M. An Eulerian-Lagrangian discontinuous Galerkin method for transient advection-diffusion equations. *Numer. Meth. Part. Diff. Eqs.*, **23**:1343–1367, 2007.

Ward, A.L., White, M.D., Freeman, E.J., and Zhang, Z.F. *STOMP Subsurface Transport Over Multiple Phase. Addendum: Sparse Vegetation Evapotranspiration Model for the Water-Air-Energy Operational Mode.* PNNL-15465, Pacific Northwest National Laboratory, Richland, Washington, 2005.

Ward, C.H., Cherry, J.A., and Scalf, M.R. *Subsurface Restoration.* CRC Press, 500 p., 1997.

Ward, D.S., Reeves, M., and Duda, L.E. Verification and field comparison of the Sandia Waste-Isolation Flow and Transport Model (SWIFT). NUREG/CR-3316, SAND-83-1154, Sandia National Laboratories, Albuquerque, New Mexico, 168 p., 1984.

Ward, S.H. *Geotechnical and Environmental Geophysics*, Soc. Exploration Geophysicists, Inv. in Geophysics, n. 5, vols. I, II, III., 1990.

Warrick, A.W., Islas, A., and Lomen, D.O. An analytical solution to Richards equation for time-varying infiltration. *Water Resour. Res.*, **27**:763–766, 1991.

Weber, W.J. *Physicochemical processes for Water Quality.* Wiley, New York, 640 p., 1972.

Wendland, H. Error estimates for interpolation by compactly supported radial basis functions of minimal degree. *J. Approx. Theory*, **93**:258–272, 1998.

Whitaker, S. Diffusion and dispersion in porous media. *AIChE J.*, **13**:420–427, 1967.

Whitaker, S. Flow in porous media. 1. A theoretical derivation of Darcy's law. *Transp. Porous Media*, **1**:3–25, 1986a.

Whitaker, S. Flow in porous media. 2. The governing equations for immiscible, two-phase flow. *Transp. Porous Media*, **1**:105–125, 1986b.

White, M.D. and Oostrom, M. *STOMP Subsurface Transport Over Multiple Phase: Theory Guide.* PNNL-11216 (UC-2010), Pacific Northwest National Laboratory, Richland, Washington, 2000.

White, M.D. and Oostrom, M. *STOMP Subsurface Transport Over Multiple Phase: User's Guide.* PNNL-15782 (UC-2010), Pacific Northwest National Laboratory, Richland, Washington, 2006.

White, W.N. *A Method of Estimating Ground Water Supplies Based on Discharge by Plants and Evaporation from Soil.* U.S. Geological Survey Water Supply Paper 659, 1932.

Wiener, A. *The Role of Water in Development.* McGraw-Hill, New York, 483 p., 1972.

Williams, G.A., Miller, C.T., and Kelley, C.T. Transformation approaches for simulating flow in variably saturated porous media. *Water Resour. Res.*, **36**:923–934, 2000.

Williams, J.R., Nicks, A.D., and Arnold, J.G. Simulator for water resources in rural basins. *J. Hyd. Eng., ASCE*, **111**:970–986, 1985.

Wilson, J.L. and Miller, P.J. Two-dimensional plumes in uniform groundwater flow. *J. Hyd. Div., ASCE*, **104**:503–514, 1978.

Wilson, L.G., Everett, L.G., and Cullen, S.J. *Handbook of Vadose Zone Characterization & Monitoring.* CRC Press, 752 p., 1994.

Wise, D.L., Trantolo, D.J., Cichon, E.J., Inyang, H.I., and Stottmeister, U. (Eds.) *Remediation Engineering of Contaminated Soils.* CRC Press, 1008 p., 2000a.

Wise, D.L., Trantolo, D.J., Cichon, E.J., Inyang, H.I., and Stottmeister, U. (Eds.) *Bioremediation of Contaminated Soils.* CRC Press, 903 p., 2000b.

Wodie, J.C. and Levy, T. Correction non lineaire de la Loi de Darcy. *C. R. Acad. Sci. Paris, Serie II*, 1991.

Wolery, T.J. *EQ3NR a computer program for geochemical aqueous speciation-solubility calculations: user's guide and documentation*, Lawrence Livermore National Laboratory, report no. UCRL-53414, Livermore, California, 1983.

Wong, S.M., Hon, Y.C., and Golberg, M.A. Compactly supported radial basis functions for shallow water equations. *Appl. Math. Computa.*, **127**:79–101, 2002.

Wösten, J.H.M., Pachepsky, Y.A., and Rawls, W. J. Pedotransfer functions: Bridging the gap between available basic soil data and missing soil hydraulic characteristics. *J. Hydrology*, **251**:123–150, 2001.

Wright, W.F., Schroeder, E.D., Chang, D.P.Y., and Romstad, K. Performance of a pilot-scale compost biofilter treating gasoline vapor. *J. Env. Eng., ASCE*, **123**:547–555, 1997.

Xue, Y., Sellers, J.J., Kinter, J.L., and Shukla, J. A simplified biosphere model for global climate studies. *J. Clim.*, **4**:346–364, 1991.

Yeh, G.T. *3DFEMWATER: A three-dimensional finite element model of water flow through saturated-unsaturated media.* ORNL-5567/R1, Oak Ridge National Laboratory, Oak Ridge, TN, 1987.

Yeh, G.T. *3DLEWASTE: A hybrid Lagrangian-Eulerian finite element model of waste transport through saturated-unsaturated media.* Pennsylvania State University Technical Report, Department of Civil Engineering, Pennsylvania State University, University Park, PA, 1990.

Yeh, G.T. and Tripathi, V.S. A Lagrangian-Eulerian approach to modeling hydrogechemical transport of multi-component systems. *Proc. Int. Conf. Groundwater Contaminant: Use of Models in Decision-Making in the European Year of Environment*, Martin-Nijhoff, 1987.

Yeh, G.T. and Tripathi, V.S. Critical evaluation of recent developments in hydrogeochemical transport models of reactive multicomponents. *Water Resour. Res.*, **31**:93–108, 1989.

Yeh, G.T. and Tripathi, V.S. A model for simulating transport of reactive multispecies components: Model development and demonstration. *Water Resour. Res.*, **27**:3075–3094, 1991.

Yeh, T.C.J. One-dimensional steady-state infiltration in heterogeneous soils. *Water Resour. Res.*, **25**:2149–2158, 1989.

Yeung, A.T. Contaminant extractability by electrokinetics. *Environ. Eng. Sci.*, **23**:202–224, 2006.

Yih, C.S. A transformation for free-surface flow in porous media. *Phys. Fluids*, **1**:20–24, 1964.

Yukselen, Y. and Kaya, A. Comparison of methods for determining specific surface area of soils. *J. Geotech. Geoenviron. Eng., ASCE*, **132**:931–936, 2006.

Yuster, S.T. Theoretical considerations of multiphase flow in idealized capillary systems. *Proc. 3rd World Petroleum Congress*, **II**:437–445, 1951.

Zeitoun, D.G. and Braester, C. A Neumann expansion approach to flow through heterogeneous formations. *Stoch. Hydrol. Hydraul.*, **5**:207–226, 1991.

Zhang, D. *Stochastic Methods for Flow in Porous Media, Coping with Uncertainties.* Academic Press, 2002.

Zhang, F., Yeh, G.-T., Parker, J.C., Brooks, S.C., Pace, M.N., Kim, Y.-J., Jardine, P.M., and Watson, D.B. A reaction-based paradigm to model reactive chemical transport in groundwater with general kinetic and equilibrium reactions. *J. Contam. Hydrol.*, **92**:10–32, 2007.

Zheng, C. *MT3D, A Modular Three-Dimensional Transport Model for Simulation of Advection, Dispersion and Chemical Reactions of Contaminants in Groundwater Systems*, Report to the U.S. Environmental Protection Agency, 170 p., 1990.

Zheng, C. *MT3DMS v5.2 Supplemental User's Guide*, Technical Report to the U.S. Army Engineer Research and Development Center, Department of Geological Sciences, University of Alabama, 24 p., 2006.

Zheng, C. and Bennett, C.D. *Applied Contaminant Transport Modeling, Theory and Practice.* Van Nostrand Reinhold, 440 p., 1995.

Zheng, C. and Wang, P.P. *MT3DMS, A Modular Three-Dimensional Multi-Species Transport Model for Simulation of Advection, Dispersion and Chemical Reactions of Contaminants in Groundwater Systems; Documentation and User's Guide.* U.S. Army Engineer Research and Development Center Contract Report SERDP-99-1, Vicksburg, MS, 202 p., 1999.

Zhou, Q. Modeling seawater intrusion in coastal aquifers. Ph.D. thesis, Technion-Israel Institute of Technology, 1999.

Zhou, Q., Bear, J., and Bensabat J. Saltwater upconing and decay beneath a well pumping above an interface zone. *Transp. Porous Media*, **61**:337–363, 2005.

Zhou, Q., Bensabat, J., and Bear, J. Accurate calculation of specific discharge in heterogeneous porous media. *Water Resour. Res.*, **37**:3057–3069, 2001.

Zhu, J. A low-diffusive and oscillation-free convection scheme. *Comm. Appl. Numer. Meth.*, **7**:225–232, 1991.

Zienkiewicz, O.C. *The Finite Element Method in Engineering Science*, 3rd ed. McGraw-Hill, 787 p., 1977.

Index

Printed in the USA
CPSIA information can be obtained
at www.ICGtesting.com
CBHW050109240824
13659CB00003B/11

9 789402 404777